Methodologies and Traditional Applications

Volume I

Chapman & Hall/CRC
Computer and Information Science Series

Series Editor: Sartaj Sahni

For more information about this series please visit:
https://www.crcpress.com/Chapman--HallCRC-Computer-and-Information-Science-Series/book-series/CHCOMINFSCI

Handbook of Approximation Algorithms and Metaheuristics, Second Edition

Methodologies and Traditional Applications

Volume I

Edited by

Teofilo F. Gonzalez

CRC Press
Taylor & Francis Group
Boca Raton London New York

CRC Press is an imprint of the
Taylor & Francis Group, an **informa** business

A CHAPMAN & HALL BOOK

CRC Press
Taylor & Francis Group
6000 Broken Sound Parkway NW, Suite 300
Boca Raton, FL 33487-2742

© 2018 by Taylor & Francis Group, LLC
CRC Press is an imprint of Taylor & Francis Group, an Informa business

International Standard Book Number-13: 978-1-4987-7011-8 (Hardback)

Visit the Taylor & Francis Web site at
http://www.taylorandfrancis.com

and the CRC Press Web site at
http://www.crcpress.com

Printed and bound in the United States of America by Sheridan

To my wife Dorothy, and our children:
Jeanmarie, Alexis, Julia, Teofilo, and Paolo.

Contents

SECTION II   Local Search, Neural Networks, and Metaheuristics

SECTION III   Multiobjective Optimization, Sensitivity Analysis, and Stability

SECTION IV   Traditional Applications

Preface

More than half a century ago the research community began analyzing formally the quality of the solutions generated by heuristics. The heuristics with guaranteed performance bounds eventually became known as approximation algorithms. The idea behind approximation algorithms was to develop procedures to generate provable near-optimal solutions to optimization problems that could not be solved efficiently by the computational techniques available at that time. With the advent of the theory of NP-completeness in the early 1970s, approximation algorithms became more prominent as the need to generate near optimal solutions for NP-hard optimization problems became the most important avenue for dealing with computational intractability. As it was established in the 1970s, for some problems one could generate near optimal solutions quickly, while for other problems it was established that generating provably good suboptimal solutions was as difficult as generating optimal ones. Other approaches based on probabilistic analysis and randomized algorithms became popular in the 1980s. The introduction of new techniques to solve linear programming problems started a new wave for developing approximation algorithms that matured and saw tremendous growth in the 1990s. To deal with the inapproximable problems, in a practical sense, there were a few techniques introduced in the 1980s and 1990s. These methodologies have been referred to as metaheuristics and may be viewed as problem independent methodologies that can be applied to sets of problems. There has been a tremendous amount of research in metaheuristics during the past three decades. During the last 25 years or so, approximation algorithms have attracted considerably more attention. This was a result of a stronger inapproximability methodology that could be applied to a wider range of problems and the development of new approximation algorithms. In the last decade there has been an explosion of new applications arising from most disciplines.

As we have witnessed, there has been tremendous growth in areas of approximation algorithms and metaheuristics. The second edition of this handbook includes new chapters, updated chapters and chapters with traditional content that did not warrant an update. For this second edition we have partitioned the handbook into two volumes. Volume 1 covers methodologies and traditional applications. Volume 2 covers contemporary and emerging applications. More specifically volume 1 discusses the different methodologies to design approximation algorithms and metaheuristics, as well as the application of these methodologies to traditional combinatorial optimization problems. Volume 2 discusses application of these methodologies to classical problems in computational geometry and graphs theory, as well as in large-scale and emerging application areas. Chapter 1 in both of these volumes presents an overview of approximation algorithms and metaheuristics as well as as an overview of both volumes of this handbook.

It has been a decade since the first edition and our authors have experienced all sorts of different transitions. Several authors expanded their families while writing chapters for the first edition. The babies born at that time are now more than ten years old! A few of the authors and the editor have now retired from their day to day obligations, but continue to be active in research. A couple of the authors became presidents of universities while others now hold prestigious chaired positions or high level positions at their institutions. But sadly, Rajeev Motwani and Ivan Stojmenovic, well-known researchers and authors of first edition chapters, passed away. Since their chapters did not changed significantly, they remain as co-authors of their chapters. Also, Imreh Csanád, a new author for the second edition, passed away in 2017. They are all missed greatly by their families, friends, and the entire research community.

We have collected in this volume a large amount of material with the goal of making it as complete as possible. We apologize in advance for any omissions and would like to invite all of you to propose new chapters for future editions of this handbook. Our research area will continue to grow and we are confident that the following words from an old song "The best is yet to come, you ain't seen nothing yet. . ." applies to our research area. We look forward to the next decade in which new challenges and opportunities await the new, as well as the established, researchers. We look forward to working on problems arising in new emerging applications and editing the third edition of this handbook.

I like to acknowledge the University of California, Santa Barbara (UCSB) for providing me the time and support needed to develop the first and second editions of this handbook for the past 12 years. I also want to thank my wife, Dorothy, our daughters, Jeanmarie, Alexis and Julia, and our sons, Teofilo and Paolo for their moral support, love, encouragement, understanding, and patience, throughout the project and my tenure at UCSB.

Teofilo F Gonzalez
Professor Emeritus of Computer Science
University of California, Santa Barbara

Contributors

Ravindra K. Ahuja
Optym, Gainesville, Florida

Enrique Alba
Departamento de Lenguajes y Ciencias
de la Computación
University of Málaga
Málaga, Spain

Giorgio Ausiello
Dipartimento di Informatica e Sistemistica
Università degli Studi di Roma "La Sapienza"
Roma, Italy

Roberto Battiti
Department of Computer Science and
Telecommunications
University of Trento
Trento, Italy

J. Benton
NASA Ames Research Center
Moffett Field, California

Christian Blum
ALBCOM, Dept. Llenguatges
i Sistemes Informátics
Universitat Politècnica de Catalunya
Barcelona, Spain

Hans-Joachim Böckenhauer
Department of Computer Science
ETH Zürich
Zürich, Switzerland

Vincenzo Bonifaci
Istituto di Analisi dei Sistemi ed Informatica
Consiglio Nazionale delle Ricerche
Roma, Italy

Mauro Brunato
Department of Computer Science and
Telecommunications
University of Trento
Trento, Italy

Niv Buchbinder
School of Mathematical Sciences
Tel Aviv University
Tel Aviv, Israel

Peter Cappello
Department of Computer Science
University of California
Santa Barbara, California

Christopher James Coakley
Department of Computer Science
University of California
Santa Barbara, California

János Csirik
Department of Computer Science
University of Szeged
Szeged, Hungary

Bhaskar DasGupta
Department of Computer Science
University of Illinois at Chicago
Chicago, Illinois

Minh Do
NASA Ames Research Center
Moffett Field, California

Marco Dorigo
Institut de Recherches Interdisciplinaires
et de Développements en
Intelligence Artificielle (IRIDIA)
Université Libre de Bruxelles
Brussels, Belgium

György Dósa
Mathematical Department
University of Pannonia
Veszprém, Hungary

Ding-Zhu Du
Department of Computer Science
University of Texas at Dallas
Richardson, Texas

Devdatt Dubhashi
Computer Science Department
Chalmers University
Goteborg, Sweden

Özlem Ergun
Department of Mechanical and
    Industrial Engineering
Northeastern University
Boston, Massachusetts

Leah Epstein
Department of Mathematics
University of Haifa
Haifa, Israel

Guy Even
School of Electrical Engineering
Tel Aviv University
Tel Aviv, Israel

Moran Feldman
Department of Mathematics and
    Computer Science
The Open University of Israel
Raanana, Israel

David Fernández-Baca
Department of Computer Science
Iowa State University
Ames, Iowa

Jeremy Frank
NASA Ames Research Center
Moffett Field, California

Daya Ram Gaur
Department of Mathematics and
    Computer Science
University of Lethbridge
Lethbridge, Canada

Fred Glover
OptTek Systems
Boulder, Colorado

Teofilo F. Gonzalez
Department of Computer Science
University of California
Santa Barbara, California

Fabrizio Grandoni
Dalle Molle Institute for Artificial
    Intelligence Research
Manno, Switzerland

Magnús M. Halldórsson
ICE-TCS, School of Computer Science
Reykjavik University
Reykjavik, Iceland

Hideki Hashimoto
Department of Logistics and
    Information Engineering
Tokyo University of Marine
    Science and Technology
Tokyo, Japan

Holger H. Hoos
Leiden Institute of Advanced
    Computer Science
Universiteit Leiden
Leiden, the Netherlands

Juraj Hromkovič
Department of Computer Science
ETH Zürich
Zürich, Switzerland

Yannan Hu
Department of Mathematical Informatics
Nagoya University
Nagoya, Japan

Toshihide Ibaraki
The Kyoto College of Graduate
    Studies for Informatics
Kyoto, Japan

Shinji Imahori
Department of Information and
    System Engineering
Chuo University
Tokyo, Japan

Csanád Imreh
Department of Computer Science
University of Szeged
Szeged, Hungary

Andrew B. Kahng
CSE and ECS Departments
University of California, San Diego
La Jolla, California

Samir Khuller
Department of Computer Science
University of Maryland
College Park, Maryland

Dennis Komm
Department of Computer Science
ETH Zürich
Zürich, Switzerland

Guy Kortsarz
Department of Computer Science
Rutgers University
Camden, New Jersey

Ramesh Krishnamurti
School of Computer Science
Simon Fraser University
Burnaby, Canada

Manuel Laguna
Leeds School of Business
University of Colorado
Boulder, Colorado

Stefano Leonardi
Dipartimento di Informatica e Sistemistica
Universitá degli Studi di Roma "La Sapienza"
Roma, Italy

Guillermo Leguizamón
Departamento de Informática
Universidad Nacional de San Luis
San Luis, Argentina

Derong Liu
Department of Electrical &
    Computer Engineering
University of Illinois at Chicago
Chicago, Illinois

Ion Măndoiu
Computer Science and
    Engineering Department
University of Connecticut
Storrs, Connecticut

Alberto Marchetti-Spaccamela
Dipartimento di Informatica e Sistemistica
Università degli Studi di Roma "La Sapienza"
Roma, Italy

Rafael Martí
Facultad de Mathemáticas
Universidad de Valencia
Valencia, Spain

Rajeev Motwani
Department of Computer Science
Stanford University
Stanford, California

Hiroshi Nagamochi
Department of Applied
    Mathematics and Physics
Kyoto University
Kyoto, Japan

Sotiris Nikoletseas
Computer Engineering &
    Informatics Department
Patras University
Patras, Greece

Liadan O'Callaghan
Google Inc
Mountain View, California

James B. Orlin
Massachusetts Institute of Technology
Cambridge, Massachusetts

Alessandro Panconesi
Computer Science Department
Università degli Studi di Roma "La Sapienza"
Roma, Italy

Luís Paquete
DEI/CISUC
University of Coimbra
Coimbra, Portugal

Vangelis Th. Paschos
LAMSADE and Université Paris-Dauphine
Paris, France

Abraham P. Punnen
Department of Mathematics
Simon Fraser University
Burnaby, Canada

Balaji Raghavachari
Department of Computer Science
University of Texas at Dallas
Richardson, Texas

Dror Rawitz
Faculty of Engineering
Bar Ilan University
Ramat Gan, Israel

Sartaj Sahni
Department of Computer Science
University of Florida
Gainesville, Florida

Celso S. Sakuraba
Department of Production Engineering
Federal University of Sergipe
Sao Cristovao, Brazil

Nasim Samei
Department of Computer Science
The University of Western Ontario
London, Canada

Sebastian Seibert
Department of Computer Science
RWTH Aachen University
Aachen, Germany

Hadas Shachnai
Computer Science Department
The Technion
Haifa, Israel

Hava T. Siegelmann
Computer Science Department
University of Massachusetts
Amherst, Massachusetts

Krzysztof Socha
Institut de Recherches Interdisciplinaires
    et de Développements en Intelligence
    Artificielle (IRIDIA)
Université libre de Bruxelles
Brussels, Belgium

Roberto Solis-Oba
Department of Computer Science
The University of Western Ontario
London, Canada

Paul Spirakis
Computer Engineering &
    Informatics Department
Patras University
Patras, Greece

Thomas Stützle
Institut de Recherches Interdisciplinaires
    et de Développements en Intelligence
    Artificielle (IRIDIA)
Université Libre de Bruxelles
Brussels, Belgium

Tami Tamir
School of Computer Science
The Interdisciplinary Center
Herzliya, Israel

Rob van Stee
Computer Science Department
University of Leicester
Leicester, United Kingdom

Balaji Venkatachalam
Department of Computer Science
Iowa State University
Ames, Iowa

Wei Wu
Department of Computer Science and
    Mathematical Informatics
Nagoya University
Nagoya, Japan

Weili Wu
Department of Computer Science
University of Texas at Dallas
Richardson, Texas

Mutsunori Yagiura
Department of Mathematical Informatics
Nagoya University
Nagoya, Japan

Neal E. Young
Department of Computer Science
University of California at Riverside
Riverside, California

Alexander Zelikovsky
Computer Science Department
Georgia State University
Atlanta, Georgia

An Zhu
Google Inc
Mountain View, California

1

Introduction, Overview, and Notation

Teofilo F. Gonzalez

1.1 Introduction

Approximation algorithms were formally introduced in the 1960s to generate near-optimal solutions to optimization problems that could not be solved efficiently by the computational techniques available at that time. With the advent of the theory of NP-completeness in the early 1970s, the area became more prominent as the need to generate near-optimal solutions for NP-hard optimization problems became the most important avenue for dealing with computational intractability. As it was established in the 1970s, for some problems it is possible to generate near-optimal solutions quickly, whereas for other problems generating provably good suboptimal solutions is as difficult as generating optimal ones. Computational approaches based on probabilistic analysis and randomized algorithms became popular in the 1980s. The introduction of new techniques to solve linear programming problems started a new wave of approximation algorithms that matured and saw tremendous growth in the 1990s. There were a few techniques introduced in the 1980s and 1990s to deal, in the practical sense, with inapproximable problems. These methodologies have been referred to as metaheuristics and include Simulated Annealing (SA), Ant Colony Optimization (ACO), Evolutionary Computation (EC), Tabu Search (TS), Memetic Algorithms (MAs), and so on. Other previously established methodologies such as local search, backtracking, and branch-and-bound were also explored at that time. There has been a tremendous amount of research in metaheuristics during the past three decades. These techniques have been evaluated experimentally and have demonstrated their usefulness for solving problems that are arising in practice. During the last 25 years or so, approximation algorithms have attracted considerably more attention. This was a result of a stronger inapproximability methodology that could be applied to a wider range of problems and the development of new approximation algorithms for problems arising in established and emerging application areas. Polynomial Time Approximation

Schemes (PTASs) were introduced in the 1960s and the more powerful Fully Polynomial Approximation Schemes (FPTASs) were introduced in the 1970s. Asymptotic PTAS (APTAS) and Asymptotic FPTAS (AFPTAS), and Fully Polynomial Randomized Approximation Schemes (FPRASs) were introduced later on.

Today approximation algorithms enjoy a stature comparable to that of algorithms in general and the area of metaheuristics has established itself as an important research area. The new stature is a byproduct of a natural expansion of research into more practical areas where solutions to real-world problems are expected, as well as by the higher level of sophistication required to design and analyze these new procedures. The goal of approximation algorithms and metaheuristics is to provide the best possible solutions and to guarantee that such solutions satisfy certain criteria. This two-volume handbook houses these two approaches and thus covers all the aspects of approximations. We hope it will serve you as a valuable reference for approximation methodologies and applications.

Approximation algorithms and metaheuristics have been developed to solve a wide variety of problems. A good portion of these algorithms have only theoretical value due to the fact that their time complexity is a high-order polynomial or they have a huge constant associated with their time complexity bound. However, these results are important because they establish what is possible, and it may be that in the near future these algorithms will be transformed into practical ones. Other approximation algorithms do not suffer from this pitfall, but some were designed for problems with limited applicability. However, the remaining approximation algorithms have real-world applications. Given this, there is a huge number of important application areas, including new emerging ones, where approximation algorithms and metaheuristics have barely penetrated and we believe there is an enormous potential for their use. Our goal is to collect a wide portion of the approximation algorithms and metaheuristics in as many areas as possible, as well as to introduce and explain in detail the different methodologies used to design these algorithms.

1.2 Overview

Our overview in this section is devoted mainly to the earlier years. The individual chapters in the two volumes discuss in detail the recent research accomplishments in different subareas. This section will also serve as an overview of both volumes of this handbook. Chapter 2 discusses some of the basic methodologies and applies them to classical problems.

Even before the 1960s researchers in applied mathematics and graph theory had established upper and lower bounds for certain properties of graphs. For example, bounds had been established for the chromatic number, achromatic number, chromatic index, maximum clique, maximum independent set, and so on. Some of these results could be seen as the precursors of approximation algorithms. By the 1960s it was understood that there were problems that could be solved efficiently, whereas for other problems all the known algorithms required exponential time in the worst case. Heuristics were being developed to find quick solutions to problems that appeared to be computationally difficult to solve. Researchers were experimenting with heuristics, branch-and-bound procedures, and iterative improvement frameworks and were evaluating their performance when solving actual problem instances. There were many claims being made, not all of which could be substantiated, about the performance of the procedures being developed to generate optimal and suboptimal solutions to combinatorial optimization problems.

Half a century ago (1966), Ronald L. Graham [1] formally introduced approximation algorithms. He analyzed the performance of list schedules for scheduling tasks on identical machines, a fundamental problem in scheduling theory.

> *Problem*: Scheduling tasks on identical machines.
> *Instance*: Set of n tasks $(T_1, T_2, \ldots, T_n)$ with processing time requirements $t_1, t_2, \ldots, t_n$, partial order C defined over the set of tasks to enforce task dependencies, and a set of m identical machines.

Objective: Construct a schedule with minimum makespan. A *schedule* is an assignment of tasks to time intervals on the machines in such a way that (1) each task T_i is processed continuously for t_i units of time by one of the machines; (2) each machine processes at most one task at a time; and (3) the precedence constraints are satisfied (i.e., machines cannot commence the processing of a task until all of its predecessors have been completed). The *makespan* of a schedule is the time at which all the machines have completed processing the tasks.

The *list-scheduling* procedure is given an ordering of the tasks specified by a list L. Then the procedure finds the earliest time t when a machine is idle, and an unassigned task is available (i.e., all its predecessor tasks have been completed). It assigns the leftmost available task in the list L to an idle machine at time t and this step is repeated until all the tasks have been scheduled.

The main result in Reference 1 was proving that for every problem instance I, the schedule generated by this policy has a makespan that is bounded above by $(2 - 1/m)$ times the optimal makespan for the instance. This is called the *approximation ratio* or *approximation factor* for the algorithm. We also say that the algorithm is a $(2 - 1/m)$-approximation algorithm. This criterion for measuring the quality of the solutions generated by an algorithm remains as one of the most important ones in use today. The second contribution in Reference 1 was showing that the approximation ratio $(2 - 1/m)$ is the best possible for list schedules, that is, the analysis of the approximation ratio for this algorithm cannot be improved. This was established by presenting problem instances (for all m and $n \geq 2m - 1$) and lists for which the schedule generated by the procedure has a makespan equal to $2 - 1/m$ times the optimal makespan for the instance. A restricted version of the list-scheduling algorithm is analyzed in detail in Chapter 2.

The third important aspect of the results in Reference 1 was showing that list scheduling may have anomalies. To explain this we need to define some terms. The makespan of the list schedule for instance I using list L is denoted by $f_L(I)$. Suppose that instance I' is a slightly modified version of instance I. The modification is such that we intuitively expect that $f_L(I') \leq f_L(I)$. But this is not always true, so there is an anomaly. For example, suppose that I' is I, except that I' has an additional machine. Intuitively $f_L(I') \leq f_L(I)$ because with one additional machine, tasks should be finished earlier or at worst at the same time as when there is one fewer machine. But this is not always the case for list schedules, there are problem instances and lists for which $f_L(I') > f_L(I)$. This is called an *anomaly*. Our expectation would be valid if list scheduling would generate minimum makespan schedules, but we have a procedure that generates suboptimal solutions. Such guarantees are not always possible in this type of environment. List schedules suffer from other anomalies, for example, relaxing the precedence constraints or decreasing the execution time of the tasks. In both cases, one would expect schedules with smaller or the same makespans. But, that is not always the case. Chapter 2 presents problem instances where anomalies occur. The main reason for discussing anomalies now is that even today, numerous papers are being published and systems are being deployed where "common sense"-based procedures are being introduced without any analytical justification and/or thorough experimental validation. Anomalies show that since we live for the most part in a "suboptimal world," the effect of our decisions is not always the intended one (unintended consequences). One can design approximation algorithms that do not suffer from certain types of anomalies but probably not for all possible ones.

Other classical problems with numerous applications are the traveling salesperson, Steiner tree, and spanning tree problems, which will be formally defined later on. Even before the 1960s there were several well-known polynomial time algorithms to construct minimum weight spanning trees for edge-weighted graphs [2]. These simple greedy algorithms have low-order polynomial time complexity bounds. It was well known at that time that the same type of procedures does not generate an optimal tour for the traveling salesperson problem (TSP) and does not construct optimal Steiner trees. However, in 1968 E. F. Moore (as discussed in Reference 3) showed that for any set of points P in metric space $L_M \leq L_T \leq 2L_S$, where L_M, L_T, and L_S are the total weight of a minimum weight spanning tree, a minimum weight tour (solution) for the TSP, and minimum weight Steiner tree for P, respectively. Since every spanning tree is a Steiner tree, the above-mentioned bounds show that when using a minimum weight spanning tree to approximate the Steiner tree we have a solution (Steiner tree) whose weight is at most twice the weight of an optimal Steiner tree. In other words, any algorithm that generates a minimum weight spanning tree is a 2-approximation algorithm for the Steiner tree problem. Furthermore, this approximation algorithm takes no more time than an algorithm that constructs a minimum weight spanning tree for edge weighted graphs [2], as such an algorithm can be used to construct an optimal spanning tree for a set of points in metric space. The above-mentioned bound is established by defining a transformation from any minimum weight Steiner tree into a TSP tour with weight at most $2L_S$. Therefore, $L_T \leq 2L_S$ [3]. Then by observing that the deletion of an edge in an optimum tour for the TSP problem results in a spanning tree, it follows that $L_M < L_T$. Chapter 3 discusses this approximation algorithm and its analysis in more detail. The Steiner ratio is defined as L_S/L_M. The earlier arguments show that the Steiner ratio is at least $\frac{1}{2}$. Gilbert and Pollak [3] conjectured that the Steiner ratio in the Euclidean plane equals $\frac{\sqrt{3}}{2}$ (the 0.86603 . . . conjecture). A proof of this conjecture and improved approximation algorithms for the Steiner tree problem are discussed in Chapter 36.

The above-mentioned constructive proof can be applied to a minimum weight spanning tree to generate a tour for the TSP problem. The construction takes polynomial time and results in a 2-approximation algorithm for the TSP problem. This approximation algorithm for the TSP is also referred to as the *double spanning tree algorithm* and is discussed in Chapters 3 and 27. Improved approximation algorithms for the TSP and algorithms for its generalizations are discussed in Chapters 3, 27, 34, 35, and Volume 2, Chapter 2. The approximation algorithm for the Steiner tree problem just discussed is explained in Chapter 3, and improved approximation algorithms and applications are discussed in Chapters 36, 37, and Volume 2, Chapter 2. Volume 2, Chapter 14 discusses approximation algorithms for computationally intractable variations of the spanning tree problem.

In 1969, Graham [4] studied the problem of scheduling tasks on identical machines but restricted to independent tasks, that is, the set of precedence constraints is empty. He analyzed the Largest Processing Time first (LPT) scheduling rule, which is list scheduling where the list of tasks L is arranged in nonincreasing order of their processing requirements. His elegant proof established that the LPT procedure generates a schedule with makespan at most $\frac{4}{3} - \frac{1}{3m}$ times the makespan of an optimal schedule, that is, the LPT scheduling algorithm has a $\frac{4}{3} - \frac{1}{3m}$ approximation ratio. He also showed that the analysis is best possible for all m and $n \geq 2m+1$. For $n \leq 2m$ tasks the approximation ratio is smaller and under some conditions, LPT generates an optimal makespan schedule. Graham [4], following a suggestion by D. Kleitman and D. Knuth, considered list schedules where the first portion of the list L consists of k tasks (without loss of generality, assume k is a multiple of m) with the longest processing times arranged by their starting times in an optimal schedule for these k tasks (only). Then the list L has the remaining $n - k$ tasks in any order. The approximation ratio for this list schedule using list L is $1 + \frac{m-1}{m+k}$. An optimal schedule for the longest k tasks can be constructed in $O(n + m^k)$ time by a straight forward branch and bound algorithm. In other words, this algorithm has an approximation ratio $1 + \epsilon$ and time complexity $O(n + m^{(m-1-\epsilon m)/\epsilon})$. For any fixed constants m and ϵ, the algorithm constructs in polynomial (linear) time with respect to n, a schedule with makespan at most $1 + \epsilon$ times the optimal makespan. Note that for a fixed constant m the time complexity is polynomial with respect to n, but it is not polynomial with respect to $1/\epsilon$. This was the first algorithm of its kind and later on, it was called Polynomial Time Approximation Scheme (PTAS). Chapter 8 discusses different PTASs. Additional PTAS appear in Chapter 36 and Volume 2, Chapters 2 and 5. The proof techniques presented in References 1, 4 are outlined in Chapter 2 and have been extended to apply to other problems. There is an extensive body of literature for approximation algorithms and metaheuristics for scheduling problems. Chapters 38, 39, and Volume 2, Chapter 25 discuss interesting approximation algorithms and heuristics for scheduling problems. The scheduling handbook [5] is an excellent source for scheduling algorithms, models, and performance analysis.

The development of NP-completeness theory in the early 1970s by Cook [6], Karp [7], and others formally introduced the notion that there is a large class of decision problems (the answer to these problems is a simple yes or no) that are computationally equivalent. This means that either every problem in this class has a polynomial time algorithm that solves it, or none of them do. Furthermore, this question is the same as the $P = NP$ question, a classical open problem in computational complexity. This question is to determine whether or not the set of languages recognized in polynomial time by deterministic Turing machines is the same as the set of languages recognized in polynomial time by nondeterministic Turing machines. The conjecture has been that $P \neq NP$, and thus the hardest problems in NP would not be solvable in polynomial time. The computationally equivalent decision problems in this class are called *NP-complete* problems. The scheduling on identical machines problem discussed earlier is an optimization problem. Its corresponding decision problem has its input augmented by an integer value B, and the yes–no question is to determine whether or not there is a schedule with makespan at most B. An optimization problem whose corresponding decision problem is NP-complete is called an *NP-hard* problem. Therefore, scheduling tasks on identical machines is an NP-hard problem. The TSP and the Steiner tree problem are also NP-hard problems. The minimum weight spanning tree problem can be solved in polynomial time and it is not an NP-hard problem under the assumption that $P \neq NP$. The next section discusses NP-completeness in more detail. There is a long list of practical problems arising in many different fields of study that are known to be NP-hard problems [8]. Because of this, the need to cope with these computationally intractable problems was recognized earlier on. Since then approximation algorithms became a central area of research activity. Approximation algorithms offered a way to circumvent computational intractability by paying a price when it comes to the quality of the solutions generated. But a solution can be generated quickly. In other words and another language, "no te fijes en lo bien, fijate en lo rápido". Words used to describe my golf playing ability when I was growing up.

In the early 1970s, Garey et al. [9] as well as Johnson [10,11] developed the first set of polynomial time approximation algorithms for the bin packing problem. The analysis of the approximation ratio for these algorithms is asymptotic, which is different from those for the scheduling problems discussed earlier. We will define this notion precisely in the next section, but the idea is that the ratio holds when the value of an optimal solution is greater than some constant. Research on the bin packing problem and its variants has attracted very talented investigators who have generated more than 1000 papers, most of which deal with approximations. This work has been driven by numerous applications in engineering and information sciences (see Chapters 28, 29, 30, and 31).

Johnson [12] developed polynomial time algorithms for the sum of subsets, max satisfiability, set cover, graph coloring, and max clique problems. The algorithms for the first two problems have a constant ratio approximation, but for the other problems the approximation ratio is $\ln n$ and n^ϵ. Sahni [13,14] developed a PTAS for the knapsack problem. Rosenkrantz et al. [15] developed several constant ratio approximation algorithms for the TSP that satisfy the triangle inequality (or simply, defined over metric graphs). This version of the problem is defined over edge weighted complete graphs, rather than for points in metric space as in Reference 3. These algorithms have an approximation ratio of two.

Sahni and Gonzalez [16] showed that there were a few NP-hard optimization problems for which the existence of a constant ratio polynomial time approximation algorithm implies the existence of a polynomial time algorithm to generate an optimal solution. In other words, complexity of generating a constant ratio approximation and an optimal solution are computationally equivalent problems. For these problems, the approximation problem is NP-hard or simply inapproximable (under the assumption that $P \neq NP$). Later on, this notion was extended to mean that there is no polynomial time algorithm with approximation ratio r for a problem under some complexity theoretic hypothesis. The approximation ratio r is called the *inapproximability ratio* (see Chapter 17 in the first edition of this handbook).

The k-min-cluster problem is one of these inapproximable problems. Given an edge-weighted undirected graph, the k-min-cluster problem is to partition the set of vertices into k sets so as to minimize the sum of the weight of the edges with endpoints in the same set. The k-maxcut problem is defined as the k-min-cluster problem, except that the objective is to maximize the sum of the weight of the edges with endpoints in different sets. Even though these two problems have exactly the same set of feasible and optimal solutions, there is a linear time algorithm for the k-maxcut problem that generates k-cuts with weight at least $\frac{k-1}{k}$ times the weight of an optimal k-cut [16], whereas approximating the k-min-cluster problem is a computationally intractable problem. The former problem has the property that a near-optimal solution may be obtained as long as partial decisions are made optimally, whereas for the k-min-cluster an optimal partial decision may turn out to force a terrible overall solution. For the k-min-cluster problem if one makes a mistake at some iteration, one will end up with a solution that is far from optimal. Whereas for the k-maxcut problem one can make many mistakes and still end up with a near-optimal solution. A similar situation arises when you make a mistake an exam where almost everyone receives a perfect score, versus a course where the average score is about 50% of the points.

Another interesting problem whose approximation problem is NP-hard is the TSP problem [16]. This is not exactly the same version of the TSP problem discussed earlier, which we said has several constant ratio polynomial time approximation algorithms. Given an edge-weighted undirected graph, the TSP is to find a least weight tour, that is, to find a least weight (simple) path that starts at vertex 1, visits each vertex in the graph *exactly* once, and ends at vertex 1. The weight of a path is the sum of the weight of its edges. The weights of the edges are unrelated and the approximation problem is NP-hard. The version of the TSP problem studied in Reference 15 is limited to metric graphs, that is, the graph is complete (all the edges are present) and the set of edge weights satisfies the triangle inequality (which means that the weight of the edge joining vertex i and j is less than or equal to the weight of any path from vertex i to vertex j). This version of the TSP problem is equivalent to the one studied by E. F. Moore [3]. The approximation algorithms given in References 3, 15 can be easily adapted to provide a constant ratio approximation to the version of the TSP problem where the tour is defined as visiting each vertex in the graph *at least* once. Since Moore's approximation algorithms for the metric Steiner tree and metric TSP are based on the same idea, one would expect that the Steiner tree problem defined over arbitrarily weighted graphs is NP-hard to approximate. However, this is not the case. Moore's algorithm [3] can be modified to be a 2-approximation algorithm for this more general Steiner tree problem.

As pointed out in Reference 17, Levner and Gens [18] added a couple of problems to the list of problems that are NP-hard to approximate. Garey and Johnson [19] show that the max clique problem has the property that if for some constant r there is a polynomial time r-approximation algorithm, then there is a polynomial time r'-approximation for any constant r' such that $0 < r' < 1$. Since that time researchers have tried many different algorithms for the clique problem, none of which were constant ratio approximation algorithms, and it was conjectured that none existed under the assumption that $P \neq NP$. This conjecture has been proved.

A PTAS is said to be an FPTAS if its time complexity is polynomial with respect to n (the problem size) and $1/\epsilon$. The first FPTAS was developed by Ibarra and Kim [20] for the knapsack problem. Sahni [21] developed three different techniques based on rounding, interval partitioning, and separation to construct FPTAS for sequencing and scheduling problems. These techniques have been extended to other problems and are discussed in Chapter 9. Horowitz and Sahni [22] developed FPTAS for scheduling on processors with different processing speeds. Reference 17 discusses a simple $O(n^3/\epsilon)$ FPTAS for the knapsack problem developed by Babat [23,24]. Lawler [25] developed techniques for maximum speed-up FPTAS for the knapsack and related problems. Chapter 9 presents different methodologies to design FPTAS. Garey and Johnson [26] showed that if any problem in a class of NP-hard optimization problems that satisfy certain properties has a FPTAS, then $P = NP$. The properties are that the objective function value of every feasible solution is a positive integer, and the problem is *strongly* NP-hard. A problem is strongly NP-hard if the problem is NP-hard even when the magnitude of the maximum number in the input is bounded by a polynomial on the input length. For example, the TSP problem is strongly NP-hard, whereas the knapsack problem is not, under the assumption that $P \neq NP$ (see Chapter 9).

Lin and Kernighan [27] developed elaborate heuristics that established experimentally that instances of the TSP with up to 110 cities can be solved to optimality with 95% confidence in $O(n^2)$ time. This was an iterative improvement procedure applied to a set of randomly selected feasible solutions. The process was to perform k pairs of link (edge) interchanges that improved the length of the tour. However, Papadimitriou and Steiglitz [28] showed that for the TSP no local optimum of an efficiently searchable neighborhood can be within a constant factor of the optimal value unless $P = NP$. Since then there has been quite a bit of research activity in this area. Deterministic and stochastic local search in efficiently searchable, as well as in very large neighborhoods, are discussed in Chapters 16, 17, 18, and 19. Chapter 13 discusses issues relating to the empirical evaluation of approximation algorithms and metaheuristics.

Perhaps the best known approximation algorithm for the TSP defined over metric graphs is the one by Christofides [29]. The approximation ratio for this algorithm is $\frac{3}{2}$, which is smaller than the approximation ratio of 2 for the algorithms reported in References 3, 15. However, looking at the bigger picture that includes the time complexity of the approximation algorithms, Christofides algorithm is not of the same order as the ones given in References 3, 15. Therefore, neither approximation algorithm dominates the other as one has a smaller time complexity bound, whereas the other (Christofides algorithm) has a smaller worst case approximation ratio.

Ausiello et al. [30] introduced the differential ratio, which is another way of measuring the quality of the solutions generated by approximation algorithms. Differential ratio destroys the artificial dissymmetry between "equivalent" minimization and maximization problems (e.g., the k-maxcut and the k-min cluster discussed earlier) when it comes to approximation. This ratio uses the difference between the worst possible solution minus the solution generated by the algorithm, divided by the difference between the worst solution minus the best solution. Cornuejols et al. [31] also discussed a variation of differential ratio approximations. They wanted the ratio to satisfy the following property: "A modification of the data that adds a constant to the objective function value should also leave the error measure unchanged." That is, the "error" by the approximation algorithm should be the same as before. Differential ratio and its extensions are discussed in Chapter 15, along with other similar notions [30]. Ausiello et al. [30] introduced *reductions that preserve approximability*. Since then there have been several new types of approximation preserving reductions. The main advantage of these reductions is that they enable us to define large classes of optimization problems that behave in the same way with respect to approximation. Informally, the class of NP Optimization (**NPO**) problems, is the set of all optimization problems Π which can be "recognized" in polynomial time (see Chapter 14 for a formal definition). An **NPO** problem Π is said to be in **APX**, if it has a constant approximation ratio polynomial time algorithm. The class **PTAS** consists of all **NPO** problems which have PTAS. The class **FPTAS** is defined similarly. Other classes, **Poly-APX**, **Log-APX**, and **Exp-APX**, have also been defined (see Chapter 14).

One of the main accomplishments at the end of the 1970s was the development of a polynomial time algorithm for Linear Programming (LP) problems by Khachiyan [32]. This result had a tremendous impact on approximation algorithm research and started a new wave of approximation algorithms. Two subsequent research accomplishments were at least as significant as Khachiyan's [32] result. The first one was a faster polynomial time algorithm for solving linear programming problems developed by Karmakar [33]. The other major accomplishment was the work of Grötschel et al. [34,35]. They showed that it is possible to solve a linear programming problem with an exponential number of constraints (with respect to the number of variables) in time which is polynomial in the number of variables and the number of bits used to describe the input, given a *separation oracle* plus a bounding ball and a lower bound on the volume of the feasible solution space. Given a solution, the separation oracle determines in polynomial time whether or not the solution is feasible, and if it is not it finds a constraint that is violated. Chapter 10 gives an example of the use of this approach. Important developments have taken place during the past 30 years. The books [35,36] are excellent references for linear programming theory, algorithms, and applications.

Because of the above-mentioned results, the approach of formulating the solution to an NP-hard problem as an integer linear programming problem and so solving the corresponding linear programming problem became very popular. This approach is discussed in Chapter 2. Once a fractional solution is obtained, one uses rounding to obtain a feasible solution to the original NP-hard problem. The rounding may be deterministic or randomized, and it may be very complex (meta-rounding). LP rounding is discussed in Chapters 2, 4, 7, 8, 10, 11, and Volume 2, Chapters 8 and 11.

Independently, Johnson [12] and Lovász [37] developed efficient algorithms for the set cover with approximation ratio of $1 + \ln d$, where d is the maximum number of elements in each set. Chvátal [38] extended this result to the weighted set cover problem. Subsequently, Hochbaum [39] developed an algorithm with approximation ratio f, where f is the maximum number of sets containing any of the elements in the set. This result is normally inferior to the one by Chvátal [38], but it is more attractive for the weighted vertex cover problem, which is a restricted version of the weighted set cover. For this subproblem it is a 2-approximation algorithm. A few months after Hochbaum's initial result,[*] Bar-Yehuda and Even [40] developed a primal-dual algorithm with the same approximation ratio as the one in Reference 39. The algorithm in Reference 40 does not require the solution of an LP problem, as in the case of the algorithm in Reference 39, and its time complexity is linear. But it uses linear programming theory to establish this result. This was the first primal-dual approximation algorithm, though some previous algorithms may also be viewed as falling into this category. An application of the primal-dual approach as well as related ones are discussed in Chapter 2. Chapters 4, 34, and Volume 2, Chapter 23 discuss several primal-dual approximation algorithms. Chapter 12 discusses "distributed" primal-dual algorithms. These algorithms make decisions by using only "local" information.

In the mid 1980s Bar-Yehuda and Even [41] developed a new framework parallel to the primal-dual methods. They called it *local ratio*; it is simple and requires no prior knowledge of linear programming. In Chapter 2 we explain the basics of this approach, and Chapter 6 and Reference 42 covers extensively this technique as well as its extensions.

Raghavan and Thompson [43] were the first to apply randomized rounding to relaxations of linear programming problems to generate solutions to the problem being approximated. This field has grown tremendously. LP randomized rounding is discussed in Chapters 2, 4, 7, 10, 11, and Volume 2, Chapter 11, and deterministic rounding is discussed in Chapters 2, 7, 8, 10, and Volume 2, Chapters 8 and 11. A disadvantage of LP-rounding is that a linear programming problem needs to be solved. This takes polynomial time with respect to the input length, but in this case it means the number of bits needed to represent the input. In contrast, algorithms based on the primal-dual approach are for the most part faster, since they take polynomial time with respect to the number of "objects" in the input. However, the LP-rounding approach can be applied to a much larger class of problems and it is more robust since the technique is more likely to be applicable after changing the objective function and/or constraints for a problem.

The first Asymptomatic PTAS (APTAS) was developed by Fernandez de la Vega and Lueker [44] for the bin packing problem. The first Asymptomatic FPTAS (AFPTAS) for the same problem was developed by Karmakar and Karp [45]. These approaches are discussed in Chapter 15. FPRASs are discussed in Chapter 11.

In the 1980s, new approximation algorithms were developed as well as PTAS and FPTAS based on different approaches. These results are reported throughout the handbook. One difference was the application of approximation algorithm to other areas of research activity (very large-scale integration [VLSI], bioinformatics, network problems) as well as other problems in established areas.

In the late 1980s Papadimitriou and Yannakakis [46] defined **MAXSNP** as a subclass of **NPO**. These problems can be approximated within a constant factor and have a nice logical characterization. They showed that if MAX3SAT, vertex cover, MAXCUT, and some other problems in the class could be approximated in polynomial time with an arbitrary precision, then all **MAXSNP** problems would. This fact was established by using *approximation preserving* reductions (see Chapter 14). In the 1990s Arora et al. [47], using complex arguments (see Chapter 17 in the 1st edition of this handbook), showed that MAX3SAT is hard to approximate within a factor of $1 + \epsilon$ for some $\epsilon > 0$ unless $P = NP$. Thus, all problems in **MAXSNP** do not admit a PTAS unless $P = NP$. This work led to major developments in the

---

[*] Here we are referring to the time when these results appeared as technical reports. Note that from the journal publication dates, the order is reversed. You will find throughout the chapters similar patterns. To add to the confusion, a large number of papers have also been published in conference proceedings. Since it would be very complex to include the dates when the initial technical report and conference proceedings were published, we only include the latest publication date. Please keep this in mind when you read the chapters and, in general, the computer science literature.

area of approximation algorithms, including inapproximability results for other problems, a bloom of approximation preserving reductions, discovery of new inapproximability classes, and construction of approximation algorithms achieving optimal or near-optimal ratios.

Feige et al. [48] showed that the clique problem could not be approximated to within some constant value. Applying the previous results in Reference 26 it showed that the clique problem is inapproximable to within any constant. Feige [49] showed that set cover is inapproximable within $\ln n$. Other inapproximable results appear in References 50, 51. Chapter 17 in the first edition of this handbook discusses all of this work in detail.

There are many other very interesting results that have been published in the past 25 years. Goemans and Williamson [52] developed improved approximation algorithms for the maxcut and satisfiability problems using *semidefinite programming* (SDP). This seminal work opened a new venue for the design of approximation algorithms. Chapter 8 in the first edition of this handbook discusses this work as well as developments in this area. Goemans and Williamson [53] also developed powerful techniques for designing approximation algorithms based on the primal-dual approach. The dual-fitting and factor revealing approach is used in Reference 54. Techniques and extensions of these approaches are discussed in Chapters 4, 12, 34, and Volume 2, Chapter 23.

This concludes our overview of Section I of this volume. Section 1.2.1 presents an overview of Section II of this volume dealing with local search, artificial neural nets and metaheuristics. Section 1.2.2 presents an overview of multiobjective optimization, reoptimization, sensitivity analysis and stability all of which are in Section III of this volume. In the last couple of decades we have seen approximation algorithms being applied to traditional combinatorial optimization problems as well as problems arising in other areas of research activity. These areas include: VLSI design automation, networks (wired, sensor, and wireless), bioinformatics, game theory, computational geometry, and graph problems. In Sections 1.2.3 through 1.2.5 we elaborate further on these applications. Section 1.2.3 overviews traditional application covered in Section IV of this volume. Sections 1.2.4 and 1.2.5 overview contemporary and emerging application which are covered in Volume 2.

1.2.1 Local Search, Artificial Neural Networks, and Metaheuristics

Local search techniques have a long history; they range from simple constructive and iterative improvement algorithms to rather complex methodologies that require significant fine-tuning, such as evolutionary algorithms (EAs) or SA. Local search is perhaps one of the most natural ways to attempt to find an optimal or suboptimal solution to an optimization problem. The idea of local search is simple: Start from a solution and improve it by making local changes until no further progress is possible. Deterministic local search algorithms are discussed in Chapter 16. Chapter 17 covers stochastic local search algorithms. These are local search algorithms that make use of randomized decisions, for example, in the context of generating initial solutions or when determining search steps. When the neighborhood to search for the next solution is very large, finding the best neighbor to move to is many times an NP-hard problem. Therefore, an approximation solution is needed at this step. In Chapter 18 the issues related to very large-scale neighborhood search are discussed from the theoretical, algorithmic, and applications point of view.

Reactive Search advocates the use of simple subsymbolic machine learning to automate the parameter tuning process and make it an integral (and fully documented) part of the algorithm. Parameters are normally tuned through a feedback loop that many times depends on the user input. Reactive search attempts to mechanize this process. Chapter 19 discusses issues arising during this process.

Artificial neural networks have been proposed as a tool for machine learning and many results have been obtained regarding their application to practical problems in robotics control, vision, pattern recognition, grammatical inferences, and other areas. Recently, neural networks have found many applications in the forefront of Artificial Intelligence (AI) and Machine Learning (ML). For example, Google's open-source deep learning neural network tools as well as the Cloud, have been a catalyst for the development of these new applications.* Recently artificial neural networks have been used to improve energy utilization during certain periods in Google's massive data centers with impressive results. The whole process has been fully automated without the use of training data. Once trained (automatically or manually), the neural network will compute an input/output mapping which, if the training data was representative enough, will closely match the unknown rule which produced the original data. Neural networks are discussed in Chapter 20 and may be viewed as heuristics to solve a large class of problems.

The work of Lin and Kernighan [27] sparked the study of modern heuristics, which have evolved and are now called *metaheuristics*. The term metaheuristics was coined by Glover [55] in 1986 and in general means "to find beyond in an upper level." The most popular metaheuristics include: TS, SA, ACO, EC, iterated local search (ILC), MAs, plus many others that keep up popping-up every year. One of the motivations for the study of metaheuristics is that it was recognized early on that constant ratio polynomial time approximation algorithms are not likely to exist for a large class of practical problems [16]. Metaheuristics do not guarantee that near-optimal solutions will be found quickly for all problem instances. However, these complex programs do find near-optimal solutions for many problem instances that arise in practice. These procedures have wide range of applicability, which is their most appealing aspect.

There are many ways of viewing metaheuristics. Some are single point while others are population based. In the former case one solution is modified over and over again until the algorithm terminates. Whereas in the latter case a set of solutions is carried throughout the execution of the algorithm. Some metaheuristics have a fixed neighborhood where moves can be made, whereas others have variable neighborhoods throughout the execution of the procedure. Metaheuristic algorithms may use memory to influence future moves and some are memoryless. Metaheuristics may be nature-inspired or algorithmic based. But no matter how they work, they have all been used successfully to solve many practical problems. In what follows we discuss several metaheuristics.

The term *tabu search* was coined by Glover [55]. TS is based on *adaptive memory* and *responsive exploration*. The former allows for an effective and efficient search of the solution space. The latter is used to guide the search process by imposing restraints and inducements based on the information collected. Intensification and diversification are controlled by the information collected, rather than by a random process. Chapter 21 discusses many different aspects of TS as well as problems to which it has been applied. Most recently, applications in the field of quantum computing as well an open-source hybrid quantum solver for D-wave systems have emerged. These developments have placed TS at the forefront of the field of quantum computing.

In the early 1980s Kirkpatrick et al. [56] and independently Černý [57] introduced SA as a randomized local search algorithm to solve combinatorial optimization problems. SA is a local search algorithm, which means that it starts with an initial solution and then searches through the solution space by iteratively generating a new solution that is "near" to it. But, sometimes the moves are to a worse solution to escape local optimal solutions. This method is based on statistical mechanics (Metropolis algorithm). It was heavily inspired by an analogy between the physical annealing process of solids and the problem of solving large combinatorial optimization problems. Chapter 25 in the 1st edition of this handbook discusses this approach in detail.

EC is a metaphor for building, applying, and studying algorithms based on Darwinian principles of natural selection. Algorithms that are based on evolutionary principles are called EAs. They are inspired by nature's capability to evolve living beings well adapted to their environment. There have been a variety of slightly different EAs proposed over the years. Three different strands of EAs were developed independently of each other over time. These are *evolutionary programming* (*EP*) introduced by Fogel [58] and Fogel et al. [59], *evolutionary strategies* (*ESs*) proposed by Rechenberg [60], and *genetic algorithms* (*GAs*) initiated by Holland [61]. GAs are mainly applied to solve discrete problems. *Genetic programming* (*GP*)

---

* https://www.tensorflow.org/ and https://aiexperiments.withgoogle.com

and *scatter search* (*SS*) are more recent members of the EA family. EAs can be understood from a unified point of view with respect to their main components and the way they explore the search space. EC is discussed in Chapter 22.

Chapter 23 presents an overview of ACO—a metaheuristic inspired by the behavior of real ants. ACO was proposed by Dorigo et al. [62] in the early 1990s as a method for solving hard combinatorial optimization problems. ACO algorithms may be considered to be part of *swarm intelligence*, the research field that studies algorithms inspired by the observation of the behavior of *swarms*. Swarm intelligence algorithms are made up of simple individuals that cooperate through self-organization.

MAs were introduced by Moscato [63] in the late 1980s to denote a family of metaheuristics, which can be characterized as the hybridization of different algorithmic approaches for a given problem. It is a population-based approach in which a set of cooperating and competing agents are engaged in periods of individual improvement of the solutions while they sporadically interact. An important component is *problem and instance-dependent knowledge*, which is used to speed-up the search process. A complete description is given in Chapter 27 of the 1st edition of this handbook.

1.2.2 Multiobjective Optimization, Reoptimization, Sensitivity Analysis, and Stability

Chapter 24 discusses *multiobjective combinatorial optimization*. This is important in practice since quite often a decision is rarely made with only one criterion. There are many examples of such applications in the areas of transportation, communication, biology, finance, and also computer science. Chapter 24 covers stochastic local search algorithms for multiobjective optimization problems.

Chapter 25 discusses *reoptimization* which tries to address the question: Given an optimal or nearly optimal solution to some instance of an NP-hard optimization problem and a small local change is applied to the instance, can we use the knowledge of the old solution to facilitate computing a reasonable solution for the new locally modified instance? As pointed out in Chapter 25, we should not expect major results for optimal solutions to NP-hard problems, but there are some interesting results for approximations. Sensitivity analysis is the dual problem, meaning that given an optimal or suboptimal solution, find all the set of related instances for which the solution remains optimal or near-optimal.

Chapter 26 covers *sensitivity analysis*, which has been around for more than 40 years. The aim is to study how variations affect the optimal solution value. In particular, parametric analysis studies problems whose structure is fixed, but where cost coefficients vary continuously as a function of one or more parameters. This is important when selecting the model parameters in optimization problems. On the other hand, Chapter 27 considers a newer area which is called *stability*. By this we mean how the complexity of a problem depends on a parameter whose variation alters the space of allowable instances.

1.2.3 Traditional Applications

We have used the label "traditional applications" to refer to more established combinatorial optimization problems. Some of these application can be categorized differently and vice-versa. The problems studied in this part of the handbook fall into the following categories: bin packing, packing, traveling salesperson, Steiner tree, scheduling, planning, generalized assignment, linear ordering, and submodular functions maximization. Let us briefly discuss these categories.

One of the fundamental problems in approximations is the bin packing problem. Chapter 28 discusses online and offline algorithms for one-dimensional bin packing. Chapters 29 and 30 discuss variants of the bin packing problem. This include variations that fall into the following type of problems: the number of items packed is maximized while keeping the number of bins fixed; there is a bound on the number of items that can be packed in each bin; dynamic bin packing, where each item has an arrival and departure time; the item sizes are not known, but the ordering of the weights is known; items may be fragmented while packing them into fixed capacity bins, but certain items cannot be assigned to the same bin; bin stretching; variable sized bin packing problem; the bin covering problem; black and white bin packing; bin packing with rejection; batched bin packing; maximal resource bin packing; and bin packing with fragile items.

Chapter 31 discusses several ways to generalize the bin packing problem to more dimensions. Two- and three-dimensional strip packing, bin packing in dimensions two and higher, vector packing, and several other variations are discussed. Cutting and packing problems with important applications in the wood, glass, steel and leather industries, as well as in LSI and VLSI design, newspaper paging, and container and truck loading are discussed in Chapter 32. For several decades, cutting and packing problems have attracted the attention of researchers in various areas including operations research, computer science, and manufacturing. Chapter 33 survey heuristics, metaheuristics, and exact algorithms for two-dimensional packing of general shapes. These problems have many practical applications in various industries such as the garment, shoe, and shipbuilding industries and many variants have been considered in the literature.

Very interesting approximation algorithms for the prize collecting traveling salesperson problem is studied in Chapter 34. In this problem a salesperson has to collect a certain amount of prizes (the quota) by visiting cities. A known prize can be collected in every city. Chapter 35 discusses branch and bound algorithms for the TSP problem. These algorithms have been implemented to run in a multicomputer environment. A general software tool for running branch and bound algorithms in a distributed environment is discussed. This framework may be used for almost any divide-and-conquer computation. With minor adjustments this tool can take any algorithm defined as a computation over directed acyclic graph, where the nodes refer to computations and the edges specify a precedence relation between computations, and run in a distributed environment.

Approximation algorithms for the Steiner tree problem are discussed in Chapter 36. This problem has applications in several research areas. One of this area is VLSI physical design. In Chapter 37 practical approximations for a restricted Steiner tree problem are discussed.

Chapter 38 surveys problems at the intersection of two scientific fields: graph theory and scheduling. These problems can either be viewed as scheduling dependent jobs where jobs have resource requirements, or as graph coloring minimization involving different objective functions. Applications include: wire minimization in VLSI design, minimizing the distance traveled by a robot moving in a warehouse, session scheduling on a path, and resource constrained scheduling.

Automated planning consists of finding a sequence of actions that transforms an initial state into one of the goal states. Planning is widely applicable and has been used in such diverse application domains as spacecraft control, planetary rover operations, automated nursing aides, image processing, business process generation, computer security, and automated manufacturing. Chapter 39 discusses approximation algorithms and heuristics for problems falling into this category.

Chapter 40 presents heuristics and metaheuristics for the generalized assignment problem. This problem is a natural generalization of combinatorial optimization problems including bipartite matching, knapsack and bin packing problems, and has many important applications in flexible manufacturing systems, facility location, and vehicle routing problems. Computational evaluation of the different procedures is discussed.

The linear ordering problem is discussed in Chapter 41. Versions of this problem were initially studied back in 1938 and 1958 in Economics. Exact algorithms, constructive heuristics, local search algorithms, and metaheuristics as well as computational results are discussed extensively in this chapter.

Chapter 42 discusses approximation algorithms and metaheuristics for submodular function maximization. These problems play a major role in combinatorial optimization. A few examples of these functions include: cut functions of graphs and hypergraphs, rank functions of matroids, and covering functions.

1.2.4 Computational Geometry and Graph Applications

The problems falling into this category have applications in several fields of study, but can be viewed as computational geometry and graph problems. The problems studied in this part of the handbook fall into the following categories: connectivity problems, design and evaluation of geometric networks, pair decomposition, covering with unit balls, minimum edge length partitions, automatic placement of labels in maps and drawings, finding corridors, clustering, maximum planar subgraphs, disjoint path problems, k-connected subgraph problems, node connectivity in survivable network problems, optimum communication spanning trees, activation network design problems, graph coloring, algorithms for a special type of graphs, and facility dispersion.

Volume 2, Chapter 2 examines approximation schemes for various geometric minimum-cost k-connectivity problems and for geometric survivability problems, giving a detailed tutorial of the novel techniques developed for these algorithms.

Geometric networks arise in many applications. Road networks, railway networks, telecommunication, pattern matching, bioinformatics—any collection of objects in space that have some connections between them can be modeled as a geometric network. Volume 2, Chapter 3 considers the problem of designing a "good" network and the dual problem, that is, evaluating how "good" a given network is. Volume 2, Chapter 4 presents an overview of several proximity problems that can be solved efficiently using the well-separated pair decomposition (WSPD). A WSPD may be regarded as a "small" set edges that approximates the dense complete Euclidean graph.

Volume 2, Chapter 5 surveys approximation algorithms for covering problems with unit balls. This problem has many applications including: finding locations for emergency facilities, placing wireless sensors or antennas to cover targets, and image processing. Approximation algorithms for minimum edge length partitions of rectangles with interior points are discussed in Volume 2, Chapter 6. This problem has applications in the area of Computer-Aided Design (CAD) of integrated circuits and systems.

Automatic placement of labels in maps and drawings is discussed in Volume 2, Chapter 7. These problems have applications in information visualization, cartography, geographic information systems, graph drawing, and so on. The chapter discusses different methodologies that have been used to provide solutions to these important problems. Volume 2, Chapter 8 discusses approximation algorithms for finding corridors in rectangular partitions. These problems have applications in VLSI, finding corridors in floorplans, and so on. Approximation algorithms and heuristics are discussed in this chapter and results of empirical evaluations are presented.

Clustering is a very important problem that has a long list of applications. Classical algorithms for clustering are discussed and analyzed in Volume 2, Chapter 9. These problems include k-median, k-center, and k-means problems in metric and Euclidean space.

Volume 2, Chapter 10 discusses the problem of finding a planar subgraph of maximum weight in a given graph. Problems of this form have applications in circuit layout, facility layout, and graph drawing. Finding disjoint paths in graphs is a problem that has attracted considerable attention from at least three perspectives: graph theory, VLSI design, and network routing/flow. The corresponding literature is extensive. Volume 2, Chapter 11 explores offline approximation algorithms for problems on general graphs as influenced from the network flow perspective.

Volume 2, Chapter 12 discusses approximation algorithms and methodologies for the k-connected subgraph problem. The problems discussed include directed and undirected graphs, as well as general, metric, and special weight functions. A survey of approximation algorithms and the hardness of approximations for survivable networks problems are discussed in Volume 2, Chapter 13. These problems include the minimum cost spanning tree, traveling salesperson, Steiner tree, Steiner forest, and their directed variants.

Besides numerous network design applications, spanning trees also play important roles in several newly established research areas, such as biological sequence alignments and evolutionary tree construction. Volume 2, Chapter 14 explores the problem of designing approximation algorithms for spanning tree problems under different objective functions. It focuses on approximation algorithms for constructing efficient communication spanning trees.

Volume 2, Chapter 15 discusses the activation network design problem where the goal is to select a "cheap" graph that satisfies some property G, meaning that the graph belongs to a family G of subgraphs of a given graph G. Many properties can be characterized by degree demands or pairwise connectivity demands.

Stochastic local search algorithms for the classical graph coloring problem are discussed in Volume 2, Chapter 16. This problem arises in many real-life applications such as register allocation, air traffic flow management, frequency assignment, light wavelengths assignment in optical networks, or timetabling. Volume 2, Chapter 17 discusses ACO for solving the maximum disjoint paths problems. This problem has many applications including the establishment of routes for connection requests between physically separated network endpoints.

Volume 2, Chapter 18 discusses efficient approximation algorithms for classical problems defined over random intersection graphs. These problems are inapproximable ones when defined over arbitrary graphs.

Facility dispersion problems are covered in Volume 2, Chapter 19. Dispersion problems arise in a number of applications, such as locating obnoxious facilities, choosing sites for business franchises, and selecting dissimilar solutions in multiobjective optimization. The facility location problem that model the placement of "desirable" facilities such as warehouses, hospitals, and fire stations, is discussed in Chapter 39 in the 1st edition of this handbook. This chapter covers approximation algorithms referred to as "dual fitting and factor revealing."

1.2.5 Large-Scale and Emerging Applications

The problems arise in the areas of wireless and sensor networks, multicasting, topology control, multimedia, data broadcast and aggregation, data analysis, computational biology, alignment problems, human genomics, VLSI placement, wavelets and streams, color quantization, digital reputation, influence maximization, and community detection. These may be referred to as "emerging" applications and normally involve large-scale problems instances. Some of these problems also fall in the other application areas.

Volume 2, Chapter 20 describes existing multicast routing protocols for ad hoc and sensor networks, and analyzes the issue of computing minimum cost multicast trees. The multicast routing problem, and approximation algorithms for mobile ad hoc networks (MANETs) and wireless sensor networks (WSNs) are presented. These algorithms offer better performance than Steiner trees.

Since flat networks do not scale, it is important to overlay a virtual infrastructure on a physical network. The design of the virtual infrastructure should be general enough so that it can be leveraged by a multitude of different protocols. Volume 2, Chapter 21 proposes a novel clustering scheme based on a number of properties of diameter-2 graphs. Extensive simulation results have shown the effectiveness of the clustering scheme when compared to other schemes proposed in the literature.

Ad hoc networks are formed by collections of nodes which communicate with each other through radio propagation. Topology control problems in such networks deal with the assignment of power values to the nodes so that the power assignment leads to a graph topology satisfying some specified properties. The problem is to minimize a specified function of the powers assigned to the nodes. Volume 2, Chapter 22 discusses some known approximation algorithms for this type of problems. The focus is on approximation algorithms with proven performance guarantees.

Recent progress in audio, video, and data storage technologies has given rise to a host of high-bandwidth real-time applications such as video conferencing. These applications require quality of service (QoS) guarantees from the underlying networks. Thus, multicast routing algorithms, which manage network resources efficiently and satisfy the QoS requirements, have come under increased scrutiny in recent years. Volume 2, Chapter 23 considers the problem of finding an optimal multicast tree with certain special characteristics. This problem is a generalization of the classical Steiner tree problem.

Scalability is especially critical for peer-to-peer systems. The basic idea of peer-to-peer systems is to have an open self-organizing system of peers that does not rely on any central server and where peers can join and leave at will. This has the benefit that individuals can cooperate without fees or an investment in additional high-performance hardware. Also, peer-to-peer systems can make use of the tremendous amount of resources (such as computation and storage) that otherwise sit idle on individual computers when they are not in use by their owners. Volume 2, Chapter 24 seeks ways of implementing join, leave, and route operations so that for any sequence of join, leave, and route requests can be executed quickly; the degree, diameter, and stretch factor of the resulting network are as small as possible; and the *expansion* of the resulting network is as large as possible. This is a multiobjective optimization problems for which they try to find good approximate solutions.

Scheduling problems modeling the broadcasting of data items over wireless channels are discussed in Volume 2, Chapter 25. The chapter covers exact and heuristic solutions for different versions of this problem.

Sensor networks are deployed to monitor a seemingly list of events in a wide rage of applications domains. By performing data analysis many patterns can be identified in sensor networks. Data analysis is based on data aggregation that must be performed at every node, but this is complicated by the fact that sensors deteriorate, often dramatically, over time. Volume 2, Chapter 26 discusses different strategies for aggregation time-discounted information in sensor networks. A related problem is studied in Volume 2, Chapter 27 where a limited number of storage nodes need to be placed in a wireless sensor network. In this chapter approximation and exact algorithm for this problem are discussed.

Volume 2, Chapter 28 considers two problems from computational biology, namely, primer selection and planted motif search. The closest string and the closest substring problems are closely related to the planted motif search problem. Representative approximation algorithms for these problems are discussed.

There are interesting algorithmic issues that arise when length constraints are taken into account in the formulation of a variety of problems on string similarity, particularly in the problems related to local alignment. Volume 2, Chapter 29 discusses these types of problems, which have their roots and most striking applications in computational biology. Volume 2, Chapter 30 discusses approximation algorithms for the selection of robust tag single nucleotide polymorphisms (SNPs). This is a problem in human genomics that arises in the current experimental environment.

VLSI has produced some of the largest combinatorial optimization problems ever considered. Placement is one of the most difficult of these. Placement problems with over 10 million variables and constraints are not unusual, and problem sizes continue to grow. Realistic objectives and constraints for placement incorporate complex models of signal timing, power consumption, wiring routability, manufacturability, noise, temperature, and so on. Volume 2, Chapter 31 considers VLSI placement algorithms.

Over the last decade the size of data seen by a computational problem have grown immensely. There appears to be more web pages than human beings, and web pages have been successfully indexed. Routers generate huge traffic logs, in the order of terabytes, in a short time. The same explosion of data is felt in observational sciences because our capabilities of measurement have grown significantly. Volume 2, Chapter 32 considers a processing mode where inputs item are not explicitly stored and the algorithm just passes over the data once.

Volume 2, Chapter 33 considers the problem of approximating "colors." Several algorithmic methodologies are presented and evaluated experimentally. These algorithms include some clustering approximation algorithms with different weights for the three dimensions.

Volume 2, Chapter 34 discusses a glowworm swarm optimization (GSO) algorithm for multimodal function optimization. This is a metaheuristic that is used in this chapter for odor source localization in an artificial olfaction system. This has applications to detect toxic gas leaks, fire origins of forest fires, leak point determination in pressurized systems, chemical discharge in water bodies, detection of mines and explosives, and so on.

A virtual community can be defined as a group of people sharing a common interest or goal who interact over a virtual medium, most commonly the Internet. Virtual communities are characterized by an absence of face-to-face interaction between participants which makes the task of measuring the trustworthiness of other participants harder than in non-virtual communities. This is due to the anonymity provided by the Internet, coupled with the loss of audiovisual cues that help in the establishment of trust. As a result, digital reputation management systems are an invaluable tool for measuring trust in virtual communities. Volume 2, Chapter 35 discusses various system which can be used to generate a good solution to this problem.

Volume 2, Chapter 36 continues with online social networks. This chapter discusses the use of social networks to detect the nodes that most influence the network. This has important applications, especially for advertising. Volume 2, Chapter 37 discusses community detection in online social networks. This problem is in some sense a clustering problem. Extracting the community structure and leveraging it to predict patterns in a dynamic network is an extremely important problem.

1.3 Definitions and Notation

One can use many different criteria to judge approximation algorithms and heuristics. For example, we could use the quality of the solution generated, and the time and space complexity needed to generate it. One may measure the criteria in different ways, for example, we could use the worst case, average case, median case, and so on. The evaluation could be analytical or experimental. Additional criteria include the following: characterization of data sets where the algorithm performs very well or very poorly; comparison with other algorithms using benchmarks or data sets arising in practice; tightness of bounds (quality of solution, time and space complexity); and the value of the constants associated with the time complexity bound including the ones for the lower order terms. For some researchers the most important aspect of an approximation algorithm is that it is complex to analyze, but for others it is more important that the algorithm be complex and involve the use of sophisticated data structures. For researchers working on problems directly applicable to the "real world," experimental evaluation, or evaluation on benchmarks is the more important criterion. Clearly, there is a wide variety of criteria one can use to evaluate approximation algorithms. The chapters in this handbook discuss different criteria to evaluate approximation algorithms.

For any given optimization problem P, let A_1, A_2, ... be the set of current algorithms that generate a feasible solution for each instance of problem P. Suppose that we select a set of criteria C and a way to measure it that we feel is the most important. How can we decide which algorithm is best for problem P with respect to C? We may visualize every algorithm as a point in multidimensional space. Now the approach used to compare feasible solutions for multiobjective function problems (see Chapter 24) can also be used in this case to label some of the algorithms as current Pareto optimal with respect to C. Algorithm A is said to be *dominated* by algorithm B with respect to C, if for each criteria $c \in C$ algorithm B is "not worse" than A, and for at least one criteria $c \in C$ algorithm B is "better" than A. An algorithm is said to be a *current Pareto optimal* algorithm with respect to C if none of the current algorithms dominates it.

In the next subsections we define time and space complexity, NP-completeness, and different ways to measure the quality of the solutions generated by the algorithms.

1.3.1 Time and Space Complexity

There are many different ways one can use to judge algorithms. The main ones we use are the time and space required to solve the problem. This is normally expressed in terms of n, the input size. It can be evaluated empirically or analytically. For the analytic evaluation we use the time and space complexity of the algorithm. Informally, this is a way to express the time the algorithm takes to solve a problem of size n and the amount of space needed to run the algorithm.

It is clear that almost all algorithms take different time to execute with different data sets even when the input size is the same. If you code it and run it on a computer you will see more variation depending on the different hardware and software installed in the system. It is impossible to characterize exactly the time and space required by an algorithm. We need a short cut. The approach that has been taken is to count the number of "operations" performed by the algorithm in terms of the input size. "Operations" is not an exact term and refers to a set of "instructions" which is independent of the problem size being solved. Then we just need to count the total number of operations.

Counting the number of operations exactly, is very complex for a large number of algorithms. So we just take into consideration the "highest" order term. This is the O notation.

> *Big oh Notation:* A (positive) function $f(n)$ is said to be $O(g(n))$ if there exist two constants $c \geq 1$ and $n_0 \geq 1$ such that $f(n) \leq c \cdot g(n)$ for all $n \geq n_0$.

The function $g(n)$ must include at least one term that is as large as the highest order term. For example, if $f(n) = n^3 + 20n^2$, then $g(n) = n^3$. Setting $n_0 = 1$ and $c = 21$ shows that $f(n)$ is $O(n^3)$. Note that $f(n)$ is also $O(n^4)$, but we like $g(n)$ to be the function with the smallest possible growth for $f(n)$. The function $f(n)$ cannot be $O(n^2)$ because it is impossible to find constants c and n_0 such that $n^3 + 20n^2 \leq cn^2$ for all $n \geq n_0$.

Time and space complexity are normally expressed using the O notation and describes the growth rate of the algorithm in terms of the problem size. Normally, the problem size is the number of vertices and edges in a graph, or the number of tasks and machines in a scheduling problem, and so on. But it can also be the number of bits used to represent the input.

When comparing two algorithms expressed in O notation we have to be careful because the constants c and n_0 are hidden. For large n, the algorithm with the smallest growth rate is the better one, but that might only hold for huge values of n. When two algorithms have similar constants c and n_0, the algorithm with the smallest growth function has a smaller running time. The book [2] discusses in detail the O notation as well as other notation.

1.3.2 NP-Completeness

Before the 1970s, researchers were aware that some problems could be computationally solved by algorithms of (low-order) polynomial time complexity ($O(n)$, $O(n^2)$, $O(n^3)$, etc.), whereas other problems had exponential time complexity, for example $O(2^n)$ and $O(n!)$. It was clear that even for small values of n, exponential time complexity equates to computational intractability if the algorithm actually performs an exponential number of operations for some inputs. The convention of computational tractability being equated to polynomial time complexity does not really fit well in practice, as an algorithm with time complexity $O(n^{100})$ is not really tractable if it actually performs n^{100} operations. But even under this relaxation of "tractability" there is a large class of problems that do not seem to have computational tractable algorithms for their solution.

We have been discussing optimization problems. But NP-completeness is defined for decision problems. A decision problem is simply one whose answer is "yes" or "no." The scheduling on identical machines problems discussed earlier is an optimization problem. Its corresponding decision problem has its input augmented by an integer value B and the yes–no question is to determine whether or not there is a schedule with makespan at most B. Every optimization problem has a corresponding decision problem. Since the solution of an optimization problem can be used directly to solve the decision problem, we say that the optimization problem is at least as hard to solve as the decision problem. If we show that the decision problem is a computationally intractable problem, then the corresponding optimization problem is also intractable.

The development of NP-completeness theory in the early 1970s by Cook [6], Karp [7], and others formally introduced the notion that there is a large class of decision problems that are computationally equivalent. By this we mean that either every problem in this class has a polynomial time algorithm that solves it, or none of them do. Furthermore, this question is the same as the $P = NP$ question, an open problem in computational complexity. This question is to determine whether or not the set of languages recognized in polynomial time by deterministic Turing machines is the same as the set of languages recognized in polynomial time by nondeterministic Turing machines. The conjecture has been that $P \neq NP$, and thus these problems would not have polynomial time algorithms for their solution. The decision problems in this class of problems are called *NP-complete* problems. Optimization problems whose corresponding decision problem is NP-complete are called *NP-hard* problems.

Scheduling tasks on identical machines is an NP-hard problem. The TSP and Steiner tree problem are also NP-hard problems. The minimum weight spanning tree problem can be solved in polynomial and it is not an NP-hard problem, under the assumption that $P \neq NP$. There is a long list of practical problems arising in many different fields of study that are known to be NP-hard problems. In fact almost all the optimization problems discussed in this handbook are NP-hard problems. The book [8] is an excellent source for information about NP-complete and NP-hard problems.

One establishes that a problem Q is an NP-complete problem by showing that the problem is in NP and giving a polynomial time transformation from an NP-complete problem to the problem Q.

A problem is said to be in NP if one can show that a yes answer to it can be verified in polynomial time. For the scheduling problem defined earlier you may think of this as providing a procedure that given any instance of the problem and an assignment of tasks to machines, the algorithm verifies in polynomial time, with respect to the problem instance size, if the assignment is a schedule and its makespan has length at most B. This is equivalent to the task a grader or teaching assistant (TA) performs when grading a question of the form "Does the following instance of the scheduling problem have a schedule with makespan at most 300? If so, give a schedule." Just verifying that the "answer" is correct is a simple problem. But solving a problem instance with 10,000 tasks and 20 machines seems much harder than simply grading it. In our over simplification it seems that $P \neq NP$. Polynomial time verification of an yes answer does not seem to imply polynomial time solvability. It is interesting to note that undergraduate students pay to take courses and exams, but the graders or TAs are paid to grade the exams, which seems to be a computationally simpler task! Though, professors just write the exam questions and are paid significantly more than the TAs. Recently, I noticed that emeriti professors do not even do that and they are still paid!

A polynomial time transformation from decision problem P_1 to decision problem P_2 is an algorithm that takes as input any instance I of problem P_1 and constructs an instance of $f(I)$ of P_2. The algorithm must take polynomial time with respect to the instance I. The transformation must be such that $f(I)$ is a yes-instance of P_2 if, and only if, I is a yes-instance of P_1.

The implication of a polynomial transformation $P_1 \alpha P_2$ is that if P_2 can be solved in polynomial time, then so can P_1, and if P_1 cannot be solved in polynomial time, then P_2 cannot be solved in polynomial time.

Consider the partition problem. We are given n items $1, 2, \ldots, n$. Item j has size $s(j)$. The problem is to determine whether or not the set of items can be partitioned into two sets such that the sum of the size of the items in one set equals the sum of the size of the items in the other set. Now let us polynomially transform the partition problem to the problem of scheduling tasks on identical machines. Given an instance I of partition, we define the instance $f(I)$ as follows. There are n tasks and $m = 2$ machines. Task i represents item i and its processing time is $s(i)$. All the tasks are independent and $B = \sum_{i=1}^{i=n} s(i)/2$. Clearly, $f(I)$ has schedule with maskespan B if, and only if, the instance I has a partition.

A decision problem is said to be *strongly NP-complete* if the problem is NP-complete even when all the "numbers" in the problem instance are less than or equal to $p(n)$, where p is a polynomial and n is the "size" of the problem instance. Partition is not NP-complete in the strong sense (under the assumption that $P \neq NP$) because there is a polynomial time dynamic programming algorithm for its solution (see Chapter 9). The book by Garey and Johnson [8] is an excellent source for NP-Completeness information.

Resolving the $P = NP$? question is one of the most important ones in the field of computer science. The Clay Institute of Mathematics[*] has offered a $1,000,000 reward for answering this question. The reward

---

[*] http://www.claymath.org/

will be given whether one proves that $P = NP$ or $P \neq NP$. No matter what, the person or people solving this problem for the first time will most likely be hired for life at the one of the most prestigious research labs or universities. However, if someone proves that $P = NP$ and the algorithm(s) to solve NP-complete problems are really efficient, then programming will be simplified considerably. To solve an NP-complete problem you just need to write a procedure to check a yes-answer. All the theoretical machinery developed so far will then be used to produce a fast algorithm to solve any NP-complete problem. If this ever happens, the best strategy would be to buy a huge farm and fill it with computers that will run implementations of such algorithms solving practical instances of NP-hard problems. Such person would become the first "trillionaire" and will be able to hire Bill to clean his/her windows, Mark to clean his/her face and books, Jeff to do his/her shopping, and so on. But, to be honest, the probability of this event is very close to zero, and most likely zero.

1.3.3 Performance Evaluation of Algorithms

The main criterion used to compare approximation algorithms has been the quality of the solution generated. Let us consider the different ways to compare the quality of the solutions generated when measuring the worst case. That is the criterion discussed in Section 1.2.

For some problems it is very hard to judge the quality of the solution generated. For example, approximating colors can only be judged by viewing the resulting images and that is subjective (see Vol. 2, Chap. 33). Volume 2, Chapter 35 covers digital reputation schemes. Here again it is difficult to judge the quality of the solution generated. Problems in the application areas of bioinformatics and VLSI CAD fall into this category because, in general, these are problems with multiobjective objective functions.

In what follows, we concentrate on problems where it is possible to judge the quality of the solution generated. At this point, we need to introduce additional notation. Let P be an optimization problem and A be an algorithm that generates a feasible solution for every instance I of problem P. We use $\hat{f}_A(I)$ to denote the objective function value of the solution generated by algorithm A for instance I. We drop A and use $\hat{f}(I)$ when it is clear which algorithm is being used. Let $f^*(I)$ be the objective function value of an optimal solution for instance I. Note that normally we do not know the value of $f^*(I)$ exactly, but we have bounds which should be as tight as possible.

Let G be an undirected graph that represents a set of cities (vertices) and roads (edges) between a pair of cities. Every edge has a positive number called the weight (or cost) and represents the cost of driving (gas plus tolls) between the pair of cities it joins. A *shortest path* from vertex s to vertex t in G is a path from s to t (st-path) such that the sum of the weight of the edges in it is the "least" possible among all possible st-paths. There are well-known algorithms that solve this shortest path problem in polynomial time [2]. Let A be an algorithm that generates a feasible solution (st-path) for every instance I of problem P. If for every instance I algorithm A generates an st-path such that

$$\hat{f}(I) \leq f^*(I) + c,$$

where c is some fixed constant, then A is said to be an *absolute* approximation algorithm for problem P with (additive) approximation bound c. Ideally, we would like to design a linear (or at least polynomial) time approximation algorithm with the smallest possible approximation bound. It is not difficult to see that this is not a good way of measuring the quality of a solution. Suppose that we have a graph G and we are running an absolute approximation algorithm for the shortest path problem concurrently in two different countries with the edge weight expressed in the local currency. Furthermore, assume that there is a large exchange rate between the two currencies. Any approximation algorithm solving the weak currency instance will have a much harder time finding a solution within the bound of c, than when solving the strong currency instance. We can take this to the extreme. We now claim that the above-mentioned absolute approximation algorithm A can be used to generate an optimal solution for any problem instance within the same time complexity bound.

The argument is simple. Given any instance I of the shortest path problem, we construct an instance I_{c+1} using the same graph, but every edge weight is multiplied by $c+1$. Clearly, $f^*(I_{c+1}) = (c+1)f^*(I)$. The st-path for I_{c+1} constructed by the algorithm is also an st-path in I and its weight is $\hat{f}(I) = \hat{f}(I_{c+1})/(c+1)$. Since $\hat{f}(I_{c+1}) \le f^*(I_{c+1}) + c$, then by substituting the above bounds we know that

$$\hat{f}(I) = \frac{\hat{f}(I_{c+1})}{c+1} \le \frac{f^*(I_{c+1})}{c+1} + \frac{c}{c+1} = f^*(I) + \frac{c}{c+1}.$$

Since all the edges have integer weights and $c/(c+1)$ is less than one, it follows that the algorithm finds an optimal solution to the problem. In other words, for the shortest path problem any algorithm that generates a solution with (additive) approximation bound c can be used to generate an optimal solution within the same time complexity bound. This same property can be established for almost all NP-hard optimization problems. Because of this the use of absolute approximation has never been given a serious consideration.

Sahni [14] defines as an ϵ-approximation algorithm for problem P an algorithm that generates a feasible solution for every problem instance I of P such that

$$\left| \frac{\hat{f}(I) - f^*(I)}{f^*(I)} \right| \le \epsilon.$$

It is assumed that $f^*(I) > 0$. For a minimization problem $\epsilon > 0$ and for a maximization problem $0 < \epsilon < 1$. In both cases ϵ represents the percentage of error. The algorithm is called an ϵ-approximation algorithm and the solution is said to be an ϵ-approximate solution. Graham's list scheduling algorithm [1] is a $1 - 1/n$ approximation algorithm, and the Sahni and Gonzalez [16] algorithm for the k-maxcut problem is a $\frac{1}{k}$-approximation algorithm (see Section 1.2). Note that this notation is different from the one discussed in Section 1.2. The difference is 1 unit, that is, the ϵ in this notation corresponds to $1 + \epsilon$ in the other.

Johnson [12] used a slightly different, but equivalent notation. He uses the approximation ratio ρ to mean that for every problem instance I of P the algorithm satisfies $\frac{\hat{f}(I)}{f^*(I)} \le \rho$ for minimization problems, and $\frac{f^*(I)}{\hat{f}(I)} \le \rho$ for maximization problems. The one for minimization problems is the same as the one given in Reference 1. The value for ρ is always greater than one, and the closer to one, the better the solution generated by the algorithm. One refers to ρ as the *approximation ratio* and the algorithm is a ρ-approximation algorithm. The list scheduling algorithm in the previous section is a $(2 - \frac{1}{m})$-approximation algorithm and the algorithm for the k-maxcut problem is a $(\frac{k}{k-1})$-approximation algorithm. Sometimes $1/\rho$ is used as the approximation ratio for maximization problems. Using this notation, the algorithm for the k-maxcut problem in the previous section is a $1 - \frac{1}{k}$-approximation algorithm.

All the above-mentioned forms are in use today. The most popular ones are ρ for minimization and $1/\rho$ for maximization. These are referred to as approximation ratios or approximation factors. We refer to all these algorithms as ϵ-approximation algorithms. The point to remember is that one needs to be aware of the differences and be alert when reading the literature. In the previous discussion we make ϵ and ρ look as if they are fixed constants. But, they can be made dependent on the size of the problem instance I. For example, it may be $\ln n$, or n^{ϵ} for some problems, where n is some parameter of the problem that depends on I, for example, the number of nodes in the input graph, and ϵ depends on the algorithm being used to generate the solutions. But it is most desirable that ϵ is a small constant.

Normally, one prefers an algorithm with a smaller approximation ratio. However, it is not always the case that an algorithm with a smaller approximation ratio always generates solutions closer to optimal than one with a larger approximation ratio. The main reason is that the notation is for the worst case ratio and the worst case does not always occur. But there are other reasons too. For example, the bound for the optimal solution value used in the analysis of two different algorithms may be different. Let P be the shortest path minimization problem and let A be an algorithm with approximation ratio 2. In this case we use d as the lower bound for $f^*(I)$, where d is some parameter of the problem instance. Algorithm B is a 1.5-approximation algorithm, but $f^*(I)$ used to establish it is the exact optimal solution value. Suppose that for problem instance I the value of d is 5 and $f^*(I) = 8$. Algorithm A will generate a path with weight at most 10, whereas algorithm B will generate one with weight at most $1.5 \times 8 = 12$. So the solution generated by Algorithm B may be worse than the one generated by A even if both algorithms generate the worst values for the instance. One can argue that the average "error" makes more sense than worst case. The problem is how to define and establish bounds for average "error." There are many other pitfalls when using worst case ratios. It is important to keep in mind all of this when making comparisons between algorithms. In practice one may run several different approximation algorithms concurrently and output the best of the solutions. This has the disadvantage that the running time of this compound algorithm will be the one for the slowest algorithm.

There are a few problems for which the worst case approximation ratio applies only to problem instances where the value of the optimal solution is small. One such problem is the bin packing problem discussed in Section 1.2. Informally, ρ_A^∞ is the smallest constant such that there exists a constant $K < \infty$ for which

$$\hat{f}(I) \le \rho_A^\infty f^*(I) + K.$$

The *asymptotic approximation ratio* is the multiplicative constant and it hides the additive constant K. This is most useful when K is small. Chapter 28 discusses this notation formally. The asymptotic notation is mainly used for bin packing and some of its variants.

Ausiello et al. [30] introduced the *differential ratio*. Informally, an algorithm is said to be a δ differential ratio approximation algorithm if for every instance I of P

$$\frac{\omega(I) - \hat{f}(I)}{\omega(I) - f^*(I)} \le \delta,$$

where $\omega(I)$ is the value of a worst solution for instance I. The worst case for the TSP over a complete graph is well defined, but for a scheduling problem, its value would be ∞ as one can keep on leaving machines idle. For problems when the worst solution is clear, differential ratio has some interesting properties for the complexity of the approximation problems. Chapter 15 discusses differential ratio approximation and its variations.

As said before there are many different criteria to compare algorithms. What if we use both the approximation ratio and time complexity? For example, the approximation algorithms in Reference 15 and the one in Reference 29 are current Pareto optimal with respect to this criteria for the TSP defined over metric graphs. Neither of the algorithms dominates the others in both time complexity and approximation ratio. The same can be said about the simple linear time (and very fast) approximation algorithm for the k-maxcut problem in Reference 16 and the complex one given in Reference 52 or the more recent ones that apply for all k.

The best algorithm to use also depends on the instance being solved. It makes a difference whether we are dealing with an instance of the TSP with optimal tour cost equal to one billion dollars and one with optimal cost equal to just a few pennies. Though, it also depends on the number of such instances being solved.

More elaborate approximation algorithms have been developed that generate a solution for any fixed constant ϵ. Formally, a *PTAS* for problem P is an algorithm A that given any fixed constant $\epsilon > 0$ it constructs a solution to every instance I problem P such that $|\frac{\hat{f}(I) - f^*(I)}{f^*(I)}| \le \epsilon$ in polynomial time with respect to the length of the instance I. Note that the time complexity may be exponential with respect to $1/\epsilon$. For example, the time complexity could be $O(n^{(1/\epsilon)})$ or $O(n + 4^{O(1/\epsilon)})$. Equivalent PTAS are also defined using different notation, for example based on $\frac{\hat{f}(I)}{f^*(I)} \le 1 + \epsilon$ for minimization problems.

One would like to design PTAS for all problems, but that is not possible unless $P = NP$. Clearly, with respect to approximation ration the PTAS is better than the ϵ-approximation algorithms for some ϵ. But their main draw back is that they are not practical because the time complexity is exponential on $1/\epsilon$. This does not preclude the existence of a practical PTAS for a "natural" occurring problem. However, a PTAS establishes that a problem can be approximated for all fixed constants. Different types of PTAS are discussed in Chapter 8. Additional PTAS are presented in Chapter 36 and Volume 2, Chapter 2.

A PTAS is said to be a FPTAS if its time complexity is polynomial with respect to n (the problem size) and $1/\epsilon$. For reasonable values of ϵ most FPTASs have practical running times. Different methodologies for designing FPTAS are discussed in Chapter 9.

Approximation schemes based on asymptotic approximation and on randomized algorithms have been developed. Chapter 10 discusses asymptotic approximation schemes and Chapter 11 discusses randomized approximation schemes.

References

1. Graham, R.L., Bounds for certain multiprocessing anomalies, *Bell Syst. Tech. J.*, 45, 1563, 1966.
2. Sahni, S., *Data Structures, Algorithms, and Applications in C++*, 2nd ed., Silicon Press, Summit, NJ, 2005.
3. Gilbert, E.N. and Pollak, H.O., Steiner minimal trees, *SIAM J. Appl. Math.*, 16(1), 1, 1968.
4. Graham, R.L., Bounds on multiprocessing timing anomalies, *SIAM J. Appl. Math.*, 17, 263, 1969.
5. Leung, J.Y.T., Ed., *Handbook of Scheduling: Algorithms, Models, and Performance Analysis*, Chapman & Hall/CRC, Boca Raton, FL, 2004.
6. Cook, S.A., The complexity of theorem-proving procedures, in *Proceedings of STOC'71*, 1971, p. 151.
7. Karp, R.M., Reducibility among combinatorial problems, in Miller, R.E. and Thatcher, J.W., Eds., *Complexity of Computer Computations*, Plenum Press, New York, 1972, p. 85.
8. Garey, M.R. and Johnson, D.S., *Computers and Intractability: A Guide to the Theory of NP-Completeness*, W. H. Freeman and Company, New York, 1979.
9. Garey, M.R., Graham, R.L., and Ullman, J.D., Worst-case analysis of memory allocation algorithms, in *Proceedings of STOC*, ACM, 1972, p. 143.
10. Johnson, D.S., *Near-Optimal Bin Packing Algorithms*, PhD Thesis, Massachusetts Institute of Technology, Department of Mathematics, Cambridge, MA, 1973.
11. Johnson, D.S., Fast algorithms for bin packing, *JCSS*, 8, 272, 1974.
12. Johnson, D.S., Approximation algorithms for combinatorial problems, *JCSS*, 9, 256, 1974.
13. Sahni, S., On the knapsack and other computationally related problems, PhD Thesis, Cornell University, 1973.
14. Sahni, S., Approximate algorithms for the 0/1 knapsack problem, *JACM*, 22(1), 115, 1975.
15. Rosenkrantz, R., Stearns, R., and Lewis, L., An analysis of several heuristics for the traveling salesman problem. *SIAM J. Comput.*, 6(3), 563, 1977.
16. Sahni, S. and Gonzalez, T., P-complete approximation problems, *JACM*, 23, 555, 1976.
17. Gens, G.V. and Levner, E., Complexity of approximation algorithms for combinatorial problems: A survey, *SIGACT News*, 12, 52, 1980.
18. Levner, E. and Gens, G.V., *Discrete Optimization Problems and Efficient Approximation Algorithms*, Central Economic and Mathematics Institute, Moscow, Russia, 1978 (in Russian).
19. Garey, M.R. and Johnson, D.S., The complexity of near-optimal graph coloring, *SIAM J. Comput.*, 4, 397, 1975.
20. Ibarra, O. and Kim, C., Fast approximation algorithms for the knapsack and sum of subset problems, *JACM*, 22(4), 463, 1975.
21. Sahni, S., Algorithms for scheduling independent tasks, *JACM*, 23(1), 116, 1976.

22. Horowitz, E. and Sahni, S., Exact and approximate algorithms for scheduling nonidentical processors, *JACM*, 23(2), 317, 1976.
23. Babat, L.G., Approximate computation of linear functions on vertices of the unit N-dimensional cube, in *Studies in Discrete Optimization,* Fridman, A.A., Ed., Nauka, Moscow, Russia, 1976 (in Russian).
24. Babat, L.G., A fixed-charge problem, *Izv. Akad. Nauk SSR, Techn, Kibernet.*, 3, 25, 1978 (in Russian).
25. Lawler, E., Fast approximation algorithms for knapsack problems, *Math. Oper. Res.*, 4, 339, 1979.
26. Garey, M.R. and Johnson, D.S., Strong NP-completeness results: Motivations, examples, and implications, *JACM*, 25, 499, 1978.
27. Lin, S. and Kernighan, B.W., An effective heuristic algorithm for the traveling salesman problem, *Oper. Res.*, 21(2), 498, 1973.
28. Papadimitriou, C.H. and Steiglitz, K., On the complexity of local search for the traveling salesman problem, *SIAM J. Comput.*, 6, 76, 1977.
29. Christofides, N., Worst-case analysis of a new heuristic for the traveling salesman problem. Technical Report 338 Grad School of Industrial Administration, CMU, 1976.
30. Ausiello, G., D'Atri, A., and Protasi, M., On the structure of combinatorial problems and structure preserving reductions, in *Proceedings of ICALP'77*, LNCS, 52, Springer-Verlag, 1977, p. 45.
31. Cornuejols, G., Fisher, M.L., and Nemhauser, G.L., Location of bank accounts to optimize float: An analytic study of exact and approximate algorithms, *Manag. Sci.*, 23(8), 789, 1977.
32. Khachiyan, L.G., A polynomial algorithms for the linear programming problem, *Dokl. Akad. Nauk SSSR*, 244(5), 1979 (in Russian).
33. Karmakar, N., A new polynomial-time algorithm for linear programming, *Combinatorica*, 4, 373, 1984.
34. Grötschel, M., Lovász, L., and Schrijver, A., The ellipsoid method and its consequences in combinatorial optimization, *Combinatorica*, 1, 169, 1981.
35. Schrijver, A., *Theory of Linear and Integer Programming*, Wiley-Interscience Series in Discrete Mathematics and Optimization, John Wiley, New York, 1998.
36. Vanderbei, R.J., *Linear Programming Foundations and Extensions,* Series: International Series in Operations Research & Management Science, No. 196, 4th Edition, Springer, Berlin, Germany, 2014.
37. Lovász, L., On the ratio of optimal integral and fractional covers, *Disc. Math.*, 13, 383, 1975.
38. Chvátal, V., A greedy heuristic for the set-covering problem, *Math. Oper. Res.*, 4(3), 233, 1979.
39. Hochbaum, D.S., Approximation algorithms for set covering and vertex covering problems, *SIAM J. Comput.*, 11, 555, 1982.
40. Bar-Yehuda, R. and Even, S., A linear time approximation algorithm for the weighted vertex cover problem, *J. Algorithms*, 2, 198, 1981.
41. Bar-Yehuda, R. and Even, S., A local-ratio theorem for approximating the weighted set cover problem, *Ann. Disc. Math.*, 25, 27, 1985.
42. Bar-Yehuda, R. and Bendel, K., Local ratio: A unified framework for approximation algorithms, *ACM Comput. Surv.*, 36(4), 422, 2004.
43. Raghavan, R. and Thompson, C., Randomized rounding: A technique for provably good algorithms and algorithmic proof, *Combinatorica*, 7, 365, 1987.
44. Fernandez de la Vega, W. and Lueker, G.S., Bin packing can be solved within $1 + \epsilon$ in linear time, *Combinatorica*, 1, 349, 1981.
45. Karmakar, N. and Karp, R.M., An efficient approximation scheme for the one-dimensional bin packing problem, in *Proceedings of FOCS*, 1982, p. 312.
46. Papadimitriou, C.H. and Yannakakis, M., Optimization, approximation and complexity classes, *J. Comput. Syst. Sci.*, 43, 425, 1991.

47. Arora, S., Lund, C., Motwani, R., Sudan, M., and Szegedy, M., Proof verification and hardness of approximation problems, in *Proceedings of FOCS*, 1992.
48. Feige, U., Goldwasser, S., Lovasz, L., Safra, S., and Szegedy, M., Interactive proofs and the hardness of approximating cliques, *JACM*, 43, 268, 1996.
49. Feige, U., A threshold of ln n for approximating set cover, *JACM*, 45(4), 634, 1998. (Prelim. version in STOC'96.)
50. Engebretsen, L. and Holmerin, J., Towards optimal lower bounds for clique and chromatic number, *TCS*, 299, 537, 2003.
51. Hastad, J., Some optimal inapproximability results, *JACM*, 48, 2001. (Prelim. version in STOC'97.)
52. Goemans, M.X. and Williamson, D.P., Improved approximation algorithms for maximum cut and satisfiability problems using semidefinite programming, *JACM*, 42(6), 1115, 1995.
53. Goemans, M.X. and Williamson, D.P., A general approximation technique for constrained forest problems, *SIAM J. Comput.*, 24(2), 296, 1995.
54. Jain, K., Mahdian, M., Markakis, E., Saberi, A., and Vazirani, V.V., Approximation algorithms for facility location via dual fitting with factor-revealing LP, *JACM*, 50, 795, 2003.
55. Glover, F., Future paths for integer programming and links to artificial intelligence, *Computers Oper. Res.*, 13, 533, 1986.
56. Kirkpatrick, S., Gelatt Jr. C.D., and Vecchi, M.P., Optimization by simulated annealing, *Science*, 220, 671, 1983.
57. Černý, V., Thermodynamical approach to the traveling salesman problem: An efficient simulation algorithm, *J. Optim. Theory Appl.*, 45, 41, 1985.
58. Fogel, L.J., Toward inductive inference automata, in *Proceedings of the International Federation for Information Processing Congress*, 1962, p. 395.
59. Fogel, L.J., Owens, A.J., and Walsh, M.J., *Artificial Intelligence through Simulated Evolution*, Wiley, New York, 1966.
60. Rechenberg, I., *Evolutionsstrategie: Optimierung technischer Systeme nach Prinzipien der biologischen Evolution*, Frommann-Holzboog, Stuttgart, Germany, 1973.
61. Holland, J.H., *Adaption in Natural and Artificial Systems*, The University of Michigan Press, Ann Harbor, MI, 1975.
62. Dorigo, M., Maniezzo, V., and Colorni, A., Positive feedback as a search strategy, Technical Report 91-016, Dipartimento di Elettronica, Politecnico di Milano, Italy, 1991.
63. Moscato, P., On genetic crossover operators for relative order preservation, C3P Report 778, California Institute of Technology, 1989.

I

Basic Methodologies

2

Basic Methodologies and Applications

Teofilo F. Gonzalez

2.1 Introduction

In Chapter 1, we present an overview of approximation algorithms and metaheuristics as well as an overview for both volumes of this handbook. In this chapter, we discuss in more detail the basic methodologies and apply them to classic problems. These methodologies include the four major ones: restriction (algorithmic and structural), relaxation (numerical and structural), rounding, and transformation. Greedy methods fall in the class of restriction methods (structural), linear programming (LP)-rounding fall under relaxation (numerical) and α-vectors, local ratio and primal-dual fall under problem transformation. We also discuss in more detail inapproximability and show that the "original" version of the traveling salesperson problem (TSP) is constant ratio inapproximable.

2.2 Restriction

Chapter 3 discusses *restriction*, which is one of the most basic techniques to design approximation algorithms. The idea is to generate a solution to a given problem P by providing an optimal or suboptimal solution to a subproblem of P. A subproblem of a problem P means restricting the solution space for P by disallowing a subset of the feasible solutions. Restriction may be structural or algorithmic. In the case of structural restriction, the idea is to impose some structure on the set of feasible solutions (the solution space), which can be exploited by an efficient algorithm that solves the problem optimally or suboptimally. In the case of algorithmic restriction, an algorithm restricts the set of feasible solutions that may be found. The problem in the restricted solution spaces is referred to as the *subproblem*. For this approach to be effective the subproblem must have the property that, for every problem instance, its optimal or suboptimal solution has an objective function value that is "close" to the optimal one for the original problem. The most common approach is to solve the subproblem once, but there are algorithms where more than one subproblem is solved and then the best of the solutions computed is the solution generated. Chapter 3 discusses this methodology and shows how to apply it to several problems.

Approximation algorithms based on this approach are discussed in Volume 1, Chapters 31, 32, 36, and Volume 2, Chapters 6 and 25. Let us now discuss a scheduling application in detail. This is the scheduling problem initially studied by Graham [1,2].

2.2.1 Scheduling

A set of n tasks denoted by $T_1, T_2, \ldots, T_n$ with processing time requirements $t_1, t_2, \ldots, t_n$ have to be processed by a set of m identical machines. A partial order C is defined over the set of tasks to enforce a set of precedence constraints or task dependencies. The partial order specifies that a machine cannot commence the processing of a task until all of its predecessors have been completed. Each task T_i has to be processed for t_i units of time by one of the machines. A (nonpreemptive) schedule is an assignment of tasks to time intervals on the machines in such a way that (1) each task T_i is processed continuously for t_i units of time by one of the machines; (2) each machine processes at most one task at a time; and (3) the precedence constraints are satisfied. The *makespan* of a schedule is the latest time at which a task is being processed. The scheduling problem discussed in this section is to construct a minimum makespan schedule for a set of partially order tasks to be processed by a set of identical machines. Several restricted versions of this scheduling problem have been shown to be NP-hard [3].

Example 2.1 *The number of tasks, n, is 8 and the number of machines, m, is 3. The processing time requirements for the tasks and the precedence constraints are given in Figure 2.1, where a directed graph is used to represent the task dependencies. Vertices represent tasks and the directed edges represent task dependencies. The integers next to the vertices represent the task processing requirements. Figure 2.2 depicts two schedules for this problem instance.*

In the next subsection, we present a simple algorithm based on restriction to generate provable good solutions to this scheduling problem. The solution space is restricted to schedules without forced "idle time,"

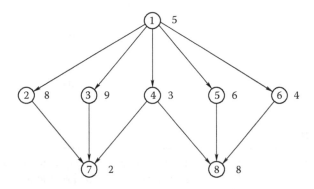

FIGURE 2.1   Precedence constraints and processing time requirements for Example 2.1.

	5		14	18	26
	1	3		6	8
	5	4			
	2		7		

(a)

	5	8		17	19
	1	4	3		7
	5		8		
	6		2		

(b)

FIGURE 2.2   Schedule (a) is a feasible schedule for Example 2.1. Schedule (b) is an optimal schedule for Example 2.1.

that is, each feasible schedule does not have idle time from the time at which all the predecessors of task T_i (in C) are completed to the time when the processing of task T_i begins, for all i. In other words, procrastination is not allowed. You may call this scheduling strategy "best effort" or "busy beaver."

2.2.2 Partially Ordered Tasks

Let us further restrict the scheduling policy to construct a schedule from time zero till all tasks have been assigned. The scheduling policy is: whenever a machine becomes idle we assign one of the unassigned tasks that is ready to commence execution (i.e., we have completed all the tasks' predecessors). Any scheduling policy in this category can be referred to as a *no-additional-delay* scheduling policy. The simplest version of this scheduling policy is to Assign Any of the Tasks (AATs) ready to be processed. A schedule generated by this policy is called an *AAT* schedule. These schedules are similar to the list schedule [1] discussed in Chapter 1. The difference is that list schedules have an ordered list of tasks which is used to break ties. The analysis for both types of algorithms is the same since the list could be any ordering of the tasks.

In Figure 2.2, we give two possible AAT schedules. The two schedules were obtained by breaking ties differently. The schedule in Figure 2.2b is a minimum makespan schedule. The reason for this is that the machines can only process one of the tasks T_1, T_5, and T_8 at a time, because of the precedence constraints and the makespan is $t_1 + t_5 + t_8$.

Figure 2.2 suggests that an optimal schedule can be generated by just finding a clever method to break ties. Unfortunately one cannot prove that this is always the case because there are problem instances for which all minimum makespan schedules are not AAT schedules, that is, in a minimum makespan schedule at least one machine procrastinates at some point in time. All the ATT schedules for the problem instance given in Figure 2.3 (the number inside the circles indicate the task index and the ones outside indicate the task time) are similar to the one given in Figure 2.4a and have makespan equal to 19. But a minimum makespan schedule has finish time 15 (Figure 2.4b).

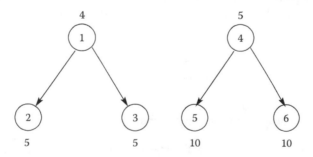

FIGURE 2.3 Problem instance.

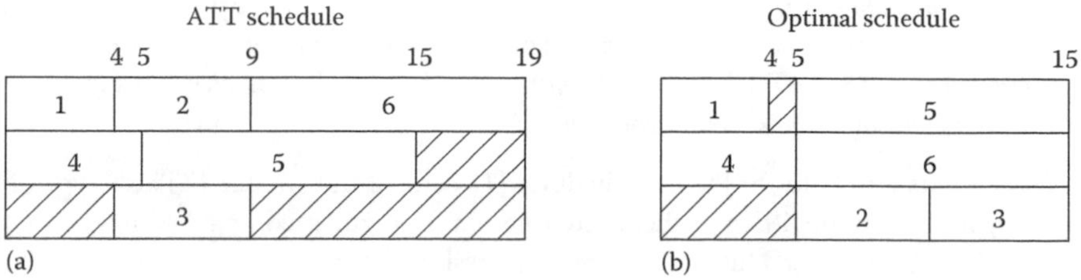

FIGURE 2.4 (a) AAT schedule. (b) Minimum makespan schedule.

The makespan time of an AAT schedule is never greater than $2 - \frac{1}{m}$ times the one of an optimal schedule for the instance. This is expressed by

$$\frac{\hat{f}_I}{f_I^*} \leq 2 - \frac{1}{m},$$

where:

$\hat{f}_I$ is the makespan of any possible AAT schedule for problem instance I

f_I^* is the makespan of an optimal schedule for I

We establish this property in the following theorem.

Theorem 2.1 *For every instance I of the identical machine scheduling problem and every AAT schedule,* $\frac{\hat{f}_I}{f_I^*} \leq 2 - \frac{1}{m}$.

Proof. Let S be any AAT schedule for problem instance I with makespan $\hat{f}_I$. By construction of the AAT schedules, it cannot be that at some time $0 \leq t \leq \hat{f}_I$ all machines are idle. In other words, at each point in time at least one task is being processed by a machine. Let i_1 be the index of a task that finishes at time $\hat{f}_I$. For $j = 2, 3, \ldots$, if task $T_{i_{j-1}}$ has at least one predecessor in C, then define i_j as the index of a task with latest finishing time that is a predecessor (in C) of task $T_{i_{j-1}}$. We refer to these tasks as a *chain* of tasks and let k be the number of tasks in the chain. By the definition of task T_{i_j} and AAT schedules, it cannot be that there is an idle machine from the time when task T_{i_j} completes its processing to the time when task $T_{i_{j-1}}$ begins processing. Therefore, a machine can only be idle when another machine is executing a task in the chain. From these two observations, we know that

$$m\hat{f}_I \leq (m-1) \sum_{j=1}^{k} t_{i_j} + \sum_{j=1}^{n} t_j$$

Since no machine can process more than one task at a time, and since not two tasks, one of which precedes the other in C, can be processed concurrently, we know that an optimal makespan schedule satisfies

$$f_I^* \geq \frac{1}{m} \sum_{j=1}^{n} t_j, \quad \text{and} \quad f_I^* \geq \sum_{j=1}^{k} t_{i_j}$$

Substituting in the above-mentioned inequality, we know that $\frac{\hat{f}_I}{f_I^*} \leq 2 - \frac{1}{m}$. ∎

The natural question to ask is whether or not the approximation ratio $2 - \frac{1}{m}$ is the best possible for AAT schedules. The answer to this question is affirmative, and a problem instance for which this bound is tight is given in Example 2.2.

Example 2.2 *There are $2m - 1$ independent tasks. The first $m - 1$ tasks have processing time requirement $m - 1$, the next $m - 1$ tasks have processing time requirement one, and the last task has processing time requirement equal to m. An AAT schedule with makespan time $2m - 1$ is given in Figure 2.5a, and in Figure 2.5b we give a minimum makespan schedule.*

Note that these results also hold for the list schedules [1] defined in Chapter 1. These type of schedules are generated by a no-additional-delay scheduling rule that is augmented by a list that is used to decide which of the ready to be processed tasks is the one to be assigned next.

Let us now consider the case when ties (among tasks that are ready) are broken in favor of the task with smallest index (T_i is selected before T_j if both tasks are ready to be processed and $i < j$). The problem

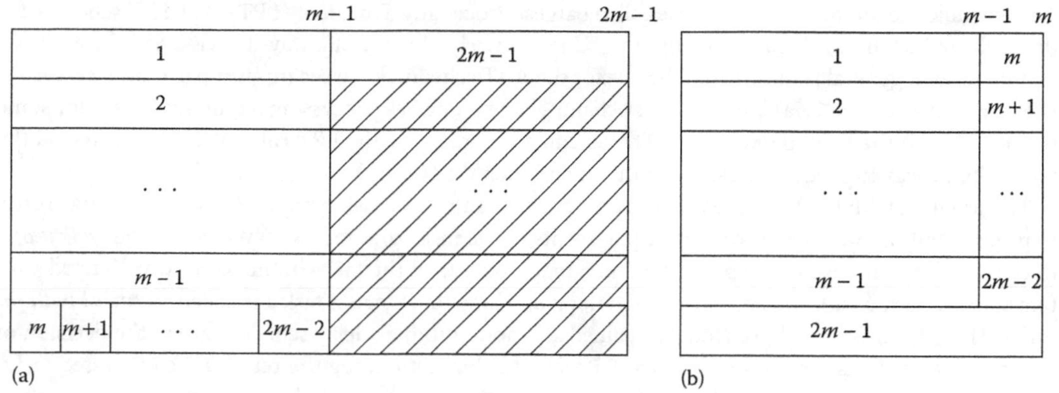

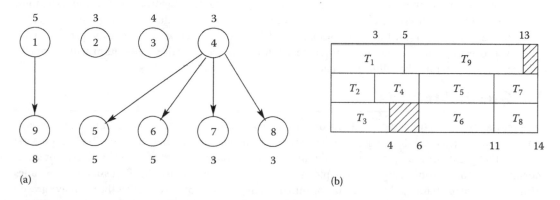

FIGURE 2.5   (a) AAT schedule. (b) Optimal schedule for Example 2.2.

FIGURE 2.6   (a) Problem instance with anomalous behavior. (b) AAT Schedule with tie breaking list.

instance I_A given in Figure 2.6 has three machines and eight tasks. Our scheduling procedure (augmented with the tie breaking rule) generates a schedule with makespan 14. In Chapter 1, we say that list schedules (which are this type of schedules) have anomalies, that is, the scheduling rule generates unexpected schedules for slightly different problem instances. To verify this, apply the scheduling algorithm to instance I_A, but now there are four machines instead of three. One would expect a schedule for this new instance to have makespan at most 14. But you can easily verify that is not the case. The makespan of the schedule with four machines is 16. Our expectation was based on the fact that $f_4^* \leq f_3^*$, where f_m^* is the minimum makespan of the problem instance when scheduling on m machines. But, our procedure does not always generates an optimal schedule. The procedure generates for the most part suboptimal schedules.

Now apply the scheduling algorithm to the instance I_A where every task has a processing requirement decreased by one unit. One would again expect a schedule for this new instance to have makespan at most 14, but you can easily verify that is not the case. Also, suppose we apply the scheduling algorithm to the problem instance I_A without the precedence constraints from task T_4 to task T_5 and task T_4 to task T_6. As before one would expect a schedule for this instance to have makespan at most 14, but that is not the case. These are anomalies. Approximation algorithms suffer from this type of anomalous behavior. We need to be aware of this fact when using approximation algorithms. Almost all of the time we use suboptimal algorithms, so we need to be aware of the fact that anomalies arise in this setting.

As in the case of Example 2.2, the worst case behavior arises when the task with the longest processing requirement is being processed while the rest of the machines are idle. Can a better approximation bound be established for the case when ties are broken in favor of a task with longest processing time?

The schedules generated by this rule are called Largest Processing Time first (LPT). Any LPT schedule for the problem instance in Figure 2.5 is optimal. Unfortunately, this is not always the case and the approximation ratio in general is the same as the one for the AAT schedules. To see this just partition task $2m - 1$ in Example 2.2 (Figure 2.5a) into a two-task chain. The first one has processing requirement of ϵ, for some $0 < \epsilon < 1$, and the second one $m - \epsilon$. The schedule generated by the LPT rule will schedule first all the tasks with processing requirement grater than 1 and then the two tasks in the chain.

The problem with the LPT rule is that it only considers the processing requirements of the tasks ready to process, but ignores the processing requirements of the tasks that follow it. We define the *weight of a directed path* as the sum of the processing time requirements of the tasks in the path. Any directed path that starts at task t with maximum weight among all paths that start at task t is called a *critical path for task t*. The critical-path (CP) schedule is defined as a no-additional-delay schedule where the decision of which task to process next is a task whose CP weight is longest among the ready to be processed tasks. The CP schedule is optimal for the problem instance that was generated by replacing the last task in Example 2.2 by two tasks. However, Graham constructed problem instances for which the makespan of the CP schedule is $2 - 1/m$ times the length of an optimal schedule.

It is not known whether or not a polynomial time algorithm exists with a smaller approximation ratio even when the processing time requirements for all the tasks are identical and $m \geq 3$. There is a polynomial time algorithm that generates an optimal schedule when $m = 2$, but the problem with different processing times is NP-hard. In the next subsection, we present an algorithm with a smaller approximation ratio for scheduling independent task.

2.3 Greedy Methods

Another traditional method to generate suboptimal solutions is to apply greedy algorithms. The idea is to generate a solution by making a sequence of irrevocable decisions. Each of these decisions is a best possible choice at that point, for example, select an edge of least weight, select the vertex of highest degree, or select the task with longest processing time. Chapter 4 discusses greedy methods. The discussion also includes primal-dual approximations algorithms falling into this category. Chapter 5 discusses the recursive greedy method. This methodology is for the case when making the best possible decision is an NP-hard problem. A large portion of the bin-packing algorithms are greedy algorithms. Bin packing and its variants are discussed in Chapters 28 through 31. Other greedy methods appear in Volume 1, Chapters 32, 41, and Volume 2, Chapters 8, 14, and 19. Let us now discuss the LPT scheduling rule for scheduling independent tasks on identical machines.

2.3.1 Independent Tasks

Another version of this scheduling problem that has received considerable attention is when the tasks are independent, that is, the partial order between the tasks is empty. Graham's [2] elegant analysis for LPT scheduling has become a classic. In fact, quite a few subsequent exact and approximation scheduling algorithms followed the same approach.

First, we analyze the LPT scheduling rule. For this case, there is only one possible schedule, modulo the relabeling of the tasks with identical processing time. We call this a "greedy method" because of the ordering of the tasks with respect to their processing requirements. This tends to generate schedules where the shortest task ends up being processed last, and the resulting schedule tends to have near-optimal makespan. However as we shall see, one may obtain the same approximation ratio by just scheduling the tasks using a list where the $2m$ task with longest processing times appear first (in sorted order) and the remaining tasks appear next in any order. This approach could be called "limited greedy." We discuss other approximation algorithms for this problem after presenting the analysis for LPT schedules.

Let I be any problem instance with n independent tasks and m identical machines. We use $\hat{f}_I$ to represent the makespan for the LPT schedule for I, and f_I^* as the one for an optimal schedule. In the next theorem, we establish the approximation ratio for LPT schedules.

Theorem 2.2 *For every scheduling problem instance I with n independent tasks and m identical machines, every LPT schedule satisfies* $\frac{\hat{f}_I}{f_I^*} \leq \frac{4}{3} - \frac{1}{3m}$.

Proof. It is clear that LPT schedules are optimal for $m = 1$. Assume that $m \geq 2$. The proof is by contradiction. Suppose the above bound does not hold. Let I be a problem instance with the least number of tasks for which $\frac{\hat{f}_I}{f_I^*} > \frac{4}{3} - \frac{1}{3m}$. Let n be the number of tasks in I, m be the number of machines and assume that $t_1 \geq t_2 \geq ... \geq t_n$. Let k be the smallest index of a task that finishes at time $\hat{f}_I$. It cannot be that $k < n$, as otherwise the problem instance $T_1, T_2, ..., T_k$ is also a counter example and it has fewer tasks than instance I, but by assumption problem instance I is a counter example with the least number of tasks. Therefore, k must be equal to n.

By the definition of LPT schedules, we know that there cannot be idle time before task T_n begins execution. Therefore,

$$\sum_{i=1}^{n} t_i + (m-1)t_n \geq m\hat{f}_I.$$

This is equivalent to

$$\hat{f}_I \leq \frac{1}{m} \sum_{i=1}^{n} t_i + (1 - \frac{1}{m})t_n.$$

Since each machine cannot process more than one task at a time, we know that $f_I^* \geq \sum_{i=1}^{n} t_i/m$. Combining these two bounds we have

$$\frac{\hat{f}_I}{f_I^*} \leq 1 + (1 - \frac{1}{m})\frac{t_n}{f_I^*}.$$

Since I is a counter example for the theorem, this bound must be greater that $\frac{4}{3} - \frac{1}{3m}$. Simplifying we know that $f_I^* < 3t_n$. Since t_n is the task with shortest processing time requirement it must be that in an optimal schedule for instance I, none of the machines can process three or more tasks. Therefore, the number of tasks n is at most $2m$.

For problem instance I, let S^* be an optimal schedule with least $\sum f_i^2$, where f_i is the makespan in S^* for machine i. Assume without loss of generality that the tasks assigned to each machine are arranged from largest to shortest with respect to their processing times. All machines have at most two tasks, as S^* is an optimal schedule for I which by definition is a counter example for the theorem.

Let i and j be two machines in schedule S^* such that $f_i > f_j$, machine i has two tasks and machine j has at least one task. Let a and b be the task index for the last task processed by machine i and j, respectively. It cannot be that $t_a > t_b$, as otherwise applying the interchange (Type I) given in Figure 2.7a results in an optimal schedule with smaller $\sum f_i^2$. This contradicts the fact that S^* is an optimal schedule with least $\sum f_i^2$. Let i and j be two machines in schedule S^* such that machine i has two tasks. Let a be the task index

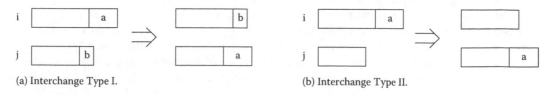

(a) Interchange Type I. (b) Interchange Type II.

FIGURE 2.7   Schedule transformations.

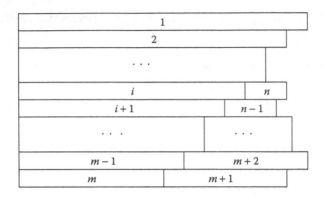

FIGURE 2.8   Optimal schedule.

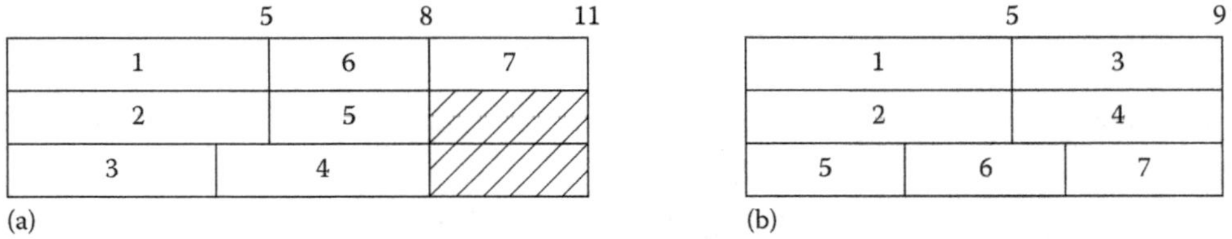

FIGURE 2.9   (a) LPT schedule. (b) Optimal schedule.

for the last task processed by machine i. It cannot be that $f_i - t_a > f_j$ as otherwise applying the interchange (Type II) given in Figure 2.7b results in an optimal schedule with smaller $\sum f_i^2$. This contradicts the fact that S^* is an optimal schedule with least $\sum f_i^2$.

Since the transformations given in Figures 2.7a and b cannot apply, the schedule S^* must be of the form shown in Figure 2.8 after renaming the machines, that is, machine i is assigned task T_i (if $i \leq n$) and task T_{2m-i+1} (if $2m - i + 1 \leq n$). But this schedule is an LPT schedule and $\hat{f} = f^*$. Therefore, there cannot be any counter examples to the theorem. This completes the proof of the theorem. ∎

For all m there are problem instances for which the ratio given by Theorem 2.2 is tight. In Figure 2.9, we give one of such problem instance for three machines.

The important properties needed to prove Theorem 2.2 are that the longest $2m$ tasks need to be scheduled via LPT, and either the schedule will be optimal for the $2m$ task or at least three tasks will be assigned to a machine. The first m tasks, the ones with longest processing time, will be assigned to one machine each, so the order in which they are assigned is not really important. The next set of m tasks need to be assigned from longest to shortest processing times as in the LPT schedule. The remaining tasks can be assigned in any order as long as whenever a machine finishes a task the next task in the list is assigned to that machine. Any list schedule whose list follows the above-mentioned ordering can be shown to have makespan at most $\frac{4}{3} - \frac{1}{3m}$ times the one of an optimal schedule. These type of schedules forms a restriction on the solution space.

It is interesting to note that the problem of scheduling $2m$ independent tasks is an NP-hard problem. However, in polynomial time we can find out if there is an optimal schedule in which each machine has at most two tasks. And this is all that is needed to establish the $\frac{4}{3} - \frac{1}{3m}$ approximation ratio. One of the first avenues of research explored was to see if the same approach would hold for the longest $3m$ tasks. That is, design a polynomial time algorithm that finds an optimal schedule in which each machine has at most three tasks. If such algorithm exists, we could use it to generate schedules that are within $\frac{5}{4} - \frac{1}{4m}$ times the makespan of an optimal schedule. This does not seem possible as Garey and Johnson [3] established that this problem is an NP-hard problem.

Other approximation algorithms with improved performance were subsequently developed. Coffman et al. [4] introduced the multifit (MF) approach. A k attempt MF approach is denoted by MF_k. The MF_k procedure performs k binary search steps to find the smallest capacity c such that all the tasks can be packed into a set of m bins when packing using first fit with the tasks sorted in nondecreasing order of their processing times. The tasks assigned to bin i correspond to machine i and c is the makespan of the schedule. The approximation ratio has been shown to be $1.22 + 2^{-k}$ and the time complexity of the algorithm is $O(n \log n + kn \log m)$. Friesen [5] subsequently improved it to $1.2 + 2^{-k}$ followed by Friesen and Langston [6] to $\frac{72}{61} + \frac{1}{2^k}$. The time complexity remained unchanged. However, the latter algorithm has a very large constant associated with the Big "oh" bound.

Following a suggestion by D. Kleitman and D.E. Knuth, Graham [2] was lead to consider the following scheduling strategy. For any $k \geq 1$, an optimal schedule for the longest k tasks is constructed and then the remaining tasks are scheduled in any order using the no-additional-delay policy. Graham showed that this algorithm has an approximation ratio $O(1 + \frac{1 - 1/m}{1 + \lceil k/m \rceil})$ and takes $O(n^{O(1/k)})$ time when there is a fixed number of machines. This was the first polynomial time approximation scheme (PTAS) for any problem. This PTAS, as well the ones for other problems, is explained in more detail in Chapter 8. Fully polynomial time approximation schemes are not possible for this problem unless $P = NP$ [3].

2.4 Relaxation: Linear Programming Approach

Let us now consider the minimum weight vertex cover, which is a fundamental problem in the study of approximation algorithms. This problem is defined as follows.

Problem: Minimum weight vertex cover.

Instance: Given a vertex weighted undirected graph G with the set of vertices $V = \{v_1, v_2, \ldots, v_n\}$, edges $E = \{e_1, e_2, \ldots, e_m\}$, and a positive real number (weight) w_i assigned to each vertex v_i.

Objective: Find a minimum weight vertex cover, that is, a subset of vertices $C \subseteq V$ such that every edge is incident to at least one vertex in C. The weight of the vertex cover C is the sum of the weight of the vertices in C.

It is well known that the minimum weight vertex cover problem is an NP-hard problem. Now consider the following simple greedy algorithm to generate a vertex cover. Assume without loss of generality that the graph G does not have isolated vertices, that is, vertices without any edges. An edge is said to be *uncovered* with respect to a set of vertices C if both of its endpoints are vertices in $V \backslash C$, that is, if both endpoints are in V, but not in C.

```
Algorithm Min-Weight(G)
    Let C = ∅;
    while there is an uncovered edge do
        Let U be the set of vertices adjacent to at least one uncovered edge;
        Add to C a least weight vertex in set U;
    endwhile
end
```

Algorithm `Min-Weight` is not a constant ratio approximation algorithm for the vertex cover problem. Consider the family of star graphs $\mathcal{K}$ each with $l + 1$ nodes, l edges, the center vertex having weight k and the l leaves having weight 1, for any positive integers $k \geq 2$ and $l \geq 3$. For each of these graphs Algorithm `Min-Weight` generates a vertex cover that includes all the leaves in the graph and the weight of the cover is l. For all graphs in $\mathcal{K}$ with $k = 2$, an optimal cover has weight 2 and includes only the center vertex. Therefore, Algorithm `Min-Weight` has an approximation ratio of at least $l/2$, which cannot be bounded above by any fixed constant.

Algorithm `Max-Weight` is identical to Algorithm `Min-Weight`, but instead of selecting the vertex in set U with least weight, it selects one with largest weight. Clearly, this algorithm constructs an optimal cover for the graphs identified earlier where Algorithm `Min-Weight` performs badly. For every graph in $\mathcal{K}$, this algorithm selects as its vertex cover the center vertex which has weight k. Now for all graphs in $\mathcal{K}$ with $l = 2$, an optimal cover consists of both leaf vertices and it has weight 2. Therefore, the approximation ratio for Algorithm `Max-Weight` is at least $k/2$, which cannot be bounded above by any fixed constant.

All of the graphs identified earlier, where one of the algorithms performs badly, have the property that the other algorithm constructs an optimal solution. A compound algorithm that runs both algorithms and then selects the better of the two vertex covers may be a constant ratio algorithm for the vertex cover problem. However, this compound algorithm can also be easily fooled by just using a graph consisting of two stars, where each of the individual algorithms failed to produce good solutions. Therefore, this compound algorithm fails to generate constant ratio approximate solutions. One may now argue that we could partition the graph into connected components and apply both algorithm to each component. For these "two star" graphs, the new compound algorithm will generate an optimal solution. But in general, this new approach fails to produce a constant ratio approximate solution for all possible graphs. Adding an edge between the center vertex in the "two star" graphs gives rise to problem instances for which the new compound algorithm fails to provide a constant ratio approximate solution.

A more clever approach is a modified version of Algorithm `Min-Weight`, where instead of selecting a vertex of least possible weight in set U, one selects a vertex v in set U with least $w(v)/u(v)$, where $u(v)$ is the number of uncovered edges incident to vertex v. This seems to be a better strategy because when vertex v is added to C it covers $u(v)$ edges at a total cost of $w(v)$. So the cost (weight) per edge of $w(v)/u(v)$ is incurred when covering the uncovered edges incident to vertex v. This strategy solves optimally the star graphs in $\mathcal{K}$ defined earlier. However, even when all the weights are equal, one can show that this is not a constant ratio approximation algorithm for the weighted vertex cover problem. In fact the approximation ratio for this algorithm is about $\log n$. Instances with a simple recursive structure that asymptotically achieve this bound as the number of vertices increases can be easily constructed. Chapter 3 gives an example on how to construct problem instances where an approximation algorithm fails to produce a good solution. Chapter 6 discusses local ratio-based algorithm that uses a variation of $w(v)/u(v)$ approach for the vertex cover. The algorithm is a 2-approximation algorithm.

Other approaches to solve the problem can also be shown to fail to provide a constant ratio approximation algorithm for the weighted vertex cover. What type of algorithm can be used to guarantee a constant ratio solution to this problem? Let us try another approach.

Another way to view the minimum weight vertex cover is by defining a 0/1 variable x_i for each vertex v_i in the graph. The 0/1 vector X defines a subset of vertices C as follows. Vertex v_i is in C if, and only if, $x_i = 1$. The set of vertices C defined by X is a vertex cover if, and only if, for every edge $\{i, j\}$ in the graph $x_i + x_j \geq 1$. The vertex cover problem is expressed as an instance of the 0/1 integer linear programming (ILP) as follows:

$$\text{minimize} \quad \sum_{i \in V} w_i x_i \tag{2.1}$$

$$\text{subject to} \quad x_i + x_j \geq 1 \quad \forall \{i, j\} \in E, \tag{2.2}$$

$$x_i \in \{0, 1\} \quad \forall i \in V. \tag{2.3}$$

The 0/1 ILP is also an NP-hard problem. So this formulation does not solve the problem. But we will use it to generate a suboptimal solution.

An important methodology for designing approximation algorithms is relaxation. In this case one relaxes the integer constraints for the x_i values. That is, we replace Constraint (2.3) by $0 \leq x_i \leq 1$ (or simply $x_i \geq 0$, which in this case is equivalent). This means that we are augmenting the solution space by adding solutions that are not feasible for the original problem. This approach will at least provide us with what appears to be a good lower bound for the value of an optimal solution of the original problem, since every feasible solution to the original problem is a feasible solution to the relaxed problem (but the converse is not true). This relaxed problem is an instance of the LP problem, which can be solved in polynomial time. Let X^* be an optimal solution to the LP problem. Clearly, X^* might not be a vertex cover as the x_i^* values may be noninteger. The previous interpretation for the X^* values has been lost because it does not make sense to talk about a fractional part of a vertex being part of a vertex cover. To circumvent this situation, we need to use the X^* vector to construct a 0/1 vector $\hat{X}$ that represents a vertex cover. In order for a vector $\hat{X}$ to represent a vertex cover, it needs to satisfy Inequality (2.2), for every edge $e_k = \{i, j\} \in E$. Clearly, the inequalities holds for X^*. This means that for each edge $e_k = \{i, j\} \in E$ at least one of x_i^* or x_j^* has value at least greater than or equal to $\frac{1}{2}$. So the vector $\hat{X}$ defined from X^* as $\hat{x}_i = 1$ if $x_i^* \geq \frac{1}{2}$ (rounding up) and $\hat{x}_i = 0$ if $x_i^* < \frac{1}{2}$ (rounding down) represents a vertex cover. Furthermore, because of the rounding up the objective function value for the vertex cover $\hat{X}$ is at most $2 \sum w_i x_i^*$. Since $\sum w_i x_i^*$ value is a lower bound for an optimal solution to the weighted vertex cover problem, we know that this procedure generates a vertex cover whose weight is at most twice the weight of an optimal cover, that is, it is a 2-approximation algorithm. This process is called (deterministic) *LP rounding*. Volume 1, Chapters 8, 10, and Volume 2, Chapters 8 and 11 discuss and apply this methodology to other problems.

Another way to round is via randomization, which means in this case that we flip a biased coin (with respect to x_i^* and perhaps other factors) to decide the value for $\hat{x}_i$. The probability that $\hat{X}$ is a vertex cover and its expected weight can be computed. By repeating this randomization process several times, one can show that a cover with weight at most twice the optimal one will be generated with very high probability. In this case, it is clear that randomization is not needed. However, for other problems it is justified. Volume 1, Chapters 4, 7, 10, 11, and Volume 2, Chapter 11 discuss LP randomized rounding.

The above-mentioned rounding methods have the disadvantage that an LP problem needs to be solved. Experimental evaluations over several decades have shown that the Simplex method solves quickly (in poly time) LP problems. But the worst case time complexity is exponential with respect to the problem size. In Chapter 1, we discuss the Ellipsoid algorithm and more recent ones that solve LP problems. Even though these algorithms have polynomial time complexity, there is a term that depends on the number of bits needed to represent the input. Much progress has been made in speeding-up these procedures, but the algorithms are not competitive with typical $O(n \log n)$ for greedy algorithms.

Let us now discuss another approximation algorithm for the minimum vertex cover problem that it is "independent" of LP, and then we discuss a local ratio and a primal-dual approach to this problem.

We call this approach the α-*vector* approach. For every vertex $i \in V$, we define $\delta(i)$ as the set of edges incident to vertex i. Let $\alpha = (\alpha_1, \alpha_2, \ldots, \alpha_m)$ be any vector of m nonnegative real values, where $m = |E|$ is the number of edges in the graph. For all k, multiply the kth edge inequality by α_k.

$$\alpha_k x_i + \alpha_k x_j \geq \alpha_k \quad \forall e_k \in E. \tag{2.4}$$

The total sum of these inequalities can be expressed as

$$\sum_{i \in V} \sum_{e_k \in \delta(i)} \alpha_k x_i \geq \sum_{e_k \in E} \alpha_k. \tag{2.5}$$

Define $\beta_i = \sum_{e_k \in \delta(i)} \alpha_k$ for every vertex $i \in V$. In other words, β_i is the sum of the α values of all the edges incident to vertex i. Substituting in the above-mentioned inequality we know that

$$\sum_{i \in V} \beta_i x_i \geq \sum_{e_k \in E} \alpha_k. \tag{2.6}$$

Suppose that the α vector is such that $w_i \geq \beta_i$ for all i. Then it follows that

$$\sum_{i \in V} w_i x_i \geq \sum_{i \in V} \beta_i x_i \geq \sum_{e_k \in E} \alpha_k. \tag{2.7}$$

In other words, any vector α such that the resulting vector β computed from it satisfies $w_i \geq \beta_i$ provides us with the lower bound $\sum_{e_k \in E} \alpha_k$ for every vector X that represents a vertex cover. In other words, if we assign a positive weight to each edge in such a way that the sum of the weight of the edges incident to each vertex i is at most w_i, then the sum of the weights of the edges is a lower bound for an optimal solution. We need to point out that if instead of positive weights we can assign nonnegative weights. When the weight is zero, means that the edge is ignored for the lower bound. But as we shall see in the next paragraph, the way we assign the α values and the way we select the vertices in the cover, guarantees that we construct a vertex cover for the original graph.

This is a powerful lower bound. To get maximum strength, we need to find a vector α such that $\sum_{e_k \in E} \alpha_k$ is maximum. But finding this vector is as hard as solving the LP problem described before. What if we find a maximal vector α, that is, a vector that cannot possibly be increased in any of its components? This is a simpler task. It is just matter of starting with an α vector with all entries being zero and then increasing one of its components until it is no longer possible to do so. We keep on doing this until there are no edges whose α value can be increased. In this maximal solution, we know that for each edge in the graph at least one of its endpoints has the property that $\beta_i = w_i$, as otherwise the maximality of α is contradicted. Define the vector $\hat{X}$ from the α vector as follows: $x_i = 1$ if $\beta_i = w_i$, and $x_i = 0$, otherwise. Clearly, $\hat{X}$ represents a vertex cover because for every edge in the graph we know that for at least one of its vertices has $\beta_i = w_i$. What is the weight of the vertex cover represented by $\hat{X}$? We know that $\sum w_i \hat{x}_i = \sum \beta_i \hat{x}_i \leq 2 \sum \alpha_k$ because each α_k can contribute its value to at most two β_is. Therefore, we have a simple 2-approximation algorithm for the weighted vertex cover problem. Furthermore, the procedure to construct the vertex cover takes linear time with respect to the number of vertices and edges in the graph.

This algorithm was initially developed by Bar-Yehuda and Even [7] using the LP relaxation and its dual. It is called the *primal-dual* approach. It will be discussed later in this section. The above-mentioned algorithm can be proven to be a 2-approximation algorithm without using this ILP formulation. That is, the same result can be established by just using simple combinatorial arguments [8].

Another related approach, called *local ratio*, was developed by Bar-Yehuda and Even [9]. It is extensively covered in Chapter 6. Initially each vertex is assigned a cost which is simply its weight and it is referred to as the *remaining cost*. The edges in the graph are labeled $\{e_1, e_2, \ldots, e_m\}$. The algorithm considers one edge at a time using this ordering. When the kth edge $e_k = \{i, j\}$ is considered, we define γ_k as the minimum of the remaining cost of vertex i and vertex j. The edge makes a down payment of γ_k to each of its two endpoints and each of the two vertices has its remaining cost decreased by γ_k. The procedure stops when we have considered all the edges. All the vertices whose current cost is zero have been paid for completely and they are yours to keep as the vertices in the cover generated by the algorithm. The remaining ones have not been paid for and there are "no refunds" (not even if you talk to the store manager). The weight of all the vertices in the cover generated by the procedure is at most twice $\sum_{e_k \in E} \gamma_k$, which is simply the sum of the down payments made. What is the weight of an optimal vertex cover? The claim is it is greater or equal to $\sum_{e_k \in E} \gamma_k$. The reason is simple. Consider the first step when we introduce γ_1 for edge e_1. Let I_0 be the initial problem instance and I_1 be the resulting instance after deleting edge e_1 and reducing the cost of the two endpoints of edge e_1 by γ_1. One can easily prove that $f^*(I_0) \geq f^*(I_1) + \gamma_1$, and inductively that $f^*(I_0) = \sum_{e_k \in E} \gamma_k$ [10]. The algorithm is a 2-approximation algorithm for the weighted vertex cover. The approach is called *local ratio* because at each step one adds at most $2\gamma_k$ to the value of the solution generated and one accounts for γ_k value of an optimal solution. We say that this method falls under the category of problem "transformation," since at each step the problem is reduced to another one of the same type but one fewer edges and two slightly different vertex weights. This local ratio approach has been successfully applied to quite a few problems. The nice feature is that it is very simple to understand and does not require any LP background. Chapter 6 discusses this approach extensively.

The primal-dual approach is similar to the previous ones, but it uses the foundations of LP theory. The LP relaxation problem is

$$\text{minimize} \quad \sum_{i \in V} w_i x_i \tag{2.8}$$

$$\text{subject to} \quad x_i + x_j \geq 1 \quad \forall e_k = \{i, j\} \in E, \tag{2.9}$$

$$x_i \geq 0 \quad \forall i \in V. \tag{2.10}$$

The LP problem is called the *primal* problem. The corresponding dual problem is

$$\text{maximize} \quad \sum_{e_k \in E} y_k \tag{2.11}$$

$$\text{subject to} \quad \sum_{e_k \in \delta(i)} y_k \leq w_i \quad \forall i \in V, \tag{2.12}$$

$$y_k \geq 0 \quad \forall e_k \in E. \tag{2.13}$$

As you can see the Y vector is simply the α vector defined before, and the dual is to find a Y vector with maximum $\sum_{i \in V} y_i$. LP theory [11,12] states that any feasible solution X to the primal problem and any feasible solution Y to the dual problem are such that

$$\sum_{e_k \in E} y_k \leq \sum_{i \in V} w_i x_i.$$

This is called *weak duality*. *Strong duality* states that

$$\sum_{e_k \in E} y_i^* = \sum_{i \in V} w_i x_i^*.$$

where:
   X^* is an optimal solution to the primal problem
   Y^* is an optimal solution to the dual problem

Note that the dual variables are multiplied by weights which are the right-hand side of the constraints in the primal problem. In this case, all of them are one.

The primal-dual approach is based on the weak duality property. The idea is to first construct a feasible solution to the dual problem. That solution will give us a lower bound for the value of an optimal vertex cover because of weak duality. Then we use this solution to construct a solution to the primal problem. The idea is that the difference of the objective function value between the primal and dual solutions we constructed is "small." In this case, we construct a maximal vector Y (as we did with the α vector earlier). We note that since the Y vector is maximal, then for at least one of the endpoints (say i) of every edge must satisfy Inequality (2.12) tight, that is, $\sum_{e_k \in \delta(i)} y_k = w_i$. Now define vector X with $x_i = 1$ iff Inequality (2.12) is tight in the dual solution. Clearly, X represent a feasible solution to the primal problem, and its objective function value is at most $2 \sum_k y_k$. It then follows by weak duality that an optimal weighted vertex cover has value at least $\sum_k y_k$, and we have a 2-approximation algorithm for the weighted vertex cover. It is simple to see that the algorithm takes linear time (with respect to the number of vertices and edges in the graph) to solve the problem.

There are other ways to construct a solution to the dual problem. For example, by increasing the weight of all edges uniformly until a vertex becomes saturated and repeating this process with the remaining edges. This produces the same worst case result, but the time complexity becomes $O(mn)$ instead of linear. In Chapters 4 and 12, another method is discussed for finding a solution to the dual problem. Note the difference in the time required to construct the solution. Chapter 12 discusses a "distributed" version of this algorithm. This algorithm makes decisions using only "local" information. Volume 1, Chapter 34 and

Volume 2, Chapter 23 discuss several approximation algorithms based on variations of the primal-dual approach. Some of these methods are not exactly primal-dual, but may be viewed this way.

LP has also been used as a tool to compute the approximation ratio of some algorithms. This type of research may eventually be called the *automatic analysis of approximation algorithms*. Chapter 3 discusses an early approach to compute the approximation ratio of an algorithms that generates 256 different solutions and then selects the best one as its solution. Chapter 39 of the first edition of this handbook discusses a more recent approach. In the former case, a set of LP needed to be solved. Once this was computed, it gave the necessary insight on how prove it analytically. In the latter case, one just formulates the problem and finds bounds for the value of an optimal solution to the LP problem.

2.5 Inapproximability

Sahni and Gonzalez [13] established back in 1973 that constant ratio approximation algorithms exist for some problems only if $P = NP$. In other words, finding a suboptimal solution to some problems is as hard as finding an optimal solution. Any polynomial time algorithm that generates k-approximate solutions can be used to find an optimal solution to the problem in polynomial time! One of these problems is the classical version of TSP defined in Chapter 1, not the one defined over metric space or metric graphs. To prove this result, we show that an NP-complete problem, called the Hamiltonian cycle (HC) problem, can be solved in polynomial time if there is a polynomial time algorithm for the TSP that generates a k-approximate solutions, for any fixed constant k. The *HC* problem is given an undirected graph, $G = (V, E)$, determine whether on not the graph has a HC. A HC for an undirected graph G is a path that starts at vertex 1, visits each vertex *exactly* once, and ends at vertex 1.

To prove this result, a polynomial transformation (Chapter 1, [3]) is used. Let $G = (V, E)$ be any instance of the HC problem with $n = |V|$. Now construct an instance $G' = (V', E', W')$ of the TSP as follows. The graph G' has n vertices and it is complete (all the edges are present). The edge $\{i, j\}$ in E' has weight 1 if the edge $\{i, j\}$ is in E, and weight Z otherwise. The value of Z is $(k - 1)n + 2 > 1$. It will be clear later on why it was defined this way. If the graph G has a HC, then we know that the graph G' has a tour with cost n. On the other hand, if G does not have a HC, then all tours for the graph G' have cost greater than or equal to $n - 1 + Z$. A k-approximate solution when $f^*(G') = n$ must have weight at most $\hat{f}(G') \leq kf^*(G') = kn$. When G does not have a HC, the best possible tour that could be found by the approximation is one with weight at least $n - 1 + Z = kn + 1$. Therefore, if the approximation algorithm returns a tour with length at most kn, then G has a HC; otherwise, the tour returned has length $> kn$ and G does not have a HC. Since the algorithm takes polynomial time with respect to the number of vertices and edges in the graph, it then follows that the algorithm solves in polynomial time the HC problem. So we say that the TSP problem is inapproximable with respect to any constant ratio. It is inapproximable in the sense that a polynomial time constant ratio approximation algorithm implies the solution of a computational complexity question. In this case, it is the $P = NP$ question.

In the last 25 years, there have been new inapproximability results. These results have been for constant, $\ln n$, and n^ϵ approximation ratios. The techniques to establish some of these results are quite complex, but an important component continues to be reducibility. Chapter 17 in the first edition of this handbook discusses all of this work in detail.

References

1. Graham, R.L., Bounds for certain multiprocessing anomalies, *Bell Syst. Tech. J.*, 45, 1563, 1966.
2. Graham, R.L., Bounds on multiprocessing timing anomalies, *SIAM J. Appl. Math.*, 17, 263, 1969.
3. Garey, M.R. and Johnson, D.S., *Computers and Intractability: A Guide to the Theory of NP-Completeness*, W. H. Freeman, San Francisco, CA, 1979.
4. Coffman Jr. E.G., Garey, M.R., and Johnson, D.S., An application of bin-packing to multiprocessor scheduling, *SIAM J. Comput.*, 7, 1, 1978.

5. Friesen, D.K., Tighter bounds for the multifit processor scheduling algorithm, *SIAM J. Comput.*, 13, 170, 1984.
6. Friesen, D.K. and Langston, M.A., Bounds for multifit scheduling on uniform processors, *SIAM J. Comput.*, 12, 60, 1983.
7. Bar-Yehuda, R. and Even, S., A linear time approximation algorithm for the weighted vertex cover problem, *J. Algorithms*, 2, 198, 1981.
8. Gonzalez, T.F., A simple LP-free approximation algorithm for the minimum web pages vertex cover problem, *Inf. Proc. Lett.*, 54(3), 129, 1995.
9. Bar-Yehuda, R. and Even, S., A local-ratio theorem for approximating the weighted set cover problem, *Ann. Disc. Math.*, 25, 27, 1985.
10. Bar-Yehuda, R. and Bendel, K., Local ratio: A unified framework for approximation algorithms, *ACM Comput. Surv.*, 36(4), 422, 2004.
11. Schrijver, A., *Theory of Linear and Integer Programming*, Wiley-Interscience Series in Discrete Mathematics and Optimization, John Wiley & Sons, Chichester, UK, 1998.
12. Vanderbei, R.J., *Linear Programming Foundations and Extensions,* Series: International Series in Operations Research & Management Science, Number 196, 4th Edition, Springer, New York, 2014.
13. Sahni, S. and Gonzalez, T., P-complete approximation problems, *JACM*, 23, 555, 1976.

3

Restriction Methods

Teofilo F. Gonzalez

3.1 Introduction

Restriction is one of the most basic techniques to design approximation algorithms. The idea is to generate a solution to a given problem P by providing an optimal or suboptimal solution to a subproblem of P. A subproblem of a problem P means restricting the solution space for P by disallowing a subset of the feasible solutions. Restriction may be structural or algorithmic. In the case of structural restriction, the idea is to impose some structure on the set of feasible solutions (the solution space), which can be exploited by an efficient algorithm that solves the problem optimally or suboptimally. In the case of algorithmic restriction, an algorithm restricts the set of feasible solutions that may be found. The problem in the restricted solution spaces is referred as the *subproblem*. For this approach to be effective the subproblem must have the property that for every problem instance, its optimal or suboptimal solution should have an objective function value that is "close" to the optimal one for the original problem. The most common approach is to solve the subproblem once but there are algorithms where more than one subproblem is solved, and then the best of the solutions computed is the solution generated. In this chapter we discuss this methodology and show how to apply it to several problems. Approximation algorithms based on this approach are discussed in Volume 1, Chapters 31, 32, 36, and Volume 2, Chapters 6 and 15.

This approach is in a sense the opposite of "relaxation," that is, augmenting the feasible solution space by including previously infeasible solutions. In this case, one needs to solve a superproblem of P. An approximation algorithm for P solves the superproblem (optimally or suboptimally) and then transforms such solution to one that is feasible for P. Approximation algorithms based on the linear programming methodology fall under this category. There are many different conversion techniques including rounding, randomized rounding, and so on. Chapters 4, 7, and 11 discuss this approach in detail. Approximation algorithms based on both restriction and relaxation exist (Vol. 2, Chap. 8). These algorithms first restrict the solution space and then relaxes it. The resulting solution space is different from the original one, as there are some infeasible solutions in the original problem that are now a part of the solution space, and some feasible solution in the original problem is no longer part of the solution space.

In this chapter we discuss several approximation algorithms based on restriction. When designing algorithms of this type the first question is: how to select one of the many subproblem(s) to provide a good approximation for a given problem? One would like to select a subproblem that "works best." But what do we mean by a subproblem that works best? The one that works best could be a subproblem that results in an approximation algorithm with smallest possible approximation ratio, or it could be a subproblem whose solution can be computed the fastest, or one may use some other criteria, for example, any of the ones discussed in Chapter 1. Perhaps "works best" should be with respect to a combination of different criteria. But even when using the approximation ratio as the only evaluation criteria for an algorithm, it is not at all clear how to select a subproblem that can be solved quickly and from which a best possible solution could be generated. These are the two most important properties when choosing a subproblem to solve. By studying several algorithms based on restriction, one learns why it works for these cases and then it becomes easier to find ways to approximate other problems.

The problems that we discuss in this chapter to illustrate "restriction" are Steiner tree, the traveling salesperson, covering points by squares, rectangular partitions, and routing multiterminal nets. The Traveling Salesperson Problem (TSP), and Steiner tree problem are classical problems in combinatorial optimization. The algorithms that we discuss for the TSP are among the best known approximation algorithms for combinatorial problems.

A closely related approach to restriction is *transformation-restriction*. The idea is to transform the problem instance to a restricted instance of the same problem. The difference is that the restricted problem instance is not a subproblem of original problem instance as in the case of restriction, but it is a "simpler" problem of the same type. In Section 3.5 we present algorithms based on this approach for routing multiterminal nets and embedding hyperedges in a cycle. The fully polynomial time approximation scheme for the knapsack problem based on rounding discussed in Chapter 9 is based on transformation-restriction. In Section 3.8 we summarize the chapter and briefly discuss other algorithms based on restriction for path problems arising in computational geometry.

3.2 Steiner Tree

The Steiner tree problem is a classical problem in combinatorial optimization. Let's define the Steiner tree problem over an edge weighted complete metric graph $G = (V, E, W)$, where V is the set of n vertices, E is the set of $m = \frac{n^2 - n}{2}$ edges, and $W : E \rightarrow R^+$ is the weight function for the edges. Since the graph is metric the set of weights satisfies the triangle inequality, that is, for every pair of vertices i, j, the weight $W(i, j)$ is less than or equal to the sum of the weight of the edges in any path from vertex i to vertex j. The Steiner tree problem consists of a metric graph $G = (V, E, W)$ and a subset of vertices $T \subseteq V$. The problem is to find a tree that includes all the vertices in T plus some other vertices in the graph such that the sum of the weight of the edges in the tree is least possible. The Steiner tree problem is an NP-hard problem.

When $T = V$ the problem is called the minimum weight (cost) spanning tree problem. By the 1960s there were several well-known polynomial time algorithms to construct a minimum weight spanning tree for edge weighted graphs [1]. These simple greedy algorithms have low-order polynomial time complexity bounds.

Given an instance of the metric graph Steiner tree problem $(G = (V, E, W), T)$ one may construct a minimum weight spanning tree for the subgraph $G' = (T, E', W')$, where E' and W' includes only the edges joining vertices in T. Clearly this minimum weight spanning tree is a restricted version of the Steiner tree problem and it seems a natural way to approximate the Steiner tree problem. This approach was analyzed in 1968 by E. F. Moore (see Reference 2) for the Steiner tree problem defined in metric space (e.g., 2D). The metric graph problem we just defined, includes only a subset of all the possible points in metric space. E. F. Moore presented an elegant proof of the fact that in metric space (and also for metric graphs) $L_M \leq L_T \leq 2L_S$, where L_M, L_T, and L_S are the weight of a minimum weight spanning tree, a minimum weight tour (solution) for the TSP, and minimum weight Steiner tree for any set of points P, respectively. We defined the TSP in Chapter 2. Since every spanning tree for T is a Steiner tree for T, the above bounds show that when using a minimum weight spanning tree to approximate the Steiner tree results in a solution whose weight is at most twice the weight of an optimal Steiner tree. In other words, any algorithm that generates a minimum weight spanning tree is a 2-approximation algorithm for the Steiner tree problem. Furthermore, this approximation algorithm takes the same time as an algorithm that constructs a minimum weight spanning trees for edge-weighted graphs [1], since such an algorithm can be used to construct an optimal spanning tree for a set of points in metric space. The above bound is established by defining a transformation from any minimum weight Steiner tree into a TSP tour in such a way that $L_T \leq 2L_S$ [2]. Then by observing that deleting an edge from an optimum tour to the TSP problem results in a spanning tree, shows that $L_M \leq L_T$. The proof is identical to the one given in the next section where we show this result, but starting from a minimum weight spanning tree. So where is restriction in this case? The algorithm finds the best possible solution that does not include Steiner points, that is, points in $V \setminus T$. In other words the restriction is to find the best tree without allowing Steiner points.

Chapter 36 also discusses several constant-ratio approximation algorithms for Steiner trees, including ones where the graph is not metric.

3.3 Traveling Salesperson

The TSP has been studied for many decades [3]. There are many variations of this problem. One of the simplest versions of the problem consists of an edge-weighted complete graph and the problem is to find a minimum weight tour that starts and ends at vertex one and visits every vertex *exactly* once. The weight of a tour is the sum of the weight of the edges in the tour. Sahni and Gonzalez [4] (Chapter 1) show that the constant-ratio approximation problem is NP-hard, that is, if for any constant c there is a polynomial time algorithm with approximation ratio c, then $P = NP$. In this section we discuss approximation algorithms for the TSP defined over complete metric graphs. These algorithms are among the best known approximation algorithms for any problem. The "double minimum weight spanning tree" (DMWST) approximation algorithm that we discuss in this section is widely known, and it is based on the constructive proof for the approximation algorithm discussed in the previous section developed for the Steiner tree problem by E. F. Moore. Additional constant-ratio approximation algorithms for this version of the TSP problem were developed by Rosenkrantz et al. [5]. These algorithms as well the DMWST algorithm have an approximation ratio of $2 - 1/n$ and take $O(n^2)$ time. Since the graph is complete, the time complexity is linear with respect to the number of edges in the graph. After presenting this result we discuss the improved approximation algorithm by Christofides [6]. This algorithm has a smaller approximation ratio, but its time complexity grows faster than that of the previous algorithms.

In the literature you will find that the TSP is also defined with tours visiting each vertex *at least* once. We now show that the "at most once" version over metric graphs and the "exactly once" version over metric graphs are equivalent problems. Consider any optimal tour R where some vertices are visited more than once. Let vertex i be a vertex visited more than once. Let vertices j and k be visited just before and just after vertex i. Delete from the tour the edges $\{j, i\}$ and $\{i, k\}$ and add edge $\{j, k\}$. Because the graph is metric the tour weight will stay the same or decrease. If it decreases, then it contradicts the optimality of R. So the weight of the tour must be the same as before. After applying this transformation until it is no longer possible we obtain a tour R' in which every vertex is visited exactly once and the weight of R' is identical to that R. Since every tour that visits every vertex exactly once also visits every vertex at least once, it follows that both versions of the problem for metric graphs have the same optimal tour weight, that is, both problems are equivalent. Since for the TSP defined over metric graphs both versions of the problem are equivalent, for convenience we use the definition of tours to visit each vertex at least once.

Now suppose that you have an optimal tour S for an instance I of the "at least once" version of the TSP. Applying the above-mentioned transformation we obtain an optimal tour S' in which every vertex is visited exactly once. Deleting an edge from the tour results in a spanning tree. Therefore, the weight of a minimum weight spanning tree is a lower bound for the weight of an optimal tour. The questions are: How good of a lower bound is it? How can one construct a tour from a spanning tree?

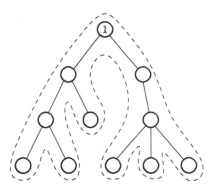

FIGURE 3.1   Spanning tree (solid lines) and tour constructed (dashed lines).

How can we find a tour from a spanning tree T? Just draw the spanning tree in the plane with a vertex as its root and construct a tour by visiting each edge in the tree T twice as illustrated in Figure 3.1. A more formal approach is to construct an *Euler circuit* in the multigraph (graph with multiple edges between between vertices) consisting of two copies of the edges in T. An Euler tour (or circuit) is a path that starts and ends at the same vertex and visits every edge in the multigraph once. An Euler tour always exists for the multigraphs we have defined because these multigraphs are connected and all their nodes are of even degree (the number of edges incident to each vertex is even). These multigraphs are called *Eulerian*, and an Euler tour can be constructed in linear time with respect to the number of nodes and edges in the multigraph [7].

The approximation algorithm, which we refer to as *DMCST*, constructs a minimum cost spanning tree, makes a copy of all the edges in the tree, and then generates a tour from this tree with weight at most equal to twice the cost of a minimum cost spanning tree. We established before that an optimal tour has weight greater than the weight of a minimum cost spanning tree, it then follows that the weight of the tour that the DMCST algorithm generates is at most twice the weight of an optimal tour for G. Therefore, algorithm DMCST generates 2-approximate solution. Actually the ratio is $2 - 1/n$, which can be established when the edge deleted for an optimal tour to obtain a spanning tree is the one with largest weight. The time complexity of the algorithm is bounded by the time complexity for generating a minimum cost spanning tree, since an Euler tour can be constructed in linear time with respect to the number of edges in the spanning tree. We formalize these results in the following theorem:

Theorem 3.1   *For the metric TSP problem algorithm DMCST generates a tour with weight at most $(2-1/n)$ times the weight of an optimal tour. The time complexity of the algorithm is $O(n^2)$ time, which is linear time with respect to the number of edges in the graph.*

Proof. The proof for the approximation ratio follows from the above-mentioned discussion. As Fredman and Tarjan [8] point out, implementing Prim's minimum cost spanning tree algorithm by using Fibonacci heaps results in a minimum cost spanning tree algorithm that takes $O(n \log n + m)$ time. Since the graph is complete, the time complexity is $O(n^2)$ which is linear with respect to the number of edges in the graph.   ∎

This constructive proof is based on Moore's proof for the approximation algorithm for the Steiner tree problem. The difference is that he starts from an optimal Steiner tree.

So where is the restriction in the above-mentioned algorithms? We are actually *restricting* tours for the TSP to traverse the least possible number of *different* edges, though a tour may traverse some of these edges more than once. The minimum number of different edges needed from G is $n - 1$ and they must form a spanning tree. It is therefore advantageous to select a spanning tree of least possible total weight.

That justifies the use of a minimum weight spanning tree. This is another way to think about the design of the DMWST algorithm.

Christofides [6] modified the previous approach so that the tours generated have total weight within 1.5 times the weight of an optimal tour. However, the currently fastest implementation of this procedure takes $O(n^3)$ time. His modification is very simple. First, observe that there are many different ways to transform a spanning tree into an Eulerian multigraph. All possible augmentations must include at least one edge incident to every odd degree vertex in the spanning tree. Let N be the set of odd-degree vertices in the spanning tree. Christofides' idea is to transform the spanning tree into an Eulerian multigraph by adding the least number of edges with the least possible total weight. He showed that such set of edges is a minimum cost complete matching on the graph G_N induced by the set of vertices N in G. A *matching* is a subset of the edges in a multigraph, no two of which are incident upon the same vertex. A matching is *complete* if every node has an edge in the matching incident to it, and the weight of a matching is the sum of the weight of the edges in it. A minimum cost complete matching can be constructed in polynomial time. The edges in the complete matching plus the ones in the spanning tree form an Eulerian multigraph, and Christofides' algorithm generates as its solution an Euler tour of this multigraph.

To establish the 1.5 approximation ratio we observe that an optimal tour can be transformed without increasing its total weight into another tour that visits only the vertices in N because the graph is metric. One can partition the edges in this reduced tour into two sets such that each set is a complete matching for the restricted graph N. One set contains the even numbered edges in the tour and the other set the odd numbered edges. Since a minimum cost complete matching for G_N has total cost at most equal to either of the previous two matchings, it then follows that the minimum cost complete matching has total cost at most half of the weight of an optimal tour. Therefore, the edges in the the tour constructed by Christofides' algorithm have weight at most 1.5 times the weight of an optimal tour. The time complexity for Christofides' algorithm is $O(n^3)$ and it is dominated by the time required to construct a minimum cost complete matching [9,10]. We formalize this result in the following theorem whose proof is omitted as it follows the above discussion.

Theorem 3.2 [6] *For the metric TSP, Christofides' algorithm generates a tour with weight at most 1.5 times the weight of an optimal tour. The time complexity of the algorithm is $O(n^3)$.*

This approach used by this algorithm is similar to the one employed by Edmonds and Johnson [11] for the *Chinese postman problem*. Given an edge weighted connected undirected graph, the Chinese postman problem is to construct a minimum weight cycle, which contains every edge in the graph possibly with repeated edges. The currently best algorithm to solve this problem takes $O(n^3)$ time, and it uses shortest paths and minimum weighted matching algorithms. There are asymptotically faster algorithms when graphs are sparse and weight of the edges are integers.

3.4 Covering Points by Squares

Given a set of n points, $P = \{(x_1, y_1), (x_2, y_2), \ldots (x_n, y_n)\}$, in two-dimensional space and an integer D, the *covering by squares in 2D* (CS_2) problem is to find the least number of $D \times D$ squares to cover P. The CS_2 problem as well as the problem of covering by disks have been shown to be NP-hard [12]. Approximation algorithm for these problems as well as their generalizations to multidimensional space have been developed [13,14]. All these problems find applications in several areas [12,15,16]. The most popular application is to find the least number of emergency facilities such that every potential patient lives at a distance at most D from at most one facility. This application corresponds to covering by the least number of disks with radius D.

We discuss in this section a simple approximation algorithm based on restriction for the CS_2 problem. Assume without loss of generality that $x_i \geq 0$ and $y_i \geq 0$ and that at least one of the points has x-coordinate value of zero. Define the function $I_x(P_i) = \lfloor x_i/D \rfloor$, used to identify vertical bands of points. For $l \geq 0$ and k consists of all the points with $I_x(P_i) = k$.

The restriction to the solution space is to only allow feasible solutions where each square covers points from only one band. Note that an optimal solution to the CS_2 problem does not necessarily satisfy this property. For example, the instance with $P_1 = (0.1, 1.0)$, $P_2 = (0.1, 2.0)$, $P_3 = (1.1, 0.9)$, $P_4 = (1.1, 2.1)$, and $D = 1$ has two squares in optimal cover. The first square covers points P_1 and P_3, and the second square covers P_2 and P_4. However, an optimal cover for the points in band 0 (i.e., P_1 and P_2) is 1 square and the one for the points in band 1 (i.e., P_3 and P_4) is 2 squares. So an optimal cover to the restricted problem has three squares, but an optimal cover for the CS_2 problem has two squares.

One reason for restricting the solution space in this way is that an optimal cover for any given band can be easily generated by a greedy procedure in $O(n \log n)$ time [14]. A greedy approach places a square as high as possible provided it includes the bottommost point in the band as well as all other points in the band at a vertical distance at most 1 from a bottommost point. All the points covered by this square are removed and the procedure is repeated until all the points have been covered. One can easily show that this is an optimal cover by transforming any optimal solution for the band, without increasing the number of squares, to the cover generated by the greedy algorithm. By using elaborate data structures Gonzalez [14] showed that the greedy algorithm can be implemented to take $O(n \log s)$ time, where s is the number of squares in an optimal solution. Actually a method that uses considerable more space can be used to solve the problem in $O(n)$ time [14].

The solution generated by our algorithm for the whole problem is the union of the covers for each of the bands generated by the greedy method. Let $\hat{f} = E + O$ be the total number of squares, where $E(O)$ is the number of square for the even (odd) numbered bands. We claim that an optimal solution to the CS_2 problem has at least $max\{E, O\}$ squares. This follows from the fact that an optimal solution for the even (odd) numbered bands is $E(O)$ because it is not possible for a square to cover points from two different even (odd) numbered bands. Therefore, $\frac{\hat{f_I}}{f_I^*} \leq 2$, where f_I^* is the number of squares in an optimal solution for problem instance I. This result is formalized in the following theorem whose proof is omitted as it from the previous discussion.

Theorem 3.3 *For the CS_2 problem the above-mentioned procedure generates a cover such that $\frac{\hat{f_I}}{f_I^*} \leq 2$ in $O(n \log s)$ time, where s is the number of squares in an optimal solution.*

A polynomial time approximation scheme for the generalization of the CS_2 to d dimensions (the CS_d problem) is discussed in Volume 1, Chapter 8 and Volume 2, Chapter 5. The idea is to generate a set of solutions by shifting the bands by different amounts and then selecting as the solution the best cover computed by the algorithm. This approach is called *shifting* and it was introduced by Hochbaum and Maass [13] and Baker [17].

3.5 Optimal Rectangular Partitions

The minimum edge-length rectangular partition (RG_P) problem has applications in the area of computer aided-design of integrated circuits and systems. Given a rectangle R with interior points P, the RG_P problem is to introduce a set of interior lines segments with least total length such that every point in P is in at least one of the partitioning line segments, and R is partitioned into rectangles. Figure 3.2a shows a problem instance I and Figure 3.2b shows an optimal rectangular partition for the problem instance I.

A rectangular partition E is said to have a *guillotine cut* if one of the vertical or horizontal line segments partitions the rectangle into two rectangles. A rectangular partition E is said to be a *guillotine partition* if either E is empty, or E has a guillotine cut and each of the two resulting rectangular partitions is a guillotine partition.

Finding an optimal rectangular partition is an NP-hard problem [18]. However, an optimal guillotine partition can be constructed in polynomial time. Therefore, it is natural to restrict the solution space to guillotine partitions when approximating rectangular partitions.

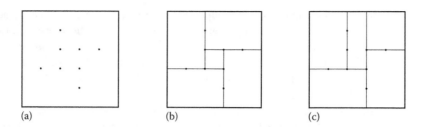

FIGURE 3.2 (a) Instance I of the RG_P problem. (b) Rectangular partition for the instance I. (c) Guillotine partition the instance I.

In Volume 2, Chapter 6 we prove that an optimal guillotine partition has total edge length, that is at most twice the length of an optimal rectangular partition. Gonzalez and Zheng [19] presented a complex proof that shows that bound is just 1.75. In Volume 2, Chapter 6 we also explain the basic ideas behind the proof of the approximation ratio of 1.75. This approach has been extended to the multidimensional version of this problem by Gonzalez et al. [20].

An optimal guillotine partition can be constructed in $O(n^5)$ time via dynamic programming. When n is large this approach is not practical. Gonzalez et al. [21] showed that suboptimal guillotine partitions that can be constructed in $O(n \log n)$ time generate solutions with total edge length at most four times the length of an optimal rectangular partition. As in the case of optimal guillotine partitions, this result has been extended to the multidimensional version of the problem [21]. Clearly, none of the methods dominates the other when considering both the approximation ratio and the time complexity bound.

Chapter 36 discusses how more general guillotine cuts can be used to develop a PTAS for the TSP in two-dimensional space. Volume 2, Chapter 2 discusses this approach for the TSP and Steiner tree problems.

3.6 Routing Multiterminal Nets

Let R be a rectangle whose sides lie on the two-dimensional integer grid. A subset of grid points on the boundary of R that do not include the corners of R is denoted by S, and its grid points are called *terminal points*. Let n be the number of terminal points, that is, the cardinality of set S, and let $N_1, N_2, \ldots, N_m$ be a partition of S such that each set N_i includes at least two terminal points. Each set N_i is called a *net* and the problem is to make all the terminal points electrically common by introducing a set wire segments in two layers. Terminal points from different nets should not be made electrically common. The wire segment must be along the grid lines outside R with at most one wire segment assigned to each grid edge. All vertical wires are in one layer and all the horizontal wires are in another layer. A *via* (or hole) is used to connect a vertical wire to a horizontal wire that have a common point. Note that each grid point may only belong to one net. In other words, when viewing the two layers from the top, dog-leg on one layer (wires from two nets bending at a grid point) are not allowed. The main reasons are that dog-legs would complicate the layer assignment without improving the layout area.

The Multiterminal net routing Around a Rectangle (*MAR*) problem is given a rectangle R and a set of nets, find a layout, subject to the constraints defined earlier, that fits inside a rectangle with least possible area. Constructing a layout in this case reduces to just finding the wire segments for each net along the grid lines (without dog-legs) outside R, since the layer assignment is straight forward.

Developing a constant-ratio approximation algorithm for this problem is complex because the objective function depends on the product of two values, rather than just one value as in most other problems. Gonzalez and Lee [22] developed a linear time algorithm for the *MAR* problem when every net consists of two terminal points. It is conjectured that the problem becomes NP-hard when the nets have three terminal points each. Gonzalez and Lee [22,23] developed constant-ratio approximation algorithms for the *MAR* problem [23,24]. The approximation ratios for these algorithms are 1.69 [23] and 1.6 [24].

The approach is to partition the set of nets into groups and then route each group of nets independently of each other. Some of the groups are routed optimally. Since the analysis of the approximation ratio for these algorithms is complex, in this section we only analyze the case when the nets contain one terminal point on the top side of R, and one or more terminal points on the bottom side of R. The set of these nets is called N_{TB}. The algorithm to route the N_{TB} nets is based on restriction and it is interesting. Readers interested in additional details are referred to papers [23,24].

Let n_{TB} be the number of N_{TB} nets. Let E be an optimal area layout for all the nets and let D be E except that the set of nets in N_{TB} are all connected by a path that crosses the left side of R. In this case the layout for the nets N_{TB} is restricted (only paths that cross the left side of R are allowed). We use $H_E(TB)$ ($H_D(TB)$) to denote the height of the layout $E(D)$ on the top plus the corresponding height on the bottom side of R. To simplify the analysis, lets assume that every net in N_{TB} is connected in E by a path that either crosses the left or right (but not both) sides of R. Gonzalez and Lee [24] explain how to modify the analysis when some of these nets are connected by paths that cross both the left and right sides of R.

By *reversing* the connecting path for a net in N_{TB} we mean to connect the net by a path that crosses the opposite side of R, that is, if it crossed the left side of R it will now cross the right side, or vice versa. When we reverse the connecting path for a net the height on the top side plus the bottom side of R increases by at most two. We say that connecting paths for two N_{TB} nets *cross on the top side* of R when their contribution to the total height of the assignment is two for at least one point in between the two terminal points. When we *interchange* the connecting paths for two N_{TB} nets that cross on the top side of R we mean reversing both connecting paths. An interchange increases by at most two the height on the top side plus the bottom side of R.

We transform E to D by reversals in order to quantify the difference in heights between E and D. The largest increase in height is when all the N_{TB} nets are connected in E by paths that cross the right side of R. In this case we need to reverse all the connecting paths for the N_{TB} nets, so $H_D(TB) \leq H_E(TB) + 2n_{TB}$. When one plugs this in the analysis for the whole problem it results in an algorithm with an approximation ratio greater than 2.

A better approach is to use the following restriction: All the connecting paths for the N_{TB} nets are identical, and either they cross the left or the right side of R. In this case we construct two different layouts. Let layout D_l (D_r) be E except that all the nets in N_{TB} are connected by a path crossing the left (right) side of R. Let M be a minimum area layout between D_l and D_r. In E let l (r) be the number of N_{TB} nets connected by a path crossing the left (right) side of R. By reversing the minimum of $\{l, r\}$ paths it is possible to transform E to D_l or D_r. Therefore, $H_M(TB) \leq H_E(TB) + n_{TB}$, which is better by 50% than for the assignment D defined before.

By trying more alternatives one can obtain better solutions. Let us partition the set of nets N_{TB} into two groups, S_l and S_r. The set S_l contains the $\frac{n_{TB}}{2}$ nets in N_{TB} whose terminal point on the top side of R is closest to the left side of R, and set S_r contains the remaining ones. For $i, j \in \{l, r\}$ let D_{ij} be E except that all the nets in S_l are connected by paths that cross the "i" side of R and all the nets in S_r are connected by paths that cross the "j" side of R. Let P be a minimum area layout among D_{ll}, D_{lr}, D_{rl}, and D_{rr}. Let $l_1(r_1)$ be the number of nets in S_l connected by a path that crosses the left side of R. We define L_2 and r_2 similarly, but using the set S_r. We establish in the following lemma that $H_P(TB) \leq H_E(TB) + \frac{3}{4}n_{TB}$.

Lemma 3.1 *Let P and E be the assignments defined previously. Then $H_P(TB) \leq H_E(TB) + \frac{3}{4}n_{TB}$.*

Proof. The proof is by contradiction. Suppose that $H_P(TB) > H_E(TB) + \frac{3}{4}n_{TB}$. There are two cases depending on the values of r_1 and l_2.

Case 1: $r_1 < l_2$.

To transform assignment E to D_{lr} we need to interchange l_2 connecting paths that cross on the top side of R and reverse $r_1 - l_2$ connecting paths. Therefore, $H_{D_{lr}}(TB) \leq H_E(TB) + 2r_1$. Since $H_{D_{lr}}(TB) \geq H_P(TB) > H_E(TB) + \frac{3}{4}n_{TB}$, we know that $2r_1 > \frac{3}{4}n_{TB}$ which is equivalent to $r_1 > \frac{3}{8}n_{TB}$. Since $r_1 + l_1 = \frac{1}{2}n_{TB}$, we know that $l_1 < \frac{1}{8}n_{TB}$.

To transform assignment E to D_{rr} we need to reverse the connecting paths of $l_1 + l_2$ connecting paths. Therefore, $H_{D_{rr}}(TB) \leq H_E(TB) + 2l_1 + 2l_2$. Since $H_{D_{rr}}(TB) \geq H_P(TB) > H_E(TB) + \frac{3}{4}n_{TB}$, we know that $l_1 + l_2 > \frac{3}{8}n_{TB}$. Applying the same argument to assignment D_{rl}, we know $l_1 + r_2 > \frac{3}{8}n_{TB}$. Adding these two last inequalities and substituting the fact that $l_2 + r_2 = \frac{1}{2}n_{TB}$, we know that $l_1 > \frac{1}{8}n_{TB}$. Which contradicts our previous finding that $l_1 < \frac{1}{8}n_{TB}$.

Case 2: $r_1 \geq l_2$.

A contradiction in this case can be obtained applying similar arguments to assignments D_{ll} and D_{lr}. ∎

For three groups, rather than two, Gonzalez and Lee [23] showed that $H_P(TB) \leq H_E(TB) + \frac{2}{3}n_{TB}$, where P is the best of the eight assignments. This is enough to prove the approximation ratio of 1.69 for the *MAR* problem. If instead of three groups one uses six, one can prove $H_P(TB) \leq H_E(TB) + 0.6n_{TB}$, where P is the best of the 64 assignments generated. In this case, approximation ratio for the whole *MAR* problem is 1.6. Interestingly, partitioning into more groups results in smaller bounds for this group, but does not reduce the approximation ratio for the *MAR* problem because the routing of other nets becomes the bottleneck. We state Gonzalez and Lee's theorem without a proof. Readers interested in the proof are referred to Reference 24.

Theorem 3.4 *For the MAR problem the procedure given in Reference 24 generates a layout with area at most 1.6 times the area of an optimal layout in $O(nm)$ time.*

An interesting observation is that the proof that the bound $H_P(TB) \leq H_E(TB) + (1.6)n_{TB}$ holds can be carried out automatically by solving a set of linear programming problems. The linear programming problems find the ratios for l_i and r_i such that the minimum increase from E to one of the layouts is maximized. Note that some of the "natural" constraints of the problem are in terms $\max\{r_1, l_2\}$ which makes the solution space nonconvex. However by replacing it with inequalities of the form $r_1 \leq l_2$ and $r_1 > l_2$ we partition the optimization region into several convex regions. By solving a set of linear programming problems (one for each convex region) the maximum possible increase can be computed.

3.7 Variations on Restriction

A closely related approach to restriction is to generate a solution by solving a restricted problem instance constructed from the original instance. We call this approach *transformation-restriction*. For example, consider the routing multiterminal nets around a rectangle discussed in Section 3.6. Remember that there are n terminal points and m nets. Suppose that we break every net i with k_i points into k_i nets with two terminal points each. The k nets consist of adjacent terminal points of the net. In order for these k_i nets to have different terminal points we make a copy of each terminal point at half integer points next to the old ones. Note that a new grid needs to be redefined to include the half integer points without introducing more horizontal (vertical) routing tracks above or below (to the left or right) of R. Figure 3.3b shows the details. The resulting two terminal net problem can be solved in linear time using the optimal algorithm developed by Gonzalez and Lee [22]. A solution to this problem can be easily transformed into a solution to the original problem after deleting the added terminal points as well as some superfluous connections. This algorithm generates a layout whose total area is at most 4 times the area of an optimal layout. Furthermore, the layout can be constructed in $O(n)$ time. With respect to the approximation ratio Gonzalez and Lee's algorithms [23,24] are better, but these algorithms take $O(nm)$ time, whereas the simple algorithm just described takes linear time.

3.7.1 Embedding Hyperedges in a Cycle

In this subsection we present an approximation algorithm for embedding hyperedges in a cycle so as to minimize the congestion (EHCMC). This problem has applications in the area of design automation and

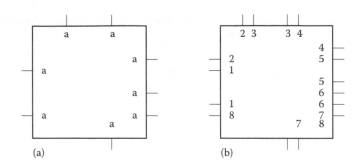

FIGURE 3.3   (a) Net with k terminal points. (b) Resulting k 2-terminal nets.

parallel computing. As input we are given a hypergraph $G = (V, H)$, where $V = \{v_1, v_2, \ldots, v_n\}$ is the set vertices and $H = \{h_1, h_2, \ldots h_m\}$ is the set of hyperedges (or subsets with at least two elements of the set V). Traversing the vertices $v_1, v_2, \ldots, v_n$ the clockwise direction forms a cycle which we call C. Let v_t and v_s be two vertices in h_i such that v_s is the next vertex in h_i in clockwise direction from v_t. Then the pair (v_s, v_t) for hyperedge h_i defines the connecting path for h_i that begins at vertex v_s then proceeds in the clockwise direction along the cycle until reaching vertex v_t. Every edge e in the cycle that is visited by the connecting path formed by pair (v_s, v_t) is said to be *covered* by the connecting path. The EHCMC problem consists of finding a connecting path c_i for every hyperedge h_i such that the maximum congestion of an edge in C is least possible, where the congestion of an edge e in cycle C is the number of connecting paths that include edge e.

Ganley and Cohoon [25] showed that when the maximum congestion is bounded by a fixed constant k, the EHCMC problem is solvable in polynomial time. But, the problem is NP-hard when there is no constant bound for k. Frank et al. [26] showed that when the hypergraph is a graph the EHCMC problem can be solved in polynomial time. We call this problem the embedding edges in a cycle to minimize congestion (EECMC). In this section we present the simple linear time algorithm with an approximation ratio of two for the EHCMC problem developed by Gonzalez [27].

The algorithm based on *transformation-restriction* for this problem is simple and uses the same approach as in the previous subsection. This general approach also works for other routing problems. A hyperedge with k vertices $x_1, x_2, \ldots, x_k$, appearing in that order around the cycle C is decomposed into the following k edges $\{x_1, x_2\}, \{x_2, x_3\} \ldots \{x_{k-1}, x_k\}, \{x_k, x_1\}$. Note that in this case we do not need to add additional vertices as in the previous subsection because a vertex may be part of several hyperedges. The decomposition transforms the problem into an instance of the EECMC problem, which can be solved by the algorithm given in Reference 26. From this embedding we can construct an embedding to the original problem instance after deleting some superfluous edges in the embedding. The resulting embedding can be easily shown to have congestion of at most twice the one in an optimal solution X. This is because there is a solution S to the EECMC problem instance in which every connecting path Y in X can be mapped to a set of connecting paths in S with the property that if the connecting path Y contributes one unit to the congestion of an edge e, then the set of connecting paths in S contribute two units to the congestion of edge e. Furthermore, each connecting path in S appears in one mapping. The time complexity of the algorithm is $O(n)$.

3.8   Concluding Remarks

We have seen several approximation algorithms based on restriction. As we have seen the restricted problem may be solved optimally or suboptimally as in Section 3.5. One generates solutions closer to optimal, whereas the other generates the solution faster. These are many more algorithms based on this technique. For example, some computational geometry problems where the objective function is in terms of distance have been approximated via restriction [28–31]. These type of problems allow feasible solutions to be any set of points. A restricted problem allows a set of points (called *artificial points*) to be part of the feasible solution space. The more artificial points, the smaller the approximation ratio of the algorithm; however, the algorithm takes longer to generate a solution.

There are problems for which it is not know whether or not there is a constant-ratio approximation algorithm. However, heuristics based on restriction are used to generate good solutions in practice. One such problems is discussed in Volume 2, Chapter 25.

A closely related approach to restriction is *transformation-restriction*. The idea is to transform the problem instance to a restricted instance of the same problem. The difference is that the restricted problem instance is not a subproblem of original problem instance as in the case of restriction. We showed how this approached is applied to a couple of problems.

Approximation algorithms that are based on restriction and relaxation exist. These algorithms first restrict the solution space and then relaxes it resulting in a solution space that is different from the original one. Gonzalez and Gonzalez [32] have applied this approach successfully to the minimum edge length corridor problem (Vol. 2, Chap. 8).

References

1. Sahni, S., *Data Structures, Algorithms, and Applications in C++*, 2nd ed., Silicon Press, Summit, NJ, 2005.
2. Gilbert, E.N. and Pollak, H.O., Steiner minimal trees, *SIAM Journal of Applied Mathematics*, 16(1), 1, 1968.
3. Lawler, E.L., Lenstra, J.K., Rinnooy Kan, A.H.G., and Shmoys, D.B., Eds., *The Traveling Salesman Problem: A Guided Tour of Combinatorial Optimization*, John Wiley & Sons, New York, 1985.
4. Sahni, S. and Gonzalez, T., P-complete approximation problems, *JACM*, 23, 555, 1976.
5. Rosenkrantz, R., Stearns, R., and Lewis, L., An analysis of several heuristics for the traveling salesman problem. *SIAM Journal on Computing*, 6(3), 563, 1977.
6. Christofides, N., Worst-case analysis of a new heuristic for the traveling salesman problem. Technical Report 338. Grad School of Industrial Administration, CMU, Pittsburgh, PA, 1976.
7. Weiss, M.A., *Data Structures & Algorithms in C++*, 2nd ed., Addison-Wesley, Reading, MA, 1999.
8. Fredman, M. and Tarjan, R.E., Fibonacci heaps and their uses in improved network optimization algorithms, *JACM*, 34(3), 596, 1987.
9. Gabow H.N., A scaling algorithm for weighted matching on general graphs, in *Proceedings of FOCS*, IEEE Press, New York p. 90, 1985.
10. Lawler, E.L., *Combinatorial Optimization: Networks and Matroids*, Holt, Rinehart and Winston, New York, 1976.
11. Edmonds, J. and Johnson, E.L., Matching Euler tours and the Chinese postman, *Mathematical Programming*, 5, 88, 1973.
12. Flower, R.J., Paterson, M.S., and Tanimoto, S.L., Optimal packing and covering in the plane are NP-complete, *Information Processing Letters*, 12, 133, 1981.
13. Hochbaum D.S. and Maass, W., Approximation schemes for covering and packing problems in image processing and VLSI, *JACM*, 32(1), 130, 1985.
14. Gonzalez, T.F., Covering a set of points in multidimensional space, *Information Processing Letters*, 40, 181, 1991.
15. Tanimoto, S.L., Covering and indexing an image subset, in *Proceedings of the IEEE Conference on Pattern Recognition and Image Processing*, IEEE Press, New York, p. 239, 1979.
16. Tanimoto, S.L. and Fowler, R.J., Covering image subsets with patches, in *Proceedings of the 5th International Conference on Pattern Recognition*, IEEE Press, New York, p. 835, 1980.

17. Baker, B.S., Approximation algorithms for NP-complete problems on planar graphs, in *Proceedings of the 24th Annual IEEE Symposium on Foundations of Computer Science* (Tucson, AZ, November 7–9), IEEE, New York, p. 133, 1983.

18. Lingas, A., Pinter, R.Y., Rivest, R.L., and Shamir, A., Minimum edge length partitioning of rectilinear polygons, in *Proceedings of the 20th Allerton Conference on Communication, Control, and Computing*, Monticello, IL, 1982.

19. Gonzalez, T.F. and Zheng, S.Q., Improved bounds for rectangular and guillotine partitions, *Journal of Symbolic Computation*, 7, 591, 1989.

20. Gonzalez, T.F., Razzazi, M., Shing, M., and Zheng, S.Q., On optimal *d*-guillotine partitions approximating hyperrectangular partitions, *Computational Geometry: Theory and Applications*, 4(1), 1, 1994.

21. Gonzalez, T.F., Razzazi, M., and Zheng, S.Q., An efficient divide-and-conquer algorithm for partitioning into d-boxes, *International Journal of Computational Geometry & Applications*, 3(4), 417, 1993.

22. Gonzalez, T.F. and Lee, S.L., A linear time algorithm for optimal wiring around a rectangle, *JACM*, 35(4), 810, 1988.

23. Gonzalez, T.F. and Lee, S.L., Routing multiterminal nets around a rectangle, *IEEE Transactions on Computers*, 35(6), 543, 1986.

24. Gonzalez, T.F. and Lee, S.L., A 1.60 approximation algorithm for routing multiterminal nets around a rectangle, *SIAM Journal on Computing*, 16(4), 669, 1987.

25. Ganley, J.L., and Cohoon, J.P., Minimum-congestion hypergraph embedding in a cycle, *IEEE Transactions on Computers*, 46(5), 600, 1997.

26. Frank, A., Nishizeki, T., Saito, N., Suzuki, H., and Tardos, E., Algorithms for routing around a rectangle, *Discrete Applied Mathematics*, 40(3), 363, 1992.

27. Gonzalez, T.F., Improved approximation algorithms for embedding hypergraphs in a cycle, *Information Processing Letters*, 67, 267, 1998.

28. Papadimitriou, C.H., An algorithm for shortest path motion in three dimensions. *Information Processing Letters*, 20, 259, 1985.

29. Choi, J., Sellen, J., and Yap, C.K., Approximate Euclidean shortest path in 3-space, *International Journal of Computational Geometry and Applications*, 7(4), 271, 1997.

30. Choi, J., Sellen, J., and Yap C.K., Approximate Euclidean shortest path in 3-space, in *Proceedings of the 10th Annual Symposium on Computational Geometry*, ACM, New York, p. 41, 1994.

31. Bhosle, A.M. and Gonzalez, T.F., Exact and approximation algorithms for finding an optimal bridge connecting two simple polygons, *International Journal on Computational Geometry and Applications*, 15(6), 609, 2005.

32. Gonzalez, A. and Gonzalez, T.F., Approximating Corridors and Tours via Restriction and Relaxation Techniques, *ACM Transactions on Algorithms*, 6(3), Article number 56, 1–36, 2010.

4

Greedy Methods

Samir Khuller

Balaji Raghavachari

Neal E. Young

4.1 Introduction

Greedy algorithms can be used to solve many optimization problems exactly and efficiently. Examples include classical problems such as finding minimum spanning trees and scheduling unit length jobs with profits and deadlines. These problems are special cases of finding a maximum- or minimum-weight basis of a *matroid*. This well-studied problem can be solved exactly and efficiently by a simple greedy algorithm [1,2].

Greedy methods are also useful for designing efficient *approximation* algorithms for intractable (i.e., NP-hard) combinatorial problems. Such algorithms find solutions that may be suboptimal, but still satisfy some performance guarantee. For a minimization problem, an algorithm has *approximation ratio* α, if, for every instance I, the algorithm delivers a solution whose cost is at most $\alpha \times \text{OPT}(I)$, where $\text{OPT}(I)$ is the cost of an optimal solution for instance I. An α-approximation algorithm is a polynomial-time algorithm with an approximation ratio of α.

In this chapter we survey several NP-hard problems that can be approximately solved via greedy algorithms. For a couple of fundamental problems we sketch the proof of the approximation ratio. For most of the other problems that we survey, we give brief descriptions of the algorithms and citations to the articles where these results were reported.

4.2 Set Cover

We start with SET COVER, perhaps one of the most elementary of the NP-hard problems. The problem is defined as follows: The input is a set $X = \{x_1, x_2, \ldots, x_n\}$ of elements and a collection of sets $\mathcal{S} = \{S_1, S_2, \ldots S_m\}$ whose union is X. Each set S_i has a weight of $w(S_i)$. A *set cover* is a subset $\mathcal{S}' \subseteq \mathcal{S}$ such that $\bigcup_{S_j \in S'} S_j = X$. Our goal is to find a set cover $S' \subseteq S$ so as to minimize $w(S') = \sum_{S_i \in S'} w(S_i)$. In other words, we wish to choose a minimum-weight collection of subsets that covers all the elements.

Intuitively, for a given weight, one prefers to choose a set that covers the most elements. This suggests the following algorithm: start with an empty collection of sets, then repeatedly add sets to the collection, each time adding a set that minimizes the cost per newly covered element (i.e., the set that minimizes the weight of the set divided by the number of its elements that are not yet in any set in the collection.)

4.2.1 Algorithm for Set Cover

Next we prove that this algorithm has approximation ratio $H(|S_{max}|)$, where S_{max} is the largest set in S and H is the harmonic function, defined as $H(d) = \sum_{i=1}^{d} 1/i$. For simplicity, assume each set has weight 1.

We use the following charging scheme: *when the algorithm adds a set S to the collection, let u denote the number of not-yet-covered elements in S and charge $1/u$ to each of those elements.* Clearly the weight of the chosen sets is at most the total amount charged. To finish, we observe that the total amount charged is at most OPT $\times H(|S_{max}|)$. To see why this is so, let $S^* = \{e_s, e_{s-1}, \ldots, e_1\}$ be any set in OPT. Assume that when the greedy algorithm chooses sets to add to its collection, it covers the elements in S^* in the order given (each e_i is covered by the time e_{i-1} is). When the charge for an element e_i is computed (i.e., when the greedy algorithm chooses a set S containing e_i for the first time) at least i elements $(e_i, e_{i-1}, e_{i-2}, \ldots, e_1)$ in S^* are not yet covered. Since the greedily chosen set S contains at least as many not-yet-covered elements as S^*, the charge to e_i is at most $1/i$. Thus, the total charge to elements in S^* is at most

$$\frac{1}{s} + \frac{1}{s-1} + \cdots + \frac{1}{2} + 1 = H(s) \leq H(|S_{max}|).$$

Thus, the total charge to elements covered by OPT is at most OPT $\times H(|S_{max}|)$. Since every element is covered by OPT, this means that the total charge altogether is at most OPT $\times H(|S_{max}|)$. This implies that the greedy algorithm is an $H(|S_{max}|)$-approximation algorithm.

These results were first reported in the mid-1970s [3–6]. Since then, it has been proven that no polynomial-time approximation algorithm for set cover has a significantly better approximation ratio unless P = NP [7].

The algorithm and approximation ratio extend to a fairly general class of problems called *minimizing a linear function subject to a submodular constraint*. This problem generalizes set cover as follows: instead of asking for a set cover, we ask for a collection of sets C such that some function $f(C) \geq f(X)$. The function $f(C)$ should be increasing as we add sets to C and it should have the following property: if $C \subset C'$, then, for any set $S, f(C' \cup \{S'\}) - f(C') \leq f(C \cup \{S\}) - f(C)$. In terms of the greedy algorithm, this means that adding a set S to the collection now increases f at least as much as adding it later. (For set cover, take $f(C)$ to be the number of elements covered by sets in C.) See Reference 8 for details.

4.2.2 Shortest Superstring Problem

We consider an application of the set cover problem, SHORTEST SUPERSTRING problem. Given an alphabet Σ, and a collection of n strings $S = \{s_1, \ldots, s_n\}$, where each s_i is a string from the alphabet Σ, find a shortest string s that contains each s_i as a substring. There are several constant-factor approximation algorithms for this problem [9]; here we simply want to illustrate how to reduce this problem to the set cover problem. The reduction is such that an optimal solution to the set cover problem has weight at most twice the length of a shortest superstring.

For each $s_i, s_j \in S$ and for each value $0 < k < \min(|s_i|, |s_j|)$, we first check to see if the last k symbols of s_i are identical to the first k symbols of s_j. If so, we define a new string β_{ijk} obtained by concatenating s_i with s_j^k, the string obtained from s_j by deleting the first k characters of s_j. Let C be the set of strings β_{ijk}. For a string π we define $S(\pi) = \{s \in S | s$ is a substring of $\pi\}$. The underlying set of elements of the set cover is S. The specified subsets of S are the sets $S(\pi)$ for each $\pi \in S \cup C$. The weight of each set $S(\pi)$ is $|\pi|$, the length of the string.

We can now apply the greedy set cover algorithm to find a collection of sets $S(\pi_i)$ and then simply concatenate the strings π_i to find a superstring. The approximation factor of this algorithm can be shown to be $2H(n)$.

4.3 Steiner Trees

The STEINER TREE problem is defined as follows. Given an edge-weighted graph $G = (V, E)$ and a set of terminals $S \subset V$, find a minimum-weight tree that includes all the nodes in S. (When $S = V$, then this is the problem of finding a minimum-weight *spanning* tree. There are several very fast greedy algorithms that can be used to solve this problem optimally.) The Steiner tree problem is NP-hard and several greedy algorithms have been designed that give a factor 2 approximation [10,11]. We briefly describe the idea behind one of the methods. Let $T_1 = \{s_1\}$ (an arbitrarily chosen terminal from S). At each step T_{i+1} is computed from T_i as follows: attach the vertex from $S - T_i$ that is the "closest" to T_i by a path to T_i and call the newly added special vertex s_{i+1}. Thus, T_i always contains the vertices $s_1, s_2, \ldots, s_i$. It is clear that the solution produces a Steiner tree. It is possible to prove that the weight of this tree is at most twice the weight of an optimal Steiner tree.

Zelikovsky [12] developed a greedy algorithm with an approximation ratio of $11/6$. This bound has been further improved subsequently, but by using more complex methods.

A generalization of Steiner trees called NODE-WEIGHTED STEINER TREES is defined as follows. Given a node-weighted graph $G = (V, E)$ and a set of terminals $S \subset V$, find a minimum-weight tree that includes all the nodes in S. Here, the weight of a tree is the sum of the weights of its nodes. It can be shown that this problem is at least as hard as the Set-Cover problem to approximate [13]. Interestingly, this problem is solved via a greedy algorithm similar to the one for the Set Cover problem with costs. We define a "spider" as a tree on ℓ terminals where there is at most one vertex with degree more than 2. Each leaf in the tree corresponds to a terminal. The weight of the spider is simply the weight of the nodes in the spider. The algorithm at each step greedily picks a spider with minimum ratio of weight to number of terminals in it. It collapses all the terminals spanned by the spider into a single vertex, makes this new vertex a terminal and repeats until one one terminal remains. The approximation guarantee of this algorithm is $2 \ln |S|$. Further improvements appear in Reference 14. For more on the Steiner tree problem, see the book by Hwang et al. [15].

4.4 *K*-Centers

The K-CENTER problem is a fundamental facility location problem and is defined as follows: given an edge-weighted graph $G = (V, E)$ find a subset $S \subseteq V$ of size at most K such that each vertex in V is close to some vertex in S. More formally, the objective function is defined as follows:

$$\min_{S \subseteq V} \max_{u \in V} \min_{v \in S} d(u, v)$$

where d is the distance function. For example, one may wish to install K fire stations and minimize the maximum distance (response time) from a location to its closest fire station.

Gonzalez [16] describes a very simple greedy algorithm for the basic K-center problem and proves that it gives an approximation factor of 2. The algorithm works as follows. Initially pick any node v_0 as a center and add it to the set C. Then for $i = 1$ to K do the following: in iteration i, for every node $v \in V$, compute its distance $d^i(v, C) = \min_{c \in C} d(v, c)$ to the set C. Let v_i be a node that is farthest away from C, that is, a node for which $d^i(v_i, C) = \max_{v \in V} d(v, C)$. Add v_i to C. Return the nodes $v_0, v_1, \ldots, v_{K-1}$ as the solution.

The above-mentioned greedy algorithm is a 2-approximation for the K-center problem. First note that the radius of our solution is $d^K(v_K, C)$, since by definition v_K is the node that is farthest away from our set of centers. Now consider the set of nodes $v_0, v_1, \ldots, v_K$. Since this set has cardinality $K + 1$, at least two of these nodes, say v_i and v_j, must be covered by the same center c in the optimal solution. Assume without loss of generality that $i < j$. Let R^* denote the radius of the optimal solution. Observe that the distance from each node to the set C does not increase as the algorithm progresses. Therefore $d^K(v_K, C) \leq d^j(v_K, C)$. Also we must have $d^j(v_K, C) \leq d^j(v_j, C)$, otherwise we would not have selected node v_j in iteration j. Therefore

$$d(c, v_i) + d(c, v_j) \geq d(v_i, v_j) \geq d^j(v_j, C) \geq d^K(v_K, C)$$

by the triangle inequality and the fact that v_i is in the set C at iteration j. But since $d(c, v_i)$ and $d(c, v_j)$ are both at most R^*, we have the radius of our solution $= d^K(v_K, C) \leq 2R^*$.

4.5 Connected Dominating Sets

The connected dominating set (CDS) problem is defined as follows: given a graph $G = (V, E)$, find a minimum size subset S of vertices, such that the subgraph induced by S is connected and S forms a dominating set in G. Recall that a dominating set is one in which each vertex is either in the dominating set, or adjacent to some vertex in the dominating set. The CDS problem is known to be *NP*-hard.

We describe a greedy algorithm for this problem [17]. The algorithm runs in two phases. At the start of the first phase all nodes are colored white. Each time we include a vertex in the dominating set, we color it black. Nodes that are dominated are colored gray (once they are adjacent to a black node). In the first phase the algorithm picks a node at each step and colors it black, coloring all adjacent white nodes gray. A *piece* is defined as a white node or a black connected component. *At each step we pick a node to color black that gives the maximum (non-zero) reduction in the number of pieces.*

It is easy to show that at the end of this phase if no vertex gives a non-zero reduction to the number of pieces, then there are no white nodes left.

In the second phase, we have a collection of black connected components that we need to connect. Recursively connect pairs of black components by choosing a chain of vertices, until there is one black connected component. Our final solution is the set of black vertices that form the connected component.

Key property: At the end of the first phase if there is more than one black component, then there is always a pair of black components that can be connected by choosing a chain of two vertices.

It can be shown that the CDS found by the algorithm is of size at most $(\ln \Delta + 3) \cdot |OPT_{CDS}|$, where Δ is the maximum degree of a node.

Let a_i be the number of pieces left after the ith iteration, and $a_0 = n$. Since a node can connect up to Δ pieces, $|OPT_{CDS}| \geq \frac{a_0}{\Delta}$. (This is true if the optimal solution has at least two nodes.) Consider the $i + 1$st iteration. An optimal solution can connect a_i pieces. Hence the greedy procedure is guaranteed to pick a node which connects at least $\left\lceil \frac{a_i}{|OPT_{CDS}|} \right\rceil$ pieces. Thus the number of pieces will reduce by at least $\left\lceil \frac{a_i}{|OPT_{CDS}|} \right\rceil - 1$. This gives us the recurrence relation,

$$a_{i+1} \leq a_i - \left\lceil \frac{a_i}{|OPT_{CDS}|} \right\rceil + 1 \leq a_i \left(1 - \frac{1}{|OPT_{CDS}|} \right) + 1.$$

Its solution is,

$$a_{i+1} \leq a_0 \left(1 - \frac{1}{|OPT_{CDS}|} \right)^i + \sum_{j=i}^{i-1} \left(1 - \frac{1}{|OPT_{CDS}|} \right)^j.$$

Notice after $|OPT_{CDS}| \cdot \ln \frac{a_0}{|OPT_{CDS}|}$ iterations, the number of pieces left is less than $2 \cdot |OPT_{CDS}|$. After this, for each node we choose, we will decrease the number of pieces by at least one until the number of black components is at most $|OPT_{CDS}|$, thus at most $|OPT_{CDS}|$ more vertices are picked. So after $|OPT_{CDS}| \cdot \ln \frac{a_0}{|OPT_{CDS}|} + |OPT_{CDS}|$ iterations at most $|OPT_{CDS}|$ pieces are left to connect. We connect the remaining pieces choosing chains of at most two vertices in the second phase. The total number of nodes chosen is at most $|OPT_{CDS}| \cdot \ln \frac{a_0}{|OPT_{CDS}|} + |OPT_{CDS}| + 2|OPT_{CDS}|$, and since $\Delta \geq \frac{a_0}{|OPT_{CDS}|}$, the solution found has at most $|OPT_{CDS}| \cdot (\ln \Delta + 3)$ nodes.

4.6 Scheduling

We consider the following simple scheduling problem [18]. There are k identical machines. We are given a collection of n jobs. Job J_i is specified by the following vector: (r_i, d_i, p_i, w_i). The job has a *release* time of r_i, a *deadline* of d_i and a *processing time* of p_i. The weight of the job is w_i. Our goal is to schedule a subset of the jobs such that each job starts after its release time and is completed by its deadline. If S is a subset of jobs that are scheduled, then the total profit due to set S is $\sum_{J_i \in S} w_i$. We do not get any profit if the job is not completed by its deadline. Our objective is to find a maximum-profit subset of jobs that can be scheduled on the k machines. The jobs are scheduled on one machine, with no preemption. In other words, if job J_i starts on machine j at time s_i, then $r_i \leq s_i$ and $s_i + p + i \leq d_i$. Moreover, each machine can be executing at most one job at any point of time.

A number of algorithms for the problem are based on linear program (LP) rounding [18]. A special case of interest is when all jobs have unit weight (or identical weight). In this case, we simply wish to maximize the number of scheduled jobs. The following greedy algorithm has the property that it schedules a set of jobs such that the total number of scheduled jobs is at least ρ_k times the number of jobs in an optimal schedule. Here $\rho_k = 1 - \frac{1}{(1+\frac{1}{k})^k}$. Observe that when $k = 1$, then $\rho_k = \frac{1}{2}$ and this bound is tight for the greedy algorithm.

The algorithm considers each machine in turn and finds a maximal set of jobs to schedule for the machine; it removes these jobs from the collection of remaining jobs, then recurses on the remaining set of jobs. Now we discuss how a maximal set of jobs is chosen for a single machine. The idea is to pick a job that can be finished as quickly as possible. After we pick this job, we schedule it, starting it at the earliest possible time. Making this choice might force us to reject several other jobs. We then consider starting a job after the end of the last scheduled job, and again pick one that we can finish at the earliest possible time. In this way, we construct the schedule for a single machine.

4.7 Minimum-Degree Spanning Trees (MDSTs)

In this problem, the input is a graph $G = (V, E)$, with non-negative weights $w : E \mapsto R^+$ on its edges. We are also given an integer $d > 1$. The objective of the problem is to find a minimum-weight spanning tree of G in which the degree of every node is at most d. It is a generalization of the Hamiltonian path problem, and is therefore NP-hard. It is known that the problem is not approximable to any ratio unless $P = NP$ or the approximation algorithm is allowed to output a tree whose degree is larger than d. Approximation algorithms try to find a tree whose degree is as close to d as possible, but whose weight is not much more than an optimal degree-d tree.

Greedy algorithms usually select one edge at a time, and once an edge is chosen, that decision is never revoked and the edge is part of the output. Here we add a subset S of edges at a time (e.g., a spanning forest), where S is chosen to minimize a relaxed version of the objective function. We get an iterative solution and the output is a union of the edges selected in each of the steps. This approach typically provides a logarithmic approximation. For MDST, the algorithm finds a tree of degree $O(d \log n)$, whose weight is within $O(\log n)$ of an optimal degree-d tree, where the graph has n vertices. The ideas have appeared in References 19, 20. Such algorithms in which two objectives (degree and weight) are approximated are called *bicriteria* approximation algorithms.

A minimum-weight subgraph in which each node has degree at most d and at least 1 can be computed using algorithms for matching. Except for possibly being disconnected, this subgraph satisfies the other properties of an MDST: degree constraints and weight at most OPT. A greedy algorithm for MDST works by repeatedly finding d-forests, where each d-forest is chosen to connect the connected components left from the previous stages. The number of components decreases by a constant factor in each stage, and, in $O(\log n)$ stages, we get a tree of degree at most $d \log n$.

4.8 Maximum-Weight b-Matchings

In this problem, we are interested in computing a maximum-weight subgraph of a given graph G in which each node has degree at most b. The classical matching problem is a b-matching with $b = 1$. This problem can be solved optimally in polynomial time, but the algorithms take about $O(n^3)$ time. We discuss a $1/2$-approximation algorithm that runs in $O(bE + E \log E)$ time. The edges are sorted by weight, with the heaviest edges considered first. Start with an empty forest as the initial solution. When an edge is considered, we see if adding it to the solution violates the degree bound of its end vertices. If not, we add it to our solution. Intuitively, each edge of our solution can displace at most 2 edges of an optimal solution, one incident to each of its end vertices, but of smaller weight.

4.9 Primal-Dual Methods

In this section we study a powerful technique, namely the primal-dual method, for designing approximation algorithms [21]. Duality provides a systematic approach for bounding OPT, a key task in proving any approximation ratio. The approach underlies many approximation algorithms. In this section, we illustrate the basic method via a simple example.

A closely related method, one that we don't explore here, is the "local-ratio" method developed by Bar-Yehuda [22]. It seems that most problems that have been solved by the primal-dual method appear amenable to attack by the local-ratio method as well.

We use as our example another fundamental NP-hard problem, the VERTEX COVER problem. Given a graph $G = (V, E)$ with weights on the vertices given by $w(v)$, we wish to find a minimum-weight vertex cover. A vertex cover is a subset of vertices, $S \subseteq V$, such that for each edge $(u, v) \in E$, either $u \in S$ or $v \in S$ or both. This problem is equivalent to the special case of the set cover problem where each set contains exactly two elements.

We describe a 2-approximation algorithm. First, write an integer linear program (ILP) for this problem. For each vertex v in the given graph, the program has a binary variable $x_v \in \{0, 1\}$. Over this space of variables, the problem is to find

$$\min \left\{ \sum_{v \in V} w(v) x_v \; : \; x_u + x_v \geq 1 \; (\forall (u, v) \in E) \right\}.$$

It is easy to see that an optimal solution to this integer program gives an optimal solution to the original vertex cover problem. Thus, the integer program is NP-hard to solve. Instead of solving it directly, we relax ILP to a linear program (LP), which is to optimize the same objective function over the same set of constraints, but with real-valued variables $x_v \in [0, 1]$.

Each LP has a dual. Let $N(v)$ denote the neighbor set of v. The dual of LP has a variable $y_{(u,v)} \geq 0$ for each edge $(u, v) \in E$. Over this space of variables, the dual of LP is to find

$$\max \left\{ \sum_{(u,v) \in E} y_{u,v} \; : \; \sum_{u \in N(v)} y_{(u,v)} \leq w(v) \; (\forall v \in V) \right\}.$$

The key properties of these programs are the following:

1. Weak duality: the cost of any feasible solution to the dual is a lower bound on the cost of any feasible solution to LP. Consequently, the cost of any feasible solution to the dual is a lower bound on the cost of any feasible solution to ILP.
2. If we can find feasible solutions for ILP and the dual, where the cost of our solution to ILP is at most α times the cost of our solution to the dual, then our solution to ILP has cost at most α OPT.

One way to get an approximate solution is to solve the vertex cover LP optimally (e.g., using a network flow algorithm [23]), and then round the obtained fractional solution to an integral solution. Here we describe a different algorithm—a greedy algorithm that computes solutions to both ILP and the dual. The solutions are not necessarily optimal, but will have costs within a factor of 2.

The dual solution is obtained by the following simple heuristic: *Initialize all dual variables to 0, then simultaneously and uniformly raise all dual variables, except those dual variables that occur in constraints that are currently tight. Stop when all constraints are tight.* The solution to ILP is obtained as follows: *Compute the above-mentioned dual solution. When the constraint for a vertex v becomes tight, add v to the cover.* (Thus, the vertices in the cover are those whose constraints are tight.)

The constraint for vertex v is tight if $\sum_{u \in N(v)} y_{(u,v)} = w(v)$. When we start to raise the dual variables, the sum increases at a rate equal to the degree of the vertex. Thus, the first vertices to be added are those minimizing $\frac{w(v)}{d(v)}$. These vertices and their edges are effectively deleted from the graph, and the process continues.

The algorithm returns a vertex cover because, in the end, for each edge (u, v) at least one of the two vertex constraints is tight. By weak duality, to see that the cost of the cover is at most 2OPT, it suffices to see that the cost of the cover S is at most twice the cost of the dual solution. This is true because each node's weight can be charged to the dual variables corresponding to the incident edges, and each such dual variable is charged at most twice:

$$\sum_{v \in S} w(v) = \sum_{v \in S} \sum_{u \in N(v)} y_{(u,v)} \leq 2 \sum_{(u,v) \in E} y_{(u,v)}.$$

The above-mentioned equality follows because $w(v) = \sum_{u \in N(v)} y_{(u,v)}$ for each vertex added to the cover. The inequality follows because each dual variable $y_{(u,v)}$ occurs at most twice in the sum.

To implement the algorithm, it suffices to keep track of the current degree $D(v)$ of each vertex v, as well as the slack $W(v)$ remaining in the constraint for v. In fact, with a little bit of effort the reader can see that the following pseudocode implements the algorithm described earlier, without explicitly keeping track of dual variables. This algorithm was first described by Clarkson [24]:

GREEDY-VERTEX-COVER(G, S)
1   **for all** $v \in V$ **do** $W(v) \leftarrow w(v); D(v) \leftarrow deg(v)$
2    $S \leftarrow \emptyset$
3   **while** $E \neq \emptyset$ **do**
4   Find $v \in V$ for which $\frac{W(v)}{D(v)}$ is minimized.
5   **for all** $u \in N(v)$ **do**
6            $E \leftarrow E \setminus (u, v)$
7            $W(u) \leftarrow W(u) - \frac{W(v)}{D(v)}$ and $D(u) \leftarrow D(u) - 1$
8   **end**
9        $S \leftarrow S \cup \{v\}$ and $V \leftarrow V \setminus \{v\}$
10   **end**

More sophisticated applications of the primal-dual method require more sophisticated proofs. In some cases, the algorithm starts with a greedy phase, but then has a final round in which some previously added elements are discarded. The key idea is to develop the primal solution hand in hand with the dual solution in a way that allows the cost of the primal solution to be "charged" to the cost of the dual.

Because the vertex cover problem is a special case of the set cover problem, it is also possible to solve the problem using the greedy set cover algorithm. This gives an approximation ratio of at most $H(|V|)$, and in fact there are vertex cover instances for which that greedy algorithm produces a solution of cost $\Omega(H(|V|))$ OPT. The greedy algorithm described earlier is almost the same; it differs only in that it modifies the weights of the neighbors of the chosen vertices as it proceeds. This slight modification yields a significantly better approximation ratio.

4.10 Greedy Algorithms via the Probabilistic Method

In their book on the probabilistic method, Alon et al. [25] describe probabilistic proofs as follows:

> In order to prove the existence of a combinatorial structure with certain properties, we construct an appropriate probability space and show that a randomly chosen element in the space has the desired properties with positive probability.

The *method of conditional probabilities* is used to convert those proofs into efficient algorithms [26].

For some problems, elementary probabilistic arguments easily prove that good solutions exist. In some cases (especially when the proofs are based on iterated random sampling) the probabilistic proof can be converted into a greedy algorithm. This is a fairly general approach for designing greedy algorithms. In this section we give some examples.

4.10.1 Max Cut

Given a graph $G = (V, E)$, the MAX-CUT problem is to partition the vertices into two sets S and $\overline{S}$ so as to maximize the number of edges "cut" (crossing between the two sets). The problem is NP-hard.

Consider the following randomized algorithm: *For each vertex, choose the vertex to be in S or $\overline{S}$ independently with probability 1/2.* We claim this is a 1/2-approximation algorithm, in expectation. To see why, note that the probability that any given edge is cut is 1/2. Thus, by linearity of expectation, in expectation $|E|/2$ edges are cut. Clearly an optimal solution cuts at most twice this many edges.

Next we apply the *method of conditional probabilities* [25,26] to convert this randomized algorithm into a deterministic one. We replace each random choice made by the algorithm by a deterministic choice that does "as well" in a precise sense. Specifically, we modify the algorithm to to maintain the following invariant:

> After each step, if we were to take the remaining choices randomly, then the expected number of edges cut in the end would be at least $|E|/2$.

Suppose decisions have been made for vertices $V_t = \{v_1, v_2, \ldots, v_t\}$, but not yet for vertex v_{t+1}. Let S_t denote the vertices in V_t chosen to be in S. Let $\overline{S}_t = V_t - S_t$ denote the vertices in V_t chosen to be in $\overline{S}$. Given these decisions, the status of each edge in $V_t \times V_t$ is known, whereas the rest still have a 1/2 probability of being cut. Let $x_t = |E \cap (S_t \times \overline{S}_t)|$ denote the number of those edges that will definitely cross the cut. Let $e_t = |E - V_t \times V_t|$ denote the number of edges which are not yet determined. Then, given the decisions made so far, the expected number of edges that would be cut if all remaining choices were to be taken randomly would be

$$\phi_t \doteq x_t + e_t/2.$$

The x_t term counts the edges cut so far, whereas the $e_t/2$ term counts the e_t edges with at least one undecided endpoint: each of those edges will be cut with probability 1/2.

Our goal is to replace the random decisions for the vertices with deterministic decisions that guarantee $\phi_{t+1} \geq \phi_t$ at each step. If we can do this, then we will have $|E|/2 = \phi_0 \leq \phi_1 \leq \cdots \leq \phi_n$, and, since ϕ_n is the number of edges finally cut, this will ensure that at least $|E|/2$ edges are cut.

Consider deciding whether the vertex v_{t+1} goes into S_{t+1} or $\overline{S}_{t+1}$. Let s be the number of v_{t+1}'s neighbors in S_t. Let $\overline{s}$ be the number of v_{t+1}'s neighbors in $\overline{S}_{t+1}$. By calculation,

$$\phi_{t+1} - \phi_t = \begin{cases} s/2 - \bar{s}/2 & \text{if } v_{t+1} \text{ is added to } \bar{S}_{t+1} \\ \bar{s}/2 - s/2 & \text{otherwise.} \end{cases}$$

Thus, the following strategy ensures $\phi_{t+1} \geq \phi_t$: *if* $s \leq \bar{s}$, *then put* v_{t+1} *in* S_{t+1}; *otherwise put* v_t *in* $\bar{S}_{t+1}$. By doing this at each step, the algorithm guarantees that $\phi_n \geq \phi_{n-1} \geq \cdots \geq |E|/2$.

We have derived the following greedy algorithm: *Start with* $S = \bar{S} = \emptyset$. *Consider the vertices in turn. For each vertex* v, *put the vertex* v *in* S *or* $\bar{S}$, *whichever has fewer of* v's *neighbors.* We know from the derivation that this is a 1/2-approximation algorithm.

4.10.2 Independent Set

Although the application of the method of conditional probabilities is somewhat technical, it is routine, in the sense that it follows a similar form in every case. Here is another example.

The problem of finding a MAXIMUM INDEPENDENT SET in a graph $G = (V, E)$ is one of the most basic problems in graph theory. An independent set is defined as a subset S of vertices such that there are no edges between any pair of vertices in S. The problem is NP-hard. Turán's theorem states the following: *Any graph* G *with* n *nodes and average degree* d *has an independent set* I *of size at least* $n/(d+1)$. Next we sketch a classic proof of the theorem using the probabilistic method. Then we apply the method of conditional probabilities to derive a greedy algorithm.

Let $\widehat{N}(v) = N(v) \cup \{v\}$ denote the neighbor set of v, including v. Consider this randomized algorithm: *Start with* $I = \emptyset$. *Consider the vertices in random order. When considering* v, *add it to* I *if* $\widehat{N}(v) \cap I = \emptyset$.

For a vertex v to be added to I, it suffices for v to be considered before any of its neighbors. This happens with probability $|\widehat{N}(v)|^{-1}$. Thus, by linearity of expectation, the expected number of vertices added to I is at least

$$\sum_v |\widehat{N}(v)|^{-1}.$$

A standard convexity argument shows this is at least $n/(d+1)$, completing the proof of Turán's theorem.

Now we apply the method of conditional probabilities. Suppose the first t vertices $V_t = \{v_1, v_2, \ldots, v_t\}$ have been considered. Let $I_t = V_t \cap I$ denote those that have been added to I. Let $R_t = V \setminus (V_t \cup \widehat{N}(I_t))$ denote the remaining vertices that might still be added to I and let $\widehat{N}_t(v) = \widehat{N}(v) \cap R_t$ denote the neighbors of v that might still be added. If the remaining vertices were to be chosen in random order, the expected number of vertices in I by the end would be at least

$$\phi_t \doteq |I_t| + \sum_{v \in R_t} |\widehat{N}_t(v)|^{-1}.$$

We want the algorithm to choose vertex v_{t+1} to ensure $\phi_{t+1} \geq \phi_t$. To do this, it suffices to choose the vertex $w \in R_t$ minimizing $|\widehat{N}_t(w)|$, then

$$\phi_{t+1} - \phi_t \geq 1 - \sum_{v \in \widehat{N}_t(w)} |\widehat{N}_t(v)|^{-1} \geq 1 - \sum_{v \in \widehat{N}_t(w)} |\widehat{N}_t(w)|^{-1} = 0.$$

This gives us the following greedy algorithm: *Start with* $I = \emptyset$. *Repeat until no vertices remain: choose a vertex* v *of minimum degree in the remaining graph; add* v *to* I *and delete* v *and all of its neighbors from the graph. Finally, return* I. It follows from the derivation that this algorithm ensures $n/(d+1) \leq \phi_0 \leq \phi_1 \leq \cdots \leq \phi_n$, so that the algorithm returns an independent set of size at least $n/(d+1)$, where d is the average degree of the graph.

As an exercise, the reader can give a different derivation leading to the following greedy algorithm (with the same performance guarantee): *Order the vertices by increasing degree, breaking ties arbitrarily. Let* I *consist of those vertices that precede all their neighbors in the ordering.*

4.10.3 Unweighted Set Cover

Next we illustrate the method on the set cover problem.

We start with a randomized rounding scheme that uses iterated random sampling to round a fractional set cover (a solution to the relaxed problem) to a true set cover. We prove an approximation ratio for the randomized algorithm, then apply the method of conditional probabilities to derive a deterministic greedy algorithm.

We emphasize that, in applying the method of conditional probabilities, we remove the explicit dependence of the algorithm on the fractional set solution. Thus, the final algorithm does not in fact require first solving the relaxed problem.

Recall the definition of the set cover problem from the beginning of the chapter. For this section, we will assume all weights $w(S_i)$ are 1.

Consider the following relaxation of the problem: assign a value $z_i \in [0, 1]$ to each set S_i so as to minimize $\sum_i z_i$ subject to the constraint that, for every element x_j, $\sum_{i:x_j \in S_i} z_i \geq 1$. We call a z meeting these constraints a *fractional set cover*.

The optimal set cover gives one possible solution to the relaxed problem, but there may be other fractional set covers that give a smaller objective function value. However, not too much smaller. We claim the following: *Let z be any fractional set cover. Then there exists an actual set cover C of size at most $T = \lceil \ln(n)|z| \rceil$, where $|z| = \sum_i z_i$.*

To prove this, consider the following randomized algorithm: given z, *draw T sets at random from the distribution p defined by $p(S_i) = z_i/|z|$*. With non-zero probability, this random experiment yields a set cover. Here is why. A calculation shows that, with each draw, the chance that any given element e is covered is at least $1/|z|$. Thus, the expected number of elements left uncovered after T draws is at most

$$n(1 - 1/|z|)^T \; < \; n \exp(-T/|z|) \; \leq \; 1.$$

Since on average less than 1 element is left uncovered, it must be that some outcome of the random experiment covers all elements.

Next we apply the method of conditional probabilities. Suppose that t sets have been chosen so far, and let n_t denote the number of elements not yet covered. Then the conditional expectation of the number of elements left uncovered at the end is at most

$$\phi_t \; \doteq \; n_t(1 - 1/|z|)^{T-t}.$$

We want the algorithm to choose each set to ensure $\phi_t \leq \phi_{t-1}$, so that in the end $\phi_T \leq \phi_0 < 1$ and the chosen sets form a cover.

Suppose the first t sets have been chosen, so that ϕ_t is known. A calculation shows that, if the next set is chosen at random according to the distribution p, then $E[\phi_{t+1}] \leq \phi_t$. Thus, choosing the next set to *minimize ϕ_{t+1} will ensure $\phi_{t+1} \leq \phi_t$*. By inspection, choosing the set to minimize ϕ_{t+1} is the same as choosing the set to minimize n_{t+1}.

We have derived the following greedy algorithm: *Repeat T times: add a set to the collection so as to minimize the number of elements remaining uncovered.* In fact, it suffices to do the following: *Repeat until all elements are covered: add a set to the collection so as to minimize the number of elements remaining uncovered.* (This suffices because we know from the derivation that a cover will be found within T rounds.)

We have proven the following fact: *The above-mentioned greedy algorithm returns a cover of size at most* $\min_z \lceil \ln(n)|z| \rceil$, *where z ranges over all fractional set covers.* Since the minimum-size set cover OPT corresponds to a z with $|z| = |\text{OPT}|$, we have the following corollary: *The above-mentioned greedy algorithm returns a cover of size at most* $\lceil \ln(n)\text{OPT} \rceil$.

This algorithm can be generalized to weighted set cover, and slightly stronger performance guarantees can be shown [3–6]. This particular greedy approach applies to a general class of problems called "minimizing a linear function subject to a submodular constraint" [8].

Comment: In many cases, applying the method of conditional probabilities will not yield a greedy algorithm, because the conditional expectation ϕ_t will depend on the fractional solution in a nontrivial way. In that case, the derandomized algorithm will first have to compute the fractional solution (typically by solving an LP). That is Raghavan and Thompson's standard method of *randomized rounding* Reference 27. The variant we see here was first observed in Reference 28. Roughly, to get a greedy algorithm, we should apply the method of conditional probabilities to a probabilistic proof based on *repeated random sampling* from the distribution defined by the fractional optimum.

4.10.4 Lagrangian Relaxation for Fractional Set Cover

The algorithms described previously fall naturally into a larger and technically more complicated class of algorithms called *Lagrangian relaxation algorithms*. Typically, such an algorithm is used to find a structure meeting in a given a set of constraints. The algorithm constructs a solution in small steps. Each step is made so as to minimize (or keep from increasing) a *penalty function* that approximates some of the underlying constraints. Finally, the algorithm returns a solution that approximately meets the underlying constraints.

These algorithms typically have a greedy outer loop. In each iteration, they solve a subproblem that is simpler than the original problem. For example, a multicommodity flow algorithm may solve a sequence of shortest-path subproblems, routing small amounts of flow along paths chosen to minimize the sum of edge penalties that grow exponentially with the current flow on the edge.

Historical examples include algorithms by von Neumann, Ford and Fulkerson, Dantzig-Wolfe decomposition, Benders' decomposition, and Held and Karp. In 1990, Shahrokhi and Matula proved a polynomial time bound for such an algorithm for multicommodity flow. This sparked a long line of work generalizing and strengthening this result [29–31]. See the recent text by Bienstock [32]. These works focus mainly on *packing and covering problems*—LPs and ILPs with non-negative coefficients.

As a rule, the problems in question can also be solved by standard linear programming algorithms such as the simplex algorithm, the ellipsoid algorithm, or interior-point algorithms. The primary motivation for studying Lagrangian relaxation algorithms has been that, similar to other greedy algorithms, they can often be implemented without explicitly constructing the full underlying problem. This can make them substantially faster.

As an example, here is a Lagrangian relaxation algorithm for fractional set cover (given an instance of the set cover problem, find a fractional set cover z of minimum size $|z| = \sum_i z_i$; see the previous subsection for definitions). Given a set cover instance and $\varepsilon \in [0, 1/2]$, the algorithm returns a fractional set cover of size at most $1 + O(\varepsilon)$ times the optimum:

---

1. Let $N = 2\ln(n)/\varepsilon^2$, where n is the number of elements.
2. Repeat until all elements are sufficiently covered ($\min_j c(j) \geq N$):
3. Choose a set S_i maximizing $\sum_{x_j \in S_i}(1 - \varepsilon)^{c(j)}$, where $c(j)$ denotes the number of times any set containing element x_j has been chosen so far.
4. Return z, where z_i is the number of times S_i was chosen divided by N.

The naive implementation of this algorithm runs in $O(nM \log(n)/\varepsilon^2)$ time, where $M = \sum_i |S_i|$ is the size of the input. With appropriate modifications, the algorithm can be implemented to run in $O(M \log(n)/\varepsilon^2)$ time.

For readers who are interested, we sketch how this algorithm may be derived using the probabilistic framework. To begin, we imagine that we have in hand any fractional set cover z^*, to which we apply the following randomized algorithm: *Define probability distribution p on the sets by $p(S_i) = z_i^*/|z^*|$. Draw sets* *randomly according to p until every element has been covered (in a drawn set) at least $N = 2\ln(n)/\varepsilon^2$ times. Return z, where z_i is the number of times set S_i was drawn, divided by N.* (The reader should keep in mind that the dependence on z^* will be removed when we apply the method of conditional probabilities.)

Claim 4.1 *With non-zero probability the algorithm returns a fractional set cover of size at most $(1 + O(\varepsilon))|z^*|$.*

Next we prove the claim. Let $T = |z^*|N/(1 - \varepsilon)$. We will prove that, with non-zero probability, within T draws each set will be covered at least N times. This will prove the claim because then the size of z is at most $T/N = |z^*|/(1 - \varepsilon)$.

Fix a given element x_j. With each draw, the chance that x_j is covered is at least $1/|z^*|$. Thus, the expected number of times x_j is covered in T draws is at least $T/|z^*| = N/(1 - \varepsilon)$. By a standard Chernoff bound, the probability that x_j is covered less than N times in T rounds is at most $\exp(-\varepsilon^2 N/2(1 - \varepsilon)) < 1/n$.

By linearity of expectation, the expected number of elements that are covered less than N times in T rounds is less than 1. Thus, with non-zero probability, all elements are covered at least N times in T rounds.

This proves the claim. Next we apply the method of conditional probabilities to derive a greedy algorithm.

Let X_{jt} be an indicator variable for the event that x_j is covered in round t, so that for any j the X_{jt}'s are independent with $E[X_{jt}] \geq 1/|z^*|$. Let $\mu = N/(1 - \varepsilon)$. The proof of the Chernoff bound bounds $\Pr[\sum_t X_{jt} \leq (1 - \varepsilon)\mu]$ by the expectation of the following quantity:

$$\frac{(1 - \varepsilon)^{\sum_t X_{jt}}}{(1 - \varepsilon)^{(1-\varepsilon)\mu}} = \frac{(1 - \varepsilon)^{\sum_t X_{jt}}}{(1 - \varepsilon)^N}.$$

Thus, the proof of our aforementioned claim implicitly bounds the probability of failure by the expectation of

$$\phi = \sum_j \frac{(1 - \varepsilon)^{\sum_t X_{jt}}}{(1 - \varepsilon)^N}.$$

Furthermore, the proof shows that the expectation of this quantity is less than 1.

To apply the method of conditional probabilities, we will *choose each set to keep the conditional expectation of the above quantity ϕ below 1.*

After the first t sets have been drawn, the random variables X_{js} for $s \leq t$ are determined, whereas X_{js} for $s > t$ are not yet determined. Using the inequalities from the proof of the Chernoff bound, the conditional expectation of ϕ given the choices for the first t sets is at most

$$\phi_t \doteq \sum_j \frac{\prod_{s \leq t}(1 - \varepsilon)^{X_{js}} \times \prod_{s > t}(1 - \varepsilon/|z^*|)}{(1 - \varepsilon)^N}.$$

This quantity is initially less than 1, so it suffices to *choose each set to ensure $\phi_{t+1} \leq \phi_t$.* If the $t + 1$'st set is chosen randomly according to p, then $E[\phi_{t+1}] \leq \phi_t$. Thus, to ensure $\phi_{t+1} \leq \phi_t$, it suffices to choose the set to minimize ϕ_{t+1}. By a straightforward calculation, this is the same as choosing the set S_i to maximize $\sum_{x_j \in S_i}(1 - \varepsilon)^{\sum_{s \leq t} X_{jt}}$. This gives us the algorithm in question (at the top of this section). From the derivation, we know the following fact: *The above-mentioned algorithm returns a fractional set cover of size at most $(1 + O(\varepsilon)) \min_{z^*} |z^*|$, where z^* ranges over all the fractional set covers.*

4.11 Conclusion

In this chapter we surveyed a collection of problems and described simple greedy algorithms for several of these problems. In several cases, the greedy algorithms described do not represent the state of the art for these problems. The reader is referred to other chapters in this handbook to read in more detail about the specific problems and the techniques that yield the best worst-case approximation guarantees. In many instances, the performance of greedy algorithms may be better than their worst-case bounds suggest. This and their simplicity make them important in practice.

For some problems (e.g., set cover) it is known that a greedy algorithm gives the best possible approximation ratio unless $NP \subset \text{DTIME}(n^{\log \log n})$. But for some problems no such intractability results are yet known. In these cases, instead of proving hardness of approximation for all polynomial-time algorithms, one may try something easier: to prove that no *greedy* algorithm gives a good approximation. Of course this requires a formal definition of the class of algorithms. (A similar approach has been fruitful in competitive analysis of online algorithms.) Such a formal study of greedy algorithms with an eye towards lower bound results has been the subject of several recent papers [33].

For additional information on combinatorial optimization, the reader is referred to books by Papadimitriou and Steiglitz [2], Cook et al. [34], and a series of three books by Schrijver [35]. For more on approximation algorithms, there is a book by Vazirani [23], lecture notes by Motwani [36], and a book edited by Hochbaum [37]. There is a chapter on greedy algorithms in several textbooks, such as Kleinberg and Tardos [38] and Cormen et al. [39]. More on randomized algorithms can be found in a book by Motwani and Raghavan [40], and a survey by Shmoys [41].

References

1. Lawler, E., *Combinatorial Optimization: Networks and Matroids*, Holt, Rinehart and Wilson, New York, 1976.
2. Papadimitriou, C.H. and Steiglitz, K., *Combinatorial Optimization*, Prentice-Hall, Englewood Cliffs, NJ, 1982.
3. Johnson, D.S., Approximation algorithms for combinatorial problems, *JCSS*, 9, 256, 1974.
4. Stein, S.K., Two combinatorial covering theorems, *J. Comb. Theor. A*, 16, 391, 1974.
5. Lovász, L., On the ratio of optimal integral and fractional covers, *Disc. Math.*, 13, 383, 1975.
6. Chvátal, V., A greedy heuristic for the set-covering problem, *Math. Oper. Res.*, 4(3), 233, 1979.
7. Raz, R. and Safra, S., A sub-constant error-probability low-degree test, and a sub-constant error-probability PCP characterization of NP, in *Proceedings of STOC*, ACM, New York, 1997, p. 475.
8. Nemhauser, G.L. and Wolsey, L.A., *Integer and Combinatorial Optimization*, John Wiley & Sons, New York, 1988.
9. Blum, A., Jiang, T., Li, M., Tromp, J., and Yannakakis, M., Linear approximation of shortest superstrings, *JACM*, 41(4), 630, 1994.
10. Markowsky, G., Kou, L., and Berman, L., An fast algorithm for Steiner trees, *Acta Informatica*, 15, 141, 1981.
11. Takahashi, H. and Matsuyama, A., An approximate solution for the Steiner problem in graphs, *Math. Japonica*, 24(6), 573, 1980.
12. Zelikovsky, A., An 11/6-approximation algorithm for the network Steiner problem, *Algorithmica*, 9(5), 463, 1993.
13. Klein, P.N. and Ravi, R., A nearly best-possible approximation algorithm for node-weighted Steiner trees, *J. Algorithms*, 19(1), 104, 1995.
14. Guha, S. and Khuller, S., Improved methods for approximating node weighted Steiner trees and connected dominating sets, *Inf. Comput.*, 150(1), 57, 1999.
15. Hwang, F.K., Richards, D.S., and Winter, P., *The Steiner Tree Problem*, Number 53 in Annals of Discrete Mathematics, Elsevier Science Publishers B. V., Amsterdam, the Netherlands, 1992.

16. Gonzalez, T.F., Clustering to minimize the maximum intercluster distance, *Theor. Comput. Sci.*, 38, 293, 1985.

17. Guha, S. and Khuller, S., Approximation algorithms for connected dominating sets, *Algorithmica*, 20(4), 374, 1998.

18. Bar-Noy, A., Guha, S., Naor, J., and Schieber, B., Approximating the throughput of multiple machines in real-time scheduling, *SIAM J. Comput.*, 31(2), 331, 2001.

19. Fürer, M. and Raghavachari, B., An NC approximation algorithm for the minimum-degree spanning tree problem, in *Proceedings of the 28th Annual Allerton Conference on Communication, Control and Computing*, University of Illinois, IL, 1990, p. 274.

20. Ravi, R., Marathe, M.V., Ravi, S.S., Rosenkrantz, D.J., and Hunt III, H.B., Approximation algorithms for degree-constrained minimum-cost network-design problems, *Algorithmica*, 31(1), 58, 2001.

21. Goemans, M.X. and Williamson, D.P., A general approximation technique for constrained forest problems, *SIAM J. Comput.*, 24(2), 296, 1995.

22. Bar-Yehuda, R., One for the price of two: A unified approach for approximating covering problems. *Algorithmica*, 27(2), 131, 2000.

23. Vazirani, V.V., *Approximation Algorithms*, Springer-Verlag New York, Inc., New York, 2001.

24. Clarkson, K., A modification of the greedy algorithm for vertex cover, *Information Processing Letters*, 16, 23, 1983.

25. Alon, N., Spencer, J.H., and Erdős, P., *The Probabilistic Method*, Wiley-Interscience Series in Discrete Mathematics and Optimization, John Wiley & Sons, Chichester, UK, 1992.

26. Raghavan, P., Probabilistic construction of deterministic algorithms approximating packing integer programs, *JCSS*, 37(2), 130, 1988.

27. Raghavan, P. and Thompson, C., Randomized rounding: A technique for provably good algorithms and algorithmic proofs, *Combinatorica*, 7, 365, 1987.

28. Young, N.E., Randomized rounding without solving the linear program, in *Proceedings of SODA*, San Francisco, CA, 1995, p. 170.

29. Plotkin, S.A., Shmoys, D.B., and Tardos, É., Fast approximation algorithms for fractional packing and covering problems, *Math. Oper. Res.*, 20(2), 257, 1995.

30. Grigoriadis, M.D. and Khachiyan, L.G., Fast approximation schemes for convex programs with many blocks and coupling constraints, Technical Report DCS-TR-273, Rutgers University Computer Science Department, New Brunswick, NJ, 1991.

31. Young, N.E., Sequential and parallel algorithms for mixed packing and covering, in *Proceedings of IEEE FOCS*, 2001, p. 538.

32. Bienstock, D., *Potential Function Methods for Approximately Solving Linear Programming Problems: Theory and Practice*, Kluwer Academic Publishers, Boston, MA, 2002.

33. Borodin, A., Nielsen, M., and Rackoff, C., (Incremental) priority algorithms. *Algorithmica*, 37, 295, 2003.

34. Cook, W.J., Cunningham, W.H., Pulleyblank, W.R., and Schrijver, A., *Combinatorial Optimization*, John Wiley, New York, 1997.

35. Schrijver, A., *Combinatorial Optimization—Polyhedra and Efficiency, Volume A: Paths, Flows, Matchings, Volume B: Matroids, Trees, Stable Sets, Volume C: Disjoint Paths, Hypergraphs*, volume 24 of *Algorithms and Combinatorics*, Springer-Verlag, Berlin, Germany, 2003.

36. Motwani, R., Lecture notes on approximation algorithms, Technical Report, Stanford University, Stanford, CA, 1992.

37. Hochbaum, D., Ed., *Approximation Algorithms for NP-hard Problems*, PWS Publishing Company, Boston, MA, 1997.

38. Kleinberg, J. and Tardos, É., *Algorithm Design*, Addison-Wesley, Boston, MA, 2005.

39. Cormen, T.H., Leiserson, C.E., Rivest, R.L., and Stein C., *Introduction to Algorithms*, MIT Press, Cambridge, MA, 2001.
40. Motwani, R. and Raghavan, P., *Randomized Algorithms*, Cambridge University Press, New York 1997.
41. Shmoys, D.B., Computing near-optimal solutions to combinatorial optimization problems, in *Combinatorial Optimization*, Cook, W., Lovasz, L., and Seymour, P.D., Eds., AMS, Providence, PI, p. 355, 1995.

5

Recursive Greedy Methods

Guy Even

5.1 Introduction

Greedy algorithms are often the first algorithm that one considers for various optimization problems, and, in particular, covering problems. The idea is very simple: try to build a solution incrementally by augmenting a partial solution. In each iteration, select the "best" augmentation according to a simple criterion. The term greedy is used because the most common criterion is to select an augmentation that minimizes the ratio of "cost" to "advantage." We refer to the cost-to-advantage ratio of an augmentation as the *density* of the augmentation.

In the Set-Cover (SC) problem, every set S has a weight (or cost) $w(S)$. The "advantage" of a set S with respect to a partial cover $\{S_1, \ldots, S_k\}$ is the number of new elements covered by S, that is, $|S \backslash (S_1 \cup \ldots \cup S_k)|$. In each iteration, a set with a minimum density is selected and added to the partial solution until all the elements are covered. In the SC problem, it is easy to find an augmentation with minimum density simply by recomputing the density of every set in every iteration.

In this chapter we consider problems for which it is NP-hard to find an augmentation with minimum density. From a covering point of view, this means that there are exponentially many sets. However, these sets are succinctly represented using a structure with polynomial complexity. For example, the sets can be paths or trees in a graph. In such problems, applying the greedy algorithm is a nontrivial task. One way to deal with such a difficulty, is to try to approximate a minimum density augmentation. Interestingly, the augmentation itself is computed using a greedy algorithm, and this is why the algorithm is called the recursive greedy algorithm.

The recursive greedy algorithm was presented by Zelikovsky [1] and Kortsarz and Peleg [2]. In Reference 1, the directed Steiner tree (DST) problem in acyclic graphs was considered. In the DST problem the input consists of a directed graph $G = (V, E)$ with edge weights $w(e)$, a subset $X \subseteq V$ of terminals, and a root $r \in V$. The goal is to find a minimum weight subgraph that contains directed paths from r to every terminal in X. In Reference 2, the bounded diameter Steiner tree (BDST) problem was considered. In the BDST problem, the input consists of an undirected graph $G = (V, E)$ with edge costs

$w(e)$, a subset of terminals $X \subseteq V$, and a diameter parameter d. The goal is to find a minimum weight tree that spans X with diameter bounded by d. In both papers, it is proved that, for every $\varepsilon > 0$, the recursive greedy algorithm achieves an $O(|X|^{\varepsilon})$ approximation ratio in polynomial time. The recursive greedy algorithm is still the only nontrivial approximation algorithm known for these problems.

The presentation of the recursive greedy algorithm was simplified and its analysis was perfected by Charikar et al. [3]. In Reference 3, the recursive greedy algorithm was used for the DST problem. The improved analysis gave a poly-logarithmic approximation ratio in quasi-polynomial time (i.e., running time is $O(n^{c \log n})$, for a constant c).

The recursive greedy algorithm is a combinatorial algorithm (i.e., no linear programming or high precision arithmetic is used). The algorithm's description is simple and short. The analysis captures the intuition regarding the segments during which the greedy approach performs well. The running time of the algorithm is exponential in the depth of the recursion, and hence, reducing its running time is an important issue.

We present modifications of the recursive greedy algorithm that enable reducing the running time. Unfortunately, these modifications apply only to the restricted case in which the graph is a tree. We demonstrate these methods on the group Steiner (GS) problem [4] and its restriction to trees [5]. Following Reference 6, we show that for the GS problem over trees, the recursive greedy algorithm can be modified to give a poly-logarithmic approximation ratio in polynomial time. Better poly-logarithmic approximation algorithms were developed for the GS problem, however, these algorithms rely on linear programming [5,7].

Organization: In Section 5.2, we review the greedy algorithm for the SC problem. In Section 5.3, we present three versions of DST problems. We present simple reductions that allow us to focus on only one version. Section 5.4 constitutes the heart of this chapter; in it the recursive greedy algorithm and its analysis are presented. In Section 5.5, we consider the GS problem over trees. We outline modifications of the recursive greedy algorithm that enable a poly-logarithmic approximation ratio in polynomial time. We conclude in Section 5.6 with open problems.

5.2 A Review of the Greedy Algorithm

In this section we review the greedy algorithm for the Set-Cover (SC) problem and its analysis.

In the Set-Cover (SC) problem we are given a a set of elements, denoted by $U = \{1, \dots, n\}$ and a collection $\mathcal{R}$ of subsets of U. Each subset $S \in \mathcal{R}$ is also given a nonnegative weight $w(S)$. A subset $\mathcal{C} \subseteq \mathcal{R}$ is an SC if $\bigcup_{S' \in \mathcal{C}} S' = \{1, \dots, n\}$. The *weight* of a subset of $\mathcal{R}$ is simply the sum of the weights of the sets in $\mathcal{R}$. The goal in the SC problem to find a cover of minimum weight. We often refer to a subset of $\mathcal{R}$ that is not a cover as a *partial cover*.

The greedy algorithm starts with an empty partial cover. A cover is constructed by iteratively asking an *oracle* for a set to be added to the partial cover. This means that no backtracking takes place; every set that is added to the partial cover is kept until a cover is obtained. The oracle looks for a set with the lowest *residual density*, defined as follows:

Definition 5.1 *Given a partial cover $\mathcal{C}$, the residual density of a set S is the ratio*

$$\rho_{\mathcal{C}}(S) \triangleq \frac{w(S)}{|S \setminus \bigcup_{S' \in \mathcal{C}} S'|}.$$

Note that the residual density is nondecreasing (and may even increase) as the greedy algorithm accumulates sets. The performance guarantee of the greedy algorithm is summarized in the following theorem (see Chapter 4 on Greedy Methods):

Theorem 5.1 *The greedy algorithm computes a cover whose cost is at most $(1 + \ln n) \cdot w(C^*)$, where C^* is a minimum weight cover.*

There are two main question that we wish to ask about the greedy algorithm:

Question 1: What happens if the oracle is approximate? Namely, the oracle does not return a set with minimum residual density, but a set whose residual density is at most α times the minimum residual density? How does such an approximate oracle affect the approximation ratio of the greedy algorithm? In particular we are interested in the case that α is not constant (e.g., α depends on the number of uncovered elements). We note that in the SC problem, an exact oracle is easy to implement. But we will see a generalization of the SC problem in which the task of an exact oracle is NP-hard, and hence we will need to consider an approximate oracle.

Question 2: What happens if we stop the execution of the greedy algorithm before a complete cover is obtained? Suppose that we stop the greedy algorithm when the partial cover covers $\beta \cdot n$ elements in U. Can we bound the weight of the partial cover? We note that one reason for stopping the greedy algorithm before it ends is that we simply run out of "budget" and cannot "pay" for additional sets.

The following lemma helps answer both questions raised earlier. Let x denote the number of elements that are not covered by the partial cover. We say that the oracle is $\alpha(x)$-approximate if the residual density of the set it finds is at most $\alpha(x)$ times the minimum residual density.

Lemma 5.1 ([3]) *Suppose that the oracle of the greedy algorithm is $\alpha(x)$-approximate and that $\alpha(x)/x$ is a nonincreasing function. Let C_i denote partial cover accumulated by the greedy algorithm after adding i sets. Then,*

$$\frac{w(C_i)}{w(C^*)} \leq \int_{n-|\bigcup_{S' \in C_i} S'|}^{n} \frac{\alpha(x)}{x} dx.$$

Proof. The proof is by induction on n. When $n = 1$, the algorithm simply returns a set S such that $w(S) \leq \alpha(1) \cdot w(C^*)$. Since $\alpha(x)/x$ is nonincreasing, we conclude that $\alpha(1) \leq \int_0^1 \frac{\alpha(x)}{x} dx$, and the induction basis follows.

The induction step for $n > 1$ is proved as follows: Let $C_i = \{S_1, \ldots, S_i\}$. When the oracle computes S_1, its density satisfies: $w(S_1)/|S_1| \leq \alpha(n) \cdot w(C^*)/n$. Hence, $w(S_1) \leq |S_1| \cdot \frac{\alpha(n)}{n} \cdot w(C^*)$. Since $\alpha(x)/x$ is nonincreasing, $|S_1| \cdot \frac{\alpha(n)}{n} \leq \int_{n-|S_1|}^{n} \frac{\alpha(x)}{x} dx$. We conclude that

$$w(S_1) \leq \int_{n-|S_1|}^{n} \frac{\alpha(x)}{x} dx \cdot w(C^*). \tag{5.1}$$

Now consider the residual set system over the set of elements $\{1, \ldots, n\} \setminus S_1$ with the sets $S' = S \setminus S_1$. We keep the set weights unchanged, that is, $w(S') = w(S)$. The collection $\{S'_2, \ldots, S'_i\}$ is the output of the greedy algorithm when given this residual set system. Let $n' = |S'_2 \cup \cdots \cup S'_i|$. Since C^* induces a cover of the residual set with the same weight as $w(C^*)$, the induction hypothesis implies that

$$w(S'_2) + \cdots + w(S'_i) \leq \int_{n-(n'+|S_1|)}^{n-|S_1|} \frac{\alpha(x)}{x} dx \cdot w(C^*). \tag{5.2}$$

The lemma follows now by adding Equations 5.1 and 5.2. ∎

We remark that for a full cover, since $\int_0^1 dx/x$ is not bounded, one could bound the ratio by $\alpha(1) + \int_1^n \frac{\alpha(x)}{x} dx$. Note that for an exact oracle $\alpha(x) = 1$, this modification of Lemma 5.1 implies Theorem 5.1.

Lemma 5.1 shows that the greedy algorithm works also with approximate oracles. If $\alpha(x) = O(\log x)$, then the approximation ratio of the greedy algorithm is simply $O(\alpha(n) \cdot \log n)$. But, for example, if $\alpha(x) = x^\varepsilon$, then the lemma "saves" a factor of $\log n$ and shows that the approximation ratio is $\frac{1}{\varepsilon} \cdot n^\varepsilon$. So this settles the first question.

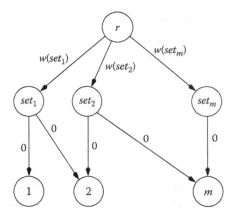

FIGURE 5.1   Reduction of SC instance to DST instance.

Lemma 5.1 also helps settle the second question. In fact, it proves that the greedy algorithm (with an exact oracle) is a bi-criteria algorithm in the following sense.

Claim 5.1 *If the greedy algorithm is stopped when $\beta \cdot n$ elements are covered, then the cost of the partial cover is bounded by $\ln\left(\frac{1}{1-\beta}\right) \cdot w(C^*)$.*

The greedy algorithm surly does well with the first set it selects, but what can we say about the remaining selections? Claim 5.1 quantifies how well the greedy algorithm does as a function of the portion of the covered elements. For example, if $\beta = 1 - 1/e$, then the partial cover computed by the greedy algorithm weighs no more than $w(C^*)$. (We ignore here the knapsack-like issue of how to cover "exactly" $\beta \cdot n$ elements, and assume that, when we stopped the greedy algorithm, the partial cover covers $\beta \cdot n$ elements.) The lesson to be remembered here is that the greedy algorithm performs "reasonably well" as long as "few" elements have been covered.

The DST problem is a generalization of the SC problem. In fact, every SC instance can be represented as a DST instance over a layered directed graph with three vertex layers (Figure 5.1). The top layer contains only a root, the middle layer contains a vertex for every set, and the bottom layer contains a vertex for every element. The weight of an edge from the root to a set is simply the weight of the set. The weight of all edges from sets to elements are zero. The best approximation algorithm for SC is the greedy algorithm. What form could a greedy algorithm have for the DST problem?

5.3 Directed Steiner Problems

In this section we present three versions of DST problems. We present simple reductions that allow us to focus on only the last version.

Notation and terminology: We denote the vertex set and edge set of a graph G by $V(G)$ and $E(G)$, respectively. An *arborescence* T rooted at r is a directed graph such that: (i) the underlying graph of T is a tree (i.e., if edge directions are ignored in T, then T is a tree) and (ii) there is a directed path in T from the root r to every node in T. If an arborescence T is a subgraph of G, then we say that T *covers* (or *spans*) a subset of vertices X if $X \subseteq V(T)$. If edges have weights $w(e)$, then the weight of a subgraph G' is simply $\sum_{e \in E(G')} w(e)$. We denote by T_v the subgraph of T that is induced all the vertices reachable from v (including v).

5.3.1 The Problems

The DST problem: In the DST problem the input consists of a directed graph G, a set of terminals $X \subseteq V(G)$, positive edge weights $w(e)$, and a root $r \in V(G)$. An arborescence T rooted at r is a DST if it spans the set of terminals X. The goal in the DST problem is to find a minimum weight directed Steiner tree.

The k-DST problem: Following Reference 3, we consider a version of the DST problem, called k-DST, in which only part of the terminals must be covered. In the k-DST problem, there is an additional parameter k, often called the *demand*. An arborescence T rooted at r is a k-*partial* directed Steiner tree (k-DST) if $|V(T) \cap X| \geq k$. The goal in the k-DST problem is to find a minimum weight k-partial directed Steiner tree. We denote the weight of an optimal k-partial directed Steiner tree by $DS^*(G, X, k)$. (Formally, the root r should be a parameter, but we omit it to shorten notation.) We encode DST instances as k-DST instances simply by setting $k = |X|$.

The ℓ-shallow k-DST problem: Following Reference 2, we consider a version of the k-DST problem in which the length of the paths from the root to the terminals is bounded by a parameter ℓ. A rooted arborescence in which every node is at most ℓ edges away from the root is called an ℓ-*layered tree*. (Note that we count the number of layers of edges; the number of layers of nodes is $\ell + 1$.) In the ℓ-shallow k-DST problem the goal is to compute a minimum k-DST among all ℓ-layered trees.

5.3.2 Reductions

Obviously, the k-DST problem is a generalization of the DST problem. Similarly, the ℓ-shallow k-DST problem is a generalization of the k-DST problem (i.e., simply set $\ell = |V| - 1$). The only nontrivial approximation algorithm we know is for the ℓ-shallow k-DST problem; this approximation algorithm is a recursive greedy algorithm. Since its running time is exponential in ℓ, we need to consider reductions that result with as small as possible values of ℓ.

For this purpose we consider two well-known transformations: transitive closure and layering. We now define each of these transformations.

Transitive closure: The *transitive closure* of G is a directed graph $TC(G)$ over the same vertex set. For every $u, v \in V$, the pair (u, v) is an edge in $E(TC(G))$ if there is a directed path from u to v in G. The weight $w'(u, v)$ of an edge in $E(TC(G))$ is the minimum weight of a path in G from u to v.

The weight of an optimal k-DST is not affected by applying transitive closure namely,

$$DS^*(G, X, k) = DS^*(TC(G), X, k). \tag{5.3}$$

This means that replacing G by its transitive closure does not change the weight of an optimal k-DST. Hence we may assume that G is transitively closed, that is, $G = TC(G)$.

Layering: Let ℓ denote a positive integer. We reduce the directed graph G into an ℓ-layered directed acyclic graph LG_ℓ as follows (Figure 5.2). The vertex set $V(LG_\ell)$ is simply $V(G) \times \{0, \ldots, \ell\}$. The jth layer in $V(LG_\ell)$ is the subset of vertices $V(G) \times \{j\}$. We refer to $V(G) \times \{0\}$ as the *bottom layer* and to $V(G) \times \{\ell\}$ as the *top layer*. The graph LG_ℓ is layered in the sense that $E(LG_\ell)$ contains only edges from the $V(G) \times \{j + 1\}$ to $V(G) \times \{j\}$, for $j < \ell$. The edge set $E(LG_\ell)$ contains two types of edges: *regular* edges and *parallel* edges. For every $(u, v) \in E(G)$ and every $j < \ell$, there is a regular edge $(u, j + 1) \to (v, j) \in E(LG_\ell)$. For every $u \in V$ and every $j < \ell$, there is a parallel edge $(u, j + 1) \to (u, j) \in E(LG_\ell)$. All parallel edges have zero weight. The weight of a regular edge is inherited from the original edge, namely, $w((u, j + 1) \to (v, j)) = w(u, v)$. The set of terminals X' in $V(LG_\ell)$ is simply $X \times \{0\}$, namely, images of terminals in the bottom layer. The root in LG_ℓ is the node (r, ℓ). The following observation shows that we can restrict our attention to layered graphs.

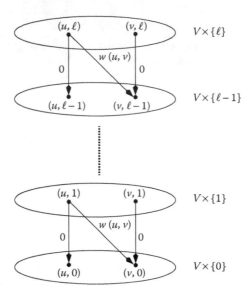

FIGURE 5.2 Layering of a directed graph G. Only parallel edges incident to images of $u, v \in V(G)$ and regular edges corresponding to $(u, v) \in E(G)$ are depicted.

Observation 5.1 *There is a weight preserving and terminal preserving correspondence between ℓ-layered r-rooted trees in G and (r, ℓ)-rooted trees in LG_ℓ. In particular, $w(LT_\ell^*) = DS^*(LG_\ell, X', k)$, where LT_ℓ^* denotes a minimum weight k-DST among all ℓ-layered trees.*

Observation 5.1 implies that if we wish to approximate LT_ℓ^*, then we may apply layering and assume that the input graph is an ℓ-layered acyclic graph in which the root is in the top layer and all the terminals are in the bottom layer.

Limiting the number of layers: As we pointed out, the running time of the recursive greedy algorithm is exponential in the number of layers. It is therefore crucial to be able to bound the number of layers. The following lemma bounds the penalty incurred by limiting the number of layers in the Steiner tree. The proof of the lemma appears in Appendix 5.6 and uses notation introduced in Section 5.4. (A slightly stronger version appears in Reference 8, with the ratio $2^{1-1/\ell} \cdot \ell \cdot k^{1/\ell}$.)

Lemma 5.2 ([1] corrected in [8]) *If G is transitively closed, then $w(LT_\ell^*) \leq \frac{\ell}{2} \cdot k^{2/\ell} \cdot DS^*(G, X, k)$.*

It follows that an α-approximate algorithm for an ℓ-shallow k-DST is also an $\alpha\beta$-approximation algorithm for k-DST, where $\beta = \frac{\ell}{2} \cdot k^{2/\ell}$. We now focus on the development of an approximation algorithm for the ℓ-shallow k-DST problem.

5.4 A Recursive Greedy Algorithm for ℓ-Shallow k-DST

In this section we present a recursive greedy algorithm for the ℓ-shallow k-DST problem. Based on the layering transformation, we assume that the input graph is an ℓ-layered acyclic directed graph G. The set of terminals, denoted by X, is contained in the bottom layer. The root, denoted by r, belongs to the top layer.

5.4.1 Motivation

We now try to extend the greedy algorithm to the ℓ-shallow k-DST problem. Suppose we have a directed tree $T \subseteq G$ that is rooted at r. This tree only covers part of the terminals. Now we wish to augment T so that it covers more terminals. In other words, we are looking for an r-rooted augmenting tree T_{aug} to be added to the T. We follow the minimum density heuristic, and define the residual density of T_{aug} by

$$\rho_T(T_{aug}) \triangleq \frac{w(T_{aug})}{|(T_{aug} \cap X) \setminus (T \cap X)|}.$$

All we need now is an algorithm that finds an augmenting tree with the minimum residual density. Unfortunately, this problem is by itself NP-hard! Consider the following reduction: Let G denote the above-mentioned 2-layered DST instance to represent a Set-Cover instance. Add a layer with a single node r' that is connected to the root r of G. The weight of the edge (r', r) should be large (say, n times the sum of the weights of the sets). It is easy to see that every minimum density subtree must span all the terminals. Hence, every minimum density subtree induces a minimum weight set cover, and finding a minimum density subtree in a 3-layered graph is already NP-hard. We show in Section 5.4.3 that for two or less layers, one can find a minimum density augmenting tree in polynomial time.

We already showed that the greedy algorithm works well also with an approximate oracle. So we try to approximate a subtree with minimum residual density. The problem is how to do it? The answer is by applying a greedy algorithm recursively!

Consider an ℓ-layered directed graph and a root r. The algorithm finds an low density ℓ-layered augmenting tree by accumulating low density $(\ell - 1)$-layered augmenting trees that hang from the children of r. These trees are found by augmenting low density trees that hang from grandchildren of r, and so on. We now formally describe the algorithm.

5.4.2 The Recursive Greedy Algorithm

Notation: We denote the number of terminals in a subgraph G' by $k(G')$ (i.e., $k(G') = |X \cap V(G')|$). Similarly, for a set of vertices U, $k(U) = |X \cap U|$. We denote the set of vertices reachable in G from u by $desc(u)$. We denote the layer of a vertex u by $layer(u)$ (e.g., if u is a terminal, then $layer(u) = 0$).

Description: A listing of the algorithm $DS(u, k, X)$ appears as Algorithm 5.1. The stopping condition is when u belongs to the bottom layer or when the number of uncovered terminals reachable from u is less than the demand k (i.e., the instance is infeasible). In either case, the algorithm simply returns the root $\{r\}$.

ALGORITHM 5.1   DS(u, k, X)—A recursive greedy algorithm for the directed Steiner tree problem. The graph is layered and all the vertices in the bottom layer are terminals. The set of terminals is denoted by X. We are searching for a tree rooted at u that covers k terminals

1. **stopping condition: if** $layer(u) = 0$ or $k(desc(u)) < k$ **then** return $(\{u\})$.
2. **initialize:** $T \leftarrow \{u\}$; $X^{res} \leftarrow X$;.
3. **while** $k(T) < k$ **do**
4.   **recurse:** for every $v \in children(u)$ and every $k' \leq \min\{k - k(T), |desc(v) \cap X^{res}|\}$

$$T_{v,k'} \leftarrow DS(v, k', X^{res}).$$

5.   **select:** Let T_{aug} be a lowest residual density tree among the trees $T_{v,k'} \cup \{(u, v)\}$, where $v \in children(u)$ and $k' \leq k - k(T)$.
6.   **augment & update:** $T \leftarrow T \cup T_{aug}$; $X^{res} \leftarrow X^{res} \setminus V(T_{aug})$.
7. **end while**
8. **return** (T).

The algorithm maintains a partial cover T that is initialized to the single vertex u. The augmenting tree T_{aug} is selected as the best tree found by the recursive calls to the children of u (together with the edge from u to its child). Note that the recursive calls are applied to all the children of u and all the possible demands k'. After T_{aug} is added to the partial solution, the terminals covered by T_{aug} are erased from the set of terminals so that the recursive calls will not attempt to cover terminals again. Once the demand is met, namely, k terminals are covered, the accumulated cover T is returned.

The algorithm is invoked with the root r, the demand k, and the set of terminals X. Note that if the instance is feasible (namely, at least k terminals are reachable from the root), then the algorithm never encounters infeasible sub-instances during its execution.

5.4.3 Analysis

Minimum residual density subtree: Consider a partial solution T rooted at u accumulated by the algorithm. A tree T' rooted at u is a *candidate tree* for augmentation if: (i) every vertex $v \in V(T')$ in the bottom layer of G is in X^{res} (i.e., T' covers only new terminals) and (ii) $0 < k(T') \leq k - k(T)$ (i.e., T' does not cover more terminals than the residual demand). We denote by T'_u a tree with minimum residual density among all the candidate trees.

We leave the proof of the following lemma as an exercise.

Lemma 5.3 *Assume that $w_i, k_i > 0$, for every $0 \leq i \leq n$. Then,* $\min_i \frac{w_i}{k_i} \leq \frac{\sum_i w_i}{\sum_i k_i} \leq \max_i \frac{w_i}{k_i}$.

Corollary 5.1 *If u is not a terminal, then we may assume that u has a single child in T'_u.*

Proof. We show that we could pick a candidate tree with minimum residual density in which u has a single child. Suppose that u has more than one child in T'_u. To every edge $e_j = (u, v_j) \in E(T'_u)$ we match a subtree A_{e_j} of T'_u. The subtree A_{e_j} contains u, the edge (u, v_j), and the subtree of T'_u hanging from v_j. The subtrees $\{A_{e_j}\}_{e_j}$ form an edge-disjoint decomposition of T'_u. Let $w_j = w(A_{e_j})$ and $k_j = k(A_{e_j} \setminus T)$. Since u is not a terminal, the subtrees $\{A_{e_j}\}_{e_j}$ partition the terminals in $V(T'_u)$, and $k(T'_u) = \sum_j k_j$. Similarly, $w(T'_u) = \sum_j w_j$. By Lemma 5.3, it follows that one of the trees A_{e_j} has a residual density that is not greater than the residual density of T'_u. Use this minimum residual density subtree instead of T'_u, and the corollary follows. ∎

Density: Note that edge weights are nonnegative and already covered terminals do not help in reducing the residual density. Therefore, every augmenting tree T_{aug} covers only new terminals and does not contain terminals already covered by T. It follows that every terminal in T_{aug} belongs to X^{res} and, therefore, $k(T_{aug}) = |T_{aug} \cap X^{res}|$. We may assume that the same holds for T'_u; namely, T'_u does not contain already covered terminals. Therefore, where possible, we ignore the "context" T in the definition of the residual density and simply refer to *density*, that is, the density of a tree T' is $\rho(T') = w(T')/|V(T') \cap X|$.

Notation and terminology: A *directed star* is a 1-layered rooted directed graph (i.e., there is a center out of which directed edges emanate to the leaves). We abbreviate and refer to a directed star simply as a star. A *flower* is a 2-layered rooted graph in which the root has a single child.

Bounding the density of augmenting trees: When $layer(u) = 1$, if u has least k terminal neighbors, then the algorithm returns a star centered at u. The number of edges emanating from r in the star equals k, and these k edges are the k lightest edges emanating from r to terminals. It is easy to see that in this case the algorithm returns an optimal k-DST.

The analysis of the algorithm is based on the following claim that bounds the ratio between the densities of the augmenting tree and T'_u.

Claim 5.2 ([3]) *If $layer(u) \geq 2$, then, in every iteration of the while loop in an execution of $DS(u, k)$, the subtree T_{aug} satisfies:*

$$\rho(T_{aug}) \leq (layer(u) - 1) \cdot \rho(T'_u)$$

Proof. The proof is by induction on $layer(u)$. Suppose that $layer(u) = 2$. By Corollary 5.1, T'_u is a flower that consists of a star S_v centered at a neighbor v of u, the node u, and the edge (u, v). Moreover, S_v contains the $k(T'_u)$ closest terminals to v. When the algorithm computes T_{aug} it considers all stars centered at children v' of u consisting of the $k' \leq k - k(T)$ closest terminals to v'. In particular, it considers the star S_v together with the edge (u, v). Hence, $\rho(T_{aug}) \leq \rho(T'_u)$, as required.

We now prove the induction step for $layer(u) > 2$. Let $i = layer(u)$. The setting is as follows: during an execution of $DS(u, X)$, a partial cover T has been accumulated, and now an augmenting tree T_{aug} is computed. Our goal is to bound the density of T_{aug}.

By Corollary 5.1, u has a single child in T'_u. Denote this child by u'. Let $B_{u'}$ denote the subtree of T'_u that hangs from u' (i.e., $B_{u'} = T'_u \setminus \{u, (u, u')\}$). Let $k' = k(T'_u)$.

We now analyze the selection of T_{aug} while bearing in mind the existence of the "hidden candidate" T'_u that covers k' terminals. Consider the tree $T_{u',k'}$ computed by the recursive call $DS(u', k', X^{res})$. We would like to argue that $T_{u',k'}$ should be a good candidate. Unfortunately, that might not be true! However, recall that the greedy algorithm does "well" as long as "few" terminals are covered. So we wish to show that a "small prefix" of $T_{u',k'}$ is indeed a good candidate. We now formalize this intuition.

The tree $T_{u',k'}$ is also constructed by a sequence of augmenting trees, denoted by $\{A_j\}_j$. Namely, $T_{u',k'} = \bigcup_j A_j$. We identify the smallest index ℓ for which the union of augmentations $A_1 \cup \cdots \cup A_\ell$ covers at least $k'/(i-1)$ terminals (recall that $i = layer(u)$). Formally,

$$k\left(\bigcup_{j=1}^{\ell-1} A_j\right) < \frac{k'}{(i-1)} \leq k\left(\bigcup_{j=1}^{\ell} A_j\right).$$

Our goal is to prove the following two facts. Fact (1): Let $k'' = k\left(\bigcup_{j=1}^{\ell} A_j\right)$, then the candidate tree $T_{u',k''} = DS(u', k'', X^{res})$ equals the prefix $\bigcup_{j=1}^{\ell} A_j$. Fact (2): The density of $T_{u',k''}$ is small, that is, $\rho(T_{u',k''}) \leq (i-1) \cdot \rho(B_{u'})$.

The first fact is a "simulation argument" since it claims that the union of the first ℓ augmentations computed in the course of the construction of $T_{u',k'}$ is actually one of the candidate trees computed by the algorithm. This simulation argument holds because, as long as the augmentations do not meet the demand, the same prefix of augmentations is computed. Note that k'' is the formalization of "few" terminals (compared to k'). Using $k'/(i-1)$ as a measure for a few terminals does not work because the simulation argument would fail.

The second fact states that the density of the candidate $T_{u',k''}$ is smaller than $(i-1) \cdot \rho(B_{u'})$. Note that $B_{u'}$ and $A_1 \cup \cdots \cup A_{\ell-1}$ may share terminals (in fact, we would "like" the algorithm to "imitate" $B_{u'}$ as much as possible). Hence, the residual density of $B_{u'}$ may increase as a result of adding the trees $A_1, \ldots, A_{\ell-1}$. However, since $k(A_1 \cup \cdots \cup A_{\ell-1}) < k'/(i-1)$, it follows that even after accumulating $A_1 \cup \cdots \cup A_{\ell-1}$, the residual density of $B_{u'}$ does not grow by much. Formally, the residual density of $B_{u'}$ after accumulating $A_1 \cup \cdots A_{\ell-1}$ is bounded as follows:

$$\begin{aligned} \rho_{(T \cup A_1 \cup \cdots \cup A_{\ell-1})}(B_{u'}) &= \frac{w(B_{u'})}{k' - k(A_1 \cup \cdots A_{\ell-1})} \\ &\leq \frac{w(B_{u'})}{k' \cdot (1 - \frac{1}{i-1})} \\ &= \left(\frac{i-1}{i-2}\right) \cdot \rho(B_{u'}). \end{aligned} \tag{5.4}$$

We now apply the induction hypothesis to the augmenting trees A_j (for $j \leq \ell$), and bound their residual densities by $(layer(u') - 1)$ times the "deteriorated" density of $B_{u'}$. Formally, the induction hypothesis implies that when A_j is selected as an augmentation tree its density satisfies:

$$\rho(A_j) \le (i-2) \cdot \rho_{(T \cup A_1 \cdots \cup A_{j-1})}(B_{u'})$$
$$\le (i-1) \cdot \rho(B_{u'}) \qquad\qquad \text{(by Eq. 5.4)}$$

By Lemma 5.3, $\rho(\bigcup_{j=1}^{\ell} A_j) \le \max_{j=1..\ell} \rho(A_j)$. Hence $\rho(T_{u',k''}) \le (i-1) \cdot \rho(B_{u'})$, and the second fact follows:

To complete the proof, we need to deal with the addition of the edge (u, u').

$$\rho(\{(u,u')\} \cup T_{u',k''}) = \frac{w(u,u') + w(T_{u',k''})}{k''}$$
$$\le \frac{w(u,u')}{k'} \cdot (i-1) + \rho(T_{u',k''}) \qquad \left(\text{since } k'' \ge \frac{k'}{i-1}\right)$$
$$\le (i-1) \cdot \rho(\{(u,u')\} \cup B_{u'}) \qquad\qquad \text{(by fact (2))}$$
$$= (i-1) \cdot \rho(T_u').$$

The claim follows since $\{(u,u')\} \cup T_{u',k''}$ is only one of the candidates considered for the augmenting tree T_{aug} and hence $\rho(T_{aug}) \le \rho(\{(u,u')\} \cup T_{u',k''})$.   ∎

Approximation ratio: The approximation ratio follows immediately from Lemma 5.1.

Claim 5.3 *Suppose that G is ℓ-layered. Then, the approximation ratio of Algorithm $DS(r,k,X)$ is $O(\ell \cdot \log k)$.*

Running time: For each augmenting tree, Algorithm $DS(u,k,X)$ invokes at most $n \cdot k$ recursive calls from children of u. Each augmentation tree covers at least one new terminal, so there are at most k augmenting trees. Hence, there are at most $n \cdot k^2$ recursive calls from the children of u. Let $time(\ell)$ denote the running time of $DS(u,k,X)$, where $\ell = layer(u)$. Then the following recurrence holds: $time(\ell) \le (n \cdot k^2) \cdot time(\ell-1)$. We conclude that the running time is $O(n^{\ell} \cdot k^{2\ell})$.

5.4.4 Discussion

Approximation of k-DST: The approximation algorithm is presented for ℓ-layered acyclic graphs. In Section 5.3.2, we presented a reduction from the k-DST problem to the ℓ-shallow k-DST problem. The reduction is based on layering and its outcome is an ℓ-layered acyclic graph. We obtain the following approximation result from this reduction:

Theorem 5.2 ([3]) *For every ℓ, there an $O(\ell^3 \cdot k^{2/\ell})$-approximation algorithm for the k-DST problem with running time $O(k^{2\ell} \cdot n^{\ell})$.*

Proof. The preprocessing time is dominated by the running time of $DS(r,k,X)$ on the graph after it is transitively closed and layered into ℓ layers.

Let R^* denote an minimum residual density augmenting tree in the transitive closure of the graph (without the layering). Let T'_{k^*} denote a minimum residual subtree rooted at u in the layered graph among the candidate trees that cover $k(R^*)$ terminals. By Lemma 5.2, $w(T'_{k^*}) \le \ell/2 \cdot k(R^*)^{\ell/2} \cdot w(R^*)$, and hence, $\rho(T'_{k^*}) \le \ell/2 \cdot k(R^*)^{\ell/2} \cdot \rho(R^*)$. Since $\rho(T'_u) \le \rho(T'_{k^*})$, by Claim 5.2 it follows that $\rho(T_{aug}) \le (\ell-1) \cdot \ell/2 \cdot k^{2/\ell} \cdot \rho(R^*)$.

We now apply Lemma 5.1. Note that $\int \frac{x^{2/\ell}}{x} dx = \frac{\ell}{2} \cdot x^{2/\ell}$. Hence, $w(T) = O(\ell^3 \cdot k^{2/\ell})$, where T is the tree returned by the algorithm, and the theorem follows.   ∎

We conclude with the following result:

Corollary 5.2 *For every constant $\varepsilon > 0$, there exists a polynomial time $O(k^{1/\varepsilon})$-approximation algorithm for the k-DST problem. There exists a quasi-polynomial time $O(\log^3 k)$-approximation algorithm for the k-DST problem.*

Proof. Substitute $\ell = 2/\varepsilon$ and $\ell = \log k$ in Theorem 5.2. ■

Preprocessing: Computing the transitive closure of the input graph is necessary for the correctness of the approximation ratio. Recall that Lemma 5.2 holds only if G is transitively closed.

Layering, on the other hand, is used to simplify the presentation. Namely, the algorithm can be described without layering [2,3]. The advantage of using layering is that it enables a unified presentation of the algorithm (i.e., there is no need to deal differently with 1-layered trees). In addition, the layered graph is acyclic, so we need not consider multiple "visits" of the same node. Finally, for a given node u, we know from its layer what the recursion level is (i.e., the recursion level is $\ell - layer(u)$) and what the height of the tree we are looking for is (i.e., current height is $layer(u)$).

Suggestions for improvements: One might try to reduce the running time by not repeating computations associated with the computations of candidate trees. For example, when computing the candidate $T_{v,k-k(T)}$ the algorithm computes a sequence of augmenting trees that is used to build also other candidates rooted at v that cover fewer terminals (we relied on this phenomenon in the simulation argument used in the proof of Claim 5.2) . However, such improvements do not seem to reduce the asymptotic running time; namely, the running time would still be exponential in the number of layers and the basis would still be polynomial. We discuss other ways to reduce the running time in the next section.

Another suggestion to improve the algorithm is to zero the weight of edges when they are added to the partial cover T [1]. Unfortunately, we do not know how to take advantage of such a modification in the analysis and, therefore, keep the edge weights unchanged even after we pay for them.

5.5 Improving the Running Time

In this section we consider a setting in which the recursive greedy algorithm can be modified to obtain a poly-logarithmic approximation ratio in polynomial time. The setting is with a problem called the Group Steiner (GS) problem, and only part of the modifications are applicable also to the k-DST problem. (Recall that the problem of finding a polynomial time poly-logarithmic approximation algorithm for k-DST is still open.)

Motivation: We saw that the running time of the recursive greedy algorithm is $O((nk^2)^{\ell})$, where k is the demand (i.e., number of terminals that need to be covered), the degree of a vertex is $n - 1$ (since transitive closure was applied, the graph is complete), and ℓ is the bound on the number of layers we allow in the k-DST.

To obtain polynomial running times, we first modify the algorithm and preprocess the input so that its running time is $\log(n)^{O(\ell)}$. We then set $\ell = \log n/ \log \log n$. Note that

$$(\log n)^{\frac{\log n}{\log \log n}} = n.$$

Hence, a polynomial running time is obtained!

Four modifications are required to make this idea work:

1. Bound the number of layers - we already saw that the penalty incurred by limiting the number of layers can be bounded. In fact, according to Lemma 5.2, the penalty incurred by $\ell = \log n/ \log \log n$ is poly-logarithmic (since $\ell \cdot k^{2/\ell} = (\log n)^{O(1)}$).

2. Degree reduction - we must reduce the maximum degree so that it is poly-logarithmic, otherwise too many recursive calls are invoked. Preprocessing of GS instances over trees achieves such a reduction in the degree.

3. Avoiding small augmenting trees - we must reduce the number of iterations of the while-loop. The number of iterations can be bounded by $(log n)^c$ if we require that every augmenting tree must cover at least a poly-logarithmic fraction of the residual demand.
4. Geometric search - we must reduce the number of recursive calls. Hence, instead of considering all demands below the residual demand, we consider only demands that are powers of $(1 + \varepsilon)$.

The GS problem over trees: We now present a setting where all four modifications can be implemented. In the GS problem over trees the input consists of: (i) an undirected tree T rooted at r with nonnegative edge edges $w(e)$ and (ii) groups $g_i \subseteq V(T)$ of terminals. A subtree $T' \subseteq T$ rooted at r covers k groups if $V(T')$ intersects at least k groups. We refer to a subtree that covers k groups as a k-GS tree. The goal is to find a minimum weight k-GS tree.

We denote the number of vertices by n and the number of groups by m. For simplicity, assume that every terminal is leaf of T and that every leaf of T is a terminal. In addition, we assume that the groups g_i are disjoint. Note that the assumption that the groups are disjoint implies that $\sum_{i=1}^{m} |g_i| \le n$.

Bounding the number of layers: Lemma 5.2 applies also to GS instances over trees, provided that the transitive closure is used. Before transitive closure is used, we direct the edges from the node closer to the root to the node farther away from the root. As mentioned earlier, limiting the number of layers to $\ell = \log n / \log \log n$ incurs a poly-logarithmic penalty.

However, there is a problem with bounding the number of layers according to Lemma 5.2. The problem is that we need to transitively close the tree. This implies that we lose the tree topology and end up with an directed acyclic graph instead. Unfortunately, we only know how to reduce the maximum degree of trees, not of directed acyclic graphs. Hence, we need to develop a different reduction that keeps the tree topology.

In Reference 6, a height reduction for trees is presented. This reduction replaces T by an ℓ-layered tree T'. The penalty incurred by this reduction is $O(n^{c/\ell})$, where c is a constant. The details of this reduction appear in Reference 6.

Reducing the maximum degree: We now sketch how to preprocess the tree T to obtain a tree $\nu(T)$ such that: (i) There is a weight preserving correspondence between k-GS trees in T and in $\nu(T)$. (ii) The maximum number of children of a vertex in $\nu(T)$ is bounded by an integer $\beta \ge 3$. (iii) The number of layers in $\nu(T)$ is bounded by the number of layers in T plus $\lfloor \log_{\beta/2} n \rfloor$. We set $\beta = \lceil \log n \rceil$, and obtain the required reduction.

We define a node $v \in V(T)$ to be β-*heavy* if the number of terminals that are descendents of v is at least n/β; otherwise v is β-*light*.

Given a tree T rooted at u and a parameter β, the tree $\nu(T)$ is constructed recursively as follows: If u is a leaf, then the algorithm returns u. Otherwise, the star induced by u and its children is locally transformed as follows: Let $v_1, v_2, \ldots, v_k$ denote the children of u.

1. Edges between u and β-heavy children v_i of u are not changed.
2. The β-light children of u are grouped arbitrarily into minimal bunches such that each bunch (except perhaps for the last) is β-heavy. Note that the number of leaves in each bunch (except perhaps for the last bunch) is in the half closed interval $[n_u/\beta, 2n_u/\beta)$. For every bunch B, a new node b is created. An edge (u, b) is added as well as edges between b and the children of u in the bunch B. The edge weights are set as follows: (a) $w(u, b) \leftarrow 0$ and (b) $w(b, v_i) \leftarrow w(u, v_i)$.

After the local transformation, let $v'_1, v'_2, \ldots, v'_j$ be the new children of u. Some of these children are the original children and some are the new vertices introduced in the bunching. The tree $\nu(T)$ is obtained by recursively processing the subtrees $T_{v'_i}$, for $1 \le i \le j$, in essence replacing $T_{v'_i}$ by $\rho(T_{v'_i})$.

The maximum number of children after processing is at most β because the subtrees $\{T_{v'_i}\}_i$ partition the nodes of $V(T_u) - \{u\}$ and each tree except, perhaps one, is β-heavy. The recursion is applied to each subtree $T_{v'_i}$, and hence $\nu(T)$ will satisfies the degree requirement, as claimed. The weight preserving correspondence between k-GS trees in T and in $\nu(T)$ follows from the fact that the "shared" edges (u, b) that were created for bunching together β-light children of u have zero weight.

We now bound the height of $\nu(T)$. Consider a path p in $\nu(T)$ from the root r to a leaf v. All we need to show is that p contains at most $\log_{\beta/2} n$ new nodes (i.e., nodes corresponding to bunches of β-light vertices). However, the number of terminals hanging from a node along p decreases by a factor of $\beta/2$ every time we traverse such a new node, and the bound on the height of $\nu(T)$ follows.

The modified algorithm: We now present the modified recursive greedy algorithm for GS over trees. A listing of the modified recursive greedy algorithm appears as Algorithm 5.2.

ALGORITHM 5.2   Modified-GS(u, k, $\mathcal{G}$)—Modified Recursive Greedy Algorithm for k-GS over Trees

1. **stopping condition: if** u is a leaf **then** return $(\{u\})$.
2. **Initialize:** $cover \leftarrow \{u\}$ and $\mathcal{G}^{res} \leftarrow \mathcal{G}$.
3. **while** $k(cover) < k$ **do**
4.   **recurse:** for every $v \in \text{children}(u)$ and
   for every k' power of $(1 + \lambda)$ in $\underline{[\gamma_r \cdot (k - k(cover)), k - k(cover)]}$

$$T_{v,k'} \leftarrow \text{Modified-GS}(v, k', \mathcal{G}^{res}).$$

5.   **select:** (pick the lowest density tree)

$$T_{aug} \leftarrow \text{MIN-DENSITY}\left\{ T_{v,k'} \cup \{(u, v)\} \right\}.$$

6.   **augment & update:** $cover \leftarrow cover \cup T_{aug}$; $\mathcal{G}^{res} \leftarrow \mathcal{G}^{res} \setminus \{g_i : T_{aug} \text{ intersects } g_i\}$.
7.   **keep** $k/h(T_u)$**-cover: if** first time $\underline{k(cover) \geq k/h(T_u)}$ **then** $cover_h \leftarrow cover$.
8. **end while**
9. **return** (lowest density tree $\in \{cover, cover_h\}$).

The following notation is used in the algorithm. The input is a rooted undirected tree T which does not appear as a parameter of the input. Instead, a node u is given, and we consider the subtree of T that hangs from u. We denote this subtree by T_u. The partial cover accumulated by the algorithm is denoted by $cover$. The set of groups of terminals is denoted by $\mathcal{G}$. The set of groups of terminals not covered by $cover$ is denoted by $\mathcal{G}^{res}$. The number of groups covered by $cover$ is denoted by $k(cover)$. The height of a tree T_u is the maximum number of edges along a path from u to a leaf in T_u. We denote the height of T_u by $h(T_u)$.

Two parameters λ and γ_v appear in the algorithm. The parameter λ is set to equal $1/h(T)$. The parameter γ_v satisfies $1/\gamma_v = |children(v)| \cdot (1 + 1/\lambda) \cdot (1 + \lambda)$.

Lines that are significantly modified (compared to Algorithm 5.1) are underlined. In line 4, two modifications take place. First, the smallest demand is not one, but a poly-logarithmic fraction of the residual demand (under the assumption that the maximum degree and the height is poly-logarithmic). Second, only demands that are powers of $(1 + \lambda)$ are considered. In line 7, the algorithm also stores the partial cover that first covers at least $1/h(T_u)$ of the initial demand k. This change is important for the simulation argument in the proof. Since the algorithm does not consider all the demands, we need to consider also the partial cover that the simulation argument points to. Finally, in line 9, we return the partial cover with the best density among $cover$ and $cover_h$. Again, this selection is required for the simulation argument.

Note that modified-GS(u, k, $\mathcal{G}$) may return now a cover that covers less than k groups. If this happens in the topmost call, then one needs to iterate until a k-GS cover is accumulated.

The following claim is proved in Reference 6. It is analogous to Claim 5.2 and is proved by rewriting the proof while taking into account error terms that are caused by the modifications. Due to lack of space, we omit the proof.

Claim 5.4 ([6]) *The density of every augmenting tree T_{aug} satisfies:*

$$\rho(T_{aug}) \leq (1 + \lambda)^{2h(T_u)} \cdot h(T_u) \cdot \rho(T'_u).$$

The following theorem is proved in Reference 6. The assumptions on the height and maximum degree are justified by the reduction discussed earlier.

Theorem 5.3 *Algorithm modified-GS(r, k, G) is a poly-logarithmic approximation algorithm with polynomial running time for GS instances over trees with logarithmic maximum degree and $O(\log n / \log \log n)$ height.*

5.6 Discussion

In this chapter we presented the recursive greedy algorithm and its analysis. The algorithm is designed for problems in which finding a minimum density augmentation of a partial solution is an NP-hard problem. The main advantages of the algorithm are its simplicity and the fact that it is a combinatorial algorithm. The analysis of the approximation ratio of the recursive greedy algorithm is nontrivial and succeeds in bounding the density of the augmentations.

The recursive greedy algorithm has not been highlighted as a general method, but rather as an algorithm for Steiner tree problems. We believe that it can be used to approximate other problems as well.

Open Problems: The quasi-polynomial time $O(\log^3 k)$-approximation algorithm for DST raises the question of finding a polynomial time algorithm with poly-logarithmic approximation ratio for DST. In particular, the question is whether the running time of the recursive greedy algorithm for DST can be reduced by modifications or preprocessing.

Appendix A. Proof of Lemma 5.2

We prove that given a k-DST T in a transitive closed directed graph G, there exists a k-DST T' such that: (i) T' is ℓ-layered and (ii) $w(T') \leq \frac{\ell}{2} \cdot k^{2/\ell} \cdot w(T)$. The proof uses notation introduced in Section 5.4.

Notation: Consider a rooted tree T. The subtree of T that consists of the vertices hanging from v is denoted by T_v. Let $\alpha = k^{2/\ell}$. We say that a node $v \in V(T)$ is α-*heavy* if $k(T_v) \geq k(T)/\alpha$. A node v is α-*light* if $k(T_v) < k(T)/\alpha$. A node v is *minimally α-heavy* if v is α-heavy and all its children are α-light. A node v is *maximally α-light* if v is α-light and its parent is α-heavy.

Promotion: We now describe an operation called promotion of a node (and hence the subtree hanging from the node). Let G denote a directed graph that is transitively closed. Let T denote a rooted tree at r that is a subgraph of G. *Promotion* of $v \in V(T)$ is the construction of the rooted tree T' over the same vertex set with the edge set: $E(T') \stackrel{\triangle}{=} E(T) \cup \{(r, v)\} \setminus \{(p(v), v)\}$. The promotion of v simply makes v a child of the root.

Height reduction: The height reduction procedure is listed as Algorithm 5.3. The algorithm iteratively promotes minimally α-heavy nodes that are not children of the root, until every α-heavy node is a child of the root. The algorithm then proceeds with recursive calls for every maximally α-light node. There are two types of maximally α-light nodes: (1) children of promoted nodes and (2) α-light children of the root (that have not been promoted).

ALGORITHM 5.3   HR(T, r, α)—A Recursive Height Reduction
Algorithm. T Is a Tree Rooted at r, and $\alpha > 1$ Is a Parameter

1. **stopping condition:** if $V(T) = \{r\}$ **then return** ($\{r\}$).
2. $T' \leftarrow T$.
3. **while** $\exists v \in V(T') : v$ is α-heavy & $dist(r, v) > 1$ **do**
4. $T' \leftarrow promote(T', v)$
5. **end while**
6. **for all** minimally α- heavy nodes $v \in V(T')$ **do**
7. $T' \leftarrow$ tree obtained from T' after replacing T'_v by HR(T'_v, v, α).
8. **end for**
9. **return** (T').

The analysis of the algorithm is as follows: Let $h_\alpha(k(T))$ denote an upper bound on the height of the returned tree as a function of the number of terminals in T. The recursion is applied only to α-light trees that are one or two edges away from the current root. It follows that $h_\alpha(k(T))$ satisfies the recurrence

$$h_\alpha(k') \leq h_\alpha(k'/\alpha) + 2.$$

Therefore, $h_\alpha(k') \leq 2 \cdot \log_\alpha k'$.

Bounding the weight: We now bound the weight of the tree T' returned by the height reduction algorithm. Note that every edge $e' \in E(T')$ corresponds to a path $path(e') \in T$. We say that an edge $e \in E(T)$ is *charged* by an edge $e' \in E(T')$ if $e \in path(e')$. If we can prove that every edge $e \in E(T)$ is charged at most β times, then $w(T') \leq \beta \cdot w(T)$.

We now prove that every edge $e \in E(T)$ is charged at most $\alpha \cdot \log_\alpha k(T)$ times. It suffices to show that every edge is charged at most α times in each level of the recursion. Since the number of terminals reduces by a factor of at least α in each level of the recursion, the recursion depth is bounded by $\log_\alpha k(T)$. Hence, the bound on the number of times that an edge is charged follows.

Consider an edge $e \in E(T)$ and one level of the recursion. During this level of the recursion, α-heavy nodes are promoted. The subtrees hanging from the promoted nodes are disjoint. Since every such subtree contains at least $k(T)/\alpha$ terminals, it follows that the number of promoted subtrees is at most α. Hence, the number of new edges $(r, v) \in E(T')$ from the root r to a promoted node v is at most α. Each such new edge charges every edge in $E(T)$ at most once, and hence every edge in $E(T)$ is charged at most α times in each recursive call. Note also that the recursive calls in the same level of the recursion are applied to disjoint subtrees. Hence, for every edge $e \in E(T)$, the recursive calls that charge e belong to a single path in the recursion tree.

We conclude that the recursion depth is bounded by $\log_\alpha k(T)$ and an edge is charged at most α times in each recursive call. Set $\ell = 2 \cdot \log_\alpha k(T)$, and then $\alpha \cdot \log_\alpha k(T) = \frac{\ell}{2} \cdot k^{2/\ell}$. The lemma follows. ∎

Acknowledgments

I would like to thank Guy Kortsarz for introducing me to the recursive greedy algorithm and sharing his understanding of this algorithm with me. Guy also volunteered to read a draft. I thank Chandra Chekuri for many discussions related to this chapter. Lotem Kaplan listened and read drafts and helped me in the search for simpler explanations. Thanks to the Max-Planck-Institut für Informatik where I had the opportunity to finish writing the chapter. Special thanks to Kurt Mehlhorn and his group for carefully listening to a talk about this chapter.

References

1. Zelikovsky, A., A series of approximation algorithms for the acyclic directed Steiner tree problem, *Algorithmica*, 18, 99, 1997.
2. Kortsarz, G. and Peleg, D., Approximating the weight of shallow Steiner trees, *Discrete Applied Mathematics*, 93, 265, 1999.
3. Charikar, M., Chekuri, C., Cheung, T., Dai, Z., Goel, A., Guha, S., and Li, M., Approximation algorithms for directed Steiner problems, *Journal of Algorithms*, 33, 73, 1999.
4. Reich, G. and Widmayer, P., Beyond Steiner's problem: A VLSI oriented generalization, in *Proceedings of Graph-Theoretic Concepts in Computer Science (WG-89)*, LNCS, 411, p. 196, 1990.
5. Garg, N., Konjevod, G., and Ravi, R., A polylogarithmic approximation algorithm for the group Steiner tree problem, *Journal of Algorithms*, 37, 66, 2000. Preliminary version in Proceedings of SODA, 253, 1998.
6. Chekuri, C., Even, G., and Kortsarz, G., A greedy approximation algorithm for the group Steiner problem, *Discrete Applied Mathematics*, 154(1), 15, 2006.
7. Zosin, L. and Khuller, S., On directed Steiner trees, in *Proceedings of SODA*, 59, 2002.
8. Helvig, C.H., Robins, G., and Zelikovsky, A., Improved approximation scheme for the group Steiner problem, *Networks*, 37(1), 8, 2001.

6

Local Ratio

Dror Rawitz

6.1 Introduction

The *local ratio technique* has been used to design approximation algorithms for optimization problems since its initiation in the 1980s by Bar-Yehuda and Even [1]. One of the main features of the technique is its simplicity and elegance. A local ratio algorithm is typically intuitive and easy to understand, and its analysis usually reveals the makings of the algorithm, namely it is easy to identify the specific key property or properties of the problem at hand that were instrumental is designing the algorithm. Initially, the simplicity of the technique leads to believe that it has very limited use. However, over the years it has evolved into a powerful technique with broad applicability.

Many approximation algorithms are based on the following simple theme. We describe it for minimization problems, but it could be easily described for maximization problems as well. The first step is to identify a lower bound B on the optimum that can be computed efficiently. Then, this lower bound is used for the computation of a feasible solution whose value W is at most r times value of the lower bound. The approximation ratio of the algorithm is r, since $W \leq r \cdot B \leq r \cdot \text{OPT}$, where OPT is the optimum. For instance, the 2-approximation algorithm for (unweighted) VERTEX COVER by Gavril [2] relies on the computation of a maximal matching whose size is the desired lower bound; the 2-approximation algorithm for the TRAVELING SALESMAN PROBLEM employs a *minimum spanning tree* whose weight is a lower bound on the optimal tour. Moreover, algorithms that are based on linear programming use either an optimal fractional solution or a dual solution whose value is the above-mentioned lower bound.

Roughly speaking, local ratio algorithms use a variation of the above-mentioned theme. Instead of coming up with a single lower bound B on the optimum such that $W \leq r \cdot B$, the algorithm induces a break down of the weight W into a series of k nonnegative weights $W_1, \ldots, W_k$, where $W = \sum_i W_i$, and compares them to k lower bounds $B_1, \ldots, B_k$, where $B = \sum_i B_i$ is a lower bound on the optimum. More specifically, the algorithm shows that $W_i \leq r \cdot B_i$, for every i. The solution is r-approximate, since $W = \sum_i W_i \leq r \sum_i B_i = r \cdot B \leq r \cdot \text{OPT}$. The tool that allows the breaking down of weights and bounds is the *Local Ratio Lemma*.

6.1.1 Local Ratio Lemma

The Local Ratio Lemma applies to minimization (maximization) problems that can be formulated as follows: Given a set of *feasibility constraints* $\mathcal{F}$, a *solution vector* $x \in \mathbb{R}^n$ satisfying the constraints in $\mathcal{F}$ is called a *feasible solution*. Given $\mathcal{F}$ and a *weight vector* $w \in \mathbb{R}^n$, the problem is to find a feasible solution x that minimizes (maximizes) the scalar product $w \cdot x$. The most common type of optimization problem that can be formulated as earlier consists of instances in which the input contains a set of n weighted items and a specification of feasibility constraints on subsets of items. A solution is a subset of items that can be described using an incidence vector $x \in \{0, 1\}^n$. In this case the cost (profit) of a feasible solution is the total weight of the elements it comprises.

The Local Ratio Lemma for minimization problems was first used by Bar-Yehuda and Even [1] who invented the local ratio technique. The maximization version of the lemma was used by Bar-Noy et al. [3] much later in the first paper that used local ratio to design an algorithm for a maximization problem.

Lemma 6.1 (Local Ratio [1,3]) *Let $\mathcal{F}$ be a set of feasibility constraints on vectors in $\mathbb{R}^n$. Let $w, w_1, w_2 \in \mathbb{R}^n$ be weight functions such that $w = w_1 + w_2$. Let $x \in \mathbb{R}^n$ be a feasible solution (with respect to $\mathcal{F}$) that is r-approximate with respect to w_1 and with respect to w_2. Then, x is also r-approximate with respect to w.*

Proof. Let x^*, x_1^*, and x_2^* be optimal solutions with respect to w, w_1, and w_2, respectively.

In the minimization case, $w_1 \cdot x_1^* \leq w_1 \cdot x^*$ and $w_2 \cdot x_2^* \leq w_2 \cdot x^*$, and therefore

$$w \cdot x = w_1 \cdot x + w_2 \cdot x \leq r(w_1 \cdot x_1^*) + r(w_2 \cdot x_2^*) \leq r(w_1 \cdot x^*) + r(w_2 \cdot x^*) = r(w \cdot x^*).$$

In the maximization case, $w_1 \cdot x_1^* \geq w_1 \cdot x^*$ and $w_2 \cdot x_2^* \geq w_2 \cdot x^*$, and it follows that

$$w \cdot x = w_1 \cdot x + w_2 \cdot x \geq \frac{w_1 \cdot x_1^*}{r} + \frac{w_2 \cdot x_2^*}{r} \geq \frac{w_1 \cdot x^*}{r} + \frac{w_2 \cdot x^*}{r} = \frac{w \cdot x^*}{r}.$$

The lemma follows: ∎

A typical local ratio algorithm is recursive, and the corresponding analysis is done by induction. (This is not crucial, but usually it is rather convenient.) At the recursive base, a solution is found, since the instance is "easy" in some sense (e.g., a zero-cost solution can be found). In each recursive step, the algorithm defines a weight function w_1 such that at least one item with non-zero weight with respect to w gets zero weight with respect to $w - w_1$. In minimization problems the algorithm also makes sure that $w - w_1$ still assigns nonnegative weights to all items. Given the reduced weights, the algorithm then modifies the instance and recursively solves the modified instance with the weight function $w - w_1$. The returned solution is fixed so it is a "good" approximate solution with respect to both w_1 and $w - w_1$. By the Local Ratio Lemma, this solution is guaranteed to be a good approximation with respect to w as well.

Several issues need to be resolved:

1. What is the set of instances that is considered "easy" enough to be handled in the recursion base?
2. How do we come up with a weight function w_1?

3. How should we modify the instance?
4. What is a "good" approximate solution, and how do we fix solutions?

The answers to these questions are interrelated and they depend on the specific problem as shown in the sequel.

6.1.2 Overview

This chapter is intended to be an introductory survey of local ratio that contains a few answers to the above-mentioned questions and several applications that demonstrate the usage of the technique. We note that the algorithms that are given in this chapter may not have the best approximation ratio. Also, full details are given only when they are relatively simple.

Section 6.2 discusses the most basic version of local ratio, as it was presented in the first local ratio paper [1]. In Section 6.3 we describe a local ratio framework for covering problems whose scope is much wider than the scope of the basic version. A framework for packing problems is given in Section 6.4. We consider several variations of the technique that do not fall within either framework in Section 6.5. A variant of local ratio, called *Fractional Local Ratio*, is covered by Section 6.6. We conclude in Section 6.7 where we mention several related topics that are not covered by this chapter.

6.2 Vertex Cover and Full Effectiveness

The local ratio technique was originally invented as a tool to design approximation algorithms for the VERTEX COVER problem. Hence it is only natural to use this problem as a first example. Indeed, we present several local ratio 2-approximation algorithm for VERTEX COVER. We also consider an extension of VERTEX COVER IN HYPERGRAPHS (also known as HITTING SET or SET COVER) in which each edge must be covered multiple times. That is, an edge e must be covered by at least $d(e)$ vertices. We describe a local ratio δ_{max}-approximation algorithm, where $\delta_{max} = \max_e \delta_e$, and $\delta_e = |e| - d(e) + 1$.

6.2.1 Vertex Cover

In VERTEX COVER we are given a simple graph $G = (V, E)$ with a nonnegative weight function w on the vertices, and the goal is to find a minimum weight subset $C \subseteq V$ of vertices that cover all edges, namely such that $e \cap C \neq \emptyset$ for every edge $e \in E$.

As mentioned earlier a local ratio step is based on finding a set of "good" solutions and a weight function w_1 such that any such solution is r-approximate with respect to w_1. A convenient option is to consider the set of all possible solutions.

Definition 6.1 *Given a set of feasibility constraints $\mathcal{F}$ on vectors in $\mathbb{R}^n$ and a number $r \geq 1$, a weight function $w \in \mathbb{R}^n$ is called* fully r-effective *if there exists a number $b \geq 0$ such that $b \leq w \cdot x \leq r \cdot b$, for every feasible solution x.*

Consider the following generic algorithm for Vertex Cover (Algorithn VC), which is based on the previous definition. We use $E(u)$ to denote the set of edges that are incident to u, that is, $E(u) \stackrel{\triangle}{=} \{(u, v) : (u, v) \in E\}$. The initial call is $\mathbf{VC}(V, E, w)$. We note that it may be that there exists a vertex $u \in V$ such that $w(u) = 0$. In this case we can use the weight function $w_1 = 0$, which is considered fully r-effective (in this case $b = 0$).

Observe that each recursive call is made after the removal of a vertex, hence there are at most $|V|$ recursive calls. It is not hard to verify that the running time of Algorithm **VC** is polynomial, if a weight function w_1 can be computed efficiently.

We show that the algorithm computes r-approximate solutions, assuming that w_1 can be computed in each recursive call.

ALGORITHM 6.1   VC(V, E, w)

1: **if** $E = \emptyset$ **then return** $\emptyset$
2: Let w_1 be a fully r-effective weight function, such that $0 \leq w_1(v) \leq w(v)$, for every v, and
   $w_1(u) = w(u)$, for some $u \in V$.
3: **return** $\{u\} \cup$ **VC**$(V \setminus \{u\}, E \setminus E(u), w - w_1)$

Lemma 6.2 *Suppose that there exists a fully r-effective weight function w_1, such that $0 \leq w_1(v) \leq w(v)$, for every v, and $w_1(u) = w(u)$, for some $u \in V$, for every graph G and weight function w. Then, Algorithm* **VC** *is an r-approximation algorithm for* VERTEX COVER.

Proof. We prove the lemma by induction on the recursion. For the recursive base, observe that $\emptyset$ is an optimal solution for a graph without edges. As for the inductive step, let $U' = $ **VC**$(V \setminus \{u\}, E \setminus E(u), w - w_1)$. We know that U' is r-approximate with respect to $G' = (V \setminus \{u\}, E \setminus E(u))$ and $w_2 = w - w_1$ due to the inductive hypothesis. Since $w_2(u) = 0$, it follows that $U = U' \cup \{u\}$ is r-approximate with respect to G and w_2. Moreover, since w_1 is fully r-effective, U is r-approximate with respect to G and w_1. It follows, due to the Local Ratio Lemma, that U is r-approximate with respect to G and w. ∎

It remains to present fully r-effective weight functions that satisfies the requirement of Lemma 6.2. One possibility is to select an arbitrary edge $e \in E$, and to define

$$
w_1(v) = \varepsilon \cdot \begin{cases} 1 & v \in e, \\ 0 & \text{otherwise.} \end{cases} \tag{6.1}
$$

where $\varepsilon = \min_{v \in e} w(v)$. The edge must be covered, and thus $\varepsilon \leq w_1 \cdot x \leq 2\varepsilon$, for every solution x. It follows that w_1 is fully 2-effective. This weight function leads to the linear time 2-approximation algorithm that was presented by Bar-Yehuda and Even [4] and analyzed using local ratio in Reference 1.

The above-mentioned weight function can be generalized as follows: Choose a vertex u whose neighborhood $N(u)$ is nonempty and a subset $S \subseteq N(u)$, and define

$$
w_1(v) = \varepsilon \cdot \begin{cases} |S| & v = u, \\ 1 & v \in S, \\ 0 & \text{otherwise,} \end{cases}
$$

where $\varepsilon = \min \{w(v) : v \in N(u)\} \cup \{w(u)/|S|\}$. In this case, w_1 is fully 2-effective, since $|S|\varepsilon \leq w_1 \cdot x \leq 2|S|\varepsilon$, for every solution x. Clarkson's algorithm [5] is obtained by choosing a vertex u that minimizes $w(u)/\deg(u)$, where $\deg(u) = |N(u)|$ is the degree of u, and $S = N(u)$ in each recursive call of Algorithm **VC**.

Another fully 2-effective weight functions is $w_1(u) = \varepsilon \cdot \deg(u)$, where $\varepsilon = \min_v \frac{w(v)}{\deg(v)}$. Observe that $|E|\varepsilon \leq w_1 \cdot x \leq 2|E|\varepsilon$, since all edges must be covered. Such a weight function is called *homogeneous*, since it gives all covering items the same cost-effectiveness, that is all items are equally good from a greedy point of view. Actually, all the above-mentioned weight functions are homogeneous with respect to a subgraph of the input graph.

Finally, observe that Algorithm **VC** consists of an alternating sequence of weight subtractions and instance modifications. Each instance modification removes a zero weight vertex from the graph, and this vertex is later added to the solution. It is possible to first perform weight subtractions, and when no more subtractions are possible to use the set of zero weight vertices as a solution. In fact, the first local ratio algorithm [1] was presented in this way.

6.2.2 Multicoverage

Now consider the following extension of VERTEX COVER IN HYPERGRAPHS in which each edge $e \in E$ has a covering demand $d(e)$. That is, a feasible solution is a subset C of the vertices such that $|e \cap C| \geq d(e)$, for every edge e. We assume that $d(e) \leq |e|$, for every e, since otherwise no solution exists (and this can be identified in linear time). We describe a local ratio $\delta_{\max}$-approximation algorithm, where $\delta_{\max} = \max_e \delta_e$, and $\delta_e = |e| - d(e) + 1$.

We use the weight function w_1 defined in Equation (6.1).

Lemma 6.3 *Given an instance of* VERTEX COVER IN HYPERGRAPHS *with edge demands and an arbitrary edge e, the weight function w_1 from Equation (6.1) is fully $\frac{|e|}{d(e)}$-effective.*

Proof. Clearly, $w_1(C) \geq \varepsilon \cdot d(e)$, for every solution C. On the other hand, $\sum_{v \in V} w_1(v) = \varepsilon |e|$. ∎

The following algorithm uses the weight function w_1 defined in Equation (6.1).

ALGORITHM 6.2   VCH(*V, E, d, w*)

1: **if** $E = \emptyset$ **then return** $\emptyset$
2: Let w_1 be the weight function defined in Equation (6.1), and let u be a vertex such that $w_1(u) = w(u)$
3: Define $d'(e) \leftarrow \begin{cases} d(e) - 1 & u \in e, \\ d(e) & \text{otherwise.} \end{cases}$
4: $E_0 \leftarrow \{e : d'(e) = 0\}$
5: **return** $\{u\} \cup \mathbf{VC}(V \setminus \{u\}, E \setminus E_0, d', w - w_1)$

Since each recursive call removes a vertex, there are $|V|$ recursive calls. Hence, the running time of the algorithm is polynomial.

Lemma 6.4 *The function δ remains unchanged throughout the execution of Algorithm* **VCH**.

Proof. We prove the lemma by induction. The base is the original instance. In the inductive step, consider a recursive call. The current instance satisfies the claim due to the inductive hypothesis. A new instance is obtained by removing a vertex u from all edges containing it, thus decreasing their size by 1. In d' the demands of edges containing u are also decreased by 1. Finally, all edges satisfying $d'(e) = 0$ are removed. The lemma follows. ∎

It remains to prove that the algorithms is a $\delta_{\max}$-approximation algorithm.

Theorem 6.1 *Algorithm* **VCH** *computes $\delta_{\max}$-approximate solutions.*

Proof. We prove the theorem by induction on the recursion. For the recursive base, observe that $\emptyset$ is an optimal solution for a graph without edges. As for the inductive step, let U' be the solution returned by the recursive call. U' is $\delta_{\max}$-approximate with respect to $G' = (V \setminus \{u\}, E \setminus E_0)$, d', and $w_2 = w - w_1$ due to the inductive hypothesis. Since $w_2(u) = 0$, it follows that $U = U' \cup \{u\}$ is r-approximate with respect to G, d, and w_2. By Lemma 6.3 we have that w_1 is fully $\frac{|e|}{d(e)}$-effective. Furthermore, $d(e) \geq 1$ and by Lemma 6.4 we have that $|e| = \delta(e) + d(e) - 1$. It follows that $\frac{|e|}{d(e)} = \frac{\delta(e) + d(e) - 1}{d(e)} \leq \delta(e)$. Hence, U is $\delta_{\max}$-approximate with respect to G, d, and w due to the Local Ratio Lemma. ∎

Algorithm **VCH** is the local ratio interpretation of a variant of the primal-dual algorithm given by Hall and Hochbaum [6]. Also, when $d(e) = 1$, for every e, Algorithm **VCH** becomes the linear time local ratio Δ_E-approximation algorithm for HITTING SET by Bar-Yehuda and Even [1], where $\Delta_E = \max_{e \in E} |e|$.

6.3 Covering Problems

In the previous section we considered covering problems that require total coverage (e.g., all edges must be covered). The local ratio algorithm were based on identifying a structure (e.g., an edge) such that a large fraction of its items must participate in any cover (e.g., one of the end-points of an edge). However, there are covering problems that do not require total coverage, and therefore it is impossible to identify such a structure. A good example is the MINIMUM KNAPSACK problem. In this problem the instance consists of n items, where item i has a length $\ell(i)$ and a weight $w(i)$, and a length lower bound L. A feasible solution is a subset of items whose total length is at least L, and the goal is to find a feasible solution of minimum weight. Clearly, any specific item need not participate in all possible covers. Bafna et al. [7] extended the local ratio technique by incorporating the notion of *minimiality with respect to set inclusion* into it. The crux is that nonessential items are removed from the cover, and this leads to a lower upper bound on the cost of a cover. Following [7], Fujito [8] gave a local ratio unified approximation algorithm for node-deletion problems with nontrivial and hereditary graph properties that computes *minimal solutions*. Bar-Yehuda [9] presented a generic algorithm for covering problems. A slightly different framework was given by Bar-Yehuda and Rawitz [10]. In the sequel we present a generic local ratio approximation algorithm for covering problems that is a variant of the one from [10].

6.3.1 Framework for Covering Problems

Before presenting the covering framework, we need the following definition: A *covering instance* $\mathcal{I} = (S, f, w)$ consists of a set S of covering items, and a weight function w on the covering items. We are also given a function $f : 2^S \rightarrow \{0, 1\}$ such that $f(U) = 1$ if and only if U is a cover. Notice that f must be *monotone*, namely $U \subseteq U'$ implies $f(U) \leq f(U')$. We assume that $f(S) = 1$, since otherwise there are no feasible solutions. For example, in the MINIMUM KNAPSACK problem $S = \{1, \ldots, n\}$, and we have that $f(U) = 1$ if and only if $\sum_{i \in U} \ell(i) \geq L$. Given a subset $Z \subseteq S$, a subset $U \subseteq S \setminus Z$ is called a *completion* of Z, if $f(Z \cup U) = 1$. We also define $f_Z(U) = f(Z \cup U)$. Notice that the *residual instance* $(S \setminus Z, w, f_Z)$ that is obtained by assuming that the items in Z were already added to the solution is also a covering instance.

As mentioned earlier local ratio is based on an upper-bound/lower-bound theme (Definition 6.1). We can obtain a lower bound on the weight of any feasible solution with respect to w_1 by assigning a positive weight to every item (or at least to many items). However, this may result in a very high upper bound on the weight of a feasible solution. We get a better upper bound by focusing on minimal solutions.

Definition 6.2 *Given a covering problem (S, f, w) and a number $r \geq 1$, a weight function $w \in \mathbb{R}^n$ is said to be r-effective if there exists a number $b \geq 0$ such that $b \leq wx \leq r \cdot b$ for all minimal solutions x.*

When a local ratio algorithm uses an r-effective weight function, it basically signs a contact saying that an r-approximate solution would be attainable, as long as the computed solution is minimal. Therefore, the algorithm must fix the solution such that it will be minimal. We note that one can also view minimality as an additional feasibility constraint. However, since minimality is nonmonotone, it is common to treat minimality as a tool to solve the problem and not as a constraint.

Algorithm **Cover** is a generic local ratio algorithm for covering problems. It works as follows. It maintains a set Z of items that are temporarily assumed to be taken into the solution. If Z is already feasible then the algorithm returns the empty set. Otherwise, it computes an r-effective weight function w_1 with respect to $(S \setminus Z, f_Z, w)$ such that $0 \leq w_1(v) \leq w(v)$, for every $v \in V$, and $w_1(u) = w(u)$, for some $u \in S \setminus Z$. The item u is called a *pivot*. (Note that $w_1 = 0$ is possible when $w(u) = 0$.) Then the algorithm recursively solves the problem for $Z \cup \{u\}$ and $w - w_1$. The item u is added to the solution returned by the recursive call, only if the solution is not a feasible completion of Z. The initial call is **Cover**$(\emptyset, w)$.

Algorithm **Cover** has at most n recursive calls. Thus, its running time is polynomial, if a weight function w_1 can be computed efficiently.

We show that the algorithm computes r-approximate solutions, assuming that an r-effective weight function w_1 can be computed in each recursive call.

ALGORITHM 6.3   Cover(Z, w)

1: **if** $f_Z(\emptyset) = 1$ **then return** $\emptyset$
2: Let w_1 be an r-effective weight function with respect to f_Z, such that $0 \le w_1(v) \le w(v)$, for every
    $v \in V$, and $w_1(u) = w(u)$, for some $u \in S \setminus Z$.
3: $U \leftarrow$ **Cover**$(Z \cup \{u\}, w - w_1)$
4: **if** $f_Z(U) = 0$ **then** $U \leftarrow U \cup \{u\}$
5: **return** U

Theorem 6.2 *Let S be a set of items, and let $f : 2^S \rightarrow \{0, 1\}$ be a monotone function. Suppose that there exists an r-effective weight function w_1 with respect to f_Z, such that $0 \le w_1(v) \le w(v)$, for every v, and $w_1(u) = w(u)$, for some $u \in S \setminus Z$, for every $Z \subseteq S$ and a weight function w. Then, Algorithm **Cover** computes r-approximate solutions.*

Proof. We prove that U is an r-approximate minimal completion of Z by induction on the recursion. For the recursive base, observe that $\emptyset$ is a minimal and an optimal competition of Z. As for the inductive step, let $U' = $ **Cover**$(Z, w - w_1)$. We know that U' is a minimal r-approximate completion of $Z \cup \{u\}$ with respect to $w_2 = w - w_1$ due to the inductive hypothesis. Since the pivot u is added to U only if U is not a completion of Z, we have that U is minimal at the end of the recursive call. Since $w_2(u) = 0$, it follows that U is also r-approximate with respect to w_2. Moreover, since w_1 is r-effective, U is an r-approximate completion of Z with respect to w_1. It follows, due to the Local Ratio Lemma, that U is a minimal r-approximate completion of Z with respect to w. ∎

As we shall see in the sequel, residual instances are typically instances of the same problem. Hence, instead of presenting an r-effective weight function w_1 for every instance and a subset Z, it is enough to provide an r-effective weight function for every possible instance.

6.3.2 Minimum Knapsack

We demonstrate the generic algorithm for covering problems by presenting a 2-approximation algorithm for MINIMUM KNAPSACK. We note that a fully polynomial time approximation scheme for MINIMUM KNAPSACK can be designed using standard techniques.

Given Algorithm **Cover** and Theorem 6.2 we need only show how to construct an r-effective weight function w_1 with respect to any subset Z, for some $r \ge 1$. First, observe that the function f that corresponds to MINIMUM KNAPSACK is defined as follows: $f(U) = 1$ if and only if $\sum_{i \in U} \ell(i) \ge L$. Given a subset Z, the residual instance is in fact a MINIMUM KNAPSACK instance whose covering requirement is $L^Z = P - \sum_{j \in Z} \ell(j)$. Thus instead of presenting a weight function for every subset Z, we simply present a weight function for every possible instance of MINIMUM KNAPSACK.

Let $\bar{\ell}(i) = \min\{\ell(i), L\}$ be the *covering potential* of item i, and consider the following weight function that gives each covering item a weight which is proportional to its covering potential: $w_1(i) = \varepsilon \cdot \bar{\ell}(i)$, where $\varepsilon = \min_{i \in Z} \frac{w(i)}{\bar{\ell}(i)}$. Observe that w_1 is homogeneous, since it gives all items the same cost-effectiveness.

Lemma 6.5 *Given a MINIMUM KNAPSACK instance, the weight function w_1 is 2-effective.*

Proof. Let U be a feasible solution. If U contains an item i such that $\bar{\ell}(i) = L$, then $w_1(U) \ge \varepsilon \bar{\ell}(i) = \varepsilon L$. Otherwise, $\bar{\ell}(i) = \ell(i)$, for every $i \in U$, and we have that

$$w_1(U) = \varepsilon \sum_{i \in U} \bar{\ell}(i) = \varepsilon \sum_{i \in U} \ell(i) \ge \varepsilon L.$$

On the other hand, let U be a minimal solution and let i be an arbitrary item in U. In this case

$$w_1(U) = \varepsilon \sum_{j \in U} \bar{\ell}(j) = \varepsilon \left[\sum_{j \in U \setminus \{i\}} \bar{\ell}(j) + \bar{\ell}(i) \right] < \varepsilon[L + L] = 2\varepsilon L. \qquad \blacksquare$$

A 2-approximation algorithm for MINIMUM KNAPSACK follows from Theorem 6.2 and Lemma 6.5.

The local ratio 2-approximation algorithm was originally given by Bar-Yehuda [11]. A primal-dual version of this algorithm that uses the so called *knapsack inequalities* was given by Williamson and Shmoys [12, Chapter 7]. In fact the local ratio 2-approximation algorithm [11] and the equivalence between local ratio and the primal-dual schema [10] lead to the knapsack inequalities which are induced by all possible residual instances (i.e., all possible values of Z).

6.3.3 Partial Vertex Cover

In the PARTIAL VERTEX COVER problem the input consists of a simple graph $G = (V, E)$, a length function ℓ on the edges, a covering requirement L, and a weight function w on the vertices. The goal is to find a minimum-weight set of vertices U that covers a total length of at least L, that is, such that $\sum_{e \cap U \neq \emptyset} \ell(e) \geq L$. PARTIAL VERTEX COVER generalizes VERTEX COVER, since in the latter $\ell(e) = 1$, for every $e \in E$, and $L = |E|$. It also generalized MINIMUM KNAPSACK, since each item i can be represented by an edge $e_i = (v_i, v_i')$, where $\ell(e_i) = \ell_i$, $w(v_i) = w(v_i') = w(i)$. It follows that PARTIAL VERTEX COVER can be viewed as a combination of VERTEX COVER and MINIMUM KNAPSACK.

In this section we describe a 2-approximation algorithm for PARTIAL VERTEX COVER by Bar-Yehuda [11]. The algorithm is based on a weight function which is a combination of two homogeneous weight functions we have seen earlier: the first is the degree-proportional weight function for VERTEX COVER and the other is the one that was given for MINIMUM KNAPSACK.

The monotone function f that corresponds to PARTIAL VERTEX COVER is defined as follows: $f(U) = 1$ if and only if $\ell(U) \geq L$, where $\ell(U) = \sum_{e \cap U \neq \emptyset} \ell(e)$. Given a subset $Z \subseteq V$, the residual problem is a PARTIAL VERTEX COVER instance, where the graph is $G' = (V \setminus Z, E \cap (V \setminus Z \times V \setminus Z))$, and the covering requirement is $L^Z = L - \ell(U)$.

Let $\ell(u) = \ell(\{u\})$, and define $\bar{\ell}(u) = \min\{\ell(u), L\}$. Consider the following homogeneous weight function: $w_1(u) = \varepsilon \cdot \bar{\ell}(u)$, where $\varepsilon = \min_u \frac{w(u)}{\bar{\ell}(u)}$.

Lemma 6.6 *Given a* PARTIAL VERTEX COVER *instance, the weight function w_1 is 2-effective.*

Proof. Let U be a feasible solution. If U contains a vertex whose weight is εL, then $w_1(U) \geq \varepsilon L$. Otherwise, U consists solely of vertices whose length is strictly smaller than L, and therefore

$$w_1(U) = \varepsilon \sum_{u \in U} \bar{\ell}(u) = \varepsilon \sum_{u \in U} \ell(u) \geq \varepsilon L.$$

Next we prove that any minimal solution U satisfies $w_1(U) \leq 2\varepsilon L$. If $|U| = 1$, then $w_1(u) = \varepsilon L$, and we are done. Otherwise, we prove the claim by showing that $\ell(U) \leq 2L$. Notice that an edge that is covered by U may be covered twice or once by vertices in U. Moreover, since U is minimal, each vertex $u \in U$ covers at least one edge by itself, that is, $E(u) \setminus E(U \setminus \{u\}) \neq \emptyset$. Let $\bar{\ell}(u) = \sum_{e \in E(u) \setminus E(U \setminus \{u\})} \ell(e)$ be the length of edges that u covers by itself, and let $u_0 = \text{argmin}_{u \in U} \bar{\ell}(u)$. Since U contains more that one vertex, we have that $\bar{\ell}(u_0) \leq \sum_{u \in U \setminus \{u_0\}} \bar{\ell}(u)$. By the minimality of U, we have that $\ell(U) - \bar{\ell}(u_0) < L$, and therefore,

$$\sum_{u \in U} \ell(u) = 2 \left(\ell(U) - \sum_{u \in U} \bar{\ell}(u) \right) + \sum_{u \in U} \bar{\ell}(u) \leq 2 \left(L - \sum_{u \in U \setminus \{u_0\}} \bar{\ell}(u) \right) + 2 \sum_{u \in U \setminus \{u_0\}} \bar{\ell}(u) = 2L.$$

as required. $\blacksquare$

Since w_1 is 2-effective, Algorithm **Cover** can be used as a 2-approximation algorithm for PARTIAL VERTEX COVER due to Theorem 6.2.

Bshouty and Burroughs [13] obtained the first 2-approximation algorithm for PARTIAL VERTEX COVER. Bar-Yehuda [11] used local ratio to obtain a $\max\{\Delta_E, 2\}$-approximation algorithm for PARTIAL VERTEX COVER in hypergraphs, where $\Delta_E = \max_{e \in E}|e|$. (Recall that the problem becomes MINIMUM KNAPSACK when $\Delta_E = 1$.)

6.3.4 Feedback Vertex Set

Given a simple graph $G = (V, E)$ a set of vertices $F \subseteq V$ is called a *feedback vertex set* if its removal from G leaves an acyclic graph (or a forest). Given a weight function w on the vertices of G, the goal in the FEEDBACK VERTEX SET problem is to find a minimum-weight feedback vertex set. FEEDBACK VERTEX SET is a covering problem, where the cycles of G are to be covered by vertices. In this section we show how to use Algorithm **Cover** to obtain a 2-approximation algorithm for FEEDBACK VERTEX SET.

The monotone function for FEEDBACK VERTEX SET is defined as follows: $f(U) = 1$ if and only if the graph $G' = (V \setminus U, E \cap (V \setminus U \times V \setminus U))$ is acyclic. Consider a residual instance with respect to a subset $Z \subseteq V$. After removing Z from G, we obtain an instance of FEEDBACK VERTEX SET. Moreover, observe that zero degree vertices can be ignored and hence can be discarded. In addition, we can also iteratively remove vertices with degree one. It follows that we may assume that all vertices have degree at least 2. Now consider the following a weight function: $w_1(u) = \varepsilon \cdot \deg(u)$, where $\varepsilon = \min_u \frac{w(u)}{\deg(u)}$.

Lemma 6.7 *Let* (G, w) *be a* FEEDBACK VERTEX SET *instance, in which* $\deg(v) \geq 2$, *for every* $v \in V$. *Then the weight function* w_1 *is 2-effective.*

Proof. Given a set of vertices U, let us denote $\deg(U) = \sum_{u \in U} \deg(u)$. We prove the lemma by showing that b such that (i) $\deg(F) \geq |E| - (|V| - |F^*|)$, for every solution F and (ii) $\deg(F) \leq 2(|E| - (|V| - |F^*|))$, for every minimal solution F, where F^* is a minimum cardinality feedback vertex set.

Consider a feasible solution F. The removal of F from G leaves a forest on $|V| - |F|$ vertices that contains less than $|V| - |F|$ edges. Hence, at least $|E| - (|V| - |F|)$ where deleted. Since each of these edges is incident on some vertex in F, we have that $\deg(F) > |E| - (|V| - |F|) \geq |E| - (|V| - |F^*|)$.

Let F be a minimal feasible solution. Since F is minimal, for every $u \in F$, there exists a cycle C_u containing u, but no other vertex from F. Let $V(C_u)$ be the vertex set C_u, and let $V' = \bigcup_{u \in F}(V(C_u) \setminus \{u\})$. Let G' be the subgraph of G induced by V'. Each connected component of G' must contain $V(C_u) \setminus \{u\}$, for some u. Moreover, C_u must contain a vertex u' from F^*, which means that either $u' = u$ or $u' \in V(C_u) \setminus \{u\}$. Hence, there are at most $|F^* \cap F|$ connected components that do not contain vertices from F^*, and at most $|F^* \setminus F|$ connected components containing vertices from F^*. It follows that there are at most $|F^*|$ connected components and at least $|V'| - |F^*|$ edges in G'. Recall that each vertex $u \in F$ has two edges with one endpoint incident on a vertex in V', and therefore $\deg(V') \geq 2(|V'| - |F^*|) + 2|F|$. Hence

$$
\begin{aligned}
\deg(F) &= \deg(V) - \deg(V') - \deg(V \setminus (V' \cup F)) \\
&\leq \deg(V) - 2(|V'| - |F^*|) - 2|F| - \deg(V \setminus (V' \cup F)) \\
&\leq \deg(V) - 2(|V'| - |F^*|) - 2|F| - 2(|V| - |V'| - |F|) \\
&= 2(|E| - (|V| - |F^*|)),
\end{aligned}
$$

where the second inequality is because the degree of every vertex is at least two. ∎

There are several proofs of the 2-effectiveness of the degree weights or similar weight functions [7,14–16]. The 2-approximation algorithm presented earlier is a local ratio interpretation of the algorithm from Reference 14. The proof of Lemma 6.7 is taken from Reference 17.

6.3.5 Steiner Tree

Given a graph $G = (V, E)$, and a set $T \subseteq V$ of *terminals*, a *Steiner tree* is a subtree of G that spans all the vertices in T. The Steiner tree may contain nonterminal vertices, also known as *Steiner vertices*. Given a nonnegative weight function w on the edges, the STEINER TREE problem is to find a minimum weight Steiner tree, where the weight of a tree is the total weight of its edges. In this section we utilize Algorithm **Cover** to obtain a local ratio $(2 - \frac{1}{|T|})$-approximation algorithm for the STEINER TREE problem.

Let (G, T, w) be a STEINER TREE instance, the corresponding monotone function may be defined as follows: $f(E') = 1$, for $E' \subseteq E$, if and only if $E' \cap E(S, V \setminus S) \neq \emptyset$, for every $S \subseteq V$, such that $S \cap T \neq \emptyset$ and $(V \setminus S) \cap T \neq \emptyset$. In word, the set E' intersect with the edges of any cut $(S, V \setminus)$ that splits T.

A residual instance is induced by an edge subset $Z \subseteq E$. We show that this instance is, in fact, a STEINER TREE instance. Since our goal is to connect terminals, we can treat each connected component in $G' = (V, Z)$ as a super-vertex. More formally, we define a graph $G^Z = (V^Z, E^Z)$, where a vertex $v^Z \in V^Z$ represents a connected component in G' and there is an edge $(v^Z, u^Z) \in E^Z$ if there is an edge between the corresponding components in G', where $w(v^Z, u^Z) = \min \{w(u, v) : v \in V(v^Z), u \in V(u^Z)\}$. Moreover, $T^Z = \{v^Z : T \cap V(v^Z) \neq \emptyset\}$.

Given a STEINER TREE instance (G, T, w), we define the following weight function: $w_1(e) = \varepsilon \cdot \tau(e)$, where $\tau(e) = |e \cap T|$ is the *terminal degree* of e, and $\varepsilon = \min_e \frac{w(e)}{\tau(e)}$.

Lemma 6.8 *Let (G, T, w) be a STEINER TREE instance. Then the weight function w_1 is $(2 - \frac{2}{|T|})$-effective. Moreover, if $T = V$, then w_1 is 1-effective.*

Proof. We prove the first part of the lemma by showing that (i) $|T| \leq \sum_{e \in F} \tau(e)$, for every solution F and (ii) $\sum_{e \in F} \tau(e) \leq 2|T| - 2$, for any minimal solution F.

The first inequality holds, since every terminal in T must be incident to at least one edge in F, for any feasible solution F.

Next consider a minimal Steiner tree F. Observe that since F is minimal, every leaf in the tree is a terminal. Choose an arbitrary terminal r as the root of the Steiner tree. We use a charging scheme argument. Place a total of $2|T| - 2$ coins on the terminals in T, two coins per terminal in $T \setminus \{r\}$. We show how to redistribute the coins such that each edge $e \in F$ receives at least $\tau(e)$ coins. Each nonroot terminal gives its first coin to the first edge in his path to the root. It give its second coin to the first edge on the way to the root that is incident on another terminal. This is well defined, since the last edge on the path in incident to the root. Notice that both coins can be given to the same edge. Consider and edge $(u, v) \in F$, where u is the vertex that is closer to the root r. If v is a terminal (u, v) gets a coin from v. Since F is minimal, (u, v) must be on the path from some terminal to r. Let t be such a terminal that is closest to r. (Observe that it may be that $t = v$.) If u is a terminal, (u, v) gets a coin from t. It follows that at least $\tau(e)$ coins are placed on every edge $e \in F$.

Finally, if $T = V$, it must be that $|F| = |V| - 1$, for any minimal solution F, and we have that $\sum_{e \in F} \tau(e) = 2(|V| - 1)$.   ∎

By Theorem 6.2 and Lemma 6.8 we have a $(2 - \frac{2}{|T|})$-approximation algorithm for the STEINER TREE problem. In particular, if $T = \{s, t\}$, we get a 1-approximation algorithm for the SHORTEST PATH problem, which can be seen as a bidirectional version of Dijkstra's algorithm [18]. If $T = V$, we get a local ratio analysis of Kruskal's MINIMUM SPANNING TREE algorithm [19].

The GENERALIZED STEINER TREE problem in an extension of STEINER TREE in which the input contains terminal pairs that should be connected. (In this case the solution may be a forest.) Agrawal et al. [20] presented a 2-approximation algorithm for this problem, and Goemans and Willimanson [21] provided primal-dual approximation framework for network design problems that simulates the algorithm from [20] in the case of GENERALIZED STEINER TREE. The framework of Goemans and Willimanson [21] can be explained using local ratio [16] and the algorithm that is presented in this section is a simplification that only applies to STEINER TREE.

6.4 Packing Problems

In this section we discuss the local ratio approach towards packing problems. More specifically, we provide a generic local ratio algorithm for packing problems and demonstrate it on several problems.

6.4.1 Framework for Packing Problems

A general packing instance $\mathcal{I} = (S, f, w)$ consists of a set S of items to be packed, and a weight function w on the items. There is also a monotone function $f : 2^S \rightarrow \{0, 1\}$ such that $f(U) = 1$ if and only if the items in U can be packed together. We assume that $f(\emptyset) = 1$, since otherwise no feasible solution exists. For example, in the INDEPENDENT SET problem we are given a graph G and weight function on the vertices, and the goal is to find a maximum weight *independent set*, where an independent set $U \subseteq V$ satisfies $E \cap (U \times U) = \emptyset$. In this case $S = V$ and $f(U) = 1$ if and only if U is an independent set.

Before describing the generic algorithm we present the maximization version of Definition 6.2. First, notice that fully r-effectiveness is unreasonable in the context of packing problems, since the empty set constitute a feasible solution. A natural choice for a property would be *maximality with respect to set inclusion*. We shall use a slightly weaker property as explained in the following.

Given a packing instance $\mathcal{I} = (S, w, f)$ and an item $u \in S$, a solution U is called *u-maximal*, if either $u \in U$ or $f(U \cap \{u\}) = 0$.

Definition 6.3 *Given a packing problem (S, w, f), an item $u \in S$, and a number $r \geq 1$, a weight function $w \in \mathbb{R}^n$ is called r-effective if there exists a number $b \geq 0$ such that $\frac{b}{r} \leq wx \leq b$ for all u-maximal solutions.*

The local ratio algorithm for packing problem is not unlike the algorithm for covering problems. It is recursive, and each recursive call involves an r-effective weight function. However, there are also several differences. First, we allow negative coefficients in the weight function $w - w_1$. In addition all items whose weight is nonpositive are removed from the instance in the recursive call. Another difference is that the algorithm constructs *u-maximal* solutions rather than minimal ones, where u is an item whose weight becomes zero during the current recursive call. As in the covering case, u is called a *pivot*. Accordingly, when the recursive call returns a solution, the algorithm turns it into a u-maximal solution.

Algorithm **Pack** works as follows: It maintains a set P of items that are candidates to enter the solution. Hence, if $P = \emptyset$ the algorithm returns the empty set (recursion base). Otherwise, it computes an r-effective weight function w_1 with respect to P such that $w_1(u) = w(u)$, for some item $u \in P$. Then it recursively solves the problem for items with positive weights and $w - w_1$. The item u is added to the solution returned by the recursive call if the solution remains feasible with u. The initial call is **Pack**(S, w).

ALGORITHM 6.4   Pack(P, w)

1: **if** $P = \emptyset$ **then return** $\emptyset$
2: Let $u \in P$, and let w_1 be a u-maximal r-effective weight function (with respect to P), such that
   $w_1(u) = w(u)$.
3: $N \leftarrow \{v \in N : w(v) - w_1(v) \leq 0\}$
4: $U \leftarrow$ **pack**$(P \setminus N, w - w_1)$
5: **if** $f(U \cap \{u\}) = 1$ **then** $U \leftarrow U \cup \{u\}$
6: **return** U

Algorithm **Pack** has at most n recursive calls. Thus, its running time is polynomial, if a weight function w_1 can be computed efficiently.

We show that the algorithm computes r-approximate solutions, assuming that the weight function w_1 can be computed in each recursive call.

Theorem 6.3 *Let (S, f, w) be a packing instance. Suppose that there exists $u \in P$ and an r-effective weight function w_1 with respect to P, such that $w_1(u) = w(u)$, for every P and a weight function w. Then, Algorithm* **Pack** *is an r-approximation algorithm.*

Proof. We prove that U is an r-approximate solution by induction on the recursion. For the recursive base, observe that $\emptyset$ is optimal. As for the inductive step, let $U' = $ **Pack**$(P \setminus N, w - w_1)$. We know that U' is r-approximate with respect to $P \setminus N$ and $w_2 = w - w_1$ due to the inductive hypothesis. Since $w_2(v) \leq 0$, for every $v \in N$, it follows that U is r-approximate with respect P and w_2. The pivot u is added to U if possible, and therefore U is u-maximal at the end of the recursive call. Furthermore, since w_1 is r-effective with respect to u-maximality, U is r-approximate with respect to w_1. It follows, due to the Local Ratio Lemma, that U is r-approximate with respect to w as well. ∎

6.4.2 Independent Set

In this section provide applications of Algorithm **Pack** involving the INDEPENDENT SET problem. More specifically, we present a Δ-approximation algorithm for graphs whose maximum degree is at most Δ and an exact optimization algorithm for interval graphs.

A graph $G = (V, E)$ is said to have a bounded degree Δ if $\max_{v \in V} \deg(v) \leq \Delta$. Given such a graph, choose an arbitrary vertex u and define the following weight function:

$$w_1(v) = w(u) \cdot \begin{cases} 1 & v \in N(v) \cup \{v\}, \\ 0 & \text{otherwise.} \end{cases} \tag{6.2}$$

We claim that w_1 is Δ-effective with respect to u-maximality.

Lemma 6.9 *Let (G, w) be an INDEPENDENT SET instance, where G has bounded degree Δ. Also, let u be a vertex in G. Then the weight function w_1 is Δ-effective.*

Proof. Clearly, $w_1(U) \leq \Delta \cdot w(u)$, for every solution U. On the other hand, any u-maximal solution U contains either u or one of its neighbors, and therefore $w_1(U) \geq w(u)$. ∎

Theorem 6.3 and Lemma 6.9 lead to a Δ-approximation algorithm for INDEPENDENT SET in graphs with bounded degree Δ. The algorithm works for any selection of pivots. However if we choose the pivot u to be a minimum weight vertex, we get the greedy algorithm, in which one picks the maximum weight vertex and eliminates its neighbors. We note that the same algorithm also works for Δ-degenerate graphs,[*] provided that the pivot is a vertex whose degree is at most Δ.

Next, we deal with INDEPENDENT SET in interval graphs. An interval graph is the intersection graph of a set of intervals on the real line. That is, each vertex corresponds to an interval, and there is an edge between two vertices if their corresponding intervals intersect. We use Algorithm **Pack** to design an exact optimization (1-approximation) algorithm for this problem. We assume that the interval representation of the graph G is given in the input (Such a representation can be computed in linear time [22]).

Lemma 6.10 *Let (G, w) be an INDEPENDENT SET instance, where G is an interval graph, and let u be the vertex that corresponds to an interval with the leftmost right end-point. Then the weight function w_1 defined in (6.2) is 1-effective.*

Proof. $w_1(U) \leq w(u)$ for any feasible solution u, since u and its neighbors form a clique. On the other hand, any u-maximal solution U contains either u or one of its neighbors, and therefore $w_1(U) \geq w(u)$. ∎

---

[*] A graph is G called *k-degenerate*, if there exists a vertex of degree at most k in every subgraph of G.

We note that the same algorithm would work for INDEPENDENT SET in chordal graphs, provided that the pivots are chosen according to a perfect elimination ordering.

6.4.3 Interval Scheduling

In this section we utilize Algorithm Pack to design a local ratio algorithm for INTERVAL SCHEDULING. We also consider a special case called BANDWIDTH ALLOCATION PROBLEM (BAP). These algorithms in this section were originally given by Bar-Noy et al. [3].

An INTERVAL SCHEDULING instance consists of a set of *activities*, each requiring the utilization of a given *resource* whose capacity is normalized to 1. The activities are specified as a collection of sets $A_1, \ldots, A_m$, where the set A_j contains all possible *incarnations* of activity j. An incarnation $I \in A_j$ is associated with the following parameters: (i) a half-open time interval $[s(I), e(I))$, where $s(I)$ and $e(I)$ are the *start-time* and *end-time* of the incarnation; (ii) a demand $d(I) \in (0, 1]$; and (iii) a weight $w(I) \geq 0$. A feasible solution, or a *schedule* is a set of incarnations that satisfies the following conditions: (i) it contains at most one incarnation per activity and (ii) for all time instants t, the total demand at t is at most 1. The goal in INTERVAL SCHEDULING is to find a maximum weight schedule.

Instead of presenting an algorithm for INTERVAL SCHEDULING, we first consider a special case—the BAP. In this problem each activity A_j has only one incarnation denoted I_j. We use the term *job* to refer to both the activity its incarnation. Given a BAP instance, we partition the jobs into two subset: $L = \{j : d(I_j) > \frac{1}{2}\}$ and $S = \{j : d(I_j) \leq \frac{1}{2}\}$. Jobs in L with demand greater than $\frac{1}{2}$ are considered *large*, and the jobs in S are considered *small*. We solve the problem on L and on S and take the solution with maximum profit. Observe that a feasible solution cannot contain two large jobs whose intervals intersect. It follows that a solution for L must be a set of pairwise nonintersecting jobs. Hence we get an instance of INDEPENDENT SET in interval graphs, and this problem can be solved in polynomial time (as shown in the previous section).

It remains to cope with solving BAP on instances contains small demands. We do so using Algorithm **Pack** and the following weight function:

$$w_1(I_j) = w(I_k) \cdot \begin{cases} 1 & j = k, \\ \frac{d(I_j)}{1 - d(I_k)} & j \neq k, I_j \cap I_k \neq \emptyset, \\ 0 & \text{otherwise}, \end{cases}$$

where k is the job whose interval has the leftmost right end-point.

Lemma 6.11 *Given a BAP instance containing small demands, let k be the job whose interval has the leftmost right end-point. Then the weight function w_1 is 2-effective with respect to k-maximality.*

Proof. First, observe that all jobs that receive a positive weight by w_1 contain the right end-point of I_k. Let $\mathcal{J}$ be a feasible solution. If $I_k \notin \mathcal{J}$, then

$$w_1(\mathcal{J}) = w(I_k) \cdot \sum_{I_j \in \mathcal{J} : I_j \cap I_k \neq \emptyset} \frac{d(I_j)}{1 - d(I_k)} \leq w(I_k) \cdot \frac{1}{1 - d(I_k)} \leq 2w(I_k),$$

Since $d(I_k) \leq \frac{1}{2}$. If $I_k \in \mathcal{J}$, then

$$w_1(\mathcal{J}) = w(I_k) + w(I_k) \cdot \sum_{I_j \in \mathcal{J} \setminus \{I_k\} : I_j \cap I_k \neq \emptyset} \frac{d(I_j)}{1 - d(I_k)} \leq w(I_k) \left(1 + \frac{1 - d(I_k)}{1 - d(I_k)} \right) = 2w(I_k).$$

On the other hand, suppose that $\mathcal{J}$ is k-maximal. If $I_k \in \mathcal{J}$, then $w_1(\mathcal{J}) = w(I_k)$. Otherwise, the total demand on the right end-point of I_k is larger than $1 - d(I_k)$, which means that

$$w_1(\mathcal{J}) = w(I_k) \cdot \sum_{I_j \in \mathcal{J}: I_j \cap I_k \neq \emptyset} \frac{d(I_j)}{1 - d(I_k)} > w(I_k)\frac{1 - d(I_k)}{1 - d(I_k)} = w(I_k).$$

The lemma follows. ∎

Theorem 6.3 and Lemma 6.11 lead to a 2-approximation algorithm for instances of BAP containing small demands, and recall that we have an algorithm for solving the problem with large demands. Now consider an optimal solution. Either at least two thirds of the weight comes from small demands or not. In the former case, a 2-approximate solution for the set S of small jobs is 3-approximate. In the latter case, the solution for the set L of large jobs is 3-approximate. Hence, the solution with maximum weight is 3-approximate. We note that BAP admits a $(2 + \varepsilon)$-approximation algorithm, for any $\varepsilon > 0$ [23–25].

In INTERVAL SCHEDULING we need to cope with the multiple incarnations of each activity. We first deal with the case of large demands. Consider the following weight function:

$$w_1(J) = w(I) \cdot \begin{cases} 1 & J \in \mathcal{A}(I) \cup \mathcal{I}(I), \\ 0 & \text{otherwise,} \end{cases}$$

where I is the incarnation with the leftmost right end-point, $\mathcal{A}(I)$ is the activity of I, and $\mathcal{I}(I)$ is the set of incarnations intersecting I, that is, $\mathcal{I}(I) = \{J : I \cap J \neq \emptyset\}$.

Lemma 6.12 *Assume that we are given an* INTERVAL SCHEDULING *instance containing large demands, and let I be the incarnation with the leftmost right end-point. Then the weight function w_1 is 2-effective with respect to I-maximality.*

Proof. Any feasible solution may contain at most two incarnations from $\mathcal{A}(I) \cup \mathcal{I}(I)$, whereas an I-maximal solution contains at least one. ∎

We define the following weight function for the case of small demands:

$$w_1(J) = w(I) \cdot \begin{cases} 1 & J \in \mathcal{A}(I), \\ \frac{d(J)}{1-d(I)} & J \neq I, I \in \mathcal{I}(I), \\ 0 & \text{otherwise,} \end{cases}$$

where I is the incarnation with the leftmost right end-point.

The proof of the following lemma can be obtained by combining the proofs of Lemmas 6.11 and 6.12.

Lemma 6.13 *Assume that we are given an* INTERVAL SCHEDULING *instance containing large demands, and let I be the incarnation with the leftmost right end-point. Then the weight function w_1 is 3-effective with respect to I-maximality.*

The 2-approximation algorithm for large demands and the 3-approximation algorithm for small demands yield a 5-approximation algorithm for INTERVAL SCHEDULING.

6.5 Technique Variations

In this section we consider applications of the local ratio technique that do not fall within the covering and packing frameworks that were given in Sections 6.3 and 6.4.

As mentioned before, a typical local ratio algorithm is recursive. Each such algorithm can be divided into four primitives: (i) the recursion base, (ii) the construction of a weight function w_1, (iii) the way that an instance is modified before a recursive call, and (iv) the way in which the solution returned by a recursive call is fixed. In this section we present local ratio algorithms in which one (or more) of the primitives extends the primitive that was used by Algorithms **Cover** and **Pack**.

6.5.1 Nontrivial Recursion Base

In the previous sections the base case was trivial. In the covering framework the base case occurs when the empty set is feasible, and in the packing framework, the base case is reached when there is nothing left to pack. In this section we consider cases in which the base case is nontrivial, namely the base case is reached when the instance satisfies a certain property, and then the local ratio algorithm invokes an approximation algorithm that works for inputs that satisfy the above-mentioned property. Therefore, when the recursion base is nontrivial, the local ratio phase of the algorithm can be seen a "clean-up" phase.

A *multiple interval* is the union of a finite number of disjoint intervals on the real line. A *multiple interval graph* is an intersection graph of a family of multiple intervals. Such a graph is called a *t-interval graph* if each multiple interval in the family is the union of at most t intervals. We present a $(2 - \frac{1}{t})$-approximation algorithm for VERTEX COVER in t-interval graphs that is taken from [26]. We assume that a corresponding family of t-intervals is given as part of the input (As opposed to the case of $t = 1$, it is NP-hard to determine whether a given graph is t-interval, for $t \geq 2$ [27]).

The general outline of the algorithm is as follows. We first perform a local ratio clean-up phase on G to remove all cliques of size at least three. It follows that after this phase there are no three pairwise intersecting t-intervals in the corresponding family of t-intervals. We call t-interval graphs with this property *flat*. Flat t-interval graphs are convenient for our purposes, since they have bounded *degeneracy* (every induced subgraph has a vertex of bounded degree), and therefore we can apply known techniques for solving VERTEX COVER in such graphs. For now we use an α-approximation algorithm for flat t-interval graphs called **Flat-Cover**.

We first show that w_1 is fully $\frac{3}{2}$-effective.

ALGORITHM 6.5   VC-*t*-Interval($G = (V, E), w$)

1: **if** G is flat **then return** **Flat-Cover**(G, w)
2: Select a clique C in G such that $|V(C)| \geq 3$, and let $\varepsilon = \min_{v \in V(C)} w(v)$
3: Define $w_1(v) = \varepsilon \cdot \begin{cases} 1 & v \in V(C), \\ 0 & \text{otherwise.} \end{cases}$
4: **return** $\{u\} \cup$ **VC-*t*-interval**$(V \setminus \{u\}, E \setminus E(u), w - w_1)$

Lemma 6.14 *Given a* VERTEX COVER *instance* (G, w), *and a clique* C, *the weight function* w_1 *is fully* $\frac{3}{2}$-*effective.*

Proof. Since any cover must contain at least $|V(C)| - 1$ vertices from any clique C, it follows that w_1 is fully $(\frac{|V(C)|}{|V(C)|-1})$-effective. ∎

Next we bound the approximation ratio of Algorithm **VC-*t*-interval**.

Lemma 6.15 *Suppose that Algorithm* **Flat-Cover** *is an* α-*approximation algorithm for* VERTEX COVER *in flat t-interval graphs. Then Algorithm* **VC-*t*-interval** *is a* $\max\{\alpha, \frac{3}{2}\}$-*approximation algorithm for* VERTEX COVER *in t-interval graphs.*

Proof. We prove the lemma by induction on the recursion. At the recursion base the solution returned by Algorithm **Flat-Cover** is α-approximate. For the induction step, let U' the solution that was returned by the recursive call, and let $U = U' \cup \{u\}$. By the inductive hypothesis, U' is $\max\{\alpha, \frac{3}{2}\}$-approximate for $G' = (V \setminus \{u\}, E \setminus E(u))$ with respect to $w - w_1$. U is feasible for G, since u covers $E(u)$. Also, since $(w - w_1)(u) = 0$,

U is max$\{\alpha, \frac{3}{2}\}$-approximate for G with respect to $w - w_1$. In addition U is $\frac{3}{2}$-approximate with respect to w_1 due to Lemma 6.14. Applying the Local Ratio Theorem, we obtain that U is max$\{\alpha, \frac{3}{2}\}$-approximate with respect w.   ■

It remains to deal with flat t-interval graphs. The next lemma bounds the degeneracy of flat t-interval graphs.

Lemma 6.16 ([26]) *Flat t-interval graphs are $(2t - 1)$-degenerate.*

Proof. Let G be a flat t-interval graph, and consider the corresponding t-interval family. G contains an edge between v and u if their t-intervals intersect. Observe that in this case either one of v's intervals contains the right end-point of one of u's intervals, or one of u's intervals contains the right end-point of one of v's intervals. Moreover, no three intervals intersect, since G is flat. Hence, $|E|$ is bounded by the number of right end-points that are contained in other intervals. Since each t-interval is composed of at most t intervals, there are at most nt interval in total. Hence $|E| \leq nt - 1$, and it follows that there exists a vertex u such that $\deg(u) < 2t$.   ■

Hochbaum [28] presented a $(2 - 2/k)$-approximation algorithm for VERTEX COVER in graphs that can be colored with k colors. Since flat t-interval graphs are $(2t - 1)$-degenerate, they can be colored using $2t$ colors. It follows that using Hochbaum's algorithm, we can find an $(2 - 1/t)$-approximate cover.

We note that Bar-Yahuda et al. [29] used local ratio to eliminate large cliques in order to approximate VERTEX COVER in rectangle intersection graphs. Actually, a nontrivial recursion base was used in the first local ratio paper [1] in order to design a $(2 - \frac{\log\log n}{2\log n})$-approximation algorithm for VERTEX COVER. Consider a odd cycle C in G, and define

$$w_1(v) = \varepsilon \cdot \begin{cases} 1 & v \in V(C) \\ 0 & \text{otherwise,} \end{cases}$$

where $\varepsilon = \min_{v \in C} w(v)$. w_1 is fully $\frac{2|C|}{|C|+1}$-effective, since any cover must contain at least $\frac{|C|+1}{2}$ vertices from $V(C)$. This weight function cannot lead to a better than 2 ratio by itself, since it may contain odd cycles of length $\Omega(n)$. However, Bar-Yehuda and Even [1] used it eliminate short odd cycles, and then utilized, at the recursion base, an algorithm for VERTEX COVER in graphs without short odd cycles.

6.5.2 Instance Modification

The framework for covering contains a relatively simple instance modification procedure: add the pivot u to Z. This procedure relies on the monotonicity of the set of feasible solutions: it is never a bad idea to add zero weight item to the solution (if one ignores minimality). The maximization framework behaves almost as a mirror image. The instance modification procedure removes items whose weights are nonpositive, since nonpositive items can always be left out. However, this approach may not work in the nonmonotone case. Adding a zero weight item to Z or removing an item may result in an instance with no feasible solutions, or with very expensive (cheap) solutions. One possible way to cope with this issue is to go back to full effectiveness.

We demonstrate this on the MINIMUM 2-SATISFIABILITY problem (MIN-2SAT), in which the input consists of a 2CNF formula φ containing m clauses on the variables $x_1, \ldots, x_n$ and a nonnegative weight function w on the variables. A feasible solution is a truth assignment $x \in \{0, 1\}^n$ that satisfies φ, and its weight is $\sum_{i=1}^{n} w(i)x_i$. The goal is to find minimum weight truth assignment that satisfies φ or to determine that no such assignment exists. Gusfield and Pitt [30] presented an $O(mn)$ time 2-approximation algorithm for MIN-2SAT, but they did not use local ratio arguments explicitly. Nevertheless, a variant of their algorithm can be explained using local ratio (see Reference 31 for more details).

Consider a variable x_i. An assignment to x_i may have implied assignments. For example, if φ contains the clause $\overline{x_i} \wedge x_j$, then $x_i = 1$ implies $x_j = 1$. Let $T_b(x_i)$ be the set of literals that must be assigned 1, if $x_i = b$ (note that $x_i \in T_1(x_i)$). We note that the set $T_b(x_i)$ can be computed efficiently using constraint propagation. Observe that it is possible that $T_b(x_i) \cap T_{-b}(x_i) \neq \emptyset$. Also, it may be that $x_j, \overline{x_j} \in T_b(x_i)$, for some $b \in \{0, 1\}$, and in this case we say that assigning b to x_i leads to a contradiction. Furthermore, it can be shown [32] that if assigning a value b to x_i does not lead to a contradiction, then the formula φ' that is obtained by the assigning 1 to the literals in $T_b(x_i)$ is satisfiable if and only if φ is satisfiable.

Now assume that both assignments to x_i do not lead to a contradiction, and consider the following weight function:

$$w_1(x_j) = \begin{cases} \varepsilon_j & x_j \in T_0(x_i) \cup T_1(x_i), \\ 0 & \text{otherwise,} \end{cases}$$

where $\sum_{x_j \in T_0(x_i) \setminus T_1(x_i)} \varepsilon_j = \sum_{x_j \in T_1(x_i) \setminus T_0(x_i)} \varepsilon_j$. The function w_1 is fully 2-effective, since any assignment weighs at least $\beta = \min\left\{\sum_{x_j \in T_0(x_i)} w(x_j), \sum_{x_j \in T_1(x_i)} w(x_j)\right\}$ and at most 2β. Assume without loss of generality that $\beta = \sum_{x_j \in T_0(x_i)} w(x_j)$. If we assign $\varepsilon_j = w(x_j)$, for every $x_j \in T_0(x_i)$, the assignment of 1 to the literals in $T_1(x_i)$ is free of charge.

The local ratio algorithm is described as follows. At the recursion base, if the formula is empty, then we return an empty assignment. If the formula φ is not empty, we choose a variable x_i, and create a new formula φ' by taking the cheaper or the only possible assignment $x_i = b$. (If both are impossible, then φ is not satisfiable.) Then, we recursively solve the problem on φ' and $w - w_1$. When a satisfying assignment for φ' is returned, we augment it with the assignment implied by $x_i = b$.

We note that MIN-2SAT can also be 2-approximated using a reduction to VERTEX COVER [33, pp. 131–132]. We also note that Bar-Yehuda and Rawitz [31] have presented a local ratio 2-approximation algorithm for the *two variables per constraint integer programming* problem (2VIP) that generalizes min-2SAT. In this problem the variables have a bounded domain which is not necessarily $\{0, 1\}$.

6.5.3 Effectiveness with Respect to a Property

Algorithm **Cover** uses a weight function which is effective with respect to minimality. Hence, on return from the recursive call Algorithm **Cover** adds the pivot to the computed solution only if feasibility requires, thus making sure that the solution is minimal. Similarly, Algorithm **Pack** adds the pivot u to the solution if it can, since the weight function is effective with respect to u-maximality. In this section we discuss the case where the effectiveness is with respect to a property $\mathcal{P}$ which is not maximality or minimality. In such cases, we need to fix the solution that was computed by the recursive call such that it will satisfy $\mathcal{P}$.

The following extends Definitions 6.1 and 6.2:

Definition 6.4 *Given a covering problem (S, f, w), a property $\mathcal{P}$, and a number $r \geq 1$, a weight function $w \in \mathbb{R}^n$ is said to be r-effective with respect to a property $\mathcal{P}$ if there exists a number $b \geq 0$ such that $b \leq wx \leq r \cdot b$ for all solutions x that satisfy $\mathcal{P}$.*

In the BANDWIDTH TRADING problem the input consists of a set M of machine types and a set J of jobs. Each machine type $i \in M$ has a time interval I_i during which a machine of type i is available and a weight $w(i)$ that represent the price one as to pay for a machine of time i. Each job j has a time interval I_j during which it must be processed. A feasible solution, or a *schedule*, is an assignment S such that $S(i, k)$ is the set of jobs that are assigned to machine k of type i such that: (i) $\{S(i, k)\}_{i,k}$ is a partition of J; (ii) if $j \in S(i, k)$, for some k, then $I_j \subseteq I_i$; and (iii) if $j, j' \in S(i, k)$, then $I_j \cap I_{j'} = \emptyset$. The goal is to find a minimum weight schedule, where the weight of a schedule is $\sum_i w(i) \left| \{k : S(i, k) \neq \emptyset\} \right|$.

We present a local ratio 3-approximation algorithm for BANDWIDTH TRADING that relies on a property other than *minimality*. A primal-dual version of this algorithm was given by Bhatia et al. [34] (see also [16]).

Given a time t, let $J_t = \{j : t \in I_j\}$ and let $M_t = \{i : t \in I_i\}$. Let τ be a time that maximizes the size of J_t, that is, $\tau = \text{argmax}_t |J_t|$. We use the following weight function:

$$w_1 = \begin{cases} \varepsilon & i \in M_\tau, \\ 0 & i \notin M_\tau. \end{cases}$$

where $\varepsilon = \min \{w(i) : i \in M_\tau\}$.

We say that a schedule S is τ-*sensible* if it uses at most $3|J_\tau|$ machines from M_τ.

Lemma 6.17 w_1 *is 3-effective with respect to τ-sensibility.*

Proof. Clearly, $w(S) \geq \varepsilon|J_\tau|$, for any schedule S. On the other hand, $w(S) \leq 3\varepsilon|J_\tau|$, for any τ-sensible schedule S. ∎

Consider the following local ratio algorithm, called **BT**. The algorithm uses a procedure called **Transform** that transform S' into a schedule S for J such that (i) S is τ-sensible and (ii) for every $i \in M \setminus \{i_0\}$, S does not use more machines of type i than S'. The existence of Procedure **Transform** was proven in Reference 34.

ALGORITHM 6.6   BT(M, J, w)

1: **if** $J = \emptyset$ **then return** $\emptyset$
2: $\tau \leftarrow \text{argmax}_t |J_t|$
3: $i_0 \leftarrow \text{argmin}_i \{w(i) : i \in M_\tau\}$
4: $J' \leftarrow \{j : I_j \subseteq I_{i_0}\}$
5: $S' \leftarrow$ **BT**($M \setminus \{i_0\}, J \setminus J', w - w_1$)
6: $S \leftarrow$ **Transform**(S')
7: **return** S

Theorem 6.4 *Algorithm **BT** is a 3-approximation algorithm for* BANDWIDTH TRADING.

Proof. We prove the theorem by induction on the recursion. In the base case ($J = \emptyset$), the schedule returned is optimal. For the inductive step, S' is 3-approximate with respect to $(M \setminus \{i_0\}, J \setminus J', w - w_1)$ due to the inductive hypothesis. Since $w_1(i) = w(i)$, S is 3-approximate with respect to $(M, J, w - w_1)$. Also, since S is τ-sensible it is 3-approximate with respect to w_1 due to Lemma 6.17. Therefore, by the Local Ratio Theorem, S is 3-approximate with respect to w. ∎

There are other local ratio algorithm that use properties other than *minimality* or *u-maximality* (when u is the pivot). Amzallag et al. [35] gave two local ratio approximation algorithms for a resource allocation problem, called the CELL SELECTION problem, in which clients may receive service from multiple severs. One of these algorithms relies on a certain kind of u-maximal solutions, where the capacity of several servers must be utilized before using the rest of the servers. Another example is CAPACITATED VERTEX COVER (soft capacities). The local ratio version of the primal-dual 2-approximation algorithm for this problem [36] relies on the following property: solutions that are computed by the algorithm (see Reference 16 for more details).

6.5.4 Nonuniform Weight Functions

Most local ratio applications utilize a single type of weight function. However, one may use several different weight function types within the same algorithm.

As an example, we revisit FEEDBACK VERTEX SET. The local ratio 2-approximation algorithm for this problem that was given in Section 6.3.4 is based on a single type of weight function: the weight of a vertex is proportional to its degree. Bafna et al. [7] presented a local ratio 2-approximation algorithm that uses two types of weight functions. We describe it as an application of Algorithm **Cover**. First, if the graph contains a *semi-disjoint* cycle* C, the algorithm uses the following weight function:

$$w_1(v) = \begin{cases} \varepsilon & v \in C, \\ 0 & \text{otherwise,} \end{cases}$$

where $\varepsilon = \min_{v \in C} w(v)$. The weight function w_1 is 1-effective since a minimal solution contains exactly one vertex from C. If the graph does not contain semi-disjoint cycles, it uses the weight function $w_1(v) = \varepsilon \cdot (\deg(v) - 1)$, where $\varepsilon = \min_{v \in V} \frac{w(v)}{\deg(v)-1}$. A proof that w_1 is 2-effective can be found in Reference 7 (see also [15]).

Mestre [37] presented a $(1 + \phi)$-approximation algorithm the DATA MIGRATION problem, where $\phi = \frac{1+\sqrt{5}}{2}$ is the golden ratio. To obtain this result he took the idea of using different weight function types (see Section 6.5.4) a step further by presenting an *adaptive* version of local ratio in which the weight function w_1 is computed in order to minimize its effectiveness. More specifically, the problem of finding a weight function with minimum effectiveness over a given support is cast as a linear program (LP), and such an LP is solved in each recursive step of the algorithm in order to obtain the weight function w_1. The upper bound on the approximation ratio is obtained by finding an upper bound on the effectiveness of the weight function that is computed by the LP solver.

6.5.5 Negative Weights

In most local ratio applications the coefficients of the weight function w_1 are nonnegative, but negative weights can also be used. We demonstrate the use of negative weights by providing a local ratio analysis of the Ford and Fulkerson algorithm [38]. This analysis that appeared in Reference 39 was the first to use negative weights in the context of local ratio.

The input in the MINIMUM s, t-CUT problem consists of a directed graph $G = (V, E)$, a source s, a destination t, and a nonnegative weight function w on the edges. An s, t-cut is a partition of V into two subsets S and T such that $s \in S$ and $t \in T$, and the weight of a cut (S, T) is $\sum_{v \in S, u \in T} w(v, u)$. The goal is to find an s, t-cut of minimum weight.

Now consider a path P from s to t, and define the following weight function:

$$w_1(v, u) = \begin{cases} \varepsilon_P & (v, u) \in P, \\ -\varepsilon_P & (u, v) \in P, \\ 0 & \text{otherwise,} \end{cases}$$

where $\varepsilon_P = \min_{(v,u) \in P} w(v, u)$.

Lemma 6.18 *Given a* MINIMUM s, t-CUT *instance, w_1 is fully 1-effective.*

Proof. Let (S, T) be an s, t-cut. The weight of (S, T) is $w_1(S, T) = \varepsilon_P \cdot |(S \times T) \cap P| + (-\varepsilon_P) \cdot |(T \times S) \cap P|$. Since P is a path from s to t, it follows that $|(S \times T) \cap P| - |(T \times S) \cap P| = 1$, and thus $w_1(S, T) = \varepsilon$. ∎

Given Lemma 6.18 the Ford and Fulkerson algorithm can be described as follows: as long as there is an augmenting path P with $\varepsilon_P > 0$, we subtract w_1 from w, and otherwise return the set of nodes that are reachable from s using nonzero weight edges.

---

* A cycle C is *semi-disjoint* if there exists a vertex $x \in C$ such that $\deg(u) = 2$ for every vertex $u \in C \setminus \{x\}$.

Bar-Yehuda and Rawitz [39] provided local ratio analyses that rely on negative coefficients for the optimality of the solutions that are obtained by two additional exact optimization algorithms: Kuhn's algorithm for the ASSIGNMENT problem [40] and Goldberg's preflow-push algorithm [41].

6.6 Fractional Local Ratio

As we saw, a typical local ratio r-approximation algorithm is recursive, and an r-effective weight function w_1 is constructed in each recursive call. Observe that *effectiveness* is local in the sense that w_1 is said to r-effective with respect to a property $\mathcal{P}$, if a solution that satisfies $\mathcal{P}$ is r-approximate compared to the optimal solution with respect to w_1. This means that the constructed solution is compared to a different optimal solution at different recursive calls. Bar-Yehuda et al. [42] came up with a more global approach: all intermediate solutions are compared to a specific solution $\tilde{x}$. Clearly, $\tilde{x}$ cannot be an optimal solution of the original instance, therefore they settled for a super optimal nonfeasible solution. More specifically, they used an optimal fraction solution to a linear programming relaxation (LP relaxation). Hence, the name *fractional local ratio*.

Fractional local ratio compares solutions to a vector $\tilde{x}$, thus it is convenient to use the following definition. Since we focus on maximization problems in this section, the definition is given for maximization problems.

Definition 6.5 *Let $w \in \mathbb{R}^n$ be a weight function and let $r \geq 1$. Also, $x, \tilde{x} \in \mathbb{R}^n$. x is said to be r-approximate relative to $\tilde{x}$ (with respect to w) if $w \cdot x \geq (w \cdot \tilde{x})/r$.*

Fractional local ratio is based on a *fractional* version of the Local Ratio Lemma.

Lemma 6.19 (Fractional Local Ratio [42]) *Let $w, w_1, w_2 \in \mathbb{R}^n$ be weight functions such that $w = w_1 + w_2$. Let $\tilde{x}, x \in \mathbb{R}^n$ such that x is r-approximate relative to $\tilde{x}$ with respect to w_1 and with respect to w_2. Then, x is r-approximate relative to $\tilde{x}$ with respect to w as well.*

Proof. $w \cdot x = w_1 \cdot x + w_2 \cdot x \geq (w_1 \cdot \tilde{x})/r + (w_2 \cdot \tilde{x})/r = (w \cdot \tilde{x})/r$. ∎

The first step of a fraction local ratio algorithm is the computation of the solution $\tilde{x}$. The rest of the algorithm is quite similar to a regular local ratio algorithm. However, in each recursive call, the construction of the weight function w_1 is based on $\tilde{x}$. Also, the corresponding analysis compares the weight of the solution returned with the weight of $\tilde{x}$, namely it is based on the Fractional Local Ratio Lemma.

The next definition reformulates the notion of effectiveness in the context of fractional local ratio (again, for maximization problems).

Definition 6.6 *Given a set of feasibility constraints $\mathcal{F}$ on vectors in $\mathbb{R}^n$, a solution $\tilde{x}$, and a number $r \geq 1$, a weight function $w \in \mathbb{R}^n$ is said to be r-effective with respect to property $\mathcal{P}$ and compared to $\tilde{x}$ if $w \cdot x \geq w \cdot \tilde{x}/r$ for all feasible solutions x that satisfy $\mathcal{P}$.*

6.6.1 Independent Set in Bounded Degree Graphs

In this section we describe a fractional local ratio $\frac{1}{2}(\Delta + 1)$-approximation algorithm for INDEPENDENT SET in graph with bounded degree $\Delta > 1$. (Recall that a local ratio Δ-approximation algorithm was given in section 6.4.2.) We note that another $\frac{1}{2}(\Delta + 1)$-approximation algorithm for INDEPENDENT SET in graph with bounded degree $\Delta > 1$ was given by Hochbaum [28].

The INDEPENDENT SET problem can be formulated using the following linear integer program:

$$
\begin{aligned}
\max \quad & \sum_{v \in V} w(v) \cdot x(v) \\
\text{s.t.} \quad & x(v) + x(u) \leq 1 && \forall (v, u) \in E \\
& x(v) \in \{0, 1\} && \forall v \in V
\end{aligned}
\tag{6.3}
$$

An LP relaxation is obtained by replacing the second set of constraints by: $x(v) \in [0, 1]$, for every $v \in V$.

We need the follow property of fractional solutions of (6.3).

Lemma 6.20 *Let $\tilde{x}$ be a feasible fraction solution of the LP relaxation of (6.3). Then there exists a vertex u satisfying $\tilde{x}(u) + \sum_{v \in N(u)} \tilde{x}(v) \le \frac{1}{2}(\Delta + 1)$.*

Proof. Let $u = \text{argmax}_{u \in V} \tilde{x}(v)$. If u is an isolated vertex, then we are done. Otherwise, observe that if $\tilde{x}(u) \le \frac{1}{2}$, then $\tilde{x}(v) \le \tilde{x}(u) \le \frac{1}{2}$, for every $v \in N(u)$, and if $\tilde{x}(u) > \frac{1}{2}$, then $\tilde{x}(v) \le 1 - \tilde{x}(u) \le \frac{1}{2}$, for every $v \in N(u)$. Let $v' \in N(u)$. We have that

$$\tilde{x}(u) + \sum_{v \in N(u)} \tilde{x}(v) \le \tilde{x}(u) + \tilde{x}(v') + \sum_{v \in N(u) \setminus \{v'\}} \tilde{x}(v) \le 1 + (\Delta - 1)\tilde{x}(u) \le \frac{1}{2}(\Delta + 1).$$

∎

The fractional local ratio algorithm first computes an optimal solution $\tilde{x}$ of the LP relaxation of (6.3). Then it uses Algorithm **Pack** with the weight function w_1 from Equation (6.2). However, now we need to choose the pivot u carefully. More specifically, the pivot would be a vertex u minimizing the expression $\tilde{x}(u) + \sum_{v \in N(u)} \tilde{x}(v)$. Lemma 6.20 ensures that $\tilde{x}(u) + \sum_{v \in N(u)} \tilde{x}(v) \le \frac{1}{2}(\Delta + 1)$ for the chosen pivot u.

Lemma 6.21 *Let (G, w) be an INDEPENDENT SET instance, where G has bounded degree Δ. Also, let $\tilde{x}$ be an feasible fractional solution. Then the weight function w_1 is $\frac{1}{2}(\Delta + 1)$-effective with respect to u-maximality and compared to $\tilde{x}$.*

Proof. Any u-maximal solution U contains either u or one of its neighbors, and therefore $w_1(U) \ge w(u)$. One the other hand, by Lemma 6.20 we have that

$$w_1\tilde{x} = w_1(u)\tilde{x}(u) + \sum_{v \in N(u)} w_1(v)\tilde{x}(v) = w_1(u)\left(\tilde{x}(u) + \sum_{v \in N(u)} \tilde{x}(v)\right) \le w_1(u) \cdot \frac{1}{2}(\Delta + 1),$$

and the lemma follows. ∎

The proof of the following theorem is similar to the proof of Theorem 6.3.

Theorem 6.5 *There exists a $\frac{1}{2}(\Delta + 1)$-approximation algorithm for INDEPENDENT SET in graphs with bounded degree $\Delta > 1$.*

Proof. Let $\tilde{x}$ be be an optimal fractional solution. Also, let x denote the incidence vector of the computed solution U. In order to compare a solution x to $\tilde{x}$, we assume that x and all intermediate weight functions are of size $|V|$. We prove that x is $\frac{1}{2}(\Delta + 1)$-approximate relative to $\tilde{x}$ with respect to w by induction on the recursion. If follows that the computed solution is a $\frac{1}{2}(\Delta + 1)$-approximate for the original instance.

For the recursive base, observe that $wx = 0$ whereas $w\tilde{x} \le 0$, and therefore $wx \ge \frac{1}{2}(\Delta + 1)\tilde{x}$. As for the inductive step, let x' be the incidence vector of $U' = \text{Pack}(P \setminus N, w - w_1)$. By the induction hypothesis we have that x' is $\frac{1}{2}(\Delta + 1)$-approximate relative to $\tilde{x}$ with respect to $w - w_1$. Since $w_1(u) = w(u)$ (u is the pivot), x is also $\frac{1}{2}(\Delta + 1)$-approximate relative to $\tilde{x}$ with respect to $w - w_1$. Next, since x is u-maximal at the end of the recursive call, Lemma 6.20 implies that x is $\frac{1}{2}(\Delta + 1)$-approximate relative to $\tilde{x}$ with respect to w_1. It follows, due to the Fractional Local Ratio Lemma, that x is r-approximate relative to $\tilde{x}$ with respect to w as well. ∎

6.6.2 Independent Set in t-Interval Graphs

The first fractional local ratio algorithm was a $2t$-approximation algorithm for INDEPENDENT SET in t-interval graphs [42].

We start with a linear integer program for INDEPENDENT SET in t-interval graphs:

$$
\begin{aligned}
\max \quad & \sum_{v \in V} w(v) \cdot x(v) \\
\text{s.t.} \quad & \sum_{v:p \in I(v)} x(v) \leq 1 \quad \forall p \in R \\
& x(v) \in \{0,1\} \quad \forall v \in V
\end{aligned}
\tag{6.4}
$$

where R is the set of right end-points of intervals, and $I(v)$ is the t-interval that corresponds to v. (We assume that a t-interval realization is provided in the input.) As before, an LP relaxation is obtained by replacing the second set of constraints by: $x(v) \in [0,1]$, for every $v \in V$.

The algorithm for the INDEPENDENT SET in t-interval graphs is almost identical to the one given in the previous section. The following lemma replaces Lemma 6.20.

Lemma 6.22 *Let x be a solution of the LP relaxation of (6.4). Then there exists a vertex u satisfying $x(u) + \sum_{v \in N(u)} x(v) \leq 2t$.*

Proof. Let $N[v] = N(v) \cup \{v\}$. In order to prove this lemma, it is enough to show that

$$
\sum_{v} \sum_{u \in N[v]} x(v)x(u) = \sum_{v} x(v) \sum_{u \in N[v]} x(u) \leq 2t \cdot \sum_{v} x(v).
$$

First, observe that if the t intervals of v_1 and v_2 intersect, then $v_1 \in N[v_2]$ and $v_2 \in N[v_1]$. Therefore, in this case, the term $x(v_1)x(v_2)$ appears twice on the LHS. Furthermore, either the exists a point $p \in R(v_1)$ such that $p \in I(v_1)$, or there exists a point $p \in R(v_2)$ such that $p \in I(v_2)$. It follows that

$$
\sum_{v} \sum_{u \in N[v]} x(v)x(u) \leq 2 \cdot \sum_{v} \sum_{p \in R(v)} \sum_{u:p \in I(u)} x(v)x(u).
$$

Since x is a feasible solution of the LP relaxation of (6.4), for every v and $p \in R(v)$ we have that

$$
\sum_{u:p \in I(u)} x(v)x(u) = x(v) \sum_{u:p \in I(u)} x(u) \leq x(v),
$$

Hence,

$$
\sum_{v} \sum_{u \in N[v]} x(v)x(u) \leq 2 \cdot \sum_{v} \sum_{p \in R(v)} x(v) = 2t \cdot \sum_{v} x(v),
$$

and we are done.   ∎

This leads to the following result:

Theorem 6.6 *There exists a 2t-approximation algorithm for* INDEPENDENT SET *in t-interval graphs.*

The previous result was extended in several directions. Bar-Yehuda and Rawitz [43] designed fractional local ratio $O(t)$-approximation algorithm for several variant of the t-INTERVAL SCHEDULING problem, in which one is asked to pack t-intervals with demands (à la BAP). Hermelin and Rawitz [44] presented a fractional local ratio algorithm for INDEPENDENT SET in t-subtree graphs, which are the intersection graphs of t-subtrees, where a sub-tree is the union of t disjoint subtrees of some tree. This graph class generalizes both the class of t-interval graphs and the class of chordal graphs. Canzar et al. [45] studied a special case of INDEPENDENT SET in 2-interval graphs. They presented two fractional local ratio algorithms: a 3-approximation algorithm with a trivial recursive base and 2-approximation algorithm whose recursive base takes advantage of the structure of basic feasible solutions of an LP relaxation.

Lewin-Eytan et al. [46] used fractional local ratio to design an approximation algorithm for INDEPENDENT SET in intersection graphs of axis parallel rectangles, which is another multi-dimensional extension of interval graphs. Rawitz [47] extended their result to packing rectangles with demands.

6.7 Additional Topics

This chapter is meant to be an introductory survey of the local ratio technique. In this section we mention several advanced topics that are related to local ratio and are not covered by the chapter.

Connection to primal-dual schema: The *primal-dual schema* is a common method for designing approximation algorithm. Such an algorithm relies on relaxing the dual complementary slackness conditions, while enforcing the primal conditions. Bar-Yahuda and Rawitz [10] showed that the local ratio technique and the primal-dual schema are equivalent. The equivalence is based on showing that an effective weight function with respect to a property $\mathcal{P}$ is equivalent to a *strong* valid inequality (in terms of Reference 48) with respect to a property $\mathcal{P}$. The equivalence is constructive in the sense that it can be used to transform a primal-dual algorithm into a local ratio algorithm, and vice versa. The equivalence holds for all variants of local ratio showed in this chapter, and it is likely to hold to extensions of both techniques. For example, a fractional version of the primal-dual schema, which is equivalent to fractional local ratio, was presented in Reference 43.

Lower bounds and the stack model: Borodin et al. [49] defined a syntactic computational model called the *stack model* that captures standard local ratio algorithms for covering and packing problems (i.e., those that are described in Sections 6.3 and 6.4). They presented several inapproximability results that apply to algorithms that fall within the stack model. Since the model is defined without any complexity limitations, the bounds are independent of the question whether $P = NP$.

Distributed local ratio algorithms: Local ratio algorithms sometimes admit efficient distributed versions. Patt-Shamir et al. [50] presented a distributed version of a local ratio approximation algorithm for the CELL SELECTION problem [35]. The distributed implementation terminates in $O(\varepsilon^{-2}\mathrm{polylog}(n))$ rounds and increases the approximation ratio by a multiplicative factor of $(1 + \varepsilon)$, for any $\varepsilon \in (0, 1)$. Also, Bar-Yehuda et al. [51] recently presented a distributed local ratio $(2+\varepsilon)$-approximation algorithm for VERTEX COVER, which terminates in $O(\frac{\log \Delta}{\varepsilon \log \log \Delta})$ rounds, for any $\varepsilon > 0$, which is at most $O(1)$. One of the main challenges in using local ratio in a distributed setting is that one cannot use a single pivot, if one hopes to get a poly-logarithmic running time. One way to solve this issue is to perform several/many weight subtractions in parallel.

Acknowledgment

The author thanks Reuven Bar-Yehuda for helpful discussions and suggestions.

References

1. R. Bar-Yehuda and S. Even. A local-ratio theorem for approximating the weighted vertex cover problem. *Annals of Discrete Mathematics*, 25:27–46, 1985.
2. M. R. Garey and D. S. Johnson. *Computers and Intractability: A Guide to the Theory of NP-Completeness*. W. H. Freeman and Company, New York, 1979.
3. A. Bar-Noy, R. Bar-Yehuda, A. Freund, J. Naor, and B. Shieber. A unified approach to approximating resource allocation and scheduling. *Journal of the ACM*, 48(5):1069–1090, 2001.
4. R. Bar-Yehuda and S. Even. A linear time approximation algorithm for the weighted vertex cover problem. *Journal of Algorithms*, 2:198–203, 1981.
5. K. L. Clarkson. A modification of the greedy algorithm for vertex cover. *Information Processing Letters*, 16(1):23–25, 1983.
6. N. G. Hall and D. S. Hochbaum. A fast approximation algorithm for the multicovering problem. *Discrete Applied Mathematics*, 15(1):35–40, 1986.
7. V. Bafna, P. Berman, and T. Fujito. A 2-approximation algorithm for the undirected feedback vertex set problem. *SIAM Journal on Discrete Mathematics*, 12(3):289–297, 1999.

8. T. Fujito. A unified approximation algorithm for node-deletion problems. *Discrete Applied Mathematics and Combinatorial Operations Research and Computer Science*, 86:213–231, 1998.

9. R. Bar-Yehuda. One for the price of two: A unified approach for approximating covering problems. *Algorithmica*, 27(2):131–144, 2000.

10. R. Bar-Yehuda and D. Rawitz. On the equivalence between the primal-dual schema and the local ratio technique. *SIAM Journal on Discrete Mathematics*, 19(3):762–797, 2005.

11. R. Bar-Yehuda. Using homogeneous weights for approximating the partial cover problem. *Journal of Algorithms*, 39:137–144, 2001.

12. D. P. Williamson and D. B. Shmoys. *The Design of Approximation Algorithms*. Cambridge University Press, New York, 2011.

13. N. H. Bshouty and L. Burroughs. Massaging a linear programming solution to give a 2-approximation for a generalization of the vertex cover problem. In *15th Annual Symposium on Theoretical Aspects of Computer Science*, volume 1373 of *LNCS*, pp. 298–308. Springer, Berlin, Germany, 1998.

14. A. Becker and D. Geiger. Optimization of Pearl's method of conditioning and greedy-like approximation algorithms for the vertex feedback set problem. *Artificial Intelligence*, 83(1):167–188, 1996.

15. F. A. Chudak, M. X. Goemans, D. S. Hochbaum, and D. P. Williamson. A primal-dual interpretation of recent 2-approximation algorithms for the feedback vertex set problem in undirected graphs. *Operations Research Letters*, 22:111–118, 1998.

16. R. Bar-Yehuda, K. Bendel, A. Freund, and D. Rawitz. Local ratio: A unified framework for approximation algorithms (in memoriam: Shimon even 1935–2004). *ACM Computing Surveys*, 36(4):422–463, 2004.

17. R. Bar-Yehuda, K. Bendel, A. Freund, and D. Rawitz. The local ratio technique and its application to scheduling and resource allocation problems. In M. C. Golumbic and I. B.-A. Hartman, Eds., *Graph Theory, Combinatorics, and Algorithms: Interdisciplinary Applications*, volume 34 of Operations Research & Computer Science, pp. 107–143. Springer-Verlag, New York, 2005.

18. E. W. Dijkstra. A note on two problems in connexion with graphs. *Numerische Mathematik*, 1:269–271, 1959.

19. J. Kruskal. On the shortest spanning subtree of a graph and the traveling salesman problem. *The American Mathematical Society*, 7:48–50, 1956.

20. A. Agrawal, P. Klein, and R. Ravi. When trees collide: An approximation algorithm for the generalized Steiner problem on networks. *SIAM Journal on Computing*, 24(3):440–456, 1995.

21. M. X. Goemans and D. P. Williamson. A general approximation technique for constrained forest problems. *SIAM Journal on Computing*, 24(2):296–317, 1995.

22. K. S. Booth and G. S. Lueker. Testing for the consecutive ones property, interval graphs, and graph planarity using pq-tree algorithms. *Journal of Computer and System Sciences*, 13(3):335–379, 1976.

23. G. Călinescu, A. Chakrabarti, H. J. Karloff, and Y. Rabani. An improved approximation algorithm for resource allocation. *ACM Transactions on Algorithms*, 7(4):48, 2011.

24. C. Chekuri, M. Mydlarz, and F. B. Shepherd. Multicommodity demand flow in a tree and packing integer programs. *ACM Transactions on Algorithms*, 3(3), 2007.

25. A. Anagnostopoulos, F. Grandoni, S. Leonardi, and A. Wiese. A mazing $(2 + \varepsilon)$ approximation for unsplittable flow on a path. In *25th Annual ACM-SIAM Symposium on Discrete Algorithms*, pp. 26–41, 2014.

26. A. Butman, D. Hermelin, M. Lewenstein, and D. Rawitz. Optimization problems in multiple-interval graphs. *ACM Transactions on Algorithms*, 6(2):40, 2010.

27. D. B. West and D. B. Shmoys. Recognizing graphs with fixed interval number is NP-complete. *Discrete Applied Mathematics*, 8:295–305, 1984.

28. D. S. Hochbaum. Efficient bounds for the stable set, vertex cover and set packing problems. *Discrete Applied Mathematics*, 6:243–254, 1983.

29. R. Bar-Yehuda, D. Hermelin, and D. Rawitz. Minimum vertex cover in rectangle graphs. *Computational Geometry*, 44(6–7):356–364, 2011.

30. D. Gusfield and L. Pitt. A bounded approximation for the minimum cost 2-SAT problem. *Algorithmica*, 8:103–117, 1992.

31. R. Bar-Yehuda and D. Rawitz. Efficient algorithms for bounded integer programs with two variables per constraint. *Algorithmica*, 29(4):595–609, 2001.

32. S. Even, A. Itai, and A. Shamir. On the complexity of timetable and multi-commodity flow problems. *SIAM Journal on Computing*, 5(4):691–703, 1976.

33. D. S. Hochbaum, Ed. *Approximation Algorithms for NP-Hard Problem*. PWS Publishing Company, Boston, MA, 1997.

34. R. Bhatia, J. Chuzhoy, A. Freund, and J. Naor. Algorithmic aspects of bandwidth trading. *ACM Transactions on Algorithms*, 3(1):1–19, 2007.

35. D. Amzallag, R. Bar-Yehuda, D. Raz, and G. Scalosub. Cell selection in 4g cellular networks. *IEEE Transactions on Mobile Computing*, 12(7):1443–1455, 2013.

36. S. Guha, R. Hassin, S. Khuller, and E. Or. Capacitated vertex covering. *Journal of Algorithms*, 48(1):257–270, 2003.

37. J. Mestre. Adaptive local ratio. *SIAM Journal on Computing*, 39(7):3038–3057, 2010.

38. L. R. Ford and D. R. Fulkerson. Maximal flow through a network. *Canadian Journal of Mathematics*, 8:399–404, 1956.

39. R. Bar-Yehuda and D. Rawitz. Local ratio with negative weights. *Operations Research Letters*, 32(6):540–546, 2004.

40. H. W. Kuhn. The Hungarian method of solving the assignment problem. *Naval Research Logistics Quarterly*, 2:83–97, 1955.

41. A. V. Goldberg. *Efficient graph algorithms for sequential and parallel computers*. PhD thesis, Department of Electrical Engineering and Computer Science, MIT, Cambridge, MA, 1987.

42. R. Bar-Yehuda, M. M. Halldórsson, J. Naor, H. Shachnai, and I. Shapira. Scheduling split intervals. *SIAM Journal on Computing*, 36(1):1–15, 2006.

43. R. Bar-Yehuda and D. Rawitz. Using fractional primal-dual to schedule split intervals with demands. *Discrete Optimization*, 3(4):275–287, 2006.

44. D. Hermelin and D. Rawitz. Optimization problems in multiple subtree graphs. *Discrete Applied Mathematics*, 159(7):588–594, 2011.

45. S. Canzar, K. M. Elbassioni, G. W. Klau, and J. Mestre. On tree-constrained matchings and generalizations. *Algorithmica*, 71(1):98–119, 2015.

46. L. Lewin-Eytan, J. Naor, and A. Orda. Admission control in networks with advance reservations. *Algorithmica*, 40(4):293–304, 2004.

47. D. Rawitz. Admission control with advance reservations in simple networks. *Journal of Discrete Algorithms*, 5(3):491–500, 2007.

48. D. Bertsimas and C.-P. Teo. From valid inequalities to heuristics: A unified view of primal-dual approximation algorithms in covering problems. *Operations Research*, 46(4):503–514, 1998.

49. A. Borodin, D. Cashman, and A. Magen. How well can primal-dual and local-ratio algorithms perform? *ACM Transactions on Algorithms*, 7(3):29, 2011.

50. B. Patt-Shamir, D. Rawitz, and G. Scalosub. Distributed approximation of cellular coverage. *Journal on Parallel and Distributed Computing*, 72(3):402–408, 2012.

51. R. Bar-Yehuda, K. Censor-Hillel, and G. Schwartzman. A distributed $(2 + \varepsilon)$-approximation for vertex cover in $O(\log \Delta / \varepsilon \log \log \Delta)$ rounds. In *35th ACM Symposium on Principles of Distributed Computing. Journal of the ACM*, 64(3):23:1–23:11, 2017.

LP Rounding
and Extensions

Daya Ram Gaur

Ramesh Krishnamurti

7.1 Introduction

Many combinatorial optimization problems can be cast as integer linear programming problems. A linear programming relaxation of an integer program provides a natural lower bound (in case of minimization problems) on the value of the optimal integral solution. An optimal solution to the linear programming relaxation may not necessarily be integral. If there exists a procedure to obtain an integral solution "close" to the fractional solution then we have an approximation algorithm. This process of obtaining the integral solution from the fractional one is referred to as "rounding." Our goal is to present an ensemble of rounding techniques (which is by no means complete) that have enjoyed some success. On occasion, for detailed correctness proofs, we refer the reader to the original paper.

Rounding techniques can be broadly divided into two categories; those which round variables non-deterministically (also called randomized rounding), and those which round variables deterministically. The randomized rounding techniques presented typically yield solutions whose expected value is bounded. At times the rounding steps can be made deterministic (derandomized) by using the method of conditional expectation due to Erdős and Selfridge [1]. We refer the reader to Alon and Spencer [2, Chapter 15] for the method of conditional expectation. Both randomized as well as deterministic rounding can be further classified into techniques which round the variables independently, and those which round the variables in groups (dependently). Our presentation is along similar lines; in Section 7.2 we discuss nondeterministic rounding techniques due to Raghavan and Thompson [3], Goemans and Williamson [4], Bertsimas et al. [5], Goemans and Williamson [6], and Arora et al. [7]. We discuss deterministic rounding techniques due to Lin and Vitter [8], Jain [9], Ageev and Sviridenko [10], and Gaur et al. [11] in Section 7.3. Finally, we conclude with a discussion. For other applications of rounding we refer the reader to the books by Hochbaum [12] and Vazirani [13].

Next, we define the performance ratio of an approximation algorithm. Associated with every instance $\mathcal{I}$ of an NP-optimization problem $\mathcal{P}$ is a nonempty set of feasible solutions $\mathcal{S}$. To each solution $S \in \mathcal{S}$ we assign a number called its value. For a minimization (maximization) problem, the goal is to determine the solution with the minimum (maximum) value. The solution with the minimum (maximum) value is denoted $OPT(\mathcal{I})$ or simply OPT when there is no ambiguity. Let A be an algorithm whose running time is bounded by a polynomial in the length of the input. $ALG(\mathcal{I})$ (or simply ALG) denotes the value of the solution returned by algorithm A on problem $\mathcal{P}$. For a minimization (maximization) problem the performance ratio of A is defined as the $\alpha = \max_{\mathcal{I}}(ALG(\mathcal{I})/OPT(\mathcal{I}))$ ($\alpha = \min_{\mathcal{I}}(ALG(\mathcal{I})/OPT(\mathcal{I}))$). For minimization (maximization) problems $\alpha \geq 1$ ($\alpha \leq 1$). The other commonly used convention is to define $\alpha = \min_{\mathcal{I}}(OPT(\mathcal{I})/ALG(\mathcal{I}))$ for a minimization problem (in which case, $\alpha \leq 1$).

7.2 Nondeterministic Rounding

7.2.1 Independent Rounding

In this section we illustrate the technique due to Raghavan and Thompson [3]. They developed the first constant factor approximation algorithm for the minimum width routing problem in two dimensions. Here we illustrate their technique on the Set Cover problem, basing our presentation on Vazirani [13]. Given a collection $S = \{S_1, S_2, \ldots, S_m\}$ of subsets of some universe $U = \{1, 2, \ldots, n\}$, the problem is to determine the minimum number of sets from S that cover all the elements of U. Let x_j be the variable associated with set S_j. In what follows IP is the integer program and LP is the corresponding linear programming relaxation for the set cover problem. The first constraint in the IP ensures that each element $i \in U$ is covered by some set in S, and the second constraint stipulates that the sets are picked integrally.

$$\begin{aligned} \text{IP: minimize} && \textstyle\sum_{j \in [1,m]} x_j && \text{LP: minimize} && \textstyle\sum_{j \in [1,m]} x_j \\ \text{subject to: } && \textstyle\sum_{j : i \in S_j} x_j \geq 1 \quad \forall i \in U && \text{subject to: } && \textstyle\sum_{j : i \in S_j} x_j \geq 1 \quad \forall i \in U \\ && x_j \in \{0, 1\} \quad \forall i \in U && && 0 \leq x_j \leq 1 \quad \forall i \in U \end{aligned}$$

Let x^* be the optimal solution to the linear programming relaxation earlier. In each iteration, we round each variable x_j to 1 with probability x_j^* and to 0 with probability $1 - x_j^*$. Each set S_j for which $x_j = 1$ is picked in the solution. The probability that element $i \in U$ is not covered in an iteration is $\prod_{j : i \in S_j} (1 - x_j^*)$. If the element $i \in U$ occurs in the k sets $S_{i_1}, S_{i_2}, \ldots, S_{i_k}$, the values $x_{i_1}^*, x_{i_2}^*, \ldots, x_{i_k}^*$, are constrained by the inequality $\sum_{j=1}^{k} x_{i_j}^* \geq 1$, since the element $i \in U$ is covered by the optimal LP solution. The probability $\prod_{i_j}^{k} (1 - x_{i_j}^*)$ is then minimized when each value $x_{i_j}^*$ takes the value $1/k$. Thus, the probability that element $i \in U$ is not covered in an iteration is at least $(1 - 1/k)^k \geq 1/e$. Thus the probability that $i \in U$ is not covered after $c \log n$ iterations is $(1/e)^{c \log n} \leq 1/(4n)$ for some constant c. Equivalently, the probability that the solution computed after $c \log n$ iterations is not a valid cover is at most $\sum_{i=1}^{n} 1/(4n) = 1/4$. Furthermore, the expected number of sets in the solution computed is $(\sum_{j \in [1,m]} x_j^*) c \log n$. The probability that number of sets is more than 4 times this expected value is at most $1/4$ (follows from the Markov inequality). Therefore, with probability at least $1/2$, the algorithm returns a cover with cost at most $(\sum_{j \in [1,m]} x_j^*) 4c \log n$, implying the performance ratio is $O(\log n)$. Srinivasan [14], observing that the constraints in the set cover problem are positively correlated, showed that that performance ratio of the randomized rounding algorithm is $\log(|U|/OPT) + O(\log \log(|U|/OPT)) + O(1)$.

Next, we consider an interesting idea due to Goemans and Williamson [4], in which two randomized rounding algorithms are run on each problem instance, and the better of the two is returned as the solution. This technique is used for the maximum satisfiability problem to obtain a 3/4 approximation algorithm, though each algorithm by itself does not provide a 3/4 approximation ratio. In the weighted version of the maximum satisfiability problem, we are given a boolean formula in conjunctive normal form with weights on the clauses, and the goal is to determine an assignment of values (true/false) to the literals, such that the sum of weights of the clauses satisfied is maximized. The simpler rounding algorithm uses a purely randomized rounding, where each variable is set to true (false) with probability 1/2. If a clause j has k literals, then the probability that this clause is not satisfied is $1/2^k$ (corresponding to the situation when each of the k variables is set to 0). Thus, the probability that the clause is satisfied equals $1 - (1/2^k)$. To illustrate the second rounding algorithm (using linear programming) for this problem, we let C_j^+ denote the unnegated literals in the *jth* clause, and C_j^- the negated literals in the *jth* clause in formula C. The integer program IP, and the corresponding linear programming relaxation LP for the problem are given in the following:

$$\text{IP: maximize} \sum_{j \in C} w_j z_j \qquad\qquad \text{LP: maximize} \sum_{j \in C} w_j z_j$$

$$\text{subject to:} \qquad\qquad\qquad \text{subject to:}$$

$$\sum_{i \in C_j^+} x_i + \sum_{i \in C_j^-} (1 - x_i) \geq z_j \quad \forall j \in C \qquad \sum_{i \in C_j^+} x_i + \sum_{i \in C_j^-} (1 - x_i) \geq z_j \quad \forall j \in C$$

$$z_j, x_i \in \{0, 1\} \quad \forall i, j \qquad\qquad 0 \leq z_j, x_i \leq 1 \quad \forall i, j$$

Let x^*, z^* be the optimal solution to the linear programming relaxation LP. The rounding sets literal i to true with probability x_i^* (without loss of generality we assume clause j contains only positive literals). The probability that clause j is satisfied after this rounding is $1 - \prod_{i \in C_j^+}(1 - x_i^*)$. If clause j contains k literals, $x_{i_1}, x_{i_2}, \ldots, x_{i_k}$, the values $x_{i_1}^*, x_{i_2}^*, \ldots, x_{i_k}^*$, are constrained by the inequality $\sum_{j=1}^{k} x_{i_j}^* \geq z_j^*$, a constraint in the LP formulation. The probability $\prod_{i_j}^{k}(1 - x_{i_j}^*)$ is then maximized when each value $x_{i_j}^*$ takes the value z_j^*/k. Thus, the probability that clause j is not satisfied after the rounding is at most $(1 - z_j^*/k)^k$. Thus the probability that clause j is satisfied after the rounding is at least $1 - (1 - z_j^*/k)^k$. Observing that $1 - (1 - z_j^*/k)^k \geq z_j^*(1 - (1 - 1/k)^k)$ for $0 \leq z_j^* \leq 1$ (due to concavity), the probability that clause j is satisfied is at least $z_j^*(1 - (1 - 1/k)^k)$.

The bound of 3/4 follows from the fact that for each clause j, $\max\{(1 - 1/2^k), z_j^*(1 - (1 - 1/k)^k)\} \geq \max\{z_j^*(1 - 1/2^k), z_j^*(1 - (1 - 1/k)^k)\} \geq \frac{z_j^*}{2}\{(1 - 1/2^k) + (1 - (1 - 1/k)^k)\} \geq 3/4 z_j^*$ for every positive integer k.

7.2.2 Dependent Rounding

7.2.2.1 Simultaneous Rounding

The idea of simultaneously rounding a set of variables, was used by Bertsimas et al. [5] to establish the integrality of several well known polytopes. In particular they established the integrality of the polytopes associated with the minimum $s - t$ cut, p-median on a cycle, uncapacitated lot sizing, and boolean optimization. Using this technique, Bertsimas et al. [5] established a bound of $2(1 - 1/2^k)$ for the minimum k-sat problem. A bound of 2 is established for the feasible cut problem, by showing it is equivalent to vertex cover, which is approximable within a factor of 2 Yu and Cheriyan [15]. Here we illustrate the technique due to Bertsimas et al. [5] on the feasible cut problem. This technique is particularly interesting as the analysis of the performance ratio is considerably simplified.

Given a graph $G = (V, E)$ with weights on the edges, M a set of pairs of nodes in G, and a source vertex s. The problem is to determine a cut of minimum weight with the additional constraints that s belongs to the cut, but for any pair $(i, j) \in M$, both i and j are not in the cut. The integer program IP for the feasible cut problem, and the corresponding linear programming relaxation LP are given in the following:

$$\text{IP: minimize} \quad \sum_{(i,j) \in E} c_{ij} x_{ij}$$
$$\text{subject to:} \quad x_{ij} \geq y_i - y_j \quad \forall (i, j) \in E$$
$$x_{ij} \geq y_j - y_i \quad \forall (i, j) \in E$$
$$y_i + y_j \leq 1 \quad \forall (i, j) \in M$$
$$y_s = 1$$
$$x_{ij}, y_j \in \{0, 1\} \quad \forall i, j$$

$$\text{LP: minimize} \quad \sum_{(i,j) \in E} c_{ij} x_{ij}$$
$$\text{subject to:} \quad x_{ij} \geq y_i - y_j \quad \forall (i, j) \in E$$
$$x_{ij} \geq y_j - y_i \quad \forall (i, j) \in E$$
$$y_i + y_j \leq 1 \quad \forall (i, j) \in M$$
$$y_s = 1$$
$$0 \leq x_{ij}, y_j \leq 1 \quad \forall i, j$$

In this technique the variables are rounded simultaneously with respect to a random variable. Let U be a random value in $[1/2, 1]$ generated uniformly. Given an optimal solution (x^*, y^*) to the linear program LP, construct the cut as follows: if $y_i^* < U$ then $y_i = 0$, and if $y_i^* > U$ then $y_i = 1$. The rounding operation gives a feasible cut, since for each $(i, j) \in M$ at most one of y_i^*, y_j^* is greater than $1/2$. Let Z_{IP} be the value of the optimal solution to IP, Z_{LP} the value of the optimal solution to LP, and $E(Z_R)$ the expected value of the solution obtained after rounding.

Theorem 7.1 *Minimum feasible cut can be approximated within a factor of 2.*

Proof. Clearly, $Z_{IP} \leq E(Z_R)$. We show that $E(Z_R) \leq 2Z_{LP}$. If $E(x_{ij}) \leq 2x_{ij}^*$ for all i, j, then by linearity of expectation the result holds. Without loss of generality assume that $y_i^* \leq y_j^*$. If $y_j^* \leq 1/2$ then $E(x_{ij}) = 0$. If $y_i^* \leq 1/2 \leq y_j^*$, then $E(x_{ij}) = P(U \in [1/2, y_j^*]) = 2(y_j^* - 1/2) \leq 2(y_j^* - y_i^*)$. If $y_i^* \geq 1/2$, then $E(x_{ij}) = 2(y_j^* - y_i^*)$. This implies that $E(x_{ij}) \leq 2(y_j^* - y_i^*) \leq 2x_{ij}^*$. ∎

7.2.2.2 Rounding against a Hyperplane

The first substantial improvement for the Max Cut problem was made by Goemans and Williamson [6], who presented a 0.87856 factor approximation based on semidefinite programming. The above-mentioned bound is also applicable to the Max 2-Sat problem. They also gave a 0.7584 factor approximation algorithm for the Max Sat problem. Here we outline their technique for the Max Cut problem. Given a graph $G = (V, E)$ with weights on the edges, the objective is to partition the vertices of G such that the sum of weights of the cut edges is maximized. The problem is formulated first as a quadratic (nonlinear) program, and a relaxation of the quadratic program is defined in which each variable corresponds to a vector. An optimal solution to this relaxed nonlinear program is then computed. Given a random hyperplane, the vertices are partitioned into two sets, corresponding to points above and below the hyperplane. This partition has the desired bound. For details of the proof and the algorithm for computing the optimal solution to the relaxed program VP, shown below, we refer the reader to Vazirani [13] and Chapter 8 on the 1st edition of this handbook. Next, we describe their formulations and the randomization procedure.

$$\text{QP: maximize } 1/2 \sum_{(i,j) \in E} w_{ij}(1 - y_i y_j) \qquad \text{VP: maximize } 1/2 \sum_{(i,j) \in E} w_{ij}(1 - v_i v_j)$$
$$\text{subject to: } \quad y_i^2 = 1 \quad \forall i \in V \qquad\qquad \text{subject to: } \quad v_i \cdot v_i = 1 \quad \forall i \in V$$
$$y_i \in \mathcal{Z} \quad \forall i \in V \qquad\qquad\qquad\qquad v_i \in \mathcal{R}^n \quad \forall i \in V$$

Let r be a uniformly distributed vector in unit sphere S_{n-1}, then $S = \{i : v_i \cdot r \geq 0\}$ and $V \setminus S$ are the two sets defining the partition.

7.2.2.3 Extensions

Next we outline some extensions of the basic rounding technique. In all these techniques the variables are rounded randomly in a somewhat dependent fashion. First we consider the assignment problem in the presence of covering constraints. Given a complete bipartite graph $G = (A \cup B, E)$, with $|A| = |B|$, and weights on the edges. The objective is to find a matching of minimum weight that satisfies the covering constraints. The integer program IP and the linear programming relaxation LP for the assignment problem are given in the following:

$$\text{IP: minimize } \quad \sum_{(i,j) \in E} c_{ij} x_{ij} \qquad\qquad \text{LP: minimize } \quad \sum_{(i,j) \in E} c_{ij} x_{ij}$$
$$\text{subject to: } \quad \sum_{j \in B} x_{ij} = 1 \quad \forall i \in A \qquad \text{subject to: } \quad \sum_{j \in B} x_{ij} = 1 \quad \forall i \in A$$
$$\sum_{i \in A} x_{ij} = 1 \quad \forall j \in B \qquad\qquad\qquad \sum_{i \in A} x_{ij} = 1 \quad \forall j \in B$$
$$\sum_{i \in A, j \in B} a_{ij}^k x_{ij} \geq b^k \quad \forall k \in [1, K] \qquad \sum_{i \in A, j \in B} a_{ij}^k x_{ij} \geq b^k \quad \forall k \in [1, K]$$
$$x_{ij}, y_j \in \{0, 1\} \quad \forall i \in A, j \in B \qquad 0 \leq x_{ij}, y_j \leq 1 \quad \forall i \in A, j \in B$$

In the absence of the covering constraint $\left(\sum_{i \in A, j \in B} a_{ij}^k x_{ij} \geq b^k\right)$ the polytope associated with the IP is integral. But in the presence of the covering constraints we can only guarantee a fractional optimal solution to the LP in polynomial time. One possibility is to obtain an integral solution by rounding [3] the optimal fractional solution. One major difficulty with independent rounding in the presence of equality constraints is that the probability that the constraint is satisfied could be as low as $1/e$ (consider the case when all the x_{ij}s have the same value $1/|A|$). Therefore, the expected number of equality constraints satisfied in one rounding iteration is low. However the covering constraints are satisfied almost approximately. Arora et al. [7] developed a randomized rounding technique that obtains an integral solution (from the fractional solution) which satisfies $(|A| - o(|A|))$ equality constraints and all the covering constraints are satisfied almost approximately $\left(\sum_{i \in A, j \in B} a_{ij}^k x_{ij} \geq b^k - O(\sqrt{|A|} \max\{a_{ij}^k\})\right)$. Next, we describe their rounding algorithm for the case when the all fractional values are constants. For the rounding in the general case, and for the proofs we refer the reader to the original paper. Let x^* be the optimal fractional solution. The algorithm first constructs a multigraph from the bipartite graph as follows: for each edge in G, toss a biased coin (with probability of head x_{ij}^*) $\Theta(\log^3(n))$ times. If heads show up a times then the multigraph has a copies of edge (i, j). Multigraph is a union of paths and cycles of length $O(\sqrt{n})$. (if not then we have to delete $O(\sqrt{n})$ edges). Now these paths and cycles are further divided into $\Theta(\sqrt{n})$ groups of size $O(\sqrt{n})$ each. Within each group, either all the edges of A are picked or all the edges of B are picked, and the decision is equally likely. Using a generalization of this technique Arora et al. [7] were able to demonstrate polynomial time approximation schemes for dense instances of minimum linear arrangement problem, minimum cut linear arrangement problem, maximum acyclic subgraph problem, and the betweenness problem.

Next we briefly mention some other techniques. Srinivasan [16] developed a rounding technique based on distributions on level sets, and established better approximation ratios for low-congestion multipath routing problem, and the maximum coverage version of set cover problem. Gandhi et al. [17] developed a new rounding scheme based on the pipage rounding method of Ageev and Sviridenko [10] (see Section 7.3.4), and the level set based method of Srinivasan [16] to obtain better approximation algorithms for the throughput maximization problem in broadcast scheduling, the delay minimization problem in broadcast scheduling, and the capacitated vertex cover problem. Another dependent rounding technique has been developed by Doerr [18], with applications to digital half toning. Doerr [19] developed another dependent randomized rounding technique that respects cardinality constraints.

7.3 Deterministic Rounding

7.3.1 Scaling

Scaling is an important technique that has been applied to covering problems such as Vertex Cover to obtain a simple 2 factor approximation. Our presentation is based on Hochbaum [12, Chapter 3]. Given that it is still not known whether vertex cover admits an approximation ratio strictly better (by a constant) than 2, scaling seems to be a powerful technique. Given a graph $G = (V, E)$ with weights on the vertices. The objective is to determine a minimum weight set $S \subset V$ such the every edge has at least one end point in S. In what follows below is the integer program IP, and the corresponding linear programming relaxation LP.

IP: minimize $\sum_{i \in V} w_i x_i$
    subject to: $x_i + x_j \geq 1$ $\forall (i, j) \in E$
                  $x_i \in \{0, 1\}$ $\forall i \in V$

LP: minimize $\sum_{i \in V} w_i x_i$
    subject to: $x_i + x_j \geq 1$ $\forall (i, j) \in E$
                 $0 \leq x_i \leq 1$ $\forall i \in V$

Let x^* be the optimal solution to the linear program LP. Let S be the set of vertices j such that $x_j^* \geq 1/2$. S is a cover because for each edge (i, j) either x_i or x_j is $\geq 1/2$, and the weight of S is at most $2 \sum_{i \in V} w_i x_i^*$.

Interestingly, the algorithm by Gonzalez [20] is the only factor 2 approximation algorithm for vertex cover whose proof does not rely on the theory of linear programming.

7.3.2 Filter and Round

Sahni and Gonzalez [21] showed that for certain problems including the p-median problem, the tree pruning problem, and the generalized assignment problem, finding an α-approximate solution is NP-hard. In light of the previous result, the next best thing is to find an α-approximate solution with the minimum number of constraint violations. Lin and Vitter [8] gave such approximation algorithms for the problems mentioned earlier. For the generalized assignment problem, we refer the reader to Chapter 40 by Yagiura and Ibaraki. Here we will illustrate their technique on the p-median problem. Our presentation is based on Lin and Vitter [8]. Given a complete graph G on n vertices with weights on the edges, and an integer p, the problem is to determine p vertices (medians) so that the sum of the distances from each vertex to its closest median is minimized. The integer program IP and the corresponding linear programming relaxation LP are given in the following.

$$
\begin{aligned}
\text{IP: minimize} \quad & \sum_{i,j \in V} c_{ij} x_{ij} \\
\text{subject to: } & \sum_{j \in V} x_{ij} = 1 \quad \forall i \in V \\
& x_{ij} \leq y_j \quad \forall i,j \in V \\
& \sum_{j \in V} y_j = p \\
& x_{ij}, y_j \in \{0,1\} \quad \forall i,j
\end{aligned}
\qquad
\begin{aligned}
\text{LP: minimize} \quad & \sum_{i,j \in V} c_{ij} x_{ij} \\
\text{subject to: } & \sum_{j \in V} x_{ij} = 1 \quad \forall i \in V \\
& x_{ij} \leq y_j \quad \forall i,j \in V \\
& \sum_{j \in V} y_j = p \\
& 0 \leq x_{ij}, y_j \leq 1 \quad \forall i,j
\end{aligned}
$$

Given an optimal solution x^*, y^* to the LP, we obtain an integer program called a filtered program (FP) by setting some variables in x to 0. The FP has the property that any integral feasible solution is at most $1 + \alpha$ times the value of the optimal solution to LP. First, a fractional feasible solution to FP is constructed from x^*, y^*. A feasible integral solution to FP is then obtained using either randomized rounding or some greedy rounding. Here we illustrate a deterministic (greedy) rounding method. We assume certain lemmas to illustrate the technique. For the proof of the lemmas we refer the reader to the original paper by Lin and Vitter [8].

Lemma 7.1 *Given y, the optimal values for x can be computed for the linear programming problem LP.*

Given an optimal solution x^*, y^* to the LP, for a vertex $i \in V$, let V_i be the set of vertices j such that $c_{ij} \leq (1 + \alpha) \sum_{j \in V} c_{ij} x_{ij}^*$. The FP and the reduced filtered program (RFP) necessary to compute the solution to FP by Lemma 7.1 are:

$$
\begin{aligned}
\text{FP: minimize} \quad & L \\
\text{subject to: } & \sum_{j \in V_i} x_{ij} = 1 \quad \forall i \in V \\
& x_{ij} \leq L y_j \quad \forall i,j \in V \\
& \sum_{j \in V} y_j = p \\
& x_{ij} = 0 \quad \forall i \in V, j \in V \setminus V_i \\
& x_{ij}, y_j \in \{0,1\} \quad \forall i,j
\end{aligned}
\qquad
\begin{aligned}
\text{RFP: minimize} \quad & \sum_{j \in V} y_j \\
\text{subject to: } & \sum_{j \in V_i} y_j \geq 1 \quad \forall i \in V \\
& y_j \in \{0,1\} \quad \forall i,j
\end{aligned}
$$

L corresponds to the factor by which the covering constraints are violated. The following lemma holds by construction.

Lemma 7.2 *Any feasible (integral) solution to FP has value at most $(1 + \alpha)$ times the value of the optimal solution to the linear programming relaxation LP.*

It is the case that $\sum_{j \in V_i} y_j^* \geq \alpha/(1 + \alpha)$. Therefore, a feasible fractional solution to RFP with value $(1 + 1/\alpha)p$ can be constructed (by assigning $y_j = y_j^*(1 + \alpha)/\alpha$). RFP is nothing but set cover and a $\log n$ approximate integral solution can be constructed using the greedy heuristic of Chvátal [22]. Therefore, by Lemma 7.2 we have a $(1 + \alpha)p \log n$ approximate constraint violations with value at most $(1 + \alpha)$ times the value of the optimal solution to the integer program.

7.3.3 Iterated Rounding

The technique of iterated rounding was introduced by Jain [9], who gave a 2-factor approximation algorithm for the generalized steiner network problem. Consider the problem of finding a minimum-cost edge-induced subgraph of a graph that contains a pre-specified number of edges from each cut. Formally, $G = (V, E)$ is a graph with weights on the edges. Also given is a function $f : 2^V \rightarrow \mathcal{Z}$. The problem is to determine a minimum-weight set of edges such that for every R subset of V, the number of edges in $\delta(R) \geq f(R)$, where $\delta(R)$ are the edges in the cut defined by the vertices in R. The integer program IP and the corresponding linear programming relaxation LP are given in the following:

IP: minimize $\sum_{e \in E} w_e x_e$

subject to: $\sum_{e \in \delta(R)} \geq f(R) \quad \forall S \subseteq V$

$x_e \in \{0, 1\} \quad \forall i \in E$

LP: minimize $\sum_{e \in E} w_e x_e$

subject to: $\sum_{e \in \delta(R)} \geq f(R) \quad \forall S \subseteq V$

$0 \leq x_e \leq 1 \quad \forall i \in E$

Note that both the above-mentioned programs contain exponentially many constraints. Jain [9] gives a separation oracle for the linear programming relaxation. Using this separation oracle an optimal solution can be computed in polynomial time Grötschel et al. [23]. Furthermore, Jain establishes the following:

Theorem 7.2 *Any basic feasible solution to the linear programming relaxation has at least one variable with value* $\geq 1/2$.

Based on the previous theorem, one can construct a solution as follows: find an optimal solution (basic) to the LP, include all the edges with value $\geq 1/2$ in the solution, recursively solve the subproblem obtained by deleting the edges included in the solution.

7.3.4 Pipage Rounding

Pipage rounding was developed by Ageev and Sviridenko [10], who applied it to the maximum coverage problem, hypergraph maximum k-cut with given sizes of parts, and scheduling on unrelated parallel machines. They showed that maximum coverage problem can be approximated within $1 - (1 - 1/k)^k$ where k is the maximum size of any subset, thereby improving the previous bound of $1 - 1/e$ due to Cornuejols et al. [24]. For the hypergraph max k-cut they obtained a bound of $1 - (1 - 1/r)^r - 1/r^r$ where r is the cardinality of the smallest edge in the hypergraph. For the scheduling problem on unrelated machines, they consider an additional constraint on the number of jobs that a given machine can process and obtained the bound of 3/2. A similar bound was also established by Skutella [25] in the absence of cardinality constraints. For the case of two machines the current best bound is 1.2752 due to Skutella [25], obtained by rounding the semidefinite programming relaxation using the dependent rounding technique of Goemans and Williamson [6]. Ageev et al. [26] obtained a 1/2 approximation algorithm for the max dicut problem with given sizes of parts by a refined application of the pipage rounding. Recently, Galluccio and Nobili [27] have improved the approximation ratio from 3/4 to $1 - 1/2q$ for the maximum coverage problem when all the sets are of size 2, where every clique in a clique cover of the input graph has size at least q. Note that $q \geq 2$. This problem is also known as the maximum vertex cover problem. Pipage rounding is especially suited to problems involving assignment and cardinality constraints.

Our description of the pipage rounding is based on Ageev and Sviridenko [10]. The idea is to deterministically round a fractional solution to an integral solution, whilst ensuring that the objective function value does not decrease in the rounding process. If the starting fractional solution was at least c times the optimal fractional solution, then the pipage rounding will guarantee a c-approximation algorithm. The rounding process converts a fractional solution into another fractional solution with less number of nonintegral components. The "δ-convexity" of the objective functions guarantees that the objective function value does not decrease in the rounding process.

Let $G = (V, E)$ be a bipartite graph with capacities c_v on the vertices. Let $f(X)$ be a polynomially computable function defined on the values $X = \{x_e : e \in E\}$ assigned to the edges of G. Consider the following integer program IP whose solution is an assignment of $0, 1$ to the edges that maximizes $f(X)$ subject to the capacity constraints, and its linear programming relaxation LP:

$$\text{IP: maximize} \qquad f(X) \qquad\qquad \text{LP: maximize} \qquad f(X)$$
$$\text{subject to: } \sum_{e \in N(v)} x_e \leq c_v \quad \forall v \in V \qquad\qquad \text{subject to: } \sum_{e \in N(v)} x_e \leq c_v \quad \forall v \in V$$
$$x_e \in \{0, 1\} \quad \forall e \in E \qquad\qquad\qquad 0 \leq x_e \leq 1 \quad \forall e \in E$$

We do not assume that the optimal solution to the LP is computable in polynomial time. Given a fractional solution X, let $G(X)$ be the subgraph induced by the edges that are assigned a nonintegral value in X. $G(X)$ either contains a path P, or a cycle C. Let $P_o(C_o)$ be the odd indexed edges in $P(C)$. Similarly, $P_e(C_e)$ the set of even indexed edges in $P(C)$. Given $P(C)$, let $lb = \min\{\min\{x_e : e \in P_o(C_o)\}, \min\{1 - x_e : e \in P_e(C_e)\}\}$. Similarly, define $ub = \min\{\min\{1 - x_e : e \in P_o(C_o)\}, \min\{x_e : e \in P_e(C_e)\}\}$. f is said to be δ-convex with respect to $\delta \in [lb, ub]$ if for each fractional solution and all paths and cycles it is convex in δ. Given δ-convexity the maximum of f in $[lb, ub]$ is attained at one of the end points. Pipage rounding amounts to either successively adding and deleting ub, or successively deleting and adding lb, from the values assigned to the edges in $P(C)$. This process yields a solution with a reduced number of nonintegral components. Let us examine the case when all the capacities are 1, and f computes the sum of the values assigned to the edges. In this case, the solution to the IP corresponds to a maximum matching, and the solution to the linear program corresponds to the maximum fractional matching. The pipage rounding (as can be readily verified) in this case converts the fractional matching into an integral matching of same or larger size. To compute an α-approximation it remains to find a function g that approximates f within α such that maximum of g can be computed in polynomial time, subject to the constraints in the LP.

We illustrate the application of pipage rounding to the maximum coverage problem, where we are given a collection S of weighted subsets of ground set I, and an integer k. The goal is to determine $X \subseteq I$ of cardinality k such that the sum of the weights of the sets in S that intersect with X is maximized. Associated with each element $i \in I$ is a variable x_i, and associated with each element S_j of S is a variable z_j. An integer program for the maximum coverage problem is given in the following:

$$\text{IP: maximize} \qquad \sum_{j=1}^{m} w_j z_j$$
$$\text{subject to: } \sum_{i \in S_j} x_i \geq z_j, \quad \forall S_j \in S$$
$$\sum_{i=1}^{n} x_i = k$$
$$x_i \in \{0, 1\} \quad \forall i \in I$$

The above-mentioned objective function in IP can be replaced with $f = \sum_{j=1}^{m} w_j (1 - \prod_{i \in S_j} (1 - x_i))$ as it has the same value over all integral vectors x. Replace f by $g = \sum_{j=1}^{m} w_j \min\{1, \sum_{i \in S_j} x_i\}$. It can be shown that f and g are δ-convex and g approximates f within a factor of $1 - (1 - 1/k)^k$, where k is the cardinality of the largest element of S. Furthermore, fractional optimal solution to g as the objective, subject to the constraints in IP can be computed in polynomial time.

7.3.5 Decompose and Round

We next describe a deterministic technique due to Gaur et al. [11]. This technique is applicable to geometric covering problems, and can be thought of as an extension of the scaling technique. We consider covering problems of the form min cx, subject to $Ax \geq b$, $x \in \{0,1\}^n$, where A is an $m \times n$ matrix with $0,1$ entries, and c, x are vectors of dimension n, and b is a vector with m entries. The geometry of the problem under consideration imposes a structure on A, and this helps us in the application of the scaling technique. We begin with a few definitions. Let $C = \{1, \ldots, n\}$ be the set of indices of the columns in A and $R = \{1, \ldots, m\}$ the set of indices of the rows in A. Let $\mathcal{R} = \{R_1, R_2, \ldots, R_k\}$ be a partition of R, and by $\mathcal{C} = \{C_1, C_2, \ldots, C_k\}$ a partition of the columns of A. $A(R_i, C_j)$ is the matrix obtained from A by removing the columns in $C \setminus C_j$, and the rows in $R \setminus R_i$. A matrix A is *totally unimodular* if the determinant of every square submatrix of A is ± 1. We say A is *partially unimodular* with respect to $\mathcal{C}$ and $\mathcal{R}$ if for all $C_i \in \mathcal{C}, R_j \in \mathcal{R}$, $A(R_j, C_i)$ is totally unimodular. For a partially unimodular matrix A, $|\mathcal{C}| = |\mathcal{R}|$ is also known as the *partial-width* of A. It is well known that if M is block structured and if all the blocks are totally unimodular then M is totally unimodular. This fact, with a suitable reordering of the rows and columns, implies the following:

Lemma 7.3 *Let A_D be the matrix whose ith diagonal block corresponds $A(R_i, C_i)$ (all other entries are 0) then A_D is totally unimodular, if A is partially unimodular with respect to $\mathcal{R}, \mathcal{C}$.*

We next describe the rectangle stabbing problem, and show that its coefficient matrix is partially unimodular and has partial width 2. A $\log n$ factor approximation for the rectangle stabbing problem is due to Hassin and Megiddo [28]. Given a set of axis-aligned rectangles (in 2D), the problem is to determine the minimum number of axis-parallel lines that are needed to stab all the rectangles. Let H be the set of horizontal lines going through the horizontal edges of the rectangles, V be the set of vertical lines going through the vertical edges of the rectangles, and R be the set of all rectangles. Let $H_r(V_r)$ be the set of lines from $H(V)$ that intersect rectangle $r \in R$. In what follows is the integer program IP and the corresponding linear programming relaxation LP.

IP: minimize $\quad \sum_{i \in H} h_i + \sum_{j \in V} v_j \qquad$ LP: minimize $\quad \sum_{i \in H} h_i + \sum_{j \in V} v_j$

subject to: $\sum_{i:i \in H_r} h_i + \sum_{j:j \in V_r} v_j \geq 1 \quad \forall r \in R \qquad$ subject to: $\sum_{i:i \in H_r} h_i + \sum_{j:j \in V_r} v_j \geq 1 \quad \forall r \in R$

$\qquad\qquad h_i, v_j \in \{0,1\} \quad \forall i \in H, j \in V \qquad\qquad\qquad 0 \leq h_i, v_j \leq 1 \quad \forall i \in H, j \in V$

Let A be the coefficient matrix corresponding to the programs earlier.

Lemma 7.4 *A is partially unimodular with respect to $\mathcal{C} = \{H, V\}$ and $\mathcal{R} = \{R_h, R_v\}$ as computed as follows.*

Given an optimal solution h^*, v^* to the linear programming relaxation LP, we construct a partition $\mathcal{R} = \{R_h, R_v = R \setminus R_h\}$ of the rows of A as follows: R_h is the set of all the rectangles r such that $\sum_{i:i \in H_r} h_i \geq 1/2$. Let A_D be the block diagonal matrix whose blocks are $A(R_h, H)$ and $A(R_v, V)$. $A(R_h, H)$ and $A(R_v, V)$ are totally unimodular as the columns can be reordered so that each row has consecutive ones property. By Lemma 7.3, A_D is totally unimodular. Consider the program min cx subject to $A_D x \geq 1$, $x \in \{0,1\}^n$. Conforti et al. [29] showed that the polytope associated with A_D is integral, hence the optimal integral solution has the same value as the optimal fractional solution. Note that $(2h^*, 2v^*)$ is feasible in the previous problem. Therefore, the performance ratio is 2 as $ALG \leq (2h^*, 2v^*)$ and $OPT \geq (h^*, v^*)$. Furthermore, the addition of capacity constraints on H and V does not affect the performance ratio. These results can be generalized for arbitrary weights on the lines and requirements on the rectangles in d dimensions. For recent results on the rectangle stabbing problem with soft capacities see Even et al. [30]. The case when rectangles have zero height has been studied extensively by Kovaleva and Spieksma, see Chapter 37 of the 1st edition of this handbook.

Handbook of Approximation Algorithms and Metaheuristics

A brief comment about the technique is in order. Every matrix A is partially unimodular with respect to the following partitions. $C_1, C_2, \ldots, C_n$ where C_i is the *ith* column in A. Let x^* be the optimal solution to the LP. Consider the following partition of rows: rectangle r belongs to set R_i in the partition if $x_i^* A[r, i] = \max_{j \in [1..n]} \{x_j^* A[r, j]\}$. A_D can now be constructed from the blocks $A(R_i, C_i)$. Once again, by Lemma 7.3 A_D is totally unimodular as each $A(R_i, C_i)$ is a column vector with all ones (the determinant for every square submatrix is 1) and totally unimodular. Let τ be the maximum number of non-zero entries in a row of A. The performance ratio using the algorithm and the previous argument is $1/\tau$. This is similar to the bound obtained for the set cover problem using the scaling technique. In this sense our approach can be viewed as a generalization of the scaling technique.

The arguments outlined in the preceding paragraphs lead to the following theorem.

Theorem 7.3 *Given a covering problem of the form min cx, subject to $Ax \geq b$, $x \in \{0, 1\}^n$, if A is partially unimodular and has partial-width $1/\alpha$, there exists an approximation algorithm with performance ratio α.*

In light of the preceding theorem it is natural to study algorithms (exact and approximation) for determining the minimum cardinality partitions with respect to which A is partially unimodular. We are not aware of any existing results and pose the determination of minimum partial-width as an interesting open problem, with application to the theory of approximation algorithms.

Next, we consider an application of the rectangle stabbing problem to a load balancing problem that arises in the context of scheduling on multiprocessor systems. In the rectilinear partitioning problem, the input is a matrix of integers, and the problem is to partition the matrix using h horizontal lines and v vertical lines, such that the load inside each rectangle (formed by two consecutive horizontal and vertical lines) is minimized, where the load of a rectangle is defined to be the sum of entries in the rectangle. Given an instance of the rectilinear partitioning problem we construct an instance of the rectangle stabbing problem as follows: let L be the minimum load (we can obtain this using binary search), all the submatrices with load in excess of L correspond to rectangles in the rectangle stabbing problem. Note that if all the rectangles are stabbed then the load is at most L. As we only have a 2-factor approximation algorithm for the rectangle stabbing problem, the number of lines returned can be twice the number of lines stipulated. Therefore, a solution to the rectilinear partitioning problem is obtained by removing every second line (horizontal as well as vertical). In the process of removing the alternate lines, a new rectangle is formed whose load is at most $4L$. Therefore, the performance ratio is 4.

7.4 Discussion

Numerous techniques have been developed over the last two decades to convert an optimal fractional solution (to the linear programming relaxation of an integer program) to an approximate integral solution. These techniques can be divided into two broad categories; those that use randomized strategies and ones that use deterministic strategies. Most of the randomized strategies can be made deterministic (at the expense of increase running time) using the method of conditional expectation. The applicability of the strategies is most evident in the context of packing and covering types of problems. Some success has been obtained in application of these techniques in presence of cardinality constraints.

References

1. Erdős, P. and Selfridge, J.L., On a combinatorial game, *J. Comb. Theor. Ser. A*, 14, 298, 1973.
2. Alon, N. and Spencer, J.H., *The probabilistic method*, Wiley-Interscience, New York, 2000.
3. Raghavan, P. and Thompson, C.D., Randomized rounding: A technique for provably good algorithms and algorithmic proofs, *Combinatorica*, 7(4), 365, 1987.
4. Goemans, M.X. and Williamson, D.P., New $\frac{3}{4}$-approximation algorithms for the maximum satisfiability problem, *SIAM J. Disc. Math.*, 7(4), 656, 1994.

5. Bertsimas, D., Teo, C., and Vohra, R., On dependent randomized rounding algorithms, *Oper. Res. Lett.*, 24(3), 105, 1999.
6. Goemans, M.X. and Williamson, D.P., Improved approximation algorithms for maximum cut and satisfiability problems using semidefinite programming, *JACM*, 42(6), 1115, 1995.
7. Arora, S., Frieze, A., and Kaplan, H., A new rounding procedure for the assignment problem with applications to dense graph arrangement problems, *Math. Prog. Ser. A*, 92(1), 1, 2002.
8. Lin, J.H. and Vitter, J.S., Approximation algorithms for geometric median problems, *Inf. Proc. Lett.*, 44(5), 245, 1992.
9. Jain, K., A factor 2 approximation algorithm for the generalized Steiner network problem, *Combinatorica*, 21(1), 39, 2001.
10. Ageev, A.A. and Sviridenko, M.I., Pipage rounding: A new method of constructing algorithms with proven performance guarantee, *J. Comb. Optim.*, 8(3), 307, 2004.
11. Gaur, D.R., Ibaraki, T., and Krishnamurti, R., Constant ratio approximation algorithms for the rectangle stabbing problem and the rectilinear partitioning problem, *J. Algor.*, 43(1), 138, 2002.
12. Hochbaum, D.S., Ed., *Approximation algorithms for NP-hard problems*, PWS Publishing Co., Boston, MA, 1997.
13. Vazirani, V.V., *Approximation Algorithms*, Springer-Verlag, Berlin, Germany, 2001.
14. Srinivasan, A., Improved approximation guarantees for packing and covering integer programs, *SIAM J. Comput.*, 29(2), 648, 1999.
15. Yu, B. and Cheriyan, J., Approximation algorithms for feasible cut and multicut problems, in *Proceedings of ESA*, LNCS, Springer, Berlin, Germany, p. 979, 1995.
16. Srinivasan, A., Distributions on level-sets with applications to approximation algorithms, in *Proceedings of FOCS*, IEEE, p. 588, 2001.
17. Gandhi, R., Khuller, S., Parthasarathy, S., and Srinivasan, A., Dependent rounding in bipartite graphs, in *Proceedings of FOCS*, IEEE, p. 323, 2002.
18. Doerr, B., Nonindependent randomized rounding and an application to digital halftoning, *SIAM J. Comput.*, 34(2), 299, 2005.
19. Doerr, B., Roundings respecting hard constraints, in *Proceedings of STACS*, Springer-Verlag, Berlin, Germany, p. 617, 2005.
20. Gonzalez, T.F., A simple LP-free approximation algorithm for the minimum weight vertex cover problem, *Inform. Proc. Lett.*, 54(3), 129, 1995.
21. Sahni, S. and Gonzalez, T.F., P-complete approximation problems, *JACM*, 23(3), 555, 1976.
22. Chvátal, V., A greedy heuristic for the set-covering problem, *Math. Oper. Res.*, 4(3), 233, 1979.
23. Grötschel, M., Lovász, L., and Schrijver, A., *Geometric algorithms and combinatorial optimization*, of *Algorithms and Combinatorics*, Springer-Verlag, Berlin, Germany, 2nd ed., 1993.
24. Cornuejols, G., Fisher, M.L., and Nemhauser, G.L., Location of bank accounts to optimize float: An analytic study exact and approximate algorithms, *Manag. Sci.*, 23, 789, 1977.
25. Skutella, M., Convex quadratic and semidefinite programming relaxations in scheduling, *JACM*, 48(2), 206, 2001.
26. Ageev, A., Hassin, R., and Sviridenko, M., An 0.5-approximation algorithm for MAX DICUT with given sizes of parts, *SIAM J. Disc. Math.*, 14(2), 246, 2001.
27. Galluccio, A. and Nobili, P. Improved approximation of maximum vertex cover, *Oper. Res. Lett.* 34, 77–84, 2006.
28. Hassin, R. and Megiddo, N., Approximation algorithms for hitting objects with straight lines, *Disc. Appl. Math.*, 30(1), 29, 1991.
29. Conforti, M., Cornuéjols, G., and Truemper, K., From totally unimodular to balanced $0, \pm 1$ matrices: A family of integer polytopes, *Math. Oper. Res.*, 19(1), 21, 1994.
30. Even, G., Rawitz, D., and Shahar, S., Approximation of rectangle stabbing with soft capacities, in *Workshop on Interdisciplinary Applications of Graph Theory, Combinatorics, and Algorithms*, 2005.

8

Polynomial Time Approximation Schemes

Hadas Shachnai

Tami Tamir

8.1 Introduction

Let Π be an NP-hard optimization problem, and let $\mathcal{A}$ be an approximation algorithm for Π. Given an instance I of Π, denote by $A(I)$ the objective value obtained by $\mathcal{A}$ for I, and by $OPT(I)$ the optimal objective value. The approximation ratio of $\mathcal{A}$ for the instance I is $R_A(I) = A(I)/OPT(I)$ when Π is a minimization problem, and $R_A(I) = OPT(I)/A(I)$ when Π is a maximization problem; thus, $R_A(I) \geq 1$.

A *polynomial time approximation scheme (PTAS)* is an algorithm that takes as input an additional parameter, $\varepsilon > 0$, which determines the desired approximation ratio. As ε approaches zero, the approximation ratio gets arbitrarily close to 1. The time complexity of the scheme is polynomial in the input size, but may be exponential in $1/\varepsilon$. This gives a clear trade-off between running time and quality of approximation. Formally,

Definition 8.1 *An approximation scheme for an optimization problem Π is an algorithm $\mathcal{A}$ that takes as input an instance I of Π and an error bound ε, runs in time polynomial in $|I|$ and has approximation ratio $R_A(I, \varepsilon) \leq 1 + \varepsilon$. In fact, such an algorithm $\mathcal{A}$ is a family of algorithms A_ε such that, for any instance I, $R_{A_\varepsilon}(I) \leq 1 + \varepsilon$.*

The approximation algorithm $\mathcal{A}$ may be deterministic or randomized. In the latter case, the result is a *randomized approximation scheme (RPTAS)*.

Definition 8.2 *A* randomized approximation scheme *for an optimization problem* Π *is a family of algorithms* A_ε *that run in time polynomial in* $|I|$ *and have, for any instance I, expected* approximation ratio $EXP[R_{A_\varepsilon}(I)] \leq 1 + \varepsilon$.

In some approximation schemes, an additive constant k, whose value is independent of I and ε, is added to the approximation ratio. Asymptotically, this constant is negligible; thus, such a scheme is called an *asymptotic PTAS (APTAS)*.

Definition 8.3 *An* asymptotic approximation scheme *for a minimization problem* Π *is a family of algorithms* A_ε *that run in time polynomial in* $|I|$, *such that for some constant k and any instance I*, $A_\varepsilon(I) \leq (1 + \varepsilon)OPT(I) + k$.

A similar definition applies for a maximization problem Π. We refer the reader to Chapter 10 in this book for a detailed study of such schemes.

Some approximation algorithms output a solution for a *relaxed instance* of the problem. For example, in packing problems, an algorithm may pack the items in bins whose sizes are slightly larger than the original. The objective value is achieved relative to the relaxed instance. This type of algorithm is called a *dual* approximation algorithm [1], or approximation with *resource augmentation* [2]. A *dual approximation scheme* is a family of algorithms A_ε that run in time polynomial in $|I|$, such that for any instance I, $R_A(I, \varepsilon) \leq 1 + \varepsilon$, and $A(I)$ is achieved for resources augmented by factor of $(1 + \varepsilon)$.

Depending on the function $f(|I|, 1/\varepsilon)$, which gives the running time of the scheme, some schemes are classified as *quasi-polynomial* and others as *efficient* or *fully polynomial*. In particular, let $n = |I|$ be the input size. Then, when the running time is $O(n^{polylog(n)})$, we get a *quasi-PTAS (QPTAS)* [3–6]; when the running time can be written as a polynomial in $|I|$ times some function $f(1/\varepsilon)$, we have an *efficient-PTAS (EPTAS)*. And, when the running time is polynomial in both $|I|$ and $1/\varepsilon$, we have an FPTAS.

There is wide literature on approximation schemes for NP-hard problems. Many of these works present PTASs for certain subclasses of instances of problems, which are in general extremely hard to solve. While some of the proposed schemes may have running times which render them inefficient in practice, these works essentially help identify the class of problems that admit PTAS. There have been some studies also towards characterizing this class of problems [7]. We focus here on techniques that have been commonly used for developing PTASs.

We refer the reader also to the comprehensive surveys on approximation algorithms by Motwani [8], and on scheduling, by Karger et al. [9], from which we borrowed some of the classic examples in this chapter. A detailed exposition of approximation schemes for some fundamental optimization problems can be found, for example, in [10–13].

8.2 Partial Enumeration

8.2.1 Extending Partial Small-Size Solutions

There are two main techniques based on extending partial small-size solutions. The first technique exploits our ability to solve the problem *optimally* on a constant-size subset of the instance. Thus, initially, such a constant-size subset is selected. This subset contains the most "significant" elements in the instance. We identify elements as *significant* depending on the problem at hand. The problem is solved optimally for this subset. This can be done by exhaustive search, since there is only a constant number of elements to consider. Next, this optimal partial solution is extended into a complete one, using some heuristic which has a bounded approximation ratio.

In the second technique, none of the elements is initially identified as "significant"; instead, *all* partial solutions of constant-size are considered, and each is extended to a complete solution using some heuristic. The best extension is selected to be the output of the scheme.

The time complexity analysis of such PTASs is based on the fact that the number of possible subsets, or solutions that are considered, is exponential in the (constant) size of these subsets. The step in which the constant-size partial solution is extended is usually based on some greedy rule that may require sorting, and is polynomial. The parameter ε specifying the required approximation ratio of $(1 + \varepsilon)$, determines the size k of the partial solution to which an exponential exhaustive search is applied. This implies that the running time of such schemes is exponential in $1/\varepsilon$.

8.2.1.1 Extending an Optimal Solution for a Single Subset

We present the first technique in the context of a classical scheduling problem, namely, the problem of finding the *minimum makespan (MM)* (or, overall completion time) of a schedule of n jobs on m identical machines. A main idea in the PTAS of Graham [14] is to schedule first optimally the k longest jobs and then schedule, using some heuristic, the remaining jobs. Formally, the input for the MM problem consists of n jobs and m identical machines. The goal is to schedule the jobs non preemptively on the machines in a way that minimizes the maximum completion time of any job in the schedule.

Denote by $p_1, \ldots, p_n$ the processing times of the jobs. Assume that $n > m$, and that the processing times are sorted in non increasing order, that is, for all $i < j$, $p_i \geq p_j$. A well-known heuristic for the makespan problem is the *longest processing time (LPT) rule*, which selects the longest unscheduled job in the sorted list and assigns it to a machine that currently has the minimum load. The PTAS combines an optimal schedule of the longest k jobs with the LPT rule, applied to the remaining jobs.

Formally, for any $k \in [0, n]$, the algorithm A_k is defined as follows:

1. Schedule optimally, with no intended idles, the first k jobs.
2. Add the remaining jobs greedily using the LPT rule.

Theorem 8.1 *Let $A_k(I)$ denote the makespan achieved by A_k on an instance I, and let $OPT(I)$ denote the MM of I, then*

$$A_k(I) \leq OPT(I) \left(1 + \frac{1 - \frac{1}{m}}{1 + \lfloor k/m \rfloor} \right).$$

Proof. Let T denote the makespan of an optimal schedule of the first k jobs. Clearly, T is a lower bound for $OPT(I)$, thus, if the makespan is not increased in the second step, that is, $A_k(I) = T$, then A_k is optimal for I. Otherwise, the makespan of the schedule is larger than T. Let j be the job to determine the makespan (the one which completes last). By the definition of LPT, this implies that all the machines were busy when job j started its execution (otherwise, job j could start earlier). Since the optimal schedule from step 1 has no intended idles, all the machines are busy during the time interval $[0; A_k(I) - p_j]$.

Let $P = \sum_{j=1}^{n} p_j$ be the total processing time of the n jobs. By the above, $P \geq m(A_k(I) - p_j) + p_j$. Also, since the jobs are sorted in nonincreasing order of processing times, we have that $p_j \leq p_{k+1}$, and therefore $P \geq mA_k(I) - (m - 1)p_{k+1}$. A lower bound for the optimal solution is the makespan of a schedule in which the load on the m machines is perfectly balanced; thus, $OPT(I) \geq P/m$, which implies that $A_k(I) \leq OPT(I) + (1 - \frac{1}{m})p_{k+1}$.

In order to bound $A_k(I)$ in terms of $OPT(I)$, we need to bound p_{k+1} in terms of $OPT(I)$. To obtain such a bound, consider the $k + 1$ longest jobs. In an optimal schedule, some machine is assigned at least $\lceil (k + 1)/m \rceil \geq 1 + \lfloor k/m \rfloor$ of these jobs. Since each of these jobs has processing time at least p_{k+1}, we conclude that $OPT(I) \geq (1 + \lfloor k/m \rfloor)p_{k+1}$, which implies that $p_{k+1} \leq OPT(I)/(1 + \lfloor k/m \rfloor)$. It follows that

$$A_k(I) \leq OPT(I) \left(1 + \frac{(1 - \frac{1}{m})}{(1 + \lfloor k/m \rfloor)} \right). \qquad \blacksquare$$

To observe that the above-mentioned family of algorithms is a PTAS, we relate the value of k to $(1 + \varepsilon)$, the desired approximation ratio. Given $\varepsilon > 0$, let $k = \lceil \frac{1-\varepsilon}{\varepsilon} m \rceil$. It is easy to verify that the corresponding algorithm A_k achieves approximation ratio at most $(1 + \varepsilon)$. Thus, we conclude that for a fixed m, there is a PTAS for the MM problem.

Note that for any fixed k, an optimal schedule of the first k jobs can be found in $O(m^k)$ steps. Applying the LPT rule takes additional $O(n\log n)$. For A_ε, we get that the running time of the scheme is $O(m^{m/\varepsilon})$, that is, exponential in m (i.e., assumed to be constant) and $1/\varepsilon$. This demonstrates the basic property of approximation schemes: a clear trade-off between running time and the quality of approximation.

8.2.1.2 Extend All Possible Solutions for Small Subsets

The second technique, of considering all possible subsets, is illustrated in an early PTAS of Sahni for the *knapsack problem* [15]. An instance of the knapsack problem consists of n items, each having a specified size and a profit, and a single knapsack, having size B. Denote by $s_i \geq 0, p_i \geq 0$ the size and profit associated with item i. The goal is to find a subset of the items such that the total size of the subset does not exceed the knapsack capacity, and the total profit associated with the items is maximized.

The PTAS in [15] is based on considering all $O(kn^k)$ possible subsets of size at most k, where k is some fixed constant. Each of these subsets is extended to a larger feasible subset by adding more items to the knapsack, using some greedy rule. The best extension among these $O(kn^k)$ candidates is selected to be the output of the scheme. Formally, for any $k \in [0, n]$, the algorithm A_k is defined as follows:

1. (Preprocessing) Sort the items in non increasing order of their profit densities, p_i/s_i.
2. For each feasible subset of at most k items,
   a. Pack the subset in the knapsack.
   b. Add to the knapsack items in the sorted list one by one, while there is enough available capacity.
3. Select among the packings generated in Step 2, a packing that maximizes the profit.

Theorem 8.2 *Let $P(A_k)$ denote the profit achieved by A_k, and let $P(OPT)$ denote the optimal profit, then*

$$OPT(A_k) \leq P(A) \left(1 + \frac{1}{k}\right).$$

Proof. Let OPT be any optimal solution. If $|OPT| \leq k$ we are done, since the subset OPT will be considered in some iteration of Step 2. Otherwise, let $H = \{a_1, a_2, \ldots, a_k\}$ be the set of k most profitable items in OPT. There exists an iteration of A_k in which H is considered. We show that the profit gained by A_k in this iteration yields the statement of the theorem. Consider the list $L_1 = OPT \setminus H = \{a_{k+1}, \ldots, a_x\}$ of the remaining items of OPT, in the order they are considered by A_k. Recall that, at some point, A_k will try H as the initial set of k packed items. The algorithm will then add greedily items, as long as the capacity constraint allows. If all the items are packed, A_k is clearly optimal; otherwise, at some point there is not enough space for the next item. Let m be the index of the first item in L_1 which is not packed in the knapsack by A_k, that is, the items $a_{k+1}, \ldots, a_{m-1}$ are packed. The item a_m is not packed because B_e, the remaining empty space at this point, is smaller than s_m. The greedy algorithm packed into the knapsack only items with profit density at least p_m/s_m. At the time that a_m is dropped, the knapsack contains the items from H, the items $a_{k+1}, \ldots, a_{m-1}$ and some items which are not in OPT.

Let G denote the items packed in the knapsack so far by the greedy stage of A_k. All of these items have profit density at least p_m/s_m. In particular, the items in $G \setminus OPT$ that have total size $\Delta = B - (B_e + \sum_{i=1}^{m-1} s_i)$ all have profit density at least p_m/s_m. Thus, the total profit of the items in G is $P(G) \geq \sum_{i=k+1}^{m-1} p_i + \Delta \frac{p_m}{s_m}$. We conclude that the total profit of the items in OPT is

$$P(OPT) = \sum_{i=1}^{k} p_i + \sum_{i=k+1}^{m-1} p_i + \sum_{i=m}^{|OPT|} p_i$$

$$\leq P(H) + \left(P(G) - \Delta \frac{p_m}{s_m}\right) + \left(B - \sum_{i=1}^{m-1} s_i\right) \frac{p_m}{s_m}$$

$$= P(H) + P(G) + B_e \frac{p_m}{s_m} < P(H \cup G) + p_m$$

Since A_k packs at least $H \cup G$, we get that $P(A_k) \geq P(H) + P(G)$, which implies that $P(OPT) - P(A_k) < p_m$. Given that there are at least k items with a profit at least as large as a_m (those selected to H), we conclude that $s_m \leq S(OPT)/(k+1)$. This gives the approximation ratio. ∎

Assuming a single preprocessing step, in which the items are sorted by their profit densities, each subset is extended to a maximal packing in time $O(n)$. Since there are $O(kn^k)$ possible subsets to consider, the total running time of the scheme is $O(kn^{k+1})$.

To obtain a PTAS for the knapsack problem, let A_ε be the algorithm A_k with $k = \lceil 1/\varepsilon \rceil$. By the above, the approximation ratio is at most $1 + \varepsilon$, and the running time of A_ε is $O(\frac{1}{\varepsilon} n^{1 + \frac{1}{\varepsilon}})$.

The technique of choosing the best among a small number of partial packings was applied also to variants of multidimensional packing. A detailed example is given in Section 8.3.2.

8.2.2 Applying Enumeration to a Compacted Instance

In this section we present the technique of applying exhaustive enumeration to a modified instance, in which we have a more compact representation of the input. Approximation schemes that are based on this approach consist of three steps:

1. The instance I is modified to a simpler instance, I'. The parameter ε determines how rough I' is compared to I. The smaller ε the more refined is I'.
2. The problem is solved optimally on I'.
3. An approximate solution for I is induced from the optimal solution for I'.

The challenge is to modify I in the first step into an instance I' that is simple enough to be solved in polynomial time, yet not too different from the original instance I, so that we can use an exact solution for I' to derive an approximate solution for I.

The use of this technique usually involves partitioning the input into *significant* and *nonsignificant* elements. The partition depends on the problem at hand. For example, it is natural to distinguish between *long* and *short* jobs in scheduling problems, and between *big* and *small*, or *high-profit* and *low-profit* elements, in packing problems. For a given instance, the distinction between the two types of elements usually depends on the input parameters (including ε), and on the optimal solution value.

In some cases, the transformation from I to I' only involves grouping of the nonsignificant elements. Each group of such elements thus forms a single significant element in I'. As a result, the instance I' consists of a small number of significant elements. More details and an example for this type of transformation are given in Section 8.2.2.1.

In other cases, all the elements, or only the more significant ones, are transformed into a set of elements with a small number of distinct values. This approach is described and demonstrated in Section 8.2.2.2.

8.2.2.1 Grouping Subsets of Elements

We illustrate the technique with the PTAS of Sahni [16] for the MM problem on two identical machines. The input consists of n jobs with processing times $p_1, \ldots, p_n$. The goal is to schedule the jobs on two identical parallel machines in a way that minimizes the latest completion time. In other words, we seek a schedule which balances the load on the two machines as much as possible.

Let $P = \sum_{j=1}^{n} p_j$ denote the total processing time of all jobs, and let p_{max} denote the longest processing time of a job. Let $C = \max(P/2, p_{max})$. Note that C is a lower bound on the MM (i.e., $OPT \geq C$), since $P/2$ is the schedule length if the load is perfectly balanced between the two machines, and since some machine must process the longest job.

The first step of the scheme is to modify the instance I into a simplified instance I'. This modification depends on the value of C and on the parameter ε. Given I, ε, partition the jobs into small jobs—of length at most εC, and big jobs—of length larger than εC. Let P_S denote the total length of small jobs. The modified instance I' consists of the big jobs in I together with $\lfloor P_S/(\varepsilon C) \rfloor$ jobs of length εC.

Next, we need to solve optimally the MM problem for the instance I'. Note that all jobs in I' have length at least εC and their total size is at most P, the total processing time of the jobs in the original instance, since the small jobs in I are replaced in I' by jobs of length εC with total length at most P_S. Therefore, the number of jobs in I' is at most *the constant* $P/\varepsilon C \leq 2/\varepsilon$. An optimal schedule of a constant number of jobs can be found by exhaustive search over all $O(2^{2/\varepsilon})$ possible schedules. This constant number is independent of n, but grows exponentially with ε, as we expect from our PTAS.

Finally, we need to transform the optimal schedule of I' into a feasible schedule of I. Note that, for the makespan objective, we are only concerned about the partition of the jobs between the machines, whereas the order in which the jobs are scheduled on each machine can be arbitrary. Denote by $OPT(I')$ the length of the optimal schedule for I'. To obtain a schedule of I, each of the big jobs is scheduled on the same machine as in the optimal schedule for I'. The small jobs are scheduled greedily in an arbitrary order on the first machine until, for the first time, the total load on the first machine is at least $OPT(I')$. The remaining small jobs are scheduled on the second machine. Clearly, the overflow on the first machine is at most εC (maximal length of a small job). Also, since the total number of (εC)-jobs was defined to be $\lfloor P_S/(\varepsilon C) \rfloor$, the overflow on the second machine is also bounded by εC. Therefore, the resulting makespan in the schedule of I is at most $OPT(I') + \varepsilon C$.

To complete the analysis we need to relate $OPT(I')$ to $OPT(I)$.

Claim 8.1 $OPT(I') \leq (1 + \varepsilon)OPT(I)$.

Proof. Given a schedule of I, in particular an optimal one, a schedule for I' can be derived by replacing— on each machine separately—the small jobs with jobs of size εC, with at least the same total size. Recall that the number of (εC)-jobs in I' is $\lfloor P_S/(\varepsilon C) \rfloor$. Regardless of the partition of the small jobs in I between the two machines, the result of this replacement is a feasible schedule of I' whose makespan is at most $OPT(I) + \varepsilon C$. Since $OPT(I) \geq C$, the statement of the claim holds.   ∎

Back to our scheme, we showed that the optimal schedule of I' is transformed into a feasible schedule of I whose makespan is at most $OPT(I') + \varepsilon C$. By Claim 8.1, this value is at most $(1 + \varepsilon)OPT(I) + \varepsilon C \leq (1 + 2\varepsilon)OPT(I)$. By selecting $\varepsilon' = \varepsilon/2$, and running the scheme with ε', we get the desired ratio of $(1 + \varepsilon)$.

The previous scheme can be extended to any constant number of machines. For *arbitrary* number of machines, a more complex PTAS exists: the scheme of [1], which requires reducing the number of distinct values in the input, is given in the next section.

8.2.2.2 Reducing the Number of Distinct Values in the Input

Any optimization problem can be solved optimally in polynomial, or even constant time, if the input size is some constant. For many optimization problems, an efficient algorithm exists if the input size is arbitrary but the number of *distinct values* in the input is some constant. Alternatively, the problem can be solved by a pseudo-polynomial-time algorithm (e.g., by dynamic programming), whose running time depends on the instance parameters, and is therefore polynomial only if the parameter values are polynomial in the problem size.

The idea behind the technique that we describe in the following is to transform the elements (or sometimes, only the significant elements) in the instance I into an instance I' in which the number of distinct values is fixed, or to scale the values according to the input size. The problem is then solved on I', and the solution for I' is transformed into a solution for the original instance. The nonsignificant elements, which are sometimes omitted from I', are added later to the solution, using some heuristic. The parameter ε determines the (constant) number of distinct values contained in I': the smaller ε, the larger is the number of distinct values. The following are two main approaches for determining the values in I'.

1. *Rounding*: The values in I' form an arithmetic series in which the difference between elements is a function of ε. For example, multiples of $\varepsilon^2 T$, for some value T. In this approach, the gap between any two values bounds the difference between the original value of an element in I and the value of

the corresponding element in I'. Note that the number of elements whose values are rounded to a single value in I' can be arbitrary.

2. *Shifting:* The values in I' are a subset of the values in I, selected such that the distribution on the number of values in I that are shifted to a single value in I' is uniform. On the other hand, in contrast to the rounding approach, there is no bound on the difference between the value of an element in I and its value in I'. For example, partition the elements into $\lceil 1/\varepsilon^2 \rceil$ groups, each having at most $\lfloor n/\varepsilon^2 \rfloor$ elements, and fix the values in each group to be (say) the minimal value of any element in the group.

In both approaches, the approximation ratio is guaranteed to be $(1+\varepsilon)$ if I' is close enough to I. Formally, an optimal solution for I' induces a solution for I whose value is larger/smaller by a factor of at most $(1+\varepsilon)$. Another factor of $(1 + \varepsilon)$ may be added to the approximation ratio due to the nonsignificant items—in case they are handled separately.

We demonstrate this technique with the classic PTAS of Hochbaum and Shmoys [1] for the MM problem on parallel machines. The input for the problem is a set of n jobs having processing times $p_1, \ldots, p_n$, and m identical machines; the goal is to schedule the jobs on the machines in a way that minimizes the latest completion time of any job. The number of machines, m, can be arbitrarily large (otherwise, a simpler PTAS exists; see Section 8.2.2.1).

First, note that the MM problem is closely related to the *bin packing* (BP) problem. The input for BP is a collection of items whose sizes are in $(0, 1]$. The goal is to pack all items using a minimal number of bins. Formally, let $I = \{p_1 \ldots p_n\}$ be the sizes in a set of n items, where $0 < p_j \leq 1$. The goal is to find a collection of subsets $U = \{B_1, B_2, \ldots, B_k\}$ which forms a disjoint partition of I, such that for all i, $1 \leq i \leq k$, $\sum_{j \in B_i} p_j \leq 1$, and the number of bins, k, is minimized.

The exact solutions of MM and BP relate in the following way. It is possible to schedule all the jobs in an MM instance on m machines with makespan C_{max} if and only if it is possible to pack all the items in a BP instance, where the size of item j is p_j/C_{max}, in m bins. The relation between the optimal solutions does not remain valid for approximations. In particular, BP admits an asymptotic FPTAS (see Vol. 2, Chap. 15) whereas MM does not. However, this relation can be used to develop a PTAS for MM.

Let $OPT_{BP}(I)$ be the number of bins in an optimal solution of BP, and let $OPT_{MM}(I) = C_{max}$ be an optimal solution for MM. Denote by $\frac{I}{d}$ the BP input in which all the values are divided by d. We already argued that:

$$OPT_{BP}\left(\frac{I}{d}\right) \leq m \quad \Leftrightarrow \quad OPT_{MM}(I, m) \leq d.$$

We define a *dual* approximation scheme for BP. For an input I, we seek a solution with at most OPT_{BP} bins, where each bin is filled to capacity at most $1 + \varepsilon$. In other words, we relax the bin capacity constraint by a factor of $1 + \varepsilon$. Let $dual_\varepsilon(I)$ be such an algorithm, and let $DUAL_\varepsilon(I)$ be the number of bins in the corresponding packing.

Theorem 8.3 *If there exists a dual approximation algorithm for BP, then there is a PTAS for the MM problem.*

Proof. The PTAS performs a binary search to find OPT_{MM}. In order to bound the range in which the optimal makespan is searched, two lower bounds and one upper bound for this value are used. The lower bounds are the length of the longest job, and the load on each machine when the total load is perfectly balanced. That is, let $SIZE(I, m) = max\{\frac{1}{m} \sum p_i, p_{max}\}$, then $OPT_{MM} \geq SIZE(I, m)$. The upper bound uses the fact that the simple *List Scheduling* algorithm attains a 2-approximation ratio [14], therefore $OPT_{MM} \leq 2SIZE(I, m)$.

Now it is possible to perform a binary search to find OPT_{MM}. Instead of checking whether $OPT_{MM} < d$, the algorithm checks whether $DUAL_\varepsilon(\frac{I}{d}) < m$.

$upper = 2SIZE(I, m)$

$lower = SIZE(I, m)$

repeat until $lower = upper$

    $d = (lower + upper)/2$

    call $dual_\varepsilon(\frac{I}{d})$

    if $DUAL_\varepsilon(\frac{I}{d}) > m$

        $lower \leftarrow d$

    else

        $upper \leftarrow d$

$d^\star \leftarrow upper$

return $dual_\varepsilon(\frac{I}{d^\star})$

Initially, $OPT_{MM}(I, m) \leq upper \Rightarrow OPT_{BP}(\frac{I}{upper}) \leq m$. Since $dual_\varepsilon$ is a relaxation of BP,

$$DUAL_\varepsilon \left(\frac{I}{upper} \right) \leq OPT_{BP} \left(\frac{I}{upper} \right).$$

This implies that $DUAL_\varepsilon(\frac{I}{upper}) \leq m$. By the update rule, the above remains true during the execution of the loop. However,

$$DUAL_\varepsilon \left(\frac{I}{upper} \right) \leq m \Rightarrow OPT_{MM}(I, m) \leq (1 + \varepsilon)upper,$$

and thus $(1 + \varepsilon)upper$ remains an upper bound on $OPT_{MM}(I, m)$ throughout the search. Similarly, before the loop, $OPT_{MM}(I, m) \geq lower$, which remains true since $DUAL_\varepsilon(\frac{I}{lower}) \geq m$ is an invariant of the loop, and

$$OPT_{BP} \left(\frac{I}{lower} \right) \geq DUAL_\varepsilon \left(\frac{I}{lower} \right) \geq m \Rightarrow OPT_{MM}(I, m) \geq lower.$$

Thus, the solution value is bounded above by

$$(1 + \varepsilon) \cdot d^\star = (1 + \varepsilon) \cdot upper = (1 + \varepsilon) \cdot lower \leq (1 + \varepsilon)OPT_{MM}(I, m).$$

In practice, assume that we stop the binary search after k iterations. At this time, it is guaranteed that $upper - lower \leq 2^{-k}SIZE(I, m) \leq 2^{-k}OPT_{MM}(I, m)$, and the value of the solution is bounded above by $(1 + \varepsilon) \cdot d^\star = (1 + \varepsilon) \cdot upper \leq (1 + \varepsilon) \cdot (lower + 2^{-k}OPT_{MM}(I, m)) \leq (1 + \varepsilon)(1 + 2^{-k})OPT_{MM}(I, m)$.

By choosing $k = O(\log \frac{1}{\epsilon})$, and taking in the scheme $\varepsilon' = \varepsilon/3$, we obtain a $(1 + \varepsilon)$-approximation. ∎

We now describe the $dual_\varepsilon$ approximation scheme for BP. This scheme uses the rounding and grouping technique.

Theorem 8.4 *There exists an* $O\left(n^{\lceil \frac{1}{\varepsilon^2} \rceil}\right)$*-time dual approximation scheme for BP.*

Proof. Recall that, for a given $\varepsilon > 0$, the dual approximation scheme needs to find a packing of all items using at most OPT_{BP} bins, such that the total size of the items packed in each bin is at most $1 + \varepsilon$. The basic idea is to omit first the "small" items and then round the sizes of the "big" items; this yields an instance in which the number of distinct item sizes is fixed. We can now solve the problem exactly using dynamic programming, and the solution induces a solution for the original instance, where each bin is filled up to capacity $1 + \varepsilon$.

The first observation is that small items, whose sizes are less that ε, can be initially omitted. The problem will be solved for big items only and the small items will be added later on greedily, in the following manner: if there is a bin filled with items of total size less than 1, small items are added to it; otherwise, a new bin is opened. If no new bin is opened, then, clearly, no more than the optimum number of bins is used (as the dual PTAS uses the minimal number of bins for the big items). If new bins were added, then all original bins are filled to capacity at least 1, and all the new bins (except maybe the last one) are also filled to capacity at least 1. This is optimal since $OPT(I) \geq \lceil \sum p_i \rceil \geq DUAL_\varepsilon(I)$. We conclude that, without loss of generality, all items are of size $\varepsilon \leq p_i \leq 1$. Divide the range $[\varepsilon, 1]$ into intervals of size ε^2. This gives $S = \lceil \frac{1}{\varepsilon^2} \rceil$ intervals. Denote by l_i the endpoints of the intervals and let b_i be the number of elements whose sizes are in the interval $(l_i, l_{i+1}]$.

We now examine a packed bin. Since the minimal item size is ε, the bin can contain at most $\lfloor \frac{1}{\varepsilon} \rfloor$ items. Denote by X_i the number of items in the bin whose sizes are in the interval $(l_i, l_{i+1}]$. X_i is in the range $[0, \lfloor \frac{1}{\varepsilon} \rfloor)$. Let the vector $(X_1, \ldots, X_S)$ denote the *configuration* of the bin. The number of feasible configurations is bounded above by $\lfloor \frac{1}{\varepsilon} \rfloor^S$. A configuration is *feasible* if and only if $\sum_{i=1}^S X_i l_i \leq 1$.

For any bin B whose packing forms a feasible configuration, the total size of the items in the bin is bounded by

$$\sum_{j \in B} p_j \leq \sum_{j \in B} X_j l_{j+1} \leq \sum_{j \in B} X_j (l_j + \varepsilon^2) \leq 1 + \varepsilon^2 \sum_{j \in B} X_j \leq 1 + \varepsilon^2 \cdot \frac{1}{\epsilon} \leq 1 + \varepsilon.$$

Therefore, it is sufficient to solve the instance with all item sizes rounded down to sizes in $\{l_1, \ldots, l_S\}$.

Finally, we describe a dynamic programming algorithm which solves the BP problem exactly when the number of distinct item sizes is fixed. Let $BINS(b_1, b_2, \ldots, b_S)$ be the minimal number of bins required to pack b_1 items of size l_1, b_2 items of size l_2, $\ldots$, and b_S items of size l_S. Let C denote the set of all feasible configurations. Observe that, by a standard dynamic programming recursion,

$$BINS(b_1, b_2, \ldots, b_S) = 1 + min_C BINS(b_1 - X_1, b_2 - X_2, \ldots, b_S - X_S).$$

We minimize over all possible vectors $(X_1, X_2, \ldots, X_S)$ that correspond to a feasible packing of the "first" bin (counted by the constant 1), and the best way to pack the remaining items (this is the recursive call). Thus, the dynamic programming procedure builds a table of size n^S, where the calculation of each entry requires $O(\lfloor \frac{1}{\epsilon} \rfloor^S)$.

This yields a running time of

$$O\left(n^S \cdot \lfloor \frac{1}{\epsilon} \rfloor^S\right) = O\left(\left(\frac{n}{\varepsilon}\right)^{\lceil \frac{1}{\varepsilon^2} \rceil}\right) = O\left(n^{\lceil \frac{1}{\varepsilon^2} \rceil}\right). \qquad \blacksquare$$

The technique of applying enumeration to a compacted instance through grouping/rounding has been extensively used in PTASs for scheduling problems [17–19]. A common approach for compacting the instance is to reduce the input parameters to *poly-bounded*, that is, parameters whose values can be bounded as function of the input size. This approach is used, for example, in the PTAS of Chekuri and Khanna for preemptive weighted flow time [4] (see also the survey paper [9]).

8.2.3 More on Grouping and Shifting

In the following we outline two extensions of the techniques described in this section.

8.2.3.1 Randomized Grouping

In some cases, we need to define a partition of the input elements to groups $(I_1, \ldots, I_k)$, using for each element x a parameter of the problem, $q(x)$, such that the elements in two groups I_j and I_{j+1} differ in their $q(x)$ value by roughly a factor of α, for some $\alpha > 1$. When such partition is infeasible, we can use randomization to achieve an *expected* separation between groups. For a parameter $\alpha > 1$, the following randomized geometric grouping technique yields an expected separation that is logarithmic in α. This technique extends the deterministic geometric rounding technique described in Section 8.2.2.2. Initially, pick a number $r \in [1, \alpha]$ at random, by a probability distribution having the density function $f(y) = 1/y \ln \alpha$. An element x with the value $q(x)$ belongs to the group I_j if $q(x) \in [r\alpha^j, r\alpha^{j+1})$. Thus, the index of the group to which x belongs, denoted by $g(x)$, is a random variable which can take two possible values: $\lfloor \log_\alpha q(x) \rfloor$ or $\lfloor \log_\alpha q(x) \rfloor + 1$. It can be shown that for a fixed α, the number of distinct partitions induced by the random choices of r is at most the number of elements in the input. This enables to easily derandomize algorithms that use randomized geometric grouping. The technique was applied, for example, by Chekuri and Khanna [4] in a PTAS for preemptive weighted flow time.

8.2.3.2 Oblivious Shifting

While applying the standard shifting technique (as described in Section 8.2.2.2) requires knowing the initial input parameters, it is possible to apply shifting also when not all values are known a-priori. In *oblivious shifting*, the input size is initially known, and the scheme starts by defining the number of values in the resulting instance, but the actual shifted values are revealed at a later stage, by optimizing on these values, considering the constraints of the problem. The technique can be used for defining a "good" compacted instance from a partial solution for the problem, which can then enable to obtain a complete solution for the problem efficiently.

For example, a variant of the BP problem, in which items may be *fragmented*, is solved in [20] in two steps. Given the input, we need to determine the set of items that will be fragmented, as well as the fragment sizes in a feasible approximate solution. Since the possible number of fragment sizes is large, a compact vector of fragments is generated, which contains a bounded number of *unknown* shifted fragment sizes. The actual sizes of the shifted fragments are determined by solving a linear program which attempts to find a feasible packing of these fragments. A detailed description is given in [20].

8.3 Rounding Linear Programs

In this section we discuss approximations obtained using linear programming (LP) relaxation for a given optimization problem. We refer the reader to Chapter 7 in this book for further background on linear programming and rounding linear programs. Most generally, the technique is based on solving a linear programming relaxation of the problem, for which an exact or approximate solution can be obtained efficiently. This solution is then rounded, thus yielding an approximate integral solution. The (fractional) solution obtained for the LP needs to have some nice properties that would allow rounding to be not too harmful, in terms of ε, the accuracy parameter of the scheme. One such property of a linear program, which is commonly used, is the existence of a small *basic solution*. We illustrate the usage of this property in the following, with examples from vector scheduling (VS) and covering integer programs. A linear program has a *small* basic solution, if there exists an optimal solution in which the number of non-zero variables is small as function of the input size and ε. For such a solution, the error incurred by rounding can be bounded, such that the resulting integral solution is within factor of $1 + \varepsilon$ from the optimal. A natural example is the class of linear programs in which either the number of variables or the number of constraints is some fixed constant. For such programs, there exists a basic solution in which the number of non-zero variables is fixed; however, depending on the problem, and in particular, on the value of an optimal solution for the LP, a basic solution can be "small," even if the number of non-zero variables is relatively large, for example, $\Omega(\varepsilon n)$, where n is the number of variables.

LP rounding can be combined with the techniques described in Section 8.2. In Section 8.3.1 we show the usage of LP rounding for a given subset of input elements satisfying certain properties. In Section 8.3.2 we show how LP rounding can be combined with the selection of *all* possible (small) subsets.

8.3.1 Solving LP for a Subset of Elements

As mentioned in Section 8.2.1, in many problems, an approximation scheme can be obtained by partitioning a set of input elements to subsets, and solving the problem for each subset separately. For some subsets, a good solution can be obtained by rounding an LP relaxation of the problem.

In certain assignment problems, we can find an almost integral basic solution for an LP, for part of the input, since the relation between the number of variables and nontrivial constraints in the linear programming relaxation, combined with the assignment requirement of the problem, imply that only few variables can get fractional values. This essential property is used, for example, in the PTAS of Chekuri and Khanna for the VS problem [21]. The VS problem is to schedule d-dimensional jobs on m identical machines, such that the maximum load over all dimensions and over all machines is minimized. Formally, an instance I of VS consists of n jobs, $J_1, \ldots, J_n$, where J_j is associated with a rational d-dimensional vector $(p_j^1, \ldots, p_j^d)$, and m machines. We need to assign the jobs to the machines, that is, schedule a subset of the jobs, A_i, on machine i, $1 \leq i \leq m$, such that $\max_{1 \leq i \leq m} \max_{1 \leq h \leq d} \sum_{J_j \in A_i} p_j^h$ is minimized.

Note that in the special case where $d = 1$ we get the MM problem (see in Section 8.2.2.2). The PTAS of [21] for the VS problem, where d is fixed, applies a nontrivial generalization of the PTAS of Hochbaum and Shmoys for the case $d = 1$ [1]. The scheme is based on a primal-dual approach, in which the primal problem is VS and the dual problem is vector packing. Thus, the machines are viewed as d-dimensional bins, and the schedule length—as bin capacity (or height). W.l.o.g., we may assume that the optimal schedule has the value 1. Given an $\varepsilon > 0$ and a correct guess of the optimal value, we describe an algorithm A_ε that returns a schedule of height at most $1 + \varepsilon$ in the following. Arriving at correct guess involves a binary search for the optimal value (which can be done in polynomial time; see below).

Let $\delta = \varepsilon/d$ be a parameter. The scheme starts with a preprocessing step, which enables to bound the ratio of the largest coordinate to the smallest non-zero coordinate in any input vector. Specifically, let $\| J_j \|_\infty = \max_{1 \leq h \leq d} p_j^h$ be the ℓ_∞ norm of J_j, $1 \leq j \leq n$, then, for any J_j, and any $1 \leq h \leq d$, if $p_j^h \leq \delta \| J_j \|_\infty$, we set $p_j^h = 0$. As shown in [21], any valid schedule for the resulting modified instance, I', yields a valid solution for the original instance, I, whose height is at most $(1 + \varepsilon)$ times that of I'.

We consider from now on only transformed instances. The scheme proceeds by partitioning the jobs to the sets L (*large*) and S (*small*). The set L consists of all vectors whose ℓ_∞ norm is greater than δ, and S contains the remaining vectors. The algorithm A_ε packs first the large jobs and then the small jobs. Note, that while in the case of $d = 1$ these packings are done independently, for $d \geq 2$, we need to consider the interaction between these two sets. Similar to the scheme of Hochbaum and Shmoys [1], a valid schedule is found for the jobs by guessing a configuration. In particular, let the d-tuple $(a_1, \ldots, a_d)$ $0 \leq a_h \leq \lceil 1/\varepsilon \rceil$, $1 \leq h \leq d$, denote a *capacity configuration*, that is, the way some bin is filled. Since $d \geq 2$ is a constant, the possible number of capacity configurations, given by $W = (1 + \lceil 1/\varepsilon \rceil)^d$, is also a constant. Then, by numbering the capacity configurations, we describe by a W-tuple $M = (m_1, \ldots, m_W)$ the number of bins having capacity configuration w, where $1 \leq w \leq W$. The possible number of *bin configurations* is then $O(m^W)$. This allows to guess a bin configuration which yields the desired $(1 + \varepsilon)$-approximate solution in polynomial time.

We say that a packing of vectors in a bin *respects* a capacity configuration $(a_1, \ldots, a_d)$ if the height of the packing is smaller than εa_h for any $1 \leq h \leq d$. Given a capacity configuration $(a_1, \ldots, a_d)$, we define the *empty capacity configuration* to be the d-tuple $(\bar{a}_1, \ldots, \bar{a}_d)$, where $\bar{a}_h = \lceil 1/\varepsilon \rceil + 1 - a_h$, for $1 \leq h \leq d$. For a given bin configuration, M, we denote by $\bar{M}$ the bin configuration obtained by taking for each of the bins in M the corresponding empty capacity configuration.

The scheme performs the following two steps for each possible bin configuration, M: (i) decides whether vectors in L can be packed respecting M and (ii) decides whether vectors in S can be packed respecting $\bar{M}$. Given that we have guessed the correct bin configuration M, both steps will succeed, and we get a packing of height at most $1 + \varepsilon$.

We now describe how the scheme packs the large and the small vectors. The vectors in L are packed using rounding and dynamic programming. In particular, since by definition, any entry in a vector in L has the value δ^2 or larger, we use geometric rounding, that is, for each vector J_j, and any entry p_j^h, $1 \leq h \leq d$, p_j^h is rounded down to the nearest value of the form $\delta^2(1 + \varepsilon)^t$, for $0 \leq t \leq \lceil \frac{2}{\varepsilon} \log 1/\delta \rceil$. Denote the resulting set of vectors L', and the modified instance I'. The vectors in L' can be partitioned into

$$q = \left(1 + \left\lceil \frac{2}{\varepsilon} \log 1/\delta \right\rceil \right)^d \tag{8.1}$$

classes. The proofs of the next lemmas are given in [21].

Lemma 8.1 *Given a solution for I', replacing each vector in L' by the corresponding vector in L results in a valid solution for I whose height is at most $1 + \varepsilon$ times that of I'.*

Lemma 8.2 *Given a correct guess of a bin configuration M, there exists an algorithm which finds a packing of the vectors in L' that respects M, and whose running time is $O((d/\delta)^q mn^q)$, where q is given in (8.1).*

The small vectors are packed using a linear programming relaxation and careful rounding. Renumber the vectors in S by $1, \ldots, |S|$. Let $x_{ji} \in \{0, 1\}$ be an indicator variable for the assignment of the vector J_j to machine i, $1 \leq j \leq n, 1 \leq i \leq m$. In the LP relaxation $x_{ji} \geq 0$. We solve the following linear program.

$$\text{(LP)} \quad \sum_{J_j \in S} p_j^h x_{ji} \leq b_i^h \quad 1 \leq i \leq m, \ 1 \leq h \leq d \tag{8.2}$$

$$\sum_{i=1}^{m} x_{ji} = 1 \quad 1 \leq j \leq |S| \tag{8.3}$$

$$x_{ji} \geq 0 \quad 1 \leq j \leq n, \ 1 \leq i \leq m \tag{8.4}$$

The constraints (8.2) guarantee that the packing does not exceed a given height bound in any dimension (i.e., the available height after packing the large vectors). The constraints (8.3) reflect the requirement that each vector is assigned to one machine. A key property of the LP, which enables to obtain an integral solution that is close to the fractional, is given in the next result.

Lemma 8.3 *In any basic feasible solution for LP, at most $d \cdot m$ vectors are assigned (fractionally) to more than one machine.*

Proof. Recall that the number of non-zero variables, in any *basic* solution for a linear program, is bounded by the number of tight constraints in some optimal solution (since nontight constraints can be omitted). Since the number of nontrivial constraints (i.e., constraints other than $x_{ji} \geq 0$) is ($|S| + d \cdot m$), it follows that the number of strictly positive variables in any basic solution is at most ($|S| + d \cdot m$). Since each vector is assigned to at least one machine, the number of vectors which are fractionally assigned to more than one machine is at most $d \cdot m$. ∎

The previous type of argument was first made and exploited by Potts [22] in the context of parallel machine scheduling. It was later applied for other problems, such as job shop scheduling [23].

Thus, we solve the above-mentioned program and obtain a basic solution. Denote by S' the set of vectors which are assigned fractionally to two machines or more. Since $|S'| \leq d \cdot m$, we can partition the set S' to subsets of size at most d each, and schedule the i-th set to the i-th machine. Since $\| J_j \|_\infty \leq \delta = \varepsilon/d$, for all $J_j \in S'$, the total height of the machines is violated at most by ε in any dimension. We can therefore summarize in the following theorem.

Theorem 8.5 *For any $\varepsilon > 0$, there is a $(1 + \varepsilon)$-approximation algorithm for VS whose running time is $(nd/\varepsilon)^{O(f)}$, where $f = O((\frac{\ln(d/\varepsilon)}{\varepsilon})^d)$.*

Proof. By the previous discussion, given the correct guess of the optimal value, the scheme yields a schedule of value (height) at most $1 + O(\varepsilon)$ the optimal. We need to find a packing of the vectors in L and S, for each bin configuration M. The running time for a single configuration is dominated by the packing of L, and since the number of configurations is $m^W = O(n^{O(1/\varepsilon^d)})$, we get the running time from Lemma 8.2. The value of an optimal schedule can be guessed, within factor $1 + \varepsilon$, by obtaining first a $(d + 1)$-approximate solution. This can be done by applying an approximation algorithm for resource constrained scheduling due to [24]. ∎

8.3.2 LP Rounding Combined with Enumeration

As described in Section 8.2.1, a common technique for obtaining a PTAS is to extend all possible solutions for small subsets of elements. This technique can be combined with LP rounding as follows. Repeatedly select a small subset of input elements, $S_g \subseteq I$, to be the basis for an approximate solution; solve an LP for the remaining elements, $I \setminus S_g$. Select the subset S_g which gives the best solution. We exemplify the usage of the technique to obtain a PTAS for *covering integer programs with multiplicity constraints (CIP)*. In this core problem, we must fill up an R-dimensional bin by selecting (with bounded number of repetitions) from a set of n R-dimensional items, such that the overall cost is minimized. Formally, let $A = \{a_{ji}\}$ denote the sizes of the items in the R dimensions, $1 \leq j \leq R, 1 \leq i \leq n$; the cost of item i is $c_i \geq 0$. Let x_i denote the number of copies selected from item i, $1 \leq i \leq n$. We seek an n-vector $\mathbf{x}$ of nonnegative integers, which minimizes $c^T\mathbf{x}$, subject to the R constraints given by $A\mathbf{x} \geq b$, where $b_j \geq 0$ is the size of the bin in dimension j. In addition, we have multiplicity constraints for the vector $\mathbf{x}$, given by $\mathbf{x} \leq d$, where $d \in \{1, 2, \ldots\}^n$.

Covering integer programs form a large subclass of integer programs encompassing such NP-hard problems as minimum knapsack and set cover. This implies the hardness of CIP in fixed dimension (i.e., where R is a fixed constant). For general instances, the hardness of approximation results for set cover carry over to CIP. Comprehensive surveys of known results for CIP and CIP_∞, where the multiplicity constraints are omitted, are given in [25] and in [26] (see also in [27]).

We describe in the following a PTAS for CIP in fixed dimension. The scheme presented in [25] builds on the classic LP-based scheme due to Frieze and Clarke for the R-dimensional knapsack problem [28]. Consider an instance of CIP in fixed dimension, R. We want to minimize $\sum_{i=1}^{n} c_i x_i$ subject to the constraints $\sum_{i=1}^{n} a_{ij} x_i \geq b_j$ for $j = 1, \ldots, R$, and $x_i \in \{0, 1, \ldots d_i\}$ for $i = 1 \ldots, n$.

Assume that we know the optimal cost, C, for the CIP instance. The scheme of [25] uses a reduction to the binary *minimum R-dimensional multiple choice knapsack (R-MMCK)* problem. For some $R \geq 1$, an instance of binary R-MMCK consists of a single R-dimensional knapsack, of size b_j in the j-th dimension, and m sets of items. Each item has an R-dimensional size and is associated with a cost. The goal is to pack a subset of items, by selecting at most one item from each set, such that the total size of the packed items in dimension j is at least b_j, $1 \leq j \leq R$, and the overall cost is minimized.

Given the value of C, the parameter ε and a CIP instance with bounded multiplicity, the scheme constructs an R-MMCK instance in which the knapsack capacities in the R dimensions are b_j, $1 \leq j \leq R$. Also, there are n sets of items denoted by A^i, $1 \leq i \leq n$. Let $\hat{K}^i$ be the integer value satisfying $d_i c_i \in [\hat{K}^i \varepsilon C/n, (\hat{K}^i + 1)\varepsilon C/n)$, then the number of items in A^i is $K^i = \min(\hat{K}^i, \lfloor n/\varepsilon \rfloor)$. The set A^i represents all possible values which x_i can take in the solution for CIP. In particular, the k-th item in A^i, denoted by (i, k), represents the assignment of a value in $[0, d_i]$ to x_i, such that $c(i, k)$, the total cost incurred by item i is in $[k\varepsilon C/n, (k + 1)\varepsilon C/n)$. This total cost is rounded down to the nearest integral multiple of $\varepsilon C/n$; thus, $c(i, k) = k\varepsilon C/n$. The size of the item (i, k) in dimension j, $1 \leq j \leq R$, is given by $s_j(i, k) = a_{ij}$.

Given an instance of R-MMCK, guess a partial solution, given by a small size set, S; these items have the maximal costs in some optimal solution. The size of S is a fixed constant, namely, $|S| = h = \lfloor \frac{2R(1+\varepsilon)}{\varepsilon} \rfloor$. The set S will be extended to an *approximate* solution, by solving a linear program for the remaining items. The value of h is chosen such that the resulting solution is guaranteed to be within $1 + \varepsilon$ from the optimal, as computed in the following. Let $E(S)$ be the subset of items with costs that are larger than the minimal cost of any item in S, that is, $E(S) = \{(i, k) \notin S \mid c(i, k) > c_{min}(S)\}$, where $c_{min}(S) = \min_{(i,k) \in S} c(i, k)$. Select all the items $(i, k) \in S$, and eliminate from the instance all the items $(i, k) \in E(S)$ and the sets A^i from which an item has been selected. In the next step we find an optimal *basic solution* for the following linear program, in which $x_{i,k}$ is an indicator variable for the selection of the item (i, k).

$$(LP(S)) \quad \text{minimize} \quad \sum_{i=1}^{n} \sum_{k=1}^{K^i} x_{i,k} \cdot c(i, k)$$

$$\text{subject to}: \quad \sum_{k=1}^{K^i} x_{i,k} \leq 1 \quad \text{for } i = 1, \dots, n,$$

$$\sum_{i=1}^{n} \sum_{k=1}^{K^i} s_j(i, k) x_{i,k} \geq b_j \quad \text{for } j = 1, \dots, R$$

$$0 \leq x_{i,k} \leq 1 \quad \text{for } (i, k) \notin S \cup E(S)$$

$$x_{i,k} = 1 \quad \text{for } (i, k) \in S$$

$$x_{i,k} = 0 \quad \text{for } (i, k) \in E(S)$$

Given an optimal fractional solution for the above-mentioned program, we get an integral solution as follows. For any i, $1 \leq i \leq n$, let $k_{max} = k_{max}(i)$ be the maximal value of $1 \leq k \leq K^i$ such that $x_{i,k} > 0$, then we set $x_{i,k_{max}} = 1$ and, for any other item in A^i, $x_{i,k} = 0$. Finally, we return to the CIP instance and assign to x_i the maximum value for which the total (rounded down) cost for item i is $c(i, k_{max})$.

The next three lemmas show that the scheme yields a $(1 + \varepsilon)$-approximation to the optimal cost, and that the resulting integral solution is feasible.

Lemma 8.4 *If there exists an optimal (integral) solution for CIP with cost C, then the integral solution obtained from the rounding for R-MMCK has the cost $\hat{z} \leq (1 + \varepsilon)C$.*

Proof. Let $\mathbf{x}^*$ be an optimal (fractional) solution for the linear program LP(S), and let S^* be the corresponding subset of items, that is, $S^* = \{(i, k) \mid x_{i,k}^* = 1\}$. If $|S^*| < h$ then we are done: in some iteration, the scheme will try S^*; otherwise, let $S^* = \{(i_1, k_1), \dots, (i_g, k_g)\}$, such that $c(i_1, k_1) \geq \dots \geq c(i_g, k_g)$, for some $g > h$. Let $S_h^* = \{(i_1, k_1), \dots, (i_h, k_h)\}$, and $\sigma = \sum_{t=1}^{h} c(i_t, k_t)$. Then, for any item $(i, k) \notin (S_h^* \cup E(S_h^*))$, we have $c(i, k) \leq \sigma/h$. Let $z^*, \hat{z}$ denote the optimal (integral) solution and the solution output by the scheme for the R-MMCK instance, respectively. Denote by $\mathbf{x}^B(S_h^*), \mathbf{x}^I(S_h^*)$ the basic and integral solutions of LP(S) as computed by the scheme, for the initial guess S_h^*.

By the above-mentioned rounding method, for any $1 \leq i \leq n$, the cost of the item selected from A^i is $c(i, k_{max})$. Let F denote the set of items for which the basic variable was a fraction, that is, $F = \{(i, k) \mid x_{i,k}^B(S_h^*) < 1\}$, and let $\delta = \sum_{(i,k) \in F} c(i, k)$.

Then, we get that

$$z^* \geq \sum_{i=1}^{n} \sum_{k=1}^{K^i} c(i, k) x_{i,k}^B(S_h^*)$$

$$\geq \sum_{i=1}^{n} \sum_{k=1}^{K^i} c(i, k) x_{i,k}^I(S_h^*) - \delta.$$

Recall that in any *basic* solution for a linear program, the number of non-zero variables is bounded by the number of tight constraints in some optimal solution. Assume that in the optimal (fractional) solution of $LP(S_h^*)$ there are L tight constraints, where $0 \leq L \leq n + R$. Then in the basic solution $\mathbf{x}^B(S_h^*)$, at most L variables can be strictly positive. Thus, at least $L - 2R$ variables get an integral value (i.e., 1), and $|F| \leq 2R$. Note that for any $(i, k) \in F$, $c(i, k) \leq \sigma/h$, since $F \cap (S_h^* \cup E(S_h^*)) = \emptyset$. Hence, we get that $z^* \geq \hat{z} + \frac{2R\sigma}{h} \geq \hat{z} + \frac{2R\hat{z}}{h} \geq \frac{\hat{z}}{1+\varepsilon}$. ∎

The next two lemmas follow from the rounding method used by the scheme.

Lemma 8.5 *The scheme yields a feasible solution for the CIP instance.*

Lemma 8.6 *The cost of the integral solution for the CIP instance is at most $\hat{z} + \varepsilon C$.*

Note that C can be guessed in polynomial time within factor $(1 + \varepsilon)$, using binary search over the range $(0, \sum_{i=1}^{n} d_i c_i)$. Thus, combining the above-mentioned lemmas we get:

Theorem 8.6 *There is a PTAS for CIP in fixed dimension.*

Consider now the special case where the multiplicity constraints are omitted; that is, each variable x_i can get any nonnegative (integral) value. For this special case, we can use a linear programming formulation in which the number of constraints is R, which is fixed. A PTAS for this problem can be derived from the scheme of Chandra et al. [29] for integer multidimensional knapsack. Drawing from recent results for CIPs, we describe the PTAS of [25] in the following, which improves the running time in [29] by using a fast approximation scheme for solving the linear program.

A Scheme for CIP_∞ The scheme, called *multi-dimensional cover with parameter ε (MDC$_\varepsilon$)*, proceeds in the following steps:

(i) For a given $\varepsilon \in (0, 1)$, let $\delta = \lceil R \cdot ((1/\varepsilon) - 1) \rceil$.
(ii) Renumber the items by $1, \dots, n$, such that $c_1 \geq c_2 \geq \cdots \geq c_n$.
(iii) Denote by Ω the set of integer vectors $\mathbf{x} = (x_1, \dots, x_n)$ satisfying $x_i \geq 0$, and $\sum_{i=1}^{n} x_i \leq \delta$. For any vector $\mathbf{x} \in \Omega$: Let $d \geq 1$ be the maximal integer i for which $x_i \neq 0$. Find a $(1 + \varepsilon)$-approximation to the optimal (fractional) solution of the following linear program.

$$(LP') \quad \text{minimize} \quad \sum_{i=d+1}^{n} c_i z_i$$

$$\text{subject to :} \quad \sum_{i=d+1}^{n} a_{ij} z_i \geq b_j - \sum_{i=1}^{n} a_{ij} x_i \quad \text{for } j = 1, \dots, R \qquad (8.5)$$

$$z_i \geq 0, \quad \text{for } i = d+1, \dots, n$$

The constraints (8.5) reflect the fact that we need to fill in each dimension j at least the capacity $b_j - \sum_{i=1}^{n} a_{ij} x_i$, once we obtained the vector $\mathbf{x}$.

Let $\hat{z}_i$, $d + 1 \leq i \leq n$, be a $(1 + \varepsilon)$-approximate solution for LP'. We take $\lceil \hat{z}_i \rceil$ as the integral solution. Denote by $C_{MDC}(\mathbf{x}) = \sum_{i=d+1}^{n} c_i \lceil \hat{z}_i \rceil$ the value obtained from the rounded solution, and let $c(\mathbf{x}) = \sum_{i=1}^{n} c_i x_i$.

(iv) Select the vector $\mathbf{x}^*$ for which $C_{MDC_\varepsilon}(\mathbf{x}^*) = \min_{\mathbf{x}}(c(\mathbf{x}) + C_{MDC}(\mathbf{x}))$.

We now show that MDC_ε is a PTAS for CIP_∞. Let C_o be the cost of an optimal integral solution for the CIP_∞ instance.

Theorem 8.7 *MDC_ε is a PTAS for CIP_∞ which satisfies the following. (i) If $C_o \neq 0, \infty$ then $C_{MDC_\varepsilon}/C_o < 1 + \varepsilon$. (ii) The running time of algorithm MDC_ε is $O(n^{\lceil R/\varepsilon \rceil} \cdot \frac{1}{\varepsilon^2} \log C)$, where $C = \max_{1 \leq i \leq n} c_i$ is the maximal cost of any item, and its space complexity is $O(n)$.*

The next lemma is required to prove the theorem.

Lemma 8.7 *For any $\varepsilon > 0$, a $(1 + \varepsilon)$-approximation to the optimal solution for LP' can be found in $O(1/\varepsilon^2 R \log(C \cdot R))$ steps.*

Proof. For a system of inequalities as given in LP', there is a solution in which at most R variables get non-zero values. This follows from the fact that the number of nontrivial constraints is R. Hence, it suffices to solve LP' for the $\binom{n-d}{R}$ possible subsets of R variables, out of $(z_{d+1}, \ldots, z_n)$. This can be done in polynomial time since R is fixed. Now, for each subset of R variables, we have an instance of the *fractional covering problem*, for which we can use a fast approximation scheme [30] to obtain a $(1+\varepsilon)$-approximate solution. ∎

Proof of Theorem 8.7. For showing (*i*), assume that the optimal (integral) solution for the CIP_∞ instance is obtained by the vector $\mathbf{y} = (y_1, \ldots, y_n)$. If $\sum_{i=1}^n y_i \leq \delta$ then $C_{MDC_\varepsilon} = C_o$, since in this case $\mathbf{y}$ is a valid solution, and $\mathbf{y} \in \Omega$, therefore, in some iteration MDC_ε will examine $\mathbf{y}$. Suppose that $\sum_{i=1}^n y_i > \delta$, then we define the vector $\mathbf{x} = (y_1, \ldots, y_{d-1}, x_d, 0, \ldots, 0)$, such that $y_1 + \cdots + y_{d-1} + x_d = \delta$. (Note that $x_d \neq 0$.) Let $\tilde{C}_o(\mathbf{x}) = \sum_{i=d+1}^n c_i \hat{z}_i$ be the approximate fractional solution for LP'. We have that $\mathbf{x} \in \Omega$, therefore

$$C_{MDC}(\mathbf{x}) - \tilde{C}_o(\mathbf{x}) \leq Rc_d, \tag{8.6}$$

Let $C_o(\mathbf{x})$ be the optimal fractional solution for LP' with the vector $\mathbf{x}$. Note that C_o, the optimal (integral) solution for CIP_∞, satisfies

$$C_o > c(\mathbf{x}) + C_o(\mathbf{x}), \tag{8.7}$$

since $C_o(\mathbf{x})$ is a lower bound for the cost incurred by the integral values $y_{d+1}, \ldots, y_n$. In addition,

$$c(\mathbf{x}) + C_{MDC}(\mathbf{x}) \geq C_{MDC_\varepsilon}. \tag{8.8}$$

Hence, we get that

$$
\begin{aligned}
\frac{C_o}{C_{MDC_\varepsilon}} &\geq \frac{c(\mathbf{x}) + C_o(\mathbf{x})}{c(\mathbf{x}) + C_{MDC}(\mathbf{x})} > 1 - \frac{C_{MDC}(\mathbf{x}) - C_o(\mathbf{x})}{c(\mathbf{x}) + C_{MDC}(\mathbf{x}) - C_o(\mathbf{x})} \\
&\geq 1 - \frac{C_{MDC}(\mathbf{x}) - \tilde{C}_o(\mathbf{x})(1 - \varepsilon)}{c(\mathbf{x}) + C_{MDC}(\mathbf{x}) - \tilde{C}_o(\mathbf{x})} \\
&\geq (1 - \varepsilon)\left(1 - \frac{C_{MDC}(\mathbf{x}) - \tilde{C}_o(\mathbf{x})}{c(\mathbf{x}) + C_{MDC}(\mathbf{x}) - \tilde{C}_o(\mathbf{x})}\right) \\
&\geq (1 - \varepsilon)\left(1 - \frac{C_{MDC}(\mathbf{x}) - \tilde{C}_o(\mathbf{x})}{\delta c_d + C_{MDC}(\mathbf{x}) - \tilde{C}_o(\mathbf{x})}\right)
\end{aligned}
$$

The first inequality follows from (8.7) and (8.8), and the third inequality follows from the fact that $\tilde{C}_o(\mathbf{x})(1 - \varepsilon) \leq C_o(\mathbf{x}) \leq \tilde{C}_o(\mathbf{x})$. The last inequality follows from the fact that $c(\mathbf{x}) \geq \delta c_d$.

Using (8.6), we get that $\frac{C_o}{C_{MDC_\varepsilon}} \geq (1 - \varepsilon)1 - Rc_d/(\delta c_d + Rc_d) \geq (1 - \varepsilon)^2$. Taking in the scheme $\tilde{\varepsilon} = \varepsilon/2$, we get the statement in (*i*).

Next, we show (*ii*). Note that $|\Omega| = O(n^\delta)$ since the number of possible choices of n nonnegative integers, whose sum is at most δ is bounded by $\binom{n+\delta}{\delta}$. Now, given a vector $\mathbf{x} \in \Omega$, we can compute $C_{MDC}(\mathbf{x})$ in $O(n^R)$ steps since at most R variables out of $z_{d+1}, \ldots, z_n$ can have non-zero values. Multiplying by the complexity of the FPTAS for fractional covering, as given in Lemma 8.7, we get the statement of the theorem. ∎

Enumeration is combined with LP rounding also in the PTAS of Caprara et al. [31] for the *knapsack problem with cardinalities constraints*, and in a PTAS for the *multiple knapsack* problem due to Chekuri and Khanna [32], among others. The scheme of [31] is based on the scheme of Frieze and Clarke [28], with the running time improved by factor of n, the number of items. The scheme of [28] is the basis also for PTASs for other variants of the knapsack problem. (A comprehensive survey is given in [33]; see also in [34].)

8.4 Approximation Schemes for Geometric Problems

In this section we present approximation techniques that are specialized for geometric optimization problems. For a complete description of these techniques we refer the reader to the survey by Arora [35], Chapter 11 in [23], and Chapters 8 and 9.3.3 in [10]. A typical input for a geometric problem is a set of elements in the space (such as points in the plane); the goal is to connect or pack these elements in a way that minimizes the resources used (e.g., total length of connecting lines, or total number of covering objects).

8.4.1 Randomized Dissection

We present the techniques used in the PTAS of Arora [36] for the Euclidean traveling salesman problem (TSP) in the following. In the classical TSP problem, given are nonnegative edge weights for the complete graph K_n, and the goal is to find a tour of minimum cost, where a tour refers to a cycle of length n. In other words, the goal is to find an ordering of the nodes such that the total cost of the edges along the path visiting all nodes according to this ordering is minimal. In general, TSP is NP-hard in the strong sense, and it cannot be approximated within any multiplicative factor $c > 1$, unless P = NP. The PTAS of Arora considers the relaxed problem of *Euclidean* TSP. The input is a set of n points in $\Re^d$, and the edge weights are the Euclidean (ℓ_2) distances between them.

The idea of the PTAS is to dissect the plane into squares, and to look (using dynamic programming) for a tour that crosses the resulting grid lines only at specific points, denoted *portals*. The parameter ε of the PTAS determines the depth of the recursive dissection, as well as the density of the portals. A smaller ε results in more portals and a finer dissection, which lead to a less restricted tour and a larger dynamic programming instance. Randomization is used to determine an initial shift of the grid lines.

A dissection of a square is a recursive partitioning into squares. It can be viewed as a tree of squares whose root is the square we started with. Each square in the tree is partitioned into four equal squares, which are its children. The leaves are squares of a small sidelength—determined by the parameter ε of the PTAS.

The location of the grid lines is determined *randomly* as follows. Given a set of n points in $\Re^2$, enclose the points in a minimum bounding square. Let ℓ be the side of this square. Let $p \in \Re^2$ be the lower left endpoint of the bounding box. Enclose the bounding box inside a larger square, denoted the *enclosing box* of sidelength $L = 2\ell$, and position the enclosing box such that p has distance a from the left edge and b from the lower edge, where $a, b \leq \ell$ are chosen randomly. The randomized dissection is the dissection of this enclosing box. Note that the randomness is used only to determine the placement of the enclosing box (and its accompanying dissection).

We now describe the PTAS in [36] for the Euclidean TSP problem, which uses the earlier randomized dissection. Formally, for every $\varepsilon > 0$, this PTAS finds a $(1 + \varepsilon)$-approximation to Euclidean TSP.

First, perform randomized dissection to the bounding box of the n points. Recall that L is the side of the enclosing box. The recursive procedure of subdividing the squares stops when the side lengths of the squares becomes less than $L\varepsilon/8n$, or when each square at the last level contains at most one point. We may assume (by scaling) that L is a power of 2 and that the sides of squares at the last level are unit length. Thus, at most $\log L$ iterations are required, and $L \leq 8n/\varepsilon$. When there is more than one point in a unit square, consolidate them into one new "bigger" point. Any tour for the resulting set of points can be augmented to a tour for the original set of points with an increase in length bounded by $\sqrt{2}nL\varepsilon/8n$, which is negligible, since $L \leq OPT/2$. Henceforth, we shall assume that there is at most one point per unit square.

The *level* of a square in the dissection is its depth in the recursive dissection tree; the root square has level 0. We also assign a level from 0 to $\log(L-1)$ to each horizontal and vertical grid line that participates in the dissection. The horizontal (resp., vertical) line that divides the enclosing box into two has level 0. Similarly, the 2^i horizontal and 2^i vertical lines that divide the level i squares into level $i+1$ squares have level i. The following property of a randomized dissection is used: Any fixed vertical grid line that intersects the bounding box of the instance has probability $\frac{2^i}{\ell} = \frac{2^{i+1}}{L}$ to be a line at level i.

Next, the location of the portals is determined. Let $m = \frac{1}{\varepsilon} \log L$. The parameter m is the *portal parameter* which determines the density of the points the path can pass through. A level i line has $2^{i+1}m$ equally spaced portals. In addition, we also refer to the corners of each square as a portal. Since a level i line has 2^{i+1} level $i+1$ squares touching it, it follows that each side of the square has at most $m+2$ portals (m regular portals plus the 2 corners), and a total of at most $4m+4$ portals on its boundary. A *portal-respecting tour* is one that, whenever it crosses a grid line, does so at a portal.

Finally, dynamic programming is used to find the optimum portal-respecting tour in time $2^{O(m)}L \log L$. Since $m = O(\log n/\varepsilon)$, we get a total running time of $n^{O(1/\varepsilon)}$. The dynamic programming as well as the complete analysis of bounding the PTAS error and the time complexity are given in [35].

Note that since the PTAS uses randomization, the error of the PTAS is a random variable. Formally, let *OPT* denote the cost of the optimum salesman tour and $OPT_{a,b,m}$ denote the cost of the best portal-respecting tour when the portal parameter is m and the random shifts are a, b, then,

Theorem 8.8 *The expectation (over the choices of a, b) of $OPT_{a,b,m} - OPT$ is at most $2 \log L/m OPT$, where L is the sidelength of the enclosing box.*

As mentioned in the survey of Arora [35], this method of dissection can be used to develop PTASs for other geometric optimization problems such as minimum Steiner tree, facility location with capacities and demands, and Euclidean min-cost k-connected subgraph.

Another class of geometric optimization problem is the class of *clustering* problems, such as metric max-cut and k-median. In recent research on clustering problems, a core idea in the design of approximation schemes is to use random sampling of data points from a biased distribution, which depends on the pairwise distances. This technique is used, for example, in the PTAS of Fernandez de la Vega and Kenyon for metric max-cut [37], and in the work of Indyk on metric 2-clustering [38]. For more details on the technique and its applications, we refer the reader to [39].

8.4.2 Shifted Plane Partitions

The *shifting* technique that is applied to geometric problems is based on selecting the best solution over a (polynomial size) set of feasible solutions. Each candidate feasible solution is obtained using a divide-and-conquer approach, in which the plane is partitioned into disjoint areas (strips). The technique can be applied to geometric problems such as square packing or covering with disks, which arise in VLSI design, image processing and many other important areas. A common goal in these problems is to cover or pack elements (e.g., points in the plane) into a minimal number of objects (e.g., squares of given size).

Recall that each candidate solution is obtained by using divide-and-conquer approach, in which the plane is partitioned into strips. A solution for the original problem is formed by taking the union of the solutions for these strips. Consecutive solutions refer to consecutive partitions of the plane into strips, which differ from each other by *shifting the partitioning bars*, using the shifting parameter. The smaller the shifting parameter, the larger is the number of candidate solutions to be considered, and the better resulting approximation.

We illustrate the shifting technique for the problem of covering n points in the 2-dimensional plane. The complete analysis is given in [10,40]. Assume that the n points are enclosed in an area I. The goal is to cover these points with a minimal number of disks of diameter D. Denote by ℓ the shifting parameter. The area I is divided into vertical strips of width D. Each set of ℓ consecutive strips are grouped together to form strips of width ℓD. Note that there are ℓ different ways to determine this grouping—and they can derive from each other by shifting the partitioning bars to the right over distance D. Denote the ℓ distinct partitions obtained this way by $S_1, S_2, \ldots, S_\ell$.

Let $\mathcal{A}$ be an algorithm to solve the covering problem on strips of width at most ℓD. The algorithm $\mathcal{A}$ can be used to generate a solution for a given partition S_j. We apply $\mathcal{A}$ to each strip in S_j and then union the sets of disks used. The *shift algorithm*, s_A, defined for a given $\mathcal{A}$, uses $\mathcal{A}$ to solve the problem for the ℓ possible partitions and selects the solution that requires minimum number of disks.

The following lemma gives the performance ratio of s_A (denoted r_{s_A}) as function of ℓ and the performance ratio of $\mathcal{A}$ (denoted r_A):

Lemma 8.8

$$r_{s_A} \leq r_A \left(1 + \frac{1}{\ell}\right).$$

The algorithm $\mathcal{A}$ may itself be derived from an application of the shifting technique. In our example, in order to solve the covering problem on a strip of width ℓD, the strip is cut into squares of size $\ell D \times \ell D$, for which an optimal solution can be found by exhaustive search.

We note that the above-mentioned shifting technique can be used to derive PTASs for several other problems, including minimum vertex-cover and maximum independent-set in planar graphs [41]. The idea is that a planar graph can be decomposed into components of bounded outer-planarity. The solution for each component can be found using dynamic programming. The shifting idea is to remove one "layer" from the graph in each iteration. This removal guarantees that the number of cross-cluster edges is small, so by considering the union of the local cluster solutions one can get a good approximation for the original problem.

8.5 Reapproximation Schemes

In previous sections, we have seen techniques for developing approximation schemes to solve NP-hard optimization problems. Many real-life applications involve systems that change dynamically over time [42–44]. Thus, throughout the continuous operation of such a system, it is required to compute solutions for new problem instances, derived from previous instances. Since the transition from one solution to another incurs some cost, a natural goal is to have the solution for the new instance close to the original one (under a certain distance measure). In this section we consider *reapproximation schemes* − developed for such scenarios.

8.5.1 Reoptimization Model

We use a model for combinatorial reoptimization introduced in [42] (see also Chapter 25 in this book for a detailed discussion of reoptimization) in the following. Let Π be an optimization problem, and I_0 an input for Π. Denote by $\mathcal{C}_{I_0} = \{C_{I_0}^1, C_{I_0}^2, \ldots\}$ the set of configurations corresponding to the solution space of Π for the input I_0. Each configuration $C_{I_0}^j \in \mathcal{C}_{I_0}$ has some value $val(C_{I_0}^j)$. In the reoptimization problem, $R(\Pi)$, we are given a configuration $C_{I_0}^j \in \mathcal{C}_{I_0}$ of an initial instance I_0, and a new instance I derived from I_0 by admissible operations, for example, addition or removal of elements, changes in element parameters, and so on. For any element $i \in I$ and configuration $C_I^k \in \mathcal{C}_I$, we are given the *transition cost* of i when moving from the initial configuration $C_{I_0}^j$ to the feasible configuration C_I^k of the new instance. Denote this transition cost by $\delta(i, C_{I_0}^j, C_I^k)$. Practically, the transition cost of i is not given as a function of two configurations, but rather as a function of the state of i in the initial configuration and its possible states in any new configuration. This keeps the input description more compact. The primary goal is to find an optimal solution for I. Among all configurations having an optimal $val(C_I^k)$ value, we seek a configuration C_I^* for which the total transition cost, given by $\sum_{i \in I} \delta(i, C_{I_0}^j, C_I^*)$ is minimized.

Definition 8.4 *An algorithm $\mathcal{A}$ is an (r, ρ)-reapproximation algorithm for $R(\Pi)$, for $\rho, r \geq 1$, if it achieves a ρ-approximation for the optimization problem Π, while incurring a transition cost that is at most r times the minimum cost required for solving Π optimally on I.*

The concept of approximation scheme, as given in Definition 8.1, is extended for reoptimization problems as follows.

Definition 8.5 *A polynomial time reapproximation scheme (PTRS) for $R(\Pi)$ is an algorithm that, given the inputs I_0 and I for $R(\Pi)$ and parameters $\varepsilon_1, \varepsilon_2 \geq 0$, yields a $(1 + \varepsilon_1, 1 + \varepsilon_2)$-reapproximation for $R(\Pi)$, in time polynomial in $|I_0|$ and $|I|$.*

8.5.2 Combining Enumeration and Minimum Cost Matching

As shown earlier, applying enumeration can lead to a PTAS for a given instance of optimization problem. However, developing a PTRS involves both finding an approximate solution for the underlying optimization problem Π, and minimizing the transition cost to this solution. A natural approach, for example, in solving packing and scheduling reoptimization problems, is to enumerate over a polynomial size set of solutions for I, while optimizing on the total transition cost to each solution, using the solution for I_0. This can be done by applying matching techniques. We demonstrate the use of this approach in the PTRS of Mordecai [45] for $R(\Pi_{ID})$, where Π_{ID} is makespan minimization on identical machines (see Section 8.2).

Let $I_0 = (\mathcal{M}_0, \mathcal{J}_0)$ be an instance of Π_{ID}, where $m_0 = |\mathcal{M}_0|$ and $n_0 = |\mathcal{J}_0|$ are the number of machines and the number of jobs, respectively. Also, let $\sigma_0 : \mathcal{J}_0 \to \mathcal{M}_0$ be an initial assignment of jobs to the machines in I_0. We denote by $I = (\mathcal{M}, \mathcal{J})$ a new instance derived from I_0 by admissible operations, for example, addition or removal of jobs and/or machines. Let $m = |\mathcal{M}|$ and $n = |\mathcal{J}|$ be the number of machines and the number of jobs in I, respectively. For any job $j \in \mathcal{J}$ and a feasible assignment $\sigma : \mathcal{J} \to \mathcal{M}$, we are given the transition cost of j when moving from the initial assignment σ_0 to σ. We denote this transition cost by $c_{\sigma_0}(j, \sigma)$. The goal is to find an optimal assignment for I, for which the total transition cost, given by $\sum_{j \in \mathcal{J}} c_{\sigma_0}(j, \sigma)$, is minimized. We assume that $c_{\sigma_0}(j, \sigma) \in \{0, 1\}$ in the following. Specifically, $c_{\sigma_0}(j, \sigma) = 0$ if $\sigma(j) = \sigma_0(j)$ and $c_{\sigma_0}(j, \sigma) = 1$ otherwise. Thus, the problem can be viewed as makespan minimization using a minimum number of job migrations.

The reapproximation scheme uses a *relaxed* packing of items in bins, where the items correspond to jobs, and the bins represent the machines.

Definition 8.6 *Given a set of bins, each of capacity $K > 0$, and a set of items packed in the bins, we say that the packing is ε-relaxed, for some $\varepsilon > 0$, if the total size of items assigned to each bin is at most $(1 + \varepsilon)K$.*

The PTRS of [45] accepts as input the instances I_0 and I, the initial assignment of jobs to the machines, σ_0, and an error bound $\varepsilon > 0$. The scheme proceeds as follows. Let $C_{max}^*(I)$ denote the MM for the instance I. Initially, the scheme calls as a subroutine a PTAS [10,46] for the new instance I, to obtain a solution of makespan $T \leq (1 + \varepsilon_0)C_{max}^*(I)$, for some $\varepsilon_0 < \varepsilon$. Then, the instance I is split into *large* and *small* jobs. The processing times of the large jobs are rounded down. This enables to define a polynomial size collection of feasible configurations of the rounded jobs on the machines in I, such that the load on each machine does not exceed T. The scheme then iterates over this collection to find a set of configurations that minimizes the transition cost from the initial assignment. It is shown in [45] that once a cost effective set of configurations is selected, the rounded jobs can be assigned to the machines with their original processing times, along with the remaining small jobs, such that the resulting makespan is at most $(1 + \varepsilon)C_{max}^*(I)$.

ALGORITHM 8.1 $\mathcal{A}_{ID}(I_0, I, \sigma_0, \varepsilon)$

1. Use a PTAS for makespan minimization on identical machines to find $T \le (1 + \varepsilon_0)C^*_{max}(I)$, where $\varepsilon_0 = \frac{\varepsilon}{4}$.

2. Associate each machine with a bin of unit capacity. Let $\alpha_j = \frac{p_j}{T}$ for all $j \in \mathcal{J}$. Consider job j as an item of size $\alpha_j \in (0, 1]$.

3. An item $j \in \mathcal{J}$ is *small* if it has a size at most ε_0; otherwise, item j is *large*.

4. Round down the sizes of the large items to the nearest multiple of ε_0^2. Denote the rounded sizes $\bar{\alpha}_j$, for every large item j.

5. For any feasible configuration $C = \{C^1, \ldots, C^m\}$ of the large items on the machines in I do:

    i. Let $\widehat{m} = \max\{m_0, m\}$. Construct a complete bipartite graph $G = (U, V, E)$, in which $V = \{1, \ldots, \widehat{m}\}$, and $U = \{C^1, C^2, \ldots, C^{\widehat{m}}\}$. Each vertex $i \in V$ corresponds to the initial configuration of machine i, given by $C_0^i = \{j \in \mathcal{J}_0 : \sigma_0(j) = i\}$, for $1 \le i \le m_0$; if $m_0 < m$, set $C_0^i = \emptyset$ for $m_0 + 1 \le i \le \widehat{m}$. If $m_0 > m$, set $C^i = \emptyset$ for all $m + 1 \le i \le \widehat{m}$. Define costs on the edges, $c(i, C^k)$, for all $1 \le i, k \le \widehat{m}$, as follows.

        a. Initialize $c(i, C^k) = 0$.

        b. Add to $c(i, C^k)$ the cost of the large items in C^k which are not contained in the initial configuration C_0^i.

        c. If $C^k = \emptyset$ add to $c(i, C^k)$ the total cost of the small items in C_0^i;
           else
           add to C^k the small items contained in C_0^i; then, omit the largest small items until the total size of items in C^k does not exceed 1. Add to $c(i, C^k)$ the cost of the omitted items.

    ii. Find a minimum cost perfect matching in the bipartite graph.

    iii. Add to the solution the omitted small items and the new small items (which did not appear in I_0) using *relaxed* First-Fit (see the following).

6. Choose the solution of minimum cost, and return the corresponding schedule of the jobs in I on the machines.

We give a detailed description of the scheme, $\mathcal{A}_{ID}$ in the following. For the case where $m < m_0$, we assume w.l.o.g. that the omitted machines are $m + 1, m + 2, \ldots, m_0$.

Theorem 8.9 *For any $\varepsilon > 0$, $\mathcal{A}_{ID}$ yields in polynomial time a $(1, 1 + \varepsilon)$-reapproximation for $R(\Pi_{ID})$.*

We prove the theorem using the next results.

Observation 8.1 *Let $I = (\mathcal{M}, \mathcal{J})$ be an instance of Π_{ID}, for which the MM is $C^*_{max} = C^*_{max}(I)$. For some $T \ge C^*_{max}$, let $\alpha_j = \frac{p_j}{T}, \forall j \in \mathcal{J}$; then $\sum_{j \in \mathcal{J}} \alpha_j \le m$.*

Proof. We note that the MM satisfies $C^*_{max} \ge \frac{\sum_{j \in \mathcal{J}} p_j}{m}$. Since $T \ge C^*_{max}$, we have that $\alpha_j = \frac{p_j}{T} \le \frac{p_j}{C^*_{max}}$ for all $j \in \mathcal{J}$, and the claim follows. ∎

Lemma 8.9 *Let $\widetilde{C}$ be a feasible assignment of rounded large items to the m bins (i.e., machines), given by the configuration $C = \{C^1, \ldots, C^m\}$, to which we add in each bin the small items that were not omitted in Step 5 of $\mathcal{A}_{ID}$. Then $\widetilde{C}$ can be expanded in polynomial time to an ε_0-relaxed packing of all items in I.*

Proof. Let $C = \{C^1, C^2, \ldots, C^m\}$ be a feasible configuration of the large rounded items, and let $S^1, S^2, \ldots, S^m$ be the subsets of small items added in Step 5 to the bins, to form $\widetilde{C}$. Then $\widetilde{C}$ yields a feasible packing, that is, for all $1 \le i \le m$,

$$\sum_{j \in C^i} \bar{\alpha}_j + \sum_{j \in S^i} \alpha_j \le 1.$$

Note that the number of large items in bin i is bounded by $\lfloor 1/\varepsilon_0 \rfloor$, since $\bar{\alpha}_j \geq \varepsilon_0$ for all $j \in C^i$. Also, since $\alpha_j - \bar{\alpha}_j \leq \varepsilon_0^2$, for any large item j, we have that

$$\sum_{j \in C^i \cup S^i} \alpha_j \leq \sum_{j \in C^i} (\bar{\alpha}_j + \varepsilon_0^2) + \sum_{j \in S^i} \alpha_j \leq \sum_{j \in C^i} \bar{\alpha}_j + \left\lfloor \frac{1}{\varepsilon_0} \right\rfloor \varepsilon_0^2 + \sum_{j \in S^i} \alpha_j \leq 1 + \varepsilon_0.$$

Hence, the packing of $\widetilde{C}$ is ε_0-relaxed. Now, we show that the packing remains ε_0-relaxed after we add the small items that were omitted in Step 5, using first-fit. Consider the next unpacked item. We apply first-fit in the following *relaxed* manner. Starting from bin 1, we seek the first bin in which the item can be added, such that the size of the items packed in this bin is at most $1 + \varepsilon_0$. Let r_i be the total size of items packed in bin i after adding the small items. Assume that, after applying first-fit, some small items remain unpacked (i.e., none of the bins can accommodate these items). Renumber the items such that the unpacked items are $\{\ell, \ldots, n\}$. Then, we have that $(1 + \varepsilon_0) - r_i < \alpha_j$ for all $\ell \leq j \leq n$ and $i = 1, \ldots, m$. It follows, that

$$(1 + \varepsilon_0)m - \sum_{i=1}^{m} r_i < m\alpha_j \qquad \forall \ell \leq j \leq n. \tag{8.9}$$

By the definition of r_i, and since all items up to $j = \ell - 1$ are packed,

$$\sum_{i=1}^{m} r_i = \sum_{j=1}^{\ell-1} \alpha_j \tag{8.10}$$

By Observation 8.1,

$$\sum_{j=1}^{\ell-1} \alpha_j \leq m - \sum_{j=\ell}^{n} \alpha_j.$$

Then, from (8.10),

$$m - \sum_{j=\ell}^{n} \alpha_j \geq \sum_{i=1}^{m} r_i. \tag{8.11}$$

By (8.9) and (8.11), it follows that for $\ell \leq j \leq n$,

$$m(1 + \varepsilon_0) - \left(m - \sum_{j=\ell}^{n} \alpha_j \right) \leq m(1 + \varepsilon_0) - \sum_{i=1}^{m} r_i < m\alpha_j,$$

or,

$$m\varepsilon_0 + \sum_{j=\ell}^{n} \alpha_j < m\alpha_j. \tag{8.12}$$

From (8.12), and since $\alpha_j \leq \varepsilon_0$ for $j = \ell, \ldots, n$, we have that $m\varepsilon_0 + \sum_{j=\ell}^{n} \alpha_j < m\varepsilon_0$; thus, $\sum_{j=\ell}^{n} \alpha_j < 0$. A contradiction, since $\alpha_j \geq 0$ for all j.

   Hence, the above-mentioned relaxed implementation of First-Fit packs all the remaining small items and yields a relaxed-packing of the original instance. ∎

Let OPT be an optimal solution for the instance I of Π_{ID}, and let $C = \{C^1, \ldots, C^m\}$ be the configuration of large items derived from OPT. Let ALG_C be the solution the scheme finds in Step 5, for configuration C. Denote by $Cost(OPT)$ and $Cost(ALG_C)$ the transition costs incurred by OPT and ALG_C, respectively.

Lemma 8.10 *The cost incurred by ALG_C satisfies $Cost(ALG_C) \leq Cost(OPT)$.*

Proof. Assume that $Cost(OPT) < Cost(ALG_C)$. We note that the cost for the large items is the same in the two solutions; therefore, any difference is due to small items. In the scheme, the cost charged for these items is either the cost of omitting a small item from its original bin (and moving it to another bin), or the cost of adding to a bin a new small item (which did not appear in I_0). The cost for new small items has to be the same in both solutions. Thus, from the unit transition costs, we conclude that the number of omitted small items in OPT is smaller than in ALG_C. In other words, the number of small items that are packed in their original bin in OPT is greater than this number in ALG_C. Hence, there exists a bin $1 \leq i \leq m$ for which this holds. Consider the small items packed in bin i in each solution. Denote by $\beta_1, \ldots, \beta_s$ the sizes of the small items packed in bin i, in OPT and in ALG_C. Also, let $\alpha_1^A, \ldots, \alpha_k^A$ and $\alpha_1^O, \ldots, \alpha_\ell^O$ denote the sizes of two disjoint sets of small items packed in bin i in ALG_C and in OPT, respectively. Assume w.l.o.g. that

$$\alpha_1^O = min\{\alpha_1^O, \ldots, \alpha_\ell^O\}. \tag{8.13}$$

$\mathcal{A}_{\text{ID}}$ chooses to omit from bin i the largest small items first; thus,

$$\alpha_1^O \geq max\{\alpha_1^A, \ldots, \alpha_k^A\}. \tag{8.14}$$

As $C_{max}^* \leq T$, we have that

$$\sum_{j \in C^i} \alpha_j + \sum_{j=1}^{s} \beta_j + \sum_{j=1}^{\ell} \alpha_j^O \leq \frac{C_{max}^*}{T} \leq 1 \tag{8.15}$$

Also, it holds that

$$\sum_{j \in C^i} \alpha_j + \sum_{j=1}^{s} \beta_j + \sum_{j=1}^{k} \alpha_j^A \leq 1 + \varepsilon_0.$$

On the other hand, the scheme chose to omit the item of size α_1^O. Hence,

$$\sum_{j \in C^i} \alpha_j + \sum_{j=1}^{s} \beta_j + \sum_{j=1}^{k} \alpha_j^A + \alpha_1^O > 1 + \varepsilon_0. \tag{8.16}$$

By the above-mentioned discussion, we also have that

$$\sum_{j \in C^i} \alpha_j + \sum_{j=1}^{s} \beta_j + \sum_{j=1}^{k} \alpha_j^A + \alpha_1^O \leq \sum_{j \in C^i} \alpha_j + \sum_{j=1}^{s} \beta_j + k\alpha_1^O + \alpha_1^O$$

$$\leq \sum_{j \in C^i} \alpha_j + \sum_{j=1}^{s} \beta_j + \ell\alpha_1^O$$

$$\leq \sum_{j \in C^i} \alpha_j + \sum_{j=1}^{s} \beta_j + \sum_{j=1}^{\ell} \alpha_j^O$$

$$\leq 1$$

The first and the third inequalities follow from (8.13) and (8.14); the second inequality holds since $k < \ell$, and the last inequality follows from (8.15). This yields a contradiction to (8.16), implying the statement of the lemma.   ∎

Lemma 8.11 [47] *Let $G = (V, U, E)$ be a bipartite graph with $|V| = |U| = n$, and let $c : E \to \mathbf{R}$ be a cost function on the edges. A minimum cost perfect matching, that is, a perfect matching $M \subseteq E$ for which $\sum_{e \in M} c(e)$ is minimized, can be found in $O(n^3)$ time.*

Proof of Theorem 8.9. Let $C = \{C^1, \ldots, C^m\}$ be the configuration of large items derived from an optimal solution, *OPT*. Let ALG_C be the packing obtained for C in Step 5 of the scheme. By Lemma 8.10, we have

$$Cost\,(ALG_C) \leq Cost\,(OPT).$$

Let *ALG* be the solution the scheme outputs. Since

$$Cost(ALG) = min \left\{ Cost(ALG_C) : C \text{ is a valid configuration} \right\},$$

we have that $Cost(ALG) \leq Cost(OPT)$. Now, we note that if *ALG* is transformed into a schedule of the jobs on the machines (using the original processing times), we get a solution of makespan at most $(1 + \varepsilon_0)T$. Since $T \leq (1 + \varepsilon_0)C^*_{max}$, we have a $(1 + \varepsilon_0)^2 \leq 1 + \varepsilon$ approximation to the MM.

It is easy to verify that the scheme has polynomial running time, using Lemma 8.11 and the polynomial number of valid configurations for the large items. We conclude that $\mathcal{A}_{ID}$ is a PTRS for $R(\Pi_{ID})$.   ∎

8.5.3 The Layered Graph Approach

In the *layered graph* approach, the stages in the solution process for a given problem instance are represented through the layers of a directed graph, whose size is polynomial in the input size. The nodes in the graph represent *states* of the algorithm, and the (weighted) edges reflect the cost of moving between states. To find a solution of small cost, we seek the lightest path from a given "initial" state and a "success" state. Hochbaum and Shmoys [48] used a layered graph to develop a PTAS for makespan minimization on *uniform* machines. Denote the problem by Π_{UN}. The scheme is based on solving an instance of the dual BP problem. We demonstrate the use of the approach in [45] for developing a PTRS for $R(\Pi_{UN})$.

The scheme of [48] constructs a layered directed graph, with two nodes designated "initial" and "success," such that there exists a path from "initial" to "success" in the graph if and only if there is a schedule of makespan at most $1 + \varepsilon$ times the MM, for a given $\varepsilon > 0$. This path defines the configuration of *medium* and *large* items in the dual BP problem on each machine; it also guarantees that the *small* items can be packed. For the reoptimization version of the problem, $R(\Pi_{UN})$, we assume as before that $c_{\sigma_0}(j, \sigma) \in \{0, 1\}$, for all $j \in \mathcal{J}$.

The PTRS of [45] adds suitable costs on the edges of the layered graph. It is shown that a feasible solution of optimal cost can be obtained by finding the lightest path from "initial" to "success" in the layered graph. The scheme also guarantees that if there is a path from "initial" to "success" in the graph, there is sufficient space in the bins to add the remaining items (e.g., using first-fit), in the bins in which they are considered small. This way, the small items can be packed in their original bins (as long as we do not exceed the capacity), and the incurred cost is due to the packing of other items.

We start with an overview of the scheme of [48]. Given a set of n jobs and m uniform machines, with the speeds $\{s_1, \ldots, s_m\}$, assume w.l.o.g. that $s_1 = \max_{1 \leq i \leq m} s_i$. Each machine $1 \leq i \leq m$ is associated with a bin of size s_i. Normalize the bin sizes by s_1; let $\{1 = \tilde{s}_1, \tilde{s}_2, \ldots, \tilde{s}_m\}$ be the resulting sizes. Consider the jobs as items whose sizes are in $(0, 1]$ (by scaling the job processing times, using an upper bound on the MM).

Round down the item sizes in $\left(\varepsilon^{k+1}, \varepsilon^k\right]$ to the nearest multiple of ε^{k+2}, for some integer $k \geq 0$. For a bin of size $\tilde{s}_i \in \left(\varepsilon^{k+1}, \varepsilon^k\right]$ define:

- Items of sizes in $\left(\varepsilon^{k+1}, \varepsilon^k\right]$ are *large* for the bin.
- Items of sizes in $\left(\varepsilon^{k+2}, \varepsilon^{k+1}\right]$ are *medium* for the bin.
- Items of sizes less than or equal to ε^{k+2} are *small* for the bin.

For convenience, the interval $\left(\varepsilon^{k+1}, \varepsilon^k\right]$ is referred as interval k. For items in interval k, a bin is *large* if it is in interval k, *huge* if it is in interval $k - 1$, and *enormous* if it is in intervals $0, \ldots, k - 2$.

A directed layered graph, $G_L = (V_L, E_L)$, is then constructed, where each node is labeled with a state vector describing the remaining items to be packed as *large* or *medium* items. The graph consists of stages, where stage k specifies the large and medium pack of bins in the interval $\left(\varepsilon^{k+1}, \varepsilon^k\right]$. Each layer within a stage corresponds to packing a bin in the corresponding interval. Both the bins within a stage and the stages are sorted in decreasing order of bin size. The state vector associated with each node is of the form $(L; M; V_1, V_2, V)$, where L and M are vectors, each describing a distribution of items in the subintervals of $\left(\varepsilon^{k+1}, \varepsilon^k\right]$ and $\left(\varepsilon^{k+2}, \varepsilon^{k+1}\right]$, respectively. There are two nodes designated "initial" and "success," such that "initial" is connected to the initial state vectors of the first stage, and every final state vector of the final stage is connected to "success." A path from "initial" to "success" in G_L specifies a packing of the rounded medium and large items for every bin.

We note that after packing the large and medium items in bins in interval k, we must allow for the packing of the remaining items in interval k that will be packed as *small*. These items must be packed in enormous bins for them; therefore, we need to have sufficient unused capacity in the enormous bins to contain the total size of these unpacked items. This is represented by the value V_1 in the state vector; it keeps the slack (or, unused capacity) in the partial packing of the enormous bins with large and medium items. For stages corresponding to intervals greater than k, we also need to have the unused capacity in the huge and large bins, and this is the role of V_2 and V, respectively.

The packing of items as small in a valid solution is reflected by the update-edges on the corresponding path from "initial" to "success." The update-edges connect bins in stage k to bins in stage $(k + 1)$; the stage vector of the end-node of each edge contains the updated value of V_1, the total available capacity for small items, after the packing of all items that are small for bins in stages $0, 1, \ldots, k$. Thus, there is at most one outgoing update-edge from each node in G_L. Since we must represent the possible values in a compact way, we consider the sizes of the items that will be packed as small items into this as-yet-unused capacity. For V_1, items in interval k are small, and thus all items to be packed into this unused capacity have rounded sizes that are multiples of ε^{k+2}. Hence, it suffices to represent V_1 as an integer multiple of ε^{k+2}. Similarly, V_2 and V will be represented as integer multiples of ε^{k+3} and ε^{k+4}, respectively.

The following lemmas, due to [48], will be useful in analyzing the PTRS of [45].

Lemma 8.12 *Given $\varepsilon > 0$, the layered graph G_L has $O(2m \cdot n^{2/\varepsilon^2+3} \cdot 1/\varepsilon^6)$ nodes and $O(2m(n/\varepsilon^2)^{(2/\varepsilon^2)+3})$ edges. The number of nodes in each layer, which is the number of state vectors corresponding to packing the bins of that layer, is $O(n^{2/\varepsilon^2}(n/\varepsilon^2)^3)$.*

Lemma 8.13 *For any $\varepsilon > 0$, there is a one to one correspondence between paths from "initial" to "success" in the layered graph G_L and ε-relaxed packings.*

Lemma 8.14 *Given an ε-relaxed packing of rounded item sizes, for some $\varepsilon > 0$, restoring the item sizes to their original size yields a $(2\varepsilon + \varepsilon^2)$-relaxed packing of the original items.*

We now turn to describe the PTRS for $R(\Pi_{\mathrm{UN}})$. Let $C^*_{max}(I)$ denote the MM for an instance I of Π_{UN}. A pseudocode of the scheme is given in Algorithm 8.2.

ALGORITHM 8.2 $\mathcal{A}_{\text{UN}}(I_0, I, \sigma_0, \varepsilon)$

1. Use a PTAS for makespan minimization on uniform machines to find $T < (1 + \varepsilon)C^*_{max}(I)$.
2. Let $\frac{s_1}{s_m} \leq b$, for some constant $b > 1$. Scale the job processing times by $s_1 \cdot T$, and associate each machine i with a bin of capacity $\frac{s_i}{s_1} \leq 1$. Consider the jobs as items of sizes in $(0, 1]$. Denote the bin sizes by $\frac{1}{b} \leq \tilde{s}_m \leq \tilde{s}_{m-1} \leq \ldots \leq \tilde{s}_1 = 1$, and let the scaled item sizes be $\alpha_1, \alpha_2, \ldots, \alpha_n$.
3. Round down the item sizes $\alpha_j \in \left(\varepsilon^{k+1}, \varepsilon^k \right]$ to the nearest multiple of ε^{k+2}.
4. Construct the directed layered graph G_L, and define edge costs as follows. Consider the edge e_ℓ^k connecting a node of the $(\ell - 1)$-th layer to a node in the ℓ-th layer in the stage that corresponds to interval k; then, e_ℓ^k describes the large and medium pack of the ℓ-th bin of this stage. Let $C_{0\ell}^k$ be the set of items packed in this bin in the initial solution. For all $\ell, k \geq 1$, fix the cost on e_ℓ^k to be the number of large and medium items in $C_{0\ell}^k$ that are not packed in the bin by e_ℓ^k. All other edges in G_L are assigned the cost zero.
5. For every choice of exactly one update-edge at the end of each stage do:
   i. Find the lightest path from "initial" to "success" in the graph (if one exists). Define the corresponding partial solution consisting of items that are packed as large or medium.
   ii. Add the remaining items greedily to the enormous bins for them: for each bin i, let S_0^i be the set of remaining items that belong to bin i in σ_0 and that are small for this bin. Start packing the items in S_0^i in bin i, in nondecreasing order of item sizes, until the total size of packed items exceeds for the first time the bin capacity. Pack all the remaining items, as small, using first-fit.
6. Return the solution of minimum cost.

8.5.3.1 Analysis

Theorem 8.10 *For any $\varepsilon > 0$, $\mathcal{A}_{\text{UN}}$ yields in polynomial time a $(1, 1 + \varepsilon)$-reapproximation for $R(\Pi_{\text{UN}})$.*

We need for the proof the next lemmas.

Lemma 8.15 *Given the partial packing of medium and large items (after Step 5(i) in $\mathcal{A}_{\text{UN}}$), packing the remaining items greedily as small, in Step 5(ii), incurs the minimum cost for these items.*

Proof. For the partial packing of medium and large items, let R be the set of remaining items. We show that packing the items in R in enormous bins for them, by the greedy algorithm in Step 5(ii), results in the minimum cost for packing R. The greedy algorithm first packs every bin with its original items that are small. It sorts these items in nondecreasing order by sizes and packs them in their original bin in this order, until the bin capacity is exceeded for the first time (or all items are packed). This way, we guarantee that each bin contains the maximum number of small items originally packed in it. Thus, the number of items packed as small—not in their original bin—is minimized. This also implies a minimum packing cost for R. ∎

Lemma 8.16 *Let OPT be an optimal solution for an instance I of Π_{UN}. Then, the cost of OPT is at least the cost of the solution obtained by $\mathcal{A}_{\text{UN}}$.*

Proof. Let $C = \{C^1, \ldots, C^m\}$ be the configuration of large and medium items derived from OPT. Since C is optimal, in particular, it is a feasible configuration of large and medium items, which also leaves sufficient space for packing the remaining items as small, without exceeding the capacity of the bins. Thus, by Lemma 8.13, there is a path from "initial" to "success" in the layered graph, such that C is the large and medium pack derived from it.

This path also contains exactly one update-edge after each stage. Therefore, $\mathcal{A}_{\text{UN}}$, in particular, considers this choice of update-edges, for which it finds a lightest path in the graph. By Lemma 8.15, the cost of the solution output by the algorithm is at most the cost of the lightest path plus the minimum cost of packing all the remaining items as small, which is at most the cost of *OPT*. ∎

Lemma 8.17 *For any partial solution of large and medium rounded items, derived from a path from "initial" to "success" in G_L, all the remaining rounded items can be packed in enormous bins for them, using the greedy algorithm in Step 5(ii), such that the load on bin i is at most $\tilde{s}_i(1 + \varepsilon + \varepsilon^3)$.*

Proof. As shown in [48], the small-pack phase, which is done after each stage, is always successfully completed, since there is an update-edge if and only if there is sufficient total slack to accommodate all items to be packed as small. Using a similar argument, $\mathcal{A}_{UN}$ is able to pack all the remaining small items after packing the large and medium items. Consider the rounded items packed into a bin of size $\tilde{s}_i$, such that $\tilde{s}_i$ is in interval k. Suppose that small item j, when added to the bin, exhausts the usable slack. We note that item j is of size less than or equal to ε^{k+2}. Before item j was added, the total size of rounded items in the bin did not exceed $\tilde{s}_i + \varepsilon^{k+4}$ (recall that the usable slack is rounded up to a multiple of ε^{k+4}). Hence, the bin contains items of total (rounded) size at most $\tilde{s}_i + \varepsilon^{k+4} + \varepsilon^{k+2}$. Let B_i denote the set of items packed in bin i, then the sum of the rounded item sizes in B_i is at most $\tilde{s}_i + \varepsilon^{k+4} + \varepsilon^{k+2}$, which is at most $\tilde{s}_i(1 + \varepsilon + \varepsilon^3)$, since $\tilde{s}_i \geq \varepsilon^{k+1}$. ∎

Lemma 8.18 *The number of different choices for update-edges, one after each stage, is at most $O((n^{2/\varepsilon^2}(n/\varepsilon^2)^3)^S)$, where S is the number of stages in G_L.*

Proof. Any two consecutive stages are connected by $O(n^{2/\varepsilon^2}(n/\varepsilon^2)^3)$ update-edges, same as the number of state vectors in each layer (by Lemma 8.12). Therefore, the number of different choices of update-edges, one in each stage, is $O((n^{2/\varepsilon^2}(n/\varepsilon^2)^3)^S)$. ∎

Lemma 8.19 *The number of stages S in G_L depends only on ε and on b.*

Proof. The number of stages in G_L is the number of intervals k that contain at least one bin. Since the smallest bin size satisfies $\tilde{s}_m \geq \frac{1}{b}$, S is bounded by $min\{k \in \mathbb{N} : \varepsilon^{k+1} < \frac{1}{b}\}$. ∎

Proof of Theorem 8.10. By Lemma 8.16, the best solution among all lightest paths (one for each choice of a set of update-edges), we have a solution whose transition cost is at most the optimum. By Lemma 8.17, this solution is also an $(\varepsilon + \varepsilon^3)$-relaxed packing of the rounded items, and by Lemma 8.14, restoring the items to their original sizes, we get a $(2(\varepsilon + \varepsilon^3) + (\varepsilon + \varepsilon^3)^2)$-relaxed packing. Hence, by running the scheme with a suitable error bound, for example, taking $\varepsilon' = \varepsilon/8$, we have that $\mathcal{A}_{UN}$ yields a $(1, 1 + \varepsilon)$-reapproximation for $R(\Pi_{UN})$.

We note that $\mathcal{A}_{UN}$ runs in polynomial time. The graph $G_L = (V_L, E_L)$ can be constructed in time linear in the graph size. And, by Lemma 8.12, we have that $V_L = O(2m \cdot n^{2/\varepsilon^2+3} \cdot 1/\varepsilon^6)$, and $E_L = O(2m(n/\varepsilon^2)^{(2/\varepsilon^2)+3})$. Using Lemma 8.18, the number of different choices of update-edges is $O((n^{2/\varepsilon^2}(n/\varepsilon^2)^3)^S)$, where S is the number of stages in the graph. By Lemma 8.19, S depends only on ε and b. Therefore, the complexity of finding a lightest path from "initial" to "success" and then greedily packing the small items, for every choice of a subset of update-edges, is polynomial in the input size. ∎

8.6 Concluding Remarks

There are many other approximation schemes that hinge on the techniques described in this chapter. We mention a few of them. Wu et al. [49] applied enumeration on a compacted instance while developing a PTAS for minimum routing cost spanning trees. Golubchik et al. [50] apply enumeration to a *structured* instance in solving the problem of data placement on disks (see also [51]). The technique of extending solutions for small subsets is applied, for example, by Sevastianov and Woeginger [17] to makespan minimization in open shop scheduling, and by Khuller et al. [52] to the problem of broadcasting in heterogeneous networks. Kenyon et al. [53] used a nontrivial combination of grouping with periodic scheduling to obtain a PTAS for data broadcast. Jansen et al. [54] combined LP rounding and enumeration to derive a PTAS for job shop scheduling problems (see also [55]).

The layered graph approach was used to develop PTASs for several scheduling problems [56,57]. Epstein and Sgall [58] used the approach to obtain PTASs for a wide class of objective functions, in scheduling on uniformly related machines.

As mentioned in Section 8.4, some techniques are specialized for certain types of problems. For graph problems, several PTASs exploit the density of the input graph [59]; other PTASs rely on properties of *planar graphs* [60,61]. Recent works on graph optimization problems include a PTAS for k-Consensus Clustering [62], for fractional multicommodity flow [63], and for unsplittable flow [64]. Hassin and Levin [65] developed an EPTAS for constrained *minimum spanning trees (MSTs)*, using Lagrangian relaxation and matroid intersection.

We mention some of the recent results for geometric problems. Awasthi et al. [66] gave a PTAS for k-median and k-means clustering for the class of graphs in which the $(k-1)$-means optimum is higher than the k-means optimum by a factor $(1 + \alpha)$ for some constant $\alpha > 0$. The PTAS of [67] for the *container selection problem* relies on geometric compacting of the instance, using a constant number of rays through the origin of $\mathbb{R}^d$, for some constant $d \geq 2$.

Techniques that are based on clustering and graph partitioning were introduced for geometric problems [6,68,69], or for planar graphs. Such techniques were used, for example, for computing a Steiner forest [70,71], and for developing a bicriteria PTAS for *minimum bisection* [72]. Given an undirected graph with edge costs and node weights, the minimum bisection problem seeks a partition of the nodes into two parts of equal weight, such that the total cost of the edges crossing the partition is minimized. The paper [72] gives a bicriteria PTAS for bisection on planar graphs. Specifically, let W be the total weight of all nodes in a planar graph G. For any constant $\varepsilon > 0$, the PTAS outputs a bipartition of the nodes, such that each part weighs at most $W/2 + \varepsilon$, and the total cost of edges crossing the partition is at most $(1 + \varepsilon)$ times the total cost of an optimal bisection.

Arora's framework of PTAS for Euclidean TSP (see Section 8.4) was extended by Talwar [73], who gave a Quasi-PTAS for TSP in metric spaces with bounded doubling dimension. Bartal et al. [74] improved this to a PTAS, by decomposing the metric space into *sparse* subsets, where the weight of the MST of each subset is within a constant factor of its diameter. A similar approach was used in [75] for obtaining a linear time approximation scheme for Euclidean TSP in the RAM model. The construction was extended in [76] to derive a PTAS for metric TSP with neighborhoods.

Some variants of TSP are also known to admit PTASs. One such variant is the *traveling repairman problem (TRP)*. An instance of TRP is given by points in a metric space, and a feasible solution is a path starting at a given origin r that visits each of the points. The completion time of a point v is the distance from r to v on the selected path. The objective is to minimize the total completion time of the points. In [77], a PTAS is given for TRP in the Euclidean plane and on weighted trees. Das and Mathieu [78] presented a Quasi-PTAS for Euclidean *capacitated vehicle routing*. In this variant of TSP, we are given the locations of customers and depots, along with a vehicle of capacity k. The objective is to find a minimum length collection of tours covering all customers, such that each tour starts and ends at a depot and visits at most k customers.

We have mentioned in Sections 8.2.3 and 8.4 some techniques used in *randomized* approximation schemes. A detailed exposition of such schemes for counting problems is given in Chapter 11 in [79] and Chapter 11 in the 1st edition of this handbook. Benczúr and Karger presented in [80] randomized approximation schemes for cuts and flows in capacitated graphs. Efraimidis and Spirakis used in [81] the technique of *filtered randomized rounding* in developing randomized approximation schemes for scheduling unrelated parallel machines.

Finally, for some NP-hard optimization problems, which are not known to admit a PTAS, there have been recent studies of *parameterized approximation schemes* [82–84].

References

1. D.S. Hochbaum and D.B. Shmoys. Using dual approximation algorithms for scheduling problems: Practical and theoretical results. *J. ACM*, 34(1), 144–162, 1987.
2. N. Bansal and M. Sviridenko. Two-dimensional bin packing with one dimensional resource augmentation. *Disc. Optim.*, 4(2), 143–153, 2007.
3. S. Arora and G. Karakostas. Approximation schemes for minimum latency problems. *SIAM J. Comput.*, 32(5), 1317–1337, 2003.
4. C. Chekuri and S. Khanna. Approximation schemes for preemptive weighted flow time. In *Proceedings of STOC*, pp. 297–305, 2002.
5. N. Bansal, A. Chakrabarti, A. Epstein, and B. Schieber. A quasi-PTAS for unsplittable flow on line graphs. In *Proceedings of STOC*, pp. 721–729.
6. A. Adamaszek and A. Wiese. Approximation schemes for maximum weight independent set of rectangles. In *Proceedings of FOCS*, pp. 400–409, 2013.
7. S. Khanna and R. Motwani. Towards a syntactic characterization of PTAS. *Proceedings of STOC*, pp. 329–337, 1996.
8. R. Motwani. Lecture notes on approximation algorithms. Technical Report, Department of Computer Science, Stanford University, Stanford, CA, 1992.
9. D. Karger, C. Stein, and J. Wein. Scheduling algorithms. In *CRC Handbook on Algorithms*, 1997.
10. D.S. Hochbaum. *Approximation Algorithms for NP-Hard Problems*. PWS Publishing Co., Boston, MA, 1996.
11. P. Schuurman and G.J. Woeginger. Approximation schemes—A tutorial. In *Lectures on Scheduling*, R.H. Möehring, C.N. Potts, A.S. Schulz, G.J. Woeginger, and L.A. Wolsey, 2000.
12. D.P. Williamson and D.B. Shmoys. *The Design of Approximation Algorithms*. Cambridge University Press, New York, 2011.
13. V. Vazirani. *Approximation Algorithms*. Springer Science & Business Media, Berlin, Germany, 2003.
14. R.L. Graham. Bounds for certain multiprocessing anomalies. *Bell Syst. Tech. J.*, 45, 1563–1581, 1966.
15. S. Sahni. Approximate algorithms for the 0/1 knapsack problem. *J. ACM*, 22, 115–124, 1975.
16. S. Sahni. Algorithms for scheduling independent tasks. *J. ACM*, 23, 555–565, 1976.
17. S.V. Sevastianov and G.J. Woeginger. Makespan minimization in open shops: A polynomial time approximation scheme. *Math. Prog.*, 82(1–2), 191–198, 1998.
18. K. Jansen and M. Sviridenko. Polynomial time approximation schemes for the multiprocessor open and flow shop scheduling problem. In *Proceedings of STACS*, pp. 455–465, 2000.
19. F.N. Afrati, E. Bampis, C. Chekuri, D.R. Karger, C. Kenyon, S. Khanna, I. Milis, M. Queyranne, M. Skutella, C. Stein, and M. Sviridenko. Approximation schemes for minimizing average weighted completion time with release dates. In *Proceedings of FOCS*, pp. 32–44, 1999.
20. H. Shachnai, T. Tamir, and O. Yehezkely. Approximation schemes for packing with item fragmentation. *Theor. Comput. Syst.*, 43, 81–99, 2008.
21. C. Chekuri and S. Khanna. On multidimensional packing problems. *SIAM J. Comput.*, 33(4), 837–851, 2004.
22. C.N. Potts. Analysis of a linear programming heuristic for scheduling unrelated parallel machines. *Disc. Appl. Math.*, 10, 155–164, 1985.
23. K. Jansen, R. Solis-Oba, and M. Sviridenko. Makespan minimization in job shops: A linear time approximation scheme. *SIAM J. Disc. Math.*, 16(2), 288–300, 2003.
24. M.R. Garey and R.L. Graham. Bounds for multiprocessor scheduling with resource constraints. *SIAM J. Comput.*, 4(2), 187–200, 1975.

25. A. Kulik, H. Shachnai, O. Shmueli, and R. Sayegh. Approximation schemes for deal splitting and covering integer programs with multiplicity constraints. *Theor. Comput. Sci.*, 412(52), 7087–7098, 2011.

26. S.G. Kolliopoulos. Approximating covering integer programs with multiplicity constraints. *Disc. Appl. Math.*, 129, 2–3, 461–473, 2003.

27. S.G. Kolliopoulos and N.E. Young. Tight approximation results for general covering integer programs. In *Proceedings of FOCS*, pp. 522–528, 2001.

28. A.M. Frieze and M.R.B. Clarke. Approximation algorithms for the m-dimensional 0-1 knapsack problem: Worst-case and probabilistic analyses. *Eur. J. Oper. Res.*, 15(1), 100–109, 1984.

29. A.K. Chandra, D.S. Hirschberg, and C.K. Wong. Approximate algorithms for some generalized knapsack problems. *Theor. Comput. Sci.*, 3, 293–304, 1976.

30. L. Fleischer. A fast approximation scheme for fractional covering problems with variable upper bounds. In *Proceedings of SODA*, pp. 994–1003, 2004.

31. A. Caprara, H. Kellerer, U. Pferschy, and D. Pisinger. Approximation algorithms for knapsack problems with cardinality constraints. *Eur. J. Oper. Res.*, 123, 333–345, 2000.

32. C. Chekuri and S. Khanna. A PTAS for the multiple knapsack problem. In *Proceedings of SODA*, pp. 213–222, 2000.

33. H. Kellerer, U. Pferschy, and D. Pisinger. *Knapsack Problems*. Springer, Berlin, Germany, 2004.

34. H. Shachnai and T. Tamir. Approximation schemes for generalized 2-dimensional vector packing with application to data placement. *J. Disc. Algor.*, 10(1), 35–48, 2012.

35. S. Arora. Approximation schemes for NP-hard geometric optimization problems: A survey. *Math. Prog.*, 97, 43–69, 2003.

36. S. Arora. Polynomial time approximation schemes for Euclidean traveling salesman and other geometric problems. *J. ACM*, 45(5), 753–782, 1998.

37. W. Fernandez de la Vega and C. Kenyon. A randomized approximation scheme for metric MAX-CUT. In *Proceedings of the 39th IEEE Symposium on Foundations of Computer Science*, pp. 468–471, 1998.

38. P. Indyk. A sublinear time approximation scheme for clustering in metric spaces. In *Proceedings of FOCS*, 1999.

39. W. Fernandez de la Vega, M. Karpinski, C. Kenyon, and Y. Rabani. Approximation schemes for clustering problems. In *Proceedings of STOC*, 2003.

40. D.S. Hochbaum and W. Maass. Approximation schemes for covering and packing problems in image processing and VLSI. *J. ACM*, 32(1), 130–136, 1985.

41. B.S. Baker. Approximation algorithms for NP-complete problems on planar graphs. *J. ACM*, 41(1), 153–180, 1994.

42. H. Shachnai, G. Tamir, and T. Tamir. A theory and algorithms for combinatorial reoptimization. In *Proceedings of LATIN*, 2012.

43. G. Baram and T. Tamir. Reoptimization of the minimum total flow-time scheduling problem. *Sustainable Computing: Informatics and Systems*, 4, 241–251, 2014.

44. B. Hiller, T. Klug, and A. Tuchscherer. An exact reoptimization algorithm for the scheduling of elevator groups. *Flexible Services and Manufacturing Journal*, 26(4), 585–608, 2014.

45. Y. Mordechai. Optimization and reoptimization in scheduling problems. M.Sc Thesis, Computer Science Department, Technion, Haifa, Israel, 2015.

46. N. Alon, Y. Azar, G. J. Woeginger, and T. Yadid. Approximation schemes for scheduling on parallel machines. *J. Scheduling* 1(1), 55–66, 1998.

47. J. Edmonds and R.M. Karp. Theoretical improvements in algorithmic efficiency for network flow problems. *J. ACM*, 19(2), 248–264, 1972.

48. D.S. Hochbaum and D.B. Shmoys. A polynomial approximation scheme for scheduling on uniform processors: Using the dual approximation approach. *SIAM J. Comput.*, 17(3), 539–551, 1988.

49. B.Y. Wu, G. Lancia, V. Bafna, K.M. Chao, R. Ravi, and C.Y. Tang. A polynomial-time approximation scheme for minimum routing cost spanning trees. *SIAM J. Comput.*, 29(3), 761–778, 2000.

50. L. Golubchik, S. Khanna, S. Khuller, R. Thurimella, and A. Zhu. Approximation algorithms for data placement on parallel disks. *ACM Trans. Algor.*, 5(4), 34, 2009.

51. S. Kashyap and S. Khuller. Algorithms for non-uniform size data placement on parallel disks. In *FST & TCS*, 2003.

52. S. Khuller, Y. Kim, and G. Woeginger. A polynomial time approximation scheme for broadcasting in heterogeneous networks. In *Proceedings of APPROX*, 2004.

53. C. Kenyon, N. Schabanel, and N.E. Young. Polynomial-time approximation scheme for data broadcast. *CoRR cs.DS/0205012*, 2002.

54. K. Jansen, M. Mastrolilli, and R. Solis-Oba. Approximation schemes for job shop scheduling problems with controllable processing times. *Eur. J. Oper. Res.*, 167(2), 297–319, 2005.

55. K. Jansen, and M. Mastrolilli. Approximation schemes for parallel machine scheduling problems with controllable processing times. *Comput. Oper. Res.*, 31(10), 1565–1581, 2004.

56. Y. Azar and L. Epstein. Approximation schemes for covering and scheduling in related machines. In *International Workshop on Approximation Algorithms for Combinatorial Optimization (APPROX)*, 1998.

57. H. Shachnai and T. Tamir. Polynomial time approximation schemes for class-constrained packing problems. *J. Scheduling*, 4(6), 313–338, 2001.

58. L. Epstein and J. Sgall. Approximation schemes for scheduling on uniformly related and identical parallel machines. *Algorithmica*, 39(1), 43–57, 2004.

59. S. Arora, D. Karger, and M. Karpinski. Polynomial time approximation schemes for dense instances of NP-hard problems. In *Proceedings of STOC*, 1995.

60. M.M. Halldórsson and G. Kortsarz. Tools for multicoloring with applications to planar graphs and partial k-trees. *J. Algorithms*, 42(2), 334–366, 2002.

61. E.D. Demaine and M. Hajiaghayi. Bidimensionality: New connections between FPT algorithms and PTASs. In *Proceedings of SODA*, pp. 590–601, 2005.

62. T. Coleman and A. Wirth. A polynomial time approximation scheme for k-consensus clustering. In *Proceedings of SODA*, pp. 729–740, 2010.

63. A. Madry. Faster approximation schemes for fractional multicommodity flow problems via dynamic graph algorithms. In *Proceedings of STOC*, pp. 121–130, 2010.

64. J. Batra, N. Garg, A. Kumar, T. Mömke, and A. Wiese. New approximation schemes for unsplittable flow on a path. In *Proceedings of SODA*, pp. 47–58, 2015.

65. R. Hassin and A. Levin. An efficient polynomial time approximation scheme for the constrained minimum spanning tree problem using matroid intersection. *SIAM J. Comput.*, 33(2), 261–268, 2004.

66. P. Awasthi, A. Blum, and O. Sheffet. Stability yields a PTAS for k-median and k-means clustering. In *Proceedings of FOCS*, pp. 309–318, 2010.

67. V. Nagarajan, K.K. Sarpatwar, B. Schieber, H. Shachnai, and J.L. Wolf. The container selection problem. In *Proceedings of APPROX-RANDOM*, pp. 416–434, 2015.

68. A. Adamaszek and A. Wiese. A QPTAS for maximum weight independent set of polygons with polylogarithmically many vertices. In *Proceedings of SODA*, pp. 645–656, 2014.

69. S. Bandyapadhyay, S. Bhowmick, and K. Varadarajan. Approximation schemes for partitioning: Convex decomposition and surface approximation. In *Proceedings of SODA*, pp. 1457–1470, 2015.

70. M. Bateni, M. Hajiaghayi, and D. Marx. Approximation schemes for Steiner forest on planar graphs and graphs of bounded treewidth. *J. ACM*, 58(5), 1–37, 2011.

71. D. Eisenstat, P. Klein, and C. Mathieu. An efficient polynomial-time approximation scheme for Steiner forest in planar graphs. In *Proceedings of SODA*, pp. 626–638, 2012.

72. K. Fox, P. Klein, and S. Mozes. A polynomial-time bicriteria approximation scheme for planar bisection. In *Proceedings of STOC*, pp. 841–850, 2015.
73. K. Talwar. Bypassing the embedding: Algorithms for low dimensional metrics. In *Proceedings of STOC*, pp. 281–290, 2004.
74. Y. Bartal, L. Gottlieb, and R. Krauthgamer. The traveling salesman problem: Low-dimensionality implies a polynomial time approximation scheme. In *Proceedings of STOC*, pp. 663–672, 2012.
75. Y. Bartal and L. Gottlieb. A linear time approximation scheme for Euclidean TSP. In *Proceedings of FOCS*, pp. 663–672, 2013.
76. T.H.H. Chan and S.H.C. Jiang. Reducing curse of dimensionality: Improved PTAS for TSP (with neighborhoods) in doubling metrics. In *Proceedings of SODA*, pp. 754–765, 2016.
77. R. Sitters. Polynomial time approximation schemes for the traveling repairman and other minimum latency problems. In *Proceedings of SODA*, pp. 604–616, 2014.
78. A. Das and C. Mathieu. A quasi-polynomial time approximation scheme for euclidean capacitated vehicle routing. In *Proceedings of SODA*, pp. 390–403, 2010.
79. R. Motwani and P. Raghavan. *Randomized Algorithms*, Cambridge University Press, New York, 1995.
80. A.A. Benczúr and D.R. Karger. Randomized approximation schemes for cuts and flows in capacitated graphs. Technical Report, MIT, July 2002.
81. P.S. Efraimidis and P.G. Spirakis. Randomized approximation schemes for scheduling unrelated parallel machines. *Electronic Colloquium on Computational Complexity (ECCC)*, 7(7), 2000.
82. D. Marx. Parameterized complexity and approximation algorithms. *Comput. J.*, 51(1), 60–78, 2008.
83. K. Jansen. Parameterized approximation scheme for the multiple knapsack problem. *SIAM J. Comput.*, 39(4), 1392–1412, 2009.
84. M.R. Fellows, D. Hermelin, F. Rosamond, and H. Shachnai. Tractable parameterizations for the minimum linear arrangement problem. *ACM Trans. Comput. Theor. (TOCT)*, 8(2), 6, 2016.

9

Rounding, Interval Partitioning and Separation

Sartaj Sahni

9.1 Introduction

This chapter reviews the three general methods—rounding, interval partitioning, and separation—proposed by Sahni [1] to transform pseudopolynomial-time algorithms into fully polynomial-time approximation schemes. The three methods, which generally apply to dynamic-programming and enumeration-type pseudopolynomial time algorithms, are illustrated using the 0/1-knapsack and multiconstrained shortest paths problems. Both of these problems are known to be NP-hard and both are solvable in pseudopolynomial time using either dynamic programming or enumeration.

9.2 Rounding

The rounding method of Reference 1 is known also by the names digit truncation and scaling. The key idea in the rounding method is to reduce the magnitude of some or all of the numbers in an instance so that the pseudopolynomial-time algorithm actually runs in polynomial time on the reduced instance. The amount by which each number is reduced is such that the optimal solution for the reduced instance is an ϵ-approximate solution for the original instance.

Rounding up, rounding down, and random rounding are three possible strategies to construct the reduced instance. In each, we employ a rounding factor $\delta(n, \epsilon)$, where n is a measure of the problem size. For convenience, we abbreviate $\delta(n, \epsilon)$ as δ. When rounding up, each number α (for convenience, we assume that all numbers in all instances are positive) that is to be rounded is replaced by $\lceil \alpha/\delta \rceil$ and when rounding down, α is replaced by $\lfloor \alpha/\delta \rfloor$. In random rounding, we round up with probability equal to the fractional part of α/δ and round down with probability equal to 1—the fractional part of α/δ. So, for example, if $\alpha = 7$ and $\delta = 4$, α is replaced by (or reduced to) 2 when rounding up and by 1 when rounding down. In random rounding, α is replaced by 2 with probability 0.75 and by 1 with probability 0.25. Random rounding is typically implemented using a uniform random number generator that generates real numbers in the range $[0, 1)$. The decision on whether to round up or down is made by generating a random number. If the generated number is $\leq$ the fractional part of α/δ, we round up; otherwise, we round down.*

As an example of the application of rounding, consider the 0/1-knapsack problem, which is known to be NP-hard [3]. In the 0/1-knapsack problem, we wish to pack a knapsack (bag or sack) with a capacity of c. From a list of n items/objects, we must select the items that are to be packed into the knapsack. Each item i has a weight w_i and a profit p_i. We assume that all weights and profits are positive integers. In a feasible knapsack packing, the sum of the weights of the packed objects does not exceed the knapsack capacity c, which also is assumed to be a positive integer. Since an item with weight more than c cannot be in any feasible packing, we may assume that $w_i \leq c$ for all i. An optimal packing is a feasible packing with maximum profit. The problem formulation is

$$\text{maximize } \sum_{i=1}^{n} p_i x_i$$

subject to the constraints

$$\sum_{i=1}^{n} w_i x_i \leq c \text{ and } x_i \in \{0, 1\}, \ 1 \leq i \leq n$$

In this formulation we are to find the values of x_i. When $x_i = 1$ means that object i is packed into the knapsack, and $x_i = 0$ means that object i is not packed.

For the instance $n = 5$, $(p_1, \cdots, p_5) = (w_1, \cdots, w_5) = \{1, 2, 4, 8, 16\}$ and $c = 27$, the optimal solution is $X = (x_1, x_2, \cdots, x_5) = (1, 1, 0, 1, 1)$, which corresponds to packing items 1, 2, 4, and 5 into the knapsack. This solution uses all the knapsack capacity and yields a profit of 27. With each feasible packing, we associate a profit and weight pair (P, W), where P is the sum of the profits of the items in the packing and $W \leq c$ is the sum of their weights. For example, a packing that generates a profit of 15 and uses 20 units of capacity is represented by the pair (15,20). P is the profit or value of the packing (P, W) and W is its weight.

Several of the standard algorithm design methods of Reference 3—backtracking, branch and bound, dynamic programming, and divide and conquer, for example, may be applied to the knapsack problem. Backtracking and branch and bound result in algorithms whose complexity is $O(2^n)$ and dynamic programming results in a pseudopolynomial time algorithm whose complexity is $O(\min\{2^n, n\tilde{F}, nc\})$, where $\tilde{F}$ is the value of the optimal solution [4]. A pseudopolynomial-time algorithm with this same complexity also may be arrived at using an enumerative approach. By coupling a divide and conquer step to this enumerative algorithm, we obtain a pseudopolynomial-time algorithm whose complexity is $O(\min\{2^{n/2}, n\tilde{F}, nc\})$ [4].

Let $(P1, W1)$ and $(P2, W2)$ represent two different feasible packings of items selected from the first i items. Tuple $(P1, W1)$ *dominates* $(P2, W2)$ iff either $P1 \geq P2$ and $W1 < W2$ or $P1 > P2$ and $W1 = W2$. The enumerative algorithm for the 0/1-knapsack problem constructs a list of (or enumerates) the profit and weight pairs that correspond to all possible nondominated feasible packings. This list is constructed

---

* There is a similar rounding, but quite different, method for approximation algorithms—randomized rounding—due to Raghavan and Thompson [2]. In randomized rounding, we start with an integer linear program formulation; relax the integer constraints to real number constraints; solve the resulting linear program; transform the noninteger values in the obtained solution to the linear program to integers using the random rounding strategy stated earlier.

incrementally. Let S_i be the list of nondominated profit and weight pairs for all possible feasible packings chosen from the first i items. We start with the list $S_0 = \{(0,0)\}$, and construct $S_1, S_2, \cdots, S_n$ in this order. Note that each S_i, $i > 0$ may be constructed from S_{i-1} using the equality

$$S_i = S_{i-1} \oplus \{(a + p_i, b + w_i) | (a, b) \in S_{i-1} \text{ and } b + w_i \leq c\} \tag{9.1}$$

where $\oplus$ denotes a union in which dominated pairs are eliminated. Equation 9.1 simply states that the nondominated pairs obtainable from the first i items are a subset of those obtainable from the first $i - 1$ items (these have $x_i = 0$) plus those obtainable from feasible packings of the first i items that necessarily include w_i (i.e., $x_i = 1$). The subset is identified by eliminating dominated pairs. Trying Equation 9.1 out on the above $n = 5$ instance, we get (since $P = W$ for every pair (P, W) in this example, we represent each pair by a single number)

$$\begin{aligned}
S_0 &= \{0\} \\
S_1 &= \{0\} \oplus \{1\} = \{0, 1\} \\
S_2 &= \{0, 1\} \oplus \{2, 3\} = \{0, 1, 2, 3\} \\
S_3 &= \{0, 1, 2, 3\} \oplus \{4, 5, 6, 7\} = \{0, 1, 2, 3, 4, 5, 6, 7\} \\
S_4 &= \{0, \cdots, 7\} \oplus \{8, \cdots, 15\} = \{0, \cdots, 15\} \\
S_5 &= \{0, \cdots, 15\} \oplus \{16, \cdots, 27\} = \{0, \cdots, 27\}
\end{aligned}$$

For the case $n = 4$, $(p_1, \cdots, p_4) = (w_1, \cdots, w_4) = \{1, 1, 8, 8\}$, and $c = 17$, we get

$$\begin{aligned}
S_0 &= \{0\} \\
S_1 &= \{0\} \oplus \{1\} = \{0, 1\} \\
S_2 &= \{0, 1\} \oplus \{1, 2\} = \{0, 1, 2\} \\
S_3 &= \{0, 1, 2\} \oplus \{8, 9, 10\} = \{0, 1, 2, 8, 9, 10\} \\
S_4 &= \{0, 1, 2, 8, 9, 10\} \oplus \{8, 9, 10, 16, 17\} = \{0, 1, 2, 8, 9, 10, 16, 17\}
\end{aligned}$$

The solution to the knapsack instance may be determined from the S_is using the procedure of Figure 9.1.

For our $n = 5$ instance with $c = 27$, (P, W) is determined to be (27,27) in Step 1. In Step 2, x_5 is set to 1 as $(27, 27) \notin S_4$ and P and W are updated to 11. Then x_4 is set to 1 as $(11, 11) \notin S_3$ and P and W are updated to 3. Next, x_3 is set to 0 as $3 \in S_2$. x_2 and x_1 are set to 1 in the remaining 2 iterations of the **for** loop.

The S_is may be implemented as sorted linear lists (note that the dominance rule ensures that if S_i is in ascending order of P, S_i is also in ascending order of W; also, no two pairs of S_i may have the same P or the same W value). The set S_i may be computed from S_{i-1} in $O(|S_{i-1}|)$ time using Equation 9.1. The time

Step 1: [Determine solution value]

Determine the pair $(P, W) \in S_n$ with maximum profit value. The value of an optimal packing is P.

Step 2: [Determine x_is]

    for $(i = n; i > 0; i--)$

        if $((P, W) \notin S_{i-1})$ $\{x_i = 1; P{-} = p_i; W{-} = w_i;\}$

        else $x_i = 0;$

FIGURE 9.1 Procedure to determine x_is from the S_is.

to compute all S_is is, therefore, $\sum_{1 \le i \le n} |S_{i-1}|$. (Note that in S_n we need only compute the pair with maximum profit. When the S_is are in ascending order of profit, this maximum pair may be determined easily.) From Equation 9.1 it follows that $|S_i| \le 2^i$ (this also follows from the observation that there are 2^i different subsets of i items). Also, since the w_is and p_is are positive integers and S_i has only nondominated pairs, $|S_i| \le \min\{\tilde{F}, c\} + 1$. Hence, the time needed to generate the S_is is $O(\min\{2^n, n\tilde{F}, nc\})$. If the sorted linear lists are array lists, each S_i may be searched for (P, W) in $O(\log |S_i|)$ time. In this case the complexity of the procedure to determine the x_is from the S_is is $O(n * \min\{n, \log \tilde{F}, \log c\})$. This may be reduced to $O(n)$ by retaining with each $(P, W) \in S_i$ a pointer to the pair (P, W) or $(P - p_i, W - w_i)$, that is, in S_{i-1} (note that at least one of these pairs must be in S_{i-1}). These pointers are added to the members of S_i at the time S_i is constructed using Equation 9.1. The inclusion of these pointers doesn't change the asymptotic complexity of the procedure to compute the S_is.

The enumerative pseudopolynomial-time algorithm just described for the knapsack problem may be transformed into a fully polynomial time approximation scheme by suitably rounding down the p_is. Suppose we round using the rounding factor δ to obtain the reduced instance with $p_i' = \lfloor p_i/\delta \rfloor$ and $w_i' = w_i$, $1 \le i \le n$ and $c' = c$. The time to solve the reduced instance is $O(n\tilde{F}')$, where $\tilde{F}'$ is the value of the optimal solution to the reduced problem (we assume the reduction is sufficient so that $n\tilde{F}' < \min\{2^n, nc'\}$). Notice that the original and reduced instances have the same feasible packings; only the profit associated with each feasible packing is different. A feasible packing has a smaller profit in the reduced instance than in the original instance.

Consider any feasible packing $(x_1, \cdots, x_n)$. Since $p_i' * \delta \le p_i < (p_i' + 1) * \delta$,

$$\delta * \sum_i p_i' x_i \le \sum_i p_i x_i < \delta * \sum_i (p_i' + 1) x_i \tag{9.2}$$

So,

$$\delta \tilde{F}' \le \tilde{F} < \delta(\tilde{F}' + n) \tag{9.3}$$

Suppose we use the just described rounding strategy on our $n = 4$ example with $(p_1, \cdots, p_4) = (w_1, \cdots, w_4) = (1, 1, 8, 8)$, $c = 17$ and $\delta = 3$. We obtain $(p_1', \cdots, p_4') = (0, 0, 2, 2)$, $(w_1', \cdots, w_4') = (1, 1, 8, 8)$, and $c' = 17$. One of the optimal solutions for the reduced instance has $(x_1, x_2, x_3, x_4) = (0, 0, 1, 1)$ and the value of this solution is $p_3' + p_4' = 4$. In the original instance, the solution $(0, 0, 1, 1)$ has value 16. Note that many different knapsack instances round to the same reduced instance. For example, $(p_1, \cdots, p_4) = (2, 1, 6, 7)$, $(w_1, \cdots, w_4) = (1, 1, 8, 8)$, and $c = 17$ (using $\delta = 3$). The value of the solution $(0, 0, 1, 1)$, for this original instance, is 13. From Equation 9.2, regardless of the original instance, the value of $(0, 0, 1, 1)$ must be at least $\delta * \sum p_i' x_i = 12$ and cannot equal or exceed $\delta * \sum (p_i' + 1) x_i = 18$.

To ensure that every optimal solution to the reduced instance also defines an ϵ-approximate solution for the original instance, we must select δ carefully. Let $\hat{F}$ be the value, in the original instance, of the optimal solution for the reduced instance. From Equations 9.2 and 9.3, we obtain

$$\hat{F} \ge \delta \tilde{F}' > \tilde{F} - n\delta$$

So, $(\tilde{F} - \hat{F}) < n\delta$ and $(\tilde{F} - \hat{F})/\tilde{F} < n\delta/\tilde{F}$. To guarantee that the optimal solution for the reduced instance is an ϵ-approximate solution for the original instance, we require $n\delta/\tilde{F} \le \epsilon$ or $\delta \le \epsilon\tilde{F}/n$. Since the reduced instance has smaller p_i' values and hence smaller complexity when δ is larger, we would like to use

$$\delta = \epsilon\tilde{F}/n$$

With this choice of δ, $\tilde{F}' \le \tilde{F}/\delta = n/\epsilon$ (Equation 9.3). So, $|S_i| \le n/\epsilon + 1$ and the complexity of the enumerative algorithm becomes $O(n^2/\epsilon)$. In other words, the enumerative algorithm becomes a fully polynomial time approximation scheme for the 0/1-knapsack problem! Unfortunately, this choice of δ is problematic as we cannot easily compute $\tilde{F}$. Since any $\delta \le \epsilon \tilde{F}/n$ guarantees ϵ-approximate solutions, we may use

$$\delta = \epsilon LB/n$$

where $LB \le \tilde{F}$ is a lower bound on the value of the optimal solution. Let $Pmax = \max_i\{p_i\}$ be the maximum profit value. Since $w_i \le c$ for all i (by assumption), $Pmax \le \tilde{F}$, and $LB = Pmax$ is a lower bound on $\tilde{F}$. So, using $\delta = \epsilon Pmax/n$ guarantees ϵ-approximate solutions. Since $\tilde{F} \le nPmax$, $\tilde{F}' \le nPmax/\delta = n^2/\epsilon$ and the complexity of the enumerative algorithm becomes $O(n^3/\epsilon)$.

An alternative way to determine a lower bound for $\tilde{F}$ is to sort the knapsack items into nondecreasing order of profit density p_i/w_i and pack the items into the knapsack in density order up to and including the first item that causes the knapsack to be either filled or over filled. Note that if there is no such first item, all items can be packed into the knapsack and this packing represents the optimal solution. Also note that if the stated packing strategy fills the knapsack completely, it represents an optimal packing. So, assume that the capacity is exceeded. Let $\bar{F}$ be the value of this packing that overfills the knapsack. In Reference 5, it is shown that $\bar{F}/2 \le \tilde{F} \le \bar{F}$. So, using $\delta = \epsilon\bar{F}/(2n)$ as the rounding factor, guarantees ϵ-approximate solutions. Since $\tilde{F} \le \bar{F}$, $\tilde{F}' \le \bar{F}/\delta$ and, for the reduced instance, $|S_i| \le \bar{F}/\delta + 1 = 2n/\epsilon + 1$. For the reduced instance, the complexity of the enumerative algorithm is, therefore, $O(n^2/\epsilon)$ and we get a fully polynomial time ϵ-approximation scheme of this complexity.

9.3 Interval Partitioning

Unlike rounding, which reduces an instance to one that is easier to solve using a known pseudopolynomial time algorithm, in interval partitioning we work with the unreduced (original) instance. In interval partitioning, we partition the solution space into buckets or intervals and for each interval, we retain only one of the (feasible) solutions (or partial solutions) that fall into it.

For the 0/1-knapsack problem, for example, each pair $(P, W) \in S_i$, $i \le n$ represents a feasible solution. We may partition the solution space based on the profit value of the pair (P, W). If we partition using an interval size of δ, then the intervals are $[0, \delta)$, $[\delta, 2\delta)$, $[2\delta, 3\delta)$, and so on. When two or more solutions fall into the same interval, all but one of them is eliminated. Specifically, we eliminate all but the one with least weight. Let S_i' be the list of (P, W) pairs for all possible feasible packings chosen from the first i items subject to the interval partitioning constraint that S_i' has at most 1 (P, W) pair in each interval. We begin with $S_0' = \{(0, 0)\}$ and compute S_i' from S_{i-1}' using Equation 9.4

$$S_i' = S_{i-1}' \odot \{(a + p_i, b + w_i)|(a, b) \in S_{i-1}' \text{ and } b + w_i \le c\} \tag{9.4}$$

where $\odot$ denotes a union in which only the least weight pair from each interval is retained. The maximum profit pair in S_n' is used as the approximately optimal solution. The x_is for this pair are obtained using the procedure of Figure 9.1. with S_i replaced by S_i'.

Consider the 0/1-knapsack instance $n = 5$, $(p_1, \cdots, p_5) = (w_1, \cdots, w_5) = \{1, 2, 4, 8, 16\}$ and $c = 27$ that was first considered in Section 9.2. Suppose we work with an interval size $\delta = 2$. The intervals are $[0, 2)$, $[2, 4)$, $[4, 6)$, and so on. The S_i's are as follows:

$$S_0' = \{0\}$$
$$S_1' = \{0\} \odot \{1\} = \{0\}$$
$$S_2' = \{0\} \odot \{2\} = \{0, 2\}$$
$$S_3' = \{0, 2\} \odot \{4, 6\} = \{0, 2, 4, 6\}$$
$$S_4' = \{0, 2, 4, 6\} \odot \{8, 10, 12, 14\} = \{0, 2, 4, 6, 8, 10, 12, 14\}$$
$$S_5' = \{0, 2, 4, \cdots, 14\} \odot \{16, 18, 20, \cdots, 26\} = \{0, 2, 4, \cdots, 26\}$$

The maximum profit pair in S_4 is (26,26). For this instance, therefore, the best solution found using interval partitioning with $\delta = 2$ has a profit 1 less than that of the optimal.

Consider the instance $n = 6$, $(p_1, \cdots, p_6) = (w_1, \cdots, w_6) = (1, 2, 5, 6, 8, 9)$, and $c = 27$. Suppose we use $\delta = 3$. The intervals are [0, 3), [3, 6), [6, 9), and so on. The S_i's are:

$$S_0' = \{0\}$$
$$S_1' = \{0\} \odot \{1\} = \{0\}$$
$$S_2' = \{0\} \odot \{2\} = \{0\}$$
$$S_3' = \{0\} \odot \{5\} = \{0, 5\}$$
$$S_4' = \{0, 5\} \odot \{6, 11\} = \{0, 5, 6, 11\}$$
$$S_5' = \{0, 5, 6, 11\} \odot \{8, 13, 14, 19\} = \{0, 5, 6, 11, 13, 19\}$$
$$S_6' = \{0, 5, 6, 11, 13, 19\} \odot \{9, 14, 15, 20, 22\} = \{0, 5, 6, 9, 13, 15, 19, 22\}$$

The profit of the best solution found for this instance is 22; the profit for the optimal solution is 27. Note that if c were 28 instead of 27,

$$S_6' = \{0, 5, 6, 11, 13, 19\} \odot \{9, 14, 15, 20, 22, 28\} = \{0, 5, 6, 9, 13, 15, 19, 22, 28\}$$

and we would have found the optimal solution.

Let $\check{F}$ be the value of the solution found by interval partitioning. It is easy to see that $\tilde{F} < \check{F} + n\delta$. So, $(\tilde{F} - \check{F})/\tilde{F} < n\delta/\tilde{F}$. To guarantee that the solution found using interval partitioning is an ϵ-approximate solution, we require $n\delta/\tilde{F} \leq \epsilon$. For this, we must choose δ so that $\delta \leq \epsilon\tilde{F}/n$. Since $\tilde{F}$ is hard to compute, we opt to select δ as in Section 9.2. Both the choices $\delta = \epsilon Pmax/n$ and $\delta = \epsilon\bar{F}/(2n)$ guarantee that the solution generated using interval partitioning is ϵ-approximate. When $\delta = \epsilon Pmax/n$, the number of intervals is $\tilde{F}/\delta + 1 \leq nPmax/\delta + 1 = n^2/\epsilon + 1$ and the complexity of the (modified) enumerative algorithm is $O(n^3/\epsilon)$. When $\delta = \epsilon\bar{F}/(2n)$, the number of intervals is $\tilde{F}/\delta + 1 \leq \bar{F}/\delta + 1 = 2n/\epsilon + 1$ and the complexity is $O(n^2/\epsilon)$.

9.4 Separation

An examination of our $n = 6$ example of Section 9.3 reveals that interval partitioning misses some opportunities to reduce the size of an S_i' while yet preserving the relationship $\tilde{F} \leq \hat{F} + n\delta$, which is necessary to ensure an ϵ-approximate solution. For example, in S_4' we have two solutions, one with value 5 and the other with value 6. Although these are within δ of each other, they fall into two different intervals and so neither is eliminated. In the separation method, we ensure that the value of retained solutions differs by more than δ.

For the 0/1-knapsack problem, let S_i'' be the list of (P, W) pairs for all possible feasible packings chosen from the first i items subject to the separation constraint that no two pairs of S_i'' have value within δ of each other. We begin with $S_0'' = \{(0,0)\}$ and compute S_i'' from S_{i-1}'' using Equation 9.5

$$S_i'' = S_{i-1}'' \otimes \{(a + p_i, b + w_i)|(a, b) \in S_{i-1}'' \text{ and } b + w_i \leq c\} \tag{9.5}$$

where $\otimes$ denotes a union that implements the separation constraint. More precisely, suppose that

$$T = S_{i-1}'' \oplus \{(a + p_i, b + w_i)|(a, b) \in S_{i-1}'' \text{ and } b + w_i \leq c\}$$

Let (P_i, W_i), $1 \leq i \leq |T|$ be the pairs in T in ascending order of profit (and hence of weight). The set S_i'' is obtained from T using the code of Figure 9.2.

$$S_i'' = \{(P_1, W_1)\};$$
$$P_{prev} = P_1;$$
$$\textbf{for } (\textbf{int } i = 1; i <= |T|; i++)$$
$$\quad \textbf{if } (P_i > P_{prev} + \delta) \ \{S_i'' = S_i'' \cup \{(P_i, W_i)\}; \ P_{prev} = P_i; \}$$

FIGURE 9.2   Computing S_i'' from T.

The maximum profit pair in S_n'' is used as the approximately optimal solution.
Consider the $n = 6$ example of Section 9.3. $S_i'' = S_i'$, $0 \le i \le 3$. The remaining S_i'''s are

$$S_0'' = \{0\}$$
$$S_1'' = \{0\} \otimes \{1\} = \{0\}$$
$$S_2'' = \{0\} \otimes \{2\} = \{0\}$$
$$S_3'' = \{0\} \otimes \{5\} = \{0, 5\}$$
$$S_4'' = \{0, 5\} \otimes \{6, 11\} = \{0, 5, 11\}$$
$$S_5'' = \{0, 5, 11\} \otimes \{8, 13, 19\} = \{0, 5, 11, 19\}$$
$$S_6'' = \{0, 5, 11, 19\} \otimes \{9, 14, 20\} = \{0, 5, 9, 14, 19\}$$

The profit of the best solution found for this instance is 19; the profit for the optimal solution is 27. We could have produced a slightly better solution by noting that we can replace the computation of S_n'' by a step in which we determine the maximum profit pair in

$$S_{n-1}'' \oplus \{(a + p_i, b + w_i) | (a, b) \in S_{n-1}'' \text{ and } b + w_i \le c\}$$

For our example, this pair has value 20.

Let $\check{F}$ be the value of the solution found by the separation method. It is easy to see that $\tilde{F} \le \check{F} + n\delta$. So, $\delta \le \epsilon \tilde{F}/n$ ensures that an ϵ-approximate solution is found. As was the case in Section 9.3, for the knapsack problem, the choices $\delta = \epsilon Pmax/n$ and $\delta = \epsilon \tilde{F}/(2n)$ guarantee that the solution generated using separation is ϵ-approximate. When $\delta = \epsilon Pmax/n$, the complexity of the (modified) enumerative algorithm is $O(n^3/\epsilon)$ and when $\delta = \epsilon \tilde{F}/(2n)$, the complexity is $O(n^2/\epsilon)$.

Intuitively, we may expect that using the same δ value, $|S_i''| \le |S_i'|$ for all i. Although this relationship holds for the $n = 6$ example considered previously, the relationship doesn't always hold. For example, consider the knapsack instance $n = 5$, $(p_1, \cdots, p_5) = (w_1, \cdots, w_5) = (30, 10, 51, 51, 51)$, $c = 186$ and $\delta = 20$. Using interval partitioning, we get

$$S_0' = \{0\}$$
$$S_1' = \{0\} \odot \{30\} = \{0, 30\}$$
$$S_2' = \{0, 30\} \odot \{10, 40\} = \{0, 30, 40\}$$
$$S_3' = \{0, 30, 40\} \odot \{51, 81, 91\} = \{0, 30, 40, 81\}$$
$$S_4' = \{0, 30, 40, 81\} \odot \{51, 81, 91, 132\} = \{0, 30, 40, 81, 132\}$$
$$S_5' = \{0, 30, 40, 81, 132\} \odot \{51, 81, 91, 132, 183\} = \{0, 30, 40, 81, 132, 183\}$$

and using separation, we get

$$S_0'' = \{0\}$$
$$S_1'' = \{0\} \otimes \{30\} = \{0, 30\}$$
$$S_2'' = \{0, 30\} \otimes \{10, 40\} = \{0, 30\}$$
$$S_3'' = \{0, 30\} \otimes \{51, 81\} = \{0, 30, 51, 81\}$$
$$S_4'' = \{0, 30, 51, 81\} \otimes \{51, 81, 102, 132\}$$
$$= \{0, 30, 51, 81, 102, 132\}$$
$$S_5'' = \{0, 30, 51, 81, 102, 132\} \otimes \{51, 81, 102, 132, 153, 183\}$$
$$= \{0, 30, 51, 81, 102, 132, 153, 183\}$$

9.5 0/1-Knapsack Problem Revisited

In Sections 9.2 through 9.4, we saw how to apply the generic rounding, interval partitioning, and separation methods to the 0/1-knapsack problem and obtain an ϵ-approximate fully polynomial time approximation scheme for this problem. The complexity of the approximation scheme is either $O(n^3/\epsilon)$ or $O(n^2/\epsilon)$, depending on the choice of δ. By tailoring the approximation method to the application, we can, at times, reduce the complexity of the approximation scheme. Ibarra and Kim [5], for example, combine rounding and interval partitioning to arrive at an $O(n \log n - (\log \epsilon)/\epsilon^4)$ ϵ-approximate fully polynomial time approximation scheme for 0/1-knapsack problem. Figure 9.3 gives their algorithm. The correctness proof and complexity analysis can be found in Reference 5.

Step 1: [Determine δ]
Sort the n items into nondecreasing order of profit density p_i/w_i.
Let $\bar{F}$ be as in Section ??.
Let $\delta = \epsilon^2 \bar{F}/9$.

Step 2: [Partition Items]
Let *small* be the items with $p_i \leq \epsilon \bar{F}/3$.
Let *big* be the remaining items.

Step 3: [Rounding]
Let *big'* be obtained from *big* by rounding down the profits using the rounding factor δ.
For each rounded-down profit p, retain up to $9/(\epsilon^2 p)$ items of least weight.
Let *big''* be the resulting item set.
Let m be the number of items in *big''*.

Step 4: [Interval Partitioning]
Use interval partitioning on *big''* and determine S_m'.

Step 5: [Augmentation]
Augment each $(P, W) \in S_m'$ by adding in items from *small* in order of nondecreasing density so as not to exceed the capacity of the knapsack.
Select the augmentation that yields maximum profit as the approximate solution.

FIGURE 9.3   Fully polynomial time ϵ-approximation scheme. (From Ibarra, O. and Kim, C., *JACM*, 22, 463, 1975.)

9.6 Multiconstrained Shortest Paths

9.6.1 Notation

Assume that a communication network is represented by a weighted directed graph $G = (V, E)$, where V is the set of network vertices or nodes and E is the set of network links or edges. We use n and e, respectively, to denote the number of nodes and links in the network, that is, $n = |V|$ and $e = |E|$. We assume that each link (u, v) of the network has $k > 1$ nonnegative weights $w_i(u, v)$, $1 \leq i \leq k$. These weights, for example, may represent link cost, delay, delay-jitter, and so on. The notation $w(u, v)$ is used to denote the vector $(w_1(u, v), \cdots, w_k(u, v))$, which gives the k weights associated with the edge (u, v). Let p be a path in the network. We use $w_i(p)$ to denote the sum of the w_is of the edges on the path p.

$$w_i(p) = \sum_{(u,v) \in p} w_i(u, v)$$

By definition, $w(p) = (w_1(p), \cdots, w_k(p))$.

In the *multiconstrained path* (*k-MCP*) problem, we are to find a path p from a specified source vertex s to a specified destination vertex d such that

$$w_i(p) \leq c_i, 1 \leq i \leq k \tag{9.6}$$

The c_is are specified quality of service (QoS) constraints. Note that Equation 9.6 is equivalent to $w(p) \leq c$, where $c = (c_1, \cdots, c_k)$. A *feasible path* is any path that satisfies Equation 9.6.

The *restricted shortest path* (*k-RSP*) problem is a related optimization problem in which we are to find a path p from s to d that minimizes $w_1(p)$ subject to

$$w_i(p) \leq c_i, 2 \leq i \leq k$$

An algorithm is an ϵ-*approximation algorithm* (or simply, an *approximation algorithm*) for *k-MCP* iff the algorithm generates a source to destination path p that satisfies Equation 9.6 whenever the network has a source to destination path p' that satisfies

$$w_i(p) \leq \epsilon * c_i, 1 \leq i \leq k \tag{9.7}$$

where ϵ is a constant between 0 and 1.

Both the *k-MCP* and *k-RSP* problems for $k > 1$ are known to be NP-hard [6] and several pseudopolynomial time algorithms, heuristics, and approximation algorithms have been proposed [7–9]. Jaffe [10] has proposed a polynomial time approximation algorithm for *2-MCP*. This algorithm, which uses a shortest path algorithm such as Dijkstra's [11], replaces the two weights on each edge by a linear combination of these two weights. The algorithm is expected to perform well when the two weights are positively correlated. Chen and Nahrstedt [12] use rounding to arrive at a polynomial time approximation algorithm for *k-MCP*. Korkmaz and Krunz [13] propose a randomized heuristic that employs two phases. In the first phase a shortest path from each vertx of V to the destination vertex d is computed for each of the k weights as well as for a linear combination of all k weights. The second phase performs a randomized breadth-first search for a solution to the *k-MCP* problem. Yuan [14] has proposed two heuristics for *k-MCP*—limited granularity and limited path. By properly selecting the parameters for the limited granularity heuristic (LGH), this heuristic becomes an ϵ-approximation algorithm for *k-MCP*.

The papers [15–19] use rounding (up, down, and random) and interval partitioning to arrive at fully polynomial time approximation schemes for *k-RSP*. Song and Sahni [20] use rounding (up), interval partitioning, and separation to develop fully polynomial time approximation schemes for *k-MCP*. We focus on the work of Reference 20 and this section is derived from Reference 20.

9.6.2 Extended Bellman-Ford Algorithm

This is an extension of the well-known dynamic programming algorithm due to Bellman and Ford that is used to find shortest paths in weighted graphs [11]. The original Bellman-Ford algorithm was proposed for graphs in which each edge has a single weight. The extension allows for multiple weights (e.g., cost, delay, delay-jitter, and so on).

Let u and v be two vertices in an instance of k-MCP. Let p and q be two different u to v paths. Path p is *dominated* by path q iff $w(q) \leq w(p)$ (i.e., $w_i(q) \leq w_i(p)$, $1 \leq i \leq k$).

In its pure form, the Bellman-Ford algorithm works in $n - 1$ (n is the number of vertices in the graph) rounds numbered 1 through $n - 1$. In round 1, the algorithm implicitly enumerates one-edge paths from the source vertex; then, in round 2, those with two edges are enumerated; and so on until finally paths with $n - 1$ edges are enumerated. Since no simple path has more than $n - 1$ edges, by the end of round $n - 1$, all simple paths have been (implicitly) enumerated. The enumeration of paths that have $i + 1$ edges is accomplished by considering all one-edge extensions of the enumerated i-edge paths. During the implicit enumeration, suboptimal paths (i.e., paths that are dominated by others) are eliminated. Suppose we have two paths p and q to vertex u and that p is dominated by q. If path p can be extended to a path that satisfies Equation 9.6, then so also can q. Hence there is no need to retain p for further enumeration by path extension. Actual implementations rarely follow the pure Bellman-Ford paradigm and enumerate some paths of length more than $i + 1$ in round i.

Figure 9.4 gives the version of the extended Bellman-Ford algorithm employed by Reference 20. This version is very similar to the version used by Yuan et al. [14,21]. $PATH(u)$ is a set of paths from the source s to vertex u. $PATH(u)$ never contains two paths p and q for which $w(p) \leq w(q)$. Lines 12 through 14 initialize $PATH(u)$ for all vertices u. The **for** loop of lines 16 through 20 attempts to implement the pure form of the extended Bellman-Ford algorithm and performs the required $n - 1$ rounds (there is a provision to terminate in fewer rounds in case the previous round added a path to no $PATH(u)$). The

```
Relax(u, v)
1.    for each new p ∈ PATH(u) such that w(p) + w(u, v) ≤ c do
2.        if (v = d) return TRUE;
3.        Flag = TRUE;
4.        for each q ∈ PATH(v) do
5.            if (w(q) ≤ w(p) + w(u, v))
6.                Flag = FALSE; Break; // exit inner for loop
7.            if ((w(p) + w(u, v)) ≤ w(q))
8.                remove q from PATH(v);
9.        if (Flag == TRUE)
10.           insert p||(u, v) into PATH(v); Change = TRUE;
11.   return FALSE;

Extended Bellman-Ford(G, c, s, d)
12.   for i = 0 to n − 1 do
13.       PATH(i) = NULL;
14.   PATH(s) = {s};
15.   Result = FALSE;
16.   for round = 1 to n − 1 do
17.       Change = FALSE;
18.       for each edge (u, v) ∈ E do
19.           if (Relax(u,v)) return "YES";
20.       if (Change == FALSE) return "NO";
21.   return "NO";
```

FIGURE 9.4 Extended Bellman-Ford Algorithm for k-MCP.

method $Relax(u, v)$ extends the new[*] paths in $PATH(u)$ by appending the edge (u, v). Feasible extended paths (i.e., those that satisfy the k constraints of Equation 9.6) are examined further. If v is the destination, the algorithm terminates as we have found a feasible source to destination path. Let the extended path $p||(u, v)$ be r. The inner **for** loop (lines 4–8) removes from $PATH(v)$ all paths that are dominated by r (lines 7 and 8). This loop also verifies that r isn't dominated by a path in $PATH(v)$ (lines 5 and 6). Notice that if r is dominated by or equal to a path in $PATH(v)$, r cannot dominate a path in $PATH(v)$. Finally, in lines 9 and 10, r is added to $PATH(v)$ only if it isn't dominated by or equal to any path in $PATH(v)$.

To see that the algorithm of Figure 9.4 is not a faithful implementation of the pure form of the Bellman-Ford algorithm, consider any iteration of the **for** loop of lines 16–20 (i.e., consider one round) and suppose that edge (u, v) is considered before edge (v, w) in the **for** loop of lines 13–14. Following the consideration of (u, v), $PATH(v)$ possibly contains paths with *round* edges. So, when (v, w) is considered, $Relax$ extends the paths in $PATH(v)$ by a single edge (v, w) thereby permitting a path of length *round* + 1 to be included in $PATH(w)$. This lack of faithfullness in implementation of the pure Bellman-Ford algorithm doesn't affect the correctness of the algorithm and, in fact, agrees with the traditional implementation of the Bellman-Ford algorithm for the case when each edge has a single weight (i.e., $k = 1$) [11].

Another implementation point worth mentioning is that although we have defined $PATH(u)$ to be a set of paths from the source to vertex u, it is more efficient to implement $PATH(u)$ to be the the set of weights (or more accurately, weight vectors $w()$) of these paths. This, in fact, is how the algorithm is implemented in Reference 14.

9.6.3 Rounding

Let $\delta_i = c_i * (1 - \epsilon)/n, 2 \le i \le k$. Suppose we replace each $w_i(u, v)$ with the weight

$$w_i'(u, v) = \lceil \frac{w_i(u, v)}{\delta_i} \rceil * \delta_i$$

Let p be a path such that satisfies Equation 9.7. Then,

$$w_i'(p) < w_i(p) + n\delta_i \le \epsilon c_i + (1 - \epsilon)c_i = c_i$$

So, algorithm *extended Bellman-Ford* of Figure 9.4 when run with the edge weights $w_i(u, v)$ replaced by the weights $w_i'(u, v), 2 \le i \le k$ will find a feasible path (either p or some other feasible path). In an implementation of the rounding method, we actually replace each $w_i(u, v), 2 \le i \le k$ by

$$w_i''(u, v) = \lceil \frac{w_i(u, v)}{\delta_i} \rceil$$

and each c_i by $\lfloor c_i/\delta_i \rfloor, 2 \le i \le k$. From the computation stand point, using the w_i's is equivalent to using the w_i''s.

Let $S = (n/(1 - \epsilon))^{k-1}$. In the w_i''s formulation, it is easy to see that $|PATH(u)| \le S$. Hence the complexity of *extended Bellman-Ford* when the w_i'' (equivalently, w_i') weights are used is $O(neS^2)$ and we have a fully polynomial time approximation scheme for k-MCP. For the case $k = 2$, the complexity is $O(neS)$ if we employ the merge strategy of Horowitz and Sahni [4] to implement $Relax$ (i.e., maintain $PATH(u)$ in ascending order of w_1; extend the new paths in 1 step; then merge these extensions with $PATH(v)$ in another step).

---

[*] A path is new iff it hasn't been the subject of a similar extension attempt on a previous round.

$RelaxIPH(u, v)$

1.   **for** each new $p \in PATH(u)$ such that $w(p) + w(u, v) \le c$ **do**
2.   **if** $(v = d)$ **return** TRUE;
3.   Let $r = p||(u, v)$;
4.   Let $q \in PATH(v)$ such that r and q fall in the same bucket;
5.   **if** (there is no such q)
6.   Add r to $PATH(v)$; Change = TRUE;
7.   **else if** $(w_1(r) < w_1(q))$
8.   Replace q by r in $PATH(v)$; Change = TRUE;
9.   **return** FALSE;

FIGURE 9.5   Relax method for interval partitioning heuristic (IPH).

9.6.4 Interval Partitioning and Separation

In interval partitioning, we partition the space of $[w_2(p), w_3(p), \cdots, w_k(p)]$ values into buckets of size $[\delta_2, \delta_3, \cdots, \delta_k]$. $PATH(u)$ is maintained so as to have at most one path in each bucket. When a *Relax* step attempts to put a second path into a bucket, only the path with the smaller w_1 value is retained. When the δ_is are chosen as in Section 9.6.3, we get a fully polynomial time approximation scheme. By choosing larger values for the δ_is, we lose the guarantee of an ϵ-approximate solution but we reduce the run time. We use the term *interval partitioning heuristic* (IPH) to refer to the interval partitioning algorithm in which the δ_is are chosen arbitrarily.

Figure 9.5 gives the relax method used by IPH. The driver *extended Bellman-Ford* is unchanged. By choosing the number of buckets (equivalently, the bucket size) as in Section 9.6.3, we get a fully polynomial time ϵ-approximation scheme. The proof of this claim is quite similar to that of the proof provided in Section 9.6.3.

Theorem 9.1 *IPH is an ϵ-approximation algorithm for k-MCP when the bucket size is chosen as in Section 9.6.3.*

In the separation method, $PATH(u)$ is such that no two paths of $PATH(v)$ are within $\delta_i/2$ of their w_i values for $2 \le i \le k$. So, if we attempt to add to $PATH(v)$ a path q such that $w_i(p) + \delta_i/2 \le w_i(q) \le w_i(p) + \delta_i/2$, $2 \le i \le k$, where $p \in PATH(v)$, then only the path with the smaller w_1 value is retained.

Since separation comes with greater implementation overheads than associated with interval partitioning [20] focuses on of the interval partitioning method for k-MCP.

9.6.5 The Heuristics of Yuan [14]

The LGH of Yuan [14] combines the interval partitioning and rounding methods. $PATH(v)$ is represented as a $(k - 1)$-dimensional array with each array position representing a bucket of size $[s_2, s_3, \cdots, s_k]$. As in the pure form of interval partitioning, each bucket can have at most one path. However, unlike interval partitioning, the exact w_i values of the retained path are not stored. Instead, the w_i values, $2 \le i \le k$ are rounded up to the maximum possible for the bucket; the smallest w_1 value of the paths that fall into a bucket is stored in the bucket. Note that because of the rounding of the w_i values, $2 \le i \le k$, we do not store these values along with the path; they may be computed as needed from the bucket indexes.

We may regard the LGH as one with delayed rounding; the rounding done at the outset when the traditional rounding method is used, is delayed to the time a path is actually constructed. By incorporating buckets, we eliminate the need to store the w_i, $2 \le i \le k$ values stored explicitly with each path when either the rounding or interval partitioning methods are used. Although there is a reduction in space (by a factor of k) on a per path basis, the array of buckets used to implement each $PATH(u)$ needs $\prod_{2 \le i \le k} c_i/s_i$ space, whereas when the w_is are explicitly stored, the space requirements can be reduced to $O(k *$ total number of paths stored). The time complexity of LGH is $O(ne \prod_{2 \le i \le k} c_i/s_i)$.

Note that when $s_i = \delta_i$, $2 \le i \le k$, the LGH becomes an ϵ-approximation algorithm.

The limited path heuristic (LPH) of Yuan [14] limits the size of $PATH(v)$ to be X, where X is a specified parameter. It differs from *Extended Bellman-Ford* (Figure 9.4) only in that line 9 is changed to **if** (*Flag* == *True* && |$PATH(v)$| < X). With this modification, the complexity of *extended Bellman-Ford* becomes $O(neX^2)$. The success of LPH hinges on the expectation that the first X nondominated paths, to vertex v, found by *extended Bellman-Ford* are more likely to lead to a feasible path to the destination than subsequent paths to v. In a pure implementation of the Bellman-Ford method (which Figure 9.4 is not), this expectation may be justified with the expectation that paths to nondestination vertices with a smaller number of edges (these are found first in a pure Bellman-Ford algorithm) are more likely to lead to a feasible path to the destination than those with a larger number of edges.

9.6.6 Generalized Limited Path Heuristic (GLPH)

LPH limits the number of paths in $PATH(u)$ to be at most X. In GLPH, the constraint on the number of paths is

$$\sum_{u \in V, u \neq s} |PATH(u)| \leq (n-1) * X$$

While both LPH and GLPH place the same limit on the total number of paths retained (i.e., $(n-1) * X$), LPH accomplishes this by explicitly restricting the number of paths in each $PATH(u)$, $u \neq s$ to be no more than X.

To ensure a performance at least as good as that of LPH, GLPH ensures that each $PATH(u)$ maintains a superset of the $PATH(u)$ maintained by LPH. So, GLPH permits the size of a $PATH(u)$ to exceed X so long as the sum of the sizes is no more than $(n-1) * X$. When the sum of the sizes equals $(n-1) * X$, we continue to add paths to those $PATH(u)$s that have fewer than X paths. However, each such addition is accompanied by the removal of a path that would not be in any $PATH(v)$ of LPH.

9.6.7 Hybrid Interval Partitioning Heuristics (HIPHs)

Although IPH becomes an ϵ-approximation algorithm when the bucket size is chosen appropriately, LPH is expected to perform well on many real-world networks because we expect paths with a small number of edges to be more likely to lead to feasible source-destination paths than those with a large number of edges. In this section we describe four hybrid heuristics: HIPH1, HIPH2, HIPH3, and HIPH4.

HIPH1 and HIPH2 combine IPH and LPH into a unified heuristic that has the merits of both. HIPH1 maintains two sets of paths for each vertex $u \in V$. The first set $PATH(u)$ is limited to have at most X paths. This set is a faithful replica of $PATH(u)$ as maintained by LPH. The second set, $ipPATH(u)$ uses interval partitioning to store additional paths found to vertex u. For the source vertex s, $PATH(s) = \{s\}$ and $ipPATH(s) = \emptyset$. Figure 9.6 gives the new relax method employed by HIPH1. It is easy to see that if on entry to *RelaxHIPH1*, $PATH(u)$ as maintained by HIPH1 is the same as that maintained by the relax method of LPH, then on exit, $PATH(v)$ is the same for both HIPH1 and LPH. Since both heuristics start with the same $PATH(u)$ for all u, both maintain the same $PATH(u)$ sets throughout. Hence HIPH1 produces a feasible solution whenever LPH does. Further, because HIPH1 maintains additional paths in $ipPATH()$, it has the potential to find feasible source-to-destination paths even when LPH fails to do so. It is easy also to see that when bucket size is selected as in Section 9.6.3, HIPH1 is an ϵ-approximation algorithm.

Theorem 9.2 *HIPH1 is an ϵ-approximation algorithm for k-MCP when the bucket size for ipPATH() is chosen as in Section 9.6.3. Further, for any given X, HIPH1 finds a feasible source-to-destination path whenever LPH finds such a path.*

HIPH2 is quite similar to HIPH1. In HIPH1 the extension $r = p||(u, v)$ of a path $p \in ipPATH(u)$ can be stored only in $ipPATH(v)$. In HIPH2, however, this extension is stored in $PATH(v)$ whenever

$RelaxHIPH1(u,v)$
1.   **for** each new $p \in PATH(u)$ such that $w(p) + w(u,v) \leq c$ **do**
2.   **if** $(v = d)$ **return** TRUE;
3.   Flag = TRUE;
4.   **for** each $q \in PATH(v)$ **do**
5.   **if** $(w(p) + w(u,v) \geq w(q))$
6.   Flag = FALSE; Break; // exit **for** loop
7.   **if** $((w(p) + w(u,v)) < w(q))$
8.   remove q from $PATH(v)$;
9.   **if** $(Flag == TRUE)$
10.   **if** $(|PATH(v)| < X)$
11.   insert $p||(u,v)$ into $PATH(v)$; Change = TRUE;
12.   **else**
13.   **do** lines 3-8 of $RelaxIPH$ using $ipPATH$ in place of $PATH$;
14.   // Relax using $ipPATH$ in place of $PATH$
15.   **return** $RelaxIPH(u,v)$;

FIGURE 9.6   Relax method for HIPH1.

$|PATH(v)| < X$. When $|PATH(v)| = X$, lines 4–8 of *RelaxIPH* are applied (using $ipPATH(v)$ in place of $PATH(v)$) to determine the fate of r. With this change, $PATH(u)$ as maintained by LPH may not be the same as that maintained by HIPH2. However, by choosing the bucket size for $ipPATH(u)$ as in Section 9.6.3, HIPH2 becomes an ϵ-approximation algorithm.

Theorem 9.3 *HIPH2 is an ϵ-approximation algorithm for k-MCP when the bucket size for ipPATH() is chosen as in Section 9.6.3.*

HIPH3 and HIPH4 are the GLPH analogs of HIPH1 and HIPH2; that is they are based on GLPH rather than LPH.

Theorem 9.4 *HIPH3 and HIPH4 are ϵ-approximation algorithms for k-MCP when the bucket size for ipPATH() is chosen as in Section 9.6.3.*

9.6.8 Performance Evaluation

The *existence ratio* (ER) and *competitive ratio* (CR) are defined, respectively, by Yuan [14] to be the number of routing requests satisfied by the extended Bellman-Ford algorithm divided by the total number of routing requests and the number of routing requests satisfied by a heuristic divided by the number satisfied by the extended Bellman-Ford algorithm. For example, if we make 500 routing requests, 100 of which are satisfiable, the ER is $100/500 = 0.2$. If LPH is able to find a feasible path for 80 of the 100 requests for which such a path exists, the CR of LPH is $80/100 = 0.8$.

Song and Sahni [20] report on an extensive simulation study involving mesh [14], power-law [22], and augmented directed chain (ADC) [20] networks. Figure 9.7 gives the smallest of the tested X values for which the CR becomes 1.0. For the case when $k = 2$, X is the bound placed on $|PATH(u)|$ and $|ipPATH(u)|$. In particular, for LGH, X is the number of positions in the 1-dimensional array used to represent each $PATH(u)$ and for IPH, X is the number of intervals for each $PATH(u)$. GLPH working on a network with n vertices is able to store at most $X * (n - 1)$ paths, which is the maximum number of paths in all $PATH(u)$ lists of LPH. For the hybrid heuristics HIPH1 and HIPH2, $|PATH(u)| \leq X$ and $|ipPATH(u)| \leq X$. For HIPH3 and HIPH4, $\sum |PATH(u)| \leq X * (n - 1)$ and $|ipPATH(u)| \leq X$. Note that since every heuristic other than LGH stores both w_1 and w_2 for each path while LGH stores only w_1, the worst-case space requirements of LGH for any X are one-half that for LPH and GLPH and one-fourth that for HIPH1 through HIPH4. In Figure 9.7, X values labeled with a $*$ indicate that the CR becomes almost 1.0, more precisely, larger than 0.99. So, for example, the entry $8 * /16$ for GLPH, HIPH3, and HIPH4 working on 16×16 unbiased meshes means that these heuristics achieved a CR very close to 1.0

DataSet	Algorithm							
	LGH	LPH	IPH	GLPH	HIPH1	HIPH2	HIPH3	HIPH4
8×8 mesh, $k = 2$, unbiased	-	8*	-	4	8	8	4	4
16×16 mesh, $k = 2$, unbiased	-	-	-	8*/16	16*	16*	8*/16	8*/16
8×8 mesh, $k = 2$, biased	-	-	-	8	8*	4*/8	2*/4	2*/4
16×16 mesh, $k = 2$, biased	-	-	1	16	16*	1	8	1
Power–law, $k = 2$, unbiased	-	4*/8	16*	2	4*/8	4	1*/2	1*/2
Power–law, $k = 2$, biased	-	4*/8	16*	2	4*/8	4	2	2
ADC, $k = 2$	-	16	1*	8	16	1*/8	8	1*/8

FIGURE 9.7 Smallest X at which CR becomes 1.0. (From Song, M. and S. Sahni, S., *IEEE Trans. Comput.*, 55, 603–617, 2006.)

when $X = 8$ and a CR of 1.0 when $X = 16$. The - in the entry for 16×16 unbiased meshes for LGH means that the CR for LGH did not become close to 1.0 for any of the tested X values.

9.6.9 Summary

All the studied *k-MCP* heuristics, with the exception of GLPH, become ϵ-approximation schemes when the bucket size is chosen as in Section 9.6.3. Although GLPH has the same bound on total memory as does the LPH of Reference 14, GLPH provides better CR; in fact, GLPH finds a feasible path whenever LPH does and is able to find feasible solutions for several instances on which LPH fails to do so. The IPH heuristic achieves significantly better CRs than are achieved by the LGH of Reference 14. LPH and GLPH do well on graphs in which there is at least one feasible path that has a small number of edges. On ADCs that do not have such feasible paths, LPH and GLPH provide miserable performance [20]. The hybrid heuristics HIPH1 through HIPH4 combine the merits of IPH (ϵ-approximation when bucket size is chosen properly) and LPH and GLPH (guaranteed success when the graph has a feasible path with few edges). Of the 4 hybrid heuristics, HIPH4 performed best in the experiments of Reference 20.

References

1. Sahni, S., General techniques for combinatorial approximation, *Operations Research*, 25(6), 920, 1977.
2. Raghavan, R. and Thompson, C., Randomized rounding: A technique for provably good algorithms and algorithmic proof, *Combinatorica*, 7, 365, 1987.
3. Horowtiz, E., Sahni, S., and Rajasekeran, S., *Fundamentals of Computer Algorithms*, W. H. Freeman, New York, 1998.
4. Horowitz, E. and Sahni, S., Computing partitions with applications to the knapsack problem, *JACM*, 21(2), 277, 1974.
5. Ibarra, O. and Kim, C., Fast approximation algorithms for the knapsack and sum of subset problems, *JACM*, 22(4), 463, 1975.
6. Garey, M. and Johnson, D., *Computers and Intractability: A Guide to the Theory of NP-Completeness*, W. H. Freeman, San Francisco, CA, 1979.
7. Chen, S. and Nahrstedt, K., An overview of quality-of-service routing for the next generation high-speed networks: Problems and solutions, *IEEE Network Magazine*, 12, 64, 1998.
8. Kuipers, F., Korkmaz, T., and Krunz, M., An overview of constraint-based path selection algorithms for QoS routing, *IEEE Communications Magazine*, 40(12), 2002.

9. Younis, O. and Fahmy, S., Constraint-based routing in the Internet: Basic principles and recent research, *IEEE Communications Surveys and Tutorials*, 5(1), 2, 2003.
10. Jaffe, J. M., Algorithms for finding paths with multiple constraints, *Networks*, 14, 95, 1984.
11. Sahni, S., *Data Structures, Algorithms, and Applications in C++*, 2nd ed., Silicon Press, Summit, NJ, 2005.
12. Chen, S. and Nahrstedt, K., On finding multi-constrained paths, *IEEE International Conference on Communications (ICC'98)*, June 1991.
13. Korkmaz, T. and Krunz, M., A randomized algorithm for finding a path subject to multiple QoS requirements, *Computer Networks*, 36, 251, 2001.
14. Yuan, X., Heuristic algorithms for multiconstrained quality of service routing, *IEEE/ACM Transactions on Networking*, 10(2), 244, 2002.
15. Chen, S., Song, M., and Sahni, S., Two techniques for fast computation of constrained shortest paths, In *Proceedings of IEEE GLOBECOM 2004*, 2004.
16. Goel, A., Ramakrishnan, K. G., Kataria, D., and Logothetis, D., Efficient computation of delay-sensitive routes from one source to all destinations, In *IEEE INFOCOM'01*, 2001.
17. Hassin, R., Approximation schemes for the restricted shortest path problem, *Mathematics of Operations Research*, 17(1), 36, 1992.
18. Korkmaz, T. and Krunz, M., Multi-constrained optimal path selection, In *IEEE INFOCOM'01*, April 2001.
19. Lorenz, D. H. and Raz, D., A simple efficient approximation scheme for the restricted shortest path problem, *Operations Research Letters*, 28, 213, 2001.
20. Song, M. and Sahni, S., Approximation algorithms for multiconstrained quality-of-service routing, *IEEE Transactions on Computers*, 55, 603–617, 2006.
21. Widyono, R., The design and evaluation of routing algorithms for real-time channels, TR-94-024, Internaltional Computer Science Institute, UC Berkeley, 1994.
22. Faloutsos, M., Faloutsos, P., and Faloutsos, C., On power-law relationships of the Internet topology, In *ACM Proceedings of SIGCOMM '99*, 1999.
23. Sahni, S., Approximate algorithms for the 0/1 knapsack problem, *JACM*, 22(1), 115, 1975.
24. Sahni, S., Algorithms for scheduling independent tasks, *JACM*, 23(1), 116, 1976.

10

Asymptotic Polynomial Time Approximation Schemes

Rajeev Motwani

Liadan O'Callaghan

An Zhu

Tribute to Rajeev Motwani (March 26, 1962–June 5, 2009)

Rajeev Motwani's vast and various research, and his exquisite course preparation and teaching, inspired his many students, fellow researchers, and friends. His knack for identifying the ideal research problems and research collaborators for each student, and his adeptness at nudging us along the path of inquiry, unlocked for us the beauty of theoretical computer science and the tremendous satisfaction of making our own discoveries. He was so generous with his knowledge and insight, and produced such far-reaching and practical results, that his influence is felt widely throughout the technological world, guaranteeing him a legacy even among those not lucky enough to have known him.

10.1 Introduction

We illustrate the concept of asymptotic (fully) polynomial-time approximation schemes (APTASs, AFPTASs) via a study of the bin packing problem. We discuss in detail an APTAS due to Vega and Lueker [1], and an AFPTAS due to Karmakar and Karp [2]. Many of the algorithmic and analytical techniques described in this chapter can be applied elsewhere in the development and study of other polynomial-time approximation schemes (PTASs). We conclude with a brief survey of other bin packing-related results and other examples of APTAS and AFPTAS.

We first introduce the classic bin packing problem, which is $\mathcal{NP}$-complete. Informally, we are given a collection of items of sizes between 0 and 1. We are required to pack them into bins of unit size so as to minimize the number of bins used. Thus, we have the following minimization problem.

BIN PACKING (**BP**):

- *Instances*: $I = \{s_1, s_2, \dots s_n\}$, such that $\forall i, s_i \in [0,1]$.
- *Solutions*: A collection of subsets $\sigma = \{B_1, B_2, \dots B_k\}$, which is a disjoint partition of I, such that $\forall i, B_i \subset I$ and $\sum_{j \in B_i} s_j \leq 1$.
- *Value*: The value of a solution is the number of bins used, or $f(\sigma) = |\sigma| = k$.

BIN PACKING is a perfect illustration of why sometimes the absolute performance ratio is not the best possible definition of the performance guarantee for an approximation algorithm. Recall that the absolute performance ratio, a.k.a. the approximation ratio, of an algorithm A for a minimization problem is defined as:

$$R_A = \inf\{r \mid R_A(I) = \frac{A(I)}{OPT(I)} < r, \forall I\},$$

where $A(I)$ and $OPT(I)$ denote the value of algorithm A's solution and the optimal solution for instance I, respectively.[*] Note that the problem of deciding if an instance of BIN PACKING has a solution with 2 bins is $\mathcal{N}P$-complete—this is exactly the PARTITION problem [3]. This implies no algorithm can guarantee an approximation ratio better than 3/2 for BIN PACKING. Consequently, no approximation schemes, PTAS or FPTAS [4], exist for BIN PACKING.

The hardness of 3/2 comes from the fact that we cannot decide between 2 bins or 3 bins, a difference of 1 bin only. It is the small value of the optimum solution that makes the approximation ratio appear to be large; the approximation ratio is misleading, since on larger instances the ratio could still be bounded by a small constant. Therefore, we introduce the asymptotic performance ratio:

Definition 10.1 *The* **asymptotic performance ratio**, R_A^∞, *of an approximation algorithm A for an optimization problem is*

$$R_A^\infty = \inf\{r \mid \exists N_0, R_A(I) \leq r \text{ for all } I \text{ with } OPT(I) \geq N_0\}$$

For BIN PACKING, the 3/2-hardness result does not preclude the existence of *asymptotic* approximation schemes, which give an approximation factor that approaches 1 in the limit:

Definition 10.2 *An APTAS is a family of algorithms $\{A_\epsilon \mid \epsilon > 0\}$ such that each A_ϵ runs in time polynomial in the length of the input and $R_{A_\epsilon}^\infty \leq 1 + \epsilon$.*

Definition 10.3 *An AFPTAS is a family of algorithms $\{A_\epsilon \mid \epsilon > 0\}$ such that each A_ϵ runs in time polynomial in the length of the input and $1/\epsilon$, whereas $R_{A_\epsilon}^\infty \leq 1 + \epsilon$.*

In this chapter we present two algorithms, an APTAS and an AFPTAS, due to Vega and Lueker [1] and Karmakar and Karp [2], respectively, for BIN PACKING. The algorithmic and analytic tools demonstrated here are widely applicable to the study and development of approximation schemes. Some of the techniques, such as interval partitioning, have been applied to similar problems such as multiprocessor scheduling, knapsack [3], and various packing-related problems and their generalizations. Other techniques are more general and apply in a broader range of problem settings; for instance, linear programming is a very powerful tool and has been used with enormous success throughout operations research, management science, and theoretical computer science.

The rest of this chapter is organized as follows: Section 10.2 presents a summary of the techniques used in the two algorithms; Section 10.3 presents the APTAS; Section 10.4 presents the AFPTAS; finally, Section 10.5 summarizes some other results related to BIN PACKING, and lists some other examples of APTAS and AFPTAS.

---

[*] $R_A(I)$, the absolute performance ratio of algorithm A on an input instance I, is defined as $OPT(I)/A(I)$ for maximization problems. Such definition ensures that $R_A(I) \geq 1$ always.

10.2 Summary of Algorithms and Techniques

The first result we present is due to Vega and Lueker [1], who provided an APTAS for BIN PACKING that runs in *linear time* and has $A_\epsilon(I) \leq (1 + \epsilon) \cdot OPT(I) + 1$. To be more specific, the running time is linear in the size of the input instance I but is severely exponential in ϵ. Note that the reason this scheme is an APTAS, and not a PTAS, is the additive error term of 1 in the approximation bound. The basic techniques used in this result may be summarized as follows:

- Separate handling of "small" items.
- Discretization via interval partitioning or linear grouping.
- Rounding of "fractional" solutions.

We then present the modification of this result due to Karmakar and Karp [2] which leads to an AFPTAS for BIN PACKING. They give an approximation scheme with a performance guarantee similar to the one described earlier, with running time improved to $O\left(\frac{n \log n}{\epsilon^8}\right)$.

We now derive the results described earlier. Our presentation combines the methods of Vega and Lueker with those of Karmakar and Karp, as the two techniques share many of the same basic tools. The general approach used in both techniques is as follows: We first define a restricted version of the problem in which all items are of at least some minimum size, and the item sizes can only take on a few distinct values. This new version of BIN PACKING turns out to be reasonably easy to solve. Then we provide a two-step reduction from the original problem instance to a restricted problem instance. The first step is to pull out the "small" items; it is shown that given any packing of the remaining items, the small items can be added back in without a significant increase in the number of bins used. The second step is to divide the item sizes into m intervals, and replace all items in the same interval by items of the same size. It turns out that this "linear grouping" affects the value of the optimal solution only marginally. In the next two sections, we consider each of these ingredients in turn and finally show how they can be combined to produce an APTAS and then an AFPTAS.

10.3 Asymptotic Polynomial-Time Approximation Scheme

Definition 10.4 *For any instance $I = \{s_1, \ldots, s_n\}$, let $SIZE(I) = \sum_{i=1}^n s_i$ denote the total size of the n items.*

Recall that $OPT(I)$ denotes the value of the optimal solution, that is, the minimum number of unit size bins needed to pack the items. We now give two inequalities relating these quantities.

Lemma 10.1 $SIZE(I) \leq OPT(I) \leq |I| = n$.

Proof. In the optimal solution, at best each bin is filled to its maximum capacity, that is, 1. Thus the total number of bins needed is at least $SIZE(I)/1$, proving $SIZE(I) \leq OPT(I)$. Since each item is of size between 0 and 1, putting each item in a separate bin is clearly a feasible (if not optimal) solution, proving $OPT(I) \leq |I| = n$. ∎

Lemma 10.2 $OPT(I) \leq 2 \cdot SIZE(I) + 1$.

Proof. Prove by contradiction. Suppose this is not the case, that is, there exists an instance I, where $OPT(I) > 2 \cdot SIZE(I) + 1$. Then in the optimal solution, there must exist at least two bins that are at least half empty. Otherwise, we have at least $OPT(I) - 1$ number of bins that are at least half full, that is, $(OPT(I) - 1)/2 \leq SIZE(I)$, which contradicts our initial assumption. Now the fact that we have two bins at least half empty contradicts the assumption that we have an optimal solution. We could have easily combined the two bins into one, and reduce the number of bins used by 1. Thus our initial assumption must be false, proving the lemma. ∎

We will represent an instance I as an *ordered* list of items $I = s_1 s_2 \ldots s_n$ such that $1 \geq s_1 \geq s_2 \geq \cdots \geq s_n > 0$.

Definition 10.5 *Let $I_1 = x_1 x_2 \ldots x_n$ and $I_2 = y_1 y_2 \ldots y_n$ be two instances of equal cardinality. The instance I_1 is said to* **dominate** *the instance I_2, or $I_1 \succeq I_2$, if it is the case that $x_i \geq y_i$, for all i.*

The following lemma follows from the fact that any feasible packing of I_1 gives a feasible packing of I_2, using the same number of bins.

Lemma 10.3 *Let I_1 and I_2 be two instances of equal cardinality such that $I_1 \succeq I_2$. Then, $SIZE(I_1) \geq SIZE(I_2)$ and $OPT(I_1) \geq OPT(I_2)$.*

We define a restricted version of BIN PACKING as follows. Suppose that the item sizes in I take on only m distinct values. Now the instance I can be represented as a multiset of items which are drawn from these m types of items.

Definition 10.6 *Suppose that we are given m distinct item sizes $V = \{v_1, \ldots, v_m\}$, such that $1 \geq v_1 > v_2 > \cdots > v_m > 0$, and an instance I of items whose sizes are drawn only from V. Then, we can represent I as multiset $M_I = \{n_1 : v_1, n_2 : v_2, \ldots, n_m : v_m\}$, where n_i is a nonnegative integer denoting the number of items in I of size v_i.*

It follows that $|M_I| = \sum_{i=1}^{m} n_i = n$, $SIZE(M_I) = \sum_{i=1}^{m} n_i v_i = SIZE(I)$ and $OPT(M_I) = OPT(I)$. We now define RBP, the restricted version of BIN PACKING.

Definition 10.7 *For all $0 < \delta < 1$ and positive integers m, the problem $RBP[\delta, m]$ is defined as* BIN PACKING *restricted to instances where the item sizes take on at most m distinct values and the size of each item is at least δ.*

Next we show how to approximately solve *RBP* via a linear programming formulation.

10.3.1 Restricted Bin Packing

Assume that δ and m are fixed independently of the input size n. The input instance for $RBP[\delta, m]$ is a multiset $M = \{n_1 : v_1, n_2 : v_2, \ldots, n_m : v_m\}$ such that $1 \geq v_1 > v_2 > \cdots > v_m \geq \delta$. Let $n = |M| = \sum_{i=1}^{n} n_i$. In the following discussion we will assume that the underlying set V for M is fixed. Note that, given M, it is trivial to determine V and verify that M is a valid instance of $RBP[\delta, m]$.

Consider a packing of some subset of the items in M into a unit size bin. We can denote this packing by a multiset $\{b_1 : v_1, b_2 : v_2, \ldots, b_m : v_m\}$ such that b_i is the number of items of size v_i that are packed into the bin. More concisely, having fixed V, we can denote the packing by the m-vector $B = (b_1, \ldots, b_m)$ of nonnegative integers. We will say that two bins packed with items from M are of the same *type* if the corresponding packing vectors are identical:

Definition 10.8 *A* **bin type** *T is an m-vector $(T_1, \ldots, T_m)$ of nonnegative integers such that $\sum_{i=1}^{m} T_i v_i \leq 1$.*

Having fixed the set V, the collection of possible bin types is fully determined and is finite, because each T_i in T must take on an integer value from 0 to $\lfloor 1/v_i \rfloor$. Let $T^1, \ldots, T^q$ denote the set of all legal bin types with respect to V. Here q, the number of distinct types, is a function of δ and m. We bound the value of q in the following lemma:

Lemma 10.4 *Let $k = \lfloor \frac{1}{\delta} \rfloor$. Then,*

$$q(\delta, m) \leq \binom{m + k}{k}$$

Proof. Each type vector $T^t = (T^t_1, \ldots, T^t_m)$ has the property that, for all i, $T^t_i \geq 0$ and $\sum_{i=1}^{m} T^t_i v_i \leq 1$. It follows that $\sum_{i=1}^{m} T^t_i \leq k$, since we have a lower bound of δ on the values v_i in V. Thus, each type vector corresponds to a way of choosing m nonnegative integers whose sum is at most k. This is the same as choosing $m + 1$ nonnegative integers whose sum is exactly k. The number of such choices is an upper bound on the value of q. A standard counting argument now gives the desired bound. ∎

Consider an arbitrary feasible solution x to an instance M of $RBP[\delta, m]$. Each packed bin in this solution can be classified as belonging to one of the $q(\delta, m)$ possible types of packed bins. The solution x can therefore be specified completely by a vector giving the number of bins of each of the q types.

Definition 10.9 *A feasible solution x to an instance M of $RBP[\delta, m]$ is a q-vector of nonnegative integers, say $x = (x_1, \ldots, x_q)$, where x_t denotes the number of bins of type T^t used in x.*

Notice that not all q-vectors correspond to a feasible solution. A feasible solution must guarantee, for each i, that exactly n_i items of size v_i are packed in the various copies of the bin types. The feasibility condition can be phrased as a series of linear equations as follows:

$$\forall i \in \{1, \ldots, m\}, \ \sum_{t=1}^{q} x_t T^t_i = n_i$$

Let the matrix A be a $q \times m$ matrix whose t^{th} row is the type vector T^t, and $\vec{n} = (n_1, \ldots, n_m)$ denote the multiplicities of the various item sizes in the input instance M. Then the above-mentioned set of equations can be concisely expressed as $\vec{x}.A = \vec{n}$. The number of bins used in the solution x is simply $\vec{x}.\vec{1} = \sum_{t=1}^{q} x_t$, where $\vec{1}$ denotes all-ones vector. In fact, we have proved the following lemma:

Lemma 10.5 *The optimal solution to an instance M of $RBP[\delta, m]$ is exactly the solution to the following integer linear program, $ILP(M)$*

$$\begin{aligned} minimize \quad & \vec{x}.\vec{1} \\ subject\ to \quad & \\ & \vec{x} \geq 0 \\ & \vec{x}.A \geq \vec{n} \end{aligned}$$

We have replaced the equations by inequalities, but, since a packing of a superset of M can always be converted into a packing of M using the same number of bins, the validity of the lemma is unaffected. It is also worth noting that the matrix A is determined completely by the underlying set V; the vector $\vec{n}$, however, is not determined *a priori* but depends on the instance M.

How easy is it to obtain this integer program? Note that the number of constraints in $ILP(M)$ is exponentially large in terms of δ and m. However, we are going to assume that both δ and m are constants which are fixed independently of the length of the input, which is n. Thus, $ILP(M)$ can be obtained in time linear in n, given any instance M of cardinality n.

How about solving ILP? Recall that the integer programming problem is $\mathcal{N}P$-complete in general [3]. However, there is an algorithm due to Lenstra [5–7] which solves any ILP in time linear in the number of constraints, provided the number of variables is fixed. This is exactly the situation in ILP: the number of variables q is fixed independent of n, as is the number of constraints, which is $q + m$. Thus, we can solve ILP exactly in time independent of n. (A more efficient algorithm for approximately solving ILP will be described in a later section.) The following theorem results. Here $f(\delta, m)$ is some constant which depends only on δ and m.

Theorem 10.1 *Any instance of $RBP[\delta, m]$ can be solved in time $O(n + f(\delta, m))$.*

10.3.2 Eliminating Small Items

We now present the second ingredient of the APTAS devised by Vega and Lueker: the separate handling of small items. It is shown that if we have a packing of all items except those whose sizes are bounded from above by δ, then it is possible to add the small items back in without much increase in the number of bins. This fact is summarized in the following lemma; the rest of this subsection is devoted to the proof of this lemma.

Lemma 10.6 *Fix some constant $\delta \in (0, \frac{1}{2}]$. Let I be an instance of* BIN PACKING *and suppose that all items of size greater than δ have been packed into β bins. Then it is possible to find in linear time a packing for I which uses at most* $\max\{\beta, (1 + 2\delta) \cdot OPT(I) + 1\}$ *bins.*

Proof. The basic idea is to start with the packing of the "large" items and to use the greedy algorithm first fit (FF) to pack the "small" items into the empty space in the β bins.

FF is a classic bin-packing algorithm of historical importance, as we shall see later. The algorithm is as follows. We are given the set of items in an arbitrary order, and we start with zero bins. For each item in the list, we consider the existing bins (if any) in order and place the item in the first bin that can accommodate it. If no existing bin can accommodate it, we make a new bin after all the existing ones, and put the item in the new bin.

To use FF to add the small items into an existing packing of the large ones, we can start by numbering the β bins in an arbitrary fashion, and also ordering the small items arbitrarily. Then we run FF as usual using this ordering to decide where each small item will be placed. If at some point the small items do not fit into any of the currently available bins, a new bin is initiated.

In the best case, the small items can all be greedily packed into the β bins which were open initially. Clearly, the lemma is valid in that case. Suppose now that some new bins were required for the small items. We claim that at the end of the entire process each of the bins used for packing I has at most δ empty space in it, with the possible exception of at most one bin. To see why this claim holds, note that at the moment when the first new bin was started, each of the original bins must have had at most δ free space. Next, observe that whenever another new bin was opened, no earlier bin could have had more than δ free space. Therefore, at every moment, at most one bin had more than δ free space.

Let $\beta' > \beta$ be the total number of bins used by *FF*. We are guaranteed that all the bins, except one, are at least $1 - \delta$ full. This implies that $SIZE(I) \geq (1 - \delta)(\beta' - 1)$. But we know that $SIZE(I) \leq OPT(I)$, implying that

$$\beta' \leq \frac{1}{1 - \delta} OPT(I) + 1 \leq (1 + 2\delta) \cdot OPT(I) + 1$$

and we have the desired result. ∎

10.3.3 Linear Grouping

The final ingredient needed for the APTAS is called interval partitioning or linear grouping. This is a technique for converting an instance I of BIN PACKING into an instance M of $RBP[\delta, m]$, for an appropriate choice of δ and m, without changing the value of the optimal solution too much. Let us assume for now that all the items in I are of size at least δ, for some choice of $\delta \in (0, \frac{1}{2}]$. All that remains is to show how to obtain an instance where the item sizes take on only m different values. First, let us fix some parameter k, a nonnegative integer to be specified later. We now show how to convert an instance of $RBP[\delta, n]$ into an instance of $RBP[\delta, m]$, for $m = \lfloor n/k \rfloor$.

Definition 10.10 *Given an instance I of RBP$[\delta, n]$ and a parameter k, let $m = \lfloor n/k \rfloor$. Define the groups of items $G_i = s_{(i-1)k+1} \ldots s_{ik}$, for $i = 1, \ldots, m$, and let $G_{m+1} = s_{mk+1} \ldots s_n$.*

Here the group G_1 contains the k largest items in I, G_2 contains the next k largest items, and so on. The following fact is an easy consequence of these definitions.

Fact 10.1 $G_1 \succeq G_2 \succeq \cdots \succeq G_m$.

From each group G_i we can obtain a new group of items H_i by increasing the size of each item in G_i to that of the largest item in that group. The following fact is also obvious.

Definition 10.11 *Let $v_i = s_{(i-1)k+1}$ be the largest item in group G_i. Then the group H_i is a group of $|G_i|$ items, each of size v_i. In other words, $H_i = v_i v_i \ldots v_i$ and $|H_i| = |G_i|$.*

Fact 10.2 $H_1 \succeq G_1 \succeq H_2 \succeq G_2 \succeq \cdots \succeq H_m \succeq G_m$ *and* $H_{m+1} \succeq G_{m+1}$.

The entire point of these definitions is to obtain two instances of $RBP[\delta, m]$ such that their optimal solutions bracket the optimal solution for I. These instances are defined as follows.

Definition 10.12 *Let the instance $I_{LO} = H_2 H_3 \ldots H_{m+1}$ and $I_{HI} = H_1 H_2 H_3 \ldots H_{m+1}$.*

Note that I_{LO} is an instance of $RBP[\delta, m]$. Moreover, it is easy to see that $I \preceq I_{HI}$. We now present some properties of these three instances.

Lemma 10.7

$$OPT(I_{LO}) \leq OPT(I) \leq OPT(I_{HI}) \leq OPT(I_{LO}) + k$$
$$SIZE(I_{LO}) \leq SIZE(I) \leq SIZE(I_{HI}) \leq SIZE(I_{LO}) + k$$

Proof. First, observe that

$$I_{LO} = H_2 H_3 \ldots H_m H_{m+1} \preceq G_1 G_2 \ldots G_{m-1} X,$$

where X is any set of $|H_{m+1}|$ items from G_m. The right hand side of this inequality is a subset of I, and so, from Lemma 10.3, $OPT(I_{LO}) \leq OPT(I)$ and $SIZE(I_{LO}) \leq SIZE(I)$. Similarly, since $I \preceq I_{HI}$, $OPT(I) \leq OPT(I_{HI})$, and $SIZE(I) \leq SIZE(I_{HI})$.

Observe now that $I_{HI} = H_1 I_{LO}$. Given any packing of I_{LO}, we can obtain a packing of I_{HI} which uses at most k extra bins. (Just pack each item in H_1 in a separate bin.) This implies that $OPT(I_{HI}) \leq OPT(I_{LO}) + k$ and $SIZE(I_{HI}) = SIZE(I_{LO}) + SIZE(H_1) \leq SIZE(I_{LO}) + k$. ∎

It is worth noting that the result presented in this lemma is constructive. It is possible in $O(n \log n)$ time to construct the instances I_{LO} and I_{HI}, and given an optimal packing of I_{LO} it is possible to construct a packing of I that meets the guarantee of the above-mentioned lemma. To construct I_{LO} and I_{HI}, it is necessary only to sort the items and perform the linear grouping. (Actually, one ingredient is still unspecified, namely the value of k; this will be given in the next section.) Given a packing of I_{LO}, we can assign all elements in $I \setminus G_1$ to bins according to the assignments of the corresponding members of I_{LO}; finally, each member of G_1 can get its own bin.

10.3.4 APTAS for Bin Packing

We now put together all these ingredients and obtain the APTAS. The algorithm A_ϵ, for any $\epsilon \in (0, 1]$, takes as input an instance I of BIN PACKING consisting of n items.

ALGORITHM 10.1 A_ϵ

Input: Instance I consisting of n item sizes $\{s_1, \ldots, s_n\}$.
Output: A packing into unit-sized bins.
1. $\delta \leftarrow \frac{\epsilon}{2}$
2. Set aside all items of size smaller than δ, obtaining an instance J of $RBP[\delta, n']$ with $n' = |J|$.
3. $k \leftarrow \left\lceil \frac{\epsilon^2}{2} n' \right\rceil$
4. Perform linear grouping on J with parameter k. Let J_{LO} be the resulting instance of $RBP[\delta, m]$ and
     $J_{HI} = H_1 \cup J_{LO}$, with $|H_1| = k$ and $m = \left\lfloor \frac{n'}{k} \right\rfloor$.
5. Pack J_{LO} optimally using Lenstra's algorithm on $ILP(J_{LO})$.
6. Pack the k items in H_1 into at most k bins.
7. Obtain a packing of J using the same number of bins as in steps 5 and 6, by replacing each item in
     J_{HI} by the corresponding (smaller) item in J.
8. Using FF, pack all the small items set aside in step 2, using new bins only if necessary.

How many bins does A_ϵ use in the worst case? Observe that we have packed the items in J_{HI}, hence the items in J, into at most $OPT(J_{LO}) + k$ bins. Consider the now the value of k in terms of the optimal solution. Since all items have size at least $\epsilon/2$ in J, it must be the case that $SIZE(J) \geq \epsilon n'/2$. This implies that

$$ k \leq \frac{\epsilon^2 n'}{2} + 1 \leq \epsilon \cdot SIZE(J) + 1 \leq \epsilon \cdot OPT(J) + 1 $$

Using Lemma 10.7, we obtain that J is packed into a number of bins not exceeding

$$ OPT(J_{LO}) + k \leq OPT(J) + \epsilon \cdot OPT(J) + 1 \leq (1 + \epsilon) \cdot OPT(J) + 1 $$

Finally, Lemma 10.6 implies that, whereas packing the small items at the last step, we use a number of bins not exceeding

$$ \max\{(1 + \epsilon) \cdot OPT(J) + 1, (1 + \epsilon) \cdot OPT(I) + 1\} \leq (1 + \epsilon) \cdot OPT(I) + 1 $$

since $OPT(J) \leq OPT(I)$. We have obtained the following theorem.

Theorem 10.2 *The algorithm A_ϵ finds a packing of I into at most $(1 + \epsilon) \cdot OPT(I) + 1$ bins in time $c(\epsilon)n \log n$, where $c(\epsilon)$ is a constant depending only on ϵ.*

For the running time, note that the only really expensive step in the algorithm is the one where we solve ILP using Lenstra's algorithm. As we observed earlier, that this requires time linear in n, although it may be severely exponential in δ and m, which are functions of ϵ.

10.4 Asymptotic Fully Polynomial-Time Approximation Scheme

Our next goal is to convert the preceding APTAS into an AFPTAS. The reason that the above-mentioned scheme is not fully polynomial is the use of the algorithm for integer linear programming which requires time exponential in $1/\epsilon$. We now describe a technique for getting rid of this step via the construction of a "fractional" solution to the restricted bin-packing problem, and a "rounding" of this to a feasible solution which is not very far from optimal. This is based on the ideas due to Karmakar and Karp.

10.4.1 Fractional Bin Packing and Rounding

Consider again the problem $RBP[\delta, m]$. By the preceding discussion, any instance I of this problem can be formulated as $ILP(I)$.

$$
\begin{aligned}
\text{minimize} \quad & \vec{x}.\vec{1} \\
\text{subject to} \quad & \\
& \vec{x} \geq 0 \\
& \vec{x}.A = \vec{n}
\end{aligned}
$$

Notice that we are stating the last constraint as we originally did: as an equality. Recall that A is a $q \times m$ matrix, $\vec{x}$ is a q-vector and $\vec{n}$ is an m-vector. The bin types matrix A, as well as $\vec{n}$, are determined by the instance I.

Consider now the linear programming relaxation of $ILP(I)$. This system $LP(I)$ is exactly the same as $ILP(I)$, except that we now relax the requirement that $\vec{x}$ be an integer vector. Recall that $SIZE(I)$ is the total size of the items in I, and that $OPT(I)$ is the value of the optimal solution to $ILP(I)$ as well as the smallest number of bins into which the items of I can be packed.

Definition 10.13 *$LIN(I)$ is the value of the optimal solution to $LP(I)$, the linear programming relaxation of $ILP(I)$.*

What does a noninteger solution to $LP(I)$ mean? The value of x_i is a real number which denotes the number of bins of type T^i which are used in the optimal packing. One may interpret this as saying that items can be "broken up" into fractional parts, and these fractional parts can then be packed into fractional bins. This in general would give us a solution of value $SIZE(I)$, but keep in mind that the constraints in $LP(I)$ do not allow any arbitrary "fractionalization." The constraints require that in any fractional bin, the items packed therein must be the same fraction of the original items. Thus, this solution does capture some of the features of the original problem. We will refer to the solution of $LP(I)$ as a *fractional bin packing*.

To analyze the relationship between the fractional and integral solutions to any instance we will have to use some basic facts from the theory of linear programming. The uninitiated reader is referred to any standard text-book for a more complete treatment; for example, see the book by Papadimitriou and Steiglitz [8].

Consider the system of linear equations implicit in the constraint* $\vec{x}.A = \vec{n}$. Here we have m linear equations in q variables, where q is much larger than m. This is an under-constrained system of equations. Let us assume that $rank(A) = m$; it is easy to modify the following analysis when $rank(A) < m$. Assume, without loss of generality, that the first m rows of A form a basis, that is, they are linearly independent. The following are standard observations from linear programming theory.

Definition 10.14 *A **basic feasible solution** to LP is a solution $\vec{x}^*$ such that only the entries corresponding to the basis of A are non-zero. In other words, $x_i^* = 0$ for all $i > m$.*

Fact 10.3 *Every LP has an **optimal** solution which is a basic feasible solution.*

We can now derive the following lemma which relates $LIN(I)$ to both $SIZE(I)$ and $OPT(I)$.

Lemma 10.8 *For all instances I of $RBP[\delta, m]$,*

$$
SIZE(I) \leq LIN(I) \leq OPT(I) \leq LIN(I) + \frac{m+1}{2}
$$

---

* We will ignore the nonnegativity constraints for now as they do not bear upon the following discussion.

Proof. To prove the first inequality, we note that $SIZE(I) = \sum_{j=1}^{m} n_j v_j = \sum_{j=1}^{m} (\vec{x} \cdot A_j) v_j$, where we use A_j to mean the jth column of A. This sum is equal to $\sum_{i=1}^{q} x_i (\sum_{j=1}^{m} a_{ij} v_j)$. Note that for all $1 \leq i \leq q$, $\sum_{j=1}^{m} a_{ij} v_j \leq 1$ is the total size accounted for by the ith bin type and is therefore at most 1. It follows that $SIZE(I) \leq \sum_{i=1}^{q} x_i = LIN(I)$. The second inequality follows from the observation that an optimal solution to $ILP(I)$ is also a feasible solution to $LP(I)$.

To see the last inequality, fix I and let $\vec{y}$ be some basic optimal solution to $LP(I)$. Since $\vec{y}$ has at most m non-zero entries, it uses only m different types of bins. Rounding up the value of each component of $\vec{y}$ will increase the number of bins by at most m, and will yield a solution to ILP. The bound promised in the lemma is slightly stronger and may be observed as follows. Define the vectors $\vec{w}$ and $\vec{z}$ in the following way:

$$\forall i, \quad w_i = \lfloor y_i \rfloor$$
$$\forall i, \quad z_i = y_i - w_i$$

The vector $\vec{w}$ is the integer part of the solution and $\vec{z}$ is the fractional part. Let J denote the instance of $RBP[\delta, m]$ that consists of the items not packed in the (integral) solution specified by $\vec{w}$. (Note that J is, indeed, a legal instance of $RBP[\delta, m]$, that is, all items occur in integral quantities, because in $\vec{w}$, all bin types, and therefore all items, occur in integral quantities.) The vector $\vec{z}$ gives a fractional packing of the items in J, such that each of the m bin types is used a number of times which is a fraction less than 1.

Just as $SIZE(I) \leq LIN(I)$, a similar argument implies that

$$SIZE(J) \leq LIN(J)$$

By Lemma 10.2 we know that

$$OPT(J) \leq 2 \cdot SIZE(J) + 1$$

It is also obvious that $OPT(J) \leq \sum_{i=1}^{m} z_i \leq m$, since rounding each non-zero z_i up to 1 gives a feasible packing of J. Thus,

$$OPT(J) \leq \min\{m, 2 \cdot SIZE(J) + 1\}$$
$$\leq (m + 2 \cdot SIZE(J) + 1)/2$$
$$= SIZE(J) + \frac{m+1}{2}$$

We will now bound $OPT(I)$ in terms of $LIN(I)$ and m.

$$OPT(I) \leq OPT(I - J) + OPT(J)$$
$$\leq \sum_{i=1}^{m} w_i + \left(SIZE(J) + \frac{m+1}{2} \right)$$
$$\leq \sum_{i=1}^{m} w_i + LIN(J) + \frac{m+1}{2}$$
$$\leq \sum_{i=1}^{m} w_i + \sum_{i=1}^{m} z_i + \frac{m+1}{2}$$
$$= LIN(I) + \frac{m+1}{2}$$

The first inequality follows from the fact that independent integer packings of $I - J$ and J can be combined to form an integer packing of I. The second and third follow from facts proved earlier, and the fact that $\vec{w}$

is a feasible solution to the $RBP[\delta, m]$ instance I. The fourth holds because $\vec{z}$ is a feasible fractional packing of J. Finally, the equality holds by the optimality of $\vec{y}$ as a solution to $LIN(I)$. ∎

It is not very hard to see that all of the above is constructive. More precisely, given the solution to $LP(I)$, we can construct in linear time a solution to I such that the bound from the above-mentioned theorem is met: We take an optimal basic solution $\vec{y}$ and break it into $\vec{w}$ and $\vec{z}$ as described, and define J as earlier. We find an integral solution for J either by rounding up each non-zero entry to 1 or by using FF, whichever produces a better solution. We then put together the solution given by $\vec{w}$ and that found for J.

The only problem is that it is not obvious that we can solve the linear program in fully polynomial time, even though there exist polynomial-time algorithms for linear programming [9], unlike the general problem of integer programming. The reason is that the number of variables is still exponential in $1/\epsilon$. All we have achieved is that we no longer need to solve an integer program.

Karmakar and Karp show how to get around this problem by resorting to the Ellipsoid method of Grötschel et al. [6,7,10]. In this method, it is possible to solve a linear program with an exponential number of constraints in time which is polynomial in the number of variables and the number sizes, given a *separation oracle*. A separation oracle takes any proposed solution vector $\vec{x}$ and either guarantees that it is a feasible solution, or provides any one constraint which is violated by it. Karmakar and Karp gave an efficient construction of a separation oracle for $LP(I)$. This would result in a polynomial-time algorithm for $LP(I)$ if it had a small number of variables, even if it has an exponential number of constraints. Since our situation is exactly the reverse, that is, we have a small number of constraints and an exponential number of variables, we will consider the *dual linear program* for $LP(I)$, which has the desired features of a small number of variables. By linear program duality, its optimal solution corresponds exactly to the optimal solution of $LP(I)$.

One important detail is that that it is impossible to solve $LP(I)$ exactly in fully polynomial time. However, it can be solved within an additive error of 1 in fully polynomial time. Moreover, the implementation of the separation oracle is in itself an approximation algorithm. The idea behind this method is due to Gilmore and Gomory [11] who observed that, in the case of an infeasible proposed solution, a violated constraint can be computed via the solution of a knapsack problem. Since this problem is $\mathcal{N}P$-complete, one must resort to the use of an approximation scheme for KNAPSACK [3], and so the solution of the dual is not exact but a close approximation. Karmakar and Karp used this approximate solution to the dual to obtain an approximate lower bound on the optimal value of the original problem. Having devised the procedure for efficiently computing an approximate lower bound, they then construct an approximate solution.

This algorithm is rather formidable and the details are omitted as it is outside the scope of this discussion. The following theorem results.

Theorem 10.3 *There is a fully polynomial-time algorithm A for solving an instance I of $RBP[\delta, m]$ such that $A(I) \leq LIN(I) + \frac{m+1}{2} + 1$.*

10.4.2 AFPTAS for Bin Packing

We are now ready to present the AFPTAS for BIN PACKING. We will need the following variant of Lemma 10.7.

Lemma 10.9 *Using the linear grouping scheme on an instance I of $RBP[\delta, n]$, we obtain an instance I_{LO} of $RBP[\delta, m]$ and a group H_1 such that, for $I_{HI} = H_1 I_{LO}$,*

$$LIN(I_{LO}) \leq LIN(I) \leq LIN(I_{HI}) \leq LIN(I_{LO}) + k$$

Proof. The proof is almost identical to that of Lemma 10.7. Recall that $m = \lfloor n/k \rfloor$. Take the original instance I, and define $G_1, \ldots, G_{m+1}, H_1, \ldots, H_m, I_{LO}$, and I_{HI} as before. From Lemma 10.3 the first two inequalities follow. The third follows from the fact that, given a solution to I_{LO}, we can solve I_{HI} by putting all members of $I_{HI} \cup I_{LO}$ in the bins assigned by the given solution, and then putting each member of H_1 in a bin by itself.   ∎

The basic idea behind the AFPTAS of Karmakar and Karp is very similar to that used in the APTAS. We first eliminate all the small items, and then apply linear grouping to the remaining items. The resulting instance of $RBP[\delta, m]$ is then formulated as an *ILP*, and the solution to the corresponding relaxation *LP* is computed using the Ellipsoid method. The fractional solution is then rounded to an integer solution. The small items are then added into the resulting packing exactly as before.

ALGORITHM 10.2 A_ϵ

Input: Instance I consisting of n item sizes $\{s_1, \ldots, s_n\}$.
Output: A packing into unit-sized bins.
1. $\delta \leftarrow \frac{\epsilon}{2}$.
2. Set aside all items of size smaller than δ, obtaining an instance J of $RBP[\delta, n']$ with $n' = |J|$.
3. $k \leftarrow \left\lceil \frac{\epsilon^2 n'}{2} \right\rceil$
4. Perform linear grouping on J with parameter k. Let J_{LO} be the resulting instance of $RBP[\delta, m]$ and
   $J_{HI} = H_1 \cup J_{LO}$, with $|H_1| = k$ and $m = \left\lfloor \frac{n'}{k} \right\rfloor$.
5. Pack the k items in H_1 into at most k bins.
6. Pack J_{LO} using the Ellipsoid method and rounding the resulting fractional solution.
7. Obtain a packing of J using the same number of bins as used for J_{HI}, by replacing each item
   in J_{HI} by the corresponding (smaller) item in J.
8. Using *FF*, pack all the small items set aside in step 2, using new bins only if necessary.

Theorem 10.4 *The approximation scheme $\{A_\epsilon : \epsilon > 0\}$ is an AFPTAS for* BIN PACKING *such that*

$$A_\epsilon(I) \leq (1 + \epsilon) \cdot OPT(I) + \frac{1}{\epsilon^2} + 3$$

Proof. The running time is dominated by the time required to solve the linear program, and we are guaranteed that this is fully polynomial.

By Lemma 10.8, the number of bins used to pack the items in J_{LO} is at most

$$(LIN(J_{LO}) + 1) + \frac{m+1}{2} \leq OPT(I) + \frac{1}{\epsilon^2} + 2$$

given the preceding lemmas and the choice of m. The number of bins used to pack the items in H_1 is at most k, which in turn can be bounded as follows using the observation that $OPT(J) \geq SIZE(J) \geq \epsilon n'/2$.

$$k \leq \left\lceil \frac{n'\epsilon^2}{2} \right\rceil \leq \epsilon \cdot OPT(J) + 1 \leq \epsilon \cdot OPT(I) + 1$$

Thus, the total number of bins used to pack the items in J cannot exceed

$$(1 + \epsilon) \cdot OPT(I) + \frac{1}{\epsilon^2} + 3$$

Lemma 10.6 guarantees that the small items can be added without an increase in the number of bins, and so the desired result follows.   ∎

10.5 Related Results

We conclude the chapter by presenting a literature survey on topics related to BIN PACKING and asymptotic approximation schemes.

BIN PACKING is a classic problem in theoretical computer science; the algorithms proposed for this problem, and the analysis of these algorithms, employ a wide variety of techniques. In the foregoing discussion, we used the fact that the FF algorithm has an asymptotic worst-case performance ratio of 2, but this is not the best bound. Ullman [12] proved an asymptotic worst-case performance bound of 17/10 for this algorithm, and subsequent papers [13–15] reduced the additive constant term from 3 to 1 or less eventually. FF is not the only algorithm considered for BIN PACKING. Many other online algorithms, semi-online algorithms, and offline algorithms have been proposed and their worst-case and average-case behavior studied extensively. We refer the reader to survey articles by Coffman, Garey and Johnson [16,17], Coffman, Csirik and Woeginger [18], and Chapters 28, 29, and 30 in this handbook for further details.

There are several commonly considered variants of the basic bin packing problem, all of which are $\mathcal{N}P$-complete. In most of these cases, it is reasonably easy to come up with bounded-ratio approximations. These variants can be classified under four main headings: packings in which the number of items per bin is bounded, packings in which certain items cannot be packed into the same bin, packings in which there are constraints (e.g., partial orders) on the way in which the items are packed, and dynamic packings in which items may be added and deleted. These variants are discussed in Chapters 29 and 30 in this handbook.

There are also some generalizations of the basic packing problem, many of which are covered in the three survey papers and chapters mentioned earlier. While some generalizations do not admit APTAS or AFPTAS, several approximation schemes have been found successfully, generally based on the ideas described earlier. Here we focus on three generalizations that admit APTAS and AFPTAS: packings into variable-sized bins, multidimensional bin packing, and BIN COVERING, the dual of BIN PACKING.

Murgolo shows an approximation scheme for the case of variable-sized bins [19]. For multidimensional bin packing, APTAS have recently been found for packing d-dimensional* cubes into the minimum number of unit cubes by Bansal and Sviridenko [20] and Correa and Kenyon [21], independently. Interestingly, the problem of packing (2-dimensional) rectangles into squares does not admit APTAS or AFPTAS [20]. However, for a more restricted version, namely, a two-stage packing of the rectangles, Caprara et al. show an AFPTAS [22]. The dual problem of BIN PACKING is BIN COVERING, in which we want to maximize the total number of bins used but must fill each bin to at least a certain capacity. Jansen and Solis-Oba show an AFPTAS for BIN COVERING [23].

BIN PACKING is not the only problem that admits APTAS and AFPTAS. Raghavan and Thompson give an APTAS for the 0-1 multicommodity flow problem [24]. Their approaches include probabilistic rounding of fractional linear-programming solutions. Cohen and Nakibli show an APTAS for a somewhat related problem, the n-hub shortest-path routing problem [25]. The goal is to minimize the overloading of links in a directed network with pairwise source-sink flows, by setting an n-hub route for each source-sink pair. This APTAS also uses probabilistic rounding. Aingworth et al. [26] show an AFPTAS for pricing Asian options on the lattice, using discretization to reduce the number of possible option values.

There are other problems that admit *absolute* approximation algorithms, that is, algorithms guaranteed to produce solutions whose costs are at most an additive constant away from the optimal. In contrast to APTAS and AFPTAS, whose approximation ratios approach a value arbitrarily *close* to 1 as the optimal cost grows, these algorithms have an asymptotic performance ratio *equal* to 1; that is, as the optimal cost grows, the approximation ratio of an absolute approximation algorithm approaches 1 itself. Examples of problems admitting absolute approximations include minimum edge coloring [27] and minimum degree spanning tree [28], where the approximate solution is guaranteed to exceed the optimal solution by at most 1. The techniques used in these algorithms, however, differ from the ones discussed in this chapter. A variation of Karmakar and Karp's ideas leads to a stronger result for BIN PACKING,

---

* Here d is assumed to be a fixed constant.

which is the construction of an approximation algorithm A that is fully polynomial and has the performance guarantee $A(I) \leq OPT(I) + O(\log^2 OPT(I))$. One is tempted to believe that there also exists an absolute approximation algorithm for BIN PACKING, that is, an algorithm that runs in polynomial time and guarantees that $A(I) \leq OPT(I) + O(1)$. The existence of such an algorithm is still an open question.

References

1. Fernandez de la Vega, W. and Lueker, G.S., Bin packing can be solved within $1 + \epsilon$ in linear time, *Combinatorica*, 1, 349, 1981.
2. Karmakar, N. and Karp, R.M., An efficient approximation scheme for the one-dimensional bin packing problem, in *Proceedings of FOCS*, The Computer Society Press, Piscataway, NJ, 1982, p. 312.
3. Garey, M.R. and Johnson, D.S., *Computers and Intractability: A Guide to the Theory of NP-Completeness*, W. H. Freeman and Company, New York, 1979.
4. Ausiello, G., Crescenzi, P., Gambosi, G., Kann, V., Marchetti-Spaccamela, A., and Protasi, M., *Complexity and Approximation: Combinatorial Optimization Problems and their Approximability Properties*, Springer-Verlag, Berlin, Germany, 1999.
5. Lenstra, H.W., Integer programming with a fixed number of variables, *Mathematics of Operations Research*, 8, 538, 1983.
6. Grötschel, M., Lovász, L., and Schrijver, A., *Geometric Algorithms and Combinatorial Optimization*, Springer-Verlag, Berlin, Germany, 1987.
7. Schrijver, A., *Theory of Linear and Integer Programming*, John Wiley & Sons, Chichester, UK, 1986.
8. Papadimitriou, C.H. and Steiglitz, K., *Combinatorial Optimization: Algorithms and Complexity*, Prentice Hall, Englewood Cliffs, NJ, 1982.
9. Karmakar, N., A new polynomial-time algorithm for linear programming, *Combinatorica*, 4, 373, 1984.
10. Grötschel, M., Lovász, L., and Schrijver, A., The ellipsoid method and its consequences in combinatorial optimization, *Combinatorica*, 1, 169, 1981.
11. Gilmore, P.C. and Gomory, R.E., A linear programming approach to the cutting-stock problem, *Operations Research*, 9, 849, 1961.
12. Ullman, J.D., The performance of a memory allocation algorithm, Technical Report 100, Princeton University, Princeton, NJ, 1971.
13. Garey, M.R., Graham, R.L., and Ullman, J.D., Worst-case analysis of memory allocation algorithms, in *Proceedings of STOC*, ACM, New York, 1972, p. 143.
14. Johnson, D.S., Demers, A., Ullman, J.D., Garey, M.R., and Graham, R.L., Worst-case performance bounds for simple one-dimensional packing algorithms, *SIAM Journal on Computing*, 3, 299, 1974.
15. Garey, M.R., Graham, R.L., Johnson, D.S., and Yao, A.C., Resource constrained scheduling as generalized bin packing, *Journal of Combinatorial Theory*, series A, 21, 257, 1976.
16. Coffman, E.G., Garey, M.R., and Johnson, D.S., Approximation algorithms for bin packing – An updated survey, in *Algorithm Design for Computer System Design*, Ausiello, G., Lucertini, M., and Serafini, P., Eds., Springer-Verlag, New York, 1984.
17. Coffman, E.G., Garey, M.R., and Johnson, D.S., Approximation algorithms for bin packing – A survey, in *Approximation Algorithms for NP-hard Problems*, Hochbaum, D.S., Ed., PWS Publishing Co., Boston, MA, 1996.
18. Coffman, E.G., Csirik, J., and Woeginger, G.J., Approximate solutions to bin packing problems, Technical Report Woe-29, Institut für Mathematik B, TU Graz, Graz, Austria, 1999.
19. Murgolo, F.D., An efficient approximation scheme for variable-sized bin packing, *SIAM Journal on Computing*, 16, 149, 1987.

20. Bansal, N. and Sviridenko, M., New approximability and inapproximability results for 2-dimensional bin packing, in *Proceedings of SODA*, Society for Industrial and Applied Mathematics, Philadelphia, PA, 2004, p. 196.

21. Correa, J.R. and Kenyon, C., Approximation schemes for multidimensional packing, in *Proceedings of SODA*, Society for Industrial and Applied Mathematics, Philadelphia, PA, 2004, p. 186.

22. Caprara, A., Lodi, A., and Monaci, M., Fast approximation schemes for two-stage, two-dimensional bin packing, *Mathematics of Operations Research*, 30, 150, 2005.

23. Jansen, K. and Solis-Oba, R., An asymptotic fully polynomial time approximation scheme for bin covering, *Theoretical Computer Science*, 306, 543, 2003.

24. Raghavan, P. and Thompson, C.D., Randomized rounding: A technique for provably good algorithms and algorithmic proofs, *Combinatorica*, 7, 365, 1987.

25. Cohen, R. and Nakibli, G., On the computational complexity and effectiveness of N-hub shortest-path routing, in *Proceedings of the 23rd Conference of the IEEE Communications Society*, IEEE, Piscataway, NJ, 2004, p. 694.

26. Aingworth, D., Motwani, R., and Oldham, J., Accurate approximations for Asian options, in *Proceedings of SODA*, Society for Industrial and Applied Mathematics, Philadelphia, PA, 2000, p. 891.

27. Vizing, V.G., On an estimate of the chromatic class of a p-graph (in Russian), *Diskret. Analiz.*, 3, 25, 1964.

28. Fürer, M. and Raghavachari, B., Approximating the minimum degree spanning tree to within one from the optimal degree, *Journal of Algorithms*, 17, 409, 1994.

11

Randomized Approximation Techniques

Sotiris Nikoletseas

Paul Spirakis

11.1 Introduction

Randomization (i.e., the use of random choice as an algorithmic step) is one of the most interesting tools in designing efficient algorithms. A remarkable property of randomized algorithms is their structural simplicity. In fact, in several cases, although the known deterministic algorithms are quite involved, the randomized ones are simpler and much easier to code.

This happens also in the case of approximation algorithms to NP-hard problems. In fact, it is exactly the area of efficient approximations where the value of randomization has been demonstrated via at least two general techniques. The first of them is the notorious randomized rounding method, which provides an unexpected association between the optimal solutions of 0/1 integer linear programs (ILPs) and their linear programming (LP) relaxations. Randomized rounding is a way to return from the fractional optimal values of the LP relaxation (which can be efficiently computed) to a good integral solution, whose expected cost is the cost of the fractional solution! We demonstrate this method here via an example application to the optimization version of the NP-complete set cover problem, and we also comment on its use (as a random projection technique) in approximations via semidefinite programs.

The second technique is used in approximately counting the number of solutions to #P-complete problems. Most of these techniques are built around the Markov chain Monte Carlo (MCMC) method. It essentially states that the time required for a Markov chain to mix (to approach its steady state) is an approximate estimator of the size of the state space of the chain (i.e., the number of the combinatorial objects that we wish to count). If the Markov chain is *rapidly mixing* (i.e., it converges in polynomial time) then we can also count the size of the state space approximately in polynomial time. We demonstrate this second approach here via an application to approximate counting of a special kind of colorings of the vertices of a graph.

The main drawback of the use of randomization in approximations is that it may only derive good results in expectation (or sometimes with high probability). This means that in certain inputs a randomized approximation technique may take a lot of time (if we want it not to fail) or may even fail. In certain cases it is possible to convert a randomized approach to a deterministic one via a *derandomization* technique (e.g., either by making the random choices dependent on each other and thus reduce the amount of randomness to the limit of allowing a deterministic brute-force search of the probability space, or by the use of *conditional probabilities*). We do not discuss derandomization here, as its application has been quite limited and as our purpose is to let the reader appreciate the simplicity and generality of the randomized methods.

11.2 Optimization and Randomized Rounding

11.2.1 Introduction

NP-hard optimization problems are not known to allow finding optimal solutions efficiently. Their combinatorial structure is elaborate and sometimes quite cryptic in the general sense.

Many NP-hard optimization problems can be coded as ILPs. In fact, in quite a lot of them, the values of the integer variables involved are only 0 and 1. We are then speaking of 0/1 integer programming problems (or 0/1 ILP). An example here is hard problems involving Boolean solutions.

Relaxing the integer constraints of the form "$x_i \in \{0, 1\}$" to the linear inequalities "$0 \le x_i \le 1$," converts a 0/1 ILP into a linear programming (LP) problem. Nowadays, it is known that LP optimization problems can be solved in polynomial time (via the ellipsoid or interior point methods). This strong similarity between 0/1 ILP and LP allows us to design efficient approximation algorithms for the hard problem at hand.

A feasible solution to the LP-relaxation can be thought of as a *fractional solution* to the original problem. The set of feasible solutions of a system of linear inequalities is known to build a *polytope* (a convex, multidimensional object, a *polyhedron*, like a diamond). To search for an optimum with respect to a linear function in a polytope is not so hard, as it has been proved that the optimum is located in some vertex of the polytope. However, in the case of an NP-hard problem, we cannot expect the polyhedron defining the set of feasible solutions to have integer vertices. Thus, our task is to somehow transform the optimal solution of the LP-relaxation into a near-optimal *integer* solution.

A basic technique for obtaining approximation algorithms using LP is what we call *LP-rounding*: that is, solve the (relaxed) linear program and then convert the fractional solutions obtained (e.g., $x_i = 2/3$) to an integral solution (e.g., here $x_i = 1$ seems more reasonable than $x_i = 0$) making sure that the cost of the solution does not increase much in the process.

A natural idea for rounding an optimal fractional solution is to view the fractions as *probabilities*. Then we can "flip coins" with these probabilities as biases, and round accordingly. So, the case "$x_i = 2/3$," obtained via LP, now leads to an experiment where "$x_i = 1$ with probability 2/3 and 0 else." This idea is called *randomized rounding*. In the sequel, we present the method via an application to the set cover problem. This application is also demonstrated in the book of Vazirani [1]. We try to be more thorough here and provide details.

11.2.2 The Set Cover Problem

The set cover problem is one of the oldest known NP-complete problems (it generalizes the vertex cover problem).

> *Problem Set Cover:*
> Given is a universal set U of n elements and also a collection of subsets of U, $S = \{S_1, S_2, \ldots, S_k\}$.
>   Given is also a cost function $c : S \to Q^+$.
> We seek to find a minimum cost subcollection of S that covers all the elements of U.

Note here that the cost of a subcollection of S, for example, $F = \{S_{i_1}, \ldots, S_{i_\lambda}\}$ is $\sum_{j=1}^{\lambda} c(S_{i_j})$. Also note that any feasible answer to the problem requires covering all of U, that is, if F is a feasible answer, then we demand that $\cup_{j=1}^{\lambda} S_{i_j} = U$. Define the *frequency* of an element of U to be the number of sets it is in.

Let us denote by f the frequency of the most frequent element. The various known approximation algorithms for set cover achieve one of the two approximation factors $O(\log n)$ or f. The special case with $f = 2$ is, basically, the vertex cover problem in graphs [2].

11.2.3 The Set Cover as an Integer Program

To formulate the set cover problem as a problem in 0/1 ILP, let us assign a variable $x(S_i)$ for each set $S_i \in S$. This variable will have the value 1 iff the set S_i is selected to be in the set cover (and will have the value 0 else).

Clearly, for each element $\alpha \in U$ we want it to be covered, that is, we want it to be in at least one of the picked sets. In other words, we want, for each $\alpha \in U$ that, at least one of the sets containing it is picked by our candidate algorithm. These considerations give the following ILP programs:

> *Set Cover ILP*
> Minimize $\sum_{S_i \in S} c(S_i) x(S_i)$
> subject to:
>
> $\forall \alpha \in U, \qquad \sum_{S_i : \alpha \in S_i} x(S_i) \geq 1$
> and $x(S_i) \in \{0, 1\}, \qquad$ for all $S_i \in S$.

The LP relaxation of this integer program can be obtained by replacing each "$x(S_i) \in \{0, 1\}$" with "$0 \leq x(S_i) \leq 1$." The reader can easily see that the upper bound on $x(S_i)$ is redundant here. So we get the following linear program:

> *Set Cover Relaxation*
> Minimize $\sum_{S_i \in S} c(S_i) x(S_i)$
> subject to:
>
> (1) $\forall \alpha \in U, \qquad \sum_{S_i : \alpha \in S_i} x(S_i) \geq 1$
> (2) $\forall S_i \in S, \qquad x(S_i) \geq 0.$

Note 1: A solution to the above-mentioned LP is a "fractional" set cover.

Note 2: A fractional set cover may be *cheaper* than the optimal (integral) set cover! To see this, let $U = \{\alpha_1, \alpha_2, \alpha_3\}$ and $S = \{S_1, S_2, S_3\}$ with $S_1 = \{\alpha_1, \alpha_3\}$, $S_2 = \{\alpha_2, \alpha_3\}$ and $S_3 = \{\alpha_3, \alpha_1\}$. Let $c(S_i) = 1, i = 1, 2, 3$. Any integral cover must pick 2 sets, for a cost of 2. On the other hand, if we pick each set with $x(S_i) = 1/2$, we satisfy all the constraints and the cost of the fractional cover obtained is 3/2.

Note 3: The LP dual [3] of the set cover relaxation is a "packing" LP: The dual tries to pack "material" into elements, trying to maximize the total amount packed, but no set must be "overpacked" (i.e., the total amount of the material packed into the elements of the set should not exceed its cost). The duality of covering-packing is a basic remark and has given lots of approximation results.

11.2.4 A Randomized Rounding Approximation to Set Cover

Let $x(S_i) = p_i, i = 1, \ldots, k$ be an *optimal* solution to the set cover relaxation program. Such a solution can be found in polynomial time.

Now, for each $S_i \in S$, select S_i with probability p_i, *independently* of the other selections.

Note: We can do it via choosing (independently for each i) k values $\gamma_1, \ldots, \gamma_k$ randomly and uniformly from the interval $[0, 1]$. Then, for $i = 1$ to k, if $\gamma_i \in [0, p_i]$, we select S_i, otherwise we do not.

Let F be the collection of the selected sets via the experiment.

The expected cost of F is

$$E(cost(F)) = \sum_{S_i \in S} c(S_i) \cdot Prob\{S_i \text{ is selected}\}$$

That is,

$$E(cost(F)) = \sum_{S_i \in S} c(S_i)p_i.$$

But $\{p_i, i = 1, \ldots, k\}$ is the optimal solution to the set cover relaxation program, hence $\sum_{S_i \in S} c(S_i)p_i$ is the optimal (minimum) value. Let us denote it by OPT_R (optimal for the relaxation).

Now let us examine whether F is a cover. For an $\alpha \in U$, suppose that α occurs in λ sets of S. W.l.o.g., let the probabilities of these sets in the optimal solution of the relaxation be $p_1, \ldots, p_\lambda$. As all the constraints are satisfied by the optimal solution we get

$$p_1 + \ldots + p_\lambda \geq 1. \tag{11.1}$$

But

$$Prob\{\alpha \text{ is covered by } F\} = 1 - \prod_{i=1}^{\lambda}(1 - p_i).$$

Because of (11.1), the above-mentioned expression becomes minimum when $p_i = \cdots = p_\lambda = \frac{1}{\lambda}$, so

$$Prob\{\alpha \text{ is covered by } F\} \geq 1 - \left(1 - \frac{1}{\lambda}\right)^\lambda \geq 1 - \frac{1}{e}$$

where $e = 2.73$ is the basis of the natural logarithms. The previous analysis holds for any $\alpha \in U$.

We now repeat the part of the experiment where we pick the collection F, independently each time. Let us pick the collections $F_1, F_2, \ldots, F_t$. Let $\tilde{F} = \cup_{i=1}^{t} F_i$. So, for all $\alpha \in U$

$$Prob\{\alpha \text{ is not covered by } \tilde{F}\} \leq \left(\frac{1}{e}\right)^t.$$

By summing over all $\alpha \in U$, we have

$$Prob\{\tilde{F} \text{ is not a cover}\} \leq n\left(\frac{1}{e}\right)^t. \tag{11.2}$$

By selecting $t = \log \xi n$ (with $\xi \geq 4$ a constant) we eventually get

$$Prob\{\tilde{F} \text{ is not a cover}\} \leq n\frac{1}{4n} = \frac{1}{4}. \tag{11.3}$$

Having established that $\tilde{F}$ is a cover with constant probability let us see its cost. Clearly

$$E(c(\tilde{F})) \leq OPT_R \cdot \log \xi n.$$

Thus, by the Markov inequality $(Prob\{X \geq mE[X]\} \leq 1/m,$ for $x \geq 0)$ we get

$$Prob\{c(\tilde{F}) \geq 4OPT_R \cdot \log \xi n\} \leq \frac{1}{4}. \tag{11.4}$$

Let A be the (undesirable) event: "$\tilde{F}$ is not a valid cover or $c(\tilde{F})$ is at least $4OPT_R \log \xi n$."

$$Prob(A) \leq \frac{1}{4} + \frac{1}{4} = \frac{1}{2}. \tag{11.5}$$

Notice that, given $\tilde{F}$, we can verify in polynomial time whether the negation of A holds. If it holds (this happens with probability $\geq 1/2$) then we have an $\tilde{F}$ which is (a) a valid set cover, (b) with cost at most $4 \log \xi n$ times above the OPT_R.

Let OPT be the optimal cost of the integer program. Clearly $OPT_R \leq OPT$ hence, when $\overline{A}$ holds, we have found a valid cover with an approximation ratio (w.r.t. the cost)

$$R = \frac{c(\tilde{F})}{OPT} \leq \frac{c(\tilde{F})}{OPT_R} \leq 4 \log \xi n.$$

Now, if A happens to hold, then we repeat the entire algorithm. The expected number of repetitions needed to get a valid cover with $R = \Theta(\log n)$ is then at most 2.

We summarize all this in the following.

Theorem 11.1 *The method of randomized rounding gives us in expected polynomial time a valid set cover with a cost approximation ratio $R = \Theta(\log n)$.*

Note: The algorithm presented here never errs (as we repeat it if $\tilde{F}$ is not a valid cover of small cost). The penalty is time but it is small as the number of repetitions follows a geometric distribution.

11.2.5 A Remark in the Analysis

In the analysis of the last section 11.2.4, we established namely

$$Prob\{\tilde{F} \text{ is not a cover}\} \leq n \left(\frac{1}{e} \right)^t$$

where t is the number of collections selected independently. By using $t = \xi \log n$ with $\xi \geq 2$ we get

$$Prob\{\tilde{F} \text{ is not a cover}\} \leq ne^{-2 \log n} \leq \frac{1}{n}$$

with an expected cover cost of $E(c(\tilde{F})) \leq OPT_R \cdot \xi \log n$, that is, $E(c(\tilde{F})) \leq OPT \cdot 2 \log n$.

If we are satisfied with a good *expected* cost, we can stop here. We get a valid cover with probability at least $1 - \frac{1}{n}$ in one repetition and the expected cost is $\Theta(OPT \cdot \log n)$.

11.2.6 Literature Notes

For more information about set cover approximation via randomized rounding, see the excellent book of V.V. Vazirani, Ch. 14. For a more advanced randomized rounding method for set cover see Reference 4. A quite similar method can be applied to the MAX SAT problem (see Reference 1, Chapter 16 or Reference 5, Chapter 7). Randomized rounding (actually the random projection method) has been also used together with semidefinite programming to give an efficient approximation to the MAX CUT problem and its variations (see Reference 1, Chapter 26) or the seminal work of Goemans and Williamson [6] who introduced the use of semidefinite programs in approximation algorithms.

11.3 Approximate Counting Using the Markov Chain Monte Carlo Method

The MCMC method is a development of the classic, well-known Monte Carlo method for approximately estimating measures and quantities whose exact computation is a difficult task. In fact, the Monte Carlo method expresses the quantity under evaluation (say x) as the expected value $x = E(X)$ of a random variable X whose samples can be estimated efficiently. By taking the mean of a sufficiently large set of samples, an approximate estimation of the quantity of interest can be obtained.

Jerrum [7] illustrates the use of the Monte Carlo method by a simple example: the estimation of the area of the region of the unit square defined by a system of polynomial inequalities. To do this, points of the unit square are randomly uniformly sampled, that is, a point is chosen uniformly at random (u.a.r.) and then it is tested whether it belongs to the region of interest (i.e., whether it satisfies or not all inequalities in the system). The probability that a randomly chosen point belongs to the area under investigation (i.e., the expectation of a random variable indicating whether the chosen point satisfies all inequalities in the system or not) is then an estimate of the area of the region of interest. By performing a sufficiently long sequence of such trials and taking their sample mean, an approximate estimation is obtained. More complex examples are the estimation of a size of a tree by sampling paths from its root to a leaf [8] and the estimation of the permanent of a 0,1-matrix [9].

It is, however, not always possible to get such samples of the random variable used. The Markov chain simulation can then be employed. The main idea of the MCMC method is to construct, for a random variable X, a Markov chain whose state space is (or includes) the range of X. The Markov chain constructed should be ergodic, that is, it converges to a stationary distribution π, and this stationary distribution matches the probability distribution of the random variable X. The desired samples can then be (indirectly) obtained by simulating the Markov chain for sufficiently many steps T, from any fixed initial state, and by taking the final state reached. If T is large enough, the Markov chain gets very close to stationarity and, thus, the distribution of the samples obtained in this way is very close to the probability distribution of the random variable X; the obtained samples are thus close to perfect and the approximation error will be negligible.

The estimation of a sufficiently large time T is important for the efficiency of the simulation. In contrast to the classical theory of stochastic process that only studies the asymptotic convergence to the stationarity, the MCMC method investigates the nonasymptotic speed of convergence and thus the computational efficiency in practical applications of the simulation. The efficiency of an algorithm using the method depends on how small is the number of simulation steps T. In efficient algorithmic uses of the MCMC method with provable performance guarantees (not just heuristic applications) we require T to be small, that is, very much smaller than the size of the state space of the simulated space. In other words, we want the Markov chain to get close to stationarity after a very short random walk on its state space. We call this time the "mixing time" of the chain and we say that an efficiently converging chain is "rapidly mixing."

Proving satisfactory upper bounds for the mixing time of the simulated Markov chain is in fact the most interesting (nontrivial) point in the application of the MCMC method. Several analytical tools have relatively recently been devised, including the "canonical path" argument, the "conductance" argument, and the "coupling" method. We here choose to illustrate the application of the "coupling" method in a particular approximate counting problem, the problem of counting radiocolorings of a graph. For the other two methods, the reader can consult References 7, 10.

The approximate counting problem is a general computing task, that of estimating the number of elements in a combinatorial space. Several interesting counting problems turn out to be complete for the complexity class #P of counting problems, and thus efficient approximation techniques become essential. Furthermore, the problem of approximate counting is closely related to the problem of random sampling of combinatorial structures, that is, generating the elements of a very large combinatorial space randomly according to some probability distribution. Combinatorial sampling problems have major computational applications, including (besides approximate counting) applications in statistical physics and in combinatorial optimization.

11.3.1 Radiocolorings of Graphs

An interesting variation of graph coloring is the k-coloring problem of graphs, defined as follows ($D(u, v)$ denotes the distance of vertices u and v in a graph G):

Definition 11.1 k-coloring problem [11] *Given a graph $G(V, E)$ find a function $\phi : V \rightarrow \{1, \ldots, \infty\}$ such that $\forall u, v \in V$, $x \in \{0, 1, \ldots, k\}$: if $D(u, v) = k - x + 1$ then $|\phi_u - \phi_v| \geq x$. This function is called a k-coloring of G. Let $|\phi(V)| = \lambda$. Then λ is the* number of colors *that ϕ actually uses (it is usually called* order *of G under ϕ). The number $v = max_{v \in V}\phi(v) - min_{u \in V}\phi(u) + 1$ is usually called the* span *of G under ϕ.*

The problem of k-coloring graphs is well motivated by practical considerations and algorithmic applications in modern networks. In fact, k-coloring is a discrete version of the frequency assignment problem (FAP) in wireless networks. FAP aims at assigning frequencies to transmitters exploiting frequency reuse while keeping signal interference to acceptable levels. The interference between transmitters are modeled by an interference graph $G(V, E)$, where V ($|V| = n$) corresponds to the set of transmitters and E represents distance constraints (e.g., if two neighbor nodes in G get the same or close frequencies then this causes unacceptable levels of interference). In most real-life cases the network topology formed has some special properties, for example, G is a lattice network or a planar graph. The FAP is usually modeled by variations of the graph coloring problem. The set of colors represents the available frequencies. In addition, each color in a particular assignment gets an integer value which has to satisfy certain inequalities compared to the values of colors of nearby nodes in G (frequency-distance constraints). The FAP has been considered in, for example, References 12–14. *Planar* interference graphs, have been studied in References 15, 16.

We have studied the case of k-coloring problem where $k = 2$, called the radiocoloring problem (RCP).

Definition 11.2 Radiocoloring Problem: *Given a graph $G(V, E)$ find a function $\Phi : V \rightarrow N^*$ such that $|\Phi(u) - \Phi(v)| \geq 2$ if $D(u, v) = 1$ and $|\Phi(u) - \Phi(v)| \geq 1$ if $D(u, v) = 2$. The least possible number λ (order) needed to radiocolor G is denoted by $X_{order}(G)$. The least possible number $v = max_{v \in V}\Phi(v) - min_{u \in V}\Phi(u) + 1$ (span) needed for the radiocoloring of G is denoted as $X_{span}(G)$.*

Real networks reserve bandwidth (range of frequencies) rather than distinct frequencies. In this case, an assignment seeks to use as small range of frequencies as possible. It is sometimes desirable to use as few distinct frequencies of a given bandwidth (span) as possible, as the unused frequencies are available for other use. Such optimization versions of the RCP are defined as follows:

Definition 11.3 Min span RCP: *The optimization version of the RCP that tries to minimize the span. The optimal span is called X_{span}.*

Definition 11.4 Min order RCP: *The optimization version of the RCP that tries to minimize the order. The optimal order is called X_{order}.*

Fotakis et al. [17] provide an $O(n\Delta)$ algorithm that *approximates the minimum order of RCP, X_{order}, of a planar graph G by a constant ratio which tends to 2 as the maximum degree Δ of G increases.*

We study here the problem of *estimating the number of different radiocolorings* of a planar graph G. This is a #P-complete problem. We employ here standard techniques of rapidly mixing Markov chains and the *new method of coupling* for proving *rapid convergence* [18] and we present a *fully polynomial randomized approximation scheme* (FPRAS) for estimating the number of radiocolorings with λ colors for a planar graph G, when $\lambda \geq 4\Delta + 50$.

Results on radiocoloring other types (periodic, hierarchical) of graphs can be found in References 19–21.

11.3.2 Outline of Our Approach

Let G be a *planar* graph of maximum degree $\Delta = \Delta(G)$ on vertex set $V = \{0, 1, \ldots, n - 1\}$ and C be a set of λ colors. Let $\Phi : V \rightarrow C$ be a (proper) radiocoloring assignment of the vertices of G. Such a radiocoloring always exists if $\lambda \geq 2\Delta + 25$ and can be found by the $O(n\Delta)$ time algorithm provided in Reference 17.

Consider the Markov chain (X_t) whose state space $R = R_\lambda(G)$ is the set of all radiocolorings of G with λ colors and whose transition probabilities from state (radiocoloring) X_t are modeled by:

1. choose a vertex $v \in V$ and a color $c \in C$ u.a.r.,
2. recolor vertex v with color c. If the resulting coloring X' is a *valid* radiocoloring assignment then let $X_{t+1} = X'$ else $X_{t+1} = X_t$.

The above-mentioned procedure is similar to the "Glauber dynamics" of an antiferromagnetic Potts model at zero temperature, and was used in Reference 18 to estimate the number of proper colorings of any low degree graph with k colors.

The Markov chain (X_t), which we refer to in the sequel as $M(G, \lambda)$, is *ergodic* (as we show below), provided $\lambda \geq 2\Delta + 26$, in which case its stationary distribution is *uniform* over R. We show here that $M(G, \lambda)$ is *rapidly mixing*, that is, converges, in time polynomial in n, to a close approximation of the stationary distribution, provided that $\lambda \geq 2(2\Delta + 25)$. This can be used to get an FPRAS for the number of radiocolorings of a planar graph G with λ colors, in the case where $\lambda \geq 4\Delta + 50$.

11.3.3 The Ergodicity of the Markov Chain $M(G, \lambda)$

For $t \in N$ let $P^t : R^2 \rightarrow [0, 1]$ denote the t-step transition probabilities of the Markov chain $M(G, \lambda)$ so that $P^t(x, y) = \Pr\{X_t = y | X_0 = x\}, \forall x, y \in R$. It is easy to verify that $M(G, \lambda)$ is (a) *irreducible* and (b) *aperiodic*. The irreducibility of $M(G, \lambda)$ follows from the observation that any radiocoloring x may be transformed to any other radiocoloring y by sequentially assigning new colors to the vertices V in ascending sequence; before assigning a new color c to vertex v it is necessary to recolor all vertices $u > v$ that have color c. If we assume that $\lambda \geq 2\Delta + 26$ colors are given, removing the color c from this set, we are left with $\geq 2\Delta + 25$ for the coloring of the rest of the graph. The algorithm presented in Reference 17 shows that the remaining graph can by radiocolored with a set of colors of this size. Hence, color c can be assigned to v.

Aperiodicity follows from the fact that the loop probabilities are $P(x, x) \neq 0, \forall x \in R$.

Thus, the finite Markov chain $M(G, \lambda)$ is *ergodic*, that is, it has a stationary distribution $\pi : R \rightarrow [0, 1]$ such that $\lim_{t \rightarrow \infty} P^t(x, y) = \pi(y), \forall x, y \in R$. Now if $\pi' : R \rightarrow [0, 1]$ is any function satisfying "local balance," that is, $\pi'(x)P(x, y) = \pi'(y)P(y, x)$ then if $\sum_{x \in R} \pi'(x) = 1$ it follows that π' is indeed the stationary distribution. In our case $P(y, x) = P(x, y)$, thus the stationary distribution of $M(G, \lambda)$ is *uniform*.

11.3.4 Rapid Mixing

The efficiency of any approach similar to sample radiocolorings crucially depends on the rate of convergence of $M(G, \lambda)$ to stationarity. There are various ways to define closeness to stationarity but all are essentially equivalent in this case and we will use the "variation distance" at time t with respect to initial vertex x

$$\delta_x(t) = \max_{S \subseteq R} \left| P^t(x, S) - \pi(S) \right| = \frac{1}{2} \sum_{y \in R} \left| P^t(x, y) - \pi(y) \right|$$

where $P^t(x, S) = \sum_{y \in S} P^t(x, y)$ and $\pi(S) = \sum_{x \in S} \pi(x)$.

Note that this is a *uniform bound* over all events $S \subseteq R$ of the difference of probabilities of event S under the stationary and t-step distributions.

The *rate of convergence to stationarity* from initial vertex x is

$$\tau_x(\epsilon) = \min\{t : \delta_x(t') \leq \epsilon, \forall t' \geq t\}$$

Our strategy is to use the coupling method, that is, construct a coupling for $M = M(G, \lambda)$, that is, a stochastic process (X_t, Y_t) on $R \times R$ such that each of the processes (X_t), (Y_t), considered in isolation, is a faithful copy of M. We will arrange a joint probability space for (X_t), (Y_t) so that, far from being independent, the two processes tend to *couple* so that $X_t = Y_t$ for t large enough. If coupling can occur rapidly (independently of the initial states X_0, Y_0), we can infer that M is rapidly mixing, because the variation distance of M from the stationary distribution is bounded above by the probability that (X_t) and (Y_t) *have not coupled* by time t.

The key result we use here is the *coupling lemma* (see Reference 22, Chapter 4 by M. Jerrum), which apparently makes its first explicit appearance in the work of Aldous [23] Lemma 3.6 (see also Diaconis [24], Chapter 4, Lemma 5).

Lemma 11.1 *Suppose that M is a countable, ergodic Markov chain with transition probabilities $P(\cdot, \cdot)$ and let $((X_t, Y_t), t \in \mathbb{N})$ be a coupling of M. Suppose further that $t : (0, 1] \rightarrow \mathbb{N}$ is a function such that $\Pr(X_{t(\epsilon)} \neq Y_{t(\epsilon)}) \leq \epsilon$, $\forall \epsilon \in (0, 1]$, uniformly over the choice of initial state (X_0, Y_0). Then the mixing time $\tau(\epsilon)$ of M is bounded above by $t(\epsilon)$.* ◇

The transition $(X_t, Y_t) \rightarrow (X_{t+1}, Y_{t+1})$ in the coupling is defined by the following experiment:

1. Select $v \in V$ u.a.r.
2. Compute a permutation $g(G, X_t, Y_t)$ of C according to a procedure that follows.
3. Choose a color $c \in C$ u.a.r.
4. In the radiocoloring X_t (respectively Y_t) recolor vertex v with color c (respectively $g(c)$) to get a new radiocoloring X' (respectively Y').
5. If X' (respectively Y') is a (valid) radiocoloring then $X_{t+1} = X'$ (respectively $Y_{t+1} = Y'$), else let $X_{t+1} = X_t$ (respectively $Y_{t+1} = Y_t$).

Note that, whatever a procedure is used to select the permutation g, the distribution of $g(c)$ is *uniform*, thus (X_t) and (Y_t) are both faithful copies of M.

We now remark that any set of vertices $F \subseteq V$ can have the same color in the graph G^2 only if they can have the same color in some radiocoloring of G. Thus, given a proper coloring of G^2 with λ' colors, we can construct a proper radiocoloring of G by giving the values (new colors) $1, 3, \ldots, 2\lambda' - 1$ in the color classes of G^2. Note that this transformation preserves the number of colors (but not the span).

Now let $A = A_t \subseteq V$ be the set of vertices on which the colorings of G^2 implied by X_t, Y_t agree and $Dim = D_t \subseteq V$ be the set on which they disagree. Let $d'(v)$ be the number of edges incident at v in G^2 that have one point in A and one in Dim. Clearly, if m' is the number of edges of G^2 spanning A, D, we get $\sum_{v \in A} d'(v) = \sum_{v \in D} d'(v) = m'$.

The procedure to compute $g(G, X_t, Y_t)$ is as follows:

(a) If $v \in D$ then g is the identity.
(b) If $v \in A$ then proceed as follows: Denote by N the set of neighbors of v in G^2. Define $C_x \subseteq C$ to be the set of all colors c, such that some vertex in N receives c in radiocoloring Y_t but no vertex in N receives c in radiocoloring Y_t. Let C_y be defined as C_x with the roles of X_t, Y_t interchanged. Observe $C_x \cap C_y = \emptyset$ and $|C_x|, |C_y| \leq d'(v)$. Let, w.l.o.g., $|C_x| \leq |C_y|$. Choose any subset $C'_y \subseteq C_y$ with $|C'_y| \leq |C_x|$ and let $C_x = \{c_1, \ldots, c_r\}, C'_y = \{c'_1, \ldots, c'_r\}$ be enumerations of C_x, C_y coming from the orderings of X_t, Y_t. Finally, let g be the permutation $(c_1, c'_1), \ldots, (c_r, c'_r)$ which interchanges the color sets $C_x, C_{y'}$ and leaves all other colors fixed.

It is clear that $|D_{t+1}| - |D_t| \in \{-1, 0, 1\}$.

(i) Consider first the probability that $|D_{t+1}| = |D_t| + 1$. For this event to occur, the vertex v selected in step (1) of the procedure for g must lie in A and hence we follow (b). If the new radiocolorings are to disagree at vertex v then the color c selected in line (3) must be an element of C_y. But $|C_y| \leq d'(v)$ hence

$$\Pr\{|D_{t+1}| = |D_t| + 1\} \leq \frac{1}{n} \sum_{v \in A} \frac{d'(v)}{\lambda} = \frac{m'}{\lambda \cdot n} \tag{11.6}$$

(ii) Now consider the probability that $|D_{t+1}| = |D_t| - 1$. For this to occur, the vertex v must lie in *Dim* and hence the permutation g selected in line (2) is the identity. For X_{t+1}, Y_{t+1} to agree at v, it is enough that color c selected in step (3) is different from all the colors that X_t, Y_t imply for the neighbors of v in G^2. The number of colors c that satisfy this, is (by our previous results) at least $\lambda - 2(2\Delta + 25) + d'(v)$ hence

$$\Pr\{|D_{t+1}| = |D_t| - 1\} \geq \frac{1}{n} \sum_{v \in D} \frac{\lambda - 2(2\Delta + 25) + d'(v)}{\lambda}$$

$$\geq \frac{\lambda - 2(2\Delta + 25)}{\lambda n} |D| + \frac{m'}{\lambda n} \tag{11.7}$$

Define now $\alpha = \frac{\lambda - 2(2\Delta + 25)}{\lambda n}$ and $\beta = \frac{m'}{\lambda n}$. So

$$\Pr\{|D_{t+1}| = |D_t| + 1\} \leq \beta$$

and $\Pr\{|D_{t+1}| = |D_t| - 1\} \geq \alpha |D_t| + \beta$

Given $\alpha > 0$, that is, $\lambda > 2(2\Delta + 25)$, from Equations (11.6) and (11.7), we get

$$E(|D_{t+1}|) \leq \beta(|D_t| + 1) + (\alpha|D_t| + \beta)(|D_t| - 1) + (1 - \alpha|D_t| - 2\beta)|D_t|$$

$$= (1 - \alpha)|D_t|$$

Thus, from Bayes, we get $E(|D_{t+1}|) \leq (1 - \alpha)^t |D_0| \leq n(1 - \alpha)^t$
and as $|D_t|$ is a nonnegative random variable, we get, by Markov inequality, that

$$\Pr\{D_t \neq 0\} \leq n(1 - \alpha)^t \leq ne^{-\alpha t}$$

So, we note that $\forall \epsilon > 0$, $\Pr\{D_t \neq \emptyset\} \leq \epsilon$ provided $t \geq \frac{1}{\alpha} \ln\left(\frac{n}{\epsilon}\right)$ thus proving

Theorem 11.2 *Let G be a planar graph of maximum degree Δ on n vertices. Assuming $\lambda \geq 2(2\Delta + 25)$ the convergence time $\tau(\epsilon)$ of the Markov chain $M(G, \lambda)$ is bounded above by,*

$$\tau_x(\epsilon) \leq \frac{\lambda}{\lambda - 2(2\Delta + 25)} \, n \ln\left(\frac{n}{\epsilon}\right)$$

regardless of the initial state x.   ∎

11.3.5 An FPRAS for Radiocolorings with λ Colors

We first provide the following definition:

Definition 11.5 *A randomized approximation scheme for radiocolorings with λ colors of a planar graph G is a probabilistic algorithm that takes as input the graph G and an error bound $\epsilon > 0$ and outputs a number Y (a random variable) such that*

$$\Pr\left\{(1 - \epsilon)\, |R_\lambda(G)| \leq Y \leq (1 + \epsilon)\, |R_\lambda(G)|\right\} \geq \frac{3}{4}$$

Such a scheme is said to be fully polynomial *if it runs in time polynomial in n and* ϵ^{-1}. *We abbreviate such schemes to* FPRAS.

The technique we employ is as in Reference 18 and is fairly standard in the area. By using it we get the following theorem:

Theorem 11.3 *There is an FPARS for the number of radiocolorings of a planar graph G with* λ *colors, provided that* $\lambda > 2(2\Delta + 25)$, *where* Δ *is the maximum degree of G.*

Proof. Recall that $R_\lambda(G)$ is the set of all radiocolorings of G with λ colors. Let m be the number of edges in G and let

$$G = G_m \supseteq G_{m-1} \supseteq \cdots \supseteq G_1 \supseteq G_0$$

be any sequence of graphs where G_{i-1} is obtained by G_i by removing a single edge. We can always erase an edge whose one node is of degree at most 5 in G_i. Clearly

$$|R_\lambda(G)| = \frac{|R_\lambda(G_m)|}{|R_\lambda(G_{m-1})|} \cdot \frac{|R_\lambda(G_{m-1})|}{|R_\lambda(G_{m-2})|} \cdots \frac{|R_\lambda(G_1)|}{|R_\lambda(G_0)|} \cdot |R_\lambda(G_0)|$$

But $|R_\lambda(G_0)| = \lambda^n$ for all kinds of colorings. The standard strategy is to estimate the ratio

$$\rho_i = \frac{|R_\lambda(G_i)|}{|R_\lambda(G_{i-1})|}$$

for each $i, 1 \leq i \leq m$.

Suppose that graphs G_i, G_{i-1} differ in the edge $\{u, v\}$ which is present in G_i but not in G_{i-1}. Clearly, $R_\lambda(G_i) \subseteq R_\lambda(G_{i-1})$. Any radiocoloring in $R_\lambda(G_{i-1}) \setminus R_\lambda(G_i)$ assigns either the same color to u, v or the color values of u, v differ by only 1. Let $deg(v) \leq 5$ in G_i. So, we now have to recolor u with one of at least $\lambda - (2\Delta + 25)$, that is, at least $2\Delta + 25$, colors (from Section 5 of Reference 18). Each radiocoloring of $R_\lambda(G_i)$ can be obtained in at most one way by our algorithm of the previous section as the result of such a perturbation, thus

$$\frac{1}{2} \leq \frac{2\Delta + 25}{2(\Delta + 1) + 25} \leq \rho_i < 1 \tag{11.8}$$

To avoid trivialities, assume $0 < \epsilon \leq 1, n \geq 3$ and $\Delta > 2$.

Let $Z_i \in \{0, 1\}$ be the random variable obtained by simulating the Markov chain $M(G_{i-1}, \lambda)$ from any certain fixed initial state for

$$T = \frac{\lambda}{\lambda - 2(2\Delta + 25)} \, n \ln\left(\frac{4nm}{\epsilon}\right)$$

steps and returning to 1 if the final state is a member of $R_\lambda(G_i)$ and 0 else. Let $\mu_i = E(Z_i)$. By our theorem of rapid mixing, we have

$$\rho_i - \frac{\epsilon}{4m} \leq \mu_i \leq \rho_i + \frac{\epsilon}{4m}$$

and by Equation (11.8), we get

$$\left(1 - \frac{\epsilon}{2m}\right)\rho_i \leq \mu_i \leq \left(1 + \frac{\epsilon}{2m}\right)\rho_i$$

As our estimator for $|R_\lambda(G)|$ we use

$$Y = \lambda^n Z_1 Z_2 \cdots Z_m$$

Note that $E(Y) = \lambda^n \mu_1 \mu_2 \cdots \mu_m$. But

$$Var(Y) \leq \frac{Var(Z_1 Z_2 \cdots Z_m)}{(\mu_1 \mu_2 \cdots \mu_m)^2} = \prod_{i=1}^{m} \left(1 + \frac{Var(Z_i)}{\mu_i^2}\right) - 1$$

By using standard techniques (as in Reference 18) one can easily show that Y satisfies the requirements for an FPRAS for the number of radiocolorings of graph G with λ colors $|R_\lambda(G)|$. ∎

References

1. Vazirani, V.V., *Approximation Algorithms,* Springer, Berlin, Germany, 2001.
2. Garey, M.R. and Johnson, D.S., *Computers and Intractability: A Guide to the Theory of NP-Completeness,* W. H. Freeman and Co., New York, 1979.
3. Papadimitriou, C.H. and Steiglitz, K., *Combinatorial Optimization: Algorithms and Complexity,* Prentice Hall, Englewood Cliffs, NJ, 1982.
4. Srinivason, A., Improved approximations of packing and covering problems, in *Proceedings of Symposium on Theory of Computing,* 2005.
5. Hromkovic, J., *Design and Analysis of Randomized Algorithms,* EATCS Series, Springer-Verlag, New York, 2005.
6. Goemans, M.X. and Williamson, D.P., Improved approximation algorithms for maximum cut and satisfiability problems using semidefinite programming, *JACM,* 42, 1115, 1995.
7. Jerrum, M., Mathematical foundations of the Markov chain Monte Carlo method, in *Probabilistic Methods for Algorithmic Discrete Mathematics,* Habib, M., McDiarmid, C., Ramirez-Alfonsin, J., and Reed, B. (Eds.), Springer Verlag, New York, p. 116, 1998.
8. Knuth, D.E., Estimating the efficiency of backtrack problems, *Mathematics of Computation,* 29, 121, 1975.
9. Rusmussen, L.E., Approximating the permanent: A simple approach, *Random Structures and Algorithms,* 5, 349, 1994.
10. Jerrum, M. and Sinclair, A., The Markov chain Monte Carlo method: An approach to approximate counting and integration, in *Approximation Algorithms for NP-hard Problems,* Hochbaum, D.S. (Ed.), PWS Publishing Company, Boston, MA, p. 482, 1997.
11. Hale, W.K., Frequency assignment: Theory and applications, in *Proceedings of the IEEE,* 68(12), 1497, 1980.
12. Griggs, J. and Liu, D., Minimum span channel assignments, in *Recent Advances in Radio Channel Assignments,* in the *Ninth SIAM Conference on Discrete Mathematics,* Toronto, Canada, 1998.
13. Fotakis, D., Pantziou, G., Pentaris, G., and Spirakis, P., Frequency assignment in mobile and radio networks, *Networks in Distributed Computing, DIMACS Series in Discrete Mathematics and Theoretical Computer Science, AMS,* 45, 73, 1999.
14. Katsela, I. and Nagshineh, M., Channel assignment schemes for cellular mobile telecommunication systems, *IEEE Personal Communication Complexity,* p. 1070, 1996.
15. Ramanathan, S. and Loyd, E.R., The complexity of distance2-coloring, in *Proceedings of the 4th International Conference of Computing and Information,* Springer, New York, p. 71, 1992.
16. Bertossi, A.A. and Bonuccelli, M.A., Code assignment for hidden terminal interference avoidance in multihop packet radio networks, *IEEE/ACM Transactions on Networking,* 3(4), 441, 1995.
17. Fotakis, D., Nikoletseas, S., Papadopoulou, V., and Spirakis, P., Radiocolorings in planar graphs: Complexity and approximations, *Theoretical Computer Science Journal,* 340(3), 205, 2005. Also, in *Proceedings of the 25th International Symposium on Mathematical Foundations of Computer Science (MFCS 2000),* LNCS, Vol. 1893, Springer Verlag, Berlin, Germany, p. 363, 2000.

18. Jerrum, M., A very simple algorithm for estimating the number of k-colourings of a low degree graph, *Random Structures and Algorithms, 7*, 157, 1994.
19. Andreou, M., Fotakis, D., Nikoletseas, S., Papadopoulou, V., and Spirakis, P., On radiocoloring hierarchically specified planar graphs: PSPACE-completeness and approximations, in *Proceedings of the 27th International Symposium on Mathematical Foundations of Computer Science (MFCS)*, LNCS, Vol. 2420, Springer Verlag, Berlin, Germany, p. 81, 2002.
20. Fotakis, D., Nikoletseas, S., Papadopoulou, V., and Spirakis, P., Radiocolorings in periodic planar graphs: PSPACE-completeness and efficient approximations for the optimal range of frequencies, in *Proceedings of the 28th International Workshop on Graph-Theoretic Concepts in Computer Science*, LNCS, Vol. 2573, Springer, Berlin, Germany, p. 223, 2002.
21. Fotakis, D., Nikoletseas, S., Papadopoulou, V., and Spirakis, P., Hardness results and efficient approximations for frequency assignment problems: Radio labelling and radio coloring, *Journal of Computers and Artificial Intelligence (CAI)*, 20(2), 121, 2001.
22. Habib, M., McDiarmid, C., Ramirez-Alfonsin, J., and Reed, B. (Eds.), *Probabilistic Methods for Algorithmic Discrete Mathematics*, Springer Verlag, Berlin, Germany, 1998.
23. Aldous, D., Random walks in finite groups and rapidly mixing Markov chains, in *Seminaire de Probabilites XVII 1981/82*, Dold, A. and Eckmann, B. (Eds.), Springer, Lecture Notes in Mathematics, Vol. 986, Springer, Berlin, Germany, p. 243, 1982.
24. Diaconis, P., *Group Representations in Probability and Statistics*, Institute of Mathematical Statistics, Hayward, CA, 1988.

12

Distributed Approximation Algorithms via LP-Duality and Randomization

Devdatt Dubhashi

Fabrizio Grandoni

Alessandro Panconesi

12.1 Introduction

The spread of computer networks, from sensor networks to the Internet, creates an ever growing need for efficient *distributed* algorithms. In such scenarios, familiar combinatorial structures such as spanning trees and dominating sets are often useful for a variety of tasks. Others, like maximal independent sets, turn out to be a very useful primitive for computing other structures. In a distributed setting, where transmission of messages can be orders of magnitude slower than local computation, the expensive resource is communication. Therefore, the running time of an algorithm is given by the number of communication rounds that are needed by the algorithm. This will be made precise in the following text.

In what follows we will survey a few problems and their solutions in a distributed setting: Dominating sets; edge and vertex colorings; matchings; vertex covers, and minimum spanning trees (MSTs). These problems were chosen for a variety of reasons: They are fundamental combinatorial structures; computing them is useful in distributed settings; and they serve to illustrate some interesting techniques and methods.

Randomization, whose virtues are well known to people coping with parallel and distributed algorithms, will be a recurrent theme. In fact, only rarely it has been possible to develop deterministic distributed algorithms for nontrivial combinatorial optimization problem. Here, in the section on vertex covers, we will discuss a novel and promising approach based on the primal-dual methodology to develop efficient, distributed deterministic algorithms. One of the main uses of randomization in distributed scenarios is to break the symmetry. This is well-illustrated in Section 12.2. discussing dominating sets. Often the analysis of simple randomized protocols requires deep results from probability theory. This will be illustrated in Section 12.3, where martingale methods are used to analyze some simple, and yet almost optimal, distributed algorithms for edge coloring. The area of distributed algorithms for graph problems is perhaps unique in complexity theory because it is possible to derive several nontrivial absolute lower-bounds (that is, not relying on special complexity assumptions, such as P $\neq$ NP). This will be discussed in Section 12.6.

Let us then define the computation model. We have a message-passing, synchronous network: vertices are processors, edges are communication links, and the network is synchronous. Communication proceeds in synchronous *rounds*: in each round, every vertex sends messages to its neighbors, receives messages from its neighbors, and does some amount of local computation. It is also assumed that each vertex has a unique identifier. In the case of randomized algorithms each node of the network has access to its own source of random bits. *In this model, the running time is the number of communication rounds.* This will be our notion of "time."

Although we place no limits on the amount of local computation, the algorithms we describe perform polynomial-time local computations only. Under the assumption that local computations are polynomial-time several of the algorithms that we describe are "state-of-the-art," in the sense that their approximation guarantee is the same, or comparable, to that obtainable in a centralized setting. It is remarkable that this can be achieved in a distributed setting.

The model is in some sense orthogonal to the PRAM model [1] for parallel computation where a set of polynomially-many, synchronous processors access a shared memory. There, communication is free: any two processors can communicate in constant time via the shared memory. In the distributed model, in contrast, messages are routed through the network and therefore the cost of sending a message is at least proportional to the length of the shortest path between the two nodes. On the other hand, local computation is inexpensive, whereas this is the expensive resource in the PRAM model.

Note that there is a trivial universal algorithm that always works: The network elects a leader which then collects the entire topology of the network, computes the answers, and notifies them to the other nodes. This will take a time proportional to the diameter of the network, which can be as large as n, the number of nodes. In general we will be looking for algorithms that take poly-logarithmically, in n, many communication rounds, regardless of the diameter of the network. Such algorithms will be called *efficient*.

Note the challenge here: if a protocol runs for t rounds then each processor can receive messages from nodes at distance at most t. For small values of t this means that the network is computing a global function of itself by relying on local information alone.

12.2 Small Dominating Sets

In this section we study the minimum dominating set (MDS) problem. The advent of wireless networks gives a new significance to the problem as (connected) dominating sets are the structure of choice to set up the routing infrastructure of such ad-hoc networks, the so-called backbone (see, e.g., Reference 2 and references therein). In the sequel we describe a nice algorithm from Reference 3 for computing small dominating sets. The algorithm is in essence an elegant parallelization of the well-known greedy heuristic for set cover [4,5]. Randomness is a key ingredient in the parallelization. The algorithm computes, on any input graph, a dominating set of size at most $O(\log \Delta)opt$, whereas customary Δ denotes the maximum degree of the graph and *opt* is the smallest size of a dominating set in the input graph. By "computing a dominating set" we mean that at the end of the protocol every vertex decides whether it is in the dominating set or not. The algorithm was originally developed for the PRAM model but, as we will show, it can be implemented distributively. It is noteworthy that the approximation bound is essentially the "best possible" under the assumption that every node performs a polynomially-bounded computation during every round. "Best possible" means that an $o(\log n)$-approximation would imply that P = NP [6], whereas a $(c \ln n)$-approximation, for a constant $c < 1$, would imply that NP could be solved exactly by means of slightly super-polynomial algorithms [7,8].

We shall then describe a surprisingly simple deterministic algorithm that, building on top of the dominating set algorithm, computes a "best possible" connected dominating set, in $O(\log n)$ additional communication rounds [9].

There are other nice algorithms to compute dominating sets efficiently in a distributed setting. The algorithm in Reference 10 is a somewhat different parallelization of the greedy algorithm, whereas Reference 11 explores an interesting trade-off between the number of rounds of the algorithm and the quality of the approximation that it achieves. This chapter makes use of linear programming (LP) based methods, an issue that we will explore in Section 12.5.

12.2.1 Greedy

Let us start by reviewing the well-known greedy heuristic for set cover. Greedy repeatedly picks the set of minimum unit cost, creating a new instance after every choice by removing the points just covered. More formally, let (X, F, c) be a set cover instance where X is a ground set of elements and $F := \{S_i : S_i \subseteq X, i \in [m]\}$ is a family of nonempty subsets of X with positive costs $c(S) > 0$. The goal is to select a subfamily of minimum cost that covers the ground set. The cost of a subfamily is the sum of the costs of each set in the subfamily.

Dominating set is a special case of set cover. A graph G with positive weights $c(u)$, $u \in V(G)$, can be viewed as a set system $\{S_u : u \in V(G)\}$ with $S_u := N(u) \cup \{u\}$, where $N(u)$ is the set of neighbors of u and $c(S_u) := c(u)$.

Given a set cover instance $I := (X, F, c)$, let $c(e) := \min_{e \in S \in F} \frac{c(S)}{|S|}$, be the *cost* of the element $e \in X$. This is the cheapest way to cover e where we do the accounting in the following natural way: when we pick a set, its cost is distributed equally to all elements it covers. An algorithm A may pick a certain set S' at this stage, then in this accounting scheme, each element $e \in S'$ pays the *price* $p(e) := \frac{c(S')}{|S'|}$. Once set s' is picked, we create a new instance I' with ground set $X' := X - \hat{S}$ and set system F' whose sets are defined as: $S'_i := S_i - \hat{S}$. The new costs coincide with the old ones: $c(S') = c(S)$, for all $S \in F'$. The algorithm continues in the same fashion until all elements are covered.

Greedy selects a set $\hat{S}$ at each stage that realizes the minimum unit cost, that is, $p(e) = c(e)$ at each stage. In other words, greedy repeatedly selects the set that guarantees the smallest unit price. For the discussion to follow concerning the distributed version of the algorithm it is important to notice that each element e is assigned a price tag $p(e)$ only once, at the time when it is covered by greedy. For a subset $A \subseteq X$, let $g(A) := \sum_{e \in A} p(e)$. Then $g(X)$, the sum of the unit prices, is the total cost incurred by greedy. The crux of the analysis is the next lemma.

Lemma 12.1 *For any set S, $g(S) \le H_{|S|} c(S)$ where $H_k := 1 + \frac{1}{2} + \frac{1}{3} + \ldots + \frac{1}{k}$ is the k-th harmonic number.*

Proof. Sort the elements of S according to the time when they are covered by greedy, breaking ties arbitrarily. Let $e_1, e_2, \ldots, e_k$ be this numbering. When greedy covers e_i it must be that $p(e_i) \le \frac{c(S)}{k-i}$. The claim follows. ∎

Clearly we have $g(A \cup B) \le g(A) + g(B)$. Denoting with C^* an optimal cover, we have, by Lemma 12.1,

$$g(X) = g\left(\cup_{S \in C^*} S\right) \le \sum_{S \in C^*} g(S) \le \sum_{S \in C^*} H_{|S|} c(S) \le \max_S H_{|S|} \sum_{S \in C^*} c(S) \le \max_S H_{|S|} opt$$

It is well known that $\log k \le H_k \le \log k + 1$. In the case of dominating set the bound becomes

$$g(X) \le H_{\Delta+1} opt = O(\log \Delta) opt$$

where Δ is the maximum degree of the graph.

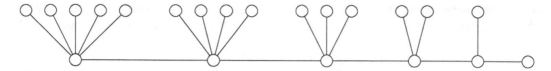

FIGURE 12.1   Example of lower bound graph for $k = 6$. The number of nodes is $n = k(k+1)/2 = \Theta(k^2)$. The bottom nodes are selected by greedy, one by one from left to right. The number of rounds is $k - 1$.

12.2.2 Greedy Hordes

We now proceed to parallelize greedy. Figure 12.1 shows that the number of steps taken by greedy can be $\Omega(\sqrt{n})$. The problem lies in the fact that at any stage there is just one candidate set that gives the minimum unit cost $\hat{c}$. It is to get around this problem that we introduce the following notion. A *candidate* is any set S such that

$$\hat{c} \leq \frac{c(S)}{|S|} \leq 2\hat{c}. \tag{12.1}$$

Let us modify greedy in such a way that, at any step, it selects any set satisfying this condition. With this modification the solution computed by greedy will still be at most $O(\log n)opt$ as the algorithm pays at most twice the smallest unit price the overall we lose only a factor of two in the approximation.

Suppose now that the algorithm is modified in such a way that it adds to the solution all candidates satisfying (12.1). With this modification the graphs of Figure 12.1 will be covered in $O(\log n)$ steps. But as the example of the clique shows (all the nodes are selected) this increase in speed destroys the approximation guarantee. This is because the key requirement of the sequential greedy procedure is violated. In the sequential procedure the price $p(e)$ is paid only once, at the time when e is covered. If we do things in parallel we need to keep two conflicting requirements in mind: picking too many sets at once can destroy the approximation guarantee but picking too few can result in slow progress. And we must come up with a charging scheme to distribute the costs among the elements in a manner similar to the sequential case.

Rajagopalan and Vazirani solved this problem by devising a scheme that picks enough sets to make progress but at the same time retains the parsimonious accounting of costs like in the sequential version. Specifically, for every set S selected by greedy, the cost $c(S)$ will be distributed among the elements of a subset $T \subset S$ of at least $|S|/4$ elements. Crucially, the elements of T will be charged only once. If we can do this then we will lose another factor of four in the approximation guarantee with respect to greedy, all in all losing a factor of eight.

The scheme works as follows: line up the candidate sets satisfying (12.1) on one side and all the elements on the other. The elements are thought of as *voters* and cast their vote for one of the candidate sets containing them by an *election*. An election is conducted as follows:

- A random permutation of the candidates is computed.
- Among all the candidate sets that contain it, each voter votes for that set which has the lowest number in the permutation.
- A candidate is elected if it obtains at least $\frac{1}{4}$ of the votes of its electorate. Elected candidates enter the set cover being constructed.

The cost of the set can now be distributed equally among the elements that voted for it, that is, at least a quarter of the elements.

Let us now describe the distributed implementation of this scheme in the specific case of the set system corresponding to the dominating set problem. During the execution nodes can be in four different states:

- They can be *free*. Initially all vertices are free.
- They can be *dominated*.

- They can be *dominators*. Dominators are added to the dominating set and removed from the graph.
- They can be *out*. Vertices are out when they are dominated and have no free neighbors. These vertices are removed from the graph as they can play no useful role.

The algorithm is a sequence of $\log \Delta$ *phases* during which the following invariant is maintained, with high probability. At the beginning of phase i, $i = 1, 2, \ldots, \log \Delta$, the maximum degree of the graph is at most $\Delta/2.^{i-1}$ The candidates during phase i are all those vertices whose degree is in the interval $(\Delta/2^i, \Delta/2^{i-1}]$, that is, they satisfy condition in (12.1). Note that candidates can be free or dominated vertices. The voters are those free nodes that are adjacent to a candidate. This naturally defines a bipartite graph with candidates on one side, voters on the other, and edges that represent domination relationships. Each phase consists of a series of $O(\log n)$ elections. A free vertex can appear on both sides, as a free vertex can dominate itself. We shall refer to the neighbors of a candidate c in the bipartite graph as the *electorate* of c, and to the neighbors of a voter v as the *pool* of v. Election are carried out and elected candidates enter the dominating set.

Step 1 of each election seem to require global synchronization but a random permutation can be generated if the value of n is known. If each element picks a random number between 1 and n^k then with probability $1 - 1/n^{k-1}$ all choices will be distinct. Thus, the probability that there is a collision is negligible during the entire execution of the algorithm.

After every election nodes are removed for two different reasons. Elected nodes disappear from the candidate side of the bipartition, whereas their neighbors disappear from the other side, as they are no more free. In the analysis we will show that after one election the expected number of edges that disappear from the bipartite graph is a constant fraction of the total. This automatically implies that the total number of elections to remove all edges from the graph is $O(\log n)$ with overwhelming probability. More precisely, for any $c > 0$ there is $\alpha > 0$ such that, the probability that the bipartite graph is nonempty after $\alpha \log n$ elections is at most n^{-c} [12,13]. It follows that α can be chosen in such a way that the probability that some phase does not end successfully is negligible.

A voter v is *influential* for a candidate c if at least $\frac{3}{4}$ of the voters in c's electorate have degree no greater than that of v. Let $d(v)$ denote the degree of v.

Lemma 12.2 *For any two voters v and w, $d(v) \geq d(w)$, in c's electorate, $\Pr[w \text{ votes } c \mid v \text{ votes } c] \geq \frac{1}{2}$.*

Proof. Let N_b denote the number of neighbors that v and w have in common, let N_v the number of neighbors of v that are not neighbors of w, and let N_w be the number of neighbors of w that are not neighbors of v. Then,

$$\Pr[w \text{ votes } c \mid v \text{ votes } c] = \frac{\Pr[w \text{ votes } c, v \text{ votes } c]}{\Pr[v \text{ votes } c]} = \frac{N_v + N_b}{N_v + N_b + N_w} \geq \frac{1}{2}. \qquad \blacksquare$$

Lemma 12.3 *Let v be an influential voter for c. Then, $\Pr[c \text{ is elected} \mid v \text{ votes } c] \geq \frac{1}{6}$.*

Proof. Let $X := (\# \text{ votes for } c)$ and $Y := c - X$ where, with abuse of notation we use c to denote the size of c's electorate. Then, by Lemma 12.2

$$E[X \mid v \text{ votes } c] \geq \sum_{w: \, d(w) \leq d(v)} \Pr[w \text{ votes } c \mid v \text{ votes } c] \geq \frac{3}{8}c.$$

Applying Markov's inequality to Y we get,

$$\Pr[c \text{ not elected} \mid v \text{ votes } c] = \Pr[X < c/4 \mid v \text{ votes } c] = \Pr[Y \geq 3c/4 \mid v \text{ votes } c]$$
$$\leq \frac{4 \, E[Y \mid v \text{ votes } c]}{3c} = \frac{4(c - E[X \mid v \text{ votes } c])}{3c} \leq \frac{5}{6}.$$

The claim follows. $\blacksquare$

Lemma 12.4 *Fix a phase and let m denote the total number of edges in the bipartite graph at any stage in this phase. Let X denote the number of edges removed from the bipartite graph after one election. Then,* $E[X] \geq \frac{m}{24}$.

Proof. An edge vc is *good* if v is influential for c. By definition, at least $\frac{1}{4}$ of the edges are good. Then,

$$E[X] = \sum_{vc} \Pr[c \text{ is elected}, v \text{ votes } c]d(v)$$

$$\geq \sum_{vc \text{ good}} \Pr[c \text{ is elected}, v \text{ votes } c]d(v)$$

$$\geq \sum_{vc \text{ good}} \Pr[v \text{ votes } c]\,\Pr[c \text{ is elected} \mid v \text{ votes } c]d(v)$$

$$= \sum_{vc \text{ good}} \Pr[c \text{ is elected} \mid v \text{ votes } c]$$

$$\geq \frac{m}{24}, \quad \text{by Lemma 12.3.} \qquad \blacksquare$$

As remarked, this lemma implies that, with high probability, $O(\log n)$ rounds are sufficient for every phase. The resulting running time is $O(\log n \log \Delta)$ communication rounds, whereas the approximation guarantee is $O(\log \Delta)$. Vertices must know n to compute a permutation and to run the correct number of elections, and they must know Δ to decide whether they are candidates at the current phase. Alternatively, if only the value of n is known, the algorithm can execute $O(\log n)$ phases, for a total of $O(\log^2 n)$ many rounds.

12.2.3 Small Connected Dominating Sets

In this section we develop an efficient distributed algorithm for computing "best possible" connected dominating sets. Again, by this we mean that the protocol computes a connected dominating set of size at most $O(\log \Delta)$ times the optimum. Nowadays, connected dominating sets are quite relevant from the application point of view as they are the solution of choice for setting up the backbones of self-organizing networks such as ad hoc and sensor networks (see Reference 2 and references therein). A backbone is a subnetwork that is in charge of administering the traffic inside a network.

What is remarkable from the algorithmic point of view is that connectivity is a strong global property, and yet we will be able to obtain it by means of a distributed algorithm that relies on local information alone. The overall strategy can be summarized as follows:

- Compute a small dominating set.
- Connect it up using a sparse spanning network.

We saw in Section 12.2.2 how to take care of Step 1. To connect up a dominating set we can proceed as follows. Let D be the dominating set in the graph G created after Step 1. Consider an auxiliary graph H with vertex set D and where any two $u, v \in D$ that are at distance $1, 2,$ or 3 in G are connected by an edge in H. It is easy to see that H is connected if G is (which we assume). Every edge in H corresponds to a path with $0, 1,$ or 2 vertices in G. If we inserted all such vertices we would still have a dominating set, as adding vertices can only improve domination. The resulting set would however be too large in general, as H can have as many as $|D|^2$ edges, each contributing with 2 vertices. The best way to connect D up would be to compute a spanning tree T. If we could do this, adding to D all vertices lying on paths corresponding to the edges of T, we would obtain the desired approximation since $E(T) = |D| - 1$ and recalling that $|D|$ is a $O(\log \Delta)$-approximation. Therefore, denoting with D^* and C^* an optimal dominating and connected dominating set, respectively, we would have (with some abuse of notation) that $|D \cup V(T)| \leq 3|D| \leq O(\log \Delta)|D^*| \leq O(\log \Delta)|C^*|$.

The problem however is that, as we discuss in Section 12.6.1, computing a spanning tree takes time $\Omega(\sqrt{n})$. In what follows we show a very simple algorithm that computes, in $O(\log |V(G)|)$ many communication rounds, a network $S \subset H$ such that (a) S is connected, (b) $|E(S)| = O(|D|)$, and (c) $V(S) = D$. In other words, S is a sparse connected network that spans the whole of D with linearly many edges. If we can compute such an S than we will have a connected dominating set of size at most $O(\log \Delta)$ times the optimum. S will not be acyclic but this is actually a positive thing as it makes S more resilient to failures. In fault-prone environments such as ad hoc and sensor networks this kind of redundancy is actually very useful. The key to computing S is given by the following lemma (see, e.g., Reference 14 Lemma 15.3.1). Recall that the *girth* of a graph G is the length of the shortest cycle in G.

Lemma 12.5 *Let* $G = (V, E)$ *be a graph of girth* g, *and let* $m := |E|$ *and* $n := |V|$. *Then,* $m \leq n + n^{.1 + 2/(g-1)}$

Proof. Assume $g = 2k + 1$ and let $d := \frac{m}{n}$. Consider the following procedure. As long as there is a vertex whose degree is less than d, remove it. Every time we remove a vertex the new minimum degree is at least as large as the old one. Therefore this procedure ends with a graph whose minimum degree is at least d. Pick now any vertex in this graph and start a breadth first search. This generates a tree in which the root has at least d children and every other node has at least $d - 1$ children. Moreover, assigning level 0 to the root, this tree is a real tree up to and including level $k - 1$, that is, no two vertices of this BFS exploration coincide up to that level. Therefore,

$$n \geq 1 + d + d(d - 1) + \ldots + d(d - 1)^{k-1} \geq (d - 1)^k.$$

Recalling the definition of d, the claim follows. The proof for the case $g = 2k$ is analogous. ∎

Note that if $g = 2 \log n + 1$ then $m \leq 3n$. Define a cycle to be *small* if it is of length at most $2 \log n + 1$. The following amazingly simple protocol removes all small cycles while, crucially, preserving connectivity:

- If an edge is the smallest in some small cycle, it is deleted.

Assume that every edge in the graph has a unique identifier. An edge is smaller than an another edge if its identifier is smaller than that of the other edge. It is clear that every small cycle is destroyed. The next lemma shows that connectivity is preserved.

Lemma 12.6 *The above-mentioned protocol preserves connectivity.*

Proof. Sort the edges by increasing ID's and consider the following sequential procedure. At the beginning all edges are present in the graph. At step i edge e_i is considered. If e_i is in a small cycle then it is removed. This breaks all small cycles and preserves connectivity, as an edge is removed only when there is another path connecting its endpoints. The claim follows by observing that the sequential procedure and the distributed protocol remove the same set of edges. ∎

To implement the protocol we only need to determine the small cycles to which an edge belong. This can be done by a breadth-first-search (BFS) of depth $O(\log n)$ starting from every vertex. If edges do not have distinct ID's to start with they can be generated by selecting a random number in the range $[m^3]$ which ensures that with all ID's are distinct with overwhelming probability. This requires the value of n or m to be known. This sparsification technique appears to be quite effective in practice [2].

12.3 Coloring: The Extraordinary Career of a Trivial Algorithm

Consider the following sequential greedy algorithm to color the vertices of an input graph with $\Delta + 1$ colors, where Δ is the maximum degree: pick a vertex, give it a color not assigned to any of its neighbors; repeat until all vertices are colored. In general, Δ can be quite far from the optimal value $\chi(G)$ but it should not be forgotten that the chromatic number is one of the most difficult combinatorial problems to approximate [15–17].

In this section we will see how efficient distributed implementations of this simple algorithm lead to surprisingly strong results for vertex and especially edge coloring. Consider first the following distributed implementation. Each vertex u is initially given a list of colors $L_u := \{1, 2, \ldots, \Delta + 1\}$. Computation proceeds in rounds, until the graph is colored. One round is as follows: each uncolored vertex u picks a tentative color $t_u \in L_u$; if no neighboring vertex has chosen the same tentative color, t_u becomes the final color of u, and u stops. Otherwise L_u is updated by removing from it all colors assigned to neighbors of u at the current round. We shall refer to this as the *trivial algorithm*. It is apparent that the algorithm is distributed.

The trivial algorithm is clearly correct. An elementary, but nontrivial analysis shows that the probability that an uncolored vertex colors itself in one round is at least $\frac{1}{4}$ [18]. As we discussed in the previous section, this implies that the algorithm will color the entire network within $O(\log n)$ communication rounds, with high probability.

The following slight generalization is easier to analyze. At the beginning of every round, uncolored vertices are *asleep* and wake up with probability p. The vertices that wake up execute the round exactly as described before. At the end of the round, uncolored vertices go back to sleep. Said it differently, the previous algorithm is obtained by setting $p = 1$. In the sequel we will refer to this generalization as the (generalized) trivial algorithm. Luby analyzed this algorithm for $p = \frac{1}{2}$ [19]. Heuristically it is not hard to see why the algorithm makes progress in this case. Assume u is awake. The expected number of neighbors of u that wake up is $d(u)/2 \leq |L_u|/2$.

In the worst case, these neighbors will pick different colors and all these colors will be in L_u. Even then, u will have probability at least $\frac{1}{2}$ to pick a color that creates no conflict. Thus, with probability $\frac{1}{2}$ a vertex wakes up and, given this, with probability at least $\frac{1}{2}$ it colors itself. The next proposition formalizes this heuristic argument.

Proposition 12.1 *When $p = \frac{1}{2}$ the probability that an uncolored vertex colors itself in one round is at least $\frac{1}{4}$.*

Proof. Let t_u denote the tentative color choice of a vertex u.

$$\Pr[u \text{ does not color} \mid u \text{ wakes up}] = \Pr[\exists v \in N(u)\ t_u = t_v \mid u \text{ wakes up}]$$

$$\leq \sum_{v \in N(u)} \Pr[t_u = t_v \mid u \text{ wakes up}]$$

$$= \sum_{v \in N(u)} \Pr[t_u = t_v \mid u \text{ and } v \text{ wake up}] \Pr[v \text{ wakes up}]$$

$$= \sum_{v \in N(u)} \frac{|L_u \cap L_v|}{|L_v||L_u|} \frac{1}{2} \leq \sum_{v \in N(u)} \frac{1}{|L_u|} \frac{1}{2} \leq \frac{1}{2}.$$

Therefore,

$$\Pr[u \text{ colors itself}] = \Pr[u \text{ colors itself} \mid u \text{ wakes up}] \Pr[u \text{ wakes up}] \geq \frac{1}{4}. \qquad \blacksquare$$

Note that the trivial algorithm works just as well if the lists are initialized as $L_u := \{1, 2, \ldots, d(u) + 1\}$, for all $u \in V(G)$, for any value of $p > 0$. Interestingly, in practice with $p = 1$ the trivial algorithm is much faster than Luby's one. In fact, experimentally, the speed of the algorithm increases regularly and monotonically as p tends to 1 [20].

In the distributed model we can simulate the trivial algorithm for the line graph with constant-time overhead. In this case the algorithm will be executed by the edges rather than the vertices, each edge e having its own list L_e. In this fashion we can compute edge colorings that are approximated by a factor of 2 (as $2\Delta - 1$ colors are used). It is a challenging open problem whether an $O(\Delta)$-approximation can be computed deterministically in the distributed model. The best known result so far is an $O(\Delta \log n)$-approximation [21]. But the real surprise is that the trivial algorithm computes near-optimal edge colorings!

Vizing's theorem shows that every graph G can be edge colored sequentially in polynomial time with Δ or $\Delta + 1$ colors [22]. The proof is in fact a polynomial time sequential algorithm for achieving a $\Delta + 1$ coloring. Thus edge coloring can be well approximated. It is a very challenging open problem whether colorings as good as these can be computed fast in a distributed model.

If the edge lists L_e's are initialized to contain just a bit more than Δ colors, say $|L_e| = (1 + \epsilon)\Delta$ for all e, then the trivial algorithm will edge color the graph within $O(\log n)$ communication rounds. Here ϵ can be any fixed, positive constant. Some lists can run out of colors and, consequently, the algorithm can fail, but this happens with a probability that goes to zero as n, the number of vertices, grows. All this is true, provided that the minimum degree $\delta(G)$ is large enough, that is, $\delta(G) \gg \log n$ [23,24]. For Δ-regular graphs the condition becomes $\Delta \gg \log n$.

In fact, the trivial algorithm has in store more surprises. If the input graph is Δ-regular and has no triangles, it colors the vertices of the graph using only $O(\Delta / \log \Delta)$ colors. This is in general optimal, as there are infinite families of triangle-free graphs that need these many colors [25]. Again, the algorithm fails with negligible probability, provided that $\Delta \gg \log n$. For the algorithm to work, the value of p must be set to a value that depends on the round: small initially, it grows quickly to 1 [26].

The condition $\Delta \gg \log n$ appears repeatedly. The reason is that these algorithms are based on powerful martingale inequalities and this condition is needed to make them work. These probabilistic inequalities are the subject of the next section.

12.3.1 Coloring with Martingales

Let $f(X_1, \ldots, X_n)$ be a function for which we can compute $\mathsf{E}[f]$, and let the X_i's be independent. Assume moreover that the following Lipshitz condition (with respect to the Hamming distance) holds:

$$|f(X) - f(Y)| \leq c_i \tag{12.2}$$

whenever $X := (x_1, \ldots, x_n)$ and $Y := (y_1, \ldots, y_n)$ differ only in the i-th coordinate. Then, f is sharply concentrated around its mean:

$$\Pr[|f - \mathsf{E}[f]| > t] \leq 2e^{-2t^2 / \sum_i c_i^2}. \tag{12.3}$$

This is the simplest of a series of powerful concentration inequalities dubbed the method of bounded differences (MOBDs) [27]. The method is based on martingale inequalities (we refer the reader to the thorough and quite accessible treatment in Reference 13). In words, if a function does not depend too much on any coordinate then it is almost constant.

To appreciate the power and ease of use of (12.3) we derive the well-known Chernoff-Hoeffding bound (see, among others, Reference 13, 28, 29). This bound states that if $X := \sum_{i=1}^n X_i$ is the sum of independent, binary random variables $X_i \in \{0, 1\}$, then X is concentrated around its mean: $\Pr[|X - \mathsf{E}[X]| > t] \leq 2e^{-2t^2/n}$. This captures the well-known fact that if a fair coin is flipped many times we expect HEADS to occur roughly 50% of the time, and this bound gives precise probability estimates of deviating from the mean. This bound can be recovered from (12.3) simply by defining $f := X$ and by noticing that condition 12.2 holds with $c_i = 1$.

We now apply the MOBD to the analysis of the trivial algorithm in a simplified setting. Let us assume that the network is a triangle-free, d-regular graph. We analyze what happens to the degree of a node after the first round. The probability with which an edge colors itself is $\left(1 - \frac{1}{d}\right)^{2d-2} \sim \frac{1}{e^2}$. Therefore, denoting with f the new degree of vertex u, we have that $\mathsf{E}[f] = \Theta(d)$. At first blush it may seem that the value of f depends on the tentative color choices of $\Theta(d^2)$ edges: those incident on u and the edges incident on them. But it is possible to express f as a function of $2d$ variables only, as follows. For every $v \in N(u)$ consider the bundle of $d - 1$ edges incident on v that are not incident on u, and treat this bundle as a single random variable, denoted as B_v. B_v is a random vector with $d - 1$ components, each specifying the tentative color choice of an edge incident on v (except uv). Furthermore, for every edge $e = uv$, let X_e denote e's color choice. Thus, f depends on d variables of type X_e and on d variables of type B_v. What is the effect of these variables on f? If we change the value of a fixed X_e, and keep all remaining variables the same, this color change can affect at most two edges (one of which is e itself). The resulting c_e is 2. The cumulative effect of the first d variables of type X_e is therefore $4d$.

Note now, that as the network is triangle-free, changing the value of a bundle B_v can only affect the edge uv the bundle is incident to. Thus the effect of changing B_v while keeping everything else fixed, is 1. Summing up we get a total effect of $\sum_i c_i^2 = 5d$. Plugging in this value in (12.3), for $t = \epsilon d$, where $1 > \epsilon > 0$ we get, $\Pr[|f - \mathsf{E}[f]| > \epsilon d] \leq 2e.^{-2\epsilon^2 d/5}$ We can see here why it is important to have $d \gg \log n$. With this condition, the bound is strong enough to hold for all vertices and all rounds simultaneously. In fact a value $d = \Theta(\log n)$ would seem to be enough, but the error terms accumulate as the algorithm progresses. To counter this cumulative effect, we must have $d \gg \log n$.

This establishes that the graph stays almost regular after one round (and in fact at all times), with high probability. For the full analysis one has to keep track of several other quantities besides vertex degrees, such as the size of the color lists. While the full analysis of the algorithm is beyond the scope of this survey, this simple example already clarifies some of the issues. For instance, if the graph is not triangle-free, then the effect of a bundle can be much greater than 1. To cope with this, more powerful inequalities, and a more sophisticated analysis, are needed [13,23,24,30]. We remark that in general these inequalities do not even require the variables X_i to be independent. In fact, only the following bounded difference condition is required,

$$|\mathsf{E}[f|X_1, \ldots, X_{i-1}, X_i = a] - |\mathsf{E}[f|X_1, \ldots, X_{i-1}, X_i = b]| \leq c_i.$$

If this condition holds for all possible choices of a and b, and for all i, then Equation 12.3 follows. What is behind this somewhat contrived definition is the fact that the sequence $Y_i := \mathsf{E}[f|X_1, \ldots, X_{i-1}, X_i]$ is a martingale (the so-called Doob martingale). A martingale is simply a sequence of random variables $Z_0, Z_1, \ldots, Z_n$ such that $\mathsf{E}[Z_i|Z_0, \ldots, Z_{i-1}] = Z_i$, for $i = 1, 2, \ldots, n$. A typical example of a martingale is a uniform random walk in the integer lattice, where a particle can move left, right, up or down with equal probability. If Z_i denotes the distance of the particle from the origin, the expected distance after one step stays put. A close relative of the Chernoff-Hoeffding bound, known as Azuma's inequality, states that if a martingale sequence $Z_0, Z_1, \ldots, Z_n$ satisfies the bounded difference condition $|Z_i - Z_{i-1}| \leq c_i$ for $i = 1, 2, \ldots, n$, then it is unlikely that Z_n is far from Z_0:

$$\Pr[|Z_n - Z_0| > t] \leq 2e^{-2t^2/\sum_i c_i^2}. \tag{12.4}$$

In other words, if a martingale sequence does not make big jumps, then it is unlikely to stray afar from its starting point. This is true for the random walk; it is very unlikely that after n steps the particle will be far from the origin. Note that for a Doob martingale $Y_0 = \mathsf{E}[f]$ and $Y_n = f$, so that Equation 12.4 becomes Equation 12.3.

To see the usefulness of this more awkward formulation, let us drop the assumption that the network is triangle-free and analyze again what happens to the vertex degrees, following the analysis from Reference 30. As observed this introduces the problem that the effect of bundles can be very large: changing the value of B_v can affect the new degree by as much as $d - 1$. We will therefore accept the fact that the new degree of a vertex is a function of $\Theta(d^2)$ variables, but we will be able to bound the effect of edges at distance one from the vertex. Fix a vertex v and let $N^1(v)$ denote the set of "direct" edges—that is, the edges incident on v—and let $N^2(v)$ denote the set of "indirect edges," that is, the edges incident on a neighbor of v. Let $N^{1,2}(v) := N^1(v) \bigcup N^2(v)$. Finally, let $T := (T_{e_1}, \ldots, T_{e_m})$, $m = |E(G)|$, be the random vector specifying the tentative color choices of the edges in the graph G. With this notation, the number of edges successfully colored at vertex v is a function $f(T_e, e \in N^{1,2}(v))$ (to study f or the new degree is the same: if f is concentrated so is the new degree).

Let us number the variables so that the direct edges are numbered *after* the indirect edges (this will be important for the calculations to follow). We need to compute

$$\lambda_k := |E[f \mid T_{k-1}, T_k = c_k] - E[f \mid T_{k-1}, T_k = c'_k]|. \tag{12.5}$$

We decompose f as a sum to ease the computations later. Introduce the indicator functions $f_e, e \in E$: $f_e(c)$ is 1 if edge e is successfully colored in coloring c, and 0 otherwise. Then $f = \sum_{v \in e} f_e$. Hence we are reduced, by linearity of expectation, to computing for each $e \in N^1(v)$, $|\Pr[f_e = 1 \mid T_{k-1}, T_k = c_k] - \Pr[f_e = 1 \mid T_{k-1}, T_k = c'_k]|$.

To compute a good bound for λ_k in (12.5), we shall lock together two distributions Y and Y'. Y is distributed as T conditioned on $T_{k-1}, T_k = c_k$, and Y', whereas Y' distributed as T conditioned on $T_{k-1}, T_k = c'_k$. We can think of Y' as identically equal to Y except that $Y'_k = c'_k$. Such a pairing (Y, Y') is called a *coupling* of the two different distributions $[T|T_{k-1}, T_k = c_k]$ and $[T|T_{k-1}, T_k = c'_k]$. It is easily seen that by the independence of all tentative colors, the marginal distributions of Y and Y' are exactly the two conditioned distributions $[T \mid T_{k-1}, T_k = c_k]$ and $[T \mid T_{k-1}, T_k = c'_k]$, respectively. Now let us compute $|E[f(Y) - f(Y')]|$.

First, let us consider the case when $e_1, \ldots, e_k \in N^2(v)$, that is, only the choices of indirect edges are exposed. Let $e_k = (w, z)$, where w is a neighbor of v. Then for a direct edge $e \neq vw$, $f_e(y) = f_e(y')$ because in the joint distribution space, y and y' agree on all edges incident on e. So we only need to compute $|E[f_{vw}(Y) - f_{vw}(Y')]|$. To bound this simply, we observe first that $f_{vw}(y) - f_{vw}(y') \in [-1, 1]$ and second that $f_{vw}(y) = f_{vw}(y')$ unless $y_{vw} = c_k$ or $y_{vw} = c'_k$. Thus we can conclude that $E[f_{vw}(Y) - f_{vw}(Y')] \leq \Pr[Y_e = c_k \vee Y_e = c'_k] \leq \frac{2}{d}$.

In fact one can do a tighter analysis using the same observations. Let us denote $f_e(y, y_{w,z} = c_1, y_e = c_2)$ by $f_e(c_1, c_2)$. Note that $f_{vw}(c_k, c_k) = 0$ and similarly $f_{vw}(c'_k, c'_k) = 0$. Hence

$$E[f_e(Y) - f_e(Y') \mid z] = (f_{vw}(c_k, c_k) - f_{vw}(c'_k, c_k))\Pr[Y_e = c_k] + (f_{vw}(c_k, c'_k) - f_{vw}(c'_k, c'_k))\Pr[Y_e = c'_k]$$

$$= (f_{vw}(c_k, c'_k) - f_{vw}(c'_k, c_k))\frac{1}{d}.$$

(Here we used the fact that the distribution of colors around v is unaffected by the conditioning around z and that each color is equally likely.) Hence $|E[f_e(Y) - f_e(Y')]| \leq \frac{1}{d}$.

Now let us consider the case when $e_k \in N^1(v)$, that is, choices of all indirect edges and of some direct edges have been exposed. In this case, we merely observe that f is Lipshitz with constant 2: $|f(y) - f(y')| \leq 2$ whenever y and y' differ in only one coordinate. Hence we can easily conclude that $|E[f(Y) - f(Y')]| \leq 2$.

Overall, $\lambda_k \leq 1/d$ for an edge $e_k \in N^2(v)$, and $\lambda_k \leq 2$ for an edge $e_k \in N^1(v)$. Therefore we get

$$\sum_k \lambda_k^2 = \sum_{e \in N^2(v)} \frac{1}{d^2} + \sum_{e \in N^1(v)} 4 \leq 4d + 1.$$

We thus arrive at the following sharp concentration result by plugging into Equation 12.3: Let v be an arbitrary vertex and let f be the number of edges successfully colored around v in one stage of the trivial algorithm. Then,

$$\Pr[|f - E[f]| > t] \leq 2 \exp\left(-\frac{t^2}{2d + \frac{1}{2}}\right).$$

As $E[f] = \Theta(d)$, this is a very strong bound.

12.4 Matchings

Maximum matching is probably one of the best studied problems in computer science: given a weighted undirected graph $G = (V, E)$, compute a subset of pairwise nonincident edges (*matching*) of maximum cost. For the sake of simplicity, we will focus on the the cardinality version of the problem, where all the edges have weight one.

It is not hard to show that a maximum matching cannot be computed efficiently (i.e., in poly-logarithmic time) in a distributed setting.

Lemma 12.7 *Any distributed maximum matching algorithm requires $\Omega(n)$ rounds.*

Proof. Consider the following *mailing problem*: let P be a path of $n = 2k + 1$ nodes, and let ℓ and r be the left and right endpoints of the path, respectively. Let moreover c be the *central* node of the path. Nodes ℓ and r receive the same input bit b, and the problem is to forward b to the central node c. Clearly this process takes at least k rounds.

Now assume by contradiction that there exists a $o(n)$ distributed maximum matching protocol $\mathcal{M}$. We can use $\mathcal{M}$ to solve the above-mentioned mailing problem in the following way. All the nodes run $\mathcal{M}$ on the auxiliary graph $P(b)$ obtained from P by removing the edge incident to ℓ if $b = 1$, and the edge incident to r otherwise. If $b = 1$ ($b = 0$), the edge on the left (right) of v must belong to the (unique) maximum matching. This way c can derive the value of the input bit b in $o(n) = o(k)$ rounds, which is a contradiction. ∎

Fischer et al. [31] described a parallel algorithm to compute a near-optimal matching in arbitrary graphs. Their algorithm can be easily turned into a distributed protocol to compute a $k/(k + 1)$-approximate solution in poly-logarithmic time, for any fixed positive integer $k > 0$. A crucial step in the algorithm by Fischer et al. is computing (distributively) a maximal independent set. As this sub-problem is rather interesting by itself in the distributed case, in Section 12.4.1 we will sketch how it can be solved efficiently. In Section 12.4.2 we will describe and analyze the algorithm by Fischer et al.

12.4.1 Distributed Maximal Independent Set

Recall that an *independent set* of a graph is a subset of pairwise nonadjacent nodes. No deterministic protocol is currently known for the problem. Indeed, this is one of the main open problems in distributed algorithms. Luby [32] and independently Alon et al. [33] gave the first distributed randomized algorithms to compute a maximal independent set. Here we will focus on Luby's result, as described in Kozen's book [34].

As the algorithm by Fischer et al., Luby's algorithm was originally thought for a parallel setting, but it can be easily turned into a distributed algorithm. It is worth to notice that transforming an efficient parallel algorithm into an efficient distributed algorithm is not always trivial. For example, there is a deterministic parallel version of Luby's algorithm, whereas, as aforementioned, no efficient deterministic distributed algorithm is known for the maximal independent set problem.

Luby's algorithm works in stages. In each stage one (not necessarily maximal) independent set I is computed, and the nodes I are removed from the graph together with all their neighbors. All the edges incident to deleted nodes are also removed. The algorithm ends when no node is left. At the end of the algorithm a maximal independent set is given by the union of the independent sets I computed in the different stages.

It remains to describe how each independent set I is computed. Each node v in the (current) graph independently becomes a *candidate* with probability $\frac{1}{2d(v)}$. Then, for any two adjacent candidates, the one of lower degree is discarded from S (ties can be broken arbitrarily). The remaining candidates form the set I.

Each stage can be trivially implemented with a constant number of communication rounds. The expected number of rounds is $O(\log n)$. More precisely, in each stage at least a constant expected fraction of the (remaining) edges are removed from the graph.

A crucial idea in Luby's analysis is the notion of good nodes: a node v is *good* if at least one third of its neighbors have degree not larger than v. In particular, this implies

$$\sum_{u \in N(v)} \frac{1}{2d(u)} \geq \frac{1}{6}. \tag{12.6}$$

Otherwise v is *bad*. Though there might be few good nodes in a given graph, the edges incident to at least one good node are a lot. Let us call an edge *good* if it is incident to at least one good node, and *bad* otherwise. The following lemma holds.

Lemma 12.8 *At least one half of the edges are good.*

Proof. Direct all the edges toward the endpoint of higher degree, breaking ties arbitrarily. Consider any bad edge e directed toward a given (bad) node v. By definition of bad nodes, the out-degree of v is at least twice its own in-degree. Thus we can uniquely map e into a pair of edges (either bad or good) leaving v. Therefore the edges are at least twice as many as the bad edges. ∎

Thus it is sufficient to show that in a given stage each good node is removed from the graph with constant positive probability.

Lemma 12.9 *Consider a node v in a given stage. Node v belongs to I with probability $\frac{1}{4d(v)}$.*

Proof. Let $L(v) = \{u \in N(v) \mid d(u) \geq d(v)\}$ be the neighbors of v of degree not smaller than $d(v)$. Then

$$Pr(v \notin I \mid v \in S) \leq \sum_{u \in L(v)} Pr(u \in S \mid v \in S) = \sum_{u \in L(v)} Pr(u \in S) \leq \sum_{u \in L(v)} \frac{1}{2d(u)} \leq \sum_{u \in L(v)} \frac{1}{2d(v)} \leq \frac{1}{2}.$$

Hence $Pr(v \in I) = Pr(v \in I \mid v \in S) \, Pr(v \in S) \geq \frac{1}{2} \frac{1}{2d(v)} = \frac{1}{4d(v)}$. ∎

Lemma 12.10 *Let v be a good node in a given stage. Node v is discarded in the stage considered with probability at least $1/36$.*

Proof. We will show that $v \in N(I) = \cup_{u \in I} N(u)$ with probability at least $1/36$. The claim follows. If v has a neighbor u of degree at most 2, by Lemma 12.9, $Pr(v \in N(I)) \geq Pr(u \in I) \geq \frac{1}{4d(u)} = \frac{1}{8}$.

Now assume all the neighbors of v have degree 3 or larger. It follows that, for every neighbor u of v, $\frac{1}{2d(u)} \leq \frac{1}{6}$. Hence by Equation 12.6 there exists a subset $M(v)$ of neighbors of v such that $\frac{1}{6} \leq \sum_{u \in M(v)} \frac{1}{2d(u)} \leq \frac{1}{3}$. Thus

$$Pr(v \in N(I)) \geq Pr(\exists u \in M(v) \cap I)$$
$$\geq \sum_{u \in M(v)} Pr(u \in I) - \sum_{u,w \in M(v), u \neq w} Pr(u \in I \wedge w \in I)$$
$$\geq \sum_{u \in M(v)} \frac{1}{4d(u)} - \sum_{u,w \in M(v), u \neq w} Pr(u \in S \wedge w \in S)$$
$$\geq \sum_{u \in M(v)} \frac{1}{4d(u)} - \sum_{u,w \in M(v), u \neq w} Pr(u \in S)Pr(w \in S)$$
$$\geq \sum_{u \in M(v)} \frac{1}{4d(u)} - \sum_{u \in M(v)} \sum_{w \in M(v)} \frac{1}{2d(u)} \frac{1}{2d(w)}$$
$$= \left(\frac{1}{2} - \sum_{w \in M(v)} \frac{1}{2d(w)} \right) \sum_{u \in M(v)} \frac{1}{2d(u)} \geq \left(\frac{1}{2} - \frac{1}{3} \right) \frac{1}{6} = \frac{1}{36}. \quad ∎$$

12.4.2 The Distributed Maximum Matching Algorithm

Consider an arbitrary matching M of a graph $G = (V, E)$. A node is *matched* if it is the endpoint of some edge in M, and *free* otherwise. An *augmenting path* P with respect to M is a path (of odd length) whose endpoints are free and whose edges are alternatively inside and outside M. The reason of the name is that we can obtain a matching M' of cardinality $|M| + 1$ from M, by removing from M all the edges which are also in P, and by adding to M the remaining edges of P (in other words, M' is the symmetric difference $M \oplus P$ of M and P).

The algorithm by Fischer et al. is based on the following two lemmas by Hopcroft and Karp [35]. Let two paths be *independent* if they are node-disjoint. Note that a matching can be augmented along several augmenting paths simultaneously, provided that such paths are independent.

Lemma 12.11 *If a matching is augmented along a maximal set of independent shortest augmenting paths, then the shortest augmenting paths length grows.*

Lemma 12.12 *Suppose a matching M does not admit augmenting paths of length $2k - 1$ or smaller. Then the size of M is at least a fraction $\frac{k}{k+1}$ of the maximum matching size.*

Proof. Let M^* be a maximum matching. The symmetric difference $M' = M \oplus M^*$ contains $|M^*| - |M|$ independent augmenting paths with respect to M. As each of these paths contains at least k edges of M, $|M^*| - |M| \leq |M|/k$. The claim follows. ∎

We are now ready to describe and analyze the approximate maximum matching algorithm by Fischer et al. The algorithm proceeds in stages. In each stage i, $i \in \{1, 2, \ldots, k\}$, the algorithm computes a maximal independent set P_i of augmenting paths of length $2i - 1$ with respect to the current matching M. Then M is augmented according to P_i. Stage i can be implemented by simulating Luby's algorithm on the auxiliary graph induced by the augmenting paths considered, where the nodes are the paths and the edges are the pairs of nonindependent paths. In particular, Luby's algorithm takes $O(\log n^{2i})$ rounds in expectation in the auxiliary graph, where each such round can be simulated within $O(i)$ rounds in the original graph. Note that, by Lemma 12.11, at the end of stage i there are no augmenting paths of length $2i - 1$ or smaller. It follows from Lemma 12.12 that at the end of the k-th stage the matching computed is $\frac{k}{k+1}$-approximate. The total expected number of rounds is trivially $O(k^3 \log n)$. The following theorem summarizes the above-mentioned discussion.

Theorem 12.1 *For every integer $k > 0$, there is a distributed algorithm which computes a matching of cardinality at least $\frac{k}{k+1}$ times the maximum matching cardinality within $O(k^3 \log n)$ communication rounds in expectation.*

Wattenhofer and Wattenhofer [36] gave a $O(\log^2 n)$ randomized algorithm to compute a constant approximation in the weighted case. In the deterministic case weaker results are available. This is mainly due to the fact that we are not able to compute maximal independent sets deterministically. Hańćkowiak et al. [37,38] described an efficient distributed deterministic algorithm to compute a maximal matching. Recall that any maximal matching is a 2-approximation for the maximum matching problem. Recently a 1.5 deterministic distributed approximation algorithm was described in Reference 39.

12.5 LP-Based Distributed Algorithms

It might come as a surprise that LP-based methods find their application in a distributed setting. In this section we describe some primal-dual algorithms for vertex cover problems that give "state-of-the-art" approximations. In general it seems that the primal-dual method, one of the most successful techniques in approximation algorithms, when applied to graph algorithms exhibits "local" properties that makes it amenable to a distributed implementation. The best way to explain what we mean is to work out an example.

We will illustrate the method by considering the *vertex cover* problem: given an undirected graph $G = (V, E)$, with positive weights $\{c(v)\}_{v \in V}$, compute a minimum cost subset V' of nodes such that each edge is incident to at least one node in V'. This *NP*-hard problem is approximable within 2 [40], and not approximable within 1.1666 unless P = NP [41]. In the centralized case there is a primal-dual 2-approximation algorithm. The distributed implementation we give yields a $2 + \epsilon$ approximation, where ϵ can be any fixed constant greater than zero. The number of communication rounds of the algorithm is $O(\log n \, \log \frac{1}{\epsilon})$.

The sequential primal-dual algorithm works as follows. We formulate the problem as an integer program (IP):

$$\min \quad \sum_{v \in V} c(v) \cdot x_v \tag{IP}$$

$$\text{s.t} \quad x_v + x_u \geq 1 \qquad\qquad \forall e = (u, v) \in E \tag{12.7}$$

$$x_v \in \{0, 1\} \qquad\qquad \forall v \in V \tag{12.8}$$

The binary indicator variable x_v, for each $v \in V$, takes value one if $v \in V'$, and zero otherwise.

We now let (LP) be the standard LP relaxation obtained from (IP) by replacing the constraints (12.8) by $x_v \geq 0$ for all $v \in V$. In the linear-programming dual of (LP) we associate a variable α_e with constraint (12.7) for every $e \in E$. The linear programming dual (D) of (LP) is then

$$\max \quad \sum_{e \in E} \alpha_e \tag{D}$$

$$\text{s.t} \quad \sum_{e = (u,v) \in E} \alpha_e \leq c(v) \qquad\qquad \forall v \in V \tag{12.9}$$

$$\alpha_e \geq 0 \qquad\qquad \forall e \in E \tag{12.10}$$

The starting primal and dual solutions are obtained by setting to zero all the variables x_v and α_e. Observe that the dual solution is feasible while the primal one is not. We describe the algorithm as a continuous process. We let all the variables α_e grow at uniform speed. As soon as one constraint of type (12.9) is satisfied with equality (it becomes *tight*), we set the corresponding variable x_v to one, and we freeze the values α_e of the edges incident to v. The α-values of frozen edges do not grow more, so that the constraint considered remains tight. The process continues until all edges are frozen. When this happens the primal solution becomes feasible. In order to prove the latter claim, assume by contradiction that this is not the case. But then there is an edge $e = uv$ which is not covered, that is, $x_u = x_v = 0$. This means that the constraints corresponding to u and v are not tight and α_e can continue to grow, a contradiction.

Thus the set $V' := \{u : x_u = 1\}$ is a cover. Its cost is upper-bounded by twice the cost of the dual solution:

$$\sum_{v \in V} c(v) x_v = \sum_{v \in V'} c(v) \leq \sum_{v \in V'} \sum_{e = (u,v) \in E} \alpha_e \leq 2 \sum_{e \in E} \alpha_e.$$

Thus the solution computed is 2-approximate by weak duality.

The continuous above-mentioned process can be easily turned into a discrete one. Let $c'(v)$ be the difference between the right-hand side and the left-hand side of constraints in (12.9) in a given instant of time (*residual weight*):

$$c'(v) = c(v) - \sum_{e = (u,v) \in E} \alpha_e.$$

Let, moreover, $d'(v)$ be the current number of nonfrozen (*active*) edges incident to v. The idea is to raise in each step the dual value α_e of all the active edges by the minimum over all nodes v such that $x_v = 0$ of the quantity $c'(v)/d'(v)$. This way, in each step at least one extra node enters the vertex cover.

There is a simple-minded way to turn the above-mentioned algorithm into a distributed algorithm: each node v maintains the quantities $c'(v)$ and $d'(v)$. A node is *active* if $c'(v) > 0$ and $d'(v) > 0$, that is if v and at least one of its neighbors are not part of the vertex cover. In each round each active node v sends a *proposal* $c'(v)/d'(v)$ to all its active neighbors. Then it decreases $c'(v)$ by the minimum of all the proposals sent and received. If $c'(v)$ becomes zero, v enters the vertex cover. Otherwise, if $d'(v)$ becomes zero, v halts as all its neighbors already belong to the vertex cover.

The main drawback of this approach is that it is very slow. In fact, it may happen that in each step a unique node enters the vertex cover, thus leading to a linear number of rounds. Khuller et al. [42] showed how to circumvent this problem by loosing something in the approximation. Here we will present a simplified version of their algorithm and analysis (which was originally thought for weighted set cover in a parallel setting). The idea is to slightly relax the condition for a node v to enter the vertex cover: it is sufficient that the residual weight $c'(v)$ falls below $\epsilon c(v)$, for a given (small) constant $\epsilon > 0$.

Theorem 12.2 *The above-mentioned algorithm computes a $\frac{2}{1-\epsilon}$-approximate vertex cover within $O(\log n \, \log \frac{1}{\epsilon})$ rounds.*

Proof. The bound on the approximation easily follows by adapting the analysis of the primal-dual centralized approximation algorithm:

$$(1 - \epsilon)\, apx = \sum_{v \in V'} (1 - \epsilon)\, c(v) \leq \sum_{v \in V'} \sum_{e=(u,v) \in E} \alpha_e \leq 2 \sum_{e \in E} \alpha_e \leq 2\, opt.$$

To bound the number of rounds we use a variant of the notion of good nodes introduced in Section 12.4.1. Consider the graph induced by the active nodes in a given round, and call the corresponding edges *active*. Let us direct all the active edges toward the endpoint which makes the smallest proposal. A node is *good* if its in-degree is at least one third of its (total) degree. By basically the same argument as in Section 12.4.1, at least one half of the edges are incident to good nodes. Moreover, the residual weight of a node which is good in a given round decreases by at least one third in the round considered. As a consequence, a node can be good in at most $\log_{3/2} \frac{1}{\epsilon}$ rounds (after those many rounds it must enter the vertex cover).

We will show next that the total number of active edges halves every $O(\log \frac{1}{\epsilon})$ rounds by means of a potential function argument. It follows that the total number of rounds is $O(\log m \, \log \frac{1}{\epsilon}) = O(\log n \, \log \frac{1}{\epsilon})$. Let us associate $2 \log_{3/2} \frac{1}{\epsilon}$ credits to each edge, and thus $2m \log_{3/2} \frac{1}{\epsilon}$ credits to the whole graph. When a node v is good in a given step, we remove one credit from each edge incident to it. Observe that an active edge e in a given round must have at least 2 credits left. This is because otherwise one of the endpoints of e would already belong to the vertex cover, and thus e could not be active. By m_j we denote the number of active edges in round j. Recall that in each round at least one half of the edges are incident to a good node, and such edges loose at least one credit each in the round considered. Thus the total number of credits in round j decreases by a quantity g_j which satisfies $g_j \geq m_j/2$. Consider an arbitrary round i, and let k be the smallest integer such that $m_{i+k} < m_i/2$ (or $i+k$ is the last round). Is is sufficient to show that $k = O(\log \frac{1}{\epsilon})$. In each round $j, j \in \{i, i+1, \ldots, i+k-1\}$, the number of edges satisfies $m_j \geq m_i/2$. The total number of credits at the beginning of round i is at most $2m_i \log_{3/2} \frac{1}{\epsilon}$, and the algorithm halts when no credit is left. Therefore

$$2m_i \log_{3/2} \frac{1}{\epsilon} \geq \sum_{j=i}^{i+k-1} g_j \geq \sum_{j=i}^{i+k-1} \frac{m_j}{2} \geq \sum_{j=i}^{i+k-1} \frac{m_i}{4} = k\frac{m_i}{4} \quad \Rightarrow \quad k \leq 8 \log_{3/2} \frac{1}{\epsilon} = O\left(\log \frac{1}{\epsilon}\right). \quad \blacksquare$$

By choosing $\epsilon = 1/(nC + 1)$, where C is the maximum weight, the algorithm by Khuller et al. computes a 2-approximate vertex cover within $O(\log n \, \log(nC))$ rounds. Recently, Grandoni et al. [43] showed how

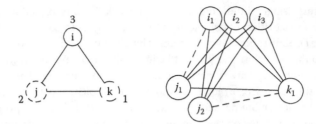

FIGURE 12.2 A weighted graph G (on the left) with the corresponding auxiliary graph $\widetilde{G}$. A maximal matching M of $\widetilde{G}$ is indicated via dashed lines. The nodes of G such that all the corresponding nodes in $\widetilde{G}$ are matched form a 2-approximate vertex cover.

to achieve the same task in $O(\log(nC))$ rounds by means of randomization. They reduce the problem to the computation of a maximal matching in an auxiliary graph of nC nodes (to have an idea of the reduction, see Figure 12.2). Such matching can be computed in $O(\log(nC))$ rounds via the randomized, distributed maximal matching algorithm by Israeli and Itai [44]. The authors also show how to keep small the message size and the local computation time by computing the matching *implicitly*.

The *capacitated* vertex cover problem is the generalization of the vertex cover problem where each node v can cover only a limited number $b(v) \le d(v)$ of edges incident to it. Grandoni et al. [45] showed how to compute within $O(\frac{\log nC}{\epsilon})$ rounds an $(2 + \epsilon)$-approximate solution, if any, which violates the capacity constraints by a factor at most $(4 + \epsilon)$. They also proved that any distributed constant approximation algorithm must violate the capacity constraints by a factor at least 2. This, together with the known lower bounds on the approximation of (classical) vertex cover, shows that their algorithm is the best possible modulo constants. The algorithm by Grandoni et al. builds up on a primal-dual centralized algorithm developed for the purpose, which computes a 2-approximation with a factor 2 violation of the capacity constraints. Turning such primal-dual algorithm into a distributed protocol is far more involved than in the case of classical vertex cover.

12.6 What Can and Cannot Be Computed Locally?

This fundamental question in distributed computing was posed by Naor and Stockmeyer [46]. Here, "locally" means that the nodes of the network use information available locally from a neighborhood that can be reached in time much smaller than the size of the network. For many natural distributed network problems such as leader election and consensus the parameter determining the time complexity is not the number of vertices, but the network *diameter D* which is the maximum distance (number of hops) between any two nodes [47]. A natural question is whether other fundamental primitives can be computed in $O(D)$ time in a distributed setting. If the model allows messages of unbounded size, then there is a trivial affirmative answer to this question: collect all the information at one vertex, solve the problem locally and then transmit the result to all vertices. The problem is therefore only interesting in the more realistic model where we assume that each link can transmit only B bits in any time step (B is usually taken to be a constant or $O(\log n)$).

A landmark negative result in this direction was that of Linial [48], which investigated the time complexity of various global functions of a graph computed in a distributed setting. Suppose that n processors are arranged in a ring and can communicate only with their immediate neighbors. Linial showed that a 3-coloring of the n-cycle requires time $\Omega(\log^* n)$. This result was extended to randomized algorithms by Naor [49]: any probabilistic algorithm for 3-coloring the ring must take a least $\frac{1}{2}\log^* n - 2$ rounds, otherwise the probability that all processors are colored legally is less than $\frac{1}{2}$. The bound is tight (up to a constant factor) in light of the deterministic algorithms of Cole and Vishkin [50].

There has been surprisingly little continuation of work in this direction until fairly recently. Garay et al. [51] gave an algorithm of complexity $O(D + \sqrt{n} \log n)$ to compute a MST of a graph on n vertices with diameter D. Similar bounds were attained by other methods, but none managed to break the $\sqrt{n}$ barrier, leading to the suspicion that it might be impossible to compute the MST in time $o(\sqrt{n})$ and so this problem is fundamentally harder than the other paradigm problems. The issue was finally settled in Reference 52 who showed a $\Omega(\sqrt{n})$ lower bound on the problem (up-to log factors). Subsequently, Elkin [53] improved the lower bound and also extended it to distributed approximation algorithms. Kuhn et al. [54] gave lower bounds on the complexity of computing the minimum vertex cover (MVC) and the MDS of a graph: in k communication rounds, the MVC and MDS can only be approximated to factors of $\Omega(n^{ck^2}/k)$ and $\Omega(\Delta^{1/k}/k)$ (where Δ is the maximum degree of the graph). Thus, the number of rounds required to reach a constant or even a poly-log approximation is at least $\Omega(\sqrt{\log n / \log \log n})$ and $\Omega(\log \Delta / \log \log \Delta)$. The same lower bounds also apply to the construction of maximal matchings and maximal independent sets via a simple reduction.

12.6.1 A Case Study: MST

Here, we give a self-contained exposition of the lower bound for the MST problem due to References 52, 53. We will give the full proof of a bound somewhat weaker than the optimal result of Elkin to convey the underlying ideas more clearly. The basic idea is easy to explain using the example of Peleg and Rubinovich [52], see Figure 12.3. The network consists of m^2 country road and one highway. Each country road has m toll stations and between every two successive toll station are m towns. The highway has m toll stations with no towns in between. Each toll station number i on each country road is connected to the corresponding highway toll station. The left end of country road i is labelled s_i and its right end r_i. The left end of the highway is labelled s and the right end r. This is the basic underlying graph. Note that there are $\Theta(m^4)$ vertices and the diameter is $\Theta(m)$.

As for the weights, every edge along the highway or on the country roads has weight 0. The roads connecting the toll stations on the country roads to the corresponding toll stations on the highway have weight ∞ *except for the first and last toll stations*. The toll station connections on the right end between each r_i and r are all 1. At the left end, between each s_i and s, they take either the value 0 or ∞.

What does the MST of this network look like? First, we may as well include the edges along the highway and each path as these have zero cost. Also the intermediate connecting edges have weight ∞ and so are excluded. That leaves us with the connecting edges on the left and on the right. The choice here depends on the weights on the left connecting edges. There are m connecting edges from the left vertex s. If the edge (s, s_i) has weight ∞, then we must exclude this and include the matching connection (r_i, r) at the right end. On the other hand, if edge (s, s_i) has weight 0, then we must include this and exclude the corresponding edge (r_i, r) at the right to r. Thus there are m^2 decisions made at s depending on the weights of the corresponding edges, and these decision must be conveyed to r to pick the corresponding complementary edges. How quickly can these m^2 bits be conveyed from s to r? Clearly it would take a

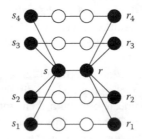

FIGURE 12.3 MST lower bound graph for $m = 2$. The black nodes are the toll stations and the white nodes are the towns.

very long to route along the country roads, and so one must use the highway edges instead. Each highway edge can forward only B bits at any time step. So, heuristically, transporting the m^2 bits takes $\Omega(m^3/B)$ steps.

To make this heuristic argument formal, Peleg and Rubinovich introduced a *mailing problem* to be solved on a given network. In the above-mentioned example, the *sender s* has m^2 bits that need to be transported to the *receiver r*. At each step one can forward B bits along any edge. How many steps do we need to correctly route the m^2 bits from the sender to the receiver? It is easy to see that there is a reduction from the mailing problem to that of computing the MST: for each of the input bits at s, set the weights on the connecting edges accordingly: the weight (s, s_i) is ∞ if the input bit i is 1 and 0 otherwise. Now compute the MST. Then, if vertex r notices that the edge (r_i, r) is picked in the MST, it decodes i as 1 and as 0 otherwise. This will correctly solve the mailing problem, due to the structure of the MST discussed earlier. Thus, a lower bound on the mailing problem implies the same lower bound on the MST problem.

In fact by a slight change, the correspondence can be extended from exact to approximation algorithms. Elkin [53] introduced the *corrupted mail* problem. Here there are Γ bits at the sender exactly $\alpha\Gamma$ of which are 1's, where α and Γ are parameters. In the aforementioned example, $\Gamma = m^2$. The receiver should get Γ bits delivered to it, but these are allowed to be somewhat corrupted. The restrictions are (a) any input bit that was 1 must be transmitted correctly without corruption and (b) the total number of 1's delivered can be at most $\beta\Gamma$ where $\beta \geq \alpha$ is another parameter. Consider solving the (α, β) corrupted mail problem on the Peleg-Rubinovich example. As in the reduction before, the vertex s sets the weights on the left connections according to its input and so exactly $\alpha\Gamma$ connections have weight ∞, and the rest 0. The optimal MST has weight exactly $\alpha\Gamma$ obtained by picking the corresponding right connections. Now, instead of the optimal MST, suppose we apply a protocol to compute a β/α approximation. This approximate MST can have weight at most $\beta\Gamma$, and it must include the connection edges at r paired with the infinite weight edges at s. Thus, if r sets its bits as before corresponding to which of its connections are in the approximate MST, we get a correct protocol for the (α, β) corrupted mail problem. Thus a lower bound for the (α, β) corrupted mail problem implies the same lower bound for a $\frac{\beta}{\alpha}$ approximate MST.

We are thus left with the task of proving a lower bound for the corrupted mail problem. Let the *state* $\psi(v, t)$ of a vertex v at some time t denote the sequence of messages it has received up to this time. Consider the start vertex s at time 0: this can be in any of $\binom{\Gamma}{\alpha\Gamma}$ states corresponding to the input it receives. At this time, on the other hand, the vertex r (and indeed, any other vertex) is in a fixed state (having received no messages at all). As time progresses and messages are passed, the set of possible states that other vertices are in expands. Eventually, the set of possible states that vertex r is in must be large enough to accommodate the output corresponding to all the possible inputs at s. Each possible state of r with at most $\beta\Gamma$ 1s can be the correct answer to at most $\binom{\beta\Gamma}{\alpha\Gamma}$ input configurations at s. Hence, the set of output states at r must be at least $\binom{\Gamma}{\alpha\Gamma}/\binom{\beta\Gamma}{\alpha\Gamma} \geq (1/e\beta)^{\alpha\Gamma}$.

Now, we will argue that it must take a long time for any protocol, before enough messages arrive at r for the set of its possible states to have this size. Consider the *tail sets* $T_i, i \geq 1$ which consist of the tail of each country road from vertex i until the end, and the corresponding fragment of the highway consisting of the vertices $h_{\lceil i/m \rceil m}$ until h_{m^2}. Also, set $T_0 := V \setminus \{h_0\}$. For a subset of vertices U, let $\mathcal{C}(U, t)$ denote set of all possible vectors of states of the vertices in U at time t, and let $\rho(U, t) := |\mathcal{C}(U, t)|$. Note that $\rho(T_0, 0) = 1$ although $\rho(\{s\}, 0) = \binom{\Gamma}{\alpha\Gamma}$.

We now focus on how set of configurations of the tail sets T_i grow in time. Fix a configuration $C \in \mathcal{C}(T_t, t)$. How many configurations in $\mathcal{C}(T_{t+1}, t+1)$ can this branch into? The tail set T_{t+1} is connected to the rest of the graph by one highway edge f and by m^2 path edges. Each of the path edges carries a unique message determined by the state of the left end point in configuration C. The state of the left end point of the highway edge f is not determined by C and hence there could be a number of possible messages that could be relayed along it. However, because of the restriction that at most B bits can be transmitted along an edge at any time step, the total number of possible behaviors observable on edge f at this time step is at most $2^B + 1$. Thus the configuration C can branch off into at most $2^B + 1$ possible configurations

$C' \in \mathcal{C}(T_{t+1}, t + 1)$. Thus we have argued that for $0 \leq t < m^2$, $\rho(T_{t+1}, t + 1) \leq (2^B + 1)\rho(T_t, t)$. By induction, this implies that for $0 \leq t < m^2$, $\rho(T_t, t) \leq (2^B + 1)^t$. Thus finally, we have, that if t^* is the time at which the protocol ends, then either $t^* \geq m^2$, or $(1/e\beta)^{\alpha\Gamma} \leq \rho(\{r\}, t^*) \leq \rho(T_{t^*}, t^*) \leq (2^B + 1)^{t^*}$. Hence, $t^* \geq \min(m^2, \alpha\Gamma \log(\frac{1}{e\beta})/(B + 1))$.

Recalling that $\Gamma = m^2$ in our specific graph, and taking β to be a constant such that $\beta e < 1$, $t^* = \Omega(\alpha m^2/B)$, or in terms of the number of vertices $n = \Theta(m^4)$ of the graph, $t^* = \Omega(\alpha\sqrt{n}/B)$. If we have a $H := \beta/\alpha$ approximation algorithm for the MST, this implies that $t^* = \Omega(\sqrt{n}/HB)$, implying the trade-off $t^*H = \Omega(\sqrt{n}/B)$ between time and approximation. Elkin [53] improves the lower bound for t^* to $t^* = \Omega\left(\sqrt{n/B}/H\right)$, implying the time-approximation tradeoff $t^{*2}H = \Omega\left(\sqrt{n/B}\right)$, and gives a protocol achieving this tradeoff.

12.6.2 The Role of Randomization in Distributed Computing

Does randomization help in a distributed setting? This is a fundamental open question in distributed computing. For some of the problems discussed, such as 3-coloring on a ring, we have noted that matching lower bounds hold for randomized algorithms. By the usual application of Yao's minimax theorem, Elkin's lower bound also applies to randomized algorithms. For the problem of computing maximal matchings and maximal independent sets, there are simple randomized algorithms, whereas the result of Kuhn et al. [54] shows a super-poly-log lower bound for deterministic algorithms. A classification of problems by the degree to which randomization helps is an interesting open problem.

References

1. JaJa, J., *An Introduction to Parallel Algorithms*, Addison-Wesley, Reading, MA, 1992.
2. Basagni, S., Mastrogiovanni, M., Panconesi, A., and Petrioli, C., Localized protocols for ad hoc clustering and backbone formation: A performance comparison, *IEEE Trans. Paral. Dist. Sys.*, 17(4), 292, 2006.
3. Rajagopalan, S. and Vazirani, V.V., Primal-dual RNC approximation algorithms for set cover and covering integer programs, *SIAM J. Comput.*, 28(2), 525(electronic), 1999.
4. Vazirani, V.V., *Approximation Algorithms*, Springer-Verlag, Berlin, Germany, 2001.
5. Ausiello, G., Crescenzi, P., Gambosi, G., Kann, V., Marchetti-Spaccamela, A., and Protasi, M., *Complexity and Approximation*, Springer-Verlag, Berlin, Germany, 1999.
6. Raz, R. and Safra, S., A sub-constant error-probability low-degree test, and a sub-constant error-probability PCP characterization of NP, in *Proceedings of the Twenty-Ninth Annual (ACM) Symposium on Theory of Computing*, pp. 475–484, El Paso, TX, 1997.
7. Arora, S. and Sudan, M., Improved low-degree testing and its applications, *Combinatorica*, 23(3), 365, 2003.
8. Feige, U., A threshold of ln n for approximating set cover, *JACM*, 45(4), 634, 1998.
9. Dubhashi, D., Mei, A., Panconesi, A., Radhakrishnan, J., and Srinivasan, A., Fast distributed algorithms for (weakly) connected dominating sets and linear-size skeletons, in *Proceedings of the Fourteenth Annual (ACM-SIAM) Symposium on Discrete Algorithms*, pp. 717–724, Baltimore, MD, 2003.
10. Jia, L., Rajaraman, R., and Suel, T., An efficient distributed algorithm for constructing small dominating sets, *Dist. Comput.*, 15, 193, 2002.
11. Kuhn, F. and Wattenhofer, R., Constant-time distributed dominating set approximation, *Dist. Comput.*, 17(4), 303, 2005.
12. Karp, R.M., Probabilistic recurrence relations, *JACM*, 41(6), 1136, 1994.
13. Dubhashi, D. and Panconesi, A., *Concentration of Measure for the Analysis of Randomised Algorithms*, Cambridge University Press, New York, 2005.

14. Matoušek, J., *Lectures on Discrete Geometry*, Graduate Texts in Mathematics, Springer, New York, p. 212, 2002.
15. Bellare, M., Goldreich, O., and Sudan, M., Free bits, PCPS and non-approximability—towards tight results, *SIAM J. Comput.*, 27, 804, 1998.
16. Feige, U. and Kilian, J., Zero knowledge and the chromatic number, *JCSS*, 57, 187, 1998.
17. Halldòrsson, M.M., A still better performance guarantee for approximate graph coloring, *Inf. Proc. Lett.*, 45, 19, 1993.
18. Johansson, O., Simple distributed $\delta + 1$-coloring of graphs, *Inf. Proc. Lett.*, 70(5), 229, 1999.
19. Luby, M., Removing randomness in parallel computation without a processor penalty, *JCSS*, 47(2), 250, 1993.
20. Finocchi, I., Panconesi, A., and Silvestri, R., An experimental study of simple, distributed vertex colouring algorithms, in *Proceedings of Symposium on Discrete Algorithms*, 2002.
21. Czygrinow, A., Hańćkowiak, M., and Karoński, M., Distributed O(Delta log(n))-edge-coloring algorithm, in *Proceedings of the 9th European Symposium on Algorithms*, pp. 345–355, Aarhus, Denmark, 2001.
22. Bollobas, B., *Graph Theory: An Introductory Course*, Springer-Verlag, Berlin, Germany, 1979.
23. Dubhashi, D., Grable, D., and Panconesi, A., Nearly-optimal, distributed edge-colouring via the nibble method, *Theor. Comp. Sci.*, 203, 225, 1998.
24. Grable, D. and Panconesi, A., Nearly optimal distributed edge colouring in O(log log n) rounds, *Random Struct. Algor.*, 10(3), 385, 1997.
25. Bollobas, B., Chromatic number, girth and maximal degree, *SIAM J. Disc. Math.*, 24, 311, 1978.
26. Grable, D. and Panconesi, A., Fast distributed algorithms for brooks-vizing colourings, in *Proceedings of the Ninth Annual (ACM-SIAM) Symposium on Discrete Algorithms*, pp. 473–480, San Francisco, CA, 1998.
27. McDiarmid, C., Concentration, in *Probabilistic Methods for Algorithmic Discrete Mathematics: Algorithms, Combinations*, Habib, M., McDiarmid, C., Ramirez-Alfonsin, J., and Reed, B. (Eds.), Springer, Berlin, Germany, Vol. 16, p. 195, 1998.
28. Molloy, M. and Reed, B., A bound on the strong chromatic index of a graph, *J. Comb. Theo.*, 31, 303–311, 1999.
29. Mitzenmacher, M. and Upfal, E., *Probability and Computing*, Cambridge University Press, New York, 2005.
30. Dubhashi, D.P., Martingales and locality in distributed computing, in *Foundations of Software Technology and Theoretical Computer Science*, Ramesh, S. and Sivakumar, G. (Eds.), Springer, Berlin, Germany, p. 174, 1998.
31. Fischer, T., Goldberg, A.V., Haglin, D.J., and Plotkin, S., Approximating matchings in parallel, *Inf. Proc. Lett.*, 46, 115, 1993.
32. Luby, M., A simple parallel algorithm for the maximal independent set problem, in *Proceedings of the Seventeenth Annual (ACM) Symposium on Theory of Computing*, pp. 1–10, Providence, RI, 1985.
33. Alon, N., Babai, L., and Itai, A., A fast and simple randomized parallel algorithm for the maximal independent set problem, *J. Algor.*, 7, 567, 1986.
34. Kozen, D., *The Design and Analysis of Algorithms*, Springer-Verlag, New York, 1992.
35. Hopcroft, J.E. and Karp, R.M., An $n^{5/2}$ algorithm for maximum matching in bipartite graphs, *SIAM J. Comput.*, 2, 225, 1973.
36. Wattenhofer, M. and Wattenhofer, R., Distributed weighted matching, in *Proceedings of the Eighteenth International Symposium on Distributed Computing*, pp. 335–348, Amsterdam, the Netherlands, 2004.
37. Hańćkowiak, M., Karoński, M., and Panconesi, A., On the distributed complexity of computing maximal matchings, in *Proceedings of Symposium on Discrete Algorithms*, p. 219, 1998.

38. Hańćkowiak, M., Karoński, M., and Panconesi, A., On the distributed complexity of computing maximal matchings, *SIAM J. Disc. Math.*, 15(1), 41, 2001.
39. Czygrinow, A., Hańćkowiak, M., and Szymanska, E., A fast distributed algorithm for approximating the maximum matching, in *Proceedings of the Twelfth Annual European Symposium on Algorithms*, pp. 252–263, Bergen, Norway, 2004.
40. Monien, B. and Speckenmeyer, E., Ramsey numbers and an approximation algorithm for the vertex cover problem, *Acta Informatica*, 22, 115, 1985.
41. Håstad, J., Some optimal inapproximability results, in *Proceedings of the Twenty-Ninth Annual (ACM) Symposium on Theory of Computing*, pp. 1–10, El Paso, TX, 1997.
42. Khuller, S., Vishkin, U., and Young, N., A primal-dual parallel approximation technique applied to weighted set and vertex cover, *J. Algor.*, 17(2), 280, 1994.
43. Grandoni, F., Könemann, J., and Panconesi, A., Distributed weighted vertex cover via maximal matchings, *ACM Trans. Algorithms*, 5(11), 1–12, 2008.
44. Israeli, A. and Itai, A., A fast and simple randomized parallel algorithm for maximal matching, *Inf. Proc. Lett.*, 22, 77, 1986.
45. Grandoni, F., Könemann, J., Panconesi, A., and Sozio, M., A primal-dual bicriteria distributed algorithm for capacitated vertex cover. *SIAM J. Comput.*, 38(3), 825–840, 2008.
46. Naor, M. and Stockmeyer, L., What can be computed locally? *SIAM J. Comput.*, 24(6), 1259, 1995.
47. Peleg, D., *Distributed Computing: A Locality-Sensitive Approach*, SIAM Monographs on Discrete Mathematics and Applications, SIAM, Philadelphia, PA, p. 5, 2000.
48. Linial, N., Locality in distributed graph algorithms, *SIAM J. Comput.*, 21(1), 193, 1992.
49. Naor, M., A lower bound on probabilistic algorithms for distributive ring coloring, *SIAM J. Disc. Math.*, 4(3), 409, 1991.
50. Cole, R. and Vishkin, U., Deterministic coin tossing with applications to optimal parallel list ranking, *Inf. Control*, 70(1), 32, 1986.
51. Garay, J.A., Kutten, S., and Peleg, D., A sublinear time distributed algorithm for minimum-weight spanning trees, *SIAM J. Comput.*, 27(1), 302, 1998.
52. Peleg, D. and Rubinovich, V., A near-tight lower bound on the time complexity of distributed minimum-weight spanning tree construction, *SIAM J. Comput.*, 30(5), 1427, 2000.
53. Elkin, M., Unconditional lower bounds on the time-approximation tradeoffs for the distributed minimum spanning tree problem, in *Proceedings of the Thirty-Sixth Annual (ACM) Symposium on Theory of Computing*, pp. 331–340, Chicago, IL, 2004.
54. Kuhn, F., Moscibroda, T., and Wattenhofer, R., What cannot be computed locally! in *Proceedings of the International Symposium on Distributed Computing*, 2004.

13

Empirical Analysis of Randomised Algorithms

Holger H. Hoos

Thomas Stützle

13.1 Introduction

Heuristic algorithms are often difficult to analyse theoretically; this holds in particular for advanced, randomised algorithms that perform well in practice, such as high-performance stochastic local search (SLS) procedures (also known as metaheuristics) [1]. Furthermore, for various reasons, the practical applicability of the theoretical results that can be achieved is often very limited. Some theoretical results are obtained under idealised assumptions that do not hold in practical situations—as is the case, for example, for the well-known convergence result for simulated annealing [2]. Also, most complexity results apply to worst-case behaviour, and average-case results, which are fewer and typically much harder to prove, are often based on instance distributions that are unlikely to be encountered in practice. Finally, theoretical bounds on the run-times of heuristic algorithms are typically asymptotic and do not reflect the actual behaviour accurately enough for many purposes, in particular, for comparative performance analyses. For these reasons, researchers (and practitioners) typically use empirical methods when analysing or evaluating heuristic algorithms.

In many ways, the issues and considerations arising in the empirical analysis of algorithmic behaviour are quite similar to those commonly encountered in experimental studies in biology, physics or any other empirical science. Fundamentally, to investigate a complex phenomenon of interest, the classical scientific cycle of observation, hypothesis, prediction, experiment is followed to obtain a model that explains the phenomenon. Different from natural phenomenae, algorithms are completely specified and mathematically defined at the lowest level; still, in many cases, this knowledge is insufficient for theoretically deriving all relevant aspects of their behaviour. In this situation, empirical approaches, based on computational experiments, are often not only the sole way of assessing a given algorithm, but also have the potential to provide insights into practically relevant aspects of algorithmic behaviour that appear well beyond the reach of theoretical analysis.

Some general goals are common to all empirical studies: *Reproducibility* ensures that experiments can be repeated with the same outcome; it requires that all relevant experimental conditions and protocols are specified clearly and in sufficient detail. In the empirical analysis of algorithmic behaviour, reproducibility is greatly facilitated by the fact that actual computations can in principle be replicated exactly. However, complications can arise when dealing with randomised algorithms or randomly generated input data, in which case statistical significance and sample sizes can become critical issues (despite the fact that typically, pseudo-random number generators are used to implement random processes). *Comparability* with past and future related results ensures that empirical results are useful in the context of larger scientific endeavors. To achieve this goal, experiments have to be designed in such a way that their results can be meaningfully compared to those from relevant previous work and facilitate comparisons with related results expected from future experiments. Finally, perhaps the main goal of any empirical study is to gain *insight and understanding*; this implies that experiments should be designed in such a way that their outcome is likely to shed light on important, previously open questions regarding the phenomenon of interest. In the empirical analysis of algorithms, in many cases these questions are of the form "Algorithm A has property X," and in particular, "Algorithm A performs better than Algorithm B."

13.2 Decision Algorithms

Many computational problems take the form of *decision problems*, in which solutions are characterised by a set of logical conditions. As an example, consider the following decision variant of the travelling salesman problem (TSP): given an edge-weighted graph and a real number b, does there exist a Hamiltonian cycle (i.e., a round trip that visits every vertex exactly once) with total weight at most b? Other well-known examples of decisions problems include the propositional satisfiability problem (SAT), the graph colouring problem and certain types of scheduling problems.

A *decision algorithm* is an algorithm that takes as an input an instance of a given decision problem and determines whether the instance is *soluble*, that is, whether it has a solution. In most cases, if a solution is found, that solution is also returned by the algorithm. Note that this notion of a decision algorithm includes algorithms that may be incomplete, that is, may fail to return a correct result within bounded time, or even incorrect, that is, sometimes return erroneous results. In the following, we will focus on decision algorithms that are correct, but incomplete; this captures most heuristic decision algorithms, including, for example, almost all SLS algorithms for SAT.

13.2.1 Analysis on Single Instances

The primary performance metric for complete (and correct) decision algorithms is typically *run-time*, that is, the time required for solving a given problem instance. For incomplete algorithms, it may happen that, although the given problem instance is soluble, a solution cannot be found. (In this case, the algorithm may not terminate, or signal failure, e.g., by returning "no solution found.") Obviously, such cases need to be noted; by further analysing them, valuable insights into weaknesses of the algorithm (or errors in its implementation) can be obtained.

Run-time is typically measured in terms of CPU time (rather than wall-clock time) to minimise the impact of other processes that are running concurrently (e.g., system processes). Obviously, CPU time measurements are always based on a concrete implementation and run-time environment, that is, machine and operating system; to facilitate reproducibility and comparability, a specification of the run-time environment (comprising at least the processor type and model, clock speed and amount of RAM, as well as the operating system, including version number) should be given along with any CPU time result.

It is often desirable to further abstract from details of the implementation and run-time environment, especially in the context of comparative performance studies. This can be achieved using *operation counts*, which reflect the number of elementary operations that are considered to contribute significantly towards an algorithm's performance, and *cost models*, which relate the cost (typically in terms of run-time per execution) of these operations relative to each other or absolute in terms of CPU-time for a given implementation and run-time environment [3]. For SLS algorithms, a commonly used operation count is the number of local search steps. When measuring performance in terms of operation counts, care should be taken to select elementary operations that are constant or close to constant cost per step within and between runs of the algorithm on the same instance. In this situation, operation counts and CPU-time measurements are related to each other by scaling with a constant factor that only depends on the given problem instance. Using operation counts and an associated cost model rather than CPU-time measurements as the basis for empirical studies often gives a clearer and more detailed picture of algorithmic performance.

While performance analysis of deterministic decision algorithms on a single problem instance consists of a simple run-time measurement, matters are slightly more involved if the algorithm under consideration is randomised. In that case, the run-time of an algorithm A applied to a given problem instance π corresponds to a random variable $RT_{A,\pi}$; the probability distribution of $RT_{A,\pi}$ is called the *run-time distribution (RTD) of A on π*. Clearly, the run-time behaviour of an algorithm A on a given problem instance π is completely and precisely characterised by the respective RTD. Furthermore, this RTD can be estimated based on run-time measurements obtained from multiple independent runs of A on π. For sufficiently high numbers of runs, the *empirical RTDs* thus obtained approximate the underlying theoretical RTD arbitrarily accurately. In practice, empirical RTDs based on 20–100 runs are sufficient for most purposes (this will be further discussed later in this chapter).

Graphical representations of empirical RTDs are often useful; plots of the respective cumulative distribution functions (CDFs) are easily obtained [1] and, unlike histograms, show the underlying data in full detail. They also make it easy to read quantiles and quantile ratios (such as the median and quartile ratio) directly off the plots; these basic descriptive statistics provide the basis for quantitative analyses and many statistical tests, which are discussed later. Compared to averages and empirical standard deviations, medians and quantile ratios have the advantage of being less sensitive with respect to outliers. Given the fact that the RTDs of many randomised heuristic algorithms show very large variability, the stability of basic descriptive statistics can become an important consideration. For the same reason, empirical RTDs are often best presented in the form of semi-log or loglog plots. Figure 13.1 shows an example of a typical empirical RTD plot.

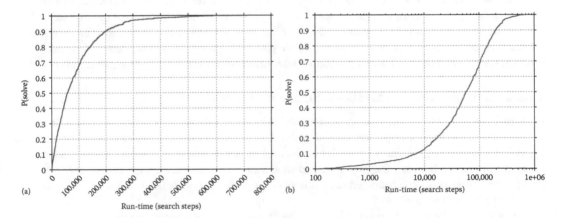

FIGURE 13.1   (a) Example of an empirical RTD of an SLS algorithm for SAT applied to a hard problem instance and (b) semi-log plot of the same RTD.

13.2.2 Analysis on Instance Ensembles

Typically, the behaviour of heuristic algorithms is analysed on a set or ensemble of instances. The selection of such benchmark sets is an important factor in the design of an empirical study, and the use of inadequate benchmark sets can lead to questionable results and misleading conclusions. Although the criteria for benchmark selection depend significantly on the problem domain under consideration, on the hypotheses and goals of the empirical study and on the algorithms being analysed, there are some general principles and guidelines, which can be summarised as follows (for more details, see Reference 1): Benchmark sets should contain a diverse collection of problem instances, ideally including instances from real-world applications as well as artificially crafted and randomly generated instances; the instances should typically be intrinsically hard or difficult to solve for a broad range of algorithms. Furthermore, to facilitate the reproducibility of empirical analyses and the comparability of results between studies, it is important to use established benchmark sets (in particular, those available from public benchmark libraries, such as ORLIB [4], TSPLIB [5] or SATLIB [6]), and to make newly created test-sets available to other researchers.

The basic approach to the empirical evaluation of an algorithm on a given ensemble of problem instances is to perform the same type of analysis described in the previous section on each individual instance. For small ensembles, it is often possible to analyse and report the results of this analysis for all instances, for example, in the form of tables or multiple RTD plots. When dealing with bigger ensembles, such as benchmark sets obtained from random instance generators, it becomes important to characterise the performance of a given algorithm on individual instances as well as across the entire ensemble. The latter can be achieved by aggregating the results obtained on all individual instances into a so-called *search cost distribution (SCD)*. For a deterministic algorithm applied to a given benchmark set, the empirical SCD is obtained from the run-time measurements on each individual problem instance. Analogously to RTDs, SCDs are typically best analysed qualitatively by means of CDF plots and quantitatively by means of basic descriptive statistics, such as quantiles and quantile ratios. For randomised decision algorithms, SCDs can be computed based on the median (or mean) run-times for each individual instance; this means that each point in the SCD plot corresponds to a statistic of an entire RTD. It is often appropriate to also analyse in more detail a small set of RTDs that have been carefully selected in such a way that they representatively illustrate the variation in algorithm behaviour across the ensemble.

In many cases, it is also of considerable interest to investigate the dependence of algorithmic performance on certain instance features, such as problem size. This is often done by studying the correlation between the feature value for a given problem instance and the corresponding run-time (or RTD) across the ensemble, for example, by means of simple correlation plots or using appropriate statistics, such as the Pearson correlation coefficient, and possibly also significance tests. The issues faced in this context are very similar to those arising in the comparative analysis of multiple algorithms on instance ensembles and will be further discussed in Section 13.2.4. In terms of qualitative analyses, choosing an appropriate graphical representation, such as a semi-logarithmic plot for the functional dependence of mean cost on problem size, is often the key for easily detecting interesting behaviour (e.g., exponential scaling).

13.2.3 Comparative Analysis on Single Instances

In many empirical studies, the main goal is to establish the superiority of one heuristic algorithm over another. The most basic form of this type of analysis is the comparative analysis between two decision algorithms on a single problem instance. If both algorithms are deterministic, this amounts to a straightforward comparison between the respective run-time measurements. Clearly, in the case of incomplete algorithms or prematurely terminated runs, it needs to be noted if one or both algorithm failed to solve the given problem instance.

If at least one of the algorithms is randomised, the situation is slightly more complicated. Intuitively, an algorithm A shows superior performance compared to another algorithm B on a given problem instance π if for no run-time, A has a lower solution probability than B, and there are some run-times for which

TABLE 13.1 Upper Bounds on the Performance Differences Detectable by the Mann-Whitney U-Test for Various Sample Sizes (Number of Runs Per RTD); m_1/m_2 Denotes the Ratio between the Medians of the Two Given RTDs. (The Values in This Table Have Been Obtained Using a Standard Procedure Based on Adjusting the Statistical Power of the Two-Sample t-Test to the Mann-Whitney U-Test Using a Worst-Case Pitman Asymptotic Relative Efficiency (ARE) Value of 0.864.)

Sign. Level 0.05, Power 0.95		Sign. Level 0.01, Power 0.99	
Sample Size	m_1/m_2	Sample Size	m_1/m_2
3010	1.1	5565	1.1
1000	1.18	1000	1.24
122	1.5	225	1.5
100	1.6	100	1.8
32	2	58	2
10	3	10	3.9

the solution probability of A is higher than that of B. In that case, we say that A *probabilistically dominates* B on π [1]. A probabilistic domination relation holds between two decision algorithms on a given problem instance if, and only if, their respective RTDs do not cross each other. This provides a simple method for graphically checking probabilistic domination between two SLS algorithms on individual problem instances. The concept of probabilistic domination also applies to situations where one of A and B is deterministic, since in terms of analysing run-time behaviour, deterministic decision algorithms can be seen as special cases of randomised decision algorithms that have degenerate RTDs whose CDFs are simple step functions. In situations where a probabilistic domination relation does not hold, that is, there is a cross-over between the respective RTD graphs, which of the two given algorithms is preferable in terms of higher solution probability depends on the time the algorithms are allowed to run.

Statistical tests can be used to assess the significance of empirically measured performance differences between randomised algorithms. In particular, the *Mann-Whitney U-test* (or, equivalently, the *Wilcoxon rank sum test*) can be used to determine whether the medians of two given RTDs are equal [7]; a rejection of this null hypothesis indicates significant performance differences. The widely used t-test compares the means of two populations, but it requires the assumption that the given samples are normally distributed with identical variance—an assumption which is usually not met when analysing individual RTDs. The more specific hypothesis whether the theoretical RTDs of two decision algorithms are identical can be tested using the *Kolmogorov-Smirnov test for two independent samples* [7].

An important question arising in comparative performance analyses of randomised algorithms is that of sample size: How many independent runs should be performed when measuring the respective empirical RTDs? Generally, the ability of statistical tests to correctly distinguish situations in which the given null hypothesis is correct from those where it is incorrect crucially depends on sample size. This is illustrated in Table 13.1, which shows the performance differences between two given RTDs that can be detected by the Mann-Whitney U-test for standard significance levels and power values in dependence of sample size. (The significance level and power value indicate the maximum probabilities that the test incorrectly rejects or accepts the null hypothesis that the medians of the given RTDs are equal, respectively.)

In cases where probabilistic domination does not hold, the previously mentioned statistical tests are still applicable. However, they do not capture interesting and potentially important performance differences, that can be easily seen from the respective RTD graphs. Such an example is depicted in Figure 13.2.

13.2.4 Comparative Analysis on Instance Ensembles

Comparative performance analyses of two decision algorithms on ensembles of problem instances are based on the same data used in the comparative analysis on the respective single instances. When dealing

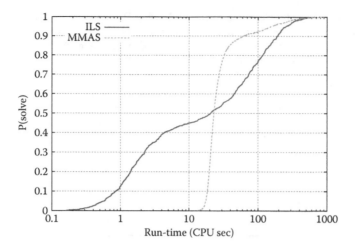

FIGURE 13.2   RTDs for two SLS algorithms for the TSP that for a given benchmark instance are required to find an optimal solution. Between 20 and 30 CPU seconds the two RTDs cross over.

with two deterministic decision algorithms, A and B, this results in pairs of run-times for each problem instance. In many cases, particularly when evaluating algorithms on big and diverse benchmark sets, there will be cases where A performs better than B and vice versa. In such situations it can be useful to use statistical tests to assess the significance of the observed performance differences; this is particularly the case for benchmark sets obtained from random instance generators. The *binomial sign test* as well as the *Wilcoxon matched pairs signed-rank test* determine whether the median of the paired differences is statistically significantly different from zero, indicating that one algorithm performs better than the other [7]. The Wilcoxon test is more sensitive, but requires the assumption that the distribution of the paired differences is symmetric. The well-known *t-test for two dependent samples* requires assumptions on the normality and homogeneity of variance of the underlying distributions of search cost over the given instance ensembles, which are typically not satisfied when dealing with the run-times of heuristic algorithms.

If one or both of the given algorithms are randomised, the same tests can be applied to RTD statistics, such as the median (or mean) run-time. However, this approach does not capture qualitative differences in performance, particularly as given in cases where there is no probabilistic domination of one algorithm over the other, and may suffer from inaccuracies due to a lack of statistical stability of the underlying RTD statistics. Therefore, additional analyses should be performed. In particular, the statistical significance of the performance differences (such as median run-time) on each individual problem instance should be investigated using an appropriate test (such as the Mann-Whitney U-test). Furthermore, for each instance it should be checked whether a probabilistic domination relation holds; based on this information, the given instance ensemble can be partitioned into three subsets: (i) those instances on which A probabilistically dominates B, (ii) those on which B probabilistically dominates A, and (iii) those for which probabilistic domination is not observed. The relative sizes and contents of these partitions give a rather realistic and detailed picture of the algorithms' relative performance on the given set of instances.

Particularly for large instance ensembles, it is often useful to study the correlation between the performance of algorithms A and B across the given set of instances. This type of analysis can help to expose (and ultimately, remedy) weaknesses of an algorithm and to refine claims about its relative superiority for certain types of problem instances. For qualitative analyses of performance correlation, scatter plots can be used in which each instance is represented by one point, whose coordinates correspond to the performance of A and B applied to that instance. Performance measures used in this context are typically run-time in case of deterministic algorithms, and RTD statistics, such as the median run-time, otherwise.

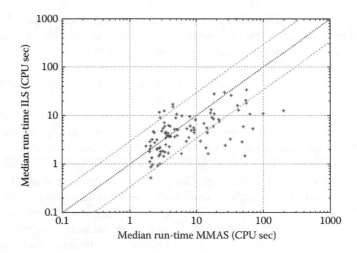

FIGURE 13.3 Correlation between the median run-times required by two high-performance SLS algorithms for finding the optimal solution to a set of 100 TSP instances of 300 cities each; each median was measured across 10 runs per algorithm. The band between the two outer lines indicates performance differences that cannot be assumed to be statistically significant for the given sample size of the underlying RTDs.

It should be noted that in the case of randomised algorithms, statistical instability of RTD statistics due to sampling error limits the accuracy of performance measurements. An example if such an analysis is shown in Figure 13.3.

Quantitatively, the correlation can be summarised using the empirical correlation coefficient. When the nature of an observed performance correlation seems to be regular (e.g., a roughly linear trend in the scatter plot), a simple regression analysis can be used to model the corresponding relationship in the algorithms' performance. It is often useful to perform correlation analyses on log-transformed data; this facilitates capturing general polynomial relationships.

To test the statistic significance of an observed performance correlation, nonparametric tests, such as Spearman's rank order test or Kendall's tau test, can be employed [7]. These tests determine whether there is a significant monotonic relationship in the performance data. They are preferable over tests based on Pearson's product-moment correlation coefficient, which require the assumption that the two random variables underlying the performance data stem from a bivariate normal distribution. (This assumption is often violated when dealing with run-times of heuristic algorithms over instance ensembles.)

13.3 Optimisation Algorithms

In many situations, the objective of a computational problem is to find a solution that is optimal with respect to some measure of quality or cost. An example of such an *optimisation problem* is the widely studied TSP: given an edge-weighted graph G, find a Hamiltonian cycle with minimal total weight, that is, a shortest round trip that visits every vertex of G exactly once. Another example is MAX-SAT, the optimisation variant of the SAT problem, where the objective is to find an assignment of truth values to the propositional variables in a given conjunctive normal form (CNF) formula F such that a maximal number of clauses in F are simultaneously satisfied.

The measure to be optimised in an optimisation problem is called the *objective function*, and the term *solution quality* is used to refer to the objective function value of a given candidate solution. In most cases, solution qualities take the form of real numbers, and the goal is to find a candidate with either minimal or maximal solution quality. Optimisation problems can include additional logical conditions that any solution needs to satisfy to be deemed *valid* or *feasible*. In the case of the TSP, such a logical condition states that to be considered a valid solution, a path in the given graph must be a Hamiltonian cycle. Logical conditions can always be integrated into the objective function in such a way that valid solutions are characterised by objective function values that exceed a specific threshold in solution quality.

An *optimisation algorithm* is an algorithm that takes as an input an instance of a given optimisation problem and returns a valid solution (or may determine that no valid solution exists). Optimisation algorithms that are theoretically guaranteed to find an optimal solution for any soluble problem instance within bounded time are called *complete* or *exact*; algorithms that are guaranteed to always return a solution that is within a specific constant factor of an optimal solution are called *approximation algorithms*.

When evaluating the performance of optimisation algorithms (theoretically or empirically), it is often useful to study the ratio between the solution quality achieved by the algorithm, q, and the optimal solution quality for the given problem instance, q^*. This performance measure is called the *approximation ratio*; formally, to be uniformly applicable to minimisation and maximisation problems, it is defined as $r := \max\{q/q^*, q^*/q\}$. When used in the empirical analysis of optimisation algorithms, solution qualities are often expressed in percent deviation from the optimum; this measure of *relative solution quality* is defined as $q' := (r - 1) \times 100$. For most heuristic optimisation algorithms, in particular for those based on SLS methods, there is a tradeoff between run-time and solution quality: the longer the algorithm is run, the better solutions are produced. The characterisation of this tradeoff is of significant importance in the empirical analysis of optimisation algorithms.

13.3.1 Analysis on Single Instances

As in the case of decision algorithms, the empirical analysis of a deterministic optimisation algorithm on a single problem instance is rather straightforward, and many of the same considerations (particularly with respect to measuring run-times and failure to produce valid solutions) apply. The run-time/solution quality trade-off is characterised by the development of solution quality over time (SQT), in the form of a so-called *SQT curve*; these represent for each point in time t the quality of the best solution seen up to time t (the so-called *incumbent solution*) and are hence always monotone.

A slightly more complicated situation arises when dealing with randomised optimisation algorithms. Following the same approach as for randomised decision algorithms, run-time is considered a random variable; in addition, a second random variable is used to capture solution quality, and the joint probability distribution of these two random variables characterises the behaviour of the algorithm on a given problem instance precisely and completely. For a given algorithm and problem instance, this probability distribution is called the *bivariate RTD of A on π* [1]; it can be visualised in the form of a cumulative distribution surface, each point of which represents the probability that A applied to π reaches (or exceeds) a certain solution quality bound within a certain amount of time (Figure 13.4).

Empirical bivariate RTDs can be easily determined from multiple *solution quality traces*, each of which represents the development of SQT for a single run of the algorithm on the given problem instance. A solution quality trace usually consists of pairs (t, q) for each point in time t at which an improvement in the incumbent solution, that is, a new best solution quality q within the current run, has been achieved. As in the case of the (univariate) RTDs for decision algorithms, a sufficient number of independent runs (i.e., solution quality traces) on any given problem instance is required for measuring reasonably accurate empirical bivariate RTDs; obviously, the same holds for basic descriptive RTD statistics on the solution quality obtained within a given run-time, or the run-time required for reaching a given solution quality.

Multivariate probability distributions are more difficult to handle than univariate distributions. Therefore, rather than working directly with bivariate RTDs, it is often preferable to focus on the (univariate) distributions of the run-time required for reaching a given solution quality threshold instead. These *qualified RTDs* (QRTDs) are the marginals of a given bivariate RTD for a specific bound on solution quality; intuitively, they correspond to cross-sections of the respective two-dimensional cumulative RTD graph for fixed solution quality values (Figure 13.4). QRTDs directly characterise the ability of an SLS algorithm for an optimisation problem to solve the associated decision problems for the given solution quality

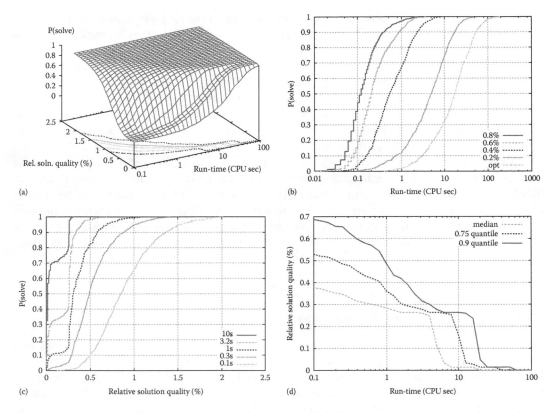

FIGURE 13.4 (a) Bivariate RTD for an SLS algorithm applied to a TSP benchmark instance; the other plots give different views on the same distribution; (b) QRTDs for various relative solution quality bounds (percentage deviation from optimum); (c) SQDs for various run-time bounds (in CPU seconds); and (d) SQT curves for various SQD quantiles.

bound. They are particularly useful for analysing an algorithm's ability to find optimal, close-to-optimal or feasible solutions and can be studied using exactly the same techniques as those applied to the (univariate) RTDs of decision algorithms. A detailed picture of the behaviour of a randomised optimisation algorithm on a single problem instance can be obtained by analysing series of qualified RTDs for increasingly tight solution quality thresholds. The solution quality bounds used in an QRTD analysis are typically derived from knowledge of optimal solutions or bounds on the optimal solution quality; the latter case includes bounds obtained from long runs of heuristic optimisation algorithms.

Another commonly used way of studying the behaviour of randomised optimisation algorithms on a given problem instance is to analyse the distribution of solution qualities obtained over multiple independent runs with a fixed time-bound. Technically, these so-called *solution quality distributions (SQDs)* are the marginals of the underlying bivariate RTDs for a fixed run-time bound. They correspond to cross-sections of the two-dimensional cumulative RTD graph for fixed run-time values; in this sense, they are orthogonal to QRTDs (Figure 13.4). Again, these univariate distributions can be studied using essentially the same techniques as for analysing the RTDs of decision algorithms.

Closely related to SQDs are the *asymptotic SQDs* obtained in the limit for arbitrarily long run-times. For complete and probabilistically approximately complete optimisation algorithms, which are guaranteed to find an optimal solution to any given problem instance with arbitrarily high probability given sufficiently long run-time, the asymptotic SQDs are degenerate distributions whose probability mass is completely concentrated on the optimal solution quality of the given problem instance. When dealing with randomised optimisation algorithms with an algorithm-dependent termination criterion, such as randomised iterative improvement methods that terminate upon reaching a local minimum, it is often also useful to study *termination time distributions (TTDs)*, which characterise the distribution of the time until termination over multiple independent runs.

Finally, the SQT curves described earlier in the context of characterising run-time/solution quality trade-offs for deterministic optimisation algorithms can be generalised to randomised algorithms. This is done by replacing the uniquely defined solution quality values obtained by a deterministic algorithm for any given run-time bound by statistics of the respective SQDs in the randomised case. Although historically, this type of analysis has most commonly used SQT curves based on mean solution quality values, it is often preferable to use SQTs that reflect the development of SQD quantiles (such as the median) over time, since these tend to be statistically more stable than means. SQTs based on SQD quantiles also offer the advantage that they directly correspond to horizontal sections or contour lines of the underlying bivariate RTD surfaces. Combinations of such SQTs can be very useful for summarising certain aspects of a complete bivariate RTD; they are particularly well-suited for analysing trade-offs between run-time and solution quality (Figure 13.4). However, the investigation of individual SQTs offers a fairly limited view of an optimisation algorithm's run-time behaviour in which important details can be easily missed and should therefore be complemented with other approaches, such as QRTD or SQD analysis. All these analyses can be carried out on the same set of solution quality traces collected over multiple indendent runs of the algorithm.

13.3.2 Comparative Analysis on Single Instances

The basic approach used for the comparative analysis of two (or more) optimisation algorithms on a single problem instance is analogous to that for decision algorithms. Often, a fixed target solution quality is used in this context, in which case the analysis involves the QRTDs of the algorithms with respect to that solution quality bound. Alternatively, a bound on run-time can be used, and the respective SQDs can be compared using the same methods as in the case of RTDs for decision algorithms. (It may be noted that the SQDs of high-performance algorithms for high run-times typically have much lower variance than QRTDs.)

Both of these methods do not take into account trade-offs between run-time and solution quality. To capture such trade-offs, it is useful to extend the concept of probabilistic domination introduced earlier for decision algorithms. We first note that in the case of two deterministic optimisation algorithms, A and B, this is straight forward: A dominates B on a given problem instance π if A gives consistently better solution quality than B for any run-time. This implies that the respective SQT curves do not cross each other. In the case of crossing SQTs, which of the two algorithm is preferable in terms of solution quality achieved depends on the time the algorithms are allowed to run.

When generalised to randomised algorithms, this leads to the concept of *probabilistic domination*. Analogous to the case of randomised decision algorithms, probabilistic domination between two randomised optimisation algorithms holds if, and only if, their (bivariate) RTDs do not cross each other. Note that this implies that there is no cross-over between any SQDs for the same run-time bound, or between any QRTDs for any solution quality bound. In practice, probabilistic domination can be tested based on a series of QRTDs for different solution quality bounds (or SQDs for various run-time bounds). This does not require substantial experimental overhead, since the solution quality traces underlying empirical QRTDs for the best solution quality bound also contain all the information for QRTDs for lower-quality bounds. When probabilistic domination does not hold, the run-time/solution quality trade-offs between the given algorithms can be characterised using the same data. In many cases, the results from empirical performance comparisons between randomised optimisation algorithms can be conveniently summarised using SQT curves over multiple SQD statistics (e.g., median and additional quantiles) in combination with SQD plots for selected run-times.

13.3.3 Analysis on Instance Ensembles

The considerations arising when extending the analyses described in the previous sections to ensembles of problem instances are essentially the same as in the case of decision convenient (see Sections 13.2.2 and 13.2.4). It is convenient (and in some special cases sufficient) to perform the analysis for a single solution quality or run-time bound, in which case the methodology is analogous to that for decision algorithms. However, in most cases, run-time/solution quality trade-offs need to be considered. This can be achieved by analysing SCDs or performance correlations for multiple solution quality or run-time bounds in addition to a more detailed analysis for carefully selected individual instances.

In the analysis of optimisation algorithms on instance ensembles, it is typically much preferable to use relative rather than absolute solution qualities. This introduces a slight complication when dealing with benchmark instances for which (provably) optimal solution qualities are unknown. To deal with such instances, theoretically or empirically determined bounds on the optimal solution quality, including best solution qualities achieved by high-performance heuristic algorithms, are often used. In this context, particularly when conducting performance comparisons related to the ability of various algorithms to find optimal or close-to-optimal solutions, it is very important to ensure that the bounds used in lieu of provably optimal solutions are as tight as possible.

13.4 Advanced RTD-Based Analysis

The measurement of RTDs for decision and optimisation problems can serve not only as a first step in the descriptive and comparative analysis of algorithm behaviour, as shown in the previous sections, but it can also form the basis of more advanced analysis techniques, for example, for examining scaling behaviour or performance robustness with respect to an algorithm's parameter settings. In what follows, we briefly outline such types of analyses; while our discussion is focused on RTDs for decision algorithms or, equivalently, on QRTDs for optimisation algorithms, many of its aspects can be extended in a straightforward way to the analysis of SQDs for optimisation algorithms.

13.4.1 Scaling with Instance Size

An important question is how an algorithm's performance scales with the size of the given problem instance. One possibility for studying scaling behaviour is to base the analysis on individual instances of various sizes. However, since there is often very substantial variation in run-time between instances of the same size, scaling studies are better based on ensembles of instances for each size. Then, the set of techniques discussed in the previous section can be applied by first measuring RTDs on individual instances; next, SCDs can be derived from appropriately chosen statistics of these RTDs, as discussed in Section 13.2.2; and finally, various statistics of these SCDs can be analysed in dependence of instance size.

As a first step, it is often useful to analyse this scaling data graphically. In this context, the use of semi-log or loglog plots can be very helpful: in particular, exponential scaling of mean or median search cost is reflected in a linear relationship between between instance size and the logarithm of run-time, whereas a linear relationship between the logarithms of both, instance size and run-time is indicative of polynomial scaling. To analyse scaling behaviour in more detail, function fitting techniques, such as statistical regression, can be used. A simple example of an empirical scaling analysis is given in Figure 13.5.

Additional support for observed or conjectured scaling behaviour can be obtained by interpolation experiments, where for instance sizes that are in the range of the previously analysed instance ensembles additional data points are measured, or extrapolation experiments, where an empirically fitted scaling function is used to predict the SCD statistics for larger instance sizes and deviations from the predicted values are analysed to possibly further refine the hypothesis on the scaling behaviour.

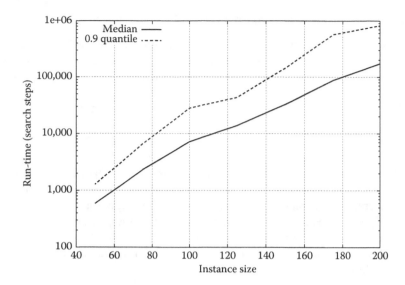

FIGURE 13.5   Scaling of the median and the 0.9 percentile for the search cost of solving SAT-encoded graph coloring instances with an SLS algorithm. Both statistics show clear signs of exponential scaling.

13.4.2 Impact of Parameter Settings

Many heuristic algorithms have one or more parameters that control their behaviour; as an example, consider the tabu tenure parameter in tabu search, a well-known SLS method (see also Chapter 21). The settings of such control parameters often have a significant yet theoretically poorly understood impact on the performance of the respective algorithm, which can be empirically studied by analysing the variation of an algorithm's RTD (or RTD statistics) in response to changes in its parameter settings. Often, the data required for this type of parameter sensitivity analysis is readily available from experiments conducted in order to optimise parameter settings for achieving peak-performance.

It should be noted that in the case of randomised algorithms, the variation of run-time for a fixed parameterisation and problem instance often depends on the parameter setting and should therefore be studied. For many SLS algorithms, suboptimal parameters can cause search stagnation and extremely high variability in run-time; in such situations, larger sample sizes may be required for obtaining reasonably accurate estimates of RTD statistics. Furthermore, for many heuristic algorithms with multiple parameters, the effects of various parameters are typically not independent, and experimental design techniques have to be employed for studying the nature and strength of these parameter dependencies.

Another important aspect of investigating parameter-dependent algorithmic performance deals with consistency across instance ensembles, that is, with the question to which degree the impact of parameter settings is similar across the instances in a given ensemble. One way of approaching this issue is to treat different parameterisations like different algorithms, and to use the methods for comparative performance analysis on instance ensembles from Section 13.2.4 (in particular, correlation analysis of RTD statistics). Consistency of performance-optimising parameter settings is often of particular interest. When consistent behaviour across an ensemble is not observed, it may still be possible to relate aspects of parameter-dependent run-time behaviour to specific characteristics of the instances. Such characteristics could be of purely syntactic nature (such as instance size or clauses/variables ratio for SAT instances) or they may be based on some deeper semantic properties (such as search space features in the case of SLS algorithms).

The need for manually tuning parameters can cause problems in practical applications of heuristic algorithms as well as in their empirical analysis. In particular, comparative performance analyses can yield misleading results when parameter settings have been tuned unevenly (i.e., more effort has been spent in optimising parameter settings for one of the algorithms). To alleviate these problems, automatic tuning techniques have been proposed [8,9]. Furthermore, mechanisms for adapting parameter values while solving a given problem instance have been used with considerable success, in particular in the context of reactive search methods [10].

13.4.3 Stagnation Detection

Intuitively, a randomised heuristic decision algorithm shows stagnation behaviour if for long runs, the probability of finding a solution can be improved by restarting the algorithm at some appropriately chosen cut-off time. For search algorithms, this effect may be due to the inability of the algorithm to trade effectively exploration of the search space and exploitation of the search experience, and may be related to the algorithm being stuck in specific areas of the search space.

Interestingly, it is quite easy to detect such stagnation behaviour from an empirical RTD. It is easy to see that only for RTDs that are identical to an exponential distribution, a well-known probability distribution from statistics, such restarts do not result in any performance loss or improvement [11] (essentially, this is due to the memory-less property of the exponential distribution). This insight provides the basis for detecting stagnation situations by comparing empirical RTDs of a given algorithm to exponential distributions. Stagnation behaviour is present if there is an exponential distribution that meets the empirical RTD from below but never crosses it. This situation is illustrated in Figure 13.6a; the arrows indicate the optimal cuttoff time for a static restart strategy, which can also be determined from the RTD.

In general, the detection of stagnation situations using the RTD-based methodology can be a key element in the systematic development of randomised heuristic algorithms; for example, in the case of SLS algorithms, the occurrence of search stagnation often indicates the need for additional or stronger diversification mechanisms. (For further details, see Chapter 4 of Reference 1.)

13.4.4 Functional Characterisation of Empirical RTDs

It is often useful (though not always possible) to characterise empirical RTDs by means of simple mathematical functions. For example, the RTDs of many high-performance SLS algorithms are well approximated by exponential distributions [11]. Such characterisations are not only useful in the context of stagnation analysis (as explained in the previous section), but also provide detailed and often very accurate summarisations of an algorithm's run-time behaviour. Furthermore, they can help in gaining insights into an algorithm's properties by providing a basis for modelling its behaviour mathematically.

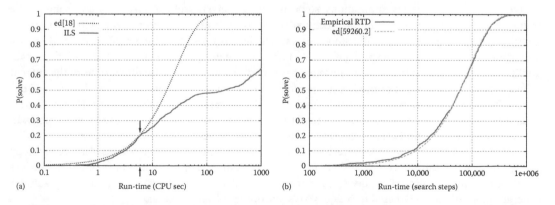

FIGURE 13.6 (a) Empirical QRTD of an iterated local search algorithm for finding the optimal solution of TSPLIB instance pcb442 (ILS); comparison with an exponential distribution ($ed[m] = 1 - 2^{-run\text{-}time/m}$) reveals severe stagnation behaviour. (b) Best fit of an empirical RTD by an exponential distribution. The fit passes a χ^2 goodness-of-fit test at a significance level of $\alpha = 0.05$.

In the context of functional RTD characterisations, it is particularly appealing to model empirical RTDs using parameterised continuous probability distributions known from statistics. This can be done using standard fitting techniques to determine suitable parameter values; the quality of the resulting approximations can be evaluated using goodness-of-fit tests, such as the χ^2 test or the Kolmogorov-Smirnov test [7]. (For an illustration, see Figure 13.6b.) The same methods can be used for functionally characterising other empirical data, such as SQDs or SCDs.

When dealing with large instance ensembles, the fitting and testing process needs to be automated. This way, more general hypotheses regarding an algorithm's run-time behaviour can be investigated empirically. Like any empirical approach, this method cannot be used for proving universal results on an algorithms' behaviour on an entire (infinite) class of problem instances, but it can be very useful in formulating, refining or falsifying hypotheses on such results.

13.5 Extensions

Most empirical analyses of heuristic algorithms in the literature focus on "classical" $\mathcal{N}P$-hard problems. It is clear, however, that sound empirical methodologies are equally important when tackling conceptually more involved types of problems, such as multi-objective, dynamic, or stochastic optimisation problems.

Multi-objective problems: In multi-objective problems, several, typically conflicting optimisation criteria need to be considered simultaneously. For these problems, a common goal is to identify the set of *pareto-optimal solutions* [12], that is, solutions for which there exists no alternative that is strictly better with respect to all optimisation criteria. Such multi-objective problems arise in many engineering and business applications, and heuristic algorithms are widely used for solving them [13,14]. The behaviour of these algorithms can be analysed empirically using a suitably generalised notion of multi-variate RTDs. Since the dimensionality of the RTDs to be measured in this case is equal to the number of objective functions plus one, data collection and analysis are considerably more complex than in the case of single-objective optimisation problems. While we are not aware of any studies based on these multi-variate RTDs, the marginal distributions obtained when keeping the computation time fixed have received considerable attention. The analysis of these so-called *attainment functions* as been proposed by Fonseca et al. [15] and has been acknowledged as one of the few approaches for a correct analysis of the performance of randomised algorithms for multi-objective optimisation [16].

Dynamic problems: In many applications, some aspects of a given problem instance may change while trying to find or implement a solution. Such dynamic problems are encountered, for example, in many distribution problems, where traffic situations can change as a result of congested or blocked routes. Two common goals in dynamic problems are to minimise the delay in recovering solutions (of a certain quality) after a change in the problem instance has occurred and to miminise disruptions of the current solution, that is, the amount of modifications required to adapt the current solution to the changed situation. The empirical analysis of heuristic (and in particular, randomised) algorithms for both of these situations can be handled using relatively straight-forward extensions of the RTD-based methodology. In the case of dynamic optimisation problems, trade-offs between solution quality and the amount of disruption can be studied using the same techniques as for static multi-objective problems. Also, particularly for dynamic optimisation problems where changes occur rather frequently, it can be useful to analyse the development of solution quality (or, for randomised algorithms, SQDs) over time, using suitable generalisations of the RTD-based techniques for static optimisation problems.

Stochastic problems: In some practical applications, important properties of solutions are subject to statistical variation. For many stochastic optimisation problems, variations in the quality of a given solution are caused by random changes (or uncertainty) in solution components that are characterised in the form of probability distributions; for example, in stochastic routing problems, the costs associated with using certain connections may be specified by Gaussian distributions. A typical goal when solving stochastic optimisation problems is to find a solution with optimal *expected quality*. In some cases, the expected quality of a solution can be determined analytically, and algorithms for such problems can be analysed using the same empirical methods as described for conventional deterministic problems. In other cases, approximation or sampling methods have to be used for estimating the quality of candidate solutions. While in principle, the techniques described in this chapter can be extended to these cases, empirical analyses (as well as algorithm development) are more involved; for example, when measuring empirical SQDs, a trade-off arises between the number of algorithm runs and the number of samples used to estimate the quality of incumbent solutions.

13.6 Further Reading

The use of principled and advanced techniques for the empirical analysis of deterministic and randomised heuristic algorithms is gaining increasing acceptance amongst researchers and practitioners. In this chapter, we have described the analysis of RTDs as a core technique for the empirical investigation and characterisation of randomised algorithms [17]. While RTDs have been previously reported in the literature [18–20], they have typically been used for purely descriptive purposes or in the context of investigating the parallelisation speed-up achievable by performing multiple independent runs of a sequential algorithm. A more detailed description of the RTD-based methodology is given in Chapter 4 of Reference 1. RTD-based methods are now being used increasingly widely for the empirical study of a broad range of SLS algorithms for numerous combinatorial problems [21–29].

SQDs of randomised heuristic optimisation algorithms have been occasionally reported in the literature; they have been used, for example, to obtain results on the scaling of SLS behaviour [30]. SQDs can also be used for estimating optimal solution qualities for combinatorial optimisation problems [31,32]. SCDs over ensembles of problem instances have been measured and characterised for deterministic, complete algorithms for binary CSP and SAT [33,34].

There is a growing body of work on general issues in empirical algorithmics. Several articles provide guidelines for the experimental study of mathematical programming software [35,36] and heuristic algorithms [37], with the aim of increasing the reproducibility of results. General guidelines for the experimental analysis of algorithms have also been proposed by McGeoch and Moret [38–40]. Johnson gives an overview of guidelines and potential pitfalls in empirical algorithmics research [41]. A more scientific approach to experimental studies of algorithms in optimisation has been advocated by Hooker [42,43], who emphasised the need for formulating and empirically investigating hypotheses about algorithm properties and behaviour rather than limiting the experimental study of algorithms to simplistic performance comparisons.

At the core of any empirical approach to investigating the behaviour and performance of randomised algorithms are statistical methods. Cohen's book [44] provides a good introduction to empirical methods in computing science with an emphasis on algorithms and applications in artificial intelligence. The handbook by Sheskin [7] is an excellent source for detailed information on statistical tests and their application, whereas Siegel et al. [45] and Conover [46] provide more specialised introductions to nonparametric statistics. For an introduction to the important topic of experimental design and data analysis we refer to the books of Dean and Voss [47] and Montgomery [48].

Acknowledgments

HH acknowledges support provided by the Natural Sciences and Engineering Research Council of Canada (NSERC) under Discovery Grant 238788-05; TS acknowledges support received from the Belgian Fonds National de la Recherche Scientifique, of which he is a research associate.

References

1. Hoos, H.H. and Stützle, T., *Stochastic Local Search—Foundations and Applications*, Morgan Kaufmann Publishers, San Francisco, CA, 2004.
2. Hajek, B., Cooling schedules for optimal annealing, *Math. Oper. Res.*, 13(2), 311, 1988.
3. Ahuja, R.K. and Orlin, J.B., Use of representative operation counts in computational testing of algorithms, *INFORMS J. Comput.*, 8(3), 318, 1996.
4. Beasley, J.E., OR-Library, http://mscmga.ms.ic.ac.uk/info.html, 2003.
5. Reinelt, G., TSPLIB, http://www.iwr.uni-heidelberg.de/groups/comopt/software/TSPLIB95, 2003.
6. Hoos, H.H. and Stützle, T., SATLIB—The Satisfiability Library, http://www.satlib.org, 2003.
7. Sheskin, D.J., *Handbook of Parametric and Nonparametric Statistical Procedures*, CRC Press, Boca Raton, FL, 2nd ed., 2000.
8. Birattari, M., Stützle, T., Paquete, L., and Varrentrapp, K., A racing algorithm for configuring metaheuristics, in *Proceedings of the Genetic and Evolutionary Computation Conference*, Morgan Kaufmann Publishers Inc., San Francisco, CA, p. 11, 2002.
9. Coy, S.P., Golden, B.L., Runger, G.C., and Wasil, E.A., Using experimental design to find effective parameter settings for heuristics, *J. Heuristics*, 7(1), 77, 2001.
10. Battiti, R., Reactive search: Toward self-tuning heuristics, in *Modern Heuristic Search Methods*, Rayward-Smith, V.J. (Ed.), John Wiley & Sons, Chichester, UK, p. 61, 1996.
11. Hoos, H.H. and Stützle, T., Characterising the behaviour of stochastic local search, *Arti. Intell.*, 112(1–2), 213, 1999.
12. Steuer, R.E., *Multiple Criteria Optimization: Theory, Computation and Application*, John Wiley & Sons, New York, 1986.
13. Gandibleux, X., Sevaux, M., Sörensen, K., and T'kindt, V., Eds., *Metaheuristics for Multiobjective Optimisation*, LNEMS, Springer, Berlin, Germany, p. 535, 2004.
14. Coello, C.A. and Lamont, G.B. Eds., *Applications of Multi-Objective Evolutionary Algorithms*, World Scientific, Singapore, 2004.
15. Fonseca, V.G.A., Fonseca, C.M., and Hall, A., Inferential performance assessment of stochastic optimizers and the attainment function, in *Evolutionary Multi-criterion Optimization*, Zitzler, E., Deb, K., Thiele, L., Coello, C.A.C., and Corne, D.W. (Eds.), LNCS, Springer, Heidelberg, Germany, Vol. 1993, pp. 213–225, 2001.
16. Zitzler, E., Thiele, L., Laumanns, M., Fonseca, C.M., and Grunert da Fonseca, V., Performance assessment of multiobjective optimizers: An analysis and review, *IEEE Trans. Evolut. Comput.*, 7(2), 117, 2003.
17. Hoos, H.H. and Stützle, T., Evaluating Las Vegas algorithms—pitfalls and remedies, in *Proceedings of the Fourteenth Conference on Uncertainty in Artificial Intelligence*, Morgan Kaufmann Publishers Inc., San Francisco, CA, p. 238, 1998.
18. Battiti, R. and Tecchiolli, G., Parallel biased search for combinatorial optimization: Genetic algorithms and TABU, *Microproc. Microsys.*, 16(7), 351, 1992.
19. Taillard, É.D., Robust taboo search for the quadratic assignment problem, *Paral. Comput.*, 17(4–5), 443, 1991.
20. ten Eikelder, H.M.M., Verhoeven, M.G.A., Vossen, T.V.M., and Aarts, E.H.L., A probabilistic analysis of local search, in *Metaheuristics: Theory & Applications*, Osman, I.H. and Kelly, J.P. (Eds.), Kluwer Academic Publishers, Boston, MA, p. 605, 1996.
21. Aiex, R.M., Resende, M.G.C., and Ribeiro, C.C., Probability distribution of solution time in GRASP: An experimental investigation, *J. Heuristics*, 8(3), 343, 2002.
22. Aiex, R.M., Pardalos, P.M., Resende, M.G.C., and Toraldo, G., GRASP with path relinking for three-index assignment, *INFORMS J. Comput.*, 17, 224, 2005.

23. Braziunas, D. and Boutilier, C., Stochastic local search for POMDP controllers, in *Proceedings of the National Conference on Artificial Intelligence*, AAAI Press ©2004, p. 690, 2004.

24. Hoos, H.H. and Boutilier, C., Solving combinatorial auctions using stochastic local search, in *Proceedings of the National Conference on Artificial Intelligence*, p. 22, 2000.

25. Hoos, H.H. and Stützle, T., Local search algorithms for SAT: An empirical evaluation, *J. Automated Reasoning*, 24(4), 421, 2000.

26. Shmygelska, A. and Hoos, H.H., An ant colony optimisation algorithm for the 2D and 3D hydrophobic polar protein folding problem, *BMC Bioinform.*, 6(30), 85, 2005.

27. Stützle, T. and Hoos, H.H., Analysing the run-time behaviour of iterated local search for the travelling salesman problem, in *Essays and Surveys on Metaheuristics*, Hansen, P. and Ribeiro, C.C. (Eds.), Springer Science+Business Media, New York, p. 589, 2001.

28. Stützle, T., Iterated local search for the quadratic assignment problem, *Eur. J. Oper. Res.*, 174(3), 1519–1539, 2006.

29. Watson, J.-P., Whitley, L.D., and Howe, A.E., Linking search space structure, run-time dynamics, and problem difficulty: A step towards demystifying tabu search, *J. Arti. Intell. Res.*, 24, 221, 2005.

30. Schreiber, G.R. and Martin, O.C., Cut size statistics of graph bisection heuristics, *SIAM J. Opt.*, 10(1), 231, 1999.

31. Dannenbring, D.G., Procedures for estimating optimal solution values for large combinatorial problems, *Management Sci.*, 23(12), 1273, 1977.

32. Golden, B.L. and Steward, W., Empirical analysis of heuristics, in *The Traveling Salesman Problem*, Lawler, E.L., Lenstra, J.K., Rinnooy Kan, A.H.G., and Shmoys, D.B. (Eds.), John Wiley & Sons, Chichester, UK, p. 207, 1985.

33. Kwan, A.C.M., Validity of normality assumption in CSP research, in *PRICAI: Topics in Artificial Intelligence*, Foo, N.Y. and Goebel, R. (Eds.), LNCS, Springer, Berlin, Germany, Vol. 1114, p. 253, 1996.

34. Frost, D., Rish, I., and Vila, L., Summarizing CSP hardness with continuous probability distributions, in *Proceedings of National Conference Artificial Intelligence*, AAAI Press ©1997, p. 327, 1997.

35. Crowder, H., Dembo, R., and Mulvey, J., On reporting computational experiments with mathematical software, *ACM Trans. Math. Soft.*, 5(2), 193, 1979.

36. Jackson, R., Boggs, P., Nash, S., and Powell, S., Report of the ad-hoc committee to revise the guidelines for reporting computational experiments in mathematical programming, *Math. Prog.*, 49, 413, 1990.

37. Barr, R.S., Golden, B.L., Kelly, J.P., Resende, M.G.C., and Stewart, W.R., Designing and reporting on computational experiments with heuristic methods, *J. Heuristics*, 1(1), 9, 1995.

38. McGeoch, C.C., Toward an experimental method for algorithm simulation, *INFORMS J. Comput.*, 8(1), 1, 1996.

39. McGeoch, C.C. and Moret, B.M.E., How to present a paper on experimental work with algorithms, *SIGACT News*, 30(4), 85, 1999.

40. Moret, B.M.E., Towards a discipline of experimental algorithmics, in *Data Structures, Near Neighbor Searches, and Methodology: Fifth and Sixth DIMACS Implementation Challenges*, Goldwasser, M.H., Johnson, D.S., and McGeoch, C.C. (Eds.), AMS, Providence, RI, p. 197, 2002.

41. Johnson, D.S., A theoretician's guide to the experimental analysis of algorithms, in *Data Structures, Near Neighbor Searches, and Methodology: Fifth and Sixth DIMACS Implementation Challenges*, Goldwasser, M.H., Johnson, D.S., and McGeoch, C.C. (Eds.), AMS, Providence, RI, p. 215, 2002.

42. Hooker, J.N., Needed: An empirical science of algorithms, *Oper. Res.*, 42(2), 201, 1994.

43. Hooker, J.N., Testing heuristics: We have it all wrong, *J. Heuristics*, 1(1), 33, 1996.

44. Cohen, P.R., *Empirical Methods for Artificial Intelligence*, MIT Press, Cambridge, MA, 1995.
45. Siegel, S., Castellan Jr., N.J., and Castellan, N.J., *Nonparametric Statistics for the Behavioral Sciences*, McGraw-Hill, New York, 2nd ed., 2000.
46. Conover, W.J., *Practical Nonparametric Statistics*, John Wiley & Sons, New York, 3rd ed., 1999.
47. Dean, A. and Voss, D., *Design and Analysis of Experiments*, Springer Verlag, Berlin, Germany, 2000.
48. Montgomery, D.C., *Design and Analysis of Experiments*, John Wiley & Sons, New York, 5th ed., 2000.

Reductions That Preserve Approximability

Giorgio Ausiello

Vangelis Th. Paschos

14.1 Introduction

The technique of transforming a problem into another in such a way that the solution of the latter entails, somehow, the solution of the former, is a classical mathematical technique that has found wide application in computer science since the seminal works of Cook [1] and Karp [2] who introduced particular kinds of transformations (called *reductions*) with the aim of studying the computational complexity of combinatorial decision problems. The interesting aspect of a reduction between two problems consists in its twofold application: on one side it allows to transfer positive results (resolution techniques) from one problem to the other and, on the other side, it may also be used for deriving negative (hardness) results. In fact, as a consequence of such seminal work, by making use of a specific kind of reduction, the polynomial-time *Karp-reducibility*, it has been possible to establish a complexity partial order among decision problems, which, for example, allows us to state that, modulo polynomial-time transformations, the SATISFIABILITY problem is as hard as thousands of other combinatorial decision problems, even though the precise complexity level of all these problems is still unknown.

Strictly associated with the notion of reducibility is the notion of completeness. Problems that are complete in a complexity class via a given reducibility are, in a sense, the hardest problems of such class. Besides, given two complexity classes **C** and **C'** $\subseteq$ **C**, if a problem Π is complete in **C** via reductions that belong (preserve membership) to **C'**, to establish whether **C'** $\subset$ **C** it is "enough" to assess the actual complexity of Π (informally we say that Π is a candidate to separate **C** and **C'**).

In this chapter we will show that an important role is played by reductions also in the field of approximation of hard combinatorial optimization problems. In this context the kind of reductions that will be applied are called *approximation preserving reductions*. Intuitively, in the most simple case, an approximation preserving reduction consists of two mappings f and g: f maps an instance x of problem Π into an instance $f(x)$ of problem Π', g maps back a feasible solution y of Π' into a feasible solution $g(y)$ of Π with the property that $g(y)$ is an approximate solution of problem Π whose quality is almost as good as the quality of the solution y for problem Π'. Clearly, again in this case, the role of an approximation preserving reduction is twofold: on one side it allows to transfer an approximation algorithm from problem Π' to problem Π; on the other side, if we know that problem Π cannot be approximated beyond a given threshold, such limitation applies also to problem Π'.

Various kinds of approximation preserving reducibilities will be introduced in this chapter, and we will show how they can be exploited in a positive way to transform solution heuristics from a problem to another and how, on the contrary, they may help in proving negative, inapproximability results.

It is well known that **NP**-hard combinatorial optimization problems behave in a very different way with respect to approximability and can be classified accordingly. Although for some problems there exist polynomial-time approximation algorithms that provide solutions with a constant approximation ratio w.r.t. the optimum solution, for some other problems even a remotely approximate solution is computationally hard to achieve. Analogously to what happens in the case of the complexity of decision problems, approximation preserving reductions allow to establish a partial order among optimization problems in terms of approximability properties, independently from the actual level of approximation that for such problems can be achieved (and that in some cases is still undefined). Approximation preserving reductions can also be used to define complete problems that play an important role in the study of possible separations between approximation classes. The discovery that a problem is complete in a given approximation class provides a useful insight in understanding what makes a problem not only computationally hard but also resilient to approximate solutions.

As a final remark on the importance of approximation preserving reductions, let us observe that such reductions require that some correspondence between the combinatorial structure of two problems is established. This is not the case for reductions between decision problems. For example, in such case, we see that all **NP**-complete decision problems turn out to be mutually interreducible by means of polynomial-time reduction, whereas when we consider the corresponding optimization problems, the different approximability properties come to evidence. As a consequence, we can say that approximation preserving reductions are also useful tools to analyze the deep relation existing between combinatorial structure of problems and the hardness of approximation.

The rest of this chapter is organized as follows. The next section is devoted to basic definitions and preliminary results concerning reductions among combinatorial optimization problems. In Section 14.3 we provide the first, simple example of approximation preserving reducibility, namely the *linear reducibility*, that while not as powerful as the reducibilities that will be presented in the sequel is widely used in practice. In Section 14.4 we introduce the reducibility that, historically, has been the first to be introduced, *the strict reducibility* and we discuss the first completeness results based on reductions of such kind. Next, in Section 14.5 we introduce AP-reducibility, and in Section 14.6 we discuss more extensive completeness results in approximation classes. In Section 14.7, we present a new reducibility, called FT-reducibility, that allows to prove the polynomial-time approximation scheme (**PTAS**)-completeness of natural **NPO** problems. Finally, in Section 14.8 we present other reductions with the specific aim of proving further inapproximability results. The last two sections of the chapter contain conclusions and references.

In this chapter we assume that the reader is familiar with the basic notions of computational complexity regarding both decision problems and combinatorial optimization problems, as they are defined in Chapter 1.

14.2 Basic Definitions

Before introducing the first examples of reductions between optimization problems, let us recall the definitions of the basic notions of approximation theory and of the most important classes of optimization problems, characterized in terms of their approximability properties. First of all we introduce the class **NPO** that is the equivalent, for optimization problems, of the class of decision problems **NP**.

Definition 14.1 *An **NP** optimization problem, **NPO**, Π is defined as a four-tuple $(\mathcal{I}, Sol, m, \text{goal})$ such that:*

- *$\mathcal{I}$ is the set of instances of Π and it can be recognized in polynomial time;*
- *given $x \in \mathcal{I}$, $Sol(x)$ denotes the set of feasible solutions of x; for any $y \in Sol(x)$, $|y|$ (the size of y) is polynomial in $|x|$ (the size of x); given any x and any y polynomial in $|x|$, one can decide in polynomial time if $y \in Sol(x)$;*
- *given $x \in \mathcal{I}$ and $y \in Sol(x)$, $m(x, y)$ denotes the value of y and can be computed in polynomial time;*
- *goal $\in \{min, max\}$ indicates the type of optimization problem.*

Given an **NPO** problem $\Pi = (\mathcal{I}, Sol, m, \text{goal})$ an optimum solution of an instance x of Π is usually denoted $y^*(x)$ and its measure $m(x, y^*(x))$ is denoted by $\text{opt}(x)$.

Definition 14.2 *Given an **NPO** problem $\Pi = (\mathcal{I}, Sol, m, \text{goal})$, an approximation algorithm A is an algorithm that given an instance x of Π returns a feasible solution $y \in Sol(x)$. If A runs in polynomial time with respect to $|x|$, A is called a polynomial-time approximation algorithm for Π.*

The quality of the solution given by an approximation algorithm A for a given instance x is usually measured as the ratio $\rho_A(x)$, *approximation ratio*, between the value of the approximate solution, $m(x, A(x))$, and the value of the optimum solution $\text{opt}(x)$. For minimization problems, therefore, the approximation ratio is in $[1, \infty)$, whereas for maximization problems it is in $[0, 1]$.

Definition 14.3 *An **NPO** problem Π belongs to the class **APX** if there exist a polynomial-time approximation algorithm A and a value $r \in \mathbb{Q}$ such that, given any instance x of Π, $\rho_A(x) \leq r$ (resp., $\rho_A(x) \geq r$) if Π is a minimization problem (resp., a maximization problem). In such case A is called an r-approximation algorithm.*

Examples of combinatorial optimization problems belonging to the class **APX** are MAX SATISFIABILITY, MIN VERTEX COVER, and MIN EUCLIDEAN TSP.

In some cases, a stronger form of approximability for **NPO** problems can be obtained by a PTAS, that is a family of algorithms A_r such that, given any ratio $r \in \mathbb{Q}$, the algorithm A_r is an r-approximation algorithm whose running time is bounded by a suitable polynomial p as a function of $|x|$.

Definition 14.4 *An **NPO** problem Π belongs to the class **PTAS** if there exists a PTAS A_r such that, given any $r \in \mathbb{Q}$, $r \neq 1$, and any instance x of Π, $\rho_{A_r}(x) \leq r$ (resp., $\rho_{A_r}(x) \geq r$) if Π is a minimization problem (resp., a maximization problem).*

Among the above-mentioned problems in **APX**, the problem MIN EUCLIDEAN TSP can be approximated by means of a PTAS and hence belongs to the class **PTAS**. Moreover, other examples of combinatorial optimization problems belonging to the class **PTAS** are MIN PARTITIONING and MAX INDEPENDENT SET ON PLANAR GRAPHS.

Finally, a stronger form of approximation scheme can be used for particular problems in **PTAS**, for example, MAX KNAPSACK or MIN KNAPSACK. In such cases, in fact, the running time of the algorithm A_r is uniformly polynomial in r as made precise in the following definition:

Definition 14.5 *An **NPO** problem Π belongs to the class fully polynomial-time approximation scheme (FPTAS) if there exists a PTAS A_r such that, given any $r \in \mathbb{Q}$, $r \neq 1$, and any instance x of Π, $\rho_{A_r}(x) \leq r$ (resp., $\rho_{A_r}(x) \geq r$) if Π is a minimization problem (resp., a maximization problem) and, furthermore, there exists a two variate polynomial q such that the running time of $A_r(x)$ is bounded by $q(x, 1/(r-1))$ (resp., $q(x, 1/(1-r))$ in case of maximization problems.*

It is worth to remember that under the hypothesis that $P \neq NP$ all the above-mentioned classes form a strict hierarchy, that is $\textbf{FPTAS} \subset \textbf{PTAS} \subset \textbf{APX} \subset \textbf{NPO}$.

Let us note that there exist also other notorious approximability classes, as **Poly-APX**, **Log-APX**, **Exp-APX**, the classes of problems approximable within ratios that are, respectively, polynomials (or inverse of polynomials if goal = max), logarithms (or inverse of logarithms), and exponentials (or inverse of exponentials) of the size of the input. The best studied among them is the class **Poly-APX**. Despite their interest, for sake of conciseness, these classes are not dealt in this chapter.

When the problem of characterizing approximation algorithms for hard optimization problems was tackled, the need arose for a suitable notion of reduction that could be applied to optimization problems to study their approximability properties ([3]):

> What is it that makes algorithms for different problems behave in the same way? Is there some stronger kind of reducibility than the simple polynomial reducibility that will explain these results, or are they due to some structural similarity between the problems as we define them?

Approximation preserving reductions provide an answer to the above-mentioned question. Such reductions have an important role when we wish to assess the approximability properties of an **NPO** optimization problem and locate its position in the approximation hierarchy. In such case, in fact, if we can establish a relationship between the given problem and other known optimization problems, we can derive both positive information on the existence of approximation algorithms (or approximation schemes) for the new problem or, on the other side, negative information, showing intrinsic limitations to approximability. With respect to reductions between decision problems, reductions between optimization problems have to be more elaborate. Such reductions, in fact, have to map both instances and solutions of the two problems, and they have to preserve, so to say, the optimization structure of the two problems.

The first examples of reducibility among optimization problems were introduced by Ausiello et al. [4,5] and by Paz and Moran [6]. In particular in Reference 5 the notion of *structure preserving reducibility* is introduced and for the first time the completeness of MAX WSAT (weighted-vertex SAT) in the class of **NPO** problems is proved. Still it took a few more years until suitable notions of approximation preserving reducibilities were introduced by Orponen and Mannila [7]. In particular, their paper presented the strict reduction (see Section 14.4) and provided the first examples of natural problems that are complete under approximation preserving reductions: (MIN WSAT, MIN 0-1 LINEAR PROGRAMMING, and MIN TSP).

Before introducing specific examples of approximation preserving reduction in the next sections, let us explain more formally how reductions between optimization problems can be defined, starting from the notion of *basic reducibility* (called R-reducibility in the following, denoted by $\leq_R$) that underlays most of the reducibilities that will be introduced later.

Definition 14.6 *Let Π_1 and Π_2 be two **NPO** maximization problems. Then we say that $\Pi_1 \leq_R \Pi_2$ if there exists two polynomial-time computable functions f, g that satisfy the following properties:*

- $f : \mathcal{I}_{\Pi_1} \rightarrow \mathcal{I}_{\Pi_2}$ *such that $\forall x_1 \in \mathcal{I}_{\Pi_1}, f(x_1) \in \mathcal{I}_{\Pi_2}$; in other words, given an instance x_1 in Π_1, f allows to build an instance $x_2 = f(x_1)$ in Π_2;*
- $g : \mathcal{I}_{\Pi_1} \times \mathrm{Sol}_{\Pi_2} \rightarrow \mathrm{Sol}_{\Pi_1}$ *such that, $\forall(x_1, y_2) \in (\mathcal{I}_{\Pi_1} \times \mathrm{Sol}_{\Pi_2}(f(x_1)))$, $g(x_1, y_2) \in \mathrm{Sol}_{\Pi_1}(x_1)$; in other words, starting from a solution y_2 of the instance x_2, g determines a solution $y_1 = g(x_1, y_2)$ of the initial instance x_1.*

As we informally said in the introduction the aim of an approximation preserving reduction is to guarantee that if we achieve a certain degree of approximation in the solution of problem Π_2, then a suitable degree of approximation is reached for problem Π_1. As we will see, the various notions of approximation preserving reducibilities that will be introduced in the following, essentially differ in the mapping that is established between the approximation ratios of the two problems.

Before closing this section, let us introduce the notion of *closure* of a class of problems under a given type of reducibility. In what follows, given two **NPO** problems Π and Π', and a reducibility X, we will generally use the notation $\Pi \leq_X \Pi'$ to indicate that Π reduces to Π' via reduction of type X.

Definition 14.7 *Let C be a class of **NPO** problems and X a reducibility. Then, the closure $\overline{C}^X$ of C under X is defined as: $\overline{C}^X = \{\Pi \in NPO : \exists \Pi' \in C, \Pi \leq_X \Pi'\}$.*

14.3 The Linear Reducibility

The first kind of approximation preserving reducibility that we want to show is a very natural and simple transformation among problems that consists in two linear mappings, one between the values of the optimum solutions of the two problems and one between the errors of the corresponding approximate solutions: the *linear reducibility* (L-reducibility, denoted by $\leq_L$).

Definition 14.8 *Let Π_1 and Π_2 be two problems in **NPO**. Then, we say that $\Pi_1 \leq_L \Pi_2$, if there exist two functions f and g (basic reduction) and two constants $\alpha_1 > 0$ and $\alpha_2 > 0$ such that $\forall x \in \mathcal{I}_{\Pi_1}$ and $\forall y' \in \mathrm{Sol}_{\Pi_2}(f(x))$:*

- $\mathrm{opt}_{\Pi_2}(f(x)) \leq \alpha_1 \mathrm{opt}_{\Pi_1}(x)$;
- $|m_{\Pi_1}(x, g(y')) - \mathrm{opt}_{\Pi_1}(x)| \leq \alpha_2 |m_{\Pi_2}(f(x), y') - \mathrm{opt}_{\Pi_2}(f(x))|$.

This type of reducibility has been introduced in Reference 8 and has played an important role in the characterization of the hardness of approximation. In fact, it is easy to observe that the following property holds.

Fact 14.1 *Given two problems Π and Π', if $\Pi \leq_L \Pi'$ and $\Pi' \in PTAS$, then $\Pi \in PTAS$. In other words, the L-reduction preserves membership in PTAS.*

Example 14.1 MAX 3-SAT $\leq_L$ MAX 2-SAT. *Let us consider an instance ϕ with m clauses (w.l.o.g., let us assume that all clauses consist of exactly three literals); let l_i^1, l_i^2, and l_i^3, be the three literals of the i-th clause, $i = 1, \ldots, m$. To any clause we associate the ten following new clauses, each one consisting of at most two literals: l_i^1, l_i^2, l_i^3, l_i^4, $\bar{l}_i^1 \vee \bar{l}_i^2$, $\bar{l}_i^1 \vee \bar{l}_i^3$, $\bar{l}_i^2 \vee \bar{l}_i^3$, $l_i^1 \vee \bar{l}_i^4$, $l_i^2 \vee \bar{l}_i^4$, $l_i^3 \vee \bar{l}_i^4$, where l_i^4 is a new variable. Let C_i' be the conjunction of the ten clauses derived from clause C_i. The formula $\phi' = f(\phi)$ is the conjunction of all clauses C_i', $i = 1, \ldots, m$, that is, $\phi' = f(\phi) = \wedge_{i=1}^{m} C_i'$ and it is an instance of MAX 2-SAT.*

*It is easy to see that all truth assignments for ϕ' satisfy at most seven clauses in any C_i'. On the other side, for any truth assignment for ϕ satisfying C_i, the following truth assignment for l_i^4 is such that the extended truth assignment satisfies exactly seven clauses in C_i': if exactly one (resp., all) of the variables l_i^1, l_i^2, l_i^3 is (resp., are) set to **true**, then l_i^4 is set to **false** (resp., **true**); otherwise (exactly one literal in C_i is set to **false**), l_i^4 can be indifferently **true** or **false**. Finally, if C_i is not satisfied (l_i^1, l_i^2, and l_i^3 are all set to **false**), no truth assignment for l_i^4 can satisfy more than six clauses of C_i' whereas six are guaranteed by setting l_i^4 to **false**. This implies that $\mathrm{opt}(\phi') = 6m + \mathrm{opt}(\phi) \leq 13 \, \mathrm{opt}(\phi)$ (as $m \leq 2 \, \mathrm{opt}(\phi)$, see Lemma 14.2 in Section 14.6.2).*

Given a truth assignment for ϕ', we consider its restriction $\tau = g(\phi, \tau')$ on the variables of ϕ; for such assignment τ we have: $m(\phi, \tau) \geq m(\phi', \tau') - 6m$. Then, $\mathrm{opt}(\phi) - m(\phi, \tau) = \mathrm{opt}(\phi') - 6m - m(\phi, \tau) \leq \mathrm{opt}(\phi') - m(\phi', \tau')$. This means that the reduction we have defined is an L-reduction with $\alpha_1 = 13$ and $\alpha_2 = 1$.

L-reductions provide a simple way to prove hardness of approximability. An immediate consequence of the above-mentioned reduction and of Fact 14.1 is that as MAX 3-SAT does not allow a PTAS so does MAX 2-SAT. The same technique can be used to show the nonexistence of PTAS for a large class of optimization problems, among other MAX CUT, MAX INDEPENDENT SET-B (i.e., MAX INDEPENDENT SET on graphs with bounded degree), MIN VERTEX COVER, and so on.

Before closing this section, let us observe that the set of ten 2-SAT clauses that we have used in Example 14.1 for constructing the 2-SAT formula $\phi' = f(\phi)$ is strongly related to the bound on approximability established in the example. The proof of the result is based on the fact that at least six out of the ten clauses can always be satisfied whereas exactly seven out of ten can be satisfied, if and only if the original 3-SAT clause is satisfied. A combinatorial structure of this kind, which allows to transfer (in)approximability results from a problem to another, is called a *gadget* [9]. The role of gadgets in approximation preserving reductions will be discussed further in Section 14.8.

14.4 Strict Reducibility and Complete Problems in NPO

As we informally said in the introduction, an important characteristic of an approximation preserving reduction from a problem Π_1 to a problem Π_2 is that the solution y_1 of problem Π_1 produced by the mapping g should be at least as good as the original solution y_2 of problem Π_2. This property is not necessarily true for any approximation preserving reduction (it is easy to observe that, e.g., L-reductions do not always satisfy it) but it is true for the most natural reductions that have been introduced in the early phase of approximation studies: the *strict reductions* [7].

In the following, we present the strict reducibility (S-reducibility, denoted by $\leq_S$) referring to minimization problems, but the definition can be trivially extended to all types of optimization problems.

Definition 14.9 *Let Π_1 and Π_2 be two **NPO** minimization problems. Then, we say that $\Pi_1 \leq_S \Pi_2$ if there exist two polynomial-time computable functions f, g that satisfy the following properties:*

- *f and g are defined as in a basic reduction;*
- *$\forall x \in \mathcal{I}_{\Pi_1}, \forall y \in \mathrm{Sol}_{\Pi_2}(f(x)), \rho_{\Pi_2}(f(x), y) \geq \rho_{\Pi_1}(x, g(x, y)).$*

It is easy to observe that the S-reducibility preserves both membership in **APX** and in **PTAS**.

Proposition 14.1 *Given two minimization problems Π_1 and Π_2, if $\Pi_1 \leq_S \Pi_2$ and $\Pi_2 \in$ **APX** (resp., $\Pi_2 \in$ **PTAS**), then $\Pi_1 \in$ **APX** (resp., $\Pi_2 \in$ **PTAS**).*

Example 14.2 *Consider the* MIN WEIGHTED VERTEX COVER *problem in which the weights of vertices are bounded by a polynomial $p(n)$ and let us prove that this problem S-reduces to the unweighted* MIN VERTEX COVER *problem. Let us consider an instance $(G(V, E), \vec{w})$ of the former and let us see how it can be transformed into an instance $G'(V', E')$ of the latter. We proceed as follows: for any vertex $v_i \in V$, with weight w_i, we construct an independent set W_i of w_i new vertices in V'; next, for any edge $(v_i, v_j) \in E$, we construct a complete bipartite graph among the vertices of the independent sets W_i and W_j in G'. This transformation is clearly polynomial as the resulting graph G' has $\sum_{i=1}^{n} w_i \leq np(n)$ vertices.*

Let us now consider a cover C' of G' and, w.l.o.g., let us assume it is minimal w.r.t. inclusion (in case it is not, we can easily delete vertices until we reach a minimal cover). We claim that at this point C' has the form: $\cup_{j=1}^{\ell} W_{i_j}$, that is, there is an ℓ such that C' consists of ℓ independent sets W_i. Suppose that the claim is not true. Let us consider an independent set W_k that is only partially included in C' (i.e., a nonempty portion W_k' of it belongs to C'). Let us also consider all independent sets W_p that are entirely or partially included in C' and moreover are connected by edges to the vertices of W_k. Two cases may arise: (i) all considered sets W_p have their vertices included in C'; in this case the existence of W_k' would contradict the minimality of C' and (ii) among the considered sets W_p there is at least one set W_q out of which only a non-empty portion W_q' is included in C'; in this case, as the subgraph of G' induced by $W_k \cup W_q$ is a complete bipartite graph, the edges connecting the vertices of $W_p \setminus W_p'$ with the vertices of $W_q \setminus W_q'$ are not covered by C' and this would contradict the assumption that C' is a cover of G'. As a consequence, the size of C' satisfies $|C'| = \sum_{j=1}^{\ell} w_{i_j}$ and the function g of the reduction can then be defined as follows: if C' is a cover of G' and if W_i, $i = 1, \ldots, \ell$, are the independent sets that form C', then a cover C for G contains all corresponding vertices $v_1, \ldots, v_\ell$ of V. Clearly g can be computed in polynomial time.

From these premises we can immediately infer that the same approximation ratio that is guaranteed for A on G' is also guaranteed by g on G. The shown reduction is hence an S-reduction.

An immediate corollary of the strict reduction shown in the example is that the approximation ratio 2 for MIN VERTEX COVER (that we know can be achieved by various approximation techniques, see Reference 10) also holds for the weighted version of the problem, dealt in Example 14.2.

The S-reducibility is indeed a very strong type of reducibility: in fact, it requires a strong similarity between two optimization problems and it is not easy to find problems that exhibit such similarity. The interest for the S-reducibility arises mainly from the fact that by making use of reductions of this kind, Orponen and Mannila have identified the first optimization problem that is complete in the class of **NPO** minimization problems: the problem MIN WSAT. Let us consider a Boolean formula in conjunctive normal form ϕ over n variables $x_1, \ldots, x_n$ and m clauses. Any variable x_i has a positive weight $w_i = w(x_i)$. Let us assume that the truth assignment that puts all variables to **true** is feasible, even if it does not satisfy ϕ. Besides, let us assume that t_i is equal to 1 if τ assigns value **true** to the i-th variable and 0 otherwise. We want to determine the truth assignment τ of ϕ that minimizes: $\sum_{i=1}^{n} w_i t_i$. The problem MAX WSAT can be defined in similar terms. In this case, we assume that the truth assignment that puts all variables to **false** is feasible and we want to determine the truth assignment τ that maximizes: $\sum_{i=1}^{n} w_i t_i$. In the variants MIN W3-SAT and MAX W3-SAT, we consider that all clauses contain exactly 3 literals.

The fact that MIN WSAT is complete in the class of **NPO** minimization problems under S-reductions implies that this problem does not allow any constant ratio approximation (unless $\mathbf{P} = \mathbf{NP}$) ([5–7]). In fact, due to the properties of S-reductions, if a problem which is complete in the class of **NPO** minimization problems was approximable then all **NPO** minimization problems would. As it is already known that some minimization problems in **NPO** do not allow any constant ratio approximation algorithm (namely MIN TSP on general graphs), then we can deduce that (unless $\mathbf{P} = \mathbf{NP}$) no complete problem in the class of **NPO** minimization problems allows any constant ratio approximation algorithm.

Theorem 14.1 MIN WSAT *is complete in the class of minimization problems belonging to* **NPO** *under S-reductions.*

Proof. The proof is based on a modification of Cook's proof of the **NP**-completeness of SAT [1]. Let us consider a minimization problem $\Pi \in \mathbf{NPO}$, the polynomial p provides the bounds relative to problem Π (see Definition 14.1) and an instance x of Π. The following nondeterministic Turing machine M (with two output tapes T_1 and T_2) generates all feasible solutions $y \in \mathrm{Sol}(x)$ together with their values:

- generate y, such that $|y| \leq p(|x|)$;
- if $y \notin \mathrm{Sol}(x)$, then reject; otherwise, write y on output tape T_1, $m(x, y)$ on output tape T_2 and accept.

Let us now consider the reduction that is currently used in the proof of Cook's theorem [11] and remember that such reduction produces a propositional formula in conjunctive normal form that is satisfied if and only if the computation of the Turing machine accepts. Let ϕ_x be such formula and $x_n, x_{n-1}, \ldots, x_0$ the variables of ϕ_x that correspond to the cells of tape T_2 where M writes the value $m(x, y)$ in binary (w.l.o.g., we can assume such cells to be consecutive), such that a satisfying assignment of ϕ_x, x_i is **true** if and only if the $(n-i)$-th bit of $m(x, y)$ is equal to 1. Given an instance x of Π the function f of the S-reduction provides an instance of MIN WSAT consisting of the pair (ϕ_x, ψ) where $\psi(x) = \psi(x_i) = 2^i$, for $i = 0, \ldots, n$ and $\psi(x) = 0$, for any other variable x in ϕ_x.

The function g of the S-reduction is defined as follows. For any instance x of Π and any solution $\tau' \in \mathrm{Sol}(f(x))$ (i.e., any truth assignment τ' that satisfies the formula ϕ_x (for simplicity we only consider the case in which the formula ϕ_x is satisfiable)), we recover from ϕ_x the representation of the solution y written on tape T_1. Besides, we have that $m(x, g(x, \tau')) = \sum_{\tau'(x_i)=\mathbf{true}} 2^i = m((\phi_x, \psi), \tau')$, whereby $\tau'(x_i)$ we indicate the value of variable x_i according to the assignment τ'. As a consequence, $m(x, g(x, \tau')) = m(f(x), \tau')$ and henceforth $r(x, g(x, \tau')) = r(f(x), \tau')$, and the described reduction is an S-reduction. ■

After having established that MIN WSAT is complete for **NPO** minimization problems under the S-reducibility we can then proceed to find other complete problems in this class.

Let us consider the following definition of the MIN 0-1 LINEAR PROGRAMMING problem (the problem MAX 0-1 LINEAR PROGRAMMING can be defined analogously). We consider a matrix $A \in \mathbb{Z}^{m \times n}$ and two vectors $\vec{b} \in \mathbb{Z}^m$ and $\vec{w} \in \mathbb{N}^n$. We want to determine a vector $\vec{y} \in \{0, 1\}^n$ that verifies $A\vec{y} \geq \vec{b}$ and minimizes the quantity $\vec{w} \cdot \vec{y}$.

Clearly MIN 0-1 LINEAR PROGRAMMING is an **NPO** minimization problem. The reduction from MIN WSAT to MIN 0-1 LINEAR PROGRAMMING is a simple modification of the standard reduction among the corresponding decision problems. Suppose that the following instance of MIN 0-1 LINEAR PROGRAMMING, consisting of a matrix $A \in \mathbb{Z}^{m \times n}$ and two vectors $\vec{b} \in \mathbb{Z}^m$ and $\vec{w} \in \mathbb{N}^n$, is the image $f(x)$ of an instance x of MIN WSAT and suppose that $\vec{y}$ is a feasible solution of $f(x)$ whose value is $m(f(x), \vec{y}) = \vec{w} \cdot \vec{y}$. Then, $g(x, \vec{y})$ is a feasible solution of x, that is a truth assignment τ, whose value is $m(x, \tau) = \sum_{i=1}^{n} w_i t_i$ where t_i is equal to 1 if τ assigns value **true** to the i-th variable and 0 otherwise. As we have $\sum_{i=1}^{n} w_i t_i = \vec{w} \cdot \vec{y}$ it is easy to see that the reduction (f, g, c), where c is the identity function, is an S-reduction* and, as a consequence, MIN 0-1 LINEAR PROGRAMMING is also complete in the class of **NPO** minimization problems.

It is not difficult to prove that an analogous result holds for maximization problems, that is, MAX WSAT is complete under S-reductions in the class of **NPO** maximization problems.

At this point of the chapter we still do not have the technical instruments to establish a more powerful result, that is, to identify problems that are complete under S-reductions for the entire class of **NPO** problems. To prove such result we need to introduce a more involved kind of reducibility, the AP-reducibility (see Section 14.5). In fact, by means of AP-reductions MAX WSAT can itself be reduced to MIN WSAT and vice versa [12] and therefore it can be shown that (under AP-reductions) both problems are indeed **NPO**-complete.

14.5 AP-Reducibility

After the seminal paper by Orponen and Mannila [7] research on approximation preserving reducibility was further developed [13–15]; nevertheless, the beginning of the structural theory of approximability of optimization problems can be traced back to the fundamental paper by Crescenzi and Panconesi [16] where reducibilities preserving membership in **APX** (A-reducibility), **PTAS** (P-reducibility), and **FPTAS** (F-reducibility) were studied and complete problems for each of the three kinds of reducibilities were shown, respectively in **NPO**, **APX**, and **PTAS**. Unfortunately the problems that are proved complete in **APX** and in **PTAS** in this paper are quite artificial.

Along a different line of research, during the same years, the study of logical properties of optimization problems has led Papadimitriou and Yannakakis ([8]) to the syntactic characterization of an important class of approximable problems, the class **Max-SNP**. Completeness in **Max-SNP** has been defined in terms of L-reductions (see Section 14.3) and natural complete problems (e.g., MAX 3-SAT, MAX 2-SAT, MIN VERTEX COVER-B) have been found. The relevance of such approach is related to the fact that it is possible to prove that **Max-SNP**-complete problems do not allow PTAS (unless **P** = **NP**).

The two approaches have been reconciled by Khanna et al. [17], where the closure of syntactically defined classes with respect to an approximation preserving reduction were proved equal to the more familiar computationally defined classes. As a consequence of this result any **Max-SNP**-completeness result appeared in the literature can be interpreted as an **APX**-completeness result. In this paper a new type of reducibility is introduced, the E-reducibility. With respect to the L-reducibility, in the E-reducibility the constant α_1 is replaced by a polynomial $p(|x|)$. This reducibility is fairly powerful as it allows to prove that MAX 3-SAT is complete for **APX-PB** (the class of problems in **APX** whose values are bounded by a polynomial in the size of the instance) such as MAX 3-SAT. On the other side, it remains somewhat restricted because it does not allow the transformation of **PTAS** problems (such as MAX KNAPSACK) into problems belonging to **APX-PB**.

The final answer to the problem of finding the suitable kind of reducibility (powerful enough to establish completeness results both in **NPO** and **APX**) is the AP-reducibility introduced by Crescenzi et al. [12].

---

* Note that, in this case, the reduction is also a linear reductions with $\alpha_1 = \alpha_2 = 1$.

In fact, the types of reducibility that we have introduced so far (linear and strict reducibilities) suffer from various limitations. In particular, we have seen that strict reductions allow us to prove the completeness of MIN WSAT in the class of **NPO** minimization problems but are not powerful enough to allow the identification of problems that are complete for the entire class **NPO**. Besides, both linear and strict reductions, in different ways, impose strong constraints on the values of the solutions of the problems among which the reduction is established.

In this section, we provide the definition of the AP-reducibility (denoted by $\leq_{AP}$) and we illustrate its properties. Completeness results in **NPO** and in **APX** based on AP-reductions are shown in Section 14.6.

Definition 14.10 *Let* Π_1 *and* Π_2 *be two minimization* **NPO** *problems. An* AP-*reduction between* Π_1 *and* Π_2 *is a triple* (f, g, α), *where* f *and* g *are functions and* α *is a constant, such that, for any* $x \in \mathcal{I}_{\Pi_1}$ *and* $r > 1$:

- $f(x, r) \in \mathcal{I}_{\Pi_2}$ *is computable in time* $t_f(|x|, r)$ *polynomial in* $|x|$ *for a fixed* r; $t_f(n, \cdot)$ *is nonincreasing;*
- *for any* $y \in \text{Sol}_{\Pi_2}(f(x, r))$, $g(x, y, r) \in \text{Sol}_{\Pi_1}(x)$ *is computable in time* $t_g(|x|, y, r)$ *that is polynomial both in* $|x|$ *and in* $|y|$ *for an fixed* r; $t_g(n, n, \cdot)$ *is nonincreasing;*
- *for any* $y \in \text{Sol}_{\Pi_2}(f(x, r))$, $\rho_{\Pi_2}(f(x, r), y) \leq r$ *implies* $\rho_{\Pi_1}(x, g(x, y, r)) \leq 1 + \alpha(r - 1)$.

It is worth to underline the main differences of AP-reductions with respect to the reductions introduced until now. In first place, with respect to L-reductions, the constraint that the optimum values of the two problems are linearly related has been dropped. In second place, with respect to the S-reductions, we allow a weaker relationship to hold between the approximation ratios achieved for the two problems. Besides, an important condition that is needed in the proof of **APX**-completeness is that, in AP-reductions, the two functions f and g may depend on the approximation ratio r. Such extension is somewhat natural as there is no reason to ignore the quality of the solution we are looking for, when reducing one optimization problem to another and it plays a crucial role in the completeness proofs. On the other side, as in many applications such knowledge is not required, whenever functions f and g do not use the dependency on r, we will avoid specifying this dependency. In other words, we will write $f(x)$ and $g(x, y)$ instead of $f(x, r)$ and $g(x, y, r)$, respectively.

Proposition 14.2 *Given two minimization problems* Π_1 *and* Π_2, *if* $\Pi_1 \leq_{AP} \Pi_2$ *and* $\Pi_2 \in$ **APX** *(resp.,* $\Pi_2 \in$ **PTAS***), then* $\Pi_1 \in$ **APX** *(resp.,* $\Pi_1 \in$ **PTAS***).*

As a last remark, let us observe that the S-reducibility is a particular case of AP-reducibility, corresponding to the case in which $\alpha = 1$. More generally, the AP-reducibility is sufficiently broad to encompass almost all known approximation preserving reducibilities while maintaining the property of establishing a linear relation between performance ratios: this is important to preserve membership in all approximation classes.

14.6 NPO-Completeness and APX-Completeness

14.6.1 NPO-Completeness

In the preceding section, we have announced that by means of a suitable type of reduction we can transform an instance of MAX WSAT into an instance of MIN WSAT. This can now be obtained by making use of AP-reductions. By combining this result with Theorem 14.1 and with the corresponding result concerning the completeness of MAX WSAT in the class of **NPO**-maximization problems, we can assemble the complete proof that MIN WSAT is complete for the entire class **NPO** under AP-reductions. The inverse reduction, from MIN WSAT to MAX WSAT can be shown in a similar way, leading to the proof that also MAX WSAT is complete for the entire class **NPO** under AP-reductions.

Theorem 14.2 MAX WSAT *can be* AP-*reduced to* MIN WSAT *and vice versa.*

Sketch of the proof: The proof works as follows. First, a simple reduction can be defined that transforms a given instance ϕ of MAX WSAT into an instance ϕ' of MIN WSAT with α depending on r. Such reduction can then be modified into a real AP-reduction in which α is a constant, not depending on r, whereas, of course, the functions f and g will depend on r. We limit ourselves to describing the first step. The complete proof can be found in Reference 18.

Let ϕ be the formula produced in the reduction proving the completeness of MAX WSAT for the class of **NPO** maximization problems. Then, $f(\phi)$ be the formula $\phi \wedge \alpha_1 \wedge \cdots \wedge \alpha_s$ where α_i is $z_i \equiv (\bar{v}_1 \wedge \cdots \wedge \bar{v}_{i-1} \wedge v_i)$, $z_1, \ldots, z_s$ are new variables with weights $w(z_i) = 2^i$ for $i = 1, \ldots, s$, and all other variables (even the v-variables) have zero weight. If τ is a satisfying truth assignment for $f(\phi)$, let $g(\phi, \tau)$ be the restriction of τ to the variables that occur in ϕ. This assignment clearly satisfies ϕ. Note that exactly one among the z-variables is **true** in any satisfying truth assignment of $f(\phi)$. If all z-variables were **false**, then all v-variables would be **false**, which is not allowed. On the other hand, it is clearly not possible that two z-variables are **true**. Hence, for any feasible solution τ of $f(\phi)$, we have that $m(f(\phi), \tau) = 2^i$, for some i with $1 \leq i \leq s$. This finally implies that: $2^s/m(f(\phi), \tau) \leq m(\phi, g(\phi, \tau)) < 2.2^s/m(f(\phi), \tau)$. This is in particular true for the optimal solution (observe that any satisfying truth assignment for ϕ can be easily extended to a satisfying truth assignment for $f(\tau)$). Thus, after some easy algebra, the performance ratio of $g(\phi, \tau)$ with respect to ϕ verifies: $r(\phi, g(\phi, \tau)) > 1/(2r(f(\phi), \tau))$.

The reduction satisfies the approximation preserving condition with a factor $\alpha = (2r - 1)/(r - 1)$. To obtain a factor α not depending on r the reduction can be modified by introducing 2^k more variables for a suitable integer k. ∎

Other problems that have been shown **NPO**-complete are MIN (MAX) W3-SAT and MIN TSP ([7]). As it has been observed before, as a consequence of their **NPO**-completeness under approximation preserving reductions, for all these problems does not exist any r-approximate algorithm with constant r, unless $\mathbf{P} = \mathbf{NP}$.

14.6.2 APX-Completeness

As it has been mentioned earlier, the existence of an **APX**-complete problem has already been shown in Reference 16 (see also Reference 19) but the problem that is proved complete in such framework is a rather artificial version of MAX WSAT. The reduction used in such result is called P-reduction. Unfortunately no natural problem has been proved complete in **APX** using the same approach. In this section, we prove the **APX**-completeness under AP-reduction of a natural and popular problem: MAX 3-SAT. The proof is crucially based on the following two lemmas (whose proofs are not provided in this paper).

The first lemma is proved in Reference 20 and is based on a powerful algebraic technique for the representation of propositional formulæ [18], whereas the second one states a well-known property of propositional formulæ and is proved in References 3, 18.

Lemma 14.1 *There is a constant $\epsilon > 0$ and two functions f_s and g_s such that, given any propositional formula ϕ in conjunctive normal form, the formula $\psi = f_s(\phi)$ is a conjunctive normal form formula with at most three literals per clause that satisfies the following property: for any truth assignment T' satisfying at least a portion $1 - \epsilon$ of the maximum number of satisfiable clauses in ψ, $g_s(\phi, T')$ satisfies ϕ if and only if ϕ is satisfiable.*

Lemma 14.2 *Given a propositional formula in conjunctive normal form, at least one half of its clauses can always be satisfied.*

Theorem 14.3 MAX 3-SAT *is **APX**-complete.*

Sketch of the proof: As it has been done in the case of the proofs of **NPO**-completeness, we split the proof in two parts. First, we show that MAX 3-SAT is complete in the class of **APX** maximization problems and then we show that any **APX** minimization problem can be reduced to an **APX** maximization problem. To make the proof easier, we adopt the convention used in Reference 18. The approximation ratio of a maximization problem in this context will be defined as the ratio between the value of the optimum solution $\text{opt}(x)$ and the value of the approximate solution $m(x, A(x))$. For both maximization and minimization problems, therefore, the approximation ratio is in $[1, \infty)$. Let us first observe that MAX 3-SAT $\in$ **APX** as it can be approximated up to the ratio 0.8006 ([9]).

Now we can sketch the proof that MAX 3-SAT is hard for the class of maximization problems in **APX**. Let us consider a maximization problem $\Pi \in$ **APX**. Let A_Π be a polynomial-time r_Π-approximation algorithm for Π. To construct an AP-reduction, let us define the parameter α as follows: $\alpha = 2(r_\Pi \log r_\Pi + r_\Pi - 1) \times ((1 + \epsilon)/\epsilon)$ where ϵ is the constant of Lemma 14.1. Let us now chose $r > 1$ and consider the following two cases: $1 + \alpha(r - 1) \geq r_\Pi$ and $1 + \alpha(r - 1) < r_\Pi$.

In the case $1 + \alpha(r - 1) \geq r_\Pi$, given any instance x of Π and given any truth assignment τ for MAX 3-SAT, we trivially define: $f(x, r)$ to be the empty formula and $g(x, \tau, r) = A_\Pi(x)$. It can easily be seen that $r(x, g(x, \tau, r)) \leq r_\Pi \leq 1 + \alpha(r - 1)$ and the reduction is an AP-reduction.

Let us then consider the case $1 + \alpha(r - 1) < r_\Pi$ and let us define $r_n = 1 + \alpha(r - 1)$; then, $r = ((r_n - 1)/\alpha) + 1$. If we define $k = \lceil \log_{r_n} r_\Pi \rceil$, we can partition the interval $[m(x, A_\Pi(x)), r_\Pi m(x, A_\Pi(x))]$ in the following k subintervals: $[m(x, A_\Pi(x)), r_n m(x, A_\Pi(x))]$, $[r_n^i m(x, A_\Pi(x)), r_n^{i+1} m(x, A_\Pi(x))]$, $i = 1, \ldots, k - 2$, $[r_n^{k-1} m(x, A_\Pi(x)), r_\Pi m(x, A_\Pi(x))]$. Then we have $m(x, A_\Pi(x)) \leq \text{opt}(x) \leq r_\Pi m(x, A_\Pi(x)) \leq r_n^k m(x, A_\Pi(x))$, that is, the optimum value of instance x of Π belongs to one of the subintervals.

Note that by, definition, $k < (r_\Pi \log r_\Pi + r_\Pi - 1)/(r_n - 1)$ and by making use of the definitions of α, r and k, we obtain: $r < (\epsilon/(2k(1 + \epsilon))) + 1$.

For any $i = 0, 1, \ldots, k - 1$, let us consider an instance x of Π and the following nondeterministic algorithm where p is the polynomial that bounds the value of all feasible solutions of Π:

- guess a candidate solution y with value at most $p(|x|)$;
- if $y \in \text{Sol}_\Pi(x)$ and $m_\Pi(x, y) \leq r_n^{i+1} m(x, A_\Pi(x))$, then return *yes*, otherwise return *no*.

Applying once again the technique of Theorem 14.1, we can construct k propositional formulæ $\phi_0, \phi_1, \ldots, \phi_{k-1}$ such that for any truth assignment τ_i satisfying ϕ_i, $i = 0, 1, \ldots, k-1$, in polynomial time we can build a feasible solution y of the instance x with $m_\Pi(x, y) \geq r_n^i m(x, A_\Pi(x))$.

Hence, the instance ψ of MAX 3-SAT that we consider is the following: $\psi = f(x, r) = \bigwedge_{i=0}^{k-1} f_s(\phi_i)$, where f_s is the function defined in Lemma 14.1; w.l.o.g., we can suppose that all formulæ $f_s(\phi_i)$, $i = 0, \ldots, k - 1$, contain the same number of clauses.

Denote by T a satisfying truth assignment of ψ achieving approximation ratio r and by r_i the approximation ratio guaranteed by τ over $f_s(\phi_i)$. By Lemma 14.2 we get: $m(r_i - 1)/(2r_i) \leq \text{opt}(\psi) - m(\psi, T) \leq km(r - 1)/r$. Using this expression for $i = 0, \ldots, k - 1$, we have $m(r_i - 1)/2r_i \leq km(r - 1)/r$, which implies $1 - (2k(r - 1)/r) \leq 1/r_i$ and, finally, $r_i \leq 1 + \epsilon$.

Using again Lemma 14.1, we derive that, for $i = 0, \ldots, k - 1$, the truth assignment $\tau_i = g_s(\phi_i, \tau)$ (where g_s is as defined in Lemma 14.1) satisfies ϕ_i if and only if ϕ_i is satisfiable. Let us call i^* the largest i for which τ_i satisfies ϕ_i; then, $r_n^{i^*} m(x, A_\Pi(x)) \leq \text{opt}_\Pi(x) \leq r_n^{i^*+1} m(x, A_\Pi(x))$. Starting from τ_{i^*}, we can then construct a solution y for Π whose value is at least $r_n^{i^*} m(x, A_\Pi(x))$. This means that y guarantees an approximation ratio r_n. In other words, $r(x, y) \leq r_n = 1 + \alpha(r - 1)$ and the reduction (f, g, α) that we have just defined (where g consists in applying g_s, determining i^* and constructing y starting from τ_{i^*}) is an AP-reduction.

As Π is any maximization problem in **APX**, the completeness of MAX 3-SAT for the class of maximization problems in **APX** follows.

We now turn to the second part of the theorem. In fact, we still have to prove that all minimization problems in **APX** can be AP-reduced to maximization problems and, henceforth, to MAX 3-SAT.

Let us consider a minimization problem $\Pi \in$ **APX** and an algorithm A with approximation ratio r for Π; let $k = \lceil r \rceil$. We can construct a maximization problem $\Pi' \in$ **APX** and prove that $\Pi \leq_{\text{AP}} \Pi'$.

The two problems have the same instances and the same feasible solutions, whereas the objective function of Π' is defined as follows: given an instance x and a feasible solution y of x, $m_{\Pi'}(x, y) = (k + 1)$ $m_\Pi(x, A(x)) - km_\Pi(x, y)$, if $m_\Pi(x, y) \leq m_\Pi(x, A(x))$, $m_{\Pi'}(x, y) = m_\Pi(x, A(x))$, otherwise.

Clearly, $m_\Pi(x, A(x)) \leq \mathrm{opt}_{\Pi'}(x) \leq (k + 1)m_\Pi(x, A(x))$ and, by definition of Π', the algorithm A is also an approximation algorithm for this problem with approximation ratio $k + 1$; therefore $\Pi' \in$ **APX**. The reduction from Π to Π' can now be defined as follows: for any instance x of Π, $f(x) = x$; for any instance x of Π and for any solution y of instance $f(x)$ of Π', $g(x, y) = y$, if $m_\Pi(x, y) \leq m_\Pi(x, A(x))$, $g(x, y) = A(x)$, otherwise; $\alpha = k + 1$. Note that f and g do not depend on the approximation ratio r.

We now show that the reduction we have just defined is an AP-reduction. Let y be an r'-approximate solution of $f(x)$; we have to show that the ratio $r_\Pi(x, g(x, y))$ of the solution $g(x, y)$ of the instance x of Π is smaller than, or equal to, $1 + \alpha(r' - 1)$. We have the following two cases: $m_\Pi(x, y) \leq m_\Pi(x, A(x))$ and $m_\Pi(x, y) > m_\Pi(x, A(x))$.

In the case $m_\Pi(x, y) \leq m_\Pi(x, A(x))$, we can derive: $m_\Pi(x, y) \leq (1 + \alpha(r' - 1))\,\mathrm{opt}_\Pi(x)$. In other words: $r_\Pi(x, g(x, y)) = r_\Pi(x, y) \leq 1 + \alpha(r' - 1)$.

In the case $m_\Pi(x, y) > m_\Pi(x, A(x))$, as $\alpha \geq 1$, we have: $r_\Pi(x, g(x, y)) = r_\Pi(x, A(x)) = r_{\Pi'}(x, y) \leq r' \leq 1 + \alpha(r' - 1)$.

In conclusion, all minimization problems in **APX** can be AP-reduced to maximization problems in **APX** and all maximization problems in **APX** can be AP-reduced to MAX 3-SAT. As the AP-reduction is transitive the **APX**-completeness of MAX 3-SAT is proved. ∎

14.6.3 Negative Results Based on APX-Completeness

Similarly to what we saw for completeness in **NPO**, also completeness in **APX** implies negative results in terms of approximability of optimization problems. In fact, if we could prove that an **APX**-complete problem admits a PTAS, then so would all problems in **APX**. On the other side, it is well known that unless **P** = **NP** there are problems in **APX** that do not admit a PTAS (one example for all, MIN SCHEDULING ON IDENTICAL MACHINES, see Reference 18), therefore, under the same complexity theoretic hypothesis, no **APX**-complete problem admits a PTAS.

As a consequence of the results in the previous subsection, we can therefore assert that, unless **P** = **NP**, MAX 3-SAT does not admit a PTAS, neither do all other optimization problems that have been shown **APX**-complete: (MAX 2-SAT, MIN VERTEX COVER, MAX CUT, MIN METRIC TSP, etc.).

Note that the inapproximability of MAX 3-SAT has been proved by Arora et al. ([20]) in a breakthrough paper by means of sophisticated techniques based on the concept of *probabilistically checkable proofs*, without any reference to the notion of **APX**-completeness. This fact, though, does not diminish the relevance of approximation preserving reductions and the related completeness notion. In fact, most results that state the nonexistence of PTAS for **APX** optimization problems have been proved starting from MAX 3-SAT, via approximation preserving reductions that allow to carry over the inapproximability results from one problem to another. In second place, it is worth noting that the structure of approximation classes with respect to approximation preserving reductions is richer than it appears from this chapter. For example, beside complete problems, other classes of problems can be defined inside approximation classes, identifying the so called *intermediate* problems [18].

14.7 FT-Reducibility

As we have already pointed out in Section 14.5, **PTAS**-completeness has been studied in Reference 16 under the so-called F-reduction, preserving membership in **FPTAS**. Under this type of reducibility, a single problem, a rather artificial version of MAX WSAT has been shown **PTAS**-complete. In fact, F-reducibility is quite restrictive as it mainly preserves optimality, henceforth, existence of a **PTAS**-complete polynomially bounded problem is very unlikely.

In Reference 21, a more "flexible" type of reducibility, called FT-reducibility has been introduced. It is formally defined as follows.

Definition 14.11 *Let Π and Π' be two maximization integer-valued problems. Then, Π FT-reduces to Π' (denoted by $\Pi \leq_{FT} \Pi'$) if, for any $\epsilon > 0$, there exist an oracle $\bigcirc_\alpha^{\Pi'}$ for Π' and an algorithm A_ϵ calling $\bigcirc_\alpha^{\Pi'}$ such that:*

- *$\bigcirc_\alpha^{\Pi'}$ produces, for any $\alpha \in]0, 1]$ and for any instance x' of Π', a feasible solution $\bigcirc_\alpha^{\Pi'}(x')$ of x' that is an $(1 - \alpha)$-approximation;*
- *for any instance x of Π, $y = A_\epsilon(\bigcirc_\alpha^{\Pi'}, x) \in \text{Sol}(x)$; furthermore the approximation ratio of y is at least $(1 - \epsilon)$;*
- *if $\bigcirc_\alpha^{\Pi'}(\cdot)$ runs in time polynomial in both $|f(x)|$ and $1/\alpha$, then $A_\epsilon(\bigcirc_\alpha^{\Pi'}(f(x)), x)$ is polynomial in both $|x|$ and $1/\epsilon$.*

For the case where at least one between Π and Π' is a minimization problem it suffices to replace $1 - \epsilon$ or/and $1 - \alpha$ by $1 + \epsilon$ or/and $1 + \alpha$, respectively.

As one can see from Definition 14.11, FT-reduction is somewhat different from the other ones considered in this chapter and, in any case, it is not conformal to Definition 14.6. In fact, it resembles a Turing-reduction. Clearly, FT-reduction transforms a fully polynomial-time approximation schema for Π' into a fully polynomial-time approximation schema for Π, that is, it preserves membership in **FPTAS**. Note also that the F-reduction, as it is defined in Reference 16, is a special case of the FT-reduction, as the latter explicitly allows multiple calls to oracle $\bigcirc$ whereas for the former this fact is not explicit.

Theorem 14.4 *Let Π' be an NP-hard problem in NPO. If $\Pi' \in$ NPO-PB (the class of problems in NPO whose values are bounded by a polynomial in the size of the instance), then any NPO problem FT-reduces to Π'. Consequently, (i) $\overline{PTAS}^{FT} =$ NPO and (ii) any NP-hard polynomially bounded problem in PTAS is PTAS-complete under FT-reductions.*

Sketch of the proof: We first prove the following claim: *if an NPO problem Π' is NP-hard, then any NPO problem Turing-reduces [18] to Π'.*

To prove this claim, let Π be an **NPO** problem and q be a polynomial such that $|y| \leq q(|x|)$, for any instance x of Π and for any feasible solution y of x. Assume that the encoding $n(y)$ of y is binary. Then $0 \leq n(y) \leq 2^{q(|x|)} - 1$. We consider problem $\hat{\Pi}$, which is the same as Π up to its value that is defined by $m_{\hat{\Pi}}(x, y) = 2^{q(|x|)+1} m_\Pi(x, y) + n(y)$. If $m_{\hat{\Pi}}(x, y_1) \geq m_{\hat{\Pi}}(x, y_2)$, then $m_\Pi(x, y_1) \geq m_\Pi(x, y_2)$. So, if a solution y is optimal for x, with respect to $\hat{\Pi}$, it is so with respect to Π. Remark now that $\hat{\Pi}$ and its evaluation version $\hat{\Pi}_e$ are equivalent as given the value of an optimal solution y, one can determine $n(y)$ (hence y) by computing the remainder of the division of this value by $2^{q(|x|)+1}$. As Π' is **NP**-hard, it can be shown that one can solve the evaluation problem $\hat{\Pi}_e$, henceforth $\hat{\Pi}$ if one can solve, the (constructive) problem Π' and the claim is proved.

We now prove the following claim: *let $\Pi' \in$ NPO-PB; then, any NPO problem Turing-reducible to Π' is also FT-reducible to Π'.*

To prove this second claim, let Π be an **NPO** problem and suppose that there exists a Turing-reduction between Π and Π'. Let $\bigcirc_\alpha^{\Pi'}$ be as in Definition 14.11. Moreover, let p be a polynomial such that for any instance x' of Π' and for any feasible solution y' of x', $m(x', y') \leq p(|x'|)$. Let x be an instance of Π. The Turing-reduction claimed gives an algorithm solving Π using an oracle for Π'. Consider now this algorithm where we use, for any query to the oracle with the instance x' of Π', the approximate oracle $\bigcirc_\alpha^{\Pi'}(x')$, with $\alpha = 1/(p(|x'|) + 1)$. This algorithm is polynomial and produces an optimal solution, as a solution y' being an $(1 - (1/(p(|x'|) + 1)))$-approximation for x' is an optimal one. So, the claim is proved.

From the combination of the above-mentioned claims the theorem is easily derived. ∎

Observe finally that MAX PLANAR INDEPENDENT SET and MIN PLANAR VERTEX COVER are in both **PTAS** ([22]) and **NPO-PB**. So, the following theorem concludes this section.

Theorem 14.5 MAX PLANAR INDEPENDENT SET *and* MIN PLANAR VERTEX COVER *are* **PTAS**-*complete under* FT-*reductions.*

14.8 Gadgets, Reductions, and Inapproximability Results

As it has been pointed out already in Section 14.3, in the context of approximation preserving reductions we call *gadget* a combinatorial structure, which allows to transfer approximability (or inapproximability) results from a problem to another. A classical example is the set of ten 2-SAT clauses that we have used in Example 14.1 for constructing the 2-SAT formula starting from a 3-SAT formula. Although gadgets are used in the seminal work of Karp on reductions among combinatorial problems, the study of gadgets has been started in References 9, 23; from the former derive most of the results discussed in this section.

To understand the role of gadgets in approximation preserving reductions, let us first go back to linear reductions and see what are the implications on the approximation ratio of two problems Π and Π', deriving from the fact that $\Pi \leq_L \Pi'$. Suppose Π and Π' are minimization problems, f, g, α_1, and α_2 are the functions and constants that define the linear reduction, x is an instance of problem Π, $f(x)$ is the instance of problem Π' determined by the reduction, y is a solution of $f(x)$. Then, the following relationship holds between the approximation ratios of Π and Π': $r_\Pi(x, g(x,y)) \leq 1 + \alpha_1\alpha_2(r_{\Pi'}(f(x), y) - 1)$, and, therefore, we have that $r_{\Pi'} \leq 1 + (r-1)/(\alpha_1\alpha_2)$ implies $r_\Pi \leq r$.

In the particular case of the reduction between MAX 3-SAT and MAX 2-SAT, we have $\alpha_1\alpha_2 = 13$ and, therefore, we can infer the following results on the approximability upper bounds and lower bounds of the two problems, which may be proved by a simple calculation:

- As it is known that MAX 2-SAT can be approximated with the ratio 0.931 ([24]), then MAX 3-SAT can be approximated with ratio 0.103;
- As it is known that MAX 3-SAT cannot be approximated beyond the threshold 7/8, then MAX 2-SAT cannot be approximated beyond the threshold 103/104.

Although better bounds are now known for these problems [25], it is important to observe that the above given bounds may be straightforwardly derived from the linear reduction between the two problems and are useful to show the role of gadgets. In such reduction, the structure of the gadget is crucial (it determines the value α_1) and it is clear that better bounds could be achieved if the reduction could make use of "smaller" gadgets. In fact, in Reference 9, by cleverly constructing a more sophisticated type of gadget (in which, in particular, clauses have real weights), the authors derive a 0.801 approximation algorithm for MAX 3-SAT, improving on previously known bounds.

Based on Reference 23, in Reference 9 the notion of α-gadget (i.e., gadget with performance α) is abstracted and formalized with reference to reductions among constraint satisfaction problems. In the same paper, it is shown that, under suitable circumstances, the search for (possibly optimum) gadgets to be used in approximation preserving reductions, can be pursued in a systematic way by means of a computer program. An example of the results that may be achieved in this way is the following.

Let PC_0 and PC_1 be the families of constraints over three binary variables defined as: $PC_i(a, b, c) = 1$, if $a \oplus b \oplus c = i$, $PC_i(a, b, c) = 0$, otherwise, and let DICUT be the family of constraints corresponding to directed cuts in a graph. There exists optimum 6.5 gadgets (automatically derived by the computer program) reducing PC_0 and PC_1 to DICUT. As a consequence, for any $\epsilon > 0$, MAX DICUT is hard to approximate to within $12/13 + \epsilon$.

14.9 Conclusion

A large number of other approximation preserving reductions among optimization problems, beside those introduced in this chapter, have been introduced throughout the years. Here we have reported only the major developments. Other overviews of the world of approximation preserving reductions can be found in References 12, 26.

As we have already pointed out in Section 14.2, we have not dealt in this chapter with approximability classes beyond **APX**, even if intensive studies have been performed, mainly for **Poly-APX**. In Reference 17, completeness results are established, under the E-reduction, for **Poly-APX-PB** (the class of problems in **Poly-APX** whose values are bounded by a polynomial in the size of the instance). Indeed, as we have already discussed in Section 14.5, use of restrictive reductions as the E-reducibility, where the functions f and g do not depend on any parameter ϵ seems very unlikely to be able to handle **Poly-APX**-completeness. As it is shown in Reference 21 (see also Chapter 15), completeness for the whole **Poly-APX** can be handled, for instance, by using PTAS-reduction, a further relaxation of the AP-reduction where the dependence between the approximation ratios of Π and Π' is not restricted to be linear ([27]). Under PTAS-reduction, MAX INDEPENDENT SET is **Poly-APX**-complete ([21]).

Before concluding, it is worth noting that a structural development (based on the definition of approximability classes, approximation preserving reductions, and completeness results), analogous to the one that has been carried on for the classical approach to the theory of approximation, has been elaborated also for the differential approach (see Chapter 15 for a survey). In References 21, 28 the approximability classes **DAPX**, **Poly-DAPX**, and **DPTAS** are introduced; suitable approximation preserving reductions are defined; and complete problems in **NPO**, **DAPX**, **Poly-DAPX**, and **DPTAS**, under such kind of reductions, are shown.

References

1. Cook, S.A., The complexity of theorem-proving procedures, in *Proceedings of Third ACM Symposium on Theory of Computing*, p. 151, 1971.
2. Karp, R.M., Reducibility among combinatorial problems, in R.E. Miller and J.W. Thatcher (Eds.), *Complexity of Computer Computations*, Plenum Press, New York, p. 85, 1972.
3. Johnson, D.S., Approximation algorithms for combinatorial problems, *J. Comput. System Sci.*, 9, 256, 1974.
4. Ausiello, G., D'Atri, A., and Protasi, M., Structure preserving reductions among convex optimization problems, *J. Comp. Sys. Sci.*, 21, 136, 1980.
5. Ausiello, G., D'Atri, A., and Protasi, M., Lattice-theoretical ordering properties for NP-complete optimization problems, *Fundamenta Informaticæ*, 4, 83, 1981.
6. Paz, A. and Moran, S., Non deterministic polynomial optimization problems and their approximations, *Theor. Comput. Sci.*, 15, 251, 1981.
7. Orponen, P. and Mannila, H., On approximation preserving reductions: Complete problems and robust measures, Technical Report C-1987-28, Department of Computer Science, University of Helsinki, Finland, 1987.
8. Papadimitriou, C.H. and Yannakakis, M., Optimization, approximation and complexity classes, *J. Comput. Sys. Sci.*, 43, 425, 1991.
9. Trevisan, L., Sorkin, G.B., Sudan, M., and Williamson, D.P., Gadgets, approximation, and linear programming, *SIAM J. Comput.*, 29(6), 2074, 2000.
10. Garey, M.R. and Johnson, D.S., *Computers and Intractability: A Guide to the Theory of NP-completeness*, W. H. Freeman, San Francisco, CA, 1979.
11. Papadimitriou, C.H., *Computational Complexity*, Addison-Wesley, Boston, MA, 1994.
12. Crescenzi, P., Kann, V., Silvestri, R., and Trevisan, L., Structure in approximation classes, *SIAM J. Comput.*, 28(5), 1759, 1999.
13. Simon, H.U., Continuous reductions among combinatorial optimization problems, *Acta Informatica*, 26, 771, 1989.
14. Simon, H.U., On approximate solutions for combinatorial optimization problems, *SIAM J. Disc. Math.*, 3(2), 294, 1990.
15. Krentel, M.W., The complexity of optimization problems, *J. Comput. System Sci.*, 36, 490, 1988.

16. Crescenzi, P. and Panconesi, A., Completeness in approximation classes, *Info. Comput.*, 93(2), 241, 1991.
17. Khanna, S., Motwani, R., Sudan, M., and Vazirani, U., On syntactic versus computational views of approximability, *SIAM J. Comput.*, 28, 164, 1998.
18. Ausiello, G., Crescenzi, P., Gambosi, G., Kann, V., Marchetti-Spaccamela, A., and Protasi, M., *Complexity and Approximation. Combinatorial Optimization Problems and Their Approximability Properties*, Springer, Berlin, Germany, 1999.
19. Ausiello, G., Crescenzi, P., and Protasi, M., Approximate solutions of NP optimization problems, *Theor. Comput. Sci.*, 150, 1, 1995.
20. Arora, S., Lund, C., Motwani, R., Sudan, M., and Szegedy, M., Proof verification and intractability of approximation problems, in *Proceedings of the 33rd IEEE Annual Symposium on Foundations of Computer Science*, p. 14, 1992.
21. Bazgan, C., Escoffier, B., and Paschos, V.Th., Completeness in standard and differential approximation classes: Poly-(D)APX- and (D)PTAS-completeness, *Theor. Comput. Sci.*, 339, 272, 2005.
22. Baker, B.S., Approximation algorithms for NP-complete problems on planar graphs, *J. Assoc. Comput. Mach.*, 41(1), 153, 1994.
23. Bellare, M., Goldreich, O., and Sudan, M., Free bits and non-approximability—towards tight results, *SIAM J. Comput.*, 27(3), 804, 1998.
24. Feige, U. and Goemans, M.X., Approximating the value of two prover proof systems, with applications to MAX 2SAT and MAX DICUT, in *Proceedings of Israel Symposium on Theory of Computing and Systems, ISTCS'95*, p. 182, 1995.
25. Karloff, H. and Zwick, U., A 7/8-approximation for MAX 3SAT?, in *Proceedings of the 38th IEEE Annual Symposium on Foundations of Computer Science*, p. 406, 1997.
26. Crescenzi, P., A short guide to approximation preserving reductions, in *Proceedings of the 12th IEEE Conference on Computational Complexity*, p. 262, 1997.
27. Crescenzi, P. and Trevisan, L., On approximation scheme preserving reducibility and its applications, *Theor. Comput. Sys.*, 33(1), 1, 2000.
28. Ausiello, G., Bazgan, C., Demange, M., and Paschos,V.Th., Completeness in differential approximation classes, *IJFCS*, 16(6), 1267, 2005.

15

Differential Ratio Approximation

Giorgio Ausiello

Vangelis Th. Paschos

15.1 Introduction

In this chapter we introduce the so-called differential approximation ratio as measure of the quality of the solutions obtained by approximation algorithms. After providing motivations and basic definitions we show examples of optimization problems for which the evaluation of approximation algorithms based on the differential ratio appears to be more meaningful than the usual approximation ratio used in the classical approach to approximation algorithms. Finally, we discuss some structural results concerning approximation classes based on the differential ratio. Throughout the chapter we make use of the notations introduced in Chapter 14. Also, given an approximation algorithm A for an **NPO** problem Π, we denote by $m_A(x, y)$, the value of the solution y computed by A on instance x of Π. When clear from the context, reference to A will be omitted. The definitions of most of the problems dealt in this chapter can be found in References 1, 2; also, for graph-theoretic notions, interested readers are referred to Reference 3.

In several cases, the commonly used approximation measure (called *standard approximation ratio* in what follows) may not be very meaningful in characterizing the quality of approximation algorithms. This happens, in particular, when the ratio of $m(x, y_w)$, the value of the worst solution for a given input x, to the value of the optimum solution $\mathrm{opt}(x)$ is already bounded (above, if goal(Π) = min, below, otherwise). Consider, for instance, the basic maximal matching algorithm for MIN VERTEX COVER that achieves approximation ratio 2. In this algorithm, given a graph $G(V, E)$, a maximal* matching M of G is computed and the endpoints of the edges in M are added in the solution for MIN VERTEX COVER. If M is perfect (almost any graph, even relatively sparse, admits a perfect matching [4]), then the whole of V will be

---

* With respect to inclusion.

included in the cover, whereas an optimum cover contains at least a half of V. So, in most cases, the absolutely worst solution (that one could compute without using any algorithm) achieves approximation ratio 2.

The above-mentioned remark is just one of the drawbacks of the standard approximation ratio. Various other drawbacks have been also observed, for instance, the artificial dissymmetry between "equivalent" minimization and maximization problems (e.g., MAX CUT and MIN CLUSTERING, see Reference 5) introduced by the standard approximation ratio. The most blatant case of such dissymmetry is the one appearing when dealing with the approximation of MIN VERTEX COVER and MAX INDEPENDENT SET (given a graph, a vertex cover is the complement of an independent set with respect to the vertex set of the graph). In other words, using linear programming vocabulary, the objective function of the former is an affine transformation of the objective function of the latter. This equivalence under such simple affine transformation does not reflect in the approximability of these problems in the classical approach: the former problem is approximable within constant ratio, in other words it is in **APX** (see Chapter 14 for definitions of approximability classes based on the standard approximation paradigm; the ones based on the differential paradigm are defined analogously in this chapter, see Section 15.5), whereas the latter is inapproximable within ratio $\Omega(n^{\epsilon-1})$, for any $\epsilon > 0$ [6]. In other words, the standard approximation ratio is unstable under affine transformations of the objective function.

To overcome these phenomena, several researchers have tried to adopt alternative approximation measures not suffering from these inconsistencies. One of them is the ratio $\delta(x, y) = (\omega(x) - m(x, y))/(\omega(x) - \text{opt}(x))$, called differential ratio in the sequel, where $\omega(x)$ is the value of a worst solution for x, called *worst value*. It will be formally dealt in the next sections. It has been used rather punctually and without following a rigorous axiomatic approach until the paper [7] where such an approach is formally defined. To our knowledge, differential ratio is introduced in Reference 8 in 1977, and in References 9–11 are, to our knowledge, the most notable cases in which this approach has been applied. It is worth noting that in Reference 11, a weak axiomatic approach is also presented.

Finally, let us note that several other authors that have also recognized the methodological problems implied by the standard ratio have proposed other alternative ratios. It is interesting to remark that, in most cases, the new ratios are very close, although with some small or less small differences, to the differential ratio. For instance, in Reference 12, for studying MAX TSP, it is proposed the ratio $d(x, y, z_r) = |\text{opt}(x) - m(x, y)|/|\text{opt}(x) - z_r|$, where z_r is a positive value computable in polynomial time, called *reference-value*. It is smaller than the value of any feasible solution of x, hence smaller than $\omega(x)$ (for a maximization problem a worst solution is the one of the smallest feasible value). The quantities $|\text{opt}(x) - m(x, y)|$ and $|\text{opt}(x) - z_r|$ are called *deviation* and *absolute deviation*, respectively. The approximation ratio $d(x, y, z_r)$ depends on both x and z_r, in other words, there exist a multitude of such ratios for an instance x of an **NPO** problem, one for any possible value of z_r. Consider a maximization problem Π and an instance x of Π. Then, $d(x, y, z_r)$ is increasing with z_r, so, $d(x, y, z_r) \leq d(x, y, \omega(x))$. In fact, in this case, for any reference value z_r: $r(x, y) \geq 1 - d(x, y, z_r) \geq 1 - d(x, y, \omega(x)) = \delta(x, y)$, where r denotes the standard approximation ratio for Π. When $\omega(x)$ is computable in polynomial time, $d(x, y, \omega(x))$ is the smallest (tightest) over all the d-ratios for x. In any case, if for a given problem, one sets $z_r = \omega(x)$, then $d(x, y, \omega(x)) = 1 - \delta(x, y)$ and both ratios have the natural interpretation of estimating the relative position of the approximate solution-value in the interval worst solution-value – optimal value.

15.2 Towards a New Measure of Approximation Paradigm

In Reference 7, it is undertaken the task of adopting, in an axiomatic way, an approximation measure founded on both intuitive and mathematical links between optimization and approximation. It is claimed there that a "consistent" ratio must be *order preserving* (i.e., the better the solution the better the approximation ratio achieved) and *stable under affine transformation of the objective function*. Furthermore, it is proved that no ratio function of two parameters—for example, m, opt—can fit this latter requirement.

Hence it is proposed what will be called *differential approximation ratio** in what follows. Problems related by affine transformations of their objective functions are called *affine equivalent*.

Consider an instance x of an **NPO** problem Π and a polynomial time approximation algorithm A for Π, the differential approximation ratio $\delta_A(x, y)$ of a solution y computed by A in x is defined by: $\delta_A(x, y) = (\omega(x) - m_A(x, y))/(\omega(x) - \text{opt}(x))$, where $\omega(x)$ is the value of a worst solution for x, called *worst value*. Note that for any goal, $\delta_A(x, y) \in [0, 1]$ and, moreover, the closer $\delta_A(x, y)$ to 1, the closer $m_A(x, y)$ to opt(x). By definition, when $\omega(x) = \text{opt}(x)$, that is, all the solutions of x have the same value, then the approximation ratio is 1. Notice that $m_A(x, y)) = \delta_A(x, y) \text{opt}(x) + (1 - \delta_A(x, y))\omega(x)$. So, differential approximation ratio measures how an approximate solution is placed in the interval between $\omega(x)$ and opt(x).

We note that the concept of the worst solution has a status similar to the optimum solution. It depends on the problem itself and is defined in a nonconstructive way, that is, independently of any algorithm that could build it. The following definition for worst solution is proposed in Reference 7.

Definition 15.1 *Given an **NPO** problem $\Pi = (\mathcal{I}, \text{Sol}, m, \text{goal})$, a worst solution of an instance x of Π is defined as an optimum solution of a new problem $\bar{\Pi} = (\mathcal{I}, \text{Sol}, m, \overline{\text{goal}})$, that is, of a **NPO** problem having the same sets of instances and of instances and of feasible solutions and the same value-function as Π but its goal is the inverse w.r.t. Π, that is, $\overline{\text{goal}} = \min$ if goal $= \max$ and vice versa.*

Example 15.1 *The worst solution for an instance of* MIN VERTEX COVER *or of* MIN COLORING *is the whole vertex-set of the input-graph, whereas for an instance of* MAX INDEPENDENT SET *the worst solution is the empty set. On the other hand, if one deals with* MAX INDEPENDENT SET *with the additional constraint that a feasible solution has to be maximal with respect to inclusion, the worst solution of an instance of this variant is a* minimum maximal independent set, *that is, an optimum solution of a very well-known combinatorial problem, the* MIN INDEPENDENT DOMINATING SET. *Also, the worst solution for* MIN TSP *is a "heaviest" Hamiltonian cycle of the input-graph, that is, an optimum solution of* MAX TSP, *whereas for* MAX TSP *the worst solution is the optimum solution of a* MIN TSP. *The same holds for the pair* MAX SAT, MIN SAT.

From Example 15.1, one can see that, although for some problems a worst solution corresponds to some trivial input-parameter and can be computed in polynomial time (i.e., for instance, the case with MIN VERTEX COVER, MAX INDEPENDENT SET, MIN COLORING, etc.), several problems exist for which determining a worst solution is as hard as determining an optimum one (as for MIN INDEPENDENT DOMINATING SET, MIN TSP, MAX TSP, MIN SAT, MAX SAT, etc.).

Remark 15.1 *Consider the pair of affine equivalent problems* MIN VERTEX COVER, MAX INDEPENDENT SET *and an input-graph $G(V, E)$ of order n. Denote by $\tau(G)$ the cardinality of a minimum vertex cover of G and by $\alpha(G)$, the stability number of G. Obviously, $\tau(G) = n - \alpha(G)$. Based on the above-mentioned discussion, the differential ratio of some vertex cover C of G is $\delta(G, C) = (n - |C|)/(n - \tau(G))$. As the set $S = V \setminus C$ is an independent set of G, its differential ratio is $\delta(G, S) = (|S| - 0)/(\alpha(G) - 0) = (n - |C|)/(n - \tau(G)) = \delta(G, C)$.*

As we have already mentioned, the differential ratio, although not systematically, has been used several times by many authors, before and after Reference 7, in various contexts going from mathematical (linear or non-linear) programming [14–16] to pure combinatorial optimization [9,10,13,17,18]. Sometimes the use of the differential approach has been disguised by considering the standard approximation ratio of affine transformations of a problem. For instance, to study differential approximation of BIN PACKING, one can deal with standard approximation of the problem of maximizing the number of unused bins; for MIN COLORING, the affinely equivalent problem is the one of maximizing the number of unused colors, for MIN SET COVER, the problem consists in maximizing the number of unused sets, and so on.

---

* This notation is suggested in Reference 7; another notation drawing the same measure is *z-approximation* suggested in Reference 13.

15.3 Differential Approximation Results for Some Optimization Problems

In general, no systematic way allows to link results obtained in standard and differential approximation paradigms when dealing with minimization problems. In other words, there is no evident transfer of positive or inapproximability results from one framework to the other one. Hence, a "good" differential approximation result does not signify anything for the behavior of the approximation algorithm studied, or of the problem itself, when dealing with the standard framework, and vice versa. Things are somewhat different for maximization problems with positive solution-values. In fact, considering an instance x of a maximization problem Π and a solution $y \in \mathrm{Sol}(x)$ that is a δ-differential approximation, we immediately get:

$$\frac{m(x,y) - \omega(x)}{\mathrm{opt}(x) - \omega(x)} \geq \delta \implies \frac{m(x,y)}{\mathrm{opt}(x)} \geq \delta + (1 - \delta)\frac{\omega(x)}{\mathrm{opt}(x)} \overset{\omega(x) \geq 0}{\implies} \frac{m(x,y)}{\mathrm{opt}(x)} \geq \delta$$

So, positive results are transferred from differential to standard approximation, whereas transfer of inapproximability thresholds is done in the opposite direction.

Fact 15.1 *Approximation of a maximization **NPO** problem Π within differential approximation ratio δ, implies its approximation within standard approximation ratio δ.*

Fact 15.1 has interesting applications. The most immediate of them deals with the case of maximization problems with worst-solution values 0. There, standard and approximation ratios coincide. In this case, the differential paradigm inherits the inapproximability thresholds of the standard one. For instance, the inapproximability of MAX INDEPENDENT SET within $n^{\epsilon-1}$, for any $\epsilon > 0$ ([6]), also holds in the differential approach.

Furthermore, as MAX INDEPENDENT SET and MIN VERTEX COVER are affine equivalent, henceforth differentially equi-approximable, the negative result for MAX INDEPENDENT SET is shared, in the differential paradigm, by MIN VERTEX COVER.

Corollary 15.1 *Both MAX INDEPENDENT SET and MIN VERTEX COVER are inapproximable within differential ratios $n^{\epsilon-1}$, for any $\epsilon > 0$, unless $P = NP$.*

Notice that differential equi-approximability of MAX INDEPENDENT SET and MIN VERTEX COVER makes that, in this framework the latter problem is not constant approximable but inherits also the positive standard approximation results of the former one ([19–21]).

In what follows in this section, we mainly focus ourselves on three well-known **NPO** problems: MIN COLORING, BIN PACKING, TSP in both minimization and maximization variants, and MIN MULTIPROCESSOR SCHEDULING. As we will see, approximabilities of MIN COLORING and MIN TSP are radically different from the standard paradigm (where these problems are very hard) to the differential one (where they become fairly well-approximable). For the first two of them, differential approximability will be introduced by means of more general problem that encompasses both MIN COLORING and BIN PACKING, namely, the MIN HEREDITARY COVER.

15.3.1 Min Hereditary Cover

Let π be a nontrivial *hereditary* property* on sets and C a ground set. A π-covering of C is a collection $\mathcal{S} = \{S_1, S_2, \ldots, S_q\}$ of subsets of C (i.e., a subset of 2^C), any of them verifying π and such that $\cup_{i=1}^{q} S_i = C$. Then, MIN HEREDITARY COVER consists, given a property π, a ground set C and a family $\mathcal{S}$ including *any*

---

* A property is hereditary if whenever it is true for some set, it is true for any of its subsets; it is nontrivial if it is true for infinitely many sets and false for infinitely many sets also.

subset of C verifying π, of determining a π-covering of minimum size. Observe that, by definition of the instances of MIN HEREDITARY COVER singletons of the ground sets are included in any of them and are always sufficient to cover C. Henceforth, for any instance x of the problem, $\omega(x) = |C|$.

It is easy to see that, given a π-covering, one can yield a π-partition (i.e., a collection S where for any $S_i, S_j \in S$, $S_i \cap S_j = \emptyset$) of the same size, by greedily removing duplications of elements of C. Henceforth, MIN HEREDITARY COVER or MIN HEREDITARY PARTITION are, in fact, the same problem. MIN HEREDITARY COVER has been introduced in Reference 22 and revisited in Reference 13 under the name MIN COVER BY INDEPENDENT SETS. Moreover, in the former paper, using a clever adaptation of the local improvement methods of Reference 23, a differential ratio 3/4 for MIN HEREDITARY COVER has been proposed. Based on Reference 24, this ratio has been carried to 289/360 by Reference 13.

A lot of well-known **NPO** problems are instantiations of MIN HEREDITARY COVER. For instance, MIN COLORING becomes an MIN HEREDITARY COVER-problem, considering as ground set the vertices of the input-graph and as set-system, the set of the independent sets* of this graph. The same holds for the partition of the covering of a graph by subgraphs that are planar, or by degree-bounded subgraphs, and so on. Furthermore, if any element of C is associated with a weight and a subset S_i of C is in S if the total weight of its members is at most 1, then one recovers BIN PACKING.

In fact, an instance of MIN HEREDITARY COVER can be seen as a virtual instance of MIN SET COVER, even if there is no need to make it always explicit. Furthermore, the following general result links MIN k-SET COVER (the restriction of MIN SET COVER to subsets of cardinality at most k) and MIN HEREDITARY COVER (see Reference 25 for its proof in the case of MIN COLORING; it can be easily seen that extension to the general MIN HEREDITARY COVER is immediate).

Theorem 15.1 *If* MIN k-SET COVER *is approximable in polynomial time within differential approximation ratio δ, then* MIN HEREDITARY COVER *is approximable in polynomial time within differential approximation ratio* $\min\{\delta, k/(k+1)\}$.

15.3.1.1 Min Coloring

MIN COLORING has been systematically studied in the differential paradigm. Subsequent papers ([18,23, 24,26–29]) have improved its differential approximation ratio from 1/2 to 289/360. This problem is also a typical example of a problem that behaves in completely different ways when dealing with the standard or the differential paradigms. Indeed, dealing with the former one, MIN COLORING is inapproximable within ratio $n^{1-\epsilon}$, for any $\epsilon > 0$, unless problems in **NP** can be solved by slightly super-polynomial deterministic algorithms [1].

As we have seen just previously, given a graph $G(V, E)$, MIN COLORING can be seen as a MIN HEREDITARY COVER-problem considering $C = V$ and taking for S the set of the independent sets of G. According to Theorem 15.1 and Reference 24, where MIN 6-SET COVER is proved approximable within differential ratio 289/360, one can derive that it is also approximable within differential ratio 289/360. Notice that any result for MIN COLORING also holds for the minimum vertex-partition (or covering) into cliques problem as an independent set in some graph G becomes a clique in the complement $\bar{G}$ of G (in other words, this problem is also an instantiation of MIN HEREDITARY COVER). Furthermore, in References 26, 27, a differential ratio preserving reduction is devised between minimum vertex-partition into cliques and minimum edge-partition (or covering) into cliques. So, as in the standard paradigm, all these three problems have identical differential approximation behavior.

Finally, it is proved in Reference 30 that MIN COLORING is **DAPX**-complete (see also Section 15.5.3.1); consequently, unless **P** = **NP**, it cannot be solved by polynomial time differential approximation schemata. This derives immediately that neither MIN HEREDITARY COVER belongs to **DPTAS**, unless **P = NP**.

---

* It is well known that the independence property is hereditary.

15.3.1.2 Bin Packing

We now deal with another very well-known **NPO** problem, the BIN PACKING. According to the above-mentioned discussion, BIN PACKING being a particular case of MIN HEREDITARY COVER, it is approximable within differential ratio 289/360. In what follows in this section, we refine this result by first presenting an approximation preserving reduction transforming any standard approximation ratio ρ into differential approximation ratio $\delta = 2 - \rho$. Then, based on this reduction we show that BIN PACKING can be solved by a polynomial time differential approximation schema ([31]); in other words, BIN PACKING $\in$ **DPTAS**. This result draws another, although less dramatical than the one in Section 15.3.1.1, difference between standard and differential approximation. In the former paradigm, BIN PACKING is solved by an *asymptotic* polynomial time approximation schema, more precisely within standard approximation ratio $1 + \epsilon + (1/\operatorname{opt}(L))$, for any $\epsilon > 0$ ([32]), but it is **NP**-hard to approximate it by a "real" polynomial time approximation schema ([2]).

Consider a list $L = \{x_1, \ldots, x_n\}$, instance of BIN PACKING, assume, without loss of generality, that items in L are rational numbers ranged in decreasing order and fix an optimum solution B^* of L. Observe that $\omega(L) = n$. For the purposes of this section, a bin i will be denoted either by b_i, or by explicit listing of the numbers placed in it; finally, any solution will be alternatively represented as union of its bins.

Theorem 15.2 *From any algorithm achieving standard approximation ratio ρ for* BIN PACKING*, can be derived an algorithm achieving differential approximation ratio $\delta = 2 - \rho$.*

Sketch of the proof: Let k^* be the number of bins in B^* that contain a single item. Then, it is easy to see that there exists an optimum solution $\bar{B}^* = \{x_1\} \cup \ldots \cup \{x_{k^*}\} \cup \bar{B}_2^*$ for L, where any bin in $\bar{B}_2^*$ contains at least two items. Furthermore, one can show that, for any optimum solution $\hat{B} = \{b_j : j = 1, \ldots, \operatorname{opt}(L)\}$ and for any set $J \subset \{1, \ldots, \operatorname{opt}(L)\}$, the solution $B_j = \{b_j \in B : j \in J\}$ is optimum for the sub-list $L_j = \cup_{j \in J} b_j$.

Consider now Algorithm SA achieving standard approximation ratio ρ for BIN PACKING, denote by SA(L) the solution computed by it, when running on an instance L (recall that L is assumed ranged in decreasing order), and run the following algorithm, denoted by DA in the sequel, which uses SA as sub-procedure:

1. for $k = 1$ to n set: $L_k = \{x_{k+1}, \ldots, x_n\}$, $B_k = \{x_1\} \cup \ldots \cup \{x_k\} \cup \text{SA}(L_k)$;
2. output $B = \operatorname{argmin}\{|B_k| : k = 0, \ldots, n - 1\}$.

Let $\bar{B}^*$ be the optimum solution claimed above. Then, $\bar{B}_2^*$ is an optimum solution for the sub-list L_{k^*}. Observe that Algorithm SA called by DA has also been executed on L_{k^*} and denote by B_{k^*} the solution so computed by DA. The solution returned in Step 2 verifies $|B| \leq |B_{k^*}|$. Finally, as any bin in $\bar{B}_2^*$ contains at least two items, $|L_{k^*}| = n - k^* \geq 2 \operatorname{opt}(L_{k^*})$. Putting all this together, we get: $\delta_{\text{DA}}(L, B) = (n - |B|)/(n - \operatorname{opt}(L)) \geq (|L_{k^*}| - |B_{k^*}|)/(|L_{k^*}| - \operatorname{opt}(L_{k^*})) \geq 2 - \rho$. ∎

In what follows, denote by SA any polynomial algorithm approximately solving BIN PACKING within (fixed) constant standard approximation ratio ρ, by ASCHEMA(ϵ) the asymptotic polynomial time standard approximation schema of Reference 32, parameterized by $\epsilon > 0$, and consider the following algorithm, DSCHEMA (L is always assumed ranged in decreasing order):

1. fix a constant $\epsilon > 0$ and set $\eta = \lfloor 2(\rho - 1 + \epsilon)/\epsilon^2 \rfloor$;
2. for $k = n - \eta + 1, \ldots, n$ build list L_{k-1} where L_{k-1} is as in Step 1 of Algorithm DA (Theorem 15.2);
3. for any list L_i computed in Step 2, perform an exhaustive search on L_i, denote by E_i the solution so computed, and set $B_i = \{\{x\} : x \in L \backslash L_i\} \cup E_i$;
4. store B, the smallest of the solutions computed in Step 3;
5. run DA both with SA and ASCHEMA($\epsilon/2$), respectively, as sub-procedures on L;
6. output the best among the three solutions computed in Steps 4 and 5.

Theorem 15.3 ([31]) *Algorithm* DSCHEMA *is a polynomial time differential approximation schema for* BIN PACKING. *So,* BIN PACKING $\in$ **DPTAS**.

Sketch of the proof: As ρ and ϵ do not depend on n, neither does η, computed at Step 1. One can then show that when dealing with a list L such that $|L_{k^*+1}| \leq \eta$, BIN PACKING can be solved in polynomial time when η is a fixed constant. One the other hand, assuming $|L_{k^*+1}| \geq \eta$, then, one can prove that, if $\mathrm{opt}(L_{k^*+1}) \leq \epsilon|L_{k^*+1}|/(\rho_{\mathrm{SA}} - 1 + \epsilon)$, the approximation ratio of algorithm DA, when calling SA as sub-procedure, is $\delta \geq 1 - \epsilon$ whereas, if $\mathrm{opt}(L_{k^*+1}) \geq \epsilon|L_{k^*+1}|/(\rho_{\mathrm{SA}} - 1 + \epsilon)$, then the approximation ratio of algorithm DA, when calling ASCHEMA($\epsilon/2$) as sub-procedure, is also $\delta \geq 1 - \epsilon$. So, when $|L_{k^*+1}| \geq 2(\rho - 1 + \epsilon)/\epsilon^2$, Step 5 of DSCHEMA achieves differential approximation ratio $1 - \epsilon$. Putting things together derives the result. ∎

Let us note that, as we will see in Section 15.5.4, BIN PACKING is **DPTAS**-complete; consequently, unless **P** $=$ **NP** it is inapproximable by fully polynomial time differential approximation schemata. Inapproximability of BIN PACKING by such schemata has independently been shown also in Reference 19.

15.3.2 Traveling Salesman Problems

MIN TSP is one of the most paradigmatic problems in combinatorial optimization and one of the hardest one to approximate. Indeed, unless **P** $=$ **NP**, no polynomial algorithm can guarantee, on an edge-weighted complete graph of size n when no restriction is imposed to the edge-weights, standard approximation ratio $O(2^{p(n)})$, for any polynomial p. As we will see in this section things are completely different when dealing with differential approximation where MIN TSP $\in$ **DAPX**. This result draws another notorious difference between the two paradigms.

Consider an edge-weighted complete graph of order n, denoted by K_n and observe that the worst MIN TSP-solution in K_n is an optimum solution for MAX TSP. Consider the following algorithm (originally proposed by Reference 33 for MAX TSP) based on a careful patching of the cycles of a minimum-weight 2-matching* of K_n:

- Compute a $M = (C_1, C_2, \ldots, C_k)$; denote by $\{v_i^j : j = 1, \ldots, k, i = 1, \ldots, |C_j|\}$, the vertex-set of C_j; if $k = 1$, return M;
- For any C_j, pick arbitrarily four consecutive vertices v_i^j, $i = 1, \ldots, 4$; if $|C_j| = 3$, $v_4^j = v_1^j$; for C_k (the last cycle of M), pick also another vertex, denoted by u that is the other neighbor of v_1^k in C_k (hence, if $|C_k| = 3$, then $u = v_3^k$ whereas if $|C_k| = 4$, then $u = v_4^k$);
- If k is even (odd), then set:
  - $R_1 = \cup_{j=1}^{k-1}\{(v_2^j, v_3^j)\} \cup \{(v_1^k, v_2^k)\}$, $A_1 = \{(v_1^k, v_3^1), (v_2^1, v_2^2)\} \cup_{j=1}^{(k-2)/2} \{(v_3^{2j}, v_3^{2j+1}), (v_2^{2j+1}, v_2^{2j+2})\}$
    $(R_1 = \cup_{j=1}^{k}\{(v_2^j, v_3^j)\}, A_1 = \{(v_2^k, v_1^1)\} \cup_{j=1}^{(k-1)/2} \{(v_2^{2j-1}, v_2^{2j}), (v_3^{2j}, v_3^{2j+1})\})$, $T_1 = (M\backslash R_1) \cup A_1$;
  - $R_2 = \cup_{j=1}^{k-1}\{(v_1^j, v_2^j)\} \cup \{(u, v_1^k)\}$, $A_2 = \{(u, v_2^1), (v_1^1, v_1^2)\} \cup_{j=1}^{(k-2)/2} \{(v_2^{2j}, v_2^{2j+1}), (v_1^{2j+1}, v_1^{2j+2})\}$
    $(R_2 = \cup_{j=1}^{k}\{(v_1^j, v_2^j)\}, A_2 = \{(v_1^k, v_2^1)\} \cup_{j=1}^{(k-1)/2} \{(v_1^{2j-1}, v_1^{2j}), (v_2^{2j}, v_2^{2j+1})\})$, $T_2 = (MM\backslash R_2) \cup A_2$;
  - $R_3 = \cup_{j=1}^{k-1}\{(v_3^j, v_4^j)\} \cup \{(v_2^k, v_3^k)\}$, $A_3 = \{(v_2^k, v_4^1), (v_3^1, v_3^2)\} \cup_{j=1}^{(k-2)/2} \{(v_4^{2j}, v_4^{2j+1}), (v_3^{2j+1}, v_3^{2j+2})\}$
    $(R_3 = \cup_{j=1}^{k}\{(v_3^j, v_4^j)\}, A_3 = \{(v_3^k, v_4^1)\} \cup_{j=1}^{(k-1)/2} \{(v_3^{2j-1}, v_3^{2j}), (v_4^{2j}, v_4^{2j+1})\})$, $T_3 = (MM\backslash R_3) \cup A_3$;
- Output T the best among T_1, T_2 and T_3.

As it is proved in Reference 33, the set $(M\backslash \cup_{i+1}^{3} R_i) \cup_{i+1}^{3} A_i$ is a feasible solution for MIN TSP, the value of which is a lower bound for $\omega(K_n)$; furthermore, $m(K_n, T) \leq (\sum_{i=1}^{3} m(K_n, T_i))/3$. Then, a smart analysis,

---

* A minimum total weight partial subgraph of K_n any vertex of which has degree at most 2; this computation is polynomial, see, for example Reference 34; in other words a 2-matching is a collection of paths and cycles, but when dealing with complete graphs a 2-matching can be considered as a collection of cycles.

leads to the following theorem (the same result has been obtained, by a different algorithm working also for negative edge-weights, in Reference 13).

Theorem 15.4 ([33]) MIN TSP *is differentially 2/3-approximable.*

Notice that MIN TSP, MAX TSP, MIN METRIC TSP, and MAX METRIC TSP are all affine equivalent (see Reference 35 for the proof; for the two former problems, just replace weight $d(i,j)$ of edge (v_i, v_j) by $M - d(i,j)$, where M is some number greater than the maximum edge weight). Hence, the following theorem holds:

Theorem 15.5 MIN TSP, MAX TSP, MIN METRICN TSP *and* MAX METRIC TSP *are differentially 2/3-approximable.*

A very famous restrictive version of MIN METRIC TSP is the MIN TSP12, where edge-weights are all either 1, or 2. In Reference 36, it is proved that this version (as well as, obviously, MAX TSP12) is approximable within differential ratio 3/4.

15.3.3 Min Multiprocessor Scheduling

We now deal with a classical scheduling problem, the MIN MULTIPROCESSOR SCHEDULING ([37]), where we are given n tasks $t_1, \ldots, t_n$ with (execution) time lengths $l(t_j)$, $j = 1, \ldots, n$, polynomial with n, that have to be executed on m processors, and the objective is to partition these tasks on the processors in such a way that the occupancy of the busiest processor is minimized. Observe that the worst solution is the one where all the tasks are executed in the same processor; so, given an instance x of MIN MULTIPROCESSOR SCHEDULING, $\omega(x) = \sum_{j=1}^{n} l(t_j)$. A solution y of this problem will be represented as a vector in $\{0, 1\}^{mn}$, the non-zero components y_j^i of which correspond to the assignment of task j to processor i.

Consider a simple local search algorithm that starts from some solution and improves upon any change of the assignment of a single task from one processor to another one. Then the following result can be obtained ([38]):

Theorem 15.6 MIN MULTIPROCESSOR SCHEDULING *is approximable within differential ratio $m/(m+1)$.*

Sketch of the proof: Assume that both tasks and processors are ranged with decreasing lengths and occupancies, respectively. Denote by $l(p_i)$, the total occupancy of processor p_i, $i = 1, \ldots, m$. Then, $\text{opt}(x) \geq l(t_1)$ and $l(p_1) = \sum_{j=1}^{n} y_j^1 l(t_j) = \max_{i=1,\ldots,m}\{l(p_i) = \sum_{j=1}^{n} y_j^i l(t_j)\}$. Denote, w.l.o.g., by $1, \ldots, q$, the indices of the tasks assigned to p_1. As y is a local optimum, it verifies, for $i = 2, \ldots, m, j = 1, \ldots q$: $l(t_j) + l(p_i) \geq l(p_1)$. We can assume $q \geq 2$ (on the contrary y is optimum). Then, adding the preceding expression for $j = 1, \ldots, q$, we get: $l(p_i) \geq l(p_1)/2$. Also, adding $l(p_1)$ with the preceding expression for $l(p_i, i = 2, \ldots, m$, we obtain: $\omega(x) \geq (m + 1)l(p_1)/2$. Putting all this together we finally get $m(x, y) = l(p_1) \leq (m \, \text{opt}(x)/(m + 1)) + (\omega(x)/(m + 1))$. ∎

15.4 Asymptotic Differential Approximation Ratio

In any approximation paradigm, the notion of asymptotic approximation (dealing, informally, with a class of "interesting" instances) is pertinent. In the standard paradigm, the asymptotic approximation ratio is defined on the hypothesis that the interesting (from an approximation point of view) instances of the simple problems are the ones whose values of the optimum solutions tend to ∞ (because, in the opposite case,[*] these problems, called *simple* ([39]), are polynomial). In the differential approximation framework on the contrary, the size (or the value) of the optimum solution is not always a pertinent hardness criterion (see Reference 40 for several examples about this claim). Henceforth, in Reference 40, another

---

[*] The case where optimum values are bounded by fixed constants.

hardness criterion, *the number $\sigma(x)$ of the feasible values of x*, has been used to introduce the *asymptotic differential approximation ratio*. Under this criterion, the asymptotic differential approximation ratio of an algorithm A is defined as

$$\delta_A^\infty(x, y) = \lim_{k \to \infty} \inf_{\substack{x \\ \sigma(x) \geq k}} \left\{ \frac{\omega(x) - m(x, y)}{\omega(x) - \mathrm{opt}(I)} \right\} \tag{15.1}$$

Let us note that $\sigma(x)$ is motivated by, and generalizes, the notion of the *structure of the instance* introduced in Reference 9. We also notice that the condition $\sigma(x) \geq k$ characterizing "the sequence of unbounded instances" (or "limit instances") cannot be polynomially verified.* But in practice, for a given problem, it is possible to directly interpret condition $\sigma(x) \geq k$ by means of the parameters $\omega(x)$ and $\mathrm{opt}(x)$ (note that $\sigma(x)$ is not a function of these values). For example, for numerous cases of discrete problems, it is possible to determine, for any instance x, a step $\pi(x)$ defined as the least variation between two feasible values of x. For example, for BIN PACKING, $\pi(x) = 1$. Then, $\sigma(x) \leq ((\omega(x) - \mathrm{opt}(x))/\pi(x)) + 1$. Therefore, from (15.1):

$$\delta_A^\infty(x, y) \geq \lim_{k \to \infty} \inf_{\substack{x \\ \frac{\omega(x) - \mathrm{opt}(x)}{\pi(x)} \geq k-1}} \left\{ \frac{\omega(x) - m(x, y)}{\omega(x) - \mathrm{opt}(x)} \right\}$$

Whenever π can be determined, condition $(\omega(x) - \mathrm{opt}(x))/\pi(x) \geq k - 1$ can be easier to evaluate than $\sigma(x) \geq k$, and in this case, the former condition is used (this is not senseless as we try to bound below the ratio).

The adoption of $\sigma(x)$ as hardness criterion can be motivated by considering a class of problems, called *radial problems* in Reference 40, that includes many well-known combinatorial optimization problems, as BIN PACKING, MAX INDEPENDENT SET, MIN VERTEX COVER, MIN COLORING, and so on. Informally, a problem Π is radial if, given an instance x of Π and a feasible solution y for x, one can, in polynomial time, on the one hand, deteriorate y as much as one wants (up to finally obtain a worst-value solution) and, on the other hand, greedily improve y to obtain (always in polynomial time) a sub-optimal solution (eventually the optimum one).

Definition 15.2 *A problem $\Pi = (\mathcal{I}, \mathrm{Sol}, m, \mathrm{goal})$ is radial if there exists three polynomial algorithms ξ, ψ, and ϕ such that for any $x \in \mathcal{I}$:*

1. *ξ computes a feasible solution $y^{(0)}$ for x;*
2. *for any feasible solution y of x strictly better (in the sense of the value) than $y^{(0)}$, algorithm ϕ computes a feasible solution $\phi(y)$ (if any) with $m(x, \phi(y))$ strictly worse than $m(x, y)$;*
3. *for any feasible solution y of x with value strictly better than $m(x, y^{(0)})$, there exists $k \in \mathbb{N}$ such that $\phi^k(y) = y^{(0)}$ (where ϕ^k denotes the k-times iteration of ϕ);*
4. *for a solution y such that, either $y = y^{(0)}$, or y is any feasible solution of x with value strictly better than $m(x, y^{(0)})$, $\psi(y)$ computes the set of ancestors of y, defined by $\psi(y) = \phi^{-1}(\{y\}) = \{z : \phi(z) = y\}$ (this set being eventually empty).*

Let us note that the class of radial problems includes, in particular, the well-known class of *hereditary* problems for which any subset of a feasible solution remains feasible. In fact, for an hereditary (maximization) problem, a feasible solutions y is a subset of the input-data, for any instances x, $y^{(0)} = \emptyset$ and for any other feasible solution y, $\phi(y)$ is just obtained from y by removing a component of y. The hereditary notion deals with problems for which a feasible solution is a subset of the input-data, whereas the radial notion allows problems for which solutions are also second-order structures of the input-data.

---

* The same holds for the condition $\mathrm{opt}(x) \geq k$ induced by the hardness criterion in the standard paradigm.

Proposition 15.1 ([40]) *Let κ be a fixed constant and consider a radial problem Π such that, for any instance x of Π of size n, $\sigma(x) \leq \kappa$. Then, Π is polynomial time solvable.*

15.5 Structure in Differential Approximation Classes

What has been discussed in the previous sections makes clear that the entire theory of approximation, that tries to characterize and to classify problems with respect to their approximability hardness, can be redone in the differential paradigm. There exist problems having several differential approximability levels and inapproximability bounds. What follows further confirms this claim. It will be shown that the approximation paradigm we deal with, allows to devise its proper tools and to use them to design an entire structure for the approximability classes involved.

15.5.1 Differential NPO-Completeness

Obviously, the *strict* reduction of Reference 41 (see also Chapter 14), can be identically defined in the framework of the differential approximation; for clarity, we denote this derivation of the strict reduction by D-reduction. Two **NPO** problems will be called D-equivalent if there exist D-reductions from any of them to the other one.

Theorem 3.1 in Reference 41 (where the differential approximation ratio is mentioned as possible way of estimating the performance of an algorithm), based on an extension of Cook's proof ([42]) of SAT **NP**-completeness to optimization problems, works also when the differential ratio is dealt instead the standard one. Furthermore, solution *triv*, as defined in Reference 41 is indeed a worst solution for MIN WSAT. On the other hand, the following proposition holds:

Proposition 15.2 ([43]) MAX WSAT *and* MIN WSAT *are D-equivalent.*

Sketch of the proof: With any clause $\ell_1 \vee \ldots \vee \ell_t$ of an instance ϕ of MAX WSAT, we associate in the instance ϕ' of MIN WSAT the clause $\bar{\ell}_1 \vee \ldots \vee \bar{\ell}_t$. Then, if an assignment y satisfies the instance ϕ, the complement y' of y satisfies ϕ', and vice-versa. So, $m(\phi, y) = \sum_{i=1}^{n} w(x_i) - m(\phi', y')$, for any y'. Thus, $\delta(\phi, y) = \delta(\phi', y')$. The reduction from MIN WSAT to MAX WSAT is completely analogous. ∎

In a completely analogous way, as in Proposition 15.2, it can be proved that MIN 0-1 INTEGER PROGRAMMING and MAX 0-1 INTEGER PROGRAMMING are also D-equivalent. Putting all the above together the following holds:

Theorem 15.7 MAX WSAT, MIN WSAT, MIN 0-1 INTEGER PROGRAMMING, *and* MAX 0-1 INTEGER PROGRAMMING *are* **NPO***-complete under D-reducibility.*

15.5.2 The Class 0-DAPX

Informally, the class **0-DAPX** is the class of **NPO** problems for which the differential ratio of any polynomial time algorithm is *equal* to 0. In other words, for any such algorithm, there exists an instance in which it will compute its worst solution. Such situation draws the worst case for the differential approximability of a problem. Class **0-DAPX** is defined in Reference 43 by means of a reduction, called G-reduction. It can be seen as a particular kind of the GAP-reduction ([1,44,45]).

Definition 15.3 *A problem Π is said to be G-reducible to a problem Π', if there exists a polynomial reduction that transforms any δ-differential approximation algorithm for Π', $\delta > 0$ into an optimum (exact) algorithm for Π.*

Let Π be an **NP**-complete decision problem and Π' an **NPO** problem. The underlying idea for $\Pi \leq_G \Pi'$ in definition 15.3 is, starting from an instance of Π, to construct instances for Π' that have only two distinct feasible values and to prove that any differential δ-approximation for Π', $\delta > 0$, could distinguish between positive instances and negative instances for Π. Note finally that the G-reduction generalizes both the D-reduction of Section 15.5.1 and the strict reduction of Reference 41.

Definition 15.4 *0-DAPX is the class of NPO problems Π' for which there exists an NP-complete problem Π G-reducible to Π'. A problem is said to 0-DAPX-hard, if any problem in 0-DAPX G-reduces to it.*

An obvious consequence of Definition 15.4 is that *0-DAPX is the class of NPO problems Π for which approximation within any differential approximation ratio $\delta > 0$ would entail P = NP.*

Proposition 15.3 ([46]) MIN INDEPENDENT DOMINATING SET $\in$ *0-DAPX*.

Sketch of the proof: Given an instance ϕ of SAT with n variables $x_1, \ldots, x_n$ and m clauses $C_1, \ldots, C_m$, construct a graph G, instance of MIN INDEPENDENT DOMINATING SET associating with any positive literal x_i a vertex u_i and with any negative literal $\bar{x}_i$ a vertex v_i. For $i = 1, \ldots, n$, draw edges (u_i, v_i). For any clause C_j, add in G a vertex w_j and an edge between w_j and any vertex corresponding to a literal contained in C_j. Finally, add edges in G to obtain a complete graph on $w_1, \ldots, w_m$. An independent set of G contains at most $n+1$ vertices. An independent dominating set containing the vertices corresponding to true literals of a non satisfiable assignment and one vertex corresponding to a clause not satisfied by this assignment, is a worst solution of G of size $n + 1$. If ϕ is satisfiable then opt$(G) = n$. If ϕ is not satisfiable then opt$(G) = n + 1$. So, any independent dominating set of G has cardinality either n, or $n + 1$. $\blacksquare$

By analogous reductions, restricted versions of optimum-weighted satisfiability problems are proved **0-DAPX** in Reference 47.

Finally, the following relationship between **NPO** and **0-DAPX** holds:

Theorem 15.8 ([43]) *Under D-reducibility, NPO-complete = 0-DAPX-complete $\subseteq$ 0-DAPX.*

If, instead of D, a stronger reducibility is considered, for instance, by allowing f and/or g to be multivalued in the strict reduction, then, under this type of reducibility, it can be proved that **NPO**-complete = **0-DAPX** ([43]).

15.5.3 DAPX- and Poly-DAPX-Completeness

In this section we address the problem of completeness in the classes **DAPX** and **Poly-DAPX**. For this purpose, we first introduce a differential approximation schemata preserving reducibility, originally presented in Reference 43, called DPTAS-reducibility.

Definition 15.5 *Given two NPO problems Π and Π', Π DPTAS-reduces to Π' if there exist a (possibly) multi-valued function $f = (f_1, f_2, \ldots, f_h)$, where h is bounded by a polynomial in the input-length, and two functions g and c, computable in polynomial time, such that:*

- *for any $x \in \mathcal{I}_\Pi$, for any $\epsilon \in (0, 1) \cap \mathbb{Q}$, $f(x, \epsilon) \subseteq \mathcal{I}_{\Pi'}$;*
- *for any $x \in \mathcal{I}_\Pi$, for any $\epsilon \in (0, 1) \cap \mathbb{Q}$, for any $x' \in f(x, \epsilon)$, for any $y \in$ sol$_{\Pi'}(x')$, $g(x, y, \epsilon) \in$ sol$_\Pi(x)$;*
- *$c : (0, 1) \cap \mathbb{Q} \to (0, 1) \cap \mathbb{Q}$;*
- *for any $x \in \mathcal{I}_\Pi$, for any $\epsilon \in (0, 1) \cap \mathbb{Q}$, for any $y \in \cup_{i=1}^h$ sol$_{\Pi'}(f_i(x, \epsilon))$, $\exists j \leq h$ such that $\delta_{\Pi'}(f_j(x, \epsilon), y) \geq 1 - c(\epsilon)$ implies $\delta_\Pi(x, g(x, y, \epsilon)) \geq 1 - \epsilon$.*

15.5.3.1 DAPX-Completeness

If one restricts herself/himself to problems with polynomially computable worst solutions, then things are rather simple. Indeed, given such a problem $\Pi \in$ **DAPX**, it is affine equivalent to a problem Π' defined on the same set of instances and with the same set of solutions but, for any solution y of an instance x of Π, the measure for solution y with respect to Π' is defined as $m_{\Pi'}(x, y) = m_\Pi(x, y) - \omega(x)$. Affine equivalence of Π and Π' ensures that $\Pi' \in$ **DAPX**; furthermore, $\omega_{\Pi'}(x) = 0$. As, for the latter problem, standard and differential approximation ratios coincide, it follows that $\Pi' \in$ **APX**. MAX INDEPENDENT SET is **APX**-complete under PTAS-reducibility ([48]), a particular kind of the AP-reducibility seen in Chapter 14. So, Π' PTAS-reduces to MAX INDEPENDENT SET. Putting together affine equivalence between Π and Π', PTAS-reducibility between Π' and MAX INDEPENDENT SET, and taking into account that composition of these two reductions is an instantiation of DPTAS-reduction, we conclude the **DAPX**-completeness of MAX INDEPENDENT SET.

However, things become much more complicated, if one takes into account problems with nonpolynomially computable worst solutions. In this case, one needs more sophisticated techniques and arguments. We informally describe here the basic ideas and the proof-schema in Reference 43. It is first shown that any **DAPX** problem Π is reducible to MAX WSAT-B by a reduction transforming a polynomial time approximations schema for MAX WSAT-B into a polynomial time differential approximation schema for Π. For simplicity, denote this reduction by S-D. Next, a particular **APX**-complete problem Π' is considered, say MAX INDEPENDENT SET-B. MAX WSAT-B, that is in **APX**, is PTAS-reducible to MAX INDEPENDENT SET-B. MAX INDEPENDENT SET-B is both in **APX** and in **DAPX** and, moreover, standard and differential approximation ratios coincide for it; this coincidence draws a trivial reduction called ID-reduction. It trivially transforms a differential polynomial time approximation schema into a standard polynomial time approximation schema. The composition of the three reductions specified (i.e., the S-D-reduction from Π to MAX WSAT-B, the PTAS-reduction from MAX WSAT-B to MAX INDEPENDENT SET-B and the ID-reduction) is a DPTAS-reduction transforming a polynomial time differential approximation schema for MAX INDEPENDENT SET-B into a polynomial time differential approximation schema for Π, that is, MAX INDEPENDENT SET-B is **DAPX**-complete under DPTAS-reducibility.

Also, by standard reductions that turn out to be DPTAS-reductions also, the following can be proved ([30,43]).

Theorem 15.9 MAX INDEPENDENT SET-B, MIN VERTEX COVER-B, *for fixed B*, MAX k-SET PACKING, MIN k-SET COVER, *for fixed k, and* MIN COLORING *are* **DAPX**-*complete under DPTAS-reducibility.*

15.5.3.2 Poly-DAPX-Completeness

Recall that a maximization problem $\Pi \in$ **NPO** is *canonically hard for* **Poly-APX** ([49]), if and only if there exist a polynomially computable transformation T from 3SAT to Π, two constants n_0 and c and a function F, hard for **Poly**,[*] such that, given an instance x of 3SAT on $n \geq n_0$ variables and a number $N \geq n^c$, the instance $x' = T(x, N)$ belongs to $\mathcal{I}_\Pi$ and verifies the three following properties: (i) if x is satisfiable, then $\mathrm{opt}(x') = N$; (ii) if x is not satisfiable, then $\mathrm{opt}(x') = N/F(N)$; and (iii) given a solution $y \in \mathrm{sol}_\Pi(x')$ such that $m(x', y) > N/F(N)$, one can polynomially determine a truth assignment satisfying x.

Based on DPTAS-reducibility and the notion of canonical hardness, the following is proved in Reference 30.

Theorem 15.10 *If a (maximization) problem* $\Pi \in$ **NPO** *is canonically hard for* **Poly-APX**, *then any problem in* **Poly-DAPX** *DPTAS-reduces to* Π.

As it is shown in Reference 49, MAX INDEPENDENT SET is canonically hard for **Poly-APX**. Furthermore, MIN VERTEX COVER is affine equivalent to MAX INDEPENDENT SET. Henceforth, use of Theorem 15.10, immediately derives the following result:

Theorem 15.11 MAX INDEPENDENT SET *and* MIN VERTEX COVER *are complete for* **Poly-DAPX** *under DPTAS-reducibility.*

---

[*] The set of functions from $\mathbb{N}$ to $\mathbb{N}$ bounded by a polynomial; a function $f \in$ **Poly** is hard for **Poly**, if and only if there exists three constants k, c and n_0 such that, for any $n \geq n_0$, $f(n) \leq kF(n^c)$.

15.5.4 DPTAS-Completeness

Completeness in **DPTAS** is tackled by means of a kind of reducibility preserving membership in **DFPTAS**, which is called DFT-reducibility in Reference 30. This type of reducibility is the differential counterpart of the FT-reducibility introduced in Section 14.7 of Chapter 14 and can be defined in an exactly similar way. Based on DFT-reducibility, the following theorem holds ([30]; its proof is very similar to Theorem 7.4 in Chapter 14). Before stating it, we need to introduce the class of diameter polynomially bounded problems that is a subclass of the radial problems seen in Section 15.4. An **NPO** problem Π is *diameter polynomially bounded* if and only if, for any $x \in \mathcal{I}_\Pi$, $|\operatorname{opt}(x) - \omega(x)| \le q(|x|)$. The class of diameter polynomially bounded **NPO** problems will be denoted by **NPO-DPB**.

Theorem 15.12 ([30]) *Let Π' be an **NP**-hard problem **NPO-DPB**. Then, any problem in **NPO** is DFT-reducible to Π'. Consequently, (i) the closure of **DPTAS** under DFT-reductions is the whole **NPO** and (ii) any **NP**-hard problem in **NPO-DPB** $\cap$ **DPTAS** is **DPTAS**-complete under DFT-reductions.*

Consider now MIN PLANAR VERTEX COVER, MAX PLANAR INDEPENDENT SET, and BIN PACKING. They are all **NP**-hard and in **NPO-DPB**. Furthermore, they are all in **DPTAS** (for the first two problems, this is derived by the inclusion of MAX PLANAR INDEPENDENT SET in **PTAS** proved in Reference 50; for the third one, revisit Section 15.3.1.2). So, the following theorem holds and concludes this section ([30]):

Theorem 15.13 MAX PLANAR INDEPENDENT SET, MIN PLANAR VERTEX COVER, *and* BIN PACKING *are* **DPTAS**-*complete under DFT-reducibility.*

15.6 Discussion and Final Remarks

As we have already claimed in the beginning of Section 15.5, the entire theory of approximation can be reformulated in the differential paradigm. This paradigm has the diversity of the standard one, it has a nonempty scientific content and, to our opinion, it represents in some sense a kind of revival for the domain of the polynomial approximation.

Since the work in Reference 7, a great number of paradigmatic combinatorial optimization problems has been studied in the framework of the differential approximation. For instance, KNAPSACK has been studied in Reference 7 and revisited in Reference 13. MAX CUT, MIN CLUSTER, STACKER CRANE, MIN DOMINATING SET, MIN DISJOINT CYCLE COVER, and MAX ACYCLIC SUBGRAPH have been dealt in Reference 13. MIN FEEDBACK ARC SET is also studied in Reference 38 together with MIN FEEDBACK NODE SET. MIN VERTEX COVER and MAX INDEPENDENT SET are studied in References 7, 13. MIN COLORING is dealt in References 18, 23, 24, 26–30, whereas MIN WEIGHTED COLORING (where the input is a vertex-weighted graph and the weight of a color is the weight of the heaviest of its vertices) is studied in Reference 51 (see also Reference 52). MIN INDEPENDENT DOMINATING SET is dealt in Reference 46. BIN PACKING is studied in References 27, 31, 40, 53. MIN SET COVER, under several assumptions on its worst value, is dealt in References 7, 13, 54, whereas MIN WEIGHTED SET COVER is dealt in References 27, 54. MIN TSP and MAX TSP, as well as, several famous variants of them, MIN METRIC TSP, MAX METRIC TSP, MIN TSPab (the most famous restrictive case of this problem is MIN TSP12), and MAX TSPab are studied in References 13, 33, 35, 36, 55. STEINER TREE problems under several assumptions on the form of the input-graph and on the edge-weights are dealt in Reference 56. Finally, several optimum satisfiability and constraint satisfaction problems (as MAX SAT, MAX E2SAT, MAX 3SAT, MAX E3SAT, MAX EkSAT, MIN SAT, MIN kSAT, MIN EkSAT, MIN 2SAT, and their corresponding constraint satisfaction versions) are studied in Reference 57.

Dealing with structural aspects of approximation, besides the existing approximability classes (defined rather upon combinatorial arguments) two logical classes have been very notorious in the standard paradigm. These are **Max-NP** and **Max-SNP**, originally introduced in Reference 58 (see also Chapters 14 and 17). Their definitions, independent from any approximation ratio consideration, make that they can identically be considered also in differential approximation. In the standard paradigm, the following strict inclusions hold: **PTAS** ⊂ **Max-SNP** ⊂ **APX** and **MAX-NP** ⊂ **APX**. As it is proved in Reference 57, MAX SAT ∉ **DAPX**, unless **P = NP**. This draws an important structural difference in the landscape of approximation classes in the two paradigms, as an immediate corollary of this result is that **MAX-NP** ⊄ **DAPX**. Position of **Max-SNP** in the differential landscape is not known yet. It is conjectured, however, that **MAX-SNP** ⊄ **DAPX**. In any case, formal relationships of **Max-SNP** and **Max-NP** with the other differential approximability classes deserve further study.

References

1. Ausiello, G., Crescenzi, P., Gambosi, G., Kann, V., Marchetti-Spaccamela, A., and Protasi, M., *Complexity and Approximation. Combinatorial Optimization Problems and their Approximability Properties*, Springer, Berlin, Germany, 1999.
2. Garey, M.R. and Johnson, D.S., *Computers and Intractability. A Guide to the Theory of NP-completeness*, W. H. Freeman, San Francisco, CA, 1979.
3. Berge, C., *Graphs and Hypergraphs*, North Holland, Amsterdam, the Netherlands, 1973.
4. Bollobás, B., *Random Graphs*, Academic Press, London, UK, 1985.
5. Sahni, S. and Gonzalez, T., P-complete approximation problems, *JACM*, 23, 555, 1976.
6. Håstad, J., Clique is hard to approximate within $n^{1-\epsilon}$, *Acta Mathematica*, 182, 105, 1999.
7. Demange, D. and Paschos, V.Th., On an approximation measure founded on the links between optimization and polynomial approximation theory, *Theor. Comput. Sci.*, 158, 117, 1996.
8. Ausiello, G., D'Atri, A., and Protasi, M., On the structure of combinatorial problems and structure preserving reductions, in *Proceedings of 4th International Colloquium on Automata, Languages, and Programming*, LNCS, Vol. 52, Springer-Verlag, Berlin, Germany, p. 45, 1977.
9. Ausiello, G., D'Atri, A., and Protasi, M., Structure preserving reductions among convex optimization problems, *JCSS*, 21, 136, 1980.
10. Aiello, A., Burattini, E., Furnari, M., Massarotti, A., and Ventriglia, F., Computational complexity: The problem of approximation, in Colloquia Mathematica Societatis, *Algebra, Combinatorics, and Logic in Computer Science*, Bolyai, J. (Ed.), Vol. I, Elsevier, the Netherlands, p. 51, 1986.
11. Zemel, E., Measuring the quality of approximate solutions to zero-one programming problems, *Math. Oper. Res.*, 6, 319, 1981.
12. Cornuejols, G., Fisher, M.L., and Nemhauser, G.L., Location of bank accounts to optimize float: An analytic study of exact and approximate algorithms, *Manage. Sci.*, 23(8), 789, 1977.
13. Hassin, R. and Khuller, S., z-approximations, *J. Algor.*, 41, 429, 2001.
14. Bellare, M. and Rogaway, P., The complexity of approximating a nonlinear program, *Math. Program.*, 69, 429, 1995.
15. Nemirovski, A.S. and Yudin, D.B., *Problem Complexity and Method Efficiency in Optimization*, Wiley, Chichester, UK, 1983.
16. Vavasis, S.A., Approximation algorithms for indefinite quadratic programming, *Math. Program.*, 57, 279, 1992.
17. Ausiello, G., D'Atri, A., and Protasi, M., Lattice-theoretical ordering properties for NP-complete optimization problems, *Fundamenta Informaticæ*, 4, 83, 1981.
18. Hassin, R. and Lahav, S., Maximizing the number of unused colors in the vertex coloring problem, *Inform. Process. Lett.*, 52, 87, 1994.
19. Demange, M. and Paschos, V.Th., Improved approximations for maximum independent set via approximation chains, *Appl. Math. Lett.*, 10(3), 105, 1997.
20. Demange M. and Paschos, V.Th., *Improved approximations for weighted and unweighted graph problems*, Theory of Computing Systems 38, 2005.
21. Halldórsson, M.M., Approximations of weighted independent set and hereditary subset problems, *J. Graph Algor. Appli.*, 4(1), 1, 2000.

22. Monnot, J., *Critical Families of Instances and Polynomial Approximation*, PhD Thesis, LAMSADE, University Paris-Dauphine, Paris, France, 1998 (in French).

23. Halldórsson, M.M., Approximating k-set cover and complementary graph coloring, in *Proceedings of International Integer Programming and Combinatorial Optimization Conference*, LNCS, Vol. 1084, Springer Verlag, Berlin, Germany, p. 118, 1996.

24. Duh, R. and Fürer, M., Approximation of k-set cover by semi-local optimization, in *Proceedings of Symposium on Theory of Computing*, p. 256, 1997.

25. Paschos, V.Th., Polynomial approximation and graph coloring, *Computing*, 70, 41, 2003.

26. Demange, M., Grisoni, P., and Paschos, V.Th., Approximation results for the minimum graph coloring problem, *Inform. Process. Lett.*, 50, 19, 1994.

27. Demange, M., Grisoni, P., and Paschos, V.Th., Differential approximation algorithms for some combinatorial optimization problems, *Theor. Comput. Sci.*, 209, 107, 1998.

28. Halldórsson, M.M., Approximating discrete collections via local improvements, in *Proceedings of Sixth Annual ACM-SIAM Symposium on Discrete Algorithms*, p. 160, 1995.

29. Tzeng, X.D. and King, G.H., Three-quarter approximation for the number of unused colors in graph coloring, *Inform. Sci.*, 114, 105, 1999.

30. Bazgan, C., Escoffier, B., and Paschos, V.Th., Completeness in standard and differential approximation classes: Poly-(D)APX- and (D)PTAS-completeness, *Theor. Comput. Sci.*, 339, 272, 2005.

31. Demange, M., Monnot, J., and Paschos, V.Th., Bridging gap between standard and differential polynomial approximation: The case of bin-packing, *Appl. Math. Lett.*, 12, 127, 1999.

32. Fernandez, D.L.V.W. and Lueker, G.S., Bin packing can be solved within $1 + \epsilon$ in linear time, *Combinatorica*, 1(4), 349, 1981.

33. Monnot, J., Differential approximation results for the traveling salesman and related problems, *Inform. Process. Lett.*, 82(5), 229, 2002.

34. Cook, W.J., Cunningham, W.H., Pulleyblank, W.R., and Schrijver, A., *Combinatorial Optimization*, John Wiley & Sons, New York, 1998.

35. Monnot, J., Paschos, V.Th., and Toulouse, S., Approximation algorithms for the traveling salesman problem, *Math. Meth. Oper. Res.*, 57(1), 387, 2003.

36. Monnot, J., Paschos, V.Th., and Toulouse, S., Differential approximation results for the traveling salesman problem with distances 1 and 2. *European J. Oper. Res.*, 145(3), 557, 2002.

37. Hochbaum, D.S. and Shmoys, D.B., Using dual approximation algorithms for scheduling problems: Theoretical and practical results, *JACM*, 34, 144, 1987.

38. Monnot, J., Paschos, V.Th., and Toulouse, S., Optima locaux garantis pour l'approximation différentielle, *Technique et Science Informatiques*, 22(3), 257, 2003.

39. Paz, A. and Moran, S., Non deterministic polynomial optimization problems and their approximations, *Theor. Comput. Sci.*, 15, 251, 1981.

40. Demange, M. and Paschos, V.Th., Asymptotic differential approximation ratio: Definitions, motivations and application to some combinatorial problems, *RAIRO Oper. Res.*, 33, 481, 1999.

41. Orponen. P. and Mannila, H., On approximation preserving reductions: Complete problems and robust measures, Technical Report C-1987-28, Department of Computer Science, University of Helsinki, Finland, 1987.

42. Cook, S.A., The complexity of theorem-proving procedures, In *Proceedings of Symposium on Theory of Computing*, p. 151, 1971.

43. Ausiello, G., Bazgan, C., Demange, M., and Paschos, V.Th., Completeness in differential approximation classes, *Inter. J. Found. Comp. Sci.*, 16(6), 1267, 2005.

44. Arora, S. and Lund, C., Hardness of approximation, In *Approximation Algorithms for NP-hard Problems*, Chapter 10, Hochbaum, D.S. (Ed.), PWS, Boston, MA, pp. 399–446, 1997.

45. Vazirani, V., *Approximation Algorithms*, Springer, Berlin, Germany, 2001.

46. Bazgan, C. and Paschos, V.Th., Differential approximation for optimal satisfiability and related problems, *European J. Oper. Res.*, 147(2), 397, 2003.

47. Paschos, V.Th., *Complexité et Approximation Polynomiale*, Hermes, Paris, France, 2004.
48. Crescenzi, P. and Trevisan, L., On approximation scheme preserving reducibility and its applications, in *Foundations of Software Technology and Theoretical Computer Science, FST-TCS*, LNCS, Vol. 880, Springer-Verlag, Berlin, Germany, p. 330, 1994.
49. Khanna, S., Motwani, R., Sudan, M., and Vazirani, U., On syntactic versus computational views of approximability, *SIAM J. Comput.*, 28, 164, 1998.
50. Baker, B.S., Approximation algorithms for NP-complete problems on planar graphs, *JACM*, 41(1), 153, 1994.
51. Demange, M., de Werra, D., Monnot, J., and Paschos, V.Th., Weighted node coloring: When stable sets are expensive, in *Proceedings of the 28th International Workshop on Graph Theoretical Concepts in Computer Science, WG'02*, Kučera, L. (Ed.), LNCS, Vol. 2573, Springer-Verlag, Berlin, Germany, p. 114, 2002.
52. Demange, M., de Werra, D., Monnot, J., and Paschos, V.Th., Time slot scheduling of compatible jobs, Cahier du LAMSADE 182, LAMSADE, Université Paris-Dauphine, Paris, France, 2001. Available on http://www.lamsade.dauphine.fr/cahdoc.html#cahiers.
53. Demange, M., Monnot, J., and Paschos, V.Th., Maximizing the number of unused bins, *Found. Comput. Decision Sci.*, 26(2), 169, 2001.
54. Bazgan, C., Monnot, J., Paschos, V.Th., and Serrière, F., On the differential approximation of MIN SET COVER, *Theor. Comput. Sci.*, 332, 497, 2005.
55. Toulouse, S., *Approximation polynomiale: Optima locaux et rapport différentiel*, Thèse de doctorat, LAMSADE, Université Paris-Dauphine, Paris, France, 2001.
56. Demange, M., Monnot, J., and Paschos, V.Th., Differential approximation results for the steiner tree problem, *Appl. Math. Lett.*, 16, 733, 2003.
57. Escoffier, B. and Paschos, V.Th., Differential approximation of MIN SAT, MAX SAT and related problems, *European J. Oper. Res.*, 181(2), 620–633, 2007.
58. Papadimitriou, C.H. and Yannakakis, M., Optimization, approximation and complexity classes. *JCSS*, 43, 425, 1991.

II

Local Search, Neural Networks, and Metaheuristics

16

Local Search

Roberto Solis-Oba[*]

Nasim Samei

16.1 Introduction

Typically, a combinatorial optimization problem involves a set E of elements (called *ground set*) and the goal is to arrange, group, order, or select a subset of elements from E that optimizes a given objective function. Classical examples of combinatorial optimization problems include the minimum spanning tree problem, the shortest paths problem, and the traveling salesman problem [1].

Local search is perhaps one of the most natural ways to attempt to solve a combinatorial optimization problem. The idea of local search is simple: Given a (possibly not very good) solution s for a combinatorial optimization problem, try to improve the value of the solution by making "local changes" to s. A local change might involve adding elements from the ground set to s, removing elements from s, changing the way in which elements are grouped in s, or changing the order of the elements in s. If an improvement can be achieved in this manner, then a new solution s' is obtained. This process is continued until no further improvement can be obtained.

Local search has been successfully used to find good solutions for a large number of complex problems, including the seminal traveling salesman problem [1]. The empirical performance of local search algorithms has been extensively studied for a large number of problems in scheduling, Very Large Scale Integration (VLSI) design, network design, distributed planning and production control, and many other fields [2–6]. Most of these studies concluded that local search is a good method for efficiently computing near-optimum solutions to problems of realistic sizes [2,3,6,7]. In this chapter we explore the use of local search in the design of approximation algorithms with provable performance guarantee for NP-hard combinatorial optimization problems.

The idea of local search might be better understood by considering an example. In the *multiprocessor scheduling problem* the goal is to schedule a set $J = \{j_1, j_2, \ldots, j_n\}$ of jobs into a group $M = \{M_1, M_2, \ldots, M_m\}$ of m identical machines so that the completion time of the last job, also called the

---

[*] Research partially supported by the Natural Sciences and Engineering Research Council of Canada, grant 04667-2015.

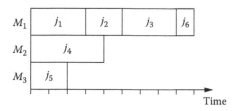

FIGURE 16.1 A schedule for *J*.

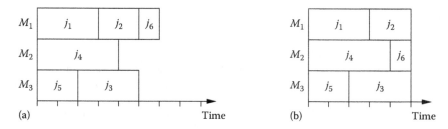

FIGURE 16.2 (a) Local improvement for the solution in Figure 16.1. (b) Local improvement for the solution in Figure 16.1.

makespan of the schedule, is minimized. Each job j_i has a processing time p_i, and every machine can process only one job at a time. Furthermore, we assume that the processing of a job cannot be interrupted (i.e., preemptions are not allowed).

Let us consider a specific instance of the multiprocessor scheduling problem. Let $J = \{j_1, j_2, j_3, j_4, j_5, j_6\}$ with processing times $p_1 = 3, p_2 = 2, p_3 = 3, p_4 = 4, p_5 = 2$, and $p_6 = 1$, and $M = \{M_1, M_2, M_3\}$. A solution for this instance of the multiprocessor scheduling problem is shown in Figure 16.1. Machine M_1 processes jobs j_1, j_2, j_3, and j_6 with total processing time $3 + 2 + 3 + 1 = 9$. This is the *load* of machine M_1. The loads of M_1 and M_3 are 4 and 2, respectively. The makespan or length of this schedule is equal to the completion time of the last job, and it is also equal to the maximum machine load, namely 9.

We can make local changes to this solution to try to improve the makespan. For example, we could move a job from one machine with maximum load to a machine with minimum load. If, say, we move j_3 from M_1 to M_3, we get the solution shown in Figure 16.2a with makespan 8. As this solution is better than the first one, we keep it and try to further improve it. Now we can move j_6 to M_3 to get the solution depicted in Figure 16.2b with makespan 5.

This last solution cannot be improved by moving any of the jobs to a different machine. In fact, one can show that this is an optimum solution for the problem as the total processing time of all the jobs is 15 and, thus, 3 machines need at least $15/3 = 5$ units of time to process them all.

16.1.1 Local Search and Combinatorial Optimization

Formally, a combinatorial optimization problem Π consists of a collection of instances (S, c). For each instance (S, c), S is the set of *feasible solutions* and it consists of a family of subsets from a finite ground set E. The second component, c of an instance is an objective function $c : S \rightarrow \mathbb{Q}$, where $\mathbb{Q}$ is the set of rational numbers. The goal of a combinatorial optimization problem is to find a solution $s^* \in S$ with minimum or maximum objective value, that is,

$$c(s^*) = \text{optimum } c(s),$$
$$s \in S$$

where optimum is either min or max.

A *neighborhood function* $\mathcal{N} : \mathcal{S} \rightarrow 2^{\mathcal{S}}$ specifies for each solution $s \in \mathcal{S}$ a subset $\mathcal{N}(S)$ of neighbors of s, or solutions that are "close" to s. The local search algorithm that we informally described in the previous section is called the *iterative improvement* algorithm:

Algorithm IterativeImprovement$(\mathcal{S}, \mathcal{N}, c)$:
In: Set $\mathcal{S}$ of feasible solutions, neighborhood function $\mathcal{N}$, and objective function c.
Out: A local optimum solution $s \in \mathcal{S}$ with respect to $\mathcal{N}$ and c.

1. Compute an initial feasible solution $s \in \mathcal{S}$.
2. **while** $\mathcal{N}(s)$ contains a solution of better cost that s **do** {
3.   Choose a solution $s' \in \mathcal{N}(s)$ with better value $c(s')$ than $c(s)$.
4.   Set $s \leftarrow s'$.
   }
5. **Output** s.

The final solution s computed by the algorithm has the best possible value among all the solutions in its neighborhood $\mathcal{N}(s)$:

$$c(s) = \operatorname*{optimum}_{s' \in \mathcal{N}(s)} c(s'). \tag{16.1}$$

Therefore, this solution s is called a *local optimum* with respect to $\mathcal{N}$ and c. The set of local optimum solutions in the feasible solution set $\mathcal{S}$ with respect to a neighborhood function $\mathcal{N}$ and objective function c is denoted as $\mathcal{L}_{\mathcal{N}c}(\mathcal{S})$. A local optimum solution in general differs from a global optimum solution s^*. To see this, let us consider the same instance of the above-mentioned multiprocessor scheduling problem. For a feasible schedule s, the objective function $c(s)$ gives the makespan of the schedule. Let us use the same neighborhood function defined before, that is, $\mathcal{N}(s)$ includes all solutions that can be obtained from s by moving a single job from a machine with maximum load to one with minimum load. This neighborhood function is called the *jump* of *move* neighborhood [8]. Let the initial solution be as shown in Figure 16.3.

Note that as M_1 and M_2 have maximum load, this solution is locally optimum as moving a single job cannot decrease the makespan. The makespan of this local optimum solution is 6, whereas the global optimum solution of Figure 16.2b has makespan 5.

Given a combinatorial optimization problem Π with instances $(\mathcal{S}, c)$ and a local search algorithm A that uses neighborhood function $\mathcal{N}$, we define the *locality gap* α_A of A as the largest possible ratio between the value of a local optimum solution and a global optimum one

$$\alpha_A = \max_{(\mathcal{S},c) \in \Pi} \left\{ \max_{s \in \mathcal{L}_{\mathcal{N}c}(\mathcal{S})} \left\{ \frac{c(s)}{c(s^*)}, \frac{c(s^*)}{c(s)} \right\} : c(s^*) = \operatorname*{optimum}_{s' \in \mathcal{S}} c(s') \right\}. \tag{16.2}$$

If an algorithm A can compute in polynomial time a local optimum solution with locality gap α_A for a combinatorial optimization problem Π, then A is, said to be an approximation algorithm for Π with approximation ratio α_A.

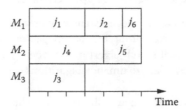

FIGURE 16.3 A local optimum solution with respect to the jump neighborhood and makespan objective function.

16.1.2 The Complexity of Computing Local Optimum Solutions

There is a large number of combinatorial optimization problems and natural neighborhood functions for them, for which we do not know any polynomial time algorithm for computing local optimum solutions. There has been a lot of research on characterizing the class of problems that admit polynomial time algorithms for finding local optimum solutions. One of the most notable works in this area is the research by Johnson et al. [9], who introduced the complexity class PLS of polynomial-time local search problems.

This class includes all those problems and associated neighborhood functions which admit polynomial time algorithms able to decide whether a given feasible solution is locally optimum and, if not, compute a better solution in its neighborhood. There is a reduction among problems in the class PLS which defines a subclass of complete PLS problems. It is unknown whether there exist polynomial time algorithms for computing local optimum solutions for PLS-complete problems.

If we look closely at the above-mentioned IterativeImprovement algorithm, we note that its time complexity is dominated by the number of iterations of the while loop and by the time needed to search the neighborhood for a better solution. Given a combinatorial optimization problem, it is possible to scale the objective function c so it yields integer values for all feasible solutions. Then, the Iterative Improvement algorithm will terminate in a (pseudo)-polynomial number of iterations, as each iteration improves the value of the solution by an integral amount.

This observation led Orlin et al. [10] to study approximate locally optimum solutions: Given a value $\varepsilon > 0$, a solution s for an instance $(\mathcal{S}, c)$ of a combinatorial minimization problem is an ε-*local optimum* with respect to the neighborhood function $\mathcal{N}$ if

$$c(s) - c(s') \leq \varepsilon c(s') \text{ for all } s' \in \mathcal{N}(s).$$

Approximate locally optimum solutions can be defined in a similar manner for maximization problems. In Reference 10 it is shown that every combinatorial optimization problem with a neighborhood function that can be efficiently searched has a fully polynomial time algorithm for computing ε-local optimum solutions. This is a very interesting result, as we show in Section 16.4, ε-local optimum solutions might be shown to be nearly global optimum.

Rardin and Sudit [11] introduced paroid search, a generic local optimization algorithm, that under very general conditions is guaranteed to produce in polynomial time a local optimum solution for any combinatorial optimization problem with a *paroid* structure. A large number of combinatorial optimization problems have a paroid structure including maximum matching, maximum satisfiability, traveling salesman and the knapsack problem. For a discussion on paroids the reader is referred to Reference 12.

Ausiello and Protasi [13] introduced the class guaranteed local optimal (GLO) of combinatorial optimization problems with the property that any local optimum solution with respect to certain neighborhood functions has value within a constant factor of that of an optimum solution. Class GLO can be used to characterize all combinatorial optimization problems whose optimum solutions can be approximated in polynomial time within a constant factor of the optimum.

16.1.3 Local Search in the Design of Approximation Algorithms

Despite its simplicity, local search has not been extensively used to design approximation algorithms. Among the reasons for this are that computing the locality gap of a local search algorithm is not easy; also for many problems, natural local search algorithms have very large locality gaps leading to poor approximation algorithms. A survey of approximation algorithms designed using local search was written by Angel [14].

Consider, for example, the above-mentioned multiprocessor scheduling problem and the jump neighborhood function. Let us consider an instance of the problem where J consist of an even number $n = km$ of jobs with unit processing times for some integer value $k > 0$, and let M consist of m machines.

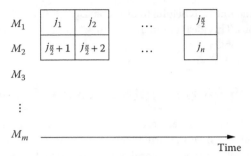

FIGURE 16.4 Instance showing a large locality gap for the jump neighborhood.

The solution shown in Figure 16.4 (half of the jobs are processed on machine M_1 and the other half on M_2) is a local optimum solution under the jump neighborhood as moving a single job cannot decrease the makespan. An optimum solution, however, distributes the jobs evenly among all machines, and so it has makespan k. The locality gap of algorithm IterativeImprovement with the jump neighborhood function is then at least $\frac{km/2}{k} = \frac{m}{2}$.

The problem with a local search algorithm based on the jump neighborhood is that it might get "trapped" in a local optimum solution of value far away from the optimum. If a neighborhood function is such that a local search algorithm always finds a global optimum solution, regardless of the initial solution, such a neighborhood function is called *exact*. For neighborhood functions that are not exact, getting trapped in a local optimum is a big problem, hence various local search techniques have been designed that can move a local search algorithm away from a local optimum solution. Variable-depth search, tabu search, simulated annealing, and genetic algorithms are among these techniques [15]. In this chapter we consider only iterative improvement algorithms, as more complex local search techniques are much harder to analyze. In the sequel "local search" will mean iterative improvement.

When designing an approximation algorithm using local search, if the locality gap of the algorithm for the selected neighborhood function is too large, we can try to improve it by selecting a different neighborhood function, and sometimes (as we show in the next section) by modifying the cost function.

Let us consider again the multiprocessor scheduling problem. Two other neighborhood functions that have been used for designing local search algorithms for this problem are the *swap* and *push* neighborhoods [8]. In the swap neighborhood, two jobs j_i and j_k on different machines are swapped (thus, interchanging their machine allocations). The push neighborhood allows moving a job j_i from its current machine M_j to a different one, M_h, and then recursively moving from M_h those jobs with processing times smaller than that of j_i. It is not hard to find instances showing that, like the jump neighborhood, these two neighborhood functions have a large locality gap.

We could also define a *k-jump* neighborhood function, where up to k jobs are selected and moved to other machines. By selecting a sufficiently large value for k (e.g., by choosing $k = n$) we guarantee that the neighborhood function is exact and, therefore, that a local search algorithm based on such a neighborhood function will always find optimum solutions. The problem with this function is that given a solution s, the size of its neighborhood $|\mathcal{N}(s)|$ is exponential in k. Therefore, Step 3 of algorithm IterativeImprovement might require time that is exponential in k. The main challenge when designing a good local search approximation algorithm is to select a neighborhood function with a small locality gap that yields small enough neighborhoods so that deciding whether the current solution is a local optimum or finding a better solution can be done efficiently.

Computing the locality gap of a local search algorithm is, in general, not easy. We need to make use of the structural properties of local optimum solutions and relate them to the properties of global optimum solutions. In the next sections we describe local search approximation algorithms for NP-hard problems to illustrate this process. Sometimes it is helpful to use equation (16.1) to relate the value of local optimum and global optimum solutions. This idea is used in Section 4.1 of this chapter to design an approximation algorithm for the k-median problem.

16.2 A Local Search for the Multiprocessor Scheduling Problem

As the example in Figure 16.4 shows, a local search algorithm for the multiprocessor scheduling problem based on the jump neighborhood might return a solution that is much worse than the optimum. This happens when the algorithm gets trapped in a local optimum where, for example, several machines have the maximum load and the rest of them are idle. As in this case at least two machines have maximum load, moving a single job will not decrease the makespan of the schedule. However, by moving a job from a maximum load machine to an idle one, the number of machines with largest load will decrease. If we continue moving jobs away from machines with maximum load, eventually we might be able to decrease the makespan of the solution.

To force the IterativeImprovement algorithm to move jobs away from machines with largest load, let us define a new objective function c' that assigns to every schedule s a pair $(c(s), n_\ell(s))$, where $c(s)$ is the makespan of s and $n_\ell(s)$ is the number of machines with load $c(s)$. Given two neighboring solutions s and s', we let $c'(s) < c'(s)$ if $c(s) < c(s')$ or if $c(s) = c(s')$ and $n_\ell(s) < n_\ell(s')$.

Given a schedule s, its neighborhood $\mathcal{N}(s)$ has size at most $O(mn)$ as a neighbor of s can be obtained by moving any of the n jobs to any of the $m - 1$ machines which do not process it in s. Therefore, algorithm IterativeImprovement requires $O(n^2)$ iterations to find a local optimum solution. Thus, the time complexity of IterativeImprovement with the jump neighborhood is $O(mn^3)$.

We use the arguments presented in Reference 16 to bound the number of iterations of the algorithm. Let $C_{\max}(i)$ and $C_{\min}(i)$ denote, respectively, the maximum and minimum machine loads at the beginning of the i-th iteration. Let $\Delta(i) = C_{\max}(i) - C_{\min}(i)$. As every iteration moves a job from a machine with maximum load to a machine with minimum load, then $C_{\max}(i)$ is a monotone nonincreasing function and $C_{\min}(i)$ is a monotone nondecreasing function. Therefore, $\Delta(i)$ is also monotone nonincreasing.

We will bound the number of iterations by bounding the maximum number of times that a job can be moved to other machines. Consider a job j_i that is moved to some machine M_k during the r-th iteration. This means that $C_{\min}(r) = \ell(M_k, r)$, where $\ell(M_k, r)$ is the load of M_k at the beginning of the r-th iteration. Then, assume that j_i is moved from M_k to some other machine at a later iteration $q > r$; thus, $C_{\max}(q) = \ell(M_k, q)$.

We need to consider two cases:

- Assume that no job is moved to machine M_k between iterations $r + 1$ and q. Then,

$$C_{\max}(q) = \ell(M_k, q) \leq \ell(M_k, r) + p_i = C_{\min}(r) + p_i \leq C_{\min}(q) + p_i.$$

The last inequality, $C_{\min}(r) + p_i \leq C_{\min}(q) + p_i$, follows as $C_{\min}(i)$ is nondecreasing. Therefore,

$$p_i \geq C_{\max}(q) - C_{\min}(q) = \Delta(q).$$

This implies that job j_i cannot be moved during the q-th iteration as by moving j_i to a different machine the value of the objective function c' will not decrease. To see this note that if job j_i is moved from M_k to another machine M_h then M_h will have load $p_i + \ell(M_h, q) \geq C_{\min}(q) + p_i \geq C_{\max}(q)$. Hence, the makespan of the new schedule will be at least $C_{\max}(q)$ and the number of machines with maximum load will remain the same.

- Therefore, at least one job needs to be moved to M_k during iterations $r+1, \ldots, q$. Let j_h be the last one of these jobs, and let j_h be moved to M_k during iteration u. Then, $C_{\min}(u) = \ell(M_k, u)$ and

$$C_{\max}(q) = \ell(M_k, q) \leq \ell(M_k, u) + p_h = C_{\min}(u) + p_h \leq C_{\min}(q) + p_h.$$

Hence,

$$p_h \geq C_{\max}(q) - C_{\min}(q) = \Delta(q).$$

As argued earlier, this implies that job j_h will not move during the following iterations.

From the above-mentioned arguments, we conclude that a job j_i can move more than once, only if between any two consecutive moves at least one other job j_h gets fixed on some machine from which it will not move any further. As at most $n - 1$ job movements can use j_h as the job that gets fixed, then the total number of job movements (and, thus, the total number of iterations of the algorithm) is $O(n^2)$.

Theorem 16.1 *The IterativeImprovement algorithm with the jump neighborhood function and cost function c' always finds a schedule s with makespan at most $2 - \frac{1}{m}$ times the optimum one.*

Proof. The proof that we give here for this theorem is essentially the same as Graham's proof for the performance ratio of his list scheduling algorithm [17]. Consider an instance (S, c) of the multiprocessor scheduling problem. The feasible set S is formed by all possible ways of scheduling the jobs J on the machines M. Let s^* be an optimum solution for this instance and let s be the local optimum solution computed by IterativeImprovement with the jump neighborhood and objective function c' (Figure 16.5). Let M_i be a machine with maximum load $c(s)$ and let j_r be the last job processed by M_i.

Let j_r start processing at time t, so $c(s) = t + p_r$. As s is a local optimum solution, then every machine has load at least t. To see this, observe that if some machine M_k has load smaller than t, then moving j_r to M_k would decrease the cost $c'(s)$ of the solution. As every machine's load is at least t, then

$$t \leq \frac{1}{m} \left(\sum_{j_i \in J} p_i - p_r \right) \leq c(s^*) - \frac{1}{m} p_r,$$

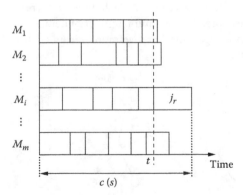

FIGURE 16.5 Solution computed by IterativeImprovement.

as no schedule can process all jobs in J in time smaller than $\frac{1}{m}\sum_{j_i\in J}p_i$. Therefore,

$$
\begin{aligned}
c(s) = t + p_r &\leq c(s^*) - \frac{1}{m}p_r + p_r \\
&= c(s^*) + \left(1 - \frac{1}{m}\right)p_r \\
&\leq c(s^*) + \left(1 - \frac{1}{m}\right)c(s^*) \quad \text{as } p_r \leq c(s^*) \\
&= c(s^*)\left(2 - \frac{1}{m}\right).
\end{aligned}
$$

■

16.3 Finding Spanning Trees with Many Leaves

In this section we describe the local search algorithm of Lu and Ravi [18] for the *maximum leaves spanning tree problem*: Given an undirected graph $G = (V, E)$, find a spanning tree of G with maximum number of leaves. This problem is known to be NP-hard and MAX SNP-complete [19]. The algorithm of Lu and Ravi is simply the IterativeImprovement algorithm with the *exchange neighborhood function*, described as follows. The feasibility space S for this problem includes all spanning trees of the input graph G. For any spanning tree $s \in S$, the objective function $c(s)$ gives the number of leaves in s.

Consider a spanning tree s. The exchange neighborhood of s is formed by all spanning trees that differ from s in a single edge, that is,

$$
\mathcal{N}(s) = \{s' : s' \text{ is a spanning tree of G and } |s \cap s'| = n - 2\},
$$

where n is the number of vertices of G. Here we view a spanning tree s as a collection of edges. Let $(u, v) \in E \setminus s$ be an edge that does not belong to the spanning tree s. Let s_{uv} be the unique path in s between vertices u and v. Note that if we add edge (u, v) to s we create a unique cycle, and by removing from this cycle any edge in s_{uv} we create a new spanning tree (Figure 16.6). As the path s_{uv} can have at most $n - 1$ edges, then the neighborhood $S(s)$ of s has size at most $m(n - 1)$, where m is the number of edges in the graph. As this neighborhood function has polynomial size, Step 3 of algorithm IterativeImprovement can be performed in $O(mn)$ time.

Furthermore, each iteration of the algorithm increases the value of the solution s by at least 1. The initial solution s_0 must have at least 2 leaves, so $c(s_0) \geq 2$. Each iteration increases the value of the solution, and $c(s)$ cannot exceed $n - 1$. Therefore, the maximum number of iterations that the while loop can perform is $n - 3$. The time complexity of algorithm IterativeImprovement is, then, $O(mn^2)$. Now, let us compute the locality gap of the algorithm.

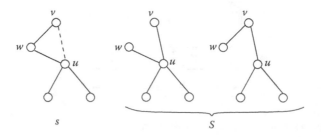

FIGURE 16.6 Neighborhood of spanning tree s. Edge (u, v) does not belong to s.

Theorem 16.2 *The IterativeImprovement algorithm with the exchange neighborhood function has a locality gap of 10.*

To prove the theorem we need to recall some basic properties of spanning trees. Let T be a spanning tree of a given graph $G = (V, E)$. A path in T containing only nodes of degree 2 in T is called a *2-path*. For the rest of this section we use the convention that when referring to a tree T, the *degree* of a vertex u is the degree of u in T (not the degree of u in the graph G). So, for example, if u is a leaf of T we will say that the degree of u is 1. We assume that some node of maximum degree is the root of the tree. Given a node u of a tree T, every node in the path from u to the root of T is called an ancestor of u.

Property 16.1 *The number of nodes $N_3(T)$ of degree at least 3 in a spanning tree T of G is at most the number of leaves in T minus 2, that is,*

$$N_3(T) \leq c(T) - 2.$$

Proof. A spanning tree has maximum number of vertices of degree at least 3 when it is a *proper binary tree*. In a proper (also called full) binary tree each internal node has two children and it is known that in a proper binary tree T' the number of leaves $c(T')$ is one more than the number of internal nodes. As in T' all internal nodes, except the root, have degree 3, then $c(T') = 1 +$ number of internal nodes of $T' = 1 + 1 + N_3(T) = 2 + N_3(T)$ for any binary tree T', and so

$$N_3(T) \leq c(T) - 2$$

for any spanning tree T of G. ∎

Property 16.2 *The number $P_2(T)$ of 2-paths in any tree T is at most twice the number of leaves of T minus 3, that is,*

$$P_2(T) \leq 2c(T) - 3.$$

Proof. Note that if the tree does not have any vertices of degree larger than 2 then T is a path, so $P_2(T) = 1$ and $c(T) = 2$.

A 2-path in T either connects a leaf or a vertex of degree at least 3 with its unique nearest ancestor of degree at least 3. Thus, every leaf and every node of degree at least 3 (except the root) has associated a unique 2-path, and so

$$P_2(T) \leq c(T) + N_3(T) - 1 \leq 2c(T) - 3.$$

The last inequality follows from Property 16.1. ∎

The last ingredient that we need to prove Theorem 16.2 is the following lemma that relates the number of leaves in any spanning tree with the number of 2-paths in a local optimum tree s selected by algorithm IterativeImprovement.

Lemma 16.1 *Let T be a spanning tree of graph $G = (V, E)$ and let s be the tree selected by IterativeImprovement. Let p be a 2-path of s. At most 4 vertices in p can be leaves in T.*

Proof. We prove the lemma by contradiction. Assume that there is a 2-path p in s such that at least 5 of its vertices are leaves in T. Let w_1, w_2, w_3, w_4, and w_5 be five of the leaves of T that belong to p, as shown in Figure 16.7. Note that if two vertices u_i, u_j of p are not adjacent in p, then there cannot be an edge between them in G. This is because if such an edge exists, then adding (u_i, u_j) to s forms a unique cycle (Figure 16.8); removing any edge of this cycle, other than (u_i, u_j), would increase the number of leaves of s by 2, contradicting the assumption that s is a local optimum solution with respect to the exchange neighborhood.

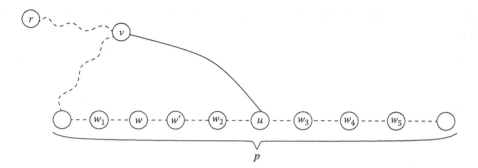

FIGURE 16.7   2-Path p of s. Edge (u, v) does not belong to s.

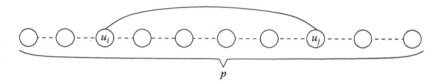

FIGURE 16.8   Cycle formed by the inclusion of edge (u_i, u_j) to p.

Let us pick any vertex r not in p as the root of T. Let (u, v) be the first edge in the path from w_3 to r that does not belong to p (Figure 16.7). Note that by the above-mentioned argument, u and v cannot both belong to p. Without loss of generality let v not belong to p. Assume that the path s_{vw_3} in s from v to w_3 goes through vertices w_1 and w_2 (the other case, when such a path goes through w_4 and w_5 is similar). Adding edge (u, v) to s creates a unique cycle. Also, observe that u is an internal vertex in s, and thus, by adding (u, v) to s we decrease the number of leaves of s by at most 1. If we now remove from p any edge (w, w') in the path from w_1 to w_2 we create a new spanning tree s' in which w and w' are leaves, and so $c(s') \geq c(s) + 1$, contradicting the assumption that s is a local optimum solution. ■

Now, we are ready to prove the theorem and to show that the IterativeImprovement algorithm with the exchange neighborhood has approximation ratio at most 10. Let s^* be a spanning tree of G with maximum number of leaves. Every leaf of s^* must be either:

(a) A leaf of s; the total number of these leaves is $c(s)$, or
(b) A vertex of degree at least 3 in s; there are at most $N_3(s) \leq c(s) - 2$ of these leaves by Property 16.1, or
(c) A vertex in a 2-path of s; by Lemma 16.1 and Property 16.2, the number of these leaves is at most $4P_2(s) \leq 8c(s) - 12$.
      Combining (a)–(c), the number of leaves in s^* is

$$c(s^*) \leq c(s) + c(s) - 2 + 8c(s) - 12 \leq 10c(s).$$

Therefore,

$$\frac{c(s^*)}{c(s)} \leq 10.$$

In Reference 18 it is shown that by using more complex arguments, it can be proven that the locality gap of the algorithm is at most 5. Furthermore, by using a neighborhood function that allows the simultaneous exchange of two tree edges with two nontree edges, Lu and Ravi [18] show that the IterativeImprovement algorithm has locality gap 3.

16.4 Clustering Problems

In a clustering problem we are given a weighted graph $G = (V, E)$ and the goal is to partition the vertices of G into groups or clusters so that a certain objective function is optimized. Two classical clustering problems are the k-median problem and the facility location problem. In the k-median problem it is desired to partition the set of vertices V into k clusters, each with a distinguished vertex called a *center* or *median*, so that the sum of distances from the vertices to their clusters' medians is minimized.

Consider, for example, the weighted graph in Figure 16.9. If $k = 2$, then the two medians in an optimum solution are vertices 3 and 6. These two medians define a partition of the set of vertices as each vertex must be in the same cluster as its nearest median (ties are broken arbitrarily); thus, the clusters are $\{1, 2, 3, 4, 8\}$ and $\{5, 6, 7, 9\}$. The sum of distances from the vertices to their nearest medians is $1 + 1 + 2 + 3 + 1 + 2 + 3 = 13$. A related problem is the *k-means* [20] problem where the goal is to minimize the sum of squared distances from the vertices to the medians.

In the facility location problem each vertex u of the graph has associated a cost $f(u)$ and the goal is to select a set F of vertices (and, thus, to partition G into $|F|$ clusters) so that the total cost $\sum_{u \in F} f(u)$ of the vertices in F plus the sum of distances from the vertices in $V \backslash F$ to F is minimized. Consider again the same graph of Figure 16.9. If the cost $f(u)$ of each vertex u is 3, then an optimum solution for the corresponding facility location problem is $F = \{3, 6\}$, and the cost of this solution is $3 + 3 + 13 = 19$.

Local search algorithms have been recently used to design good approximation algorithms for a variety of clustering problems including the k-median, k-means, facility location and k-facility location problems [20–25]. The k-facility location problem is a generalization of the k-median and facility location problems where the goal is to select a set F of k vertices so that the total cost of the vertices in F plus the sum of distances from the vertices in $V \backslash F$ to F is minimized. In the next section we describe the algorithm of Arya et al. [26] for the metric version of the k-median problem.

16.4.1 Local Search Algorithm for the k-Median Problem

Consider a weighted graph $G = (V, E)$ in which the edge weight function $d : E \to \mathbb{R}$ satisfies the triangle inequality, that is, for any three vertices $u, v, w \in V$, $d(u, v) + d(v, w) \geq d(u, w)$. The *metric k-median problem* is the k-median problem restricted to weighted functions that satisfy the triangle inequality. In this section we describe the local search algorithm of Arya et al. [26] for the metric k-median problem.

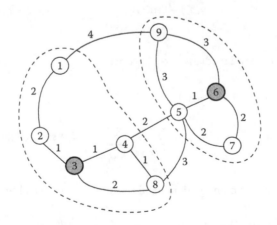

FIGURE 16.9 Clustering the vertices around vertices 3 and 6.

As a feasible solution for the k-median problem is a set $s \subseteq V$ of k vertices, the feasible set $\mathcal{S}$ consists of all subsets of k vertices. A natural way of defining the neighborhood of a solution s is through the use of the *swap operation*. A swap operation replaces one vertex in s with a vertex in $V \setminus s$, so $\mathcal{N}(s) = \{s' : s' = (\mathcal{S} \setminus \{u\}) \cup \{v\}$ for every $u \in \mathcal{S}, v \in V \setminus s\}$. For notational simplicity we denote the set $(\mathcal{S} \setminus \{u\}) \cup \{v\}$ as $\mathcal{S} - u + v$. The objective function $c(s) = \sum_{u \in V} d(u, s)$ gives the sum of distances from the vertices to the medians in the solution s. The distance $d(u, s)$ from a vertex u to the set s is defined as the distance from u to its closest median in the set s.

The size of the neighborhood $\mathcal{N}(s)$ of a solution s is $k(n - k)$. However, the number of iterations of the algorithm might not be polynomial in the size of the input. We show in the next section how to overcome this problem.

Theorem 16.3 *The IterativeImprovement algorithm with the swap neighborhood has a locality gap of 5.*

Let $s = \{s_1, s_2, \ldots, s_k\}$ be the solution computed by the algorithm. This set partitions the vertices V into k clusters $V_1, V_2, \ldots, V_k$, where all vertices in cluster V_i are closer to s_i than to any other median in s (ties are broken arbitrarily). Let $s^* = \{s_1^*, s_2^*, \ldots, s_k^*\}$ be an optimum solution, and let $V_1^*, V_2^*, \ldots, V_k^*$ be the partition induced by s^*. For any vertex $u \in V_i$ let $V_{\sigma(u)}$ denote its cluster in s and let $s_{\sigma(u)}$ be the median in cluster $V_{\sigma(u)}$. Then, the value $c(s)$ of solution s is $c(s) = \sum_{u \in V} d(u, s_{\sigma(u)})$.

Note that as s is a local optimum solution, then no swap operation can improve its value, that is, for any pair of vertices $s_i \in s, v \in V \setminus s$,

$$c(s - s_i + v) \geq c(s). \tag{16.3}$$

To prove Theorem 16.3, first we pair each median $s_i^* \in s^*$ with some median $\rho(s_i^*) \in s$ in such a way that no vertex in s is paired with more than two vertices of s^*. The pairing relation ρ is specified later. We then swap each $s_i^* \in s^*$ with its corresponding vertex $p(s_i^*)$; by inequality (16.3), for every $s_i^* \in s^*$,

$$c(s - \rho(s_i^*) + s_i^*) - c(s) \geq 0 \tag{16.4}$$

and this inequality holds regardless of whether s_i^* is in s or not. We show now that inequality (16.4) can be used to bound $c(s)$ in terms of $c(s^*)$.

Observe that $c(s - \rho(s_i^*) + s_i^*)$ and $c(s)$ differ only by the contributions made by those vertices in $\rho(s_i^*)$'s cluster and s_i^*'s cluster to the values of the two solutions, as shown in Figure 16.10. Therefore,

$$0 \leq c(s - \rho(s_i^*) + s_i^*) - c(s) \leq \sum_{u \in V_i^*} (d(u, s^*) - d(u, s)) + \sum_{v \in V_{\sigma(\rho(s_i^*))} \setminus V_i^*} (d(v, s - \rho(s_i^*) + s_i^*) - d(v, s)). \tag{16.5}$$

Adding inequalities (16.5) over all medians $s_i^* \in s^*$, we get

$$0 \leq \sum_{s_i^* \in s^*} (c(s - \rho(s_i^*) + s_i^*) - c(s))$$

$$\leq \sum_{s_i^* \in s^*} \sum_{u \in V_i^*} (d(u, s^*) - d(u, s)) + \sum_{s_i^* \in s^*} \sum_{v \in V_{\sigma(\rho(s_i^*))} \setminus V_i^*} (d(v, s - \rho(s_i^*) + s_i^*) - d(v, s)). \tag{16.6}$$

Let us look closely at the first term on the right hand side of this last inequality. As $\cup_{i=1}^k V_i^* = V$, then

$$\sum_{s_i^* \in s^*} \sum_{u \in V_i^*} (d(u, s^*) - d(u, s)) = \sum_{u \in V} d(u, s^*) - \sum_{u \in V} d(u, s) = c(s^*) - c(s). \tag{16.7}$$

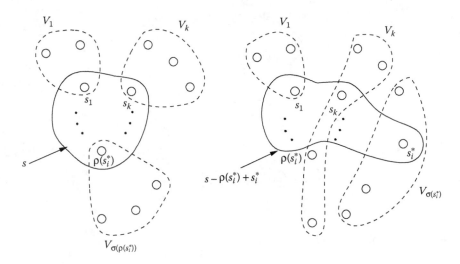

FIGURE 16.10 Solutions s and $s - \rho(s_i^*) + s_i^*$.

The second term in the last inequality of (16.6) is more complicated, so we will spend most of the rest of this section showing that

$$\sum_{s_i^* \in s^*} \sum_{v \in V_{\sigma(\rho(s_i^*))} \setminus V_i^*} (d(v, s - \rho(s_i^*) + s_i^*) - d(v, s)) \leq 4c(s^*). \tag{16.8}$$

Combining equations (16.6)–(16.8) we get,

$$0 \leq c(s^*) - c(s) + 4c(s^*) = 5c(s^*) - c(s).$$

From which it follows that

$$\frac{c(s)}{c(s^*)} \leq 5,$$

as Theorem 16.3 claims. It just remains to prove inequality (16.8). The proof we present here for the validity of this inequality is an elegant argument from Reference 27 that exploits the fact that the edge weights satisfy the triangle inequality. First, let us define the relation ρ. For each $s_i^* \in s^*$ let $\alpha(s_i^*) \in s$ be a closest vertex to s_i^* in s, breaking ties arbitrarily. Notice that α might map 0, 1 or more than 1 vertices from s^* to each vertex $s_i \in s$. Let $R_0 \subseteq s$ ($R_1, R_2 \subseteq s$) be the set of vertices from s with the property that α maps 0 (1, or more than 1, respectively) vertices from s^* to each $u \in R_0$ ($u \in R_1$, $u \in R_2$, respectively). Note that by the way in which sets R_0, R_1, R_2 are defined,

$$|R_0| + |R_1| + |R_2| = k$$

and

$$2|R_2| \leq k - |R_1|$$

as each vertex in R_2 is mapped to at least two of the $k - |R_1|$ vertices in s^* not mapped by α to R_1. Combining these two inequalities we get,

$$|R_0| = k - |R_1| - |R_2| \geq k - |R_1| - \frac{1}{2}k + \frac{1}{2}|R_1| = \frac{1}{2}(k - |R_1|),$$

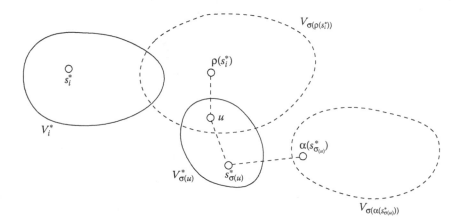

FIGURE 16.11   Re-assigning vertices from $V_{\sigma(\rho(s_i^*))}$.

so

$$2|R_0| \geq k - |R_1|. \tag{16.9}$$

We are ready to define the pairing relationship ρ:

  i) For each $s_i^* \in s^*$ such that $\alpha(s_i^*) \in R_1$, define $\rho(s_i^*) = \alpha(s_i^*)$
 ii) For the remaining vertices $s_i^* \in s^*$, define $\rho(s_i^*) = s_j$, where s_j is selected so that $s_j \in R_0$ and s_j is mapped to at most 2 vertices in s^*. This can be done by inequality (16.9).

Consider a pair $(\rho(s_i^*), s_i^*)$, and the swap operation that exchanges these two vertices transforming the solution s into $s - \rho(s_i^*) + s_i^*$. As Figure 16.10 shows, this swap modifies some of the clusters by re-assigning some of the vertices in $V_{\sigma(\rho(s_i^*))}$ and V_i^*. To show that inequality (16.8) holds consider the following clustering for $s - \rho(s_i^*) + s_i^*$. Vertices in $\bigcup_{s_j \neq \rho(s_i^*)} V_j \setminus V_i^*$ are partitioned into $k - 1$ clusters exactly as in s. Vertices in V_i^* form a cluster with median s_i^*. Finally, vertices in $V_{\sigma(\rho(s_i^*))} \setminus V_i^*$ are distributed among the other clusters of s as follows.

Let $u \in V_{\sigma(\rho(s_i^*))} \setminus V_i^*$. Let $V_{\sigma(u)}^*$ be the cluster in s^* containing u and $s_{\sigma(u)}^*$ be its median. Then, vertex u is assigned to the cluster $V_{\sigma(\alpha(s_{\sigma(u)}^*))}$ of s with median $\alpha(s_{\sigma(u)}^*)$; see Figure 16.11.

Notice that $\alpha(s_{\sigma(u)}^*) \neq \rho(s_i^*)$ as otherwise $\rho(s_i^*) = \alpha(s_{\sigma(u)}^*)$ and $\rho(s_i^*) = \alpha(s_i^*)$ so $\rho(s_i^*)$ would belong to R_2 and then $\rho(s_i^*)$ would not have been paired with s_i^* as no vertex from R_2 belongs to a pair of ρ.

By re-assigning vertex u from cluster $V_{\sigma(\rho(s_i^*))}$ to cluster $V_{\sigma(\alpha(s_{\sigma(u)}^*))}$ the change in cost for the solution is

$$
\begin{aligned}
d(u, s - \rho(s_i^*) + s_i^*) - d(u, s) &\leq d(u, s_{\sigma(u)}^*) + d(s_{\sigma(u)}^*, \alpha(s_{\sigma(u)}^*)) - d(u, \rho(s_i^*)) \\
&\leq d(u, s_{\sigma(u)}^*) + d(s_{\sigma(u)}^*, \rho(s_i^*)) - d(u, \rho(s_i^*)) \\
&\leq d(u, s_{\sigma(u)}^*) + d(u, s_{\sigma(u)}^*) \\
&= 2d(u, s^*),
\end{aligned}
$$

where the second inequality holds because $\alpha(s_{\sigma(u)}^*)$ is the closest vertex in s to $s_{\sigma(u)}^*$ and the first and third inequalities hold by the triangle inequality. Using the above-mentioned inequality, the left hand side of

(16.8) becomes

$$\sum_{s_i^* \in s^*} \sum_{v \in V_{\sigma(\rho(s_i^*))} \setminus V_i^*} (d(v, s - \rho(s_i^*) + s_i^*) - d(v, s)) \le 2 \sum_{s_i^* \in s^*} \sum_{v \in V_{\sigma(\rho(s_i^*))} \setminus V_i^*} d(v, s^*) \le$$

$$2 \sum_{s_i^* \in s^*} \sum_{v \in V_{\sigma(\rho(s_i^*))}} d(v, s^*) \le 2(2 \sum_{s_i \in s} \sum_{v \in V_i} d(v, s^*)) \le 4c(s^*)$$

where the second to last inequality holds because a vertex s appears in at most two pairs of ρ.

16.4.2 Near Local-Optimum Solutions

Even when the locality gap of the algorithm in the previous section is at most 5, it is not a polynomial time approximation algorithm, as the number of iterations needed by IterativeImprovement to find a local optimum solution might be super polynomial. This is because the algorithm might potentially select a large number of subsets of k vertices in the input graph G as intermediate solutions.

To fix this problem we can change the stopping condition of the IterativeImprovement algorithm so it finishes as soon as it finds an *approximate local optimum solution*, that is, when the solution s is such that every neighboring solution $s' \in \mathcal{N}(s)$ has value

$$c(s') > (1 - \varepsilon)c(s)$$

for some accuracy value $\varepsilon > 0$. By changing in this manner the condition of the while loop in Step 2 of the algorithm, we ensure that in each iteration the value of the solution decreases by at least a factor of $1 - \varepsilon$. Therefore, the maximum number of iterations is $\log(c(s_0)/c(s^*))/\log(\frac{1}{1-\varepsilon})$ where s_0 is the initial solution selected in Step 1; this number of iterations is polynomial in the size of the input.

If we change the algorithm as described earlier, the analysis also needs to change, as now condition in (16.4) does not hold, but the following weaker one does:

$$c(s - \rho(s_i^*) + s_i^*) \ge (1 - \varepsilon)c(s) \text{ for all } s_i^* \in s^*.$$

Using this inequality in our analysis gives only a slight worsening in the locality gap (and, thus on the approximation ratio) of the algorithm, as now

$$\frac{c(s)}{c(s^*)} \le \frac{5}{1 - 3\varepsilon}.$$

Arya et al. [26] show that by using the *p-swap* neighborhood which adds to the neighborhood $\mathcal{N}(s)$ of a solution s any subset s' of k vertices that differs from s in at most p vertices, the IterativeImprovement algorithms has locality gap $(3 + 2/p)$. To the date when this paper was written this was the best known approximation algorithm for the metric k-median problem.

16.5 k-Set Packing Problem

In the *k-set packing problem* we are given a finite set E of n elements and a collection F of m subsets of E where each subset in F contains at most k-elements. The goal is to find a maximum cardinality sub-collection of F formed by pairwise disjoint sets. In the weighted version of this problem, called W-k-set packing, each set is assigned a cost and the goal is to find a maximum cost sub-collection of sets. Both problems are *NP*-hard for all $k \ge 3$ [28].

The W-k-set packing problem is a special case of the weighted independent set problem. In the independent set problem we are given a graph $G = (V, E)$ and the goal is to find a set $A \subseteq V$ of vertices such that there is no edge between any two vertices $u, v \in A$. In the weighted version of this problem each vertex $v \in V$ is assigned a cost and the goal is to find a maximum cost independent set. We can see that any instance of the W-k-set packing problem can be reduced to an instance of the weighted independent set problem by building an intersection graph $G' = (V', E')$ where each vertex represents a set and the cost of a set is assigned to the corresponding vertex; there is an edge between two vertices S_i and S_j if and only if the corresponding sets share a common element. Therefore, finding a maximum weight k-set packing reduces to finding a maximum weight independent set in the intersection graph.

Perhaps the simplest approach for approximately solving the W-k-set packing problem is a greedy one: Start with an empty solution S, add to S a maximum cost set in F and then remove from F this set and all the sets intersecting it; repeat this step until F becomes empty. We now show that this greedy algorithm has approximation ratio k. Let $S = \{S_1, S_2, ..., S_r\}$ be the solution selected by the greedy algorithm and let $S^* = \{S_1^*, S_2^*, ..., S_p^*\}$ be an optimum solution. Sets in S are indexed in the order in which they were chosen by the algorithm. Note that at most k sets in S^* intersect S_1 (because sets in S^* are disjoint) and all of them have cost no larger than the cost of S_1. Remove these sets from S^*. Now at most k sets in S^* intersect S_2 and none of them has cost larger than S_2. Remove these sets from S^* and proceed as above-mentioned with the remaining subsets in S. As every subset in S is associated with at most k subsets in S^* of equal or smaller cost, then the cost of S^* is at most k times larger than the cost of S.

Chandra and Halldórsson [29] combined greedy and local search approaches to design an approximation algorithm for the W-k-set packing problem with locality gap $(\frac{n}{n-1})(\frac{2(k+1)}{3})$. Here we analyze a simpler version of that algorithm with locality gap $(\frac{n}{n-1})(\frac{4k+2}{5})$. The algorithm is a modification of the IterativeImprovement algorithm where the initial solution is computed using the above-mentioned greedy algorithm. The neighborhood function used, that we call the *independent neighborhood function*, is as follows. Let $s \subseteq F$ be solution formed by a collection of disjoint subsets of F and let $s_i \in S$. We define $M(s_i) = \{s_j : s_j \in F \text{ and } s_j \cap s_i \neq \emptyset\}$ to be those subsets in F that have at least one common element with s_i and let $Q(s_i) = \{F_j : F_j \subseteq M(s_i) \text{ and all subsets in } F_j \text{ are pairwise disjoint}\}$.

For $F_j \subseteq F$ let $M(F_j) = \cup_{s_i \in F_j} M(s_i)$ be the collection of subsets from F intersecting any of the subsets in F_j. Given a solution s, let $s_i \in s$, $F_j \in Q(s_i)$ and $s' = (s \cup F_j) \setminus (M(F_j) \cap s)$ be the solution obtained from s by adding F_j to s and removing the sets that intersect F_j. We define the *payoff* of s' as the ratio $c(F_j)/c(M(F_j) \cap s)$, where c is the cost function. Note that the payoff of s' can be also written as $c(s' \setminus s)/c(s \setminus s')$. Then the independent neighborhood $\mathcal{N}(s)$ of s includes all those collections $F' \subseteq F$ of pairwise disjoint sets obtained from s by

- Adding some $F_j \in Q(s_i)$ to s for some $s_i \in s$ such that $c(F_j)/c(M(F_j) \cap s) \geq 2$, and
- Removing from s all sets $s_a \notin F_j$ that intersect a set from F_j.

In other words, given a solution s its neighborhood contains any other solution with payoff at least 2 obtained by removing some set s_i from s, then adding any collection $C \subseteq F$ of pairwise disjoint sets all of which intersect s_i, and removing any sets that intersect C.

Note that every sub-collection $F_j \in Q(s_i)$ has at most k subsets in it and s has at most n subsets in it. Therefore, given a solution s it takes $O(mk + nk^2\Delta^k)$ time to explore all the neighborhood of s, where $\Delta = max_{s_i \in F}\{M(s_i)\}$ is the maximum number of sets from F that intersect any set from F.

The running time of the IterativeImprovement algorithm might not be polynomial because each iteration might improve the value of the solution by an arbitrary small amount. To deal with this problem we present an approach different from the one used in the last section to ensure that the IterativeImprovement algorithm performs a polynomial number of iterations. We round down the cost of each set to the nearest multiple of $\frac{w_{gr}}{n^2}$ where w_{gr} is the cost of the solution computed by the greedy algorithm. Since, as shown above, an optimum solution has cost at most kw_{gr} then the maximum number of iterations performed by IterativeImprovement is n^2k because each iteration increases the value of the solution by at least $\frac{w_{gr}}{n^2}$. The time complexity of the algorithm is then $O((mk + nk^2\Delta^k)n^2k)$.

Note that rounding the costs of the sets in F changes the instance of the problem. When we convert the costs back to their original values the cost of an optimum solution might increase by at most

$$n(\frac{w_{gr}}{n^2}) = \frac{w_{gr}}{n} \leq \frac{c(s^*)}{n}, \tag{16.10}$$

where $c(s^*)$ is the cost of an optimum solution with the original costs for the sets.

Theorem 16.4 *The IterativeImprovement algorithm with the independent neighborhood function has locality gap $(\frac{n}{n-1})(\frac{4k+2}{5})$.*

It will be shown that if the set costs are rounded as described earlier, the locality gap of the algorithm is $\frac{4k+2}{5}$. Therefore, by (16.10) on instances with arbitrary costs for the sets the locality gap of the IterativeImprovement algorithm is $(\frac{n}{n-1})(\frac{4k+2}{3})$.

For the rest of the section we assume that the costs of the sets are multiples of $\frac{w_{gr}}{n^2}$. Let s be the solution computed by algorithm IterativeImprovement and let s^* be an optimum solution. We define a mapping $f_s : s^* \to s$ that maps each set $s_i^* \in s^*$ to a set $s_j \in s$ of maximum cost such that $s_i^* \cap s_j \neq \emptyset$.

The inverse mapping $g_s : s \to 2^{s^*}$ is defined as $g_s(s_j) = \{s_i^* \in s^* : f_s(s_i^*) = s_j\}$ for all $s_j \in s$.

The *projection* of s^* to s is

$$\text{proj}(s) = \sum_{s_i^* \in s^*} c(f_s(s_i^*)) = \sum_{s_j \in s} |g_s(s_j)| c(s_j).$$

Lemma 16.2 $kc(s) + \text{proj}(s) \geq 2c(s^*)$.

Proof. The proof is by induction on the number of iterations of IterativeImprovement. Let s_0 be the initial solution computed by the greedy algorithm. As the greedy algorithm repeatedly chooses the largest cost set from F that does not intersect any set in the current solution, then for each $s_i^* \in s^*$ either

- s_i^* is selected to be in s_0 by the greedy algorithm, or
- Some set s_j that intersects s_i^* and has cost $c(s_j) \geq c(s_i^*)$ is added to s_0.

Therefore, $\text{proj}(s_0) \geq c(s^*)$. Also, as the approximation ratio of the greedy algorithm is k, then $c(s_0) \geq \frac{c(s^*)}{k}$. Combining these two inequalities we get $kc(s_0) + \text{proj}(s_0) \geq 2c(s^*)$. For the inductive step, assume that the lemma holds for the solution selected by the algorithm after i iterations. Let s' be the solution selected by the algorithm at the $(i + 1)$-th iteration.

Let $s_\cap^*$ contain those sets in s^* that are mapped by f_s to a set in $s \cap s'$, that is, $s_\cap^* = \{s_j^* \in s^* \mid f_s(s_j^*) \in s \cap s'\}$, and let $s_{\bar{\cap}}^* = s^* \setminus s_\cap$.

Note that for every $s_j^* \in s_\cap^*$, $c(f_{s'}(s_j^*)) \geq c(f_s(s_j^*))$ because s' might contain a set of cost larger than $c(f_s(s_j^*))$ that intersects s_j^*. Therefore,

$$
\begin{aligned}
\text{proj}(s) - \text{proj}(s') &= \sum_{s_j^* \in s^*} (c(f_s(s_j^*)) - c(f_{s'}(s_j^*))) \\
&\leq \sum_{s_j^* \in s_{\bar{\cap}}^*} (c(f_s(s_j^*)) - c(f_{s'}(s_j^*))) \\
&\leq \sum_{s_j^* \in s_{\bar{\cap}}^*} c(f_s(s_j^*)) = \sum_{s_j \in s \setminus s'} |g_s(s_j)| c(s_j) \\
&\leq k\, c(s \setminus s'), \tag{16.11}
\end{aligned}
$$

where the last inequality holds because at most k sets from s^* can be mapped by f_s to the same subset of s.

As s' is in the neighborhood of s then the payoff of s' is at least 2, that is, $c(s' \setminus s) \geq 2c(s \setminus s')$; therefore,

$$
\begin{aligned}
c(s') - c(s) &= c(s' \setminus s) - c(s \setminus s') \\
&\geq 2\,c(s \setminus s') - c(s \setminus s') = c(s \setminus s').
\end{aligned}
\tag{16.12}
$$

Combining (16.11) and (16.12) we get,

$$
k\,c(s') + \mathrm{proj}(s') \geq k\,c(s) + \mathrm{proj}(s) \geq 2c(s^*).
$$

where the last inequality follows by induction hypothesis.   ∎

Lemma 16.3 $(k + 1)c(s) \geq \frac{1}{2}c(s^*) + \mathrm{proj}(s)$

Proof. As the mapping g_s is onto, then

$$
c(s^*) = \sum_{s_j \in s} c(g_s(s_j)).
\tag{16.13}
$$

For any set $s_j \in s$, the set $s' = (s \cup g_s(s_j)) \setminus (M(g_s(s_j)) \cap s)$ contains pairwise disjoint sets, but it is not in the neighborhood of s as s is locally optimum. This means that the payoff of s' is less than 2, that is, $c(g_s(s_j)) < 2c(M(g_s(s_j)) \cap s)$ and so

$$
\sum_{s_j \in s} c(g_s(s_j)) < 2 \sum_{s_j \in s} c(M(g_s(s_j)) \cap s).
$$

By this inequality and (16.13),

$$
\begin{aligned}
\frac{1}{2}c(s^*) &< \sum_{s_j \in s} c(M(g_s(s_j)) \cap s) \\
&= \sum_{s_i \in s} \left| \{s_j \,|\, s_i \in (M(g_s(s_j) \cap s)\} \right| c(s_i).
\end{aligned}
\tag{16.14}
$$

Given a set $s_i \in s$ to determine the number of sets $s_j \in s$ for which $s_i \in M(g_s(s_j)) \cap s$ note that the mapping g_s partitions s^* into disjoint sets so s_i can belong to at most k sets $M(g_s(s_j))$ for $s_j \in s$. However, if $g_s(s_i) \neq \emptyset$ then s_i belongs to $M(g_s(s_i))$ and in addition it can belong to at most $k - |g_s(s_j)|$ other sets $M(g_s(s_j))$ for $s_j \in s$. Therefore, equation (16.14) can be re-written as

$$
\begin{aligned}
\frac{1}{2}c(s^*) &< \sum_{\substack{s_i \in s \\ |g_s(s_i)| > 0}} c(s_i) + \sum_{s_i \in s}(k - |g_s(s_i)|)c(s_i) \\
&\leq (k + 1)c(s) - \mathrm{proj}(s).
\end{aligned}
$$

∎

From Lemmas 16.2 and 16.3 we get

$$
\frac{5}{2}c(s^*) < (2k + 1)c(s),
$$

and so

$$
c(s^*) < \frac{4k + 2}{5}c(s)
$$

as required.

References

1. C.H. Papadimitriou and K. Steiglitz, *Combinatorial Optimization: Algorithms and Complexity*, Prentice-Hall, New York, 1982.
2. E.H.L. Aarts and J.K. Lenstra (Eds.), *Local Search in Combinatorial Optimization*, Wiley, Chichester, UK, 1997.
3. I.H. Osman and J.P. Kelly (Eds.), *Metaheuristics: Theory and Applications*, Kluwer Academic Publishers, Boston, MA, 1996.
4. O. Bräysy and M. Gendreau, Vehicle routing problem with time windows, Part I: Route construction and local search algorithm, *Transportation Science*, 39(1), 104–118, 2005.
5. P. Galinier and A. Hertz, A survey of local search methods for graph coloring. *Computers and Operation Research*, 33(9), 2547–2562, 2006.
6. R.A. Ahoja, O. Ergun, I.B. Orlin, and A.P. Punncu, A survey of very large-scale neighborhood search techniques, *Discrete Applied Mathematics*, 123(1–3), 75–102, 2002.
7. H.H. Hoos and T. Stutzle, Stochastic local search algorithms: An overview, *Springer Handbook of Computational Intelligence*, J. Kacprzyk and W. Pedrycz (Eds.), Springer, Berlin, Germany, pp. 1085–1105, 2015.
8. P. Schuurman and T. Vredeveld, Performance guarantees of local search for multiprocessor scheduling, *Integer Programming and Combinatorial Optimization*, Lecture Notes in Computer Science, Vol. 2081, Springer, Berlin, Germany, pp. 370–382, 2001.
9. D.S. Johnson, C.H. Papadimitriou, and M. Yannakakis, How easy is local search? *Journal of Computer and System Science*, 37, 79–100, 1988.
10. J.B. Orlin, A.P. Punnen, and A.S. Schulz, Approximate local search in combinatorial optimization, *SIAM Journal on Computing*, 33(5), 1201–1214, 2004.
11. R.L. Rardin and M. Sudit, Paroid search: Generic local combinatorial optimization, *Discrete Applied Mathematics*, 43, 155–174, 1993.
12. R.L. Rardin and M. Sudit, Paroids: A canonical format for combinatorial optimization, *Discrete Applied Mathematics*, 39, 37–56, 1992.
13. G. Ausiello and M. Protasi, Local search, reducibility and approximability of NP-optimization problems, *Information Processing Letters*, 54, 73–74, 1995.
14. E. Angel, A survey of approximation results for local search algorithms, in *Efficient Approximation and Online Algorithms*, E. Bampis, K. Jansen, and C. Kenyon (Eds.), Lecture Notes in Computer Science, Vol. 3484, Springer, Berlin, Germany, pp. 30–73, 2006.
15. I. Boussaïd, J. Lepagnot, and P. Siarry, A survey on optimization metaheuristics, *Information Sciences*, 237, 82–117, 2013.
16. P. Brucker, J. Hurink, and F. Werner, Improving local search heuristics for some scheduling problems II, *Discrete Applied Mathematics*, 72, 47–69, 1997.
17. R.L. Graham, Bounds for certain multiprocessor anomalies, *Bell System Technical Journal*, 45, 1563–1581, 1966.
18. H. Lu and R. Ravi, The power of local optimization: approximation algorithms for maximum-leaf spanning tree, *Proceedings of the Thirtieth Annual Allerton Conference on Communication, Control, and Computing*, The Printing House, Inc., pp. 533–542, 1992.
19. G. Galbiati, F. Maffioli, and A. Morzenti, A short note on the approximability of the maximum leaves spanning tree problem, *Information Processing Letters*, 52, 45–49, 1994.
20. T. Kanungo, D.M. Mount, N.S. Netanyahu, C.D. Piatko, R. Silverman, and A.Y. Wu, A local search approximation algorithm for k-means clustering, *Computational Geometry: Theory and Applications*, 28, 89–112, 2004.
21. M. Charikar and S. Guha, Improved combinatorial algorithms for the facility location and k-median problems, *Proceedings of the 40th Annual Symposium on Foundations of Computer Science*, IEEE Computer Society, pp. 378–388, 1999.

22. F. Chudak and D. Williamson, Improved approximation algorithms for capacitated facility location problems, *Proceedings of the 7th Conference on Integer Programming and Combinatorial Optimization*, Springer, pp. 9–113, 1999.

23. M. Korupolu, C.G. Plaxton, and R. Rajaraman, Analysis of a local search heuristic for facility location problems, *Journal of Algorithms*, 37, 237–254, 2000.

24. P. Zhang, A new approximation algorithm for the k-facility location problem, *Theoretical Computer Science*, 384(1), 126–135, 2007.

25. N. Samei and R. Solis-Oba, Analyses of a local search algorithm for the k-facility location problem, *RAIRO-Theoretical Informatics and Applications*, 49(4), 285–306, 2015.

26. V. Arya, N. Garg, R. Khandekar, A. Meyerson, K. Munagala, and V. Pandit, Local search heuristics for k-median and facility location problems, *SIAM Journal on Computing*, 33(3), 544–562, 2004.

27. A. Gupta and K. Tangwongsan, Simpler analyses of local search algorithms for facility location, Data Structures and Algorithms, arXiv, 0809, 2554v1, 2008.

28. M.R. Garey and D.S. Johnson, *Computers and Intractability*, Freeman, New York, 1979.

29. B. Chandra and M.M. Halldórsson, Greedy local improvement and weighted set packing approximation, *Journal of Algorithms*, 39(2), 223–240, 2001.

17

Stochastic Local Search

Holger H. Hoos

Thomas Stützle

17.1 Introduction

Stochastic local search (SLS) algorithms are among the most successful techniques for solving computationally hard problems from computing science, operations research and various application areas; examples range from propositional satisfiability, routing- and scheduling problems to genome sequence assembly, protein structure prediction and winner determination in combinatorial auctions. Because of their versatility and excellent performance in combination with the fact that efficient implementations can often be achieved relatively easily, SLS methods enjoy an ever-increasing popularity among researchers and practitioners.

Local search techniques have a long history; they range from simple constructive and iterative improvement algorithms to rather complex methods that require significant fine-tuning, such as evolutionary algorithms (EAs) or simulated annealing (SA). The key idea behind local search is to iteratively expand or improve a current candidate solution by means of small modifications. Most local search algorithms make use of randomised decisions, for example, in the context of generating initial solutions or when determining search steps, and are therefore referred to as *SLS algorithms*. (It may be noted that formally, deterministic local search algorithms can be seen as special cases of SLS algorithms, as deterministic decisions can be modelled using degenerate probability distributions.) To define an SLS algorithm, the following components have to be specified. (A formal definition can be found in Chapter 1 of the book by Hoos and Stützle [1].)

Search space: The set of *candidate solutions* (or *search positions*) for the given problem instance; candidate solutions typically consist of a number of discrete *solution components*.

Solution set: Specifies all search positions that are considered to be (feasible) solutions of the given problem instance.

Neighbourhood relation: Specifies the direct neighbours of each candidate solution *s*, that is, the search positions that can be reached from *s* in one search step.

Memory states: Used to hold information about the search mechanism beyond the search position (e.g., tabu tenure of solution components in tabu search, or temperature in SA); may consist of a single, constant state in the case of algorithms that do not use memory, such as simple iterative improvement.

Initialisation function: Specifies search initialisation in the form of a probability distribution over initial search positions and memory states.

Step function: Determines the computation of search steps by mapping each search position and memory state to a probability distribution over its neighbouring search positions and memory states.

Termination predicate: Used to decide search termination based on the current search position and memory state.

Based on these components, SLS algorithms work as illustrated in Figure 17.1. When applied to optimisation problems, whose definition comprises an objective function that specifies the quality of solutions, SLS algorithms need to keep track of the best solution encountered during the search process, the so-called *incumbent solution*. For decision problems, the search process is usually terminated as soon as a solution for the given problem instance is found, and repeated updating of an incumbent solution is not required.

Among the components underlying any SLS algorithm, the neighbourhood relation and the step function are particularly important. Typically, neighbourhood relations have to be defined in a problem-specific way, and it is often difficult to predict which of the various choices that can be made in this context will result in the best performance. However, standard types of neighbourhood relations exist; particularly widely used are the so-called *k-exchange neighbourhoods*, in which two candidate solutions are direct neighbours if, and only if, they differ in at most k solution components. As an example, consider the 2-exchange neighbourhood for the (symmetric) travelling salesman problem (TSP), under which two candidate solutions are direct neighbours if one can be obtained from the other by replacing two edges of s by two alternate edges (see Figure 17.2). Every neighbourhood relation induces a *neighbourhood graph*, whose vertex set corresponds to the given search space, and in which each pair of neighbouring search positions is connected by an edge. Many important properties of the neighbourhood relation are reflected in the neighbourhood graph; for example, k-exchange neighbourhoods induce symmetric, k-regular neighbourhood graphs.

The step function defines how the search process moves from one search state to the next, where a *search state* is a combination of a search position and a memory state. Search steps are usually defined by means of a procedure that draws a sample from the probability distribution determined by the underlying step function; similar procedural specifications are used for the initialisation and termination functions. The search process is typically guided by an *evaluation function* that is used to heuristically assess or rank candidate solutions. For combinatorial optimisation problems, the objective function (which is part of the problem definition) is often also used as an evaluation function. In the case of combinatorial decision

Stochastic Local Search:
```
       determine initial search state
       While termination criterion is not satisfied:
         perform search step
         if necessary, update incumbent solution
       return incumbent solution or report failure
```

FIGURE 17.1   General outline of an SLS algorithm.

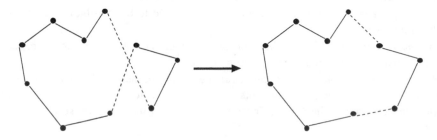

FIGURE 17.2   Example of a 2-exchange step for the symmetric TSP.

problems, on the other hand, there can be considerable freedom in choosing an appropriate evaluation function. Furthermore, some SLS methods use multiple evaluation functions or modify the evaluation function while searching. (For this reason, we do not explicitly specify a single evaluation function as a component of our formal definition of an SLS algorithm.)

The concept of SLS, as aforementioned, provides a unifying framework for many different types of algorithms. In particular, it captures constructive algorithms, whose search space contain partial solutions of a given problem, *that is*, candidate solutions that can be extended by adding solution components (such as edges in the case of the TSP). Consequently, search methods that are strongly based on constructive search procedures, such as greedy randomised adaptive search procedure (GRASP) [2,3] or ant colony optimisation (ACO) [4,5], fall under the general SLS framework. Similarly, population-based algorithms, which operate on ensembles of candidate solutions, are covered by our definition by considering search positions to consist of sets of individual candidate solutions. Note that in this case, step functions can be defined to model operations on populations, such as recombination in the case of EAs (see also Chapter 2 of Hoos and Stützle' book [1]).

Many SLS algorithms are based on generic methods for controlling and directing a simpler, subsidiary search process (such as an iterative improvement procedure, see Section 17.2), and substantial research efforts have been focused on the development and study of such general SLS methods, which are also commonly known as *metaheuristics* [6]. Although high-level search strategies and mechanisms provide an important basis for developing SLS algorithms for a broad range of problems, it is important to keep in mind that other aspects, including the choice of the search space and neighbourhood relation, as well as techniques for the efficient implementation of local search steps, are also crucial ingredients of high-performance SLS algorithms. Research in the area of SLS comprises all aspects of the design, implementation and analysis of SLS algorithms.

The remainder of this chapter provides an overview of widely used SLS methods for discrete combinatorial problems, including: iterative improvement techniques (Section 17.2); so-called "simple" SLS methods (Section 17.3), including SA and tabu search, as well as the less widely known method of dynamic local search (DLS); hybrid SLS methods (Section 17.4), such as iterated local search (ILS) and GRASP; and population-based SLS methods (Section 17.5), in particular, ACO and EAs. The chapter closes with a brief discussion of some general issues and interesting research directions in the area of SLS.

17.2 Iterative Improvement

One of the most basic SLS methods is based on the idea of iteratively improving a candidate solution of the given problem with respect to an evaluation function. More precisely, the search is started from some initial position, and in each search step, the current candidate solution s is replaced with a neighbouring candidate solution s' with better evaluation function value. The search is terminated when a *local minimum* is reached, *that is*, a candidate solution s with $g(s) \leq g(s')$ for all direct neighbours s' of s, where g is the evaluation function that is to be minimised. This SLS method is called *iterative improvement*; it is also known as *iterative descent* or *hill-climbing* (the latter is motivated by an equivalent formulation where a given evaluation function is to be maximised).

Iterative improvement algorithms find local optima of a given evaluation function. As local minima are defined with respect to a given neighbourhood relation N, it is quite obvious that the choice of N is of crucial importance for the performance of any iterative improvement procedure. Although the use of larger neighbourhoods results in better quality local optima, the time-complexity of determining improving search steps increases with neighbourhood size. For example, in the commonly used k-exchange neighbourhoods, each search position has $\mathcal{O}(n^k)$ direct neighbours (where n is the number of solution components in each candidate solution), *that is*, the neighbourhood size is exponential in k, and the same holds for the time for identifying improving neighbours in the worst case. This leads to a general trade-off between solution quality and run-time of iterative improvement algorithms. In practice, search steps with quadratic or cubic time-complexity can already lead to prohibitively high computation times when solving large problem instances.

The time-complexity of local search steps can be significantly reduced using two generic techniques. First, *caching and incremental updating techniques* can be used to significantly reduce the often considerable cost of computing from scratch the evaluation function values of all neighbours of the current search position in each search step. Instead, these values are stored and updated based on the actual effects of each search step (see also Chapter 1 of the book by Hoos and Stützle [1]). Second, the size of large neighbourhoods can be substantially reduced by excluding from consideration neighbours of the current candidate solution that are provably or likely nonimproving. Such *neighbourhood pruning techniques* have a long history; they include fixed radius searches, nearest neighbour lists and the use of so-called "don't look bits" in the context of the TSP [7], and reduced neighbourhoods in the case of the job-shop scheduling problem [8]. These techniques are essential for the design of high-performance SLS algorithms based on large neighbourhoods, but they can also lead to significant performance improvements when used in combination with smaller neighbourhoods.

Another factor that has a significant influence on the speed and performance of an iterative improvement algorithm is the mechanism used for determining the search steps—the so-called *pivoting rule* [9]. The two most widely used pivoting rules are best improvement and first improvement. *Iterative best improvement* chooses in each step one of the neighbouring search positions that results in a maximal possible improvement of the evaluation function value; ties can be broken randomly, based on the order in which the neighbourhood is examined or using secondary criteria. *Iterative first improvement* examines the neighbourhood in some pre-defined order and performs the first improving search step encountered during this inspection. Clearly, the local optima found by this method depend on the order in which the neighbourhood is scanned. Instead of using predefined, fixed orders, a random order for scanning the neighbourhood can be useful, and repeated runs of such random-order first improvement algorithms are often able to identify many different local optima, even when started from the same initial position [1].

Iterative first improvement usually requires more steps than iterative best improvement to reach local optima of comparable quality, but the individual improvement steps are typically found much faster, as in many cases, they do not need to inspect the entire neighbourhood. However, best-improvement algorithms often benefit more significantly from caching and incremental updating strategies, and as a result, are not always slower than first-improvement algorithms in reaching a local minimum. It may also be noted that to identify a candidate solution as a local optimum both, first- and best-improvement algorithms need to inspect the entire local neighbourhood; the time required for this final check (the so-called check-out time) can be reduced using don't look bits [7,10].

An interesting way of achieving a good trade-off between neighbourhood size and time-complexity of local search steps is to use multiple neighbourhood relations. *Variable neighbourhood descent (VND)*, a variant of a more general SLS method called variable neighbourhood search (VNS) [11,12], is an iterative improvement method that switches systematically between several neighbourhood relations $N_1, N_2, \ldots, N_k$, which are typically ordered according to increasing size. Starting from an initial candidate solution, VND performs iterative improvement using N_1. Once a local optimum w.r.t. N_1 is found, the search is continued in N_2. Generally, whenever a local optimum of N_i has been found, VND switches to N_{i+1}. However, as soon as an improvement has been achieved in any neighbourhood N_i (with $i > 1$), the search is continued using N_1. The idea underlying this mechanism is to use small neighbourhoods (which can be searched most efficiently) whenever possible. VND terminates when a local optimum of N_k has been found. Empirical results have shown that VND algorithms are often significantly more efficient in finding high-quality local optima than simple iterative improvement algorithms using large neighbourhoods. It may be noted that other variants of the more general approach of VNS, such as basic VNS or skewed VNS [11,12], are conceptually more closely related to ILS (see Section 17.4).

Finally, a number of SLS algorithms use *very large scale neighbourhoods* [13], whose size is often exponential in the size of the given problem instance. This type of neighbourhood usually has to be searched heuristically, which is the case in variable-depth search [14,15] and ejection chain algorithms [16].

However, there are specially structured neighbourhoods that can be searched exactly and efficiently using network-flow techniques or dynamic programming [17–21]. For an overview of these techniques we refer to Chapter 18.

17.3 "Simple" SLS Methods

Iterative improvement algorithms terminate when they reach a local optimum of the given evaluation function. In the following, we discuss several approaches that allow SLS algorithms to escape from local optima by occasionally accepting worsening search steps. The methods discussed in this section are "simple" in the sense that they are usually based on a single, fixed neighbourhood relation.

Randomised iterative improvement (RII): One of the simplest ways to allow worsening search steps is to occasionally move to a randomly chosen neighbouring search position. In RII, this idea is implemented by switching probabilistically between two types of search steps within the same neighbourhood structure: iterative improvement steps and so-called *uninformed random walk steps*, in which a neighbour is chosen uniformly at random. More precisely, in each search step, an uninformed random walk step is performed with probability w_p, and an iterative improvement step is performed otherwise (i.e., with probability $1 - w_p$). The parameter w_p is called *walk probability* or *noise parameter*. Using this mechanism, arbitrarily long sequences of random walk steps can be performed, where the probability of r consecutive random walk steps is w_p^r. Extending an iterative improvement algorithm into an RII algorithm typically requires only few lines of code and introduces only one parameter. Despite their conceptual simplicity, RII algorithms can perform quite well; for example, they have provided the basis for GSAT with random walk, a well-known algorithm for the propositional satisfiability problem (SAT) [22]. However, there are relatively few RII algorithms, perhaps because more complex SLS algorithms often achieve better performance.

Probabilistic iterative improvement (PII): Random walk steps, as performed in RII, may lead to significant deteriorations in evaluation function value. An attractive alternative is to base the acceptance of a worsening search step on the change in evaluation function caused by it; this is the key idea behind PII. At each step, PII selects a neighbouring candidate solution according to a function $p(g, s)$, which determines a probability distribution over the neighbourhood of S taking into account the evaluation function g. In practice, samples of this probability distribution can be taken as follows: first, a neighbour s' of the current search position s is selected uniformly at random; then, a decision is made whether s' is accepted as the new current search position. When minimising g, the probability of accepting s', $p_{accept}(T, s, s')$, is in many cases based on the following probability function:

$$p_{accept}(T, s, s') := \begin{cases} 1 & \text{if } g(s') < g(s) \\ \exp(\frac{g(s) - g(s')}{T}) & \text{otherwise} \end{cases} \qquad (17.1)$$

where T is a parameter that determines the degree to which worsening search steps are likely to be accepted. This so-called *Metropolis condition* is frequently used in SA algorithms (discussed in the following), and in this context, the parameter T is called *temperature*. In fact, PII with the metropolis condition is equivalent to constant-temperature SA, which has been shown to perform well if suitable temperature values are used [23,24].

Simulated annealing (SA): A natural generalisation of PII with the metropolis condition is obtained by modifying the parameter T during the search; for example, by gradually decreasing T, the search process becomes increasingly greedy. This is the key idea underlying SA, an SLS method that is inspired by the annealing process of solids, and which has been independently proposed by Kirkpatrick et al. [25] and Cerný [26].

In each step of a standard SA algorithm, first, a neighbour s' of the current candidate solution s is chosen using a *proposal mechanism* (in the simplest case, this can be a uniform random choice from the local neighbourhood of s); next, a parameterised probabilistic *acceptance criterion* (e.g., the metropolis condition—see Equation 17.1) is used to decide whether to perform a search step from s to s'. The way in which the temperature parameter T is modified during the search process is determined by the *annealing schedule*, a function that defines for each search step i a temperature value $T(i)$. Annealing schedules are usually specified in the form of an initial temperature T_0, the number of search steps performed at each temperature value (this is often chosen as a multiple of the neighbourhood size) and a temperature update scheme (e.g., geometric cooling according to $T_{k+1} := \alpha \cdot T_k$, where α is a parameter between 0 and 1). Obviously, the performance of an SA algorithm is highly dependent on its annealing schedule. SA algorithms frequently use special termination conditions, such as the *acceptance ratio* (i.e., the ratio of accepted vs proposed search steps) or the number of subsequent temperature values that have been used as the last improvement in the incumbent solution has been achieved.

SA is one of the earliest and most prominent generic SLS methods. SA algorithms have been applied to a wide range of combinatorial problems; they have also been intensely studied analytically (giving rise to some interesting results, such as proofs regarding the convergence to optimal solutions) and experimentally. Many variants of SA have been proposed and studied, such as threshold accepting, where the probabilistic acceptance criterion is replaced by a deterministic mechanism [27]. Chapter 25 in the 1st edition of this handbook discusses this approach in detail.

Tabu search (TS): Differently from the previously discussed SLS methods, tabu search makes direct and systematic use of memory to guide the search process [28]. The key idea behind *simple tabu search*, the most basic tabu search variant, is to use short-term memory to prevent a subsidiary iterative improvement procedure from returning to recently visited search positions, which allows the search process to escape from local optima of the given evaluation function. When performing iterative improvement steps, only *permissible neighbours* of the current candidate solution are considered, *that is*, neighbouring search positions that are not declared tabu. The tabu memory is updated after each search step, such that search steps cannot be undone for a fixed amount of time (*tabu tenure*). Typically, the tabu mechanism is implemented by storing with each solution component the time (in terms of step number) when it was last used in the current candidate solution, and to only permit solution components to be introduced by a search step for which the difference between this time-stamp and the current time exceeds the tabu tenure. For example, in a simple tabu search algorithm based on the 2-exchange neighbourhood for the TSP, the edges removed in a 2-exchange step may not be re-introduced for tt subsequent search steps, where tt is the tabu tenure parameter. Many simple tabu search algorithms use an *aspiration criterion*, which allow a search step to a nonpermissible neighbour if it leads to an improvement in the incumbent solution.

The performance of simple tabu search algorithms can be very impressive, but usually depends strongly on the value of the tabu tenure parameter. Several extensions have been proposed that adjust the tabu tenure while running the search process; the best known example is reactive tabu search [29] (see also Chapter 19). Furthermore, additional intermediate and long-term memory mechanisms can be used to further improve the effectiveness of simple tabu search. These mechanisms aim to either intensify the search in particular regions of the search space or to diversify the search to prevent stagnation. For a detailed description of these techniques, we refer to the book by Glover and Laguna [28]. More detailed information on tabu search can also be found in Chapter 21 of this book.

Dynamic local search (DLS): A different approach for escaping from local optima provides the basis for DLS: Rather than explicitly using worsening search steps, DLS modifies the given evaluation function whenever its subsidiary local search procedure encounters a local optimum. More precisely, DLS works as follows. Initially, a subsidiary local search algorithm, typically an iterative improvement procedure, is executed until a local optimum s is encountered. At this point, the evaluation function is modified such that s is no longer a local optimum. Then, the subsidiary local search is run again until it reaches a local optimum of the modified evaluation function, at which point, further modifications of the evaluation function are performed. These phases of local search followed by evaluation function modifications are repeated until a termination criterion is satisfied.

The modifications to the evaluation function are typically achieved by means of *penalty weights* that are associated with individual solution components. The augmented evaluation function effectively used by the subsidiary local search algorithm is then of the form

$$g_p(s) := g(s) + \sum_{i \in C(s)} penalty(i), \tag{17.2}$$

where:

$g(s) = $ the original evaluation function
$C(s) = $ the set of solution components of candidate solution s
$penalty(i) = $ the penalty of solution component i of s

At the beginning of the search process, all penalties are set to zero. DLS algorithms mainly differ in the way the penalty update is performed, for example, the usage of additive versus multiplicative increments or the selection of the solution whose penalties are increased. Typically, when a local optimum s is encountered, penalties are only increased for solution components occurring in s. For example, in guided local search (GLS), a well-known family of DLS algorithms [30,31], the following penalty update mechanism is used. When encountering a local minimum s of g_p, GLS computes a utility value $u(i) := g_i(s)/(1 + penalty(i))$ for each solution component i of s, where $g_i(s)$ is the contribution of i to the original evaluation (or objective) function. (e.g., in the case of the TSP, the solution components are typically edges of the given graph, and $g_i(s)$ is the cost of edge i.) Then, only the penalties of solution components with maximal utility are increased. Note that the term $1 + penalty(i)$ in the denominator of definition of $u(i)$ avoids overly frequent penalisation of solution components.

In addition to penalty increases in local optima, many DLS algorithms also occasionally decrease penalty values. DLS algorithms have been demonstrated to achieve state-of-the-art performance for various combinatorial problems, including SAT [32]. They are conceptually related to Lagrangian methods in continuous optimisation.

17.4 Hybrid SLS Methods

Although the previously discussed SLS methods provide the basis for state-of-the-art algorithms for many hard combinatorial problems, often, further performance improvements can be achieved by combining various SLS strategies into hybrid algorithms. In fact, RII can already be seen as a hybrid SLS method, as it combines two different types of search steps—random walk and iterative improvement steps. In the following, we briefly discuss several other hybrid SLS methods. In many cases, hybrid SLS methods combine constructive search steps and perturbative local search steps, or larger modifications of candidate solutions with the application of "simple" local search algorithms.

Iterated local search (ILS): The key idea behind this SLS method is to escape from local optima of the given evaluation function by means of a perturbation mechanism. Essentially, three components form the core of any ILS algorithm. A subsidiary local search procedure is used to efficiently find local optima; this is typically based on an iterative improvement algorithm or a "simple" SLS method. The perturbation procedure introduces a modification to a given candidate solution to allow the search process to escape from local optima. Finally, an acceptance criterion is used to decide whether the search should be continued from a newly found local optimum. Based on these components, ILS works as outlined in Figure 17.3.

When using an iterative improvement algorithm as the subsidiary local search procedure, ILS can be seen as performing a biased random walk in the space of local optima reached by this procedure. The perturbation procedure should introduce a modification that cannot be immediately reversed by the local

Iterated Local Search (ILS):
  determine initial candidate solution s
  perform subsidiary local search on s
  While termination criterion is not satisfied:
    $r := s$
    perform perturbation on s
    perform subsidiary local search on s
    based on acceptance criterion, keep s or revert to $s := r$

FIGURE 17.3   Outline of iterated local search.

search procedure, to achieve effective diversification of the search. The acceptance criterion determines the degree of search intensification; for example, if only improving candidate solutions are accepted, ILS performs randomised first-improvement search in the space of local optima. Together, the perturbation procedure and the acceptance criterion determine the balance between search intensification and diversification.

One of the major attractions of ILS stems from the fact that it is typically very easy to extend an existing implementations of a simple local search algorithm into a basic ILS algorithm. ILS algorithms define the current state-of-the-art for solving many hard combinatorial problems, the most prominent of which is the TSP [33]. Conceptually, there is a close relationship between ILS and some advanced forms of tabu search; furthermore, the previously mentioned basic and skewed VNS algorithms can be seen as special cases of ILS. For more details on ILS we refer to Lourenço et al. [34].

Iterated greedy (IG) algorithms: Greedy construction methods are at the core of many well-known approximation algorithms. They also provide the basis for IG algorithms, which can be seen as a variant of ILS in which local search and perturbation phases are replaced by construction and destruction phases. Starting from a complete candidate solution, IG alternates between phases of destructive and constructive search. During a destruction phase, some solution components are removed from the current candidate solution s (e.g., uniformly at random or heuristically, depending on their impact on the evaluation function), resulting in some partial (nonempty) candidate solution s_p. During a construction phase, solution components are added heuristically, starting from s_p, until a new complete candidate solution s' has been obtained. As in ILS, an acceptance criterion is used to decide whether to continue the search from s' or s.

Note that the initial candidate solution in IG may be generated by a construction method that may be different from the one used in subsequent construction phases. Also, a perturbative local search procedure may be used to further improve any complete candidate solutions considered during the search. In this case, IG may be seen as a variant of ILS in which a combination of a destruction and a construction phase forms the perturbation procedure. IG methods are a promising approach for solving problems for which good constructive algorithms exist. Problems for which IG algorithms reach state-of-the-art performance include set covering [35,36] and flow-shop scheduling [37].

Greedy randomised adaptive search procedures (GRASPs): A combination of constructive and perturbative local search forms the basis for GRASP [2,3], which works as follows: Using a randomised construction procedure, a complete candidate solution is generated, which is then improved using a perturbative local search procedure; this two-phase process is repeated until a termination criterion is satisfied.

The construction procedure in GRASP iteratively adds solution components that are chosen randomly from a *restricted candidate list*; the elements of this list are determined using a heuristic function, whose value for a given solution component may depend on the solution components already present in the current partial candidate solution (this is the "adaptive" aspect of the search process). GRASP has been applied to a broad range of combinatorial problems; for a detailed description of GRASP and various extensions we refer to Resende and Ribeiro [3].

17.5 Population-Based SLS Methods

Various SLS methods maintain a population of candidate solutions which are simultaneously manipulated in each search step. As previously noted, such population-based methods can be formulated within the unified SLS framework described in Section 17.1 by considering search positions that consist of sets of candidate solutions in combination with suitably defined neighbourhood relations as well as initialisation and step functions. The appeal of population-based SLS methods partly stems from the fact that the use of a population provides a straightforward means for achieving search diversification; however, compared to other SLS methods, this often comes at the cost of higher implementation effort and more parameters that need to be tuned to achieve competitive performance. The two widely known population-based methods described in the following are inspired by biological mechanisms; however, it should be noted that there are many other population-based SLS methods, such as population-based extensions of ILS, that are not motivated in this way.

Ant colony optimisation (ACO): The inspiration for ACO stems from the pheromone trail laying and following behaviour exhibited by several ant species [5]. In ACO, the (artificial) ants are randomised construction procedures that take into account (artificial) pheromone trails and heuristic information when iteratively constructing complete candidate solutions. Essentially, the pheromone trails are modelled by numerical values that are associated with solution components; these are adapted while solving a given problem instance and hence reflect the search experience of the (artificial) ant colony.

In each construction step, solution components are chosen with a probability that is proportional to their pheromone value and the heuristic information. For example, in the TSP, the first problem tackled by ACO algorithms [4], a pheromone value τ_{ij} is associated with each edge (i, j) of the given graph, and the heuristic information η_{ij} is typically defined as the inverse of the length of edge (i, j). Hence, an ant located at a vertex i would add vertex $j \in N(i)$ to its current partial tour s_p with probability

$$p_{ij} = \frac{\tau_{ij}^{\alpha} \cdot \eta_{ij}^{\beta}}{\sum_{l \in N(i)} \tau_{il}^{\alpha} \cdot \eta_{il}^{\beta}}, \tag{17.3}$$

where $N(i)$ is the feasible neighbourhood of vertex i, *that is*, the set of all vertices that have not yet been visited in s_p, and α and β are parameters that control the relative influence of pheromone values and heuristic information, respectively. This probabilistic selection mechanism was used in ant system, the first ACO algorithm [38]; the tour construction procedure applied by the artificial ants using this mechanism resembles a probabilistic variant of the nearest neighbour construction heuristic.

After all ants have constructed a complete candidate solution, many ACO algorithms apply a perturbative local search procedure (such as iterative improvement) to each complete candidate solution; it has been shown that in the context of solving $\mathcal{NP}$-hard problems, the overall performance of an ACO algorithm often depends crucially on this local search phase [5,39]. Then, the pheromone values are updated in two steps. In the first step (which models pheromone evaporation), all pheromone values are decreased, typically by multiplication with a constant factor. In the second step (which models pheromone deposit), the pheromone values of the solution components contained in one or more of the current candidate solutions are increased; the amount of increase often depends on the quality of the respective solutions. Cycles of construction (and subsidiary local search) phases followed by pheromone updates are repeated until a termination criterion is satisfied.

Several variants of ACO have been proposed that all share the same basic underlying ideas (see the book by Dorigo and Stützle [5] for an overview). The ACO metaheuristic [40] gives a general framework for these variants. In general, ACO is steadily gaining popularity and provides the basis for state-of-the-art algorithms for a number of hard combinatorial optimisation problems [5]. More information on ACO can be found in Chapter 23 of this book.

Evolutionary algorithms: One of the most prominent classes of population-based SLS methods has been inspired by concepts from biological evolution. EAs start with an initial set of candidate solutions (which can be generated randomly or by means of heuristic construction search methods), and iteratively modify this population by means of three types of operations: mutation, recombination and selection.

Typical *mutation operators* introduce small perturbations to candidate solutions in a randomised fashion; the amount of perturbation applied is usually controlled by a parameter called the *mutation rate*. *Recombination operators* generate one or more new candidate solutions, often called "offspring," by combining information from two or more "parent" candidate solutions. One of the most common types of recombination is known as *crossover*; it generates offspring by assembling partial candidate solutions from two parents. *Selection operators* are used to determine which candidate solutions from the current population and from the set of new candidate solutions obtained from mutation and recombination will form the population used in the next iteration of the search process. This choice is typically based on the value of candidate solutions under the given evaluation function (which, in the context of EAs, is commonly referred to as *fitness function*), such that better candidate solutions have a higher probability to "survive" the selection process. Selection operators are also used for choosing candidate solutions from the current population to undergo mutation or recombination.

The performance of EAs depends crucially on the choice of the evolutionary operators. In general, the performance of EAs improves if knowledge about the given problem is exploited in the operators that are applied. In fact, much research in EAs has been devoted to the design of effective mutation and crossover operators; a good example for this is the TSP [41,42]. As for ACO, substantial performance improvements can often be achieved by additionally optimising candidate solutions using a perturbative local search method, such as iterative improvement. The resulting hybrid algorithms are also known as *memetic algorithms (MAs)* [43]. For a detailed account on memetic algorithms and their performance we refer to Chapter 27 of this 1st edition of this handbook. *Scatter search* and *path relinking* are SLS methods whose roots can be traced back to the mid-1970s [44] and that only recently have regained considerable attention. They can be seen as memetic algorithms that uses special types of recombination and selection operators. For details on these methods, we refer to the work of Glover et al. [45,46].

17.6 Discussion and Research Directions

The study of SLS methods lies at the intersection of computing science, operations research, statistics and various application areas. Their successful application requires knowledge of the various SLS techniques and their properties; despite the generic nature of many SLS methods, insights into the problem to be solved are often also crucial.

Overall, the behaviour of most high-performance SLS algorithms is not well understood, and although some theoretical results exist (e.g., for SA and ACO algorithms), these are often obtained under specific assumptions that limit their practical relevance [47,48]. Still, some theoretical properties, such as empirical approximate completeness (which guarantees that when run arbitrarily long, an algorithm finds an (optimal) to any given, soluble problem instance with probability approaching one), have been shown to be useful in guiding the development of high-performance SLS algorithms Hoos [49].

Nevertheless, the analysis of SLS algorithms relies mostly on computational experiments, and advanced methods for the empirical analysis of SLS behaviour play an important role in the development of new algorithms. Empirical studies in this area are complicated by the fact that most SLS algorithms are heavily randomised; but using advanced empirical methods, such as the RTD-based approach discussed in Chapter 13, have contributed significantly to an improved understanding of SLS behaviour as well as to the development of high-performance SLS algorithms for many hard combinatorial problems [49,50]. Yet, compared to mature empirical sciences, such as physics or biology, computing science in general, and the area of algorithmics in particular, are still in the early stages of adopting and exploiting a well-founded empirical approach.

In the past, much of the successful work on SLS algorithms and generic SLS methods has relied crucially on prior experience and good intuitions. One of the major directions in SLS research concerns the development of improved practices for developing SLS algorithms and for their application to new problems. In this context, it is likely that conceptual frameworks for the design and exploration of SLS methods, as well as implementation and experimentation environments will play an important role [51,52]. Furthermore, an improved understanding of the relationship between the features of problems and problem instances on one hand, and properties and behaviour of SLS methods on the other hand will likely provide the basis for more principled approaches for the the design and application of SLS algorithms. Insights to be gained in this context are also of considerable scientific interest. Progress in this research direction is significantly leveraged by advanced search space analysis techniques, statistical methods and machine learning approaches (see, e.g., Merz and Freisleben [53] or Watson et al. [54]).

The formalisation of general principles underlying the design of successful SLS algorithms remains a major challenge. One interesting question in this context concerns the role and degree of randomisation. Although deterministic local search algorithms exist, most high-performance SLS algorithms are highly randomised. It is not clear to which extent it is generally possible to derandomise these algorithms without major losses in robustness or peak performance. Interestingly, there is evidence that in most cases, the quality of the random number source used by the algorithm is not critical and that the number of randomised decisions can often be significantly reduced [55]. However, in most application areas, the use of deterministic algorithms is not inherently preferable, and randomised algorithms have some general advantages, for example with respect to straight-forward and efficient parallelisation techniques. Therefore, although from a scientific point of view, the derandomisation of SLS algorithms poses interesting questions and challenges, its practical importance is somewhat unclear.

Finally, many challenges remain in the context of SLS methods to more complex combinatorial problems, including multi-objective, dynamic and stochastic problems. Furthermore, there appears to be considerable potential in the context of SLS methods for solving continuous optimisation problems and hybrid problems with discrete and continuous components. Overall, there is no doubt that as one of the most versatile and successful approaches for solving hard combinatorial problems, SLS methods will continue to attract the attention of researchers and practitioners from many academic and application areas.

Acknowledgments

Holger H. Hoos acknowledges support provided by the Natural Sciences and Engineering Research Council of Canada (NSERC) under Discovery Grant 238788-05; Thomas Stützle acknowledges support of the Belgian FNRS, of which he is a research associate.

References

1. Hoos, H.H. and Stützle, T., *Stochastic Local Search—Foundations and Applications*, Morgan Kaufmann Publishers, San Francisco, CA, 2004.
2. Feo, T.A. and Resende, M.G.C., A probabilistic heuristic for a computationally difficult set covering problem, *Oper. Res. Lett.*, 8(2), 67, 1989.
3. Resende, M.G.C. and Ribeiro, C.C., Greedy randomized adaptive search procedures, in *Handbook of Metaheuristics*, Glover, F. and Kochenberger, G. (Eds.), Kluwer Academic Publishers, Dordrecht, the Netherlands, p. 219, 2002.
4. Dorigo, M., Maniezzo, V., and Colorni, A., Positive feedback as a search strategy, TR 91-016, Dipartimento di Elettronica, Politecnico di Milano, Italy, 1991.
5. Dorigo, M. and Stützle, T., *Ant Colony Optimization*, MIT Press, Cambridge, MA, 2004.
6. Voß, S., Martello, S., Osman, I.H., and Roucairol, C. (Eds.), *Meta-Heuristics: Advances and Trends in Local Search Paradigms for Optimization*, Kluwer Academic Publishers, Boston, MA, 1999.

7. Bentley, J.L., Fast algorithms for geometric traveling salesman problems, *ORSA J. Comput.*, 4(4), 387, 1992.

8. Jain, A.S., Rangaswamy, B., and Meeran, S., New and "stronger" job-shop neighbourhoods: A focus on the method of Nowicki and Smutnicki, *J. Heuristics*, 6(4), 457, 2000.

9. Yannakakis, M., The analysis of local search problems and their heuristics, in *Proceedings of Symposium on Theoretical Aspects of Computer Science*, LNCS, Springer, Berlin, Germany, Vol. 415, p. 298, 1990.

10. Martin, O.C., Otto, S.W., and Felten, E.W., Large-step Markov chains for the traveling salesman problem, *Comp. Sys.*, 5(3), 299, 1991.

11. Hansen, P. and Mladenović, N., Variable neighborhood search: Principles and applications, *Eur. J. Oper. Res.*, 130(3), 449, 2001.

12. Hansen, P. and Mladenović, N., Variable neighborhood search, in *Handbook of Metaheuristics*, Glover, F. and Kochenberger, G., (Eds.), Kluwer Academic Publishers, Norwell, MA, p. 145, 2002.

13. Ahuja, R.K., Ergun, O., Orlin, J.B., and Punnen, A.P., A survey of very large-scale neighborhood search techniques, *Disc. Appl. Math.*, 123(1–3), 75, 2002.

14. Kernighan, B.W. and Lin, S., An efficient heuristic procedure for partitioning graphs, *Bell Sys. Tech. J.*, 49, 213, 1970.

15. Lin, S. and Kernighan, B.W., An effective heuristic algorithm for the traveling salesman problem, *Oper. Res.*, 21(2), 498, 1973.

16. Glover, F., Ejection chain, reference structures and alternating path methods for traveling salesman problems, *Disc. Appl. Math.*, 65(1–3), 223, 1996.

17. Thompson, P.M. and Orlin, J.B., The theory of cycle transfers, Working Paper OR 200-89, Operations Research Center, MIT, Cambridge, MA, 1989.

18. Thompson, P.M. and Psaraftis, H.N., Cyclic transfer algorithm for multivehicle routing and scheduling problems, *Oper. Res.*, 41, 935, 1993.

19. Ahuja, R.K., Orlin, J.B., and Sharma, D., Multi-exchange neighborhood structures for the capacitated minimum spanning tree problem, *Math. Prog., Ser. A*, 91, 71–97, 2001. doi:10.1007/s I 01070100234.

20. Potts, C.N. and van de Velde, S., Dynasearch: Iterative local improvement by dynamic programming; part I, the traveling salesman problem, Technical Report LPOM–9511, Faculty of Mechanical Engineering, University of Twente, Enschede, the Netherlands, 1995.

21. Congram, R.K., Potts, C.N., and van de Velde, S., An iterated dynasearch algorithm for the single-machine total weighted tardiness scheduling problem, *INFORMS J. Comput.*, 14(1), 52, 2002.

22. Selman, B., Kautz, H., and Cohen, B., Noise strategies for improving local search, in *Proceedings of the National Conference on Artificial Intelligence*, MIT Press, Cambridge, MA, p. 337, 1994.

23. Connolly, D.T., An improved annealing scheme for the QAP, *Eur. J. Oper. Res.*, 46(1), 93, 1990.

24. Fielding, M., Simulated annealing with an optimal fixed temperature, *SIAM J. Optimization*, 11(2), 289, 2000.

25. Kirkpatrick, S., Gelatt Jr., C.D., and Vecchi, M.P., Optimization by simulated annealing, *Science*, 220, 671, 1983.

26. Cerný, V., A thermodynamical approach to the traveling salesman problem, *J. Optimiz. Theor. Appl.*, 45(1), 41, 1985.

27. Dueck, G. and Scheuer, T., Threshold accepting: A general purpose optimization algorithm appearing superior to simulated annealing, *J. Comput. Phys.*, 90(1), 161, 1990.

28. Glover, F. and Laguna, M., *Tabu Search*, Kluwer Academic Publishers, Norwell, MA, 1997.

29. Battiti, R. and Tecchiolli, G., Simulated annealing and tabu search in the long run: A comparison on QAP tasks, *Computer Math. Appl.*, 28(6), 1, 1994.

30. Voudouris, C., *Guided Local Search for Combinatorial Optimization Problems*, PhD Thesis, Department of Computer Science, University of Essex, Colchester, UK, 1997.

31. Voudouris, C. and Tsang, E., Guided local search and its application to the travelling salesman problem, *Eur. J. Oper. Res.*, 113(2), 469, 1999.
32. Hutter, F., Tompkins, D.A.D., and Hoos, H.H., Scaling and probabilistic smoothing: Efficient dynamic local search for SAT, in *Principles and Practice of Constraint Programming*, Van Hentenryck, P. (Ed.), LNCS, Vol. 2470, Springer Verlag, Berlin, Germany, p. 233, 2002.
33. Johnson, D.S. and McGeoch, L.A., Experimental analysis of heuristics for the STSP, in *The Traveling Salesman Problem and its Variations*, Gutin, G. and Punnen, A. (Eds.), Kluwer Academic Publishers, Secaucus, NJ, p. 369, 2002.
34. Lourenço, H.R., Martin, O., and Stützle, T., Iterated local search, in *Handbook of Metaheuristics*, Glover, F. and Kochenberger, G. (Eds.), Kluwer Academic Publishers, Norwell, MA, p. 321, 2002.
35. Jacobs, L.W. and Brusco, M.J., A local search heuristic for large set-covering problems, *Naval Res. Logis. Quart.*, 42(7), 1129, 1995.
36. Marchiori, E. and Steenbeek, A., An evolutionary algorithm for large scale set covering problems with application to airline crew scheduling, in *Real-World Applications of Evolutionary Computing*, Cagnoni, S. et al. (Eds.), LNCS, Vol. 1803, Springer Verlag, Berlin, Germany, p. 367, 2000.
37. Ruiz, R. and Stützle, T., A simple and effective iterated greedy algorithm for the permutation flowshop scheduling problem, *Eur. J. Oper. Res.*, 177, 2033–2039, 2007.
38. Dorigo, M., Maniezzo, V., and Colorni, A., Ant system: Optimization by a colony of cooperating agents, *IEEE Trans. Sys. Man Cybernet. – Part B*, 26(1), 29, 1996.
39. Stützle, T. and Hoos, H.H., $\mathcal{MAX}$–$\mathcal{MIN}$ Ant system, *Future Gen. Comp. Sys.*, 16(8), 889, 2000.
40. Dorigo, M. and Di Caro, G., The ant colony optimization meta-heuristic, in *New Ideas in Optimization*, Corne, D., Dorigo, M., and Glover, F. (Eds.), McGraw-Hill, London, UK, p. 11, 1999.
41. Potvin, J.Y., Genetic algorithms for the traveling salesman problem, *Annals Oper. Res.*, 63, 339, 1996.
42. Merz, P. and Freisleben, B., Memetic algorithms for the traveling salesman problem, *Comp. Sys.*, 13(4), 297, 2001.
43. Moscato, P., Memetic algorithms: A short introduction, in *New Ideas in Optimization*, Corne, D., Dorigo, M., and Glover, F., (Eds.), McGraw-Hill, New York, p. 219, 1999.
44. Glover, F., Heuristics for integer programming using surrogate constraints, *Decision Sci.*, 8, 156, 1977.
45. Glover, F., Laguna, M., and Martí, R., Scatter search and path relinking: Advances and applications, in *Handbook of Metaheuristics*, Glover, F. and Kochenberger, G. (Eds.), Kluwer Academic Publishers, Norwell, MA, p. 1, 2002.
46. Laguna, M. and Martí, R., *Scatter Search: Methodology and Implementations in C*, Vol. 24, Kluwer Academic Publishers, Dordrecht, the Netherlands, 2003.
47. Hajek, B., Cooling schedules for optimal annealing, *Math. Oper. Res.*, 13(2), 311, 1988.
48. Gutjahr, W.J., ACO algorithms with guaranteed convergence to the optimal solution, *Inf. Proc. Lett.*, 82(3), 145, 2002.
49. Hoos, H.H., On the run-time behaviour of stochastic local search algorithms for SAT, in *Proceedings of the National Conference on Artificial Intelligence*, MIT Press, Cambridge, MA, p. 661, 1999.
50. Stützle, T. and Hoos, H.H., Analysing the run-time behaviour of iterated local search for the travelling salesman problem, in *Essays and Surveys on Metaheuristics*, Hansen, P. and Ribeiro, C.C. (Eds.), Kluwer Academic Publishers, Norwell, MA, p. 589, 2001.
51. Van Hentenryck, P. and Michel, L., *Constraint-Based Local Search*, MIT Press, Cambridge, UK, 2005.
52. Voß, S. and Woodruff, D.L. (Eds.), *Optimization Software Class Libraries*, Kluwer Academic Publishers, Boston, MA, 2002.

53. Merz, P. and Freisleben, B., Fitness landscapes and memetic algorithm design, in *New Ideas in Optimization*, Corne, D., Dorigo, M., and Glover, F. (Eds.), McGraw-Hill, London, UK, p. 244, 1999.

54. Watson, J.-P., Whitley, L.D., and Howe, A.E., Linking search space structure, run-time dynamics, and problem difficulty: A step towards demystifying tabu search, *J. Arti. Intelli. Res.*, 24, 221, 2005.

55. Hoos, H.H. and Tompkins, D.A.D., On the quality and quantity of random decisions in stochastic local search for SAT, in *Advances in Artificial Intelligence, Conference of the Canadian Society for Computational Studies of Intelligence*, Vol. 4013, Springer, Berlin, Germany, p. 146–148, 2006.

18

Very Large-Scale Neighborhood Search: Theory, Algorithms, and Applications

Ravindra K. Ahuja

Özlem Ergun

James B. Orlin

Abraham P. Punnen

18.1 Introduction

A combinatorial optimization problem (COP) P consists of a collection of its instances. An instance of P can be represented as an ordered pair (F, f), where F is the family of feasible solutions and f is the objective function, which is used to compare two feasible solutions. The family F is formed by subsets of a finite set $E = \{1, 2, \ldots, m\}$ called the ground set. The objective function $f : F \rightarrow Q^+ \cup \{0\}$ assigns a nonnegative cost to every feasible solution S in F. The Traveling Salesman Problem (TSP) and the Minimum Spanning Tree Problem are typical examples of COPs.

Most of the COPs of practical interest are NP-hard. Local search is one of the primary solution approaches for computing an approximate solution for such hard problems. To describe a local search algorithm formally, we need the concept of a neighborhood for each feasible solution. A *neighborhood function* for an instance (F, f) of a COP P is a mapping $N_F : F \rightarrow 2^F$. We assume that N_F does not depend on the objective function f. For convenience, we usually drop the subscript and simply write N. For a feasible solution S, $N(S)$ is called the neighborhood of S. We assume that $S \in N(S)$. For a minimization problem, a feasible solution $\bar{S} \in F$ is said to be *locally optimal* with respect to N if $f(\bar{S}) \leq f(S)$ for all $S \in N(\bar{S})$. Then the *local search problem* is to find a locally optimal solution for a given COP.

Classic neighborhood functions studied in the combinatorial optimization literature include the 2-opt and 3-opt neighborhoods for the TSP [1,2], the flip neighborhood for Max Cut [3], and the swap neighborhood for graph partitioning [3,4]. The sizes of these neighborhoods are polynomial in the problem size. However, the class of local search algorithms that we are considering here primarily concentrates on neighborhoods of very large size, often exponential in the problem size. For the TSP, this class includes variable depth neighborhoods, ejection chain neighborhoods, pyramidal tours neighborhoods, permutation tree neighborhoods, neighborhoods induced by polynomial-time solvable special cases, and so on. For partitioning problems, various multi-exchange neighborhoods studied in literature have this property.

Roughly speaking, a local search algorithm starts with an initial feasible solution and then repeatedly searches the neighborhood of the "current" solution to find better and better solutions until it reaches a locally optimal solution. Computational studies of local search algorithms and their variations have been extensively reported in the literature for various COPs (see, e.g., [2] and [5] for studies of the graph partitioning problem and the TSP, respectively). Empirically, local search heuristics appear to converge rather quickly, within low-order polynomial time. In general, the upper bound on the guaranteed number of improving moves is pseudopolynomial.

This chapter is organized as follows. In Section 18.2, we consider various applications of very large-scale neighborhood (VLSN) search algorithms and discuss in detail how to develop such algorithms using multi-exchanges for various partitioning problems. Section 18.3 deals with theoretical concepts of extended neighborhoods and linkages with domination analysis of algorithms. The results in this section reaffirm the importance of using large-scale neighborhoods in local search. In Section 18.4, we deal with performance guarantees for computing an approximation to a local minimum. This is especially relevant for VLSN search. Searching a large-scale neighborhood is sometimes NP-hard and thus approximation algorithms are used. In these cases, the algorithm may terminate at a point that is not a local minimum.

18.2 VLSN Search Applications

VLSN search algorithms have been successfully applied to solve various optimization problems of practical interest. This includes the capacitated minimum spanning tree (CMST) problem [6–8,61], vehicle routing problems [9–12], the TSP [11–16], the weapon target assignment problem [17], the generalized assignment problem [18–20], the plant location problem [21], parallel machine scheduling problems [22], airline fleet assignment problems [23–25], the quadratic assignment problem [26], pickup and delivery problems with time windows [4,27,57], the multiple knapsack problem [28], manufacturing problems [29], optimization problems in intensity-modulated radiation therapy (IMRT) treatment planning [30], school timetabling problems [31], the graph coloring problem [32], and so on. The successful design of an effective VLSN search algorithm depends on the ability to identify a good neighborhood function and the ability to design an effective exact or heuristic algorithm to search the neighborhood for an improved solution. A VLSN search algorithm can be embedded within a metaheuristic framework such as tabu search [33], genetic algorithms [34], scatter search [35], greedy randomized adaptive search procedure (GRASP) [36], and so on in order to achieve further enhances in performance. Simulated annealing may be used in principle but is not likely to be a successful approach when the size of the neighborhood is exponentially large.

Researchers have used various techniques for developing good neighborhood functions that lead to effective VLSN search algorithms. This includes multi-exchange [6,7,37,38], ejection chains [39], variable depth methods [40,41], integer programming [10], weighted matching [10,12,13,15,16], set partitioning [10], and so on. A comprehensive discussion of all these techniques is outside the scope of this chapter. Applications of ejection chains in local search are discussed in other chapters of this book and hence we will not discuss it here. To illustrate the features of a VLSN search algorithm we primarily concentrate on applications of multi-exchange neighborhoods originally developed by Thompson and Orlin [42] and Thompson and Psaraftis [38]. It is one of the successful approaches for developing VLSN search algorithms for various partitioning problems [6,7,28].

18.2.1 Partitioning Problems and Multi-Exchange Neighborhoods

A subset B of the ground set E is said to be a *feasible subset* if it satisfies a collection of prescribed conditions, called *feasibility conditions*. We assume there is a *feasibility oracle* ζ that verifies whether a subset B

of E satisfies the feasibility conditions. A partition $S = \{S_1, S_2, \ldots, S_p\}$ is said to be *feasible* if S_i is a feasible subset for each $i = 1, 2, \ldots, p$. Let $c : 2^E \to \mathbb{Q}$ be a prescribed cost function. The cost $C(S)$ of the partition $S = \{S_1, S_2, \ldots, S_p\}$ is defined as $C(S) = \sum_{i=1}^{p} c(S_i)$. *Partitioning problems* are the subclass of COP where the family of feasible solutions F is the collection of all feasible partitions of E with $C(.)$ as the objective function.

Partitioning problems are typically NP-hard even if the feasibility oracle ζ and evaluation of the cost $c(.)$ of the subsets of a partition run in polynomial time. Several well-studied COPs are special cases (or may be viewed as special cases) of partitioning problems. These include the CMST problem [6,7,37,43], the generalized assignment problem [18–20], vehicle routing problems [10,22], graph partitioning problems [4,41], the k-constrained multiple knapsack problem [28], and so on. We will discuss some of these problems in detail later. Next, we develop a VLSN search algorithm for the general partitioning problem.

The first step in the design of a VLSN search algorithm is to develop a suitable neighborhood function. Let $S = \{S_1, S_2, \ldots, S_p\}$ be a feasible partition. A *cyclic exchange* of S selects a collection $\{S_{\pi_1}, S_{\pi_2}, \ldots, S_{\pi_q}\}$ of subsets from S and moves an element say b_i from subset S_{π_i} to subset $S_{\pi_{i+1}}$ for $i = 1, 2, \ldots, q$, where $q + 1 = 1$ (Figure 18.1). Let S^* be the resulting partition. If S^* is a feasible partition, then we call the exchange a *feasible cyclic exchange*. It can be verified that

$$C(S^*) = C(S) + \sum_{i=1}^{q} \left(c\left(\{b_{l-1}\} \cup S_{\pi_i} \setminus \{b_l\}\right) - c(S_{\pi_i}) \right) \qquad (18.1)$$

where b_0 is defined to be b_q. If $q = 2$, the cyclic exchange reduces to a 2-exchange, an operation well studied for various partitioning problems [5,14,44]. In a (feasible) cyclic exchange, if we omit the move from S_{π_q} to $S_{\pi_{q+1}}(= S_{\pi_1})$ the resulting exchange is called *a (feasible) path exchange*. For $q = 2$, a path exchange reduces to the shift operation, again well studied for various partitioning problems [44].

The cyclic and path exchanges discussed earlier were introduced by Thompson and Orlin [42] and Thompson and Psaraftis [38] and subsequently used by several researchers in the context of various partitioning problems.

Given a feasible partition S, let $Z(S)$ be the collection of all feasible partitions of E that can be obtained by cyclic exchanges from S. Similarly, let $P(S)$ be the collection of feasible partitions of E that can be obtained by a path exchange from S. We call $Z(S)$ as the *cyclic exchange neighborhood* of S, and we call $P(S)$ as the *path exchange neighborhood* of S. The union $Z(S) \cup P(S)$ is called the *multi-exchange neighborhood* of S. The cardinality of the neighborhoods $Z(S)$ and $P(S)$ is often exponential. A precise estimate of this cardinality depends on various parameters including the nature of the feasibility oracle ζ. Thus $Z(S)$ and $P(S)$ qualify as very large-scale neighborhoods that can be used in the design of a VLSN search algorithm.

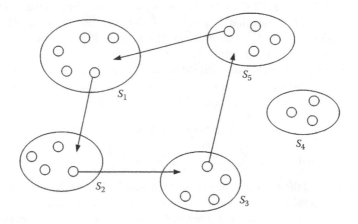

FIGURE 18.1 A cyclic exchange.

Besides the size of the neighborhood, the power of a VLSN search algorithm also depends on our ability to obtain an improved solution in the neighborhood. Given a feasible solution S, the problem of finding an improved solution in a given neighborhood is called a *local augmentation problem* [45]. We next discuss how to solve the local augmentation problem for $Z(S)$. This is achieved by (approximately) solving an optimization problem on an associated structure called the *improvement graph*.

The improvement graph associated with a feasible partition is denoted by $G(S)$. Its node set is E, which is also the ground set. Given a partition S, for each element i of E, let $\theta(i)$ denote the index of the subset in which it belongs in the partition S. That is, for $i \in S_r$, $\theta(i) = r$. The arc set $A = A(G(S))$ of $G(S)$ is constructed as follows. An arc $(i, j) \in A$ signifies that element i leaves subset $S_{\theta(i)}$ and enters subset $S_{\theta(j)}$, whereas element j leaves $S_{\theta(j)}$. So,

$$A = \left\{ (i, j) : \ i, j \in E, \ \theta(i) \neq \theta(j) \ \text{and} \ \{i\} \cup S_{\theta(j)} \backslash \{j\} \text{is a feasible subset} \right\}$$

The cost α_{ij} of arc $(i, j) \in A$ is given by

$$\alpha_{ij} = c(\{i\} \cup S_{\theta(j)} \backslash \{j\}) - c(S_{\theta(j)}). \tag{18.2}$$

A directed cycle $< i_1 - i_2 - \ldots - i_r - i_1 >$ in the improvement graph is said to be *subset disjoint* if $\theta(i_p) \neq \theta(i_q)$ for $p \neq q$. In this definition, if we replace "cycle" by "path," we get a *subset disjoint directed path*.

Lemma 18.1 [42] *There is a one-to-one cost-preserving correspondence between cyclic exchanges with respect to S and subset-disjoint directed cycles in $G(S)$.*

A negative cost subset disjoint directed path (cycle) in $G(S)$ is called a *valid path* (*cycle*). In view of Lemma 18.1, a valid cycle in $G(S)$ yields an improved partition. If $G(S)$ contains no valid cycle, then S is locally optimal with respect to $N(S)$. Unfortunately, finding such a valid cycle is NP-hard [42]. Thus the local augmentation problem for the neighborhood $N(S)$ is also NP-hard.

18.2.1.1 Local augmentation algorithm for N(S)

Ahuja et al. [7] proposed a heuristic algorithm to compute a valid cycle by modifying the label correcting algorithm for shortest paths. An exact algorithm based on dynamic programming (implicit enumeration) to solve this problem was introduced in Reference 6. We briefly discuss the algorithm, and refer the reader to Reference 6 for further details.

For any subgraph H of $G(S)$, its cost $\alpha(S)$ is given by $\alpha(S) = \sum_{ij \in P} \alpha_{ij}$, where ij is shorthand for (i, j). For a directed path P in $G(S)$, let $tail(P)$ denotes the start node, $head(P)$ denotes the end node and $label(P)$ denotes the set $\{\theta(i) : \ i \in P\}$. We say that a path P_1 dominates another path P_2 in $G(S)$ if $\alpha(P_1) < \alpha(P_2)$, $tail(P_1) = tail(P_2)$, $head(P_1) = head(P_2)$, and $label(P_1) = label(P_2)$. If P_1 dominates, P_2 and P_2 are part of a valid cycle $\mathbb{C}$, then clearly $P_1 \cup \mathbb{C} \backslash P_2$ contains a valid cycle with cost no more than $\alpha(\mathbb{C})$. For each path P, we associate a triplet $(tail(P), head(P), label(P))$ called the *key value* of P. In our search for a valid cycle in $G(S)$, among paths with the same key value, we only need to consider one with the smallest cost. The following lemma further cuts down the number of paths needed to be considered.

Lemma 18.2 [40] *If $W = < i_1 - i_2 - \ldots - i_r - i_1 >$ is a negative cost cycle, then there exists a node i_h in W such that each directed path $i_h - i_{h+1}$, $i_h - i_{h+1} - i_{h+2}, \ldots, i_h - i_{h+1} - i_{h+2} - \ldots - i_{h-1}$ (where indexes are modulo r) is a negative cost directed path.*

In view of Lemma 18.2 and the preceding discussions, we need only consider nondominated valid paths as candidates for forming valid cycles. Let $\mathcal{P}_k$ be the set of all valid non-dominated paths of length k in $G(S)$. The algorithm of Ahuja et al. [6] progressively generates $\mathcal{P}_k$ for larger values of k. The algorithm enumerates valid paths using a forward dynamic programming recursion. It first obtains all valid paths of length 1 and uses these paths to generate valid paths of length 2 and so on. This process is repeated until we have obtained all valid paths of length R for some given length R or until we find a valid cycle. From each valid path, candidate cycles are examined by connecting the head node with the tail node by an arc in $G(S)$ if such an arc exists. A formal description of the valid cycle detection algorithm is given in the following:

ALGORITHM   Valid Cycle Detection

begin

    $P := \{(i,j) \in A(G(S)): \alpha_{ij} < 0\}$;

    $k := 1$;

    $W^* := \varnothing; \alpha(W^*) = \infty$;

    while $k < R$ and $\alpha(W^*) \geq 0$ **do**

    begin

        while $P_k \neq \varnothing$ **do**

        begin

            remove a path P from P_k

            let $i := head(P)$ and $h := tail(P)$

            if(i, h) $\in A(G(S))$ **and** $\alpha(P) + \alpha_{ih} < \alpha(W^*)$ **then** $W^* := P \cup \{(i,h)\}$;

            for each $(i,j) \in A(i)$ **do**

                if $label(j) \notin label(P)$ **and** $\alpha(P) + \alpha_{ij} < 0$ **then**

                begin

                    add the path $P \cup \{(i,j)\}$ to P_{k+1}

                    if P_{k+1} contains another path with the same key as

                    $P \cup \{(i,h)\}$ **then** remove the dominated path;

                endif;

            endfor

        endwhile;

        $k = k + 1$;

    endwhile;

    return W^*;

end;

Although the above-mentioned algorithm is not polynomial, it worked very well in practice for some partitioning problems on which the algorithm was tested [6,7,28].

Path exchange as a cyclic exchange

Let us now consider how a valid path can be obtained if $G(S)$ has one. Note that by Lemma 18.2, every valid cycle of length r gives $r-1$ valid paths. Sometimes path exchanges resulting from these valid paths may be better than the best cyclic exchange. Even in the absence of a valid cycle, it is possible that $G(S)$ may have valid paths that lead to an improved solution. Let S be a partition containing p subsets, and suppose that S^* is a partition obtained from S by a cyclic exchange. In general, S^* will also contain p subsets, and they will have the same cardinalities as the original subsets of S. These are limitations of

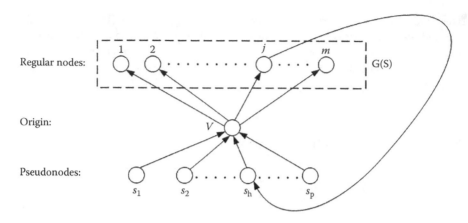

FIGURE 18.2 Construction of the improvement graph $\tilde{G}(S)$.

the cyclic exchange. One approach to overcoming this drawback is to periodically explore the possibility of splitting one or more subsets within a partition or merging two or more subsets within a partition. In the case of communication spanning tree problems, some cyclic exchange moves lead to such simple split operations in a natural way. Interestingly, path exchange moves have the natural property of the possibility of decreasing the number of subsets in a partition. By allowing the possibility of an empty set, an improvement graph can be constructed where a path exchange can even increase the number of subsets while moving from one partition to another. This variation of path exchange is called *enhanced path exchange*.

Interestingly, we now observe that path exchanges can be viewed as cyclic exchanges in a modi-fied improvement graph $\tilde{G}(S) = (\tilde{V}, \tilde{A})$, which is a supergraph of G(S). Introduce p new nodes (called *pseudonodes*) $s_1, s_2, \ldots, s_p$ and another node v, called the *origin node*. Thus the node set $\tilde{V}$ of $\tilde{G}(S)$ is $E \cup \{s_1, s_2, \ldots, s_p, v\}$. The graph $\tilde{G}(S)$ contains all arcs of G(S) along with additional arcs from the set:

$$\{(v,j) : j \in E\} \cup \{(j, s_i) : i \neq \theta(j) \text{ and } S_i \cup \{j\} \text{ is a feasible subset}\} \cup \{(s_i, v) : i = 1, 2.., p\}$$

(Figure 18.2). The cost of these additional arcs are defined as follows: $\alpha_{vj} = c(S_{\theta(j)} \backslash \{j\}) - c(S_{\theta(j)})$ for all $j \in E$, $\alpha_{s_i v} = 0$ for $i = 1, 2, \ldots, p$ and $\alpha_{js_i} = c(S_i \cup \{j\}) - c(S_i)$ for all $(j, s_i) \in \{(j, s_i) : i \neq \theta(j)$ and $S_i \cup \{j\}$ is a feasible subset}.

Note that a directed cycle $\mathbb{C}$ in $\tilde{G}(S)$ containing the node v will have exactly one pseudonode and the remaining nodes are regular. Further, these regular nodes form a path in $\mathbb{C}$ called the *regular path* associated with $\mathbb{C}$. A directed cycle $\mathbb{C}$ containing the node v in the graph $\tilde{G}(S)$ is said to be *subset disjoint* if its associated regular path is subset disjoint.

Lemma 18.3 [7] *There is a one-to-one cost preserving correspondence between path exchanges with respect to the partition S and subset-disjoint directed cycles in $\tilde{G}(S)$ containing the origin node v.*

If a valid cycle in the improvement graph contains the origin node v, it corresponds to an improving path exchange. If it does not contain v, it corresponds to an improving cyclic exchange. Moreover, the improve-ment graph $\tilde{G}(S)$ can be further modified by adding a new node and appropriate arcs so that improving enhanced path exchanges can be identified. A VLSN search algorithm based on a multi-exchange (path or cyclic) can be described as follows:

ALGORITHM   Multi-Exchange

begin

compute a feasible solution S to the partitioning problem;

construct the improvement graph $\tilde{G}(S)$;

while $\tilde{G}(S)$ contains a valid cycle **do**

obtain a valid cycle W in $\tilde{G}(S)$;

If W contains the origin node v **then**

perform a path exchange using the corresponding regular path;

else

perform a cyclic exchange corresponding to W;

endif

update S and $\tilde{G}(S)$;

endwhile;

end.

Many problem specific features can be used to simplify calculations in the VLSN search algorithm discussed earlier for the partitioning problem. The complexity of the construction of the improvement graph depends on that of the feasibility oracle and on the evaluation of the cost function $c(.)$. Also, problem-specific information may be used to update the improvement graph from iteration to iteration efficiently.

The Capacitated Minimum Spanning Tree Problem

Let G be a graph on the node set $V \cup \{0\}$ where $V = \{1, 2, \ldots, n\}$. Nodes in V are called *terminal nodes*, and the node 0 is called the *central node*. For each arc (i, j), a nonnegative cost c_{ij} is prescribed. Also for each node $i \in V$, a nonnegative demand w_i is prescribed. Let L be a given real number. For any spanning tree T of G, let $T_1, T_2, \ldots, T_{K_T}$ denote the components of T-$\{0\}$. Then the CMST problem is to

$$\text{Minimize} \sum_{ij \in T} c_{ij}$$

subject to

$$T \text{ is a spanning tree of } G$$

$$\sum_{i \in T_j} w_i \leq L, j = 1, 2, \ldots, K_T \tag{18.3}$$

For any subgraph H of G, we sometimes let $V(H)$ denote its node set and let $E(H)$ denote its edge set. We also use the notation $ij \in H$ $(i \in H)$ to represent $(i, j) \in E(H)$ $(i \in V(H))$ when there is no ambiguity. It is easy to see that for an optimal spanning tree T, the component T_i must be a minimum spanning tree of the subgraph of G induced by $V(T_i), i = 1, 2, \ldots, K_T$.

Thus CMST can be viewed as a partition problem where the ground E is the set V of nodes in G, the feasibility oracle is to verify the condition $\sum_{i \in S_j} w_i \leq L$. The cost $c(S_i)$ is the cost of the minimum spanning tree T_i of the subgraph of G induced by $S_i \cup \{0\}$. Without loss of generality, we assume that node 0 is a pendant node (node of degree 1) of T_i. Otherwise, we can decompose T_i into a number of subtrees equal to the degree of node 0 where 0 is a pendant node in each such subtree, yielding an alternative feasible solution.

TABLE 18.1   Evaluation of the Composite Neighborhood Algorithm

Problem ID	# of Nodes	# of Arcs	Capacity	Best Available Solution	Composite Neighborhood Solution
tc80-3	81	3,240	10	880	878
CM50-3	50	1,225	400	735	732
CM50-4	50	1,225	400	567	564
CM50-5	50	1,225	400	612	611
CM50-2	50	1,225	800	515	513
CM100-1	100	4,950	200	520	516
CM100-2	100	4,950	200	602	596
CM100-3	100	4,950	200	549	541
CM100-4	100	4,950	200	444	437
CM100-5	100	4,950	200	427	425
CM100-1	100	4,950	400	253	252
CM100-5	100	4,950	400	224	223
CM200-1	200	19,900	200	1037	1017
CM200-2	200	19,900	200	1230	1221
CM200-3	200	19,900	200	1367	1365
CM200-4	200	19,900	200	942	927
CM200-5	200	19,900	200	981	965
CM200-1	200	19,900	400	399	397
CM200-2	200	19,900	400	486	478
CM200-3	200	19,900	400	566	560
CM200-4	200	19,900	400	397	392
CM200-5	200	19,900	400	425	420
CM200-1	200	19,900	800	256	254
CM200-3	200	19,900	800	362	361
CM200-4	200	19,900	800	276	275
CM200-5	200	19,900	800	293	292
CM200-2	200	19,900	400	486	478
CM200-3	200	19,900	400	566	560
CM200-4	200	19,900	400	397	392
CM200-5	200	19,900	400	425	420
CM200-1	200	19,900	800	256	254
CM200-3	200	19,900	800	362	361
CM200-4	200	19,900	800	276	275
CM200-5	200	19,900	800	293	292

The algorithm multi-exchange can be used to find a heuristic solution for the CMST. For details on implementation aspects of this heuristic specifically for the CMST problem, we refer to Reference 6. Instead of using the simple node exchange, Ahuja et al. [6,7] also considered subtree exchange neighborhoods and composite neighborhoods [6]. The node exchange worked well for problems where the node weights are identical (homogeneous problems). Subtree exchanges worked well for problems with different node weights (heterogeneous problems). The composite neighborhood worked well for both class of problems and produced improved solutions for several benchmark problems. Table 18.1 (taken from [6]) summarizes these results.

18.3 Extended Neighborhoods and Domination Analysis

Let us now look at some theoretical issues related to performance guarantees of VLSN search algorithms. Glover and Punnen [46] introduced the concept of domination ratio to assess the performance guarantee of a heuristic algorithm. Let α be a heuristic algorithm for a COP P. Then the domination ratio of α, denoted by $dom(\alpha)$, is

$$\underset{(F,f)}{Inf} \mid \left\{ S \in F \; : \; f(S) \geq f(S_\alpha) \right\} \mid / |F|,$$

where S_α is the solution obtained by the heuristic α on instance (F, f). Note that $0 < dom(\alpha) \leq 1$ and $dom(\alpha) = 1$ if and only if α guarantees an optimal solution.

Identifying tight deterministic bounds on domination ratio could be difficult for many algorithms. For various recent works on domination analysis of algorithms, we refer to [15,29,47,48]. For VLSN search algorithms for a COP,

$$\underset{(F,f)}{Inf} \left\{ \frac{|N(S)|}{|F|} \; : \; S \in F \right\}$$

gives a trivial lower bound on the domination ratio, provided we can find an improving solution in $N(S)$ if exists.

We next introduce the concept of extended neighborhoods, which can be used to find improved domination ratios for a VLSN algorithm. The value of extended neighborhoods goes beyond establishing improved domination ratios. It generalizes the concept of exact neighborhoods and provides some theoretical insights on why the VLSN search algorithms work well in practice. We will address this aspect later in this section. Let us first consider some basic definitions and properties. For details of the theory of extended neighborhoods, we refer to the Reference 49.

Consider an instance $I = (F, f)$ of a COP and let N be a neighborhood function defined on it. Let L_I^N denote the collection of all locally optimal solutions of I with respect to N. Two neighborhoods N^1 and N^2 are said to be LO-equivalent for a COP P if and only if $L_I^{N^1} = L_I^{N^2}$ for all instances I of P. A neighborhood function N^* of P is called an *extended neighborhood* of N if

1. N^* and N are LO-equivalent
2. For every neighborhood function N^0 that is LO-equivalent to N, $N^0(S) \subseteq N^*(S)$ for all $S \in F$ and for all $I = (F, f)$ of P

Equivalently, N^* is the largest neighborhood that is LO-equivalent to N for P. If $N^* = F$, then N is an exact neighborhood [41]. (A neighborhood is *exact* if every local optimum is a global optimum.) In this way, the concept of an extended neighborhood generalizes the concept of exact neighborhoods.

Note that the domination ratio of any local search algorithm using the neighborhood N will be at least

$$\underset{(F,f)}{Inf} \left\{ \frac{|N^*(S)|}{|F|} \; : \; S \in F \right\}.$$

The well-known 2-opt neighborhood for the TSP on n nodes contain $n(n-1)/2$ elements, whereas the extended neighborhood of 2-opt for the TSP contains at least $(n/2)!$ elements (see 50).

A COP is said to have linear cost if $f(x) = cx$, where $c = (c_1, c_2, \ldots, c_m)$ is the vector element costs and x is the incidence vector representing a feasible solution. Note that each feasible solution S can be represented by its incidence vector $x = (x_1, x_2, \ldots, x_m)$ where

$$x_i = \begin{cases} 1 & \text{if } i \in S \\ 0 & \text{if } i \notin S. \end{cases}$$

We denote by F_x the collection of all incidence vectors of elements of F. When x is the incidence vector of S, we sometimes denote $N(S)$ by $N(x)$. Let $N_F(x) = \{x^1, x^2, \ldots, x^K\}$ for some K. Then for $i = 1$ to K, let $v^i = x^i - x$. We refer to v^i as a *neighborhood vector* and let $V_F^N(x)$ be an $m \times K$ matrix whose ith column is v^i. $V_F^N(x)$ is called the matrix of neighborhood vectors at x. The following lemma characterizes extended neighborhoods for COPs with a linear cost function.

Lemma 18.4 **[49]** *Let P be a COP with a linear cost function and let N be an associated neighborhood function. Let F be the collection of feasible solutions of any instance and let $V_F^N(x)$ be the matrix of neighborhood vectors. Then the extended neighborhood N_F^* of N_F is given by*

$$N_F^*(x) = \left\{ \hat{x} \in F_x \ : \ \hat{x} = x + V_F^N(x)\lambda; \ \lambda \geq 0 \text{ and } \lambda \in \Re^{|N_F(x)|} \right\}$$

To illustrate the characterization of extended neighborhood in Lemma 18.4, let us construct the extended neighborhood of the 2-opt neighborhood for the TSP. In the definition of COP with a linear cost function, if E is selected as the edge set of a complete graph G and if F is selected as the family of all tours in G, we get an instance of the TSP. Let the number of nodes in G be n. A tour T in G is represented as a sequence $<i_1, i_2, \ldots, i_n, i_1>$, where $(i_k, i_{k+1}) \in T$ for $k = 1, 2, \ldots, n-1$ and $(i_n, i_1) \in T$. Given a tour $T = <i_1, i_2, \ldots, i_n, i_1>$, a 2-opt move can be represented as a set $\{k, l\}$ where i_k and i_l are nonadjacent in T. The move $\{k, l\}$ represents the operations of removing edges (i_k, i_{k+1}), (i_l, i_{l+1}) from T and adding edges (i_k, i_l), (i_{k+1}, i_{l+1}) to $T-\{(i_k, i_{k+1}), (i_l, i_{l+1})\}$, where the index $n+1 = 1$. The 2-opt neighborhood of a tour T denoted by 2-opt(T) is the collection of all tours in G that can be obtained by a 2-opt move from T. Let $T^{k,l}$ be the tour obtained from T by the 2-opt move $\{k, l\}$. The neighborhood vector corresponding to this move denoted by $v_T^{k,l}$ is given by

$$v_T^{k,l}(e) = \begin{cases} -1 & \text{if } e = (i_k, i_{k+1}) \text{ or } e = (i_l, i_{l+1}) \\ 1 & \text{if } e = (i_k, i_l) \text{ or } e = (i_{k+1}, i_{l+1}) \\ 0 & \text{otherwise.} \end{cases}$$

If x is the incidence vector of T, then $x+v_T^{k,l}$ is the incidence vector of $T^{k,l}$. Thus by Lemma 18.4, we have the extended neighborhood of 2-opt(T) denoted by 2-opt*(T), which is given by

$$\left\{ x' : x' \text{ is a tour in G and } x' = x + \sum_{\{k,l\} \in 2-opt} \lambda_{kl} v_T^{k,l}, \lambda_{kl} \geq 0 \right\} \qquad (18.4)$$

Lemma 18.5 **[49]**

 (i) $|2 - opt^*(T)| \geq (\lceil n/2 \rceil - 3)!(n - 3)$
 (ii) *Finding the best tour in 2-opt*(T) is NP-hard*

By definition, 2-opt and 2-opt* have the same set of local optima but the size of 2-opt*(T) is exponentially larger than the size of 2-opt. We now show that using the same starting solution, a local search with respect to 2-opt* can terminate at a solution that cannot be reached by a local search with respect to 2-opt.

Lemma 18.6 **[49]** *Let T be any tour with at least 11 nodes. There is a cost vector c and a tour T^* such that T^* is not reachable from T by local search with respect to 2-opt, but T^* is reachable from T by a local search with respect to 2-opt*.*

Proof. Without loss of generality assume that $T = <1, 2, \ldots, n, 1>$. Let all edges of T have cost 0. Introduce edges (1, 5), (2, 6), (3, 8), and (4, 9) with cost -2 and introduce edges (6, 10) and (7, 11) with cost 1. Complete the graph by introducing the remaining edges with a large cost. (Figure 18.3).

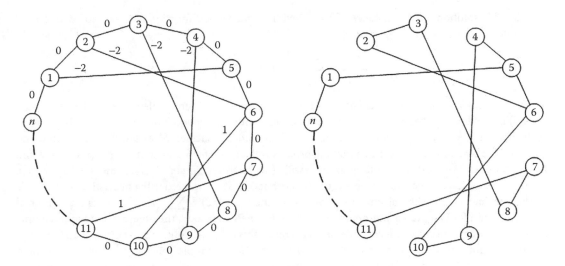

FIGURE 18.3 The graph constructed in the proof of Equation 18.6 and tour T^*.

It can be verified that $T^* = < 1, 5, 4, 9, 10, 6, 2, 3, 8, 7, 11, 12, 13, \ldots, n, 1 >$ is an optimal tour with cost -6. Since $T^* = T + v_T^{1,5} + v_T^{6,10} + v_T^{3,8}$, by Lemma 18.4, $T^* \in 2\text{-opt}^*$. T^* however is not reachable from T using improving exchanges as the moves $\{1, 5\}$ and $\{3, 8\}$ lead to local optima different from T^* and $\{6, 8\}$ is not an improving move. This completes the proof. ∎

Thus local search with respect to extended neighborhoods fundamentally offers more possibilities than does local search using the original neighborhood. This observation further strengthens our confidence in using VLSN search algorithms to obtain good quality solutions.

18.4 Approximate Local Search

One of the most important theoretical questions in VLSN search (and local search in general) is to find a good bound on the number of improvement steps required to reach termination. In addressing this important theoretical question, Johnson et al. [5] introduced the complexity class *PLS*. A COP together with a neighborhood function belongs to class PLS if

1. Its instances are polynomial-time recognizable and an initial feasible solution is efficiently computable
2. The feasibility of a proposed solution can be checked in polynomial time
3. The local optimality of a solution with respect to the neighborhood under consideration can be verified in polynomial time and if it is not locally optimal, an improving solution can be obtained in polynomial time

The class PLS has its own reduction scheme that leads to the class of PLS-complete problems. If a local optimum can be found in polynomial time for one of the PLS complete problems, then a local optimum for every other PLS-complete problem can be found in polynomial time. For example, the TSP with Lin-Kernighan neighborhood [40], graph partitioning with swap neighborhood, and MAX CUT with flip neighborhood are all PLS-complete [3,52,58–60]. It is an outstanding open question whether there is a polynomial time algorithm for finding a locally optimal solution for a PLS-complete problem.

In view of this open question, it is reasonable to ask whether it is possible to efficiently compute a solution "close" to a local optimum for a PLS-complete problem. This question is equally relevant for problems not in the class PLS, where the neighborhoods are searched by approximation algorithms. In this case, it is interesting to see how close the solution produced is to a locally optimal solution.

A feasible solution S to an instance of a COP with neighborhood function N is said to be an *ε-local optimum* [45] if

$$\frac{f(S^*) - f(S)}{f(S)} \leq \varepsilon \quad \text{for all } S \in N(S)$$

where $\varepsilon > 0$. Computing an *ε-local optimum* is relatively easy [45] for problems where the cardinality of $N(S)$ is small (i.e., polynomial) for all S. Let S be a current solution and if no solution S^* exists in $N(S)$ such that $f(S^*) < f(S)/(1 + \varepsilon)$, then S is an *ε-local optimum*. If such an S^* exists, then move to it and continue the search. It can be verified that the process will terminate in polynomial time (for fixed ε and integer cost function) and the resulting solution will be an *ε-local optimum*. This scheme is not applicable in many VLSN search algorithms where the neighborhood $N(S)$ is explored using heuristics.

We consider COPs with objective function f is of the form $f(S) = \sum_{e \in S} c_e$. If there is a polynomial time algorithm for computing an improving solution in a neighborhood, then there is a polynomial time algorithm for computing an *ε-local optimum*. This result was originally obtained by Orlin et al. [45].

The algorithm to compute an *ε-local optimum* starts with a feasible solution S^0. Then the element costs c_e for $e \in E$ are modified using a prescribed scaling procedure to generate a modified instance. Using local search on this modified problem, we look for a solution with an objective function value (with respect to the original cost) that is half that of S^0. If no such solution is found, we are at a local optimum for the modified problem and output its solution. Otherwise we replace S^0 by the solution of cost less than half and the algorithm is repeated. A formal description of the algorithm is given in the following. We assume that a local augmentation procedure IMPROVE$_N$ is available which with input a neighborhood $N(S)$ and a cost function f computes an improved solution or declare that no improved solution exists in $N(S)$.

ALGORITHM   ε-Local Search

Input: Objective function $f: 2^E \to \mathbb{N}$; subroutine IMPROVE$_N$; initial feasible
solution $S^0 \in F$; accuracy $\varepsilon > 0$.

Output: Solution $S^\varepsilon \in F$ that is an *ε*-local optimum with respect to N and f.

Step 1: $i := 0$

Step 2: $K := f(S^i)$, $q := \frac{K\varepsilon}{2m(1+\varepsilon)}$, and $c'_e := \left\lceil \frac{c_e}{q} \right\rceil q$ for $e \in E$;

Step 3: $k := 0$ and $S^{i,k} := S^i$

Step 4: **repeat**

> Call IMPROVE$_N(S^{i,k}, f')$;
> {comment: $f'(S) = \sum_{e \in S} c'_e$}
> **if** the answer is "NO", **then**
> > Let $S^{i,k+1} \in N(S^{i,k})$ such that $c'(S^{i,k+1}) < c'(S^{i,k})$; set $k := k+1$;
> > **else** $S^\varepsilon := S^{i,k}$; stop
>
> **until** $c(S^{i,k}) \leq K/2$;

Step 5: $S^{i+1} := S^{i,k}$, set $i := i+1$ and **goto** Step 2.

Lemma 18.7 [45] *The ε-local search algorithm produces an ε-local optimum*

Proof. Let S^ε be the solution produced by the algorithm, and let S be an arbitrary solution in the neighborhood $N(S^\varepsilon)$. Let K and q denote the corresponding values from the last execution of Step 2 of the algorithm. Note that

$$f(S^\varepsilon) = \sum_{e \in S^\varepsilon} c_e \leq \sum_{e \in S^\varepsilon} \left\lceil \frac{c_e}{q} \right\rceil q \leq \sum_{e \in S} \left\lceil \frac{c_e}{q} \right\rceil q \leq \sum_{e \in S} q \left(\left\lceil \frac{c_e}{q} \right\rceil + 1 \right) \leq \sum_{e \in S} c_e + mq = f(S) + mq,$$

where $m = |E|$. Here, the second inequality follows from the fact that S^ε is locally optimal with respect to f'. Together with $f(S^\varepsilon) \geq K/2$, we have $\frac{f(S^\varepsilon) - f(S)}{f(S)} \leq \frac{mq}{f(S)} \leq \frac{mq}{f(S^\varepsilon) - mq} \leq \frac{2mq}{K - 2mq} = \varepsilon$. This completes the proof. ∎

We now analyze the complexity of ε-local search algorithm as given in Reference 45. In each improving move within the local search in Step 4 of the algorithm, the objective function value (with respect to f') is decreased by at least q units. Thus, the number of calls to IMPROVE$_N$ between two consecutive iterations of Step 2 is $O(m(1 + \varepsilon)/\varepsilon) = O(m/\varepsilon)$. Step 2 is executed at most $\log f(S^0)$ times, where S^0 is the starting solution. Thus the total number of times neighborhoods searched is $O(m\varepsilon^{-1} \log f(S^0))$. Thus whenever IMPROVE$_N$ is polynomial algorithm ε-local search computes an ε-local optimum in polynomial time for fixed ε.

A strongly polynomial bound on the number of iterations was also proved in Reference 45. The proof makes use of the following lemma.

Lemma 18.8 [50] *Let $d = (d_1, \ldots, d_m)$ be a real vector and let $y_1, \ldots, y_p$ be vectors on $\{0, 1\}^m$. If for all $i = 1, \ldots, p - 1, 0 \leq dy_{i+1} \leq \frac{1}{2} dy_i$ then $p = O(m \log m)$.*

After each execution of Step 2, K is reduced at least by half. Further, K is a linear combination of c_e for $e \in E$ with coefficients 0 or 1. Lemma 18.8 implies that Step 2 of the ε-local search algorithm can be executed at most $O(m \log m)$ times. Thus IMPROVE$_N$ at Step 4 is called at most $O(\varepsilon^{-1} m^2 \log m)$ times. Let $\zeta(m, \log c_{\max})$ be the complexity of IMPROVE$_N$, $\xi(n)$ be the time needed to obtain a feasible starting solution and $K^0 = f(S^0)$, where $c_{\max} = \max\{c_e : e \in E\}$. Thus we have the following complexity result.

Lemma 18.9 [45] *The ε-local search algorithm correctly identifies an ε-locally optimal solution of an instance of a COP in $O(\xi(n) + \zeta(n, \log c_{\max}) n \varepsilon^{-1} \min\{n \log n, \log K^0\})$ time.*

If the neighborhood N is exact, then ε-local search produces an ε-optimal solution [45]. Suppose that the neighborhood N is searched approximately. That is, IMPROVE$_N$ detects an improved solution or declares that the current solution is δ-locally optimal. Even in this case an ε-local optimum can be identified in (strongly) polynomial time for any fixed ε.

18.5 Concluding Remarks

In this chapter, we have discussed techniques for developing VLSN search algorithms. Empirical and theoretical indicators are provided to substantiate our belief that VLSN search algorithms are powerful tools for obtaining high-quality solutions for hard problems in reasonable amount of computational time.

Acknowledgments

We are thankful to Temel Oncan for his assistance in preparing the manuscript. This work was partially supported by National Science Foundation (NSF) grant DMI-0217359 awarded to Ravi Ahuja, DMI-0238815 grant awarded to Özlem Ergun, NSF grant DMI-0217123 awarded to James B Orlin, and a Natural Sciences and Engineering Research Council (NSERC) discovery grant awarded to Abraham P Punnen.

References

1. Gutin, G. and Punnen, A.P., *The Traveling Salesman Problem and Its Variations*, Kluwer Academic Publishers, Dordrecht, the Netherlands, 2002.
2. Johnson, D.S., and McGeoch, L.A., Experimental analysis of heuristics for the STSP, in *The Traveling Salesman Problem and its Variations*, Gutin, G. and Punnen, A.P., Eds., Kluwer Academic Publishers, Dordrecht, the Netherlands, pp. 369–443, 2002.
3. Schaffer A.A. and Yannakakis, M., Simple local search problems that are hard to solve, *SIAM Journal on Computing*, **20**, 56, 1991.
4. Johnson, D.S., Aragon, C.R., McGeoch, L.A., and Schevon, C., Optimization by simulated annealing: An experimental evaluation; Part 1, Graph partitioning, *Operations Research*, **37**, 865, 1989.
5. Johnson, D.S., Papadimitriou, C.H., and Yannakakis, M., How easy is local search? *Journal of Computer and System Sciences*, **37**, 79, 1988.
6. Ahuja, R.K., Orlin, J.B., and Sharma, D., A composite very large-scale neighborhood structure for the capacitated minimum spanning tree problem, *Operations Research Letters*, **31**, 185, 2003.
7. Ahuja, R.K., Orlin, J.B., and Sharma, D., Multi-exchange neighborhood search structures for the capacitated minimum spanning tree problem, *Mathematical Programming*, **91**, 71, 2001.
8. Sharma, D., Cyclic exchange and related neighborhood structures for combinatorial optimization problems, PhD Thesis, Operations Research Center, MIT, Cambridge, MA, 2002.
9. Agarwal, R., Ahuja, R.K., Laporte, G., and Shen, Z.J., A composite very large-scale neighborhood search algorithm for the vehicle routing problem, in *Handbook of Scheduling: Algorithms, Models and Performance Analysis*, Leung, J.Y.-T., Ed., Chapman & Hall/CRC, Boca Raton, FL, 2003, Chapter 49.
10. De Franceschi, R., M. Fischetti, M., and Toth, P., A new ILP-based refinement heuristic for vehicle routing problems, *Mathematical Programming*, **105**, 417–499, 2005.
11. Ergun, Ö., New neighborhood search algorithms based on exponentially large neighborhoods, PhD Thesis, Operations Research Center, MIT, Cambridge, MA, 2001.
12. Ergun, Ö., Orlin, J.B., and Steele-Feldman, A., Creating very large-scale neighborhoods out of smaller ones by compounding moves, Technical Report, 2002.
13. Deineko, V.G. and Woeginger, G.J, A study of exponential neighborhoods for the traveling salesman problem and the quadratic assignment problem, *Mathematical Programming*, **87**, 519, 2000.
14. Gutin, G., On the efficiency of a local algorithm for solving the traveling salesman problem, *Automation and Remote Control*, **49**, 1514, 1988.
15. Gutin, G., Yeo, A., and Zverovitch, A., Exponential neighborhoods and domination analysis for the TSP, in *The Traveling Salesman Problem and Its Variations*, Gutin, G. and Punnen, A.P., Eds., Kluwer Academic Publishers, Dordrecht, the Netherlands, p. 223, 2002.
16. Punnen, A.P., The traveling salesman problem: New polynomial approximation algorithms and domination analysis, *Journal of Information and Optimization Sciences*, **22**, 191, 2001.
17. Sarvanov, V.I. and Doroshko, N.N., The approximate solution of the traveling salesman problem by a local algorithm that searches neighborhoods of exponential cardinality in quadratic time, *Software: Algorithms and Programs*, **31**, 8, 1981.
18. Ahuja, R.K., Kumar, A., Jha, K.C., and Orlin, J.B., Exact and heuristic algorithms for the weapon-target assignment problem, Technical Report, 2003.
19. Yagiura, M., Ibaraki, T., Glover, F., An ejection chain approach for the generalized assignment problem, *INFORMS Journal on Computing*, **16**, 133, 2004.
20. Yagiura, M., Iwasaki, S., Ibaraki, T., and Glover, F., A very large-scale neighborhood search algorithm for the multi-resource generalized assignment problem, *Discrete Optimization*, **1**, 87, 2004.
21. Yagiura, M., Ibaraki, T., and Glover, F., A path relinking approach with ejection chains for the generalized assignment problem, *European Journal of Operational Research*, **169**, 548–569, 2006.

22. Ahuja, R.K., Orlin, J.B., Pallottino, S., Scaparra, M.P., and Scutella, M.G., A multi-exchange heuristic for the single source capacitated facility location, *Management Science*, **50**, 749, 2004.
23. Agarwal, R., Ergun, Ö., Orlin, J.B., and Potts, C.N., Solving parallel machine scheduling problems with variable depth local search, Technical Report, 2004.
24. Ahuja, R.K., Goodstein, J., Liu, J., Mukherjee, A., and Orlin, J.B., A neighborhood search algorithm for the combined through and fleet assignment model with time windows, *Networks*, **44**, 160–171, 2004.
25. Ahuja, R.K., Goodstein, J., Liu, J., Mukherjee, A., Orlin, J.B., and Sharma, D., Solving multi-criteria combined through-fleet assignment model, in *Operations Research in Space and Air*, Ciriani, T.A., Fasano, G., Gliozzi, S., and Tadei, R., Eds., Kluwer Academic Publishers, Dordrecht, the Netherlands, p. 233, 2003.
26. Ahuja, R.K., Goodstein, J., Mukherjee, A., Orlin, J.B., and Sharma, D., A very large-scale neighborhood search algorithm for the combined through-fleet assignment model, Technical Report, 2001.
27. Ahuja, R.K., Jha, K.C., Orlin, J.B., and Sharma, D., Very large scale neighborhood search algorithm for the quadratic assignment problem, Technical Report, 2002.
28. Ropke, S. and Pisinger, D., An adaptive large neighborhood search heuristic for the pickup and delivery problem with time windows, Technical Report, 2004.
29. Cunha, C.B. and Ahuja, R.K., Very large-scale neighborhood search for the K-constrained multiple knapsack problem, Technical Report, 2004.
30. Huang, W., Romeijn, H.E., and Geunes, J., The continuous-time single-sourcing problem with capacity expansion opportunities, *Naval Research Logistics*, **52**, 193, 2005.
31. Ahuja, R.K., Very large-scale neighborhood search algorithms for minimizing the beam-on time and set-up time in radiation therapy treatment planning, Working Paper, Department of Industrial and Systems Engineering, University of Florida, Gainesville, FL, 2002.
32. Avella, P., D'Auria, B., Salerno, S., and Vasil'ev, I., Computational experience with very large-scale neighborhood search for high-school timetabling, Technical Report, 2004.
33. Chiarandini, M., Dumitrescu, I., and Stützle, T., Local search for the colouring graph problem. A computational study, TR AIDA-03-01, FG Intellektik, FB Informatik, TU Darmstadt, Germany, 2003.
34. Glover, F. and Laguna, M., *Tabu Search*, Kluwer Academic Press, Dordrecht, the Netherlands, 1997.
35. Michalewicz, Z., 1996, *Genetic Algorithms+Data Structure=Evolution Programs*, 3rd ed, Springer-Verlag, Berlin, Germany, 1996.
36. Rego, C. and Alidaee, B., *Metaheuristic Optimization Via Memory and Evolution*, Kluwer Academic Publishers, Dordrecht, the Netherlands, 2005.
37. Feo, T.A. and Resende, M.G.C., Greedy randomized adaptive search procedures, *Journal of Global Optimization*, **6**, 109, 1995.
38. Thompson, P.M. and Psaraftis, H.N., Cyclic transfer algorithms for multi-vehicle routing and scheduling problems, *Operations Research*, **41**, 935, 1993.
39. Tam, V. and Kwan, M.C., Adapting the large neighborhood search to effectively solve pickup and delivery problems with time windows, *International Conference on Tools with Artificial Intelligence (ICTAI)*, IEEE, 2004, p. 519.
40. Glover, F., Ejection chains, reference structures and alternating path methods for traveling salesman problems, *Discrete Applied Mathematics*, **65**, 223, 1996.
41. Lin, S. and Kernighan, B.W., An effective heuristic algorithm for the traveling salesman problem, *Operations Research*, **21**, 972, 1973.
42. Papadimitriou, C.H. and Steiglitz, K., *Combinatorial Optimization: Algorithms and Complexity*. Prentice-Hall, New Delhi, India, 1982.
43. Thompson, P.M. and Orlin, J.B., The theory of cyclic transfers, Working Paper OR200-89, Operations Research Center, MIT, Cambridge, MA, 1989.

44. Gavish, B., Topological design telecommunications networks- Local access design methods, *Annals of Operations Research*, **33**, 17, 1991.
45. Amberg, A., Domschke, W., and Voß, S., Capacitated minimum spanning trees: Algorithms using intelligent search, *Combinatorial Optimization: Theory and Practice*, **1**, 9, 1996.
46. Orlin, J.B., Punnen, A.P., and Schulz, A.S., Approximate local search in combinatorial optimization, *SIAM Journal on Computing*, **33**, 1201, 2004.
47. Glover, F. and Punnen, A.P., The traveling salesman problem: New solvable cases and linkages with the development of approximation algorithms, *Journal of the Operational Research Society*, **48**, 502, 1997.
48. Punnen, A.P. and Kabadi, S.N., Domination analysis of some heuristics for the asymmetric traveling salesman problem, *Discrete Applied Mathematics*, **119**, 117, 2002.
49. Punnen, A.P., Margot, F., and Kabadi, S.N., TSP heuristics: Domination analysis and complexity, *Algorithmica*, **35**, 111, 2003.
50. Orlin, J.B. and Sharma, D., Extended neighborhood: Definition and characterization, *Mathematical Programming*, **101**, 537, 2004.
51. Radzik, T., Parametric flows, weighted means of cuts, and fractional combinatorial optimization, in *Complexity in Numerical Optimization*, Pardalos, P., Ed., World Scientific, River Edge, NJ, 1993, p. 351.
52. Aarts, E.M.L. and Lenstra, J.K., *Local Search in Combinatorial Optimization*, John Wiley & Sons, Chichester, Germany, 1979.
53. Ahuja, R.K., Ergun, Ö., Orlin, J.B., and Punnen, A.P., A survey of very large-scale neighborhood search techniques, *Discrete Applied Mathematics*, **23**, 75, 2001.
54. Ahuja, R.K., Boland, N., and Dumitriscue, I., Exact and heuristic algorithms for the subset-disjoint minimum cost cycle problems, Working Paper, Industrial and Systems Engineering, University of Florida, Gainesville, FL, 2001.
55. Ergun, Ö. and Orlin, J.B., Very large-scale neighborhood search based on solving restricted dynamic programming recursions, Technical Report, 2004.
56. Ergun, Ö. and Orlin, J.B., Fast neighborhood search for the single machine total weighted tardiness problem, *Operations Research Letters*, **34**, 41–45.
57. Ibaraki, T., Imahori, S., Kubo, M., Masuda, T., Uno, T., and Yagiura, M., Effective local search algorithms for routing and scheduling problems with general time-window constraints, *Transportation Science*, **39**, 206, 2005.
58. Krentel, M. W., Structure in locally optimal solutions, in *Proceedings of FOCS*, IEEE, 1989, p. 216.
59. Papadimitriou, C.H., The complexity of the Lin–Kernighan heuristic for the traveling salesman problem, *SIAM Journal on Computing*, **21**, 450, 1992.
60. Papadimitriou, C.H. and Steiglitz, K., On the complexity of local search for the traveling salesman problem, *SIAM Journal on Computing*, **6**, 76, 1977.
61. Sharaiha, Y.M., Gendreau, M., Laporte, G., and Osman, I.H., A tabu search algorithm for the capacitated shortest spanning tree problem, *Networks*, **29**, 161, 1997.

19

Reactive Search: Machine Learning for Memory-Based Heuristics

Roberto Battiti

Mauro Brunato

19.1 Introduction: The Role of the User in Heuristics

Most state-of-the-art heuristics are characterized by a certain number of choices and free parameters, whose appropriate setting is a subject that raises issues of research methodology [1–3].

In some cases, these parameters are tuned through a feedback loop that *includes the user as a crucial learning component*: depending on preliminary algorithm tests some parameter values are changed by the user, and different options are tested until acceptable results are obtained. Therefore, the quality of results is not automatically transferred to different instances and the feedback loop can require a lengthy "trial and error" process every time the algorithm has to be tuned for a new application.

Parameter tuning is therefore a crucial issue both in the scientific development and in the practical use of heuristics. In some cases the role of the user as an intelligent (learning) part makes the reproducibility of heuristic results difficult and, as a consequence, the competitiveness of alternative techniques depends in a crucial way on the user's capabilities.

Reactive Search advocates the use of simple sub-symbolic machine learning to automate the parameter tuning process and make it an integral (and fully documented) part of the algorithm.

If learning is performed on line, task-dependent and local properties of the configuration space can be used by the algorithm to determine the appropriate balance between *diversification* (looking for better solutions in other zones of the configuration space) and *intensification* (exploring more intensively a small but promising part of the configuration space). In this way, a single algorithm maintains the flexibility to deal with related problems through an internal feedback loop that considers the previous history of the search.

In the following, we shall call *reaction* the act of modifying some algorithm parameters in response to the search algorithm's behavior *during* its execution, rather than between runs. Therefore, a *reactive heuristic* is a technique with the ability of tuning some important parameters during execution by means of a machine learning mechanism.

It is important to notice that such heuristics are intrinsically history-dependent; thus, the practical success of this approach in some cases raises the need of a sounder theoretical foundation of non-Markovian search techniques.

19.1.1 Machine Learning for Automation and Full Documentation

Parameter tuning is a typical "learning" process where experiments are designed in a focused way, with the support of statistical estimation (parameter identification) tools.

Because of its familiarity with algorithms, the computer science (CS) community masters a very powerful tool for describing processes so that they can be reproduced even by a (mechanical) computer. In particular, the machine learning community, with significant influx from statistics, developed in the last decades a rich variety of "design principles" that can be used to develop machine learning methods and algorithms.

It is, therefore, appropriate to consider whether some of these design principles can be profitably used in the area of parameter tuning for heuristics. The long-term goal is that of completely *eliminating the human intervention* in the tuning process. This does not imply higher unemployment rates in the CS community, on the contrary, the researcher is now loaded with a heavier task: the algorithm developer must aim at transferring his/her expertise into the algorithm itself, a task that requires the *exhaustive description of the tuning phase* in the algorithm.

Let us note that the algorithm "complexity" will increase as a result of the process, but the price is worth paying if the two following objectives are reached:

Complete and unambiguous documentation: The algorithm (and the research paper based on the algorithm) becomes self-contained, and its quality can be judged independently from the designer or specific user. This requirement is particularly important from the scientific point of view, where objective evaluations are crucial. The recent introduction of software archives (in some cases related to scientific journals) further simplifies the test and *simple re-use* of heuristic algorithms.

Automation: The time-consuming tuning phase is now substituted by an automated process. Let us note that only the final user will typically benefit from an automated tuning process. On the contrary, the algorithm designer faces a longer and harder development phase, with a possible preliminary phase of exploratory tests, followed by the above-described exhaustive documentation of the tuning process when the algorithm is presented to the scientific community.

Although formal learning frameworks do exist in the CS community (notably, the PAC learning model [4,5]) one should not reach the conclusion that these models can be simply adapted to the new context. On the contrary, the theoretical framework of computational learning theory and machine learning is very different from that of heuristics. For example, the definition of a "quality" function against which the learning algorithm has to be judged is complex. In addition, the abundance of negative results in computational learning should warn about excessive hopes.

Nonetheless, as a first step, some of the principles and methodology used in machine learning can be used in an analogic fashion to develop "reactive heuristics."

19.1.2 Asymptotic Results Are Irrelevant for Optimization

Scientists, and also final users, might feel uneasy working with non-Markovian techniques because they don't benefit from the deep and wide theoretical background that covers Markovian algorithms.

However, asymptotic convergence results of many Markovian search algorithms, such as simulated annealing (SA) [6], are often irrelevant for their application to optimization. As an example, a comparison of SA and reactive search has been presented in References 7, 8.

In any finite-time approximation one must resort to approximations of the asymptotic convergence. In SA, for instance, the "speed of convergence" to the stationary distribution is determined by the second largest eigenvalue of the transition matrix. The number of transitions is at least *quadratic* in the size of the solution space [9], which is typically exponential in n.

When using a time-varying temperature parameter, it can happen (e.g., traveling salesman problem (TSP)) that the complete enumeration of all solutions would take less time than approximating an optimal solution with arbitrary precision by SA [9].

In addition, repeated local search [10], and even random search [11], have better asymptotic results. According to Reference 9 "approximating the asymptotic behavior of SA arbitrarily closely requires a number of transitions that for most problems is typically larger than the size of the solution space [...] Thus, the SA algorithm is clearly unsuited for solving combinatorial optimization problems to optimality." Of course, practical utility of SA has been shown in many applications, in particular with fast cooling schedules, but then the asymptotic results are not directly applicable. The optimal finite-length annealing schedules obtained on specific simple problems do not always correspond to those intuitively expected from the limiting theorems [12].

19.2 Reactive Search Applied to Tabu Search (RTS)

In this section, we illustrate the potential of reactive search by installing a reaction mechanism on the prohibition period T of a tabu search (TS) [13] algorithm. For the complete description of TS, the reader is referred to Chapter 21 of this handbook.

19.2.1 Prohibition-Based Diversification: Tabu Search

The TS meta-heuristic is based on the use of *prohibition-based* techniques and "intelligent" schemes as a complement to basic heuristic algorithms such as local search, with the purpose of guiding the basic heuristic *beyond local optimality*. It is difficult to assign a precise date of birth to these principles. For example, ideas similar to those proposed in TS can be found in the *denial* strategy of Reference 14 (once common features are detected in many suboptimal solutions, they are forbidden) or in the opposite *reduction* strategy of Reference 15 (in an application to the TSP, all edges that are common to a set of local optima are fixed). In very different contexts, prohibition-like strategies can be found in *cutting planes* algorithms for solving integer problems through their linear programming relaxation (inequalities that cut off previously obtained fractional solutions are generated) and in branch and bound algorithms (subtrees are not considered if the leaves cannot correspond to better solutions). For many examples of such techniques, see the textbook of Reference 16.

The renaissance and full blossoming of "intelligent prohibition-based heuristics" starting from the late eighties is greatly due to the role of Glover in the proposal and diffusion of a rich variety of meta-heuristic tools [13,17], but see also Reference 18 for an independent seminal paper. A growing number of TS-based algorithms has been developed in the last years and applied with success to a wide selection of problems [19]. It is therefore difficult, if not impossible, to characterize a "canonical form" of TS, and classifications tend to be short-lived. Nonetheless, at least two aspects characterize many versions of TS: the fact that TS is used to complement local (neighborhood) search, and the fact that the main modifications to local search are obtained through the *prohibition* of selected moves available at the current point. TS acts to continue the search beyond the first local minimizer without wasting the work already executed, as it is the case if a new run of local search is started from a new random initial point, and to enforce appropriate amounts of diversification to avoid that the search trajectory remains confined near a given local minimizer.

In our opinion, the main competitive advantage of TS with respect to alternative heuristics based on local search such as SA [6] lies in the intelligent use of the past history of the search to influence its future steps.

For a generic search space $\mathcal{X}$, let $X^{(t)} \in \mathcal{X}$ be the current configuration and $N(X^{(t)}) \subseteq \mathcal{X}$ its neighborhood. In prohibition-based search (TS) some of the neighbors can be *prohibited*, and let the subset $N_A(X^{(t)}) \subseteq N(X^{(t)})$ contain the *allowed* ones. The general way of generating the search trajectory that we consider is given by:

$$X^{(t+1)} = \text{Best-Neighbor}\left(N_A(X^{(t)})\right) \tag{19.1}$$

$$N_A(X^{(t+1)}) = \text{Allow}\left(N(X^{(t+1)}), X^{(0)}, \ldots, X^{(t+1)}\right) \tag{19.2}$$

The set-valued function Allow selects a subset of $N(X^{(t+1)})$ in a manner that depends on the search trajectory $X^{(0)}, \ldots, X^{(t+1)}$.

This general framework allows several specializations. In many cases, the dependence of Allow on the entire search trajectory introduces too many constraints on the next move, causing the search path to avoid an otherwise promising area, or even prohibiting all neighbors. It is therefore advisable to reduce the amount of constraints by limiting the Allow function to the latest T configurations (where the parameter T is often called the *prohibition period*), so that equation (19.2) becomes

$$N_A(X^{(t+1)}) = \text{Allow}\left(N(X^{(t+1)}), X^{(t')}, \ldots, X^{(t+1)}\right), \qquad t' = \max\{0, t - T + 1\} \tag{19.3}$$

A practical example for equation (19.3) is a function Allow that forbids all moves that have been performed within the last T iterations. For instance, let us assume that the feasible search space $\mathcal{X}$ is the set of binary strings with a given length L: $\mathcal{X} = \{0, 1\}^L$ (this case shall be considered also in the example of Section 19.4). In this case, a practical neighborhood of configuration $X^{(t)}$ is given by the L configurations that differ from $X^{(t)}$ by a single entry. In such case, a simple prohibition scheme may allow a move if and only if it changes an entry that has remained fixed for the previous T iterations. In other words, after an entry has been changed, it shall remain *frozen* for the following T steps.

It is apparent that the choice of the right prohibition period T is crucial in order to balance the amount of *intensification* (small T) and *diversification* (large T).

19.2.2 Reaction on Tabu Search Parameters

Some problems arising in TS that have been investigated in reactive search papers are:

1. The determination of an appropriate prohibition T for the different tasks,
2. The robustness of the technique for a wide range of different problems,
3. The adoption of minimal computational complexity algorithms for using the search history.

The three issues are briefly discussed in the following sections, together with the reaction-based methods proposed to deal with them.

19.2.2.1 Self-Adjusted Prohibition Period

In RTS, the *prohibition period* T is determined through feedback (i.e., *reactive*) mechanisms during the search. At the beginning, we let $T = 1$ (the inverse of a given move is prohibited only at the next step). During the search, T increases only when there is *evidence* that diversification is needed, and it decreases when this evidence disappears. In detail: the evidence that diversification is needed is signalled by the repetition of previously visited configurations. For this purpose, all configurations found during the search

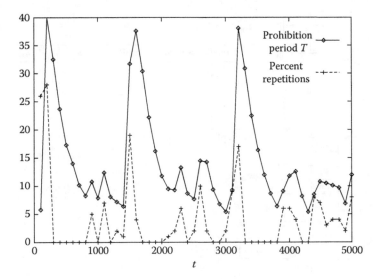

FIGURE 19.1   Dynamics of the prohibition period T on a QAP task.

are stored in memory. After a move is executed, the algorithm checks whether the current configuration has already been found and reacts accordingly (T increases if a configuration is repeated, T decreases if no repetitions occurred during a sufficiently long period).

By means of this self-adjustment algorithm, T is not fixed during the search, but it is determined in a dynamic way depending on the *local structure* of the search space. This is particularly relevant for "inhomogeneous" tasks, where the statistical properties of the search space vary widely in the different regions (in these cases a fixed T would be inappropriate).

An example of the behavior of T during the search is illustrated in Figure 19.1, for a quadratic assignment problem task [20]. T increases in an exponential way when repetitions are encountered, it decreases in a gradual manner when repetitions disappear.

19.2.2.2   The Escape Mechanism

The basic tabu mechanism based on prohibitions is not sufficient to avoid long cycles. As an example, when operating on binary strings of length L, the prohibition T must be less than the length of the string, otherwise all moves are eventually prohibited; therefore, cycles longer than $2 \times L$ are still possible. In addition, even if "limit cycles" (endless cyclic repetitions of a given set of configurations) are avoided, the first reactive mechanism is not sufficient to guarantee that the search trajectory is not confined in a limited region of the search space. A "chaotic trapping" of the trajectory in a limited portion of the search space is still possible (the analogy is with *chaotic attractors* of dynamical systems, where the trajectory is confined in a limited portion of the space, although a limit cycle is not present).

For both reasons, to increase the robustness of the algorithm a second more radical diversification step (*escape*) is needed. The *escape* phase is triggered when too many configurations are repeated too often [20]. A simple *escape* consists of a number of random steps executed starting from the current configuration (possibly with a bias toward steps that bring the trajectory away from the current search region).

With a stochastic escape, one can easily obtain the *asymptotic convergence* of RTS: in fact, in a finite search space *escape* is activated infinitely often; if the probability for a point to be reached after escaping is different from zero for all points, eventually all points will be visited–clearly including the globally optimal points. The detailed investigation of the asymptotic properties and finite-time effects of different *escape* routines to enforce long-term diversification is an open research area.

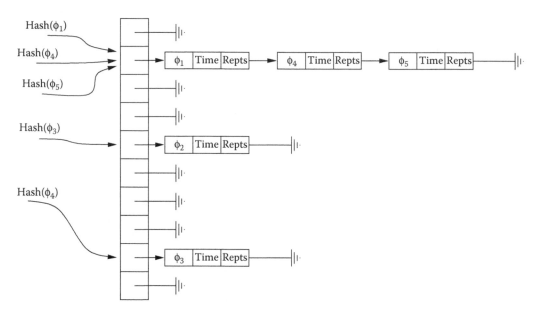

FIGURE 19.2 Open hashing scheme: items (configuration, or compressed hashed value, etc.) are stored in "buckets." The index of the bucket array is calculated from the configuration.

19.2.3 Implementation of History-Sensitive Techniques

The efficiency and competitiveness of history-based reaction mechanisms strongly depend on the detailed data structures used in the algorithms and on the consequent realization of the needed operations. Different data structures can possess widely different computational complexities so that attention should be spent on this subject before choosing a version of reactive search that is efficient on a particular problem.

Reactive-TS can be implemented through a simple list of visited configurations, or with more efficient hashing [20,21] or radix tree [20] techniques. At a finer level of detail, hashing can be realized in different ways. If the entire configuration is stored (see also Figure 19.2) an exact answer is obtained from the memory lookup operation (a repetition is reported if and only if the configuration has been visited before). On the contrary, if a "compressed" item is stored, such as a hashed value of a limited length derived from the configuration, the answer will have a limited probability of *false positives* (a repetition can be reported even if the configuration is new, because the compressed items are equal by chance–an event called "collision"). Experimentally, small collision probabilities do not have statistically significant effects on the use of Reactive-TS as heuristic tool, and hashing versions that need only a few bytes per iteration can be used.

19.2.3.1 Fast Algorithms for Using the Search History

The storage and access of the past events is executed through the well-known hashing or radix-tree techniques in a CPU time that is approximately *constant* with respect to the number of iterations. Therefore the overhead caused by the use of the history is negligible for tasks requiring a nontrivial number of operations to evaluate the cost function in the neighborhood.

An example of a memory configuration for the hashing scheme is shown in Figure 19.2. From the current configuration ϕ one obtains an index into a "bucket array." The items (configuration or hashed value or derived quantity, last time of visit, total number of repetitions) are then stored in linked lists starting from the indexed array entry. Both storage and retrieval require an approximately constant amount of time if: (i) the number of stored items is not much larger than the size of the bucket array and (ii) the *hashing function* scatters the items with a uniform probability over the different array indices.

More precisely, given a hash table with m slots that stores n elements, a load factor $\alpha = n/m$ is defined. If collisions are resolved by chaining searches take $O(1 + \alpha)$ time, on the average.

19.2.3.2 Persistent Dynamic Sets

Persistent dynamic sets are proposed to support memory–usage operations in history-sensitive heuristics in References 22, 23.

Ordinary data structures are *ephemeral* [24] because when a change is executed the previous version is destroyed. Now, in many contexts such as computational geometry, editing, implementation of very high level programming languages, and, last but not least, the context of history-based heuristics, multiple versions of a data structure must be maintained and accessed. In particular, in heuristics one is interested in *partially persistent* structures, where all versions can be accessed but only the newest version (the *live* nodes) can be modified. A review of *ad hoc* techniques for obtaining persistent data structures is given in Reference 24 that is dedicated to a systematic study of persistence, continuing the previous work of Reference 25.

Hashing combined with persistent red-black trees: The basic observation is that, because *TS* is based on local search, configuration $X^{(t+1)}$ differs from configuration $X^{(t)}$ only because of the addition or subtraction of a single index (a single bit is changed in the string). Let us define the operations INSERT(i) and DELETE(i) for inserting and deleting a given index i from the set. As aforementioned, configuration X can be considered as a set of indices in $[1, L]$ with a possible realization as a balanced red-black tree, see References 26, 27 for two seminal papers about red-black trees. The binary string can be immediately obtained from the tree by visiting it in symmetric order, in time $O(L)$. INSERT(i) and DELETE(i) require $O(\log L)$ time, whereas at most a single node of the tree is allocated or deallocated at each iteration. Rebalancing the tree after insertion or deletion can be done in $O(1)$ rotations and $O(\log L)$ color changes [28]. In addition, the amortized number of color changes per update is $O(1)$, see for example Reference 29.

Now, the reverse elimination method [13,17,30] (a technique for the storage and analysis of the ordered list of all moves performed throughout the search) is closely reminiscent of a method studied in Reference 25 to obtain partial persistence, in which the entire update sequence is stored and the desired version is rebuilt from scratch each time an access is performed, whereas a systematic study of techniques with better space-time complexities is present in References 24, 31. Let us now summarize from Reference 31 how a partially persistent red-black tree can be realized. An example of the realizations that we consider is presented in Figure 19.3.

The trivial way is that of keeping in memory all copies of the ephemeral tree (see Figure 19.3a), each copy requiring $O(L)$ space. A smarter realization is based on *path copying*, independently proposed by many researchers, see Reference 31. Only the path from the root to the nodes where changes are made is copied: a set of search trees is created, one per update, having different roots but *sharing* common subtrees. The time and space complexities for INSERT(i) and DELETE(i) are now of $O(\log L)$.

The method that we will use is a space-efficient scheme requiring only linear space proposed in Reference 31. The approach avoids copying the entire access path each time an update occurs. To this end, each node contains an additional "extra" pointer (beyond the usual left and right ones) with a time stamp.

When attempting to add a pointer to a node, if the extra pointer is available, it is used and the time of the usage is registered. If the extra pointer is already used, the node is copied, setting the initial left and right pointers of the copy to their latest values. In addition, a pointer to the copy is stored in the last parent of the copied node. If the parent has already used the extra pointer, the parent, too, is copied. Thus copying proliferates through successive ancestors until the root is copied or a node with a free extra pointer is encountered. Searching the data structure at a given time t in the past is easy: after starting from the appropriate root, if the extra pointer is used the pointer to follow from a node is determined by examining the time stamp of the extra pointer and following it iff the time stamp is not larger than t. Otherwise, if the extra pointer is not used, the normal left-right pointers are considered. Note that the pointer direction (left or right) does not have to be stored: given the search tree property it can be derived by comparing

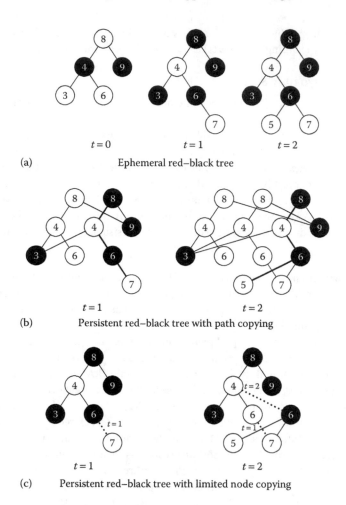

(a)   Ephemeral red–black tree

(b)   Persistent red–black tree with path copying

(c)   Persistent red–black tree with limited node copying

FIGURE 19.3 How to obtain a partially persistent red–black tree from an ephemeral one (a), containing indices 3,4,6,8,9 at $t = 0$, with subsequent insertion of 7 and 5. Path copying (b), with thick lines marking the copied part. Limited node copying (c) with dashed lines denoting the "extra" pointers with time stamp.

the indices of the children with that of the node. In addition, colors are needed only for the most recent (live) version of the tree. In Figure 19.3 null pointers are not shown, colors are correct only for the live tree (the nodes reachable from the rightmost root), extra pointers are dashed and time-stamped.

The worst-case time complexity of INSERT(i) and DELETE(i) remains of $O(\log L)$, but the important result derived in Reference 31 is that the amortized space cost per update operation is $O(1)$. Let us recall that the total amortized space cost of a sequence of updates is an upper bound on the actual number of nodes created.

Let us now consider the context of history-based heuristics. Contrary to the popular usage of persistent dynamic sets to search past versions at a specified time t, one is interested in checking whether a configuration has already been encountered in the previous history of the search, at *any* iteration.

A convenient way of realizing a data structure supporting X-SEARCH(X) is to combine *hashing* and *partially persistent dynamic sets*, see Figure 19.4. From a given configuration X an index into a "bucket array" is obtained through a hashing function, with a possible incremental evaluation in time $O(1)$. Collisions are resolved through chaining: starting from each bucket header there is a linked list containing a pointer to the appropriate root of the persistent red-black tree and satellite data needed by the search (time of configuration, number of repetitions).

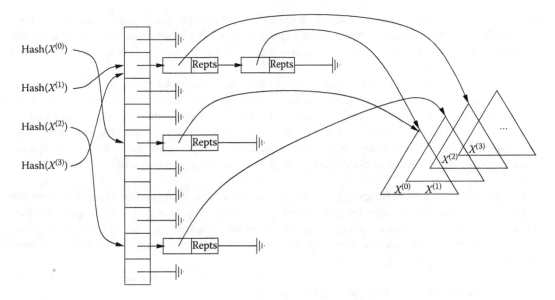

FIGURE 19.4 Open hashing scheme with persistent sets: a pointer to the appropriate root for configuration $X^{(t)}$ in the persistent search tree is stored in a linked list at a "bucket." Items on the list contain satellite data. The index of the bucket array is calculated from the configuration through a hashing function.

As soon as configuration $X^{(t)}$ is generated by the search dynamics, the corresponding persistent red-black tree is updated through INSERT(i) or DELETE(i). Let us now describe X-SEARCH($X^{(t)}$): the hashing value is computed from $X^{(t)}$ and the appropriate bucket searched. For each item in the linked list the pointer to the root of the past version of the tree is followed and the old set is compared with $X^{(t)}$. If the sets are equal, a pointer to the item on the linked list is returned. Otherwise, after the entire list has been scanned with no success, a null pointer is returned.

In the last case a new item is linked in the appropriate bucket with a pointer to the root of the live version of the tree (X-INSERT(X, t)). Otherwise, the last visit time t is updated and the repetition counter is incremented.

After collecting the above-cited complexity results, and assuming that the bucket array size is equal to the maximum number of iterations executed in the entire search, it is straightforward to conclude that each iteration of *reactive-TS* requires $O(L)$ average-case time and $O(1)$ amortized space for storing and retrieving the past configurations and for establishing prohibitions.

In fact, both the hash table and the persistent red-black tree require $O(1)$ space (amortized for the tree). The worst-case time complexity per iteration required to update the current $X^{(t)}$ is $O(\log L)$, the average-case time for searching and updating the hashing table is $O(1)$ (in detail, searches take time $O(1 + \alpha)$, α being the load factor, in our case upper bounded by 1). The time is therefore dominated by that required to compare the configuration $X^{(t)}$ with that obtained through X-SEARCH($X^{(t)}$), that is, $O(L)$ in the worst case. Because $\Omega(L)$ time is needed during the neighborhood evaluation to compute the f values, the above-mentioned complexity is optimal for the considered application to history-based heuristics.

19.3 Wanted: A Theory of History-Sensitive Heuristics

Randomized Markovian local search algorithms have enjoyed a long period of scientific and applicative excitement, in particular see the flourishing literature on SA [6,32]. SA generates a *Markov chain*: the successor of the current point is chosen stochastically, with a probability that does not depend on the previous history (standard SA does not learn). A consequence is that the "trapping" of the search trajectory in an attractor cannot be avoided: the system has no memory and cannot detect that the search is localized. Incidentally, the often cited asymptotic convergence results of SA are unfortunately irrelevant for the application of SA to optimization. In fact, repeated local search [10], and even random search [11] has better asymptotic results.

History-sensitive techniques in local search contain an internal *feedback loop* that uses the information derived from the past history to influence the future behavior. In the cited prohibition-based diversification techniques one can, for example, decide to increase the diversification when configurations are encountered again along the search trajectory [20,33]. It is of interest that state-of-the-art versions of SA incorporate "temporal memory" [34]. The non-Markovian property is a mixed blessing: it permits heuristic results that are much better in many cases, but makes the theoretical analysis of the algorithm difficult.

Therefore one has an unfortunate chasm: on one side there is an abundance of mathematical results derived from the theory of Markov processes, but their relevance to optimization is dubious, on the other side there is mounting evidence that simple *machine learning* or history-sensitive schemes can augment the performance of heuristics in a significant way, but the theoretical instruments to analyze history-sensitive heuristics are lacking.

The practical success of history-sensitive techniques should motivate new search streams in Mathematics and Computer Science for their theoretical foundation.

19.4 Applications of Reactive Search and the Maximum Clique Example

Reactive search principles have been used for example for the problem of quadratic assignment [20], training neural nets and control problems [35], vehicle-routing problems [36–38], structural acoustic control problems [39], special-purpose very large scale integration (VLSI) [40], graph partitioning [41], electric power distribution [42], maximum satisfiability [43], constraint satisfaction [44,45], optimization of continuous functions [46,47], traffic grooming in optical networks [48], maximum clique (MC) [33], real-time dispatch of trams in storage yards [49], and increasing Internet capacity [50]. Because of space limitation we consider with some detail only the application to the MC problem.

The MC problem is NP-hard, and strong negative results have been shown about its approximability [51,52]. In particular, if P $\neq$ NP, MC is *not approximable* within $n^{1/4-\epsilon}$ for any $\epsilon > 0$, n being the number of nodes in the graph [53], and it is not approximable within $n^{1-\epsilon}$ for any $\epsilon > 0$, unless coRP = NP [54].

These theoretical results stimulated a research effort to design efficient heuristics for this problem, and computational experiments to demonstrate that optimal or close approximate values can be efficiently obtained for significant families of graphs [55,56].

In particular, a new *reactive* heuristic (reactive local search or RLS) is proposed for the MC problem in [33]. The present description is a summarized version of the cited paper.

The experimental efficacy and efficiency [33] of RLS is strengthened by an analysis of the complexity of a single iteration. It is possible to show that the worst-case cost is $O(\max\{n, m\})$ where n and m are the number of nodes and edges, respectively. In practice, the cost analysis is pessimistic and the measured number of operations tends to be a small constant times the average degree of nodes in $\overline{G}$, the complement of the original graph.

19.4.1 Reactive Local Search for the Maximum Clique Problem

The RLS algorithm for the MC problem takes into account the particular neighborhood structure of MC. This is reflected in the following two facts: a single reactive mechanism is used to determine the

Global variables and data structures	
t	Time (iteration counter)
t_T	Time of last period change
S	Nodes in $V \setminus X$ adjacent to all nodes in X
$deltaS[j]$	Nodes in $V \setminus S$ adjacent to all nodes in $X \setminus \{j\}$
k_b	Cardinality of best configuration
$lastMoved[v]$	Time of last movement concerning node v

Local variables	
T	Prohibition period
t_R	Time of last restart
X	Current configuration
I_b	Best configuration
t_b	Time of best configuration

```
1.  procedure REACTIVE-LOCAL-SEARCH
2.      t ← 0 ; T ← 1 ; t_T ← 0 ; t_R ← 0
3.      X ← ∅ ; I_b ← ∅ ; k_b ← 0 ; t_b ← 0
4.      S ← V ; ∀j ∈ V deltaS[j] ← ∅
5.      ∀j ∈ V lastMoved[j] ← −∞
6.      repeat
7.          T ← HISTORY-REACTION(X, T)
8.          X ← BEST-NEIGHBOR (X)
9.          t ← t + 1
10.         if |X| > k_b
11.             I_b ← X ; k_b ← |X| ; t_b ← t
12.         if t - max {t_b, t_R} > 100k_b
13.             t_R ← t; RESTART
14.     until k_b is acceptable
15.         or maximum number of iterations reached
```

FIGURE 19.5   RLS algorithm: pseudocode description.

prohibition parameter T, and an explicit restart scheme is added so that all possible configurations will eventually be visited, even if the search space is not connected by using the basic local-search moves. Both building blocks of RLS use the past history of the search (set of visited configurations) to influence the choice.

The admissible search space $\mathcal{X}$ is the set of all cliques in a graph G defined over a vertex set V. Let us recall that a clique is a subset X of V such that all pairs of nodes in X are connected by an edge. The function to be maximized is the clique size $f(X) = |X|$, X being the current clique, and the neighborhood $M(X)$ consists of all cliques that can be obtained from X by adding or dropping a single vertex (*add* or *drop* moves).

At a given iteration, the neighborhood set $M(X)$ is partitioned into the set of *prohibited* neighbors and the set of *allowed* ones. As soon as a vertex is moved (added or removed from the current clique), changing its status is prohibited for the next T iterations; it is allowed otherwise. With a slight abuse of terminology, the terms *allowed* and *prohibited* shall also be applied to vertices.

The top-level description of RLS is shown in Figure 19.5. First (lines 2–5) the relevant variables and structures are initialized: they are the iteration counter t, the prohibition period T, the time t_T of the last change of T, the last restart time t_R, the current clique X, the largest clique I_b found so far with its size k_b, and the iteration t_b at which it is found. The set S shall be used by function BEST-NEIGHBOR and contains the set of eligible nodes to improve the current clique (initially, the current clique is empty, and no node is prohibited, so all nodes in V are eligible); the role of array $deltaS$ shall be explained in Section 19.4.1.1. Then the loop (lines 6–15) continues to be executed until a satisfactory solution is found or a limiting number of iterations is reached.

The function HISTORY-REACTION searches for the current clique X in memory, inserts it if it is a new one, and adjusts the prohibition T through feedback from the previous history of the search.

Then the best neighbor is selected and the current clique updated (line 8). The iteration counter is incremented. If a better solution is found, the new solution, its size and the time of the last improvement are saved (lines 10–11). A restart is activated after a suitable number of iterations are executed from the last improvement and from the last restart (lines 12–13).

The prohibition period T is equal to one at the beginning, because in this manner one avoids coming back to the just abandoned clique. Nonetheless, let us note that RLS behaves exactly as local search in the first phase, as long as only new vertices are added to the current clique X, and therefore prohibitions do not have any effect. The difference starts when a maximal clique with respect to set inclusion is reached and the first vertex is dropped.

```
1.  function  BEST-NEIGHBOR (X)
2.    ⌈ type ← notFound
3.    │ if {allowed nodes ∈ S} ≠ ∅
4.    │   ⌈ type ← addMove
5.    │   │ maxDegAllowed ← maximum degree in G(S)
6.    │   │ v ← random allowed w ∈ S
7.    │   ⌊    with deg_{G(S)}(w) = maxDegAllowed
8.    │ if type = notFound and X ≠ ∅
9.    │   ⌈ type ← dropMove
10.   │   │ if {allowed v ∈ X} ≠ ∅
11.   │   │   ⌈ maxDeltaS ← max_{allowed j∈X} deltaS[j]
12.   │   │   │ v ← random allowed w ∈ X
13.   │   │   ⌊    with deltaS [w] = maxDeltaS
14.   │   │ else
15.   │   ⌊   ⌊ v ← random w ∈ X
16.   │ if type = notFound
17.   │   ⌈ type ← addMove
18.   │   ⌊ v ← random w ∈ V
19.   │ INCREMENT AL-UPDATE (v, type)
20.   │ if type = addMove return X ∪ {v}
21.   ⌊   else return X \ {v}
```

Parameters	
X	Configuration to be changed

Local variables	
type	Type of move (add Move or drop Move)
v	Node to be added or removed

FIGURE 19.6   RLS algorithm: the function BEST-NEIGHBOR.

19.4.1.1 Choice of the Best Neighbor

The function BEST-NEIGHBOR is described in Figure 19.6. Given a current clique X, let us define as S the vertex set of possible additions, that is, the vertices that are adjacent to all nodes of X. Let $G(S)$ is the subgraph induced by S. Finally, if $j \in X$, $deltaS[j]$ is the number of vertices adjacent to all nodes of X but j. A vertex v is *prohibited* at iteration t iff it satisfies $lastMoved[v] \geq (t - T^{(t)})$, where $lastMoved[v]$ is the last iteration at which it has been added to or dropped from the current clique. Vector $lastMoved$ is used to determine the allowed vertices, see lines 3 and 11.

The best neighbor is chosen in stages with this overall scheme: first an allowed vertex that can be added to the current clique is searched for (lines 3–7). If none is found, an allowed vertex to drop is searched for (lines 8–13). Finally, if no allowed moves are available, a random vertex in X is dropped if X is not empty (line 15), a random vertex in V is added in the opposite case (lines 16–18).

Ties among *allowed* vertices that can be added are broken by preferring the ones with the largest degree in $G(S)$ (line 7); a random selection is executed among vertices with equal degree in $G(S)$.

Ties among *allowed* vertices that can be dropped are broken by preferring those causing the largest increase $|S^{(t+1)}| - |S^{(t)}|$ where $S^{(t)}$ is the set S at iteration t (line 13). Again, a random selection is then executed if this criterion selects more that one winner.

19.4.1.2 Reaction and Periodic Restart

The function HISTORY-REACTION is illustrated in Figure 19.7. The prohibition T is minimal at the beginning ($T = 1$), and is then determined by two competing processes: T increases when the current clique comes back to one already found, it decreases when no cliques are repeated in a suitable period. In detail: the current clique X is searched in memory, by utilizing hashing techniques (line 1). If X is found, a reference Z is returned to a data structure containing the last visit time (line 2). If the repetition interval R is sufficiently short, cycles are discouraged by increasing T (lines 5–7). If X is not found, it is stored in memory with the time t when it was encountered (line 9). If T remained constant for a number of iterations greater than $10k_b$, and therefore no clique is repeated during this interval, it is decreased (lines 10–12).

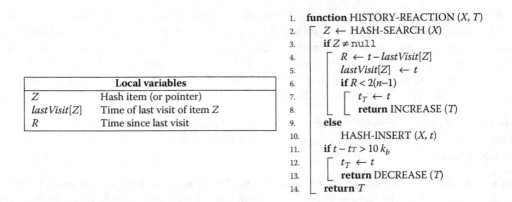

FIGURE 19.7   RLS algorithm: routine HISTORY-REACTION.

FIGURE 19.8   RLS algorithm: routine RESTART.

Increases and decreases, with a minimal change of one unit plus upper and lower bounds, are realized by the two following functions:

$$\text{INCREASE}(T) = \min\{\max\{T \cdot 1.1, T + 1\}, n - 2\}$$

$$\text{DECREASE}(T) = \max\{\min\{T \cdot 0.9, T - 1\}, 1\}$$

The routine RESTART is similar to that in Reference 57. If there are vertices that have never been part of the current clique during the search, that is, that have never been moved since the beginning of the run, one of them with maximal degree in V is randomly selected (lines 3–5 in Figure 19.8). If all vertices have already been members of X in the past, a random vertex in V is selected (line 7). Data structures are updated to reflect the situation of $X = \emptyset$, then the incremental update is applied and the vertex v is added.

19.4.2 Complexity per Iteration

The computational complexity of each iteration of RLS is the sum of a term caused by the usage and updating of reaction-related structures and a term caused by the local search part: neighborhood evaluation and generation of the next clique.

Let us first consider the reaction-related part. The overhead per iteration incurred to determine the prohibitions is $O(|M(X)|)$, $M(X)$ being the neighborhood, that for updating the last usage time of the chosen move is $O(1)$, that to check for repetitions, and to update and store the new *hashing value* of the current configuration has an average complexity of $O(1)$, if an incremental *hashing* calculation is applied.

In our case the single-iteration RLS complexity is dominated by the neighborhood evaluation. This evaluation requires an efficient update of the sets S and SMINUS plus the computation of the degrees of the vertices in the induced subgraph $G(S)$ (used in function BEST-NEIGHBOR, Figure 19.6, line 6). It is therefore crucial to consider incremental algorithms, in an effort to reduce the complexity. In our algorithm, the sets S and SMINUS are maintained by the routine INCREMENTAL-UPDATE that is used in the function BEST-NEIGHBOR and in the procedure RESTART. The limited space of this extended abstract force us to omit the detailed description of INCREMENTAL-UPDATE and of the related data structures. The following theorem is proved in the full paper:

Theorem 19.1 *The incremental algorithm for updating X, S and SMINUS during each iteration of RLS has a worst case complexity of $O(n)$. In particular, if vertex v is added to or deleted from S, the required operations are $O(\deg_{\overline{G}}(v))$.*

Let us note that the actual multiplicative constant is very small and that the algorithm tends to be faster for dense graphs where the average degree $\deg_{\overline{G}}(v)$ in the complement graph can be much smaller than n.

Finally, the computation of the vertex degrees in the induced subgraph $G(S)$ costs at most $O(m)$ by the following trivial algorithm. All the edges are inspected, if both end-points are in S, the corresponding degrees are incremented by 1. In practice the degree is not computed from scratch but it is updated incrementally with a much lesser computational effort: in fact the maximum number of nodes that enter or leave S at a given iteration is at most $\deg_{\overline{G}}(v)$, v being the just moved vertex. Therefore, the number of operations performed is at most $O(\deg_{\overline{G}}(v) \cdot |S^{(t+1)}|)$. Because the search aims at maximizing the clique X, the set S tends to be very small (at some steps empty!) after a first transient period, and the dominant factor is the same $O(\deg_{\overline{G}}(v))$ factor that appears in Theorem 19.1.

Putting together all the complexity considerations Corollary 19.1 is immediately implied:

Corollary 19.1 *The worst-case complexity of a single iteration is $O(\max\{n, m\})$.*

Experimental results of RLS on the benchmark suite for the MC problem by the organizers of the second DIMACS implementation challenge [55], are analyzed in Reference 33.

19.5 Related Reactive Approaches

Because reactive search is rooted in the automation of the algorithm tuning process by embodying the typical experimental cycle followed by heuristic algorithm designers, it is not surprising to see related ideas and principles arising in various areas. For space limitation we list in this section a limited selection of interesting related papers.

Probabilistic methods that explicitly maintain statistics about the search space by creating models of the good solutions found so far are considered for example in References 58, 59.

Implicit models of the configuration space are built by a population of searchers in genetic algorithms, where learning comes in the form of "survival of the fittest" and generation of new sampling points depends on the previous evolution, see for example Reference 60. The issue of controlling parameters of an evolutionary algorithm, including adaptive and "self-adaptive" techniques is considered in Reference 61, while fitness landscape analysis for the choice of appropriate operators in "memetic" algorithms (combining genetic algorithms with local search) is considered in Reference 62.

In the area of stochastic optimization, which considers noise in the evaluation process, memory based schemes for validation and tuning of function approximators are used in References 63, 64.

Guided local search aims at exploiting the problem and search-related information to effectively guide local search heuristics [65]. Evaluation functions for global optimization and Boolean satisfiability are learnt in Reference 66.

Learning mechanisms with biological motivations are used in ant colony optimization based on feedback, distributed computation, and the use of a constructive greedy heuristic. For example "pheromone trail" information to perform modifications on solutions for the quadratic assignment problem is considered in References 67, 68.

Dynamic local search, which increases penalties of some solution components to move the tentative solution away from a given local minimum can also be considered as a form of learning based on the previous history of the search. An example is the dynamic local search algorithm on the MAXSAT problem in Reference 69, see also Reference 70 for additional references about stochastic local search methods including adaptive versions.

In the area of computer systems management, "autonomic" systems are based on self-regulating biological systems (http://www.research.ibm.com/autonomic/) that have many points of contact with Reactive Search.

Acknowledgment

The work of R. Battiti was supported by the Russian Science Foundation through the Project entitled Global Optimization, Supercomputing Computations, and Applications under Grant 15-11-30022.

Note: Most of the Battiti's papers are available in postscript form at http://rtm.science.unitn.it/~battiti/battiti-publications.html.

References

1. Barr, R.S., Golden, B.L., Kelly, J.P., Resende, M.G.C., and Stewart, W., Designing and reporting on computational experiments with heuristic methods, *Journal of Heuristics,* 1(1), 9, 1995.
2. Hooker, J.N., Testing heuristics: We have it all wrong, *Journal of Heuristics,* 1(1), 33, 1995.
3. McGeoch, C.C., Toward an experimental method for algorithm simulation, *INFORMS Journal on Computing,* 8(1), 1, 1996.
4. Valiant, L.G., A theory of the learnable, *CACM,* 27(11), 1134, 1984.
5. Kearns, M.J. and Vazirani, U.V., *An Introduction to Computational Learning Theory,* MIT Press, Cambridge, MA, 1994.
6. Kirkpatrick, S., Gelatt Jr., C.D., and Vecchi, M.P., Optimization by simulated annealing, *Science,* 220, 671, 1983.
7. Battiti, R. and Tecchiolli, G., Simulated annealing and tabu search in the long run: A comparison on QAP tasks, *Computer and Mathematics with Applications,* 28(6), 1, 1994.
8. Battiti, R. and Tecchiolli, G., Local search with memory: Benchmarking RTS, *Operations Research Spektrum,* 17(2/3), 67, 1995.
9. Aarts, E.H.L., Korst, J.H.M., and Zwietering, P.J., Deterministic and randomized local search, in *Mathematical Perspectives on Neural Networks,* Mozer, M., Smolensky, P., and Rumelhart, D., Eds., Lawrence Erlbaum Publishers, Hillsdale, NJ, 1995.
10. Ferreira, A.G. and Zerovnik, J., Bounding the probability of success of stochastic methods for global optimization, *Computers and Mathematics Applications,* 25, 1, 1993.
11. Chiang, T.S. and Chow, Y., On the convergence rate of annealing processes, *SIAM Journal on Control and Optimization,* 26(6), 1455, 1988.
12. Strenski, P.N. and Kirkpatrick, S., Analysis of finite length annealing schedules, *Algorithmica,* 6, 346, 1991.
13. Glover, F., Tabu search—Part I, *ORSA Journal on Computing,* 1(3), 190, 1989.
14. Steiglitz, K. and Weiner, P., Algorithms for computer solution of the traveling salesman problem, in *Proceedings of the Sixth Allerton Conference on Circuit and System Theory,* Urbana, IL, 1968, p. 814.
15. Lin, S., Computer solutions of the travelling salesman problems, *Bell Systems Technical J,* 44(10), 2245, 1965.
16. Papadimitriou, C.H. and Steiglitz, K., *Combinatorial Optimization, Algorithms and Complexity,* Prentice-Hall, Upper Saddle River, NJ, 1982.

17. Glover, F., Tabu search—Part II, *ORSA Journal on Computing,* 2(1), 4, 1990.
18. Hansen, P. and Jaumard, B., Algorithms for the maximum satisfiability problem, *Computing,* 44, 279, 1990.
19. Glover, F., Tabu search: Improved solution alternatives, in *Mathematical Programming, State of the Art,* Birge, J.R. and Murty, K.G., Eds., The University of Michigan, Ann Arborm, MI, 1994, p. 64.
20. Battiti, R. and Tecchiolli, G., The reactive tabu search, *ORSA Journal on Computing,* 6(2), 126, 1994.
21. Woodruff, D.L. and Zemel, E., Hashing vectors for tabu search, *Annals of Operations Research,* 41, 123, 1993.
22. Battiti, R., Time- and space-efficient data structures for history-based heuristics, Technical Report UTM-96-478, Dipartimento di Matematica, University di Trento, Italy, 1996.
23. Battiti, R., Partially persistent dynamic sets for history-sensitive heuristics, in *Data Structures, Near Neighbor Searches, and Methodology: Fifth and Sixth DIMACS Challenges,* Johnson, D.S., Goldwasser, M.H., and McGeoch, C.C., Eds., DIMACS Series in Discrete Mathematics and Theoretical Computer Science, 59, AMS, 2002, p. 1.
24. Driscoll, J.R., Sarnak, N., Sleator, D.D., and Tarjan, R.E., Making data structures persistent, in *Proceedings of Symposium on Theory of Computing Conference,* 1986.
25. Overmars, M.H., Searching in the past II: General transforms, Technical Report, Department of Computer Science, University of Utrecht, the Netherlands, 1981.
26. Bayer, R., Symmetric binary B-trees: Data structure and maintenance algorithms, *Acta Informatica,* 1, 290, 1972.
27. Guibas, L.J. and Sedgewick, R., A dichromatic framework for balanced trees, in *Proceedings of Foundations of Computer Science (FOCS),* IEEE Computer Society, Ann Arbor, Michigan, 1978, p. 8.
28. Tarjan, R.E., Updating a balanced search tree in $O(1)$ rotations, *Information Processing Letters,* 16, 253, 1983.
29. Maier, D. and Salveter, S.C., Hysterical B-trees, *Information Processing Letters,* 12(4), 199, 1981.
30. Dammeyer, F. and Voss, S., Dynamic tabu list management using the reverse elimination method, *Annals of Operations Research,* 41, 31, 1993.
31. Sarnak, N. and Tarjan, R.E., Planar point location using persistent search trees, *CACM,* 29(7), 669, 1986.
32. Johnson, D.S., Aragon, C.R., McGeoch, L.A., and Schevon, C., Optimization by simulated annealing: An experimental evaluation; part II, graph coloring and number partitioning, *Operations Research,* 39(3), 378, 1991.
33. Battiti, R. and Protasi, M., Reactive local search for the maximum clique problem, *Algorithmica,* 29(4), 610, 2001.
34. Fox, B.L., Simulated annealing: Folklore, facts, and directions, *Monte Carlo and Quasi-Monte Carlo Methods in Scientific Computing,* Niederreiter, H. and Shiue, P.J.-S., Eds., Springer-Verlag, New York, 1995.
35. Battiti, R. and Tecchiolli, G., Training neural nets with the reactive tabu search, *IEEE Transactions on Neural Networks,* 6(5), 1185, 1995.
36. Chiang, W.C. and Russell, R.A., A reactive tabu search metaheuristic for the vehicle routing problem with time windows, *INFORMS Journal on Computing,* 9, 417, 1997.
37. Osman, I.H. and Wassan, N.A., A reactive tabu search meta-heuristic for the vehicle routing problem with back-hauls, *Journal of Scheduling,* 5(4), 287, 2002.
38. Bräysy, O., A reactive variable neighborhood search for the vehicle-routing problem with time windows, *INFORMS Journal on Computing,* 15(4), 347, 2003.
39. Kincaid, R.K. and Labal, K.E., Reactive tabu search and sensor selection in active structural acoustic control problems, *Journal of Heuristics,* 4(3), 199, 1998.

40. Anzellotti, G., Battiti, R., Lazzizzera, I., Soncini, G., Zorat, A., Sartori, A., Tecchiolli, G., and Lee, P., TOTEM: A highly parallel chip for triggering applications with inductive learning based on the reactive tabu search, *International Journal of Modern Physics C*, 6(4), 555, 1995.

41. Battiti, R. and Bertossi, A.A., Greedy, prohibition, and reactive heuristics for graph partitioning, *IEEE Transactions on Computers*, 48(4), 361, 1999.

42. Toune, S., Fudo, H., Genji, T., Fukuyama, Y., and Nakanishi, Y., Comparative study of modern heuristic algorithms to service restoration in distribution systems, *IEEE Transactions on Power Delivery*, 17(1), 173, 2002.

43. Battiti, R. and Protasi, M., Reactive search, a history-sensitive heuristic for MAX-SAT, *ACM Journal of Experimental Algorithmics*, 2, 1997.

44. Battiti, R. and Protasi, M., Reactive local search techniques for the maximum k-conjunctive constraint satisfaction problem, *Discrete Applied Mathematics*, 96–97, 3, 1999.

45. Nonobe, K. and Ibaraki, T., A tabu search approach for the constraint satisfaction problem as a general problem solver, *European Journal of Operational Research*, 106, 599, 1998.

46. Battiti, R. and Tecchiolli, G., The continuous reactive tabu search: Blending combinatorial optimization and stochastic search for global optimization, *Annals of Operations Research—Metaheuristics in Combinatorial Optimization*, 63, 153, 1996.

47. Chelouah, R. and Siarry, P., Tabu search applied to global optimization, *European Journal of Operational Research*, 123, 256, 2000.

48. Battiti, R. and Brunato, M., Reactive search for traffic grooming in WDM networks, *Evolutionary Trends of the Internet, IWDC2001*, Palazzo, S., Ed., LNCS, Vol. 2170, Springer-Verlag, Heidelberg, Germany, 2001, p. 56.

49. Winter, T. and Zimmermann, U., Real-time dispatch of trams in storage yards, *Annals of Operations Research*, 96, 287, 2000.

50. Fortz, B. and Thorup, M., Increasing Internet capacity using local search, *Computational Optimization and Applications*, 29(1), 13, 2004.

51. Ausiello, G., Crescenzi, P., and Protasi, M., Approximate solution of NP optimization problems, *Theoretical Computer Science*, 150, 1, 1995.

52. Crescenzi, P. and Kann, V., A compendium of NP optimization problems, Technical Report, 1996, electronic notes: http://www.nada.kth.se/~viggo/problemlist/compendium.html.

53. Bellare, M., Goldreich, O., and Sudan, M., Free bits, PCP and non-approximability—towards tight results, in *Proceedings of Foundations of Computer Science* (FOCS), Milwaukee, Wisconsin, 1995, p. 422.

54. Hastad, J., Clique is hard to approximate within $n^{1-\epsilon}$, in *Proceedings of Foundations of Computer Science* (FOCS), Burlington, Vermont, 1996, p. 627.

55. Johnson, D.S. and Trick, M., Eds., *Cliques, Coloring, and Satisfiability: Second DIMACS Implementation Challenge*, DIMACS Series in Discrete Mathematics and Theoretical Computer Science, Vol. 26, AMS, 1996.

56. Pardalos, P.M. and Xu, J., The maximum clique problem, *Journal of Global Optimization*, 4, 301, 1994.

57. Soriano, P. and Gendreau, M., Tabu search algorithms for the maximum clique problem, Technical Report CRT-968, Centre de Recherche sur les Transports, Universite de Montreal, Canada, 1994.

58. Pelikan, M., Goldberg, D.E., and Lobo, F., A survey of optimization by building and using probabilistic models, *Computational Optimization and Applications*, 21(1), 5, 2002.

59. Baluja, S. and Davies, S., Using optimal dependency trees for combinatorial optimization: Learning the structure of the search space, in *Proceedings of the Fourteenth International Conference on Machine Learning*, Fisher, D.H., Ed., Morgan Kaufmann Publishers, San Francisco, CA, 1997, p. 30.

60. Syswerda, G., Simulated crossover in genetic algorithms, in *Foundations of Genetic Algorithms*, Whitley, D.L., Ed., Morgan Kaufmann Publishers, San Mateo, CA, 1993, p. 239.

61. Eiben, A.E., Hinterding, R., and Michalewicz, Z., Parameter control in evolutionary algorithms, *IEEE Transactions on Evolutionary Computation,* 3(2), 124, 1999.
62. Merz, P. and Freisleben, B., Fitness landscape analysis and memetic algorithms for the quadratic assignment problem, *IEEE Transactions on Evolutionary Computation,* 4(4), 337, 2000.
63. Moore, A.W. and Schneider, J., Memory-based stochastic optimization, in *Advances in Neural Information Processing Systems,* Touretzky, D.S., Mozer, M.C., and Hasselmo, M.E., Eds., Vol. 8, MIT Press, Cambridge, MA, 1996, p. 1066.
64. Dubrawski, A. and Schneider, J., Memory based stochastic optimization for validation and tuning of function approximators, *Conference on AI and Statistics,* 6th International Workshop on Artificial Intelligence and Statistics, Fort Lauderdale, Florida, 1997.
65. Voudouris, Ch. and Tsang, E., Guided local search and its application to the traveling salesman problem, *European Journal of Operational Research,* 113, 469, 1999.
66. Boyan, J.A. and Moore, A.W., Learning evaluation functions for global optimization and boolean satisfiability, in *Proceedings of 15th National Conference on Artificial Intelligence,* Press, A., ed., Madison, Wisconsin, 1998, p. 3.
67. Dorigo, M., Maniezzo, V., and Colorni, A., Ant system: Optimization by a colony of cooperating agents, *IEEE Transactions on Systems, Man and Cybernetics,* Part B, 26(1), 29, 1996.
68. Gambardella, L.M., Taillard, E.D., and Dorigo, M., Ant colonies for the quadratic assignment problem, *Journal of the Operational Research Society,* 50(2), 167, 1999.
69. Hutter, F., Tompkins, D.A.D., and Hoos, H.H., Scaling and probabilistic smoothing: Efficient dynamic local search for SAT, in *Proceedings CP-02,* LNCS, 2470, Springer-Verlag, Heidelberg, Germany, 2002, p. 233.
70. Hoos, H.H. and Stuetzle, T., *Stochastic Local Search: Foundations and Applications,* Morgan Kaufmann, San Mateo, CA, 2005.

20

Neural Networks

Bhaskar DasGupta[*]

Derong Liu

Hava T. Siegelmann

20.1 Introduction

Artificial neural networks have been proposed as a tool for machine learning [1–4] and many results have been obtained regarding their application to practical problems in robotics control, vision, pattern recognition, grammatical inferences, and other areas [5–8]. In these roles, a neural network is trained to recognize complex associations between inputs and outputs that were presented during a *supervised* training cycle. These associations are incorporated into the weights of the network, which encode a distributed representation of the information that was contained in the input patterns. Once trained, the network will compute an input/output mapping which, if the training data was representative enough, will closely match the unknown rule that produced the original data. Massive parallelism of computation, as well as noise and fault tolerance, are often offered as justifications for the use of neural nets as learning paradigms.

Traditionally, especially in the structural complexity literature [4], feedforward circuits composed of AND, OR, NOT, or threshold gates have been thoroughly studied. However, in practice, when designing a neural net, continuous activation functions such as the *standard sigmoid* are more commonly used. This is because usual learning algorithms such as the backpropagation algorithm assumes a continuous activation function. As a result, neural nets are distinguished from those conventional circuits because they perform real-valued computation and admit efficient learning procedures. The last three decades have seen a resurgence of theoretical techniques to design and analyze the performances of neural nets (e.g., see the survey in Reference 9) as well as novel application of neural nets to various applied areas (e.g., see Reference 6 and some of the references there). Theoretical researches in computational capabilities of neural nets have given valuable insights into the mechanisms of these models.

In subsequent discussions, we distinguish between two types of neural networks, commonly known as the "feedforward" neural nets and the "recurrent" neural nets. A feedforward net consists of a number of

---

[*] Supported in part by NSF grants CCR-0206795, CCR-0208749, and IIS-0346973.

processors ("nodes" or "neurons") each of which computes a function of the type $y = \sigma \left(\sum_{i=1}^{k} a_i u_i + b \right)$ of its inputs $u_1, \ldots, u_k$. These inputs are either external (input data is fed through them) or they represent the outputs y of other nodes. No cycles are allowed in the connection graph and the output of one designated node is understood to provide the output value produced by the entire network for a given vector of input values. The possible coefficients a_i and b appearing in the different nodes are the *weights* of the network, and the functions σ appearing in the various nodes are the *node, activation,* or *gate* functions. An *architecture* specifies the interconnection structure and the σ's, but not the actual numerical values of the weights. A recurrent neural net, on the other hand, allows cycles in the connection graph, thereby allowing the model to have substantially more computational capabilities (see Section 20.3.2).

In this chapter we survey research works dealing with basic questions regarding *computational capabilities* and *learning* of neural models. There are various types of such questions that one may ask, most of them closely related and complementary to each other. We next describe a few of them *informally*.

One direction of research deals with the *representational capabilities* of neural nets, assuming *unlimited* number of neurons are available [10–16]. The origin of this type of research can be traced back to the work of the famous mathematician Kolmogorov [17], who essentially proved the first existential result on the representation capabilities of depth 2 neural nets. This type of research ignores the training question itself, asking instead if it is *at all* possible to compute or approximate arbitrary functions [11,17] or if the net can simulate, say, Turing machines [15,16]. Many of the results and proofs in this direction are nonconstructive.

Another perspective to learnability questions of neural nets takes a *numerical analysis or approximation theoretic* point of view. There one asks questions such as *how many* hidden units are necessary to well-approximate, that is to say, approximate with a small overall error, an unknown function. This type of research also ignores the training question, asking instead what is the best one could do, in this sense of overall error, if the best possible network with a given architecture were to be eventually found. Some papers along these lines are seen in References 18, 19, which dealt with single hidden layer nets, and in Reference 20, which dealt with multiple hidden layers.

Another possible line of research deals with the *sample complexity* questions, that is, the quantification of the amount of information (number of samples) needed to characterize a given unknown mapping. Some recent references to such work, establishing sample complexity results, and hence "weak learnability" in the Valiant model, for neural nets, are the papers in References 21 through 25; the first of these references deals with networks that employ hard threshold activations, the third and fourth cover continuous activation functions of a type (piecewise polynomial), and the last one provides results for networks employing the standard sigmoid activation function.

Yet another direction in which to approach theoretical questions regarding learning by neural networks originates with the work of Judd (see for instance References 26, 27, as well as the related work [28,29]). Judd was motivated by the observation that the "backpropagation" algorithm often runs very slowly, especially for high-dimensional data. Recall that this algorithm is used to find a network (i.e., find the weights, assuming a fixed architecture) that reproduces the observed data. Of course, many modifications of the vanilla "backprop" approach are possible, using more sophisticated techniques such as high-order (Newton), conjugate gradient, or sequential quadratic programming methods. However, the "curse of dimensionality" seems to arise as a computational obstruction to all these training techniques as well, when attempting to learn arbitrary data using a standard feedforward network. For the simpler case of linearly separable data, the perceptron algorithm and linear programming techniques help to find a network—with no "hidden units"—relatively fast. Thus one may ask if there exists a *fundamental barrier* to training by general feedforward networks, a barrier that is insurmountable no matter which particular algorithm one uses. (Those techniques which *adapt* the architecture to the data, such as cascade correlation or incremental techniques, would not be subject to such a barrier.)

20.2 Feedforward Neural Networks

As mentioned before, a feedforward neural net is one in which the underlying connection graph contains *no* directed cycles. More precisely, a feedforward neural net (or, in our terminology, a Γ-net) can be defined as follows.

Let Γ be a class of real-valued functions, where each function is defined on some subset of $\mathbb{R}$. A Γ-net C is an unbounded fan-in circuit whose edges and vertices are labeled by real numbers. The real number assigned to an edge (resp. vertex) is called its *weight* (resp. its *threshold*). Moreover, to each vertex v a *gate* (or *activation*) function $\gamma_v \in \Gamma$ is assigned.

The circuit C *computes* a function $f_C : \mathbb{R}^m \rightarrow \mathbb{R}$ as follows. The components of the input vector $x = (x_1, \ldots, x_m) \in \mathbb{R}^m$ are assigned to the sources of C. Let $v_1, \ldots, v_n$ be the immediate predecessors of a vertex v. The input for v is then $s_v(x) = \sum_{i=1}^{n} w_i y_i - t_v$, where w_i is the weight of the edge (v_i, v), t_v is the *threshold* of v and y_i is the value assigned to v_i. If v is not a sink, then we assign the value $\gamma_v(s_v(x))$ to v. Otherwise we assign $s_v(x)$ to v.

Without any loss of generality, one can assume that C has a single sink t. Then $f_C = s_t$ is the function computed by C.

The function class Γ quite popular in the structural-complexity literature [30–32] is the *binary threshold function* $\mathcal{H}$ defined by:

$$\mathcal{H}(x) = \begin{cases} 0 & \text{if } x \leq 0 \\ 1 & \text{if } x > 0 \end{cases}$$

However, in practice this function is not so popular because the function is discrete and hence using this gate function may pose problems in most commonly used learning algorithms like the backpropagation algorithms [3] or their variants. Also, from biological perspectives, real neurons have continuous input–output relations [33]. In practice, various continuous (or, at least locally smooth) gate functions have been used, for example, *the cosine squasher, the standard sigmoid, radial basis functions, generalized radial basis functions, piecewise-linear, polynomials, and trigonometric polynomial* functions. In particular, the standard sigmoid function $\sigma(x) = 1/(1 + e^{-x})$ is very popular.

The simplest type of feedforward neural net is the classical *perceptron*. This consists of one single neuron computing a threshold function (Figure 20.1). In other words, the perceptron P is characterized by a vector (of "weights") $\vec{c} \in \mathbb{R}^m$, and computes the inner product $\vec{c}.v + c_0 = c_1 v_1 + \ldots + c_m v_m + c_0$. Such a model has been well studied and efficient learning algorithms for it exists (e.g., Reference 34, see also Reference 35). In this chapter we will be, however, more interested in more complex *multi-layered neural nets* (Figure 20.2).

20.2.1 Approximation Properties

It is known that neural nets of only depth 2 and with arbitrarily large number of nodes can approximate any real-valued function up to any desired accuracy, using a continuous activation function such as the sigmoidal function [11,17]. However, these proofs are mostly nonconstructive and, from a practical point of view, one is more interested in designing *efficient* neural nets (i.e., roughly speaking, neural nets

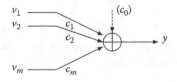

FIGURE 20.1 Classical perceptrons.

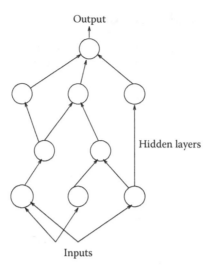

FIGURE 20.2   A feedforward neural net with three hidden layers and two inputs.

with small size and depth) to exactly or approximately compute different functions. It is also important, from a practical point of view, to understand the *size-depth tradeoff* of the complexity of feedforward nets while computing functions, as generally neural nets with more layers are more costly to simulate or implement.

Threshold circuits, *that is*, feedforward nets with threshold activation functions, have been quite well studied, and upper/lower bounds for them have been obtained while computing various Boolean functions (see, e.g., References 30–32, 36–38 among many other works). Functions of special interest have been the parity function, computing the multiplication and division of binary numbers and so forth.

However, as mentioned before, it is more common in practice to use a continuous activation function, such as the standard sigmoid function. References 39, 40, among others, considered efficient computation or approximation of various functions by feedforward circuits with continuous activation functions and also studied size-depth tradeoffs. In particular, Reference 39 showed that any polynomial of degree n with polynomially bounded coefficients can be approximated with exponential accuracy by depth 2 feedforward sigmoidal neural nets with a polynomial number of nodes. References 39, 40 also show how to simulate threshold circuits by sigmoidal circuits with a polynomial increase in size and a constant factor increase in depth. Thus, in effect, functions computed by threshold circuits can also be computed by sigmoidal circuits with not too much increase in size and depth. Maass [23] shows how to simulate nets with piecewise-linear activation functions with bounded depth, arbitrary real weights, and for Boolean inputs and outputs by a threshold net of somewhat larger size and depth with weights from $\{-1, 0, 1\}$. References 39, 40 showed that circuits composed of sufficiently smooth gate functions are capable of efficiently approximating polynomials within any degree of accuracy. Complementing these results, Reference 39 also provided nontrivial lower bounds on the size of *bounded-depth* sigmoidal nets with polynomially large weights when computing oscillatory functions. In essence, one can prove results of the following types.

Definition 20.1 ([39])[*] *Let* $\gamma : \mathbb{R} \to \mathbb{R}$ *be a function. We call* γ *nontrivially smooth with parameter* k *if and only if there exists rational numbers* α, β ($\alpha > 0$) *and an integer* k *such that* α *and* β *have logarithmic size at most* k *and* **(a)** γ *can be represented by the power series* $\sum_{i=0}^{\infty} a_i (x - \beta)^i$ *for all* $x \in [\beta - \alpha, \beta + \alpha]$.

---

[*] The notation $\gamma^{(i)}$ denotes the ith derivative of γ.

For each $i > 1$, a_i is a rational number of logarithmic size at most i^k. (**b**) *For each $i > 1$ there exists j with $i \leq j \leq i^k$ and $a_j \neq 0$.* (**c**) *For each $i > 1$, $||\gamma^{(i)}||_{[-\alpha,\alpha]} \leq 2^{i^k}$.*

Theorem 20.1 ([39])* *Assume that γ is nontrivially smooth with parameter k. Let $p(x)$ be a degree n polynomial whose coefficients are rational numbers of logarithmic size at most max. Then $p(x)$ can be ε-approximated (over the domain $[-D, D]$ with $[\beta - \alpha, \beta + \alpha] \subseteq [-D, D]$) by a $\{\gamma\}$-circuit C_p. C_p has depth 2 and size $O(n^{2k})$. The Lipschitz-bound† of C_p (over $[-D, D]^n$) is at most $c_\gamma \cdot \left(2^{max} \cdot (2 + D) \cdot \frac{1}{\varepsilon}\right)^{poly(n)}$ where the constant c_γ depends only on γ and not on p.*

Theorem 20.2 ([39]) *Let $f : [-1, 1] \to \mathbb{R}$ be a function that ε-oscillates t times‡ and let C be a Γ-circuit of depth d, size s and Lipschitz-bound 2^s over $[-1, 1]$. If C approximates f with error at most $\frac{\varepsilon}{4}$, then $s \geq t^{\Omega(1/d)}$.*

Notice that the standard sigmoid is nontrivially smooth with a constant k, and so is the case for most of the other continuous activation functions mentioned in Section 20.1. However, the simulation in Theorem 20.1 needs quadratically many nodes to simulate a polynomial by sigmoidal nets with exponential accuracy. Unfortunately, the proof of Theorem 20.2 relies on efficient simulation of a sigmoidal-circuit by a spline-circuit and hence cannot be extended to the case of arbitrary weights (i.e., the Lipschitz-bound condition cannot be dropped).

20.2.2 Backpropagation Algorithms

Basic backpropagation [41] is currently the most popular supervised learning method that is used to train multi-layer feedforward neural networks with differentiable transfer functions. It is a gradient descent algorithm in which the network weights are moved along the negative of the gradient of the performance function.

The basic backpropagation algorithm performs the following steps:

1. Forward pass: Inputs are presented and the outputs of each layer are computed.
2. Backward pass: Errors between the target and the output are computed. Then, these errors are "back-propagated" from the output to each layer until the first layer. Finally, the weights are adjusted according to the gradient descent algorithm with the derivatives obtained by backpropagation.

We will discuss in more detail a generalized version of this approach for recurrent networks, termed as the "real-time backpropagation through time (BPTT)," in Section 20.3.1. The key point of basic backpropagation is that the weights are adjusted in response to the derivatives of performance function with respect to weights, which only depend on the current pattern; the weights can be adjusted sequentially or in batch mode. More details about the basic backpropagation can be found in Reference 41. The asymptotic convergence rates of backpropagation is proved in Reference 42. Traditionally, the parity function has been used as an important benchmark for testing the efficiency of a learning algorithm. Empirical studies in Reference 43 show that the training time of a feedforward net using backpropagation while learning the parity function grows exponentially with the number of inputs, thereby rendering the learning algorithm to be very time-consuming. Unfortunately, a satisfactory theoretical justification for this behavior is yet

---

* *poly(n)* denotes a polynomial in *n*.

† The Lipschitz-bound of the net is a measure of the numerical stability of the circuit. Informally speaking, a net has a Lipschitz bound of L if all its weights and thresholds are bounded in absolute value by L.

‡ f ε-oscillates t times if and only if there are real numbers $-1 \leq x_1 < \ldots < x_{t+1} \leq 1$ such that (a) $f(x_1) = f(x_2) = \ldots = f(x_{t+1})$, (b) $|x_{i+1} - x_i| \geq \varepsilon$ for all i and (c) there are real numbers $y_1, \ldots, y_t$ such that $x_i \leq y_i \leq x_{i+1}$ and $|f(x_i) - f(y_i)| \geq \varepsilon$ for all i.

to be shown. Also, it is well known that the backpropagation algorithm may get stuck in local minima, and in fact, in general gradient descent algorithms may fail to classify correctly data that even simple perceptrons can classify correctly [44–46]. Strategies of avoiding local minima include local perturbation and simulated-annealing techniques, whereas the later problem (in absence of a local minima) can be avoided using, say, threshold least mean squares (LMS) procedures.

20.2.3 Learning Theoretic Results

Approximation results discussed in Section 20.2.1 do not necessarily translate into good learning algorithms. For example, even though a sigmoidal net has great computational power, we still need to investigate how to learn the weights of such a network from a set of examples. Learning is a very important aspect of designing efficient neural models from a practical point of view. There are a few possible approaches to tackle this issue; we describe one of those approaches next.

20.2.3.1 Vapnik-Chervonenkis (VC)-Dimension Approach

VC-dimensions (and, their suitable extensions to real valued computations) provide *information-theoretic bounds* to the sample complexities for learning problems in neural nets. We briefly (also, somewhat informally) review some (by now standard) notions regarding sample complexity, which deals with the calculation of VC dimensions as applicable for neural nets (for more details, see the books in References 47, 48, the paper in Reference 49, or the survey in Reference 9).

In the general classification problem, an input space $\mathbb{X}$ as well as a collection $\mathcal{F}$ of maps $\mathbb{X} \to \{-1, 1\}$ are assumed to have been given. (The set $\mathbb{X}$ is assumed to be either countable or an Euclidean space, and the maps in $\mathcal{F}$, the set of functions computable by the specific neural nets under consideration, are assumed to be measurable. In addition, mild regularity assumptions are made that ensure that all sets appearing below are measurable, but details are omitted as in the context of neural nets these assumptions are almost always satisfied.) Let W be the set of all sequences

$$w = (u_1, \psi(u_1)), \ldots, (u_s, \psi(u_s))$$

over all $s \geq 1$, $(u_1, \ldots, u_s) \in \mathbb{X}^s$, and $\psi \in \mathcal{F}$. An *identifier* is a map $\varphi : W \to \mathcal{F}$. The value of φ on a sequence w as aforementioned will be denoted as φ_w. The *error* of φ with respect to a probability measure P on $\mathbb{X}$, a $\psi \in \mathcal{F}$, and a sequence $(u_1, \ldots, u_s) \in \mathbb{X}^s$, is

$$\mathrm{Err}_\varphi(P, \psi, u_1, \ldots, u_s) := \mathrm{Prob}\left[\varphi_w(u) \neq \psi(u)\right]$$

(where the probability is being understood with respect to P).

The class $\mathcal{F}$ of functions is said to be (uniformly) *learnable* if there is some identifier φ with the following property: For each $\varepsilon, \delta > 0$ there is some s so that, for every probability P and every $\psi \in \mathcal{F}$,

$$\mathrm{Prob}\left[\mathrm{Err}_\varphi(P, \psi, u_1, \ldots, u_s) > \varepsilon\right] < \delta$$

(where the probability is being understood with respect to P^s on $\mathbb{X}^s$).

In the learnable case, the function $s(\varepsilon, \delta)$ that provides, for any given ε and δ, the smallest possible s as aforementioned, is called the *sample complexity* of the class $\mathcal{F}$. It can be proved that learnability is equivalent to finiteness of a combinatorial quantity called *VC dimension* ν of the class $\mathcal{F}$ in the following sense (cf. [49,50]):

$$s(\varepsilon, \delta) \leq \max\left\{\frac{8\nu}{\varepsilon} \log\left(\frac{13}{\varepsilon}\right), \frac{4}{\varepsilon} \log\left(\frac{2}{\delta}\right)\right\}$$

Moreover, lower bounds on $s(\varepsilon, \delta)$ are also known, in the following sense (cf. [49]): for $0 < \varepsilon < \frac{1}{2}$, and assuming that the collection $\mathcal{F}$ is not trivial (i.e., $\mathcal{F}$ does not consist of just one mapping or a collection of two disjoint mappings, see Reference 49 for details), we must have

$$s(\varepsilon, \delta) \geq \max\left\{\frac{1 - \varepsilon}{\varepsilon}\ln\left(\frac{1}{\delta}\right), v(1 - 2(\varepsilon(1 - \delta) + \delta))\right\}$$

The above-mentioned bounds motivate studies dealing with estimating VC dimensions of neural nets. When there is an algorithm that allows computing an identifier φ in time polynomial on the sample size, the class is said to be learnable in the probably approximately correct (PAC) sense of Valiant (cf. Reference 51). Generalizations to the learning of real-valued (as opposed to Boolean) functions computed by, say, sigmoidal neural nets, by evaluation of the "pseudo-dimension," are also possible; see the discussion in Reference 9.

It is well known that a simple perceptron with n inputs has a VC-dimension of $n + 1$ [49]. However, the VC-dimension of a threshold network with w programmable parameters is $\Theta(w \log w)$ [21,23,52,53]. Maass [23] and Goldberg and Jerrum [24], among others, investigated VC-dimensions of neural nets with continuous activations and showed polynomial bounds on neural nets with piecewise-polynomial activation functions. Even finiteness of the VC-dimension of sigmoidal neural nets was unknown for a long time, until it was showed to be finite [25]. Subsequently, an $O(w^2 n^2)$ bound on the VC-dimension of sigmoidal neural nets, where w is the number of programmable parameters and n is the number of nodes, was established [54]. Reference 55 gives a $\Omega(w^2)$ lower bound for sigmoidal nets.

20.2.3.2 The Loading (Consistency) Problem

The VC-dimensions provide information-theoretic bounds on sample complexities for learning. To design an *efficient* learning algorithm, the learner should be able to design a neural net consistent with the (polynomially many) samples it receives. This is known as the *consistency* or the *loading* problem. In other words, now we consider the *tractability* of the training problem, that is, of the question (essentially quoting Judd [27]): "Given a network architecture (interconnection graph as well as choice of activation function) and a set of training examples, does there exist a set of weights so that the network produces the correct output for all examples?"

The simplest neural network, *that is*, the perceptron, consists of one threshold neuron only. It is easily verified that the computational time of the loading problem in this case is polynomial in the size of the training set irrespective of whether the input takes continuous or discrete values. This can be achieved via a linear programming technique. Blum and Rivest [28] showed that this problem is NP-hard for a simple 3-node threshold neural net. References 56, 57 extended this result to show NP-hardness of a 3-node neural net where the activation function is a simple, saturated piecewise linear activation function, the extension was nontrivial due to the continuous part of the activation function. It was also observed in Reference 57 that the loading problem is polynomial-time if the input dimension is constant. However, the complexity of the loading problem for sigmoidal neural nets still remains an open problem, though some partial results when the net is somewhat restricted appeared in references such as Reference 58. Any NP-hardness results of the loading problems also prove hardness in the PAC learning model, due to the result in Reference 59.

Another possibility to design efficient learning algorithms is to assume that the inputs are drawn according to some particular distributions. For example, see Reference 60 for efficient learning a depth 2 threshold net with a fixed number of hidden nodes and with the output gate being an AND gate, assuming that the inputs are drawn uniformly from a n-dimensional unit ball.

20.3 Recurrent Neural Networks

As stated in the introduction, a recurrent neural net allows cycles in the connection graph. A sample recurrent neural network is illustrated in Figure 20.3.

20.3.1 Learning Recurrent Networks: Backpropagation through Time

BPTT is an approach to solve temporal differentiable optimization problems with continuous variables [61] and used most often as a training method for recurrent neural networks. In this section, we describe the method in more details.

20.3.1.1 Network Definition and Performance Measure

We will use the general expression of Werbos [41] to describe the network dynamics. Symbols y denote node inputs and outputs, whereas symbol s denote the weighted sum of node inputs. An ordered set of i, j, l, k on the weights denotes a connection from node j of layer i to node k of layer l; $w_{0,j,l,k}$ denotes connections from outside the network. The node activation function is denoted by $f(\cdot)$. The last layer of the network is denoted by M. The number of nodes in a layer l is denoted by n_l. The bias inputs to each node are handled homogeneously using the connection weights for zeroth input where inputs $y_0^{\text{ext}}(t)$ and $y_{l-1,0}(t)$ are fixed at unity for this purpose.

All the algorithms presented in the following part of the chapter are based on the following network dynamics expressed in *pseudocode* format:

for $k = 1$ to n_1 {

$$s_{1,k}(t) = \sum_{j=0}^{n^{\text{ext}}} w_{0,j,1,k}(t) y_j^{\text{ext}}(t) + \sum_{j=1}^{n_M} w_{M,j,1,k}(t) y_{M,j}(t-1) + \sum_{j=1}^{n_1} w_{1,j,1,k}(t) y_{1,j}(t-1) \tag{20.1}$$

$$y_{1,k} = f(s_{1,k}(t)) \tag{20.2}$$

}
for $l = 2$ to M {
  for $k = 1$ to n_l {

$$s_{l,k}(t) = \sum_{j=0}^{n_{l-1}} w_{l-1,j,l,k}(t) y_{l-1,j}(t) + \sum_{j=1}^{n_l} w_{l,j,l,k}(t) y_{l,j}(t-1) \tag{20.3}$$

$$y_{1,k} = f(s_{1,k}(t)) \tag{20.4}$$

  }
}

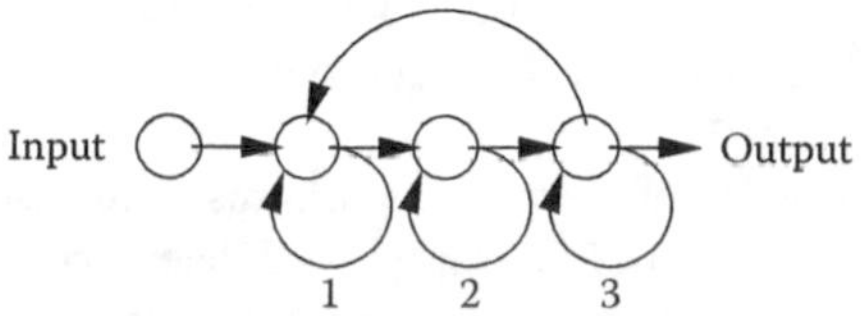

FIGURE 20.3   A simple recurrent network.

Note that the input line is not the first layer. Assume that the task to be performed by the network is sequential supervised learning task, meaning that certain of units' output values are to match specified target values at each time step. Define a time-varying $e_j(t)$:

$$e_j(t) = \begin{cases} d_j(t) - y_j(t) & \text{if } j \in \text{layer } M \\ 0 & \text{otherwise} \end{cases} \tag{20.5}$$

where $d_j(t)$ is the target of the output of the jth unit at time t and define the two performance measure functions:

$$J(t) = \frac{1}{2} \sum_{k \in M} [e_k(t)]^2 \tag{20.6}$$

$$J^{\text{total}}(t_0, t) = \sum_{\tau=t_0}^{t} J(\tau). \tag{20.7}$$

20.3.1.2 Unrolling a Network

In essence, BPTT is the algorithm that calculates derivatives of performance measure with respect to weights for a feedforward neural network that is obtained by *unrolling* the network in time. Let $\mathcal{N}$ denote the network that is to be trained to perform a desired sequential behavior. Assume that $\mathcal{N}$ has n units and that it is to run from time t_0 up through some time t. As described by Rumelhart et al., we may "unroll" this network in time to obtain a feedforward network $\mathcal{N}^*$ that has a layer for each time step in $[t_0, t]$ and n units in each layer. Each unit in $\mathcal{N}$ has a copy in each layer of $\mathcal{N}^*$, and each connection from unit j to unit i in $\mathcal{N}$ has a copy connecting unit j in layer τ to unit i in layer $\tau + 1$, for each $\tau \in [t_0, t)$. An example of this unrolling mapping is given in Figure 2 in Reference 62. The key value of this conceptualization is that it allows one to regard the problem of training a recurrent network as the corresponding problem of training a feedforward neural network with certain constraints imposed on its weights. The central result driving the BPTT approach is that to compute $\partial J(t_0, t)/\partial w_{ij}$ in $\mathcal{N}$ one simply computes the partial derivatives of $\partial J(t_0, t)$ with respect to each of the τ weights in $\mathcal{N}^*$ corresponding to w_{ij} and adds them up.

Straightforward application of this idea leads to two different algorithm, depending on whether an epochwise or continual operation approach is sought. One is real-time BPTT and the other is epochwise BPTT. We only describe the real-time BPTT due to space limitations.

20.3.1.3 Derivation of BPTT Formulation

Suppose that a differentiable function F expressed in terms of $\{y_{l,j}(\tau) | t_0 \le \tau \le t\}$, the outputs of the network over time interval $[t_0, t]$ is given. Note while F may have an *explicit* dependence on $y_{l,j}(\tau)$, it may also have an *implicit* dependence on this same value through later output values. To avoid the ambiguity in interpreting partial derivatives like $\frac{\partial F}{\partial y_{l,j}(\tau)}$, we introduce variable $y_{l,j}^*(\tau)$ such that $y_{l,j}^*(\tau) = y_{M,j}(\tau)$ for all $l = M$. Define the following:

$$\epsilon_{l,j}(\tau) = \frac{\partial F}{\partial y_{l,j}(\tau)}, \tag{20.8}$$

$$\delta_{l,j}(\tau) = \frac{\partial F}{\partial s_{l,j}(\tau)}. \tag{20.9}$$

As F depends on $y_{l,j}(\tau)$, $s_{l,k}(\tau + 1)$ and $s_{l+1,m}(\tau)$, we have:

$$\frac{\partial F}{\partial y_{l,j}(\tau)} = \frac{\partial F}{\partial y_{l,j}^*(\tau)} + \sum_{k=1}^{n_l} \frac{\partial F}{\partial s_{l,k}(\tau + 1)} \frac{\partial s_{l,k}(\tau + 1)}{\partial y_{l,j}(\tau)} + \sum_{m=1}^{n_{l+1}} \frac{\partial F}{\partial s_{l+1,m}(\tau)} \frac{\partial s_{l+1,m}(\tau)}{\partial y_{l,j}(\tau)} \tag{20.10}$$

from which we derive the following:

1. $\tau = t$. For this case,

$$\epsilon_{M,j}(\tau) = \frac{\partial F}{\partial y_{M,j}^*(\tau)} = -e_j(\tau) \tag{20.11}$$

where M means the output layer of the network and $j \in \{1, 2, \ldots, n_M\}$ and

$$\epsilon_{l,j}(\tau) = \frac{\partial F}{\partial y_{l,j}(\tau)} = \sum_{m=1}^{n_{l+1}} \frac{\partial F}{\partial s_{l+1,m}(\tau)} \frac{\partial s_{l+1,m}(\tau)}{\partial y_{l,j}(\tau)} = \sum_{m=1}^{n_{l+1}} \frac{\partial F}{\partial y_{l+1,m}(\tau)} \frac{\partial s_{l+1,m}(\tau)}{\partial y_{l,j}(\tau)} \frac{\partial s_{l+1,m}(\tau)}{\partial y_{l,j}(\tau)}$$

$$= \sum_{m=1}^{n_{l+1}} \epsilon_{l+1,m}(\tau) f^{'}(s_{l+1,m}(\tau)) w_{l,j,l+1,m} \tag{20.12}$$

where $l = 1, 2, \ldots, M - 1$ and $j = 1, 2, \ldots, n_l$, and

$$\delta_{l,j}(\tau) = \frac{\partial F}{\partial s_{l,j}(\tau)} = \frac{\partial F}{\partial y_{l,j}(\tau)} \frac{\partial y_{l,j}(\tau)}{\partial s_{l,j}(\tau)} = \epsilon_{l,j}(\tau) f^{'}(s_{l,j}(\tau)) \tag{20.13}$$

where $l = 1, 2, \ldots, M$ and $j = 1, 2, \ldots, n_l$.

2. $\tau = t - 1, \ldots, t_0$. In this case,

$$\epsilon_{M,j}(\tau) = \frac{\partial F}{\partial y_{M,j}(\tau)} = \frac{\partial F}{\partial y_{M,j}^*(\tau)} + \sum_{k=1}^{n_1} \frac{\partial F}{\partial s_{1,k}(\tau+1)} \frac{\partial s_{1,k}(\tau+1)}{\partial y_{M,j}(\tau)} + \sum_{k=1}^{n_M} \frac{\partial F}{\partial s_{M,k}(\tau+1)} \frac{\partial s_{M,k}(\tau+1)}{\partial y_{M,j}(\tau)}$$

$$= -e_j(\tau) + \sum_{k=1}^{n_1} \delta_{1,k}(\tau+1) w_{M,j,1,k} + \sum_{k=1}^{n_M} \delta_{M,k}(\tau+1) w_{M,j,M,k} \tag{20.14}$$

where M means the output layer M of network and $j \in \{1, 2, \ldots, n_M\}$,

$$\epsilon_{l,j}(\tau) = \frac{\partial F}{\partial y_{l,j}(\tau)} = \sum_{k=1}^{n_l} \frac{\partial F}{\partial s_{l,k}(\tau+1)} \frac{\partial s_{l,k}(\tau+1)}{\partial y_{l,j}(\tau)} + \sum_{m=1}^{n_{l+1}} \frac{\partial F}{\partial s_{l+1,m}(\tau)} \frac{\partial s_{l+1,m}(\tau)}{\partial y_{l,j}(\tau)}$$

$$= \sum_{k=1}^{n_l} \delta_{l,k}(\tau+1) w_{l,j,l,k} + \sum_{m=1}^{n_{l+1}} \frac{\partial F}{\partial y_{l+1,m}(\tau)} \frac{\partial y_{l+1,m}(\tau)}{\partial s_{l+1,m}(\tau)} \frac{\partial s_{l+1,m}(\tau)}{\partial y_{l,j}(\tau)}$$

$$= \sum_{m=1}^{n_l} \delta_{l,k}(\tau+1) w_{l,j,l,k} + \sum_{m=1}^{n_{l+1}} \epsilon_{l+1,m}(\tau) f^{'}(s_{l+1,m}(\tau)) w_{l,j,l+1,m} \tag{20.15}$$

where $l = 1, 2, \ldots, M - 1$ and $j = 1, 2, \ldots, n_l$ and

$$\delta_{l,j}(\tau) = \frac{\partial F}{\partial s_{l,j}(\tau)} = \frac{\partial F}{\partial y_{l,j}(\tau)} \frac{\partial y_{l,j}(\tau)}{\partial s_{l,j}(\tau)} = \epsilon_{l,j}(\tau) f^{'}(s_{l,j}(\tau)) \tag{20.16}$$

where $l = 1, 2, \ldots, M$ and $j = 1, 2, \ldots, n_l$.

In addition, for any appropriate i and j

$$\frac{\partial F}{\partial w_{i,j,l,k}} = \sum_{\tau=t_0}^{t} \frac{\partial F}{\partial w_{i,j,l,k}(\tau)} \tag{20.17}$$

and, for any τ

$$\frac{\partial F}{\partial w_{i,j,l,k}(\tau)} = \frac{\partial F}{\partial s_{l,k}(\tau)} \frac{\partial s_{l,k}(\tau)}{\partial w_{i,j,l,k}(\tau)} = \delta_{l,k}(\tau) y_{i,j}(\tau) \ or \ \delta_{l,k}(\tau) y_{i,j}(\tau - 1). \tag{20.18}$$

Combining these last two equations yields:

$$\frac{\partial F}{\partial w_{i,j,l,k}} = \sum_{\tau=t_0}^{t} \delta_{l,k}(\tau) y_{i,j}(\tau) \ or \ \delta_{l,k}(\tau) y_{i,j}(\tau - 1). \tag{20.19}$$

Equations (20.11), (20.12), (20.13), (20.14), (20.15), (20.16), and (20.19) represent the BPTT computation of $\partial F/\partial w_{i,j,l,k}$ for differentiable function F expressed in terms of the outputs of individual units in the network.

20.3.1.4 Real-Time Backpropagation through Time

In real-time BPTT, the performance measure is $J(t)$ at each time. To compute the gradient of $J(t)$ at time t, we proceed as follows. First, consider t fixed for the moment. This allows us the notational convenience of suppressing any reference to t in the following. Compute $\epsilon_{l,j}(\tau)$ and $\delta_{l,k}(\tau)$ for $\tau \in [t_0, t]$ by means of equations (20.11), (20.12), and (20.13). Equation (20.14) needs a little change as with $F = J(t)$, $e_j(\tau) = 0$; thus, for $\tau < t$,

$$\begin{aligned}
\epsilon_{M,j}(\tau) = \frac{\partial F}{\partial y_{M,j}(\tau)} &= \frac{\partial F}{\partial y_{M,j}^*(\tau)} + \sum_{k=1}^{n_1} \frac{\partial F}{\partial s_{1,k}(\tau+1)} \frac{\partial s_{1,k}(\tau+1)}{\partial y_{M,j}(\tau)} + \sum_{k=1}^{n_1} \frac{\partial F}{\partial s_{M,k}(\tau+1)} \frac{\partial s_{M,k}(\tau+1)}{\partial y_{M,j}(\tau)} \\
&= \sum_{k=1}^{n_1} \frac{\partial F}{\partial s_{1,k}(\tau+1)} \frac{\partial s_{1,k}(\tau+1)}{\partial y_{M,j}(\tau)} + \sum_{k=1}^{n_1} \frac{\partial F}{\partial s_{M,k}(\tau+1)} \frac{\partial s_{M,k}(\tau+1)}{\partial y_{M,j}(\tau)} \\
&= \sum_{k=1}^{n_1} \delta_{1,k}(\tau+1) w_{M,j,1,k} + \sum_{k=1}^{n_M} \delta_{M,k}(\tau+1) w_{M,j,M,k}
\end{aligned} \tag{20.20}$$

Thus, equations (20.11), (20.12), (20.13), (20.15), (20.16), (20.19), and (20.20) represent the real-time BPTT. The process begins by using (20.11) to determine $\epsilon_{M,j}(t)$. This step is called *injecting error*, or, to be more precise, *injecting e(t) at time t*. Then δ and ϵ are obtained for successively earlier time steps through the repeated use of equations (20.15), (20.16), and (20.20). Here $\epsilon_{l,j}(\tau)$ represents the sensitivity of the *instantaneous performance* measure $J(t)$ to small perturbations in the output of the jth unit at layer l at time τ, whereas $\delta_{l,j}(\tau)$ represents the corresponding sensitivity to small perturbations to that unit's net input at that time. Once the backpropagation computation has been performed down to time t_0, the desired gradient of instantaneous performance is computed by the following pseudocode:

for $\tau = t$ *to* t_0 {
  for $l = 2$ *to* M {
    for $k = 1$ *to* n_l {
for $j = 0$ *to* n_{l-1} {

$$\frac{\partial F}{\partial w_{l-1,j,l,k}} += \frac{\partial F}{\partial s_{l,k}(\tau)} \frac{\partial s_{l,k}(\tau)}{\partial w_{l-1,j,l,k}} = \delta_{l,k}(\tau) y_{l-1,j}(\tau) \tag{20.21}$$

}
for $j = 1$ *to* n_l {

$$\frac{\partial F}{\partial w_{l,j,l,k}} += \frac{\partial F}{\partial s_{l,k}(\tau)} \frac{\partial s_{l,k}(\tau)}{\partial w_{l,j,l,k}} = \delta_{l,k}(\tau) y_{l,j}(\tau - 1) \tag{20.22}$$

```
}
} /*k loop*/ } /*l loop*/
for k = 1 to n₁ {
for j = 1 to nᵉˣᵗ {
```

$$\frac{\partial F}{\partial w_{0,j,1,k}} + = \frac{\partial F}{\partial s_{l,k}(\tau)}\frac{\partial s_{l,k}(\tau)}{\partial w_{0,j,1,k}} = \delta_{1,k}(\tau)y_j^{ext}(\tau) \tag{20.23}$$

```
}
for j = 1 to n_M {
```

$$\frac{\partial F}{\partial w_{M,j,1,k}} + = \frac{\partial F}{\partial s_{l,k}(\tau)}\frac{\partial s_{l,k}(\tau)}{\partial w_{M,j,1,k}} = \delta_{1,k}(\tau)y_{M,j}(\tau - 1) \tag{20.24}$$

```
}
for j = 1 to n₁ {
```

$$\frac{\partial F}{\partial w_{1,j,1,k}} + = \frac{\partial F}{\partial s_{l,k}(\tau)}\frac{\partial s_{l,k}(\tau)}{\partial w_{1,j,1,k}} = \delta_{1,k}(\tau)y_{1,j}(\tau - 1) \tag{20.25}$$

```
}
} /*k loop*/
} /*τ loop*/
```

where the notation "+ =" is to indicate that the quantity on the right hand side of an expression is added to the previous value (time) of the left hand side. Thus, the sum of $\frac{\partial F}{\partial w_{i,j,l,k}}$ from t_0 to t is computed. Because this algorithm makes use of potentially unbounded history storage, it also sometimes called BPTT(∞).

20.3.2 Computational Capabilities of Discrete and Continuous Recurrent Networks

The computational power of recurrent nets is investigated in references such as 15, 16; see also Reference 63 for a thorough discussion of recurrent nets and analog computation in general. Recurrent nets include feedforward nets and thus the results for feedforward nets apply to recurrent nets as well. But recurrent nets gain considerable more computational power with increasing computation time. In the following, for the sake of concreteness, we assume that the piecewise-linear function

$$\pi(x) = \begin{cases} 0 & \text{if } x \leq 0 \\ x & \text{if } 0 \leq x \leq 1 \\ 1 & \text{if } x \geq 1 \end{cases}$$ is chosen as activation function. We concentrate on binary input and assume

that the input is provided one bit at a time.

First of all, if weights and thresholds are integers, then each node computes a bit. Recurrent net with integer weights thus turn out to be equivalent to finite automata and they recognize exactly the class of regular language over the binary alphabet $\{0, 1\}$.

The computational power increases considerably for rational weights and thresholds. For instance, a "rational" recurrent net is, up to a polynomial time computation, equivalent to a Turing machine. In particular, a network that simulates a universal Turing machine does exist and one could refer to such a network as "universal" in the Turing sense. It is important to note that the number of nodes in the simulating recurrent net is fixed (i.e., *does not grow* with increasing input length).

Irrational weights provide a further boost in computation power. If the net is allowed exponential computation time, then arbitrary Boolean functions (including noncomputable functions) are recognizable.

However, if only polynomial computation time is allowed, then nets have less power and recognize exactly the languages computable by polynomial-size Boolean circuits.

References

1. Hertz, J., Krogh, A., and Palmer, R.G., *Introduction to the Theory of Neural Computation*, Addison Wesley, Redwood City, CA, 1991.
2. Parberry, I., A primer on the complexity theory of neural networks, in *Formal Techniques in Artificial Intelligence: A Sourcebook*, Banerji, R.B. (Ed.), Elsevier Science Publishers B. V., North-Holland, Raynham, MA, p. 217, 1990.
3. Rumelhart, D.E. and McClelland, J.L., *Parallel Distributed Processing: Explorations in the Microstructure of Cognition*, Vols. 1 and 2, MIT Press, Cambridge, MA, 1986.
4. Siu, K.-Y., Roychowdhury, V., and Kailath, T., *Discrete Neural Computation: A Theoretical Foundation*, Prentice Hall, Englewood Cliffs, NJ, 1994.
5. Carpenter, G.A. and Grossberg, S., A massively parallel architecture for a self-organizing neural pattern recognition machine, *Computer Vision, Graphics, and Image Processing*, 37, 54, 1987.
6. Giles, C.E., Sun, G.Z., Chen, H.H., Lee, Y.C., and Chen, D., Higher order recurrent networks and grammatical inference, in *Advances in Neural Information Processing Systems*, 2, Touretzky, D.S. (Ed.), Morgan Kaufmann, San Mateo, CA, 1990.
7. Kawato, M., Furukawa, K., and Suzuki, R., A hierarchical neural-network model for control and learning of voluntary movement, *Biological Cybernetics*, 57, 169, 1987.
8. Widrow, B., Winter, R.G., and Baxter, R.A., Layered neural nets for pattern recognition, *IEEE Transactions on Acoustics, Speech and Signal Processing*, 36, 1109, 1988.
9. Maass, W., Perspectives of current research about the complexity of learning in neural nets, in *Theoretical Advances in Neural Computation and Learning*, Roychowdhury, V.P., Siu, K.Y., and Orlitsky, A. (Eds.), Kluwer Academic Publishers, Boston, MA, p. 295, 1994.
10. Arai, W., Mapping abilities of three-layer networks, in *Proceedings of the International Joint Conference on Neural Networks*, IEEE Press, New York, p. 419, 1989.
11. Cybenko, G., Approximation by superposition of a sigmoidal function, *Mathematics of Control, Signals, and System*, 2, 303, 1989.
12. Gallant, A.R. and White, H., There exists a neural network that does not make avoidable mistakes, in *Proceedings of the International Joint Conference on Neural Networks*, IEEE Press, New York, p. 657, 1988.
13. Mhaskar, H.N., Approximation by superposition of sigmoidal and radial basis functions, *Advances in Applied Mathematics*, 13, 350, 1992.
14. Poggio, T. and Girosi, F., A theory of networks for approximation and learning, Artificial Intelligence Memorandum, No. 1140, Massachusetts Institute of Technology, Cambridge, MA, 1989.
15. Siegelmann, H. and Sontag, E.D., Analog computation, neural networks, and circuit, *Theoretical Computer Science*, 131, 331, 1994.
16. Siegelmann, H. and Sontag, E.D., On the computational power of neural nets, *JCSS*, 50, 132, 1995.
17. Kolmogorov, A.N., On the representation of continuous functions of several variables by superposition of continuous functions of one variable and addition, *Doklady Akademii Nauk USSR*, 114, 953, 1957.
18. Barron, A.R., Approximation and estimation bounds for artificial neural networks, in *Proceedings of Conference on Learning Theory (COLT)*, Morgan Kaufmann, San Mateo, CA, p. 243, 1991.
19. Darken, C., Donahue, M., Gurvits, L., and Sontag, E.D., Rate of approximation results motivated by robust neural network learning, in *Proceedings of Conference on Learning Theory (COLT)*, p. 303, 1993.

20. DasGupta, B. and Schnitger, G., Analog versus discrete neural networks, *Neural Computation*, 8(4), 805, 1996.

21. Baum, E.B. and Haussler, D., What size net gives valid generalization? *Neural Computation*, 1, 151, 1989.

22. DasGupta, B. and Sontag, E.D., Sample complexity for learning recurrent perceptron mappings, *IEEE Transactions on Information Theory*, 42(5), 1479, 1996.

23. Maass, W., Bounds for the computational power and learning complexity of analog neural nets, in *Proceedings of Symposium on Theory of Computing*, ACM Press, New York, p. 335, 1993.

24. Goldberg, P. and Jerrum, M., Bounding the Vapnik-Chervonenkis dimension of concept classes parameterized by real numbers, in *Proceedings of Conference on Learning Theory (COLT)*, ACM Press, New York, p. 361, 1993.

25. Macintyre, A. and Sontag, E.D., Finiteness results for sigmoidal 'neural' networks, in *Proceedings of Symposium on Theory of Computing*, ACM Press, New York, p. 325, 1993.

26. Judd, J.S., On the complexity of learning shallow neural networks, *Journal of Complexity*, 4, 177, 1988.

27. Judd, J.S., *Neural Network Design and the Complexity of Learning*, MIT Press, Cambridge, MA, 1990.

28. Blum, A. and Rivest, R.L., Training a 3-node neural network is NP-complete, *Neural Networks*, 5, 117, 1992.

29. Lin, J.-H. and Vitter, J.S., Complexity results on learning by neural networks, *Machine Learning*, 6, 211, 1991.

30. Goldmann, M. and Hastad, J., On the power of small-depth threshold circuits, *Computational Complexity*, 1(2), 113, 1991.

31. Hajnal, A., Maass, W., Pudlak, P., Szegedy, M., and Turan, G., Threshold circuits of bounded depth, in *Proceedings of Foundations of Computer Science*, IEEE Press, New York, p. 99, 1987.

32. Reif, J.H., On threshold circuits and polynomial computation, in *Proceedings of the 2nd Annual Structure in Complexity Theory*, IEEE Press, New York, p. 118, 1987.

33. Hopfield, J.J., Neurons with graded response have collective computational properties like those of two-state neurons, in *Proceedings of the National Academy of Science USA*, p. 3008, 1984.

34. Minsky, M. and Papert, S., *Perceptrons*, Expanded edition, The MIT Press, Cambridge, MA, 1988.

35. Littlestone, N., Learning quickly when irrelevant attributes abound: A new linear-threshold algorithm, in *Proceedings of Foundations of Computer Science*, IEEE Press, New York, p. 68, 1987.

36. Parberry, I. and Schnitger, G., Parallel computation with threshold functions, *JCSS*, 36(3), 278, 1988.

37. Paturi, R. and Saks, M.E., On threshold circuits for parity, in *Proceedings of Foundations of Computer Science*, IEEE Press, New York, p. 397, 1990.

38. Siu, K.-Y., Roychowdhury, V., and Orlitsky, A., Lower bounds on threshold and related circuits via communication complexity, in *IEEE International Symposium on Information Theory*, 1992.

39. Dasgupta, B. and Schnitger, G., The power of approximating: A comparison of activation functions, in *Advances in Neural Information Processing Systems*, 5, Giles, C.L., Hanson, S.J., and Cowan, J.D. (Eds.), Morgan Kaufmann, San Mateo, CA, p. 615, 1993.

40. Maass, W., Schnitger, G., and Sontag, E.D., On the computational power of sigmoid versus boolean threshold circuits, in *Proceedings of Foundations of Computer Science*, IEEE Press, New York, p. 767, 1991.

41. Werbos, P.J., Backpropagation through time: What it does and how to do it, in *The Roots of Backpropagation: From Ordered Derivatives to Neural Network and Political Forecasting*, Wiley-Interscience Publication, New York, p. 269, 1994.

42. Teasuro, G., He, Y., and Ahmad, S., Asymptotic convergence of backpropagation, *Neural Computation*, 1, 382, 1989.

43. Teasuro, G. and Janssens, B., Scaling relationships in back-propagation learning, *Complex Systems,* 2, 39, 1988.

44. Brady, M., Raghavan, R., and Slawny, J., Backpropagation fails to separate where perceptrons succeed, *IEEE Transactions on Circuits and Systems,* 26, 665, 1989.

45. Shrivastave, Y. and Dasgupta, S., Convergence issues in perceptron based adaptive neural network models, in *Proceedings of the 25th Allerton Conference on Communication, Control and Computing,* University of Illinois, Urbana-Champaign, p. 1133, 1987.

46. Wittner, B.S. and Denker, J.S., Strategies for teaching layered networks classification tasks, in *Proceedings of Conference on Neural Information Processing Systems,* Anderson, D. (Ed.), American Institute of Physics, New York, 1987.

47. Vapnik, V.N., *Estimation of Dependencies Based on Empirical Data,* Springer-Verlag, Berlin, Germany, 1982.

48. Vidyasagar, M., *Learning and Generalization with Applications to Neural Networks,* Springer Verlag, London, UK, 1996.

49. Blumer, A., Ehrenfeucht, A., Haussler, D., and Warmuth, M., Learnability and the Vapnik-Chervonenkis dimension, *JACM,* 36, 929, 1989.

50. Vapnik, V.N. and Chervonenkis, A.Ja., *Theory of Pattern Recognition (in Russian),* Nauka, Moscow, 1974. (German translation: Wapnik, W.N. and Chervonenkis, A.Ja., *Theorie der Zeichenerkennung,* Berlin, Akademia-Verlag, 1979.)

51. Valiant, L.G., A theory of the learnable, *CACM,* 27, 1134, 1984.

52. Cover, T.M., Geometrical and statistical properties of linear threshold elements, Stanford PhD Thesis 1964, Stanford SEL TR 6107-1, May 1964.

53. Cover, T.M., Capacity problems for linear machines, in *Pattern Recognition,* Kanal, L. (Ed.), Thompson Book Company, Washington, DC, p. 283, 1968.

54. Karpinski, M. and Macintyre, A., Polynomial bounds for VC dimension of sigmoidal and general Pfaffian neural networks, *JCSS,* 54, 169, 1997.

55. Koiran, P. and Sontag, E.D., Neural networks with quadratic VC dimension, *JCSS,* 54, 190, 1997.

56. DasGupta, B., Siegelmann, H.T., and Sontag, E.D., On a learnability question associated to neural networks with continuous activations, in *Proceedings of Conference on Learning Theory (COLT),* ACM Press, New York, p. 47, 1994.

57. DasGupta, B., Siegelmann, H.T., and Sontag, E.D., On the complexity of training neural networks with continuous activation functions, *IEEE Transactions on Neural Networks,* 6(6), 1490, 1995.

58. Höffgen, K-U., Computational limitations on training sigmoidal neural networks, *Information Processing Letters,* 46, 269, 1993.

59. Kearns, M., Li, M., Pitt, L., and Valiant, L., On the learnability of boolean formulae, in *Proceedings of Symposium on Theory of Computing,* ACM Press, New York, p. 285, 1987.

60. Blum, A.L. and Kannan, R., Learning and intersection of k halfspaces over a uniform distribution, in *Proceedings of Foundations of Computer Science,* IEEE Press, New York, p. 312, 1993.

61. Prokhorov, D.V., Backpropagation through time and derivative adaptive critics-a common framework for comparison, in *Handbook of Learning and Approximate Dynamic Programming,* Si, J., Barto, A.G., Powell, W.B., and Wunsch, D. (Eds.), IEEE Press, New York, p. 376, 2004.

62. Willams, R.J. and Zipser, D., Gradient-based learning algorithm for recurrent networks and their computational complexity, in *Backpropagation: Theory, Architectures, and Applications,* Chauvin, Y. and Rumelhart, D.E. (Eds.), Lawrence Erlbaum Associates, Hillsdale, NJ, p. 433, 1994.

63. Siegelmann, H.T., *Neural Networks and Analog Computation: Beyond the Turing Limit,* Birkhäuser Publishers, Boston, MA, 1998.

21

Principles and Strategies of Tabu Search

Fred Glover

Manuel Laguna

Rafael Martí

21.1 Introduction

The term *tabu search* (TS) was coined in the same paper that introduced the term *metaheuristic* (Glover, 1986). TS is based on the premise that problem-solving, to qualify as intelligent, must incorporate *adaptive memory* and *responsive exploration*. The adaptive memory feature of TS allows the implementation of procedures that are capable of searching the solution space economically and effectively. As local choices are guided by information collected during the search, TS contrasts with memoryless designs that heavily rely on semi-random processes that implement a form of sampling. The emphasis on responsive exploration (and hence purpose) in tabu search, whether in a deterministic or probabilistic implementation, is derived from the supposition that a bad strategic choice can often yield more information than a good random choice.

TS can be directly applied to virtually any kind of optimization problem. We can state most of these problems in the following form, where "optimize" means to minimize or maximize:

$$\text{Optimize } f(x)$$

subject to

$$x \in X$$

The function $f(x)$ may be linear, nonlinear, or even stochastic, and the set X summarizes constraints on the vector of decision variables x. The constraints may similarly include linear, nonlinear, or stochastic inequalities and may compel all or some components of x to receive discrete values.

Although this representation is useful for discussing a number of problem-solving considerations, we emphasize that in many applications of combinatorial optimization, the problem of interest may not be easily formulated as an objective function subject to a set of constraints. The requirement $x \in X$, for example, may specify logical conditions or interconnections that would be cumbersome to formulate mathematically but may be better to be left as verbal stipulations that can be then coded as rules.

The TS technique is rapidly becoming the method of choice for designing solution procedures for hard combinatorial optimization problems. A comprehensive examination of this methodology can be found in the book by Glover and Laguna (1997), and we describe some recent developments in Section 21.8 of this chapter. Widespread successes in practical applications of optimization have spurred a rapid growth of the method as a means of identifying extremely high-quality solutions efficiently. TS methods have also been used to create hybrid procedures with other heuristic and algorithmic methods to provide improved solutions to problems in scheduling, sequencing, resource allocation, investment planning, telecommunications, and many other areas. Some of the diversity of tabu search applications is shown in Table 21.1.

The tabu search emphasis on adaptive memory makes it possible to exploit the types of strategies that underlie the best of human problem-solving instead of being confined to mimicking the processes found in lower orders of natural phenomena and behavior. The basic elements of tabu search have several important features that are summarized in Table 21.2. Tabu search is concerned with finding new and more effective ways of taking advantage of the concepts embodied in Table 21.2 and with identifying associated principles that can expand the foundations of intelligent search.

In this chapter, we will describe some key aspects of this methodology, as the use of memory structures and search strategies, and illustrate them in an implementation to solve the linear ordering problem (LOP).

21.2 Memory Structures

Tabu search begins in the same way as ordinary local or neighborhood search, proceeding iteratively from one point (solution) to another until a chosen termination criterion is satisfied. Each solution x has an associated neighborhood $N(x) \subseteq X$, and each solution $x' \in N(x)$ is reached from x by an operation called a *move*.

We may contrast TS with a simple descent method where the goal is to minimize $f(x)$. Such a method only permits moves to neighbor solutions that improve the current objective function value and ends when no improving solutions can be found. The final x obtained by a descent method is called a local optimum, since it is at least as good as or better than all solutions in its neighborhood. The evident shortcoming of a descent method is that such a local optimum in most cases will not be a global optimum, that is, it usually will not minimize $f(x)$ over all $x \in N(x)$.

Tabu search permits moves that deteriorate the current objective function value but the moves are chosen from a modified neighborhood $N^*(x)$. Short- and long-term memory structures are responsible for the specific composition of $N^*(x)$. In other words, the modified neighborhood is the result of maintaining a selective history of the states encountered during the search. In the TS strategies based on short-term considerations, $N^*(x)$ characteristically is a subset of $N(x)$, and the tabu classification serves to identify elements of $N(x)$ excluded from $N^*(x)$. In TS strategies that include longer term considerations, $N^*(x)$ may also be expanded to include solutions not ordinarily found in $N(x)$, such as solutions found and evaluated in past search, or identified as high-quality neighbors of these past solutions. Characterized in this way, TS may be viewed as a dynamic neighborhood method. This means that the neighborhood of x is not a static set but rather a set that can change according to the history of the search.

The structure of a neighborhood in tabu search differs from that used in local search in an additional manner, by embracing the types of moves used in constructive and destructive processes (where the foundations for such moves are accordingly called *constructive neighborhoods* and *destructive neighborhoods*).

TABLE 21.1   Illustrative Tabu Search Applications

Scheduling	**Telecommunications**
Flow-time cell manufacturing	Call routing
Heterogeneous processor scheduling	Bandwidth packing
Workforce planning	Hub facility location
Rostering	Path assignment
Machine scheduling	Network design for services
Flow shop scheduling	Customer discount planning
Job shop scheduling	Failure immune architecture
Sequencing and batching	Synchronous optical networks
Design	**Production, Inventory, and Investment**
Computer-aided design	Supply chain management
Fault-tolerant networks	Flexible manufacturing
Transport network design	Just-in-time production
Architectural space planning	Capacitated materials requirement planning (MRP)
Diagram coherency	Part selection
Fixed charge network design	Multi-item inventory planning
Irregular cutting problems	Volume discount acquisition
Lay-out planning	Project portfolio optimization
Logical and Artificial Intelligence	**Routing**
Maximum satisfiability	Vehicle routing
Probabilistic logic	Capacitated routing
Pattern recognition/classification	Time window routing
Data mining	Multimode routing
Clustering	Mixed fleet routing
Statistical discrimination	Traveling salesman
Neural network training	Traveling purchaser
Neural network design	Convoy scheduling
Location and Allocation	**Graph Optimization**
Multicommodity location/allocation	Graph partitioning
Quadratic assignment	Graph coloring
Quadratic semi-assignment	Clique partitioning
Multilevel generalized assignment	Maximum clique problems
Large-scale GAP problems	Maximum planner graphs
Technology	**General Combinational Optimization**
Seismic inversion	Zero-one programming
Electrical power distribution	Fixed charge optimization
Engineering structural design	Nonconvex nonlinear programming
Minimum volume ellipsoids	All-or-none networks
Space station construction	Bilevel programming
Circuit cell placement	Multiobjective discrete optimization
Off-shore oil exploration	General mixed integer optimization

Such expanded uses of the neighborhood concept reinforce a fundamental perspective of TS, which is to define neighborhoods in dynamic ways that can include serial or simultaneous consideration of multiple types of moves.

TS uses attributive memory for guiding purposes (i.e., to compute $N^*(x)$). Instead of recording full solutions, attributive memory structures are based on recording attributes. This type of memory records information about solution properties (attributes) that change in moving from one solution to another. The most common attributive memory approaches are recency-based memory and frequency-based memory. Recency, as its name suggests, keeps track of solutions attributes that have changed during the recent past. Frequency typically consists of ratios about the number of iterations that a certain attribute has changed or not (depending whether it is a transition or a residence frequency). Some examples of recency- and frequency-based memory are shown in Tables 21.3 and 21.4, respectively.

TABLE 21.2 Principal Tabu Search Features

Adaptive Memory

Selectivity (including strategic forgetting)

Abstraction and decomposition (through explicit and attributive memory)

Timing:

  Recency of events

  Frequency of events

  Differentiation between short term and long term

Quality and impact:

  Relative attractiveness of alternative choices

  Magnitude of changes in structure or constraining

  Relationships

Context:

  Regional interdependence

  Structural interdependence

  Sequential interdependence

Responsive Exploration

Strategically imposed restraints and inducements

  (*tabu conditions* and *aspiration levels*)

Concentrated focus on good regions and good solution features

  (*intensification processes*)

Characterizing and exploring promising new regions

  (*diversification processes*)

Nonmontonic search patterns

  (*strategic oscillation*)

Integrating and extending solutions

  (*path relinking*)

TABLE 21.3 Examples of Recency-Based Memory

Context	Attributes	To Record the Last Time ...
Binary problems	Variable index (i)	Variable i changed its value from 0–1 or 1–0 (depending on its current value)
Job sequencing	Job index (j)	Job j changed positions
	Job index (j) and position (p)	Job j occupied position p
	Pair of job indexes (i, j)	Job i exchange positions with job j
Graphs	Arc index (i)	Arc i was added to the current solution
		Arc i was dropped from the current solution

Characteristically, a TS process based strictly on short-term strategies may allow a solution x to be visited more than once, but it is likely that the corresponding reduced neighborhood $N^*(x)$ will be different each time. With the inclusion of longer term considerations, the likelihood of duplicating a previous neighborhood upon revisiting a solution, and more generally of making choices that repeatedly visit only a limited subset of X, is all but nonexistent.

Recency-based memory is the most common memory structure used in TS implementations. As its name suggests, this memory structure keeps track of solutions attributes that have changed during the recent past. To exploit this memory, selected attributes that occur in solutions recently visited are labeled *tabu-active*, and solutions that contain tabu-active elements, or particular combinations of these attributes, are those that become tabu. This prevents certain solutions from the recent past from belonging to $N^*(x)$ and hence from being revisited. Other solutions that share such tabu-active attributes are

TABLE 21.4   Examples of Frequency-Based Memory

Context	Residence Measure	Transition Measure
Binary problems	Number of times variable i has been assigned the value of 1.	Number of times variable i has changed values.
Job sequencing	Number of times job j has occupied position p.	Number of times job i has exchanged positions with job j.
	Average objective function value when job j occupies position p.	Number of times job j has been moved to an earlier position in the sequence.
Graphs	Number of times arc i has been part of the current solution.	Number of times arc i has been deleted from the current solution when arc j has been added.
	Average objective function value when arc i is part of the solution.	Number of times arc i has been added during improving moves.

also similarly prevented from being visited. Note that while the tabu classification strictly refers to solutions that are forbidden to be visited, by virtue of containing tabu-active attributes (or more generally by violating certain restriction based on these attributes), moves that lead to such solutions are also often referred to as being tabu.

Frequency-based memory provides a type of information that complements the information provided by recency-based memory, thus broadening the foundation for selecting preferred moves. Similar to recency, frequency often is weighted or decomposed into subclasses. Also, frequency can be integrated with recency to provide a composite structure for creating penalties and inducements that modify move evaluations.

Frequencies typically consist of ratios, whose numerators represent counts expressed in two different measures: a *transition measure*—the number of iterations where an attribute changes (enters or leaves) the solutions visited on a particular trajectory, and a *residence measure*—the number of iterations where an attribute belongs to solutions visited on a particular trajectory, or the number of instances where an attribute belongs to solutions from a particular subset. The denominators generally represent one of three types of quantities: (1) the total number of occurrences of all events represented by the numerators (such as the total number of associated iterations), (2) the sum (or average) of the numerators, and (3) the maximum numerator value. In cases where the numerators represent weighted counts, some of which may be negative, denominator (3) is expressed as an absolute value and denominator (2) is expressed as a sum of absolute values (possibly shifted by a small constant to avoid a zero denominator). The ratios produce *transition frequencies* that keep track of how often attributes change and *residence frequencies* that keep track of how often attributes are members of solutions generated. In addition to referring to such frequencies, thresholds based on the numerators alone can be useful for indicating when phases of greater diversification are appropriate.

21.3 Search Strategies

The use of recency and frequency memory in tabu search generally fulfills the function of preventing searching processes from *cycling*, that is, from endlessly executing the same sequence of moves (or more generally, from endlessly and exclusively revisiting the same set of solutions). More broadly, however, the various manifestations of these types of memory are designed to impart additional robustness or vigor to the search.

A key element of the adaptive memory framework of tabu search is to create a balance between search intensification and diversification. Intensification strategies are based on modifying choice rules to encourage move combinations and solution features historically found good. They may also initiate a return to attractive regions to search them more thoroughly. Diversification strategies, on the other hand,

> *Apply TS short term memory*
> *Apply an elite selection strategy.*
> **do** {
>   *Choose one of the elite solutions.*
>   *Resume short term memory TS from chosen*
> *solution.*
>   *Add new solutions to elite list when applicable.*
> } **while** *(iterations < limit and list not empty)*

FIGURE 21.1   Simple TS intensification approach.

seek to incorporate new attributes and attribute combinations that were not included within solutions previously generated. In one form, these strategies undertake to drive the search into regions dissimilar to those already examined. It is important to keep in mind that intensification and diversification are not mutually opposing but are rather mutually reinforcing.

Most types of intensification strategies require a means for identifying a set of elite solutions as basis for incorporating good attributes into newly created solutions. Membership in the elite set is often determined by setting a threshold that is connected to the objective function value of the best solution found during the search. A simple instance of the intensification strategy is shown in Figure 21.1. Two simple variants for elite solution selection have proved quite successful. One introduces a diversification measure to assure that the solutions recorded differ from each other by a desired degree and then erases all short-term memory before resuming from the best of the recorded solutions. The other keeps a bounded length sequential list that adds a new solution at the end only if it is better than any previously seen, and the short-term memory that accompanied this solution is also saved.

Diversification is automatically created in TS (to some extent) by short-term memory functions but is particularly reinforced by certain forms of longer term memory. TS diversification strategies are often based on modifying choice rules to bring attributes into the solutions that are infrequently used. Alternatively, they may introduce such attributes by periodically applying methods that assemble subsets of these attributes into candidate solutions for continuing the search, or by partially or fully restarting the solution process. Diversification strategies are particularly helpful when better solutions can be reached only by crossing barriers or "humps" in the solution space topology.

The incorporation of modified choice rules can be moderated by using the following penalty function:

$$MoveValue' = MoveValue + d^*Penalty.$$

This type of penalty approach is commonly used in TS, where the *Penalty* value is often a function of frequency measures such as those indicated in Table 21.2, and d is an adjustable diversification parameter. Larger d values correspond to a desire for more diversification.

21.4 Advanced Designs: Strategic Oscillation

There are many forms in which a simple tabu search implementation can be improved by adding long-term elements. In this paper, we restrict our attention to two of the most used methods, namely strategic oscillation and path relinking (PR), which constitute the core of many adaptive memory programming algorithms.

Strategic oscillation operates by orienting moves in relation to a critical level, as identified by a stage of construction or a chosen interval of functional values. Such a critical level or *oscillation boundary* often represents a point where the method would normally stop. Instead of stopping when this boundary is reached, however, the rules for selecting moves are modified to permit the region defined by the critical level to be crossed. The approach then proceeds for a specified depth beyond the oscillation boundary

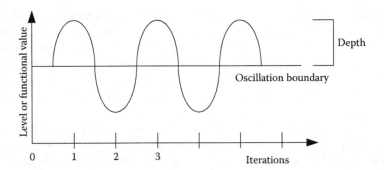

FIGURE 21.2   Strategic oscillation.

and turns around. The oscillation boundary again is approached and crossed, this time from the opposite direction, and the method proceeds to a new turning point (Figure 21.2).

The process of repeatedly approaching and crossing the critical level from different directions creates an oscillatory behavior, which gives the method its name. Control over this behavior is established by generating modified evaluations and rules of movement, depending on the region navigated and the direction of search. The possibility of retracing a prior trajectory is avoided by standard tabu search mechanisms, such as those established by the recency-based and frequency-based memory functions.

When the level or functional values in Figure 21.2 refer to degrees of feasibility and infeasibility, a vector-valued function associated with a set of problem constraints can be used to control the oscillation. In this case, controlling the search by bounding this function can be viewed as manipulating a parameterization of the selected constraint set. A preferred alternative is often to make the function a Lagrangean or surrogate constraint penalty function, avoiding vector-valued functions and allowing tradeoffs between degrees of violation of different component constraints.

Lozano et al. (2014) considered a constructive/destructive type of strategic oscillation in the context of the quadratic minimum spanning tree problem (QMSTP), where constructive steps "add" elements and destructive steps "drop" elements from the solution. As described in Glover (1977), the alternation of constructive with destructive processes, which strategically dismantle and then rebuild successive trial solutions, affords an enhancement of such traditional constructive procedures.

A special case of these constructive and destructive neighborhood ideas has been applied within a simplified method known as Iterated Greedy—IG—(Ruiz and Stützle, 2008), which generates a sequence of solutions by joining constructive and destructive phases with a greedy constructive heuristic. IG is a memory-less version of strategic oscillation that has exhibited state-of-the-art performance in some settings. Lozano et al. (2014) integrated this approach with broader strategic oscillation strategies by first proposing an adaptation of the IG methodology to the QMSTP and then extending it to include short-term memory structures to create a tabu search approach based on strategic oscillation, which outperforms the previous methods.

21.5 Advanced Designs: Path Relinking

PR, as a strategy of creating trajectories of moves passing through high-quality solutions was first proposed in connection with tabu search in Glover (1989). The approach was then elaborated in greater detail as a means of integrating TS intensification and diversification strategies, and given the name *PR*, in Glover and Laguna (1993). PR generally operates by starting from an *initiating solution* selected from a subset of high-quality solutions and by generating a path in the neighbourhood space that leads toward the other solutions in the subset, which are called *guiding solutions*. This is accomplished by selecting moves that introduce attributes contained in the guiding solutions.

PR can be considered an extension of the Combination Method of Scatter Search (Glover and Laguna, 1993; Laguna and Martí, 2003). Instead of directly producing a new solution when combining two or more original solutions, PR generates paths between and beyond the selected solutions in the neighborhood space. The character of such paths is easily specified by reference to solution attributes that are added, dropped, or otherwise modified by the moves executed. Examples of such attributes include edges and nodes of a graph, sequence positions in a schedule, vectors contained in linear programming basic solutions, and values of variables and functions of variables.

The approach may be viewed as an extreme (highly focused) instance of a strategy that seeks to incorporate attributes of high-quality solutions by creating inducements to favor these attributes in the moves selected. However, instead of using an inducement that merely encourages the inclusion of such attributes, the PR approach subordinates other considerations to the goal of choosing moves that introduce the attributes of the guiding solutions to create a "good attribute composition" in the current solution. The composition at each step is determined by choosing the best move, using customary choice criteria, from a restricted set—the set of those moves currently available that incorporate a maximum number (or a maximum weighted value) of the attributes of the guiding solutions. (Exceptions are provided by aspiration criteria, as subsequently noted.) The approach is called PR either by virtue of generating a new path between solutions previously linked by a series of moves executed during a search or by generating a path between solutions previously linked to other solutions but not to each other.

To generate the desired paths, it is only necessary to select moves that perform the following role: upon starting from an *initiating solution*, the moves must progressively introduce attributes contributed by a *guiding solution* (or reduce the distance between attributes of the initiating and guiding solutions). The roles of the initiating and guiding solutions are interchangeable; each solution can also be induced to move simultaneously toward the other as a way of generating combinations. First consider the creation of paths that join two selected solutions x' and x'', restricting attention to the part of the path that lies "between" the solutions, producing a solution sequence $x' = x(1), x(2), \ldots, x(r) = x''$. To reduce the number of options to be considered, the solution $x(i + 1)$ may be created from $x(i)$ at each step by choosing a move that minimizes the number of moves remaining to reach x''. The relinked path may encounter solutions that may not be better than the initiating or guiding solution, but that provide fertile "points of access" for reaching other, somewhat better, solutions. For this reason, it is valuable to examine neighboring solutions along a relinked path and to keep track of those of high quality that may provide a starting point for launching additional searches.

Laguna and Martí (1999) adapted PR in the context of greedy randomized adaptive search procedure (GRASP) as a form of intensification. The relinking in this context consists of finding a path between a solution found with GRASP and a chosen elite solution. Therefore, the solutions submitted to PR are drawn from a population (as in Glover, 1997) whose members may not be linked by a previous sequence of moves. Resende et al. (2010) explored the adaptation of GRASP with PR to the Max–Min Diversity Problem across different designs in which greedy, randomized, and evolutionary PR are considered in the implementation as follows:

- *Greedy path relinking*: In this method the moves in the path from a solution to another are selected in a greedy fashion, according to the objective function value.
- *Greedy randomized path relinking*: Here the method creates a candidate list with the good intermediate solutions and randomly selects among them.
- *Truncated path relinking*: In this application of PR the path between two solutions is not completed. It is applied, for example, in problems where good solutions are found close to the end points (original solutions) in the path.
- *Evolutionary path relinking*: This method iterates over the set of high-quality solutions, applying successively the relinking mechanism. It has many similarities with the scatter search methodology (Glover, 1997; Laguna and Martí, 2003).

As described in Glover and Laguna (1993) and Martí et al. (2006), we can apply different PR elements to perform more elaborated designs. Some examples are: simultaneous relinking, tunneling strategy, extrapolated relinking, multiple guiding solutions, constructive neighborhoods, or vocabulary building.

21.6 Application: The Linear Ordering Problem

Given a matrix of weights $E = \{e_{ij}\}_{m \times m}$, the LOP consists of finding a permutation p of the columns (and rows) to maximize the sum of the weights in the upper triangle. In mathematical terms, we seek to maximize:

$$C_E(p) = \sum_{i=1}^{m-1} \sum_{j=i+1}^{m} e_{p_i p_j}.$$

where p_i is the index of the column (and row) in position i in the permutation. Note that in the LOP, the permutation p provides the ordering of both the columns and the rows. Solution methods for this NP-hard problem have been proposed since 1958, when Chenery and Watanabe outlined some ideas on how to obtain solutions for this problem (Martí and Reinelt, 2011). In this section, we describe a tabu-search implementation (Laguna et al. 1999) for the LOP.

The LOP has a wide range of applications in several fields. Perhaps, the best known application occurs in the field of economics. In this application, the economy (regional or national) is first subdivided into sectors. Then, an input/output matrix is created, in which the entry (i, j) represents the flow of money from sector i to sector j. Economists are often interested in ordering the sectors so that suppliers tend to come first, followed by consumers. This is achieved by permuting the rows and columns of the matrix so that the sum of entries above the diagonal is maximized, which is the objective of the LOP.

Insertions are used as the primary mechanism to move from one solution to another in Laguna's et al. method for the LOP. *INSERT_MOVE(p_j, i)* consists of deleting p_j from its current position j to be inserted in position i (i.e., between the current sectors p_{i-1} and p_i). This operation results in the ordering p', as follows:

$$p' = \begin{cases} (p_1, \ldots, p_{i-1}, p_j, p_i, \ldots, p_{j-1}, p_{j+1}, \ldots, p_m) & \text{for } i < j \\ (p_1, \ldots, p_{j-1}, p_{j+1}, \ldots, p_i, p_j, p_{i+1}, \ldots, p_m) & \text{for } i > j \end{cases}$$

The neighborhood N consists of all permutations resulting from executing general insertion moves as:

$$N = \{p' : INSERT\_MOVE(p_j, i), \text{ for } j = 1, \ldots, m \text{ and } i = 1, 2, \ldots, j-1, j+1, \ldots, m\},$$

and N is partitioned into m N^j neighborhoods associated with each sector p_j, for $j = 1, \ldots, m$.

$$N_j = \{p' : INSERT\_MOVE(p_j, i), \ i = 1, 2, \ldots, j-1, j+1, \ldots, m\}$$

Starting from a randomly generated permutation p, the basic TS procedure alternates between an intensification and a diversification phase. An iteration of the *Intensification Phase* begins by randomly selecting a sector. The probability of selecting sector j is proportional to its weight w_j according to:

$$w_j = \sum_{i \neq j} \left(e_{ij} + e_{ji} \right)$$

The move INSERT_MOVE(p_j, i) $\in N^j$ with the largest move value is selected. (Note that this rule may result in the selection of a nonimproving move.) The move is executed even when the move value is not positive, thereby resulting in a deterioration of the current objective function value. The moved sector becomes tabu-active for *TabuTenure* iterations, and therefore it cannot be selected for insertions during this time.

The number of times that sector j has been chosen to be moved is accumulated in the value *freq(j)*. This frequency information is used for diversification purposes. The intensification phase terminates after *MaxInt* consecutive iterations without improvement. Before abandoning this phase, a local search procedure based on the same neighborhood is applied to the best solution found (during the current intensification). We denote this solution as $p^\#$, in contrast to p^* (the best solution found over the entire search). By applying this greedy procedure (without tabu restrictions), a local optimum is guaranteed as the output of the intensification phase.

The *Diversification Phase* is performed for *MaxDiv* iterations. In each iteration, a sector is randomly selected, where the probability of selecting sector j is inversely proportional to the frequency count *freq(j)*. The chosen sector is placed in the best position as determined by the move values associated with the insert moves in N^j. The procedure stops when *MaxGlo* global iterations are performed without improving $C_E(p^*)$. A global iteration is an application of the intensification phase followed by the application of the diversification phase.

An additional intensification is introduced by implementing a long-term PR phase. Specifically, the best solution found at the end of an intensification phase $p^\#$ (which not necessarily represents p^*, the best solution overall) is subjected to a relinking process. The process consists of making moves starting from $p^\#$ (the initiating solution) in the direction of a set of elite solutions (also referred to as guiding solutions). The set of elite solutions consists of the *EltSol* best solutions found during the entire search. The insertions used to move the initiating solution closer to the guiding solutions can be described as follows. For each sector p_j in the current solution:

1. Find the position i for which the absolute value of $(j-i)$ is minimized, where i is the position that p_j occupies in at least one of the guiding solutions.
2. Perform *INSERT_MOVE(p_j, i)*.

A long-term diversification phase is also implemented to complement the diversification phase in the basic procedure. The long-term diversification is applied after *MaxLong* global iterations have elapsed without improving $C_E(p^*)$. For each sector p_j, a rounded average position $\alpha(p_j)$ is calculated using the positions occupied by this sector in the set of elite solutions and the solutions visited during the last intensification phase. Then, m diversification steps are performed that insert each sector p_j in its complementary position $m - \alpha(p_j)$, that is, *INSERT_MOVE(p_j, m-$\alpha(p_j)$)* is executed for $j = 1, \ldots, m$.

After preliminary experimentation, the search parameters are set to *MaxGlo* $= 100$, *MaxLong* $= 50$, *EltSol* $= 4$, *TabuTenure* $= 2\sqrt{m}$, *MaxInt* $= m$, and *MaxDiv* $= 0.5\,m$ and *EltSol* $= 4$. In the 49 instances of the public domain LOLIB library, the method obtains the optimal solution within 1 second of computer time run on a Pentium IV at 3 Ghz. The method is also compared with a previous procedure due to Chanas and Kobylanski (1996) and a greedy procedure based on the N local search. The methods were run in a way that the best solution found was reported every 0.5 seconds. These data points were used to generate the performance graph in Figure 21.3. The superior performance of TS_LOP is made evident by Figure 21.3.

21.7 The Tabu Cycle and Conditional Probability

In this section, we describe the implementation and testing of the tabu cycle method and two variants of the conditional probability method (Laguna, 2005). These methods were originally described in Glover (1990) and again in the book by Glover and Laguna (1997) but have been largely ignored in the tabu search literature. The *tabu cycle method* is a short-term memory mechanism that is based on partitioning the elements (i.e., move attributes) of a tabu list. The methodology is general and capable of accommodating multiattribute tabu search memory, as described in Glover and Laguna (1997). In its most basic form, the

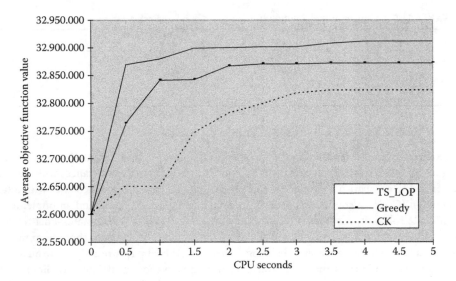

FIGURE 21.3 Performance graph.

tabu cycle method divides the short-term memory list into *TabuGroups* groups, where group k consists of elements that were added to the list between a specified range of iterations ago. Although in some variants of tabu search (e.g., probabilistic tabu search) it is common to progressively relax the tabu status of elements as they become older, the tabu cycle method, by contrast, allows the elements of some groups to fully escape their tabu status according to certain frequencies that increase with the age of the groups. The method is based on the use of iteration intervals called *tabu cycles*, which are made smaller for older groups than for younger groups (with the exception of a small buffer group). Specifically, if group k has a tabu cycle of $TC(k)$ iterations, then at each occurrence of this many iterations, on average, the elements of group k escape their tabu status and are free to be chosen. In other words, on average, group k is designated as FREE every $TC(k)$ iterations. Mechanisms and data structures that are useful for achieving this are described in Laguna (2005).

The *conditional probability method* is a variant of the tabu cycle method that chooses elements by establishing the probability that a group will be FREE on a given iteration. The probability assigned to group k may be viewed conceptually as the inverse of the tabu cycle value. That is, $P(k) = 1/TC(k)$. Analogous to the tabu cycle method, group k is FREE if all older groups likewise are FREE. The method employs a *conditional probability*, $CP(k)$, as a means of determining whether a particular group k can be designated as FREE. The conditional probability values are fixed and that at each iteration, the status of a group is determined by a probabilistic process that is not affected by previous choices. Consequently, the approach ignores the possibility that actual tabu cycle values may be far from their targets for some groups. This may happen, for example, when for a number of iterations, no elements are chosen from a particular set of FREE groups. The conditional probability method also makes use of a buffer group, for which no element is allowed to escape its tabu status.

A variant of the conditional probability method uses substitute probability values to keep the expected number of elements per iteration chosen from groups that are no older than any given group k close to $P(k)$. The substitute probabilities replace the original $P(k)$ values in the determination of the conditional probabilities. These substitute probabilities make use of cycle counts, which are also used in connection with the tabu cycle method.

Laguna (2005) used a single machine scheduling problem to test the merit of implementations of the tabu cycle method and both variants of the conditional probability method. The problem consists of minimizing the sum of the setup costs and linear delay penalties when n jobs, arriving at time zero, are to be scheduled for sequential processing on a continuously available machine. Several variants of tabu search

TABLE 21.5   Number of Best Solutions (out of 100) Found by Each Method

Problem Set	Static	Dynamic	Cycle	C-Prob	S-Prob
$n = 50$	2	50	9	31	65
$n = 100$	0	10	28	17	47
$n = 200$	0	8	37	26	29

for this problem have been reported in the literature (Laguna et al., 1991 and 1993; Glover and Laguna, 1991; Laguna and Glover, 1993). Experiments with more than 300 problem instances with up to 200 jobs were performed to compare a simple static and dynamic short-term memory schemes with a tabu cycle implementation (Cycle), a conditional probability implementation (C-Prob), and an implementation of the conditional probability method with substitute probabilities (S-Prob). The static short-term memory assigns a constant tabu tenure to all attributes during the search. The dynamic short-term memory randomly selects a tabu tenure from a specified range. Therefore, the tabu tenure assigned to an attribute in a given iteration may not be the same as the tabu tenure assigned to another attribute in a different iteration. Table 21.5 shows the number of best solutions found by each method in each set of 100 problems.

The results in Table 21.5 show the merit of the tabu cycle and the conditional probability variants as the problem size increases. In addition to these results, the S-Prob is able to find 17 new best solutions to 20 problems used for experimentation in Glover and Laguna (1991). For problems with up to 60 jobs, for which a lower bound can be computed, S-Prob produces a maximum gap of 3.56% in relation to this optimistic bound.

These results confirm that a tabu search procedure based solely on a static tabu list is not a robust method, because it is incapable of maintaining an acceptable level of diversity during the search. The dynamic short-term memory continues to be an appealing alternative, because it is easy to implement and provides a good balance between diversification and intensification. The results also show that improved outcomes are possible with the additional effort required to implement the tabu cycle or conditional probability methods.

Additional strategies identified in Glover and Laguna (1997) can be valuable for exploiting other aspects of intensification and diversification, but this example demonstrates the importance of handling short-term memory in a strategic way, especially when faced with larger and more difficult problems.

21.8 Additional Uses of Conditional Relationships, Solution Attributes, Embedded Tabu Search, Unified Models, and Learning

21.8.1 Conditional Relationships

Principles for exploiting conditional relationships in tabu search and strategic oscillation originally proposed in Glover (2000) have recently been elaborated to produce a class of methods called multiwave algorithms (Glover, 2016a), whose preliminary implementation has achieved notable success (Oualid, 2017). These principles, which are applicable both to multistart constructive methods and iterated local search methods, may be briefly summarized as follows.

Define a *boundary solution* to be a local optimum for a neighborhood search method or a completed construction for a constructive solution method, and define a solution *wave* to be a succession of moves, starting from a given initial solution (which can be a null solution for a constructive method) that lead to a boundary solution. We consider forms of constructive or neighborhood search that permit moves to be dropped (or equivalently, reversed) *en route* from an initial solution to a boundary solution. Hence, constructive search is treated within the framework of strategic oscillation, which permits destructive moves at intermediate stages and neighborhood search is treated in a similar framework, where move reversals are employed at specific junctures. The following observations motivate the use of such a framework.

Principle of Marginal Conditional Validity (MCV Principle): Starting from a given initial solution, as more moves are made, the information that permits these moves to be evaluated becomes increasingly effective for guiding the search to a high-quality boundary solution, conditional upon the decisions previously made.

This principle has several consequences, which can be stated as heuristic inferences.

Inference 1: Early decisions are more likely to be bad ones.

Inference 2: Early decisions are likely to look better than they should (i.e., receive higher evaluations than would be appropriate for achieving global optimality), once later decisions have been made.

Inference 3: The outcome of a constructive or local improvement method can often be improved by examining the resulting complete solution, where all decisions have been made, and seeing whether one of the decisions can now be advantageously replaced with a different one.

Inference 4: The outcome of a constructive or improvement method can often be improved by examining a solution at an intermediate stage, before reaching a boundary solution, and seeing whether one of the decisions can now be advantageously replaced with a different one.

These conditionality principles are reinforced by an associated set of principles based on the notion of persistent attractiveness, which is manifested in three types of situations as follows:

PA Type 1: Exhibited by a move that is attractive enough more than once during a wave to be chosen a second or third time (a move that receives a high enough evaluation after being dropped to cause it to be chosen again).

PA Type 2: Exhibited by a move that repeatedly has high evaluations during a wave, whether chosen or not, and if ultimately chosen is not selected until a step that occurs somewhat after it first appears to be attractive.

PA Type 3: Exhibited by a move that repeatedly has a high evaluation during one or more waves but which is not selected.

21.8.2 Principle of Persistent Attractiveness

(PA-1) A move of PA Type 1 allows a focus on dropping moves that will have greater impact by exempting such a move from being dropped anew once it achieves the Type 1 status.

(PA-2) A move of PA Type 2, which was chosen on a given wave offers an opportunity for producing a better solution on a new wave if it is selected at an earlier stage in the wave. (In particular, given that the move was eventually chosen, selecting it earlier affords a chance to make decisions based on this choice that were not influenced at the later point in the decision sequence when the move was chosen previously).

(PA-3) A move of PA Type 3 can be used to combine diversification with intensification by choosing at least one member of this type at a relatively early stage of a wave, assuring that the resulting solution will not duplicate a previous solution and allowing subsequent decisions to take this move into account.

These types of conditional relationships are amplified in the design of multiwave algorithms by reference to an associated principle for combining attractiveness and move influence and are conjectured to be fundamental to designing more effective uses of memory in tabu search.

21.8.3 Solution Attributes

The literature of evolutionary algorithms points out the relevance of encoding solutions properly to allow operations for combining solutions to generate offspring that can be meaningfully exploited. As emphasized in Glover and Laguna (1993), the flip side of this for neighborhood search algorithms using adaptive memory is to select particular attributes of solutions for defining tabu status (i.e., for determining conditions under which certain moves should be prohibited) to produce solutions of highest quality. Recent studies by Wang et al. (2016) and Wu et al. (2016) have documented the impact of attribute selection strategies by employing a balance between attributes consisting of hash-coded solutions (drawing on ideas of Woodruff and Zemel, 1993) and attributes that are more commonly used. Computational experiments show that such blends of attributes yield tabu search algorithms that significantly outperform all other methods previously proposed for the classes of problems tested. These outcomes motivate the exploration of other ways of generating compound solution attributes to identify attributes that likewise lead to more effective versions of tabu search.

21.8.4 Embedded Tabu Search

Numerous methods, most commonly evolutionary methods, have been joined with tabu search to produce effective hybrid (composite) methods. Such a hybrid method can be viewed in one of two ways: either as an evolutionary method that relies on tabu search as a means for generating improved solutions or as a tabu search method that incorporates evolutionary search in the role of an intensification/diversification strategy. The first of these two perspectives (e.g., Glover, 1997) has often proved more popular, giving rise to a large number of metaheuristics that "embed" tabu search within them. Prominent among these hybrid methods are Memetic Algorithms, in which tabu search is often used as a solution improvement procedure. Examples of such evolutionary/tabu search hybrids that have achieved particularly noteworthy successes appear in the work of Chen et al. (2016), Jin and Hao (2016), and Lai and Hao (2016).

21.8.5 Unified Models and Applications

A wide range of metaheuristic applications have shown to be encompassed in the quadratic unconstrained binary optimization (QUBO) model. By means of special reformulation techniques, these include many constrained optimization models as well. For example, Kochenberger et al. (2014) identified important QUBO applications in the areas of quadratic assignment, capital budgeting, task allocation, distributed computer systems, maximum diversity, symmetric and asymmetric assignments, constraint satisfaction, set partitioning, warehouse location, maximum clique and independent sets, graph coloring, graph partitioning, number partitioning, and linear ordering, among a variety of others.

A highly effective tabu search metaheuristic for solving QUBO problems has been identified in Wang et al. (2014). This has led to additional applications in the field of quantum computing, as documented in Rosenberg et al. (2016) and Mniszewski et al. (2016), culminating in an open-source hybrid quantum solver called *Qbsolve* released by D-Wave Systems (Booth et al., 2017) and reviewed in *Wired* magazine (Finley, 2017). These developments are placing tabu search at the forefront of the burgeoning field of quantum computing.

21.8.6 Tabu Search with Learning

Learning is automatically built into tabu search with its emphasis on adaptive memory. In addition, proposals have often been made for joining tabu search strategies with classical forms of learning. For example, the paper that lays the foundation for scatter search and strategic oscillation (Glover, 1977) points out the relevance of using clustering in connection with *strongly determined* and *consistent* variables, whose identification depends on frequency-based memory applied to high-quality solutions. In general, within a memory-based method, clustering can take advantage of the fact that solutions generated over time may be exploited more effectively by analyzing subsets of good solutions that exhibit common characteristics (as proposed in Glover, 1989 and in Glover and Laguna, 1997). PR provides an indirect means for accomplishing this, as observed in connection with the successful learning approach in Lai et al. (2016).

Classification, another form of classical learning, can be exploited in retrospective analysis, particularly in relation to a tabu search learning strategy called target analysis, to identify characteristics of moves that lead to elite solutions, as noted in Glover and Greenberg (1989) and Glover and Laguna (1997). In reverse, adaptive memory search methods have been utilized as a means to obtain improved learning procedures, as in neural networks (Kelly et al. 1996; Dengiz et al. 2009) and in clustering (Cao et al. 2015). Nevertheless, the exploration of classical forms of learning within metaheuristics remains a topic that has not been covered in nearly the depth that it deserves. Key observations related to this topic emerge in the context of clustering and intensification/diversification strategies, where spatially defined measures of distance may be inapplicable (Glover, 2016b).

A significant recent emphasis on learning has been launched under the banner of "Opposition-based learning (OBL)" introduced in Tizhoosh (2005). OBL has become the focus of numerous research initiatives in machine learning and metaheuristic optimization, and a variety of proposals and studies have been undertaken to exploit its underlying ideas, as noted in the survey of Xu et al. (2014). Still more recently, intensification and diversification strategies introduced in the setting of tabu search and PR have been discovered to provide a more flexible and comprehensive learning framework called diversification-based learning (Glover and Hao, 2017), which makes it possible to remedy limitations of OBL such as failing to establish a link to feasibility and being compelled to resort to randomization when its framework breaks down. Once again, this work shows the value of giving increased attention to the topic of learning in future studies.

21.9 Conclusion

The adaptive memory underpinnings of tabu search, together with the special strategies such as PR and strategic oscillation that have been introduced for exploiting this memory, have produced numerous advances for designing improved optimization procedures. These advances in turn offer numerous research opportunities for applying these approaches to the abundance of new optimization problems that are emerging in the scientific and social domains and for identifying enhanced versions of tabu search and TS hybrids for treating these applications.

Acknowledgments

This work was supported by the Spanish Ministerio de Economía y Competividad and Fondo Europeo de Desarrollo Regional (FEDER) (project TIN-2015-65460-C02-01-MINECO/FEDER).

References

Booth, M., S. Reinhardt, and A. Roy (2016) Partitioning optimization problems for hybrid classical/quantum execution, D-Wave Systems Technical Report 2017-01-09.

Cao, B., F. Glover, and C. Rego (2015) A tabu search algorithm for cohesive clustering problems, *Journal of Heuristics*, 21: 457–477. doi:10.1007/s10732-015-9285-2.

Chanas, S. and P. Kobylanski (1996) A new heuristic algorithm solving the linear ordering problem, *Computational Optimization and Applications,* 6: 191–205.

Chen, Y., J.-K. Hao, and F. Glover (2016) An evolutionary path relinking approach for the quadratic multiple knapsack problem, *Knowledge-Based Systems*, 92: 23–34.

Dengiz, B., C. Alabas-Uslu, and O. Dengiz (2009) A tabu search algorithm for the training of neural networks, *Journal of the Operational Research Society*, 60(2): 282–291.

Finley, K. (2017) Quantum computing is real, and D-Wave just open-sourced It, *Wired*, January 11, 2017, https://www.wired.com/2017/01/d-wave-turns-open-source-democratize-quantum-computing/.

Glover, F. (1977) Heuristics for integer programming using surrogate constraints, *Decision Sciences*, 8(1): 156–166.

Glover, F. (1986) Future paths for integer programming and links to artificial intelligence, *Computers and Operations Research,* 13: 533–549.

Glover, F. (1989) Tabu search, Part I, *ORSA Journal on Computing*, 1(3): 190–206.

Glover, F. (1990) Tabu search, Part II, *ORSA Journal on Computing*, 2(1): 4–32.

Glover, F. (1997) A template for scatter search and path relinking, *Artificial Evolution, Lecture Notes in Computer Science*, Vol. 1363, J.-K. Hao, E. Lutton, E. Ronald, M. Schoenauer, and D. Snyers, (Eds.), Springer, Berlin, Germany, pp. 13–54.

Glover, F. (2000) Multi-start and strategic oscillation methods—Principles to exploit adaptive memory, *Computing Tools for Modeling, Optimization and Simulation: Interfaces in Computer Science and Operations Research*, M. Laguna and J. L. Gonzales Velarde, (Eds.), Kluwer Academic Publishers, Boston, MA, pp. 1–24.

Glover, F. (2016a) Multi-Wave algorithms for metaheuristic optimization, *Journal of Heuristics*, 22(3): 331–358. doi:10.1007/s10732-016-9312-y.

Glover, F. (2016b) Pseudo-centroid clustering, *Soft Computing*, 21(22): 6571–6592, Springer, Berlin, Germany, published online http://link.springer.com/article/10.1007/s00500-016-2369-6. doi:10.1007/s00500-016-2369-6.

Glover, F. and H. Greenberg (1989) New approaches for heuristic search: A bilateral linkage with artificial intelligence, *European Journal of Operational Research*, 39(2): 119–130.

Glover, F. and J.-K. Hao (2017) Diversification-based learning in computing and optimization, *Metaheuristic International Conference* (MIC2017), Barcelona, Spain.

Glover, F. and M. Laguna (1997) *Tabu Search,* Kluwer Academic Publishers, Boston, MA.

Glover, F. and M. Laguna (1991) Target analysis to improve a tabu search method for machine scheduling, *The Arabian Journal for Science and Engineering*, 16(2B): 239–253.

Glover, F. and M. Laguna (1993) Tabu search, *Modern Heuristic Techniques for Combinatorial Problems*, C. Reeves (Ed.), Blackwell Scientific Publishing, Oxford, UK, pp. 70–150.

Jin, Y. and J.-K Hao (2016) Hybrid evolutionary search for the minimum sum coloring problem of graphs, *Information Sciences* 352–353: 15–34.

Kelly, J. P., B. Rangaswamy, and J. Xu (1996) A scatter-search-based learning algorithm for neural network training, *Journal of Heuristics*, 2(2): 129–146.

Kochenberger, G., J.-K. Hao, F. Glover, M. Lewis, Z. Lu, H. Wang, and Y. Wang (2014) The unconstrained binary quadratic programming problem: A survey, *Journal of Combinatorial Optimization*, 28(1): 58–81.

Laguna, M. (2005) Implementing and testing the tabu cycle and conditional probability methods, *Computers & Operations Research*, 33(9): 2495–2507. http://leeds-faculty.colorado.edu/laguna/articles/tabucycle.html.

Laguna, M. and F. Glover (1993) Integrating target analysis and tabu search for improved scheduling systems, *Expert Systems with Applications*, 6: 287–297.

Laguna, M., J. W. Barnes, and F. Glover (1993) Intelligent scheduling with tabu search: An application to jobs with linear delay penalties and sequence dependent setup costs and times, *Journal of Applied Intelligence*, 3: 159–172.

Laguna, M. and R. Martí (1999) GRASP and Path relinking for two layer straight-line crossing minimization, *INFORMS Journal on Computing*, 11(1): 44–52.

Laguna, M. and R. Martí (2003) *Scatter Search—Methodology and Implementations in C,* Kluwer Academic Publishers, Boston, MA.

Laguna, M., R. Martí, and V. Campos (1999) Intensification and diversification with elite tabu search solutions for the linear ordering problem, *Computers and Operations Research*, 26: 1217–1230.

Lai, X. and J.-K Hao (2016) A tabu based memetic algorithm for the max-mean dispersion problem, *Computers & Operations Research*, 72: 118–127.

Lai, X., J.-K. Hao, Z. Lü, and F. Glover (2016) A learning-based path relinking algorithm for the bandwidth coloring problem, *Engineering Applications of Artificial Intelligence*, 52: 81–91.

LOLIB (1997) http://www.iwr.uni-heildelberg.de/iwr/comopt/soft/LOLIB/LOLIB.html.

Lozano, M., F. Glover, C. García-Martinez, F. Rodríguez, and R. Martí (2014) Tabu search with SO for the quadratic mininum spanning tree, *IIE Transactions* 46 (4), pp. 414–428.

Martí, R., M. Laguna, and F. Glover (2006) Principles of scatter search, *European Journal of Operational Research*, 169(2): 359–372.

Martí, R. and G. Reinelt (2011) *The Linear Ordering Problem: Exact and Heuristic Methods in Combinatorial Optimization*, Springer-Verlag, Berlin, Germany.

Mniszewski, S. M., C. F. A. Negre, and H. Ushijima-Mwesigwa (2016) Graph partitioning using the d-wave for electronic structure problems, *ISTI D-Wave Quantum Computing Efforts Debrief*, Los Alamos National Laboratories.

Oualid, G. (2017) Multi-flux strategic oscillation method for the uncapacitated facility location problem, *18th Annual Congress of the French Society for Operational Research and Decision Support*, Metz, France.

Resende, M., R. Martí, M. Gallego, and A. Duarte (2010) GRASP and path relinking for the max-min diversity problem, *Computers and Operations Research*, 37(3): 498–508.

Rosenberg, G., M. Vazifeh, B. Woods, and E. Haber (2016) Building an Iterative Heuristic Solver for a Quantum Annealer, *1QB Information Technologies Research Paper*, 2016.

Ruiz, R. and T. Stützle (2008) A simple and effective iterated greedy algorithm for the permutation flowshop scheduling problem, *European Journal of Operational Research*, 177(3): 2033–2049.

Tizhoosh, H. R. (2005) Opposition-based learning: A new scheme for machine intelligence, in: *Proceedings of International Conference on Computational Intelligence for Modelling, Control and Automation, and International Conference on Intelligent Agents, Web Technologies and Internet Commerce* (*CIMCA/IAWTIC-2005*), pp. 695–701, 2005.

Wang, Y., J.-K Hao, F. Glover, and Z. Lu (2014) A tabu search based memetic algorithm for the maximum diversity problem, *Engineering Applications of Artificial Intelligence*, 27: 103–114.

Wang, Y., Q. Wu, and F. Glover (2016) Effective metaheuristic algorithms for the minimum differential dispersion problem, *European Journal of Operational Research*, pp. 1–15. doi:10.1016/j.ejor.2016.10.035.

Woodruff, D. and E. Zemel (1993) Hashing vectors for tabu search, *Annals of Operations Research* 41: 123–137.

Wu, Q., Y. Wang, and F. Glover (2016) Advanced algorithms for bipartite Boolean quadratic programs guided by tabu search, strategic oscillation and path relinking, submitted to the *INFORMS Journal on Computing*.

Xu, Q., L. Wang, N. Wang, X. Hei, and L. Zhao (2014) A review of opposition based learning from 2005 to 2012, *Engineering Applications of Artificial Intelligence*, 29: 1–12.

22

Evolutionary Computation

Guillermo Leguizamón

Christian Blum

Enrique Alba

22.1 Introduction

This chapter aims to give a summary of evolutionary computation (EC) techniques. The summary includes a general description of the main families of algorithms belonging to the EC field as well as their main components. Also, we describe their evolution through the last years including latest advances in constraint handling methods, parallel models and algorithms, methods for dynamic environments, and multiobjective optimization, and so on. Also included are some observations about the relationship between EC techniques and other methods for optimization, in particular metaheuristics. Finally, we describe some new trends and give a glimpse of the numerous applications of EC techniques.

Optimization problems are of high importance in industry and science. Examples of practical optimization problems include train scheduling, time tabling, shape optimization, or telecommunication network design. An optimization problem $\mathcal{P}$ can be described as a triple $(\mathcal{S}, \Omega, f)$, where

1. $\mathcal{S}$ is the search space defined over a finite set of decision variables X_i, $i = 1, \ldots, n$. In case these variables have discrete domains we deal with discrete optimization (or combinatorial optimization), and in case of continuous domains $\mathcal{P}$ is called a continuous optimization problem. Mixed variable problems also exist. Ω is a set of constraints among the variables;

2. $f : \mathcal{S} \rightarrow \mathbb{R}^+$ is the objective function that assigns a positive cost value to each element (or solution) of $\mathcal{S}$.

The goal is to find a solution $s \in \mathcal{S}$ such that $f(s) \leq f(s')$, $\forall\ s' \in \mathcal{S}$ (in case we want to minimize the objective function), or $f(s) \geq f(s')$, $\forall\ s' \in \mathcal{S}$ (in case the objective function must be maximized). In real-life problems the goal is often to optimize several objective functions at the same time. This form of optimization is labeled multiobjective optimization.

Because of the practical importance of optimization problems, many algorithms to tackle them have been developed. These algorithms can be classified as either *complete* or *approximate* algorithms. Complete algorithms are guaranteed to find for every finite size instance of an optimization problem an optimal solution in bounded time [1,2]. Yet, many practically relevant optimization problems are—due to their size or their structure—very hard to be solved to optimality. Therefore, complete methods might need a computation time too high for practical purposes. Thus, the development of approximate methods—in which we sacrifice the guarantee of finding optimal solutions for the sake of getting good solutions in a significantly reduced amount of time—has received more and more attention in the last 30 years.

A very successful strand of approximate optimization algorithms originates from a field known as EC. An EC can be regarded as a metaphor for building, applying, and studying algorithms based on Darwinian principles of natural selection. The instances of algorithms that are based on evolutionary principles are called EAs ([3]). EAs can be characterized as computational models of evolutionary processes. They are inspired by nature's capability to evolve living beings well adapted to their environment. At the core of each EA is a population of individuals. At each algorithm iteration a number of *reproduction* operators is applied to the individuals of the current population to generate the individuals of the population of the next generation. EAs might use operators called *recombination* or *crossover* to recombine two or more individuals to produce new individuals. They also might use *mutation* or *modification* operators that cause a self-adaptation of individuals. The driving force in EAs is the *selection* of individuals based on their *fitness* (which might be based on the objective function, the result of a simulation experiment, or some other kind of quality measure). Individuals with a higher fitness have a higher probability to be chosen as members of the population of the next generation (or as parents for the generation of new individuals). This corresponds to the principle of *survival of the fittest* in natural evolution. It is the capability of nature to adapt itself to a changing environment, which gave the inspiration for EAs.

There has been a variety of slightly different EAs proposed over the years. Three different strands of EAs were developed independently of each other over time. These are evolutionary programming (EP) as introduced by Fogel [4] and Fogel et al. [5], evolutionary strategies (ESs) proposed by Rechenberg [6], and genetic algorithms (GAs) initiated by Holland [7] ([8–11] for further references). EP arose from the desire to generate machine intelligence. Although EP originally was proposed to operate on discrete representations of finite state machines, most of the present variants are used for continuous optimization problems. The latter also holds for most present variants of ES, whereas GAs are mainly applied to solve discrete problems. More recently, other members of the EA family such as genetic programming (GP) and scatter search (SS) were developed. Despite this division into different strands, EAs can be understood from a unified point of view with respect to their main components and the way they explore the search space. Over the years there have been quite a few overviews and surveys about EC methods. Among those are the ones by Bäck [12], Fogel et al. [13], Kobler and Hertz [14], Spears et al. [15], and Michalewicz and Michalewicz [16]. In Reference 17 a taxonomy of EAs is proposed.

The structure of this chapter is as follows. Section 22.2 gives an introduction to EAs and covers shortly important aspects such as the solution representations and search operators. Section 22.3 analyses the evolution of EAs emphasizing some of the important recent research advances. Section 22.4 discusses and highlights the connections between EAs and other stochastic search methods, namely metaheuristics. We conclude the chapter in Section 22.5 with an overview on EAs applications and an outline of new research trends.

22.2 Evolutionary Algorithms (EAs)

The algorithm in Figure 22.1 shows the basic structure of EAs. In this algorithm, P denotes the population of individuals. These individuals are not necessarily solutions to the considered problem. They may be partial solutions, or sets of solutions, or any object that can be transformed into one or more solutions in a structured way. The set of all possible individuals is generally called the *genotype space* denoted by $\mathcal{G}$,

Evolutionary-Algorithm:

1. $P \leftarrow$ **apply** ι on $\mathcal{G}$ to generate μ individuals (the initial population);

2. **while** termination criteria not met **do**

    (a) $P' \leftarrow$ **apply** σ on P; /* selection */

    (b) $P'' \leftarrow$ **apply** ω_r on P'; $r \in \{1, \ldots, \#operators\}$; /* reproduction */

    (c) $P \leftarrow$ **apply** ψ on P and P''; /* replacement */

    endwhile

FIGURE 22.1 Pseudocode of an evolutionary algorithm.

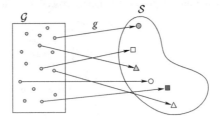

FIGURE 22.2 Mapping from the genotype space to the phenotype space.

whereas the search space of the tackled optimization problem is called the *phenotype space* denoted by $\mathcal{S}$. Hereby, the structure of the individuals is called the *solution representation, or solution encoding*. $\mathcal{G}$ and $\mathcal{S}$, respectively, constitute the domain and codomain of a function g known as the *growth* (or *expression*) function. In some cases $\mathcal{G}$ and $\mathcal{S}$ are actually equivalent, g being a trivial identity function. However, this is not the general situation. As a matter of fact, the only requirement posed on g is surjectivity, that is, for each $s \in \mathcal{S}$ there must be an individual $i \in \mathcal{G}$ such that $g(i) = s$ (Figure 22.2). Furthermore, g might be undefined for some elements in $\mathcal{G}$ or even might not be injective at all (a redundant representation).

At the start of an EA, an initial population of individuals is generated by applying a function ι to the genotype space $\mathcal{G}$. Function ι might represent a random procedure that generates individuals at random, or it might represent a heuristic seeding procedure. Then, at each iteration (also called generation) of the algorithm, the following three major operations are performed: First, a set of individuals P' are selected from the current population P by applying a function σ to P. The selection process is based on the individuals' *fitness* value, which is a measure of how good the solution represented by an individual is for the problem being considered. The fitness value is in general tightly coupled to the objective function. It is the main source for guiding the search process. Second, a population P'' of offspring is generated from P' by the application of *reproduction* operators, that is, the application of a function ω_r to P'. Finally, the replacement function ψ is applied to the current population P and the set of offspring individuals P'' to generate the population for the next generation. To choose the individuals for the next population exclusively from the offspring is called *generational replacement*. In some schemes, such as *elitist strategies*, successive generations overlap to some degree, that is, some portion of the previous generation is retained in the new population. The fraction of new individuals at each generation is called the *generational gap* [18]. In *steady state* selection, only a few individuals are replaced in each iteration: usually a small number of the least fit individuals are replaced by offspring.

The aforementioned process is repeated until certain termination criteria (usually reaching a maximum number of iterations, or reaching a time limit) are satisfied. Most EAs deal with populations of constant size. However, it is also possible to have a variable population size. In case of a continuously shrinking population size, the situation in which only one individual is left in the population (or no crossover partners can be found for any member of the population) might also be one of the stopping conditions of the algorithm. For a graphical presentation of an algorithm iteration see Figure 22.3.

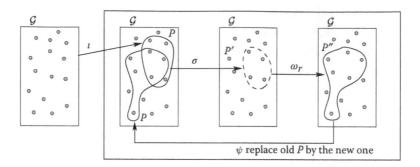

FIGURE 22.3 Illustration of the evolutionary process from the perspective of the genotype space $\mathcal{G}$.

FIGURE 22.4 The different steps in the process of applying an EA to an optimization problem.

Every possible instantiation of this general framework gives rise to different EAs. In fact, it is possible to distinguish among different EA families by considering some guidelines on how to perform this instantiation. However, before we outline the main EA families, we first deal in more detail with some important aspects that have to be taken into account when applying EAs. Figure 22.4 shows a typical line of decisions with which a practitioner is faced when applying an EA.* The first step consists in finding an appropriate *problem formulation*, that is, a concise and manageable description of the problem to be tackled (e.g., a sentence, a mathematical formula, a mathematical program, etc.). Having decided on the problem formulation, the choice of the solution representation (which defines the EA search space), and the development of the reproduction operators are crucial for the success of the algorithm.

22.2.1 Solution Representation (i.e., the Individuals)

Most commonly used in EAs is the representation of solutions as bit-strings or as permutations of n integer numbers. However, also real value encodings, tree-structures or other complex structures are possible. See Figure 22.5 for some examples. In GAs, an individual usually consists of one or more so-called chromosomes. Chromosomes are strings of smaller units termed genes. The different values a gene can take are called the alleles for that gene. Holland's schema analysis [7] and Radcliffe's generalization to formae [19] are examples of how theory can help to guide representation choices.

22.2.2 Reproduction Operators

The chosen solution encoding determines the complexity and size of the EA search space. To explore this search space it is mandatory to design a set of reproduction operators (the number and type of these operators will depend on the solution encoding and the particular EA instance). In general, we distinguish between *recombination* operators (i.e., N-ary operators) and *mutation* operators (i.e., unary operators). In most cases it is possible to recombine all individuals with each other. However, sometimes this is not the case. In general, a neighborhood function $\mathcal{N}_{\mathcal{EC}} : \mathcal{G} \rightarrow 2^{\mathcal{G}}$ assigns to each individual $i \in \mathcal{G}$ a set of individuals $\mathcal{N}_{\mathcal{EC}}(i) \subseteq \mathcal{G}$ whose members are permitted to act as recombination partners for i to create offspring. If an individual can be recombined with any other individual (as, e.g., in the simple

---

* Note that this is similar when applying an approximate optimization technique other than EC.

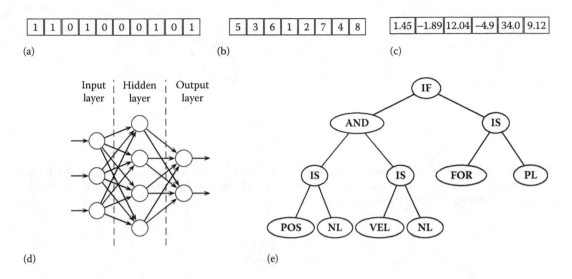

| 1 | 1 | 0 | 1 | 0 | 0 | 0 | 1 | 0 | 1 |

(a)

| 5 | 3 | 6 | 1 | 2 | 7 | 4 | 8 |

(b)

| 1.45 | −1.89 | 12.04 | −4.9 | 34.0 | 9.12 |

(c)

(d)

Input layer   Hidden layer   Output layer

(e)

FIGURE 22.5 Five examples for solution representations. (a) Binary strings are popular, for example, to encode solutions to subset problems such as the knapsack problem. (b) The permutation representation is often used for permutation problems such as the traveling salesman problem. (c) The gray value (or real vector) representation is used, for example, for continuous optimization. (d) Graphs or trees are often represented by edge sets. (e) Trees are often used as encoding for logical rules.

GA [11]) we talk about *unstructured* populations, otherwise we talk about *structured* populations. An example of an EA that works on structured populations is the parallel genetic algorithm (PGA) proposed by Mühlenbein [20] (see Section 22.2.4 for more information on parallel models and algorithms).

The most common form of a recombination operator is two-parent crossover, in which one or two offspring are produced from two parents. But there are also recombination operators that operate on more than two individuals to create new individuals (multi-parent crossover) [21,22]. More recent developments even use population statistics for generating the individuals of the next generation. Examples are the recombination operators called gene pool recombination [23] and bit-simulated crossover [24], which make use of a probability distribution over the search space given by the current population to generate the next population. The simplest form of a mutation operator just performs a small random perturbation of an individual, introducing a kind of *noise*. More complex mutation operators include, for example, random moves or more goal-directed moves in a neighborhood of the individual.

It is important to remark that each possible operator will define its own characteristic neighborhood structure, that is, the set of individuals that might be the outcome of the operators' application. Figures 22.6 and 22.7 show a general view of the neighborhood sizes with respect to the whole EA search space $\mathcal{G}$. In Figure 22.6 we can observe two different neighborhoods for the same individual. For example, consider $\mathcal{G}$ to consist of all binary strings of length n. A mutation operator might be allowed to flip maximally $n/2$ bits of each individual, or, for example, only $n/10$ bits. In the first case the neighborhood of the operator is much bigger than in the second case. Therefore, different mutation operator will define neighborhoods of different sizes and complexities.

Figure 22.7 shows the neighborhood of a hypothetical (binary) recombination operator. As the distance between two individuals gets smaller (from (a) over (b) to (c)) the size of the respective neighborhood for the same operator also diminishes. The neighbor size generally depends on the operator characteristics in addition to the distance between the parent individuals. The sequence from left to right can also be seen as the changing exploration capacity of the operator as the population tends to converge, that is, the individuals from the population are getting closer, which means that the EA converges.

Finally, in Table 22.1 we provide some commonly used reproduction operators for popular solution representations.

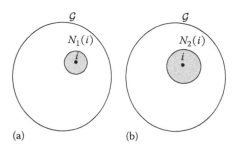

FIGURE 22.6 Graphical representation of the neighborhood $N(i)$ of a mutation operator around an individual i. The neighborhood size depends on the operator characteristics. (a) Shows the example of a mutation operator with a smaller neighborhood than the one shown in (b).

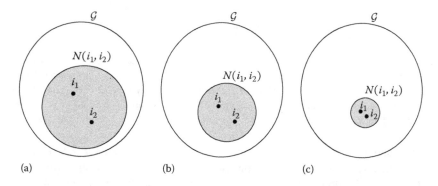

FIGURE 22.7 Graphical representation of the neighborhood $N(i_1, i_2)$ of a binary recombination operator when applied to two individuals i_1 and i_2. From (a) over (b) to (c) the distance between the two individuals changes, and therefore also the neighborhood size.

TABLE 22.1 Some Representative Solution Encodings Together with Commonly used Reproduction Operators

Solution Representation	Reproduction Operators
Binary strings	One-point crossover, uniform crossover, flip mutation, etc.
Permutations	Partially matched crossover (PMX), order crossover (OX), SWAP mutation, etc.
Gray value encoding (i.e., real vectors)	Arithmetic crossover, k-point crossover, Gauss mutation, etc.
Integer vectors	k-point crossover, little and big creep mutation, random mutation, etc.
S-expressions	Subree exchange, node function alteration, etc.

22.2.3 Exploitation versus Exploration

Of great importance for the success of an EA is the right balance between *exploitation* and *exploration*. The term exploitation refers to the use of the accumulated search experience, whereas the term exploration generally refers to the search for areas in the search space that were not visited yet. In terms of exploitation, it proved in many EA applications quite beneficial to use improvement mechanisms to increase the fitness of individuals. Although the use of a population ensures an exploration of the search space, the use of local search techniques helps to quickly identify "good" areas in the search space. Another exploitation strategy is the use of recombination operators that explicitly try to combine "good" parts of individuals (rather than, e.g., a simple one-point crossover for bit-strings). This may guide the search performed by EAs to areas of individuals with certain "good" properties. Techniques of this kind are sometimes called *linkage learning* or *building block learning* [25–28]. Furthermore, generalized recombination operators which incorporate the notion of "neighborhood search" into EC have been proposed in the literature. An example can be found in Reference 29.

One of the major difficulties of EAs (especially when applying local search) is the premature convergence toward sub-optimal solutions. The simplest mechanism to increase exploration is the use of a mutation operator. The simplest form of a mutation operator just performs a small random perturbation of an individual, introducing a kind of *noise*. To avoid premature convergence there are also a number of other ways of maintaining the population diversity. Probably the oldest strategies are *crowding* [18] and its close relative, *preselection* [30]. Newer strategies are *fitness sharing* [31] and *niching* [32]. Hereby, the reproductive fitness allocated to an individual in a population is reduced proportionally to the number of the other individuals that share the same region of the search space.

22.2.4 Families of Evolutionary Algorithms

EAs, as we know them today, were first introduced during the late 1960s and early 1970s. In these years scientists from different places in the world almost simultaneously began to transfer nature's optimization capabilities into algorithms for search and problem solving. The existence of these different primordial sources resulted in the rise of three different EA models. These classical families are:

- *Evolutionary programming*: This EA family originated from the work of Fogel et al. [5]. EP focuses on the adaption of individuals rather than on the evolution of their genetic information. This implies a much more abstract view of the evolutionary process, in which the behavior of individuals is directly modified (as opposed to manipulating its genes). This behavior is typically modeled by using complex data structures such as finite automata or graphs (see, e.g., Figure 22.5d). Traditionally, EP uses asexual reproduction (i.e., mutation) by introducing slight changes in an existing solution, and selection techniques based on direct competition among individuals.
- *Evolutionary strategies*: These techniques were initially developed by Rechenberg and Schwefel [6]. Their original goal was to create a tool for solving engineering problems. With this goal in mind, these techniques are characterized by manipulating arrays of floating-point numbers (see, e.g., Figure 22.5c). Nowadays, also exist versions of ES for discrete problems, but still their application to continuous optimization problems is predominant. As in EP, mutation is often the unique reproductive operator used in ES; however, sometimes also recombination operators are used within ES. A very important feature of ES is the utilization of self-adaptive mechanisms for controlling the application of mutation. These mechanisms are aimed at optimizing the progress of the search by evolving not only the individuals, but also some parameters for mutating these individuals (in a typical situation, an ES individual is a pair $(i, \overrightarrow{\sigma})$, where $\overrightarrow{\sigma}$ is a vector of standard deviations used to control the Gaussian mutation exerted on the individual i).
- *Genetic algorithms*: GAs are possibly the most widespread variant of EAs. They were first introduced by Holland [7]. Holland's work had a great influence on the developments in the field, to the point that some aspects of his work nearly achieved dogma status (e.g., the ubiquitous use of binary strings as solution representation, as shown in Figure 22.5a). Later, the seminal book of Goldberg [8] resulted in a widespread use of GAs for discrete optimization. The main feature of GAs is the use of a recombination (or *crossover*) operator as the primary search tool. The rationale is the assumption that different parts of the optimal solution can be independently discovered, and be later combined to create better solutions. Additionally, mutation is used, but is generally considered a secondary background operator whose purpose is merely 'to keep the pot boiling' by introducing new information in the population. However, this classical interpretation is no longer considered valid nowadays.

The above-mentioned three EA families have not developed in complete isolation from each other. On the contrary, numerous researchers have built bridges among them, and cross-fertilization was and is common. As a result of this interaction, the borders of these classical families tend to be fuzzy (the reader may refer to Reference 12 for a unified presentation of EA families), and new variants have emerged of which we cite the following ones:

- *Evolution programs*: This term is due to Michalewicz [33], and comprises those techniques that, while using the fundamental principles of GAs, evolve complex data structures, as in EP. Nowadays, it is accepted to use the acronym GA—or more generally EA—to refer to such an algorithm, leaving the term "traditional GA" to denote classical bit-string-based GAs.
- *Genetic programming*: The roots of GP can be traced back to the work of Cramer [34]. However, nowadays this technique is mostly associated with Koza, who promoted GP to its current status [35]. Essentially, GP could be viewed as an evolution program in which the structures evolved represent computer programs. Such programs are typically encoded by trees (see, e.g., Figure 22.5e). The ultimate goal of GP is the automatic design of a program for performing a certain task, formulated as a collection of (input, output) examples.
- *Memetic algorithms*: These techniques owe their name to Moscato [36]. Some widespread misconception equates MAs to EAs augmented with local search; although such an augmented EA could be indeed considered an MA, there are additional properties that define MAs (e.g., restarting procedures). In general, an MA is a problem-aware EA. This problem awareness is typically acquired by combining the EA with existing algorithms such as hill climbing, branch and bound, and so on.
- *Scatter search*: The original ideas for SS go back to Glover [37] and originated from strategies for creating composite decision rules and surrogate constraints [38]. The generation of new solutions is achieved by systematically following unifying principles for joining solutions based on generalized path constructions in Euclidean spaces. The terminology in SS is slightly different to the rest of the EAs. For example, the main population of individuals is called the *reference set*. This reference set is manipulated by the application of the following five methods: (a) diversification generation, (b) improvement, (c) reference set update, (d) subset generation, and (e) solution combination.

22.3 Advanced EA Research Topics

Because of their success in practice, EAs have received an ever growing amount of attention during the last two decades. Nowadays, the diversity of research on EAs is impressive. Besides the improvement of their theoretical understanding, the main aims of this research concern their efficiency and the discovery of new problem domains to which they can be applied. More and more advanced versions of EAs for increasingly complex problems were developed. In the following we will shortly deal with the following examples: EAs for constrained optimization problems (COPs), for dynamic optimization problems (DOPs), and for multiobjective optimization problems (MOPs). Figure 22.8 shows a time line in which—in addition to the main EA families—is indicated the start of the research activities in these areas. Furthermore, we will shortly present parallel EAs (PEAs) whose aim is to exploit the implicit parallelism that is based on the fact that EAs are population-based.

22.3.1 Constrained Problems

So far we have assumed that the outcome of a reproduction operator consists always of one ore more feasible individuals. However, sometimes reproduction operators are applied that do not guarantee that. This is the case in particular when real world problems are concerned that often include constraints in their formulation. However, early applications of EAs were designed to solve mainly unconstrained combinatorial and continuous optimization problems. As soon as the research community began to explore the application of EAs to more realistic problems, several constraint handling techniques were accordingly developed. The simplest method is to *reject* infeasible individuals. Nevertheless, for some

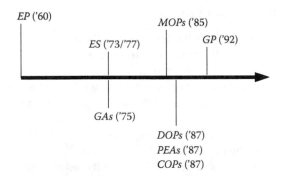

FIGURE 22.8 Time line of EA research including the main EA families and some new developments. The dates indicate the beginning of the respective research activity.

highly constrained problems (e.g., for timetabling problems) it might be very difficult to find feasible individuals. A very popular method is the incorporation of constraints into the fitness function through a penalty factor. However, this technique experiences many drawbacks due to the need of finding appropriate penalty factors. Therefore, other proposals were designed to automatically adapt the penalty factors during the evolutionary process. In addition, plenty of works have been devoted in the last years to develop improved and alternative constraint handling techniques including: advanced penalty approaches, repair of infeasible individuals, design of special representations and operators, and several versions of hybrid methods. For a comprehensive survey of constrain handling techniques see Reference 39.

22.3.2 Dynamic Problems

One of the most recent applications of EAs is the application to DOPs. These are problems that exhibit changes in the problem formulation during the algorithms' execution. These changes might concern different parts of the problem formulation. For example, Morrison [40] exclusively dealt with changes of the objective function. These problems are also called problems with a *dynamic fitness landscape*. A different example is the work of Branke [41], who adopts a more general point of view in which the problem might change either in the objective function, in the problem instance, or in terms of the problem constraints (Figure 22.9). A general definition of DOPs does not exist so far. However, Branke gives a categorization of dynamic environments with respect to the frequency, the severity, the predictability, and the cycle of the changes [41]. The central aspects of EAs that are specially designed for dynamic optimization are the introduction and maintenance of diversity in the population, and the use of implicitly or explicitly stored search history. These adaptive EAs must find a trade-off among the quality, the robustness, and the flexibility of the found solutions.

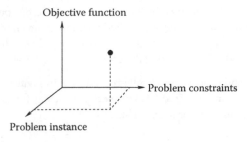

FIGURE 22.9 Dynamic environments seen as a three-dimensional field.

22.3.3 Multiobjective Problems

Multiobjective optimization is another important research field in which EAs have been successfully applied [42]. As a population-based technique, EAs seem to be particularly suitable to face this class of problems. In contrast to other traditional techniques, EAs provide an entire set of Pareto optimal solutions in a single run. During the last years a number of different EA-based approaches have been proposed to deal with multiobjective optimization. Some of the most popular ones are (according to Coello [42]): the aggregation of the objective functions (combinations of all objectives into a single one), VEGA* (an extended version of Grefenstette's GENESIS which implements an alternative selection operator), MOGA (is a scheme in which the population is ranked on the basis of nondominance; this rank is used to accordingly calculate each individuals' fitness value), NSGA (bases its behavior on successive classification and manipulation of nondominated individuals; this allows to search and quickly converge to nondominated regions), and NPGA (implements an extended version of tournament selection in which the competitors compete against a certain number of other individuals). Of course, each of these approaches has strengths and weakness. However, they offer a clear evidence of the intrinsic flexibility and robustness of EAs which makes them suitable for tackling MOPs.

22.3.4 Parallel Models and Algorithms

Even though the use of populations in EAs has many advantages, EAs often suffer from the disadvantage of consuming quite a lot of computation time. This is especially the case when the application of the fitness function is costly, or when the population size is required to be large. However, EAs are naturally prone to parallelism as the most usual operations on the individuals are mostly independent from each other. Besides that, the whole population (also called panmixia) can be geographically structured in separate subpopulations, often leading to better algorithms. The geographical structuring of a population is a form of decentralization. Decentralizing an EA by structuring the population has many advantages, but using structured algorithms may have some disadvantages too, as for example the higher complexity of implementation and analysis, added to the fact that decentralizing the algorithm is not always better. Research on EAs that exploit these aspects, called parallel evolutionary algorithms (PEAs) [44], give often evidence of higher efficiency, larger diversity maintenance, higher availability of memory and CPU resources, and multisolution capabilities of these algorithms. In the following we shortly describe two important models of PEAs:

- In coarse-grained or distributed EAs (dEAs) the population is partitioned into several subpopulations (islands). Each island is treated independently, and there are exchanges of information among them.
- In fine-grained or cellular evolutionary algorithms (cEAs) the individuals are placed on a toroidal N-dimensional grid (where $N = 1, 2, 3$ is used in practice), with one individual per grid location (this location is often referred to as a cell, the fine-grained approach being also known as cellular). Every individual has a neighborhood and can only be recombined with individuals from this neighborhood. The main difference of a cEA with respect to a panmistic EA is its decentralized selection, as the reproductive loop is performed inside each of the numerous pools. Hereby, each individual has its own pool composed of neighboring individuals, and at the same time, this individual belongs to many other pools. Usually, a 2D structure with overlapped neighborhoods is used to provide a smooth diffusion of good solutions along the grid.

The relation between dEAs and cEAs is graphically represented in Figure 22.10. Structured EAs usually outperform standard EAs (i.e., unstructured EAs) numerically. A way for further improving PEAs lies in parallelized implementations and further decentralization.

---

* VEGA was probably the first approach proposed by Schaffer [43] explicitly aimed at solving MOPs.

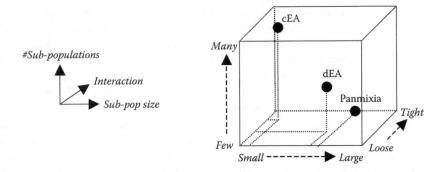

FIGURE 22.10 The structured-population genetic algorithm cube.

22.4 EAs in the Context of Metaheuristics

In a wider context, EAs belong to the class of metaheuristic algorithms, which are search algorithms for approximate optimization. The term *metaheuristic*, first introduced in Reference 45, derives from the composition of two Greek words. *Heuristic* derives from the verb *heuriskein* ($\epsilon \upsilon \rho \iota \sigma \kappa \epsilon \iota \nu$) which means "to find," whereas the suffix *meta* means "beyond, in an upper level." Before this term was widely adopted, metaheuristics were often called *modern heuristics* [46]. Except for EAs, this class of algorithms includes[*]—but is not restricted to—ant colony optimization (ACO), iterated local search (ILS), simulated annealing (SA), and tabu search (TS).

Metaheuristics can be characterized as high level strategies for exploring search spaces by defining general rules that are later customized to the problem at hands. Of great importance hereby is that a dynamic balance is achieved between *diversification* and *intensification*. The term diversification generally refers to the exploration of the search space, whereas the term intensification refers to the exploitation of the accumulated search experience. These terms stem from the TS field [47] and it is important to clarify that the terms *exploration* and *exploitation* are used instead in EC [48] (as outlined before). The balance between diversification and intensification is important, on one side to quickly identify regions in the search space with high quality solutions, and on the other side not to waste too much time in regions of the search space which are either already explored or which do not provide high quality solutions. Blum and Roli elaborated on the importance of the two concepts in their survey on metaheuristics [49].

The search strategies of different metaheuristics are highly dependent on the philosophy of the metaheuristic itself. There are several different philosophies apparent in the existing metaheuristics. Some of them can be seen as "intelligent" extensions of local search algorithms. The goal of this kind of metaheuristic is to escape from local minima to proceed in the exploration of the search space and to move on to find other hopefully better local minima. This is, for example, the case in TS, ILS, variable neighborhood search (VNS), and SA. These metaheuristics (also called trajectory methods) work on one or several neighborhood structure(s) imposed on the search space. We can find a different philosophy in algorithms such as ACO and EAs. They incorporate a learning component in the sense that they implicitly or explicitly try to learn correlations between decision variables to identify high quality areas in the search space. This kind of metaheuristic performs, in a sense, a biased sampling of the search space. For instance, in EAs this is achieved by recombination of solutions and in ACO by sampling the search space at each iteration according to a probability distribution.

An important research direction is the hybridization of EAs with other techniques for optimization, and in particular with other metaheuristics. One of the most popular ways of hybridization concerns the use of trajectory methods in EAs. Many of the successful applications of EC and make use of this feature.

---

[*] In alphabetical order.

The reason for that becomes apparent when analyzing the respective strengths of trajectory methods and EAs. The power of EAs is certainly based on the concept of recombining solutions to obtain new ones. This allows to make guided steps in the search space which are usually "larger" than the steps done by trajectory methods. In other words, a solution resulting from a recombination in EAs is usually more "different" from the parents than, say, a predecessor solution to a successor solution (obtained by applying a move) in TS. We also have "large" steps in trajectory methods such as ILS and VNS, but in these methods the steps are usually not guided (these steps are rather called "kick move" or "perturbation" indicating the lack of guidance). It is interesting to note that in population-based methods there are mechanisms in which elite solutions from the search history influence the search process in the hope of finding better solutions in-between those elite solutions and the current solutions. In EAs this is often obtained by keeping the best (or a number of the best) solution(s) found since the beginning of the respective run of the algorithm in the population. This is called a *steady state* evolution process. SS performs a steady state process by definition. In contrast, the strength of trajectory methods is rather to be found in the more structured way in which they explore a promising region in the search space. In this way the danger of being close to good solutions but "missing" them is not as high as in population-based methods such as EAs.

In summary, population-based methods are better in identifying promising areas in the search space, whereas trajectory methods are better in exploring promising areas in the search space. Thus, EA hybrids that in some way manage to combine the advantage of population-based methods with the strength of trajectory methods are often very successful.

22.5 Applications and New Trends

EAs have been applied to most optimization problems (discrete, continuous, as well as multiobjective). In the classical area of combinatorial optimization, for example, EAs have been applied to *NP*-hard problems such as the travelling salesman problem, the multiple knapsack problem, number partitioning, max independent set, and graph coloring, among others. Other nonclassical—yet important—combinatorial optimization (CO) problems to which EAs have been applied are scheduling (in many variants), timetabling, vehicle routing, quadratic assignment, placement problems, and transportation problems. Telecommunications is also a field that has witnessed successful applications of EAs. For example, EAs have been applied to the placement of antennas and converters, frequency assignment, digital data network design, predicting bandwidth demands in ATM networks, error code design, and so on. EAs have been actively used in electronics and engineering as well. For example, work has been done in structure optimization, aeronautic design, power planning, circuit design, computer-aided design, analogue network synthesis, and service restoration, among other areas. For an extensive collection of references to EC applications we refer to Reference 3.

In addition to the above-mentioned areas of applications there is an increasing interest in new areas for which EAs seems to be a good choice. Recent successes were obtained, for example, in the rapidly growing bioinformatics area (see, e.g., References 50–52), in software engineering [53], in the banking sector [54,55], in computer imagery [56], in drug design [57], in clustering [58], in cryptology [59], in games [60,61], in elevator groups [62], and in evolvable hardware [63].

The research field of EC is vast, indeed, and continuously under development. This is indicated by a significant number of international conferences, such as

- The *International Congress on Evolutionary Computation (CEC)*
- The *Genetic and Evolutionary Computation Conference (GECCO)*
- The *European Conference on Evolutionary Computation in Combinatorial Optimization (EvoCOP)*
- The *Metaheuristics International Conference (MIC)*
- *International Conference on Parallel Problems Solving in Nature (PPSN)*, among others

Furthermore, there exists a number of periodical journals, such as *Evolutionary Computation, IEEE Transactions on Evolutionary Computation, and so on*, that are devoted specifically to this area. The interested reader may also query any bibliographical database or web search engine for "evolutionary algorithm application" to get an idea of the vast number of problems that have been tackled with EAs.

22.6 Conclusion

In this chapter we gave an introduction into the working of EAs. We presented an overview of the main families of EAs as well as a description of their main components from a unifying perspective. Furthermore, we dealt with advanced research directions in EC such as parallel models and algorithms, and the application of EAS to constrained, dynamic, and multiobjective problems. We have put EAs into perspective by describing them as a member of the class of metaheuristic algorithms, highlighting hereby the strengths of EAs in contrast to the strengths of other metaheuristics. Finally, we shortly dealt with typical as well as with more recent EA applications. The rich diversity of EA applications gives a flavor of the popularity and the generality of these techniques.

Acknowledgments

This work has been partially funded by the Ministry of Science and Technology (MCYT) and Regional Development European Found (FEDER) under contract TIC2002-04498-C05-02 (the TRACER project) http://tracer.lcc.uma.es.

References

1. Nemhauser, G.L. and Wolsey, A.L., *Integer and Combinatorial Optimization*, John Wiley & Sons, New York, 1988.
2. Papadimitriou, C.H. and Steiglitz, K., *Combinatorial Optimization - Algorithms and Complexity*, Prentice Hall, Englewood Cliffs, NJ, 1982.
3. Bäck, T., Fogel, D.B., and Machalewicz, Z. (Eds.), *Handbook of Evolutionary Computation*, Institute of Physics Publishing, Philadelphia, PA, 2000.
4. Fogel, L.J., Toward inductive inference automata, in *Proceedings of the International Federation for Information Processing Congress*, North-Holland, Groningen, the Netherlands, 1962, p. 395.
5. Fogel, L.J., Owens, A.J., and Walsh, M.J., *Artificial Intelligence through Simulated Evolution*. John Wiley & Sons, New York, 1966.
6. Rechenberg, I., *Evolutionsstrategie: Optimierung technischer Systeme nach Prinzipien der biologischen Evolution*, Frommann-Holzboog, Stuttgart, Germany, 1973.
7. Holland, J.H., *Adaption in Natural and Artificial Systems*, The University of Michigan Press, Ann Harbor, MI, 1975.
8. Goldberg, D.E., *Genetic Algorithms in Search, Optimization and Machine Learning*, Addison Wesley, Reading, MA, 1989.
9. Mitchell, M., *An Introduction to Genetic Algorithms*, MIT Press, Cambridge, MA, 1998.
10. Reeves, C.R. and Rowe, J.E., *Genetic Algorithms: Principles and Perspectives. A Guide to GA Theory*, Kluwer Academic Publishers, Boston, MA, 2002.
11. Vose, M.D., *The Simple Genetic Algorithm: Foundations and Theory*, MIT Press, Cambridge, MA, 1999.
12. Bäck, T., *Evolutionary Algorithms in Theory and Practice*, Oxford University Press, New York, 1996.
13. Fogel, D.B., An introduction to simulated evolutionary optimization, *IEEE Transactions on Neural Networks*, 5(1), 3, 1994.

14. Hertz, A. and Kobler, D., A framework for the description of evolutionary algorithms, *European Journal of Operational Research*, 126, 1, 2000.

15. Spears, W.M., De Jong, K.A., Bäck, T., Fogel, D.B., and de Garis, H., An overview of evolutionary computation, in *Proceedings of the European Conference on Machine Learning (ECML-93)*, Brazdil, P.B., (Ed.), Vol. 667, Vienna, Austria, 1993, p. 442.

16. Michalewicz, Z. and Michalewicz, M., Evolutionary computation techniques and their applications, in *Proceedings of the IEEE International Conference on Intelligent Processing Systems*, Beijing, China, 1997, p. 14.

17. Calégary, P., Coray, G., Hertz, A., Kobler, D., and Kuonen, P., A taxonomy of evolutionary algorithms in combinatorial optimization, *Journal of Heuristics*, 5, 145, 1999.

18. DeJong, K.A., An Analysis of the Behavior of a Class of Genetic Adaptive Systems, PhD Thesis, University of Michigan, Ann Arbor, MI, 1975. Dissertation Abstracts International 36(10), 5140B, University Microfilms Number 76-9381.

19. Radcliffe, N.J., Forma analysis and random respectful recombination, in *Proceedings of the Fourth International Conference on Genetic Algorithms, ICGA 1991*, Morgan Kaufmann Publishers, San Mateo, CA, 1991, p. 222.

20. Mühlenbein, H., Evolution in time and space – the parallel genetic algorithm, in *Foundations of Genetic Algorithms*, Rawlins, J.E., (Ed.), Morgan Kaufmann, San Mateo, CA, 1991.

21. Bersini, H. and Seront, G., In search of a good evolution-optimization crossover, in *Proceedings of PPSN-II, Second International Conference on Parallel Problem Solving from Nature*, Männer, R., and Manderick, B., (Eds.), Elsevier, Amsterdam, the Netherlands, 1992, p. 479.

22. Eiben, A.E., Raué, P.-E., and Ruttkay, Z., Genetic algorithms with multi-parent recombination, in *Proceedings of the 3rd Conference on Parallel Problem Solving from Nature*, Davidor, Y., Schwefel, H.-P., and Manner, R., (Eds.), LNCS, Vol. 866, Springer, Berlin, Heidelberg, New York, 1994, p. 78.

23. Mühlenbein, H. and Voigt, H.-M., Gene pool recombination in genetic algorithms, in *Proceedings of the Metaheuristics Conference*, Osman, I.H. and Kelly, J.P., (Eds.), Kluwer Academic Publishers, Norwell, MA, 1995, pp. 53–62.

24. Syswerda, G., Simulated Crossover in Genetic Algorithms, in *Proceedings of the Second Workshop on Foundations of Genetic Algorithms*, Whitley, L.D., (Ed.), San Mateo, CA, 1993, p. 239.

25. Goldberg, D.E., Deb, K., and Korb, B., Don't worry, be messy, in *Proceedings of the Fourth International Conference on Genetic Algorithms*, La Jolla, CA, 1991.

26. Harik, G., Linkage learning via probabilistic modeling in the ECGA, Technical Report No. 99010, IlliGAL, University of Illinois, Champaign, IL, 1999.

27. van Kemenade, C.H.M., Explicit filtering of building blocks for genetic algorithms, in *Proceedings of the 4th Conference on Parallel Problem Solving from Nature – PPSN IV*, Voigt, H.-M., Ebeling, W., Rechenberg, I., and Schwefel, H.-P., (Eds.), LNCS, Vol. 1141, Springer, Berlin, Germany, 1996, p. 494.

28. Watson, R.A., Hornby, G.S., and Pollack, J.B., Modeling building-block interdependency, in *Late Breaking Papers at the Genetic Programming 1998 Conference*, Koza, J.R., (Ed.), University of Wisconsin, Madison, WI, 1998.

29. Rayward-Smith, V.J., A unified approach to tabu search, simulated annealing and genetic algorithms, in *Applications of Modern Heuristics*, Rayward-Smith, V.J., (Ed.), Alfred Waller Limited, Oxfordshire, UK, 1994.

30. Cavicchio, D.J., Adaptive Search Using Simulated Evolution, PhD Thesis, University of Michigan, Ann Arbor, MI, 1970.

31. Goldberg, D.E. and Richardson, J., Genetic algorithms with sharing for multimodal function optimization, in *Genetic Algorithms and their Applications*, Grefenstette, J.J., (Ed.), Lawrence Erlbaum Associates, Hillsdale, NJ, 1987, p. 41.

32. Mahfoud, S.W., *Niching Methods for Genetic Algorithms*, PhD Thesis, University of Illinois at Urbana-Champaign, Urbana, IL, 1995.

33. Michalewicz, A., *Genetic Algorithms + Data Structures = Evolution Programs*, Springer-Verlag, Berlin, Germany, 3rd ed., 1998.

34. Cramer, M.L., A representation for the adaptive generation of simple sequential programs, in *Proceedings of the First International Conference on Genetic Algorithms*, Grefenstette, J.J., (Ed.), Lawrence Erlbaum Associates, Hillsdale, NJ, 1985.

35. Koza, J.R., *Genetic Programming*, MIT Press, Cambridge, MA, 1992.

36. Moscato, P. and Cotta, C., A gentle introduction to memetic algorithms, in *Handbook of Metaheuristics*, Glover, F. and Kochenberger, G., (Eds.), Kluwer Academic Publishers, Boston, MA, 2003, p. 105.

37. Glover, F., *Parametric Combinations of Local Job Shop Rules*, ONR Research Memorandum No. 117, GSIA, Carnegie Mellon University, Pittsburgh, PA, 1963.

38. Laguna, M. and Martí, R., *Scatter Search. Methodology and Implementations in C*, Kluwer Academic Publishers, Boston, MA, 2003.

39. Coello Coello, C.A., A comprehensive survey of evolutionary-based multiobjective optimization techniques, *Knowledge and Information Systems*, 1(3), 269, 1999.

40. Morrison, R.W., *Designing Evolutionary Algorithms for Dynamic Environments*, Springer, Heidelberg, Germany, 2004.

41. Branke, J., *Evolutionary Optimization in Dynamic Problems*, Kluwer Academic Publishers, Boston, MA, 2002.

42. Coello Coello, C.A., An updated survey of GA-based multiobjective optimization techniques, *ACM Computing Surveys*, 32(2), 109, 2000.

43. Schaffer, J.D., Multiple objective optimization with vector evaluated genetic algorithms, in *Genetic Algorithms and Their Applications: First International Conference on Genetic Algorithms*, Grefensette, J.J., (Ed.), Lawrence Erlbaum, Hillsdale, NJ, 1985, p. 93.

44. Alba, E. and Tomassini, M., Parallelism and evolutionary algorithms, *IEEE Transactions on Evolutionary Computation*, 6(5), 443, 2002.

45. Glover, F., Future paths for integer programming and links to artificial intelligence, *Computers & Operations Research*, 13, 533, 1986.

46. Reeves, C.R. (Ed.), *Modern Heuristic Techniques for Combinatorial Problems*, Blackwell Scientific Publishing, Oxford, UK, 1993.

47. Glover, F. and Laguna, M., *Tabu Search*, Kluwer Academic Publishers, Norwell, MA, 1997.

48. Eiben, A.E. and Schippers, C.A., On evolutionary exploration and exploitation, *Fundamenta Informaticae*, 35, 1, 1998.

49. Blum, C. and Roli, A., Metaheuristics in combinatorial optimization: Overview and conceptual comparison, *ACM Computing Surveys*, 35(3), 268, 2003.

50. Fogel, G.B., Porto, V.W., Weekes, D.G., Fogel, D.B., Griffey, R.H., McNeil, J.A., Lesnik, E., Ecker, D.J., and Sampath, R., Discovery of RNA structural elements using evolutionary computation, *Nucleic Acids Research*, 30(23), 5310, 2002.

51. Seehuus, R., Tveit, A., and Edsberg, O., Discovering biological motifs with genetic programming: Comparing linear and tree-based representations for unaligned protein sequences, in *Proceedings of the Genetic and Evolutionary Computation Conference (GECCO-2005)*, Vol. 1, Washington, DC, Vol. 1, 2005, p. 401.

52. Rowland, J., On genetic programming and knowledge discovery in transcriptome data, in *Proceedings of the Congress on Evolutionary Computation (CEC'05)*, Vol. 1, Portand, OR, 2004, p. 158.

53. Briand, L.C., Labiche, Y., and Shousha, M., Stress testing real-time systems with genetic algorithms, in *Proceedings of the Genetic and Evolutionary Computation Conference (GECCO-2005)*, Vol. 1, Washington, DC, 2005, p. 1021.

54. Budynek, J., Bonabeau, E., and Shargel, B., Evolving computer intrusion scripts for vulnerability assessment and log analysis, in *Proceedings of the Genetic and Evolutionary Computation Conference (GECCO-2005)*, Washington, DC, 2005, p. 1905.

55. Wagner, N., Michalewicz, Z., Khouja, M., and McGregor, R.R.N., Time series forecasting for dynamic environments: The dyfor genetic program model, *IEEE Transactions on Evolutionary Computation*, 11(4), 433–452, 2007.

56. Grasemann, U. and Miikkulainen, R., Effective image compression using evolved wavelets, in *Proceedings of the Genetic and Evolutionary Computation Conference (GECCO-2005)*, Vol. 2, Washington, DC, 2005, p. 1961.

57. Lameijer, E.W., Ijzerman, A., Kok, J., and Bäck, T., The molecule evaluator: An interactive evolutionary algorithm for designing drug molecules, in *Proceedings of the Genetic and Evolutionary Computation Conference (GECCO-2005)*, Washington, DC, 2005, p. 1969.

58. Sheng, W. and Liu, X., A hybrid algorithm for K-medoid clustering of large data sets, in *Proceedings of the Congress on Evolutionary Computation (CEC'05)*, Vol. 1, Portand, OR, 2004, p. 77.

59. Seredynski, M. and Bouvry, P., Block cipher base on reversible cellular automata, in *Proceedings of the Congress on Evolutionary Computation (CEC'05)*, Portand, OR, 2004, p. 2138.

60. Denzinger, J., Chan, B., Gates, D., Loose, K., and Buchanan, J., Evolutionary behavior testing of commercial computer games, in *Proceedings of the Congress on Evolutionary Computation (CEC'05)*, Vol. 1, Portand, OR, 2004, p. 125.

61. Fogel, D.B., Evolving strategies in blackjack, in *Proceedings of the Congress on Evolutionary Computation*, Portand, OR, 2004, p. 1427.

62. Eguchi, T., Hirasawa, K., Hu, J., and Markon, S., Elevator group supervisory control systems using genetic network programming, in *Proceedings of the Congress on Evolutionary Computation (CEC'05)*, Portand, OR, 2004, p. 1661.

63. Sipper, M., Sanchez, E., Mange, D., Tomassini, M., Pérez-Uribe, A., and Stauffer, A., A phylogenetic, ontogenetic, and epigenetic view of bio-inspired hardware systems, *IEEE Transactions on Evolutionary Computation*, 1(1), 83, 1997.

23

An Introduction to Ant Colony Optimization

Marco Dorigo

Krzysztof Socha

23.1 Introduction

This chapter presents an overview of ant colony optimization (ACO)—a metaheuristic inspired by the behavior of real ants. ACO was proposed by Dorigo et al. [1–3] as a method for solving hard combinatorial optimization problems (COPs).

ACO algorithms may be considered to be part of *swarm intelligence*, that is, the research field that studies algorithms inspired by the observation of the behavior of *swarms*. Swarm intelligence algorithms are made up of simple individuals that cooperate through self-organization, that is, without any form of central control over the swarm members. A detailed overview of the self-organization principles exploited by these algorithms, as well as examples from biology, can be found in Reference 4. Many swarm intelligence algorithms have been proposed in the literature. For an overview of the field of swarm intelligence, we refer the interested reader to Reference 5.

This chapter, which is dedicated to present a concise overview of ACO, is organized as follows. Section 23.2 presents the biological phenomenon that provided the original inspiration. Section 23.3 presents a formal description of the ACO metaheuristic. Section 23.4 overviews the most popular variants of ACO and gives examples of their application. Section 23.5 shows current research directions, and Section 23.6 summarizes and concludes the chapter.

23.2 From Biology to Algorithms

ACO was inspired by the observation of the behavior of real ants. In this section, we present a number of observations made in experiments with real ants, and then we show how these observations inspired the design of the ACO metaheuristic.

23.2.1 Ants

One of the first researchers to investigate the social behavior of insects was the French entomologist Pierre-Paul Grassé. In the 40s and 50s of the twentieth century, he was observing the behavior of termites—in particular, the *Bellicositermes natalensis* and *Cubitermes* species. He discovered [6] that these insects are capable to react to what he called "significant stimuli," signals that activate a genetically encoded reaction. He observed [7] that the effects of these reactions can act as new significant stimuli for both the insect that produced them and for the other insects in the colony. Grassé used the term *stigmergy* [7] to describe this particular type of indirect communication in which the "workers are stimulated by the performance they have achieved."

The two main characteristics of stigmergy that differentiate it from other means of communication are:

- The physical, nonsymbolic nature of the information released by the communicating insects, which corresponds to a modification of physical environmental states visited by the insects
- The local nature of the released information, which can only be accessed by those insects that visit the place where it was released (or its immediate neighborhood)

Examples of stigmergy can be observed in colonies of ants. In many ant species, ants walking to, and from, a food source deposit on the ground a substance called *pheromone*. Other ants are able to smell this pheromone, and its presence influences the choice of their path—that is, they tend to follow strong pheromone concentrations. The pheromone deposited on the ground forms a *pheromone trail*, which allows the ants to find good sources of food that have been previously identified by other ants.

Some researchers investigated experimentally this pheromone laying and following behavior to better understand it and to be able to quantify it. Deneubourg et al. [8] set up an experiment called a "binary bridge experiment." They used *Linepithema humile* ants (also known as Argentine ants). The ants' nest was connected to a food source by two bridges of equal length. The ants could freely choose which bridge to use when searching for food and bringing it back to the nest. Their behavior was then observed over a period of time.

In this experiment, initially there is no pheromone on the two bridges. The ants start exploring the surroundings of the nest and eventually cross one of the bridges and reach the food source. When walking to the food source and back, the ants deposit pheromone on the bridge they use. Initially, each ant randomly chooses one of the bridges. However, because of random fluctuations, after some time there will be more pheromone deposited on one of the bridges than on the other. Because ants tend to prefer in probability to follow a stronger pheromone trail, the bridge that has more pheromone will attract more ants. This in turn makes the pheromone trail grow stronger, until the colony of ants converges towards the use of a same bridge.*

This colony level behavior, based on autocatalysis, that is, on the exploitation of positive feedback, can be exploited by ants to find the shortest path between a food source and their nest. This was demonstrated in another experiment conducted by Goss et al. [9], in which the two bridges were not of the same length: one was significantly longer than the other. In this case, the stochastic fluctuations in the initial choice of a bridge were much reduced as a second mechanism played an important role: those ants choosing by chance the shorter bridge were also the first to reach the nest and when returning to the nest they chose the shorter bridge with higher probability as it had a stronger pheromone trail. Therefore, the ants—thanks to the pheromone following and depositing mechanism—quickly converged to the use of the shorter bridge.

In the next section we explain how these experiments and findings were used to develop optimization algorithms.

---

* Deneubourg et al. conducted several experiments, and results show that each of the two bridges was used in about 50% of the cases.

23.2.2 Algorithms

Stimulated by the interesting results of the experiments described in the previous section, Goss et al. [9] developed a model to explain the behavior observed in the binary bridge experiment. Assuming that after t time units as the start of the experiment m_1 ants had used the first bridge and m_2 the second one, the probability p_1 for the $(m + 1)$-th ant to choose the first bridge can be given by:

$$p_{1(m+1)} = \frac{(m_1 + k)^h}{(m_1 + k)^h + (m_2 + k)^h},$$ (23.1)

where parameters k and h are needed to fit the model to the experimental data. The probability that the same $(m + 1)$-th ant chooses the second bridge is $p_{2(m+1)} = 1 - p_{1(m+1)}$. Monte Carlo simulations, run to test whether the model corresponds to the real data [10], showed very good fit for $k \approx 20$ and $h \approx 2$.

This basic model, which explains the behavior of real ants, may be used as an inspiration to design artificial ants that solve optimization problems defined in a similar way. In the above-described *ant foraging behavior* example, stigmergic communication happens via the pheromone that ants deposit on the ground. Analogously, artificial ants may simulate pheromone laying by modifying appropriate pheromone variables associated with problem states they visit while building solutions to the optimization problem. Also, according to the stigmergic communication model, the artificial ants would have only local access to these pheromone variables.

Therefore, the main characteristics of stigmergy mentioned in the previous section can be extended to artificial agents by:

- Associating state variables with different problem states
- Giving the agents only local access to these variables

Another important aspect of real ants' foraging behavior that may be exploited by artificial ants is the coupling between the autocatalytic mechanism and the *implicit evaluation* of solutions. By implicit solution evaluation, we mean the fact that shorter paths (which correspond to lower cost solutions in the case of artificial ants) are completed earlier than longer ones, and therefore they receive pheromone reinforcement quicker. Implicit solution evaluation coupled with autocatalysis can be very effective: the shorter the path, the sooner the pheromone is deposited, and the more ants use the shorter path. If appropriately used, it can be a powerful mechanism in population-based optimization algorithms (e.g., in evolutionary algorithms [11,12] autocatalysis is implemented by the selection/reproduction mechanism).

Stigmergy, together with implicit solution evaluation and autocatalytic behavior, gave rise to ACO. The basic idea of ACO follows very closely the biological inspiration. Therefore, there are many similarities between real and artificial ants. Both real and artificial ant colonies are composed of a population of individuals that work together to achieve a certain goal. A colony is a population of simple, independent, asynchronous agents that cooperate to find a good *solution* to the problem at hand. In the case of real ants, the problem is to find the food, in the case of artificial ants, it is to find a good solution to a given optimization problem. A single ant (either a real or an artificial one) is able to find a solution to its problem, but only cooperation among many individuals through stigmergy enables them to find *good* solutions.

In the case of real ants, they deposit and react to a chemical substance called *pheromone*. Real ants simply deposit it on the ground while walking. Artificial ants live in a *virtual* world, hence they only modify numeric values (called for analogy *artificial pheromones*) associated with different problem states. A sequence of pheromone values associated with problem states is called *artificial pheromone trail*. In ACO, the artificial pheromone trails are the sole means of communication among the ants. A mechanism analogous to the evaporation of the physical pheromone in real ant colonies allows the artificial ants to *forget* the past history and focus on new promising search directions.

Just like real ants, artificial ants create their solutions sequentially by moving from one problem state to another. Real ants simply walk, choosing a direction based on local pheromone concentrations and a stochastic decision policy. Artificial ants also create solutions step-by-step, moving through available problem states and making stochastic decisions at each step.

There are, however, some important differences between real and artificial ants:

- Artificial ants live in a discrete world—they move sequentially through a finite set of problem states.
- The pheromone update (i.e., pheromone depositing and evaporation) is not accomplished in exactly the same way by artificial ants as by real ones. Sometimes the pheromone update is done only by some of the artificial ants, and often *only after* a solution has been constructed.
- Some implementations of artificial ants use additional mechanisms that do not exist in the case of real ants. Examples include look-ahead, local search, backtracking, etc.

23.3 The Ant Colony Optimization Metaheuristic

ACO has been formalized into a combinatorial optimization metaheuristic by Dorigo et al. [13,14] and has since been used to tackle many COPs.

Given a COP, the first step for the application of ACO to its solution consists in defining an adequate model. This is then used to define the central component of ACO: the pheromone model. The model of a COP may be defined as follows:

Definition 23.1 *A model $P = (\mathbf{S}, \Omega, f)$ of a COP consists of:*

- *A search space $\mathbf{S}$ defined over a finite set of discrete decision variables and a set Ω of constraints among the variables*
- *An objective function $f : \mathbf{S} \to \mathbb{R}_0^+$ to be minimized*[*]

The search space $\mathbf{S}$ is defined as follows: Given is a set of discrete variables X_i, $i = 1, ..., n$, with values $v_i^j \in \mathbf{D}_i = \{v_i^1, ..., v_i^{|D_i|}\}$. A variable instantiation, that is, the assignment of a value v_i^j to a variable X_i, is denoted by $X_i \leftarrow v_i^j$. A solution $s \in \mathbf{S}$—that is, a complete assignment in which each decision variable has a value assigned—that satisfies all the constraints in the set Ω, is a feasible solution of the given COP. If the set Ω is empty, P is called an unconstrained problem model, otherwise it is said to be constrained. A solution $s^* \in \mathbf{S}$ is called a global optimum if and only if: $f(s^*) \leq f(s) \ \forall_{s \in \mathbf{S}}$. The set of all globally optimal solutions is denoted by $\mathbf{S}^* \subseteq \mathbf{S}$. Solving a COP requires finding at least one $s^* \in \mathbf{S}^*$.

The model of a COP is used to derive the pheromone model used by ACO. First, an instantiated decision variable $X_i = v_i^j$ (i.e., a variable X_i with a value v_i^j assigned from its domain $\mathbf{D}_i$), is called a *solution component* and denoted by c_{ij}. The set of all possible solution components is denoted by $\mathbf{C}$. A pheromone trail parameter T_{ij} is then associated with each component c_{ij}. The set of all pheromone trail parameters is denoted by $\mathbf{T}$. The value of a pheromone trail parameter T_{ij} is denoted by τ_{ij} (and called pheromone value).[†] This pheromone value is then used and updated by the ACO algorithm during the search. It allows modeling the probability distribution of different components of the solution.

In ACO, artificial ants build a solution to a COP by traversing the so-called *construction graph*, $G_C(\mathbf{V}, \mathbf{E})$. The fully connected construction graph consists of a set of vertexes $\mathbf{V}$ and a set of edges $\mathbf{E}$. The set of components $\mathbf{C}$ may be associated either with the set of vertexes $\mathbf{V}$ of the graph G_C, or with the set of its edges $\mathbf{E}$. The ants move from vertex to vertex along the edges of the graph, incrementally building a *partial solution*. Additionally, the ants deposit a certain amount of pheromone on the components, that is, either on the vertexes or on the edges that they traverse. The amount $\Delta \tau$ of pheromone deposited

---

[*] Note that minimizing over an objective function f is the same as maximizing over $-f$. Therefore, every COP can be described as a minimization problem.

[†] Note that pheromone values are in general a function of the algorithm's iteration $t : \tau_{ij} = \tau_{ij}(t)$.

ALGORITHM 23.1   Ant Colony Optimization Metaheuristic

Set parameters, initialize pheromone trails
while termination conditions not met **do**
    ConstructAntSolutions
    ApplyLocalSearch   {optional}
    UpdatePheromones
end while

---

may depend on the quality of the solution found. Subsequent ants utilize the pheromone information as a guide towards more promising regions of the search space.

The ACO metaheuristic is shown in Algorithm 23.1. It consists of an initialization step and a loop over three algorithmic components. A single iteration of the loop consists of constructing solutions by all ants, their (optional) improvement with the use of a local search algorithm, and an update of the pheromones. In the following, we explain these three algorithmic components in more detail.

ConstructAntSolutions: A set of m artificial ants construct solutions from elements of a finite set of available solution components $\mathbf{C} = \{c_{ij}\}$, $i = 1, ..., n$, $j = 1, ..., |\mathbf{D}_i|$. A solution construction starts with an empty partial solution $s^p = \emptyset$. Then, at each construction step, the current partial solution s^p is extended by adding a feasible solution component from the set of feasible neighbors $\mathbf{N}(s^p) \subseteq \mathbf{C}$. The process of constructing solutions can be regarded as a path on the construction graph $G_C = (\mathbf{V}, \mathbf{E})$. The allowed paths in G_C are hereby implicitly defined by the solution construction mechanism that defines the set $\mathbf{N}(s^p)$ with respect to a partial solution s^p.

The choice of a solution component from $\mathbf{N}(s^p)$ is done probabilistically at each construction step. The exact rules for the probabilistic choice of solution components vary across different ACO variants. The best known rule is the one of ant system (AS) [3]:

$$p(c_{ij}|s^p) = \frac{\tau_{ij}^{\alpha} \cdot \eta(c_{ij})^{\beta}}{\sum_{c_{il} \in \mathbf{N}(s^p)} \tau_{il}^{\alpha} \cdot \eta(c_{il})^{\beta}}, \quad \forall c_{ij} \in \mathbf{N}(s^p), \tag{23.2}$$

where:

τ_{ij} is the pheromone value associated with the component c_{ij}

$\eta(\cdot)$ is a function that assigns at each construction step a heuristic value to each feasible solution component $c_{ij} \in \mathbf{N}(s^p)$

The values that are given by this function are commonly called *heuristic information*. Furthermore, α and β are positive parameters, whose values determine the relative importance of pheromone versus heuristic information. Equation 23.2 is a generalization of Equation 23.1 presented in Section 23.2: ACO formalization follows closely the biological inspiration.

ApplyLocalSearch: Once solutions have been constructed, and before updating pheromones, often some optional actions may be required. These are often called *daemon actions*, and can be used to implement problem specific and/or centralized actions, which cannot be performed by single ants. The most used daemon action consists in the application of local search to the constructed solutions: the locally optimized solutions are then used to decide which pheromones to update.

UpdatePheromones: The aim of the pheromone update is to increase the pheromone values associated with good or promising solutions, and to decrease those that are associated with bad ones. Usually, this is achieved (i) by decreasing all the pheromone values through *pheromone evaporation* and (ii) by increasing the pheromone levels associated with a chosen set of good solutions $\mathbf{S}_{upd}$:

$$\tau_{ij} \leftarrow (1 - \rho) \cdot \tau_{ij} + \rho \cdot \sum_{s \in \mathbf{S}_{upd}|c_{ij} \in s} F(s), \tag{23.3}$$

where:

S_{upd} is the set of solutions that are used for the update

$\rho \in (0, 1]$ is a parameter called evaporation rate

$F : S \to \mathbb{R}_0^+$ is a function such that $f(s) < f(s') \Rightarrow F(s) \geq F(s'), \forall s \neq s' \in S$

$F(\cdot)$ is commonly called the *fitness function*.

Pheromone evaporation is needed to avoid a too rapid convergence of the algorithm. It implements a useful form of *forgetting*, favoring the exploration of new areas in the search space. Different ACO algorithms, for example, ant colony system (ACS) [15] or $\mathcal{MAX}$-$\mathcal{MIN}$ ant system ($\mathcal{MMAS}$) [16] differ in the way they update the pheromone.

Instantiations of the update rule presented in Equation 23.3 are obtained by different specifications of S_{upd}, which in many cases is a subset of $S_{iter} \cup \{s_{bs}\}$, where S_{iter} is the set of solutions that were constructed in the current iteration, and s_{bs} is the *best-so-far* solution, that is, the best solution found since the first algorithm iteration. A well-known example is the AS-update rule, that is, the update rule of AS [3], where:

$$S_{upd} \leftarrow S_{iter}. \tag{23.4}$$

An example of a pheromone update rule that is more often used in practice is the IB-update rule (where IB stands for *iteration-best*):

$$S_{upd} \leftarrow \arg \max_{s \in S_{iter}} F(s). \tag{23.5}$$

The IB-update rule introduces a much stronger bias towards the good solutions found than the AS-update rule. Although this increases the speed with which good solutions are found, it also increases the probability of premature convergence. An even stronger bias is introduced by the BS-update rule, where BS refers to the use of the best-so-far solution s_{bs}. In this case, S_{upd} is set to $\{s_{sb}\}$. In practice, ACO algorithms that use variations of the IB-update or the BS-update rules and that additionally include mechanisms to avoid premature convergence, achieve better results than those that use the AS-update rule.

23.3.1 Example: The Traveling Salesman Problem (TSP)

One of the most popular ways to illustrate how the ACO metaheuristic works, is via its application to the TSP. The TSP consists of a set of locations (cities) and a traveling salesman that has to visit all the locations once and only once. The distances between the locations are given and the task is to find a Hamiltonian tour of minimal length. The problem has been proven to be NP-hard [17].

The application of ACO to the TSP is straightforward. The moves between the locations become the solution components—that is, the move from city i to city j becomes a solution component $c_{ij} \equiv c_{ji}$. The construction graph $G_C = (V, E)$ is defined by associating the set of locations with the set V of vertices of the graph. As, in principle, it is possible to move from any city to any other one, the construction graph is fully connected and the number of vertices is equal to the number of locations defined by the problem instance. Furthermore, the lengths of the edges between the vertices are proportional to the distances between the locations represented by these vertices. The pheromone is associated with the set E of edges of the graph. An example of the resulting construction graph G_C is presented in Figure 23.1a.

The ants construct the solutions as follows. Each ant starts from a randomly selected location (vertex of the graph G_C). Then, at each construction step it moves along the edges of the graph. Each ant keeps a memory of its path through the graph, and in subsequent steps it chooses among the edges that do not lead to vertexes that it has already visited. An ant has constructed a solution once it has visited all the vertexes of the graph. At each construction step an ant chooses probabilistically the edge to follow among the available ones (those that lead to yet unvisited vertices). The exact rule depends on the implementation,

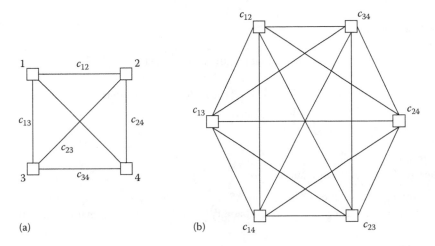

FIGURE 23.1 Example of construction graphs for a 4-city TSP when (a) components are associated with the edges of the graph and when (b) components are associated with the vertexes of the graph. Note that $c_{ij} \equiv c_{ji}$.

an example being Equation 23.2. Once all the ants have finished their tour, the pheromone on the edges is updated according to one of the possible implementations of Equation 23.3. ACO has been shown to perform quite well on the TSP [18].

It is worth noticing that it is also possible to associate the set of solution components of the TSP (or any other COP) with the set of vertices **V** rather than the set of edges **E** of the construction graph G_C. For the TSP, this would mean associating the moves between locations with the set **V** of vertices of the construction graph, and the locations with the set **E** of its edges. The corresponding example construction graph for a 4-city TSP is presented in Figure 23.1b. When using this approach, the ants' solution construction process has to be also properly modified: the ants would have to move from vertex to vertex of the construction graph choosing thereby the *connections between the cities*.

It is important to note that both ways of defining the construction graph are correct and both may be used in practice. Although for the TSP the first way seems more intuitive and was in fact used in all the applications of ACO to the TSP that we are aware of, in other cases the second way might be better suited. For example, it was the selected choice in the case of the university course timetabling problem (UCTP) [19].

23.4 Main Variants of ACO

Several variants of ACO have been proposed in the literature. Here we present the three most successful ones, AS—the first implementation of an ACO algorithm—followed by $\mathcal{MMAS}$ and ACS, together with a short list of their applications.

To illustrate the differences between them clearly, we use the example of the TSP, as described in Section 23.3.1.

23.4.1 Ant System

AS was the first ACO algorithm to be proposed in the literature [1–3]. Its main characteristic is that the pheromone values are updated by *all* the ants that have completed the tour. The pheromone update for τ_{ij}, that is, for edge joining cities i and j, is performed as follows:

$$\tau_{ij} \leftarrow (1 - \rho) \cdot \tau_{ij} + \sum_{k=1}^{m} \Delta\tau_{ij}^{k}, \quad (23.6)$$

where:

ρ is the evaporation rate

m is the number of ants

$\Delta \tau_{ij}^k$ is the quantity of pheromone per unit length laid on edge (i, j) by the k-th ant:

$$\Delta \tau_{ij}^k = \begin{cases} \frac{Q}{L_k} & \text{if ant } k \text{ used edge } (i, j) \text{ in its tour,} \\ 0 & \text{otherwise,} \end{cases} \quad (23.7)$$

where:

Q is a constant

L_k is the tour length of the k-th ant

When constructing the solutions, the ants in AS traverse a construction graph and make probabilistic decision at each vertex. The transitional probability p_{ij}^k of the k-th ant moving from city i to city j is given by:

$$p_{ij}^k = \begin{cases} \dfrac{\tau_{ij}^\alpha \cdot \eta_{ij}^\beta}{\sum_{l \in \text{allowed}_k} \tau_{il}^\alpha \cdot \eta_{il}^\beta} & \text{if } j \in \text{allowed}_k, \\ 0 & \text{otherwise,} \end{cases} \quad (23.8)$$

where allowed_k is the list of cities not yet visited by the k-th ant, and α and β are parameters that control the relative importance of the pheromone versus the heuristic information η_{ij} given by:

$$\eta_{ij} = \frac{1}{d_{ij}}, \quad (23.9)$$

where d_{ij} is the length of edge (i, j).

Several implementations of the AS algorithm have been applied to different COPs. The first and best known is the application to the TSP [1–3]. However, AS was also used successfully to tackle other combinatorial problems. The AS-QAP [20,21] algorithm was used to tackle the QAP, AS-JSP [22] for the job-shop scheduling problem (JSP), AS-VRP [23,24] for the vehicle routing problem (VRP), and AS-SCS [25,26] for the shortest common supersequence (SCS) problem.

23.4.2 $\mathcal{MAX}$-$\mathcal{MIN}$ Ant System

$\mathcal{MMAS}$ is an improvement over the original AS idea. $\mathcal{MMAS}$ was proposed by Stützle and Hoos [16] who introduced a number of changes of which the most important are the following:

- Only the best ant can update the pheromone trails
- The minimum and maximum values of the pheromone are limited

Equation 23.6 takes the following new form:

$$\tau_{ij} \leftarrow (1 - \rho) \cdot \tau_{ij} + \Delta \tau_{ij}^{\text{best}}, \quad (23.10)$$

where $\Delta \tau_{ij}^{\text{best}}$ is the pheromone update value defined by:

$$\Delta \tau_{ij}^{\text{best}} = \begin{cases} \frac{1}{L_{\text{best}}} & \text{if the best ant used edge } (i, j) \text{ in its tour,} \\ 0 & \text{otherwise,} \end{cases} \quad (23.11)$$

L_{best} is the length of the tour of the best ant. This may be (subject to the algorithm designer decision) either the best tour found in the current iteration—*iteration-best, L_{ib}*—or the best solution found since the start of the algorithm—*best-so-far, L_{bs}*—or a combination of both.

Concerning the limits on the minimal and maximal pheromone values allowed, respectively τ_{min} and τ_{max}, Stützle and Hoos suggest that they should be chosen experimentally based on the problem at hand. The maximum value τ_{max} may be calculated analytically provided that the optimum ant tour length is known. In the case of the TSP, τ_{max} is given by:

$$\tau_{max} = \frac{1}{\rho} \cdot \frac{1}{L^*},$$ (23.12)

where L^* is the length of the optimal tour. If L^* is not known, it can be approximated by L_{bs}. The minimum pheromone value τ_{min} should be chosen with caution as it has a rather strong influence on the algorithm performance. They present an analytical approach to finding this value based on the probability p_{best} that an ant constructs the best tour found so far. This is done as follows. First, it is assumed that at each construction step an ant has a constant number k of options available. Therefore, the probability that an ant makes the *right* decision (i.e., the decision that belongs to the sequence of decisions leading to the construction of the best tour found so far) at each of n steps is given by $p_{dec} = \sqrt[n]{p_{best}}$. The analytical formula they suggest for finding τ_{min} is:

$$\tau_{min} = \frac{\tau_{max} \cdot (1 - p_{dec})}{k \cdot p_{dec}}.$$ (23.13)

For more details on how to choose τ_{max} and τ_{min}, we refer to Reference 16. It is important to mention here that it has also been shown [19] that for some problems the choice of an appropriate τ_{min} value is more easily done experimentally than analytically.

The process of pheromone update in $\mathcal{MMAS}$ is concluded by verifying that all pheromone values are within the imposed limits:

$$\tau_{ij} = \begin{cases} \tau_{max} & \text{if } \tau_{ij} > \tau_{max}, \\ \tau_{min} & \text{if } \tau_{ij} < \tau_{min}. \end{cases}$$ (23.14)

$\mathcal{MMAS}$ provided a significant improvement over the basic AS performance. Although the first implementations focused on the TSP [16], it has been later applied to many other COPs such as the QAP [27] or the UCTP [19], the generalized assignment problem (GAP) [28], and the set covering problem (SCP) [29].

23.4.3 Ant Colony System

Another improvement over the original AS was ACS, introduced by Gambardella and Dorigo [15,30]. The most interesting contribution of ACS is the introduction of a *local pheromone update* in addition to the pheromone update performed at the end of the construction process (called here *offline* pheromone update).

The local pheromone update is performed by all the ants after each construction step. Each ant applies it only to the last edge traversed:

$$\tau_{ij} = (1 - \varphi) \cdot \tau_{ij} + \varphi \cdot \tau_0,$$ (23.15)

where:

$\varphi \in (0, 1]$ is the pheromone decay coefficient

τ_0 is the initial value of the pheromone

The main goal of the local update is to diversify the search performed by subsequent ants during one iteration. In fact, decreasing the pheromone concentration on the edges as they are traversed during one iteration encourages subsequent ants to choose other edges and hence to produce different solutions. This makes less likely that several ants produce identical solutions during one iteration.

The offline pheromone update, similar to $\mathcal{MMAS}$, is applied at the end of each iteration by only one ant (the one that found the best solution in the iteration). However, the update formula is slightly different:

$$\tau_{ij} \leftarrow \begin{cases} (1-\rho) \cdot \tau_{ij} + \rho \cdot \Delta\tau_{ij} & \text{if edge } (i,j) \text{ belongs to the tour of the best ant,} \\ \tau_{ij} & \text{otherwise,} \end{cases} \tag{23.16}$$

and in case of TSP, $\Delta\tau_{ij} = \frac{1}{L_{\text{best}}}$ (as in $\mathcal{MMAS}$, L_{best} can be set to either L_{ib} or L_{bs}).

Another important difference between AS and ACS is in the decision rule used by the ants during the construction process. Ants in ACS use the so-called *pseudorandom proportional* rule: the probability for an ant to move from city i to city j depends on a random variable q uniformly distributed over $[0, 1]$, and a parameter q_0; if $q \leq q_0$, then $j = \text{argmax}_{l \in N(s^p)} \{\tau_{il} \eta_{il}^\beta\}$, otherwise Equation 23.8 is used.

ACS has been initially developed for the TSP [15,30], but it was later used to tackle various COPs, including vehicle routing [31], sequential ordering [32], and timetabling [33].

23.5 Current Research Directions

Research in ACO is very active. It includes the application of ACO algorithms to new real-world optimization problems or new types of problems, such as dynamic optimization [34], multiobjective optimization [35], stochastic problems [36], or continuous and mixed-variable optimization [37]. Also, with an increasing popularity of parallel hardware architectures (multi-core processors and the grid technology), a lot of research is being done on creating parallel implementations of ACO that will be able to take advantage of the available hardware. In Section 23.5, we briefly present current research in these new areas.

23.5.1 Other Types of Problems

One of the new areas of application of ACO is dynamic optimization. This type of problems are characterized by the fact that the search space dynamically changes. Although an algorithm searches for good solutions, the conditions of the search as well as the quality of the solutions already found may change. This poses a whole new set of issues for designing successful algorithms that can deal with such situations. It becomes crucial for an algorithm to be able to adjust the search direction, following the changes of the problem being solved. Initial attempts to apply ACO to dynamic optimization problems have been quite successful [34,38,39].

Multiobjective optimization is another area of application for metaheuristics that has received increasing attention over the past years. A multiobjective optimization problem involves solving simultaneously several optimization problems with potentially conflicting objectives. For each of the objectives, a different objective function is used to assess the quality of the solutions found. Algorithms usually aim at finding the so-called *Pareto set*—that is, a set of nondominated solutions—based on the defined objective functions. In the Pareto set, no solution is worse than any other in the set, when evaluated over all the objective functions. Some ACO algorithms designed to tackle multiobjective problems have been proposed in the literature [35,40,41].

Finally, recently, researchers attempted to apply ACO algorithms to continuous optimization problems. When an algorithm designed for combinatorial optimization is used to tackle a continuous problem, the simplest approach is to divide the domain of each variable into a set of intervals. The set of intervals is finite and may be handled by the original discrete optimization algorithm. However, when the domain of the variables is large, and the required accuracy is high, this approach runs into problems. The problem size (i.e., the number of intervals) grows, and combinatorial optimization algorithms become less efficient. Also, this approach requires setting the number of intervals a priori—before the algorithm is run. In case of real-world problems, this is not always a sensible thing to do.

Because of these reasons, optimization algorithms able to handle continuous parameters natively have been developed. Recently, Socha [37,42] has extended ACO to continuous (and mixed-variable— continuous and discrete) problems. Research in this respect is ongoing and should result in new, efficient ACO implementations for continuous and mixed-variable problems.

23.5.2 Parallel ACO Implementations

Parallelization of algorithms becomes more and more an interesting and practical option for algorithm designers. ACO is, particularly, well suited for parallel implementations thanks to ants operating in an independent and asynchronous way. There have already been many attempts to propose parallel ACO algorithms. They are usually classified by their *parallel grain*, that is, the relationship between computation and communication. We can then distinguish between *coarse-grained* and *fine-grained* models. Although the former are characterized by many ants using the same CPU and rare communication between the CPUs, in the latter only few ants use each CPU and there is a lot of communication going on. An overview of the trends and strategies in designing parallel metaheuristics may be found in [43,44].

Randall and Lewis proposed a first reasonably complete classification of parallel ACO implementations [45]. Although many parallel ACO implementations have been proposed in the literature [46–51], the results are fragmented and difficult to compare. Experiments are usually of limited scale and concern different optimization problems. Also, not all parallel implementations proposed are compared with their sequential counterparts, which is an essential measure of their usefulness [51]. All these imply that more research is necessary in the area of parallelization of the ACO metaheuristic (for some recent work in this direction see Reference 52).

23.6 Conclusion

We have presented an introduction to ACO—a metaheuristic inspired by the foraging behavior of real ants. The central component of ACO is the pheromone model based on the underlying model of the problem being solved. The basic idea of ACO, which has been formalized into a metaheuristic framework, leaves many options and choices to the algorithm designer. Several variants of ACO have been already proposed, the most successful being $\mathcal{MMAS}$ and ACS.

ACO is a relatively young metaheuristic, when compared to others such as evolutionary computation, tabu search, or simulated annealing. Yet, it has proven to be quite efficient and flexible. ACO algorithms are currently state-of-the-art for solving many COPs including the SOP [32], RCPS problem [53], OSS problem [54]. For an in-depth overview of ACO, including applications, the interested reader should refer to [55].

Acknowledgments

Marco Dorigo acknowledges support from Belgian FNRS, of which he is a Research Director.

References

1. Dorigo, M., Maniezzo, V., and Colorni, A., *Positive Feedback as a Search Strategy*, Technical Report 91-016, Dipartimento di Elettronica, Politecnico di Milano, Italy, 1991.
2. Dorigo, M., *Optimization, Learning and Natural Algorithms*, PhD Thesis, Dipartimento di Elettronica, Politecnico di Milano, Milan, Italy, 1992. (In Italian.)
3. Dorigo, M., Maniezzo, V., and Colorni, A., Ant system: Optimization by a colony of cooperating agents, *IEEE Transactions on Systems, Man, and Cybernetics – Part B*, 26(1), 29, 1996.

4. Camazine, S., Deneubourg, J.-L., Franks, N., Sneyd, J., Theraulaz, G., and Bonabeau, E., *Self-Organization in Biological Systems*, Princeton University Press, Princeton, NJ, 2003.

5. Bonabeau, E., Dorigo, M., and Theraulaz, G., *Swarm Intelligence: From Natural to Artificial Systems*, Oxford University Press, New York, 1999.

6. Grassé, P.P., *Les Insectes Dans Leur Univers*, Ed. du Palais de la decouverte, Paris, France, 1946.

7. Grassé, P.P., La reconstruction du nid et les coordinations interindividuelles chez *bellicositermes natalensis* et *cubitermes* sp. la théorie de la stigmergie: Essai d'interprétation du comportement des termites constructeurs, *Insectes Sociaux*, 6, 41, 1959.

8. Deneubourg, J.-L., Aron, S., Goss, S., and Pasteels, J.-M., The self-organizing exploratory pattern of the Argentine ant, *Journal of Insect Behavior*, 3, 159, 1990.

9. Goss, S., Aron, S., Deneubourg, J., and Pasteels, J.-M., Self-organized shortcuts in the Argentine ant, *Naturwissenschaften*, 76, 579, 1989.

10. Pasteels, J.M., Deneubourg, J.-L., and Goss, S., Self-organization mechanisms in ant societies (i): Trail recruitment to newly discovered food sources, *Experientia Supplementum*, 54, 155, 1987.

11. Holland, J.H., *Adaptation in Natural and Artiffcial Systems*, University of Michigan Press, Ann Arbor, MI, 1975.

12. Fogel, D.B., *Evolutionary Computation*, IEEE Press, Piscataway, NJ, 1995.

13. Dorigo, M. and Di Caro, G., The ant colony optimization meta-heuristic, in *New Ideas in Optimization*, Corne, D., Dorigo, M., and Glover, F., (Eds.), McGraw-Hill, London, UK, 1999, p. 11.

14. Dorigo, M., Di Caro, G., and Gambardella, L.M., Ant algorithms for discrete optimization, *Artificial Life*, 5(2), 137, 1999.

15. Dorigo, M. and Gambardella, L.M., Ant colony system: A cooperative learning approach to the traveling salesman problem, *IEEE Transactions on Evolutionary Computation*, 1(1), 53, 1997.

16. Stützle, T. and Hoos, H.H., $\mathcal{MAX}$-$\mathcal{MIN}$ ant system, *Future Generation Computer Systems*, 16(8), 889, 2000.

17. Lawler, E.L., Lenstra, J.K., Rinnooy-Kan, A.H.G., and Shmoys, D.B., *The Travelling Salesman Problem*, John Wiley & Sons, New York, 1985.

18. Stützle, T. and Dorigo, M., ACO algorithms for the traveling salesman problem, in *Evolutionary Algorithms in Engineering and Computer Science*, Miettinen, K., Mäkelä, M.M., Neittaanmäki, P., and Périaux, J., (Eds.), John Wiley & Sons, Chichester, UK, 1999, p. 163.

19. Socha, K., Knowles, J., and Sampels, M., A $\mathcal{MAX}$-$\mathcal{MIN}$ ant system for the university timetabling problem, in *Proceedings of ANTS 2002 – Third International Workshop on Ant Algorithms*, LNCS, 2463, Springer-Verlag, Berlin, Germany, 2002, p. 1.

20. Maniezzo, V., Colorni, A., and Dorigo, M., The Ant System applied to the quadratic assignment problem, Technical Report IRIDIA/94-28, IRIDIA, Université Libre de Bruxelles, Bruselles, Belgium, 1994.

21. Maniezzo, V. and Colorni, A., The ant system applied to the quadratic assignment problem, *IEEE Transactions on Knowledge and Data Engineering*, 11(5), 769, 1999.

22. Colorni, A., Dorigo, M., Maniezzo, V., and Trubian, M., Ant system for job-shop scheduling, *JORBEL—Belgian Journal of Operations Research, Statistics and Computer Science*, 34(1), 39, 1994.

23. Bullnheimer, B., Hartl, R.F., and Strauss, C., Applying the ant system to the vehicle routing problem, in *Meta-Heuristics: Advances and Trends in Local Search Paradigms for Optimization*, Osman, I.H., Voß, S., Martello, S., and Roucairol, C., (Eds.), Kluwer Academic Publishers, Dordrecht, the Netherlands, 1998, p. 109.

24. Bullnheimer, B., Hartl, R.F., and Strauss, C., An improved ant system algorithm for the vehicle routing problem, *Annals of Operations Research*, 89, 312, 1999.

25. Michel, R. and Middendorf, M., An island model based ant system with lookahead for the shortest supersequence problem, in *Proceedings of PPSN-V, Fifth International Conference on Parallel Problem Solving from Nature*, Springer-Verlag, Berlin, Germany, 1998, p. 692.

26. Michel, R. and Middendorf, M., An ACO algorithm for the shortest common supersequence problem, *New Methods in Optimisation*, Corne, D., Dorigo, M., and Glover, F., (Eds.), McGraw-Hill, Boston, MA, 1999.

27. Stützle, T. and Hoos, H., The *MAX-MIN* ant system and local search for combinatorial optimization problems: Towards adaptive tools for combinatorial global optimisation, in *Meta-Heuristic, Advances and Trends in Local Search Paradigms for Optimization*, Voss, S., Martello, S., Ossmann, I.H., and Roucairol, C., (Eds.), Kluwer Academic Publishers, Boston, MA, 1998, p. 313.

28. Lourenço, H.R. and Serra, D., Adaptive approach heuristics for the generalized assignment problem, TR Economic Working Papers Series No. 304, Universitat Pompeu Fabra, Department of Economics and Management, Barcelona, Spain, 1998.

29. Lessing, L., Dumitrescu, I., and Stützle, T., A comparison between ACO algorithms for the set covering problem, in *Fourth International Workshop on Ant Algorithms and Swarm Intelligence*, LNCS, Vol. 3172, Springer-Verlag, Heidelberg, Germany, 2004, p. 1.

30. Gambardella, L.M. and Dorigo, M., Solving symmetric and asymmetric TSPs by ant colonies, in *Proceedings of the 1996 IEEE International Conference on Evolutionary Computation (ICEC'96)*, IEEE Press, 1996, p. 622.

31. Bianchi, L., Birattari, M., Chiarandini, M., Manfrin, M., Mastrolilli, M., Paquete, L., Rossi-Doria, O., and Schiavinotto, T., Metaheuristics for the vehicle routing problem with stochastic demands, in *Proceedings of Parallel Problem Solving from Nature - PPSN VIII, 8th International Conference*, LNCS, Vol. 3242, Springer-Verlag, Birmingham, UK, 2004, p. 450.

32. Gambardella, L.M. and Dorigo, M., Ant colony system hybridized with a new local search for the sequential ordering problem, *INFORMS Journal on Computing*, 12(3), 237, 2000.

33. Socha, K., Sampels, M., and Manfrin, M., Ant algorithms for the university course timetabling problem with regard to the state-of-the-art, in *Proceedings of EvoCOP 2003 – 3rd European Workshop on Evolutionary Computation in Combinatorial Optimization*, LNCS, Vol. 2611, Springer-Verlag, Heidelberg, Germany, 2003, p. 334.

34. Guntsch, M. and Middendorf, M., Applying population based ACO to dynamic optimization problems, in *Proceedings of ANTS 2002 – Third International Workshop on Ant Algorithms*, LNCS, Vol. 2463, Springer-Verlag, Bruselles, Belgium, 2002, p. 111.

35. Iredi, S., Merkle, D., and Middendorf, M., Bi-criterion optimization with multi colony ant algorithms, in *Proceedings of the Evolutionary Multi-Criterion Optimization, First International Conference (EMO'01)*, LNCS, Vol. 1993, Springer-Verlag, London, UK, 2001, p. 359.

36. Gutjahr, W.J., S-ACO: An ant-based approach to combinatorial optimization under uncertainty, in *Fourth Internatinal Workshop on Ant Algorithms and Swarm Intelligence*, LNCS, Vol. 3172, Springer-Verlag, Berlin, Germany, 2004, p. 238.

37. Socha, K., ACO for continuous and mixed-variable optimization, in *Ant Colony Optimization and Swarm Intelligence, 4th International Workshop, ANTS 2004*, LNCS, Vol. 3172, Springer-Verlag, Berlin, Germany, 2004, p. 25.

38. Di Caro, G. and Dorigo, M., AntNet: Distributed stigmergetic control for communications networks, *Journal of Artificial Intelligence Research (JAIR)*, 9, 317, 1998.

39. Guntsch, M., Middendorf, M., and Schmeck, H., An ant colony optimization approach to dynamic TSP, in *Proceedings of the Genetic and Evolutionary Computation Conference (GECCO-2001)*, Morgan Kaufmann Publishers, New York, 2001, p. 860.

40. Guntsch, M. and Middendorf, M., Solving multi-criteria optimization problems with population-based ACO, in *Proc. of Evolutionary Multi-Criterion Optimization: Second International Conference, EMO 2003*, LNCS, Vol. 2632, Springer-Verlag, Faro, Portugal, 2003, p. 464.

41. Doerner, K., Gutjahr, W., Hartl, R., Strauss, C., and Stummer, C., Pareto ant colony optimization: A metaheuristic approach to multiobjective portfolio selection, *Annals of Operations Research*, 131(1–4), 79, 2004.

42. Socha, K. and Dorigo, M., Ant Colony Optimization for Continuous Domains, Technical Report TR/IRIDIA/2005-037, IRIDIA, Université Libre de Bruxelles, Bruselles, Belgium, 2005.

43. Cung, V.-D., Martins, S.L., Ribeiro, C.C., and Roucairol, C., Strategies for the parallel implementation of metaheuristics, in *Essays and Surveys in Metaheuristics*, Ribeiro, C.C. and Hansen, P., (Eds.), Operations Research/Computer Science Interfaces, Vol. 15, Kluwer Academic Publishers, Dordrecht, the Netherlands, 2001, chap. 13.

44. Alba, E., *Parallel Metaheuristics. A New Class of Algorithms*, Wiley, Cambridge, NJ, 2005.

45. Randall, M. and Lewis, A., A parallel implementation of ant colony optimization, *Journal of Parallel and Distributed Computing*, 62(9), 1421, 2002.

46. Merkle, D. and Middendorf, M., Fast ant colony optimization on runtime reconfigurable processor arrays, *Genetic Programming and Evolvable Machines*, 3(4), 345, 2002.

47. Talbi, E.-G., Roux, O., Fonlupt, C., and Robillard, D., Parallel ant colonies for combinatorial optimization problems, in *Proceedings of the 11th IPPS/SPDP'99 Workshops Held in Conjunction with the 13th International Parallel Processing Symposium and 10th Symposium on Parallel and Distributed Processing*, Springer-Verlag, London, UK, 1999, p. 239.

48. Gambardella, L.M., Taillard, E., and Agazzi, G., MACS-VRPTW: A multiple ant colony system for vehicle routing problems with time windows, in *New Ideas in Optimization*, Corne, D., Dorigo, M., and Glover, F., (Eds.), McGraw-Hill, London, UK, 1999, p. 63.

49. Rahoual, M., Hadji, R., and Bachelet, V., Parallel ant system for the set covering problem, in *Proceedings of Ant Algorithms - Third International Workshop, ANTS 2002*, LNCS, Vol. 2463, Springer-Verlag, Heidelberg, Germany, 2002, p. 262.

50. Bullnheimer, B., Kotsis, G., and Strauß, G., Parallelization strategies for the ant system, Technical Report 8, Vienna University of Economics and Business Administration, Vienna, Austria, 1997.

51. Stützle, T., Parallelization strategies for ant colony optimization, in *Proceedings of Parallel Problem Solving from Nature - PPSN V: 5th International Conference*, Springer-Verlag, Amsterdam, the Netherlands, 1998, p. 722.

52. Manfrin, M., Birattari, M., Stützle, T., and Dorigo, M., Parallel ant colony optimization for the traveling salesman problem, Technical Report TR/IRIDIA/2006-007, IRIDIA, Université Libre de Bruxelles, Bruselles, Belgium, 2006.

53. Merkle, D., Middendorf, M., and Schmeck, H., Ant colony optimization for resource-constrained project scheduling, *IEEE Transactions on Evolutionary Computation*, 6(4), 333, 2002.

54. Blum, C., Beam-ACO – Hybridizing ant colony optimization with beam search: An application to open shop scheduling, *Computers & Operations Research*, 32(6), 1565, 2005.

55. Dorigo, M. and Stützle, T., *Ant Colony Optimization*, MIT Press, Cambridge, MA, 2004.

III

Multiobjective Optimization, Sensitivity Analysis, and Stability

24

Stochastic Local Search Algorithms for Multiobjective Combinatorial Optimization: A Review

Luís Paquete

Thomas Stützle

24.1 Introduction

Multiobjective combinatorial optimization problems (MCOPs) are combinatorial problems that involve the optimization of several, typically conflicting objectives. An MCOP arises, for example, when planning a holiday trip from a city A to some city B. Besides the minimization of the overall distance between the two sites, one may also be interested in minimizing the cost, the overall travel time, and so on. An unequivocal solution to such problems is the one which is *optimal* with respect to all objectives. But is there such a solution? The shortest tour is not necessarily the fastest nor the fastest needs to be the cheapest one—just consider tolled highways.

Which is then the optimal solution to such a multiobjective problem? It depends on the notion of *optimality*. In this chapter, we focus on the notion of Pareto optimality, which arises when the decision maker is not able to express his preferences *a priori*, simply because he is not present in the process or not able to give an *a priori* formula or ranking of the objectives. In this case, one is interested in obtaining a set of solutions that represents the *optimal trade-off* between the objectives, that is, solutions that are not worse than any other and strictly better in at least one of the objectives.

The set of available algorithms for computing high quality approximations to the Pareto optimal set has grown enormously over the recent years as witnessed by a large number of papers at international conferences and workshops [1–4], special issues of scientific journals [5–7], and numerous regular papers at multiple criteria decision making as well as at theory and algorithm conferences. The majority of these approaches are based on stochastic local search (SLS) algorithms (see Chapter 20 for more details on SLS), a trend that reflects the enormous success of these algorithms for single-objective problems.

Here, we review the main developments in the application of SLS algorithms to MCOPs. SLS techniques range from simple constructive algorithms and iterative improvement algorithms to general algorithm frameworks that can be adapted to a specific problem under consideration. These latter general-purpose SLS methods (also often called metaheuristics) include simulated annealing, tabu search, evolutionary algorithms, ant colony optimization (ACO) and many others. Although these techniques can result in rather complex algorithms already for single-objective problems, when applied to MCOPs they become even more complicated because they need to return a set of solutions instead of a single one.

For tackling MCOPs with SLS algorithms, two fundamentally different approaches can be distinguished. The first is to base the search on the component-wise ordering of the objective value vectors of solutions (or some *ranking* derived from these orderings). We will say that SLS algorithms that mainly focus on this approach follow the *component-wise acceptance criterion (CWAC) search model*. The second approach is based on the usage of parameterized scalarization methods by aggregating the objectives; the SLS algorithms following such lines use the *scalarized acceptance criterion (SAC) search model*. These two choices define, somehow, the two main *schools* for the design of SLS algorithms for MCOPs. Different choices for the remaining components of an SLS algorithm for MCOPs crucially depend on the choice taken for the search model.

In the following sections, we review available SLS algorithms in dependence of these two choices, which also makes this review different from several earlier ones [8–13]. In addition, some proposals combine these two search models. We will review these latter *hybrid* approaches separately from the others. Despite the view taken here, it is inevitable to further discuss the proposed algorithms in dependence of the analogy to known SLS methods for the single objective problems. However, as we are more interested in the differences between the main search strategies, we do not consider specific problem-dependent implementation choices in detail and also try to avoid the highly specialized jargon found in the context of various SLS methods.

24.2 Basics

The main goal of solving MCOPs in terms of Pareto optimality is to find solutions, which are not worse than any other solution and strictly better in at least one of the objectives. Let Q be the number of objectives and S be the set of all candidate solutions; then the objective function for a solution $s \in S$ to MCOPs can be defined as a mapping $\vec{f} : s \mapsto \mathbb{R}^Q$. The following orders hold for objective function vectors in $\mathbb{R}^Q$. Let $\vec{u}$ and $\vec{v}$ be vectors in $\mathbb{R}^Q$; we define the (i) *weak component-wise order* as $\vec{u} \leq \vec{v}$, that is, $u_i \leq v_i$, $i = 1, \ldots, Q$; and (ii) the *component-wise order* as $\vec{u} \prec \vec{v}$, that is, $\vec{u} \neq \vec{v}$ and $u_i \leq v_i$, $i = 1, \ldots, Q$. In the context of optimization, we denote the relation between objective function value vectors of two feasible solutions s and s' as follows: (i) if $\vec{f}(s) \prec \vec{f}(s')$, we say that $\vec{f}(s)$ *dominates* $\vec{f}(s')$; and (ii) if $\vec{f}(s) \leq \vec{f}(s')$, then $\vec{f}(s)$ *weakly dominates* $\vec{f}(s')$. In addition, we say that $\vec{f}(s)$ and $\vec{f}(s')$ are *nondominated* if $\vec{f}(s) \nprec \vec{f}(s')$ and $\vec{f}(s') \nprec \vec{f}(s)$, and they are *nonweakly dominated* if $\vec{f}(s) \nleq \vec{f}(s')$ and $\vec{f}(s') \nleq \vec{f}(s)$. Note that the latter implies that $\vec{f}(s) \neq \vec{f}(s')$. For simplification purposes, we shall use the same notation among solutions when the above-mentioned relations hold between their objective function value vectors.

As the notion of optimal solution clearly differs from the single-objective counterpart, we need to define the notion of a *Pareto global optimum solution* and a *Pareto global optimum set*: A solution $s \in S$ is a Pareto global optimum if and only if there is no $s' \in S$ such that $\vec{f}(s') \prec \vec{f}(s)$; we say that $S' \subseteq S$ is a Pareto global optimum set if and only if it contains *only* and *all* Pareto global optimum solutions. We call the image of the Pareto global optimum set in the objective space *the efficient set*. In most cases, solving an MCOP in terms of Pareto optimality would correspond to finding solutions that are representative of the efficient set.

When solving an MCOP in terms of scalarized optimality, it is assumed that the decision maker is able to *weigh* the importance of each objective; the objective function vector is then *scalarized* according to some weight vector $\vec{\lambda} = (\lambda_1, \ldots, \lambda_Q)$. We will denote the scalarized objective function value by $f_\lambda(s)$. We then say that a solution $s \in S$ is a scalarized global optimum solution if and only if there is no $s' \in S$ such that $f_\lambda(s) < f_\lambda(s')$ with respect to a given $\vec{\lambda}$. The weight vector $\vec{\lambda}$ is usually normalized such that $\sum_{q=1}^Q \lambda_q = 1$. Thus, $\vec{\lambda}$ is an element from the set of normalized weight vectors Λ given by

$$\Lambda = \{\vec{\lambda} \in R^Q : \lambda_q > 0, \sum_{q=1}^Q \lambda_q = 1, q = 1, \ldots, Q\} \tag{24.1}$$

In the case of different ranges of values between objectives, a normalization by range equalization factors must be considered [14]. The scalarization of the objective function vector is usually based on the family of weighted L_p-metrics as

$$f_\lambda(s) = \Big[\sum_{i=1}^Q \big(\lambda_i |f_i(s) - y_i|\big)^p \Big]^{1/p}, \tag{24.2}$$

where $s \in S$, $p > 0$, and $\vec{y} = y_1, \ldots, y_Q$ is the *ideal* vector, where we have $y_i = \min f_i$, $i = 1, \ldots, Q$. Settings of $p = 1$ or $p = \infty$ are most often used. When $p = 1$, we have the well-known *weighted sum* formulation given by

$$f_\lambda(s) = \sum_{i=1}^Q \lambda_i f_i(s). \tag{24.3}$$

It is well known that a scalarized global optimum solution for Equation (24.2) with $p \neq \infty$ is also a Pareto global optimum solution, either if it is a unique solution or if the components of the weight vectors are all positive [14].

Obviously, the great advantage of using a scalarized objective function is that the same SLS algorithm for solving the single-objective problem can be used for tackling the multiobjective version. When using such scalarized objective functions, the available algorithms typically change the weight vectors to generate solutions that are of high quality for different weights. A disadvantage is that, when finding optimal solutions with respect to the scalarized objective function, only *supported solutions*, that is, solutions on the convex hull of the efficient set, are obtained.

Finally, let us mention that (with few exceptions) common to all SLS algorithms for MCOPs is that they return a set of nondominated solutions. Therefore, most of these algorithms include an additional data structure that maintains a set of solutions during the search process that we call *archive* and that is returned when the algorithm is terminated at an arbitrarily chosen time. In our review, we consider that the best nondominated solutions found during the algorithm's run are maintained in the archive, if not expressed otherwise. During the search process, the algorithm needs to *update* the archive and, if nothing else is said, we assume that this update consists of (1) adding new nondominated solutions and (2) removing dominated ones.

Many of the available SLS algorithms make use of two further techniques. The first is *archive bounding*; it is used because the archive may grow very strongly and the operations for manipulating the archive become increasingly time consuming. The second are techniques for maintaining the solutions in the archive spread in the objective space, as it is assumed that clusters of solutions are not informative for the decision maker.

24.3 Component-Wise Acceptance Criterion

We first give an overview of SLS algorithms that make direct or indirect use of the component-wise ordering when deciding about the acceptance of new candidate solutions. By *direct* we mean that this decision is exclusively based on the component-wise ordering introduced in Section 24.2; by *indirect* we understand that from this component-wise ordering some *ranking* of candidate solutions is derived that is then finally used for deciding on which solutions to accept or choose for further manipulation. The algorithms that fall into this latter category are mainly evolutionary algorithms.

24.3.1 Direct Use of the Component-Wise Ordering

When designing an SLS algorithm for single-objective problems, one typically starts by implementing some form of an iterative improvement algorithm. In fact, it is relatively straightforward to apply iterative improvement algorithms also to multiobjective problems by modifying the acceptance criterion, making use of the component-wise ordering of solutions, and the use of an archive of nondominated solutions found so far. With such modifications, iterative improvement algorithms under the CWAC search model can iteratively improve the current set of candidate solutions in the archive by adding nondominated neighboring solutions to it [15]. Such an algorithm can be seeded either by one single solution that may be generated randomly, or by a a set of candidate solutions generated by, for example, an exact algorithm. Despite their wide spread for single-objective problems, iterative improvement algorithms for MCOPs were proposed only recently; nevertheless, almost all SLS algorithms that make direct usage of the component-wise ordering are such iterative improvement algorithms or extensions thereof.

Iterative improvement algorithms: Among the first such approaches is *Pareto local search* (PLS) by Paquete et al. [16,17]. PLS applies iteratively the two following steps. First, it selects randomly one candidate solution s from the archive that has not been visited and examines all neighbors of s. Second, it adds all neighbors of s that are nonweakly dominated with respect to the archive. It stops when the neighborhood of all candidate solutions has been examined. Independent of PLS, Angel et al. [18] proposed a similar approach called bicriteria local search (BLS). The main difference between PLS and BLS is that the latter examines the neighborhood of all candidate solutions in the archive, whereas PLS chooses only one and updates the archive immediately after examining a candidate solution's neighborhood. Angel et al. also proposed an extension of BLS by using an archive bounding technique that only accepts neighboring solutions whose objective function value vectors do not lie in a same partition of the objective space; in addition, also a restart version of BLS was presented. A similar idea was also proposed by Laumanns et al. [19] for a simple evolutionary multiobjective optimizer (SEMO) that examines only one randomly chosen solution in the neighborhood. A variant of SEMO [19], called fair evolutionary multiobjective optimizer (FEMO), selects the candidate solution of the archive whose neighborhood was examined least often.

An iterative improvement algorithm with a more complex acceptance criterion, called *Pareto archived evolution strategy* (PAES), was proposed by Knowles and Corne [20] that induces explicitly some dispersion among the solutions of a bounded-size archive. The acceptance criterion of PAES works as follows: a neighboring candidate solution s' is chosen randomly from the neighborhood of a current candidate solution s of the archive; if s' is dominated by any of the candidate solutions in the archive it is discarded, whereas if s' dominates candidate solutions in the archive, s' is added and the dominated candidate solutions are removed. If s' is nondominated with respect to s and to the archived candidate solutions, one of the two following possibilities is applied: (i) if the archive is not full, s' is added to the archive; (ii) if the archive is already full, s' is only added if there exists another solution s^* lying in a partition of the objective space that contains more solutions. In that case, s^* is removed from the archive. A more complex version of PAES, called M-PAES, is proposed in Reference 21.

Extensions of tabu search: An obvious further extension of the above-mentioned iterative improvement algorithms is to incorporate features of other general-purpose SLS methods. Few approaches in that direction have been proposed so far and the ones we are aware of are based on multiobjective tabu search. The central idea is to examine the neighborhood of a set of solutions, extract nondominated solutions, and accept among those only some non-tabu ones for inclusion into an archive. Examples of such an approach have been presented by Baykasoglu et al. [22,23] and by Armetano and Arroyo [24]. In the latter approach, a bounded size archive is used and special bounding and dispersion techniques are used that are based on the location of centroids of clusters of solutions in the objective space.

24.3.2 Indirect Use of the Component-Wise Ordering

Many current population-based SLS algorithms rely on a mapping of the objective function value vector of each candidate solution in the archive into a single value, a *rank*, where the lower a candidate solution's rank is, the higher are its chances of being chosen. This indirect use of the component-wise ordering is mainly made by a number of *multiobjective evolutionary algorithms* (MEAs). Here, we describe the most relevant of these MEAs with particular emphasis on the ranking procedure. All these approaches consider an archive of bounded size. An illustration of the ranking procedure used in the algorithms presented next is given in Figure 24.1; the brighter the color of an objective function value vector the better is the solution considered by the ranking procedure. Note that, as said before, we do not discuss details of these approaches such as crossover or mutation operators, as these are problem-specific.

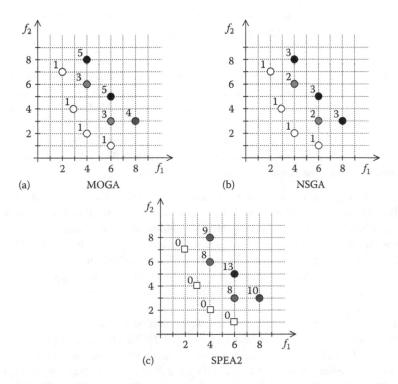

FIGURE 24.1 Graphical illustration of ranking procedures used in MOGA (a), NSGA (b) and SPEA2 (c) where the numbers at the points indicate the respective rank values (see texts for more details).

- Fonseca and Fleming [25] proposed the multiple objective genetic algorithm (MOGA), which assigns to each solution the number of solutions that weakly dominate it in the archive. Then, solutions are ranked according to those values, where ties result in ranks being averaged. A sampling algorithm chooses the next set of solutions to remain in the archive, even if some of them are dominated. Figure 24.1a shows this ranking procedure, before averaging tied ranks.
- Srinivas and Deb [26] proposed the nondominated sorting genetic algorithm (NSGA), which extended a ranking procedure initially proposed (but not tested) by Goldberg [27]: the lowest rank is assigned to the set of candidate solutions in the archive that are nondominated. These solutions are then removed and the nondominated solutions among the remaining candidate solutions are assigned the next rank level. This procedure is iterated until no candidate solution remains to be assigned a rank. Next, a sampling procedure is used for choosing the next set of solutions according to the ranking. Similarly to MOGA, also dominated solutions could be chosen. Figure 24.1b illustrates this ranking procedure. Some further improvements are found in NSGA-II [28]; these include a faster computation of the ranks and a strategy for maintaining a dispersed set of solutions called *crowding* (this technique is explained later in this section).
- Zitzler et al. [29] proposed SPEA2 that maintains two sets of solutions, one being the archive of the best nondominated solutions found so far and the second being the set of current candidate solutions that play the usual role of the population in evolutionary algorithms. Some of the solutions from the archive might be removed by a clustering algorithm if their number exceeds the maximum allowable size. The ranking procedure of SPEA2 works as follows:

  1. To each solution in the archive and in the set of current solutions is assigned the number of current solutions that are dominated by it
  2. For each solution in both sets, its rank is given by the sum of the values computed for the solutions from both sets that weakly dominate it

  Solutions, whose rank are less than one, are added to the archive. The size of the archive is maintained fixed at some value j in two ways: if the number of solutions becomes larger than j, solutions that are clustered in the objective space are removed (except those that are the best to each objective); if the archive is not full, the best current solutions are added to it until the total number of solutions reaches j. Figure 24.1c illustrates this ranking procedure, where circles correspond to the set of current solutions and squares correspond to the solutions in the archive. The ranking procedure used by SPEA2 corrects the previous version called SPEA [30], in which a solution that is dominated by another one could be assigned a lower rank.

As these approaches typically use an archive of fixed size, it is desirable that the existing solutions are dispersed in the objective space. Therefore, this aspect also affects how solutions are going to be chosen. Both MOGA and NSGA use a *fitness sharing* strategy, which decreases a solution's rank depending on how large is the number of solutions in the archive within a certain radius in the objective space. NSGA-II applies a *crowding* strategy, which modifies a solution's rank based on an average distance from the nearest nondominated solutions in the objective space. A similar approach is also used in SPEA2, where to each solution rank the inverse of the distance from the kth nearest neighbor solution in the objective space is added (k is a parameter).

Recently, some MEAs have been proposed that add a further exploration step by making direct use of the component-wise ordering. For instance, Talbi et al. [31] proposed an algorithm that starts with an MEA, whose final set of solutions is used afterwards as starting solutions for an iterative improvement similar to PLS; similar approaches are found also in Brizuela et al. [32] and Basseur et al. [33]. In Jozefowiez et al. [34], the further step after the termination of a MEA consists of a Tabu Search algorithm, with similar principles to Reference 22, that is run several times for different regions of the objective space. Finally, Morita et al. [35] proposed a more complex combination by maintaining two sets of solutions such as SPEA2. At each iteration, some solutions are chosen from each set and are recombined and mutated; the resulting solutions are then added to the current set of solutions and to the archive. Then, the CWAC step consists on the examination of the neighboring solutions to a chosen one from the archive. Finally, the archive is updated and some solutions from the current set of solutions are removed to maintain a given cardinality at the end of each iteration.

24.4 Scalarized Acceptance Criterion

The main principle underlying the SAC search model is to use the value returned by the scalarization of the objective function vector with respect to some weight vector to distinguish between *better* and *worse* solutions. Obviously, using only one weight vector is not enough for obtaining a reasonable approximation to the efficient set. Hence, most approaches consider to change the components of the weight vector while running the algorithm to attain different regions of the objective space.

In the following, we divide the approaches in *nonproprietary* and *proprietary* ones. In nonproprietary approaches, an SLS algorithm is embedded into a general framework that mainly says how an underlying SLS algorithm is applied to tackle an MCOP. In the proprietary approaches, some specific, general-purpose SLS method is enhanced by additional features or specific search strategies that make it adapted for tackling MCOPs. Several examples of such proprietary approaches are therefore discussed in dependence of the underlying general-purpose SLS method.

24.4.1 Nonproprietary Approaches

In the simplest case, a single-objective SLS algorithm could be run several, say k, times using k different weight vectors and one could for each scalarization return the best solution found by the SLS algorithm. Then, the set of objective value vectors of the k returned final solutions forms an approximation to the efficient set. To output a set of nondominated solutions, the dominated solutions from the final set are removed.

Surprisingly, such a very basic approach is very rarely used, not even for a comparison to more complex algorithms. Exceptions are found in Borges and Hansen [36] and Knowles and Corne [37], who used the set of solutions returned by such an approach to get insight into certain instance features. In the following, we describe some approaches that further extend these ideas:

- Borges [38] proposed a general framework, called CHESS, whose acceptance criterion is based on a function of the distance between a new solution and an archive; in particular, the neighboring solution that is accepted to the archive is the one that maximizes the minimum difference between each component of the objective function value vector to any solution in the archive. However, this distance does not take into account any weight vector. This general rule for the acceptance criterion can be applied to SLS methods such as simulated annealing or tabu search.

- Paquete and Stützle [39] proposed two-phase local search (TPLS), which works as follows: in a first phase, a high quality solution for one objective is obtained by some high performance SLS algorithm. Then, in the second phase, a sequence of scalarizations is solved; the initial solution of each scalarization is the one returned by the previous scalarization, the first scalarization being initialized by the solution returned from the first phase. The weight vectors that define the scalarizations in the second phase are modified according to some strategy. This strategy could consist of a random sequence of weight vectors, or of a sequence such that a small change is incurred between components of successive weight vectors. This latter strategy is the most applied one so far. For a detailed description of this algorithm see References 17, 39, and 40.

24.4.2 Proprietary Approaches

Proprietary simulated annealing: Differently from the CWAC model, many algorithms that follow the SAC model use simulated annealing principles. Here, we describe some of the most relevant ones:

- Serafini [41] proposed several ways of modifying the usual probabilistic acceptance criterion of simulated annealing for tackling multiobjective problems. Given the current solution s and a neighboring solution s', he gives guidelines that should be applied to the computation of the probability p of accepting s': if $\vec{f}(s') \prec \vec{f}(s)$, then $p = 1$; if $\vec{f}(s) \prec \vec{f}(s')$, then $p < 1$; otherwise, p depends on the value returned from a function of a parameter called "temperature" and the weighted distance between $\vec{f}(s')$ and $\vec{f}(s)$ (or between $\vec{f}(s')$ and the ideal vector). To attain more solutions, Serafini proposed to use small random variations on the components of the weight vector during the run.
- Ulungu [42] proposed MOSA, where a set of weight vectors is defined *a priori* and for each scalarization one run of a simulated annealing algorithm is done. The probabilistic acceptance criterion in MOSA follows similar principles to those proposed by Serafini [41]. Each time a neighboring solution is accepted, an archive A_λ of nondominated solutions for the current weight vector λ is updated; the final set of solutions returned by the algorithm is obtained after removing the dominated solutions from the union of the resulting sets A_λ.

Proprietary tabu search: Hansen [43] proposed MOTS that uses an archive, which is improved during the search process. In MOTS, only neighboring solutions that are non-tabu and the best with respect to a given scalarization can be added to the archive. To obtain a final set of solutions that is dispersed in the objective space, the current weight vector is updated such that neighboring solutions that are isolated in the objective space are preferably chosen. This update is done as follows: given a solution s from the archive, the i-th weight vector component increases by a fixed amount for each solution in the archive that is worse in the i-th objective; then, a *non-tabu* neighbor of s that has the best scalarized objective function value with respect to the new weight vector is chosen, added to the archive if it is nondominated with respect to all solutions in the archive and, in that case, the tabu list is updated. Further features of MOTS are described in Reference 43.

Proprietary memetic algorithms: Several memetic algorithms for MCOPs have been proposed in analogy to principles of the single-objective case. Usually, these algorithms consist of a sequence of runs of an SLS algorithm using a scalarized objective function, each one seeded by solutions that are generated by recombination and mutation procedures applied to elements of the current set of solutions. We describe the following main approaches:

- Murata and Ishibuchi [44] proposed MOGLS, which is a straightforward extension of MEAs to memetic algorithms. At each iteration, two solutions are selected from the archive by a sampling procedure that takes into account the ranking of all solutions based on a scalarization of the objective function vector with respect to a randomly generated weight vector. These two solutions are then recombined, generating a new one, which is then further improved by a local search algorithm that uses the current weight vector. Although in the first proposals only one randomly chosen solution from the neighborhood is considered at each iteration of the local search, later also analyses on the influence of the strength of the local search on the overall performance have been done [45].
- Jaszkiewicz [46] proposed another memetic algorithm also called MOGLS. This algorithm maintains two sets of solutions as SPEA2, where one set maintains the current solutions, whereas a second one corresponds to the archive of the best solutions found. At each iteration, two solutions are taken from a subset of the best current solutions with respect to a certain randomly chosen weight vector and then recombined. Next, a local search algorithm starts from this new solution using the scalarized objective function defined by the generated weight vector. The solution returned by the local search algorithm is then added to the two sets of solutions according to some acceptance rules.

24.5 Combination of Search Models

Several recently proposed algorithms combine the SAC and the CWAC search model, giving place to hybrid algorithms. Two main trends can be identified. Either the overall SLS algorithm applies two clearly distinct phases based on the two search models in sequence (sequential combination), or the SLS algorithm combines components of the two search models and iteratively changes between those in the overall search process (iterative combination).

24.5.1 Sequential Combination

All the sequential combinations we are aware of first apply algorithms based on the SAC search model. The reason for this choice may be due to the fact that optimal solutions with respect to scalarizations of the objective function vector can identify only supported solutions, possibly leaving large gaps in the final set of solutions returned; in addition, the number of solutions returned by some methods following the SAC model is limited to the number of scalarizations. Algorithms that follow the CWAC search model can easily be applied after the SAC step with the aim of filling these gaps and increasing the number of returned solutions. Note that in the SAC step of such a combination, not necessarily SLS algorithms are required. In fact, if the scalarized problem is polynomially solvable, it is more useful to use an exact algorithm; this is also the case in the first two combinations described as follows:

- Hamacher and Ruhe [47] and Andersen et al. [48] proposed an algorithm that combines the two search models for tackling the multiobjective minimum spanning tree problem. In their approaches, the SAC step consists in obtaining several supported solutions for different scalarizations (note that a minimum spanning tree can be computed in polynomial time) and in the CWAC step, candidate solutions in the neighborhood of the supported ones are added to the archive if they are nondominated by any other candidate solution.
- Gandibleux et al. [49] also proposed an algorithm to search for nondominated solutions with respect to a set of supported solutions that was previously obtained in the SAC step. (As in References 47, 48 an efficient algorithm is known to the single objective version of the MCOP tackled in that paper, the multiobjective assignment problem). The CWAC step follows principles from evolutionary algorithms: several pairs of solutions are chosen randomly from the archive and are recombined to generate a new set of solutions; the recombination operator used here takes into account some information about the components of the supported solutions. The so obtained solutions are then changed by a mutation procedure and, if the mutated solutions are not dominated by some previously calculated bounds, their neighborhood is explored and the nondominated neighbors are added to the archive. This step is repeated for a given number of iterations.
- Paquete and Stützle [39] proposed a Pareto double two-phase local search (PDTPLS) as a further extension of the TPLS approach (see page 421). For each scalarization, a solution s is returned and a CWAC step examines all neighboring solutions of s; all nondominated solutions found in this way are returned once all scalarizations have been examined.

24.5.2 Iterative Combination

Iterative combinations of the SAC and CWAC search models could be conceived in various ways and, hence, it is probably not surprising that the approaches in this category are more varied than when only sequential combinations are considered. In the following we restrict ourselves to give some examples of such combinations that range from local search algorithms like tabu search and simulated annealing to population-based algorithms like memetic algorithms or ACO algorithms. These examples illustrate the range of possibilities that are opened by such iterative combinations.

- Gandibleux et al. [50] use tabu search principles in an algorithm called MOTS. This algorithm iteratively changes between the neighborhood exploration based on a SAC search model and CWAC search model. In the SAC step, a non-tabu solution that minimizes the distance from a *local* utopian point* or a tabu but aspired solution replaces the current one. Once a new solution in the neighborhood of the current one, s, is accepted, a CWAC step adds the nondominated neighbors (or a subset thereof) of s to the archive. To favor more isolated regions of the objective space, the weight vector is updated periodically; to maintain the diversity of the search process, a new weight vector is declared tabu for a given number of iterations.

- Abdelaziz and Krichen [51] developed an algorithm that was also called MOTS. Given some solution s, the SAC step works as follows: a constructive algorithm is run several times for several scalarizations to generate a set S of solutions; for each run of the constructive algorithm, the weight vector takes into account the two worst components of the objective function value vector of s multiplied by some random values. The neighboring solutions to s are added to S, from which then the dominated solutions are removed and the resulting set S' is used to update the archive. Then, a CWAC step that follows similar principles to [22] (see also page 419) is applied for a given number of iterations. The iterative combination of the SAC and CWAC steps is repeated until no solution can be added to the archive during some iterations. A further extension, using a MEA, has been proposed in Reference 52.

- Czyzak and Jaszkiewicz [53] proposed an algorithm that further extends the simulated annealing principles for tackling MCOPs. First, a set of solutions S is generated randomly and each solution from this set is assigned a weight and added to the archive. For each candidate solution s from S, a neighboring solution s' is added to the archive if not dominated by s (CWAC step). The SAC step is then applied for determining s' as the new current solution according to probabilistic rules similar to the ones proposed by Serafini; the weight vector is changed automatically during the search process to favor the acceptance of neighboring solutions that are more *isolated* in the objective space. This iterative process is repeated for all solutions in S.

- López-Ibáñez et al. [54] proposed a combination of SPEA2 with SLS algorithms using an SAC, which combines a CWAC step, as executed by SPEA2, with a SAC step used for defining the search direction for the SLS algorithm. In that proposal, the SLS algorithm is applied to every new solution obtained from a recombination of solutions in the archive using randomly generated weights.

Recently, there has been some interest in applying ACO algorithms [55] to MCOPs. ACO algorithms are based on the repeated, construction of solutions that is stochastically biased by artificial *pheromone* trails (pheromone trails are essentially some numerical information attached to solution components that are used in the solution construction) and heuristic information on the problem under being solved. The pheromone trails are updated during the algorithm's run in dependence of the search experience. Interestingly, all applications of ACO algorithms to MCOPs that are solved under the notion of Pareto optimality combine the two search models: The solution construction is typically based on an SAC step, where weights are defined to join various types of pheromone information with respect to the various objectives, whereas the artificial ants (representing the generated solutions) that update the pheromone information are typically chosen based on a CWAC step. Some approaches are described next:

- Iredi et al. [56] proposed several extensions of ACO algorithms where the pheromone and heuristic information are associated to different weight vectors, defined according to subintervals within the range $[0, 1]$. Thus, different weight vectors direct the constructive steps towards different regions of the objective space. The artificial pheromones are then updated with the nondominated solutions found. Extensions of these algorithms were also proposed by López-Ibáñez et al. [54,57], where an SLS algorithm based on a scalarization of the objective function vector is applied after each new constructed solution.

---

* An utopian point dominates the ideal point; In MOTS, a local utopian point dominates all neighbors of the current solution.

- Dörner et al. [58] describe an ACO algorithm that uses one pheromone matrix for each objective but only one same type of heuristic information. Each solution is constructed based on the heuristic information and on the weighted aggregation of the pheromone matrices according to a randomly chosen weight vector. Some of the best nondominated solutions found are then used for updating the pheromones.

24.6 Conclusion

We have reviewed the research on SLS algorithms for tackling MCOPs with respect to the notion of Pareto optimality. The existing approaches were classified according to whether they use the SAC and CWAC search models or some combinations thereof. For each of the resulting main classes of algorithms we have shortly described the main representatives without the intention of providing a fully comprehensive enumeration of all existing proposals, which would be almost impossible, given their large number and the limited amount of space. Although we described here only the main features of available approaches, several other interesting details can be found in the original papers.

Among the possible approaches for tackling MCOPs, a recent and interesting one is to use an acceptance criterion based on solution quality indicators. Such an algorithm has been proposed by Knowles et al. [59]; it applies at each step the hypervolume indicator [30] to decide which solutions are added to the archive. A similar approach has also been followed in Reference 60, where binary performance indicators are used, that is, quality indicators that return a pair of values for each nondominated set. We remark that the performance of these algorithm strongly also depends on the performance of the underlying algorithm for computing the quality indicators and we expect further development on ways of computing these more efficiently.

There are a few important areas related to SLS algorithms for MCOPs that were not covered here. One of the most important is the empirical assessment of the performance of these algorithms. This is far from being a trivial issue since, as shown by Zitzler et al. [61], frequently used quality indicators for comparing nondominated sets obtained by SLS algorithms have severe problems. A recognized exception is the use of attainment functions. Initially proposed in Reference 62, the attainment function characterizes the performance of the SLS algorithms by describing the distribution of the outcomes. This function has shown to be a first-order moment measure of these outcomes and it can be seen as generalization of the multivariate empirical distribution function [63]. This allows the use of statistical inference and experimental design techniques to *infer* conclusions on the performance of SLS algorithms [64,65]. Some further extension of attainment functions for second-order moments can be found in Reference 66.

Finally, we remark that little is known on the dependence between the performance of SLS algorithms and certain features of the MCOPs. Exceptions can be found in Mote et al. [67] and Müller-Hannemann and Weihe [68] who identified several features of the multiobjective shortest path problem and variations that translate in a tractable number of Pareto optimal solutions, which contrasts with the known worst case for the same problem in Reference 69. In References 16, 36 it is conjectured that most efficient solutions and approximations thereof are strongly clustered in the solution space for the multiobjective traveling salesman problem with respect to a small-sized neighborhood. This means that local search algorithms using this neighborhood are very suitable for this problem, which is confirmed in Reference 39. In addition, the correlation between objectives seems to have a strong effect on the choice for the search model; in Reference 17, experimental results indicate that simple SLS algorithms based on the SAC and CWAC search model, respectively, can behave strongly differently for the multiobjective quadratic assignment problem as the correlation between objectives is changed. Therefore, an interesting future line of research is to investigate, both experimentally and analytically, which and how instance features affect the performance of SLS algorithms.

Although the research field of applying SLS algorithms to MCOPs poses many open questions and significant research issues, the set of available SLS algorithms shows that nowadays it is becoming increasingly feasible to tackle MCOPs with respect to the notion of Pareto optimality and that therefore these approaches are very likely to receive more attention for the solution of difficult real-world multiobjective problems.

Acknowledgments

This work was supported in part by the "Metaheuristics Network," a Research Training Network funded by the Improving Human Potential programme of the CEC. The information provided is the sole responsibility of the authors and does not reflect the Community's opinion. The Community is not responsible for any use that might be made of data appearing in this publication. Thomas Stützle acknowledges support of the Belgian FNRS, of which he is a research associate.

References

1. Coello, C.C., Aguirre, A.H., and Zitzler, E. (Eds.), *Evolutionary Multi-criterion Optimization*, LNCS, Guanajuato, Mexico, Vol. 3410, 2005.
2. Fonseca, C.M., Fleming, P., Zitzler, E., Deb, K., and Thiele, L. (Eds.), *Evolutionary Multi-criterion Optimization*, LNCS, Heidelberg, Germany, Vol. 2632, 2003.
3. Gandibleux, X., Sevaux, M., Sörensen, K., and T'kindt, V. (Eds.), *Metaheuristics for Multiobjective Optimisation*, LNEMS, Springer-Verlag Berlin Heidelberg, Vol. 535, 2004.
4. Zitzler, E., Deb, K., Thiele, L., Coello, C.C., and Corne, D. (Eds.), *Evolutionary Multi-criterion Optimization*, LNCS, Springer-Verlag Berlin Heidelberg, Vol. 1993, 2001.
5. Coello, C.C., Special issue on "evolutionary multiobjective optimization," *IEEE Trans. Evol. Comput.*, 7(2), 2003.
6. Gandibleux, X., Jaszkiewicz, A., Fréville, A., and Slowinski, R., Special issue on "multiple objective metaheuristics," *J. Heuristics*, 6(3), 2000.
7. Jaszkiewicz, A., Special issue on "evolutionary and local search heuristics in multiple objective optimization," *Found. Comput. Decis. Sci. J.*, 26(1), 2001.
8. Coello, C.C., A comprehensive survey of evolutionary-based multiobjective optimization techniques, *Know. Inf. Sys.*, 1, 269, 1999.
9. Coello, C.C., An updated survey of GA-based multiobjective optimization techniques, *ACM Comput. Surv.*, 32, 109, 2000.
10. Ehrgott, M. and Gandibleux, X., Approximative solution methods for combinatorial multicriteria optimization, *TOP*, 12(1), 1, 2004.
11. Fonseca, C.M. and Fleming, P., An overview of evolutionary algorithms in multiobjective algorithms, *Evolut. Comput.*, 3, 1, 1995.
12. Knowles, J. and Corne, D., Memetic algorithms for multiobjective optimization: Issues, methods and prospects, in *Recent Advances in Memetic Algorithms*, Hart, W.E., Smith, J., and Krasnogor, N., (Eds.), Springer-Verlag, New York, 2005, pp. 313–352; *Studies in Fuzziness and Soft Computing*, Vol. 166, 2004, p. 313.
13. Jones, D., Mirrazavi, S., and Tamiz, M., Multi-objective meta-heuristics: An overview of the current state-of-the-art, *Eur. J. Oper. Res.*, 137(1), 1, 2002.
14. Steuer, R.E., *Multiple Criteria Optimization: Theory, Computation and Application*, John Wiley & Sons, New York, 1986.
15. Paquete, L., Schiavinotto, T., and Stützle, T., On local optima in multiobjective combinatorial optimization problems, *Annals of Operations Research*, 156(1), 83–97, 2007.
16. Paquete, L., Chiarandini, M., and Stützle, T., Pareto local optimum sets in the biobjective traveling salesman problem: An experimental study, in *Metaheuristics for Multiobjective Optimisation*, LNEMS, Gandibleux. X. et al. (Eds.), Springer, Heidelberg, Germany, Vol. 535, 2004, p. 177.

17. Paquete, L. and Stützle, T., A study of stochastic local search algorithms for the biobjective QAP with correlated flow matrices, *Eur. J. Oper. Res.*, 169(3), 943, 2006.

18. Angel, E., Bampis, E., and Gourvés, L., A dynasearch neighborhood for the bicriteria traveling salesman problem, in *Metaheuristics for Multiobjective Optimisation*, Gandibleux, X. et al. (Eds.), LNEMS, Springer-Verlag Berlin Heidelberg, Vol. 535, 2004, p. 153.

19. Laumanns, M., Thiele, L., Zitzler, E., Welzl, E., and Deb, K., Running time analysis of multi-objective evolutionary algorithms on a simple discrete optimization problem, in *Proceedings of PPSN-VII*, LNCS, Springer-Verlag Berlin Heidelberg, Vol. 2439, 2002, p. 44.

20. Knowles, J. and Corne, D., The Pareto archived evolution strategy: A new base line algorithm for multiobjective optimisation, in *Proceedings of the Congress on Evolutionary Computation*, IEEE Service Center, Washington, DC, 1999, p. 98.

21. Knowles, J. and Corne, D., M-PAES: A memetic algorithm for multiobjective optimization, in *Proceedings of the Congress on Evolutionary Computation*, IEEE Service Center, Piscataway, NJ, Vol. 1, 2000, p. 325.

22. Baykasoglu, A., Owen, S., and Gindy, N., A taboo search based approach to find the Pareto optimal set in multiobjective optimization, *J. Eng. Opt.*, 31, 731, 1999.

23. Baykasoglu, A., Özbakir, L., and Sönmez, A., Using multiple objective tabu search and grammars to model and solve multi-objective flexible job shop scheduling problems, *J. Intell. Manuf.*, 15, 777, 2004.

24. Armetano, V.A. and Arroyo, J.E., An application of a multi-objective tabu search algorithm to a bicriteria flowshop problem, *J. Heuristics*, 10(5), 463, 2004.

25. Fonseca, C.M. and Fleming, P., Genetic algorithms for multiobjective optimization: Formulation, discussion and generalization, in *Proceedings of the International Conference on Genetic Algorithms*, Morgan Kaufmann, San Mateo, CA, 1993, p. 416.

26. Srinivas, N. and Deb, K., Multiobjective optimization using nondominated sorting in genetic algorithms, *Evol. Comput.*, 3(2), 221, 1994.

27. Goldberg, D.E., *Genetic Algorithms in Search, Optimization and Machine Learning*, Addison-Wesley Longman Publishing, Boston, MA, 1989.

28. Deb, K., Agrawal, S., Prataand, A., and Meyarivan, T., A fast and elitist multi-objective genetic algorithm: NSGA-II, *IEEE Trans. Evol. Comput.*, 6(2), 182, 2002.

29. Zitzler, E., Laumanns, M., and Thiele, L., SPEA2: Improving the strength Pareto evolutionary algorithm for multiobjective optimization, TR TIK-Report 103, Computer Engineering and Networks Laboratory, Swiss Federal Institute of Technology, Zürich, Switzerland, May 2001.

30. Zitzler, E. and Thiele, L., Multiobjective evolutionary algorithms: A comparative case study and the strength Pareto approach, *IEEE Trans. Evol. Comput.*, 4(3), 257, 1999.

31. Talbi, E.G., A hybrid evolutionary approach for multicriteria optimization problems: Application to the flow shop, in *Evolutionary Multi-criterion Optimization*, Fonseca, C.M., Fleming, P., Zitzler, E., Deb, K., and Thiele, L., (Eds.), LNCS, Springer, Berlin, Germany, Vol. 2632, 2003, p. 416.

32. Brizuela, C.A., Sannomiya, N., and Zhao, Y., Multi-objective flow-shop: Preliminary results, in *Evolutionary Multi-criterion Optimization*, Zitzler, E., Deb, K., Thiele, L., Coello, C.C., and Corne, D., (Eds.), LNCS, Springer, Berlin, Germany, Vol. 1993, 2001, p. 443.

33. Basseur, M., Seynhaeve, F., and Talbi, E.G., Design of multi-objective evolutionary algorithms: Application to the flow-shop, in *Proceedings of the Congress on Evolutionary Computation*, IEEE Press, Piscataway, NJ, Vol. 2, 2002, p. 151.

34. Jozefowiez, N., Semet, F., and Talbi, E.-G., Parallel and hybrid models for multi-objective optimization: Application to the vehicle routing problem, in *Proceedings of the International Conference on Parallel Problem Solving from Nature*, LNCS, Springer, Berlin, Germany, Vol. 2439, 2002, p. 271.

35. Morita, H., Gandibleux, X., and Katoh, N., Experimental feedback on biobjective permutation scheduling problems solved with a population heuristic, *Found. Comput. Decis. Sci.*, 26(1), 23, 2001.

36. Borges, P.C. and Hansen, M.P., A study of global convexity for a multiple objective travelling salesman problem, in *Essays and Surveys in Metaheuristics*, Ribeiro, C.C. and Hansen, P., (Eds.), Kluwer, Dordrecht, the Netherlands, 2000, p. 129.

37. Knowles, J. and Corne, D., Towards landscape analyses to inform the design of a hybrid local search for the multiobjective quadratic assignment problem, in *Soft Computing Systems: Design, Management and Applications*, Abraham, A., Ruiz-del-Solar, J., and Köppen, M., (Eds.), IOS Press, Amsterdam, the Netherlands, 2002, p. 271.

38. Borges, P., CHESS – Changing horizon efficient set search: A simple principle for multiobjective optimization, *J. Heuristics*, 6(3), 405, 2000.

39. Paquete L. and Stützle, T., A two-phase local search for the biobjective traveling salesman problem, in *Evolutionary Multi-criterion Optimization*, Fonseca, C.M., Fleming, P., Zitzler, E., Deb, K., and Thiele, L., (Eds.), LNCS, Springer, Berlin, Germany, Vol. 2632, 2003, p. 479.

40. Paquete, L., *Stochastic Local Search Algorithms for Multiobjective Combinatorial Optimization: Methods and Analysis Dissertations in Artificial Intelligence*, AKA Verlag/ IOS Press, Germany, Vol. 295, 2006.

41. Serafini, P., Simulated annealing for multiobjective optimization problems, in *Multiple Criteria Decision Making. Expand and Enrich the Domains of Thinking and Application*, Tzeng, G.H., Wang, H.F., Wen, V.P., and Yu, P.L., (Eds.), LNEMS, Springer-Verlag, Berlin, Germany, 992, p. 283.

42. Ulungu, E.L., Optimisation combinatoire multicritére: Détermination de l'ensemble des solutions efficaces et méthodes interactives, PhD Thesis, Université de Mons-Hainaut, Mons, Belgium, 1993.

43. Hansen, M.P., Tabu search for multiobjective optimisation: MOTS, in *Proceedings of the International Conference on Multiple Criteria Decision Making*, Rutgers Center for Operation Research, Piscataway, NJ, 1997.

44. Ishibuchi, H. and Murata, T., A multi-objective genetic local search algorithm and its application to flow-shop scheduling, *IEEE Trans. Sys., Man, and Cybern. – Part C*, 28(3), 392, 1998.

45. Ishibuchi, H., Yoshida, T., and Murata, T., Balance between genetic search and local search in memetic algorithms for multiobjective permutation flowshop scheduling, *IEEE Trans. Evolut. Comput.*, 7(2), 204, 2003.

46. Jaszkiewicz, A., *Genetic local search for multiple objective combinatorial optimization*, TR RA-014/98, Institute of Computer Science, Poznań University of Technology, Poznań, Poland, 1998.

47. Hamacher, H.V. and Ruhe, G., On spanning tree problems with multiple objectives, *Ann. Oper. Res.*, 52, 209, 1994.

48. Andersen, K., Jörnsten, K., and Lind, M., On bicriterion minimal spanning trees: An approximation, *Comput. Oper. Res.*, 23(12), 1171, 1996.

49. Gandibleux, X., Morita, H., and Katoh, N., Use of a genetic heritage for solving the assignment problem, in *Evolutionary Multi-criterion Opt.*, Fonseca, C.M., Fleming, P., Zitzler, E., Deb, K., and Thiele, L., (Eds.), LNCS, Springer, Berlin, Germany, Vol. 2632, 2003, p. 43.

50. Gandibleux, X., Mezdaoui, N., and Fréville, A., A tabu search procedure to solve multiobjective combinatorial optimization problems, in *Advances in Multiple Objective and Goal Programming*, Caballero, R., Ruiz, F., and Steuer, R., (Eds.), LNEMS, Springer-Verlag, Berlin, Germany, Vol. 455, 1997, p. 291.

51. Abdelaziz, F.B. and Krichen, S., A tabu search heuristic for multiobjective knapsack problems, TR RRR 28-97, Rutgers Center for Operations Research, Piscataway, NJ, 1997.

52. Abdelaziz, F.B., Chaouachi, J., and Krichen, S., A hybrid heuristic for multiobjective knapsack problems, in *Meta-Heuristics: Advances and Trends in Local Search Paradigms for Optimization*, Voss, S., Martello, S., Osman, I., and Roucairol, C., (Eds.), Kluwer Academic Publishers, Boston, MA, 1999, p. 205.

53. Czyzak, P. and Jaszkiewicz, A., Pareto simulated annealing—a metaheuristic technique for multiple objective combinatorial optimization, *J. Multi-Criteria Decis. Anal.*, 7, 34, 1998.

54. López-Ibáñez, M., Paquete, L., and Stützle, T., Hybrid population-based algorithms for the biobjective quadratic assignment problem, *JMMA*, 5(1), 2006.
55. Dorigo, M. and Stützle, T., *Ant Colony Optimization*, MIT Press, Cambridge, MA, 2004.
56. Iredi, S., Merkle, D., and Middendorf, M., Bi-criterion optimization with multi colony ant algorithms, in *Evolutionary Multi-criterion Optimization*, Zitzler, E., Deb, K., Thiele, L., Coello, C.C., and Corne, D., (Eds.), LNCS, Springer, Berlin, Germany, Vol. 1993, 2001.
57. López-Ibáñez, M., Paquete, L., and Stützle, T., On the design of ACO for the biobjective quadratic assignment problem, in *Ant Colony Optimization and Swarm Intelligence (ANTS)*, LNCS, Vol. 3172, Springer, Berlin, Heidelberg, 2004, p. 214.
58. Doerner, K., Gutjahr, W.J., Strauss, C., and Stummer, C., Pareto ant colony optimization: A metaheuristic approach to portfolio selection, *Ann. Oper. Res.*, 131, 79, 2004.
59. Knowles, J., Corne, D., and Fleischer, M., Bounded archiving using the lebesgue measure, in *Proceedings of the Congress on Evolutionary Computation*, 2003, IEEE Press, Piscataway, NJ, p. 2498.
60. Zitzler, E. and Künzli, S., Indicator-based selection in multiobjective search, in *Proceedings of the International Conference on Parallel Problem Solving from Nature*, LNCS, Springer, Heidelberg, Germany, Vol. 3242, 2004, p. 832.
61. Zitzler, E., Thiele, L., Laumanns, M., Fonseca, C.M., and Grunert da Fonseca, V., Performance assessment of multiobjective optimizers: An analysis and review, *IEEE Trans. Evol. Comput.*, 7(2), 117, 2003.
62. Grunert da Fonseca, V., Fonseca, C.M., and Hall, A., Inferential performance assessment of stochastic optimizers and the attainment function, in *Evolutionary Multi-criterion Optimization*, Zitzler, E., Deb, K., Thiele, L., Coello, C.C., and Corne, D., (Eds.), LNCS, Springer, Berlin, Germany, Vol. 1993, 2001, p. 213.
63. Grunert da Fonseca, V. and Fonseca, C.M., A link between the multivariate cumulative distribution function and the hitting function for random closed sets, *Stat. Prob. Lett.*, 57(2), 179, 2002.
64. Shaw, K.J., Fonseca, C.M., Nortcliffe, A.L., Thompson, M., Love, J., and Fleming, P.J., Assessing the performance of multiobjective genetic algorithms for optimization of a batch process scheduling problem, in *Proceedings of the 1999 Congress on Evolutionary Computation*, IEEE Press, Piscataway, NJ, Vol. 1, 1999, p. 34.
65. Paquete, L., López-Ibáñez, M., and Stützle, T., Towards an empirical analysis of SLS algorithms for multiobjective combinatorial optimization problems by experimental design, in *Proceedings of the International Conference Metaheuristics*, Vienna, Austria, 2005.
66. Fonseca, C.M., Grunert da Fonseca, V., and Paquete, L., Exploring the performance of stochastic multiobjective optimisers with the second-order attainment function, in *Evolutionary Multi-criterion Optimization*, Coello, C.C., Aguirre, A.H., and Zitzler, E., (Eds.), LNCS, Springer, Berlin, Heidelberg, Vol. 3410, 2005, p. 250.
67. Mote, J., Murthy, I., and Olson, D.L., A parametric approach to solving bicriterion shortest path problem, *Eur. J. Oper. Res.*, 53, 81, 1991.
68. Müller-Hannemann, M. and Weihe, K., Pareto shortest paths is often feasible in practice, in *International Workshop Algorithm Engineering*, LNCS, Springer, Heidelberg, Germany, Vol. 2141, 2001, p. 185.
69. Hansen, P., Bicriterion path problems, in *Multiple Criteria Decision Making Theory and Application*, Fandel, G. and Gal, T., (Eds.), LNEMS, Springer, Heidelberg, Germany, Vol. 177, 1979, p. 109.

25

Reoptimization of Hard Optimization Problems

Hans-Joachim
Böckenhauer

Juraj Hromkovič

Dennis Komm

25.1 Introduction

For many practically important optimization problems, no efficient algorithms for solving them exactly or even in an approximate way are known, and, relying on the standard complexity-theoretic assumptions such as $P \neq NP$, it is likely that this situation will never change. In a classical optimization problem, a set of instances is given and an algorithm is considered successful if it computes an exact solution or a solution with a pre-specified approximation guarantee for each of these instances without any additional prior knowledge. This viewpoint might be overly pessimistic as, in many situations, some additional knowledge is available that is not incorporated into the modeled optimization problem. A common thread of modern algorithmics is to identify such additional information and to make good use of it.

Often, one has to solve multiple instances of one optimization problem that might be somehow related. Consider the example of a timetable for some railway network. Assume that we have spent a lot of effort and resources to compute an optimal or near-optimal timetable satisfying all given requirements. Now, a small local change occurs, like, for example, the closing of a station due to construction work. This leads to a new instance of our timetable problem that is closely related to the old one. Such a situation naturally raises the question whether it is necessary to compute a new solution from scratch or whether the known old solution can be of any help. The framework of *reoptimization* tries to address this question: We are given an optimal or nearly optimal solution to some instance of a hard optimization problem, then a small local change is applied to the instance, and we ask whether we can use the knowledge of the old solution to facilitate computing a reasonable solution for the locally modified instance. It turns out that, for different problems and different kinds of local modifications, the answer to this question might be completely different. Generally speaking, we should not expect that solving the problem on the modified instance optimally can be done in polynomial time, but, in some cases, the approximability might improve a lot.

This notion of reoptimization was mentioned for the first time by Schäffter [1] in the context of a scheduling problem. Archetti et al. [2] used it for designing an approximation algorithm for the metric traveling salesman problem (ΔTSP) with an improved running time, but still the same approximation ratio as for the original problem. But the real power of the reoptimization concept lies in its potential to improve the approximation ratio compared to the original problem. This was observed for the first time by Böckenhauer et al. [3] for the ΔTSP, considering the change of one edge weight as a local modification. Independently at the same time, Ausiello et al. [4] proved similar results for TSP reoptimization under the local modification of adding or removing vertices.

In this chapter, we give an overview of known reoptimization results and some concrete examples illustrating some main proof techniques. There are different approaches in the literature for using the additional knowledge provided by related instances. A lot of work has been done on designing *dynamic data structures*. One example is to maintain a spanning tree of a graph under a sequence of insertions and deletions of vertices. Although the problem of computing a minimum spanning tree is solvable in polynomial time, a properly chosen data structure can speed up the frequent recomputations at least in an amortized way. An excellent survey of such dynamic data structures is given by Boria and Paschos [5]. Here, we focus on the reoptimization of NP-hard optimization problems.

A similar model, which we do not discuss in detail in this chapter, is the so-called *postoptimality analysis* or *sensitivity analysis* that asks a kind of a dual question: Given an instance together with an optimal solution, find the set of all instances for which this solution remains optimal. For an introduction to this concept, we refer to, for example, the papers by Libura [6], Sotskov et al. [7], Greenberg [8], or Gordeev [9].

Moreover, the term *reoptimization* is often used in the context of local-search algorithms and other heuristic approaches; as examples, see the papers by Thiongane et al. [10] or by Hiller et al. [11]. We do not discuss this approach any further in this chapter, but focus on algorithms with provable approximation guarantees.

We assume that the reader is familiar with the concepts of optimization problems and approximation algorithms as introduced in the first chapters of this book. We just briefly fix the notation we will use throughout this chapter.

Definition 25.1 (Optimization Problem) *An* optimization problem Π *is a 4-tuple* $(\mathcal{I}, \mathcal{M}, \text{measure}, \text{goal})$ *where* $\mathcal{I}$ *is the set of instances,* $\mathcal{M}$ *is a function assigning to each instance* $x \in \mathcal{I}$ *the set* $\mathcal{M}(x)$ *of feasible solutions for x, measure is a function assigning a positive real number* $\text{measure}(x, y)$ *to every instance* $x \in \mathcal{I}$ *and every feasible solution* $y \in \mathcal{M}(x)$, *and* goal $\in \{\min, \max\}$ *denotes the objective of* Π. *If* goal $= \min$, *we call* Π *a* minimization problem *and write* cost *instead of* measure; *likewise, if* goal $= \max$, Π *is called a* maximization problem *and* gain *is used instead of* measure.

Whenever x is clear from the context, we simply write measure(y) instead of measure(x, y). We consider a special class of optimization problems, namely the class NPO. Considering NP-hard NPO-problems, we observe that the hardness is a result of finding the optimal solution, and not of, for example, computing some feasible solution or its cost or gain, respectively. This truly assesses what we think of as the challenge in this case [12]. Unless P $=$ NP, if we want to obtain polynomial running time, we cannot generally guarantee the optimality of the solution computed. Our goal is, however, to bound how much more expensive or less profitable our solution is in the worst case. This is the idea behind efficient approximation algorithms.

Definition 25.2 (Approximation Algorithm) *Let* $\Pi = (\mathcal{I}, \mathcal{M}, \text{measure}, \text{goal})$ *be an optimization problem and let r be any positive (possibly constant) real-valued function of the input size. For every* $x \in \mathcal{I}$, *let* $y_{\text{opt}} \in \mathcal{M}(x)$ *denote an optimal solution for x, that is,*

$$\text{measure}(x, y_{\text{opt}}) = \text{goal}\{\text{measure}(x, y) \mid y \in \mathcal{M}(x)\}.$$

An algorithm ALG *is called* feasible *for* Π *if it computes an output* $\text{ALG}(x) \in \mathcal{M}(x)$ *for each instance* $x \in \mathcal{I}$. *A feasible algorithm* ALG *is called an r-approximation algorithm for* Π *if, for every* $x \in \mathcal{I}$,

$$\max \left\{ \frac{\text{measure}(x, \text{ALG}(x))}{\text{measure}(x, y_{\text{opt}})}, \frac{\text{measure}(x, y_{\text{opt}})}{\text{measure}(x, \text{ALG}(x))} \right\} \leq r(x).$$

NP-hard optimization problems can be classified according to what kind of approximation algorithm they admit [12,13]. In this chapter, we use the classical complexity classes for classifying optimization problems according to their approximability. We say that a problem is in APX if it admits a constant-ratio approximation in polynomial time. If it admits a $(1 + \varepsilon)$-approximation for any $\varepsilon > 0$, we distinguish approximation algorithms according to their running time: If the running time is polynomial in the input size, but possibly exponential in ε^{-1}, we call such an algorithm a *polynomial-time approximation scheme* or PTAS. If the running time is polynomial in both the input size and ε^{-1}, we speak of a *fully polynomial-time approximation scheme* or FPTAS.

25.2 Definition of Reoptimization Problems

To formalize our reoptimization setting, we first need the notion of a *local modification*. Let Π be an optimization problem with a set $\mathcal{I}$ of instances. The function $\text{lm} : \mathcal{I} \to 2^{\mathcal{I}}$ determines, for an instance x of Π, the set of all instances we consider as locally modified versions of x. The way lm is defined in detail depends on the given problem. For instance, if the optimization problem is defined on graphs, the local modification may remove or add a vertex, change the weight of a single edge, and so on. A more general definition is difficult to give and thus omitted; a possible approach is followed by Zych [14], who uses the notion of "atoms" [15]. Formally, a reoptimization problem is an optimization problem where the input contains additional information, namely a locally modified instance together with one of its optimal solutions.

Definition 25.3 (Reoptimization Problem) *Let* $\Pi = (\mathcal{I}, \mathcal{M}, \text{measure}, \text{goal})$ *be an optimization problem. For a local modification* lm, lm-Π *is called a* reoptimization version of Π *under* lm. *An instance of* lm-Π *is a triple* $x = (x_{\text{new}}, x_{\text{old}}, y_{\text{old}})$ *where* $x_{\text{old}}, x_{\text{new}} \in \mathcal{I}$ *are instances of* Π *with* $x_{\text{new}} \in \text{lm}(x_{\text{old}})$ *and* y_{old} *is an optimal solution for* x_{old}.

As an example, consider ΔTSP. An instance is a complete metric undirected weighted graph G, and the set of feasible solutions for an instance contains all Hamiltonian cycles in G; the goal is to find such a cycle of minimum weight. There are different local modifications that can be considered; as mentioned earlier, one is the increase of the weight of single edges. For this problem, we define a local modification IncEdge that maps a graph G to some graph G' such that the following holds: $G = (V, E, w)$, with $w : E \to \mathbb{R}^+$, and $G' = (V', E', w')$, whereas $w' : E' \to \mathbb{R}^+$, are complete metric undirected weighted graphs (i.e., instances of ΔTSP) such that $V = V'$, $E = E'$, and there is one edge $e \in E$ with $w'(e) > w(e)$, whereas $w'(\hat{e}) = w(\hat{e})$, for all $\hat{e} \in E \setminus \{e\}$. IncEdge-$\Delta\text{TSP} = (\mathcal{I}, \mathcal{M}, \text{measure}, \text{goal})$ denotes the corresponding reoptimization problem, where every instance $x \in \mathcal{I}$ is a triple $x = (G_{\text{new}}, G_{\text{old}}, H_{\text{old}})$ with $G_{\text{new}} \in \text{IncEdge}(G_{\text{old}})$ and $H_{\text{old}} \in \mathcal{M}(G_{\text{old}})$ is an optimal solution (i.e., a Hamiltonian cycle of minimum weight) for G_{old}.

25.3 NP-Hardness Results

Intuitively, the additional information that is given in a reoptimization setup seems to be rather powerful. Intriguingly, many reoptimization variants of NP-hard optimization problems are also NP-hard. A general approach towards proving the NP-hardness of reoptimization problems uses a sequence of reductions and can on a high level be described as follows [16]. Consider an NP-hard optimization problem Π, a local modification lm, and a resulting reoptimization problem lm-Π. Moreover, suppose we are able to transform an efficiently solvable instance x' of Π to any instance x of Π in a polynomial number of local modifications of type lm. Then, any efficient algorithm for lm-Π can be used to efficiently solve Π.

To illustrate this approach by an example, consider the shortest common superstring problem (SCSP). As an input, we are given a set of strings over some alphabet. The objective is to find a shortest string that contains all of them as a substring; for instance, $\{aab, abbbb, bbbbb, baa\}$ has the optimal solution $baabbbbb$.

Definition 25.4 (Shortest Common Superstring Problem, SCSP) *Given a substring-free set of strings* $S = \{s_1, \ldots, s_m\}$ *over an alphabet* Σ, *the feasible solutions are all strings* t *with* $|t| \leq |s_1| + \cdots + |s_m|$ *over* Σ *that contain all* s_i *with* $1 \leq i \leq m$ *as substring;* t *is called a* superstring of S. *The cost of* t *is its length* $|t|$, *and the goal is to find a superstring* t *of minimum length.*

The SCSP is known to be NP-hard [17]. Now, consider the local modification where one string is added to the input set. The corresponding reoptimization problem is denoted by AddString-SCSP. Analogously, DelString-SCSP denotes the reoptimization variant where the local modification consists in deleting a single string from the input set. We employ the above-mentioned strategy to show that these two problems are NP-hard. The following proof was first given by Bilò et al. [18].

Theorem 25.1 (Bilò et al. [18]) *The AddString-SCSP and DelString-SCSP are NP-hard.*

Proof. We first consider the AddString-SCSP. Assume for a contradiction that AddString-SCSP is solvable in polynomial time. Let S be any instance of the SCSP with strings $s_1, \ldots, s_m$ over an alphabet Σ. We construct a very simple instance S' of SCSP by setting $S' = \{s_1\}$. Then $t = s_1$ is obviously an optimal solution for S' and both S' and t can be computed from S in polynomial time. The idea is now to add all strings s_i with $2 \leq i \leq m$ from S in consecutive steps. First, we obtain an instance (S_1, S_0, t_0) of the AddString-SCSP with $t_0 = t$, $S_0 = S'$ and $S_1 = S_0 \cup \{s_2\}$. By assumption, there is an efficient algorithm ALG for the AddString-SCSP. Using ALG, we get an optimal solution t_1 for the above-mentioned instance. Next, we use ALG on (S_2, S_1, t_1) with $S_2 = S_1 \cup \{s_3\}$. This is iterated until we finally (after the $(m-1)$th involvement of ALG) obtain an optimal solution t_{m-1} for the instance $(S_{m-1}, S_{m-2}, t_{m-2})$ with $S_{m-1} = S$. This contradicts the NP-hardness of the SCSP.

Now we turn to DelString-SCSP. Again, we assume for a contradiction that DelString-SCSP is solvable in polynomial time and consider any instance S of the SCSP with strings $s_1, \ldots, s_m$ over an alphabet Σ with $|\Sigma| = \sigma$. As all strings are substring-free, we can assume $|s_i| > 1$ for all i with $1 \leq i \leq m$. We now again construct an instance S' of the SCSP. To this end, for every string s_i with $1 \leq i \leq m$, we obtain a string s_i' by the following procedure. Let s_i'' denote the string that is obtained by removing the first letter of s_i. Consider σ different symbols $X_1, \ldots, X_\sigma \notin \Sigma$. Let s_i' denote the string that is obtained by concatenating s_i'' and some X_j with $1 \leq j \leq \sigma$ such that s_i' is different from all previously constructed s_l' with $l < i$. This is always possible as S is substring-free and removing the first letter can make at most σ many substrings s_i'' identical.

The new instance S' of the SCSP contains the $2m$ strings $s_1, \ldots, s_m, s_1', \ldots, s_m'$. Note that this set is also substring-free. Let t_i' denote the string that is obtained from maximally intersecting s_i with s_i' in this order; we have $|t_i'| = |s_i| + 1 = |s_i'| + 1$. An optimal solution t for S' is obtained by concatenating all t_i' in increasing order of their indices, which leads to a total length

$$|t| = m + \sum_{i=1}^{m} |s_i'|.$$

Obviously, also in this case, both t and S' can be constructed from S in polynomial time. The idea is now to remove all strings s_i' with $1 \le i \le m$ from S' in consecutive steps. First, we obtain an instance (S_1, S_0, t_0) of the DelString-SCSP with $t_0 = t$, $S_0 = S'$, and $S_1 = S_0 \setminus \{s_1'\}$. Using an efficient algorithm ALG for the DelString-SCSP, which exists by assumption, we compute an optimal solution t_1 for the above-mentioned instance. Iterating this procedure, we finally (after the mth involvement of ALG) obtain an optimal solution t_m for the instance (S_m, S_{m-1}, t_{m-1}) with $S_m = S$, which contradicts the NP-hardness of the SCSP. ∎

As mentioned earlier, constructions that follow the same idea can be used to prove the NP-hardness of many other reoptimization problems. For some problems, however, this technique cannot be applied as easily. As an example, consider a generalization of the $\triangle$TSP where some of the vertices of the input graph are assigned real numbers; these numbers are called *deadlines* and the corresponding vertices are called *deadline vertices*. Furthermore, there is a dedicated vertex s, which is called the *start vertex*. A Hamiltonian cycle H in such a graph is a feasible solution if and only if there is a traversal of H starting at s such that, whenever we arrive at a deadline vertex, the total distance traveled so far is at most as large as the deadline of this vertex; in this case, we say that "H satisfies the deadlines."

Definition 25.5 ($\triangle$TSP with Deadlines, DL$\triangle$TSP) *Given a complete metric undirected weighted graph $G = (V, E, w)$ and a deadline triple (s, D, d), where $s \in V$ is the start vertex, $D \subseteq V$ is a set of deadline vertices, and $d : D \to \mathbb{R}^+$ is a deadline function, the feasible solutions are all Hamiltonian cycles in G that satisfy the deadlines. The goal is to compute such a Hamiltonian cycle of minimum total edge weight.*

For the DL$\triangle$TSP, there are a few additional reoptimization variants possible (compared to the $\triangle$TSP). Vertices can be added to or removed from the graph; these vertices may or may not have deadlines. Moreover, a deadline may be removed from or added to a vertex. Let us consider the addition of a deadline to a vertex and denote this reoptimization variant by AddD-DL$\triangle$TSP. Here, an approach as aforementioned does not seem promising to prove NP-hardness. To do this, we would have to modify the given graph $G = (V, E)$ with deadline vertices D and deadlines d to an efficiently solvable instance G' that, using a polynomial number of local modifications, can become the original instance. As the local modifications are essentially successively adding deadlines to V, G' has to contain fewer deadlines than G. However, as the DL$\triangle$TSP is NP-hard even if there are no deadlines at all (i.e., if we deal with the usual $\triangle$TSP), we are stuck. Therefore, we use another approach, namely a reduction from the following decision problem.

Definition 25.6 (Restricted Hamiltonian Path Problem, RHPP) *Given an undirected graph $G = (V, E)$ with two distinguished vertices $s', t' \in V$ and a Hamiltonian path P' from s' to t' in G, the question is whether there exists some Hamiltonian path in G from s' to some vertex $t'' \in V$ with $t'' \ne t'$.*

The RHPP is known to be NP-hard [12]. We now show that efficiently solving the AddD-DL$\triangle$TSP allows for an efficient algorithm for the RHPP. The following theorem even holds if the number of deadlines is a small constant.

Theorem 25.2 (Böckenhauer and Komm [19]) *The AddD-DL$\triangle$TSP is NP-hard.*

Proof. Let (G, P') with $G = (V, E)$ and $|V| = n$ be any instance of the RHPP; the Hamiltonian path P' starts at s' and ends at t'. We construct an instance of AddD-DL$\triangle$TSP as follows. First, we construct a complete metric undirected weighted graph K_G from G. To this end, all edges from E are assigned weight 1 and all newly added edges get a weight of 2 each. After that, K_G is extended to a graph G_{old} by adding

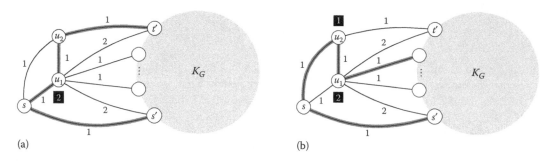

FIGURE 25.1 The two instances of the DLΔTSP as used in the proof of Theorem 25.2. (a) The old instance and H_{old}, which uses P'; (b) the new instance and H_{new}, which uses P if it exists.

three vertices s, u_1, and u_2 as shown in Figure 25.1. Here, the vertex u_1 is connected to s' and t' by edges of weight 2 and to all other vertices of K_G by edges of weight 1 each. All edges not shown have weight 2 each. The resulting graph thus only contains edges with weights either 1 or 2 and accordingly satisfies the triangle inequality. Moreover, we set $D_{\text{old}} = \{u_1\}$ and $d_{\text{old}}(u_1) = 2$. As G contains a Hamiltonian path P' from s' to t', K_G can be traversed according to P' backwards, that is, from t' to s', with total cost $n - 1$. Let Q' denote the corresponding path. It follows that $H_{\text{old}} = (s, u_1, u_2, Q', s)$ is an optimal solution of cost $n + 3$ for this instance; see Figure 25.1a. Indeed, it only uses edges of cost 1 each; thus, no other solution can be better. However, H_{old} is not necessarily the only optimal solution.

Now suppose, as a concrete local modification, we add a deadline of 1 to the vertex u_2. Hence, we obtain a new instance with $G_{\text{new}} = G_{\text{old}}$, $D_{\text{new}} = D_{\text{old}} \cup \{u_2\}$, $d_{\text{new}}(u_1) = d_{\text{old}}(u_1)$, and $d_{\text{new}}(u_2) = 1$. Observe that any feasible solution now has to start with the three vertices s, u_2, and u_1 in that order. Consequently, H_{old} becomes infeasible. Now suppose there is a second Hamiltonian path in G from s' to some vertex $t'' \neq t'$. Then K_G can be traversed by a path Q from t'' to s' with a total cost of $n - 1$. As a result, there is an optimal solution $H_{\text{new}} = (s, u_2, u_1, Q, s)$ of G_{new} of cost $n + 3$; see Figure 25.1b. Every other feasible solution, not using a Hamiltonian path from some vertex t'' to s' in G, has to use at least one edge of weight 2 and thus has a cost of at least $n + 4$. Hence, if G does not contain a Hamiltonian path from s' to some vertex t'', then there is no solution with a cost of less than $n + 4$.

As the instance $((G_{\text{new}}, (s, D_{\text{new}}, d_{\text{new}})), (G_{\text{old}}, (s, D_{\text{old}}, d_{\text{old}})), H_{\text{old}})$ can clearly be constructed from (G, P') in polynomial time, any efficient algorithm for AddD-DLΔTSP can be used to efficiently decide the given instance of the RHPP, contradicting the NP-hardness of the problem. ∎

Similar reductions can be used for other reoptimization variants of the problem, even to show APX-hardness [19,20]. In Subsection 25.4.3, we will show a linear lower bound on the approximation ratio of any algorithm for the problem if the number of deadline vertices is allowed to depend on the number of vertices.

25.4 Approximation Algorithms

Although reoptimization does not seem substantially easier with respect to NP-hardness than computing from scratch for most problems, the game changes when considering the approximation ratio that is achievable in polynomial time.

25.4.1 Improved Constant-Factor Approximations

For many reoptimization problems, algorithms with a constant approximation ratio have been designed, improving over the approximations reachable for computing without additional knowledge. A general strategy for the design of an approximation algorithm for a reoptimization problem is to output the better of two specific solutions for the new instance: One of the considered solutions is the given optimal solution for the old instance, if necessary slightly adapted to be feasible for the new instance. The other one is either computed completely from scratch, or the knowledge about the local modification is used to guess (by some brute-force search) a local part of the solution and to compute the rest from scratch. This strategy was applied for many reoptimization problems. It was for the first time described in some generality by Bilò et al. [21]. A very detailed discussion of the preconditions of a problem that make this approach applicable was given by Zych [14], based on the theory of self-reducibility.

A very nice and somewhat general application of this generic algorithm was proposed by Ausiello et al. [22], who considered *hereditary subgraph problems*. A *graph property* is called hereditary if, for any graph having this property, also all induced subgraphs have this property. We can now define a class of maximization problems in graphs that are based on hereditary properties.

Definition 25.7 (Hereditary Weighted Graph Problem) *Let $G = (V, E, w)$ be a graph with a vertex weight function $w : V \rightarrow \mathbb{N}^+$. We call a maximization problem hereditary if it consists in finding a subset $S \subseteq V$ of maximum total weight, that is, maximizing $\sum_{v \in S} w(v)$, such that the subgraph $G[S]$ of G induced by S satisfies some given hereditary property. We denote the class of all hereditary graph maximization problems by $\mathcal{H}$.*

Examples of hereditary weighted problems include finding a maximum (weighted) independent set, a maximum induced planar subgraph, or a maximum clique. For all hereditary problems, a very simple reoptimization algorithm yields a 2-approximation for the local modification of vertex addition, AddVert for short.

Theorem 25.3 (Ausiello et al. [22]) *Let $\Pi \in \mathcal{H}$ be any hereditary weighted graph problem. Then there is a polynomial-time 2-approximation algorithm for AddVert-Π.*

Proof. Let $(G_{new}, G_{old}, S_{old})$ be an instance of AddVert-Π, where $G_{new} = (V_{new}, E_{new}, w_{new})$, $G_{old} = (V_{old}, E_{old}, w_{old})$ such that $V_{new} = V_{old} \cup \{v_{new}\}$ and w_{old} and w_{new} coincide on all vertices of G_{old}. Consider the algorithm ALG that outputs the best of two solutions, namely the old optimal solution $S_1 = S_{old}$ or the solution $S_2 = \{v_{new}\}$ consisting of the new vertex only.

As Π is a hereditary problem, both solutions are feasible for the new instance. Let S_{new} be an optimal solution for the new instance. Because of the hereditary property, the weight of the sub-solution of S_{new} induced by V_{old} cannot be larger than the weight of S_{old}. Thus,

$$w_{new}(S_1) \geq w_{new}(S_{new}) - w_{new}(S_2).$$

Consequently, one of the solutions S_1 and S_2 has to have a weight of at least $w_{new}(S_{new})/2$. ∎

Of course, this very simple algorithm can be improved for many hereditary graph problems. For some problems, it is possible to improve the solution S_2 by extending it to a larger feasible solution by adding some vertices according to some problem-dependent strategy. We illustrate this approach with a reoptimization algorithm for the weighted maximum planar subgraph problem, MAXPSP for short. This is the problem of finding a maximum-weight subset $S \subseteq V$ of vertices such that the graph induced by S is planar. It is known that the MAXPSP is not approximable within a ratio of n^ε, for any $\varepsilon > 0$, where n is the size of the graph [23]. We consider the reoptimization variant AddVert-MAXPSP and present an improvement over the result from Theorem 25.3 by Boria et al. [24]. Observe that already the result from Theorem 25.3 constitutes a jump from an almost inapproximable problem to a constant approximation, that is, that the difference in hardness between the original problem and the reoptimization variant is huge.

Theorem 25.4 (Boria et al. [24]) *There is a polynomial-time 3/2-approximation algorithm for AddVert-MAXPSP.*

Proof sketch. Let $(G_{new}, G_{old}, S_{old})$ be an instance of the AddVert-MaxPSP, where $G_{new} = (V_{new}, E_{new}, w_{new})$, $G_{old} = (V_{old}, E_{old}, w_{old})$ such that $V_{new} = V_{old} \cup \{v_{new}\}$ and w_{old} and w_{new} coincide on all vertices of G_{old}. As w_{new} just extends w_{old} to all edges incident to v_{new}, we use the notation $w = w_{new}$ in the following. It is known that S_{old} can be partitioned in polynomial time into two vertex sets X_1 and X_2 that both induce an outerplanar graph, see Boria et al. [24] for details. Let, without loss of generality, $w(X_1) \geq w(X_2)$. Then, our algorithm chooses the best of the two solutions $S_1 = S_{old}$ and $S_2 = X_1 \cup \{v_{new}\}$. It can be easily shown that the union of an outerplanar graph and a single vertex is planar; thus, both S_1 and S_2 are feasible solutions for G_{new}.

Let S_{new} be an optimal solution for G_{new}. We now bound the weights of S_1 and S_2. Because of the hereditary property, the weight of S_{new} can exceed the weight of S_{old} only by the weight of v_{new}. Hence, we have

$$w(S_1) = w(S_{old}) \geq w(S_{new}) - w(v_{new}) \tag{25.1}$$

and

$$w(S_2) \geq \frac{w(S_{old})}{2} + w(v_{new}) \geq \frac{w(S_{new}) - w(v_{new})}{2} + w(v_{new}) = \frac{1}{2} \cdot w(S_{new}) + \frac{1}{2} \cdot w(v_{new}). \tag{25.2}$$

Adding (25.1) and two times (25.2) yields

$$3 \cdot \max\{w(S_1), w(S_2)\} \geq 2 \cdot w(S_{new}),$$

which immediately implies the claim. ∎

The generic approach for hereditary problems can also be adapted for the somewhat dual situation of finding a minimum-weight subset of vertices satisfying any property where, for each set $S \subseteq V$ satisfying P, also all sets S' with $S \subseteq S' \subseteq V$ satisfy P. One prominent example of such a problem is the problem of finding a minimum-weight vertex cover, where a *vertex cover* is a subset $C \subseteq V$ of vertices such that each edge of the graph is incident to at least one vertex from C. The minimum weighted vertex cover problem, MinVCP for short, is known to be NP-hard and APX-hard [25]. The best known approximation ratio for it is 2 [26]. In the following, we present a reoptimization algorithm for the AddVert-MinVCP that uses a slightly generalized variant of the above-mentioned generic strategy and yields a 3/2-approximation.

Theorem 25.5 (Ausiello et al. [22]) *There is a polynomial-time 3/2-approximation algorithm for the AddVert-MinVCP.*

Proof. Let $(G_{new}, G_{old}, S_{old})$ be an instance of the AddVert-MinVCP, where $G_{new} = (V_{new}, E_{new}, w_{new})$ and $G_{old} = (V_{old}, E_{old}, w_{old})$ such that $V_{new} = V_{old} \cup \{v_{new}\}$ and w_{old} and w_{new} coincide on all vertices from G_{old}. The algorithm considers three feasible solutions for the new instance and chooses the best of them. The first solution is $S_1 = S_{old} \cup \{v_{new}\}$, the second solution is $S_2 = S_{old} \cup N(v_{new})$, where $N(v_{new})$ is the set of neighboring vertices of v_{new} in G_{new}, and the third solution S_3 is obtained by removing v_{new} and $N(v_{new})$ from the graph G_{new}, running the 2-approximation algorithm for MinVCP from Bar-Yehuda and Even [26] on the remainder of the graph, and adding $N(v_{new})$ to the obtained solution.

Let S_{new} be an optimal solution for the new instance. For the analysis, we distinguish two cases. If S_{new} contains v_{new}, then S_1 is also an optimal solution because of the analog of the hereditary property. Thus, in the following, we consider the case that $v_{new} \notin S_{new}$. We want to show that either S_2 or S_3 is a 3/2-approximation. We observe that

$$w(S_2) \leq w(S_{old}) + w(N(v_{new})) \leq w(S_{new}) + w(N(v_{new})), \tag{25.3}$$

as the weight of the new optimal solution cannot be smaller than the one of the old optimum.

As, according to our assumption in this case, $N(v_{new}) \subseteq S_{new}$, and using the analog of the hereditary property, we can bound the weight of S_3 by

$$w(S_3) \leq 2 \cdot (w(S_{new}) - w(N(v_{new}))) + w(N(v_{new})) = 2 \cdot w(S_{new}) - w(N(v_{new})). \quad (25.4)$$

Thus, combining (25.3) and (25.4), we get

$$\min\{w(S_2), w(S_3)\} \leq \frac{1}{2} \cdot (w(S_2) + w(S_3)) \leq \frac{3}{2} \cdot w(S_{new}),$$

which proves the claim. ∎

Choosing the best of two solutions, where one resembles the given old optimal solution and the other one tries to guess a large portion of an optimal solution and approximates the remainder, is a successful strategy that cannot only be applied to graph problems. As a further example, where one can even apply this idea iteratively, we consider the SCSP.

We already mentioned that the SCSP is NP-hard; as a matter of fact, it is even known to be APX-hard [27]. Karpinski and Schied [28] showed that it cannot be approximated with a factor better than 333/332. The currently best-known algorithm is by Mucha [29] and it achieves an approximation ratio of 57/23. Let us revisit the reoptimization problem AddString-SCSP, which we proved to be NP-hard in Theorem 25.1. However, it is easy to see that a 2-approximation easily follows for this problem.

For this, and also for the subsequent algorithms, we need the notion of a *merge* of two strings. If we have two strings $s = uv$ and $t = vx$, where the suffix v of s is equal to the prefix v of t, the string uvx is called a merge of s and t. A *maximal merge* of s and t is a merge such that the length of the common part is maximized. For example, $aabababb$ is a merge of $s = aabab$ and $t = ababb$, but the maximal merge of these two strings is $aababb$. For two strings s'_1 and s'_2, let $merge(s'_1, s'_2)$ denote the maximal merge of s'_1 and s'_2. For any $l \in \mathbb{N}^+$ and strings $s'_1, \ldots, s'_l$, we define the string $merge(s'_1, \ldots, s'_l) = merge(merge(s'_1, \ldots, s'_{l-1}), s'_l)$ inductively. Note that the merge of two or more strings depends on their order.

Now consider any instance $(S_{new}, S_{old}, t_{old})$ of the AddString-SCSP with $S_{new} = S_{old} \cup \{s_{new}\}$. As the local modification consists of adding the string s_{new} to the input set, it follows for an optimal solution t_{new} of S_{new} that $|t_{new}| \geq |t_{old}|$; moreover, clearly we have $|t_{new}| \geq |s_{new}|$. Now consider the trivial algorithm that simply outputs $merge(t_{old}, s_{new})$ as a solution for S_{new}. Easily, this solution has a length of at most $|t_{old}| + |s_{new}| \leq 2 \cdot |t_{new}|$.

Bilò et al. [18] improved this algorithm, where the concrete bound on the obtained approximation ratio depends on the best approximation algorithm for the original problem. As already mentioned earlier, the improved algorithm is based on choosing the better of two solutions, where one solution is just the given old solution merged with the new string. The other solution is computed without using the given optimum for the old instance in a way we will describe later. Intuitively, this computation is based on guessing a large part of an optimal solution and approximating the remainder using a known approximation algorithm. The large part of the optimal solution is guessed in the following way: Consider the largest string in the input and guess the leftmost and rightmost string that overlap with it in the optimal solution, as well as the lengths of these overlaps. This is a successful strategy in the case that there is a long string in the input and, thus, the guessed part is relatively large in comparison with the whole optimal solution. In the other case, we will prove that the first solution is not too bad.

As the computation of the second solution does not depend on the old instance or its optimum, we will first describe it as an approximation strategy for the classical nonreoptimization SCSP. In what follows, let $S = \{s_1, \ldots, s_m\}$ be some instance of the SCSP with a fixed optimal solution t_{opt}; let n denote the total length of all strings in S, that is, $n = \sum_{i=1}^{m} |s_i|$. Suppose there is an efficient r-approximation algorithm SCSBEST for the SCSP. Let w_0 be a string in S such that $|w_0| = \alpha_0 \cdot |t_{opt}|$, for some α_0; we can think of w_0 being the longest string in S, but this is not essential for our argument. Let l_1 be the leftmost string and let r_1 be the rightmost string that overlap with w_0 in t_{opt}; if l_1 does not exist, we set $l_1 = w_0$, and if r_1 does not exist, we set $r_1 = w_0$.

For any three strings l, w, r and any numbers $x \leq \min\{|l|, |w|\}$ and $y \leq \min\{|w|, |r|\}$, we define the merge $\text{merge}_{x,y}(l, w, r)$ as the merge of l, w, and r such that l and w overlap in exactly x symbols and w and r overlap in exactly y symbols. Obviously, $\text{merge}_{x,y}(l, w, r)$ does not necessarily exist for all values of x and y if the strings l, w, and r do not match accordingly. Moreover, if l, w, and r appear in this order as substrings in t_{opt}, we denote by $\text{optmerge}(l, w, r)$ the merge of these three strings with exactly the overlaps as in t_{opt}.

Now consider the following algorithm GUESS that guesses (i.e., tries by a brute-force approach) l_1 and r_1 together with $\text{optmerge}(l_1, w_0, r_1)$. More precisely, GUESS tries all possibilities of picking l and r from S and, for each such pair (l, r), all values for x and y such that $\text{merge}_{x,y}(l, w_0, r)$ is well defined. In each of these tries, GUESS removes all strings from S that are contained as substrings in $\text{merge}_{x,y}(l, w_0, r)$ and runs SCSBEST on the resulting instance $S_1(l, r, x, y)$ yielding a string $u_1(l, r, x, y)$. The computed solution in this try is then a maximal merge of $u_1(l, r, x, y)$ and $\text{merge}_{x,y}(l, w_0, r)$, that is, $\text{merge}(u_1(l, r, x, y), \text{merge}_{x,y}(l, w_0, r))$. As all possibilities for l and r and all possible overlaps are tried, also l_1, r_1, and $\text{optmerge}(l_1, w_0, r_1)$ were tried. Thus, the minimal solution computed in one of the tries is at least as good as the solution corresponding to $\text{optmerge}(l_1, w_0, r_1)$, for which we will later be able to bound the approximation ratio.

Next, for any $k \in \mathbb{N}^+$, we extend GUESS to an algorithm GUESS_k by iterating the above-mentioned idea k times. In the second iteration, we would like $w_1 = \text{optmerge}(l_1, w_0, r_1)$ to play the role of w_0 and to guess the two outermost strings l_2 and r_2 overlapping w_1 in t_{opt} together with their overlaps with w_1. Unfortunately, in its first iteration, our algorithm cannot recognize which of its tries resulted in the merge corresponding to t_{opt}. This means that we have to take any of the candidates from the first iteration into account as possible middle strings for the second iteration. Only this way we can be sure that one of our tries really starts with w_1 and extends it to $w_2 = \text{optmerge}(l_2, w_1, r_2)$.

In general, the string w_0 is defined as earlier. For every i with $1 \leq i \leq k$, let l_i (r_i, respectively) be the leftmost (rightmost, respectively) string that overlaps with w_{i-1} in t_{opt}. Again, if l_i (r_i, respectively) does not exist, we set $l_i = w_{i-1}$ ($r_i = w_{i-1}$, respectively) as earlier. Furthermore, let $w_i = \text{optmerge}(l_i, w_{i-1}, r_i)$. GUESS_k tries all combinations of the strings in S as candidates for $l_1, \ldots, l_k$ and $r_1, \ldots, r_k$ together with all possible overlaps. This way, throughout the k iterations, among many other feasible solutions, k candidate solutions $t_1, \ldots, t_k$ are computed such that t_i with $1 \leq i \leq k$ is equal to $\text{merge}(u_i, w_i)$, where u_i is the solution returned by SCSBEST on the instance S_i that is obtained by removing all substrings of w_i from S_{i-1} (where we define $S_0 = S$). The output of GUESS_k is the best solution it encounters in its many tries, which is certainly not worse than one such candidate solution. We start by bounding the approximation ratio of GUESS_k.

Lemma 25.1 (Bilò et al. [18]) GUESS_k *computes a solution of length at most*

$$\left(1 + \frac{r^k(r-1)}{r^k - 1}(1 - \alpha_0)\right) \cdot |t_{\text{opt}}|.$$

Proof. Let r' with $0 \leq r' \leq 1$ be such that GUESS_k computes a solution with a length greater than $(1 + r')|t_{\text{opt}}|$. In what follows, we bound r' from above. As the algorithm considers all possible assignments for $l_1, \ldots, l_k$ and $r_1, \ldots r_k$ and all possible overlaps (and thus $w_0, \ldots, w_{k-1}$), we focus only on the "correct" such assignment (i.e., the one corresponding to t_{opt}). For this assignment, k candidate solutions $t_1, \ldots, t_k$ are computed, and thus $|t_i| \geq (1 + r')|t_{\text{opt}}|$, for every i with $1 \leq i \leq k$.

Recall that t_i is defined as $\text{merge}(u_i, w_i)$, and thus $|t_i| \leq |u_i| + |w_i|$. Let α_i with $1 \leq i \leq k$ be such that $|w_i| = \alpha_i |t_{\text{opt}}|$; recall that $|w_0| = \alpha_0 |t_{\text{opt}}|$ as defined earlier. Note that, if we remove all substrings from w_i from t_{opt}, this implies that we remove all strings that overlap w_{i-1} in t_{opt}. The resulting string is thus a solution for S_i of length at most $|t_{\text{opt}}| - |w_{i-1}| = (1 - \alpha_{i-1})|t_{\text{opt}}|$. Hence, also an optimal solution for S_i

is bounded by $(1 - \alpha_{i-1})|t_{\mathrm{opt}}|$. The output u_i of the r-approximative algorithm SCSBEST on S_i therefore has a length of at most $r(1 - \alpha_{i-1})|t_{\mathrm{opt}}|$ by assumption. With this, we obtain

$$(1 + r')|t_{\mathrm{opt}}| < |t_i| \leq r(1 - \alpha_{i-1})|t_{\mathrm{opt}}| + |w_i| \leq (r(1 - \alpha_{i-1}) + |\alpha_i|) \cdot |t_{\mathrm{opt}}|,$$

for all i with $1 \leq i \leq k$, which is equivalent to

$$\alpha_i > 1 + r' - r + r\alpha_{i-1}. \tag{25.5}$$

This yields

$$\alpha_k > (1 + r + r')\frac{r^k - 1}{r - 1} + r^k \alpha_0 \tag{25.6}$$

as a solution for the system of recurrences given by (25.5). Note that w_k is by definition a substring of t_{opt}, and thus $\alpha_k \leq 1$, which, together with (25.6), implies

$$r' \leq \frac{r^k(r - 1)}{r^k - 1}(1 - \alpha_0)$$

as we claimed. ∎

Next, we analyze the running time of GUESS$_k$ depending on k, the input size, and the running time of the given algorithm SCSBEST.

Lemma 25.2 (Bilò et al. [18]) *Let n be the total length of all strings in S, let m be the number of strings in S, and let $T(m, n)$ be the running time of* SCSBEST. *Then, the running time of* GUESS$_k$ *is in $O(m^{2k}n^{2k}k(mn + T(m, n)))$.*

Proof. We first compute an upper bound on the number of considered candidates in each iteration i and denote it by Cand(i). In the first iteration, $O(m^2)$ pairs (l, r) of candidates for l_1 and r_1 have to be considered, together with their possible overlaps with w_0. The length of each of these overlaps can be bounded from above by $O(n)$. Thus, overall $O(m^2 n^2)$ potential solutions have to be considered, that is, Cand(1) $= m^2 n^2$. In each further iteration $i \geq 2$, the algorithm considers all solutions, more precisely, all merges merge$_{x,y}(l, w_{i-2}, r)$ from iteration $i - 1$ as possible middle strings, that is, as candidates for w_{i-1}. For each of these middle strings, again two candidates l_i and r_i have to be guessed, together with the appropriate overlap. Thus, Cand(i) $=$ Cand($i - 1$) $\cdot m^2 n^2 = m^{2i} n^{2i}$. Summing up over all iterations, GUESS$_k$ considers at most $\sum_{i=1}^{k}$ Cand(i) $\leq k \cdot m^{2k} n^{2k}$ solution candidates.

For each of these candidate solutions, we have to find the set of uncovered strings and to run SCSBEST on them. Testing all strings from S as to whether they are covered by merge$_{x,y}(l, w_{i-2}, r)$ can be done by some standard algorithm for exact string matching in time $O(mn)$; see, for example, the textbook by Gusfield [30] for the description of such algorithms that find a single pattern of length p in a text of length t in time $O(p + t)$. The time for running SCSBEST on a set of uncovered strings can be bounded by $T(m, n)$ by assumption. Thus, the overall running time of GUESS$_k$ is in $O(m^{2k}n^{2k}k \cdot (mn + T(m, n)))$ as we claimed. ∎

Observe that GUESS$_k$ runs in polynomial time for any fixed k. Finally, we can use GUESS$_k$ to give an improved approximation algorithm for the AddString-SCSP, which runs in polynomial time, and whose approximation ratio depends on the approximation ratio r of SCSBEST. For $k \in \mathbb{N}^+$, we consider an algorithm that, for a given instance $(S_{\mathrm{new}}, S_{\mathrm{old}}, t_{\mathrm{old}})$, outputs the better solution of that computed by GUESS$_k$ on S_{new}, which we call t_{guess}, and $t_{\mathrm{merge}} = \mathrm{merge}(t_{\mathrm{old}}, s_{\mathrm{new}})$, where $s_{\mathrm{new}} \in S_{\mathrm{new}} - S_{\mathrm{old}}$ is the newly added string.

Theorem 25.6 (Bilò et al. [18]) *Let $\varepsilon > 0$. There is a polynomial-time $((2r - 1)/r - \varepsilon)$-approximation algorithm for the AddString-SCSP.*

Proof. Let α be such that $|s_{\text{new}}| = \alpha|t_{\text{new}}|$ with t_{new} being an optimal solution for S_{new}. As $|t_{\text{old}}| \leq |t_{\text{new}}|$, it follows that $|t_{\text{merge}}| = |\text{merge}(t_{\text{old}}, s_{\text{new}})| \leq (1+\alpha)|t_{\text{new}}|$. Because of Lemma 25.1, GUESS_k computes t_{guess} with $|t_{\text{guess}}| \leq (1 + r^k(r - 1))/(r^k - 1)(1 - \alpha))|t_{\text{new}}|$. As $|t_{\text{merge}}|$ grows with α whereas $|t_{\text{guess}}|$ decreases with α, the minimum of $|t_{\text{guess}}|$ and $|t_{\text{merge}}|$ (and hence the output of the algorithm) is maximal if

$$(1 + \alpha)|t_{\text{new}}| = \left(1 + \frac{r^k(r - 1)}{r^k - 1}(1 - \alpha)\right)|t_{\text{new}}|,$$

which is the case if

$$\alpha = \frac{r^{k+1} - r^k}{r^{k+1} - 1}.$$

With this value for α, we obtain an approximation ratio of at most

$$\frac{2r^{k+1} - r^k - 1}{r^{k+1} - 1}. \tag{25.7}$$

For any $\varepsilon > 0$, (25.7) can be made smaller than $(2r - 1)/r - \varepsilon$ by choosing a sufficiently large k. ∎

Bilò et al. [18] additionally gave a similar result (also based on the algorithm GUESS_k) for the DelString-SCSP; here, a slightly worse approximation ratio of $(3r - 1)/(r + 1) - \varepsilon$ can be achieved, for any $\varepsilon > 0$. Using the algorithm from Mucha [29] as SCSBEST, the following results follow immediately.

Corollary 25.1 *There is a polynomial-time 1.596-approximation algorithm for the AddString-SCSP and a polynomial-time 1.850-approximation algorithm for the DelString-SCSP.*

Many other reoptimization problems are known that admit an improved constant-factor approximation; for an overview, see Section 25.5.

25.4.2 Polynomial-Time Approximation Schemes

In subsection 25.4.1, we have seen that reoptimization allows for lowering the constant approximation ratio for many hard optimization problems. Moreover, we have seen a first example where reoptimization can even help to cross the boundaries between different approximation complexity classes: for the MaxPSP, which is not approximable within a constant, we saw a constant-factor approximation for a reoptimization variant. In this subsection, we want to give more examples where reoptimization is significantly easier than the original problem, also with respect to the standard complexity classes. More precisely, we present some examples of APX-hard problems whose reoptimization variants admit a PTAS.

We have considered weighted versions of hereditary graph problems in Theorem 25.3, where we have shown that reoptimizing them allows for a 2-approximation for the local modification of adding a vertex. Most of these hereditary problems stay APX-hard even in their unweighted variants without reoptimization, see, for example, Ausiello et al. [13] for a survey of inapproximability results. Amazingly, all of them admit a PTAS for reoptimization under vertex additions as was shown by Ausiello et al. [22].

Theorem 25.7 (Ausiello et al. [22]) *Let $\Pi \in \mathcal{H}$ be any hereditary unweighted graph problem. Then there is a PTAS for AddVert-Π.*

Proof. Let $(G_{\text{new}}, G_{\text{old}}, S_{\text{old}})$ be an instance of a hereditary unweighted maximization problem Π, where $V_{\text{new}} = V_{\text{old}} \cup \{v_{\text{new}}\}$. Let $\varepsilon > 0$ be some given number such that we want to compute a $(1 + \varepsilon)$-approximation for G_{new}. Our algorithm uses a standard approach for designing approximation schemes:

Depending on ε, we choose a natural number $k = \lceil 1/\varepsilon \rceil + 1$ and check all possible subsets of V_{new} of size at most k. Let S be (one of) the largest of these sets that satisfies the hereditary property as defined by Π. The algorithm then returns the larger one of S_{old} and S.

Let S_{new} be an optimal solution for G_{new}. We distinguish two cases. If $|S_{\text{new}}| \leq k$, then S_{new} was considered by the algorithm and thus the computed solution is optimal. So assume that $|S_{\text{new}}| > k$. We know that, as Π being a hereditary graph problem, $|S_{\text{new}}| \leq |S_{\text{old}}| + 1$, otherwise, $S_{\text{new}} - \{v_{\text{new}}\}$ would be a better solution for G_{old}. Thus,

$$\frac{|S_{\text{new}}|}{|S_{\text{old}}|} \leq \frac{|S_{\text{new}}|}{|S_{\text{new}}| - 1} \leq \frac{k}{k-1} \leq 1 + \varepsilon,$$

which proves that, in this case, S_{old} is a $(1 + \varepsilon)$-approximation.   ∎

Another APX-hard problem that admits a PTAS in its reoptimization version is a weight-bounded version of the Steiner tree problem (STP). In the *minimum Steiner tree problem*, MINSTP for short, the input consists of a connected metric undirected weighted graph $G = (V, E, w)$ where $w : E \rightarrow \mathbb{R}^+$, and a subset $T \subseteq V$ of vertices, the so-called *terminals*. The goal is to find a minimum-weight subgraph of G that contains all terminals, and possibly some of the other vertices, sometimes called *Steiner vertices*. Note that this minimal spanning subgraph will always be a tree. The MINSTP is a very prominent optimization problem; see, for example, Prömel and Steger [31] for a survey. Also its reoptimization variants have attracted a lot of attention; for a brief overview, see Section 25.5. Here, we want to consider a restricted version of the problem where $w : E \rightarrow \{1, \dots, r\}$ for some arbitrary integer r. We call this problem the r-MINSTP. It is known that even the 2-MINSTP is APX-hard [32]. We consider the local modification of increasing the terminal set, that is, of changing one nonterminal vertex of the graph into a terminal and call the resulting reoptimization problem the IncTerm-r-MINSTP. It was shown that this problem is NP-hard for any r [33], but there is a PTAS for it.

Theorem 25.8 (Böckenhauer et al. [33]) *Let $r \in \mathbb{N}^+$. There is a PTAS for the IncTerm-r-MINSTP.*

Proof. Consider an instance $((G, T_{\text{new}}), (G, T_{\text{old}}), S_{\text{old}})$ of the IncTerm-r-MINSTP such that $T_{\text{new}} = T_{\text{old}} \cup \{v_{\text{new}}\}$. Let $n = |V|$ and $\varepsilon > 0$. We describe an algorithm that computes a $(1 + \varepsilon)$-approximate Steiner tree for (G, T_{new}) in polynomial time.

Let $k = \lceil 1/\varepsilon \rceil$. There is an exact algorithm for the MINSTP, called the Dreyfus-Wagner algorithm [34], with a running time of roughly $O(3^t \cdot n^2)$, where t is the number of terminals. If $|T_{\text{new}}| \leq r \cdot k$, then our algorithm computes an optimal solution for (G, T_{new}) using the Dreyfus-Wagner algorithm. In this case, the time complexity is bounded by $O(3^{r \cdot k} \cdot n^2)$, which is polynomial in n and exponential only in the constants r and k.

Otherwise, our algorithm extends the given old optimal solution S_{old} by adding an arbitrary edge that connects v_{new} to it, if necessary, and obtains the solution S. We now prove that S is a $(1 + \varepsilon)$-approximation. As $|T_{\text{new}}| > r \cdot k$, any optimal solution S_{new} for (G, T_{new}) has a weight of at least $r \cdot k$. The weight of S can be bounded as $w(S) \leq w(S_{\text{old}}) + r$ as the added edge has a weight of at most r. Moreover, $w(S_{\text{old}}) \leq w(S_{\text{new}})$. Hence,

$$\frac{w(S)}{w(S_{\text{new}})} \leq \frac{w(S_{\text{old}}) + r}{w(S_{\text{new}})} \leq \frac{w(S_{\text{new}}) + r}{w(S_{\text{new}})} = 1 + \frac{r}{w(S_{\text{new}})} \leq 1 + \frac{r}{r \cdot k} \leq 1 + \varepsilon,$$

which proves the desired approximation ratio.   ∎

Several further reoptimization problems are known that admit a PTAS, including the maximum ΔTSP under the local modification of changing the weight of an edge [35].

25.4.3 Approximation Hardness

We have seen in the previous subsections that reoptimization problems may allow for improved approximations compared to the corresponding original problems. In this subsection, we will see that some reoptimization problems are hard to approximate and show some examples of how to prove hardness of such reoptimization problems.

Following a reduction from the RHPP along the same lines as in the proof of Theorem 25.2, also APX-hardness for different reoptimization variants of the $DL\Delta TSP$ can be proven. More specifically, two cases are distinguished depending on whether the number of deadline vertices is constant or is allowed to depend on the number of vertices of the input graphs. For the former case, Böckenhauer et al. [20] gave a lower bound of $2 - \varepsilon$, for any $\varepsilon > 0$, for the local modifications of increasing or decreasing single edge weights. Böckenhauer and Komm [36] gave the same bound for the local modifications of adding or removing vertices, or adding or removing a deadline. For the second case, it can be shown that, for a number of local modifications, no polynomial-time $(1/2 - \varepsilon)|V|$-approximation algorithm exists, for any $\varepsilon > 0$ and input graphs with $|V|$ vertices. To present the proof, we need the following lemma, which was first proven by Böckenhauer et al. [3], and which essentially shows how to force an algorithm to follow an expensive path once the local modification is applied and the (potentially existing) second Hamiltonian path for an instance of the RHPP is not found.

Lemma 25.3 (Zigzag Lemma, Böckenhauer et al. [3]) *Let $k, n, \gamma \in \mathbb{N}^+$ such that k is even and $\gamma \geq n$; let l be a constant. Let G^* be a complete metric undirected weighted graph with $n + l$ vertices and a deadline triple (s^*, D^*, d^*) such that any Hamiltonian path in G^* satisfying the deadlines (which implies starting at s^*) ends in the vertex t^*. Then, we can construct a complete metric undirected weighted graph $G_{zigzag} = (V, E)$ with $n + l + k$ vertices such that $G_{zigzag} \supset G^*$ and a deadline triple (s, D, d) such that $D \supset D^*$, d^* coincides with d on D^*, and any Hamiltonian path in G^* that reaches t^* in time $7n$ can be extended to a Hamiltonian cycle in G_{zigzag} with cost at most $7n + kn + 2\gamma$, whereas any Hamiltonian path in G^* that reaches t^* after $8n$, but before $9n$, can only be extended to a Hamiltonian cycle in G_{zigzag} with cost at least $7n + (k+1)n/2 + k\gamma$.*

Proof sketch. Consider a graph G^* as given by the lemma, and suppose we extend G^* to a graph G_{zigzag} as shown in Figure 25.2. The starting vertex s of G_{zigzag} is equal to the vertex s^* of G^*. The essential part is that the k newly added vertices have deadlines assigned to them such that they can be traversed in time $7n + kn + 2\gamma$ if an algorithm, starting from s^*, arrives at t^* at time $7n$; this is sketched in Figure 25.2a.

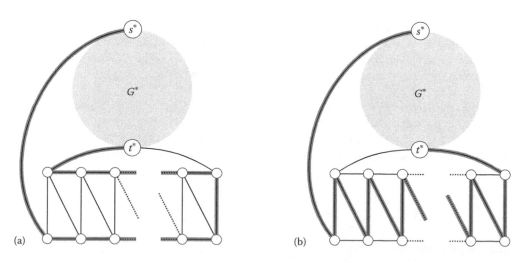

FIGURE 25.2 The zigzag construction used in the proof of Lemma 25.3. (a) Traversing G with cost $7n + kn + 2\gamma$; (b) Traversing G with cost $7n + (k+1)n/2 + k\gamma$.

However, if t^* is visited later, the newly added vertices have to be traversed using the zigzag path as shown in Figure 25.2b. The corresponding zigzag edges have a weight of γ each of which leads to a total cost of $7n + (k + 1)n/2 + k\gamma$. The edges of G_{zigzag} that are not shown in Figure 25.2 are chosen such that any solution that diverges from the two above-mentioned strategies has an even larger cost. ■

Now consider the local modification where the deadline of a deadline vertex is decreased; we call the corresponding reoptimization variant DecD-DL$\triangle$TSP. We use Lemma 25.3 to prove the subsequent theorem. Similarly to the proof of Theorem 25.2 (where a deadline was added to an existing vertex), we give a reduction from the RHPP. Here, the given graph G is first converted analogously to a complete graph K_G, which is then converted to a graph G^* that satisfies the requirements to apply the lemma. The graph G_{zigzag} then corresponds to G_{new} (together with its deadlines).

Theorem 25.9 (Böckenhauer and Komm [19]) *Let $\varepsilon > 0$. Unless* $P = NP$, *the DecD-DL$\triangle$TSP cannot be efficiently approximated with a factor of $(1/2 - \varepsilon)|V|$, where V is the set of vertices of both G_{old} and G_{new}.*

Proof. For any given $\varepsilon > 0$, we choose $k \geq (n + 7)(1 - \varepsilon)/\varepsilon$ and $\gamma \geq (k + 7)n/\varepsilon$ such that k is even and $\gamma \geq n$. Let (G, P') with $G = (V', E')$ and $|V'| = n$ be an instance of the RHPP, and let K_G be a graph constructed from G as in the proof of Theorem 25.2. From K_G, we construct a graph G^* as shown in Figure 25.3; we define $s^* = s'$. Note that every path that starts at the start vertex s^* has to end at t^*. Otherwise, at least one deadline of a vertex visited after t^* cannot be satisfied. It is easy to see that there is a Hamiltonian path $(P', u_1, u_2, u_3, u_4, u_5, t^*)$ of cost $7n$ in G^* from s^* to t^* that makes use of the Hamiltonian path P' in G (or K_G, respectively); see Figure 25.3a. We connect G^* as in Figure 25.2, yielding a graph G_{old} with vertex set V together with a deadline triple (s^*, D_{old}, d_{old}). This Hamiltonian path can thus be extended to an optimal solution H_{old} for G_{old} that satisfies the deadlines.

If the deadline of u_2 is decreased from $3n$ to $2n$, this path becomes infeasible; the corresponding graph is the new instance G_{new} with a deadline triple (s^*, D_{new}, d_{new}). If there is a Hamiltonian path P from s' to some vertex $t'' \neq t'$ in K_G of cost $n - 1$, there is a Hamiltonian path $(P, u_2, u_1, u_3, u_4, u_5, t^*)$ of

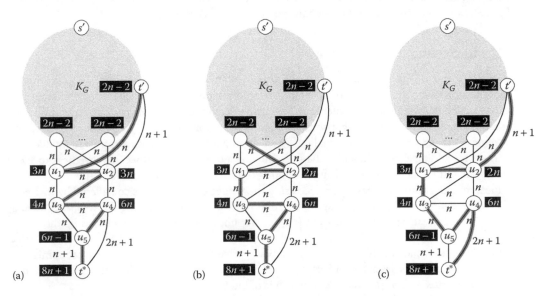

FIGURE 25.3 The graph G^* before and after the local modification and three solutions used in the proof of Theorem 25.9. (a) The subgraph of G_{old}, and a Hamiltonian path of cost $7n$, which uses P'; (b) the subgraph of G_{new}, and a Hamiltonian path of cost $7n$, which uses P if it exists; and (c) the subgraph of G_{new}, and a Hamiltonian path of cost $8n + 1$, which again uses P'.

cost $7n$ from s^* to t^* as shown in Figure 25.3b. Conversely, if there is no such path in K_G, the shortest Hamiltonian path from s^* to t^* has cost $8n + 1$. Recall that the vertices from K_G have to be visited first. As shown in Figure 25.3c, if the given path P' from s' to t' is used, this results in a Hamiltonian path $(P', u_2, u_1, u_3, u_5, u_4, t^*)$. In particular, as the edge $\{t', u_2\}$ has a weight of $n + 1$, the vertex u_4 cannot be visited before u_5. The same holds if any other Hamiltonian path in K_G is used that ends in some vertex other than t', but has cost at least n (because of using an edge of weight 2).

As a consequence of Lemma 25.3, we have that there is a Hamiltonian path in G from s' to some vertex $t'' \neq t'$ if and only if there is a Hamiltonian cycle in G_{new} of cost $7n + kn + 2\gamma$ satisfying the deadlines. If there is no such path in G, the shortest Hamiltonian cycle in G_{new} has cost $7n + (k + 1)n/2 + k\gamma$. Because of the choice of k and γ and the fact that $n + k + 7 > |V|$, we obtain

$$\frac{7n + (k+1)n/2 + k\gamma}{7n + kn + 2\gamma} > \frac{k\gamma}{(7+k)n + 2\gamma} \geq \frac{k}{2+\varepsilon} \geq \frac{1-\varepsilon}{2+\varepsilon}(n + k + 7) \geq \left(\frac{1}{2} - \varepsilon\right)|V|.$$

For a contradiction, suppose there is a polynomial-time approximation algorithm ALG for DecD-DLΔTSP with an approximation ratio better than $(1/2 - \varepsilon)|V|$. ALG outputs a solution with a cost smaller than $7n + (k + 1)n/2 + k\gamma$ if and only if there is a solution with cost $7n + kn + 2\gamma$, that immediately contradicts the NP-hardness of the RHPP. ∎

Note that, opposed to the proof of Theorem 25.2, every vertex of the graphs constructed in the proof of Theorem 25.9 (except for s^* that may be thought of as having a deadline of zero) is assigned a deadline. Moreover, observe that the same construction also works if the vertex u_2 does not have any initial deadline. Likewise, we can use the same construction if the deadline vertex u_2 is added.

25.5 Overview of Further Reoptimization Results

In the preceding sections, we have explored the virtues and limitations of reoptimization using several different problems as examples. In this section, we give a brief overview of further reoptimization problems that were studied in the literature. Other surveys on reoptimization results were given by Ausiello et al. [22,37], Boria and Paschos [5], Böckenhauer et al. [16], and Zych [14].

25.5.1 The Steiner Tree Problem

The MINSTP is one of the problems that received most attention in reoptimization. Several different local modifications were considered. For changing the set of terminals, but not the underlying graph, the first results were obtained by Böckenhauer et al. [33] who proved the NP-hardness of this reoptimization variant and gave an approximation algorithm with a ratio of 1.5 for both adding a vertex to the terminal set and removing a vertex from the terminal set. These approximation ratios were improved by Bilò et al. [38] to 1.344 for adding a terminal and 1.408 for removing a terminal. A further improvement to 1.2 in both cases, but at the expense of a much larger running time, is due to Zych and Bilò [39].

The local modification of increasing or decreasing the weight of an edge was considered by Bilò et al. [38]; they gave algorithms achieving an approximation ratio of 4/3 for increasing the weight of an edge and of 1.302 for decreasing it. Bilò and Zych [40] further improved the ratio for increasing the weight to 1.279.

Most of these results are based on adapting known approximation algorithms to the reoptimization scenario in a similar way as described in Section 25.4.1, but Zych and Bilò [39] also generalized this to an *iterative parameterized contraction technique*, which is of independent interest.

Escoffier et al. [41] considered the local modification of adding one or more vertices to the graph that can be either terminals or nonterminals or both. They gave approximation algorithms achieving ratios of 3/2 for inserting either one terminal vertex or a series of nonterminals, and of $(2 - (1/(t + 2)))$ for inserting a series of vertices containing t terminals. Their algorithms are based on the strategy of ignoring all nonterminals that were neither used in the given old solution nor newly added. For this restricted class of algorithms, they also gave some constant lower bounds, even for graphs with edge weights from $\{1, 2\}$, and for the local modification of removing a vertex.

Steiner tree reoptimization was also considered for restricted edge-weight functions. We have already seen a PTAS for edge weights from the set $\{1, \ldots, r\}$ for some constant r and for adding a vertex to the terminal set in Theorem 25.8. A similar PTAS for removing a vertex from the terminal set (but not from the graph) was presented by Böckenhauer et al. [33]. Furthermore, Böckenhauer et al. [42] considered Steiner tree reoptimization under all local modifications as described earlier for graphs where the edge weights satisfy a sharpened triangle inequality, see Chapter 27 for a formal definition. They proved NP-hardness of all variants and gave a PTAS for some of them and constant-factor approximations for the others.

Goyal and Mömke [43] studied Steiner tree reoptimization for all the above-mentioned modifications for the case that not an optimal solution for the old instance is given, but only a $(1 + \varepsilon)$-approximation, for some $\varepsilon > 0$. They proved that the resulting problem is as hard as the nonreoptimization MINSTP for adding or removing a vertex, decreasing the weight of an edge, or changing a leaf terminal into a nonterminal. Moreover, they gave improved approximation algorithms for increasing the weight of an edge, changing a nonterminal into a terminal, or changing an internal terminal into a nonterminal.

25.5.2 The TSP and Other Routing Problems

Also the TSP has been intensively studied from a reoptimization point of view. In one of the first papers on reoptimization, Archetti et al. [2] considered the ΔTSP under the local modifications of adding or removing a vertex. They proved NP-hardness and focussed on improving the running time by giving a linear-time 3/2-approximation algorithm for it. Note that the same approximation ratio can also be reached by computing from scratch using the famous Christofides algorithm [44]. Ausiello et al. [4,35] improved on this by giving a 4/3-approximation for adding a vertex. They also considered adding a series of vertices and gave an algorithm that achieves a ratio of less than 3/2 for a constant number of added vertices. Monnot [45] improved the running time for adding a single vertex.

The local modification of changing the weight of a single edge while maintaining the metricity was considered by Böckenhauer et al. [3,46]. They proved NP-hardness of this variant and gave a 1.4-approximation algorithm. This was improved to a ratio of 4/3 by Berg and Hempel [47].

TSP reoptimization was also considered for graphs satisfying a relaxed or sharpened triangle inequality, see Chapter 27 for a formal definition. Böckenhauer et al. [3,46] showed that, for changing the weight of an edge, the problem remains NP-hard for all reasonable cases of a sharpened triangle inequality and gave an algorithm improving over computing from scratch for any relaxed β-triangle inequality with $\beta < 3.348$.

The maximization version of the TSP was considered by Ausiello et al. [35] for the local modification of adding a series of vertices. They gave a 5/4-approximation for the case of general edge weights, thus improving over the ratio of 81/61 achievable by computing from scratch [48], and a PTAS for the metric case. Similar results were obtained by Berg and Hempel [47] for changing the weight of an edge.

The TSP with deadlines, as already discussed in Subsection 25.4.3, has been considered by Böckenhauer et al. [3,20] for the case of changing a deadline or the weight of an edge and by Böckenhauer and Komm [19,36] for the case of adding or removing vertices or deadlines. They proved that, in the case of an unbounded number of deadlines, reoptimization under insertion or deletion of nondeadline vertices allows for a significant improvement in the approximation ratio from linear to constant. In all other cases, it was shown that reoptimization does not help for an unbounded number of deadlines (see also Theorem 25.9).

Archetti et al. [49] considered reoptimization of the rural postman problem and designed approximation algorithms with the same or even worse ratios than for computing from scratch, but with much improved running times.

25.5.3 Other Graph Problems

Besides the TSP and the MINSTP, many other graph problems were investigated in the reoptimization scenario. Bilò et al. [21] considered the problems of finding a maximum independent set, a maximum clique, a minimum vertex cover, or a minimum dominating set under the local modification of inserting or deleting an edge. They gave a PTAS for the unweighted versions of these problems and some constant-factor approximations for the weighted case, for example, for the maximum independent set problem, which is not approximable within $O(|V|/(\log|V|)^2)$ without reoptimization [50].

The study of hereditary graph problems was started by Ausiello et al. [22] and further investigated by Boria et al. [24,51–53] for the local modifications of inserting or deleting a vertex. We have already seen in Theorem 25.7 that the unweighted versions admit a PTAS, and in Theorem 25.4 we presented a constant-factor approximation algorithm for finding a maximum planar subgraph. Further constant-factor approximations were given, for example, for finding the maximum k-colorable subgraph [53], the maximum induced split graph [53], or the maximum P_k-free subgraph [24].

Mikhailyuk [54] proved the APX-hardness of graph coloring reoptimization under the local modification of inserting a vertex of maximum degree 2.

Moreover, the minimum spanning tree reoptimization problem under the insertion or deletion of a series of vertices was considered by Boria and Paschos [55]. They designed constant-factor approximation algorithms that run in linear time in the number of vertices, that is, in sublinear time for dense graphs. An experimental analysis of these algorithms was done by Paschos and Paschos [56].

25.5.4 Covering, Packing, and Scheduling Problems

The minimum set cover problem asks for covering a given ground set by as few sets as possible from a given set system over the ground elements. It was considered under the local modification of adding or removing some elements from the sets of the set system (but not from the ground set) by Bilò et al. [21]; see also Zych [14] who proved the NP-hardness of the problem and gave some approximation algorithms for it. For the case of deleting an element, the problem was further considered by Mikhailyuk who gave an improved approximation algorithm [57] and proved the APX-hardness [54]. The inverse problem of covering as many elements as possible from the ground set with k sets was also considered by Mikhailyuk [58].

The knapsack problem was investigated by Archetti et al. [59]; they gave an improved FPTAS for the local modification of adding or removing one or more elements. Boria et al. [24] considered the bin packing problem and proved a lower bound of 3/2 on the approximation ratio under the insertion of elements. For the deletion of elements, Mikhailyuk [54] proved APX-hardness.

The first problem ever investigated in the framework of reoptimization was a scheduling problem with forbidden sets, which was considered by Schäffter [1] who gave a constant-factor approximation algorithm and a corresponding lower bound. Several other scheduling variants were addressed by Boria and Della Croce [60] who designed very simple constant-factor algorithms based on reusing the given old solution.

Finally, the SCSP, which can also be viewed as a kind of scheduling problem, was investigated by Bilò et al. [18,61]. Besides the results already mentioned in Sections 25.3 and 25.4, they also considered the deletion of a string and obtained a 13/7-approximation algorithm for this variant.

25.5.5 Satisfiability and Constraint Satisfaction

The weighted version of the maximum satisfiability problem was considered by Ausiello et al. [22] under the modification of inserting a clause. They proved that an approximation ratio of 1.23 can be reached.

The maximum satisfiability problem for a system of linear equations modulo 2 with exactly 3 variables per equation has been considered by Mikhailyuk [62] for inserting one equation. He designed a 3/2-approximation algorithm and proved a corresponding lower bound.

Mikhailyuk and Sergienko [63] considered the constraint satisfaction problem under the modification of inserting a constraint. They proved a constant-factor approximation for some special class of constraints.

25.6 Generalized Models of Reoptimization

In this section, we will take a look at some possible generalizations of the reoptimization principle. On the one hand, some of the above-mentioned hardness proofs were heavily based on the fact that, besides the given optimal solution, there exists a second, completely different optimal or near-optimal solution, which can be provably hard to find. This raises the natural question whether it is helpful for reoptimization to know not only one optimal solution, but also all optimal solutions or even some additional near-optimal solutions. In Subsection 25.6.1, we will show that, unfortunately, at least for some problems, this extra knowledge still does not help. On the other hand, in many applications such as scheduling or planning problems, it would be very helpful if the computed solution for the locally modified instance resembles the old given solution as much as possible, as a complete rescheduling can be very costly. We describe an alternative model of reoptimization in Subsection 25.6.2 that takes these considerations into account.

25.6.1 Multiple Given Solutions

So far, we dealt with very specific additional information given to us, namely one optimal solution for a related instance. In this subsection, we allow for more general additional knowledge. First, we show that there are problems where even knowing all optimal solutions for the old instance does not help. As an example, we consider the TSP on complete graphs with arbitrary edge weights and the local modification of increasing the weight of an edge, IncEdge-TSP for short. It is well known that IncEdge-TSP does not allow for any reasonable approximation algorithm.

Theorem 25.10 (Böckenhauer et al. [3,46]) *Unless $P = NP$, there is no polynomial-time algorithm that approximates IncEdge-All-Sol-TSP within a ratio of $p(n)$ for any polynomial function p, where n is the number of vertices in the instance.*

We now consider the variant of the problem where we are given an old instance together with all its optimal solutions and our goal is to find a good solution for a related instance where again the weight of a single edge is increased.

Definition 25.8 (TSP Reoptimization with All Optimal Solutions Given) *The reoptimization problem IncEdge-All-Sol-TSP takes as an input $(G_{new}, G_{old}, S_{old})$, where $G_{old} = (V, E, w_{old})$ and $G_{new} = (V, E, w_{new})$ are complete undirected weighted graphs such that w_{old} and w_{new} coincide, except for one edge $e_{change} \in E$, where $w_{old}(e_{change}) < w_{new}(e_{change})$. Moreover, S_{old} is the set of all optimal TSP solutions, that is, all Hamiltonian cycles of minimum total edge weight in G_{old}. The goal is to compute an optimal TSP solution for G_{new}.*

The following result, by Böckenhauer et al. [64], shows that also IncEdge-All-Sol-TSP is not even approximable with any polynomially bounded ratio. Note that, for measuring the size of an IncEdge-All-Sol-TSP instance, we only take into account the graph and its edge weights, but not the space needed to store the (possibly exponentially many) optimal solutions.

Theorem 25.11 (Böckenhauer et al. [64]) *Unless $P = NP$, there is no polynomial-time algorithm that approximates IncEdge-All-Sol-TSP within a ratio of 2^n, where n is the number of vertices in the instance.*

For the proof of Theorem 25.11, we use a technique invented by Papadimitriou and Steiglitz [65] for proving the hardness of local search for the TSP. This technique employs a so-called *diamond graph* as a building block.

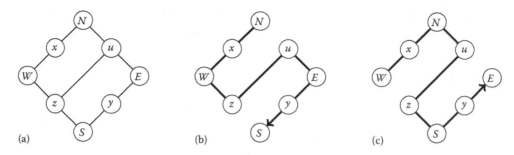

FIGURE 25.4 The diamond graph and its traversals.

Definition 25.9 (Diamond Graph) *The diamond graph $D = (V_D, E_D)$ is an undirected graph with eight vertices $V_D = \{N, S, W, E, u, x, y, z\}$ and nine edges consisting of a cycle $N, u, E, y, S, z, W, x, N$ and an additional edge $\{u, z\}$, as shown in Figure 25.4a. We call N (north), S (south), W (west), and E (east) the connecting vertices and $u, x, y,$ and z the inner vertices of D.*

The diamond graph is suited for our needs because, as an induced subgraph of a larger graph G with a Hamiltonian cycle H, it is traversed by H in exactly one of two ways, namely via a north-south traversal N, x, W, z, u, E, y, S (see Figure 25.4b) or a west-east traversal W, x, N, u, z, S, y, E (see Figure 25.4c), as was proven by Papadimitriou and Steiglitz [65]. Particularly, it is not possible for H to return to a partially visited diamond.

For the proof of Theorem 25.11, we use diamonds as gadgets in a reduction from the well-known Hamiltonian cycle problem, HCP for short. The HCP asks for the existence of a Hamiltonian cycle in a given undirected graph; it is known to be NP-hard [66,67]. We sketch the idea of the proof in the following, for a full proof, see Böckenhauer et al. [64].

Proof sketch of Theorem 25.11. We reduce the HCP to IncEdge-All-Sol-TSP. Let $G = (V, E)$ be an instance of the HCP with $|V| = k$. We first construct an undirected graph $G' = (V', E')$ from G by replacing each vertex $v_i \in V$ by a copy D_i of the diamond graph. Let $N_i, S_i, W_i,$ and E_i denote the connecting vertices of the ith diamond. Every edge $\{v_i, v_j\} \in E$ in G is transformed into two edges $\{W_i, E_j\}$ and $\{E_i, W_j\}$. Additionally, G' contains all edges connecting all north and south vertices from different diamonds.

Observe that G' contains $2^k \cdot (k-1)!$ Hamiltonian cycles that traverse all diamonds in north-south or south-north direction. In addition, G' contains a Hamiltonian cycle traversing all diamonds in west-east or east-west direction if and only if G contains a Hamiltonian cycle. No Hamiltonian cycle in G' can traverse some diamonds in north-south (or south-north, respectively) direction and others in west-east (or east-west, respectively) direction.

From G', we now construct an instance $(G_{new}, G_{old}, S_{old})$ of IncEdge-All-Sol-TSP. Let $G_{reopt} = (V', E_{reopt})$ be the complete undirected graph on the vertex set V'. To define G_{new} and G_{old}, we now define edge weight functions w_{new} and w_{old} for G_{reopt}. For all edges not present in G', we set $w_{old}(e) = w_{new}(e) = M$, where M is a very large number, for example, $M = 2^{9k}$. Except for two special edges, namely $e_{change} = \{W_1, z_1\}$ and $e_{block} = \{N_1, u_1\}$, we set $w_{old}(e) = w_{new}(e) = 1$ for all edges $e \in E'$.

In the old instance, we also set $w_{old}(e_{change}) = 1$. As optimal solutions for the old instance, we want that S_{old} contains exactly all Hamiltonian cycles traversing the diamonds in north-south or south-north direction. To make sure that these are really all optimal solutions, we make the west-east (or east-west) traversal of the first diamond slightly suboptimal by setting $w_{old}(e_{block}) = w_{new}(e_{block}) = 1 + \varepsilon$, for some arbitrarily small ε.

In the new instance, we change the weight of e_{change} to $w_{new}(e_{change}) = M$. This means that all the old optimal solutions in S_{old} become more expensive by an exponential factor. Thus, a Hamiltonian cycle of cost less than M can exist in G_{new} if and only if it uses the diamonds in west-east direction. According to our above-mentioned observation, this means that finding such a cheap Hamiltonian cycle in G_{new} implies finding a Hamiltonian cycle in G. ∎

The construction used in the proof of Theorem 25.11 is in some sense similar to lower-bound proofs for classical reoptimization problems: The optimal solution for the new instance is well hidden because it is slightly suboptimal in the old instance and thus not reported within the extra information. This raises the question whether it would be helpful to give as extra information not only all optimal solutions for the old instance, but also some of the best suboptimal ones. But again, the answer to this question is negative, at least for some problems. The result for the TSP can be generalized to the scenario where also exponentially many near-optimal solutions for the old instance are given [68]. The proof of this result is based on the construction of a generalized class of diamond graphs. Instead of going into the technical details of this construction, we present a similar result for the MINSTP, which is also by Böckenhauer et al. [68].

We consider the local modification of adding a new terminal vertex to the graph and call the resulting reoptimization problem AddTerm-STP. More specifically, the instances of the problem are of the form $((G_{new}, T_{new}), (G_{old}, T_{old}), S_{old})$, with edge-weight functions w_{new} and w_{old}, where $V_{new} = V_{old} \cup \{v_{new}\}$, $T_{new} = T_{old} \cup \{v_{new}\}$, and w_{new} and w_{old} coincide on all edges from E_{old}; again, S_{old} is an optimal solution for the old instance (G_{old}, T_{old}). This problem was considered by Böckenhauer et al. [42,69] and shown to be APX-hard. For a generalized version AddTerm-l-Sol-STP, the input is a triple $((G_{new}, T_{new}), (G_{old}, T_{old}), S_{old})$, where S_{old} contains the l best solutions for (G_{old}, T_{old}).

Theorem 25.12 (Böckenhauer et al. [68]) *The AddTerm-n^{n-2}-Sol-STP is APX-hard on graphs with n vertices.*

Proof sketch. To prove the claim, we use an approximation-preserving reduction from the AddTerm-STP. Given an instance of the AddTerm-STP, the idea of how to construct an instance of the AddTerm-n^{n-2}-Sol-STP is the following: We choose an arbitrary terminal vertex t (but not the newly added one) and transform it into a clique C of n terminals that are very close to each other compared to the distances in the rest of the graph. Then, from the given optimal solution for the old instance of the AddTerm-STP, we can construct a large set of optimal or near-optimal solutions by replacing t by some arbitrary spanning tree for C; all these solutions are then part of the instance of the AddTerm-n^{n-2}-Sol-STP. This does not change the overall structure of the graph, and thus the extra information does not help to compute an approximate solution for the new instance. ∎

25.6.2 Optimizing the Distance from the Given Solution

So far, we have analyzed to what extent the additional information given by some solution for a locally close instance helps to improve the approximability. However, in many practical applications, just finding some good approximate solution for the new instance might not be completely satisfactory. Consider again the example of a timetable for some railway network. If some local modification occurs, for example, a station temporarily closing down due to construction work, we are not only interested in computing a new timetable with good approximation ratio, but we also have to communicate it to all passengers. Thus, our goal is to have a timetable that is not only a good approximation for the new instance, but also as similar as possible to the given timetable for the old instance.

Recently, Shachnai et al. [70] introduced a model of reoptimization that incorporates this idea of viewing the given solution for the old instance not primarily as some additional information facilitating the computation of a good approximation, but as an additional constraint on the set of desirable solutions for the new instance. In this model, there are two parameters in the game, the approximation ratio achieved on the new instance, and the distance between the given old solution and the computed new solution according to some distance measure. This distance is called the *transition cost* of the computed solution. The goal is to compute a solution for the new instance satisfying some given approximation bound α and minimizing the transition cost, that is, being as close as possible to the given old solution.

To make our argument more precise, we start with formally defining the transition costs. Consider some optimization problem $\Pi = (\mathcal{I}, \mathcal{M}, \text{measure, goal})$ with some local modification lm and an instance $x = (x_{\text{new}}, x_{\text{old}}, y_{\text{old}})$ of lm-Π. As the sets of feasible solutions for x_{old} and x_{new} do not have to coincide, we also have to consider the distance between partial solutions. The exact definition of partial solutions depends upon the problem under consideration. We do not try to give a general formalization for it here. For example, for the MINSTP, the set of partial solutions consists of all subsets of vertices; for the TSP it contains, besides all Hamiltonian cycles, also all sets of vertex-disjoint paths in the graph. Let $\mathcal{M}_{\text{part}}(x)$ denote the set of all partial solutions for some instance x of Π. Let δ be a metric distance function on the set $\mathcal{M}_{\text{part}}(x_{\text{old}}) \cup \mathcal{M}_{\text{part}}(x_{\text{new}})$ that measures how costly it is to transform one partial solution s into another partial solution s'. We call $\delta(s, s')$ the transition cost from s to s'. For example, for the MINSTP, the distance function could simply count the minimum number of vertices that have to be inserted or deleted to transform one partial solution into another. We are mainly interested in the transition cost from y_{old} to some feasible solution y_{new} for the new instance x_{new}. Therefore, we use the short notation $\delta(y_{\text{new}}) = \delta(y_{\text{old}}, y_{\text{new}})$. With the use of this notation, we can now define the concept of *strong reapproximation*.[*]

Definition 25.10 (Strong Reapproximation) *Let $d, \alpha \geq 1$. An algorithm* ALG *yields a* strong (d, α)-reapproximation *for* lm-Π *if, for all instances $x = (x_{\text{new}}, x_{\text{old}}, y_{\text{old}})$ of* lm-Π, ALG(x) *is an α-approximate solution for x_{new} and*

$$\delta(\text{ALG}(x)) \leq d \cdot \min\{\delta(y_{\text{new}}) \mid y_{\text{new}} \text{ is an } \alpha\text{-approximate solution for } x_{\text{new}}\}.$$

We illustrate this concept by makespan scheduling.

Definition 25.11 (Minimum Makespan Scheduling, MINMS) *An instance of the* minimum makespan scheduling *problem* MINMS *on identical machines consists of m identical machines $M_1, \ldots, M_m$ and n jobs $J_1, \ldots, J_n$ that each have to be processed on (an arbitrary) one of these machines. The processing times of the jobs can be different, we denote them by $j_1, \ldots, j_n$. The feasible solutions are all* a *schedules, that is, assignments of jobs to machines. The* load *of a machine is the sum of processing times of all jobs assigned to this machine. The goal is to minimize the* makespan *of the schedule, that is, the maximum load over all machines.*

There is a very simple approximation algorithm for MINMS by Graham [71] that reaches an approximation ratio of $2 - 1/m$. This algorithm, called ALG$_{\text{list}}$, simply takes the jobs in some arbitrary order and places each job on one machine with the currently least load. For a nice presentation of the analysis and a hard example showing its tightness, see, for example, Albers [72]. On the other hand, MINMS is known to be strongly NP-hard [67].

As a reoptimization scenario, imagine that one of the machines breaks down (before the start of the processing) and all of its jobs have to be rescheduled to the remaining machines. As transition costs, we simply count how many jobs get moved to a different machine. We call this problem RemMachine-MINMS, and we want to show that it allows for reoptimization with optimal transition costs and an approximation ratio that is not worse than that of ALG$_{\text{list}}$. The following result is by Schieber et al. [73].

Theorem 25.13 (Schieber et al. [73]) *There is a strong $(1, 2 - 1/m)$-reapproximation algorithm for Rem-Machine-MINMS, where m is the number of machines in the new instance.*

Proof. Let $J_1, \ldots, J_n$ be the jobs of the old instance with processing times $j_1, \ldots, j_n$, to be scheduled on $m + 1$ machines. In the given optimal schedule, without loss of generality, let $J_k, \ldots, J_n$, for some $k \leq n$,

---

[*] Following the notation from Shachnai et al. [70], we use their term "reapproximation" instead of "reoptimization" in this subsection; they reserve the term "reoptimization" for the case $\alpha = 1$.

be the jobs scheduled on machine M_{m+1} that is removed in the new instance. We can further assume without loss of generality that M_{m+1} is not empty in the given optimal schedule; otherwise, it is obviously optimal for the new instance without any transition costs.

Our reapproximation algorithm is very simple, it uses ALG_{list} to reschedule the jobs from M_{m+1} according to the order given by the indices of the jobs. This obviously leads to optimal transition costs as any feasible solution has to reschedule these jobs.

We now prove that the computed solution has the claimed approximation ratio. As long as the re-scheduling does not increase the makespan, the computed solution obviously stays optimal, as using less machines cannot decrease the makespan.

If there exists some job J_l, with $k \leq l \leq n$, that increases the makespan on $M_1, \ldots, M_m$ for the first time, from this time step on, we can use the same argument as in the proof of the approximation ratio for ALG_{list} from Albers [72]: J_l gets placed on the machine with the least load from $J_1, \ldots, J_{l-1}$, say M_m. Then, the idle period of the machines $M_1, \ldots, M_{m-1}$ is at most j_l long, which can be bounded from above by the largest processing time among all jobs $J_1, \ldots, J_l$. Let $S(l)$ denote the makespan after placing J_l. Then we have

$$m \cdot S(l) \leq \sum_{i=1}^{l} j_i + (m-1) \cdot \max_{1 \leq i \leq l} j_i,$$

which can be rewritten as

$$S(l) \leq \frac{1}{m} \cdot \sum_{i=1}^{l} j_i + \left(1 - \frac{1}{m}\right) \cdot \max_{1 \leq i \leq l} j_i.$$

As both $(1/m) \cdot \sum_{i=1}^{l} j_i$ and $\max_{1 \leq i \leq l} j_i$ are lower bounds on the optimal makespan for $J_1, \ldots, J_l$ on m machines, the claimed approximation ratio follows for the partial instance consisting of $J_1, \ldots, J_l$. For any further job, there are two cases. Either rescheduling does not increase the makespan, then the approximation bound still holds as adding a job cannot decrease the optimal makespan; or it increases the makespan, then exactly the same argument as earlier shows that the approximation bound still holds for the larger subinstance. Thus, it also holds for the complete instance. ∎

Although the above-mentioned notion captures exactly our intuition about the kind of reoptimization task we would like to solve and can be successfully applied to some problems as we have just shown for MinMS, it has a serious drawback. Namely, for some problems where an α-approximation is easy to accomplish in a classical, nonreoptimization setting, computing a strong (d, α)-reapproximation can be NP-hard.

As an example, we consider the knapsack problem (KP for short). Here, the input consists of a set of n items $I_1, \ldots, I_n$ having positive weights $w_1, \ldots, w_n$ and positive values $v_1, \ldots, v_n$, and a knapsack capacity b. The goal is to choose a subset of items S such that the total weight of the chosen items does not exceed the capacity b whereas the total value of the items from S is maximized. This problem is well known to admit an FPTAS [74]. Nevertheless, obtaining a strong reapproximation in the sense of Definition 25.10 is NP-hard. The proof of the following theorem by Shachnai et al. [70] is taken from the full version of the paper [73] (where it is phrased in slightly different terms).

Theorem 25.14 (Shachnai et al. [70]) *Let $d, \alpha \geq 1$. The problem of obtaining a strong (d, α)-reapproximation of the KP under the local modification of changing the value of an item, ChangeValue-KP for short, is NP-hard.*

Proof. We use a reduction from the classical decision version of the KP, where some bound k on the value is given, and it has to be decided whether there exists a feasible solution with a value larger than k. This problem is well known to be NP-hard [66,67]. We prove how to efficiently solve it using a strong

(d, α)-reapproximation algorithm for the ChangeValue-KP, for any parameters $d, \alpha \geq 1$, thus showing that also this reapproximation problem is NP-hard. For the proof, we assume that the transition cost simply counts the number of deleted or added items.

Consider some instance $x_{\text{dec}} = (w_1, \ldots, w_n, v_1, v_n, b, k)$ of the decision version of KP, where b denotes the capacity bound of the knapsack and k denotes the desired bound on the value. For given d and α, we construct an instance $x = (x_{\text{new}}, x_{\text{old}}, y_{\text{old}})$ of the ChangeValue-KP from x_{dec} as follows: Let $x_{\text{old}} = (w_1, \ldots, w_n, w_{n+1}, v_1, \ldots, v_n, v_{n+1}, b)$ such that $w_{n+1} = b$ and $v_{n+1} > \sum_{i=1}^{n} v_i$. Then $y_{\text{old}} = \{I_{n+1}\}$ with $I_{n+1} = (w_{n+1}, v_{n+1})$ obviously is the unique optimal solution for x_{old}. Let $x_{\text{new}} = (w_1, \ldots, w_n, w_{n+1}, v_1, \ldots, v_n, v'_{n+1}, b)$, where $v'_{n+1} = k/\alpha$.

We claim that any strong (d, α)-reapproximation algorithm leaves the item I_{n+1} in the knapsack if and only if x_{dec} is a no-instance for the decision problem. Assume that x_{dec} is a no-instance. This means that all solutions for x_{dec} have a value that is at most k. Thus, y_{old} is an α-approximate solution for x_{new}, and its transition cost is 0 and thus optimal. On the other hand, if x_{dec} is a yes-instance, then there exists a solution for x_{dec}, and thus also for x_{new}, with value larger than k. This implies that y_{old} is no α-approximate solution and no optimal solution contains I_{n+1} as it fills the knapsack completely. ∎

This result shows that reapproximation in this model can be substantially harder than approximation from scratch. It turns out that reapproximation becomes easier if we compare the transition costs of our computed solution not against the optimal transition costs for computing an α-approximation, but against the transition costs of an optimal solution. Also this approach was proposed by Shachnai et al. [70]. The following definition again considers a reoptimization variant lm-Π of some optimization problem Π under a local modification lm.

Definition 25.12 (Reapproximation) *Let $d, \alpha \geq 1$. An algorithm* ALG *yields a (d, α)-reapproximation for* lm-Π *if, for all instances $x = (x_{\text{new}}, x_{\text{old}}, y_{\text{old}})$ of* lm-Π, ALG(x) *is an α-approximate solution for x_{new} and*

$$\delta(\text{ALG}(x)) \leq d \cdot \min\{\delta(y_{\text{new}}) \mid y_{\text{new}} \text{ is an optimal solution for } x_{\text{new}}\}.$$

Also with this definition, the given old solution is an additional constraint, so we cannot expect the reapproximation problem to be easier than the original optimization problem. Note that the strong reapproximation from Definition 25.10 and the reapproximation from Definition 25.12 coincide for $d = 1$, that is, for optimal transition costs.

The concept of reapproximation according to Definition 25.12 was successfully used for several problems. A special class of problems admitting an FPTAS in the classical approximation setting was considered by Shachnai et al. [70] and Schieber et al. [73], the so-called *DP-benevolent problems* as defined by Woeginger [75]. It was shown there that these problems also admit a kind of reapproximation scheme, namely a $(1 + \varepsilon_1, 1 + \varepsilon_2)$-reapproximation algorithm, for any $\varepsilon_1, \varepsilon_2 > 0$, with a fully polynomial time complexity. If the transition costs are polynomially bounded, this can even be improved to a strong $(1, 1 + \varepsilon)$-reapproximation algorithm. Moreover, the concept was applied to total flow-time scheduling [76], budgeted subset selection problems including a version of the independent set problem [77], and data placement problems [78]. An improved result for MinMS, namely a $(1, 1 + \varepsilon)$-reapproximation, is by Mordechai [79].

References

1. M.W. Schäffter. Scheduling with forbidden sets. *Discrete Applied Mathematics*, 72(1–2):155–166, 1997.
2. C. Archetti, L. Bertazzi, and M.G. Speranza. Reoptimizing the traveling salesman problem. *Networks*, 42(3):154–159, 2003.

3. H.-J. Böckenhauer, L. Forlizzi, J. Hromkovič, J. Kneis, J. Kupke, G. Proietti, and P. Widmayer. Reusing optimal TSP solutions for locally modified input instances. In: *Proceedings of the 4th IFIP International Conference on Theoretical Computer Science (IFIP TCS 2006)*. IFIP, New York, 2006, pp. 251–270.

4. G. Ausiello, B. Escoffier, J. Monnot, and V. Paschos. Reoptimization of minimum and maximum traveling salesman's tours. In: *Proceedings of the 10th Scandinavian Workshop on Algorithm Theory (SWAT 2006)*. Vol. 4059. Lecture Notes in Computer Science. Springer-Verlag, Berlin, Germany, 2006, pp. 196–207.

5. N. Boria and V.Th. Paschos. A survey on combinatorial optimization in dynamic environments. *RAIRO Operations Research*, 45(3):241–294, 2011.

6. M. Libura. Sensitivity analysis for minimum Hamiltonian path and traveling salesman problems. *Discrete Applied Mathematics*, 30:197–211, 1991.

7. Y.N. Sotskov, V.K. Leontev, and E.N. Gordeev. Some concepts of stability analysis in combinatorial optimization. *Discrete Applied Mathematics*, 58:169–190, 1995.

8. H.J. Greenberg. An annotated bibliography for post-solution analysis in mixed integer and combinatorial optimization. In: *Advances in Computational and Stochastic Optimization, Logic Programming, and Heuristic Search*. David, L. (Ed.), Woodruff. Kluwer Academic Publishers, Dordrecht, the Netherland, 1998, pp. 97–148.

9. E.N. Gordeev. Comparison of three approaches to studying stability of solutions to problems of discrete optimization and computational geometry. *Journal of Applied and Industrial Mathematics*, 9(3):358–366, 2015.

10. B. Thiongane, A. Nagih, and G. Plateau. Lagrangean heuristics combined with reoptimization for the 0–1 bidimensional knapsack problem. *Discrete Applied Mathematics*, 154:2200–2011, 2006.

11. B. Hiller, T. Klug, and J. Witzig. Reoptimization in branch-and-bound algorithms with an application to elevator control. In: *Proceedings of the 12th International Symposium on Experimental Algorithms (SEA 2013)*. Vol. 7933. Lecture Notes in Computer Science. Springer-Verlag, Rome, Italy, 2013, pp. 378–389.

12. J. Hromkovič. *Algorithmics for Hard Problems*. 2nd ed., Springer-Verlag, Berlin, Germany, 2004.

13. G. Ausiello, P. Crescenzi, G. Gambosi, V. Kann, A. Marchetti-Spaccamela, and M. Protasi. *Complexity and Approximation*. Springer-Verlag, Berlin, Germany, 1999.

14. A. Zych. Reoptimization of NP-hard problems. PhD thesis, ETH Zürich, 2012.

15. V.V. Vazirani. *Approximation Algorithms*. Springer-Verlag, Berlin, 2003.

16. H.-J. Böckenhauer, J. Hromkovič, T. Mömke, and P. Widmayer. On the hardness of reoptimization. In: *Proceedings of the 34th Conference on Current Trends in Theory and Practice of Computer Science (SOFSEM 2008)*. Vol. 4910. Lecture Notes in Computer Science, Springer-Verlag, Berlin, Germany, 2008, pp. 50–65.

17. J. Gallant, D. Maier, and J.A. Storer. On finding minimal length superstrings. *Journal of Computer and System Sciences*, 20(1):50–58, 1980.

18. D. Bilò, H.-J. Böckenhauer, D. Komm, R. Královič, T. Mömke, S. Seibert, and A. Zych. Reoptimization of the shortest common superstring problem. *Algorithmica*, 61(2):227–251, 2011.

19. H.-J. Böckenhauer and D. Komm. Reoptimization of the metric deadline TSP. In: *Proceedings of the 33rd International Symposium on Mathematical Foundations of Computer Science (MFCS 2008)*. Vol. 5162 Lecture Notes in Computer Science. Springer-Verlag, Berlin, Germany, 2008, pp. 156–167.

20. H.-J. Böckenhauer, J. Kneis, and J. Kupke. Approximation hardness of deadline-TSP reoptimization. *Theoretical Computer Science*, 410(21–23):2241–2249, 2009.

21. D. Bilò, P. Widmayer, and A. Zych. Reoptimization of weighted graph and covering problems. In: *Proceedings of the 6th International Workshop on Approximation and Online Algorithms (WAOA 2008)*. Vol. 5426. Lecture Notes in Computer Science. Springer-Verlag, Aarhus, Denmark, 2008, pp. 201–213.

22. G. Ausiello, V. Bonifaci, and B. Escoffier. Complexity and Approximation in Reoptimization. Technical Report hal-00906941, HAL archives-ouvertes.fr, 2008.

23. C. Lund and M. Yannakakis. The approximation of maximum subgraph problems. In: *Proceedings of the 20th International Colloquium on Automata, Languages and Programming (ICALP 1993)*. Vol. 700. Lecture Notes in Computer Science. Springer-Verlag, Berlin, 1993, pp. 40–51.

24. N. Boria, J. Monnot, and V.Th. Paschos. Reoptimization under vertex insertion: Max p_k-free subgraph and max planar subgraph. *Discrete Mathematics, Algorithms and Applications*, 5(2):1360004, 2013.

25. P. Crescenzi, R. Silvestri, and L. Trevisan. To Weight or not to weight: where is the question? In: *Proceedings of the 4th IEEE Israel Symposium on Theory of Computing and Systems*. New York, 1996, pp. 68–77.

26. R. Bar-Yehuda and S. Even. A linear-time approximation algorithm for the weighted vertex cover problem. *Journal of Algorithms*, 2(2):198–203, 1981.

27. V. Vassilevska. Explicit inapproximability bounds for the shortest superstring problem. In: *Proceedings of the 30th International Symposium on Mathematical Foundations of Computer Science (MFCS 2005)*. Jedrzejowicz, J. and Szepietowski, A. (Eds.), Vol. 3618. Lecture Notes in Computer Science. Springer-Verlag, Gdansk, Poland, 2005, pp. 793–800.

28. M. Karpinski and R. Schmied. Improved inapproximability results for the shortest superstring and related problems. In: *Proceedings of the Theory of Computing 2013 (CATS 2013)*. Wirth, A. (Ed.), 2013, pp. 27–36.

29. M. Mucha. Lyndon words and short superstrings. In: *CoRR*, abs/1205.6787 2013.

30. D. Gusfield. *Algorithms on Strings, Trees, and Sequences*. Cambridge University Press, New York, 1997.

31. H.J. Prömel and A. Steger. *The Steiner Tree Problem*. Vieweg, Braunschweig, 2002.

32. M.W. Bern and P.E. Plassmann. The steiner problem with edge lengths 1 and 2. *Information Processing Letters*, 32(4):171–176, 1989.

33. H.-J. Böckenhauer, J. Hromkovič, R. Královič, T. Mömke, and P. Rossmanith. Reoptimization of steiner trees: Changing the terminal set. *Theoretical Computer Science*, 410:3428–3435, 2009.

34. S.E. Dreyfus and R.A. Wagner. The steiner problem in graphs. *Networks*, 1:195–207, 1971.

35. G. Ausiello, B. Escoffier, J. Monnot, and V. Paschos. Reoptimization of minimum and maximum traveling salesman's tours. *Journal of Discrete Algorithms*, 7:453–463, 2009.

36. H.-J. Böckenhauer and D. Komm. Reoptimization of the metric deadline TSP. *Journal of Discrete Algorithms*, 8:87–100, 2010.

37. G. Ausiello, V. Bonifaci, and B. Escoffier. Complexity and approximation in reoptimization. In *Computability in Context: Computation and Logic in the Real World*. Imperial College Press, London, UK, 2011.

38. D. Bilò, H.-J. Böckenhauer, J. Hromkovič, R. Královič, T. Mömke, P. Widmayer, and A. Zych. Reoptimization of steiner trees. In: *Proceedings of the 11th Scandinavian Workshop on Algorithm Theory (SWAT 2008)*. Vol. 5124. Lecture Notes in Computer Science. Springer-Verlag, Berlin, 2008, pp. 258–269.

39. A. Zych and D. Bilò. New reoptimization techniques applied to Steiner tree problem. *Electronic Notes in Discrete Mathematics*, 37:387–392, 2011.

40. D. Bilò and A. Zych. New advances in reoptimizing the minimum Steiner tree problem. In: *Proceedings of the 37th International Symposium on Mathematical Foundations of Computer Science (MFCS 2012)*. Vol. 7464. Lecture Notes in Computer Science. Springer-Verlag, 2012, pp. 184–197.

41. B. Escoffier, M. Milanič, and V.Th. Paschos. Simple and fast reoptimizations for the Steiner tree problem. *Algorithmic Operations Research*, 4(2):86–94, 2009.

42. H.-J. Böckenhauer, K. Freiermuth, J. Hromkovič, T. Mömke, A. Sprock, and B. Steffen. Steiner tree reoptimization in graphs with sharpened triangle inequality. *Journal of Discrete Algorithms*, 11:73–86, 2012.

43. K. Goyal and T. Mömke. Robust reoptimization of Steiner trees. In: *Proceedings of the 35th IARCS Annual Conference on Foundations of Software Technology and Theoretical Computer Science (FSTTCS 2015)*, Vol. 45. Leibniz International Proceedings in Informatics (LIPIcs). Schloss Dagstuhl - Leibniz Center for Informatics, 2015, pp. 10–24.

44. N. Christofides. Worst-case analysis of a new heuristic for the travelling salesman problem. Technical Report 388, Graduate School of Industrial Administration, CMU, Pittsburgh, 1976.

45. J. Monnot. A note on the traveling salesman reoptimization problem under vertex insertion. *Information Processing Letters*, 115:435–438, 2015.

46. H.-J. Böckenhauer, L. Forlizzi, J. Hromkovič, J. Kneis, J. Kupke, G. Proietti, and P. Widmayer. On the approximability of TSP on local modifications of optimally solved instances. *Algorithmic Operations Research*, 2:83–93, 2007.

47. T. Berg and H. Hempel. Reoptimization of traveling salesperson problems: Changing single edge-weights. In: *Proceedings of the 3rd International Conference on Language and Automata Theory and Applications (LATA 2009)*. Vol. 5457. Lecture Notes in Computer Science. Springer-Verlag, Berlin, 2009, pp. 141–151.

48. Z.-Z. Chen, Y. Okamoto, and L. Wang. Improved deterministic approximation algorithms for Max TSP. *Information Processing Letters*, 95:333–342, 2005.

49. C. Archetti, G. Guastaroba, and M.G. Speranza. Reoptimizing the rural postman problem. *Computers & Operations Research*, 40:1306–1313, 2013.

50. M. Bellare, O. Goldreich, and M. Sudan. Free bits, PCPs, and nonapproximability – towards tight results. *SIAM Journal on Computing*, 27(3):804–915, 1998.

51. N. Boria, J. Monnot, and V.Th. Paschos. Reoptimization of some maximum weight induced hereditary subgraph problems. In: *Proceedings of the 10th Latin American Symposium on Theoretical Informatics (LATIN 2012)*. Vol. 7256. Lecture Notes in Computer Science. Springer-Verlag, Berlin, Germany, 2012, pp. 73–84.

52. N. Boria, J. Monnot, and V.Th. Paschos. Reoptimization of the maximum weighted P_k-free subgraph problem under vertex insertion. In: *Proceedings of the 6th International Workshop on Algorithms and Computation (WALCOM 2012)*. Vol. 7157. Lecture Notes in Computer Science. Springer-Verlag, Berlin, 2012, pp. 76–87.

53. N. Boria, J. Monnot, and V.Th. Paschos. Reoptimization of maximum weight induced hereditary subgraph problems. *Theoretical Computer Science*, 514:61–74, 2013.

54. V.A. Mikhailyuk. On the existence of polynomial-time approximation schemes for the reoptimization of discrete optimization problems. *Cybernetics and Systems Analysis*, 47(3):368–374, 2011.

55. N. Boria and V.Th. Paschos. Fast reoptimization for the minimum spanning tree problem. *Journal of Discrete Algorithms*, 8:296–310, 2010.

56. S.A. Paschos and V.Th. Paschos. Reoptimization of the minimum spanning tree. *Wiley Interdisciplinary Reviews: Computational Statistics*, 4(2):211–217, 2012.

57. V.A. Mikhailyuk. Reoptimization of set covering problems. *Cybernetics and Systems Analysis*, 46(6):879–883, 2010.

58. V.A. Mikhailyuk. On the approximation ratio threshold for the reoptimization of the maximum number of satisfied equations in linear systems over a finite field. *Cybernetics and Systems Analysis*, 48(3):335–348, 2012.

59. C. Archetti, L. Bertazzi, and M.G. Speranza. Reoptimizing the 0–1 knapsack problem. *Discrete Applied Mathematics*, 158:1879–1887, 2010.

60. N. Boria and F.D. Croce. Reoptimization in machine scheduling. *Theoretical Computer Science*, 540–541:13–26, 2014.

61. D. Bilò, H.-J. Böckenhauer, D. Komm, R. Královič, T. Mömke, S. Seibert, and A. Zych. Reoptimization of the shortest common superstring problem. In: *Proceedings of the 20th Annual Symposium on Combinatorial Pattern Matching (CPM 2009)*. Vol. 5577. Lecture Notes in Computer Science. Springer-Verlag, Berlin, Germany, 2009, pp. 78–91.

62. V.A. Mikhailyuk. Reoptimization of max k-cover: Approximation ratio threshold. *Cybernetics and Systems Analysis*, 48(2):242–248, 2012.
63. V.A. Mikhailyuk and I.V. Sergienko. Reoptimization of constraint satisfaction problems with approximation resistant predicates. *Cybernetics and Systems Analysis*, 48(1):73–85, 2012.
64. H.-J. Böckenhauer, J. Hromkovič, and A. Sprock. Knowing all optimal solutions does not help for TSP reoptimization. In *Computation, Cooperation, and Life - Essays Dedicated to Gheorghe Păun on the Occasion of His 60th Birthday*. Kelemen, J. and Kelemenová, A. (Eds.), Vol. 6610. Lecture Notes in Computer Science. Springer-Verlag, Berlin, 2011, pp. 7–15.
65. C.H. Papadimitriou and K. Steiglitz. Some examples of difficult traveling salesman problems. *Operations Research*, 26:434–443, 1978.
66. R.M. Karp. Reducibility among combinatorial problems. In: *Proceedings of a Symposium on the Complexity of Computer Computations*. Plenum Press, New York, 1972, pp. 85–103.
67. M.R. Garey and D.S. Johnson. *Computers and Intractability. A Guide to the Theory of NP-Completeness*. W. H. Freeman and Company, New York, 1979.
68. H.-J. Böckenhauer, J. Hromkovič, and A. Sprock. On the hardness of reoptimization with multiple given solutions. *Fundamenta Informaticae*, 110:59–76, 2011.
69. H.-J. Böckenhauer, K. Freiermuth, J. Hromkovič, T. Mömke, A. Sprock, and B. Steffen. The Steiner tree reoptimization problem with sharpened triangle inequality. In: *Proceedings of the 7th International Conference on Algorithms and Complexity (CIAC 2010)*. Vol. 6078. Lecture Notes in Computer Science. Springer-Verlag, Berlin, 2010, pp. 180–191.
70. H. Shachnai, G. Tamir, and T. Tamir. A theory and algorithms for combinatorial reoptimization. In: *Proceedings of the 10th Latin American Symposium on Theoretical Informatics (LATIN 2012)*. Vol. 7256. *Lecture Notes in Computer Science*. Springer-Verlag, Berlin, 2012, pp. 618–630.
71. R.L. Graham. Bounds for certain multiprocessor anomalies. *Bell System Technical Journal*, 45:1563–1581, 1966.
72. S. Albers. Online scheduling. In *Introduction to Scheduling*. Robert, Y. and Vivien, F. (Eds.), Chapman and Hall/CRC Press, Boca Raton, 2009, pp. 57–84.
73. B. Schieber, H. Shachnai, G. Tamir, and T. Tamir. A theory and algorithms for combinatorial reoptimization. Available at http://www.cs.technion.ac.il/hadas/PUB/Reopt_full.pdf, 2014 (Accessed on November 6, 2018).
74. O.H. Ibarra and C.E. Kim. Fast approximation algorithms for the knapsack and sum of subset problems. *Journal of the ACM*, 22(4):463–468, 1975.
75. G.J. Woeginger. When does a dynamic programming formulation guarantee the existence of an FPTAS? In: *Proceedings of the 10th Annual ACM-SIAM Symposium on Discrete Algorithms (SODA 1999)*. ACM/SIAM, Philadelphia, 1999, pp. 820–829.
76. G. Baram and T. Tamir. Reoptimization of the minimum total flow-time scheduling problem. *Sustainable Computing: Informatics and Systems*, 4:241–251, 2014.
77. A. Kulik, H. Shachnai, and G. Tamir. On lagrangian relaxation and reoptimization problems. Technical Report 1512.06736v1, arXiv, 2015.
78. H. Shachnai, G. Tamir, and T. Tamir. Minimal cost reconfiguration of data placement in a storage area network. *Theoretical Computer Science*, 460:42–53, 2012.
79. Y. Mordechai. Optimization and reoptimization in scheduling problems. Master's thesis, Technion – Israel Institute of Technology, 2015.

26

Sensitivity Analysis in Combinatorial Optimization*

David Fernández-Baca

Balaji Venkatachalam

26.1 Introduction

Anyone formulating an optimization problem faces uncertainty in selecting the model parameters. Transportation costs, probabilities of different events, budget constraints, among other parameters, are all unlikely to be known with any degree of accuracy. There may also be doubts about the structure of the model, such as whether to add or remove constraints. Understanding the effect of changes in parameter values and model structure on the optimum solution is the subject matter of *sensitivity analysis*. Our goal in this chapter is to give an overview of the main algorithmic ideas in sensitivity analysis for combinatorial optimization.

Research in sensitivity analysis started shortly after the development of the simplex method [1,2]. Since then, the literature has grown rapidly: as of 1997, over 1000 journal articles had been published on the subject [3]. To keep things manageable, we focus our attention on *parametric analysis*, which studies problems whose structure remains fixed, but where cost coefficients vary continuously as a function of one or more parameters. The reader interested in more information on topics not covered here has other good sources to turn to. Starting points into the literature include Geoffrion and Nauss' classical survey of the field in the late 1970s [4], the PhD Thesis of Wagelmans [5], the collection of articles [3], and the bibliography [6].

* Work partially supported by grants CCR-9988348 and EF-0334832 from the National Science Foundation.

Parametric analysis is a rich field, whose applications extend beyond the examination of alternative scenarios. In fact, parametric problems arise as auxiliary problems in other areas, such as Lagrangian relaxation and minimum-ratio optimization (see Section 26.2). Parametric analysis should not, however, be confused with the study of *parameterized complexity* [7]. The latter deals with finding efficient algorithms for problems when one of the input parameters is fixed. These parameters are typically discrete (e.g., the maximum size of a vertex cover), and the goal is to obtain the best solution, not to study its stability. More in the spirit of parametric analysis, and also beyond the scope of this chapter, is the study of *stability of approximation*, which considers how the complexity of a problem depends on a parameter whose variation alters the space of allowable instances (Chapter 27, and Reference 8). An example is analyzing how the approximability of the traveling salesman problem varies as we allow intercity distances to deviate from the triangle inequality as a function of a continuous parameter [9].

This chapter consists of five parts, in addition to this Introduction. Section 26.2 gives basic definitions and notation. Section 26.3 introduces the key issues in sensitivity analysis. Section 26.4 is devoted to general-purpose techniques for sensitivity analysis. Section 26.5 discusses selected results on sensitivity analysis of polynomially-solvable problems. Section 26.6 is devoted to sensitivity analysis of NP-hard problems.

26.2 Preliminaries

We consider combinatorial optimization problems of the form

$$Z^* = \min\{cost(x) : x \in X\}, \tag{26.1}$$

where $X \subset \mathbb{R}^n$ is the set of *feasible solutions* and *cost*, referred to as the *cost function* or the *objective function*, maps each feasible solution x to a positive real number. Problem (26.1) is called a 0/1 *problem* when $X \subseteq \{0,1\}^n$. The cost function is *linear* if $cost(x) = \sum_{i=1}^{n} c_i \cdot x_i$. The cost function is a *bottleneck* (or *min-max*) function when $cost(x) = \max_{1 \le i \le n} c_i \cdot x_i$. The c_i's are the *coefficients* of the objective function and $c = (c_1, \ldots, c_n)$ is the *coefficient vector*.

The minimum spanning tree and traveling salesman problems can be formulated as 0/1 problems with linear cost functions. In both cases, the coefficient vector consists of the edge costs and x_i is 1 or 0 depending on whether or not edge i is chosen. Bottleneck versions of both problems can also be defined. For example, in the bottleneck traveling salesman problem, the cost of a tour is determined by its costliest edge. Unless stated otherwise, our discussion will center on optimization problems with linear cost functions.

For any problem of the form (26.1), we can define a corresponding *parametric cost problem* where the cost of each feasible solution depends on a *parameter vector* λ:

$$Z(\lambda) = \min\{cost(x, \lambda) : x \in X\}. \tag{26.2}$$

Equation (26.2) defines an infinite family of optimization problems of the form (26.1), one for each fixed λ. We refer to $Z(\lambda)$ as the *optimum cost function*. An *evaluator* for problem (26.2) is an algorithm that, for any given λ, computes $Z(\lambda)$ and an optimum solution at λ.

The optimum cost function Z induces a decomposition $\mathcal{M}(Z)$ of the parameter space into maximal connected regions, such that within each of region A, there exists a feasible solution x that is optimum for every $\lambda \in A$. The *combinatorial complexity* of $\mathcal{M}(Z)$ is the size of the description of $\mathcal{M}(Z)$ as a function of the input size.

Unless stated otherwise, we consider parametric problems where the coefficients of the objective function are linear in λ; that is, $c_i(\lambda) = c_{i0} + \sum_{j=1}^{d} c_{ij}\lambda_j$ for $i = 1, \ldots, n$. In this case, $Z(\lambda)$ is a piecewise linear concave function of λ (assuming X corresponds to any reasonable combinatorial optimization problem), as it is the lower envelope of the set of cost functions of the feasible solutions. Hence, $\mathcal{M}(Z)$ subdivides the parameter space into convex polyhedral regions; its combinatorial complexity is its total number of facets of dimensions 0 through d.

An important subcase is the one where the parameter vector λ is precisely the coefficient vector c. This occurs, for example, when studying the sensitivity of a minimum spanning tree or of an optimum traveling salesman tour to variations in the edge costs. The special case where $d = 1$ is also of interest. Here, the optimum cost function Z consists of a concave sequence of straight-line segments and the points at which these segments meet are called *breakpoints*; the combinatorial complexity of $\mathcal{M}(Z)$ is proportional to the number of breakpoints.

In addition to their applications to sensitivity analysis, parametric problems arise as auxiliary problems in other contexts. Two important cases are discussed next.

Lagrangian relaxation [10,11] is a well-known technique to handle certain kinds of hard optimization problems. Specifically, consider an NP-hard problem of the form

$$z^* = \min\{cx : Ax = b, x \in X\} \tag{26.3}$$

that is polynomially-solvable in the absence of constraints $Ax = b$. For instance, the minimum spanning tree problem with degree constraints (known to be NP-hard [12]) can be expressed in the form (26.3): X is the set of spanning trees and the degree constraints are given by a system of linear equalities.

The *Lagrangian function* for problem (26.3) is

$$Z_L(\lambda) = \min\{cx + \lambda(Ax - b) : x \in X\}. \tag{26.4}$$

Observe that (26.4) defines a linear parametric problem. For example, the Lagrangian function for the degree-constrained minimum spanning tree yields a minimum spanning tree problem where edge weights are functions of a parameter vector. It can be shown that for all λ, $z^* \geq Z_L(\lambda)$. The best lower bound obtainable from (26.4) is, therefore, $\max_\lambda Z_L(\lambda)$. In practice, this can be a quite accurate estimate of the optimum solution to problem (26.3) [13,14].

Minimum ratio optimization, also known as *fractional programming*, deals with problems of the form

$$z^* = \min\{f(x)/g(x) : x \in X\}, \tag{26.5}$$

where $g(x) > 0$ for all $x \in X$. One setting for this question arises in graph problems where each edge has a cost and a profit and the goal is to find a subgraph with a certain property that minimizes the ratio of total cost to total profit. Two examples are the minimum ratio spanning tree problem [15] and the minimum ratio cycle problem [16].

Define

$$Z_R(\mu) = \min\{f(x) - \mu \cdot g(x) : x \in X\}, \tag{26.6}$$

where μ is a scalar parameter. It is well-known [15] that $z^* = \max\{\mu : Z_R(\mu) > 0\}$. Observe that $Z_R(\mu)$ is a concave piecewise linear decreasing function of μ. For the above-mentioned weighted graph problems (26.6) defines a parametric problem where edge costs are decreasing linear functions of μ.

For reasons of space, we do not treat problems with parametric constraint set in any depth here. An example is the knapsack problem with variable knapsack size, which can be expressed as

$$Z_K(\mu) = \max \left\{ \sum_{i=1}^{n} c_i x_i : \sum_{i=1}^{n} s_i x_i \leq b + \mu b', x \in \{0, 1\}^n \right\}, \tag{26.7}$$

where μ is a scalar parameter. Problems with parametric constraints generally behave quite differently from those with parametric objective function; indeed, the optimum cost function is typically not continuous, as the reader may verify for function Z_K mentioned above. An exception to this rule is linear programming with parametric right-hand side, which, because of duality, has the same behavior as linear programming with parametric objective function (see Section 26.5.1).

26.3 Issues in Sensitivity Analysis

Sensitivity analysis questions can be formulated in different ways. The first important class of problems we consider are collectively referred to as *posterior* analysis problems. Here, an optimum solution $x^{(0)}$ for some parameter vector $\lambda^{(0)} = (\lambda_1^{(0)}, \ldots, \lambda_d^{(0)}) \in \mathbb{R}^d$ is known and the question is to determine the *stability* of $x^{(0)}$ with respect to parameter variation around $\lambda^{(0)}$. Some common notions of stability are defined next.

Definition 26.1 *The* lower tolerance *and the* upper tolerance *of parameter i at $\lambda^{(0)}$ are the largest values l_i and u_i such that $x^{(0)}$ is optimum for every λ' such that $\lambda_i' \in [\lambda_i^{(0)} - l_i, \lambda_i^{(0)} + u_i]$ and $\lambda_j' = \lambda_j^{(0)}$ for $j \neq i$.*

A related problem is *ray shooting*: Given a ray ρ originating at $\lambda^{(0)}$, the question is to find the point $\lambda^* \in \rho$ farthest from $\lambda^{(0)}$ such that $x^{(0)}$ is optimum for every $\lambda \in \rho$ between $\lambda^{(0)}$ and λ^*. By suitable re-parametrization, ray shooting can be expressed as a question of computing tolerances.

Definition 26.2 *The* stability region *of $x^{(0)}$ at $\lambda^{(0)}$ is the largest connected subset F of $\mathbb{R}^d$ containing $\lambda^{(0)}$ such that $x^{(0)}$ is an optimum solution for all $\lambda \in F$.*

Observe that each region of $\mathcal{M}(Z)$ is a stability region of some feasible solution x. Thus, stability regions are convex polyhedral subsets of the parameter space. In the important special case where the parameter vector is precisely the coefficient vector, the stability regions are polyhedral cones that meet at the origin; this is because all solutions have cost 0 at the origin of the parameter space. Note also that, in the one-parameter case, computing tolerances and stability regions are equivalent problems.

Definition 26.3 *The* stability radius *of $x^{(0)}$ at $\lambda^{(0)}$ is the radius of the largest ball B such that $x^{(0)}$ is optimum for all $\lambda \in B$.*

The notion of a ball in the above-mentioned definition depends on the metric used. A common choice is the L_∞ norm, where the distance between parameter vectors λ and λ' is given by $\max_i |\lambda_i - \lambda_i'|$.

The *robustness function* [17] provides an alternative way to measure the effect of multi-parameter variation.

Definition 26.4 *The* robustness function *at $\lambda^{(0)}$, is the function $R : \mathbb{R} \to \mathbb{R}$ given by*

$$R(b) = \max \left\{ Z(\lambda^{(0)} + \delta) : \sum_{i=1}^{d} w_i \cdot \delta_i \leq b, \ \delta_i \geq 0 \text{ for } i = 1, \ldots, n \right\}, \quad (26.8)$$

where $b \geq 0$ is the budget, *$\delta = (\delta_i, \ldots, \delta_d)$ is the* increment vector, *and $w_i \geq 0$ is the* unit cost of increasing *parameter λ_i, for each $i \in \{1, \ldots, n\}$.*

Intuitively, the robustness function yields the maximum effect that a total weighted increase of the parameters within a given budget can have on the cost of the optimum solution.

The next problems do not depend on advance knowledge of an optimum solution at some parameter value; questions of this kind are traditionally referred to as *prior analysis* problems. The following notion has some relationships with computing tolerances.

Definition 26.5 *A* most vital variable *in a 0/1 combinatorial optimization problem is a variable x_i such that forcing x_i to equal zero increases the value of the optimum solution as much as possible.*

In this context of network optimization problems, such as the minimum spanning tree problem, the shortest path problem, or the traveling salesman problem, identifying most vital variables often translates into finding most vital edges (see Sections 26.5 and 26.6).

Parametric search encompasses a broad class of problems whose goal is to locate a point λ^* in the parameter space where the optimum cost function satisfies some specified property. For example, ray shooting is a parametric search problem, although it is not an instance of prior analysis. Lagrangian relaxation and minimum ratio optimization lead to two basic parametric search problems: *maximization*, that is, locating $\lambda^* = \arg\max_\lambda Z(\lambda)$, and *root-finding*, that is, solving $Z(\lambda) = 0$ for λ.

Inverse optimization is another parametric search problem. The input here is a *reference solution* $x^{(0)}$, and the problem is to locate a parameter vector $\lambda^{(0)}$ such that $x^{(0)}$ is optimal, or as close to optimum as possible, at $\lambda^{(0)}$. Examples of inverse parametric minimum spanning tree, shortest path, matching, and other optimal subgraph problems are given in Reference 18; applications to biological sequence alignment are discussed in References 19, 20.

The construction problem is to build a complete representation of $\mathcal{M}(Z)$. This is the most general problem considered here, as almost any sensitivity analysis question can be answered given $\mathcal{M}(Z)$. Clearly the time required to solve the construction problem depends heavily on the combinatorial complexity of Z.

Stability of approximate solutions and heuristics: Variants of all the issues outlined so far arise when considering approximate solutions or the behavior of heuristic algorithms. These questions are of particular interest when studying NP-hard problems (see Section 26.6).

Analogs to the notions of tolerance, stability radius, and stability region for ϵ-approximate solutions are straightforward to define: simply replace "optimum" by "ϵ-approximate" in Definitions 26.1–26.3. As observed in Reference 21, the stability region for an ϵ-approximate solution x has properties similar to those of an ordinary stability region is closed; for example, it is convex, and polyhedral. Unlike ordinary stability regions, however, the intersection between the stability regions of two different ϵ-approximate solutions, $\epsilon > 0$, can have dimension d. Note also that, for $\epsilon_1 < \epsilon_2$, the region where x is ϵ_1-approximate is contained in the region where x is ϵ_2-approximate [21].

So far we have only considered the stability of *solutions*; however, it is also natural to enquire about the sensitivity of an *algorithm* to changes in parameter values (see Section 26.6.3). Suppose A is an algorithm for some optimization problem and let $x^{(0)}$ be the solution returned by A at parameter vector $\lambda^{(0)}$. Then, the *upper tolerance* of parameter i for A at $\lambda^{(0)}$ is the largest value u_i such that $x^{(0)}$ is the solution returned by A for all λ' such that $\lambda'_i \in [\lambda_i^{(0)}, \lambda_i^{(0)} + u_i]$, and $\lambda'_j = \lambda_j^{(0)}$ for all $j \neq i$. Lower tolerances, stability regions, and radii are defined similarly.

26.4 General Techniques

General-purpose methods for certain parametric analysis problems are known. These include methods for parametric search, construction of the space decomposition, and determining the combinatorial complexity of low-dimensional problems. We survey them next.

26.4.1 Parametric Search

We review four methods of parametric search: bisection search, Newton's method, gradient descent, and Megiddo's method. We focus on one-parameter problems, where the cost of a solution is a linear function of a single scalar parameter μ and to goal is to locate a value μ^* satisfying certain properties.

Newton's method: The classic root-finding method of Newton directly applies to the problem of finding a zero of $Z(\mu)$, when Z is a concave decreasing function of μ: Assume that we know a value $\mu^{(1)}$ such that $\mu^{(1)} \leq \mu^*$, as well as the optimum solution $x^{(1)}$ at $\mu^{(1)}$. Let $\mu^{(2)}$ be the value such that $cost(x^{(1)}, \mu^{(2)}) = 0$. Now compute $\nu = Z(\mu^{(2)})$ and the optimum solution $x^{(2)}$ at $\mu^{(2)}$. Note that $\nu \geq 0$. If $\nu = 0$, stop and return $\mu^* = \mu^{(2)}$. Otherwise, set $\mu^{(1)} = \mu^{(2)}$ and $x^{(1)} = x^{(2)}$ and repeat the process. This procedure generates an increasing sequence of μ-values, each of which corresponds to a different feasible solution. Thus, the number of evaluations required by Newton's method is at most equal to the number of breakpoints of Z. This naive bound can be improved substantially in certain applications; in fact, Newton's method yields the fastest-known algorithms for several parametric search problems, including certain minimum-ratio network flow and minimum cut problems [22,23]. Newton's method can also be applied to other parametric search problems, including finding the maximizer of Z or ray shooting [24].

Bisection search: Suppose that we have an *oracle* for our parametric search problem; that is, a procedure that determines whether a given parameter value μ is less than or equal to the parameter value being sought. Suppose, additionally, that we have an interval $\mathcal{I}$ known to contain μ^*. Bisection search locates μ^* by repeatedly halving $\mathcal{I}$, taking the left or right half depending on the outcome of an oracle call at the midpoint. The search stops when $\mathcal{I}$ is small enough, according to some criterion.

Oracles can often be constructed from evaluators. For example, consider the one-parameter upper tolerance problem. Let $x^{(0)}$ be the optimum solution at $\mu^{(0)}$, and let u be the upper tolerance at $\mu^{(0)}$. Then, the parameter value being sought is $\mu^* = \mu^{(0)} + u$. The oracle must determine whether a given $\mu \geq \mu^{(0)}$ is less than or equal to μ^*. To test this, first use the evaluator to find an optimum solution $x^{(1)}$ at μ. If $cost(x^{(0)}, \mu) = cost(x^{(1)}, \mu)$, then $\mu \leq \mu^*$. Otherwise (as $x^{(1)}$ is optimum at μ), the only possibility is that $cost(x^{(0)}, \mu) < cost(x^{(1)}, \mu)$, and thus $\mu > \mu^*$.

Bisection search leads to fast algorithms for certain inverse sequence alignment problems with one or two parameters [20].

Megiddo's method: Megiddo's method [25,26] solves parametric search problems by simulating the execution of an evaluator for the underlying fixed-parameter problem to find its computation path at μ^*. This evaluator must be *piecewise linear*; that is, each value it computes must be a linear combination of the input parameters. This condition is not particularly restrictive, as many combinatorial optimization algorithms have this property; for example, most minimum spanning tree, maximum flow, and dynamic programming sequence alignment algorithms are piecewise linear. Like bisection search, Megiddo's method relies on repeated invocations of an oracle to narrow down the search range for μ^*. We have the following result.

Theorem 26.1 *Let $\mathcal{P}$ be a parametric search problem that has an oracle that runs in worst-case time b. Suppose that there exists a piecewise linear algorithm to evaluate $Z(\lambda)$ that executes t steps in the worst case. Then, $\mathcal{P}$ can be solved in time $O(t \cdot b)$. If the evaluator for $Z(\mu)$ is a piecewise linear parallel algorithm that has d parallel steps, using w processors, $\mathcal{P}$ can be solved in time $O(d \cdot b \cdot \log w + d \cdot w)$.*

The second of the above-mentioned time bounds can be improved by a factor of $\log n$ for some problems [27]. Among these are several parametric search problems related to minimum spanning trees, including finding the maximizer of Z_G and the root of Z_G, when this function is strictly decreasing (as is the case for the minimum-ratio spanning tree problem). All these questions can be answered in $O(T_{MST}(n, m) \log n)$ time, where $T_{MST}(n, m)$ is the time to compute a minimum spanning tree in an n-vertex, m-edge graph. Megiddo's method also yields efficient algorithms for the minimum ratio cycle problem [27]. Theorem 26.1 can be extended to nonlinear problems and to any fixed number of parameters [28–31].

In some cases, such as parametric minimum spanning trees on planar or dense graphs or for certain optimization problems on graphs of bounded tree-width, the poly-logarithmic slowdown of Megiddo's technique relative to the fixed-parameter problem can be eliminated entirely [32–35]. This yields parametric algorithms that are asymptotically as fast as fixed-parameter evaluators.

Gradient ascent: Gradient ascent [14,36] can be used to search for the maximizer λ^* of the optimum cost function $Z(\lambda)$. The method is iterative, generating a sequence of points that converges to λ^*. If the current point $\lambda^{(0)}$ is not minimum, the algorithm chooses the next point by moving some distance (given by some predetermined increment sequence) in the direction of a *subgradient* of $Z(\lambda)$. Informally, a vector $s \in \mathbb{R}^d$ is a subgradient of Z at $\lambda^{(0)} \in \mathbb{R}^d$ if s points in a direction along which Z is nondecreasing; that is, s plays the role a gradient plays for a differentiable function. A subgradient of Z at $\lambda^{(0)}$ can be computed using an evaluator: If $x^{(0)}$ is an optimum solution at $\lambda^{(0)}$, then, the function $cost(x^{(0)}, \lambda)$ has the form $a_0 + \sum_{i=1}^{d} a_i \lambda_i$ and the vector $(a_1, \ldots, a_d)$ is a sub-gradient at $\lambda^{(0)}$. It can be shown that $\lambda^{(0)}$ is a maximizer of Z if the 0 vector is a subgradient at $\lambda^{(0)}$ [14]. In practice this termination condition can be hard to test, as only one subgradient is computed at any point. One way to handle this is by ending the search if the function has not decreased by a certain amount after some number of iterations. Although gradient ascent is fast in practice (indeed, it is widely used in Lagrangian relaxation), it is not in general possible to establish combinatorial bounds on its running time, as the convergence time of the algorithm depends on the choice of the increment sequence [14].

26.4.2 Construction Algorithms

Constructing the space decomposition $\mathcal{M}(Z)$ induced by the optimum cost function Z requires producing all the regions of $\mathcal{M}(Z)$, along with the incidence relationships between regions, and the optimum solution associated with each region. The construction problem can be solved by evaluating Z at various points in the parameter space, producing a sequence of hyperplanes, each of which is the cost function of some optimum solution. Each new hyperplane is used to incrementally update the current estimate of Z. The technique is reminiscent of the methods used to determine the shape of a convex polyhedron through a series of hyperplane probes [37,38] (see also References 39, 40). Indeed, these results can be used to prove the following.

Theorem 26.2 *Let d denote the number of parameters, let f and v be, respectively, the number of regions and vertices of $\mathcal{M}(Z)$, and let t be the time needed to evaluate Z. Then, $\mathcal{M}(Z)$ can be constructed in $O(t \cdot (f + dv))$ time, plus the time needed to construct the lower envelope of all the score functions generated during the computation. In particular, for $d = 1$ and 2, $\mathcal{M}(Z)$ can be constructed in time $O(t \cdot f)$ and $O(t \cdot f + f^2)$, respectively.*

Lower envelope construction is dual to constructing the upper convex hull of a set of points in $\mathbb{R}^{d+1}$, a problem for which an extensive literature exists [41]. The time needed to construct a lower envelope depends heavily on the complexity of the output produced, as measured by its total number of faces of dimensions 0 through d. By the *upper bound theorem* [42], the complexity of Z is $\Theta(f^{\lfloor (d+1)/2 \rfloor})$, where f is the number of regions. Thus, for $d \geq 3$, the time needed to build the lower envelope can dominate the total computation time.

An alternative method for building $\mathcal{M}(Z)$ is *lifting*, which works by executing the fixed-parameter problem for *all* parameter values simultaneously. Another construction technique relies on ray shooting (see Section 26.3). Applications of these techniques to parametric sequence alignment are given in References 19, 43, 44.

26.4.3 Combinatorial Complexity Bounds for Small-Dimensional Problems

In some problems, the cost of a feasible solution is the weighted sum of a relatively small number of aggregate *features*. For example, in one common sequence alignment scoring scheme, whereas the number of feasible solutions (alignments) is exponentially large, the cost of a feasible solution is depends only on the total number of matches, mismatches and indels, each weighted by separate parameters [43]. In instances of this kind, Theorem 26.3, from Reference 45, is useful.

Theorem 26.3 *Consider an instance of problem (26.2) where the cost function is given by $cost(x, \lambda) = \sum_{i=1}^{d} \lambda_i \cdot x_i$ (i.e., the cost is a linear function of only the first d coordinates of a feasible solution). Suppose that there exist integers $n_1, \ldots, n_d$ such that for every $x \in X$ and each $i \in \{1, \ldots, d\}$, x_i is an integer in the set $\{0, \ldots, n_1\}$. Then, $\mathcal{M}(Z)$ has $O\left(\left(\prod_{i=1}^{d} n_i \right)^{(d-1)/(d+1)} \right)$ regions.*

Two examples of the use of Theorem 26.3 are given next; applications to sequence alignment can be found in Reference 43. Note that Theorem 26.3 applies to certain NP-hard problems as well—for an example of this, see Reference 46.

- An instance of the *stable marriage problem* has n men and n women. Each man m assigns a rank $r(m, w) \in [n]$ to every woman w, and each woman w assigns a rank $r(w, m) \in [n]$ to each man m. A feasible solution is a pairing M of men and women called a *marriage*; M is *stable* if there is no man-woman pair $(m, w) \notin M$ such that m and w prefer each other to their current mate. The cost of M is $\lambda_1 x_M + \lambda_2 y_M$, where $x_M = \sum_{(m,w) \in M} r(m, w)$, $y_M = \sum_{(m,w) \in M} r(w, m)$, and λ_1 and λ_2 weigh the relative preference of men over women in forming M. The goal is to find the stable marriage of minimum cost [47]. As x_M and y_M are $O(n^2)$, Theorem 26.3 gives a bound of $O((n^2)^{2 \cdot \frac{1}{3}}) = O(n^{4/3})$ on the number of regions of $\mathcal{M}(Z)$ [48].
- The input to the *maximum parsimony problem* [49] consists of an n-leaf rooted binary tree T and a labeling $S_1, \ldots, S_n$ of T's leaves by DNA sequences of length k. A feasible solution is a labeling of the internal nodes of T by DNA sequences of length k. Let $e = (x, y)$ be an edge of T whose endpoints are labeled by sequences X and Y and let $i \in [k]$ be an index for which $X[i] \neq Y[i]$. Then, there is a *transition* at i if either both $X[i]$ and $Y[i]$ are purines (A or G nucleotides) or both are pyrimidines (C or T nucleotides); otherwise, there is a *transversion* at i. The cost of a labeling L is $\lambda_1 v + \lambda_2 t$, where v is the total number of transversions and t is the total number of transitions over all the edges of T and all indices i. The goal is to find a labeling of minimum cost. As $0 \leq v, t \leq 2(n-1)k$, Theorem 26.3 implies that $\mathcal{M}(Z)$ has $O((nk)^{2 \cdot \frac{1}{3}}) = O(n^{2/3} k^{2/3})$ regions.

In combination with Theorem 26.2, Theorem 26.3 allows us to bound the time needed to construct $\mathcal{M}(Z)$. For instance, the time to construct the optimum score function for the maximum parsimony problem is $O(n^{5/3} k^{5/3})$, as the number of regions is $O(n^{2/3} k^{2/3})$ and each evaluation takes $O(kn)$ time [49].

26.5 Sensitivity Analysis for Polynomially-Solvable Problems

A number of sensitivity analysis algorithms are known for polynomially-solvable problems. In addition to their intrinsic interest, these results are also useful in analyzing the stability of NP-hard problems. Some of this work has already been described in Section 26.4.1; we now survey further results on the area, which rely on more problem-specific approaches. We concentrate on linear programming, matroids, shortest paths, and network flows. For graph problems, we write n and m to denote the number of vertices and edges of the input graph.

26.5.1 Linear Programming

Parametric linear programming seems to have the distinction of being the first sensitivity analysis problem to be studied [1]. Let us consider first the simplest case, where the objective function depends on a single scalar parameter: μ.

$$Z_{LP}(\mu) = \min \left\{ \sum_{i=1}^{n} (c_i + \mu c_i') x_i : \sum_{i=1}^{n} a_{ij} x_i = b_j, \text{ for } j \in \{1, \ldots, m\}, x_i \geq 0 \text{ for } i \in \{1, \ldots, n\} \right\}. \quad (26.9)$$

If problem (26.9) is feasible, then it must have an optimum basic feasible solution (bfs) [50]. As the number of such solutions is finite, function Z_{LP} is piecewise linear and concave. Its worst-case number of breakpoints is known to be exponential in n [51]. The upper and lower tolerances for an optimum bfs $x^{(0)}$ at $\mu^{(0)}$ can be computed using the *parametric simplex method* [50]. Parametric linear programming has also been studied from the perspective of interior point methods [52–54].

When the objective function depends on more than one parameter, the optimum cost function is again piecewise linear and concave; however, little is known about the complexity of the associated subdivision, other than that it must be exponential. Methods for multi-parameter sensitivity analysis were first discussed in Reference 55; see also References 56, 57.

By linear programming duality, all the results obtained for parametric linear programming apply to linear programming problems where the right-hand side of the constraint set is a function of a parameter vector [50].

26.5.2 Parametric Minimum Spanning Trees and Matroids

The minimum spanning tree problem is a special case of *matroid optimization*. A *matroid* [58] is a pair $M = (E, \mathcal{I})$ where E is a finite nonempty set of elements and $\mathcal{I}$ is a family of subsets of E, called *independent sets*, satisfying two axioms: (i) any subset of an independent set is independent and (ii) if A and B are independent, with $|A| < |B|$, then there exists some $x \in B$ such that the set $A \cup \{x\}$ is independent. The *rank* of M is the cardinality of its largest independent set. A *base* of a matroid is a maximal independent set. A *weighted matroid* is one whose elements have real-valued weights. The *matroid optimization problem* is to find a minimum-weight base, in a weighted matroid; that is, a base minimizing the sum of the element weights.

The *graphic matroid* of an undirected graph G, denoted by $M(G)$, is a matroid whose elements are the edges of G, whose independent sets are the sub-forests of G, and whose bases are the spanning forests of G. Thus, when G is connected, a minimum-weight base in $M(G)$ is a minimum spanning tree of G, and the rank of $M(G)$ is $n - 1$. For further examples of matroids see Reference 58.

Let $Z_M(\mu)$ be the function giving the weight of the optimum base of a matroid M where the weight of each element is a linear function of a scalar parameter μ. Then, $Z_M(\mu)$ has $O(mr^{1/3})$ breakpoints, where m and r are the number of elements and the rank of M, respectively [59]; this bound is tight in general [60]. The upper bound implies that the number of breakpoints for the parametric minimum spanning tree problem is $O(mn^{1/3})$. However, the best known lower bound for this special case is $\Omega(m\alpha(n))$, where α is the inverse Ackermann function [60].

The optimum cost function for the one-parameter minimum spanning tree problem can be constructed in time $O(n^{2/3} \log^{4/3} n)$ per breakpoint or in randomized expected time $O(n^{2/3} \log n)$, per breakpoint [61]. For planar graphs, the time bound can be improved to deterministic $O(n^{1/4} \log^{3/2} n)$ or randomized expected $O(n^{1/4} \log n)$ time per breakpoint; further results for minor-closed families of graphs are given in Reference 61. The latter reference also offers algorithms for *kinetic* minimum spanning tree problem, which extends the parametric minimum spanning tree problem by additionally allowing arbitrary insertions or deletions of parametrically weighted edges.

All edge tolerances of a minimum spanning tree can be computed in deterministic $O(m \log \alpha(m, n))$ time [62] (see also Reference 63), where α is the inverse Ackerman function, and in randomized $O(m)$ time [64]. A deterministic $O(n)$-time algorithm exists for planar graphs [65]. These bounds match the bounds known for solving the minimum spanning tree problem with fixed edge costs; indeed, it has been shown that any minimum spanning tree algorithm can be used to calculate minimum spanning tree edge tolerances, without an asymptotic loss in efficiency [62].

Algorithms for finding the most vital edges in minimum cost spanning trees can be found in References 66–69.

The robustness function for the minimum spanning tree problem is concave, piecewise linear, with $O(mn)$ breakpoints and can be constructed in $O(m^2 n^3 \log(n^2/m))$ time [17]. More generally, the robustness function of a weighted matroid can be constructed in $O(m^5 n^2 + m^4 n^4 \tau)$ time, where τ is the time needed to test the independence of a set of at most n elements [70].

When the number of parameters is fixed, the inverse parametric minimum spanning tree problem can be solved in randomized linear expected time, and deterministically in $O(m \log^2 n)$ worst case running time. The inverse problem can be solved in polynomial time by means of the ellipsoid method for linear programming, even when the number of parameters is large [18]. This result extends to shortest path, matching, and other "optimal subgraph" problems.

26.5.3 Shortest Paths and Related Problems

Two kinds of shortest path sensitivity analysis problems have been considered in the literature. The first deals with the sensitivity of the shortest path tree rooted at a specified source vertex s. The edge tolerance problem for this case has been studied by several authors, often alongside the tolerance problem for minimum spanning trees [63,71,72]. Thus, all edge tolerances of a shortest path tree can be computed in deterministic $O(m \log \alpha(m, n))$ time [62]; for planar graphs, the same problem can be solved in $O(n)$ time [65]. Algorithms for the most vital edge problems on shortest path trees on undirected graphs have also been developed [73,74].

The other kind of problem studies the sensitivity of the shortest path from a source s to a destination t. Ramaswamy et al. [75] have devised algorithms for two related questions. Let P^* be a shortest s-t path in an edge-weighted undirected graph G. Then, there is an $O(m + |P^*| \log |P^*|)$ algorithm for finding all upper and lower tolerances of all edges in G with respect to P^*. Ramaswamy et al. also consider the following max-min problem. Let G be an undirected graph where each edge has a real-valued capacity. The *capacity* of a path is the minimum capacity of an edge on the path. Let Q^* be the maximum capacity s-t path in G. Then, all upper and lower tolerances of the edges with respect to Q^* can be computed in $O(m + |Q^*| \log |Q^*|)$ time. These questions have applications to the pricing of edges in networks [76,77]. Note that, for the shortest path problem, there is a close relationship between computing tolerances and finding most vital edges. For instance, the most vital edge with respect to the shortest s-t path is the edge with the largest upper tolerance. In contrast, the most vital edge problem for minimum cost spanning trees is not directly related to the problem of computing edge tolerances.

Let $Z(\mu)$ be the weight of the shortest s-t path when each edge weight varies linearly as a function of a scalar parameter μ. Gusfield [78] shows that Z has $O(n^{\log n})$ breakpoints. Carstensen [79] provides evidence that this bound is tight.

26.5.4 Parametric Maximum Flow

The *maximum-flow problem* asks to find a source-to-sink flow of maximum value in a capacitated network G. A closely related question is the *minimum cut problem*, which asks to find a partition (S, T) of the vertex set—that is, a *cut*—in G such that the source is in S and the sink is in T, and such that the total capacity of the edges from S to T is minimized. By the *max-flow min-cut theorem* [80], the value of the maximum flow in G equals that of the minimum cut in G.

In the parametric maximum flow problem, the capacities are functions of a parameter vector. Not much is known about the properties of the general version of the problem, even for a single parameter; however, the following important special case has received considerable attention. A parametric flow network is *simple* if the capacity of each edge out of the source is a nondecreasing linear function of μ, the capacity of each edge into the sink is a nonincreasing function of μ, and the capacity of every other edge is constant (we assume that there are no arcs between source and sink). Simple parametric flow networks arise in diverse applications, including scheduling, stable marriage problems, finding maximum density subgraphs, and baseball elimination problems [81,82].

In simple networks the series of minimum cuts $(S_i, T_i), \ldots, (S_r, T_r)$ encountered as μ is increased have a *nesting property*; that is, $S_i \subset S_{i+1}$, for $i = 1, \ldots, r-1$ [39]. Thus, the function $Z_C(\mu)$ giving the capacity of the minimum cut has at most $n - 2$ breakpoints. Remarkably, Gallo et al. [81] show that the preflow-push algorithm [83] can be modified to construct $\mathcal{M}(Z_C)$ in $O(nm \log(n^2/m))$ time, the same time as that algorithm takes to compute a *single* maximum flow. They also show how to achieve the same running time for finding the value of μ at which $Z_C(\mu)$ is zero, as required in certain minimum-ratio applications, and for locating a maximizer of $Z_C(\mu)$. Further applications and extensions are given in Reference 82; improvements for bipartite flow networks are given in Reference 84. Instances with piecewise-linear capacities or where certain arcs not incident on s or t have varying capacities—which have applications to preemptive scheduling with release times and deadlines, where processing times are varied, at a cost—have also been studied [85].

26.6 Sensitivity Analysis of NP-Hard Problems

We now give an overview of the work on sensitivity analysis for NP-hard problems, primarily with respect to changes in the coefficient vector.

26.6.1 Complexity of Sensitivity Analysis

Unsurprisingly, several results exist indicating that analyzing the sensitivity of NP-hard problems is itself hard. Thus, Hall and Posner [86] have shown that, for a large class of NP-hard scheduling problems, performing sensitivity analysis relative to the processing times is itself NP-hard. Their results, which actually apply to the problem of recomputing the optimum solution after a specific change of parameters, strongly suggest that the computation of tolerances, stability regions, and radii is hard.

van Hoesel and Wagelmans [87] have proved a variety of hardness results for 0/1 problems with linear objective function, including the following:

- If the tolerances can be computed in polynomial time, then (under some mild assumptions) the fixed-parameter version of the problem can be solved in polynomial time as well. Thus, for instance, there is no polynomial time algorithm for determining the edge tolerances for the traveling salesman problem unless P = NP.
- If it can be checked in polynomial time whether an optimum solution $x^{(0)}$ for a given cost vector $c^{(0)}$ is also optimum for another cost vector $c^{(1)}$, then the upper and lower tolerances at $c^{(0)}$ can be computed in polynomial time. Thus, it is NP-hard to test whether the optimum solution to an NP-hard 0/1 optimization problem remains optimum after arbitrary parameter changes.
- If there is a polynomial-time ϵ-approximate algorithm for computing all the upper tolerances for a 0/1 problem, then there exists a polynomial-time algorithm to compute the upper tolerances exactly. Thus, it is NP-hard to compute the upper tolerances of the optimum solution to an NP-hard 0/1 optimization problem approximately.
- If all upper tolerances for an ϵ-approximate solution $x^{(0)}$ at $c^{(0)}$ can be computed in polynomial time, then the optimum solution at $c^{(0)}$ can be computed in polynomial time. Thus, for instance, the existence of a polynomial algorithm to determine upper tolerances for an ϵ-approximate solution to the minimum-cost knapsack problem, would imply that P = NP.

In the light, if the above-mentioned results, computing stability regions, and radii for ϵ-approximate solutions is a hard problem in general.

Sotskov et al. [21] establish a relationship between the complexity of computing exact solutions and that of multi-parameter sensitivity analysis. Let x be an ϵ-optimal solution and let $v(x)$ denote the number of 1's in x. Let $c^{(0)}$ be a coefficient vector and $x^{(0)}$ be an ϵ-approximate solution at $c^{(0)}$. Then, if the optimum cost $Z(c)$ can be calculated in $O(g(n))$ time for any coefficient vector $c \in \mathbb{R}^n$, the stability radius of x at $c^{(0)}$ can be computed in $O(2^{v(x)} n g(n))$ time. In contrast, Chakravarti and Wagelmans [88] show that the stability radius can be computed in polynomial time for problems with linear objective functions whose fixed-parameter versions can be solved in polynomial time.

Although we have limited our discussion to problems with linear objective function, we should note that several results are known for bottleneck problems. See, for example, References 89–91; the last of these references surveys the extensive treatment of the subject in the Russian literature.

Graphs of bounded tree-width: Many NP-complete graph optimization problems are polynomially-solvable on graphs of constant-bounded tree-width [92–95] (for a definition of tree-width, see References 96, 97). Indeed, parametric versions of several of these problems, where edge and/or vertex costs are linear functions of a single scalar parameter μ, are often themselves efficiently solvable [34,98]. This has some implications to the approximate parametric analysis of NP-hard problems, through Baker's [99] ingenious technique for obtaining polynomial-time approximation schemes for certain (nonparametric) problems on weighted planar graphs. Among these problems are maximum independent set, maximum tile salvage, partition into triangles, maximum H-matching, minimum vertex cover, minimum dominating set, and minimum edge dominating set.

Baker's approach is based on decomposing the input planar graph into *k-outerplanar* graphs, which are graphs of constant-bounded tree-width. This enables us to use parametric algorithms for bounded-tree-width graphs to obtain approximation algorithms for constructing the optimum cost function for of Baker's problems. More precisely, let $Z(\mu)$ denote the optimal cost function for one of Baker's problems. Then, for every fixed ϵ, there exists a polynomial-time algorithm that produces a function $Z_\epsilon(\mu)$ that has polynomially-many breakpoints and such that for each μ, $Z_\epsilon(\mu)$ is ϵ-approximate to $Z(\mu)$ [34].

26.6.2 Bounding the Stability Region

The notions of *restriction* and *relaxation* can help in obtaining bounds on the stability region. Let $cost_A$ and X_A denote the cost function and set of feasible solutions of a problem A of the form (26.2). Problem Q is a *restriction* of problem P if $X_Q \subseteq X_P$ and $cost_Q(x) \geq cost_P(x)$ for every $x \in X_Q$. Problem R is a *relaxation* of problem P if $X_R \supseteq X_P$ and $cost_R(x) \leq cost_P(x)$ for every $x \in X_P$. Note that restriction and relaxation are inverse relations; that is, R is a relaxation of P if and only if P is a restriction of R. Also, the cost of the optimum solution of any relaxation (restriction) of P is a lower (upper) bound on the cost of the optimum solution for P. These facts lead to the following elementary, but useful, observations [4]:

(GN1): If an optimum solution x^* of P is feasible for a restriction Q of P and $cost_Q(x^*) = cost_P(x^*)$, then x^* is optimum for Q.

(GN2): If $x^{(0)}$ is a feasible solution for P and $cost_P(x^{(0)})$ equals the cost of an optimum solution for some relaxation of P, then $x^{(0)}$ must be optimum for P.

Observation (GN1) allows us to make statements about the sensitivity of an optimum solution x^* to changes in the coefficient vector. For example, x^* remains optimum if we replace c by any c' such that $c'_j \geq c_j$ for every j such that $x^*_j = 0$ and $c'_j = c_j$ otherwise; that is, each such c_j has infinite upper tolerance. Similarly, for a 0/1 optimization problem, x^* remains optimum if we replace c by any c' such that $c'_j \leq c_j$ for every j such that $x_j = 1$ and $c'_j = c_j$ otherwise; that is, each such c_j has infinite lower tolerance.

Hall and Posner note that observation (GN1) can be used to study the sensitivity of scheduling problems [86]. As a simple example, an optimal schedule remains optimal if we increase the release date of a job, but not beyond the time when the job starts processing.

Observation (GN2) applies to Lagrangian relaxation (Section 26.2): Suppose that λ^* is the maximizer of the Lagrangian function (Equation (26.2)), and that x^* is optimum solution for this parametric problem at λ^*. Then, if x^* is also feasible for the original problem (26.2), x^* must be optimum for the latter problem. Thus, we can get some idea of the stability of x^* in the original (hard) problem, by analyzing the sensitivity of x^* in the relaxed (usually polynomially-solvable) problem, relying on results such as those presented in Section 26.5. Along these lines, Libura [100] presents methods for computing lower bounds on the edge tolerances for the traveling salesman problem and the minimum-weight Hamiltonian path problems, which rely on relaxations to minimum spanning tree problems. This allows one to use the results on sensitivity analysis for minimum spanning trees described in Section 26.5.2.

Sotskov et al. [21] studied the stability region of ϵ-optimal solutions and prove certain necessary and sufficient conditions to obtain lower and upper bounds on the stability radius. They considered the stability of nonpreemptive general shop scheduling with precedence constraints, when the processing times of the jobs are varied and provide conditions for the stability radius to be strictly larger than 0 or for it to be infinite. Unfortunately, verifying these conditions is not easy.

k-best solutions and sensitivity analysis: Intuitively, the boundaries of the stability region of an optimum $x^{(0)}$ at $\lambda^{(0)}$ are determined by a set of near-optimum solutions at $\lambda^{(0)}$. This observation motivated Libura et al. [101,102] to study of the connections between near-optimality and sensitivity analysis for the traveling salesman problem. Unsurprisingly, they show that it is NP-hard to find the smallest k such that the

k best solutions at $c^{(0)}$ suffice to determine the stability region of an optimum tour at $c^{(0)}$. Nevertheless, they do obtain some positive results, which we review in the following.

Suppose we are given a set S containing the k best tours for some edge cost vector $c^{(0)}$; S of course includes an optimal tour, $H^{(0)}$. Let E_1 be the set of edges that are present in $H^{(0)}$, but not in every tour in S. Then, the upper tolerances of the costs of the edges in E_1 can be computed exactly and one can obtain lower bounds on the upper tolerances of the other edges. Similarly, let E_2 be the set of edges present in some tour in S. Then, one can calculate the lower tolerances of the edges present in E_2 and lower bounds on the lower tolerances of the other edges. The authors also show how to derive upper and lower bounds for the stability radius and how to use this information to derive subsets of the stability region. The authors evaluated their results experimentally on instances from TSPLIB [103]. They observe that the number of edges for which the tolerances can be computed increases quickly for smaller values of k, and the rate of growth decreases for larger values of k.

26.6.3 Sensitivity of Heuristics

Analyzing the stability of a heuristic algorithm is conceivably easier than analyzing the stability of an exact or ϵ-approximate solution, as the heuristic need not have any sort of performance guarantee (recall the distinction between algorithm and solution stability, explained at the end of Section 26.3). For example, Ghosh et al. [104] consider the greedy heuristic for binary knapsack and subset sum problems. They show that the upper and lower tolerances of this algorithm for any parameter (the knapsack capacity, the weights and the profits of the items) can be computed in polynomial time. They also study the conditions under which the sensitivity analysis of the heuristic generates bounds for the tolerances for the optimal solutions, and the empirical behavior of the greedy output when there is a change in the problem data. Hall and Posner [86] study the behavior of greedy heuristics for two NP-hard problems: scheduling two machines to minimize the weighted completion time and scheduling two machines to minimize the makespan. They provide approaches to obtain bounds on the upper tolerances of each heuristic to changes in the processing times.

Intuitively, as a simple heuristic uses less of the input information than an exact algorithm, it may produce a poor result, but the solution should be less susceptible to parameter changes than the optimal solution. This intuition is supported by the work of Kolen et al. comparing two heuristics for scheduling jobs on identical parallel machines to minimize the makespan [105]: *shortest processing time* (SPT) and *longest processing time* (LPT). The ratios between the solution returned and the optimum are $2 - 1/m$ for SPT, where m is the number of machines, and $4/3 - 1/(3m)$ for LPT [106,107].

Suppose we vary the processing time μ of one of the jobs from 0 to ∞. For $H \in \{SPT, LPT\}$, let $Z_H(\mu)$ be the length of the makespan of the schedule returned by μ, let $B_H(n, m)$ denote the worst case number of breakpoints of Z_H, and let $A_H(n, m)$ be the number of different assignments of jobs to machines. The latter two values serve as an indication of sensitivity of H. Kolen et al. show that $Z_H(\mu)$ is a continuous piecewise linear function. They also prove that $A_{SPT}(n, m) \leq n$ and $B_{SPT}(n, m) \leq 2\lceil n/m \rceil$ and that both bounds are tight. On the other hand, $A_{LPT}(n, m) \leq 2^{n-m}$ and $B_{LPT}(n, m) \leq 2^{n-m+1}$. The first bound is tight and there exists an example for which $B_{LPT}(n, m) > 2^{(n-m)/2}$. These results support the intuition that the sensitivity of a list scheduling rule increases with the quality of the schedule produced.

References

1. Gass, S.I. and Saaty, T., The computational algorithm for the parametric objective function, *Naval Res. Logis. Quart.*, 2, 39, 1955.
2. Hoffman, H. and Jacobs, W., Smooth patterns of production, *Manag. Sci.*, 1, 86, 1954.
3. Gal, T. and Greenberg, H.J., (Eds.), *Advances in Sensitivity Analysis and Parametric Programming*, International Series in Operations Research & Management Science, Kluwer Academic Publishers, Boston, MA 1997.

4. Geoffrion, A.M. and Nauss, R., Parametric and postoptimality analysis in integer linear programming, *Manag. Sci.*, 23(5), 453, 1977.
5. Wagelmans, A., Sensitivity analysis in combinatorial optimization, Ph.D. Thesis, Econometric Institute, Erasmus University, Rotterdam, the Netherlands, 1990.
6. Greenberg, H., An annotated bibliography for post-solution analysis in mixed integer programming and combinatorial optimization, in *Advances in Computational and Stochastic Optimization, Logic Programming, and Heuristic Search*, Woodruff, D. (Ed.), Kluwer Academic Publishers, Boston, MA, 1998, p. 97. (Note: updated bibliography at http://www.cudenver.edu/~hgreenbe/aboutme/papers/mipbib.pdf).
7. Downey, R.G. and Fellows, M.R., *Parameterized Complexity*, Springer-Verlag, New York, 1999.
8. Hromkovič, J., Stability of approximation algorithms for hard optimization problems, in *SOFSEM*, LNCS, Springer, Berlin, Vol. 1725, 1999, p. 29.
9. Bender, M.A. and Chekuri, C., Performance guarantees for the TSP with a parameterized triangle inequality, *Inf. Proc. Lett.*, 73(1–2), 17, 2000.
10. Held, M. and Karp, R., The traveling salesman problem and minimum spanning trees, *Oper. Res.*, 18, 1138, 1970.
11. Held, M. and Karp, R., The traveling salesman problem and minimum spanning trees: Part II, *Math. Prog.*, 6, 6, 1971.
12. Garey, M.R. and Johnson, D.S., *Computers and Intractability: A Guide to the Theory of NP-Completeness*, W. H. Freeman and Company, Gordon and Breach, New York, 1979.
13. Fisher, M.L., The Lagrangian relaxation method for solving integer programming problems, *Manag. Sci.*, 27(1), 1, 1981.
14. Nemhauser, G.L. and Wolsey, L.A., *Integer and Combinatorial Optimization*, Wiley-Interscience Series in Discrete Mathematics and Optimization, John Wiley & Sons, New York, 1988.
15. Chandrasekaran, R., Minimal ratio spanning trees, *Networks*, 7, 335, 1977.
16. Lawler, E.L., Optimal cycles in doubly-weighted linear graphs, in *Proceedings of the International Symposium on Theory of Graphs*, Gordon and Breach, New York, 1966, p. 209.
17. Frederickson, G.N. and Solis-Oba, R., Increasing the weight of minimum spanning trees, *J. Algor.*, 33, 394, 1999.
18. Eppstein, D., Setting parameters by example, *SIAM J. Comput.*, 32(3), 643, 2003.
19. Gusfield, D. and Stelling, P., Parametric and inverse-parametric sequence alignment with XPARAL, in *Methods in Enzymology, Computer Methods for Macromolecular Sequence Analysis*, Doolittle, R.F. (Ed.), Vol. 266, Academic Press, New York, 1996, p. 481.
20. Sun, F., Fernández-Baca, D., and Yu, W., Inverse parametric sequence alignment, *J. Algor.*, 53(1), 36, 2004.
21. Sotskov, Y.N., Wagelmans, A.P.M., and Werner, F., On the calculation of the stability radius of an optimal or an approximate schedule, *Ann. Oper. Res.*, 83, 213, 1998.
22. Radzik, T., Newton's method for fractional combinatorial optimization, in *Proceedings of Foundations of Computer Science*, 1992, p. 659.
23. Radzik, T., Parametric flows, weighted means of cuts, and fractional combinatorial optimization, in *Complexity in Numerical Computations*, Pardalos, P. (Ed.), World Scientific, Singapore, 1993, p. 351.
24. Radzik, T., *Algorithms for some linear and fractional combinatorial optimization problems*, Department of Computer Science, Stanford University, Stanford, CA, 1992.
25. Megiddo, N., Combinatorial optimization with rational objective functions, *Math. Oper. Res.*, 4, 414, 1979.
26. Megiddo, N., Applying parallel computation algorithms in the design of serial algorithms, *JACM*, 30(4), 852, 1983.
27. Cole, R., Slowing down sorting networks to obtain faster sorting algorithms, *JACM*, 34(1), 200, 1987.

28. Agarwala, R. and Fernández-Baca, D., Weighted multidimensional search and its application to convex optimization, *SIAM J. Comput.*, 25, 83, 1996.

29. Cohen, E. and Megiddo, N., Maximizing concave functions in fixed dimension, in *Complexity in Numerical Optimization*, Pardalos, P.M. (Ed.), World Scientific Publishing Co, Singapore, 1993, p. 74.

30. Fernández-Baca, D., On non-linear parametric search, *Algorithmica*, 30, 1, 2001.

31. Toledo, S., Maximizing non-linear concave functions in fixed dimension, in *Complexity in Numerical Computations*, Pardalos, P. (Ed.), World Scientific Publishing Co, Singapore, 1993, p. 429.

32. Frederickson, G.N., Optimal algorithms for tree partitioning, in *Proceedings of the Symposium on Discrete Algorithms*, Society for Industrial and Applied Mathematics Philadelphia, PA, 1991, p. 168.

33. Fernández-Baca, D., Slutzki, G., and Eppstein, D., Using sparsification for parametric minimum spanning tree problems, *Nordic J. of Comput.*, 34(4), 352, 1996.

34. Fernández-Baca, D. and Slutzki, G., Optimal parametric search on graphs of bounded tree-width, *J. Algor.*, 22, 212, 1997.

35. Fernández-Baca D. and Slutzki, G., Linear-time algorithms for parametric minimum spanning tree problems on planar graphs, *Theor. Comp. Sci.*, 181, 57, 1997.

36. Press, W.H., Flannery, B.P., Teukolsky, S.A., and Vetterling, W.T., *Numerical Recipes: The Art of Scientific Computing*, Cambridge University Press, Cambridge, UK, 2nd ed., 1992.

37. Dobkin, D., Edelsbrunner, H., and Yap, C.K., Probing convex polytopes, in *Proceedings of Symposium on Theory of Computing*, ACM Press, New York, 1986, p. 424.

38. Dobkin, D., Edelsbrunner, H., and Yap, C.K., Probing convex polytopes, in *Autonomous Robot Vehicles*, Cox, I.J. and Wilfong, G.T. (Eds.), Springer-Verlag, New York, 1990, p. 328.

39. Eisner, M. and Severance, D., Mathematical techniques for efficient record segmentation in large shared databases, *JACM*, 23, 619, 1976.

40. Fernández-Baca, D. and Srinivasan, S., Constructing the minimization diagram of a two-parameter problem, *Oper. Res. Lett.*, 10, 87, 1991.

41. Seidel, R., Convex hull computations, in *Handbook of Discrete and Computational Geom.*, Goodman, J. and O'Rourke, J. (Ed.), CRC Press LLC, Boca Raton, FL, 1997, chapter 19.

42. McMullen, P. The maximum number of faces of a convex polytope, *Mathematika*, 17, 179, 1970.

43. Fernández-Baca, D. and Venkatachalam, B., Parametric sequence alignment, in *Handbook of Computational Molecular Biology*, Aluru, S. (Ed.), Vol. 2, Chapman & Hall/CRC, Boca Raton, FL, 2005, p. 271.

44. Pachter, L. and Sturmfels, B., Parametric inference for biological sequence analysis, in *Proceedings of the National Academy of Science*, 101(46), 2004, p. 16138.

45. Pachter, L. and Sturmfels, B., Tropical geometry of statistical models, *Proceedings of the National Academy of Science*, 101(46), 2004, p. 16132.

46. Fernández-Baca, D., Seppäläinen, T., and Slutzki, G., Parametric multiple sequence alignment and phylogeny construction, *J. Disc. Algor.*, 2, 271, 2004.

47. Gusfield, D. and Irving, R.W., *The Stable Marriage Problem: Structure and Algorithms*, MIT Press, Cambridge, MA, 1989.

48. Gusfield, D. and Irving, R.W., Parametric stable marriage and minimum cuts, *Inf. Proc. Lett.*, 30, 255, 1989.

49. Felsenstein, J., *Inferring Phylogenies*, Sinauer Associates, Sunderland, MA, 2003.

50. Vanderbei, R.J., *Linear Programming: Foundations and Extensions*, International Series in Operations Research and Management Science, 2nd ed., Vol. 37, Kluwer Academic Publishers, Boston, MA, 2001.

51. Murty, K., Computational complexity of parametric linear programming, *Math. Prog.*, 19, 213, 1980.

52. Adler, I. and Monteiro, R.D.C., A geometric view of parametric linear programming, *Algorithmica*, 8(2), 161, 1992.

53. Berkelaar, A., Roos, C., and Terlaky, T., The optimal set and optimal partition approach to linear and quadratic programming, in *Recent Advances in Sensitivity Analysis and Parametric Programming*, Greenberg, H. and Gal, T. (Eds.), Kluwer Academic Publishers, Boston, MA, 1997, chapter 6.

54. Yildirim, E.A. and Todd, M.J., Sensitivity analysis in linear programming and semidefinite programming using interior-point methods, *Math. Prog.*, 90(2), 229, 2001.

55. Gal, T. and Nedoma, J., Multiparametric linear programming, *Manag. Sci.*, 18, 406, 1972.

56. Borrelli, F., Bemporad, A., and Morari, M., Geometric algorithm for multiparametric linear programming, *J. Opt. Theor. Appl.*, 118(3), 515, 2003.

57. Schechter, M., Polyhedral functions and multiparametric linear programming, *J. Opt. Theor. Appl.*, 53, 269, 1987.

58. Welsh, D.J.A., *Matroid Theory*, Academic Press, New York, 1976.

59. Dey, T., Improved bounds on planar k-sets and related problems, *Disc. Comput. Geom.*, 19(3), 373, 1998.

60. Eppstein, D., Geometric lower bounds for parametric matroid optimization, *Disc. Comput. Geom.*, 20, 463, 1998.

61. Agarwal, P.K., Eppstein, D., Guibas, L.J., and Henzinger, M.R., Parametric and kinetic minimum spanning trees, in *Proceedings of Foundations of Computer Science*, 1998.

62. Pettie, S., Sensitivity analysis of minimum spanning trees in sub-inverse-Ackermann time, in *Proceedings of ISAAC*, LNCS, 3827, 2005, p. 964.

63. Tarjan, R.E., Sensitivity analysis of minimum spanning trees and shortest path problems, *Inf. Proc. Lett.*, 14(1), 30, 1982.

64. Dixon, B., Rauch, M., and Tarjan, R., Verification and sensitivity analysis of minimum spanning trees in linear time, *SIAM J. Comput.*, 21, 1184, 1994.

65. Booth, H. and Westbrook, J., A linear algorithm for analysis of minimum spanning and shortest-path trees of planar graphs, *Algorithmica*, 11, 341, 1994.

66. Hsu, L.H., Jan, R., Lee, Y., Hung, C., and Chern, M., Finding the most vital edge with respect to minimum spanning tree in weighted graphs, *Inf. Proc. Lett.*, 39, 277, 1991.

67. Hsu, L., Wang, P., and Wu, C., Parallel algorithms for finding the most vital edge with respect to minimum spanning tree, *Parallel Comput.*, 18, 1143, 1992.

68. Iwano, K. and Katoh, N., Efficient algorithms for finding the most vital edge of a minimum spanning tree, *Inf. Proc. Lett.*, 48, 211, 1993.

69. Banerjee, S. and Saxena, S., Parallel algorithm for finding the most vital edge in weighted graphs, *J. Parall. Dist. Comput.*, 46, 101, 1997.

70. Frederickson, G.N. and Solis-Oba, R., Algorithms for measuring perturbability in matroid optimization, *Combinatorica*, 18(4), 503, 1998.

71. Shier, D. and Witzgall, C., Arc tolerances in shortest path and network flow problems, *Networks*, 10(4), 277, 1980.

72. Gusfield, D., A note on arc tolerances in sparse shortest-path and network flow problems, *Networks*, 13, 191, 1983.

73. Malik, K., Mittal, A., and Gupta, S., The k most vital edges in the shortest path problem, *Oper. Res. Lett.*, 8, 223, 1989.

74. Venema, S., Shen, H., and Suraweera, F., NC algorithms for the single most vital edge problem with respect to shortest paths, *Inf. Proc. Lett.*, 60, 243, 1996.

75. Ramaswamy, R., Orlin, J.B., and Chakravarti, N., Sensitivity analysis for shortest path problems and maximum capacity path problems in undirected graphs, *Math. Prog.*, 102(2. Ser. A.), 355, 2005.

76. Hershberger, J. and Suri, S., Vickrey prices and shortest paths: What is an edge worth? in *Proceedings of Foundations of Computer Science*, IEEE Press, New York, 2001, p. 252.

77. Hershberger, J. and Suri, S., Erratum to: Vickrey prices and shortest paths: What is an edge worth?, in *Proceedings of Foundations of Computer Science*, IEEE Press, New York, 2002, p. 809.

78. Gusfield, D., Sensitivity analysis for combinatorial optimization, TR UCB/ERL M80/22, University of California, Berkeley, CA, 1980.

79. Carstensen, P., Complexity of some parametric integer and network programming problems, *Math. Prog.*, 26, 64, 1983.

80. Ahuja, R., Magnanti, T., and Orlin, J., *Network Flows: Theory, Algorithms, and Applications*, Prentice-Hall, Englewood Cliffs, NJ, 1993.

81. Gallo, G., Grigoriades, M., and Tarjan, R., A fast parametric maximum flow algorithm and applications, *SIAM J. Comput.*, 18, 30, 1989.

82. Gusfield, D. and Martel, C., A fast algorithm for the generalized parametric minimum cut problem and applications, *Algorithmica*, 7, 499, 1992.

83. Goldberg, A.V. and Tarjan, R.E., A new approach to the maximum-flow problem, *JACM*, 35(4), 921, 1988.

84. Ahuja, R., Orlin, J., Stein, C., and Tarjan, R., Improved algorithms for bipartite network flow, *SIAM J. Comput.*, 23(5), 906, 1994.

85. McCormick, S.T., Fast algorithms for parametric scheduling come from extensions to parametric maximum flow, *Oper. Res.*, 47(5), 744, 1999.

86. Hall, N.G. and Posner, M.E., Sensitivity analysis for scheduling problems, *J. Schedul.*, 7(1), 49, 2004.

87. van Hoesel, S. and Wagelmans, A., On the complexity of postoptimality analysis of 0/1 programs, *Disc. Appl. Math.*, 91(1–3), 251, 1999.

88. Chakravarti, N. and Wagelmans, A.P.M., Calculation of stability radii for combinatorial optimization problems, *Oper. Res. Lett.*, 23(1–2), 1, 1998.

89. Ramaswamy, R. and Chakravarti, N., Complexity of determining exact tolerances for min-sum and min-max combinatorial optimization problems, Working Paper WPS-247/95, Indian Institute of Management, Calcutta, India, 1995.

90. Ramaswamy, R., Chakravarti, N., and Ghosh, D., Complexity of determining exact tolerances for min-max combinatorial optimization problems, SOM Report 00A22, University of Groningen, Groningen, the Netherlands, 2000.

91. Sotskov, Y.N., Leontev, V.K., and Gordeev, E.N., Some concepts of stability analysis in combinatorial optimization, *Disc. Appl. Math.*, 58(2), 169, 1995.

92. Arnborg, S. and Proskurowski, A., Linear time algorithms for NP-hard problems restricted to partial k-trees, *Disc. Appl. Math.*, 23(1), 11, 1989.

93. Bern, M.W., Lawler, E.L., and Wong, A.L., Linear time computation of optimal subgraphs of decomposable graphs, *J. Algor.*, 8, 216, 1987.

94. Bodlaender, H.L., Treewidth: Algorithmic techniques and results, in *Proceedings of the International Symposium on Mathematical Foundations of Computer Science*, Springer-Verlag Berlin Heidelberg New York, LNCS, 1295, 1997, p. 29.

95. Courcelle, B., Graph rewriting: An algebraic and logic approach, in *Handbook of Theoretical Computer Science*, Vol. B., van Leeuwen, J. (Ed.), Elsevier, Amsterdam, the Netherlands, 1990, chapter 5.

96. Robertson, N. and Seymour, P.D., Graph minors II: Algorithmic aspects of tree-width, *J. Algor.*, 7, 309, 1986.

97. Diestel, R., *Graph Theory, Graduate Texts in Mathematics,* Vol. 173, Springer-Verlag, New York, 2005.

98. Fernández-Baca, D., and Slutzki, G., Parametric problems on graphs of bounded tree-width, *J. Algor.*, 16, 408, 1994.

99. Baker, B.S., Approximation algorithms for NP-complete problems on planar graphs, *JACM*, 41(1), 153, 1994.

100. Libura, M., Sensitivity analysis for minimum Hamiltonian path and traveling salesman problems, *Disc. Appl. Math.*, 30(2–3), 197, 1991.

101. Libura, M., van der Poort, E.S., Sierksma, G., and van der Veen, J.A.A., Stability aspects of the traveling salesman problem based on k-best solutions, *Disc. Appl. Math.*, 87(1–3), 159, 1998.

102. v. d. Poort, E.S., *Aspects of Sensitivity Analysis for the Traveling Salesman Problem*, PhD Thesis, University of Groningen, Groningen, the Netherlands, 1997.

103. Reinelt, G. and Bixby, B., TSPLIB, http://nhse.cs.rice.edu/softlib/catalog/tsplib.html.

104. Ghosh, D., Chakravarti, N., and Sierksma, G., Sensitivity analysis of the greedy heuristic for binary knapsack problems, *Eur. J. Oper. Res.*, 169(2), 340, 2006.

105. Kolen, A.W.J., Rinnooy Kan, A.H.G., van Hoesel, C.P.M., and Wagelmans, A.P.M., Sensitivity analysis of list scheduling heuristics, *Disc. Appl. Math.*, 55(2), 145, 1994.

106. Graham, R.L., Bounds for certain multiprocessing anomalies, *Bell Sys. Tech. J.*, 45, 1563, 1966.

107. Graham, R.L., Bounds on multiprocessing timing anomalies, *SIAM J. Appl. Math.*, 17(2), 416, 1969.

27

Stability of Approximation

Hans-Joachim
Böckenhauer

Juraj Hromkovič

Sebastian Seibert

27.1 Introduction

The design of approximation algorithms has evolved as one of the most successful approaches for dealing with hard optimization problems. Already a short time after introducing NP-hardness (and completeness) as a concept for proving the intractability of computational problems [1,2], the first successful approximation algorithms were proposed [3–6]. Since then, a wealth of approximation algorithms for very many practically relevant problems have been developed, as large parts of this handbook stand witness to. Overviews on approximation algorithms can be found, for example in References 7–12.

Approximation algorithms allow us to jump from exponential time complexity (unless $P = NP$) to polynomial time complexity by a small change in the requirement from searching for the optimal solution to searching for a near-optimal solution. This effect is especially strong in the case of so-called approximation schemes, where one can guarantee to find a solution whose cost can become arbitrarily close to the optimal value (Chapter 8 for more details).

Another possibility to make intractable problems tractable is to consider only a subset of input instances satisfying some special properties instead of the set of all inputs for which the problem is well-defined. A nice example is the traveling salesman problem (TSP), which has served from the beginning as an archetype of a hard problem [13,14]. TSP is not only NP-hard, but also the search for an α-approximate solution for TSP is NP-hard for every constant α. But if one considers TSP for inputs satisfying the triangle inequality (the so-called Δ-TSP), one can even design an approximation algorithm [5] with approximation ratio $\alpha = \frac{3}{2}$. The situation is still more interesting, if one considers the Euclidean TSP, where the vertices are points in the Euclidean space and the edge weights correspond to the distances between the vertices in the Euclidean metrics. The Euclidean TSP is strongly NP-hard [15], but, for every small $\varepsilon > 0$, one can design a $(1 + \varepsilon)$-approximation algorithm [16–18] with almost linear time complexity.

The fascinating observations of huge quantitative changes as mentioned earlier have lead to the proposal of considering the *stability* of approximation algorithms as introduced in Reference 19. Let us consider the following scenario. One has an optimization problem P for two sets of inputs L_1 and L_2, $L_1 \subset L_2$. For L_1, there exists a polynomial-time α-approximation algorithm A, but for L_2 there is no polynomial-time δ-approximation algorithm for any $\delta > 1$ (unless $P = NP$). We pose the following question: Is the algorithm A really useful for inputs from L_1 only? Let us consider a distance measure for any

$y \in L_2 - L_1$ as a degree of violating the specification of L_1. The goal is to prove that A achieves an approximation ratio for instances in distance k from L_1 that depends only on k and α, but not on the size of the instance itself. In case that A does not exhibit this behavior, we try to modify it and get a new approximation algorithm that works for all instances from L_2 and has an approximation ratio that depends on k only, but not on the input size. In this case, we say that the algorithm A is *(approximation) stable* according to the used distance measure.

The idea of the concept of stability of approximation algorithms is similar to the concept of stability of numerical algorithms. But instead of observing the size of the change of the output value according to a small change of the input value, we look for the size of the change of the approximation ratio according to a small change in the specification (some parameters, characteristics) of the set of problem instances considered. If the change of the approximation ratio is small for every small change in the specification of the set of problem instances, then we have a stable algorithm. If a small change in the specification of the set of problem instances causes an essential (depending on the size of the input instances) increase of the relative error, then the algorithm is unstable.

The concept of stability was implicitly used for the first time in Reference 26 for the TSP. It was formally introduced in Reference 19. It enables us to show positive results extending the applicability of known approximation algorithms. As we will see later, the concept also motivates to modify an unstable algorithm A to get a stable algorithm B that achieves the same approximation ratio on the original set of problem instances as A has, but can also be successfully used outside of this original set. This concept is useful because there are a lot of problems for which an additional assumption on the "parameters" of the problem instances leads to an essential decrease of the hardness of the problem. Such effects are the starting points for trying to partition the whole set of problem instances into a spectrum of classes according to their polynomial-time approximability.

As one can observe, this approach is similar to the concept of parameterized complexity, introduced by Downey and Fellows [20,21], in trying to overcome the troubles caused by measuring complexity and approximation ratio in the worst-case manner. The main aim of both concepts is in partitioning the set of all instances of a hard problem into infinitely many classes with respect to the hardness of particular instances. We believe that approaches similar to these lie at the core of modern algorithmics, because they provide a deeper insight into the nature of the hardness of specific problems. In many applications, we are not interested in the worst-case problem hardness, but in the hardness of forthcoming problem instances.

This chapter is organized as follows: In Section 27.2, we formally define the concept of approximation stability. In Section 27.3, we apply the concept of stability to the TSP and exhibit a partition of the set of general TSP input instances into infinitely many classes, according to their approximability in dependence of a relaxation of the triangle inequality. In Section 27.4, we then complement the picture by introducing also a partition into infinitely many approximability classes inside the metric TSP. Moreover, lower bounds are exhibited for all these classes, inside as well as outside the metric case. They serve for judging the quality of the obtained algorithms, and they show that one cannot, in principle, expect much better results. Section 27.5 concludes the chapter with a survey of other successful applications of approximation stability and a discussion of the concept.

27.2 Definition of the Stability of Approximation Algorithms

We assume that the reader is familiar with the basic concepts and notions of algorithmics and complexity theory as presented in standard textbooks such as in References 11, 13, 22–24. Next, we give a formal definition of the notion of an optimization problem, which is an extended, more detailed version [7] of the standard definition. This extended definition allows us to study the influence of the input sets on the hardness of the problem considered.

Let $\mathbb{N} = \{0, 1, 2, \ldots\}$ be the set of nonnegative integers, and let $\mathbb{R}^+$ be the set of positive reals.

Definition 27.1 *An* optimization problem U *is a 7-tuple* $U = (\Sigma_I, \Sigma_O, L, L_I, \mathcal{M}, cost, goal)$, *where*

(i) Σ_I *is an alphabet called* input alphabet,

(ii) Σ_O *is an alphabet called* output alphabet,

(iii) $L \subseteq \Sigma_I^*$ *is a language over* Σ_I *called the* language of consistent inputs,

(iv) $L_I \subseteq L$ *is a language over* Σ_I *called the* language of actual inputs,

(v) $\mathcal{M}$ *is a function from* L *to* $2^{\Sigma_O^*}$, *where, for every* $x \in L$, $\mathcal{M}(x)$ *is called the* set of feasible solutions *for the input* x,

(vi) cost *is a function, called* cost function *that, for every pair* (u, x), *where* $u \in \mathcal{M}(x)$ *for some* $x \in L$, *assigns a positive real number* $cost(u, x)$,

(vii) $goal \in \{\min, \max\}$ *is the* optimization goal.

For every $x \in L$, *we define*

$$SolOpt_U(x) = \{y \in \mathcal{M}(x) \mid cost(y) = goal\{cost(z) \mid z \in \mathcal{M}(x)\}\}$$

as the set of optimal solutions, and $Opt_U(x) = cost(y)$, *for some* $y \in SolOpt_U(x)$.

Clearly, the meaning of Σ_I, Σ_O, $\mathcal{M}$, $cost$, and $goal$ is the usual one. L may be considered as the set of consistent inputs, that is, the inputs for which the optimization problem is consistently defined. L_I is the set of inputs actually considered and only these inputs are taken into account for determining the complexity of the optimization problem U. This kind of definition is useful for considering the complexity of optimization problems parameterized according to their languages of actual inputs.

Definition 27.2 *Let* $U = (\Sigma_I, \Sigma_O, L, L_I, \mathcal{M}, cost, goal)$ *be an optimization problem. We say that an algorithm* A *is a* consistent algorithm *for* U *if, for every input* $x \in L_I$, A *computes an output* $A(x) \in \mathcal{M}(x)$. *We say that* A solves U *if, for every* $x \in L_I$, A *computes an output* $A(x)$ *from* $SolOpt_U(x)$. *The time complexity of* A *is defined as the function*

$$Time_A(n) = \max\{Time_A(x) \mid x \in L_I \cap (\Sigma_I)^n\}$$

from $\mathbb{N}$ *to* $\mathbb{N}$, *where* $Time_A(x)$ *is the length of the computation of* A *on* x.

In this chapter, we deal with the case $Time_A(n) \in O(p(n))$, for some polynomial p, only, that is, we just speak of an approximation algorithm when meaning *polynomial-time approximation algorithm*. Next, we fix our notation for some standard notions in the area of approximation algorithms.

Definition 27.3 *Let* $U = (\Sigma_I, \Sigma_O, L, L_I, \mathcal{M}, cost, goal)$ *be an optimization problem, and let* A *be a consistent algorithm for* U.

For every $x \in L_I$, *the* approximation ratio $R_A(x)$ *of* A *on* x *is defined as*

$$R_A(x) = \max\left\{\frac{cost(A(x))}{Opt_U(x)}, \frac{Opt_U(x)}{cost(A(x))}\right\}.$$

For any $n \in \mathbb{N}$, *we define the* approximation ratio of A *as*

$$R_A(n) = \max\{R_A(x) \mid x \in L_I \cap (\Sigma_I)^n\}.$$

For any positive real $\delta \geq 1$, *we say that* A *is a* δ-approximation algorithm *for* U *if* $R_A(x) \leq \delta$ *for every* $x \in L_I$. *For every function* $f: \mathbb{N} \to \mathbb{R}$, *we say that* A *is a* $f(n)$-approximation algorithm *for* U *if* $R_A(n) \leq f(n)$, *for every* $n \in \mathbb{N}$.

To define the notion of stability of approximation algorithms we need to consider something like a distance between a language L and a word outside L.

Definition 27.4 (Böckenhauer et al. [19]) *Let $U = (\Sigma_I, \Sigma_O, L, L, \mathcal{M}, cost, goal)$ and $\overline{U} = (\Sigma_I, \Sigma_O, L, L_I, \mathcal{M}, cost, goal)$ be two optimization problems with $L_I \subset L$. A distance function for U according to L_I is any function $h_L \colon L \to \mathbb{R}^+$ satisfying the property*

$$h_L(x) = 0 \text{ for every } x \in L_I.$$

We define, for any $r \in \mathbb{R}^+$,

$$Ball_{r,h}(L_I) = \{w \in L \mid h(w) \le r\}.$$

Let A be a consistent algorithm for $\overline{U}$, and let A be an ε-approximation algorithm for U for some $\varepsilon \in \mathbb{R}^+$. Let p be a positive real. We say that A is p-stable according to h if, for every real $0 \le r \le p$, there exists a $\delta_{r,\varepsilon} \in \mathbb{R}^+$ such that A is a $\delta_{r,\varepsilon}$-approximation algorithm for $U_r = (\Sigma_I, \Sigma_O, L, Ball_{r,h}(L_I), \mathcal{M}, cost, goal)$.[*]

A is stable *according to h if A is p-stable according to h for every $p \in \mathbb{R}^+$. We say that A is* unstable *according to h if A is not p-stable for any $p > 0$.*

For every positive integer r, and every function $f_r \colon \mathbb{N} \to \mathbb{R}^+$ we say that A is $(r, f_r(n))$-quasi-stable according to h if A is an $f_r(n)$-approximation algorithm for $U_r = (\Sigma_I, \Sigma_O, L, Ball_{r,h}(L_I), \mathcal{M}, cost, goal)$.

One may see that the notion of stability can be useful for answering the question how broadly a given approximation algorithm is applicable. If one is interested in negative results, then one can try to show that, for any reasonable distance measure, the considered algorithm cannot be extended to work for a much larger set of inputs than the original one. In this way, one can search for some more exact boundaries between polynomial approximability and polynomial nonapproximability.

27.3 Stable Approximation Algorithms for the TSP

To illustrate the concept of approximation stability, we consider the well-known TSP. The TSP in its general form is very hard to approximate, but if one considers complete graphs in which the triangle inequality holds, then we have a 1.5-approximation algorithm due to Christofides [5].

To extend the part of the TSP that can be considered as reasonably well solvable, it is therefore a natural idea to look at instances that "violate the triangle inequality not by much." This is formalized by the idea of the parameterized triangle inequality.

Definition 27.5 *For a fixed $\beta \ge \frac{1}{2}$, a weighted graph (G, c), where $G = (V, E)$, obeys the β-triangle inequality if and only if*

$$c(u, w) \le \beta \left(c(u, v) + c(v, w) \right) \text{ for all } u, v, w \in V.$$

In case of $\beta > 1$ we speak of a *relaxed triangle inequality* which is the the case we want to investigate in this section. The case of a *sharpened triangle inequality* is considered in Section 27.4. We call Δ_β-TSP the TSP subproblem consisting of all instances obeying the β-triangle inequality (Δ_β-inequality for short).

In terms of stability, we start from Δ-TSP $= (\Sigma_I, \Sigma_O, L, L_I, \mathcal{M}, cost, \min)$ where we assume $\Sigma_I = \Sigma_O = \{0, 1, \#\}$. L contains codes of all weight functions for edges of complete graphs, and L_I contains codes of weight functions that satisfy the triangle inequality. Let, for every $x \in L$, $G_x = (V_x, E_x, c_x)$ be the complete weighted graph coded by x.

For every TSP input outside L_I (i.e., not obeying the standard triangle inequality), we obtain a distance from L_I as follows. Let β be minimal such that the given input belongs to Δ_β-TSP, then $\beta - 1$ is the distance from L_I.

---

[*] Note that $\delta_{r,\varepsilon}$ is a constant depending on r and ε only.

Definition 27.6 *For every $x \in L$,*

$$dist(x) = \max \left\{ 0, \max \left\{ \left. \frac{c_x(\{u,v\})}{c_x(\{u,p\}) + c_x(\{p,v\})} - 1 \right| \; u,v,p \in V_x \right\} \right\}.$$

Now let us look at known algorithms for Δ-TSP. We can easily observe that these algorithms work on any TSP-instance, that is, they are consistent for $(\Sigma_I, \Sigma_O, L, L, \mathcal{M}, cost, \min)$. Only the guarantee for the approximation ratio depends on the triangle inequality. It is therefore natural to ask how these algorithms perform on other inputs.

We start with a 2-approximation algorithm that simply converts a doubled minimum spanning tree (MST) into a Hamiltonian path, see, for example, Reference 14, where also a few equivalent variants are shown.

Double Spanning Tree Algorithm

Input: A complete graph $G = (V, E)$, and a cost function $c \colon E \to \mathbb{R}^+$.
Step 1: $T := \text{MST of } (G, c)$.
Step 2: $\omega := \text{Eulerian tour in the multi-graph obtained from } T \text{ by doubling each edge.}$
Step 3:   Construct a Hamiltonian tour H of G by shortening ω (i.e., by removing all repetitions of the occurrences of every vertex in ω in one run via ω from left to right).
Output: H.

As each Hamiltonian tour becomes a spanning tree after removing any single edge, clearly, an MST is at most as expensive as an optimal TSP solution for the same graph. Consequently, ω costs at most twice as much as an optimal solution.

Although this holds for any TSP instance, only under the triangle inequality, it can be guaranteed that the "shortening" in step 3 does not increase the cost. The shortening is executed by repeatedly replacing a path $x, u_1, \ldots, u_m, y$ by the edge $\{x, y\}$ (because $u_1, \ldots, u_m$ have already occurred before in the prefix of ω). Before further pursuing that thought, let us see how the Christofides algorithm builds on the previous one.

At closer inspection, it becomes clear that the doubling of the spanning tree was needed only to get an Eulerian subgraph. Consequently, in the Christofides algorithm the idea is to find an Eulerian subgraph in a more elaborate way, namely to add only one edge each to the vertices of odd degree in T.

Christofides Algorithm

Input: A complete graph $G = (V, E)$, and a cost function $c \colon E \to \mathbb{R}^+$.
Step 1: $T := \text{MST of } (G, c)$.
Step 2: $S := \{v \in V \mid deg_T(v) \text{ is odd}\}$.
Step 3:   Compute a minimum-weight perfect matching M on S in G.
Step 4: $\omega := \text{Eulerian tour in the multi-graph } G' = (V, E(T) \cup M)$.
Step 5:   Construct a Hamiltonian tour H of G by shortening ω.
Output: H.

In addition to the above-mentioned reasoning for the double spanning tree algorithm, here we have to convince ourselves that M costs at most half as much as an optimal TSP solution. The reasoning is that, by identifying the vertices of S on an optimal solution H_{opt}, we cut the circle H_{opt} at every vertex from S. Putting the resulting paths segments alternatingly into two sets and shortening each into a single edges, we obtain two matchings M_1, M_2. Obviously, M_1, M_2 cost as least as much as M.

But from this, $cost(M) \leq \frac{1}{2} cost(H_{opt})$ follows only under the triangle inequality as we had to shorten the path segments from H_{opt} into single edges to apply our argument.

We can see how the Christofides algorithm depends on the triangle inequality even more intricately than the double spanning tree algorithm. Here, the inequality is not only used in a constructive shortening step, but also already in the argument to bound the cost of the initial component M of the solution.

When applying these two algorithms to input instances from Δ_β-TSP for $\beta > 1$, one has to expect the approximation ratio to grow at least with β as β-factors appear immediately in the "shortening" of paths. Unfortunately however, for $\beta > 1$, the approximation ratio becomes dependent also on the input size as we will see now.

Lemma 27.1 (Böckenhauer et al. [19]) *The Christofides algorithm is unstable for dist. More precisely, for every $r \in \mathbb{R}^+$, if the Christofides algorithm is $(r, f_r(n))$-quasi-stable for dist, then*

$$f_r(n) \geq n^{\log_2(1+r)}/(2 \cdot (1+r)).$$

Proof. Remember, that the distance r from Δ-TSP of a n input x is the minimal value $\beta - 1$ such that x is a Δ_β-TSP instance, that is, we may use $\beta = 1 - r$ in the following as a shorthand notation.

We construct a weighted complete graph from $Ball_{r,dist}(L_I)$ as follows (Figure 27.1). We start with a path $p_0, p_1, \ldots, p_n$ for $n = 2^k$, $k \in \mathbb{N}$, where every edge $\{p_i, p_{i+1}\}$ has weight 1. Then, we add edges $\{p_i, p_{i+2}\}$ for $i = 0, 1, \ldots, n - 2$ with weight $2 \cdot \beta$. Generally, for every $m \in \{1, \ldots, \log_2 n\}$, we define $c(\{p_i, p_{i+2^m}\}) = 2^m \cdot \beta^m$ for $i = 0, \ldots, n - 2^m$. For all other edges, we assign maximal possible weights in such a way that the constructed input is in $Ball_{r,dist}(L_I)$.

Let us have a look on the work of the Christofides algorithm on the input (G, c). There is only one MST that corresponds to the path containing all edges of weight 1 (Figure 27.1). As every path contains exactly two vertices of odd degree, the Eulerian graph constructed in step 4 is the cycle $D = p_0, p_1, p_2, \ldots, p_n, p_0$ with the n edges of weight 1 and the edge of the maximal weight $n \cdot \beta^{\log_2 n} = n^{1+\log_2 \beta}$. As the Eulerian tour is a Hamiltonian tour, the output of the Christofides algorithm is unambiguously the cycle $p_0, p_1, \ldots, p_n, p_0$ with cost $n + n\beta^{\log_2 n}$. The optimal tour for this input is

$$H_{Opt} = p_0, p_2, p_4, \ldots, p_{2i}, p_{2(i+1)}, \ldots, p_n, p_{n-1}, p_{n-3}, \ldots, p_{2i+1}, p_{2i-1}, \ldots, p_3, p_1, p_0.$$

This tour contains two edges $\{p_1, p_0\}$ and $\{p_n, p_{n-1}\}$ of weight 1 and all $n - 2$ edges of weight $2 \cdot \beta$. Thus, $cost(H_{Opt}) = 2 + 2 \cdot \beta \cdot (n - 2)$ and

$$\frac{cost(D)}{cost(H_{Opt})} = \frac{n + n \cdot \beta^{\log_2 n}}{2 + 2 \cdot \beta \cdot (n - 2)} \geq \frac{n^{1+\log_2 \beta}}{2n \cdot \beta} = \frac{n^{\log_2 \beta}}{2\beta}. \qquad \blacksquare$$

Roughly the same happens for the double spanning tree algorithm. Only the fact that the choice of the root of the MST is not fixed may allow that algorithm to gain a factor of β at best (if two edges replacing paths of about $n/2$ edges each are used).

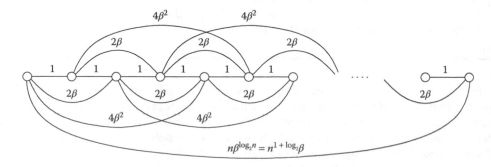

FIGURE 27.1 Sample graph showing that Christofides algorithm is unstable.

One may observe that, in this example, the double spanning tree algorithm can perform much better if, in shortening the Eulerian tour to a Hamiltonian tour, a more clever strategy is used. The Eulerian tour is in this case just a double traversal of the path that forms the MST. If, on each traversal, every second vertex is used, an optimal solution will be obtained. This idea might not work on every MST, but it is the basis for the first Δ_β-TSP algorithm by Andreae and Bandelt discussed in the following.

Before having a closer look at this algorithm, we just note that the above-mentioned example marks roughly the worst case for both algorithms.

Lemma 27.2 (Böckenhauer et al. [19]) *For every positive real number r, the Christofides algorithm and the double spanning tree algorithm are $(r, O(n^{\log_2((1+r)^2)}))$-quasi-stable for dist.*

Now we turn to the question how to modify those algorithms to get algorithms that are stable according to *dist*. As we have observed, the main problem is that shortening a path $u_1, u_2, \ldots, u_{m+1}$ to the edge $\{u_1, u_{m+1}\}$ can lead to

$$cost(\{u_1, u_{m+1}\}) = \beta^{\lceil \log_2 m \rceil} \cdot cost(u_1, u_2, \ldots, u_{m+1}).$$

Obviously, we need to limit the number of consecutive edges replaced by a single new edge by a constant. For a spanning tree being just a path, we have mentioned that it suffices to replace at most two consecutive edges. For a general spanning tree, the following result implies that replacing at most three consecutive edges suffices.

We use, for any graph $G = (V, E)$ and $k \in \mathbb{N}_{\geq 2}$, the notation

$$G^k = (V, \{\{x, y\} \mid x, y \in V, \text{ there is a path } x, P, y \text{ in } T \text{ of length at most } k\}).$$

Theorem 27.1 (Sekanina [25]) *For every tree $T = (V, E)$, the graph T^3 contains a Hamiltonian tour H.*

This means that every edge $\{u, v\}$ of H has a corresponding unique path $u, P_{u,v}, v$ in T of a length at most three. This is a helpful insight, but it still does not solve our problem completely. The remaining task is to estimate a good upper bound on the cost of the path

$$P_T(H) = u_1, P_{u_1, u_2}, u_2, P_{u_2, u_3}, u_3, \ldots, u_{n-1} P_{u_{n-1}, u_n}, u_n, P_{u_n, u_1}, u_1$$

in T that corresponds to the Hamiltonian tour $u_1, u_2, \ldots, u_n, u_1$ in T^3.

The problem is that we do not know the frequency of the occurrences of particular edges of T in $P(H)$. It may happen that the most expensive edges of T occur more frequently in $P(H)$ than the cheap edges. Observe also that $cost(T^3)$ cannot be bounded by $c \cdot cost(T)$ for any constant c independent of T because T^3 may be even a complete graph for some trees T. Thus, we need the following refinement of the above-mentioned theorem that is obtained by a slight modification of the original proof.

Lemma 27.3 (Andreae and Bandelt [26]) *Let T be a tree with $n \geq 3$ vertices, and let $\{p, q\}$ be an edge of T. Then, T^3 contains a Hamiltonian path $U = v_1, v_2, \ldots, v_n, p = v_1, v_n = q$, such that every edge of $E(T)$ occurs exactly twice in $P_T(H)$, where $H = U, p$ is a Hamiltonian tour in T^3.*

Let us call a Hamiltonian path of the type as described in the lemma *admissible*. Then we have the following refinement of the double spanning tree algorithm, developed by Andreae and Bandelt [26].

T^3-Algorithm

Input: A complete graph $G = (V, E)$, and a cost function $c \colon E \to \mathbb{R}^+$.

Step 1: $T := $ MST of (G, c).

Step 2:   Construct an admissible Hamiltonian tour H from T^3.

Output: H.

Originally, the approximation ratio of the T^3-algorithm was estimated to be $\frac{3}{2}\beta^2 + \frac{1}{2}\beta$, but, by further refining step 2, Andreae [27] could lower this bound.

Theorem 27.2 (Andreae [27]) *The refined T^3-algorithm is a $\beta^2 + \beta$-approximation algorithm for Δ_β-TSP, and it runs in time $O(n^2)$ on a graph with n vertices.*

Taking into account that T^2 does not always contain a Hamiltonian tour, if one wants to get below the β^2-factor, one needs to consider a suitable replacement for a spanning tree as starting point. This is the approach followed by Bender and Chekuri [31], building on a 2-vertex-connected spanning subgraph of G.

Theorem 27.3 (Fleischner [28]) *For every 2-vertex-connected graph S, the graph S^2 contains a Hamiltonian tour H.*

As desired, now we need to replace only two consecutive edges by one. However, as opposed to an MST, a minimal spanning 2-vertex-connected subgraph (2VCSS) of G cannot be computed in polynomial time, unless $P = NP$. Therefore, one has to use an approximation instead.

Theorem 27.4 (Penn and Shasha-Krupnik [29]) *There is a polynomial-time 2-approximation algorithm for the minimal 2-vertex-connected spanning subgraph problem.*

Again, as for the T^3-algorithm, the existence of a Hamiltonian tour is not sufficient. From a constructive version of Fleischner's proof by Lau [30], one can get a 8β-approximation. By refining this such that the resulting Hamiltonian tour uses every edge of S at most twice, one gets the following result. Again, we call such a tour *admissible*.

S^2-Algorithm

Input: A complete graph $G = (V, E)$, and a cost function $c \colon E \to \mathbb{R}^+$.
Step 1: $S := 2$-approximation of a minimal 2VCSS of (G, c).
Step 2:   Construct an admissible Hamiltonian tour H of S.
Output: H.

Theorem 27.5 (Bender and Chekuri [31]) *The S^2-algorithm is a 4β-approximation algorithm for Δ_β-TSP, and it runs in time $O(n^5)$ on a graph with n vertices.*

The impractical running time, going back to the 2VCSS computation, is clearly a drawback. Obviously, any improvement on the 2VCSS approximation would also improve the S^2-algorithm, with respect to running time or approximation ratio. However, as the algorithm by Penn and Shasha-Krupnik is already the result on several improvements in smaller and smaller steps, one cannot be too optimistic in that regard. At the current state, the S^2-algorithm surpasses the much faster T^3-algorithm only for $\beta > 3$, at which point the approximation ratio already is 12.

Another downside to both algorithms is that, for inputs close to Δ-TSP, they cannot get close to the results by the Christofides algorithm. Especially, for instances where only a few edge costs violate the triangle inequality, an algorithm based on the Christofides algorithm should in most cases get close to or reach the 1.5-approximation ratio, whereas the S^2-algorithm and the T^3-algorithm can only hope to return a 2-approximation in such cases.

The difficulty, as mentioned before, is that the Christofides algorithm uses a matching as one of its starting points, and the estimate that this does not cost more than half of a Hamiltonian tour fails if the triangle inequality does not hold. Therefore, it was even conjectured in Reference 26 that the Christofides algorithm cannot be adopted to Δ_β-TSP for $\beta > 1$.

However, there is a way around this by computing a so-called path matching.

Definition 27.7 *For a weighted graph* (G, c), $G = (V, E)$ *and an even-size set* $W \subseteq V$, *a* path matching *is a set M of paths in G such that each vertex from W is endpoint of exactly one path from M.*

A path matching of minimal cost can be computed in time $|V|^3$. First one solves the all-pairs cheapest path problem on (G, c), let $c'(u, v)$ be the cost of cheapest path between u and v. Then one computes a minimum matching M' for (G, c'), and M is obtained from M' by replacing every edge u, v by the cheapest path from u to v computed previously on (G, c).

It is easy to see that, for every TSP instance, independent of the triangle inequality, a minimum path matching costs at most half as much as an optimum TSP solution. This holds because an optimum TSP solution can simply be divided into two path matchings.

Now that this initial problem is solved, one is left with a combination of a spanning tree T and a path matching M. This is a structure for which no result similar to the ones of Sekanina and Fleischner is known. To eventually construct a Hamiltonian path without replacing more than 4 consecutive edges by a new one, one needs some additional detail work. ("PMCA" stands for path-matching and the Christofides-algorithm-based approach.)

Algorithm PMCA [19]

Input: A complete graph $G = (V, E)$, and a cost function $c \colon E \to \mathbb{R}^+$.

Step 1: $T := \text{MST of } (G, c)$.

Step 2: $S := \{v \in V \mid deg_T(v) \text{ is odd}\}$.

Step 3:   Compute a minimum-weight path matching M_1 on S in G.

Step 4:   Replace M_1 by a vertex-disjoint path matching M_2.

Step 5: $\omega_1 :=$ Eulerian tour in the multi-graph $G' = (V, E(T) \cup M_2)$.

Step 6:   Modify parts from T in ω_1 such that they form a forest of degree at most 3, obtain ω_2

Step 7:   Construct a Hamiltonian tour H of G by shortening ω_2.

Output: H.

The detail work happens in steps 4, 6, and 7.

Although any minimum-weight path matching can easily be shown to be edge-disjoint, for obtaining vertex-disjointness in step 4, a price has to be paid. At vertices used by two paths, one of them has to give way by "bridging" the two edges to and from that vertex, that is, replacing them by a new edge. This can be done such that a new edge is never replaced again later. Thus, the costs increase at most by a factor of β, that is, $cost(M_2) \le \beta \cdot cost(M_1)$.

Similarly, in step 6, we look at subpaths from ω_1 in T, that is, at those pieces we obtain after dividing ω_1 into parts from T and parts from M_2 alternatingly. At a vertex of degree ≥ 4 in T, two or more subpaths from ω_1 in T will cross, and we can bridge some of them as in step 4, replacing at most two consecutive edges from T by a new one. Thus, we have

$$cost(\omega_2) \le cost(M_2) + \beta \cdot cost(T) \le \beta \left(cost(M_1) + cost(T) \right).$$

After this preparation, in ω_2, each vertex will occur not more than twice, and bridging one of the occurrences for each such vertex is performed in step 7 again by replacing at most two consecutive edges from ω_2 by a new one. This gives the resulting estimate, where H_{opt} is an optimal TSP solution.

$$cost(H) \le \beta cost(\omega_2) \le \beta^2 \left(cost(M_1) + cost(T) \right) \le \beta^2 \cdot 1.5 cost(H_{opt}).$$

Theorem 27.6 (Böckenhauer et al. [19]) *The algorithm PMCA is a $1.5\beta^2$-approximation algorithm for Δ_β-TSP, and it runs in time $O(n^3)$ on a graph with n vertices.*

This algorithm has a practical running time, and it gives the best approximation ratio so far for $\beta \le 2$. Note that, for $\beta = 2$, the approximation ratio of the algorithms PMCA and T^3 is 6 already. Thus, algorithm PMCA covers the cases close to the kernel L_I that are the most interesting ones in practice.

Probably, it is the most prominent open question in this area whether this algorithm can be improved to ratio 1.5β, respectively, whether another approach can reach that goal. Certainly, one should not hope for anything better in view of the lower bounds discussed in the next section.

Let us finish this section by summarizing what we have seen by studying the stability for the TSP, this archetype of a hard problem. It has been very fruitful, starting from known algorithms for a kernel problem, first to identify how they depend on properties of that kernel, and then to look for modifications or similar approaches. There, the crucial point was to transform a dependency of a kernel property into one that holds for each distance from the kernel, albeit with a worse ratio, but independent of the input size. One can see that different tools may come into play, modifying the starting point, using general graph properties, or fine-tuning the algorithm's behavior at points where it did not matter for the original kernel problem.

27.4 Lower Bounds and the Situation Inside Δ-TSP

We have seen in the previous section how to use the concept of stability to partition all TSP instances into infinitely many classes according to their approximability, depending on the parameter β of a relaxed β-triangle inequality. Our aim in this section is twofold: We will first exhibit a similar partition into infinitely many classes also for the input instances inside the Δ-TSP, based on the parameter β, and we will show that the concept of stability can also be used to obtain lower bounds on the approximability for all these classes.

Recall that, for a complete edge-weighted graph $G = (V, E, c)$, we say that it satisfies a *sharpened triangle inequality*, if, for all $u, v, w \in V$,

$$c(\{u, v\}) \le \beta \cdot (c(\{u, w\}) + c(\{w, v\}))$$

holds for some $\frac{1}{2} \le \beta < 1$. Note that $\beta = \frac{1}{2}$ corresponds to the trivial case where all edge weights are equal. In Reference 32, it was shown how to estimate the approximation ratio of any approximation algorithm for Δ-TSP on instances of Δ_β-TSP for $\frac{1}{2} < \beta < 1$.

Theorem 27.7 (Böckenhauer et al. [32]) *Let A be an approximation algorithm for Δ-TSP with approximation ratio α, and let $\frac{1}{2} < \beta < 1$. Then A can be used as an approximation algorithm for Δ_β-TSP with approximation ratio $\frac{\alpha \cdot \beta^2}{\beta^2 + (\alpha - 1) \cdot (1 - \beta)^2}$.*

The proof of Theorem 27.7 is based on the following idea: For any input instance for the Δ_β-TSP, we can subtract a certain amount (depending on β and the value of the minimum edge weight) from all edge weights such that the resulting TSP instance still satisfies the triangle inequality. Furthermore, the optimal Hamiltonian tours for both instances coincide. Using the observation that, in a weighted graph obeying a sharpened β-triangle inequality, the maximum edge weight is bounded from above by $2\beta^2/(1 - \beta)$ times the minimum edge weight, we can then estimate the length of a Hamiltonian tour as computed by algorithm A as stated in Theorem 27.7.

Note that the approximation ratio guaranteed by Theorem 27.7 tends to 1 for β tending to $\frac{1}{2}$, and it tends to α for β tending to 1. Applied to the Christofides algorithm, we obtain an approximation ratio of $1 + \frac{2\beta - 1}{3\beta^2 - 2\beta + 1}$.

For $\frac{1}{2} \le \beta < \frac{2}{3}$, this result can be improved by a specific algorithm, called cycle cover algorithm, which is explained in what follows.

Theorem 27.8 (Böckenhauer et al. [32]) *For $\frac{1}{2} \le \beta < 1$, the cycle cover algorithm is a $(\frac{2}{3} + \frac{1}{3} \cdot \frac{\beta}{1-\beta})$-approximation algorithm for Δ_β-TSP.*

This algorithm is based on the idea of first computing a minimum cycle cover of the given graph, this can be done in polynomial time [33]. Then the cycles are joined together to form a Hamiltonian tour by removing the cheapest edge from every cycle and replacing it by an edge to the next cycle. One can easily observe that, in a graph satisfying a sharpened β-triangle inequality, for any two adjacent edges the costs differ by a factor of at most $\beta/(1 - \beta)$. As every cycle of the cover has at least three edges, this possible growth of the costs affects at most one-third of the total cost of the cycle cover, which gives the claimed approximation ratio.

We have seen so far that we can partition the set of TSP input instances into infinitely many classes according to their approximability. We have used the relaxed and sharpened triangle inequality to define these classes. In the following, we show that it is also possible to give an explicit lower bound on the approximability of the TSP for each of these classes, for any $\beta > \frac{1}{2}$. These lower bounds show that, unless P = NP, one can expect to obtain only gradually better results, not principally. Specifically, as we will see, no polynomial-time approximation scheme (PTAS) can exist for Δ_β-TSP for $\beta > \frac{1}{2}$, even when getting arbitrarily close to the trivial case of $\beta = \frac{1}{2}$. Also, for growing β, we will see the necessity for the approximation ratio of any polynomial-time algorithm to grow at least linearly with β.

The first inapproximability proof for the metric TSP, that is, for $\beta = 1$, goes back to Papadimitriou and Yannakakis [34], who proved that even TSP restricted to edge weights from the set $\{1, 2\}$ does not admit a PTAS.

The first explicit lower bound on the approximability of the metric TSP was given by Engebretsen [35], who proved that it is NP-hard to approximate the Δ-TSP within a factor of $\frac{5381}{5380} - \varepsilon$ for any small $\varepsilon > 0$, even in the case where all edge weights are from $\{1, 2\}$.

This result was not only improved for the metric case, but also extended to the cases of a relaxed or sharpened triangle inequality by Böckenhauer and Seibert [36], who proved the following theorem.

Theorem 27.9 (Böckenhauer and Seibert [36]) *Unless* P $\neq$ NP, *there is no polynomial-time α-approximation algorithm, if*

$$\alpha < \frac{7611 + 10\beta^2 + 5\beta}{7612 + 8\beta^2 + 4\beta} \quad \text{for the case } \frac{1}{2} < \beta \leq 1,$$

$$\alpha < \frac{3803 + 10\beta}{3804 + 8\beta} \quad \text{for the case } \beta \geq 1.$$

For the metric TSP, that is, for $\beta = 1$, this leads to a lower bound of $\frac{3813}{3812}$. The currently best known lower bound of $\frac{220}{219}$ for the metric TSP is due to Papadimitriou and Vempala [37]. However, as in Reference 37 edge weight 0 was used, this result cannot be adapted to Δ_β-TSP for $\beta < 1$.

The proof of Theorem 27.9 is based on a gap-preserving reduction from the LinEq2-2(3) problem.[*] LinEq2-2(3) is the following problem: Given a system of linear equations modulo 2 with exactly 2 variables in each equation, and with exactly three occurrences of each variable, find the maximum number of equations that can be simultaneously satisfied. This problem was proven to be not approximable in polynomial time within $\frac{332}{331} - \varepsilon$ for an arbitrarily small $\varepsilon > 0$ by Berman and Karpinski [38].

The idea of the reduction is based on building so-called gadgets. For each equation and for each variable, a small graph is built, and all these graphs are put together into a large graph that serves as the input for the TSP. In this construction, only three different edge weights are used such that, for each nonsatisfied equation in an optimal variable assignment, the optimal Hamiltonian tour through the constructed graph has to pass through one edge of highest cost. All other edges of an optimal tour are shown to be of lowest cost.

Theorem 27.9 is most interesting for the case $\beta < 1$. It shows the strong result that the TSP with sharpened triangle inequality does not admit a PTAS even if β comes arbitrarily close to the trivial case $\frac{1}{2}$.

---

[*] For an introduction into the theory of gap-preserving reductions, see Chapter 14, or for a broader treatment see Reference 39.

For the case of a relaxed triangle inequality, the result from Theorem 27.9 tends to $\frac{5}{4}$ for β tending to infinity. Bender and Chekuri [31] extended the result from Reference 34 to the case of the relaxed triangle inequality and showed a lower bound of $1 + \varepsilon\beta$ for some very small ε, not given explicitly in the proof. This result shows that the approximation ratio of any polynomial-time algorithm for the Δ_β-TSP has to grow at least linearly with β.

27.5 Discussion and Related Work

In the previous sections, we have introduced the concept of stability of approximation. Here we discuss the potential applicability and usefulness of this concept.

Using this concept, one can establish positive results of the following types:

1. An approximation algorithm or a PTAS can be successfully used for a larger set of inputs than the set usually considered.
2. We are not able to successfully apply a given approximation algorithm A (or a PTAS) for additional inputs, but one can simply modify A to get a new approximation algorithm (or a new PTAS) working for a larger set of inputs than the set of inputs of A.
3. To learn that an approximation algorithm is unstable for a distance measure could lead to the development of completely new approximation algorithms that would be stable according to the considered distance measure.

The following types of negative results may be achieved:

4. The fact that an approximation algorithm is unstable according to all "reasonable" distance measures and so that its use is really restricted to a specific set of problem instances.
5. Let $Q = (\Sigma_I, \Sigma_O, L, L_I, \mathcal{M}, cost, goal) \in NPO$ be well approximable. If, for a distance measure D and a constant r, one proves the nonexistence of any polynomial-time approximation algorithm for $Q_{r,D} = (\Sigma_I, \Sigma_O, L, Ball_{r,D}(L_I), \mathcal{M}, cost, goal)$, then this means that the problem Q is "unstable" according to D.

Thus, using the notion of stability one can search for a spectrum of the hardness of a problem according to the set of inputs. For instance, considering a hard problem such as the TSP or the maximum clique problem, one could get an infinite sequence of input languages $L_0, L_1, L_2, \ldots$ given by some distance measure, where $R_r(n)$ is the best achievable approximation ratio for the language L_r. Results of this kind can essentially contribute to understanding the nature of hardness of specific problems.

In case of the TSP with relaxed triangle inequality, this approach has been followed successfully as we have seen in the preceding sections. A variety of new and adapted algorithms have been developed after the known ones have proven to be unstable. All in all, an approximation ratio of $\min\{\beta^2 + \beta, \frac{3}{2}\beta^2, 4\beta\}$ has been reached so far, and the proven lower bounds show the β-dependency to be invariable.

Further applications of these ideas for different versions of the Hamiltonian path problem were developed by Forlizzi et al. [40], where a few stable algorithms with respect to relaxed triangle inequality were designed.

Also, we have seen how these studies gave rise to looking into subcases inside the metric one. Though not a direct application of the stability concept, the treatment here was inspired by the stability approach, and it has proven successful, too. The combined approximation ratio in this case is $\min\{\frac{2}{3} + \frac{1}{3} \cdot \frac{\beta}{1-\beta}, 1 + \frac{2\beta-1}{3\beta^2-2\beta+1}\}$.

Chandran and Ram [41], and then Bläser [42] succeeded in transferring this approach to the asymmetric TSP. Remember that no constant-factor approximation algorithm for the metric case is known there, so the combined approximation factor of $\min\{\frac{\beta}{1-\beta}, \frac{1}{1-\frac{1}{2}(\beta+\beta^2)}\}$ is quite an achievement in the asymmetrical case. Moreover, a variant of the TSP where a desired order of appearance in the tour is given for a subset of vertices was analyzed under a relaxed triangle inequality in References 43–45.

The subproblems with sharped triangle inequality were also successfully attacked for a variety of minimum connected spanning subgraph problems in References 46, 47. As these problems allow for constant approximation algorithms in their general case, a treatment analogously to the TSP could not be expected. Moreover, several graph problems have been analyzed under a sharpened triangle inequality in the context of reoptimization, see Chapter 25 for an overview.

However, the approach guided by the parameterized triangle inequality led to the development of a few new simple and fast algorithms that improve the best known approximation ratio at least for a part of the sharpened-triangle-inequality case.

Overall, we have seen that the stability of approximation approach can lead to rethinking known algorithms and developing new ones. The common goal is to provide a variety of algorithms for the whole range of a problem that deliver approximation ratios depending not on the input size, but on a suitable parameter indicating the hardness of each input instance.

References

1. S.A. Cook. The complexity of theorem proving procedures. *Proceedings of the 3rd ACM STOC*, ACM, New York, 1971, pp. 151–158.
2. R.M. Karp. Reducibility among combinatorial problems. In: R.E. Miller and J.W. Thatcher (Eds.), *Complexity of Computer Computations*. Plenum Press, New York, 1972, pp. 85–103.
3. D.S. Johnson. Approximation algorithms for combinatorial problems. *J. Comp. System Sciences* 9, (1974), 256–278.
4. L. Lovász. On the ratio of the optimal integral and functional covers. *Discrete Mathematics* 13, (1975), 383–390.
5. N. Christofides. Worst-case analysis of a new heuristic for the travelling salesman problem. Technical Report 388, Graduate School of Industrial Administration, Carnegie-Mellon University, Pittsburgh, PA, 1976.
6. O.H. Ibarra, C.E. Kim. Fast approximation algorithms for the knapsack and sum of subsets problem. *J. ACM* 22, (1975), 463–468.
7. J. Hromkovič. *Algorithmics for Hard Problems. Introduction to Combinatorial Optimization, Randomization, Approximation, and Heuristics*. 2nd ed., Springer, Berlin, Germany, 2003.
8. D.P. Williamson, D.B. Shmoys. *The Design of Approximation Algorithms*. Cambridge University Press, New York, 2011.
9. V.V. Vazirani. *Approximation Algorithms*. Springer, Berlin, Germany, 2003.
10. D.-Z. Du, K.-I. Ko, X. Hu. *Design and Analysis of Approximation Algorithms*. Springer, Berlin, Germany, 2012.
11. D.S. Hochbaum (Ed.). *Approximation Algorithms for NP-hard Problems*. PWS Publishing Company, Boston, MA, 1996.
12. G. Ausiello, P. Crescenzi, G. Gambosi, V. Kann, A. Marchetti-Spaccamela, M. Protasi. *Complexity and Approximation—Combinatorial Optimization Problems and Their Approximability Properties*. Springer, Berlin, Germany, 1999.
13. M.R. Garey, D.S. Johnson. *Computers and Intractibility. A Guide to the Theory of NP-Completeness*. W. H. Freeman and Company, New York, 1979.
14. E.L. Lawler, J.K. Lenstra, A.H.G. Rinnooy Kan, D.B. Shmoys (Eds.): *The Traveling Salesman Problem*. John Wiley & Sons, New York, 1985.
15. C.H. Papadimitriou. The Euclidean travelling salesman problem is NP-complete. *Theoretical Computer Science* 4, (1977), 237–244.
16. S. Arora. Polynomial time approximation schemes for Euclidean TSP and other geometric problems. *Proceedings of the FOCS '96 Proceedings of the 37th Annual Symposium on Foundations of Computer Science*, IEEE Computer Society Washington, DC, 1996, pp. 2–11.

17. S. Arora. Nearly linear time approximation schemes for Euclidean TSP and other geometric problems. *Proceedings of the 38th IEEE FOCS*, IEEE, New York, 1997, pp. 554–563.
18. J.S.B. Mitchell. Guillotine subdivisions approximate polygonal subdivisions: Part II—A simple polynomial-time approximation scheme for geometric k-MST, TSP and related problems. Technical Report, Department of Applied Mathematics and Statistics, Stony Brook, NY, 1996.
19. H.-J. Böckenhauer, J. Hromkovič, R. Klasing, S. Seibert, W. Unger. Towards the notion of stability of approximation for hard optimization tasks and the traveling salesman problem. *Theoretical Computer Science* 285, (2002), 3–24.
20. R.G. Downey, M.R. Fellows. Fixed-parameter tractibility and completeness I: Basic results. *SIAM Journal of Computing* 24, (1995), 873–921.
21. R.G. Downey, M.R. Fellows. *Parametrized Complexity*. Springer-Verlag, Berlin, Germany, 1999.
22. D.P. Bovet, C. Crescenzi. *Introduction to the Theory of Complexity*. Prentice-Hall, New York, 1993.
23. C.H. Papadimitriou. *Computational Complexity*. Addison-Wesley, Reading, MA, 1994.
24. J. Hromkovič. *Theoretical Computer Science*. Springer-Verlag, Berlin, Germany, 2004.
25. M. Sekanina. On an Ordering of a Set of Vertices of a Connected Graph. Publications of the Faculty of Science, University Brno, Czechoslovakia, Sovereign state, 1960, 412, pp. 137–142, 1960.
26. T. Andreae, H.-J. Bandelt. Performance guarantees for approximation algorithms depending on parameterized triangle inequalities. *SIAM Journal on Discrete Mathematics* 8, (1995), 1–16.
27. T. Andreae. On the traveling salesman problem restricted to inputs satisfying a relaxed triangle inequality. *Networks* 38, (2001), 59–67.
28. H. Fleischner. The square of every two-connected graph is Hamiltonian. *Journal of Combinatorial Theory (B)* 16, (1974), 29–34.
29. M. Penn, H. Shasha-Krupnik. Improved approximation algorithms for weighted 2- and 3-vertex connectivity augmentation problems. *Journal of Algorithms* 22, (1997), 187–196.
30. H.T. Lau. Finding EPS-graphs. *Monatshefte für Mathematik* 92, (1981), 37–40.
31. M. Bender, C. Chekuri. Performance guarantees for TSP with a parametrized triangle inequality. *Information Processing Letters* 73, (2000), 17–21.
32. H.-J. Böckenhauer, J. Hromkovič, R. Klasing, S. Seibert, W. Unger. Approximation algorithms for TSP with sharped triangle inequality. *Information Processing Letters* 75, (2000), 133–138.
33. J. Edmonds, E.L. Johnson. Matching. A well-solved class of integer linear programs. In: *Proceedings Calgary International Conference on Combinatorial Structures and Their Applications*, Gordon and Breach, New York, 1970, pp. 89–92.
34. C.H. Papadimitriou, M. Yannakakis. The traveling salesman problem with distances one and two. *Mathematics of Operations Research* 18, (1993), 1–11.
35. L. Engebretsen. An explicit lower bound for TSP with distances one and two. Extended abstract in: C. Meinel, S. Tison (Eds.), *STACS 99, Proceedings of the 16th Annual Symposium on Theoretical Aspects of Computer Science*, LNCS, 1563, Springer, Berlin, Germany, 1999, pp. 373–382. Full version in: *Electronic Colloquium on Computational Complexity* (http://www.eccc.uni-trier.de/eccc/), Revision 1 of Report No. 46 (1999).
36. H.-J. Böckenhauer, S. Seibert. Improved lower bounds on the approximability of the traveling salesman problem. *RAIRO Theoretical Informatics and Applications* 34, (2000), 213–255.
37. C.H. Papadimitriou, S. Vempala. On the approximability of the traveling salesman problem. *Combinatorica* 26(1), 2006, 101–120.
38. P. Berman, M. Karpinski. On some tighter inapproximability results. *Proceedings of the 26th International Colloquium on Automata, Languages, and Programming (ICALP'99)*, LNCS, 1644, Springer, Berlin, Germany, 1999, pp. 200–209.
39. E.W. Mayr, H.J. Prömel, A. Steger (Eds.). *Lecture on Proof Verification and Approximation Algorithms*. LNCS, 1967, Springer, Berlin, Germany, 1998.

40. L. Forlizzi, J. Hromkovič, G. Proietti, S. Seibert. On the stability of approximation for Hamiltonian path problems. *Proceedings of the SOFSEM 2005: Theory and Practice of Computer Science*, LNCS, Vol. 3381, Springer, Berlin, Germany, 2005, pp. 147–156.

41. L. Sunil Chandran, L. Shankar Ram. Approximations for ATSP with parametrized triangle inequality. *STACS 99, Proceedings of the 19th Annual Symposium on Theoretical Aspects of Computer Science*, LNCS, 2285, Springer, Berlin, Germany, 2002, pp. 227–237.

42. M. Bläser. An improved approximation algorithm for the asymmetric TSP with strengthened triangle inequality. *Proceedings of the 30th International Colloquium on Automata, Languages, and Programming (ICALP'03)*, LNCS, 2719, Springer, Berlin, Germany, 2003, pp. 157–163.

43. H.-J. Böckenhauer, J. Hromkovič, J. Kneis, J. Kupke. The parameterized approximability of TSP with deadlines. *Theory Computer System* 41(3), (2007), 431–444.

44. H.-J. Böckenhauer, T. Mömke, M. Steinová. Improved approximations for TSP with simple precedence constraints. *Journal of Discrete Algorithms* 21, (2013), 32–40.

45. H.-J. Böckenhauer, M. Steinová. Improved approximations for ordered TSP on near-metric graphs. *RAIRO Theoretical Informatics and Applications* 48(5), (2014), 479–494.

46. H.-J. Böckenhauer, D. Bongartz, J. Hromkovič, R. Klasing, G. Proietti, S. Seibert, W. Unger. On the hardness of constructing minimal 2-connected spanning subgraphs in complete graphs with sharped triangle inequality. *Theoretical Computer Science* 326, (2004), 137–153.

47. H.-J. Böckenhauer, D. Bongartz, J. Hromkovič, R. Klasing, G. Proietti, S. Seibert, W. Unger. On k-connectivity problems with sharpened triangle inequality. *Algorithms and Complexity, Proceedings of the 5th Italian Conference, CIAC 2003*, LNCS, 2653, Springer, 2003, pp. 189–200.

IV

Traditional Applications

28

Performance Guarantees for One Dimensional Bin Packing[*]

János Csirik

György Dósa

28.1 Introduction

Let $a_1, \ldots, a_n$ be a given collection of *items* with *sizes* $s(a_i) > 0$, $1 \leq i \leq n$. In mathematical terms, *bin packing* is a problem of partitioning the set $\{a_i\}$ under a sum constraint: Divide $\{a_i\}$ into a minimum number of blocks, called *bins*, such that the sum of the sizes of the items in each bin is at most a given *capacity* $C > 0$. To avoid trivialities, it is assumed that all item sizes fall in $(0, C]$. Research into bin packing and its many variants, which began some 45 years ago [1,2], continues to be driven by a countless variety of applications in the engineering and information sciences. The following examples give an idea of the scope of the applications:

- (Stock cutting) Lumber with a fixed cross section comes in a standard board C units in length. The items are demands for pieces that must be cut from such boards. The objective is to minimize the number of boards (bins) used for the pieces $\{a_i\}$, or equivalently, to minimize the *trim loss* or waste (the total board length used minus the sum of the $s(a_i)$). It is easy to see that this type of application extends to industries that supply cable, tubing, cord, tape, and so on, from standard lengths.

---

[*] This chapter includes its previous version coauthor with Edward G. Coffman, Jr. that appeared in the First Edition of the Handbook on Approximation Algorithms and Metaheuristics.

- (Television programming) Fixed duration time slots are provided between segments of entertainment programs for the use of commercials. The objective is to minimize the number of time slots (bins) that need to be devoted to a given collection of variable length commercials $\{a_i\}$.
- (Transportation) The a_i are items to be loaded onto a collection of identical transports (e.g., trucks, railway cars, airplanes) with given weight and volume limits. The objective is to minimize the number of transports (bins) needed; if only the weight limit is operative then the $s(a_i)$ denote weights, and if only the volume limit is operative then the $s(a_i)$ represent volumes.
- (Computer storage allocation) In this application, the items are files to be stored on a set of identical disks with the constraint that each file must be stored entirely on one disk; and the objective is to minimize the number of disks (bins) needed for the set of files.

In the above-mentioned applications bin packing has been presented as the *primary* combinatorial problem. In many applications, it is a secondary problem or an embedded special case. For example, capacitated vehicle routing is a classic problem in which bin packing is embedded.

Bin packing is an NP-hard problem, so a large majority of the research on bin packing focuses on the design and analysis of approximation algorithms. Apart from applications, bin packing has acquired an added, more general importance in the development of complexity theory and algorithms. It was one of a relatively few "test" problems, such as satisfiability and chromatic index, in which reducibility arguments were most often formulated in NP-completeness proofs. Bin packing has also served as a testbed for advances in algorithmics such as approximation schemes and average-case analysis.

We normalize the problem by dividing all item sizes by C and letting the bin capacity be 1. Let $A(L)$ denote the number of bins needed by algorithm A to pack the items of L. The symbol OPT stands for the number of bins used by an optimal algorithm. Define V_α as the set of all lists of items with sizes no larger than α and consider, for given k and α, the upper bound

$$R_A(k, \alpha) := \sup \left\{ \frac{A(L)}{k} : L \in V_\alpha \text{ and } \text{OPT}(L) = k \right\}.$$

The *asymptotic worst-case ratio* for algorithm A is given by

$$R_A^\infty(\alpha) := \limsup_{k \to \infty} R_A(k, \alpha).$$

A weaker, less formal, but more instructive, definition states that $R_A^\infty(\alpha)$ is the smallest constant such that there exists a constant $K < \infty$ for which

$$A(L) \leq R_A^\infty(\alpha) \cdot \text{OPT}(L) + K$$

for every list $L \in V_\alpha$; the asymptotic ratio, a multiplicative constant, hides the additive constant K. This ratio is of most interest in those applications where K is small relative to $A(L)$.

The effect of the additive constant is preserved in the *absolute worst-case ratio* for algorithm A

$$R_A(\alpha) = \sup_{L \in V_\alpha} \left\{ \frac{A(L)}{\text{OPT}(L)} \right\}. \tag{28.1}$$

For $\alpha = 1$ we use R_A^∞ and R_A.

An algorithm A is either *online* or *offline*. If online, A assigns items to bins in the order they are given in the original list, without using any knowledge about subsequent items in the list. If offline, the entire list is available to A as it computes the packing. In case of online algorithms the *absolute worst-case ratio* is sometimes called *competitive ratio*.

Online algorithms may be the only ones that can be used in certain situations, where the items arrive in sequence according to some physical process and must be assigned to a bin at arrival time. In many cases an algorithm is offline only in that it performs an initial ordering of the items before applying an online rule. An algorithm is *bounded-space* if the bins available for packing (called "open" bins) are limited in number.

A classification scheme for the bin packing problem, similar to the one used for scheduling, cutting and packing problems, is given in References 3, 4. The classification scheme applies to a large number of variations of the bin packing problem.

28.2 Online Algorithms

Let $B_1, B_2, \ldots$ denote the sequence of initially empty bins. The Next Fit (NF) algorithm was one of the first, and simplest, approximation algorithms to be analyzed: NF begins by packing $a_1, a_2, \ldots$ into B_1 until an item, say a_i, is encountered that does not fit, that is, $a_i > 1 - \sum_{1 \le j < i} a_j$. Item a_i is packed in B_2 and B_1 is *closed*, that is, no further items are packed in B_1. Bin B_2 becomes the new *open* bin and, such as B_1, is packed with items until one that does not fit is encountered, whereupon B_2 is closed; this bin-by-bin process repeats until all items are packed. NF is obviously linear-time; it is also bounded-space, as the number of open bins never exceeds 1.

Johnson's [2,5] parametric analysis of NF showed that

$$R_{NF}^{\infty}(\alpha) = \begin{cases} 2 & \text{if } \frac{1}{2} \le \alpha \le 1, \\ \frac{1}{1-\alpha} & \text{if } \alpha < \frac{1}{2}. \end{cases}$$

As observed by Fischer [6] NF uses the same number of bins for the reverse of L, $(a_n, a_{n-1}, \ldots, a_1)$, as it does for L, for every L.

An obvious drawback of NF is that it can not make use of the empty space in closed bins. When packing a new item, the First Fit (FF) algorithm tries to exploit this empty space by scanning all bins each time an item is packed; thus, bins are never closed. When packing a new item, FF puts it into the lowest indexed bin in which it fits; a new bin is started only if the current item does not fit into any nonempty bin. With an appropriate data structure, the running time of FF is $O(n \log n)$ and thus greater than the $O(n)$ running time of NF. However, FF has a much better asymptotic ratio. Johnson et al. [7] proved that, if m is a positive integer such that $1/(m+1) < \alpha \le 1/m$, then

$$R_{FF}^{\infty}(\alpha) = \begin{cases} \frac{17}{10} & \text{if } m = 1, \\ \frac{m+1}{m} & \text{if } m \ge 2. \end{cases}$$

For the case $\alpha = 1/m$, Dósa [8] proves that the maximum number of possible *FF* bins is exactly $\lfloor \frac{m+1}{m} \text{OPT} \rfloor$, if $\text{OPT} = k \cdot m + 1$; and $\lceil \frac{m+1}{m} \text{OPT} \rceil$, if $\text{OPT} = k \cdot m + r$, where $2 \le r \le m$. Proving the upper bound is relatively simple. To establish the lower bound one applies the tight lower bound for the general case given in Reference 9.

Turning back to the general case (i.e., $\alpha = 1$), a worst-case list proving that the 17/10 ratio can not be improved is quite complicated. Much simpler lists showing behavior nearly as bad, in particular a 5/3 ratio, are

$$L_{6k} = \left(\underbrace{\frac{1}{6} - 2\varepsilon, \ldots, \frac{1}{6} - 2\varepsilon}_{6k}, \underbrace{\frac{1}{3} + \varepsilon, \ldots, \frac{1}{3} + \varepsilon}_{6k}, \underbrace{\frac{1}{2} + \varepsilon, \ldots, \frac{1}{2} + \varepsilon}_{6k} \right).$$

Then $\text{OPT}(L_{6k}) = 6k$, $FF(L_{6k}) = 10k$, and the 5/3 ratio follows.

Other classical online algorithms are Best Fit (BF), Worst Fit (WF), and Almost Worst Fit (AWF). BF behaves similar to FF, except that it puts the next item into the bin in which it fits with the smallest gap left over; ties are broken arbitrarily. WF puts the next item into a nonempty bin with the largest gap, starting a new bin only if this largest gap is too small. AWF first tries to put the next item into a nonempty bin with the second largest gap; if the item does not fit there then AWF behaves similar to WF. All three variants belong to the class of so-called Any Fit (AF) algorithms: An AF algorithm scans once through the list L packing items as they are encountered. It never puts an item into an empty bin, unless it does not fit into any available partially filled bin. Similarly, an Almost Any Fit (AAF) algorithm is an AF algorithm that never puts an item into a partially filled bin with the lowest level unless there is more than one bin having this level, or unless the bin of lowest level is the only one that has enough room. Johnson [5] proved that, although the class of AF algorithms is clearly large, no algorithm in this class can improve upon FF. Moreover, all AAF algorithms have the same asymptotic ratio as FF; these statements hold even for the parametric ratios. One simple consequence is that the AAF algorithms BF and AWF have asymptotic ratios 17/10. However the asymptotic ratio of the AF algorithm WF is 2, the same as that for NF.

28.2.1 Bounded-Space Online Algorithms

An online bin packing algorithm is said to use *k-bounded-space* if, for each new item, the number of bins in which it may be packed is at most k. NF uses 1-bounded-space, whereas FF, WF, and AWF each use unbounded space. There are four very natural bounded-space bin packing algorithms that are defined via simple *packing* rules for items and simple *closing* rules for bins: A new item can always be packed into the lowest indexed bin (as in FF) or into the bin with the smallest remaining gap (as in BF). If the new item does not fit into any active bin, some active bin has to be closed; in this case one can always choose the lowest indexed bin (the First bin) or the bin with the greatest sum of item sizes (the Best bin). The corresponding four algorithms are called AFF_k, AFB_k, ABF_k, and ABB_k. Here A stands for Algorithm, the second letter denotes the packing rule (BF or FF), the third letter denotes the closing rule (Best bin or First bin), and k is the upper bound on the number of active bins. Results are as follows:

- Algorithm AFF_k. This algorithm was termed Next-k Fit by Johnson [5]. Provably tight bounds were not in hand until almost 20 years later. Csirik and Imreh [10] constructed a sequence of worst-case examples which show that $R^\infty_{AFF_k} \geq \frac{17}{10} + \frac{3}{10(k-1)}$, and Mao [11] proved a matching upper bound on $R^\infty_{AFF_k}$.
- Algorithm ABF_k. Mao [12] proved that $R^\infty_{ABF_k} = \frac{17}{10} + \frac{3}{10k}$.
- Algorithm AFB_k. Zhang [13] adapted the analysis of Mao [12] to this algorithm and proved that $R^\infty_{AFB_k} = \frac{17}{10} + \frac{3}{10(k-1)}$, that is, this algorithm has the same ratio as AFF_k.
- Algorithm ABB_k. This algorithm was investigated by Csirik and Johnson [14,15]. Compared with the other three algorithms, ABB_k uses the best packing and closing rules and it has the best asymptotic ratio; $R^\infty_{ABB_k} = \frac{17}{10}$ holds for any $k \geq 2$.

So, except for small k, the asymptotic ratios of all four algorithms are around 17/10; as they are of the AF type, they of course cannot outperform FF.

Lee and Lee [16] introduced a new class of bounded-space algorithms using bin reservation techniques. Their algorithm Harmonic$_k$ (H_k) is based on a partition of the interval $(0, 1]$ into k subintervals, where the partitioning points are $1/2, 1/3, \ldots, 1/k$. To each of these subintervals there corresponds a different active bin; items belonging to a given subinterval are packed only into the corresponding active bin. If a new item arrives that does not fit into its corresponding active bin, the bin is closed and a new bin is activated. Thus, the packing in bins containing items with sizes in one of the subintervals is a NF packing. Note that H_k is not an AF algorithm because it may close a bin even when the next item could be packed in one of the active bins (belonging to a different subinterval). In Reference 16 it was proved that

$$R^\infty_{H_k} \to h_\infty \approx 1.69103.$$

This number occurs frequently in bin packing and is defined by

$$h_\infty = \sum_{i=1}^{\infty} \frac{1}{t_i - 1}$$

TABLE 28.1 Asymptotic Ratios for Bounded-space Bin Packing Algorithms, Rounded to Five Decimal Places

k	AFF_k	ABF_k	ABB_k	$H_k \geq$	$H_k \leq$	SH_k	Champion
2	2.00000	1.85000	1.70000	2.00000	2.00000	2.00000	ABB
3	1.85000	1.80000	1.70000	1.75000	1.75000	1.75000	ABB
4	1.80000	1.77500	1.70000	1.71429	1.71429	1.72222	ABB
5	1.77500	1.76000	1.70000	1.70000	1.70000	1.70000	ABB, H, SH_k
6	1.76000	1.75000	1.70000	1.70000	1.70000	1.69444	SH_k
7	1.75000	1.74286	1.70000	1.69444	1.69444	1.69388	SH_k
8	1.74286	1.73750	1.70000	1.69377	1.69388	1.69106	SH_k
9	1.73750	1.73333	1.70000	1.69326	1.69345	1.69104	SH_k
∞	1.70000	1.70000	1.70000	1.69103	1.69103	1.69103	H, SH_k

where

$$t_{i+1} = t_i(t_i - 1) + 1, \ i \geq 1, \quad t_1 = 2. \tag{28.2}$$

Tight bounds for $R_{H_k}^{\infty}$ are not known for every value of k, so Table 28.1 gives the best upper and lower bounds currently known. The upper bounds for every value of k except 4 and 5 are from Reference 16. Tight upper bounds for $k \in \{4, 5\}$ are due to van Vliet, as are the lower bounds for $k \geq 4$ [17,18].

Woeginger [19] introduced the Simplified Harmonic (SH_k) algorithm, a modification of Harmonic with a different interval structure (basically using the sequence t_i) and with a slightly better asymptotic ratio for small values of k. He proved that SH_k only needs $O(\log k)$ open bins to achieve the performance ratio H_k.

A summary of the asymptotic ratios of the bounded-space algorithms for some small values of k is given in Table 28.1. The asymptotic ratios of all five algorithms always remain above h_{∞}. In fact, Lee and Lee [16] showed that a bounded-space algorithm cannot have an asymptotic ratio better than h_{∞}. However, at present no online bounded-space algorithm A is known for which $R_A^{\infty} = h_{\infty}$.

Csirik and Woeginger [20] compared online bounded-space algorithms that pack into bins of size $b \geq 1$, with optimal offline algorithms that pack into bins of size 1. In a decreasing scan, choose reciprocals iteratively so long as those chosen sum to no more than $1/b$. If $1/b_i$ is the i-th one chosen then $1/b = \sum_{i \geq 1} 1/b_i$, with $b_1 > b_2 > b_3 > \cdots$. The authors showed that, for every bin size $b \geq 1$, there exist online bounded-space algorithms packing into bins of size b which have a worst-case performance arbitrarily close to

$$\rho(b) := \sum_{i=1}^{\infty} \frac{1}{b_i - 1}.$$

They also showed that for every $b \geq 1$, the bound cannot be beaten by any online bounded-space bin packing algorithm.

28.2.2 Better Online Algorithms

The first online algorithm for bin packing with $R_A^{\infty} < h_{\infty}$ was Yao's [21] Revised FF (RFF). Both the definition and analysis of RFF are fairly involved. It is essentially based on FF, but similar to H_k it uses separate bins for items from the intervals $(0, 1/3]$, $(1/3, 2/5]$, $(2/5, 1/2]$, and $(1/2, 1]$, respectively. Moreover, every sixth item from the interval $(1/3, 2/5]$ is treated differently, the idea being to occasionally start a new bin with an item of this size in the hope of subsequently adding an item of size greater than 1/2 to that bin. Yao showed that

$$R_{RFF}^{\infty} = \frac{5}{3}.$$

All other known online algorithms that beat the h_∞ bound are variants of Harmonic that give special treatment to items $\geq 1/3$. Lee and Lee [16] described the Refined Harmonic (RH_K) algorithm, which was based on H_{20}. It uses the partitioning scheme of H_{20}, with the modification that the two size-intervals $(1/3, 1/2]$ and $(1/2, 1]$ are replaced by the four intervals $(1/3, y]$, $(y, 1/2]$, $(1/2, 1-y]$, and $(1-y, 1]$, where $y = 37/96$. Packing proceeds as in a Harmonic algorithm, except one now attempts to pair items whose sizes are in the first of the new intervals with items whose sizes are in the third interval, as such pairs can always fit in the same bin. Every seventh item with sizes in the first interval is handled differently. Lee and Lee proved that

$$R_{RH_{20}}^{\infty} \leq 373/228 = 1.6359\ldots.$$

Ramanan et al. [22] introduced the Modified Harmonic (MH) algorithms, which added the possibility of packing still smaller items together with items having sizes in $(1/2, 1/2 + y]$, and consequently created a more complicated algorithmic structure (as well as a different value for y – in this case $y = 265/684$). For this first variant they showed that

$$1.6156146 < R_{MH_{38}}^{\infty} \leq 1.615615\ldots,$$

and gave a general lower bound of $1.6111\ldots$ for this type of algorithm. They introduced a second variant of MF (MF-2) which divides $(1/3, 1/2]$ and $(1/2, 1]$ into more than two parts. MF-2 has an asymptotic ratio < 1.61217 and the general lower bound $1.58333\ldots$. Hu and Kahng [23] used similar principles to construct an unnamed variant for which they claimed $R_A^{\infty} \approx 1.6067$.

Richey [24] introduced the Harmonic+1 algorithm, and claimed that it has a performance ratio of 1.58872. Harmonic+1 uses a partition of $[0, 1]$, dividing it into more than 70 intervals. Its design and analysis were carried out with the help of linear programming.

Seiden [25] presented a general framework for analyzing a large class of online algorithms. This class includes Harmonic, Refined Harmonic, MH, MH-2, and Harmonic+1. He showed that all these algorithms are merely instances of a general class of algorithms which he called Super Harmonic. In his new approach, he reduced the problem of analyzing an instance of Super Harmonic to that of solving a specific knapsack instance. He developed a branch-and-bound algorithm for solving such knapsack problems, and furnished a general computer-assisted method of proving upper bounds for every algorithm that can be expressed in terms of Super Harmonic. As a result of this new technique, Seiden found a flaw in the analysis of Harmonic+1 and showed that the performance ratio of Harmonic+1 is at least 1.59217. He developed a new algorithm called Harmonic++ and showed that it has an asymptotic performance ratio of at most 1.58889. He also proved that 1.58333 is a lower bound for any Super Harmonic algorithm.

This was the best known result until 2016 when Heydrich and van Stee [26] developed a new approach. They developed an algorithm with the bound 1.5817, which is below the 1.58333 lower bound for any Super Harmonic-kind algorithm. The algorithm is very similar to Harmonic+1. It also applies an interval structure (with many dividing points), and for certain groups different types of items are paired, but there are two main differences. First, for items larger than 1/3, their exact size is considered for pairing, rather then their type. Second, to pair the largest number of items a medium item with size between 1/3 and 1/2 may be paired with an item that is larger than a half. Also, the pairing decision for medium items is delayed, and at least a certain percent of the smaller medium items will be paired with large items that may appear later. It is also shown that no algorithm that follows this approach can have a better asymptotic bound than 1.5766.

28.2.3 Lower Bounds for Online Algorithms

Consider next lower bounds on performance that the asymptotic worst-case ratios of *all* online bin packing algorithms must satisfy. Roughly speaking, one way to prove such lower bounds is to argue as follows. Suppose an online algorithm A is confronted initially with a huge set of tiny items. If A packs these tiny items very tightly, it would not be able to find an efficient packing for the larger items that might arrive later; if such items actually do arrive, A is going to lose. On the other hand, if A leaves lots of room for large items while packing the tiny items, the large items might not arrive; and in that case, A is again going to lose. To illustrate this idea let us consider a simple example involving two lists L_1 and L_2, each containing n identical items. The size of each item in list L_1 is $1/2 - \varepsilon$, and the size of each item in list L_2 is $1/2 + \varepsilon$. We will investigate the performance of an arbitrary online algorithm A on the following two lists: L_1 alone and the concatenation, $L_1 L_2$, of L_1 and L_2. The items of L_1 should be packed first. The algorithm will pack some items separately (j of them, say) and it will match up the remaining $n - j$. If we stop after L_1 has been packed then, obviously, $A(L_1)/\text{OPT}(L_1) = (n+j)/n$. If L_2 comes after L_1, then the best our algorithm can do is to add one element of L_2 to each of j bins containing a single element of L_1, and to pack separately the remaining $n - j$ elements. Clearly, $A(L_1 L_2)/\text{OPT}(L_1 L_2) = (3n - j)/(2n)$. The best choice of j is where the maximum of the two previous ratios is minimal. It is attained when $j = n/3$, implying a lower bound of 4/3 for the asymptotic worst-case performance ratio of any online algorithm A.

Yao [21] formalized this idea using three item sizes: $1/7 + \varepsilon$ as the size of small items, and $1/3 + \varepsilon$ and $1/2 + \varepsilon$ as sizes of larger items. (Note that these item sizes are simply $1/t_i + \varepsilon$, $1 \leq i \leq 3$ as given in (28.2)). Yao proved that, given such a list, the asymptotic ratio of every online bin packing algorithm A must satisfy $R_A^\infty \geq 1.5$. Brown [27] and Liang [28] independently generalized this lower bound to 1.53635 using Yao's construction for $1 \leq i \leq 5$. Ten years later [17,29] found an elegant linear programming formulation for the Brown-Liang construction. van Vliet gave an exact analysis and increased the lower bound to

$$R_A^\infty \geq 1.5401.$$

This was the best lower bound known for online bin packing for twenty years, until it was improved by Balogh et al. [30] to $248/161 \approx 1.54037$.

Galambos [31] and Galambos and Frenk [32] simplified and extended the Brown-Liang construction to the parametric case. Combining these results with the linear programming formulation of van Vliet [29] yields the lower bounds given in Table 28.2. For comparison, corresponding upper bounds for some algorithms have also been included.

Applying the above-mentioned general argument, Csirik et al. [33] proved that for lists of items in nonincreasing order, no online algorithm can have an asymptotic performance ratio smaller than 8/7. Balogh et al. [30] improved this result to $54/47 \approx 1.1489$.

Faigle et al. [34] showed that, when lists are restricted to those with only the item sizes $1/2 - \epsilon$ and $1/2 + \epsilon$, this lower bound becomes 4/3. Chandra [35] argued that all these (known) lower bounds can be carried over to *randomized* online bin packing algorithms in which the choice of an allowable packing decision can be made at random.

TABLE 28.2 Lower Bounds for the Parametric Case

α	1	1/2	1/3	1/4	1/5
Lower bound on $R_A^\infty(\alpha)$	1.540	1.389	1.291	1.229	1.188
Current champion	1.581	1.423	1.302	1.234	1.191
$R_{H_k}^\infty(\alpha)$	1.691	1.423	1.302	1.234	1.191
$R_{FF}^\infty(\alpha)$	1.700	1.500	1.333	1.250	1.200

28.3 Semi-Online Algorithms

Better bounds can be achieved when we relax the online condition by allowing the repacking of items. In these cases, we have to bound the number of times items are repacked, for otherwise, we would end up with an offline algorithm. Gambosi et al. [36] proposed two algorithms of this type. The first is based on a nonuniform partition of $(0, 1]$ into four subintervals: $I_0 = (2/3, 1], I_1 = (1/2, 2/3], I_2 = (1/3, 1/2], I_3 = (0, 1/3]$. Essentially, their first algorithm uses a Harmonic type algorithm for each interval, but it tries to fill bins packed with items from I_1 (i.e., items with sizes in I_1) up to 2/3 with already-packed I_3 items (by repacking) or with I_3 items from the remaining part of the list. This is a linear time algorithm (items can be packed at most twice) and it is quite easy to see that its asymptotic ratio is 3/2. Their second algorithm uses 6 intervals. Its time complexity is $O(n \log n)$ and the corresponding ratio is 4/3. In a subsequent paper, the same authors offered a more detailed analysis of the same set of algorithms [37].

Ivkovic and Lloyd [38] investigated dynamic bin packing, but as a side result they produced a quite complicated semi-online algorithm with an asymptotic ratio of 5/4. They also proved a lower bound for the special class of semi-online algorithms that only use *atomic* repacking moves; such a move is limited to the transfer of a single item from one bin to another. They proved that for any semi-online algorithm that performs only a bounded number of atomic repacking moves at each step, the asymptotic worst-case ratio is at least 4/3 [39].

Balogh et al. [40] improved this result to about 1.3871 and show that the same result holds for fully dynamic bin packing (where algorithms are allowed to repack a limited number of items at each step). Later on, Balogh et al. [41] present a semi-online algorithm that at each step repacks at most k elements (for a fixed positive integer k), and prove that the upper bound for the asymptotic worst-case ratio of the algorithm is a decreasing function of k, which tends to 3/2 as k goes to infinity.

One may only be willing to consider bounded-space algorithms of the semi-online type. There are such algorithms whose worst-case ratios match the limiting value, as has been shown by Galambos and Woeginger [42] and Grove [43]. Galambos and Woeginger [42] expanded on the idea of using two buffer-bins for temporarily storing some items, which originated in Reference 44.

The relaxation used by Galambos and Woeginger [42] is to allow the repacking of currently open bins, that is, items can be removed from current open bins and reassigned before packing the current item. Galambos and Woeginger presented an "online with repacking" algorithm REP$_3$ that uses a weighting function w and three open bins concurrently as follows. The current item is always packed in an open bin. Then all elements in the three open bins are repacked by the first fit decreasing (FFD) algorithm, with the result that either one bin becomes empty or at least one bin B has $w(B) := \sum_{a_i \in B} w(a_i) \geq 1$. All bins with $w(B) \geq 1$ are then closed and replaced by empty bins. The authors proved that

$$R^\infty_{REP_3} = h_\infty.$$

Grove [43] independently designed an algorithm with the same behavior using an alternative notion that he calls *lookahead*. In his algorithm, one is given a fixed *warehouse size*, W; an item a_i need not be packed until one has looked at every item a_i through a_j, for $j > i$ such that $\sum_{h=i}^{j} s(a_h) \leq W$. Allowing lookahead, an appropriate choice of parameters again gives an algorithm with an asymptotic worst-case ratio of h_∞.

Balogh and Békési [45] review semi-online bin packing research, and present a new asymptotic ratio lower bound of any online bin packing algorithm that knows in advance the optimum value. Normally a method for packing patterns is used to provide new lower bounds for online bin packing problems. This chapter extends this method by using "branching" list sequences. Then solving a relatively "simple" corresponding linear programming (LP) problem (with 42 variables, 81 constraints, and 178 non-zero entries) the lower bound of 1.30556 can be established, which is identical to the earlier bound of Epstein and Levin [46]. An improved lower bound 1.3231 can be established using a much more complicated branching structure and then solving the corresponding LP problem (with 9,632 variables, 9,784 constraints, and 72,616 non-zero entries) [45]. Standard LP solvers take about 0.14 s to solve the instance.

In Reference 47 several new lower bounds are established for the asymptotic worst-case ratio of semi-online bin packing and for a special case of batched bin packing algorithms, where the number of different item sizes is bounded by p ($p \geq 0$). Batched bin packing, first studied by Gutin et al. [48], means that the items arrive in consecutive batches, all items in a batch are available at the same time, and the number of batches is small. The new bounds are established from the solution of nonlinear optimization problems using theoretical analysis and a numerical global optimization method.

Now consider the model studied in Reference 49 where a buffer is available, and the item sizes are between α and $1/2$, where $0 < \alpha < 1/2$. For this case, NF (without buffers) has an asymptotic worst-case ratio $2/(1 + \alpha)$, and a modification of NF (with a buffer) has an asymptotic worst-case ratio at most $13/9 \approx 1.444$, if the buffer size is at least 1 [49].

28.3.1 Online Bin Packing with Advice

In this model the online constraint is relaxed and partial information about the future items is provided. This partial information is called "advice," and it is given by some offline oracle. This additional can be used to decrease the worst-case ratio. There are several different options for defining the type of the advise: "advice-with-request" and "advice-on-tape."

In what follows we discuss the latter model. Boyar et al. [50] present tight upper and lower bounds for the amount of advice an algorithm needs to achieve an optimal packing, which is approximately $n \cdot \log(\text{OPT})$, where n is the number of items and OPT is the number of bins in an offline optimal solution. An algorithm, which achieves a worst-case ratio of $3/2$, when provided with $\log n + o(\log n)$ bits of advice is analyzed. If additional advice ($2n + o(n)$ bits) is provided, it is possible to show the worst-case ratio is $4/3 + \varepsilon$. When three advice bits are provided per item, instead of two, the worst-case ratio decreases (roughly speaking) from $4/3$ to $5/4$ [51]. Finally, it is shown that "linear size" advice is needed to achieve a competitive ratio better than $9/8$.

For online bin packing with advice, Paper [52] shows that the worst-case ratio of 1 can be arbitrarily approached with only a constant number of bits of advice per request. More precisely, an online algorithm using $O(\frac{1}{\varepsilon} \log \frac{1}{\varepsilon})$ bits of advice per request and its worst-case ratio is $(1 + \varepsilon)$ is presented.

Paper [53] extends the results of Boyar et al. [50] as follows. On the positive side, a relatively simple algorithm is given that achieves a worst-case ratio arbitrarily close to $3/2$, using constant-size advice. For example, with the help of 16 bits of advice a worst-case ratio of 1.530 can be achieved. A more complex algorithm is discussed that requires only constant-size advice per request, but its worst-case ratio is below 1.5 (the ratio can be made arbitrarily close to 1.47012). On the negative side, the construction of Boyar et al. is extended to show that no online algorithm with sub-linear advice can have a worst-case ratio below $7/6$ (improving the previous bound of $9/8$).

28.4 Offline Algorithms

As unconstrained algorithms for partitioning the sets $\{a_i\}$, offline algorithms need not be thought of in terms of sequential packings of an ordered list. On the other hand, the best known offline algorithms are usually expressed in just these terms, but with an initial ordering of the list allowed. That this is a natural first approach can be seen from the bad examples for online algorithms, which were either increasing (FF) or alternating (NF) sequences of item sizes. Thus, sorting the items in decreasing order and then employing NF, FF, or BF creates three interesting candidates for simple, but effective offline algorithms; they are denoted by NFD, FFD, and BFD, with the "D" standing for "decreasing." The sorting needs $O(n \log n)$ time and so the total running time of each algorithm will be $O(n \log n)$.

Baker and Coffman [54] proved that $R_{NFD}^{\infty} = h_\infty$, and they gave a parametric bound as well: if $\alpha \in (1/(s + 1), 1/s]$, $s \geq 1$ then

$$R_{NFD}^{\infty}(\alpha) = h_\infty(\gamma_s^*).$$

where $h_\infty(\gamma_s^*) = \frac{s-1}{s} + \gamma_s$. Here γ_s is a slight generalization of the previous t_i sequence, as follows: $t_1(s) = s + 1, t_2(s) = s + 2$, and $t_{i+1}(s) = t_i(s)(t_i(s) - 1) + 1$, for $i \geq 2$. Then $\gamma_s = \sum_{i=1}^{\infty} \frac{1}{t_i(s)-1}$.

As expected, better results are achieved when using FF (or BF) after the initial sort. Johnson [2] showed that for every list L

$$\text{FFD}(L) \leq \frac{11}{9} \cdot \text{OPT}(L) + 4$$

He could also prove that this bound is tight, that is, $R_{\text{FFD}}^\infty = 11/9$. Later, Baker [55] gave a shorter and simpler proof and cut the constant to 3. Further shortening in the proof and reducing the constant to 1 was given by Reference 56. Later, even this value was cut to 7/9 by Li and Yue [57]. The constant 6/9 was conjectured by Dósa [58], and completely proved by Dósa et al. [59]. The proof is very long, and quite technical.

For the parametric case Johnson [2] showed that

$$R_{\text{FFD}}^\infty(\alpha) = \begin{cases} \frac{71}{60} & \text{for } \frac{8}{29} < \alpha \leq \frac{1}{2}, \\ \frac{7}{6} & \text{for } \frac{1}{4} < \alpha \leq \frac{8}{29}, \\ \frac{23}{20} & \text{for } \frac{1}{5} < \alpha \leq \frac{1}{4}, \end{cases}$$

and he conjectured that

$$R_{\text{FFD}}^\infty\left(\frac{1}{m}\right) = 1 + \frac{1}{m+2} - \frac{2}{m(m+1)(m+2)} = F_m$$

for all integers $m \geq 4$.

Csirik proved that this conjecture is true when m is even but is false when m is odd [60]. Defining

$$G_m = 1 + \frac{1}{m+2} - \frac{1}{m(m+1)(m+2)} = F_m + \frac{1}{m(m+1)(m+2)},$$

he was able to prove that

$$R_{\text{FFD}}^\infty\left(\frac{1}{m}\right) = \begin{cases} F_m & \text{if } m \text{ is even}, \\ G_m & \text{if } m \text{ is odd}, \end{cases}$$

for all $m \geq 5$.

Xu [61] completed the proof for arbitrary values of $\alpha \leq 1/4$. He showed that when m is even, F_m holds true for any $\alpha \in (1/(m+1), 1/m]$, whereas for m odd the interval has to be divided into two parts with

$$R_{\text{FFD}}^\infty(\alpha) = \begin{cases} F_m & \text{if } \frac{1}{m+1} < \alpha \leq d_m, \\ G_m & \text{if } d_m < \alpha \leq \frac{1}{m} \end{cases}$$

where $d_m := (m+1)^2/(m^3 + 3m^2 + m + 1)$.

The lower bound for FFD is given via the following lists:

$$L_{6k} = \left(\underbrace{\frac{1}{2} + \varepsilon, \ldots, \frac{1}{2} + \varepsilon}_{6k}, \underbrace{\frac{1}{4} + 2\varepsilon, \ldots, \frac{1}{4} + 2\varepsilon}_{6k}, \underbrace{\frac{1}{4} + \varepsilon, \ldots, \frac{1}{4} + \varepsilon}_{6k}, \underbrace{\frac{1}{4} - 2\varepsilon, \ldots, \frac{1}{4} - 2\varepsilon}_{12k} \right)$$

It is then straightforward to check that $\text{OPT}(L_{6k}) = 9k$ and $\text{FFD}(L_{6k}) = 11k$.

After sorting the items we may use an ANY FIT (AF) algorithm AF to pack the list. Unfortunately there are no exact bounds for this case. We only know [2,5,7] that

$$\frac{11}{9} \le R_{AF}^{\infty}(1) \le \frac{5}{4},$$

and

$$\frac{1}{m+2} - \frac{2}{m(m+1)(m+2)} \le R_{AF}^{\infty}(\alpha) \le \frac{1}{m+2},$$

where $m = \lfloor \frac{1}{\alpha} \rfloor$ and $\alpha < 1$.

Johnson [2] made an interesting attempt to get a better offline algorithm. He proposed the Most-k Fit (MF_k) algorithm, which first sorts the list in decreasing order. The next bin to be packed will first be filled with the largest as yet unpacked element. If the smallest item in the list does not fit in the bin, we close the bin and continue with the next bin. If the smallest item does fit, we pack at most k additional items with the least space left over. The running time is $O(n^k \log n)$. For a long time Johnson conjectured that $\lim_{k \to \infty} R_{MF_k}^{\infty} = 10/9$, but Friesen and Langston [62] provided a counterexample showing that $R_{MF_k}^{\infty} \ge 5/4$ for $k \ge 2$.

For several years FFD had the provably smallest asymptotic bound. Yao [21] proposed a complicated algorithm A with a $O(n^{10} \log n)$ running time and with $R_A^{\infty} \le 11/9 - 10^{-7}$, which proved that FFD could in fact be beaten. A much larger improvement was achieved by Garey and Johnson [63]. They proposed the Modified First Fit Decreasing (MFFD), which differs from FFD only when packing the items from $(1/6, 1/3]$, called *key* items, just after packing all items $>1/3$. MFFD attempts to pack 2 key items in every *key* bin, defined as a bin having only an item $>1/2$. The key bins are packed in largest-gap-first order. If the gap in the current key bin can accommodate the two smallest items in the set S_U of the as yet unpacked key items, then the first key item packed is the smallest in S_U and the second one packed is the one remaining in S_U that would be chosen by BF. As soon as at most one key item remains or a key bin is encountered that can not accommodate the smallest two remaining key items, MFFD reverts to FFD for the remainder of the packing. Garey and Johnson proved that

$$R_{MFFD}^{\infty} = \frac{71}{60} = 1.183333\ldots.$$

This is the best known offline algorithm. It is interesting to note that $R_{MFFD}^{\infty} = R_{FFD}^{\infty}\left(\frac{1}{2}\right)$.

In their attempt to find a better algorithm, Friesen and Langston [62] investigated the Best Two Fit (B2F) algorithm, which is basically a Most-k Fit-type algorithm with $k = 2$: B2F first packs a bin by FFD. If the bin contains more than a single item, then the list is checked to see if the smallest item in the bin could be replaced by two items that would occupy more of the bin. If so, the two largest such items are inserted in place of the smallest item. The algorithm is also simplified by requiring all items smaller than 1/6 be withheld until all larger items have been packed. FFD is used to complete the packing when only items no greater than 1/6 are left. The authors proved that

$$R_{B2F}^{\infty} = \frac{5}{4} = 1.25,$$

which is worse than those of FFD. The bad lists for this algorithm have the following B2F packing:

$$\left(\underbrace{\frac{1}{3} - 4^k \frac{\varepsilon}{2}, \frac{1}{3} - 4^k \frac{\varepsilon}{2}, \frac{1}{3} + 4^k \varepsilon}_{2 \text{ bins}}, \underbrace{\frac{1}{3} - 4^{k-1} \frac{\varepsilon}{2}, \frac{1}{3} - 4^{k-1} \frac{\varepsilon}{2}, \frac{1}{3} + 4^{k-1} \varepsilon}_{8 \text{ bins}}, \ldots, \right.$$

$$\left. \underbrace{\frac{1}{3} - 2\varepsilon, \frac{1}{3} - 2\varepsilon, \frac{1}{3} + 4\varepsilon}_{2 \cdot 4^{k-2} \text{ bins}}, \underbrace{\frac{1}{3} + \varepsilon, \frac{1}{3} + \varepsilon}_{4^{k-1} \text{ bins}} \right).$$

Fortunately the counterexamples for FFD and B2F are complementary and so the authors defined a compound algorithm (CFB) that runs both FFD and B2F on the list and takes the better result. The running time may double, but for the performance ratio we get:

$$1.164\ldots = \frac{227}{195} \leq R_{CFB}^{\infty} \leq \frac{6}{5} = 1.2.$$

The worst-case example is given by an example where the bin size is 559, and lists where the optimal packing consists of $60k$ bins each having three items of size $381, 98, 80$, of $120k$ bins each having five items of size $191, 96, 96, 96, 80$, and of $15k$ bins each having six items of size $99, 99, 99, 99, 80, 80$. Then the optimal packing clearly uses $195k$ bins, and CFB will use $227k$ bins. The exact bound for CFB has not yet been determined and so it is not known whether it is better than MFFD.

An earlier step in this direction (i.e., sequentially packing each bin as well as possible) did not entail orderings by size. Graham [64] proposed a greedy algorithm subsequently called "subset sum" (SS) algorithm in which, at each step, packing the next bin solved a knapsack problem. Surprisingly, however, the worst-case behavior of this algorithm A is rather poor:

$$R_A^{\infty} \geq \sum_{k=1}^{\infty} \frac{1}{2^k - 1} = 1.6067\ldots\,.$$

This lower bound is proved by instances L_b that have b items of size $1/2^k + \varepsilon$ for $k = 1, 2, \ldots, p$, with $0 < \varepsilon \leq 1/2^{2p}$ and for values of b that are multiples of $2^k - 1, k = 1, 2, \ldots, p$, for example, $b = a \prod_{k=1}^{p}(2^k - 1)$ for some positive integer a. Then the optimal packing of L_b has b bins, each bin consisting of one item from each size. Graham's algorithm first packs $b/(2^p - 1)$ bins, each with $2^p - 1$ items of size $1/2^p + \varepsilon$, then $b/(2^{p-1} - 1)$ bins, each with $2^{p-1} - 1$ items of size $1/2^{p-1} + \varepsilon$, and so on, ending with b bins, each with one item of size $1/2 + \varepsilon$. Summing up the lengths of the sub-packings gives the lower bound.

Determining the exact bound of sum of subsets algorithm was an open problem for decades. Caprara and Pferschy [65] proved in 2004 an upper bound on the asymptotic performance of Graham's algorithm:

$$R_{SS}^{\infty} \leq \frac{4}{3} + \ln\left(\frac{4}{3}\right) \approx 1.6210.$$

They also dealt with the parametric case, showing that, if $\frac{1}{m+1} < \alpha \leq \frac{1}{m}$, then

$$R_{SS}^{\infty}(\alpha) \geq \sum_{k=1}^{\infty} \frac{m}{(m+1)^k - 1}.$$

and

$$R_{SS}^{\infty}(\alpha) \leq \begin{cases} 2 - \frac{4m}{3(m+1)} + \ln(\frac{4}{3}) & \text{if } m \leq 2, \\ 1 + \ln\frac{m+1}{m} & \text{if } m \geq 3. \end{cases}$$

The exact worst-case bound of the subset sum algorithm was determined only until recently. It is shown in References 66, 67, that for any $1/2 < \alpha \leq 1$,

$$R_{SS}^{\infty}(\alpha) = \sum_{k=1}^{\infty} \frac{1}{2^k - 1} \approx 1.6067.$$

Note that we revisit this tight bound (i.e., 1.6067) in the section that deals with bin packing games. Epstein et al. also shows that for each integer m and $\frac{1}{m+1} < \alpha \leq \frac{1}{m}$, the SS algorithm has an approximation ratio of

$$R^\infty_{SS}(\alpha) = 1 + \sum_{k=1}^{\infty} \frac{1}{(m+1)2^k - 1}.$$

Note that the true ratio lies strictly between previous bounds given in Reference 65.

28.5 Other Worst-Case Issues

28.5.1 Special Case Optimality

One of the easiest cases is when the number of different items in the lists is bounded, that is, we have k different item sizes. Let us denote the item sizes by $s_1, s_2, \ldots, s_k$. In this case we have a smallest item size. Then let us denote the smallest integer by j so that all item sizes are $\geq 1/j$. It is quite clear that we cannot pack more than j items in one bin. So a legal packing (we call it here a configuration) of a bin can be given by a k-tuple

$$p_i = (p_{i1}, p_{i2}, \ldots, p_{ik}), \sum_{l=1}^{k} p_{il} \leq j, \sum_{l=1}^{k} s_l \cdot p_{il} \leq 1$$

where p_{ij} denotes the number of s_j items in configuration i. The total number of lists having not more than j items is $O(k^j)$, a part of them form a configuration. But then the total number of all possible packings of a list $L = (a_1, a_2, \ldots, a_n)$ where $a_i \in \{s_1, s_2, \ldots, s_k\}$ is $O(n^{k^j})$, that is, a polynomial in n. This is a brute force method—the optimum can be computed more easily in this case, as proposed by Błazewicz and Ecker [68]. Let us denote the number of items s_i in the list L by m_i, $1 \leq i \leq k$. Then the problem of finding the optimal packing of L can be formulated as the following integer programming problem:

$$\min \sum_{i=1}^{N} b_i, w.r.t \sum_{i=1}^{N} p_{ij} b_i = m_j \ (j = 1, \ldots, k)$$

and $b_i \geq 0$ are integers $(i = 1, \ldots, N)$. Here N is the number of all possible configurations and b_i gives the number of bins of configuration p_i. This matrix has k rows and $O(k^j)$ columns and can be solved in polynomial time in the number of constraints [69]. The optimal packing can then be constructed in linear time. If we do not need the exact optimum but just a good approximation of it then we can relax the integer condition from the above-mentioned inequalities and compute the optimum of the remaining problem. Then we may round up the solution, as proposed by Gilmore and Gomory [70,71].

A different approach was chosen by Hochbaum and Schmoys [72]. They investigated the classical bin packing problem with item sizes bounded from below. They found an easily solvable case: if all item sizes are $\geq 1/3$, then a relatively straightforward polynomial algorithm will furnish an optimal packing. This algorithm will pack, in the first stage, all items larger than a half separately in bins. In the second stage all items $= 1/3$ are packed by three items in bins. In the third stage the remaining items, that is, those with size $> 1/3$ and $\leq 1/2$ are paired up with items from stage one, and those items where the pairing is not possible are packed by two. The second step in this direction was to prove that if item sizes are $\geq 1/3 - \varepsilon$ for any fixed ε, $0 < \varepsilon < 1/3$, then the problem becomes strongly NP-complete. Even for this case, supposing $0 < \varepsilon \leq 1/12$, and slightly relaxing the problem allowing the bins to be over-packed a little bit, they gave a polynomial time algorithm that gives the optimal number of bins but an over-packing with at most $3\varepsilon/2$.

Leung [73] derived the limits for polynomially solvable cases. He proved that the bin packing problem remains NP-hard even if the the bin capacity is 1 and if item sizes must be drawn from the set $\{1, r, r^2, , r^3, \ldots\}$, where r is an arbitrary positive rational number < 1.

Another type of optimality result was covered by Coffman et al. [74]. They introduced the restricted lists L where the item sizes form *divisible sequences*. Now let us suppose that the bin capacity is an integer. A sequence of item sizes $s_1 > s_2 > \ldots > s_i > s_{i+1} > \ldots$ is called *divisible* if for each $i \geq 1, s_{i+1}$ exactly divides s_i. A list L is called *weakly divisible* if its items when sorted forms a divisible sequence. If, in addition, the largest item size exactly divides the bin capacity, then the list is called *strongly divisible*. In Reference 74 it was proven that when L is weakly divisible FFD gives optimal packing. If L is strongly divisible then even FF produces optimal packing.

Let us now restrict our attention to the case when there are only two different item sizes. The offline variant is defined by the values of α and β ($\beta \leq \alpha$) which are the item sizes, and two integers which are the number of items with sizes of α and β, respectively. This offline problem can be solved in polynomial time using the algorithm of McCormick et al. [75].

Gutin et al. [76] dealt with the online setting, where the item sizes are known in advance, but not their numbers. They showed that the overall best worst-case ratio for this problem is 4/3. This means that there exists a pair of sizes for which 4/3 is a lower bound for the asymptotic worst-case ratio of any algorithm. On the other hand, Reference 76 presents an algorithm with this worst-case ratio for any pair of sizes. An improved exact bound for the best possible worst-case ratio is also presented for the case when $\alpha > 1/2$.

It is worth noting that if the two sizes are not known in advance, the lower bound is larger. It is shown by the same authors in Reference 48 that the worst-case ratio of any algorithm for this model is at least 1.3871 as shown in Reference 48, therefore the advance knowledge of the sizes makes a difference.

Epstein and Levin [77] generalize the previous result for the parametric case and derive the best overall worst-case ratio as a function of k, where $1/k$ is an upper bound for both item sizes. The tight ratio is $(k + 1)^2/(k^2 + k + 1)$. They also establish that if there are two item sizes but they are not known in advance, then this ratio cannot be achieved.

28.5.2 Other Measures

Previously the competitive ratio compares the solution of online algorithms to the offline optimum. In what follows we discuss several other attempts to measure the "goodness" of online algorithms.

Kenyon [78] introduced the random order ratio:

$$RR_A = \sup_I \frac{E_\sigma[A(\sigma(I)]}{\text{OPT}(I)},$$

that is, the expected performance of an algorithm is compared to the optimum, for that input where this ratio is the largest possible. Kenyon showed that RR_{BF} is between 1.08 and 1.5. Clearly, RR_A is hard to determine exactly, at least for BF. Coffman et al. [79] introduce an alternative (slightly different) definition for random order ratio.

Boyar and Favrholdt [80] define the relative worst order ratio (WR) as follows:

$$WR_{A,B} = \max_I \frac{\max_\sigma\{A(\sigma(I))\}}{\max_\sigma\{B(\sigma(I))\}},$$

where:

    A and B be two online algorithms
    and I is an arbitrary input
    $\sigma(I)$ denotes a permutation of input I.

Then we take the worst order (for algorithm A) in the numerator and the worst order (for algorithm B) in the denominator. Note that the orderings may be different. If this ratio is always at least 1, or always at most 1, the algorithms may be compared with this measure. The relative worst order ratio can be seen as the worst-case version of the random order ratio, but the online algorithms are compared directly, rather than through the offline optimum solution.

It turns out that WR can be used to make more accurate comparisons between algorithms than when using the competitive ration. Reference 80 shows that any AF algorithm is not better than FF and no worse than WF. Moreover, it is shown that $WR_{WF,FF} = 2$ and $WR_{NF,WF} = 2$.

Epstein et al. [81] extended the results to parametric bin packing, moreover they showed that AWF is strictly worse than FF and strictly better than WF, both for standard and parametric bin packing. The authors also discuss the adaptation of the Harmonic algorithms for three versions: standard bin packing, bin covering, and open-end bin packing. For bin coloring two algorithms are compared, a greedy one and one that uses the NF strategy. It is interesting to note that the greedy algorithm has worse competitive ratio, but it is better in terms of WR.

Paper [82] compares four different measures (competitive analysis, random order ratio, worst order ratio, and one another one) and three simple algorithms for an "easy" problem which is called *baby server* problem, a simplified version of the k-server problem. The complete characterization of all three algorithms is given in all four measures. This is the first time an algorithm is proven optimal under WR analysis.

An additional measure, called differential approximation, has been introduced and used for other problems too [83,84].

28.5.3 Resource Augmentation

Resource augmentation means that the online algorithm gets some extra power (compared to the much powerful offline algorithm), it was previously proposed for scheduling and other problems. For bin packing, resource augmentation was introduced by Csirik and Woeginger [20], and here resource augmentation means that the online algorithm is allowed to use bins with capacity $b > 1$ (whereas the offline algorithm still packs into bins of unit capacity). This paper considered bounded space bin packing, and for this case the problem is completely solved: They defined a function $\rho(b)$, and showed that the worst-case ratio of the modification of the Harmonic algorithm comes arbitrary close to this bound. They also proved that no online bounded space algorithm can have better worst-case ratio.

Unbounded space resource augmented bin packing is considered in Reference 85. Here NF and WF are compared. It is well known that for standard bin packing both algorithms have tight two worst-case ratios, but WF seems to be more reasonable as never closes a bin until the list of items ends. A complete analysis is given regarding the tight asymptotic worst-case ratios of these two algorithms. According this, WF is strictly better than NF for any $1 < b < 2$, whereas they have the same ratios, that is, $1/(b-1)$, for $b > 2$. For $1 < b < 2$, the tight ratio of WF is $2b/(3b-2)$, the tight ratio of NF is quite complicated.

Epstein and van Stee [86] present general lower bounds by using improved sequences, these new bounds are nearly matching for bin sizes $b \geq 2$. It means that for $b \geq 2$, the lower bounds show that the best bounded space algorithm is very close to optimal (among unbounded space algorithms). Getting intuitions from the lower bounds new algorithms are developed, specifically for $b \in [1, 2)$. Interval $[1, 2)$ is partitioned into four subintervals and for each of them an algorithm is defined. In this way the new upper bounds improve upon the bounds from Reference 20 on the entire interval. That is, these algorithms are better than the best possible bounded space algorithm. These algorithms combine items in bins in an unusual way: the small items are packed together with large items in the same bins to achieve good performance.

28.5.4 Linear Time Algorithms

Most algorithms investigated up till now have a running time superlinear in n. It would be interesting to know what could be said about algorithms that have a linear running time. Naturally NF is one of these, but it has poor performance. Johnson quite early on [5] proposed a better linear-time algorithm, Group-X Fit Grouped (GXFG). First he defined the Group-X Fit (GXF) algorithm. Let a *schedule of intervals* be the set $X = \{x_0, x_1, \ldots, x_k\}$, where $x_0 = 0, x_k = 1$, and $x_i < x_{i+1}, 0 \le i < k$. X can be thought of as a partition of the unit interval $[0, 1]$ into the subintervals $(0, x_1], (x_1, x_2], \ldots, (x_{k-1}, 1]$. Interval $(x_{i-1}, x_i]$ will be called the X_i interval $1 \le i \le k$, and the points in X will be called breakpoints. Items from the list are classified similarly: a_j will be called an I_l item if it belongs to X_l. Note that this is a more general definition than that given for the Harmonic algorithm, the latter having $x_i = 1/i$ as breakpoints. GXF will round down the free space in open bins to the next breakpoint (if the free space is equal to a breakpoint then it is kept). The algorithm is actually a variant of BF: an I_l item will be packed in a bin with the least free space left over (if no bin has enough space for the item, a new bin will be used). The algorithm has linear running time and is online. Johnson [2,5] did show that for $m = \lfloor 1/\alpha \rfloor \ge 2$,

$$R_{GXF}^{\infty}(\alpha) = R_{FF}^{\infty}(\alpha),$$

when $\{1/(m+1), 1\} \subseteq X$. He also proved that if $m = 1$ then $1.7 \le R_{GXF}^{\infty} \le 2$ whenever $\{1/6, 1/3, 1/2\} \subseteq X$ and conjectured that the lower bound is the tight value.

The better algorithm GXFG first scans through the list and collects items from the same intervals together. The second step is just GXF which starts with items from interval X_k, followed by items from $X_{k-1}, X_{k-2}, \ldots$ Johnson also proved, for all $m = \lfloor 1/\alpha \rfloor \ge 1$, if X contains $1/(m+2), 1/(m+1), 1/m$, that

$$R_{GXFG}^{\infty}(\alpha) = \frac{m+2}{m+1}.$$

Because of the first scan GXFG is not an online algorithm. Basically it uses the idea of FFD (it first starts to pack the large elements) but without really sorting the elements.

A further improvement was achieved by Martel [87]. He used subintervals too, but packed items more freely than Johnson in his GXFG. The algorithm is the following. First we partition the items using the following five sets:

$C_1 = \{x_i \mid 2/3 < x_i \le 1\},$

$C_2 = \{x_i \mid 1/2 < x_i \le 2/3\},$

$C_3 = \{x_i \mid 1/3 < x_i \le 1/2\},$

$C_4 = \{x_i \mid 1/4 < x_i \le 1/3\},$

$C_5 = \{x_i \mid x_i \le 1/4\}.$

Let $c_i = |C_i|, i = 1, 2, \ldots, 5$ be the numbers of items in intervals. An item in C_i will be called a C_i−piece. The motivation behind this partitioning is to allow items to be packed based on the set to which they belong. Items in C_1 can only be combined with items in C_5. Items in C_2 can always be combined with an item in C_4, but never with an item in C_2. Items in C_3 can be packed together two to a bin, items in C_4 can be packed three to a bin, and items in C_5 can be packed at least four to a bin. The algorithm OffBP is based on these observations and is defined in the following.

Step 1. Form the sets $C_i, i = 1, 2, \ldots, 5$.

Step 2. Let $k = \lceil \min\{c_2, c_3\}/2 \rceil$. Split C_2 into the two sets C_2^s that contains the smallest k elements in C_2 and C_2^b, which contains the remaining elements. In a similar way, split C_3 into C_3^s and C_3^b. Pair up each element in C_2^s with an element in C_3^s. If the pair fits in a bin, create a bin containing both items. If the pair does not fit, create a bin containing only the element from C_2^s.

Step 3. Put each C_1−piece into its own bin and put each C_2^b−piece into its own bin.

Step 4. Put all C_3^b−pieces and all C_3^s−pieces which were not combined with a C_2^s−piece in Step 2 into bins, two into each bin.

Step 5. Each bin with a single C_2−piece has a C_4−piece added to it (until one runs out of C_4−pieces.)

Step 6. Any C_4−pieces not packed in Step 5 are put three to a bin into empty bins.

Step 7. Pack C_5−pieces into the bins created in Steps 1–6 using NF: for each bin we add C_5−pieces until the next C_5−piece does not fit; then go to the next bin and never consider the previous bin again.

Step 8. Any remaining C_5−pieces are packed into empty bins using NF.

Martel has shown that the performance ratio of OffBP is 4/3. He also gave hints on how the algorithm could perhaps be improved. These were followed up by Békési et al. [88]. They used the breakpoints $\{1/5, 1/4, 1/3, 3/8, 1/2, 2/3, 4/5\}$ and a much more complicated algorithm and obtained a performance bound of 5/4.

28.5.5 Randomization

Bin packing was a pioneering theme in many fields of analysis of algorithms. This is the case for randomized algorithms as well. Perhaps, AF is the first randomized algorithm that was really analyzed in detail. This is certainly a classical result for randomized algorithms.

Only a few additional results are known for randomized algorithms in bin packing. As mentioned earlier, Chandra [35] proved that the general lower bound for online algorithms hold for randomized algorithms too.

Kenyon [78] analyzed BF from a new point of view and defined a new measure of performance. This new measure is the expectation over random packing orders of a *worst-case list* of input items; packing orders are drawn uniformly at random from the set of all permutations. She defined the random-order performance-ratio RC(A) of an online algorithm A as

$$RC(A) = \limsup_{OPT(L) \to \infty} \frac{E_\sigma A(L_\sigma)}{OPT(L)},$$

where L_σ is the permuted list $(x_{\sigma(1)}, x_{\sigma(2)}, \ldots, x_{\sigma(n)})$ and the summation for the expectation is taken over all permutations $\sigma \in S_n$. She proved that for BF

$$1.08 \leq RC(BF) \leq 1.5,$$

a quite large gap between the lower and upper bound! She also conjectured that the true answer was probably close to 1.15. To estimate the lower bound she used a nice simplification: if we have only two different item sizes then the permutations can be simulated by drawing each item independently with the appropriate probabilities. So the sequence can be viewed as an unbiased random walk in the plane, which can be analyzed with help of the Markov chains. The lower bound was found with aid of the lists where we have 1/2 items with probability p and 1/3 items with probability $1 - p$. In this case, we have a Markov chain with five states:

(a) There is no open bin
(b) There is one open bin whose current content is 1/2
(c) There is one open bin whose current content is 1/3
(d) There is one open bin whose current content is 2/3
(e) There are two open bins, one containing 1/2 and one containing 2/3

A quite standard analysis of this Markov chain leads to the lower bound 1.08. The proof of the upper bound is much more complicated and uses some deeper results from probability theory.

Coffman et al. [79] investigate a simpler algorithm NF. For this, tight result is given, the random order ratio of NF is exactly 2, that is, the same as its asymptotic and absolute worst-case ratio.

28.5.6 Absolute Worst-Case Ratios

Up till now we have considered the asymptotic worst-case ratios of bin packing heuristics. For lists where OPT(L) is small a different ratio may also be helpful. This is the absolute worst-case ratio, which is defined by (28.1). It can readily be seen that bin packing is hard even for lists for which OPT(L) is small. This is true because the 2-partition problem, which is known to be NP-complete [89], can be polynomially reduced to the problem whether or not it is possible to pack all items in two bins. This also implies that no polynomial-time heuristic has an absolute worst-case ratio smaller than 1.5 for the bin packing problem, unless $P = NP$. This is obvious as such a heuristic could solve the 2-partition in polynomial time [89]. On the other hand, for the list $L = (0.4, 0.4, 0.3, 0.3, 0.3, 0.3)$ OPT(L)) $= 2$ and FFD(L) $= 3$, so FFD has the best possible absolute worst-case bound for this list. Li and Yue [57] give a sketch of a possible proof for

$$FFD(L) \leq \frac{11}{9} \cdot OPT(L) + \frac{7}{9}$$

for all lists. This gives an upper bound of 4 for lists with OPT(L) $= 3$ and bound of 5 for lists with OPT(L) $= 4$. The same bounds apply for BF.

But, what is the upper bound when OPT(L) $= 6$? Is there an input for which FFD(L) $= 8$? Not having a tight upper bound for the additive constant at hand, even this innocent-looking problem remains unsolved. Paper [90] gives a proof to exclude the case FFD(L) $= 7$ and OPT(L) $= 5$. Dósa [58] gives a table, indicating for each integer value OPT(L) $= m$, the biggest possible value for FFD(L). From this investigation it follows that the next tight additive constant for FFD is:

$$FFD(L) \leq \frac{11}{9} \cdot OPT(L) + \frac{6}{9}.$$

This bound is conjectured in Reference 58, and proved in Reference 59. The construction for the tight lower bound is a slight modification of the one for the tight asymptotic ratio; which can be found, for example, in Reference 91. The above-mentioned table is generated by letting OPT(L) in the form OPT(L) $= 9n + i$ where n is an integer and, i is in the range $2 \leq i \leq 10$ and then proving that

$$FFD(L) \leq \begin{cases} 11n + i + 1, & 2 \leq i \leq 5; \\ 11n + i + 2, & 6 \leq i \leq 10; \end{cases}$$

On the other hand, there exists a list L for which FFD(L) is not smaller.

With respect to FF (and BF), we know from Reference 92 that, for any list L,

$$FF(L) \leq \left\lceil \frac{17}{10} \cdot OPT(L) \right\rceil$$

and Reference 7 gives instances for which $FF(L)/OPT(L) = 17/10$. Simchi-Levi [93] proved that for all lists $FF(L) \leq 1.75 \cdot OPT(L)$.

Xie and Liu [94] improved the upper bound to

$$FF(L) \leq \min \left(\left\lfloor \frac{33}{19} \cdot OPT(L) \right\rfloor, \left\lceil \frac{17}{10} \cdot OPT(L) \right\rceil \right).$$

To improve $FF(L) \leq \left\lceil \frac{17}{10} \cdot OPT(L) \right\rceil \leq 1.7 \cdot OPT(L) + 0.9$ and $FF(L) \leq 1.75 \cdot OPT(L)$ either one decreases the multiplicative factor 1.75 or one eliminates the 0.9 additive constant. Xia and Tan [95] made progress in both directions, proving that $FF(L) \leq 1.7 \cdot OPT(L) + 0.7$ and $FF(L) \leq 12/7 \cdot OPT(L)$ (where $12/7 \approx 1.7143$). The latter result was also proven independently by Boyar et al. [96]. Then, the multiplicative factor was slightly decreased by Németh to $101/59 \approx 1.7119$ [97]. As we approach 1.7, there are more and more counterexamples that need to be excluded one by one; obviously, we cannot proceed this way. So we need to reduce the additive constant. Its tight value (taking into account also the lower bound) can only be $0.1 \cdot i$ for certain $i \in \{0, 1, \ldots, 7\}$, due to integrality of FF and OPT.

Sgall [98] presented a new analysis of BF algorithm, based on the "old" weighting function that was used by Johnson [2]. But the weight function is now divided into two main parts. One part is called scaled size, it is simply 6/5 times the size of an item. The other part is called "bonus," this is the difference of the weight and the scaled size. Using this new framework the proof for the asymptotic ratio becomes relatively simple and short. Using this "key" approach Dósa and Sgall [99], prove that the tight value for the additive constant is zero, that is,

$$FF(L) \leq \left\lfloor \frac{17}{10} \cdot \mathrm{OPT}(L) \right\rfloor.$$

This paper also presents tight lower bounds for almost all residual classes, whereas Reference 9 presents a brand new lower bound construction, and gives tight bounds for all residual classes. This paper also includes the proof for the tight analysis of BF algorithm, that is,

$$BF(L) \leq \left\lfloor \frac{17}{10} \cdot \mathrm{OPT}(L) \right\rfloor$$

Sgall [100] summarizes these final improvements regarding the FF and BF algorithms.

Berghammer and Reuter [101] present a linear time algorithm with *absolute* worst-case bound 3/2. Note, that this is best possible, unless $P = NP$. The 3/2 absolute bound holds for FFD and BFD too, but these algorithms take $O(n \log n)$ time. The algorithm given in Reference 101 follows the BF up to some point and then becomes more complex.

Zhang et al. [102] developed a different offline algorithm with absolute worst-case ratio 3/2 using an *extra* bin. They also defined an online algorithm (bounded-space) with a tight absolute worst-case ratio of 7/4.

Balogh et al. [103] presents another online algorithm, with a tight absolute worst-case ratio of 5/3, which is best possible. The main idea behind the algorithm (called FT = five–third) is to use FF whenever possible, but avoid creating "too many" 2-bins (i.e., bins with exactly 2 items). This is accomplished by creating "special" bins. The main problem is finding a good balance in the number of special bins, they should be not too many, and neither too few.

28.5.7 Bounds on Anomalous Behavior

As we have already seen, the two well-known algorithms FF and BF have the same asymptotic ratios. However, they can give strikingly different packings for individual lists. Examples are given in Reference 2 of lists L with arbitrarily large values of OPT(L), both such that BF(L) = (4/3) · FF(L) and FF(L) = (3/2) · BF(L). Similarly, AWF can be just as far away from FF, and examples exist such that AWF(L) = (5/4) · FF(L) and that FF(L) = (9/8) · AWF(L).

The decreasing-type algorithms FFD and BFD may be likewise compared. In Reference 2 lists are given for which BFD(L) = (10/9) · FFD(L). On the other hand, BFD can produce better packings than FFD as well: There are lists for which FFD(L) = (11/10) · BFD(L).

It has been shown that certain algorithms possess the undesirable property of sometimes performing "worse" when their inputs are made "better." Let us start from a list L and let us form a new list L' by deleting some elements of L and/or reducing the size of some elements of L. We will say in this case that L dominates L'. If an algorithm never uses more bins to pack L' than it uses to pack L we will say that the algorithm is monotonic, otherwise we will say the algorithm is nonmonotonic. Graham [64] and Johnson [2] once showed that FF is nonmonotonic. Murgolo [104] introduced a technique that allows one to completely characterize the monotonic behavior of any algorithm in a large, natural class. He provided upper bounds on nonmonotonic behavior for any *reasonable* algorithm. (A reasonable algorithm is one which never packs an item into an empty bin if it can fit into an already allocated bin.) We note that a reasonable algorithm is an AF algorithm. For example, he showed that there exist arbitrarily long lists satisfying *L dominates L'* such that $FF(L') \geq FF(L) + \frac{1}{75}FF(L)$ and this bound is tight to within a constant factor. Similar results hold for BF (with a constant of $\frac{1}{42}$) and WF (with a constant of $\frac{1}{15}$).

28.5.8 Parallel Algorithms

Anderson et al. [105] have investigated the FFD algorithm from a parallel computational view. They used the PRAM model of parallel computation of Fortune and Willie [106]. They considered a parallel algorithm to be fast if it is an $\mathcal{NC}$ algorithm, that is, if it runs in polylogarithmic time using a polynomial number of processors. The main algorithm they gave has a more reasonable bound, running in $O(\log n)$ time on an $n/\log n$ processor EREW (exclusive read, exclusive write) PRAM, and so is asymptotically optimal. A problem is called inherently sequential if it is $\mathcal{P}$-complete. This furnishes a relatively strong evidence that the problem is not in $\mathcal{NC}$, as if it were, we would have $\mathcal{P} = NC$. The interpretation of this is that deciding the value of a specific bit of the output of the algorithm is $\mathcal{P}$-complete.

The main result of Reference 106 is that the FFD heuristic is a $\mathcal{P}$-complete algorithm, and that a packing that obeys the same performance bound as FFD can be computed by a fast parallel algorithm. The algorithm packs the large items (items of size $\geq 1/6$) in the same manner as FFD and then fills in the remaining items. Now let $u_1, u_2, \dots, u_r$ be the list of remaining items, all with a size less than $1/6$. We first combine these items into chunks so that every chunk (except possibly the last) has a size between $1/24$ and $1/6$. The items with a size $1/24$ or larger are big enough and each is put into a chunk by itself. For the remaining items, the partial sums $s_k = \sum_{1 < j \leq k} u_j$ are determined and for each i, we combine the set of items $\{u_k \mid i/12 \leq s_k < (i+1)/12\}$ to form a chunk. Although the item sizes are less than $1/24$, each chunk will have a size between $1/24$ and $3/24$.

Now the bins packed using the FFD algorithm with items of size $1/6$ or larger will be filled in. We will add, in parallel, a distinct chunk to each bin filled less than $5/6$. Although the sizes are at least $1/24$, only a constant number of passes are needed. If there are any leftover items they are packed in new bins, just as in the method used in FFD packing.

The algorithm runs in $O(\log n)$ time using $n/\log n$ processors.

28.6 Bin Packing Games

There are two kinds of bin packing games: cooperative and noncooperative.

Faigle and Kern [107] defined the cooperative bin packing game in 1993 [107]. This game is an N-person game, where the set N consists of k bins (with unit capacities) and n different size items. A coalition is a subset of N, whereas N itself is the grand coalition. The objective function $v(S)$ of a coalition $S \subseteq N$ is the maximum total value of the items in coalition S which can be packed to the bins of S. The members of the coalition cooperate if there is a fair allocation of the value $v(N)$ among the individual players, in this case this allocation is called *core allocation*. A core allocation satisfies the following properties:

(i) $x(N) = v(N)$
(ii) $x(S) \geq v(S), \forall S \subseteq N$
  where $x(S) = \sum_{i \in S} x_i$. Unfortunately, in many cases the set of cores is empty. For such cases the notion of ε-core is defined (requiring (i) and (ii')), where
(ii') $x(S) \geq (1 - \varepsilon)v(S), \forall S \subseteq N$.

Faigle and Kern in 1998 [108] showed that the $1/2$-core is non-empty, more precisely the smallest ε for which the core is non-empty lies in $1/7 \leq \varepsilon_{\min} \leq 1/2$. Woeginger [109] shows that the $1/3$-core is also non-empty. Matsui [110] claims that this $\varepsilon = 1/3$ is the smallest possible value, but this claim has been disproved. Kuipers [111] shows that the 1/7-core is non-empty if any item has size above 1/3, and it is also known by Reference 108 that for every $\varepsilon > 0$, the ε-core is non-empty when there are many bins, that is, k is sufficiently large.

The computational complexity of determining whether ε-core is non-empty has been investigated. In Reference 108 it is shown that the core membership test is NP-hard. Then Matsui [110] proved that the core emptiness test is unary NP-hard even if $k = 3$, but this question remained open for $k = 2$, and was answered later by Reference 112. More precisely, the problem of deciding whether a payoff vector is a core allocation of an instance of the bin packing game is binary NP-hard if k is fixed, and is unary NP-hard if k is arbitrary. A pseudopolynomial time algorithm is also presented in this paper to determine the core membership test for fixed k.

The sequence of papers by Kern and Qiu [113–116] further elaborate on this topic. Paper [113] gives an alternative proof that the 1/3-core of the game is nonempty, and by this approach they are able to slightly improve the previous bound to $1/3 - 1/108$. It is also conjectured that $\varepsilon_{\min} = 1/7$. Paper [115] shows that the 1/4-core is non-empty for all instances. Finally, Reference 116 considers a generalization, where the bins may have different sizes. For this generalization it is shown that the 1/2-core is always non-empty; moreover, if all items have sizes above 1/3, then even the 5/12-core is always nonempty. Finally, it is shown that for any fixed ε, the ε-core is non-empty if the number of bins is sufficiently large. This last result is an extension of the main result of Reference 108 for the nonuniform case.

Now let us discuss noncooperative bin packing games, initially defined by Bilò [117]. In this game (given an input for bin packing), the set of players are the items. The pure strategy of a player is the bin in which it is packed. A bin is valid if its load (the sum of sizes of items within the bin) is at most 1. Changing the strategy of an item means that this item moves to a different (nonempty or empty) bin. Any item is charged a cost for being in a bin, this cost is proportional to the size of the item. So, if the level of a bin is l, and the size of an item is s (which is included in the level l), the item pays s/l for being in this bin. And an item moves (migrates) into another bin (the target bin) if

   (i) this item fits into the target bin, and
   (ii) this movement is beneficial for the item, that is, the item will pay less cost for being in its new bin.

This second property is equivalent to the property, that the level of the target bin will be strictly bigger after the movement of the item than the level of the item's actual bin. Some questions arise. Starting from a feasible packing, will the movement of the items be finished after some time? If so, when? And how can we characterize the stable packings, when there are no more improving steps?

The main concept for such stable stations is the idea of (pure) Nash equilibrium. An *NE* means that no player (i.e., item) can decrease his cost by changing only his own strategy unilaterally (i.e., moving into another bin). That is, if each player has chosen a strategy and no player can benefit by changing its strategy whereas the other players keep theirs unchanged, then the current packing is a Nash equilibrium. We are interested in pure Nash equilibria, where the actions of each player are chosen in a deterministic way, and discuss only this kind of *NE*. A solution which is a *NE* is often not socially optimal, where by a socially optimal solution we mean that the items are packed into the minimum number of bins, denoted as OPT. (This is the minimum total cost of the items if they would cooperate.) Considering an input, a *NE* is called "best *NE*," if it is a *NE* and the number of bins used is as small as possible. Furthermore a *NE* is called "worst *NE*," if the number of used bins is as large as possible, but the packing is a *NE*. Another important notation is the price of anarchy (*PoA*), which is defined as:

$$PoA = \lim_{N \to \infty} \sup \max_{I} \{NE(I)/OPT(I) \mid OPT(I) = N\} \tag{28.3}$$

It means that we consider the ratio of the worst *NE* and the socially optimal solution (i.e., OPT) for any input, and we take the lim sup of these ratios as the value of the optimum tends to infinite. If instead of the worst *NE* packings we use the best *NE* packings in the formula, we arrive at the price of stability (*PoS*). By definition, $PoS \leq PoA$ for any game. In many cases $PoS = 1$, but not always.

Turning back to the paper of Bilò [117], he proved that from any initial feasible packing, the sequence of selfish movements of the items always converges in a finite number of steps to a *NE*. The number of such steps (called improving steps) is at most $O(P^2)$, where P is the total size of the items. Moreover he found that the value of the *PoA* is between $8/5 = 1.6$ and $5/3 \approx 1.666$.

Paper [118] establishes that $1.6416 \leq PoA \leq 1.6575$, and shows that computing a pure *NE* takes polynomial time. This was not clear before as the number of migrations in Bilò's paper could be exponential.

Miyazawa and Vignatti [119,120] find better bounds for the number of migration steps. The former paper [119] gives an upper bound for the number of migrations if the best-response strategy is used (i.e., when a player is allowed to move only to a bin with the lowest cost for him). In this case the number of steps needed to reach a Nash equilibrium is $O(m \cdot w_{max}^2 + n \cdot w_{max})$ and $O(n \cdot k \cdot w_{max})$, where n, m, k, and w_{max} denotes, respectively, the number of items, the number of bins, the number of distinct item sizes, and the size of a largest item. In the other paper [120] a quite different model is considered, upper bounds are derived for the possible number of migration steps.

Epstein and Kleiman [121] showed that $1.6416 \leq PoA \leq 1.6428$, and also established lower and upper bounds for the strong price of anarchy (*SPoA*). This concept is used if strong Nash equilibria (*SNE*) is applied instead of the pure Nash equilibria (*NE*) in formula (28.3). A packing is a *SNE* if there exists no subset of items, where all items in the subset benefit (i.e., have a strictly smaller cost as a result of moving) from jointly moving to different bins; whereas, of course, the other items remain in their positions. Thus, comparing the worst *SNE* packings to the OPT packings, we get the strong price of anarchy (*SPoA*), whereas comparing the best *SNE* packings to the OPT packing we gain the strong price of stability (*SPoS*). The paper shows that a packing is a *SNE* iff it is produced by the subset sum algorithm for bin packing. From this consideration it follows that the *SPoA* equals the approximation ratio of the subset sum algorithm. Moreover, as a lower bound instance for the subset sum algorithm can be converted easily to a lower bound instance on the *SPoS*, it follows that for this bin packing game *SPoA* = *SPoS*. Finally, it is shown that no polynomial time algorithm exists for finding *SNE*, unless $P = NP$. Paper [66] established the exact value of the *SPoA*, which is approximately 1.6067. Interested readers are referred to Epstein's survey [122].

Let us now discuss other versions and generalizations of noncooperative bin packing games. Ma et al. [123] introduced another version where an item pays cost $1/k$, if the item is packed into a k-bin (i.e., together with other $k - 1$ items), no matter the sizes of the items, provided they fit together in the bin. This means that the movement of an item t packed in a k_1-bin B_1 (where t is included in the number of items of B_1) to a k_2-bin B_2 (where t is not included in the number of items of B_2) is beneficial iff the load of B_2 plus the size of t is at most 1 and $k_2 \geq k_1$. This game is quite different from Bilò's version. Ma et al. constructed an input L for which there is a *NE* with a cost $1.7 \cdot OPT(L)$; on the other hand, for any input the social cost of any *NE* is at most $1.7 \cdot OPT(L)$. This upper bound follows from the fact that the approximation ratio of FF is 1.7. This paper also gives a simple algorithm to pack all the items in $O(n \cdot \log n)$ time, where the resulting packing is a *NE*, and the social cost is at most $1.69103 \cdot OPT(L) + 3$. Finally, it is shown that from any initial feasible packing the packing surely converges to a *NE* in at most $O(n^2)$ steps, where n is the number of items. This result is improved to $\Theta(n^{3/2})$ by Dósa and Epstein [124].

Dósa and Epstein [125] defined a common generalization of the previous two versions. Every item has a size (as before) and also a positive weight. A packing is valid as before, but the movements are beneficial for the items using the weight of the items. An item moves into another bin if its weight divided by the total weight of the items in the target bin (including the weight of the moving item) will be smaller than the cost of the item in its actual bin. If the weight of any item is unit we get back the version of Ma et al., and if the weight of an item is the same as its size we get back the version of Bilò, but the weight can differ from both previous cases, as well. Paper [125] presents several results. It is shown that any game of this general class admits all types of equilibria mentioned above. The case of general weights is strongly related to the FF algorithm, and *PoA* = *SPoA* = *SPoS* = 1.7, but *PoS* = 1. For the special case of unit weights (i.e., Ma's version) the paper gives improved results, as follows: The *PoA* is between 1.6966 and 1.6993. Moreover, the values of *SPoA* and *SPoS* are determined exactly, their values are approximately

$SPoA \approx 1.69103$ and $SPoS \approx 1.611824$, this last value is not known to be the (asymptotic) ratio of any well-known algorithm for bin packing. The PoS equals 1 also in this case.

Recently Wang et al. [126] defined a much more general model. Given n items with positive sizes $\{a_1, a_2, \ldots, a_n\}$ and an $n \times n$ real matrix $S = [s(i,j)]$, the payoff of item i is $p_i = \sum_{j \in B_k} s(i,j)$ if i is packed into bin B_k. All bins have unit capacity. An item has an incentive to move to another bin, if its payoff will be higher in the target bin. In this model, $s(i,j)$ means how much items i and j "like" each other. The paper shows that for any symmetric matrix S, there always exists a pure NE. In the general case the PoA may be very large. So several special cases are considered, and lower and upper bounds are derived for the value of the PoA.

References

1. Garey, M.R., Graham, R.L., and Ullman, J.D., Worst-case analysis of memory allocation algorithms, *Proceedings of Symposium on Theory of Computing*, ACM, Denver, Co, 1972, p. 143.
2. Johnson, D.S., Near-optimal bin packing algorithms, PhD Thesis, Massachusetts Institute of Technology, Department of Mathematics, Cambridge, MA, 1973.
3. Coffman, Jr., E.G., and Csirik J., A classification scheme for bin packing theory, *Acta Cybern.* 18, 47, 2007.
4. Coffman, Jr., E.G., Csirik, J., Galambos, G., Martello, S., and Vigo, D., Bin packing approximation algorithms: Survey and classification, in *Handbook of Combinatorial Optimization*, Springer, New York, 2013, p. 455.
5. Johnson, D.S., Fast algorithms for bin packing, *JCSS*, 8, 272, 1974.
6. Fisher, D.C., Next-fit packs a list and its reverse into the same number of bins, *Oper. Res. Lett.*, 7, 291, 1988.
7. Johnson, D.S., Demers, A., Ullman, J.D., Garey, M.R., and Graham, R.L., Worst-case performance bounds for simple one-dimensional packing algorithms, *SIAM J. Comput.*, 3, 299, 1974.
8. Dósa, G., The tight absolute bound of First Fit in the parameterized case, *Theor. Comput. Sci.* 596, 149, 2015.
9. Dósa, G. and Sgall, J., Optimal analysis of Best Fit bin packing, *ICALP*, LNCS, vol. 8572, Springer, Heidelberg, Germany, 2014, p. 429.
10. Csirik, J. and Imreh, B., On the worst-case performance of the NkF bin-packing heuristic, *Acta Cybern.*, 9, 89, 1989.
11. Mao, W., Tight worst-case performance bounds for next-k-fit bin packing, *SIAM J. Comput.*, 22, 46, 1993.
12. Mao, W., Best-k-fit bin packing, *Computing*, 50, 265, 1993.
13. Zhang, G., Tight worst-case performance bound for AFB$_k$ bin packing, Technical Report 015, Institute of Applied Mathematics, Academia Sinica, Beijing, China, May 1994.
14. Csirik, J. and Johnson, D.S., Bounded space online bin packing: Best is better than first, *Proceedings of the Second Annual ACM-SIAM Symposium on Discrete Algorithms*, Philadelphia, PA, 1991, p. 309.
15. Csirik, J. and Johnson, D.S., Bounded space online bin packing: Best is better than first, *Algorithmica*, 31, 115, 2001.
16. Lee, C.C. and Lee, D.T., A simple online packing algorithm, *JACM*, 32, 562, 1985.
17. van Vliet, A., Lower and upper bounds for online bin packing and scheduling heuristic, PhD Thesis, Erasmus University, Rotterdam, the Netherlands, 1995.
18. van Vliet, A., On the asymptotic worst-case behavior of harmonic fit, *JACM*, 20, 113, 1996.
19. Woeginger, G.J., Improved space for bounded-space, online bin-packing, *SIAM J. Disc. Math.*, 6, 575, 1993.
20. Csirik, J. and Woeginger, G.J., Resource augmentation for online bounded space bin packing, *ICALP*, LNCS, Vol. 1853, Springer, Geneva, Switzerland, 2000, p. 296.

21. Yao, A.C., New algorithms for bin packing, *JACM*, 27, 207, 1980.
22. Ramanan, P., Brown, D.J., Lee, C.C., and Lee, D.T., On-line bin packing in linear time, *J. Algorithms*, 10, 305, 1989.
23. Hu, T.C. and Kahng, A.B., Anatomy of online bin packing, Technical Report CSE-137, Department of Computer Science and Engineering, University of California at San Diego, La Jolla, CA, 1988.
24. Richey, M.B., Improved bounds for harmonic-based bin packing algorithms, *Disc. Appl. Math.*, 34, 203, 1991.
25. Seiden, S.S., On the online bin packing problem, *JACM*, 49, 640, 2002.
26. Heydrich, S. and van Stee, R., Beating the harmonic lower bound for online bin packing, *Proceedings of the ICALP 2016*, Leibniz International Proceedings, Vol. 41, 1, 2016.
27. Brown, D.J., A lower bound for online one-dimensional bin packing algorithms, Technical Report R-864, Coordinated Science Laboratory, University of Illinois, Urbana, IL, 1979.
28. Liang, F.M., A lower bound for online bin packing, *Inf. Proc. Lett.*, 10, 76, 1980.
29. van Vliet, A., An improved lower bound for online bin packing algorithms, *Inf. Proc. Lett.*, 43, 277, 1992.
30. Balogh, J., Békési, J., and Galambos G., New lower bounds for certain classes of bin packing algorithms, *Theor. Comp. Sci.*, 440–441, 1, 2012.
31. Galambos, G., Parametric lower bound for online bin-packing, *SIAM J. Alg. Disc. Meth.*, 7, 362, 1986.
32. Galambos, G. and Frenk, J.B.G., A simple proof of LIang's lower bound for online bin packing and the extension to the parametric case, *Disc. Appl. Math.*, 41, 173, 1993.
33. Csirik, J., Galambos, G., and Turán, G., Some results on bin-packing, in *Proceedings of EURO VI*, Vienna, Austria, 1983, p. 52.
34. Faigle, U., Kern, W., and Turán, Gy., On the performance of online algorithms for partition problems, *Acta Cybern.*, 9, 107, 1989.
35. Chandra, B., Does randomization help in online bin packing? *Inf. Proc. Lett.*, 43, 15, 1992.
36. Gambosi, G., Postiglione, A., and Talamo, M., New algorithms for online bin packing, in *Algorithms and Complexity, Proceedings of the First Italian Conference*, World Scientific, Rome, Italy, 1990, p. 44.
37. Gambosi, G., Postiglione, A., and Talamo, M., Algorithms for the relaxed online bin-packing model, *SIAM J. Comput.*, 30, 1532, 2000.
38. Ivković, Z. and Lloyd, E., Fully dynamic algorithms for bin packing: Being (mostly) myopic helps, *SIAM J. Comput.*, 28, 574, 1998.
39. Ivković, Z. and Lloyd, E., A fundamental restriction on fully dynamic maintenance of bin packing, *Inf. Proc. Lett.*, 59, 229, 1996.
40. Balogh, J., Békési, J., Galambos, G., and Reinelt, G., Lower bound for the online bin packing problem, with restricted repacking, *SIAM J. Comput.*, 38(1), 398, 2008.
41. Balogh, J., Békési, J., Galambos, G., and Reinelt, G., Online bin packing with restricted repacking, *J. Comb. Opt.*, 27(1), 115, 2014.
42. Galambos, G. and Woeginger, G.J., Repacking helps in bounded space online bin packing, *Computing*, 49 (4), 329, 1993.
43. Grove, E.F., Online bin packing with lookahead, *Proceedings of the Sixth Annual ACM-SIAM Symposium on Discrete Algorithms*, San Francisco, CA, 1995, 430.
44. Galambos, G., A new heuristic for the classical bin packing problem, in *Tech. Rept. 82*, Institut für Mathematik, Universität Augsburg, Augsburg, Germany, 1985.
45. Balogh, J. and Békési, J., Semi-online bin packing: a short overview and a new lower bound, *CEJOR*, 21(4), 685, 2013.
46. Epstein, L. and Levin, A., On bin packing with conflicts, *SIAM J. Optim.*, 19, 1270, 2008.

47. Balogh, J., Békési, J., Galambos G., and Markót, M.C., Improved lower bounds for semi-online bin packing problems, *Computing* 84, 139, 2009.

48. Gutin, G., Jensen, T., and Yeo, A., Batched bin packing, *Discr. Opt.*, 2(1), 71, 2005.

49. Zheng, F., Luo, L., and Zhang, E., NF-based algorithms for online bin packing with buffer and bounded item size, *J. Comb. Opt.* 30, 360, 2015.

50. Boyar, J., Kamali, S., Larsen, K.S., and López-Ortiz, A., Online bin packing with advice, *Algorithmica*, 74, 5, 2016.

51. Zhao, X. and Shen, H., On the advice complexity of one-dimensional online bin packing, *LNCS*, 8497, 2014, p. 320.

52. Renault, M. P., Rosén, Á., and van Stee, R., Online algorithms with advice for bin packing and scheduling problems, *Theor. Comp. Sci.* 600, 155, 2015.

53. Angelopoulos, S., Dürr, C., Kamali, S., Renault, M., and Rosén, A., Online bin packing with advice of small size, *LNCS*, Victoria, BC, Vol. 9214, 2015, p. 40.

54. Baker, B.S. and Coffman, Jr. E.G., A tight asymptotic bound for next-fit-decreasing bin-packing, *SIAM J. Alg. Disc. Meth.*, 2, 147, 1981.

55. Baker, B.S., A new proof for the first-fit decreasing bin-packing algorithm, *J. Algorithms*, 6, 49, 1985.

56. Yue, M., A simple proof of the inequality FFD$(L) \leq \frac{11}{9}$OPT$(L) + 1$ $\forall L$, for the FFD bin-packing algorithm, *Acta Math. App. Sinica*, 7, 321, 1991.

57. Li, R. and Yue, M., The proof of FFD$(L) \leq 11/9$OPT$(L) + 7/9$, *Chinese Science Bulletin*, 42, 1262, 1997.

58. Dósa, G., The tight bound of First Fit Decreasing bin-packing algorithm is FFD$(I) \leq 11/9$ OPT$(I) + 6/9$, *LNCS*, Vol. 4614, 2007, p. 1.

59. Dósa, G., Li, R., Han, X., and Tuza, Z., Tight absolute bound for First Fit Decreasing bin-packing: FFD$(I) \leq 11/9$ OPT$(I) + 6/9$, *Theor. Comp. Sci.* 510, 13, 2013.

60. Csirik, J., The parametric behavior of the first-fit decreasing bin packing algorithm, *J. Algorithms*, 15, 1, 1993.

61. Xu, K., The asymptotic worst-case behavior of the FFD heuristics for small items, *J. Algorithms*, 37, 237, 2000.

62. Friesen, D.K. and Langston, M.A., Analysis of a compound bin-packing algorithm, *SIAM J. Disc. Math.*, 4, 61, 1991.

63. Garey, M.R. and Johnson, D.S., A 71/60 theorem for bin packing, *J. Complexity*, 1, 65, 1985.

64. Graham, R.L., Bounds on multiprocessing anomalies and related packing algorithms, in *Proceedings 1972 Spring Joint Computer Conference*, Atlantic City, NJ, AFIPS Press, 1972, p. 205.

65. Caprara, A. and Pferschy, U., Worst-case analysis of the subset sum algorithm for bin packing, *Oper. Res. Lett.*, 32, 159, 2004.

66. Epstein, L., Kleiman, E., and Mestre, J., Parametric packing of selfish items and the subset sum algorithm, in *The 5th International Workshop on Internet and Network Economics (WINE'09)*, 2009, p. 67.

67. Epstein, L., Kleiman, E., and Mestre, J., Parametric packing of selfish items and the subset sum algorithm, *Algorithmica* 74, 177, 2016.

68. Błazewicz, J. and Ecker, K., A linear time algorithm for restricted bin packing and scheduling problems, *Oper. Res. Lett.*, 2, 80, 1983.

69. Lenstra, Jr., H.W., Integer programming with a fixed number of variables, *Math. Oper. Res.*, 8, 538, 1983.

70. Gilmore, P.C. and Gomory, R.E., A linear programming approach to the cutting stock problem, *Oper. Res.*, 9, 839, 1961.

71. Gilmore, P.C. and Gomory, R.E., A linear programming approach to the cutting stock program—Part II, *Oper. Res.*, 11, 863, 1963.

72. Hochbaum, D.S. and Shmoys, D.B., A packing problem you can almost solve by sitting on your suitcase, *SIAM J. Alg. Disc. Meth.*, 7, 247, 1986.

73. Leung, J.Y.-T., Bin packing with restricted piece sizes, *Inf. Proc. Lett.*, 31, 145, 1989.

74. Coffman, Jr., E.G., Garey, M.R., and Johnson, D.S., Bin packing with divisible item sizes, *J. Complexity*, 3, 405, 1987.

75. McCormick, S.T., Smallwood, S.R., and Spieksma, F.C.R., A polynomial algorithm for multiprocessor scheduling with two job lengths, *Mathematics of Operations Research*, 26, 31, 2001.

76. Gutin, G., Jensen, T., and Yeo, A., Online bin packing with two item sizes, *Algorithmic Op. Res.*, 1(2), 72, 2006.

77. Epstein, L. and Levin, A., More on online bin packing with two item sizes, *Discr. Opt.* 5(4), 705, 2008.

78. Kenyon, C., Best-fit bin-packing with random order, *Proceedings of the 7th Symposium on Discrete Algorithms (SODA)*, Atlanta, GA, ACM/SIAM, 1996, p. 359.

79. Coffman, Jr., E.G., Csirik, J., Rónyai, L., and Zsbán, A., Random-order bin packing, *Discr. Appl. Math.*, 156, 2810, 2008.

80. Boyar, J. and Favrholdt, L.M., The relative worst order ratio for online algorithms, *ACM Trans. Alg.*, 3(2), Article 22, 2007.

81. Epstein, L., Favrholdt, L.M., and Kohrt, J.S., Comparing online algorithms for bin packing problems, *J. Sched.*, 15, 13, 2012.

82. Boyar, J., Irani, S., and Larsen, K.S., A comparison of performance measures for online algorithms, *Algorithmica*, 72(4), 969, 2015.

83. Demenge, M., Grison, P., and Paschos, V.Th., Differential approximation algorithms for some combinatorial optimization problems, *Theor. Comp. Sc.*, 209, 107, 1998.

84. Demenge, M., Monnot, J., and Paschos, V.Th., Bridging gap between standard and differential approximation: The case of bin-packing, *Applied Math. Lett.*, 12, 127, 1999.

85. Boyar, J., Epstein, L., and Levin, A., Tight results for Next Fit and Worst Fit with resource augmentation, *Theor. Comp. Sci.*, 411, 2572, 2010.

86. Epstein, L. and van Stee, R., Online bin packing with resource augmentation, *Discr. Opt.* 4, 322, 2007.

87. Martel, C.U., A linear time bin-packing algorithm, *Oper. Res. Lett.*, 4, 189, 1985.

88. Békési, J., Galambos, G., and Kellerer, H., A 5/4 linear time bin packing algorithm, *JCSS*, 60, 145, 2000.

89. Garey, M.R. and Johnson, D.S., *Computers and Intractability: A Guide to the Theory of NP-completeness*, W. H. Freeman, New York, 1979.

90. Zhong, W., Dósa, G., and Tan, Z., On the machine scheduling problem with job delivery coordination, *European Journal of Operational Research*, 182, 1057, 2007.

91. Coffman, Jr., E.G., Garey, M.R., and Johnson, D.S., Approximation algorithms for bin packing: A survey, in: D. Hochbaum (Ed.) *Approximation Algorithms for NP-Hard Problems*, PWS Publishing, Boston, MA, 1997.

92. Garey, M.R., Graham, R.L., Johnson, D.S., and Yao, A.C., Resource constrained scheduling as generalized bin packing, *J. Comb. Th. Ser. A*, 21, 257, 1976.

93. Simchi-Levi, D., New worst-case results for the bin packing problem, *Nav. Res. Log.*, 41, 579, 1994.

94. Xie, J. and Liu, Z., New worst-case bound of first-fit heuristic for bin packing problem (Unpublished manuscript).

95. Xia, B. and Tan, Z., Tighter bounds of the First Fit algorithm for the bin-packing problem, *Discr. Appl. Math.*, 158, 1668, 2010.

96. Boyar, J., Dósa, G., and Epstein, L., On the absolute approximation ratio for First Fit and related results. *Discr. Appl. Math.*, 160, 1914, 2012.

97. Németh, Z., A First Fit algoritmus abszolút hibájáról (in Hungarian), *Eötvös Lóránd Univ.*, Budapest, Hungary, 2013.

98. Sgall, J., A new analysis of Best Fit bin packing, *LNCS*, Vol. 7288, 2012, p. 315.
99. Dósa, G. and Sgall, J., First Fit bin packing: A tight analysis, in *Proceedings of the 30th Symposium on Theoretical Aspects of Computer Science, (STACS 2013), LIPIcs 3*, Schloss Dagstuhl, Wadern, Germany, 2013, p. 538.
100. Sgall, J., Online bin packing: Old algorithms and new results, LNCS, Vol. 8493, 2014, p. 362.
101. Berghammer, R. and Reuter, F., A linear approximation algorithm for bin packing with absolute approximation factor 3/2, *Sci. Comp. Progr.*, 48, 67, 2003.
102. Zhang, G., Cai, X., and Wong, C.K., Linear time-approximation algorithms for bin packing, *Oper. Res. Lett.*, 26, 217, 2000.
103. Balogh, J., Bekesi, J., Dósa, G., Sgall, J., and van Stee, R., The optimal absolute ratio for online bin packing, *SODA*, 26, 2015, p. 1425.
104. Murgolo, F.D., Anomalous behavior in bin packing algorithms, *Disc. Appl. Math.*, 21, 229, 1988.
105. Anderson, R.J., Mayr, E.W., and Warmuth, M.K., Parallel approximation algorithms for bin packing, *Inf. and Comp.*, 82, 262, 1989.
106. Fortune, S. and Wyllie, J., Parallelism in random access machines, in *Proceedings of Symposium on Theory of Computing*, ACM Press, San Diego, CA, 1978, p. 114.
107. Faigle, U. and Kern, W., On some approximately balanced combinatorial cooperative games, *Methods and Models of Operation Research*, 38, 141, 1993.
108. Faigle, U. and Kern, W., Approximate core allocation for binpacking games, *SIAM J. Discr. Math.*, 11, 387, 1998.
109. Woeginger, G.J., On the rate of taxation in a cooperative bin packing game, *Methods and Models of Operations Research* 42, 313, 1995.
110. Matsui, T., A minimum taxrate core allocation of bin packing games, in L.A. Petrosjan, V.V. Mazalov (Eds.), *Game Theory and Applications V*, Nova Science Publishers, Hauppauge, NY, 2000, p. 73.
111. Kuipers, J., Bin packing games, *Mathematical Methods of Operations Research*, 47, 499, 1998.
112. Liu, Z., Complexity of core allocation for the bin packing game, *Op. Res. Lett.*, 37, 225, 2009.
113. Kern, W. and Qiu, X., Improved taxation rate for bin packing games, *LNCS*, 6595, 2011, p. 175.
114. Kern, W. and Qiu, X., Integrality gap analysis for bin packing games, *Op. Res. Lett.*, 40, 360, 2012.
115. Kern, W. and Qiu, X., The 1/4-core of the uniform bin packing game is nonempty, *COCOON*, 7936, 2013, p. 41.
116. Kern, W. and Qiu, X., Note on non-uniform bin packing games, *Discr. Appl. Math.*, 165, 175, 2014.
117. Bilò, V., On the packing of selfish items, in *IPDPS'06*, IEEE, 2006, p. 9.
118. Yu, G. and Zhang, G., Bin packing of selfish items, *LNCS*, Vol. 5385, 2008, p. 446.
119. Miyazawa, F.K. and Vignatti, A.L., Convergence time to Nash equilibrium in selfish bin packing, *Electron. Notes Discrete Math.*, 35, 151, 2009.
120. Miyazawa, F.K. and Vignatti, A.L., Bounds on the convergence time of distributed selfish bin packing, *Int. J. Found. Comp. Sci.*, 22(3), 565, 2011.
121. Epstein, L. and Kleiman, E., Selfish bin packing, *Algorithmica*, 60(2), 368, 2011.
122. Epstein, L., Bin packing games with selfish items, *LNCS*, 8087, 2013, p. 8.
123. Ma, R., Dósa, G., Han, X., Ting, H., Ye, D., and Zhang, Y., A note on a selfish bin packing problem, *J. Global Opt.*, 56, 1457, 2012.
124. Dósa, G. and Epstein, L., The convergence time for selfish bin packing, *International Symposium on Algorithmic Game Theory*, Springer, Berlin, Germany, 8768, 2014, p. 37.
125. Dósa, G. and Epstein, L., Generalized selfish bin packing, *arXiv:1202.4080*, 2012.
126. Wang, Z., Han, X., Dósa, G., and Tuza, Z., Bin packing game with an interest matrix, *International Computing and Combinatorics Conference*, Springer, Cham, Germany, Vol. 9198, 2015, p. 57.

29

Variants of Classical One Dimensional Bin Packing[*]

János Csirik

Csanád Imreh[**]

29.1 Introduction

Few problems compete with the bin packing problem in having fascinated so many people for so long a time. Research into the classical bin packing problem dates back over four decades to the early seventies. In the original version, a list $L = (a_1, a_2, \ldots, a_n)$ of n items, each with a size no larger than 1, is given along with an infinite supply of unit capacity bins. The goal is to pack the list into as few bins as possible so that no bin capacity is exceeded. Because the problem is NP-hard, most research has concentrated on designing fast approximation algorithms with good performance guarantees. The studies have spanned both online and offline algorithms, and have applied both combinatorial and probabilistic analysis.

In parallel with the development of approximation algorithms for the classical problem, many variants have been proposed. Much of this research is covered in this and the next chapter, but because of the extent of the work done in this area, we cannot hope to cover every problem. Rather, we choose representative models and results, and trust that they will give the reader a flavor of the richness of the problem area.

---

[*] This chapter includes its previous version coauthor with Edward G. Coffman, Jr. and Joseph Y-T. Leung that appeared in the First Edition of the Handbook on Approximation Algorithms and Metaheuristics.

[**] Professor Imreh passed away in January 2017.

In Section 29.2, we survey the variant in which the number of items packed is maximized. In this problem the number of bins, m, is fixed and the goal is to pack a maximum number of items into the m bins. This problem was first proposed by Coffman et al. [1] in 1978.

Section 29.3 describes a variant that places a bound on the number of items that can be packed in each bin. This problem instance is identical to the classical problem, except for the additional parameter $k > 0$ that limits the number of items a bin can contain. The problem was first proposed and studied in 1975 by Krause et al. [2]; similar to the above-mentioned work in Reference 1, the problem originated in scheduling theory.

Section 29.4 covers a dynamic bin packing model, in which packings change with time; each item has an arrival and departure time that define the time interval during which an item occupies a bin. This problem was introduced by Coffman et al. [3] in 1983.

Section 29.5 surveys three variants that place constraints on the data. The first of these problems, a model studied by Liu and Sidney [4], is concerned with an ordinal data scenario: the item sizes are not known but the ordering of sizes is known. The second problem, proposed by Mandal et al. [5], allows the items to be fragmented while packing them into bins of fixed capacity. The last problem, introduced by Jansen and Öhring [6], considers the problem where certain items cannot be packed into the same bin. An undirected graph $G = (V, E)$ is used to describe these relationships, where V is a set of vertices representing the items and E is a set of edges (a_i, a_j) that specify the pairs of items that have to be packed into different bins.

In Section 29.6, we describe the bin *stretching* problem introduced by Azar and Regev [7]. The minimal number of bins needed to pack a given list of items is known and the goal is to find a good heuristic for packing the items in the same number of bins while *stretching* the required bin size as little as possible. For a given algorithm, the supremum of the ratio of its required bin capacity to the minimal value is the stretching factor. The challenge is to find a simple algorithm with a small stretching factor.

The final section surveys the main results of several interesting models. First we consider the *black-and-white* bin packing problem defined by Balogh et al. [8] where the colors of the items place restrictions on the order in which they can be packed. Then we discuss the problem defined by Dósa and He [9] where items can be withheld from the packing. The goal is to minimize the number of bins packed plus the total penalty paid for rejecting the items that were withheld. Then we survey the results on batch packing defined by Gutin et al. [10] where the items arrive in batches. The algorithm has to pack the items from a batch without any information about future batches. The section then overviews the results on maximal resource bin packing defined by Boyar et al. [11], where the objective is to use a maximal number of bins. The section closes with a model defined by Bansal et al. [12], which includes the notion of items having not only sizes but *fragility* values. The content of a bin is limited by both the fragility values and the sizes of the items it contains.

Chapter 30 covers problems in which the bins have variable sizes, a problem first studied by Friesen and Langston [13] in 1986, and has attracted considerable attention since then. The survey then turns to bin covering problems, which ask for a partition of a set of items into a maximum number of subsets such that, in every subset, the total size of the items is not less than some lower bound. The problem was first studied by Assman et al. [14] in 1984.

29.2 Maximizing the Number of Items Packed

Coffman et al. [1] introduced a bin packing variant to model processor and storage allocation problems. For example, it might be desirable to maximize the number of records stored in multiple, autonomous storage units, or to maximize the number of tasks that can be executed on multiple processors during a fixed time interval. The formal model consists of a given a set of m unit capacity bins $B_1, B_2, \ldots, B_m$ and a list of $L = (a_1, a_2, \ldots, a_n)$ of items. The goal is to pack a maximum subset of L into the bins such that no bin capacity is exceeded. It is intuitively clear that any algorithm having a reasonable worst-case performance must attempt to pack a maximum number of the smaller items. So it is natural to start by sorting the items of L into a nondecreasing order by size, and then to pack a maximum prefix of L. It is obvious that for every sublist $L' \subseteq L$ that can be packed into m bins there is a prefix of L having at least as many items that can also be packed into the m bins. In the following algorithms the item sizes L are assumed to be in nondecreasing order.

Let $\text{OPT}(L, m)$ denote the maximum number of items of L that can be packed into m bins, and let $A(L, m)$ denote the number packed by algorithm A. Coffman et al. bounded the absolute worst-case ratios $A(L, m)/\text{OPT}(L, m)$. They first considered the Smallest-Item-First (SIF) algorithm, which packs items in nondecreasing order of size, each being placed into a bin with currently the *least* occupied space, ties being broken in favor of the bin having the lowest index. The algorithm halts when it first encounters an item that is too large to fit in any bin. One can easily show that the asymptotic worst-case performance of SIF is 1/2.

The next algorithm they investigated was the First-Fit-Increasing (FFI) algorithm. They proved that, for any list packed into m bins,

$$\frac{FFI(L, m)}{\text{OPT}(L, m)} \geq \frac{3}{4},$$

and that, for every m, there exists a list that realizes this bound. As a straightforward example, consider a list beginning with m items of size $\frac{1}{2} - \varepsilon$ and ending with m items of size $\frac{1}{2} + \varepsilon$ where m is even. The optimal packing will pack all $2\,m$ into m bins, whereas FFI will pack $3\,m/2$ items.

They also derived a parametric bound in terms of k_m, the number of items in the last bin, B_m, the bin with a minimum number of items whenever FFI fails to pack all items. They proved that

$$FFI(L, m) \geq \begin{cases} \frac{mk_m+1}{m(k_m+1)}\text{OPT}(L, m) & m \leq k_m + 1, \\ \frac{k_m^2+k_m+1}{(k_m+1)^2}(\text{OPT}(L, m) - 1), & m > k_m + 1. \end{cases}$$

Both bounds are attainable.

Coffman and Leung [15] described a better algorithm for this problem, which is far more difficult to analyze. They introduced the Iterated-First-Fit Decreasing rule (FFD*) that begins by scanning L to find the maximum length prefix $L^{(1)} = (a_1, a_2, \cdots, a_t) \subseteq L$ such that $\sum_{i=1}^{t} a_i \leq m$. The algorithm then packs $L^{(1)}$ into as many bins as required (say m'), by scanning right to left and placing the next smaller item into that bin with the lowest index into which it will fit. The algorithm terminates successfully if $m' \leq m$; otherwise, the algorithm constructs $L^{(2)} \subset L^{(1)}$ by discarding the largest item in $L^{(1)}$ and repeating the process on $L^{(2)}$. Repeated shortening of the list continues until for some $j, L^{(j)}$ has been packed into $m' \leq m$ bins. They proved that, for all L and $m \geq 1$,

$$FFD^*(L, m) \geq \frac{6}{7}\text{OPT}(L, m) - \frac{18}{7},$$

and that there exist lists for each m such that FFD* will pack 7/8 the optimum number of items. As of this writing, a tight bound has yet to be found.

It is easy to see that without some restriction on the lists to be processed, all online algorithms can perform very poorly; just take lists starting with large items so that, after all the bins are filled with these items, many very small items remain. Here we need restrictions to ensure reasonably good online algorithms. In the classical problem all items have to be packed, so it seems natural to restrict the input sequences to those that can be completely packed by an optimal offline algorithm. Such sequences are called *accommodating sequences* [16].

Boyar et al. [16] considered greedy algorithms, which they called *fair* algorithms: an item is rejected only if there is not enough space to pack it. They analyzed the First Fit (FF) algorithm and proved that it has a competitive ratio of at least 5/8 on accommodating sequences. Azar et al. [17] showed that this bound is asymptotically tight, that is, the competitive ratio for accommodating sequences tends to 5/8 as $n \rightarrow \infty$. As an estimate of the rate of convergence, $5/8 + O(1/\sqrt{n})$ has been shown to bound the competitive ratio.

So-called *unrestricted* algorithms with admission control can reject an item even when there is enough space for it. Azar et al. [17] designed an unrestricted algorithm called Unfair-First-Fit (UFF), which has a competitive ratio $2/3 + \Theta(1/n)$. As every item has the same value, it seems reasonable to reject large items. So UFF examines whether the item is larger than 1/2 and whether the performance ratio would still be at least 2/3 if the item were rejected. If both conditions are satisfied, the item is rejected; otherwise, it is accepted. All accepted items are packed according to the FF rule.

The authors also showed that the competitive ratio of any online algorithm that packs accommodating sequences is no better than $0.857\ldots + O(1/n)$, even when randomized algorithms are considered. For deterministic greedy algorithms, they proved the slightly better bound, $0.809\ldots + O(1/n)$.

Additional results were presented by Boyar et al. [16] for the special case where the item sizes and bin size k are integers. They investigated special lists for which optimal offline algorithms could use no more than α times the given number of bins, with $\alpha \geq 1$ (for $\alpha = 1$ one obtains the accommodating sequences). They called these sequences α-*sequences*. Algorithm A is called c-accommodating with respect to α-sequences if $c \leq 1$ and for every α-sequence L

$$A(L) \leq c \cdot \text{OPT}(L) + b,$$

where b is a universal constant. Let

$$C_\alpha = \{c \mid A \text{ is } c\text{-accommodating with respect to } \alpha\text{-sequences}\}.$$

The *accommodating function* is defined as $\mathcal{A}(\alpha) = \sup C_\alpha$. To maximize the number of items packed, Boyar et al. proved that if $k \geq 3$, then

$$\mathcal{A}(\alpha) \leq \begin{cases} \frac{6}{7} & \text{for } \alpha = 1, \\ \frac{2}{2+(\alpha-1)(k-2)} & \text{for } 1 < \alpha \leq \frac{5}{4}, \\ \frac{8}{6+k} & \text{for } \alpha > \frac{5}{4}, \end{cases}$$

must hold. They also proved lower bounds for $\mathcal{A}(\alpha)$.

Boyar et al. [18] studied the same model restricted to the integers, in which case it is easy to see that the competitive ratio of FF is $1/k$. They introduced a new algorithm Dlog and compared it to FF under various metrics. Dlog is given m bins with $m \geq c\lfloor \log_2 k \rfloor$, for some constant c. It divides the bins into $\lceil \log_2 k \rceil$ groups $G_1, G_2, \ldots, G_{\lceil \log_2 k \rceil}$. Let $p = \left\lfloor \frac{n}{\lceil \log_2 k \rceil} \right\rfloor$ and let $s = n - p \cdot \lceil \log_2 k \rceil$. Groups $G_1, G_2, \ldots, G_s$ consist of $p+1$ bins and the other groups consist of p bins. Let $S_1 = \{x \mid \frac{k}{2} \leq x \leq k\}$ and $S_i = \{x \mid \frac{k}{2^i} \leq x \leq \frac{k}{2^{i-1}}\}$ for $2 \leq i \leq \lceil \log_2 k \rceil$. When Dlog receives an item o of size $s_o \in S_i$, it decides which group G_j of bins to pack it in by calculating

$$j = \max\{j \leq i \mid \text{there is a bin in } G_j \text{ that has room for } o\}.$$

If j exists, o is packed into G_j according to the FF rule. Otherwise, the item o is rejected. The authors proved that, for every $\alpha \geq 1$, if $m \geq c\lfloor \log_2 k \rfloor$, then the competitive ratio of Dlog on α-sequences is $\Theta\left(\frac{1}{\log k}\right)$. This is quite close to the best possible estimate, as they had already shown that any deterministic or randomized online algorithm for this problem must have a competitive ratio of less than $\frac{4}{\lceil \log_2 k \rceil}$. Additional results are given for cases where the optimum value is known in advance.

Coffman et al. [19] investigated the FFI and FFD* algorithms for lists satisfying a divisibility condition. If $s_1 > \ldots > s_j$ denote the possible item sizes with $j > 1$, then they are said to form a weakly divisible sequence if s_{i+1} divides s_i for every $i \in \{1, \ldots, j-1\}$. They proved that, under weak divisibility, FFD* produces optimal packings. If in addition s_1 divides the bin size, then strong divisibility is said to apply, in which case even FFI yields optimal packings.

29.2.1 Generalizations of the Objective Function

In the classical problems posed in terms of cost minimization, the cost associated with a given bin is defined implicitly as either zero or one, depending on whether it is empty or not, respectively. Anily et al. [20] defined a more general model, where the cost of a bin is a monotone increasing and concave function of the number of items in the bin. Worst-case ratios assessing the performance of algorithms are defined in a way similar to those for the classical problem—just replace the number of bins by the sum of the costs of all occupied bins. The authors verify that classical algorithms such as NF, FF, BF, FFD, BFD have neither finite absolute worst-case ratios nor finite asymptotic worst-case ratios for the bin packing problem with a general cost structure. However, they proved that the absolute worst-case bound for the Next-Fit-Increasing (NFI) algorithm is at most

$$\min\left\{1.75, 1.7 + \frac{2}{b_{\text{OPT}}(L)}, 1.691\ldots + \frac{3}{b_{\text{OPT}}(L)}\right\},$$

where $b_{\text{OPT}}(L)$ denotes the optimal cost for the list L. This is the best known algorithm for the problem.

29.3 Bounds on the Number of Items per Bin

Krause et al. [2] investigated a multiprocessor scheduling problem that can be recast as the simple variant of classical bin-packing where a bound $k > 1$ places a limit on the number of items in a bin. In the scheduling interpretation, the items are tasks with unit execution times and sizes given by varying storage requirements (numbers in $(0, 1]$), bins are the time units $[i, i+1)$, $i \geq 0$, assigned to one or more tasks, and k is the number of available processors. The problem is to assign tasks to time units such that (a) there are at most k per time unit, (b) the total storage required by the tasks at any time is at most 1 (the normalized storage capacity), and (c) the latest task finishing time is minimized.

In their analysis of the FF rule constrained to pack at most k items per bin, Krause et al. proved the following bounds on the asymptotic worst-case ratio with $k > 1$ a parameter,

$$\left(\frac{27}{10} - \left\lceil\frac{37}{10k}\right\rceil\right) \leq R_{FF}(k)^\infty \leq \left(\frac{27}{10} - \frac{24}{10k}\right).$$

The authors also proposed offline algorithms that improve on FF. The first modifies the first fit decreasing (FFD) rule by again requiring that at most k items be packed in a bin. They proved that, for $k > 1$ $R^\infty_{\text{FFD}}(k) = 2 - \frac{2}{k}$.

The second algorithm is called the Iterated Worst-Fit Decreasing (IWFD) rule, which starts by sorting the items of a given list L in ascending order of size. It then opens q empty bins, where

$$q = \left\lceil \max\left(\frac{n}{k}, \sum_{j=1}^{n} s(a_j)\right) \right\rceil,$$

with n being the number of items in L. Note that q is a lower bound on $\text{OPT}(L, k)$. The IWFD rule applies the largest-first (LF) rule, iteratively putting the next item into the bin having the lowest level. It breaks ties in favor of the highest bin index. If LF does not pack all items into q bins, then q is increased by 1, all bins are cleared, and the procedure is repeated. This process is iterated until all items are packed. The number of bins packed in the last iteration defines $\text{IWFD}(L, k)$. The authors proved the bounds $4/3 \leq R^\infty_{IWFD} \leq 2$, and gave more refined results for a number of special cases. As the gap in the above-mentioned result is quite large, better bounds remain an interesting open question. The authors suggest that $4/3$ is likely to be a tight bound.

Much later, Babel et al. [21] revisited this bin packing problem. They presented two heuristics for the online version, both of which improve the results by Krause et al. For increasing values of k, the asymptotic worst-case ratio tends to 2 for the first method, and it is 2 for the second algorithm. The chief feature of the first heuristic is that a new item can be packed into a bin only if the bin contains relatively few items or, if it does not, the bin together with this new item is sufficiently filled. The second algorithm defines three different bin types (depending on the bin level and the number of items already packed) and attempts to pack the next item in the first, then in the second, and finally in the third type of bins. The authors also studied special cases for small k. For $k = 2$, they defined an online algorithm with asymptotic worst-case ratio of $1 + 1/\sqrt{5} \approx 1.44721$ and they proved that no online algorithm can have a smaller asymptotic worst-case ratio than $\sqrt{2} \approx 1.41421$. For $k = 3$, they defined an extension of the online Harmonic bin packing algorithm and proved that its asymptotic worst-case ratio is 1.8. Further, they showed that no online algorithm for this case can have an asymptotic worst-case ratio smaller than 3/2.

Epstein [22] extended the online problem by considering bounded-space algorithms, which place a bound on the number of open bins available at any time. If the algorithm has the maximal number of open bins, then it can open a new bin only after permanently closing one of the currently open bins. Epstein extended the Harmonic algorithm to this more general model, and proved that the asymptotic worst-case ratio tends to $1 + h_\infty \approx 2.69103$ as both the number of classes used by the algorithm and the bound on the cardinality tend to ∞. Epstein also showed that this bound is best possible in that no smaller asymptotic worst-case ratio can be achieved. These results were extended to bins of varying sizes and to the resource augmentation model. Unbounded space algorithms for small k were also studied. For $k = 3$, an algorithm with asymptotic worst-case ratio 1.75 was presented based on a more sophisticated partition of the items than the algorithm given in Reference 21. An algorithm for larger k was also presented, it has a smaller asymptotic worst-case ratio than the best known algorithm for $k = 4, 5, 6$.

The lower bounds on the achievable asymptotic worst-case ratio of online algorithms were tightened by Fujiwara and Kobayashi [23]. For $k = 2$, they increased the lower bound to $1.427. \ldots$ For $k \geq 4$ the known lower bounds were immediate consequences of the results derived for the classical bin packing problem. Fujiwara and Kobayashi extended the lower bound technique based on packing patterns and improved the lower bounds. For $k = 4$ and $k = 5$, they increased the 4/3 bound to 3/2 and $25/17 \approx 1.47058$, respectively. They also gave improved lower bounds for $10 \leq k \leq 41$.

This work was extended by Békési et al. [24] who improved the lower bounds on the asymptotic worst-case ratio of online algorithms for $5 \leq k \leq 35$. To prove these bounds they extended the method proving the previous bounds by applying a weighting function technique. Moreover, they investigated the absolute worst-case ratio of the online algorithms and presented an algorithm that has an absolute worst-case ratio of 2 for $k \geq 3$. They analyzed in more detail the kFF algorithm and proved that it has the absolute worst-case ratios of 1.5, 11/6, and 2 for $k = 2, 3$, and 4, respectively, and a ratio larger than 2 for $k > 4$. Lower bounds for the absolute worst-case ratio were also given. For $k = 2$ the lower bound was 3/2, for $k = 3$ the lower bound was 1.75 and for $k \geq 4$ the lower bound was 2. This means that for $k = 2$ and $k \geq 4$, best possible algorithms were presented in Reference 24 under the absolute worst-case ratio metric.

Adar and Epstein [25] studied bin packing with cardinality constraints from a game theoretical point of view. The items are considered as selfish players and their goal is to find a valid packing in a bin where the total size is maximal, as the cost of the bin is divided among the items assigned to it. The authors analyzed the price of anarchy for this game.

29.3.1 Online Bin Coloring

A coloring problem defined by Shachnai and Tamir [26] generalizes the analysis of bounds on the number of items per bin. In their version every item has the same (unit) size and a color. The number of different colors is M. The bins each have a capacity B, and each has c compartments which can be identified by a unique color. An algorithm must fill the bins with items subject to both capacity constraints: the sum of the items in a bin must be no more than B and the number of compartments, each containing items of just one color, must be at most c. The goal is to minimize the number of bins used. The absolute and asymptotic worst-case ratios are defined in a similar way to the classical problem. This variant is called *class-constrained bin racking*, CCBP.

Their first result is a tight bound of 2 on the competitive ratio of any online algorithm. For special lists they obtained better results. Let $S_{c,B}(k,h)$ denote the set of all lists, where $kc < M \le (k+1)c$ and $h \cdot B < n \le (h+1) \cdot B$, where n is the number of items in the list. The competitive ratio of an algorithm A restricted to an input set S of lists will be denoted by $r_A(S)$. They showed that for any deterministic algorithm A, $1 < c < B, k \ge 0$ and $h \ge k - 1 + \max\left\{\lceil k/(c-1)\rceil, \lceil(kc+1)/B\rceil\right\}$,

$$r_A(S_{c,B}(k,h)) \le 1 + \frac{k+1-\left\lceil\frac{kc+1}{B}\right\rceil}{h+1}.$$

The authors showed that a variant of the FF rule achieves this bound. A greedy algorithm based on partitioning the items by color was shown to be efficient as well.

The general case, where items can have different sizes, was studied by Xavier and Miyazawa [27]. They introduced the color-set first-fit (CSFF) algorithm, which partitions the colors into c-element color sets. Each set has its dedicated bins. Then each item is packed FF into the bins of the set containing its color. The authors proved that CSFF has an asymptotic worst-case ratio of at most 3. They further designed an algorithm based on an online partition of the items and showed that its asymptotic worst-case ratio is 2.75.

A tighter bound, $3 - 1/c$, for the asymptotic worst-case ratio of CSFF was established by Epstein et al. [28]. Moreover, they described a general method that can be used to extend a large set of the algorithms (called *uniform*) for the classical bin packing problem to this problem. Applying this technique increases the asymptotic worst-case ratio by at most 1. Using this method and a further partition-based algorithm for smaller c, Epstein et al. proved that there exists an online algorithm with a worst-case ratio of at most 2.634... for every c. They also proved a lower bound of 1.565... on the possible asymptotic worst-case ratio of online algorithms for $k = 2$.

The next, and last, "coloring" model to be discussed is a variant called "bin coloring" introduced by Krumke et al. [29]. In this model, there are items of varying colors, each item having size 1, which have to be packed into bins of size B, where $B > 1$ is a positive integer. They considered an online, bounded-space problem, that is, one where at most m open bins are available at any give time. A further restriction is that a bin can be closed only when it becomes full. The goal is to minimize the maximum number of different colors assigned to a bin. As a performance measure, they used the absolute worst-case ratio. General bounds were derived for two heuristic algorithms.

- *The first method was the Greedyfit rule*: If upon the arrival of item a_i the color of this item is already contained in one of the currently open bins, put it into this bin. If the bin becomes full with this item, then close it. Otherwise, put the item into a bin that contains the least number of different colors (which means opening a new bin if currently fewer than m bins are partially filled). When the capacity $B \ge 2m^3 - m^2 - m + 1$, the competitive ratio of this method has been shown to be at most $min[3m - 1, B]$ but not less than $2m$.
- *The second method was the trivial algorithm Onebin, which is actually the Next Fit algorithm*: The next item is packed into the one open bin. A new bin is opened only if the previous item closed the previous bin by filling it up completely. Quite surprisingly, this algorithm has a better competitive ratio than Greedyfit, it being $min[2m - 1, B]$.

They also proved that no online algorithm can be substantially better than Onebin. Specifically, no deterministic online algorithm can have a competitive ratio better than m. This bound holds for randomized algorithms as well. If one were to adopt resource-augmentation, that is, allowing the online algorithm to use a fixed number of $m' \ge m$ open bins, the lower bound would still be $\Omega(m)$.

Contributions to the offline version of this problem were made by Lin et al. [30], who devised an algorithm that uses at most OPT(L)+1 colors. They also defined a *Maximal bin coloring* problem with a dual objective function, where the goal is to maximize the minimum number of different colors assigned to bins. They presented a greedy algorithm for the online problem, and proved that it has an asymptotic worst-case bound of 2, and that there does not exists an online algorithm with a smaller worst-case bound.

29.4 Dynamic Bin Packing

Coffman et al. [3] defined a generalization of the classical problem in which arrival and departure times are associated with each item, thus defining a sample function of a stochastic model. These times define the various time intervals during which items occupy bins. Compared to the classical "archival" model, this one is more realistic for certain applications, such as those relating to computer storage allocation.

The formal model consists of a list $L = (p_1, p_2, \ldots, p_n)$. Each element/item p_i in L corresponds to a triple (a_i, d_i, s_i), where a_i is the arrival time for p_i, d_i is its departure time, and s_i is its size. The item p_i resides in the packing for the time interval $[a_i, d_i)$, which is assumed to be strictly positive for all i. It is also assumed that the items in L are ordered in such a way that $a_1 \leq a_2 \leq \ldots \leq a_n$.

Coffman et al. analyzed the FF algorithm applied to dynamic bin packing. With this setting, FF maintains two lists of bins: a list of currently empty bins and a list of currently occupied bins. The latter list is in nondecreasing order of the times when transitions from empty to nonempty occur. The occupied bins denoted by $B_1, B_2, \ldots$ are indexed in this transition order.

For each i, as in the classical version, the FF rule attempts to pack the i-th item into the first occupied bin (the one with the smallest index) that has sufficient available space for it. If no such occupied bin exists, the item is put into an empty bin which is then appended to the list of occupied bins. The departure of the i-th item from bin B_j at time d_i causes an increase by s_i in the available space of B_j. If the bin becomes empty at that time, it is moved to the list of empty bins.

For algorithm A, let $A(L, t)$ denote the number of occupied bins in the packing of the L at time t. The performance of A applied to the list L is defined by

$$A(L) = \max_{0 \leq t \leq a_n} A(L, t).$$

This is the maximum number of occupied bins ever required by A for processing L. This measure is applied in two distinct ways to optimal packing algorithms, depending on how the items in L can be packed. The first method, $\text{OPT}_R(L)$, is the maximum number of bins ever required in dynamically packing L when the current set of items is *repacked* into the minimum number of bins each time a new item arrives. The second, $\text{OPT}_{NR}(L)$, is the maximum number of bins ever required in dynamically packing L when *no rearrangement* of items is allowed, but is otherwise packed optimally with L assumed to be known in advance. The performance ratio of an algorithm A is now defined in the following natural way. Let

$$W_{R,n}\left(A, \frac{1}{k}\right) = \sup_{|L|=n} \frac{A(L)}{\text{OPT}_R(L)},$$

where all item sizes are less than or equal to $1/k$. Then define asymptotic performance measure

$$W_R^\infty\left(A, \frac{1}{k}\right) = \limsup_{n \to \infty} W_{R,n}\left(A, \frac{1}{k}\right).$$

Similar definitions apply to $W_{NR}^\infty(A)$. Coffman et al. established the following bounds:

$$W_R^\infty\left(FF, \frac{1}{k}\right) \leq \frac{k+1}{k} + \frac{1}{k-1} \log \frac{k^2}{k^2 - k + 1}, \quad k \geq 2,$$

$$W_R^\infty(FF, 1) \le \frac{5}{2} + \frac{3}{2} \log \frac{\sqrt{13} - 1}{2} \approx 2.897.$$

They also described a slightly improved FF version for which W_R^∞ is bounded by 2.788. ... In addition, they proved that

$$W_R^\infty \left(FF, \frac{1}{k} \right) \ge \frac{k+1}{k} + \frac{1}{k^2}, \; k \ge 2,$$

$$W_R^\infty(FF, 1) \ge \frac{11}{4} = 2.75.$$

They derived a lower bound that holds for any online algorithm A:

$$W_R^\infty \left(A, \frac{1}{k} \right) \ge W_{NR}^\infty \left(A, \frac{1}{k} \right) \ge \frac{k+1}{k} + \frac{1}{k(k+1)}, \; k \ge 2.$$

$$W_R^\infty(A, 1) \ge \frac{5}{2}, \qquad W_{NR}^\infty(A, 1) \ge \frac{43}{18} \sim 2.388.$$

Note that the upper and lower bounds are rather close to each other. The largest gap occurs in

$$2.75 \le W_R^\infty(FF, 1) \le 2.897 \dots.$$

Chan et al. [31] studied the special case where each item has the same size. They focused on the model where the offline algorithm is not allowed to repack the items. They proved that the asymptotic worst-case ratio of BF and WF is 3, and that the asymptotic worst-case ratio for FF is between 2.45 and 2.4942. More generally, they proved that no online algorithm can have a smaller asymptotic worst-case ratio than 2.428. ...

For the case where each item has the same size, Han et al. [32] conducted a deeper analysis of FF. They proved that the asymptotic worst-case ratio of FF is at most 2.484 . ... They also presented improved bounds for the parametric case.

Keeping with the nonrepacking assumption, Chan et al. [33] allowed for general item sizes. They proved that no online algorithm can have a smaller asymptotic worst-case ratio than 2.5. They also investigated the problem in the resource augmentation model, where an online algorithm can use larger bins. They showed that no online algorithm using bins of size $b < 2$ can have a better asymptotic worst-case bound than $2/b$. In addition, worst-case ratios for FF, BF, and WF were obtained. Their results show that 2 is a critical bin size. If an online algorithm is allowed to use bins of size at least 2, then it can achieve optimal performance with an AF algorithm, but when the bins are smaller, there is no online algorithm that can compete with an optimal offline algorithm in the worst case. The lower bounds were improved further by Wong et al. [34] who proved that no online algorithm can have a smaller asymptotic worst-case bound than 8/3.

Li et al. [35] considered a minimum-cost objective function for dynamic bin packing, where each bin has a cost equal to the total length of the time periods during which it is nonempty. The authors studied the absolute worst-case cost ratio comparing online algorithms with offline algorithms that allow repacking. They analyzed the FF, BF, and AF algorithms and a new algorithm (called Hybrid Fit) that partitions the items by size, then packs the size classes by FF. The worst-case ratios depend on the ratio of the maximum item residence time to the minimum item residence time.

Ivković and Lloyd [36] studied the semi-online version where the online algorithm is also allowed to repack one or more items. They gave an algorithm called Mostly Myopic Packing (MMP), which is $\frac{5}{4}$-competitive and requires $\Theta(\log n)$ time per operation.

29.5 Constraints on Data

29.5.1 Bin Packing Using Semi-Ordinal Data

Liu and Sidney [4] investigated the bin packing problem under the ordinal data scenario: the item sizes are not known, but the order of the weights is known, that is, $a_1 \geq a_2 \geq \ldots \geq a_n$. Specific sizes may be obtained, and the trade-off between the amount of information obtained versus the solution quality (performance ratio) is investigated.

If the only information available about L is the ordinal assumption, then we can place only one item per bin, as all items may be larger than $1/2$. Even in this case the optimal solution may be 1 if $\sum a_i \leq 1$. So we have a worst-case ratio of n. Suppose we could purchase perfect knowledge of k sizes, where we can specify the ranks (ordinal positions) of the desired sizes. Which ordinal positions should we choose to guarantee a good feasible solution that utilizes the knowledge of these sizes? Liu and Sidney proved that the worst-case ratio of an integer ρ can be achieved by at most

$$\lfloor \ln[n(\rho - 1) + 1]/ \ln \rho \rfloor$$

exact observations, and that this worst-case performance cannot be achieved with fewer than

$$\left\lfloor \frac{\ln \frac{n(\rho-1)}{\rho^2 - \rho + 1}}{\ln \rho} \right\rfloor + 1$$

exact observations. They also gave a method for choosing a good set of indices.

29.5.2 Fragmentable Bin Packing

Mandal et al. [5] defined the following variant, called *Fragmentable Object Bin Packing*: it is permissible to fragment the items while packing them into bins of fixed capacity. Fragmentation has an additional cost associated with it, leading to the consumption of additional bin capacity. They proved that this problem is NP-hard too. Unfortunately no heuristic was analyzed from the worst-case point of view.

Menakerman and Rom [37] tackled this problem further and defined two variants. The first variant is called Bin Packing With Size-Increasing Fragmentation or BP-SIF. In this variant, when packing a fragment of an item, one unit of overhead is added to the size of every fragment. They supposed that the item sizes were integer-valued and that each bin size had an integer value U. An algorithm is said to prevent unnecessary fragmentation if it follows the following two rules:

- *No unnecessary fragmentation*: An item (or fragment of an item) is fragmented only if it is to be packed into a bin which cannot contain it. In this case the item is fragmented into two fragments. The first fragment must fill one of the bins, and the second fragment must be packed according to the packing rule of the algorithm.
- *No unnecessary bins*: An item is packed into a new bin only if it cannot fit in any of the open bins used by the algorithm (greedy).

They proved that for any algorithm that prevents unnecessary fragmentation the absolute worst-case bound is $\leq \frac{U}{U-2}$. They also showed that the following variant of the Next Fit rule, called NF_f, achieves this bound. When an item does not fit in the open bin, it is fragmented into two parts. The first part fills the open bin and the bin is closed. The second part is packed into a new bin which becomes the open bin. The Next Fit Decreasing and the Next Fit Increasing rules have basically the same bound, for large enough U. Neither the First Fit Decreasing rule nor the following iterative variant of the First Fit Decreasing rule improves the bound: the latter algorithm starts with $m = \lceil s(L)/U \rceil$ bins and tries to pack the items by the FFD rule. If every item has been packed then the algorithm stops. If some remain then m is increased by 1 and we start packing the whole list again. This is repeated until every item has been packed.

The second variant is Bin Packing With Size-Preserving Fragmentation or BP-SPF. In this version each item has a size and a cost. The items must be packed into m bins, where $s(L) \leq mU$. It is possible to fragment any item, in which case one unit is added to its cost while keeping the size fixed. The goal is to minimize the total cost. As every item must be packed and for the chosen m this is possible because the fragmentation does not increase the size we have a fixed cost for the whole list. The minimization of the cost is equivalent to the minimizing the fragmentation. The problem remains NP-hard. They proved that for any algorithm A that has no unnecessary fragmentation, the extra cost is bounded above by $m - 1$, and that NF_f achieves this bound. The appropriate variant of the FFD rule betters this bound only when U is small. If the bin size is unbounded, the worst-case behavior of the FFD rule is the maximum possible value.

LeCun et al. [38] studied the size preserving version where the goal is to minimize the number of fragments. They presented a method to reduce the variable sized bin case to the case of equal sized bins and analyzed several heuristics, the best one has absolute worst-case ratio 6/5.

29.5.3 Packing Nodes of Graphs

Jansen and Öhring [6] once looked at a constrained bin packing problem: in this case there is an undirected graph (*the conflict graph*) $G = (J, E)$ on the elements. The adjacent items $(a_i, a_j) \in E$ have to be packed into different bins, and the goal is to minimize the number of bins. Clearly, if E is an empty set, the problem is equivalent to the classical bin packing problem. On the other hand, if $s(L) \leq 1$, then the problem is to compute the chromatic number $\chi(G)$ of the conflict graph. Both special cases are NP-complete. The authors used the absolute worst-case ratios and then proved that simply using the appropriate modifications of the classical algorithms NF, FF, and FFD will not yield a constant competitive ratio. (The appropriate modification simply means that we use the heuristic, but if the item to be packed has an adjacent item in the bin being packed, then we do not pack it into this bin.) Their other algorithms were based on the composition of two algorithms – a coloring algorithm and a bin packing heuristic.

The first algorithm of this type uses an optimal coloring, that is, it finds a minimum partition of L into independent sets $U_1, U_2, \ldots, U_{\chi(G)}$, and then applies one of the NF, FF, and FFD bin packing heuristics to each independent set. They proved that in this case NF has a competitive ratio of 3, FF of 2.7, and for the FFD the bound lies between 2.691 and 2.7.

The second algorithm of this type is based on a pre-coloring method. The main step is to compute a minimum coloring of the conflict graph where the large items are separated and colored differently. Based on this method, they obtained an approximation algorithm with competitive ratio of 5/2 for graphs such as interval graphs, split graphs, cographs, and other graphs. More involved coloring methods are used to obtain algorithms with worst-case bounds of 7/3, 11/5, and 15/7, respectively. The last of these methods is a general separation method that works for cographs and partial K-trees. Applying this separation method they obtained an approximation algorithm with worst-case ratio of $2 + \varepsilon$ for these two classes of graphs. This result implies an approximation with factor $2 + \varepsilon$ for any class of graphs with a constant upper bound on the treewidth (e.g., outerplanar graphs, series parallel graphs).

Epstein and Levin [39] studied the problem further. For the offline problem, they presented an algorithm for perfect graphs with absolute worst-case ratio 2.5, an algorithm for interval graphs with absolute worst-case ratio 7/3 and an algorithm for bipartite graphs with absolute worst-case ratio 7/4. The authors also investigated the online problem for interval graphs and presented an algorithm with absolute worst-case ratio 4.7. Then they proved that no online algorithm exists with a worst-case ratio less than $155/36 \approx 4.30566$ and that no online algorithm that knows the optimal solution value can have an asymptotic worst-case ratio smaller than $47/36 \approx 1.30566$.

There are some models where the conflicts are directed, that is, there are restrictions on the order in which items can be packed into the same bin. Manyem [40] defined the online bin packing problem with Largest in Bottom (LIB) constraint. This means that no item can be packed into a bin which already contains smaller items. He proved that the extension of the NF algorithm has no constant asymptotic worst-case bound and studied the average case behavior of FF experimentally. Manyem et al. [41] proved that an AF algorithm and a bounded space Harmonic Fit algorithm have no better asymptotic worst-case ratio than 2. Later, Manyem [42] proved that an extension of FF has asymptotic worst-case bound of at most 3. Finley and Manyem [43] proved that no online algorithm can have better asymptotic worst-case ratio than 1.76. Epstein [44] improved both the lower and upper bounds significantly. She proved that no online algorithm can have asymptotic worst-case ratio smaller than 2. She analyzed an extension of the FF and showed that it has asymptotic worst-case ratio of at most 5/2. These results were also extended to the parametric case. The extensions of some standard bin packing algorithms (WF, BF, Harmonic) were also studied and she proved that none of them have constant worst-case ratio. Dósa et al. [45] proved that FF has an asymptotic worst-case bound that is at most $13/6 \approx 2.1666$. They also improved the bound for the parametric case and defined a more general model called generalized LIB constraints. In this model the restrictions on the packability of the items depend on a graph and not on the sizes of the items. Bódis [46] defined a similar model, where the conflict graph is directed and the item represented by the ending vertex of an edge cannot be packed into the bin after the item belonging to the starting vertex of the edge. For the online case, this model is equivalent to the general LIB constraint model. He defined a special case called Hanoi conflicts where each item has a Hanoi value and no item with higher Hanoi value can be packed into a bin after an item with smaller Hanoi value and analyzed some Fit algorithms for Hanoi conflicts in the special case of unit size items.

Katona [47] investigated the edge disjoint polyp packing problem. A graph is called a *p-polyp* if it consists of p simple paths of the same length and one vertex of these paths is a common vertex. The polyp packing problem is a generalization of the classical bin packing problem: how does one pack a set of paths with different lengths into a set of disjoint polyps edges? Edge disjoint packing means that the embedded path may contain a vertex more than once, but each edge only once. Katona proved that this problem is NP-complete and that FF can be modified in such a way that its asymptotic worst-case bound is between 1.6904 and 1.7.

Codenetti et al. [48] have investigated the following variant: the items to be packed are structured as the leaves of a tree. They called this problem a hierarchically *structured bin packing problem* (SBPP). The goal is to pack the items while preserving some locality properties, that is, leaves whose lowest common ancestor has low height should be packed into the same bin. This problem comes up in the area of document organization and retrieval where one is confronted with searching issues on very large structured, tree-like ontologies. The authors presented a heuristic with an asymptotic worst-case ratio of 2. For the special case where every item size is equal to 1 and the bin capacity is k, they designed a heuristic with a worst-case bound of 3/2.

Bujtás et al. [49] defined a very general model called graph bin packing which is a common generalization of bin packing with conflicts and many graph problems. In the problem the algorithm has to pack an input graph into a given host graph. The vertices of the input graph have weight, these vertices can be considered as the items. Each edge of the input graph has two values: a lower and an upper bound value. The goal is to assign each vertex of the input graph to one of the vertices of the host graph satisfying the following properties:

- The total size of the vertices assigned to a vertex of the host graph cannot be more than the capacity of the vertex
- For each edge of the input graph the distance of the host vertices of the endpoints of the edge must be between the lower and upper bounds assigned to the edge

We say that a graph is packable into the host graph if it has a valid packing. The authors presented several necessary and sufficient structural statements on packability for special classes of graphs. Moreover, some optimization problems were studied where the goal is to minimize some parameters of the used part of the host graph. Bujtás et al. [50] defined further versions of packability. A graph is called well-packable if it can be packed in an online way choosing at each step an arbitrary vertex of the input graph connected to one of the already packed vertices and assigning it to any vertex of the host graph where it does not violate the rules for the distances and the capacity. The other model is called connected-online packability. There we are looking for an online algorithm that can always find a feasible packing receiving at each step a vertex connected to one of the already packed vertices.

29.6 Changing the Bin Size

Azar and Regev [7] investigated the online version of a special case of this problem and called it bin stretching: they supposed that the optimal bin capacity is known and that the goal is to find a good heuristic, given this information. The supremum of the bin capacity ratio of an algorithm and the optimal value is called the stretching factor. They first proved that the stretching factor of any deterministic online algorithm is at least 4/3 for any number $m \geq 2$ of bins. They also gave two algorithms with a stretching factor of 5/3. To describe these algorithms we first need some additional notation. Let $c_j(B_i)$ denote the level of bin i after a_j was packed. Both algorithms use a parameter $\alpha > 0$ to classify the bins according to their levels. An appropriate choice of α will lead to an algorithm with a stretching factor of $1 + \alpha$. We will call a bin *short* if its level is at most α. Otherwise, it is *tall*. When item a_j arrives, $1 \leq j \leq n$, we define the following three disjoint sets of bins:

$$S_1^\alpha(j) = \left\{ i \in M \mid c_{j-1}(B_i) + s(a_j) \leq \alpha \right\},$$
$$S_2^\alpha(j) = \left\{ i \in M \mid c_{j-1}(B_i) \leq \alpha, \alpha < c_{j-1}(B_i) + s(a_j) \leq 1 + \alpha \right\},$$
$$S_3^\alpha(j) = \left\{ i \in M \mid c_{j-1}(B_i) > \alpha, c_{j-1}(B_i) + s(a_j) \leq 1 + \alpha \right\}.$$

The set S_1 consists of bins that are short and remain short if the current item is placed into them. The set S_2 contains bins that are short but become tall if the next item is packed. The last set S_3 consists of bins that are tall but remain below $1 + \alpha$ if the next item is packed. Note that there may be bins that are not in any of these three sets. Now the two algorithms are defined as follows.

$ALG1_\alpha$: when the next item a_j arrives

- Put the item into any bin belonging to the set S_3 or S_1, but not in an empty bin belonging to S_1 if there is a nonempty bin belonging to S_1.
- If $S_1 = S_3 = \emptyset$ then put the item into the bin belonging to S_2 with the least level.
- If $S_1 = S_2 = S_3 = \emptyset$ then report failure.

$ALG2_\alpha$: when the next item a_j arrives

- Put the item into any bin belonging to S_1.
- If $S_1 = \emptyset$ then put the item into any bin belonging to S_3.
- If $S_1 = S_3 = \emptyset$ then put the item into the bin with the least level.
- If $S_1 = S_2 = S_3 = \emptyset$ then report failure.

Note that both algorithms actually define a family of algorithms. Azar and Regev proved that both algorithms have a stretching factor of 5/3 for the best possible choice of α (which is equal to 2/3). They also presented an improved algorithm using five sets of bins. This has a stretching factor of 13/8.

We note that the bin stretching problem can be interpreted as a multiprocessor scheduling problem where the goal is to minimize the makespan and the value of the optimal solution (scaled to 1) is known in advance. Cheng et al. [51] studied the multiprocessor scheduling problem with known total processing time, and presented an algorithm with absolute worst-case ratio 1.6. In that scheduling problem the algorithm has less information than in bin stretching therefore their algorithm is also a bin stretching algorithm which has a stretching factor of 1.6. Kellerer and Kotov [52] developed a better algorithm with a stretching factor of $11/7 \approx 1.5714$. The basic idea of the algorithm is to pack the items in 2 phases. In the first phase the algorithm constructs tiny, small, medium, large, and huge bins (these classes are defined by the sum of the sizes of the items assigned to the bin) and some fraction of the bins are left empty. Then, in the second phase the remaining items are packed by an extended FF algorithm where some four-element batches of bins are handled separately. This two phase packing technique was further studied by Gabay et al. [53] and Böhm et al. [54,55]. Gabay et al. defined an algorithm with stretching factor of $26/17 \approx 1.5294$. Böhm et al. presented an algorithm with a stretching factor of $3/2$ and proved that there does not exist an algorithm based on this two-stage method with a stretching factor smaller than $3/2$.

Kellerer et al. [56] showed that for two bins the List Scheduling algorithm has a stretching factor of 2. Epstein [57] gave a tight analysis of bin stretching for two bins, where the items arrive sorted in nonincreasing size. She showed that the tight competitive ratio of this problem is $10/9$ and gave a quite complicated algorithm that achieves this bound. Gabay et al. [58] and Böhm et al. [59] studied the special case of 3 bins. Gabay et al. proved that no algorithm can have smaller stretching factor than $19/14 \approx 1.3571$ and they also extended the lower bound for 4 bins. Böhm et al. improved the lower bound to $46/33 \approx 1.363637$ and extended the bound $19/14$ to the case of 5 bins. They also presented an algorithm with stretching factor $11/8 = 1.375$ for 3 bins.

Speranza and Tuza [60] investigated a different problem. Suppose we have m bins of unit capacity and that the bin capacity can be exceeded. If the sum of the sizes of the items in the bin is larger than 1, then the cost of the bin will be this sum. Otherwise, the cost is equal to the capacity 1. The goal is to minimize the total cost of the bins. They investigated the online version of this problem. First they proved that every online algorithm has an absolute worst-case ratio of at least $7/6$. This can be demonstrated with two bins: simply make the first two elements equal to $1/3$, and for the sum of the next two items, let $a + b = 4/3$. Now, if both $1/3$ items are packed into the same bin, let $a = b = 2/3$. If they are packed into different bins, let $a = 1, b = 1/3$. They proved as well that the List Scheduling algorithm has a tight bound of $5/4$.

The authors Speranza and Tuza also presented an improved algorithm H_x. This depends on a parameter x, $0 < x < 1$. The algorithm assigns an item to the bin with the highest level under the condition that after the assignment of the item to the bin the level of the bin does not exceed $1 + x$. If the item causes excess greater than x in each open bin, it is then assigned to an empty bin—if there is any—or to the bin with the lowest level. In case of ties, the bin with the lowest index is used. They proved that

$$\frac{H_x(L)}{\text{OPT}(L)} \leq \max\left(1 + \frac{x}{1+x}, \frac{5}{4} - \frac{x^2}{4}\right).$$

The right hand side has a minimum at $x \approx 0.2956$, the minimum being ≈ 1.228. It is not known whether this bound is tight.

An offline algorithm with worst-case bound $13/12$ was given by Dell'Olmo et al. [61].

Yang and Leung [62] introduced the *Ordered Open-end Bin Packing Problem*, OOBP. Here a bin can be filled to a level exceeding 1 so long as there is a designated last item in the bin such that the removal of this item brings the bin's level back to below 1. In this problem items of size 1 play a special role, and will be called 1-items. It is quite clear that a good algorithm for this problem will fill the majority of bins to levels no less than 1, whereas for any algorithm, no bin can be filled to a level more than 2. Hence, any good algorithm will have a worst-case ratio of no more than 2. The authors showed that for any online algorithm A, $R_A^\infty \geq 1.630297$ with the 1-items, and $R_A^\infty \geq 1.415715$ without the 1-items.

They then introduced two algorithms. The first is called MXF and is defined as follows: we divide the items into four types according to their size. Type-1 items will have a size $0 < s(a_i) < 1/3$, type-2 items have size $1/3 \leq s(a_i) < 1/2$, type-3 items have size $1/2 \leq s(a_i) < 1$, and type-4 items have size 1. Accordingly, we define a type-1 open bin as an open bin containing a number of type-1 items, a type-2 open bin as a bin containing one or two type-2 items, and a type-3 open bin as a bin containing one type-3 item. We pack the items in order of their arrival. When the current item is of type-1, we pack it into the type-1 open bins using the FF rule without closing any bin. When the current item is of type-2, we pack it similarly into type-2 bins. Suppose the current item is of type-3 or type-4. We pack it into the first type-1 or type-2 open bin whose level is at least $2/3$ and close this bin. If there is no such bin, we pack the item into an open type-3 bin and then close it. If there is no such bin, we pack the item into a new bin and close it if the item is a type-4 item. The authors showed that with no 1-item present,

$$R^\infty_{MXF} \le 35/18 \approx 1.9444,$$

and they also showed that regardless of the presence of the 1-items

$$R^\infty_{MXF} \ge 25/13 \approx 1.9231.$$

Their second algorithm is the Greedy Look-Ahead Next Fit (GLANF), which is an offline algorithm. With this algorithm there is one open bin at any given moment. GLANF keeps on filling the current bin with items in their original order unless the first item is a 1-item or the addition of the next item will bring the bin's level to be at least 1. For the latter situation, GLANF makes some greedy effort in filling the current open bin to the highest possible level. They proved that, without the 1-items,

$$\frac{27}{20} \le R^\infty_{GLANF} \le \frac{3}{2},$$

and with the 1-items,

$$R^\infty_{GLANF} \ge \frac{3}{2}.$$

29.7 Additional Models

29.7.1 Black-and-White Bin Packing and Its Generalized Version

Balogh et al. [8,63] defined the black-and-white bin packing problem. The items are of two types, called black and white, and the item types must alternate in each bin, which means that two items of the same type cannot be assigned consecutively into a bin. The offline problem was defined in two different ways. In the restricted offline problem the offline algorithm knows the list of items in advance but it is not allowed to change the order of the items, it must pack them one by one. In the unrestricted offline problem, the offline algorithm is allowed to change the order of the items, it can produce any packing containing the items from the list. It is simple to prove that no online algorithm can have a constant worst-case bound when compared to an offline algorithm. The main result was an online algorithm called *Pseudo*. It packs the items into infinite sized pseudo bins considering only the colors and the items from these pseudo bins are packed into the real bins by NF. The algorithm has absolute worst-case ratio 3. The authors also studied the parametric case where the size of each item is at most $1/d$ for some integer d and proved that algorithm Pseudo has absolute worst-case ratio $1 + d/(d-1)$. They studied the straightforward extensions of some classical algorithms (NF, BF, WF) and proved that their absolute worst-case ratio is not larger than 5 and their asymptotic worst-case ratio is at least 3. It is also proved that the asymptotic worst-case ratio of any online algorithm is at least $1 + 1/(2 \ln 2) \approx 1.7213$ which shows that the problem is harder than the classical bin packing problem. Chen et al. [64] studied the online problem for the case of small items having size at most $1/2$. They developed an algorithm and proved that its absolute worst-case ratio is at most $8/3$ for this special case.

Balogh et al. [65] considered the unrestricted offline problem and designed a fast heuristic with absolute worst-case ratio 2. They also designed an asymptotic polynomial approximation scheme that can be shown to have an absolute worst-case ratio $3/2$, which has the best possible absolute worst-case ratio of any polynomial time algorithm for this problem unless P=NP.

Dósa and Epstein [66] and Böhm et al. [67] defined and studied the extension of the black-and-white bin packing to more colors. In this model the items can have colors from a color set containing c colors and again no two items of the same color can be assigned consecutively to a bin. Dósa and Epstein proved that no online algorithm can have better asymptotic worst-case ratio than 2 for this colorful bin packing model. Their result, which also holds for $c = 2$, improves on the lower bound given in Reference 8.

They studied the special case of zero sized items and proved that for $c \geq 3$ no online algorithm can have an asymptotic worst-case ratio smaller than 3/2. We note that for $c = 2$ algorithm Pseudo finds an optimal solution if the sizes of the items are zero. The authors studied the extensions of the classical algorithms FF, BF, WF, and the extension of Pseudo to $c \geq 3$ and they proved that none of these algorithm has bounded asymptotic worst-case bound. They developed a more sophisticated online algorithm and proved that it has absolute worst-case ratio of 4 for the general problem and an absolute worst-case ratio of 2 for items of size zero. Böhm et al. [67] improved the lower bound further, they proved that for $c \geq 3$ no online algorithm can have better asymptotic worst-case ratio than 2.5. They presented an algorithm which has absolute worst-case ratio of 3.5. They also studied the special case of zero size items: in this special case an algorithm is presented which has asymptotic worst-case ratio of 3/2. The algorithm uses at most $\lceil 1.5\text{OPT} \rceil$ bins, therefore it is also a good result for the absolute case. They also improved the analysis of the AF algorithm for the case of 2-colors and proved that it has absolute worst-case ratio of 3 for general item sizes and absolute worst-case ratio of $1 + d/(d-1)$ for the parametric case.

29.7.2 Bin Packing with Rejection

In the bin packing with rejection model, the items have an additional value called the penalty or rejection cost. An algorithm is allowed to reject any item, but in this case the algorithm has to pay the rejection cost. Therefore, the cost of an algorithm is the sum of all rejection costs of rejected items plus the number of unit sized bins used for packing all the other items. The problem was defined by Dósa and He in Reference 9 and some FF type algorithms were analyzed. The authors presented an online algorithm that has an absolute worst-case bound of $(3 + \sqrt{5})/2 \approx 2.618$, and a class of online algorithms that can have an asymptotic worst-case ratio arbitrary close to 7/4. It is also proved that no algorithm can have smaller absolute worst-case ratio than 2.343. For the offline problem, they present an algorithm with absolute worst-case ratio of 2 and an algorithm with asymptotic worst-case ratio of 3/2 is given. The authors extended the rejection model to its dual, the bin covering problem. Epstein [68] also studied this problem packing with rejection and presented an algorithm that is the extension of the Modified Harmonic bin packing algorithm to this more general model. It is proved that this algorithm has an asymptotic worst-case bound of 1.61562. Bounded space extending the Harmonic algorithm to this more general model, was introduced and was shown to have the same worst-case ratio as Harmonic for the standard model. This ratio tends to $h_\infty = \approx 1.69103$ as the number of item classes used by the algorithm tends to ∞. She also considered the offline problem, an asymptotic polynomial approximation scheme and a 3/2-approximation algorithm in the absolute sense were presented.

Correa and Epstein [69] defined a more general model called bin packing with controllable item sizes. In this more general model each item has a list of pairs associated with it. Each pair is a configuration of an allowed size for this item, together with a nonnegative penalty, and an item can be packed using any configuration in its list. The goal is to select a configuration for each item so that the number of unit bins plus the sum of penalties is minimized. If the list contains only two pairs where one of them has penalty 0 and the other has processing time 0, then the problem is reduced to bin packing with rejection. The authors studied online bounded space algorithms, they extended the results from the bin packing with rejection model, and presented an algorithm with an asymptotic worst-case ratio arbitrary close to h_∞. We note that the authors also investigated the case of variable sized bins and the bin covering problem with controllable item sizes.

29.7.3 Batched Bin Packing

Gutin et al. [10] defined the *batched* bin packing problem, where the items arrive in a sequence of batches. The items in each batch must be packed before next batch arrives. If each batch contains only one item then the problem reduces to online bin packing. The authors' main result for sequences of two batches is a proof that no algorithm can have an asymptotic worst-case bound better than $r \approx 1.3871$, where r

solves the equation $r/(r - 1) = 3 + \ln r/(2r - 2)$. Balogh et al. [70] considered the problem for input sequences consisting of three batches and proved that no online algorithm can have a better asymptotic worst-case bound than 1.51211. The first batch packing approximation algorithm was studied by Dósa [71] and was restricted to two batches. He analyzed an algorithm which applies FFD to both batches, packing the items into two separate sets of bins. He proved that this algorithm has the asymptotic worst-case bound $19/12 \approx 1.5833$. Based on this algorithm he defined the restrictive *disjunctive* model in which algorithms have to pack batches into disjoint sets of bins—each bin must have items from only one batch. Epstein [72] proved that, in the disjunctive model, the algorithm that applies FF to each batch has absolute worst-case ratio 2, and that no algorithm with a smaller worst-case ratio exists. Epstein also proved that an algorithm which packs the items of each of k batches optimally has an asymptotic worst-case ratio given by the k-th harmonic number H_k, and thus the estimate $1.69103\ldots$ for large k.

29.7.4 Maximal Resource Bin Packing

Boyar et al. [11] model a company/provider that allocates bins (units of a resource) to customers with each bin having a fixed cost. The provider wants to maximize income by allocating a maximum number of bins. In the on-line version of the problem an algorithm must pack items in a given sequence/list into bins under the following constraints: (1) an item must be packed without knowing the sizes of the remaining unpacked items, (2) a new bin is started only when the next item to be packed does not fit in any of the currently occupied bins, and (3) as usual, items can not be moved once they are packed. The authors showed that there is no algorithm with an asymptotic worst-case ratio greater than 2 or less than 3/2. But they were unable to exhibit an algorithm with a bound provably less than 2. Their conjecture that worst-fit achieves the 3/2 lower bound remains open.

For the offline version of the problem, the bins are packed one by one, each opened, packed, then closed once and for all. The offline assumption is that all the items yet to be packed are available for packing the next bin. A *fairness* constraint is imposed on the process of maximizing the number of bins used: A bin cannot be closed if there is at least one item remaining that fits into the empty space of the bin. Boyar et al. adapted the FFI and FFD algorithms to this problem and proved the asymptotic worst-case bounds of 6/5 and 1.7, respectively. They also proved that, under the fairness constraint, all algorithms have asymptotic worst-case ratios no more than 1.7.

Lin et al. [73] extended the FFI result to the cardinality-bounded version of maximal resource bin packing. Moreover, they defined the *lazy bin covering problem* that can be represented as a dual version of the original problem. The goal is to minimize the number of bins but the total size of the items can be larger than 1 in the bins (this assumption relates the problem to bin covering). On the other hand, each bin has to satisfy the constraint that removing any item from it the total size of the remaining items will be no more than 1. The authors proved that the extension of FFD and the extension of FFI both have an asymptotic worst-case bound of 71/60. For the online version of the problem they analyzed the NF, WF, FF, and harmonic algorithms and proved that the respective asymptotic worst-case ratios are at most 4, 3, 2, and 71/60. They also extended the results to the parametric case. Lin et al. [74] presented better algorithms for both problems. In case of maximal resource bin packing they defined a modification of the FFI algorithm and proved that it has asymptotic worst-case bound of $80/71 \approx 1.12676$. In case of lazy bin covering, they defined a modification of the FFD algorithm and proved that it has asymptotic worst-case bound of $17/15 \approx 1.13333$.

29.7.5 Bin Packing with Fragile Items

Bansal et al. [12] defined the bin packing problem with fragile items model. In this model each item has a fragility value. The items have to be assigned to a minimal number of bins where in each bin the total size of the items cannot be larger than the minimal fragility value for the bin. The main result is an algorithm with an absolute worst-case ratio at most 2. Moreover, it is proved that no polynomial time algorithm with better asymptotic worst-case ratio than $3/2$ exists unless $P = NP$. The authors defined another model, which is based on an idea similar to resource augmentation. An algorithm is called c-feasible with respect to fragility if it uses the same number of bins as the optimal solution but the total size of the items in each bin is at most c times more than the minimal fragility. The authors presented a 2-feasible algorithm. Chan et al. [75] considered the online version of the problem. They proved that no online algorithm can have better asymptotic worst-case bound than 2. Moreover, they presented results where the bounds are depending on the fragility ratio which is the ratio of the maximal and minimal fragility values. They proved that the asymptotic worst-case bound of the AF algorithm is not better than the fragility ratio, and they designed a class of algorithms which partitions the items into classes based on their size and packs the classes into separated sets of bins. If the fragility ratio is bounded by a constant, then the asymptotic worst-case ratio of this algorithm tends to 1.7 as the number of classes tends to ∞.

References

1. Coffman, Jr., E.G., Leung, J.Y-T., and Ting, D.W., Bin packing: Maximizing the number of pieces packed, *Acta Inf.*, 9, 263, 1978.
2. Krause, K.L., Shen, Y.Y., and Schwetman, H.D., Analysis of several task-scheduling algorithms for a model of multiprogramming computer systems, *JACM*, 22, 522, 1975.
3. Coffman, Jr., E.G., Garey, M.R., and Johnson, D.S., Dynamic bin packing, *SIAM J. Comput.*, 12, 227, 1983.
4. Liu, W.P. and Sidney, J.B., Bin packing using semi-ordinal data, *Oper. Res. Lett.*, 19, 101, 1996.
5. Mandal, C.A., Chakrabarti, P.P., and Ghose, S., Complexity of fragmentable object bin packing and an application, *CANDM: An Int. J.: Comput. Math. Appl.*, 35, 91, 1998.
6. Jansen, K. and Öhring, S., Approximation algorithms for time constrained scheduling, *Inf. Cont.*, 132, 85, 1997.
7. Azar, Y. and Regev, O., On-line bin-stretching, *Theor. Comp. Sci.*, 268, 17, 2001.
8. Balogh, J., Békési, J., Dósa, Gy., Kellerer, H., and Tuza, Zs., Black and white bin packing, in *Proceedings of WAOA*, LNCS, Vol. 7846, Springer, Ljubljana, Slovenia, 2012, p. 131.
9. Dósa, G.Y. and He, Y., Bin packing problems with rejection penalties and their dual problems, *Inf. Comp.*, 204(5), 795, 2006.
10. Gutin, G., Jensen, T., and Yeo, A., Batched bin packing, *Discrete Optim.*, 2(1), 71, 2005.
11. Boyar, J., Epstein, L., Favrholdt, L.M., Kohrt, J.S., Larsen, K.S., Pedersen, M.M., and Wohlk, S., The maximum resource bin packing problem, *Theor. Comp. Sci.*, 362, 127, 2006.
12. Bansal, N., Zhen, L., and Sankar, A., Bin-packing with fragile objects and frequency allocation in cellular networks, *Wireless Networks*, 15(6), 821, 2009.
13. Friesen, D.K. and Langston, M.A., Variable sized bin packing, *SIAM J. Comput.*, 15, 222, 1986.
14. Assman, S.F., Johnson, D.S., Kleitman, D.J., and Leung, J.Y-T., On a dual version of the one-dimensional bin packing problem, *J. Algorithms*, 5, 502, 1984.
15. Coffman, Jr., E.G. and Leung, J.Y-T., Combinatorial analysis of an efficient algorithm for processor and storage allocation, *SIAM J. Comput.*, 8, 202, 1979.
16. Boyar, J., Larsen, K.S., and Nielsen, M.N., The accommodation function: A generalization of the competitive ratio, *SIAM J. Comput.*, 31, 233, 2001.
17. Azar, Y., Boyar, J., Favrholdt, L.M., Larsen, K.S., and Nielsen, M.N., Fair versus unrestricted bin packing, *Algorithmica*, 34, 181, 2002.
18. Boyar, J., Favrholdt, L.M., Larsen, K.S., and Nielsen, M.N., The competitive ratio for on-line dual bin packing with restricted input sequences, *Nordic J. Comp.*, 8, 463, 2001.
19. Coffman, Jr., E.G., Garey, M.R., and Johnson, D.S., Bin packing with divisible item sizes, *J. Complexity*, 3, 405, 1987.
20. Anily, S., Bramel, J., and Simchi-Levi, D., Worst-case analysis of heuristics for the bin packing problem with general cost structures, *Oper. Res.*, 42, 287, 1994.

21. Babel, L., Chen, B., Kellerer, H., and Kotov, V., Algorithms for on-line bin-packing problems with cardinality constraints, *Disc. Appl. Math.*, 143(1–3), 238, 2004.
22. Epstein, L., Online bin packing with cardinality constraints, *SIAM J. Discrete Math.* 20(4), 1015, 2006.
23. Fujiwara, H. and Kobayashi, K., Improved lower bounds for the online bin packing problem with cardinality constraints, *J. Comb. Optim.* 29, 67, 2015.
24. Békési, J., Dósa, Gy., and Epstein, L., Bounds for online bin packing with cardinality constraints, *Inf. Comp.*, 249, 190, 2016.
25. Adar, R. and Epstein, L., Selfish bin packing with cardinality constraints, *Theor. Comp. Sci.*, 495, 66, 2013.
26. Shachnai, H. and Tamir, T., Tight bounds for online class-constrained packing, *Theor. Comp. Sci.*, 321, 103, 2004.
27. Xavier, E.C. and Miyazawa, F.K., The class constrained bin packing problem with applications to video-on-demand, *Theor. Comp. Sci.*, 393(1–3), 240, 2008.
28. Epstein, L., Imreh, Cs., and Levin, A., Class constrained bin packing revisited, *Theor. Comp. Sci.*, 411(34–36), 3073, 2010.
29. Krumke, S.O., de Paepe, W., Rambau, J., and Stougie, L., Online bin-coloring, *Theor. Comp. Sci.*, 407, 231, 2008.
30. Lin, M., Lin, Z., and Xu, J., Almost optimal solutions for bin coloring problems, *J. Comb. Optim.*, 16, 16, 2008
31. Chan, J.W.T., Lam, T.W., and Wong, P.W.H., Dynamic bin packing of unit fractions items, *Theor. Comp. Sci.*, 409, 521, 2008.
32. Han, X., Peng, C., Yec, D., Zhang, D., and Lane, Y., Dynamic bin packing with unit fraction items revisited, *Inf. Proc. Lett.*, 110, 1049, 2010.
33. Chan, J.W.T., Wong, P.W.H., and Yung, F.C.C., On dynamic bin packing: An improved lower bound and resource augmentation analysis, *Algorithmica*, 53, 172, 2009.
34. Wong, P.W.H., Yung, F.C.C., and Burcea, M., An 8/3 lower bound for online dynamic bin packing, in *Proceedings of the ISAAC*, LNCS 7676, Springer, Taipei, Taiwan, 2012, p. 44.
35. Li, Y., Tang, X., and Cai, W., Dynamic bin packing for on-demand cloud resource allocation, *IEEE Trans. Par. Distr. Sys.*, 27(1), 157, 2016.
36. Ivković, Z. and Lloyd, E., Fully dynamic algorithms for bin packing: Being (mostly) myopic helps, *SIAM J. Comput.*, 28, 574, 1998.
37. Menakerman, N. and Rom, R., Bin packing with item fragmentation, in *Proceedings of WADS*, LNCS 2125, Springer, Providence, RI, 2001, p. 313.
38. LeCun, B., Mautor, T., Quessette, F., and Weisser, M.A., Bin packing with fragmentable items: Presentation and approximations, *Theor. Comp. Sci.*, 602, 50, 2015.
39. Epstein L. and Levin A., On bin packing with conflicts, *SIAM J. Optim.*, 19(3), 1270, 2008.
40. Manyem, P., Bin packing and covering with longest items at the bottom: Online version. *ANZIAM Journal*, 43(E), 186, 2002.
41. Manyem, P., Salt, R.L., and Visser, M.S., Approximation lower bounds in online LIB bin packing and covering, *Journal of Automata, Languages and Combinatorics*, 8(4), 663, 2003.
42. Manyem, P., Uniform sized bin packing and covering: online version, in *Topics in Industrial Mathematics*, Narosa Publishing House, New Delhi, India, 2003, p. 447.
43. Finlay, L. and Manyem, P., Online LIB problems: heuristics for bin covering and lower bounds for Bin Packing, *RAIRO Rech. Opr.* 39, 163, 2005.
44. Epstein, L., On online bin packing with LIB constraints, *Nav. Res. Logist.* 56(8), 780, 2009.
45. Dósa, Gy., Tuza, Zs., and Ye, D., Bin packing with "Largest In Bottom" constraint: Tighter bounds and generalizations, *J. Comb. Optim.*, 26(3), 416, 2013.
46. Bódis, A., Bin packing with directed stackability conflicts, *Acta Univ. Sapientiae, Inf.*, 7(1), 31, 2015.
47. Katona, G.Y., Edge disjoint polyp packing, *Disc. Appl. Math.*, 78, 133, 1997.

48. Codenetti, B., De Marco, G., Leoncini, M., Montangero, M., and Santini, M., Approximation algorithms for a hierarchically structured bin packing problem, *Inf. Proc. Lett.*, 89, 215, 2004.
49. Bujtás, C.S., Dósa, G.Y., Imreh, C.S., Nagy-György, J., and Tuza, Z.S., The graph-bin packing problem, *Int. J. Found. Comp. Sci.*, 22(8), 1971, 2011.
50. Bujtás, C.S., Dósa, G.Y., Imreh, C.S., Nagy-György, J., and Tuza, Z.S., New models of graph-bin packing, *Theor. Comp. Sci.*, 640, 94, 2016.
51. Cheng, T.C.E., Kellerer, H., and Kotov, V., Semi-on-line multiprocessor scheduling with given total processing time, *Theor. Comp. Sci.*, 337(13), 134, 2005.
52. Kellerer, H. and Kotov, V., An efficient algorithm for bin stretching, *Oper. Res. Lett.*, 41(4), 343, 2013.
53. Gabay, M., Kotov, V., and Brauner, N., Online bin stretching with bunch techniques, *Theor. Comp. Sci.*, 602, 103. 2015.
54. Böhm, M., Sgall, J., van Stee, R., and Vesely, P., Better algorithms for online bin stretching, in *Proceedings of WAOA*, LNCS 8952, Springer, Patras, Greece, 2015, p. 23.
55. Böhm, M., Sgall, J., van Stee, R., and Vesely, P., The best two-phase algorithm for bin stretching, http://arxiv.org/abs/1601.08111 (Accessed on 4 November, 2017).
56. Kellerer, H., Kotov, V., Speranza, M.G., and Tuza, Zs., Semi on-line algorithms for the partition problem, *Oper. Res. Lett.*, 21, 235, 1997.
57. Epstein, L., Bin stretching revisited, *Acta Inf.*, 39, 97, 2003.
58. Gabay, M., Kotov, V., and Brauner, N., *Improved Lower Bounds for the Online Bin Stretching Problem.* HAL preprint hal-00921663, version 3, 2015.
59. Böhm, M., Sgall, J., van Stee, R., and Vesely, P., Online bin stretching with three bins, http://arxiv.org/pdf/1404.5569v3.pdf (Accessed on 4 November, 2017).
60. Speranza, M.G. and Tuza, Z.S., On-line approximation algorithms for scheduling tasks on identical machines with extendable working time, *Ann. Oper. Res.*, 86, 491, 1999.
61. Dell'Olmo, P., Kellerer, H., Speranza, M.G., and Tuza, Z.S., A 13/12 approximation algorithm for bin packing with extendable bins, *Inf. Proc. Lett.*, 65, 229, 1998.
62. Yang, J. and Leung, J.Y.-T., The ordered open-end bin packing problem, *Oper. Res.*, 51, 759, 2003.
63. Balogh, J., Békési, J., Dósa, G.Y., Epstein, L., Kellerer, H., and Tuza, Z.S., Online results for black and white bin packing, *Theory Comput. Syst.*, 56(1), 137, 2015.
64. Chen, J., Han, X., Bein, W., and Ting, H.F., Black and white bin packing revisited, in *Proceedings of COCOA*, LNCS 9486, Springer, Houston, Texas, 2015, p. 45.
65. Balogh, J., Békési, J., Dósa, G.Y., Epstein, L., Kellerer, H., Levin, A., and Tuza, Z.S., Offline black and white bin packing, *Theor. Comput. Sci.*, 596, 92, 2015.
66. Dósa, G.Y. and Epstein, L., Colorful bin packing, in *Proceedings of SWAT*, LNCS 8503, Springer, Copenhagen, Denmark, 2014, p. 170.
67. Böhm, M., Sgall, J., and Vesely, P., Online colored bin packing. In: Bampis, E., (Ed.), in *Proceedings of WAOA* LNCS 8952, Springer, Patras, Greece, 2015, p. 35.
68. Epstein, L., Bin packing with rejection revisited, *Algorithmica* 56, 505, 2010
69. Correa, J.R. and Epstein, L., Bin packing with controllable item sizes, *Inf. and Comp.*, 206, 1003, 2008.
70. Balogh, J., Békési, J., Galambos, G., Dósa, Gy., and Tan, Z., Lower bound for 3-batched binpacking, *Dis. Opt.*, 21, 14, 2016.
71. Dósa, G.Y., Batched bin packing revisited, *J. Sched.* doi:10.1007/s10951-015-0431-3
72. Epstein, L., More on batched bin packing, *Oper. Res. Lett.*, 44, 273, 2016.
73. Lin, M., Yang, Y., and Xu, J., On lazy bin covering and packing problems, *Theor. Comp. Sci.*, 411, 277, 2010.
74. Lin, M., Yang, Y., and Xu, J., Improved approximation algorithms for maximum resource bin packing and lazy bin covering problems, *Algorithmica*, 57, 232, 2010.
75. Chan, W., Chin, F.Y., Ye, D., Zhang, G., and Zhang, Y., Online bin packing of fragile objects with application in cellular networks, *J. Comb. Optim.*, 14(4), 427, 2007.

30

Variable Sized Bin Packing and Bin Covering[*]

30.1 Introduction

In this chapter we continue our survey with a focus on two variants of the one-dimensional bin packing problem: the variable sized bin packing problem, and the bin covering problem. In Section 30.2, we survey algorithms for packing into bins of different sizes, a problem first studied by Friesen and Langston [1] in 1986. In Section 30.3, we survey the bin covering problem that asks for a partition of a given set of items into a maximum number of subsets such that, in every subset, the total item size is always at least some lower bound. This problem was first studied by Assmann et al. [2] in 1984. Concluding remarks are given in Section 30.4.

30.2 Variable Sized Bin Packing

In the classical bin packing problem the measure of a packing is the number of bins used. The wasted space in the packing is an equivalent measure; a packing that minimizes one of these measures minimizes the other. In what follows, we focus on the wasted–space measure.

Interesting results are available if we relax the condition of having bin sizes fixed in advance. In a problem posed by Friesen and Langston [3], the bin size, which must be the same for all bins, is part of the solution. What is a bin size α, and a packing into bins of this size, which minimizes the wasted space? If there were no bound on the maximum bin size, the problem would be trivial: use only one bin with a size equal to the total item size. But a maximum bin size is a constraint of the problem, and we normalize to 1 for convenience. The goal is to choose a number $\alpha \leq 1$, and produce a packing of a list $L = (a_1, a_2, \ldots, a_n)$ into N bins of size α such that

$$N\alpha - \sum_{i=1}^{n} s(a_i)$$

---

[*] This chapter includes its previous version coauthor with Edward G. Coffman, Jr. and Joseph Y-T. Leung that appeared in the First Edition of the Handbook on Approximation Algorithms and Metaheuristics.

is as small as possible. The analysis compares the worst-case wasted space of approximation algorithms to OPT_{α_0}, the minimal space wasted using bin size $\alpha_0 \leq 1$. Friesen and Langston proved that if a bin packing algorithm A can guarantee a worst case ratio R for the classical bin packing problem, then an iterated version of A can be used to achieve a bound as close to R as we would like. More precisely, if a bound of $R + \epsilon$ is sought, we can generate a sequence of sizes $\alpha_1, \alpha_2, \ldots, \alpha_k$ with $\alpha_{i+1} = \alpha_i R / (R + \epsilon)$ and prove that, if $\alpha_{i+1} < \alpha_0 \leq \alpha_i$, then running algorithm A on bins of size α_i will ensure a wasted space $A_{\alpha_i}(L) < (R + \epsilon)\text{OPT}_{\alpha_0}(L)$.

The next step taken by Friesen and Langston [1] allowed more than one bin size for packing. They defined the variable-sized bin packing problem on a finite collection of k available bin sizes, $s_1 > s_2 > \ldots > s_k$, with an inexhaustible supply of each size. We adopt the normalization whereby the items $\{a_i\}$ and bins are such that the largest bin has size $s_1 = 1$ and $s(a_i) \leq 1$ for all $1 \leq i \leq n$. The goal is to pack the items into bins so that the sum of the sizes of the bins used is minimum.

The variable-sized bin packing problem is NP-hard [1], so as usual, efficient algorithms that ensure near-optimal packings comprise the design goal. Let $s(A, L)$ denote the total size of the bins used by algorithm A to pack the items of L. Then

$$R_A^\infty = \limsup_{k \to \infty} \left\{ \max \frac{s(A, L)}{s(\text{OPT}, L)} : s(\text{OPT}, L) \geq k \right\}$$

is the asymptotic worst case ratio of algorithm A.

Kinnersley and Langston [4] presented three approximation algorithms, with asymptotic worst case ratios of 2, 3/2, and 4/3, respectively. The first algorithm is a simple adaptation of the Next Fit (NF) rule. They called it NFL (Next Fit using Largest bins only), that is, NF packing only into bins of size 1. The proof of $NFL(L) < 2 \cdot \text{OPT}(L) + 1$ for any list L is quite standard. Letting ϵ denote an arbitrarily small positive real value, any instance consisting of items of size $1/2 + \epsilon$ and bins of sizes 1 and $1/2 + \epsilon$ shows that 2 is a matching asymptotic lower bound.

The second algorithm is a variant of the First Fit Decreasing (FFD) rule called FFDLR (First Fit Decreasing using Largest bins, then Repack into smallest possible bins); and it does what its name says: in the first round the elements are sorted in decreasing order and the items are packed into bins of size 1 by the FFD rule. In the second round it attempts to repack all used bins into the smallest possible bins, where only the full content of a bin can be repacked. This algorithm has an asymptotic worst-case bound of 3/2.

The third algorithm FFDLS (First Fit Decreasing using Largest bins, but Shifting as necessary) uses two rounds and also tries to repack the contents of certain bins even in the first round. It is a modification of the FFDLR rule in the sense that it uses the same method in both rounds, but in the first round when a_i is packed into bin B_j it checks whether B_j contains an item of size greater than $1/3$. If so it repacks, where possible, all the items in B_j into the smallest empty bin $B_{j'}$ that will hold them, and for which $c(B_{j'}) \geq (3/4)c(B_j)$ holds. The second round of the FFDLS rule is the same as that for FFDLR. This algorithm has an asymptotic worst-case bound of 4/3.

Note that only the first of the three algorithms is online. Kinnersley and Langston [4] also gave the following better online. First, they proved that two variants of First Fit (FFL - First Fit, using Largest possible bins and FFS - First Fit, using Smallest possible bins) both have a worst-case bound of two. They observed that the FFL rule fails in its packing of "large" items (those with a size exceeding $1/2$) and FFS fails in its packing of "small" items (those with size at most $1/2$). So they focused on a hybrid approach that we denote by FFf. Let f denote a user specified factor in the range $[1/2, 1]$. Suppose FFf has to start a new bin when it is packing a piece a_i. If a_i is a small piece, FFf starts a new bin of size 1. If a_i is a large piece, then it selects the smallest bin size in the range $[s(a_i), s(a_i)/f]$ if such a size exists, otherwise it will use bin size 1. Kinnersley and Langston proved that

$$FFf(L) \leq (1.5 + \frac{f}{2})\text{OPT}(L) + 2$$

for any list L. We should remark here that for the smallest possible f this gives an asymptotic upper bound of 1.75.

They also proved that for the parametric case, FFL gives the same performance as FF for the classical problem, if every item is at most $1/2$. Later, Zhang [5] showed that the best possible FFf algorithm (with $f = 1/2$) has a tight bound of $17/10$, the bound for First Fit in the classical bin packing problem.

Csirik [6] introduced a Harmonic type heuristic, VH_M (Variable Harmonic M) that is based on a harmonic subdivision of each bin size. Let $M > 1$ be a positive integer and let $M_j = \lceil M \cdot s(B_j) \rceil$ $(j = 1, 2, \ldots, k)$. The algorithm is defined only for those M where $M_k \geq 2$. Let us divide the intervals $(0, s_j]$ $(j = 1, 2, \ldots, k)$ into M_j parts according to the following harmonic partitioning:

$$I_{j,l} = \left(\frac{s_j}{l+1}, \frac{s_j}{l} \right], j = 1, 2, \ldots, k; \ l = 1, 2, \ldots, M_j - 1,$$

and

$$I_{j,*} = \left(0, \frac{s_j}{M_j} \right].$$

For each bin size s_j we define a weighting function as follows:

$$W_j(a_i) = \begin{cases} \frac{M_j}{M_j - 1} s(a_i) & \text{if } s(a_i) \in I_{j,*}, \\ \frac{s_j}{l} & \text{if } s(a_i) \in I_{j,l}, \ l = 1, 2, \ldots, M_j - 1, \\ \infty & \text{if } s(a_i) > s_j). \end{cases}$$

Let

$$W(a_i) = \min_{j=1,2,\ldots,k} W_j(a_i),$$

be the weight of a_i.

Now we can define the algorithm VH_M as follows:

Step 1. We assign to each item of the list a bin size; an item assigned to a bin of size s_j is called a type j item.
   a. $s_1 = 1$ is assigned to each element a_i with

$$W(a_i) = (M_1/(M_1 - 1)) \cdot s(a_i).$$

These items are called small items, and all others big items.
   b. A big item a_i will be a type-j element if j is the smallest integer such that $W(a_i) = W_j(a_i)$. A big element will be called an $I_{j,l}$ element if it is a type-j element and belongs to $I_{j,l}$.

Step 2: VH_M performs a Harmonic fit-type packing:
   a. Each big type-j element is packed by Harmonic Fit into bins of size s_j as follows. We classify these bins into $M_j - 1$ categories. Each category is designated to pack the same type of items. A bin of size s_j designated to pack $I_{j,l}$ items is called an $I_{j,l}$ bin. Clearly, each $I_{j,l}$ bin has room for exactly l items. We use a NF packing in all $I_{j,l}$ $(j = 1, 2, \ldots, k; \ l = 1, 2, \ldots, M_j - 1)$ bins, that is, after packing l items into an $I_{j,l}$ bin, we close this bin and open a new $I_{j,l}$ bin.
   b. All small items are packed in bins of size 1 by NF.

Csirik proved that, using the sequence t_i defined in the Harmonic Algorithm, the following holds: if $t_{i(k)-1} < M_k \leq t_{i(k)}$, then

$$R_{VH_M} \leq \sum_{l=1}^{i(k)-1} \frac{1}{t_l - 1} + \frac{M_k}{(M_k - 1)t_{i(k)}}.$$

He also considered the question of what performance can be achieved if we are free to choose the bin sizes. He proved that if we have just two bin sizes and the smaller bin size is optimally selected (this is 0.7) then the asymptotic worst-case ratio is 1.4.

Seiden [7] showed that Variable Harmonic is an optimal bounded-space algorithm. Seiden et al. [8] generalized the Refined Harmonic idea of the classical bin packing problem to the variable sized problem. Making use of Refined Harmonic they bettered the upper bound from 1.69103 to 1.63597. Seiden [9] once demonstrated that if we are allowed to choose the bin sizes then, with two bin sizes, the optimal performance ratio lies in the range [1.37530, 1.37532].

Epstein et al. [10] investigated in more detail the case of two different bin sizes. They presented two different algorithms for this case, both being combinations of Harmonic and Refined Harmonic. They proved that the best of these two algorithms and the Variable Harmonic yields an upper bound of $373/227 < 1.63597$. They also gave the first lower bound for this problem, showing that for two bin sizes any online algorithm has an asymptotic performance ratio of at least 1.33561.

Burkard and Zhang [11] generalized the bounded space algorithm of Csirik and Johnson [12] so that it applied to the variable bin size problem. In the latter case we have to define the closing rule slightly differently because we may choose in between bin sizes too. Hence their closing rules are:

C-VF: Close one active bin with size less than 1 if a such bin exists, otherwise use FF.

C-VB: Close one active bin with size less than 1 if a such bin exists, otherwise use BF.

They defined the opening rule in the following way: suppose a_i is a large item with size greater than 1/2. If it can be contained in a bin with size less than 1, it is called a B-item, otherwise it is called an L-item. The smallest bin that can contain a large item a_i is called an a_i−home-bin. Obviously, if a_i is an L-item, the size of the a_i−home-bin is 1. Their opening rule is: suppose the current item to be packed is a_i. If a_i is a B-item, open an a_i−home-bin and pack a_i into it. Otherwise, start a new bin of size 1 for a_i.

Burkard and Zhang proved that using this opening rule and the closing rule C-VB we will get a tight 1.7 performance bound for this algorithm for those cases where we have at least three open bins at the same time. This means that—taking into account Csirik and Johnson's result for classical bin packing—we need one more open bin for the variable-sized bin packing problem to have an algorithm with the same performance.

Chu and La [13] used a best-fit-like idea for variable-sized bin packing. They defined four algorithms: Largest object First with Least Absolute Waste (LFLAW), Largest object First with Least Relative Waste (LFLRW), Least Absolute Waste (LAW), and Least Relative Waste (LRW). Naturally, the first two are not online algorithm as they need a sorted list as input. They proved tight bounds for each algorithm, showing that

$$R^\infty_{LFLAW} = R^\infty_{LFLRW} = 2, R^\infty_{LAW} = 3 \text{ and } R^\infty_{LRW} = 2 + \ln 2.$$

Zhang [14] introduced a relaxation on the online condition: here we have all the information about the items, but we cannot preview the type of bin before it arrives. We must decide which items should be packed into the bin as it arrives. We will suppose that the largest item still fits in the smallest bin. In this case we have of course just one open bin, and we close it as we cannot pack more items into it. The goal is evidently to fill bins with the minimal total size. Zhang investigated the classical algorithms NF, First Fit, Next Fit Decreasing, and FFD and proved that each of them has an asymptotic worst-case ratio of 2. He also remarked that if there are at most $l, l \geq 1$, open bins at a time this ratio cannot be improved.

Boyar and Favrholdt [15] analyzed a version of Zhang's relaxation, under the assumption that the largest item fits in the smallest bin. Bins and items have integer sizes between 1 and some maximum size M. Whenever a bin arrives, the algorithm must keep packing items in it as long as any unpacked item still fits in the bin. They call this problem *grid scheduling* problem. Boyar and Favrholdt derived some new results for Zhang's bin packing problem too. They proved that any grid scheduling algorithm has an asymptotic worst case ratio of at least $\frac{5}{4}$ and at most 2. Any algorithm for Zhang's bin packing problem has an asymptotic worst case ratio of at least $\frac{17}{15}$, and no general upper bound applies to this problem. They proposed algorithm $FFD_{2/3}$ with an asymptotic worst case ratio of at most $\frac{13}{7}$ for both problems. This result improves Zhang's best algorithm. In addition, Boyar and Favrholdt derived some results for the relative worst-order ratio.

Boyar and Ellen [16] obtained matching lower and upper bounds for specific item and bin sizes for the previous problem, which vary depending on the ratio of the number of small jobs to the number of large jobs.

Epstein et al. [17] generalized the grid scheduling problem. At the variant of Boyar and Favrholdt any pair of jobs can be executed on the same processor. However, security reasons may prevent certain pairs of jobs from sharing a common processor. So, Epstein et al. defined a conflict graph where an edge between items i and j means that job i and j cannot be processed on the same processor. This means in bin packing formulation that these two items cannot be packed in the same bin. Epstein et al. called this problem *grid scheduling with conflicts* (GSCs). They proved the following results for GSC:

- For the semi-online version of GSC, with monotonically nondecreasing bin sizes, there is no competitive algorithm even if the conflict graph is a disjoint union of cliques. This result holds for the online problem too.
- For the semi-online version of GSC with monotonically nonincreasing bin sizes they showed a lower bound of $\frac{5}{4}$ on the asymptotic worst ratio of any algorithm, and a lower bound of $\frac{7}{6}$ on the asymptotic worst ratio of any algorithm for bipartite conflict graphs.
- Designed an asymptotically 3-competitive algorithm for the semi-online version where the item sizes are monotonically nonincreasing and an asymptotically 2-competitive algorithm for the online problem when the conflict graph is constant-colorable, that is, for k-partite graphs for any constant k.
- For the special case in which the conflict graphs is empty they gave an algorithm called Knapsack with asymptotic worst ratio between $\frac{5}{3}$ and $\frac{9}{5}$ for nonincreasing bin sizes. They also showed that its asymptotic worst ratio is at least 1.734 for nondecreasing bin sizes. For identical bin sizes it is known that the asymptotic worst ratio of Knapsack is 1.6067.
- When the conflict graph is a union of two cliques and the bin sizes are monotonically nonincreasing, they proved that a modified versions of FFD and Knapsack have asymptotic worst case ratios of exactly 2 and $\frac{3}{2}$, respectively.

Boyar and Favrholdt [18] defined a new variable-sized bin packing problem with applications in genetics. The problem is to compare k query sequences with l sequences in a database. This problem is modeled by a $k \times l$ matrix that is to be partitioned into smaller rectangles that are to be packed in variable-sized bins (that represent processors) arriving online. A set of rectangles fit in a bin, if the sum of their height and width is no more than the size of the bin. The goal is to minimize the total size of the bins used for packing the whole matrix. The connected variable-sized bin packing problem is a variation of the above-mentioned problem where the size of an item depends on the other items packed in the same bin. Boyar and Favrholdt proved that most reasonable algorithms have an asymptotic worst ratio of $\Theta(M/m)$. They also applied worst-order and bijective analysis.

Dell'Olmo and Speranza [19] introduced a different relaxation on the variable sized bin packing problem: they supposed that the bin sizes are extendible, that is, the total size of items packed into a single bin can exceed 1, if necessary. The goal is to minimize the total bin size, where the size of bin B_j is now given by $\max(c(B_j), s(B_j))$. The authors investigated two heuristics: Best Fit Decreasing for the offline case and Best Fit for the online case. They proved that BFD's performance bound is at most $2(2 - \sqrt{2})$ and conjectured that the exact bound is 8/7. For the online case they proved a tight 5/4 bound for the BF algorithm and showed that no online algorithm can have better bound than 7/6.

Ye and Zhang [20] studied a similar problem. They supposed that we have m bins with different bin sizes. They called the set of bins a collection. We have to pack the items into these bins and over-packing is allowed. They distinguished two cases. In the first case the largest item will fit into the smallest bin. They carefully analyzed the List Scheduling algorithm and gave competitive ratios for each m and each collection. The ratios differ for an even number of bins and an odd number of bins. They were able to prove that

$$R_{LS}(m) = \begin{cases} 1 + \frac{m \cdot b_{\min}}{4 \sum_{j=1}^{m} b_j}, & \text{if } m \text{ is even,} \\ 1 + \frac{(m^2 - 1) \cdot b_{\min}}{4m \sum_{j=1}^{m} b_j}, & \text{if } m \text{ is odd,} \end{cases}$$

where $b_{\min}$ is the smallest bin size. For equal bin size the ratio is 5/4 when m is even and $5/4 - 1/(4m^2)$ when m is odd. They presented an improved algorithm for $m = 2$ and $m = 3$.

In the second case the largest item does not fit into the smallest bin. A lower bound for the overall competitive ratio is given for two bins.

Kang and Par [21] investigated the special case of variable-sized bin packing where we have (weak) divisibility conditions. Two types of conditions were looked at:

- where the sizes of the items are divisible
- where the sizes of the bins are divisible, that is, $s(B_{j+1})$ exactly divides $s(B_j)$ for all $j = 1, 2, \ldots, k-1$

They also used a more general cost function: they assumed that the bin sizes are sorted in descending order, that is, $b_1 \geq b_2 \geq \ldots \geq b_k$, and bins of size b_i have a cost c_i, where the unit size cost of each bin does not increase as the bin size increases, that is, the costs and the bin sizes satisfy $\frac{c_{i_1}}{b_{i_1}} \leq \frac{c_{i_2}}{b_{i_2}}$ for all $1 \leq i_1 \leq i_2 \leq k$. They studied iterative versions of the FFD and BFD rules. The Iterative First Fit Decreasing (IFFD) algorithm works as follows: we first pack all the items into the largest size bins using the FFD algorithm, and obtain a feasible solution. We then get another solution by repacking the items in the last bin of the solution into the next largest bin using the FFD rule. We obtain another feasible solution by continuing this procedure until we have repacked every item. In this way, we obtain feasible solutions for each type of bin. Then the best solution among them is selected as the final solution. The Iterative Best Fit Decreasing (IBFD) algorithm is similarly defined, using the BFD rule instead of the FFD rule in each step. They obtained a series of results for this algorithm:

- When L has divisible item sizes and bin sizes are divisible as well, there exists an optimal solution using k_1 or $k_1 - 1$ bins of type 1, where k_1 is the number of bins used in the first step of IFFD. This means that IFFD provides an optimal solution.
- For each L, if bin sizes are divisible, then $C(IFFD(L)) \leq \frac{11}{9} \cdot C(OPT(L)) + 5\frac{2}{9}$. Here $C(A(L))$ denotes the cost of packing list L by algorithm A. The bound of 11/9 is tight.
- The general case, that is, when we have no divisibility condition, for each $L, C(IFFD(L)) \leq \frac{3}{2} \cdot C(OPT(L)) + 1$. This bound is tight.

The same results apply to IBFD.

Xing [22] investigated a special case of variable-sized bin packing where we can have over-sized items, that is, items with a size larger than the largest bin size. The bins cannot be over-packed, so we are free to divide the over-sized items such that the parts are no larger than the largest bin size. The problem is called *bin packing with over-sized (BPOS) items*. Xing also defined a special objective function for this problem: if an item a_i is over-sized and so packed into more than one bin, the extra bins do not contribute to the objective function. If we use bins $B_1, B_2, \ldots, B_m$ in the packing, then the objective function is

$$\sum_{i=1}^{m} s(B_i) - \sum_{i=1}^{n} (\lceil s(a_i) \rceil - 1).$$

Xing defined the following two-stage procedure: let us delete size $\lfloor s(a_i) \rfloor$ from item a_i – now the size of the remaining part does not exceed the largest bin size. We can easily extend a heuristic algorithm of variable-sized bin packing to a heuristic of BPOS: first we pack $\lceil s(a_i) \rceil - 1$ unit-sized bins with unit parts of a_i and then the remaining part will be packed by the variable-sized bin packing heuristic. Let TOPT(L) denote the optimal value of the BPOS using a two-stage procedure. It can readily be seen that if there is only one bin size then OPT(L) = TOPT(L) for each list. If there are several bin sizes and the largest item is packed into l bins then

$$2 - \frac{1}{l+1} \leq R^{\infty}_{TOPT} \leq 2,$$

and the ratio 2 is asymptotically tight.

Xing defined a modification of the First Fit algorithm for this problem and showed that its asymptotic worst-case ratio is 7/4. If there is a bin size between [2/3, and 3/4] then the ratio is 3/2.

Naaman and Rom [23] investigated a scheduling problem called *packet scheduling with fragmentation*. They transformed this problem into a variant of bin packing where items may be fragmented. Bins have integer capacities and item sizes are integers too. When an item is fragmented, r units of overhead are added to the size of every fragment. The goal is to pack items of maximum total size into at most m variable size bins, in other words, to maximize the bin utilization. Algorithm NF_f, a version of the NF algorithm that allows fragmentation was analyzed. The average bin size is denoted by $U = \frac{1}{m} \sum_{j=1}^{m} s(B_j)$. The asymptotic worst-case performance ratio of algorithm NF_f for this problem with variable size bins is $R^{\infty}_{NF_f} = \frac{U}{U-2r}$ for every $r \geq 0$ and $U > 4r$.

30.2.1 Number of Items

Langston [24] turned to the problem of maximizing the number of items packed into m available bins, where the bin sizes can be different. He analyzed the three heuristics given by Coffman et al. [25], and proved that both Smallest-Piece-First and First-Fit-Increasing have an asymptotic worst-case bound of 1/2. The best algorithm is the FFD* rule, for which Langston was able to prove that for all L and bin size-set B

$$n_{FFD*}(L, B) \geq \frac{2}{3} n_O(L, B) - \frac{2}{3}.$$

He conjectured that the tight bound is 8/11—this is still an open question.

Friesen and Kuhl [26] gave the best known algorithm for maximizing the number of items in a given set of variable sized bins. Their algorithm is an iterative version of the Best-2-Fit and FFD algorithms, and is similar to the compound algorithm given for the classical problem in Reference 27. This hybrid algorithm has a tight asymptotic worst-case ratio of 3/4.

Epstein and Favrholdt [28,29] investigated the online version of maximizing the number of items packed into variable-sized bins. They restricted the input sequences to be accommodating, that is, sequences that we know in advance every item can be packed by an optimal offline algorithm. They studied fair algorithms that rejects an item only if the item does not fit in the empty space of any bin. They proved that any fair algorithm has a competitive ratio of at least 1/2, and Best-Fit has a performance bound of $n/(2n - 1)$. They also showed that any fair, deterministic algorithm has a competitive ratio at most 2/3, and any fair, randomized algorithm has a competitive ratio at most 4/5.

30.3 Bin Covering

In the *packing* problems studied up to now, the goal was to partition a set of items into the *minimum* number of subsets such that, in every subset, the total size of the items does not exceed some upper bound. In a *covering* problem, the goal is to partition a set of items into the *maximum* number of subsets such that, in every subset, the total size of the items is always above some *lower* bound. Covering problems model a variety of situations encountered in business and in industry, from packing peach slices into tin cans so that each tin can contains at least its advertised net weight, to such complex problems as breaking up monopolies into smaller companies, each of which is large enough to be viable. As covering problems can be viewed as a kind of inverse or dual version of the packing problem, they are sometimes called "dual bin packing" problems in the literature. We note that the task of maximizing the number of items to be packed is sometimes also called dual bin packing problem.

In the one-dimensional bin covering problem, the goal is to pack a list $L = (a_1, a_2, \ldots, a_n)$ into a maximum number of bins of size 1 such that the contents of each bin is at least one. Let R_A^∞ denote the asymptotic worst case ratio of an approximation algorithm for bin covering ($A(L)$ and $\mathrm{OPT}(L)$ denote the numbers of bins in the packing constructed by algorithm A and an optimization algorithm, respectively). It is defined by

$$R_A^n = \min\left\{ \frac{A(L)}{\mathrm{OPT}(L)} \mid \mathrm{OPT}(L) = n \right\},$$

and

$$R_A^\infty = \liminf_{n \to \infty} R_A^n.$$

As we are dealing with a maximization problem, the larger the worst case ratio of an algorithm, the better the approximation algorithm will be. The bin covering problem was investigated for the first time in the PhD thesis of Assmann [30] and it was referred to as the *dual bin packing problem by Assmann et al.* [2].

First of all one might adapt heuristics from the classical bin packing to the bin covering problem. This was indeed done by Assmann et al. [2]. They considered the following adaptation of Next Fit, called DNF (Dual Next Fit): DNF always has a single active bin. Newly arriving items a_i are packed into the active bin until the active bin is full (i.e., it has contents of at least one). Then the active bin is closed and another (empty) bin becomes the active bin. It is not difficult to show that DNF has performance

$$R_{DNF}^\infty = \frac{1}{2},$$

as all bins packed by DNF will have a content of less than 2, whereas for the lists

$$L_{4k} = \left(\underbrace{1 - \varepsilon, \ldots, 1 - \varepsilon}_{2k \text{ items}}, \underbrace{\varepsilon, \ldots, \varepsilon}_{2k \text{ items}} \right)$$

we get $\mathrm{OPT}(L_{4k}) = 2k$ and $DNF(L_{4k}) = k$, if $\varepsilon < 1/2k$. Analogous modifications of First Fit, Best Fit, and Harmonic *do not* improve this worst case ratio of 1/2 (e.g., modifying First Fit is useless, as after filling a bin, placing further items into it does not make sense). Actually, from the worst case point of view, algorithm DNF is the best possible online bin covering algorithm, as Csirik and Totik [31] have proved that every *online bin* covering algorithm A satisfies the condition

$$R_A^\infty \leq \frac{1}{2}.$$

Christ et al. [32] compared online algorithms from a different point of view. They investigated DNF and an appropriately modified harmonic algorithm called Dual Harmonic (DH_k) when the items are drawn from special (a,b) intervals, and shown that DH_k is better than DNF. The same is true if instead of competitive ratio we use different performance measures, such as relative worst order ratio [33] or max/max ratio [34]. On the other hand, DNF is better than DH_k for the random order ratio [35] and the average case analysis [36].

A bettering of the performance ratio of 1/2 was achieved by Assmann et al. [2] by defining an *artificial upper bound* on the sum of the sizes of elements placed into the same bin. This upper bound can be regarded as the capacity of a bin and has some similarities with classical bin packing. However, after packing the items with a good heuristic for the classical bin packing problem, it might happen that in some of the bins the sum of item sizes is less than 1. Hence we have to use a second step to fill these bins. The algorithm based on the above-mentioned observation is called FFD_r (First Fit Decreasing with parameter r) and proceeds as follows:

Phase I ("Classical *FFD*"):
   a.   Presort the items in L such that $s(a_1) \geq s(a_2) \geq \ldots \geq s(a_n)$.
   b.   If there is still an unpacked element, do the following: Let a_i be the first unpacked item and let B_j be the first (leftmost) unfilled bin with a current content less than or equal to $r - s(a_i)$. If such a bin exists, place a_i in B_j, otherwise open a new empty bin and pack a_i into this bin.

Phase II (Repacking unfilled bins):
    If there is more than one open nonfilled bin, remove an item from the rightmost such bin and add it to the leftmost one.

It is clear that after Phase I there is no bin with content more than r. Furthermore, it follows from the definition in the first phase that adding an element from the rightmost open nonfilled bin will increase the total sum of sizes of elements in the leftmost bin to more than r. Finally, the time complexity of FFD_r can be seen to be $O(n \log n)$. For this algorithm the following result holds [30]: for all r, $4/3 \leq r \leq 3/2$, and every lists L,

$$FFD_r(L) \geq \frac{2}{3}(\mathrm{OPT}(L) - 1),$$

and the bound is tight.

Assmann et al. also suggested a further improvement by defining a pretty sophisticated algorithm called Iterated Lowest Fit Decreasing (ILFD). To define this heuristic we first consider the following problem: given the list L and a fixed number M of bins, what is the *maximum* possible value for the minimum bin level in a packing of L into M bins? From a good heuristic A for this problem we can derive a good approximation algorithm for the bin covering problem by iteratively applying this algorithm A. Let $A(L, M)$ stand for the minimum bin level in the packing of L generated by the heuristic A when the number of bins is fixed to be M. Now the algorithm iteratively applies A and proceeds as follows:

   ITERATED "A"
     Step 1. Let $UB = \lfloor \sum_{i=1}^{n} s(a_i) \rfloor$, $LB = 1$. (Clearly $LB \leq \mathrm{OPT}(L) \leq UB$.)
     Step 2. While $UB - LB > 1$ take $M = \lfloor (LB + UB)/2 \rfloor$ and apply heuristic A. If $A(L, M) > 1$ take $LB = M$, otherwise $UB = M$.

The resulting algorithm provides a feasible solution to the bin covering problem with LB bins. Clearly, the performance of this method depends on the choice of A. Although the problem to be solved by A is closely related to multiprocessor scheduling problems, it seems natural to use for the heuristic A the

Lowest Fit Decreasing (*LFD*) algorithm, as studied by Graham [37] and Deuermeyer et al. [38]. This algorithm proceeds as follows:

Step 1. Order L so that $s(a_1) \geq s(a_2) \geq \ldots \geq s(a_n)$ and start with M empty bins.

Step 2. When there is an unpacked item in L do the following: let a_i be the first unpacked item and let B_j be the bin with a minimum level (in case of ties, choose the rightmost one). Put a_i into B_j.

It is not difficult to verify that the time complexity of *ILFD* is $O(n \log^2 n)$. Furthermore, one can prove the following result [2]:

$$R_{ILFD}^{\infty} = \frac{3}{4}.$$

Csirik et al. [39] achieved these bounds via simpler algorithms. Their first algorithm, called SI (Simple), first sorts the items in descending order, and then packs the prefix of the list by Next Fit into the next bin until the content becomes larger than 1 when the piece following the prefix is packed. It then fills the bin with a postfix, that is, packs the smallest items as long as the content of the bin is less than 1. For this algorithm

$$R_{SI}^{\infty} = \frac{2}{3}.$$

The second algorithm, Improved Simple (ISI) is a bit more complicated. It divides list L into three sublists:

- $s(a_1) \geq s(a_2) \geq \ldots \geq s(a_p) \geq 1/2$, (X-sublist),
- $1/2 > s(a_{p+1}) \geq s(a_{p+2}) \geq \ldots \geq s(a_r) \geq 1/3$, (Y-sublist),
- $s(a_{r+1}) \geq s(a_{r+2}) \geq \ldots \geq s(a_n)$, (Z-sublist).

Now, ISI proceeds in two phases. In Phase 1, if $s(a_1) \geq s(a_{p+1}) + s(a_{p+2})$ then packs a_1 into an empty bin, otherwise packs a_{p+1} and a_{p+2}. Then it fills the just opened bin with elements from the end of the Z-sublist, that is, with $a_n, a_{n-1}, \ldots$ until the content of the bin is larger than 1, removes the packed elements, and then repeats the packing until $X \cup Y$ or Z is empty. In the second phase, if $X \cup Y$ is empty, it packs the remaining elements from the Z sublist by Next Fit. Otherwise, if Z is empty, it packs the remaining elements from the X-sublist by two, from the Y-sublist by three. The authors subsequently proved that

$$R_{ISI}^{\infty} = \frac{3}{4}.$$

The time complexity of both algorithms here is $O(n \log n)$.

Coffman et al. [40] analyzed the bin covering problem when the list has divisible item sizes. They proved that the Next Fit Decreasing rule gives an optimal packing when the list is strongly divisible. They also proved that if the list is divisible, the IFFD rule gives an optimal packing as well.

Epstein et al. [41] investigated the bin covering problem with cardinality constraints when a parameter k indicates the minimum number of items in a covered bin. They provided that for any $k > 4$ the asymptotic competitive ratio (ACR) of any algorithm is at most $\frac{1}{\frac{3}{2} - \frac{2}{k}}$. This bound holds even when the items are given in sorted nonincreasing order with respect of their size. They present an algorithm with an ACR that can be made arbitrarily close to $\frac{1}{3 - \frac{2}{k}}$. This bound is tight for $k = 2$. For the semi-online variant with nondecreasing sizes they established the tight bound of $\frac{1}{2}$ for all k.

Dósa and He [42] defined the *bin covering with rejection* problem. Each item i has a penalty p_i and items can be rejected. When an item is rejected, the penalty must be paid. The objective is to maximize the number of covered bins plus the total penalty of all rejected items. For online case it is shown that no algorithm can have an absolute worst case ratio greater than 0. They presented for the online case an algorithm with an asymptotic worst case ratio of $\frac{1}{2}$ and showed that this is the best possible. For the offline case they presented an algorithm with absolute worst case ratio of $\frac{1}{2}$.

Deuermeyer et al. [38] studied the problem of machine covering. Here we have m machines (bins) and we want to pack the list so as to maximize the minimum load for the machines. This means, in bin covering terms the bin sizes would be maximized such that the packing of the list will cover all the bins. They investigated the performance of the longest processing time (LPT) heuristic and proved that its performance is not larger than 3/4. The tight ratio of this method was found by Csirik et al. [43] to be $\frac{3m-1}{4m-2}$.

Azar and Epstein [44] investigated the online version of machine covering. They proved that there is a randomized $O(\sqrt{m}\log m)$ competitive algorithm and any randomized algorithm is at least $\Omega(\sqrt{m})$ competitive. This is in marked contrast to the competitive ratio of the best possible deterministic algorithm, which is m. For the parametric version of this problem, where the sizes may vary up to a factor F they showed that there is a randomized $O(\log F)$ competitive algorithm. For the same problem where the optimal value is known in advance they offered a deterministic $2 - \frac{1}{m}$ competitive algorithm.

Csirik et al. [45] analyzed the sum minimization version of the online bin covering problem. Given m bins (of size 1), the goal is to cover all the bins. The objective function is the sum of the size of all the items packed when the goal is first fulfilled. They considered the online problem, and two semi-online models, when the items arrive sorted by size either nondecreasing or nonincreasing order. The offline problem is considered as well. For $m = 2$ they presented the best possible online and semi-online algorithms. The absolute competitive ratios are $\frac{5}{3}$ (for arbitrary list of items), $\frac{3}{2}$ (for lists of items sorted by nondecreasing order), and $\frac{6}{5}$ (for lists of items sorted by nonincreasing order). For $m > 2$ bins they presented two algorithms for the semi-online problems: for lists of items sorted in nondecreasing order an algorithm with competitive ratio of at most 1.931215, and for the other semi-online variant a simple algorithm of ACR of $\frac{4}{3}$. Lower bounds were derived for three variants: 1.387 for items given in arbitrary order, 1.302 in nondecreasing order, and 1.111 in nonincreasing order.

Woeginger and Zhang [46] considered bin covering with variable sized bins: there are several different types $B_1, \ldots, B_k$ of bins with sizes $1 = b_1 = s(B_1) > b_2 = s(B_2) > \ldots > b_k = s(B_k)$; there is an infinite supply of bins of each size. We will denote the set of feasible bin sizes here by **B**. The problem is to cover, with a given list of items $a_i \in [0, 1]$, a set of bins with the *largest total size*. Formally: let $\text{OPT}(L, \mathbf{B})$ and $A(L, \mathbf{B})$ denote, respectively, the total size of the bin cover produced by an optimum algorithm and the total size of the bin cover produced by an approximation algorithm A for an input list L and a set **B** of bin sizes. The asymptotic worst-case $R_{A,\mathbf{B}}^{\infty}$ of algorithm A for the set **B** is defined as

$$R_{A,\mathbf{B}}^{\infty} = \lim_{s \to \infty} \inf_L \left\{ \frac{A(L, \mathbf{B})}{\text{OPT}(L, \mathbf{B})} \mid \text{OPT}(L, \mathbf{B}) \geq s \right\}.$$

Woeginger and Zhang [46] determined, for each finite set of bin sizes, the worst case ratio of the best possible online covering algorithm. Let k_1 denote the number of bin sizes in **B** that are strictly greater than 1/2. Define

$$q(\mathbf{B}) = \max \left\{ \frac{b_j}{b_{j+1}} : 1 \leq j \leq k_1 - 1 \right\} \cup \{2b_{k_1}\}.$$

Notice that $q(\mathbf{B}) > 1$. Finally, define

$$r(\mathbf{B}) = \frac{1}{q(\mathbf{B})}.$$

Woeginger and Zhang proved that for every set **B** of bin sizes, there exists an online approximation algorithm A for variable-sized bin covering with asymptotic worst case ratio $r(\mathbf{B})$. For every set **B** this result is the best possible.

Epstein [47] gave similar results for the parametric case, that is, where each item size is less than or equal to $1/m$, where m is a positive integer. Similar to Reference 46, she defined the number $r(\mathbf{B}, m)$ in the following way. For each $1 \leq i \leq k$, let $b_{i,j} = b_i/j$ and let $B_i(m)$ be the set of fractions of b_i between sizes $1/(2m)$ and $1/m$, that is, $\mathbf{B}_i(m) = \{b_{i,j} \mid 1 \leq j \leq 2m\} \cap [1/(2m), 1/m]$. We define a new set of bins by $\mathbf{C}(m) = \cup_{1 \leq i \leq k} \mathbf{B}_i(m)$. Enumerate the sizes of numbers in $\mathbf{C}(m)$, $\mathbf{C}(m) = \{c_1, \ldots, c_l\}$ where $1/m = c_1 > c_2 > \ldots > c_l = 1/(2m)$. For every element in $\mathbf{C}(m)$, recall an original bin size that caused it to be inserted into $\mathbf{C}(m)$. For c_i, let $b(c_i)$ be the smallest b_j such that there exists an integer y that satisfies $yc_i = b_j$. Now define

$$q(\mathbf{B}, m) = \max \left\{ \frac{c_i}{c_{i+1}} \mid 1 \leq i \leq l-1 \right\}.$$

Note that $1 + 1/m \geq q(\mathbf{B}, m) > 1$. Finally, define

$$r(\mathbf{B}, m) = \frac{1}{q(\mathbf{B}, m)}.$$

For every finite set of bins and integer $m \geq 1$, Epstein gave a deterministic algorithm with competitive ratio $r(\mathbf{B}, m)$. She also proved that this is the best possible for any online randomized algorithms.

Zhang [48] defined a special variable-sized bin covering problem that he called the *bin-batching* problem: in this problem, in addition to a list L, a parameter $0 \leq t \leq 1$ is also given. We are asked to group the items into batches. If a batch B has a total content $c(B)$, the gain of this batch is defined as

$$g(B) = \begin{cases} 0 & \text{if } c(B) < 1 - t, \\ c(B) & \text{if } 1 - t \leq c(B) \leq 1, \\ 1 & \text{if } c(B) > 1. \end{cases}$$

The goal is to make the total gain as large as possible. We may note that this problem reduces to the bin covering problem when $t = 0$. In bin-batching all bin sizes in $[1 - t, 1]$ are allowed, that is, we allow underpacking of bins up to a level of $1 - t$. It is clear that the problem is simple if $t \geq 1/2$, so we may assume $t < 1/2$. Zhang proved that every online algorithm for the bin-batching problem has an asymptotic worst-case ratio of at most $1/(2 - 2t)$, and that the following simple algorithm—called SA—will meet this bound: assume that the currently incoming item is a_i; if $s(a_i) \geq 1 - t$, put it into an empty bin with size $s(a_i)$ and close this bin immediately. If $s(a_i) < 1 - t$, we apply the Next Fit algorithm with bin size 1.

Woeginger and Zhang [46] looked at the special problem of having only two different bin sizes (this was done by Csirik [6] for the classical variable-sized problem). They proved that, for a set $\mathbf{B} = \{1, b\}$ with two bin sizes, there exists an online approximation algorithm A for the bin covering problem with the best possible asymptotic worst case ratio

$$R_{A,\mathbf{B}}^\infty = \begin{cases} \frac{1}{2} & \text{if } 0 < b \leq \frac{1}{2}, \\ b & \text{if } \frac{1}{2} \leq b \leq \frac{1}{\sqrt{2}}, \\ \frac{1}{2b} & \text{if } \frac{1}{\sqrt{2}} \leq b < 1. \end{cases}$$

For $\mathbf{B} = \{1, 1/\sqrt{2}\}$, the asymptotic worst-case ratio is $1/\sqrt{2}$. This is the best ratio that can be achieved with two bins.

Du et al. [49] defined the Min-Max Bin Covering Problem. The items are integers and the bin capacity is denoted by k. The problem is to partition $\{1, 2, \ldots, n\}$ into subsets S_j, satisfying for each j, $\sum_{i \in S_j} s(a_i) \geq k$, objective function is to minimize $\max_j \sum_{i \in S_j} s(a_i)$. A polynomial time algorithm with an (absolute) approximation factor of $\frac{5}{2}$ is presented.

30.4 Conclusion

In this chapter we conclude our survey of variants of the classical bin packing problem. The main topics covered include the variable sized bin packing problem and the bin covering problem. Both problems have generated tremendous interests in the past and promise to have more results in the future.

References

1. Friesen, D.K., Langston, M.A., Variable sized bin packing, *SIAM J. Comput.*, 15, 222, 1986.
2. Assmann, S.F., Johnson, D.S., Kleitman, D.J., Leung, J.Y.-T., On a dual version of the one-dimensional bin packing problem, *J. Algorithms*, 5, 502, 1984.
3. Friesen, D.K., Langston, M.A., A storage-selection problem, *Inf. Proc. Lett.*, 18, 295, 1984.
4. Kinnersley, N.G., Langston, M.A., Online variable-sized bin packing, *Disc. Appl. Math.*, 22, 143, 1988.
5. Zhang, G., Worst-case analysis of the FFH algorithm for online variable-sized bin packing, *Computing*, 56, 165, 1996.
6. Csirik, J., An on-line algorithm for variable-sized bin packing, *Acta Inf.*, 26, 697, 1989.
7. Seiden, S.S., An optimal online algorithm for bounded space variable-sized bin packing, *SIAM J. Disc. Math.*, 458, 2001.
8. Seiden, S.S., van Stee, R., Epstein, L., New bounds for variable-sized online bin packing, *SIAM J. Comput.*, 33, 455, 2003.
9. Seiden, S.S., An optimal online algorithm for bounded space variable-sized bin packing, in *Proceedings of International Colloquium on Automata, Languages and Programming* (ICALP), LNCS, Geneva, Switzerland, Vol. 1853, 2000, p. 283.
10. Epstein, L., Seiden, S., van Stee, R., New bounds for variable-sized and resource augmented online bin packing, *Proceedings of International Colloquium on Automata, Languages and Programming* (ICALP), LNCS, Malaga, Spain, Vol. 2380, 2002, p. 306.
11. Burkard, R.E., Zhang, G., Bounded space on-line variable-sized bin packing, *Acta Cybern.*, 13, 63, 1997.
12. Csirik, J., Johnson, D.S., Bounded space on-line bin packing: Best is better than first, *Algorithmica*, 31, 115, 2001.
13. Chu, C., La, R., Variable-sized bin packing: Tight absolute worst-case performance ratios for four approximation algorithms, *SIAM J. Comp.*, 30, 2069, 2001.
14. Zhang, G., A new version of on-line variable-sized bin packing, *Disc. Appl. Math.*, 72, 193, 1997.
15. Boyar, J., Favrholdt, L.M., Scheduling jobs on grid processors, *Algorithmica*, 57(4), 819, 2010.
16. Boyar, J., Ellen F., Tight bounds for an online bin packing problem, *CoRR*, abs/1404.7325, 2014.
17. Epstein, L., Favrholdt, L.M., Levin, A., Online variable-sized bin packing with conflicts, *Dis. Opt.*, 8(2), 333, 2011.
18. Boyar, J., Favrholdt, L.M., A new variable-sized bin packing problem, *J. Scheduling*, 15(3), 273, 2012.
19. Dell'Olmo, P., Speranza, M.G., Approximation algorithms for partitioning small items in unequal bins to minimize the total size, *Disc. Appl. Math.*, 94, 181, 1999.
20. Ye, D., Zhang, G., On-line extensible bin packing with unequal bin sizes, in *Proceedings of Workshop on Approximation and Online Algorithms* (WAOA), LNCS, Bergen, Norway, Vol. 2909, 2004, p. 235.
21. Kang, J., Park, S., Algorithms for the variable sized bin packing problem, *Eur. J. Oper. Res.*, 147(2), 365, 2003.
22. Xing, W., A bin packing problem with over-sized items, *Oper. Res. Lett.*, 30, 83, 2002.
23. Naaman, N., Rom, R., Packet scheduling with fragmentation, in *INFOCOM*, New York, 2002, p. 537.

24. Langston, M.A., Performance of heuristics for a computer resource allocation problem, *SIAM J. of Alg. Disc. Meth.*, 5, 154, 1984.
25. Coffman, Jr., E.G., Leung, J.Y.-T., Ting, D.W., Bin packing: Maximizing the number of pieces packed, *Acta Informatica*, 9, 263, 1978.
26. Friesen, D.K., Kuhl, F.S., Analysis of a hybrid algorithm for packing unequal bins, *SIAM J. Comput.*, 17, 23, 1988.
27. Friesen, D.K., Langston, M.A., Analysis of a compound bin-packing algorithm, *SIAM J. Disc. Math.*, 4, 61, 1991.
28. Epstein, L., Favrholdt, L.M., On-line maximizing the number of items packed in variable-sized bins, in *Proceedings of Conference on Computing and Combinatorics (COCOON)*, LNCS, Singapore, Singapore, Vol. 2387, 2002, p. 467.
29. Epstein, L., Favrholdt, L.M., On-line maximizing the number of items packed in variable-sized bins, *Acta Cybern.*, 16, 57, 2003.
30. Assmann, S.F., Problems in discrete applied mathematics, PhD Thesis, Department of Mathematics, MIT, Cambridge, MA, 1983.
31. Csirik, J., Totik, V., On-line algorithms for a dual version of bin packing, *Disc. Appl. Math.*, 21, 163, 1988.
32. Christ, M.G., Favrholdt, L.M., Larsen, K.S., Online bin covering: Expectations vs. guarantees, *Theor. Comp. Sci.*, 556, 71, 2014.
33. Boyar, J., Favrholdt, L.M., The relative worst order ratio for online algorithms, *ACM Trans. Algorithms*, 3 (2), article 22, 2007.
34. Ben-David, S., Borodin, A., A new measure for the study of on-line algorithms, *Algorithmica*, 11 (1), 73, 1994.
35. Kenyon, C., Best-fit bin-packing with random order, *Proceedings of Seventh Symposium Discrete Algorithms (SODA)*, Atlanta, Georgia, 359, 1996.
36. Coffman, Jr., E.G., Lueker, G.S., *An Introduction to the Probabilistic Analysis of Packing and Partitioning Algorithms*, Wiley & Sons, New York, 1991.
37. Graham, R.L., Bounds on multiprocessing timing anomalies, *SIAM J. Appl. Math.*, 17, 416, 1969.
38. Deuermeyer, B.L., Friesen, D.K., Langston, M.A., Scheduling to maximize the minimum processor finish time in a multiprocessor system, *SIAM J. Alg. Disc. Meth.*, 3, 190, 1982.
39. Csirik, J., Frenk, J.B.G., Labbé, M., Zhang, S., Two simple algorithms for bin covering, *Acta Cybern.*, 14, 13, 1997.
40. Coffman, Jr., E.G., Garey, M.R., Johnson, D.S., Bin packing with divisible item sizes, *J. Complexity*, 3, 405, 1987.
41. Epstein, L., Imreh, Cs., Levin, A., Bin covering with cardinality constraints, *Disc. Appl. Math.*, 161(13–14), 1973, 2013.
42. Dósa, Gy., He, Y., Bin packing problems with rejection penalties and their dual problems, *Inf. Comp.*, 204, 795, 2006.
43. Csirik, J., Kellerer, H., Woeginger, G., The exact LPT-bound for maximizing the minimum completion time, *Oper. Res. Lett.*, 11, 281, 1992.
44. Azar, Y., Epstein, L., On-line machine covering, in *European Symposium Algorithms*, LNCS, Graz, Austria, Vol. 1284, 1997, p. 23.
45. Csirik, J., Epstein, L., Imreh, C.S., Levin, A., On the sum minimization version of the online bin covering problem, *Disc. Appl. Math.*, 158(13), 1381, 2010.
46. Woeginger, G.J., Zhang, G., Optimal on-line algorithms for variable-sized bin covering, *Oper. Res. Lett.*, 25, 47, 1999.
47. Epstein, L., On-line variable sized covering, *Inf. Comp.*, 171, 294, 2001.
48. Zhang, G., An on-line bin batching problem, *Disc. Appl. Math.*, 108, 329, 2001.
49. Du, W., Eppstein, D., Goodrich, M.T., Lueker, G.S., On the approximability of geometric and geographic generalization and the min-max bin covering problem, In *WADS*, LNCS, Banff, Alberta, 5664, 2009, p. 242.

31

Multidimensional Packing Problems

Leah Epstein

Rob van Stee

31.1 Introduction

There are several ways to generalize the bin packing problem to more dimensions. We consider two- and three-dimensional strip packing, and bin packing in dimensions two and higher. Finally we consider vector packing and several other variations. In this chapter we consider only algorithmic aspects (as opposed to combinatorial aspects and game theoretical aspects).

In the most common two-dimensional version, the items are rectangles or squares, and the bins are unit squares (where items are to be packed in an nonoverlapping way). In the strip packing problem, instead of bins, we are given a strip of width 1 and unbounded height. In higher dimensions, the rectangles are replaced by boxes (or hyperboxes), the squares by cubes (or hypercubes), and the unit square by a unit cube (or a hypercube of the relevant dimension). Strip packing becomes column packing.

We start by mentioning an interesting difference between one-dimensional bin packing and its multidimensional generalizations. Although for a single dimension, offline algorithms clearly outperform online algorithms, this is not always the case in more dimensions in the sense that there are several cases where an online algorithm is or was at one point the best known approximation algorithm. Most likely, this simply reflects the fact that after many years of study, we still do not understand the multidimensional case as well as the one-dimensional case. One the other hand, some results simply cannot be generalized. For instance, for offline problems, it is known that there cannot be an asymptotic polynomial approximation scheme (APTAS) for two-dimensional bin packing [1], or for two-dimensional vector packing [2], and for online problems, there cannot be an algorithm with an asymptotic constant approximation ratio for all dimensions simultaneously [3].

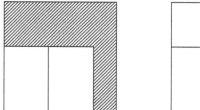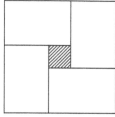

FIGURE 31.1   A comparison between the oriented and the rotatable models.

An important special case in multidimensional bin and strip packing is the case where squares or (hyper-)cubes need to be packed. For this case, often better results are known than for those of the more general case of rectangles or boxes. In particular, the offline version of packing cubes into cubes (or squares into squares) admits an APTAS [4].

As is the case for one-dimensional bin packing, much of the attention has gone to the asymptotic worst-case ratio, but in the course of this chapter we will encounter some results on the absolute ratio as well.

When packing of rectangles or boxes is considered, there are several ways to define the problem. In the oriented problem, items have a fixed orientation, and cannot be rotated (this orientation is usually axes-parallel). In the rotatable (or nonoriented) version, an item can be rotated and placed in any position such that its sides are parallel to the sides of the bin. Finally, there are mixed versions where items can be rotated in certain directions, but not all directions. One such three-dimensional model where items can be rotated to the left or to the right but the top and bottom must remain such is the "z-oriented" packing [5,6] studied by Miyazawa and Wakabayashi, also known as the "This Side Up" problem [7].

An illustration of the difference between the oriented and rotatable variants is given in Figure 31.1. In this figure we see packings of rectangles of sides $\frac{3}{5}$ and $\frac{2}{5}$. If the rectangles are oriented so that their height is $\frac{3}{5}$ and they cannot be rotated, we can pack at most two such items in one bin. However, if rotation is allowed, we can pack as much as four such rectangles together in one bin.

This chapter is organized as follows. We begin by presenting the algorithm Next Fit Decreasing Height (NFDH), which is a fundamental algorithm for two-dimensional packing problems, in Section 31.2. We then discuss results on multidimensional packing problems, in order of increasing dimension. That is, we start with strip packing in Section 31.3 and move to two-dimensional bin packing in Section 31.4. We then discuss column packing in Section 31.5 and three- and more-dimensional bin packing in Section 31.6. Finally, we mention results on vector packing in Section 31.7 and discuss several variations on multidimensional packing in Section 31.8.

31.2 Next Fit Decreasing Height

In 1968, Meir and Moser [8] introduced an algorithm for packing d-dimensional cubes into a d-dimensional hyperbox, which they called Next Fit Decreasing (NFD). This algorithm sorts the cubes by decreasing volume and packs them into layers. The authors show that if the sides of the cubes are denoted by $x_1, x_2, \ldots,$ and they are packed into a hyperbox of sides $a_1, \ldots, a_d$ where $x_1 \leq a_i$ for $i = 1, \ldots, d$, then the cubes can be packed into the hyperbox as long as their total volume is at most $x_1^d + \prod_{i=1}^{d}(a_i - x_1)$.

For $d = 2$ (packing squares into a rectangle), the algorithm works as follows. The largest square is put in the bottom left corner of the rectangle. The height of the first layer is equal to the side of this square. The next squares are put in this layer one by one, next to each other, and touching each other and the bottom of the layer, until one does not fit. At this point we define a new layer above the first layer, with height equal to the side of the first square packed into it. This continues until all squares are packed, or there is not enough room to pack some item (it does not fit into the current layer, and the last layer that

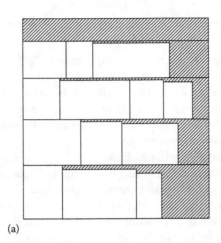
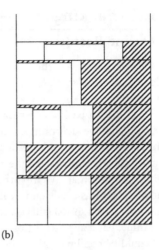

FIGURE 31.2   An illustration of a packing of NFDH (a) and of a shelf packing algorithm (b).

is left is either empty or not high enough). It is possible to define a version of this algorithm called First Fit Decreasing (FFD), where one tries to pack each square not only into the current layer. FFD tests all previous layer in the order they were created until a suitable place is found. If none is found, a new layer is used. FFD is less efficient than NFD, but it can perform better on some inputs.

This algorithm (for two dimensions) was extended to an algorithm for packing rectangles into a rectangle (or a strip) by Coffman et al. [9], which was called Next Fit Decreasing Height (NFDH). It sorts the rectangles by decreasing height and then packs them as aforementioned. Obviously, First Fit Decreasing Height (FFDH) can be used instead. They showed that if this algorithm is applied to pack rectangles into a strip (of unbounded height), then the height used to pack the rectangles is at most twice the optimal height, plus an additive constant which is equal to the height of the highest rectangle. (Thus its absolute worst-case ratio is 3.) See Figure 31.2 for an illustration for NFDH.

The proof is quite straightforward. In each level, there may be wasted space to the right of the rightmost item, and above all items except the first. The height of a level is the height of the first item in it. This item did not fit on the previous level. This implies that the total area of the items in level i plus the first item in level $i + 1$ is at least the area of level $i + 1$. That is, if we move all items in level i up to level $i + 1$, and shift the first item in level $i + 1$ to the right, then level $i + 1$ is entirely covered by items. This implies that the sum of the heights of all levels is upper bounded by twice the area of the packed items plus the height of the first level. This explains the performance guarantee bound including the additive constant, as the total area is an obvious lower bound for the optimal required height.

This fundamental algorithm was used in many later papers as a subroutine. It works especially well when all rectangles are guaranteed to have a small width (relative to the width of the strip), and this property was for instance used by Kenyon and Rémila [10] in their approximation scheme for strip packing.

Meir and Moser also showed the following important result in the same paper [8].

Theorem 31.1 *Any set of rectangles with sides at most x and total area A can be packed into any rectangle of size $a \times b$ if $a \geq x$ and $ab \geq 2A + a^2/8$. This result is the best possible (in the sense that the bound on the area cannot be decreased for at least one example).*

For packing rectangles into a unit square, this result states that any set of rectangles of total area at most $7/16$ (and sides not larger than 1) can be packed into a unit square. Moon and Moser [11] showed that for squares, it is possible to pack any set of area at most $1/2$, and online algorithms with constant (but smaller) area guarantees of 3 and $\frac{32}{11} \approx 2.9091$, were designed as well [12,13].

31.3 Strip Packing

31.3.1 Online Results

Baker and Schwarz [14] were the first to study two-dimensional online strip packing. They introduced a class of algorithms called *shelf algorithms*. A shelf algorithm uses a one-dimensional bin packing algorithm A and a parameter $\alpha \in (0, 1)$. Items are classified by height: an item is in class s if its height is in the interval $(\alpha^{s-1}, \alpha^s]$. Each class is packed in separate *shelves*, where we use A to fill a shelf and open a new shelf when necessary. Note that the algorithm A is not necessarily on-line. See Figure 31.2 for an illustration of a shelf algorithm. This partition into classes is also called *geometric rounding*.

Baker and Schwarz showed that the algorithm First Fit Shelf, which uses First Fit as a subroutine, has an asymptotic approximation ratio arbitrarily close to 1.7. Csirik and Woeginger [15] showed that by using Harmonic [16] as a subroutine, it is possible to achieve an asymptotic approximation ratio arbitrarily close to $h_\infty \approx 1.69103$ (this is the sum of a series). Moreover, they show that any shelf algorithm, online or offline, has a approximation ratio of at least h_∞. The idea of the lower bound is that items are given that could be combined nicely next to each other, but due to their heights, they are classified into different height classes and are therefore packed in separate shelves. So basically, the best thing one can do is to use a bounded space algorithm (which has a constant number of simultaneously active bins) such as Harmonic as the subroutine. Finally they mention that the one-dimensional lower bound of van Vliet [17] together with the insights of Baker et al. [18] imply a general lower bound for online algorithms of 1.5401. The upper bound was improved by Han et al. [19]. They packed items of similar width into batches of sufficiently large height. This enabled them to use a one-dimensional algorithm of a class they call "Harmonic-based algorithms" as a subroutine and achieve the same approximation ratio as the one-dimensional algorithm they use, avoiding the aforementioned problem of shelf algorithms. This means that the approximation ratio was immediately reduced to approximately 1.58889 based on Seiden [20]. As for the absolute approximation ratio, Brown et al. [21] showed a lower bound of 2 for any algorithm. They also show some lower bounds for algorithms that may sort the items.

31.3.2 Offline Results

The strip packing problem was introduced in 1980 by Baker et al. [22]. They developed the first offline approximation algorithms for this problem, and give an upper bound of 3 on the absolute performance guarantee. This bound was later improved to 2 independently by Schiermeyer [23] and by Steinberg [24], using different approaches. The barrier of 2 was first broken by Harren and van Stee [25], who gave an upper bound of 1.94. This result was then improved to $5/3 + \varepsilon$ by Harren et al. [26]. Very recently Nadiradze and Wiese [27] gave a pseudo-polynomial time algorithm that achieves a ratio of $1.4 + \varepsilon$. A straightforward reduction from Partition shows that (under standard complexity assumptions) it is not possible to obtain an algorithm whose absolute approximation ratio is strictly below $3/2$ in polynomial time.

In the same issue of *SIAM Journal on Computing* [22], Coffman et al. [9] proved upper bounds on the asymptotic asymptotic approximation ratio of several algorithms, those are bounds of 2 for NFDH, 1.7 for FFDH, and $\frac{3}{2}$ for an algorithm called Split-Fit. Also in 1980, Sleator [28] gave an algorithm with asymptotic approximation ratio 2.5, but absolute approximation ratio of 2, which is better than the absolute approximation ratio of at most 3 for Split-Fit. In 1981, Baker et al. [18] gave an offline algorithm with asymptotic worst case ratio $5/4$. Finally, Kenyon and Rémila [10] designed an asymptotic fully polynomial time approximation scheme.

This scheme uses some nice ideas, which we describe as follows.

Fractional strip packing: A fractional strip packing of L is a packing of any list L' obtained from L by subdividing some of its rectangles by horizontal cuts: each rectangle (w_i, h_i) is replaced by a sequence of rectangles $(w_i, h_i^1), (w_i, h_i^2), \ldots, (w_i, h_i^{k_i})$ such that $\sum_{j=1}^{k_i} h_i^j = h_i$.

In the case that L contains only items of m distinct widths in $(\varepsilon', 1]$, where $\varepsilon' > 0$ is some constant, it is possible to find a fractional strip packing of L which is within 1 of the optimal fractional strip packing $\text{FSP}(L)$ in polynomial time. Moreover, it is possible to turn this packing into a regular strip packing at the loss of only an additive constant $2\,m$. Denote the height of the optimal strip packing for L by $\text{OPT}(L)$. We conclude that we find a packing with height at most $\text{FSP}(L) + 1 + 2\,m \le \text{OPT}(L) + 2\,m + 1$.

Modified NFDH: This is a method for adding narrow items (items of width at most ε') to a packing of wide items such as described earlier. Such a packing leaves empty rectangles at the right hand side of the strip. Each of these rectangles is packed with narrow items using NFDH (starting with the highest narrow item in the first rectangle). When all rectangles have been used, the remaining items (if any) are packed above the packing using NFDH on the entire width of the strip.

Grouping and rounding: This method is a variation on the linear rounding defined by de la Vega and Lueker [29]. It works as follows.

We stack up the rectangles of L by order of nonincreasing widths to obtain a left-justified stack of total height $h(L)$. We define $m - 1$ threshold rectangles, where a rectangle is a threshold rectangle if its interior or lower boundary intersects some line $y = ih(L)/m$ for some $i \in \{1, \ldots, m - 1\}$. We cut these threshold rectangles along the lines $y = ih(L)/m$. This creates m groups of items that have height exactly $h(L)/m$.

First, the widths of the rectangles in the first group are rounded up to 1, and the widths of the rectangles in each subsequent group are rounded up to the widest width in that group. This defines L_+.

Second, the widths of the rectangles in each group are rounded down to the widest width of the next group (down to 0 for the last group). This defines L_-.

It is easy to find a strip packing for L_- using a reduction to fractional strip packing. Moreover, it can be seen that the stack associated with L_+ is exactly the union of a bottom part of width 1 and height $h(L)/m$ and the stack associated with L_-. Thus $\text{FSP}(L) \le \text{FSP}(L_+) = \text{FSP}(L_-) + h(L)/m$.

Partial ordering: We say that $L \le L'$ if the stack associated to L (used for the above-mentioned grouping), viewed as a region of the plane, is contained in the stack associated to L'. Note that $L \le L'$ implies that $\text{FSP}(L) \le \text{FSP}(L')$. As an example, in the above-mentioned grouping we have $L_- \le L \le L_+$.

31.3.3 Rotations

The asymptotic upper bound of 2 on the approximation ratio of NFDH and Bottom Leftmost Decreasing Width (BLDW) remain valid if orthogonal rotations are allowed, as the proofs use only area arguments. Miyazawa and Wakabayashi [6] presented an algorithm with asymptotic approximation ratio of 1.613. Epstein and van Stee [7] improved this to 3/2 using a simpler algorithm. This algorithm packs items that are wider and higher than 1/2 optimally, and packs remaining items first next to this packing (where possible) and finally on top of this packing. In this way, the resulting packing is either optimal, or at almost all heights a width of at least 2/3 is occupied by items. Finally, an asymptotic fully polynomial approximation scheme (AFPTAS) was given by Jansen and van Stee [30].

31.4 Two-Dimensional Bin Packing

We saw in Section 31.3.1 that in some cases we can use a one-dimensional bin packing algorithm as a subroutine for a strip packing algorithm, basically without a loss in (asymptotic) performance guarantee. Similarly, a two-dimensional bin packing algorithm can be used as a subroutine to create a three-dimensional strip packing algorithm, and this also holds for higher dimensions.

On the other hand, a d-dimensional strip packing algorithm can also be used to create a d-dimensional bin packing algorithm at a cost of a factor of two in the approximation ratio. The idea is to cut the packing generated by the strip packing algorithm into pieces of unit height. For each piece we do the following. Items that are completely contained in the piece are put together in one bin. Items that are partially in the next piece are put together in a second bin. See Figure 31.3.

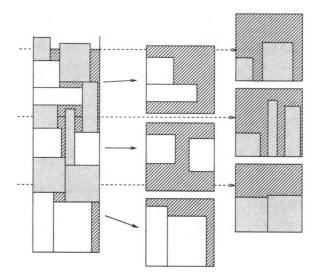

FIGURE 31.3   Converting a packing in a strip into a packing in bins.

Assume that we have a guarantee of R on the asymptotic approximation ratio of the strip packing algorithm. Then this method gives us $2R \cdot \text{OPT}(L) + C$ bins for an input L, where $\text{OPT}(L)$ is the height of the optimal strip packing. On the other hand, there cannot be a *bin* packing into less than $\text{OPT}(L)$ bins, because this packing could be trivially turned into a strip packing of height less than $\text{OPT}(L)$. This explains the factor of two loss.

31.4.1 Online Results

Coppersmith and Raghavan were the first to study the online version of this problem. They gave an online algorithm with asymptotic approximation ratio of 3.25 for $d = 2$ (and 6.25 for $d = 3$) [31]. This result was improved by Csirik et al., who presented an algorithm with approximation ratio 3.0625 [32]. In the same year, Csirik and van Vliet showed an online bin packing algorithm for arbitrary dimensions, that achieves a approximation ratio of h_∞^d, where d is the dimension [33]. Note that already for $d = 2$, this improves over the previous result, as $h_\infty^2 \approx 2.85958$. (See also [34] for $d = 2, 3$.) Seiden and van Stee [35] gave an algorithm with ratio 2.66013 for two-dimensional bin packing. Finally, Han et al. [36] gave an upper bound of 2.55 by using a Super Harmonic (instead of plain Harmonic) algorithm in the first dimension. It remains an open question as to whether Super Harmonic could be applied in both dimensions for an even better result.

Epstein and van Stee [37] introduced a new technique for packing small multidimensional items online, enabling them to achieve the asymptotic approximation ratio of h_∞^d [33] using only bounded space.

Galambos [38] was the first to give a lower bound for this problem which was higher than the best known lower bound for one-dimensional bin packing. His bound was 1.6. This was later successively improved to 1.808 by Galambos and van Vliet [39], 1.851 by van Vliet [40], and finally to 1.907 by Blitz et al. [41,42]. The gap between the upper and lower bounds remains relatively large to this day, and it is unclear how to improve either of them significantly.

Packing squares: An interesting special case is where all items are squares. Coppersmith and Raghavan [31] showed that their algorithm has an asymptotic approximation ratio of 2.6875 for this case, and gave a lower bound of 4/3 on the asymptotic approximation ratio for any algorithm. This lower bound actually holds for the more general problem of packing hypercubes. Seiden and van Stee [35] showed that the algorithm Harmonic × Harmonic, which uses the Harmonic algorithm [16] to find slices for items, and then uses the Harmonic algorithm again to find bins for slices, has an asymptotic approximation ratio of at most 2.43828. They gave a lower bound of 2.28229 asymptotic approximation ratio for bounded space algorithms.

Epstein and van Stee [43] give an algorithm with asymptotic approximation ratio at most 2.24437. Han et al. [44] claimed that the upper bound can be improved further by roughly 0.1 by using Super Harmonic in the first dimension. Finally, Epstein and van Stee [45] give bounds for the performance of their optimal bounded space algorithm from [37], showing that its approximation ratio lies between 2.3638 and 2.3692. The best known lower bound on the asymptotic approximation ratio for arbitrary online algorithms is 1.68025 [41]. The gap between the lower and the upper bounds for arbitrary online algorithms remains disappointingly large.

31.4.2 Offline Results

As mentioned at the start of this chapter, Bansal et al. [4] proved that the two-dimensional bin packing problem is APX-hard. Thus, there cannot be an asymptotic polynomial time approximation scheme for this problem. Chung et al. [46] were the first to give an approximation algorithm for this problem. It has an asymptotic approximation ratio of 2.125. As mentioned earlier, the APTAS for strip packing by Kenyon and Rémila implies a $(2 + \varepsilon)$-approximation for any $\varepsilon > 0$. In 2002, Caprara [47] gave a h_∞-approximation. The upper bound on the asymptotic approximation ratio was improved to 1.52 by Bansal et al. [48] and then to $1.5 + \varepsilon$ by Jansen and Prädel [49]. More recently, Bansal and Khan [50] gave a $(1.405 + \varepsilon)$-approximation.

Packing squares: Leung et al. [51] proved that the special case of packing squares into squares is still NP-hard (for general two-dimensional bin packing, this follows immediately from the one-dimensional case). Ferreira et al. [52] gave a 1.988-approximation for this problem, which uses as a subroutine an optimal algorithm for packing items with sides larger than 1/3. They conjecture that packing items with sides larger than 1/4 is already NP-hard. Independently of each other, Kohayakawa et al. [53] and Seiden and van Stee [35] gave a $(14/9 + \varepsilon)$-approximation ($1.5555 \cdots + \varepsilon$). However, the first result is more general in that it actually gives a $(2 - (2/3)^d + \varepsilon)$-approximation for packing d-dimensional hypercubes. The idea of both these algorithms is to find an optimal packing for large items (items with sides larger than ε) and to add the small items to this packing. Specifically, any bins in the optimal packing which contain only a single item with sides larger than 1/2 are filled with small items using the algorithm Next Fit Decreasing (NFD) from Meir and Moser (see Section 31.2). It is shown that all other bins are already "reasonably full," leading to the approximation guarantee.

In the same year, Caprara [47] gave an algorithm with approximation ratio in the interval $(1.490, 1.507)$ provided a certain conjecture holds. Two years later, Epstein and van Stee [54] gave a $(16/11 + \varepsilon)$-approximation ($1.4545 \cdots + \varepsilon$). Simultaneously and independently of each other, Bansal and Sviridenko [1] and Correa and Kenyon [55] presented an asymptotic polynomial time approximation scheme for this problem, which also works for the more general problem of packing hypercubes. That general problem was shown to be NP-hard as well by Lu et al. [56,57].

Guillotine cuts: Bansal et al. [58] considered a variant of the problem which admits an APTAS. This is rectangle packing, where the packing of each bin must be possible to achieve using guillotine cuts only. That is a sequence of edge to edge cuts, parallel to the edges of the bin. The case where the number of stages in the sequence of guillotine cuts is limited was studied by Caprara et al. [59]. They designed an APTAS for the two-stage problem. Note the shelf packing described previously actually uses two stages of guillotine cuts. Kenyon and Rémila [10] point out that their approximation scheme uses five stages of guillotine cuts. Seiden and Woeginger [60] observed that it can be readily adapted to use only three stages. They showed more generally that the asymptotic worst case ratio of strip packing versus its k-stage version is 1 for any $k > 2$, leading to an APTAS for the latter. Recently, Abed et al. [61] studied other related packing problems under guillotine cuts.

Absolute approximations: As regards the absolute performance guarantee, Zhang [62] gave an approximation algorithm with absolute worst-case ratio of 3 for two-dimensional bin packing. Van Stee [63] gave an absolute 2-approximation for the special case where squares need to be packed, which is optimal by the result of Leung et al. [51]. This result was generalized to bin packing with rotations by Harren and van Stee [64] and finally to the case without rotations by Harren et al. [65].

31.4.3 Resource Augmentation

As there cannot be an approximation scheme for general two-dimensional bin packing, several authors have looked at the possibility of resource augmentation, that is, giving the approximation algorithm slightly larger bins than the offline algorithm that it is compared to. Correa and Kenyon [55] give a dual polynomial time approximation scheme. That is, they give a polynomial time algorithm to pack rectangles into the k bins of size $1 + \varepsilon$, where these rectangles cannot be packed in less than k bins of size 1. Bansal and Sviridenko [66] showed that it is possible to achieve this even if the size of the bin is relaxed in one dimension only.

31.4.4 Rotations

For the case where rotations are allowed, Epstein [67] showed an online algorithm with asymptotic approximation ratio of 2.45. The online problem was studied before by Fujita and Hada [68]. They presented two online algorithms and claimed asymptotic approximation ratios of at most 2.6112 and 2.56411. Epstein [67] mentioned that the proof in Reference 68 only shows that the first algorithm has an asymptotic approximation ratio of at most 2.63889 and that the proof of the second algorithm is incomplete.

Two years later, Miyazawa and Wakabayashi [6] gave an offline algorithm with asymptotic approximation ratio 2.64. Epstein and van Stee [37] gave an approximation algorithm with asymptotic approximation ratio 2.25. It divides the items into types and combines them into bins such that in almost all bins, an area of 4/9 is occupied. Correa [69] adapted the dual polynomial time approximation scheme from Correa and Kenyon [55] to rotatable items. The APTAS of Jansen and van Stee [30] implies a $(2 + \varepsilon)$-approximation. This was improved to 1.52 by Bansal et al. [48] and finally to 1.5 by Jansen and Prädel [49].

31.5 Column (Three-Dimensional Strip) Packing

31.5.1 Online and Offline Results

Li and Cheng [70] were the first to consider this problem. In their paper from 1990, they showed that three-dimensional versions of NFDH and FFDH have unbounded worst-case ratio. They gave several approximation algorithms, the best of which has an asymptotic approximation ratio of 3.25. Their first algorithm sorts the items by height and then divides them into groups of area (in the first two dimensions) at most 7/16, so that they can be packed into a single layer by Theorem 31.1. They improve on this by classifying items with similar bottoms, and packing similar items together into layers. Two items have similar bottoms if both their length and their width fall into the same class when classified by the Harmonic algorithm. For the case where all items have square bottoms, the ratio improves to 2.6875.

Two years later, the same authors [71] presented an online algorithm with asymptotic approximation ratio arbitrarily close to $h_\infty^2 \approx 2.89$ for three-dimensional strip packing. At the time, there was no better *offline* approximation known. This algorithm uses the Harmonic algorithm as a subroutine in both horizontal dimensions (i.e., to find a strip for a two-dimensional item, and a place inside a strip for a one-dimensional item), and a geometric rounding for the heights. The paper actually discusses several online algorithms for this problem and only mentions the use of Harmonic in the summary section. The authors note that the improvement in the asymptotic approximation ratio compared to the approximation algorithm from their earlier paper [70] comes at the cost of a high additive constant.

In 1997, Miyazawa and Wakabayashi [72] improved the offline upper bound on the asymptotic approximation ratio to 2.66994 (2.36 for items with square bottoms). This algorithm places columns of similar items next to eachother in the strip, thus avoiding the layer structure of the previous algorithms. This result was improved to h_∞ by Bansal et al. [73] and then to 1.5 by Jansen and Prädel [74].

31.5.2 Rotations

In the case where rotations are allowed, it becomes relevant what exactly the dimensions of the strip are. In two-dimensional strip packing, this does not really play a part (unless rectangles with heights that can exceed 1 are to be packed), but in column packing, the base of the column might not be a square. However, if the base is not a square but may be an arbitrary rectangle, then having the ability to rotate items horizontally (leaving the top side unchanged) does not help, as was shown by Miyazawa and Wakabayashi [5]. The idea is that in this case it is possible to scale the input so that the smallest width of an item is still larger than the length of the base of the strip, so that no item can be rotated and still fit inside the strip. For this reason, in this section we focus on the case where the base of the strip is a square.

Epstein and van Stee [7] give an approximation algorithm with asymptotic worst-case ratio of $9/4 = 2.25$, improving on the upper bound of 2.76 by Miyazawa and Wakabayashi [6]. The special case where only rotations that leave the top side of items at the top are allowed has received more attention. It was introduced by Li and Cheng [75] as a model for a job scheduling problem in partitionable mesh connected systems. Here each job i is given by a triple (x_i, y_i, t_i), meaning that job i requires a submesh of dimensions $x_i \times y_i$ or $y_i \times x_i$ for t_i time units. They give an algorithm for minimizing the makespan (i.e., the height of the packing) which has asymptotic performance guarantee bound $4\frac{4}{7}$. This was improved to 2.543 by Miyazawa and Wakabayashi [6] and to 2.25 by Epstein and van Stee [7].

31.6 Three- and More-Dimensional Bin Packing

The best known result for packing multidimensional items into bins for $d \geq 3$ is h_∞^{d-1} [47], and this can be achieved by an online algorithm using bounded space. Clearly, this problem is APX-hard as well as it includes the two-dimensional bin packing problem as a special case [1].

Blitz et al. [41,42] gave a lower bound of 2.111 for online algorithms for $d = 3$. However, there is no good lower bound known for larger dimensions: no bound above 3 is known. It appears likely that the asymptotic performance guarantee bound of any online algorithm must grow with the dimension.

The bounded space algorithm by Epstein and van Stee [37] for this problem has a approximation ratio which is sublinear in d: it is $O(d/\log d)$ and $\Omega(\log d)$.

For $d = 3$ (i.e., online cube packing), Miyazawa and Wakabayashi [76] showed that the algorithm of Coppersmith and Raghavan [31] has an asymptotic performance guarantee bound of 3.954. Epstein and van Stee [43] give an algorithm with asymptotic approximation ratio at most 2.9421 (later claimed to be decreased by roughly 0.33 using a different algorithm [44]), and a lower bound of 1.6680. Furthermore, the same authors [45] give bounds for the performance of their bounded space algorithm from [37], showing that its approximation ratio lies between 2.95642 and 3.0672.

As was seen in Section 31.4.2, we can do even better offline. Before Bansal and Sviridenko [1] and Correa and Kenyon [55] gave their asymptotic polynomial time approximation scheme for any dimension $d \geq 2$, Miyazawa and Wakabayashi [76] gave two approximation algorithms, of which the best had an asymptotic approximation ratio of 2.6681. Soon afterwards, Kohayakawa et al. [53] presented their paper which we discussed in Section 31.4.2 as well. For $d = 3$, its asymptotic performance guarantee bound is $46/27 + \varepsilon \approx 1.7037\cdots + \varepsilon$.

31.7 Vector Packing

In this section, we discuss the nongeometric version of multidimensional bin packing. The d-dimensional "vector packing," or "vector bin packing" problem is defined as follows. The bins are instances of the "all-1" vector $(1, 1, \ldots, 1)$ of length d. Items are d-dimensional vectors, whose components are all in $[0, 1]$.

A packing is valid if the vector sum of all items assigned to one bin does not exceed the capacity of the bin (i.e., 1) in any component. As all bins are identical, the goal is to minimize the number of bins used. The problem is monotone in the dimension in the sense that an algorithm for dimension d can be used for any dimension $d' < d$, by adding $d - d'$ zero components to all items.

The problem can be seen as a scheduling problem with limited resources [77]. The machines (with correspond to bins) have fixed capacities of several resources as memory, running time, access to other computers, and so on. The items in this case are jobs that need to be run, each job requires a certain amount of each resource. Another application arises from viewing the problem as a storage allocation problem. Each bin has several qualities as volume, weight, and so on. Each item requires a certain amount of each quantity. Both applications are relevant to both offline and online environments.

For many years there were very few results on this problem. In the first paper [29] that obtained an APTAS for classical bin packing, de la Vega and Lueker gave a $(d + \varepsilon)$-approximation for the vector packing problem. This improved very slightly on some online results. These results were an upper bound of $d + 1$ on the approximation ratio of any algorithm whose output never has two bins that can be combined, given by Kou and Markowsky [78], and a tight bound on the performance guarantee of First Fit of $d + \frac{7}{10}$, given by Garey et al. [77]. Note that this is a generalization of the tight bound of $\frac{17}{10}$ for First Fit in one dimension.

Since these results were obtained, for a while there was hope that an APTAS would be found for this problem. However, Woeginger proved in Reference 2 that unless $P = NP$, there cannot be such an APTAS, already for two-dimensional vectors. Clearly, more restricted classes of vectors may still admit an APTAS. One such type of input is one where there is a total order on all vectors. In Reference 79, Caprara et al. showed that an APTAS for this problem indeed exists.

The offline result for the general case was finally improved by Chekuri and Khanna [80]. They designed an algorithm of asymptotic performance guarantee $O(\ln d)$. They proved that for an arbitrary d, it is APX-hard to approximate the problem within a factor of $d^{\frac{1}{2}-\varepsilon}$ for every fixed positive ε. This was shown using a reduction from graph coloring. The result was improved to $d^{1-\varepsilon}$ in a very recent survey by Christensen et al. [81]. Bansal et al. [48] improved the upper bound to $1 + \ln d + \varepsilon$ and Bansal et al. [82] very recently improved this to $\ln(\frac{d+1}{2}) + 1.5 + \varepsilon$. For two-dimensional vector packing, they give an asymptotic bound of $1 + \varepsilon + \ln 1.5 \approx 1.405 + \varepsilon$ and an almost tight absolute bound of $1.5 + \varepsilon$. The absolute bound improves over the result of Kellerer and Kotov [83], whose designed an algorithm for two-dimensional vector packing with absolute approximation ratio of at most 2. Yao [84] showed that no algorithm running in time $o(n \log n)$ can give better than a d-approximation.

The online result was not improved since 1976. Azar et al. [3] provided a lower bound of $\Omega(d^{1-\varepsilon})$ for this problem, holds for the case where both d and the optimal cost are functions of a parameter n, and they grow to infinity when n grows to infinity. This result is also proved via the hardness of graph coloring (this time, for online algorithms), and explains why there are no algorithms with a constant asymptotic approximation ratio in the online scenario. The last result does not give any lower bound on the approximation ratio for constant values of d. Prior to the almost linear lower bound of Reference 3, only lower bounds on the approximation ratio of online algorithms, that tend to 2 as d grows were known [41,42,85]. Recently, Balogh et al. [86] proved a lower bound of 2 on the asymptotic competitive ratio for any dimension and claimed lower bounds that are strictly higher than 2 for all dimensions.

Very recently, Azar et al. [87] gave an algorithm with constant approximation ratio slightly larger than e (for arbitrary d) for the special case of packing small vectors (vectors where all components are small relative to the size of a bin, and the asymptotic approximation ratio is a function of the upper bound on these sizes). For two dimensions, they give a tight upper bound of $4/3$ (see also [88] for the tightness of the bound of the general case).

As for variable sized packing, the online problem was studied by Epstein [89]. In this problem, the algorithm may use bins out of a given finite subset. This subset contains the standard "all-1" vector, and possibly other vectors. The cost of a bin is the sum of its components. She showed that there exists a finite set where an online algorithm can achieve approximation ratio $1 + \varepsilon$ (by defining the class of bins to be dense enough), whereas for another set (which contains except for the "all-1" bin only bins that have relatively small components), the ratio must be linear. Clearly, no matter what the set is, there exists a simple algorithm with linear approximation ratio.

Analogously to the bin covering problem, we can define the vector covering problem, where the vector sum of all vectors assigned to one bin is *at least* 1 in every component. This problem was studied by Alon et al. [90]. In this paper it was shown that the approximation ratio of any online algorithm is at least $d + \frac{1}{2}$. A linear upper bound of $2d$ is achieved by an algorithm which partitions the input into classes. The same paper contains offline results as well. An algorithm of performance guarantee $O(\log d)$ is presented as well as a simple and fast 2-approximation for $d = 2$. In Reference 91 some results on variable-sized vector covering are given. These results focus on cases where all bins are vectors of zeros and ones. The benefit of a covered bin is the sum of its nonzero components. The considered cases for the bins set are as follows. A set which consists of a single type of bin, a set of all unit vectors (all components are zero except for one), unit prefix vectors (some prefix of the vector consists of ones only) and the set of all zero-one vectors.

31.8 Variations

31.8.1 Items Appear from the Top

A "Tetris like" online model was studied in a few papers. This is similar to strip packing, however, in this model, a rectangle cannot be placed directly in its designated area, but it arrives from the top as in the Tetris game, and should be moved continuously around only in the free space until it reaches its place (Figure 31.4), and then cannot be moved again.

In Reference 92, the model was introduced by Azar and Epstein. In that paper, both the rotatable and the oriented models were studied. For the rotatable model, a 4-approximation algorithm was designed. An algorithm for squares with an asymptotic approximation ratio at most $\frac{34}{13} \approx 2.6154$ was designed by Fekete et al. [93]. The situation for the oriented problem is more difficult, as no algorithm with constant approximation ratio exists for unrestricted inputs. If the width of all items is bounded below by ϵ and/or bounded above by $1 - \epsilon$, the authors showed a lower bound of $\Omega(\sqrt{\log \frac{1}{\epsilon}})$ on the approximation ratio of any online algorithm for any deterministic or randomized algorithm. Restricting the width, they designed an $O(\log \frac{1}{\epsilon})$-approximation algorithm.

The oriented version of the problem was studied by Coffman et al. [94]. They assume a probabilistic model where item heights and widths are drawn from a uniform distribution on $[0, 1]$. They show that any online algorithm which packs n items has an asymptotic expected height of at least $0.313827n$ and design an algorithm of asymptotic expected height of $0.369764n$.

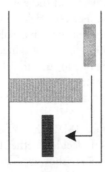

FIGURE 31.4 The process of packing an item in the "Tetris like" model.

31.8.2 Dynamic Bin Packing

A multidimensional version of a dynamic bin packing model, which was introduced in Reference 95 for the one-dimensional case, was studied by Epstein and Levy [96]. This is an online model where items do not only arrive but may also leave. Each event is an arrival or a departure of an item. Durations are not known in advance, that is, an algorithm is notified about the time that an items leaves only upon its departure. An algorithm may rearrange the locations inside bins, but the items may not migrate between bins. In Reference 96, the same problem was studied in multiple dimensions.

In two dimensions, they designed a 4.25-approximation algorithm for dynamical packing of squares, and provided a lower bound of 2.2307 on the approximation ratio. For rectangles the upper and lower bounds are 8.5754 and 3.7, respectively. For three-dimensional cubes they presented an algorithm which is a 5.37037-approximation, and a lower bound of 2.117. For three-dimensional boxes, they supplied a 35.346-approximation algorithm and a lower bound of 4.85383. For higher dimensions, they define and analyze the algorithm NFDH for the offline box packing problem. This algorithm was studied before for rectangle packing (two-dimensional only) [9], and for square and cube packing for any dimension [8,53], but not for box packing. For d-dimensional boxes they provided an upper bound of $2 \cdot 3.5^d$ and a lower bound of $d + 1$. Note that, as already mentioned in this survey, the best bound known for the regular offline multi-dimensional box packing problem is exponential as well. For d-dimensional cubes they provided an upper bound of $O\left(\frac{d}{\ln d}\right)$ and a lower bound of 2.

The constants have been improved significantly by Wong and Yung [97] (see also [98]) by using L-shape packing. This technique involves partitioning bins into parts, some of which have the shape of the letter L. In particular, their upper bound for rectangles is 7.788, their bound for boxes is 22.788, and their bound for dimension $d \geq 4$ is 3^d.

One older paper by Coffman and Gilbert [99] studies a related problem. In this problem, squares of a bounded size, which arrive and leave at various times, must be kept in a single bin. The paper gives lower bounds on the size of such a bin, so that all squares can fit. It is not allowed to rearrange the locations in the bin.

31.8.3 Packing Rectangles in a Single Rectangle

Another version is concerned with maximizing the number, area, or weight of a subset of the input rectangles, that can be packed into a larger rectangle (of given height and width). The maximization problem with respect to the number of rectangles was studied already in 1983 by Baker et al. [100]. They designed an asymptotic $\frac{4}{3}$-approximation. This offline problem was studied by Jansen and Zhang [101,102]. The first paper considered the case of weighted rectangles, and maximizing the total weight packed, whereas the second one considered unweighted rectangles, and maximizing the number of packed rectangles. The problem is considered without rotation.

In Reference 101, Jansen and Zhang proved that there exists an asymptotic FPTAS, and an absolute PTAS, for packing squares into a rectangle. For rectangles they gave an approximation algorithm with asymptotic ratio of at most two, and a simple one with an absolute ratio of $2 + \varepsilon$. In Reference 103, Jansen and Zhang gave a more complicated algorithm for the weighted case with an absolute ratio of $2 + \varepsilon$. This algorithm has higher running time than the one for the unweighted problem. The special case where the weight of each rectangle is equal to its area has also been studied. The area maximization problem was studied by Caprara and Monaci [104]. They designed an algorithm with (absolute) approximation ratio $3 + \varepsilon$.

An online version was studied by Han et al. [13]. In this version, we are given a unit square bin, rectangles arrive online, and the algorithm needs to decide whether to accept an arriving rectangle or not. The goal is again to maximize the packed area. They showed that if the algorithm is not allowed to remove rectangles accepted in the past, no algorithm with constant approximation ratio exists. This holds already for squares. It is easy to see that this holds with the following example. Take a first square which is very small, and another one which fills the bin completely. An algorithm must accept the first square and

FIGURE 31.5 The optimal packing for five unit squares.

therefore cannot accept the larger one. Next, they show that there is no algorithm with constant approximation ratio for rectangles, even if the algorithm is allowed to remove previously accepted rectangles. Therefore, the paper studies removable square packing. Before describing the results, we discuss a related paper which was used in this paper.

Januszewski and Lassak [105] studied a similar problem from the point of view of finding a threshold $\alpha \leq 1$ such that a set of squares of total area of at most α can be always packed online in a bin, without rearranging the contents of the bin. They showed that $\frac{5}{16}$ is a lower bound on α. Moreover, the considered this problem for multidimensional cubes, and showed a lower bound of $\frac{1}{2^d-1}$ for $d \geq 5$. For the packing they used a nice tool which they called bricks. A brick is a rectangle, where the ratio of the maximum between height and width to the minimum between the two remains the same after cutting the rectangle into two identical parts. Clearly, this can work if the ratio is $\sqrt{2}$.

Han et al. [13] adopted this method. They showed that any algorithm has approximation ratio of at least $\phi + 1 \approx 2.618$. They designed a matching algorithm for the case where rearranging is allowed, and a 3-approximation algorithm without rearranging. A direct consequence is that a lower bound on α for two dimensions is $\frac{1}{3}$.

Finally, another related problem is packing squares or rectangles into a square or rectangle of minimum size, where arbitrary rotations are allowed (not just over $90°$). For example, five unit squares can be packed inside a square with side $2 + \frac{1}{2}\sqrt{2}$, by placing four squares in the corners and one in the center at a $45°$ angle. See Figure 31.5. For a survey on packing equal squares into a square, see Friedman [106]. Novotný [107] showed that any set of squares with total area 1 can be packed in a rectangle of area at most 1.53 (without rotations).

References

1. N. Bansal and M. Sviridenko. New approximability and inapproximability results for 2-dimensional packing. In *Proceedings of the 15th Annual Symposium on Discrete Algorithms (SODA'04)*, pp. 189–196, ACM/SIAM, New Orleans, LA, 2004.

2. G.J. Woeginger. There is no asymptotic PTAS for two-dimensional vector packing. *Inf. Process. Lett.*, 64(6):293–297, 1997.

3. Y. Azar, I.R. Cohen, S. Kamara, and F.B. Shepherd. Tight bounds for online vector bin packing. In *Proceedings of the 45th Symposium on Theory of Computing Conference (STOC'13)*, Palo Alto, CA, pp. 961–970, 2013.

4. N. Bansal, J.R. Correa, C. Kenyon, and M. Sviridenko. Bin packing in multiple dimensions: Inapproximability results and approximation schemes. *Math. Oper. Res.*, 31(1):31–49, 2006.

5. F.K. Miyazawa and Y. Wakabayashi. Approximation algorithms for the orthogonal z-oriented 3-d packing problem. *SIAM J. Comput.*, 29(3):1008–1029, 1999.

6. F.K. Miyazawa and Y. Wakabayashi. Three-dimensional packings with rotations. *Comput. OR*, 36(10):2801–2815, 2009.

7. L. Epstein and R. van Stee. This side up! *ACM Trans. Algor.*, 2(2):228–243, 2006.

8. A. Meir and L. Moser. On packing of squares and cubes. *J. Combin. Theory*, 5:126–134, 1968.

9. E.G. Coffman, M.R. Garey, D.S. Johnson, and R.E. Tarjan. Performance bounds for level oriented two-dimensional packing algorithms. *SIAM J. Comput.*, 9:808–826, 1980.

10. C. Kenyon and E. Rémila. A near optimal solution to a two-dimensional cutting stock problem. *Math. Oper. Res.*, 25(4):645–656, 2000.

11. J. Moon and L. Moser. Some packing and covering theorems. *Colloq. Math.*, 17:103–110, 1967.

12. S.P. Fekete and H.-F. Hoffmann. Online square-into-square packing. *Algorithmica*, 1–35, 2016. doi: 10.1007/s00453-016-0114-2.

13. X. Han, K. Iwama, and G. Zhang. Online removable square packing. *Theory Comput. Syst.*, 43(1): 38–55, 2008.

14. B.S. Baker and J.S. Schwartz. Shelf algorithms for two-dimensional packing problems. *SIAM J. Comput.*, 12:508–525, 1983.

15. J. Csirik and G.J. Woeginger. Shelf algorithms for on-line strip packing. *Inform. Process. Lett.*, 63:171–175, 1997.

16. C.C. Lee and D.T. Lee. A simple online bin packing algorithm. *J. ACM*, 32:562–572, 1985.

17. A. van Vliet. An improved lower bound for online bin packing algorithms. *Inform. Process. Lett.*, 43:277–284, 1992.

18. B.S. Baker, D.J. Brown, and H.P. Katseff. A 5/4 algorithm for two-dimensional packing. *J. Algorithms*, 2:348–368, 1981.

19. X. Han, K. Iwama, D. Ye, and G. Zhang. Approximate strip packing: Revisited. *Inf. Comput.*, 249:110–120, 2016.

20. S.S. Seiden. On the online bin packing problem. *J. ACM*, 49(5):640–671, 2002.

21. D.J. Brown, B.S. Baker, and H.P. Katseff. Lower bounds for on-line two-dimensional packing algorithms. *Acta Inf.*, 18:207225, 1982.

22. B.S. Baker, E.G. Coffman, and R.L. Rivest. Orthogonal packings in two dimensions. *SIAM J. Comput.*, 9:846–855, 1980.

23. I. Schiermeyer. Reverse-fit: A 2-optimal algorithm for packing rectangles. In *Proceedings of the Second Annual European Symposium on Algorithms (ESA'94)*, Utrecht, The Netherlands, pp. 290–299, 1994.

24. A. Steinberg. A strip-packing algorithm with absolute performance bound 2. *SIAM J. Comput.*, 26(2):401–409, 1997.

25. R. Harren and R. van Stee. Improved absolute approximation ratios for two-dimensional packing problems. In I. Dinur, K. Jansen, J. Naor, and J.D.P. Rolim, (Eds.), *Proceedings Approximation, Randomization, and Combinatorial Optimization. Algorithms and Techniques, 12th International Workshop, APPROX 2009, and 13th International Workshop, RANDOM 2009*, Berkeley, CA, August 21–23, 2009. Vol. 5687, Lecture Notes in Computer Science, pp. 177–189. Springer, Berlin, Germany, 2009.

26. R. Harren, K. Jansen, L. Prädel, and R. van Stee. A $(5/3 + \epsilon)$-approximation for strip packing. *Comput. Geom.*, Arlington, VA, 47(2):248–267, 2014.

27. G. Nadiradze and A. Wiese. On approximating strip packing with a better ratio than 3/2. In *Proceedings of the 27th Annual ACM-SIAM Symposium on Discrete Algorithms, (SODA'16)*, pp. 1491–1510, 2016.

28. D.D.K.D.B. Sleator. A 2.5 times optimal algorithm for packing in two dimensions. *Inform. Process. Lett.*, 10:37–40, 1980.

29. W.F. de la Vega and G.S. Lueker. Bin packing can be solved within $1 + \varepsilon$ in linear time. *Combinatorica*, 1:349–355, 1981.

30. K. Jansen and R. van Stee. On strip packing with rotations. In *Proceedings of the 37th ACM Symposium on Theory of Computing (STOC'05)*, Baltimore, MD, pp. 755–761, 2005.

31. D. Coppersmith and P. Raghavan. Multidimensional online bin packing: Algorithms and worst case analysis. *Oper. Res. Lett.*, 8:17–20, 1989.

32. J. Csirik, J.B.G. Frenk, and M. Labbe. Two dimensional rectangle packing: On line methods and results. *Dis. Appl. Math.*, 45:197–204, 1993.

33. J. Csirik and A. van Vliet. An online algorithm for multidimensional bin packing. *Oper. Res. Lett.*, 13:149–158, 1993.

34. K. Li and K.-H. Cheng. A generalized harmonic algorithm for on-line multi-dimensional bin packing. Technical Report UH-CS-90-2, University of Houston, Houston, TX, January 1990.

35. S.S. Seiden and R. van Stee. New bounds for multi-dimensional packing. *Algorithmica*, 36(3): 261–293, 2003.

36. X. Han, F.Y.L. Chin, H.-F. Ting, G. Zhang, and Y. Zhang. A new upper bound 2.5545 on 2d online bin packing. *ACM Trans. Algor.*, 7(4):50, 2011.

37. L. Epstein and R. van Stee. Optimal online algorithms for multidimensional packing problems. *SIAM J. Comput.*, 35(2):431–448, 2005.

38. G. Galambos. A 1.6 lower bound for the two-dimensional online rectangle bin packing. *Acta Cybernet.*, 10:21–24, 1991.

39. G. Galambos and A. van Vliet. Lower bounds for 1-, 2-, and 3-dimensional online bin packing algorithms. *Computing*, 52:281–297, 1994.

40. A. van Vliet. Lower and upper bounds for online bin packing and scheduling heuristics. PhD Thesis, Erasmus University, Rotterdam, the Netherlands, 1995.

41. D. Blitz. Lower bounds on the asymptotic worst-case ratios of on-line bin packing algorithms. M.Sc. Thesis, University of Rotterdam, Rotterdam, The Netherlands, Number 114682, 1996.

42. D. Blitz, A. van Vliet, and G.J. Woeginger. Lower bounds on the asymptotic worst-case ratio of online bin packing algorithms. Unpublished manuscript, 1996.

43. L. Epstein and R. van Stee. Online square and cube packing. *Acta Inf.*, 41(9):595–606, 2005.

44. X. Han, D. Ye, and Y. Zhou. A note on online hypercube packing. *Central Europ. J. Oper. Res.*, 18(2):221–239, 2010.

45. L. Epstein and R. van Stee. Bounds for online bounded space hypercube packing. *Dis. Opt.*, 4(2):185–197, 2007.

46. F.R.K. Chung, M.R. Garey, and D.S. Johnson. On packing two-dimensional bins. *SIAM J. Alg. Dis. Meth.*, 3:66–76, 1982.

47. A. Caprara. Packing 2-dimensional bins in harmony. In *Proceedings of the 43rd Symposium on Foundations of Computer Science (FOCS'02)*. IEEE Computer Society. Vancouver, BC, pp. 490–499, 2002.

48. N. Bansal, A. Caprara, and M. Sviridenko. A new approximation method for set covering problems, with applications to multidimensional bin packing. *SIAM J. Comput.*, 39(4):1256–1278, 2009.

49. K. Jansen and L. Prädel. New approximability results for two-dimensional bin packing. *Algorithmica*, 74(1):208–269, 2016.

50. N. Bansal and A. Khan. Improved approximation algorithm for two-dimensional bin packing. In *Proceedings of the 25th Annual ACM-SIAM Symposium on Discrete Algorithms (SODA'14)*. Portland, Oregon, pp. 13–25, 2014.

51. J.Y.-T. Leung, T.W. Tam, C.S. Wong, G.H. Young, and F.Y.L. Chin. Packing squares into a square. *J. Para. Dis. Comput.*, 10:271–275, 1990.

52. C.E. Ferreira, F.K. Miyazawa, and Y. Wakabayashi. Packing squares into squares. *Pesquisa Operacional*, 19(2):223–237, 1999.

53. Y. Kohayakawa, F.K. Miyazawa, P. Raghavan, and Y. Wakabayashi. Multidimensional cube packing. *Algorithmica*, 40(3):173–187, 2004.

54. L. Epstein and R. van Stee. Optimal online bounded space multidimensional packing. In *Proceedings of 15th Annual ACM-SIAM Symposium on Discrete Algorithms (SODA'04)*, New Orleans, LA, pp. 207–216, 2004.

55. J.R. Correa and C. Kenyon. Approximation schemes for multidimensional packing. In *Proceedings of the 15th Annual Symposium on Discrete Algorithms (SODA'04)*, New Orleans, LA, pp. 179–188, 2004.

56. Y. Lu, D.Z. Chen, and J. Cha. Packing cubes into a cube in ($D > 3$)-dimensions. In D. Xu, D. Du, and D.-Z. Du, (Eds.), *Proceedings, Computing and Combinatorics: 21st International Conference, COCOON 2015*, Beijing, China, August 4–6, 2015, pp. 264–276. Springer, 2015.

57. Y. Lu, D.Z. Chen, and J. Cha. Packing cubes into a cube is NP-complete in the strong sense. *J. Comb. Optim.*, 29(1):197–215, 2015.

58. N. Bansal, A. Lodi, and M. Sviridenko. A tale of two dimensional bin packing. In *Proceedings of the 46th Annual IEEE Symposium on Foundations of Computer Science (FOCS'05)*, pp. 657–666, IEEE Computer Society Washington, DC, 2005.

59. A. Caprara, A. Lodi, and M. Monaci. Fast approximation schemes for the two-stage, two-dimensional bin packing problem. *Math. Oper. Res.*, 30:150–172, 2005.

60. S.S. Seiden and G.J. Woeginger. The two-dimensional cutting stock problem revisited. *Math. Program.*, 102(3):519–530, 2005.

61. F. Abed, P. Chalermsook, J.R. Correa, A. Karrenbauer, P. Pérez-Lantero, J.A. Soto, and A. Wiese. On guillotine cutting sequences. In *The 18th International Workshop on Approximation Algorithms for Combinatorial Optimization Problems, and the 19th International Workshop on Randomization and Computation (RANDOM-APPROX'15)*, Princeton, NJ, pp. 1–19, 2015.

62. G. Zhang. A 3-approximation algorithm for two-dimensional bin packing. *Op. Res. Lett.*, 33(2):121–126, 2005.

63. R. van Stee. An approximation algorithm for square packing. *Op. Res. Lett.*, 32(6):535–539, 2004.

64. R. Harren and R. van Stee. Packing rectangles into 2opt bins using rotations. In J. Gudmundsson, (Ed.), *Proceedings, Algorithm Theory - SWAT 2008, 11th Scandinavian Workshop on Algorithm Theory*, Gothenburg, Sweden, July 2–4, 2008, Vol. 5124, Lecture Notes in Computer Science, pages 306–318. Gothenburg, Sweden, Springer, 2008.

65. R. Harren, K. Jansen, L. Prädel, U.M. Schwarz, and R. van Stee. Two for one: Tight approximation of 2d bin packing. *Int. J. Found. Comput. Sci.*, 24(8):1299–1328, 2013.

66. N. Bansal and M. Sviridenko. Two-dimensional bin packing with one-dimensional resource augmentation. *Dis. Opt.*, 4(2):143–153, 2007.

67. L. Epstein. Two-dimensional online bin packing with rotation. *Theor. Comput. Sci.*, 411(31–33): 2899–2911, 2010.

68. S. Fujita and T. Hada. Two-dimensional on-line bin packing problem with rotatable items. *Theor. Comp. Sci.*, 289(2):939–952, 2002.

69. J.R. Correa. Resource augmentation in two-dimensional packing with orthogonal rotations. *Oper. Res. Lett.*, 34(1):85–93, 2006.

70. K. Li and K.-H. Cheng. On three-dimensional packing. *SIAM J. Comput.*, 19(5):847–867, 1990.

71. K. Li and K.-H. Cheng. Heuristic algorithms for online packing in three dimensions. *J. Algor.*, 13:589–605, 1992.

72. Flavio K. Miyazawa and Y. Wakabayashi. An algorithm for the three-dimensional packing problem with asymptotic performance analysis. *Algorithmica*, 18(1):122–144, 1997.

73. N. Bansal, X. Han, K. Iwama, M. Sviridenko, and G. Zhang. A harmonic algorithm for the 3d strip packing problem. *SIAM J. Comput.*, 42(2):579–592, 2013.

74. K. Jansen and L. Prädel. A new asymptotic approximation algorithm for 3-dimensional strip packing. In V. Geffert, B. Preneel, B. Rovan, J. Stuller, and A.M. Tjoa, (Eds.), *Prodceeings SOFSEM 2014: Theory and Practice of Computer Science - 40th International Conference on Current Trends in Theory and Practice of Computer Science*, Nový Smokovec, Slovakia, Europe, January 26–29, 2014, Vol. 8327, Lecture Notes in Computer Science, pp. 327–338. Springer, Cham, Switzerland, 2014.

75. K. Li and K.-H. Cheng. Generalized First-Fit algorithms in two and three dimensions. *Int. J. on Found. Comput. Sci.*, 2:131–150, 1990.

76. F.K. Miyazawa and Y. Wakabayashi. Cube packing. *Theor. Comp. Sci.*, 297(1–3):355–366, 2003.

77. M.R. Garey, R.L. Graham, D.S. Johnson, and A.C.C. Yao. Resource constrained scheduling as generalized bin packing. *J. Comb. Theor. (Series A)*, 21:257–298, 1976.

78. L.T. Kou and G. Markowsky. Multidimensional bin packing algorithms. *IBM J. Res. Dev.*, 21:443–448, 1977.

79. A. Caprara, H. Kellerer, and U. Pferschy. Approximation schemes for ordered vector packing problems. *Nav. Res. Logist.*, 92:58–69, 2003.

80. C. Chekuri and S. Khanna. On multidimensional packing problems. *SIAM J. Comput.*, 33(4):837–851, 2004.

81. H.I. Christensen, A. Khan, S. Pokutta, and P. Tetali. Multidimensional bin packing and other related problems: A survey. *Comput. Sci. Rev.*, 24:63–79, 2017.

82. N. Bansal, M. Eliás, and A. Khan. Improved approximation for vector bin packing. In *Proceedings of the 27th Annual ACM-SIAM Symposium on Discrete Algorithms (SODA'16)*, Arlington, VA, pp. 1561–1579, 2016.

83. H. Kellerer and V. Kotov. An approximation algorithm with absolute worst-case performance ratio 2 for two-dimensional vector packing. *Oper. Res. Lett.*, 31(1):35–41, 2003.

84. A.C.-C. Yao. New algorithms for bin packing. *J. ACM*, 27(2):207–227, 1980.

85. G. Galambos, H. Kellerer, and G.J. Woeginger. A lower bound for online vector packing algorithms. *Acta Cybernet.*, 10:23–34, 1994.

86. J. Balogh, J. Békési, G. Dósa, L. Epstein, and A. Levin. Online bin packing with cardinality constraints resolved. *CoRR*, abs/1608.06415, 2016.

87. Y. Azar, I.R. Cohen, A. Fiat, and A. Roytman. Packing small vectors. In *Proceedings of the 27th Annual ACM-SIAM Symposium on Discrete Algorithms (SODA'16)*, Arlington, VA, pp. 1511–1525, 2016.

88. Y. Azar, I.R. Cohen, and A. Roytman. Online lower bounds via duality. *CoRR*, abs/1604.01697, 2016.

89. L. Epstein. On variable sized vector packing. *Acta Cyber.*, 16:47–56, 2003.

90. N. Alon, Y. Azar, J. Csirik, L. Epstein, S.V. Sevastianov, A. Vestjens, and G.J. Woeginger. On-line and off-line approximation algorithms for vector covering problems. *Algorithmica*, 21:104–118, 1998.

91. L. Epstein. On-line variable sized covering. *Inf. Comput.*, 171(2):294–305, 2001.

92. Y. Azar and L. Epstein. On two dimensional packing. *J. Algorithms*, 25(2):290–310, 1997.

93. S.P. Fekete, T. Kamphans, and N. Schweer. Online square packing with gravity. *Algorithmica*, 68(4):1019–1044, 2014.

94. E.G. Coffman, P.J. Downey, and P.M. Winkler. Packing rectangles in a strip. *Acta Inf.*, 38(10):673–693, 2002.

95. E.G. Coffman, M.R. Garey, and D.S. Johnson. Dynamic bin packing. *SIAM J. Comput.*, 12:227–258, 1983.

96. L. Epstein and M. Levy. Dynamic multi-dimensional bin packing. *J. Dis. Algor.*, 8(4):356–372, 2010.

97. P.W.H. Wong and F.C.C. Yung. Competitive multi-dimensional dynamic bin packing via L-shape bin packing. In *Proceedings of 7th International Workshop on Approximation and Online Algorithms (WAOA'09)*, Copenhagen, Denmark, pp. 242–254, 2009.

98. M. Burcea, P.W.H. Wong, and F.C.C. Yung. Online multi-dimensional dynamic bin packing of unit-fraction items. In *Proceedings of 8th International Conference on Algorithms and Complexity (CIAC'13)*, Barcelona, Spain, pp. 85–96, 2013.

99. E.G. Coffman and E.N. Gilbert. Dynamic, first-fit packings in two or more dimensions. *Inf. Cont.*, 61(1):1–14, 1984.

100. B.S. Baker, A.R. Calderbank, E.G. Coffman, and J.C. Lagarias. Approximation algorithms for maximizing the number of squares packed into a rectangle. *SIAM J. Alg. Discr. Meth.*, 4(3), 383–397, 1983.

101. K. Jansen and G. Zhang. On rectangle packing: Maximizing benefits. In *Proceedings of the 15th Annual Symposium on Discrete Algorithms (SODA'04)*, New Orleans, LA, pp. 204–213, 2004.

102. K. Jansen and G. Zhang. Maximizing the total profit of rectangles packed into a rectangle. *Algorithmica*, 47(3):323–342, 2007.

103. K. Jansen and G. Zhang. Maximizing the number of packed rectangles. In *Proceedings of the 9th Scandinavian Workshop on Algorithm Theory (SWAT'04)*, Copenhagen, Denmark, pp. 362–371, 2004.

104. A. Caprara and M. Monaci. On the 2-dimensional knapsack problem. *Oper. Res. Lett.*, 32:5–14, 2004.

105. J. Januszewski and M. Lassak. Online packing sequences of cubes in the unit cube. *Geomet. Ded.*, 67:285–293, 1997.

106. E. Friedman. Packing unit squares in squares: A survey and new results. *Electr. J. Comb.*, 7, 2009.

107. P. Novotný. On packing of squares into a rectangle. *Archi. Math.*, 32:75–83, 1996.

32

Practical Algorithms for Two-Dimensional Packing of Rectangles

Shinji Imahori

Mutsunori Yagiura

Hiroshi Nagamochi

32.1 Introduction

Cutting and packing problems consist of placing a given set of (small) items into one or more (larger) containers without overlap so as to minimize/maximize a given objective function. This is a combinatorial optimization problem with many important applications in the wood, glass, steel and leather industries, as well as in large-scale integration (LSI) and very large-scale integration (VLSI) design, container and truck loading, and newspaper paging. For several decades, cutting and packing problems have attracted the attention of researchers in various areas including operations research, computer science and manufacturing.

Cutting and packing problems can be classified using different criteria. The dimensionality of a problem is one of such criteria, and most problems are defined over one, two, or three dimensions. In this chapter we consider two-dimensional problems. The next criterion to classify two-dimensional packing problems is the shape of items to pack. We focus on the *rectangle packing problem* in this chapter. Almost all two-dimensional packing problems are known to be NP-hard, and hence it is impossible to solve them exactly in polynomial time unless P = NP. Therefore, heuristics and metaheuristics are very important to design practical algorithms for these problems. We survey practical algorithms for the rectangle packing problem in this chapter. We also survey various schemes used to represent solutions to the rectangle packing problem, and then we introduce algorithms based on these *coding schemes*.

The remainder of this chapter is organized as follows: Section 32.2 defines the rectangle packing problem and its variations. Section 32.3 introduces coding schemes for the rectangle packing problem, which are used to represent solutions. Section 32.4 presents heuristic algorithms, from the traditional to the latest ones, for the rectangle packing problem. Many of the heuristic algorithms place items one by one at bottom-left stable positions, and in Section 32.5 we introduce techniques to enumerate all the bottom-left stable positions in a layout. Section 32.6 discusses metaheuristic algorithms for the rectangle packing problem.

32.2 Rectangle Packing Problem

We consider the following two-dimensional rectangle packing problem. We are given n items (small rectangles) $I = \{1, 2, \ldots, n\}$, each of which has width w_i and height h_i, and one or many large containers (rectangles). We are required to place the items orthogonally (an edge of each item is parallel to an edge of the container) without any overlap so as to minimize (or maximize) a given objective function. The rectangle packing problem arises in many industrial applications, often with slightly different constraints, and many variants of this problem have been considered in the literature. The following characteristics are important to classify the problems [1, 2]: type of assignment, assortment of containers, and assortment of items. We will review some specific variations of the rectangle packing problem in this section. We should mention two more important constraints for the rectangle packing problem: orientation and guillotine cut constraint. As for the orientation of the items, we usually assume that "each rectangle has a fixed orientation" or "each rectangle can be rotated by 90°." Rotation of items is usually not allowed in newspaper paging or when the items to be packed are decorated or corrugated, whereas orientation is often free in the case of plain materials. Guillotine cut constraint signifies that the items must be obtained through a sequence of edge-to-edge cuts parallel to the edges of the large container (see Figure 32.1 for an example), which is usually imposed by technical limitations of the automated cutting machines or the material.

We introduce six types of rectangle packing problems that have been actively studied. For simplicity, we define the problems assuming that each item has a fixed orientation and the guillotine cut constraint is not imposed unless otherwise stated. It is straightforward to extend our definitions for other cases where each item can be rotated by 90° and/or the guillotine cut constraint is imposed. We first consider two types of typical rectangle packing problems in which all the items must be placed in one large rectangular container that has flexible width and/or height. The problems are called *strip packing* and *area minimization*.

> *Strip packing problem*: We are given n items (small rectangles) each having width w_i and height h_i, and one large container (called a *strip*) whose width W is fixed, but its height H is variable. The objective is to minimize the height H of the strip under the constraint that all items should be packed into the strip.
>
> *Area minimization problem*: We are given n items each having width w_i and height h_i, and one large rectangular container, where both of its width W and height H are variables. The objective is to minimize the area WH of the container under the constraint that all items should be packed into the container.

Under the improved typology of cutting and packing problems by Wäscher et al. [2], the strip packing problem is a two-dimensional rectangular open dimension problem with one variable dimension and the area minimization problem is also a two-dimensional rectangular open dimension problem with two variable dimensions. In the rest of this chapter, we mainly focus on the strip packing and area minimization problems, which are formally formulated as follows:

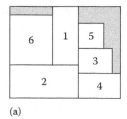

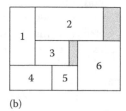

(a)   (b)

FIGURE 32.1 Examples of placements with/without guillotine cut constraint: (a) guillotine cut and (b) nonguillotine cut.

minimize the height of the strip H (or the area of the large container WH)

subject to

$$0 \le x_i \le W - w_i, \qquad \text{for all } i \in I \tag{32.1}$$

$$0 \le y_i \le H - h_i, \qquad \text{for all } i \in I \tag{32.2}$$

At least one of the next four inequalities holds for every pair i and j:

$$x_i + w_i \le x_j, \tag{32.3}$$

$$x_j + w_j \le x_i, \tag{32.4}$$

$$y_i + h_i \le y_j, \tag{32.5}$$

$$y_j + h_j \le y_i, \tag{32.6}$$

where (x_i, y_i) is the coordinate of the lower left corner of an item i. Constraints (32.1) and (32.2) mean that all items must be placed into the large container, and Constraints (32.3) to (32.6) mean that no two items overlap (i.e., each inequality signifies one of the four positional relationships: i is to the left of, right of, below, and above j).

Two other rectangle packing problems are the *two-dimensional bin packing* and *knapsack* problems, which have (many or one) fixed sized containers.

> *Two-dimensional bin packing problem*: We are given a set of items, where each item i has width w_i and height h_i, and an unlimited number of large containers (rectangular bins) having identical width W and height H. The objective is to minimize the number of rectangular bins used to place all the items.

> *Two-dimensional knapsack problem*: We are given a set I of items, where each item $i \in I$ has width w_i, height h_i and value c_i. We are also given a rectangular knapsack with fixed width W and height H. The objective is to find a subset $I' \subseteq I$ of items with the maximum total value $\sum_{i \in I'} c_i$ such that all items $i \in I'$ can be packed into the rectangular knapsack.

For the two-dimensional bin packing problem, Lodi et al. [3] proposed practical heuristic and meta-heutistic algorithms and performed computational experiments on various benchmark instances. Charalambous and Fleszar [4] presented a constructive heuristic algorithm for the problem with the guillotine cut constraint. For the two-dimensional knapsack problem, Wu et al. [5] proposed heuristic algorithms that are effective for many test instances. Egeblad and Pisinger [6] presented a heuristic approach based on the sequence pair representation, which is a coding scheme for the rectangle packing problem and is explained in Section 32.3. Leung et al. [7] proposed constructive heuristic and hybrid simulated annealing (SA) algorithms for the two-dimensional knapsack problem and performed computational experiments on 221 benchmark instances.

We should also mention the following two problems, *two-dimensional cutting stock* and *pallet loading*. For some industrial applications, such as mass production manufacturing, many small items of an identical shape or a relatively small number of shapes are packed into the containers. The following two problems are useful for modeling these situations.

> *Two-dimensional cutting stock problem*: We are given a set of items each with width w_i, height h_i, and demand d_i. We are also given an unlimited number of containers having identical width W and height H. The objective is to minimize the number of containers used to place all the items (i.e., for each i, we place d_i copies of item i into the containers).

> *Pallet loading problem*: We are given a sufficiently large number of items with identical size (w, h), and one large rectangular container with size (W, H). The objective is to place the maximum number of items into the container, where each item can be rotated by $90°$.

We note that the pallet loading problem with a fixed orientation of items is trivial to solve. Among many studies on the two-dimensional cutting stock problem, Gilmore and Gomory [8] provided one of the earliest solution method. They proposed a column generation scheme in which new cutting patterns are produced by solving a generalized knapsack problem. Some practical algorithms for the two-dimensional cutting stock problem have been proposed [9–12]. For the pallet loading problem, Morabito and Morales [13] proposed a simple but effective algorithm. Silva et al. [14] presented a comprehensive review of solution methods and computational experiments for the pallet loading problem.

The complexity of the pallet loading problem is open (this problem is not known to be in class NP, because of the compact input description), whereas the other problems we defined in this section are known to be NP-hard.

32.3 Coding Schemes for Rectangle Packing

In this section, we review coding schemes for rectangle packing problems. For simplicity, we focus on the area minimization problem. An effective search will be difficult if we directly search for a good combination of the x and y coordinates of items, as the number of solutions is uncountable and eliminating overlaps between items is not easy. To overcome this difficulty, various coding schemes have been proposed, and many algorithms for rectangle packing problems are based on some coding schemes.

A *coding scheme* consists of a set of coded solutions, a mapping from coded solutions to placements, and a *decoding algorithm* that computes for a given coded solution the corresponding placement using the mapping. The mapping is sometimes defined by the decoding algorithm. Desirable properties of a coding scheme (and a decoding algorithm) are as follows.

1. There exists a coded solution that corresponds to an optimal placement.
2. The number of all possible coded solutions is finite, where a small total number is preferable.
3. Every coded solution corresponds to a feasible placement.
4. Decoding is possible in polynomial time, where faster algorithms are more desirable.

Some of the coding schemes in the literature satisfy all the above-mentioned four properties, but others do not.

One of the most popular coding schemes is to represent a solution by a permutation of the n items, where a coded solution (i.e., a permutation of n items) specifies the placement order. The number of all possible coded solutions is $O(n!)$, which is smaller than the numbers for other coding schemes in the literature, and every permutation corresponds to a layout without overlaps. A decoding algorithm computes a placement from a given coded solution by specifying the locations of the items one by one, which defines a mapping from coded solutions to placements and is sometimes called a *placement rule*. The decoding time complexity and the existence of a coded solution that corresponds to an optimal placement depend on the decoding algorithms. In the literature, a number of decoding algorithms for permutation coding schemes have been proposed. We explain some typical decoding algorithms in Section 32.4.

We now explain different types of coding schemes for rectangle packing problems. The schemes we explain hereafter specify a positional relationship for each pair of items by a coded solution. In other words, for every pair of items, a coded solution determines one of the four inequalities (32.3) to (32.6) that must be satisfied to avoid overlap. The placement corresponding to a coded solution is the best one among those that satisfy the positional relationship specified by the coded solution.

One of the most popular coding schemes of this type is to represent a solution by an n-leaf binary tree [15]. This coding scheme can represent only slicing structures (in other words, every placement obtained by this representation always satisfies the guillotine cut constraint). The leaves of a binary tree correspond to items, and each internal node has a label "h" or "v," where h stands for horizontal and v stands for vertical. This coding scheme uses $O(n)$ space to represent a solution and the number of all possible coded solutions is $O(n! \, 2^{5n-3}/n^{1.5})$. In this scheme, one of the four positional relationships is assigned for each pair i and j of items as follows: If i is a left descendant of an internal node u with "h"

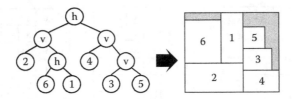

FIGURE 32.2 A binary tree τ and a layout $\pi \in \Pi_\tau$.

label and j is a right descendant of the same internal node u (i.e., u is the least common ancestor of i and j), then we must place i to the left of j (i.e., $x_i + w_i \leq x_j$). If the label of the least common ancestor is "v," then we place i below j (i.e., $y_i + h_i \leq y_j$). Figure 32.2 shows a binary tree and a placement that satisfies these constraints. For example, in Figure 32.2, there is an internal node with "h" label for which the node for item 6 is a left descendant and the node for item 1 is a right descendant, and hence item 6 is placed to the left of item 1; the node for item 4 is a left descendant and the node for item 3 is a right descendant of the least common ancestor with "v" label, and hence item 4 is placed below item 3. For a given binary tree τ, let Π_τ denote the set of all placements that satisfy the above-mentioned horizontal/vertical constraints. The placement corresponding to a coded solution τ is one of the best (i.e., the most compact) placements in Π_τ. Though Π_τ contains infinitely many placements for any τ, natural decoding algorithms for this coding scheme run in linear time of the number of items and compute one of the best placements among Π_τ. Moreover, for any placement π that satisfies the guillotine cut constraint, there exists a binary tree τ that satisfies $\pi \in \Pi_\tau$. That is, the binary tree coding scheme satisfies all of the four desirable properties of a coding scheme if the guillotine cut constraint is imposed.

Murata et al. [16] proposed a coding scheme called *sequence pair*. For the sequence pair representation, a solution is represented by a pair of permutations $\sigma = (\sigma_+, \sigma_-)$ of the n items (see Figure 32.3 for an example). A coded solution σ of this scheme specifies the positional relationship of each pair of items i and j as follows: If item i is before item j in both permutations σ_+ and σ_-, then item i must be placed to the left of j. If i is before j in σ_+ and after j in σ_-, then we place i above j. For example, in Figure 32.3, element 1 is before element 2 in both permutations, and hence item 1 is placed to the left of item 2; element 2 is before element 3 in permutation σ_+ and after element 3 in σ_-, and hence item 2 is placed above item 3. For a given pair of permutations $\sigma = (\sigma_+, \sigma_-)$, let Π_σ be the set of placements that satisfy the above-mentioned constraints. The placement corresponding to a coded solution σ is one of the best placements in Π_σ. Murata et al. [16] proposed an $O(n^2)$ time decoding algorithm to obtain one of the best placements $\pi \in \Pi_\sigma$ for a given coded solution σ. Takahashi [17] improved the time complexity of the decoding algorithm to $O(n \log n)$; Tang et al. [18] further improved it to $O(n \log \log n)$. Imahori et al. [19] and Pisinger [20] presented new decoding algorithms for a given coded solution σ. Moreover, for any feasible placement π, there exists a coded solution σ that satisfies $\pi \in \Pi_\sigma$ (such a σ can be computed in $O(n \log n)$ time [19]). That is, the sequence pair coding scheme satisfies all of the four desirable properties of a coding scheme.

Nakatake et al. [21] proposed a coding scheme called *bounded sliceline grid* (BSG). A grid of the BSG scheme consists of a set of small rooms that are separated by horizontal and vertical segments, where the

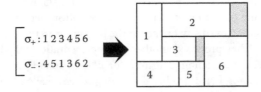

FIGURE 32.3 A pair of permutations $\sigma = (\sigma_+, \sigma_-)$ and a layout $\pi \in \Pi_\sigma$.

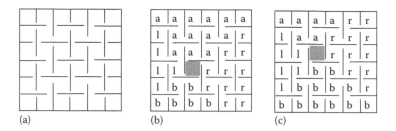

FIGURE 32.4 (a) Rooms of bounded sliceline grid representation; (b) specified positional relationships against a shaded room; (c) specified positional relationships against another shaded room.

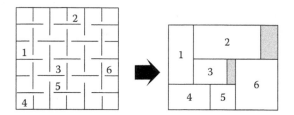

FIGURE 32.5 A solution α of bounded sliceline grid representation and a layout $\pi \in \Pi_\alpha$.

number of rooms in the horizontal and vertical directions, denoted p and q, respectively, are parameters that satisfy $pq \geq n$ (Figure 32.4a with $p = q = 6$). It introduces one of the four orthogonal relations (left-of, right-of, above and below) uniquely for each pair of rooms (Figures 32.4b,c). In these figures, a room with label l (resp., r, a, b) is left-of (resp., right-of, above, below) the shaded room. A solution is represented by an assignment of the items to rooms, where at most one item can be assigned to each room. The assigned items inherit the relations defined on the rooms. Figure 32.5 shows an example of an assignment of items to rooms and a placement that satisfies all specified constraints on positional relationships. For example, in the figure, item 3 is placed to the right of item 1, item 3 is placed below item 2, item 3 is placed above item 4. For a given assignment α, let Π_α be the set of all placements that satisfy the constraints given by α. The placement corresponding to a coded solution α is one of the best placements in Π_α. A decoding algorithm proposed by Nakatake et al. [21] runs in linear time with respect to the number of small rooms pq and can find one of the best placements $\pi \in \Pi_\alpha$ for a given coded solution α. As for the existence of a coded solution that corresponds to an optimal placement, it is known that an assignment α such as $\pi \in \Pi_\alpha$ always exists for any placement π if and only if $p \geq n$ and $q \geq n$ hold. The BSG coding scheme with parameters $p \geq n$ and $q \geq n$ satisfies all of the four desirable properties of a coding scheme.

There are many other coding schemes that describe the positional relationship of each pair of items. Guo et al. [22] proposed a tree representation called O-tree: Two ordered trees for the horizontal and vertical directions are used to represent a coded solution. This coding scheme can represent nonslicing structures and the number of all possible coded solutions is $O(n!\ 2^{2n-2}/n^{1.5})$; this is smaller than the number of all coded solutions by the binary tree representation for slicing structures. There exists a coded solution corresponding to any placement π that satisfies the bottom-left stability; that is, in the resulting placement, all items cannot be moved any further to the bottom or to the left. Chang et al. [23] extended the result by Guo et al. [22]. They proposed another tree representation called B*-tree; it is easy to implement this data structure, and a decoding algorithm for B*-tree runs in linear time with respect to the number of items. Lin and Chang [24] introduced a coding scheme called transitive closure graph-based representation (TCG), which uses a horizontal and a vertical transitive closure graphs to describe the horizontal and vertical relations for each pair of items. Sakanushi et al. [25] proposed another coding scheme called quarter-state sequence: They utilize a string of items and labels to represent a solution and their decoding algorithm also runs in linear time of the number of items.

32.4 Heuristics for Rectangle Packing

In this section, we describe heuristic algorithms for rectangle packing problems. (For simplicity, we mainly focus on the strip packing problem.) We first explain some heuristic algorithms based on the permutation coding scheme. These algorithms consist of two phases: (1) construct a permutation and (2) place the items one by one according to the permutation.

For the first phase, a standard strategy to construct a permutation is "a larger item has higher priority than a smaller one." To realize this, the items are sorted by some criteria, such as decreasing height, decreasing width or decreasing area. It is difficult to decide a priori which criterion is the best for numerous instances that arise in practice. Hence, many algorithms generate several permutations with different criteria and apply a decoding algorithm to all the generated permutations.

We then consider decoding algorithms for permutations used in the second phase. We first explain *level algorithms* in which the placement is obtained by placing items from left to right in rows forming levels (see Figure 32.6 for an example). The first level is the bottom of the container, and each subsequent level is along the horizontal line coinciding with the top of the tallest item packed on the level below.

The most popular level algorithms are the *next fit*, *first fit* and *best fit* strategies, which are extended from the algorithms for the (one-dimensional) bin packing problem. Let i ($i = 1, 2, \ldots, n$) denote the current item to be placed, and s be the level created most recently. In most cases, items are sorted by decreasing height.

- *Next fit* strategy: Item i is packed on level s left justified (i.e., place it at the left-most feasible position) if it fits. Otherwise, a new level ($s := s + 1$) is created and i is packed on it left justified.
- *First fit* strategy: We check whether or not item i fits from level 1 to level s and pack it left justified on the first level where it fits. If no level can accommodate i, it is placed on the new level $s := s + 1$ as in the next fit strategy.
- *Best fit* strategy: Item i is packed left justified on the level that minimizes the unused horizontal space among those where it fits. If no level can accommodate i, it is placed on the new level $s := s + 1$ as in the next fit and first fit strategies.

Computation time of these algorithms is $O(n)$, $O(n \log n)$, and $O(n \log n)$, respectively, if appropriately implemented. The above-mentioned strategies are illustrated through the example in Figure 32.7 (in this figure, items are sorted by decreasing height and are numbered accordingly). The resulting placements of these algorithms always satisfy the guillotine cut constraint. More precisely, they are called two-stage guillotine placement where two-stage means that all the items can be cut out in two stages (i.e., the first stage for horizontal cuts and the second stage for vertical cuts).

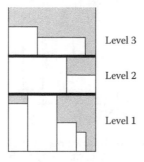

FIGURE 32.6 An example of level packing.

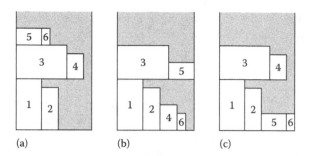

FIGURE 32.7   Three level algorithms for the strip packing problem. (a) Next fit, (b) first fit, and (c) best fit.

Another classical approach, and the most documented one, is the *bottom-left* approach. The first algorithm of this type was proposed by Baker et al. [26] in 1980, and some variants of this method have been proposed [27–29]. A common characteristic of this type of algorithms is to place items one by one at bottom-left stable positions, where a bottom-left stable position is a location where the item to place next can be placed without overlap and cannot be moved leftward or downward.

Baker et al. [26] used a bottom-left rule that places each item at the left-most point among the lowest possible positions (this position is called the *bottom-left* position). This approach is called *bottom-left-fill* (BLF) strategy in References [30, 31], which is illustrated through the example in Figure 32.8a. The "x" marks in the figure show the bottom-left stable positions. There are natural algorithms that require $O(n^3)$ time in the worst case for this strategy, and Hopper and Turton implemented one of them in their article [31]. Chazelle [27] devised an efficient algorithm that requires $O(n^2)$ time and $O(n)$ space in the worst case.

Jakobs [28] utilized another bottom-left method: For each item, it is first placed at the top right location of the container, and successive sliding moves down and to the left are applied alternately as long as possible (see the example in Figure 32.8b). This strategy is called *bottom-left* (BL) in References [30, 31] and it runs in $O(n^2)$ time, if appropriately implemented. Liu and Teng [29] also developed a bottom-left heuristic rule similar to Jakobs's algorithm. In their strategy, the downward movement has priority such that items slide leftwards only if no downward movement is possible (Figure 32.8c). This algorithm also runs in $O(n^2)$ time with appropriate implementations.

There are more algorithms that utilize a permutation to represent a solution. For example, Lodi et al. [3] proposed several decoding algorithms such as *floor ceiling, alternate directions*, and *touching perimeter*, and experimentally compared these algorithms with other decoding algorithms in the literature.

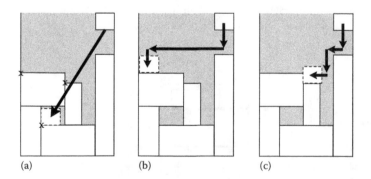

FIGURE 32.8   Three bottom-left algorithms for the strip packing problem. (a) From Baker, B.S. et al., *SIAM J. Comput.*, 9, 846, 1980; (b) From Jakobs, S., *Eur. J. Oper. Res.*, 88, 165, 1996; and (c) From Liu, D. and Teng, H., *Eur. J. Oper. Res.*, 112, 413, 1999.

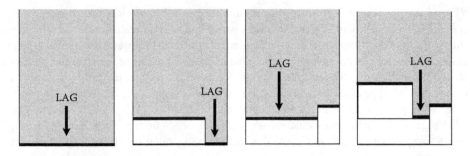

FIGURE 32.9 A heuristic algorithm for the strip packing. (From Burke, E.K. et al., *Oper. Res.*, 52, 655, 2004.)

Ortmann et al. [32] proposed four new and improved level algorithms for the strip packing problem and a bin packing problem with different-sized bins. Leung and Zhang [33] presented a fast heuristic algorithm that works well for large-scale instances of the strip packing problem. Thiebaut [34] packed one billion rectangles under ten minutes with a heuristic algorithm on a multithreaded computational environment.

We discuss a different type of heuristic algorithm called *best-fit* proposed by Burke et al. [30]. (We note that this algorithm is not a level algorithm based on the best fit strategy.) This method does not have a permutation of items to place, but it dynamically decides the next item to place during the packing stage. More precisely, it finds the lowest available gap (LAG) within the large container and then places the item that best fits there (see Figure 32.9 for an example). This enables the algorithm to make informed decisions about which item should be packed next and where it should be placed. A natural implementation of this strategy runs in $O(n^2)$ time. Imahori and Yagiura [35] presented an optimal implementation of the best-fit algorithm that requires $O(n \log n)$ time and $O(n)$ space. Aşik and Özcan [36], Verstichel et al. [37], and Wei et al. [38] proposed improved best-fit heuristic algorithms that attain better placements than the layouts obtained by the original best-fit algorithm.

32.5 Bottom-Left Stable Positions

Some construction heuristic algorithms for the rectangle packing problem place rectangles at bottom-left stable positions, and hence any layouts constructed by these algorithms (including intermediate layouts) satisfy bottom-left stability. For layouts with bottom-left stability, Chazelle [27] showed that the number of bottom-left stable positions for a rectangle to place is at most $k+1$ when the number of placed rectangles is k. He proposed an algorithm to enumerate all the bottom-left stable positions in linear time. By using this algorithm, he presented an efficient implementation of the bottom-left-fill algorithm that requires $O(n^2)$ time and $O(n)$ space in the worst case.

Another common framework of practical algorithms for two-dimensional packing problems is the improvement method, which places all the rectangles in the given area without overlap and iteratively improves the layout by some operations. Algorithms of this type often place a rectangle at a bottom-left stable position, but in this case, they may need to solve the problem of finding such a position in a layout without bottom-left stability. For this case, Healy et al. [39] showed that the number of bottom-left stable positions for a rectangle to place is $O(k)$ when k rectangles are placed in the area without overlap, and they proposed an $O(k \log k)$ time algorithm to enumerate all the bottom-left stable positions.

For some packing problems including the two-dimensional irregular packing problem, algorithms with *compaction and separation operations* were proposed [40, 41]. These algorithms generate layouts with overlap during their execution. Imahori et al. [42] considered the problem of enumerating bottom-left stable positions for a rectangle to place within a given layout that may not satisfy bottom-left stability and may have overlap between rectangles. They proposed an enumeration algorithm that runs in $O((k + p) \log k)$ time, where k is the number of placed rectangles and p is the number of bottom-left stable positions. It is noted that if the given layout has no overlap between the placed rectangles, then

$p = O(k)$ and the time complexity of this algorithm becomes $O(k \log k)$, which is the same as the result in Reference [39]. They used no-fit polygons and a sweep line to enumerate bottom-left stable positions, where no-fit polygons are widely used in packing algorithms and the sweep line technique is used for many problems in computational geometry and other areas. Their algorithm enumerates bottom-left stable positions from bottom to top (from left to right for positions with an identical y-coordinate), and it outputs the bottom-left position first in $O(k \log k)$ time.

For the same problem, Ahmadinia et al. [43] and Zhu et al. [44] gave algorithms that run in $O(k \log k + p)$ time, which are more efficient when the number of bottom-left stable positions p is large (it is known that $p = O(k^2)$). Zhu et al. [44] also considered a similar problem in the three-dimensional space.

32.6 Metaheuristics for Rectangle Packing

In the last two decades, many local search and metaheuristic algorithms for rectangle packing problems have been proposed. Dowsland [45] is one of the early researchers who implemented metaheuristics for rectangle packing problems. Her SA algorithm explores both feasible and infeasible (i.e., some items overlap) solutions. During the search, the objective is to reduce the overlapping area. Computational results for small problem instances were reported in Reference [45].

We then explain some metaheuristic algorithms based on the permutation coding scheme. Such algorithms usually consist of two phases: (1) find a good permutation using metaheuristics and (2) place the items one by one following the permutation order. Jakobs [28] proposed a metaheuristic algorithm for the strip packing problem. In this algorithm, he used a genetic algorithm (GA) to find a good permutation, and to place items, the bottom-left strategy explained in the previous section. He treated not only rectangle packing problems but also irregular packing problems and reported several computational results. Liu and Teng [29] also proposed a GA using their bottom-left type decoding algorithm. More recent applications of GAs for searching good permutations were presented in Hadjiconstantinou and Iori [46] and Burke et al. [47].

Hopper and Turton [31] compared the performance of various metaheuristics (multistart local search (MLS), SA, GA, etc.) with two decoding rules on small and large test instances. The first decoding rule is the BL heuristic proposed by Jakobs [28], and the second one is the BLF strategy proposed by Baker et al. [26]. Their results indicate that the performance of metaheuristic algorithms is strongly dependent on the decoding rule and the instance size. Moreover, it is also reported that certain computation time is needed for metaheuristics to attain better solutions than those obtained by well-designed heuristics such as BLF-DW (decreasing width) and BLF-DH (decreasing height) , which are the heuristics that apply the BLF strategy to the order of decreasing width and decreasing height, respectively.

Lesh et al. [48] proposed a stochastic search variation of the bottom-left heuristics for the strip packing problem. Their algorithm outperforms simple heuristic and metaheuristic algorithms based on the bottom-left strategy reported in the literature. Furthermore, they incorporated their algorithm in an interactive system that combines the advantages of computer speed and human reasoning. Using the interactive system, they succeeded in quickly producing significantly better solutions than their original algorithm.

Burke et al. proposed an SA enhancement [49] and a squeaky wheel optimization method [50] for the strip packing problem, which are based on the best-fit heuristic algorithm proposed in Reference [30]. Wei et al. [51] also proposed an intelligent search algorithm based on the best-fit heuristic and attained good computational results for most of the benchmark instances.

We now discuss metaheuristic algorithms based on other coding schemes. For the sequence pair representation, Murata et al. [16] proposed an SA algorithm and Imahori et al. [19] proposed an iterated local search (ILS) algorithm. To find good coded solutions they used metaheuristics in which each coded solution is evaluated with their own decoding algorithms. One of the advantages of the above-mentioned algorithms is generality: Imahori et al. [19] incorporated "spatial cost functions" into their algorithm, which were used to handle various types of rectangle packing problems and scheduling problems. Chang et al. [23] and Nakatake et al. [21] proposed SA algorithms using B*-trees and BSG, respectively. They designed algorithms that can treat various rectangle packing problems such as a problem with pre-placed items and a problem in which items can be transformed.

Imahori et al. [52] proposed an improved metaheuristic algorithm based on the sequence pair representation. Metaheuristic algorithms generate numerous number of coded solutions and evaluate all of them. Hence, the efficiency of metaheuristic algorithms strongly depends on the time complexity of decoding algorithms. In Reference [52], they proposed new decoding algorithms to efficiently evaluate all coded solutions in various neighborhoods. As a result, they attained amortized constant time to evaluate one coded solution in basic neighborhoods.

We finally mention some algorithms that search for good layouts without coding schemes. Bortfeldt [53] proposed a GA for the strip packing problem that works without any encoding of solutions. Instead of using a coding scheme, fully defined layouts are directly manipulated by specific genetic operators. Zhang et al. [54] presented another GA, and Alvarez-Valdes et al. [55] presented a greedy randomized adaptive search procedure (GRASP) for the strip packing problem.

32.7 Conclusion

In this chapter, we surveyed practical algorithms for the two-dimensional rectangle packing problem. For the rectangle packing problem, we first introduced some coding schemes (in other words, how to represent a solution) such as permutation, binary tree, sequence pair, and BSG. We then explained various heuristic and metaheuristic algorithms, most of which are based on these coding schemes. We also introduced algorithms to enumerate bottom-left stable positions in layouts with some already placed rectangles.

The survey in this chapter is by no means comprehensive, but we hope this article gives valuable information to the readers who are interested in devising practical algorithms for the rectangle packing problem. Fortunately, there are many survey papers on cutting and packing problems. Dychhoff [1] and Wäscher et al. [2] presented typologies of cutting and packing problems and categorized existing literature. Survey papers on the two-dimensional packing problems by Hopper and Turton [31], Lodi et al. [56], and Oliveira et al. [57] are useful for readers. Several chapters of this handbook are also useful for readers who are interested in related topics.

References

1. Dyckhoff, H., A typology of cutting and packing problems, *Eur. J. Oper. Res.*, 44, 145, 1990.
2. Wäscher, G., Haußner, H., and Schumann, H., An improved typology of cutting and packing problems, *Eur. J. Oper. Res.*, 183, 1109, 2007.
3. Lodi, A., Martello, S., and Vigo, D., Heuristic and metaheuristic approaches for a class of two-dimensional bin packing problems, *INFORMS J. Comput.*, 11, 345, 1999.
4. Charalambous, C. and Fleszar, K., A constructive bin-oriented heuristic for the two-dimensional bin packing problem with guillotine cuts, *Comput. Oper. Res.*, 38, 1443, 2011.
5. Wu, Y.L., Huang, W., Lau, S., Wong, C.K., and Young, G.H., An effective quasi-human based heuristic for solving the rectangle packing problem, *Eur. J. Oper. Res.*, 141, 341, 2002.
6. Egeblad, J. and Pisinger, D., Heuristic approaches for the two- and three-dimensional knapsack packing problem, *Comput. Oper. Res.*, 36, 1026, 2009.
7. Leung, S.C.H., Zhang, D., Zhou, C., and Wu, T., A hybrid simulated annealing metaheuristic algorithm for the two-dimensional knapsack packing problem, *Comput. Oper. Res.*, 39, 64, 2012.
8. Gilmore, P.C. and Gomory, R.E., Multistage cutting stock problems of two and more dimensions, *Oper. Res.*, 13, 94, 1965.

9. Alvarez-Valdés, R., Parajón, A., and Tamarit, J.M., A tabu search algorithm for large-scale guillotine (un)constrained two-dimensional cutting problems, *Comput. Oper. Res.*, 29, 925, 2002.

10. Cintra, G.F., Miyazawa, F.K., Wakabayashi, Y., and Xavier, E.C., Algorithms for two-dimensional cutting stock and strip packing problems using dynamic programming and column generation, *Eur. J. Oper. Res.*, 191, 61, 2008.

11. Silva, E., Alvelos, F., and Valério de Calvalho, J.M., An integer programming model for two- and three-stage two-dimensional cutting stock problems, *Eur. J. Oper. Res.*, 205, 699, 2010.

12. Vanderbeck, F., A nested decomposition approach to a three-stage, two-dimensional cutting-stock problem, *Management Sci.*, 47, 864, 2001.

13. Morabito, R. and Morales, S., A simple and effective recursive procedure for the manufacturer's pallet loading problem, *J. Oper. Res. Soc.*, 49, 819, 1998.

14. Sliva, E., Oliveira, J.F., and Wäscher, G., The pallet loading problem: A review of solution methods and computational experiments, *Intl. Trans. Op. Res.*, 23, 147, 2016.

15. Wong, D.F. and Liu, C.L., A new algorithm for floorplan design, in *Proceedings of the Design Automation Conference (DAC)*, 1986, p. 101.

16. Murata, H., Fujiyoshi, K., Nakatake, S., and Kajitani, Y., VLSI module placement based on rectangle-packing by the sequence-pair, *IEEE Trans. CAD*, 15, 1518, 1996.

17. Takahashi, T., An algorithm for finding a maximum-weight decreasing sequence in a permutation, motivated by rectangle packing problem, Technical Report of the IEICE, VLD96, 1996, p. 31.

18. Tang, X., Tian, R., and Wong, D.F., Fast evaluation of sequence pair in block placement by longest common subsequence computation, *IEEE Trans. CAD*, 20, 1406, 2001.

19. Imahori, S., Yagiura, M., and Ibaraki, T., Local search algorithms for the rectangle packing problem with general spatial costs, *Math. Prog.*, 97, 543, 2003.

20. Pisinger, D., Denser packings obtained in $O(n \log \log n)$ time, *INFORMS J. Comput.*, 19, 395, 2007.

21. Nakatake, S., Fujiyoshi, K., Murata, H., and Kajitani, Y., Module packing based on the BSG-structure and IC layout applications, *IEEE Trans. Comput. Aided Des.*, 17, 519, 1998.

22. Guo, P.N., Takahashi, T., Cheng, C.K., and Yoshimura, T., Floorplanning using a tree representation, *IEEE Trans. Comput. Aided Des.*, 20, 281, 2001.

23. Chang, Y.C., Chang, Y.W., Wu, G.M., and Wu. S.W., B*-trees: A new representation for non-slicing floorplans, in *Proceedings of the Design Automation Conference (DAC)*, 2000, p. 458.

24. Lin, J.M. and Chang, Y.W., TCG: A transitive closure graph-based representation for non-slicing floorplans, in *Proceedings of the Design Automation Conference (DAC)*, 2001, p. 764.

25. Sakanushi, K., Kajitani, Y., and Mehta, D.P., The quarter-state-sequence floorplan representation, *IEEE Trans. Circuits Sys.*, 50, 376, 2003.

26. Baker, B.S., Coffman, Jr., E.G., and Rivest, R.L., Orthogonal packing in two dimensions, *SIAM J. Comput.*, 9, 846, 1980.

27. Chazelle, B., The bottom-left bin-packing heuristic: An efficient implementation, *IEEE Trans. Comput.*, 32, 697, 1983.

28. Jakobs, S., On genetic algorithms for the packing of polygons, *Eur. J. Oper. Res.*, 88, 165, 1996.

29. Liu, D. and Teng, H., An improved BL-algorithm for genetic algorithm of the orthogonal packing of rectangles, *Eur. J. Oper. Res.*, 112, 413, 1999.

30. Burke, E.K., Kendall, G., and Whitwell, G., A new placement heuristic for the orthogonal stock-cutting problem, *Oper. Res.*, 52, 655, 2004.

31. Hopper, E. and Turton, B.C.H., An empirical investigation of meta-heuristic and heuristic algorithms for a 2D packing problem, *Eur. J. Oper. Res.*, 128, 34, 2001.

32. Ortmann, F.G., Ntene, N., and van Vuuren, J.H., New and improved level heuristics for the rectangular strip packing and variable-sized bin packing problems, *Eur. J. Oper. Res.*, 203, 306, 2010.

33. Leung, S.C.H. and Zhang, D., A fast layer-based heuristic for non-guillotine strip packing, *Expert Syst. Appl.*, 38, 13032, 2011.

34. Thiebaut, D., 2D-packing images on a large scale: Packing a billion rectangles under 10 minutes, *Int. J. Adv. Syst. Measure.*, 7, 80, 2014.

35. Imahori, S. and Yagiura, M., The best-fit heuristic for therectangular strip packing problem: An efficient implementation and the worst-case approximation ratio, *Comput. Oper. Res.*, 37, 325, 2010.

36. Aşik, Ö.B. and Özcan, E., Bidirectional best-fit heuristic for orthogonal rectangular strip packing, *Annals Oper. Res.*, 172, 405, 2009.

37. Verstichel, J., De Causmaecker, P., and Berghea, G.V., An improved best-fit heuristic for the orthogonal strip packing problem, *Intl. Trans. Op. Res.*, 20, 711, 2013.

38. Wei, L., Oon, W.-C., Zhu, W., and Lim, A., A skyline heuristic for the 2D rectangular packing and strip packing problems, *Eur. J. Oper. Res.*, 215, 337, 2011.

39. Healy, P., Creavin, M., and Kuusik, A., An optimal algorithm for rectangle placement, *Oper. Res. Letters*, 24, 73, 1999.

40. Imamichi, T., Yagiura, M., and Nagamochi, H., An iterated local search algorithm based on nonlinear programming for the irregular strip packing problem, *Dis. Opt.*, 6, 345, 2009.

41. Li, Z. and Milenkovic, V., Compaction and separation algorithms for non-convex polygons and their applications, *Eur. J. Oper. Res.*, 84, 539, 1995.

42. Imahori, S., Chien, Y., Tanaka, Y., and Yagiura, M., Enumerating bottom-left stable positions for rectangle placements with overlap, *J. Oper. Res. Soc. Japan*, 57, 45, 2014.

43. Ahmadinia, A., Bobda, C., Fekete, S., Teich, J., and van der Veen, J.C., Optimal free-space management and routing-conscious dynamic placement for reconfigurable devices, *IEEE Trans. Comput.*, 56, 673, 2007.

44. Zhu, W., Luo, Z., Lim, A., and Oon, W.C., A fast implementation for the 2D/3D box placement problem, *Comput. Optim. Appl.*, 63, 585, 2016.

45. Dowsland, K., Some experiments with simulated annealing techniques for packing problems, *Eur. J. Oper. Res.*, 68, 389, 1993.

46. Hadjiconstantinou, E. and Iori, M., A hybrid genetic algorithm for the two-dimensional single large container placement problem, *Eur. J. Oper. Res.*, 183, 1150, 2007.

47. Burke, E.K., Hyde, M., Kendall, G., and Woodward, J., A genetic programming hyperheuristic approach for evolving 2-D strip packing heuristics, *IEEE Trans. Evolut. Comput.*, 14, 942, 2010.

48. Lesh, N., Marks, J., McMahon, A., and Mitzenmacher, M., New heuristic and interactive approaches to 2D rectangular strip packing, *ACM J. Experimental Algorithmics*, 10, 1, 2005.

49. Burke, E.K., Kendall, G., and Whitwell, G., A simulated annealing enhancement of the best-fit heuristic for the orthogonal stock-cutting problem, *INFORMS J. Comput.*, 21, 505, 2009.

50. Burke, E.K., Hyde, M.R., and Kendall, G., A squeaky wheel optimisation methodology for two-dimensional strip packing, *Comput. Oper. Res.*, 38, 1035, 2011.

51. Wei, L., Qin, H., Cheang, B., and Xu, X., An efficient intelligent search algorithm for the two-dimensional rectangular strip packing problem, *Intl. Trans. Op. Res.*, 23, 65, 2016.

52. Imahori, S., Yagiura, M., and Ibaraki, T., Improved local search algorithms for the rectangle packing problem with general spatial costs, *Eur. J. Oper. Res.*, 167, 48, 2005.

53. Bortfeldt, A., A genetic algorithm for the two-dimensional strip packing problem with rectangular pieces, *Eur. J. Oper. Res.*, 172, 814, 2006.

54. Zhang, D.-F., Chen, S.-D., and Liu, Y.-J., An improved heuristic recursive strategy based on genetic algorithm for the strip rectangular packing problem, *Acta Automat. Sinica*, 33, 911, 2007.

55. Alvarez-Valdes, R., Parreño, F., and Tamarit, J.M., Reactive GRASP for the strip-packing problem, *Comput. Oper. Res.*, 35, 1065, 2008.

56. Lodi, A., Martello, S., and Monaci, M., Two-dimensional packing problems: A survey, *Eur. J. Oper. Res.*, 141, 241, 2002.

57. Oliveira, J.F., Neuenfeldt, A., Silva, E., and Carravilla, M.A., A survey on heuristics for the two-dimensional rectangular strip packing problem, *Pesquisa Operacional*, 36, 197, 2016.

33

Practical Algorithms for Two-Dimensional Packing of General Shapes

Yannan Hu

Hideki Hashimoto

Shinji Imahori

Mutsunori Yagiura

33.1 Introduction

Two-dimensional packing problems involve packing a set of two-dimensional items into larger rectangular containers such that no item overlaps with each other. Items to be packed can take various shapes, such as the rectilinear blocks, raster graphics, and irregular polygons. These problems are classical combinatorial optimization problems and known to be NP-hard. Many packing problems are related to real-world applications in the wood, glass, steel, and leather industries, as well as in *large-scale integration* (*LSI*) and *very-large-scale integration* (*VLSI*) design and newspaper paging. They have been studied for a long time from both theoretical and practical points of view in various areas including operations research, computer science, and manufacturing.

Cutting problems are closely related to packing problems, where a cutting problem is the problem of cutting some pieces of stock material into pieces of specified sizes so as to minimize the material waste. Although they are different from packing problems from the viewpoint of applications, their problem structures are the same from the viewpoint of optimization, and hence they are often called *cutting and packing problems*. Dyckhoff [1] and Wäscher et al. [2] gave typologies for cutting and packing problems, in which they classified the cutting and packing problems by dimensionality (i.e., one-, two-, and three-dimensional), objective functions, and shapes of items.

In this chapter, we survey heuristics, metaheuristics, and exact algorithms for two-dimensional packing problems of general shapes. Considering the geometric representation, irregular shapes are usually represented by a list of vertices (*vector format* or *vector representation*) or approximately represented in *bitmap format* (also called *raster format* or *grid approximation*). The bitmap format is a digitized representation in which shapes are represented by two-dimensional matrices of small dots (Figure 33.1b).

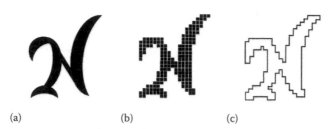

FIGURE 33.1   An arbitrarily shaped object represented as a bitmap shape: (a) orbitrary shape, (b) bitmap shape, and (c) rectilinear shape.

Regarding each dot as a square, the resulting bitmap matrix that represents a shape can be treated as a *rectilinear block*, where a rectilinear block is a polygonal block whose interior angles are either 90° or 270° (see Figure 33.1c). Figure 33.1 shows an example of representing an arbitrary shape in bitmap format and by a rectilinear block.

We focus on packing problems of irregular shapes represented in vector or bitmap format and those of rectilinear blocks. Section 33.2 defines the irregular strip packing problem and introduces some practical algorithms for irregular packing problems. Section 33.3 focuses on the problem of packing irregular shapes represented in bitmap format. Section 33.4 first defines the rectilinear block packing problem and introduces various schemes used to represent solutions to the rectilinear block packing problem and algorithms based on such coding schemes.

33.2 Irregular Packing Problems

In this section, we consider two-dimensional irregular packing problems, which have been actively studied in the last decade. Irregular packing problems have many practical applications in various industries such as the garment, shoe, and shipbuilding industries, and many variants have been considered in the literature. An irregular packing problem involves packing a set of arbitrarily shaped objects, called items, into containers, and the objective is to pack all the items into a single container as densely as possible, or pack all items into as few containers as possible.

Many of the two-dimensional packing problems are related to real-world applications and they have been studied for a long time from both theoretical and practical points of view. There are many survey papers in the literature. Hopper and Turton [3] gave a review of metaheuristics developed for two-dimensional packing problems, in which they experimentally analyzed several methods including genetic algorithms, simulated annealing, tabu search, and artificial neural networks. Bennell and Oliveira [4] also conducted a survey of irregular shape packing problems. Gomes [5] surveyed geometric representations and efficient tools for irregular packing problems. He focused on comparing representative approaches in the literature and gave different ways of representing irregular shapes.

We start by introducing in Section 33.2.1 some basic techniques including bottom-left strategy, clustering strategy, and no-fit polygons. We then show in Sections 33.2.2 and 33.2.3 various algorithms proposed for the irregular strip packing and bin packing problems and their variants.

33.2.1 Basic Techniques

There are many strategies and algorithms proposed for two-dimensional packing problems. The *bottom-left* strategy [6] and the *clustering* strategy [7] are known as the most popular strategies for these problems.

The bottom-left strategy, which was originally proposed for the rectangle strip packing problem [6], packs items into the container one by one, each at the *bottom-left position*. The bottom-left position of an item *i* to be packed next relative to a packing layout is defined as the leftmost location among the lowest

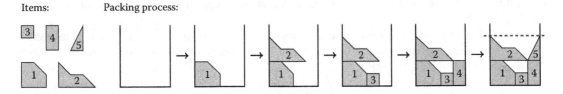

FIGURE 33.2 Bottom-left strategy.

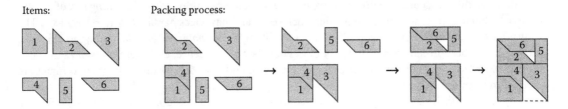

FIGURE 33.3 Clustering strategy.

bottom-left stable feasible positions, where a bottom-left stable feasible position is a location at which item i can be placed without overlap and cannot be moved leftward or downward. Note that the layout obtained by the bottom-left strategy is subject to the order it packs items, and various sequences are possible. For example, it is observed that good layouts are often obtained by packing items in the decreasing order of area or height [8]. In this chapter, we call the procedure of applying the bottom-left strategy to a given order the *bottom-left algorithm* (n.b., we assume that the choice of the sequence is outside of the algorithm and is given to it in addition to the instance data). Figure 33.2 shows an example of packing items in the order of indices according to the bottom-left strategy.

The clustering strategy repeatedly selects a small number of items from the remaining ones and packs them as compactly as possible and then treats the resulting combination of items as a new item. Figure 33.3 shows an example of the processing when items are packed according to the clustering strategy. Bennell et al. [9] proposed an efficient method for identifying a good clustering of two items.

The algorithm proposed in Reference [10] utilizes the concept of the bottom-left strategy. It first approximates every nonconvex polygon by a convex polygon that encloses it and then places them one by one at their bottom-left positions. Heuristics in References [11, 12] also take the bottom-left strategy when packing items. Recall that the bottom-left algorithm packs items one by one according to a given sequence, which indicates that different sequences lead to different packing layouts. Many approaches have been proposed for finding a good sequence [13–15].

Adamowicz and Albano [7] proposed an algorithm using the concept of the clustering strategy. Their algorithm first partitions the items into several groups, and then it generates for each group, a rectangle enclosure in which the polygons in the group are packed as compactly as possible.

Algorithms that take advantages of mathematical programming techniques for the irregular strip packing problem are proposed in References [16–18]. Bennell and Dowsland [16] proposed an algorithm that incorporates both the bottom-left algorithm and a linear-programming-based *compaction algorithm*, where a compaction algorithm translates the packed items continuously so as to minimize the height of the container. There also exists a *separation algorithm* [18], which is often used in combination with a compaction algorithm. The separation algorithm works for a packing layout in which there are some polygons having overlap with others. It translates the items in the layout continuously to make them separate with each other.

The main difference between rectangle and irregular packing problems is that the intersection test between irregular items is considerably more complicated than the case of rectangular items. To overcome this difficulty, some approximation techniques and geometric algorithms have been incorporated into packing algorithms.

The use of no-fit polygons (NFPs) is one of the most efficient geometric techniques for the intersection test between two polygons in two-dimensional space. This concept was introduced by Art [10] in 1960s, who used the term "shape envelope" to describe the positions at which two polygons can be placed without intersection. Albano and Sapuppo [13] proposed an algorithm to solve the irregular packing problem with this geometric technique, and this is the first paper in which the term "no-fit polygon" was used. This concept is also known as Minkowski sums, and it is utilized in various fields such as the motion planning problem for polygonal robots.

To describe this concept more precisely, we give below some definitions. The *bounding box* of a polygon P_j is defined as the smallest rectangle that encloses P_j and has edges parallel to the x- or y-axis. The location of P_j is described by the coordinate (x_j, y_j) of its *reference point*, where we define the reference point to be the bottom-left corner of its bounding box. For convenience, each item, as well as the container, is regarded as a set of points (including both interior and boundary points) whose coordinates are determined from the origin $O = (0, 0)$.

Let $P_j(x_j, y_j)$ be the polygon P_j placed at (x_j, y_j), which is the region occupied by P_j when its reference point is located at (x_j, y_j). The region of $P_j(x_j, y_j)$ can be represented by a *Minkowski sum*. The Minkowski sum of two sets $A \subset \mathbb{R}^2$ and $B \subset \mathbb{R}^2$ is defined as

$$A \oplus B = \{a + b \mid a \in A, b \in B\}, \tag{33.1}$$

where $a + b$ is the vector sum of a and b, and $\mathbb{R}$ is the set of real numbers. For convenience, when B consists of a single point b (i.e., $B = \{b\}$), $A \oplus \{b\}$ is denoted as $A \oplus b$. Then, a polygon P_j placed at $v_j = (x_j, y_j)$ is represented by $P_j(x_j, y_j) = P_j \oplus v_j = \{p + v_j \mid p \in P_j\}$.

An NFP is defined for an ordered pair of two polygons P_i and P_j, where the position of polygon P_i is fixed and polygon P_j can be moved. The NFP of P_j relative to P_i, $NFP(P_i, P_j)$, denotes the set of all positions of polygon P_j having an intersection with polygon P_i. For a polygon P_j, let $\text{int}(P_j)$ be the interior of P_j (i.e., the boundary of P_j is not included in $\text{int}(P_j)$). We assume that the position of polygon P_i is fixed at the origin $(0, 0)$. Then we have

$$NFP(P_i, P_j) = \{(x, y) \mid \text{int}(P_i(0, 0)) \cap P_j(x, y) \neq \emptyset\}. \tag{33.2}$$

For a point $r = (x, y)$ in the plane and a set $S \subset \mathbb{R}^2$, we define $-r = (-x, -y)$ and $-S = \{-r \mid r \in S\}$. In other words, $-S$ is obtained by reflecting S according to the origin. Then, $NFP(P_i, P_j)$ is formally defined by the Minkowski sum

$$NFP(P_i, P_j) = \text{int}(P_i) \oplus (-P_j) = \{u - w \mid u \in \text{int}(P_i), w \in P_j\}. \tag{33.3}$$

Note that $NFP(P_i, P_j)$ is an open set that consists of all points inside of a polygon except for boundary points. Let $\partial NFP(P_i, P_j)$ denote the boundary of $NFP(P_i, P_j)$, and $\text{cl}(NFP(P_i, P_j))$ denote the closure of $NFP(P_i, P_j)$, which means $\text{cl}(NFP(P_i, P_j)) = \partial NFP(P_i, P_j) \cup NFP(P_i, P_j)$. The NFP has the following important properties:

- $P_j(x_j, y_j)$ overlaps with $P_i(x_i, y_i) \iff (x_j, y_j) \in NFP(P_i, P_j) \oplus (x_i, y_i)$
- $P_j(x_j, y_j)$ touches $P_i(x_i, y_i) \iff (x_j, y_j) \in \partial NFP(P_i, P_j) \oplus (x_i, y_i)$
- $P_j(x_j, y_j)$ and $P_i(x_i, y_i)$ are separated $\iff (x_j, y_j) \notin \text{cl}(NFP(P_i, P_j)) \oplus (x_i, y_i)$

Utilizing the technique of NFP, the problem of checking whether two polygons overlap or not becomes an easier problem of checking whether a point is in a polygon or not. To achieve a feasible (i.e., without overlap) and tight layout, each item should have its reference point on the boundary of at least one NFP and not in the interior of any NFPs.

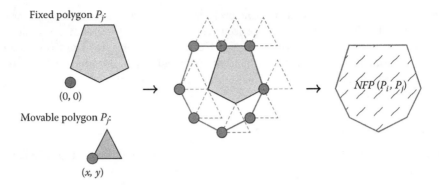

FIGURE 33.4 An example of $NFP(P_i, P_j)$.

When P_i and P_j are both convex, $\partial NFP(P_i, P_j)$ can be computed by the following simple procedure: Slide P_j around P_i having it keep touching with P_i. Then the trace of the reference point of P_j is $\partial NFP(P_i, P_j)$. Such an $NFP(P_i, P_j)$ is illustrated through the example in Figure 33.4.

There have been many different approaches to producing NFPs for nonconvex shapes, such as Minkowski sum [19], decomposition [20], and orbital approaches [21, 22].

Avnaim and Boissonnat [20] proposed a decomposition approach to produce NFPs and mathematically proved that their algorithm works correctly for the case when both polygons may not be convex and not even connected. The time complexity of their algorithm is $O(m^3 n^3 \log mn)$, where m and n are the number of vertices of the two polygons.

Burke et al. [21] proposed a robust orbital sliding approach to produce NFPs that uses standard trigonometry to physically slide one polygon around another. Their approach handles degenerate cases such as holes, interlocking concavities, and jigsaw type pieces. They then gave a new construction algorithm consisting of two geometric stages for robust NFP generation using their orbital method. The computational results on benchmark instances showed that the running time required to produce an NFP by their new approach is reasonable (within 0.23 seconds on a PC with a 2 GHz CPU) and that the resulting NFPs are consistent with those generated by the Minkowski approach in Reference [19]. They further extended the orbital sliding method in Reference [22] of calculating NFP so that it can also handle curved arcs (i.e., not only straight edges). They also showed that the resulting NFPs are successfully used to solve the irregular strip packing problem. This is the first paper in which algorithms based on the NFP technique could deal with curved arcs robustly without approximately representing them by decomposing each curved arc into line segments. Their algorithms obtained high quality (including several best-known) solutions on benchmark instances in a few minutes on a 2 GHz Intel Pentium 4 processor with 256 MB RAM.

33.2.2 Irregular Strip Packing Problem

There are a variety of objective functions for packing problems, and some typical ones are shown as follows:

- Minimize the height of the container whose width is fixed (strip packing problem),
- Minimize the area of the container whose width and height are variables (area minimization problem),
- Minimize the number of used containers whose shapes are fixed (bin packing problem),
- Maximize the total profit of the packed items in the container whose width and height are fixed (knapsack problem).

In the strip packing, area minimization, and knapsack problem, only one container is considered. On the other hand, in the bin packing problem, containers with a fixed shape (i.e., width and height are fixed) are considered and the objective is to minimize the number of used containers to pack all the given items.

Among all types of two-dimensional packing problems, the *irregular strip packing problem* has been intensively investigated. This problem is categorized as *open dimension problem* in the typology proposed by Wäscher et al. [2]. In the irregular strip packing problem, we are given a set of items of arbitrary shapes, and a rectangular container (called a strip) with fixed width and unrestricted height. The task is to pack all the items without overlap into the strip so as to minimize the height of the strip. The irregular strip packing problem is formally described as follows.

Irregular strip packing problem:

> *Instance:* A set of items of arbitrary shapes, and a rectangular strip with fixed width W and unrestricted height H.
>
> *Task:* Pack all the items without overlap into the strip so as to minimize height H.

Three cases of this problem are often considered in the literature:

1. All the items are not allowed to be rotated,
2. All the items can be rotated by a few specified degrees (e.g., $0°$ and $180°$),
3. All the items can be rotated arbitrarily.

Case 1 is assumed in this manuscript unless otherwise stated. It is often easy to apply the results for case 1, in which rotation is not allowed, to case 2, in which rotation of a few specified degrees is allowed.

In the following sections, we introduce some existing heuristic and exact methods for the irregular strip packing problem (and some for the area minimization problem), including construction heuristics, metaheuristics, and exact and heuristic algorithms based on mathematical programming techniques. For some of them, we summarize their computational results for representative benchmark instances in Tables 33.2 and 33.3 in Section 33.2.2.4.

33.2.2.1 Construction Heuristics

This section introduces some construction heuristics proposed for the irregular strip packing problem. Most of them use a permutation to represent a solution and consider effective strategies to pack the items one by one into the strip.

Albano and Sapuppo [13] proposed a construction algorithm to solve the irregular strip packing problem using NFPs. They approached the problem using a bottom-left algorithm, which utilized the NFP to reduce the geometric complexity of the packing process. Their algorithm places each item one by one at the top frontier called the leading edge of the current layout only, that is, without hole filling capabilities. See Figure 33.5 for an example of the leading edge and holes in such a layout, where the layout in the figure is rotated $90°$ clockwise, following a convention often adopted in the literature of irregular packing. Blazewicz et al. [23] presented an extension of the work performed by Albano and Sapuppo [13]. Their method is an extension of the bottom-left-fill algorithm; that is, their approach attempts to fill holes in the existing layout before attempting to place an item on the leading edge. Their algorithm utilizes the tabu search technique to produce moves from one solution to another.

Oliveira et al. [15] also tackled the irregular strip packing problem using NFPs. Their algorithm places all items one by one at a nonoverlapping position in such a way that the item to be placed touches at least one item already placed. Several criteria to choose the next item to place and its orientation were proposed (in Reference [15], they treated a problem such that each item can be rotated by some fixed degrees). Different evaluation functions were also proposed to evaluate partial solutions and to decide the position of each item. A total of 126 variants of the algorithm, generated by the complete set of combinations of criteria and evaluation functions, were computationally compared. In their computational experiments, they solved several types of test instances: test instances from a fabric cutting company and a test instance generated by Blazewicz et al. [23]. Oliveira et al. compared their results against an implementation of Albano and Sapuppo's algorithm [13], and against the results from Blazewicz et al. [23]. In some cases,

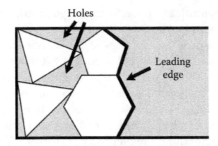

FIGURE 33.5 The leading edge and holes in a layout of irregular shapes.

their new algorithm generated better solutions than the best known solutions in the literature; more precisely, their algorithm generated solutions that ranged from 6.2% better to 4% worse than the best known results before 2000.

Bennell and Song [24] used an order of items to represent a solution of the irregular strip packing problem. They modified the placement heuristic [15] with a powerful NFP generator and evaluation criteria to decode the order. They gave a beam search algorithm to search for orders. The computational results on benchmark instances are shown in columns "BS" of Tables 33.2 and 33.3.

33.2.2.2 Metaheuristics

Some metaheuristics represent solutions by permutations (i.e., a permutation of items is a coded solution). Most of these algorithms begin with an initial permutation, which is usually randomly generated, and then they apply metaheuristics such as genetic algorithms, simulated annealing, and tabu search to improve solutions.

Jakobs [25] used an approximation scheme: for all irregular shapes, the minimum bounding rectangles of the given items are calculated and his algorithm treats these rectangles instead of the original shapes. For these shapes, a good placement is computed by his bottom-left decoding algorithm and genetic algorithm. After finding a good placement for the rectangles, the algorithm replaces rectangles with the original irregular shapes and computes a better placement of the irregular items. Jakobs also discussed an idea of clustering several shapes and finding the minimum bounding rectangle of several items.

Gomes and Oliveira [14] developed shape-ordering heuristics for an extended irregular packing algorithm similar to that given by Oliveira et al. [15]. The algorithm is improved by the introduction of the inner-fit rectangle, which is derived from the concept of NFP and represents the feasible set of points for placing a new polygon inside the container. In addition to this extension of geometric techniques, the paper introduces a 2-exchange heuristic for manipulating a permutation that specifies the order of placing items one by one. They generated some initial permutations with various criteria, such as random, decreasing order of area, and decreasing order of the longest length of items, and they improved such permutations using the 2-exchange heuristic over a number of iterations. In Reference [14], they conducted thorough computational experiments and compared the proposed algorithm with existing algorithms. They improved almost all best known solutions for well-known benchmark instances (except for an instance called SHAPE0).

Gomes and Oliveira [17] also developed a hybrid algorithm of simulated annealing and linear programming-based compaction technique to solve the irregular strip packing problem. In this algorithm, a neighborhood structure based on the exchange of items in the layout was used for simulated annealing. For a given layout (which can be neither feasible nor tight), they solved a linear programming (LP) problem to locally optimize the layout. Computational tests were conducted using 15 benchmark instances that are commonly used in the literature, and the best results published by then were improved for all instances by their new algorithm. The results obtained by their *simulated annealing hybrid algorithm* (*SAHA*) are shown in column "SAHA" of Tables 33.2 and 33.3 to compare it with other algorithms in the literature.

Burke et al. [11] proposed a heuristic algorithm for the irregular strip packing problem with new shape overlap resolution techniques applied to the given shapes directly (i.e., without reference to NFPs). In this article, they treated not only polygons (i.e., shapes with line representation) but also shapes that incorporate circular arcs and holes, and they proposed overlap resolution techniques for line and line, line and arc, and arc and arc. Items are placed one by one according to a permutation (coded solution) with the bottom-left-fill strategy, and tabu search is used to find a good permutation. They conducted computational experiments on 26 existing benchmark instances and 10 new test instances with items having circular arcs and holes. Their technique produced 25 new best solutions for the 26 existing benchmark instances; most of them were found within 5 minutes on a PC with a 2 GHz CPU.

Instead of representing a solution by a permutation, Dighe and Jakiela [26] considered a tree structure to represent a solution. They proposed an algorithm for the irregular strip packing problem, which is based on a clustering method with a tree structure. Their algorithm uses a tree structure to represent a solution: The leaves of a tree correspond to items, and clustering operations are applied to items from the leaves to the root of the tree. To find a good coded solution (i.e., tree), they utilized a genetic algorithm. Dighe and Jakiela created new test instances and conducted computational experiments on these instances.

There is also research using a similar concept to solve the irregular area minimization problem, in which both the width and height of the rectangular container are variables and the objective is to minimize the area of the container. Martins and Tsuzuki [27] considered the irregular area minimization problem. They considered the case in which all the items are rotatable and nonconvex. They proposed a construction algorithm with parameters controlled by a simulated annealing algorithm. The construction algorithm packs items one by one in a given sequence at a position in a region called the *collision free region*, originally introduced in Reference [28], which descries all feasible positions for each item relative to the current packing layout. The placement of an item is determined by two parameters (among the set of parameters controlled by simulated annealing): one specifies the rotation angle of the item and the other is used to determine its relative position to the container and to items already packed. They also considered two simple heuristic methods to define the sequence.

There are algorithms that transform the irregular strip packing problem to an *overlap minimization problem*, whose objective is to find a placement (not necessarily feasible) of given items in a container with fixed dimensions so as to minimize an overlap penalty function. Such algorithms find a layout without overlap by trying various heights of the container and solving the overlap minimization problem for each of such heights.

Imamichi et al. [29] proposed an iterated local search (ILS) based on nonlinear programming for the irregular strip packing problem. They considered an overlap minimization problem consolidating a separation algorithm based on nonlinear programming. They used the *penetration depth*, an important notion used for robotics and computer vision, to evaluate overlap penalty of two items [30]. The penetration depth of two overlapping polygons is defined to be the minimum translational distance to separate them. Their ILS also uses swap operations of items to apply big changes to the layout. They ran their algorithm 10 times for each instance with a time limit of 600 or 1200 seconds for each run on a PC with an Intel Xeon 2.8 GHz processor (NetBurst microarchitecture) and 2 GB memory. Their approach was shown to be competitive through computational results, updating the best known solutions before 2009 for several benchmark instances. We show their results on 15 benchmark instances in column "ILSQN" of Tables 33.2 and 33.3 to compare them with those of other representative algorithms in the literature.

Umetani et al. [31] proposed a heuristic algorithm called the *fast iterative translation search* (FITS) for the irregular strip packing problem. They also transformed the irregular strip packing problem to an overlap minimization problem. To compute overlap penalty of a pair of two items, they proposed an evaluation criterion, the *directional penetration depth*, which is defined as the minimum translational distance in a specified direction to separate the two items. They gave an efficient algorithm to find a position with the minimum overlap penalty for each item when it is translated in a specified direction. They proposed a local search algorithm that translates an item in horizontal and vertical directions alternately until no better position is found in either direction. This is incorporated into a variant of a guided local search approach. Their algorithm was tested on 15 well-known benchmark instances and the results were compared with those reported for four algorithms in the literature [11, 17, 29, 32]. The algorithm FITS obtained the best results for four and five instances out of 15 instances in the best and average occupation rate of 10 runs, respectively, with a time limit of 1200 seconds for each run on a PC with an Intel Xeon 2.8 GHz processor (NetBurst microarchitecture) and 2 GB memory. It also obtained results with almost equivalent efficiencies to the best results before 2009 for some instances. Its results are shown in columns "FITS" of Tables 33.2 and 33.3.

Leung et al. [33] proposed an extended local search algorithm for the irregular strip packing problem. They considered an unconstrained nonlinear programming model to minimize the overlap during the search process. They also adopted a tabu search algorithm and a compaction algorithm to make the search efficient. Their algorithm was tested on 15 instances on a PC with a Pentium 4, 2.4 GHz processor and 2 GB memory. The results are reported in columns "ELS" of Tables 33.2 and 33.3.

Elkeran [34] proposed an algorithm that hybridizes cuckoo search and guided local search for the irregular strip packing problem. He also considered an overlap minimization problem and adopted the overlap penalty function proposed in Reference [29]. To reduce the search space, he also proposed a pairwise clustering approach that discovers matched features of congruent polygons and groups them in pairs. The algorithm obtained high quality solutions for 15 benchmark instances including two proven optimal solutions. He also compared his computational results with those obtained by the algorithms in References [29, 32, 33, 35], and the results showed that the proposed algorithm obtained better solutions than those obtained by these algorithms for all the 15 benchmark instances. The computational results obtained by his algorithm, *guided cuckoo search* (*GCS*), are shown in columns "GCS" of Tables 33.2 and 33.3.

Sato et al. [35] proposed a two-level heuristic for the irregular strip packing problem. They adopted the concept of collision free region to obtain a feasible placement. Their two-level algorithm consists of two parts: the inner level applies a simulated annealing algorithm during which the height of the strip is fixed, and the external level controls the height of the strip and the initial temperature. They tested their algorithm on benchmark instances on a PC with 4 GB memory and a Core i7 860 processor whose frequency is 2.8 GHz. The results showed that their algorithm performed better than existing algorithms for some instances. We report their computational results on benchmark instances in columns "CFREFP" (collision free region exact fitting placement) of Tables 33.2 and 33.3. They then proposed in Reference [36] an algorithm that also adopted the concept of collision free region. It uses a pairwise placement strategy such that the second item is always placed at an *exact fitting* or *exact sliding position*. An exact fitting position is a place at which the item's movement is fully constrained by its surroundings, and an exact sliding position is a place at which the item's movement is constrained in all but one direction. It packs items one by one in a given sequence at such positions. Because there may exist more than one such position for an item, they proposed three deterministic heuristics to decide positions for items. It also uses a simulated annealing algorithm to find a good sequence of items.

33.2.2.3 Algorithms Using Mathematical Programming Techniques

Some heuristics and exact algorithms use mathematical programming techniques to solve the irregular strip packing problems.

Alvarez-Valdes et al. [37] proposed an exact branch-and-bound algorithm for the irregular strip packing problem. They reviewed and tested the existing integer programming (IP; also known as integer linear programming) models [11, 17, 24, 29, 38] and proposed two models based on the ideas in Reference [38], called *HS1* and *HS2*. Both HS1 and HS2 considered constraints called bound constraints to ensure that each item does not exceed the boundaries of the strip. The HS1 model considers the original simple bound constrains, whereas HS2 lifts the bound constraints using the interaction between pairs of items. They considered three branching strategies, the *Fischetti and Luzzi* (*FL*) strategy [38], a dynamic strategy, and an alternative branching strategy to branch on constraints. The branching strategies were tested on instances generated from benchmark instances and the results showed that the FL strategy performed better when the items have been ordered by nondecreasing area. The proposed models are more efficient than the algorithm in Reference [17] and *HS2* works better than *HS1*.

Santoro and Lemos [39] also proposed an exact algorithm for the irregular strip packing problem. They considered special convex shapes such that they have only 3–8 edges and their opposite edges are parallel or null. Utilizing the characteristics of the items, they proposed a mixed integer programming (MIP; also known as mixed integer linear programming (MILP)) model. The proposed model was solved by CPLEX and tested on 270 small-scale instances that were generated for this particular problem. The results showed that the MIP model can achieve optimal solutions for relatively small instances (with less than 17 simple shaped pieces) in reasonable time.

Cherri et al. [40] proposed robust MIP models. Most of the MIP models for the irregular strip packing problems are limited to the handling of nonoverlapping constraints among items. Cherri et al. proposed two robust MIP models: one states the nonoverlapping constraints based on direct trigonometry [41] and the other first divides items into convex shapes and then represents the nonoverlapping constraints based on NFPs of the convex shapes. The two approaches are robust both in terms of numeric stability and on the type of geometries they can address, considering any kind of nonconvex pieces with any type of holes inside them. The proposed models were tested on 35 benchmark instances and the results showed that the model based on NFP performed better than that based on direct trigonometry and the best exact model in the literature proposed in Reference [37]. The model based on NFP solved 77% of the tested instances to optimality in one hour and ran faster than other models on 57% of the instances whose optimal solutions were known. For instances whose optimality was not proven, it obtained feasible solutions with small gaps. The computational results on a set of instances, which were created based on real-world instances, showed that the proposed model was able to solve such instances that the shapes have holes and it obtained optimal solutions for instances with less than 10 items and feasible solutions for instances with up to 45 items.

Some algorithms are based on the idea of representing items by a set of circles. With this idea, it often becomes easy to allow continuous rotation of items.

Imamichi [42] proposed a multi-sphere scheme for the two- and three-dimensional packing problems, in which each two-dimensional (resp., three-dimensional) item is approximated by a set of circles (resp., spheres). The multi-sphere scheme consists of three procedures: approximation, local search, and global search. The approximation procedure generates a set of spheres that represents each item (including the container) so that all the spheres are contained in the interior of the item. It then transforms the strip packing problem to an overlap minimization problem as in Reference [29] and applies a local search to find a layout without overlap. The local search is based on the quasi-Newton method for solving an unconstrained nonlinear programming problem that minimizes an overlap penalty function. To find a layout without penetration and protrusions, the basic local search is incorporated in an iterated local search algorithm, in which the initial solutions for the local search are generated by perturbing good solutions obtained by then by swapping two randomly chosen items and rotating them randomly.

Jones [43] proposed an exact algorithm for a more general irregular strip packing problem in which both the container and the items can be arbitrary nonconvex polygons and the items can have holes. The key idea is to inscribe a few circles in each irregular shape and then relax the nonoverlap constraints for the shapes by replacing them with nonoverlap constraints for the inscribed circles. The relaxed problems are formulated as *quadratic programming* (QP) problems and the resulting optimal values are used as lower bounds. The upper bounds are computed by a local search approach with strict nonoverlapping constraints. For items that overlap with others in the solution of the relaxed problem, it creates new circles for them to forbid this overlapping configuration and then solves the resulting QP problem to obtain improved lower bounds. He gave a branch-and-bound algorithm combining it with some heuristics to generate initial sets of circles to represent items. He tested his algorithm on three instances made from an instance with four complicated shapes in Reference [44], one consisting of the first two, another the first three, and the other all four shapes. The algorithm proved global optimality when solving the instances with two or three shapes, and when the instance with all four shapes was considered, the algorithm found the best known solution without a guarantee of optimality.

Rocha et al. [45] considered the irregular strip packing problem with continuous rotations. They used the *circle covering* geometric representation in which each irregular shape is represented by a set of circles. Circle covering representation usually leads to high-quality solutions and with this representation, it is easy to deal with rotations. Differently from the algorithm by Jones [43], which computes lower bounds by solving a relaxation problem derived by representing irregular shapes by sets of circles that may not cover them completely, the algorithm in Reference [45] represents every item by circles in such a way that they completely cover the item. They proposed a two-phase approach. In the first phase, the algorithm divides items into groups and uses a nonlinear programming model proposed in Reference [46] to generate a layout of large items. Then, the second phase places the remaining small items into the holes of the layout obtained in the first phase and compacts all the items. The computational results showed that their algorithm obtained better solutions in about five times the running time spent by existing algorithms that use compaction methods.

33.2.2.4 Computational Results

We report computational results obtained by some heuristic and metaheuristic algorithms for 15 benchmark instances of the irregular strip packing problem in Tables 33.2 and 33.3. The characteristics of these benchmark instances are reported in Table 33.1. The columns of "Instance," "Shapes," "#Pieces," "Avg_vertices," "Rotations," and "Width" report for each instance the name, number of different polygons, total number of polygons, average number of vertices of polygons, allowable rotations of polygons in degrees, and width of the strip. In Tables 33.2 and 33.3, we show the occupation rate and computation time for each instance reported in Reference [17] (column "SAHA"), [29] (column "ILSQN"), [31] (column "FITS"), [24] (column "BS"), [33] (column "ELS"), [35] ("CFREFP"), and [34] (column "GCS"). We also give their publication years below the names of algorithms in Table 33.2.

Table 33.2 shows the occupation rates in % of layouts computed by $Area(P)/Area(strip)$, where $Area(P)$ is the total area of all polygons and $Area(strip)$ denotes the area of the strip. The rows "Avg." and "Best" show the average occupation rate and the best occupation rate of each algorithm for multiple runs. The best results in each row are marked with "*". To show the performance of each algorithm on the qualities of solutions, we also report the average values of the occupation rates of all 15 instance in row

TABLE 33.1 Information of Benchmark Instances

Instance	Shapes	#Pieces	Avg_vertices	Rotations	Width
Albano	8	24	7.25	0, 180	4900
Dagli	10	30	6.30	0, 180	60
Dighe1	16	16	3.87	0	100
Dighe2	10	10	4.70	0	100
Fu	12	12	3.58	0, 90, 180, 270	38
Jakobs1	25	25	5.60	0, 90, 180, 270	40
Jakobs2	25	25	5.36	0, 90, 180, 270	70
Mao	9	20	9.22	0, 90, 180, 270	2550
Marques	8	24	7.37	0, 90, 180, 270	104
Shapes0	4	43	8.75	0	40
Shapes1	4	43	8.75	0, 180	40
Shapes2	7	28	6.29	0, 180	15
Shirts	8	99	6.63	0, 180	40
Swim	10	48	21.9	0, 180	5752
Trousers	17	64	5.06	0, 180	79

TABLE 33.2   Occupation Rates Obtained by Seven Algorithms for Benchmark Instances

Instance		SAHA [17] 2006	ILSQN [29] 2009	FITS [31] 2009	BS [24] 2010	ELS [33] 2012	CFREFP [35] 2012	GCS [34] 2013
Albano	Avg.	84.70	87.41	87.28	–	87.38	86.31	*87.47
	Best	87.43	88.16	87.56	87.88	88.48	89.21	*89.58
Dagli	Avg.	85.38	85.80	85.71	–	86.27	*87.13	87.06
	Best	87.15	87.40	86.35	87.97	88.11	88.36	*89.51
Dighe1	Avg.	82.13	90.49	99.84	–	91.61	*100.00	*100.00
	Best	*100.00	99.89	99.89	*100.00	*100.00	*100.00	*100.00
Dighe2	Avg.	84.17	84.21	99.99	–	*100.00	*100.00	*100.00
	Best	*100.00	99.99	*100.00	*100.00	*100.00	*100.00	*100.00
Fu	Avg.	87.17	87.57	90.31	–	90.00	89.53	*90.68
	Best	90.96	90.67	91.23	90.28	91.94	91.96	*92.41
Jakobs1	Avg.	75.79	84.78	88.74	–	88.35	88.59	*88.90
	Best	78.89	86.89	89.09	85.96	*89.10	89.07	*89.10
Jakobs2	Avg.	74.66	80.50	80.26	–	80.97	80.18	*81.14
	Best	77.28	82.51	80.84	80.40	83.92	84.83	*87.73
Mao	Avg.	80.72	81.31	82.79	–	82.57	81.14	*82.93
	Best	82.54	83.44	83.73	84.07	84.33	84.23	*85.44
Marques	Avg.	86.88	86.81	88.80	–	88.32	87.70	*89.40
	Best	88.14	89.03	89.21	88.92	89.73	90.01	*90.59
Shapes0	Avg.	63.20	66.49	66.20	–	66.85	65.31	*67.26
	Best	66.50	68.44	66.50	64.35	67.63	67.59	*68.79
Shapes1	Avg.	68.63	72.83	72.60	–	*74.24	71.71	73.79
	Best	71.25	73.84	73.88	72.55	75.29	72.52	*76.73
Shapes2	Avg.	81.41	81.72	80.87	–	*82.55	80.81	82.40
	Best	83.60	84.25	81.68	81.29	84.23	83.30	*84.84
Shirts	Avg.	85.67	*88.12	86.13	–	87.20	87.18	87.59
	Best	86.79	88.78	86.92	*89.69	88.40	87.59	88.96
Swim	Avg.	72.28	*74.62	72.97	–	74.10	70.82	74.49
	Best	74.37	75.29	74.54	75.04	75.43	71.78	*75.94
Trousers	Avg.	*89.02	88.69	88.78	–	88.29	87.74	*89.02
	Best	89.96	89.79	89.40	90.38	89.63	90.07	*91.00
AVG.(ALL)	Avg.	80.12	82.82	84.75	–	84.58	84.27	*85.48
	Best	84.32	85.89	85.39	85.25	86.41	86.03	*87.37

"AVG.(ALL)." For each algorithm, the value in row "Avg." of "AVG.(ALL)," called *avg-avg-OCR*, shows the average value of the average occupation rates of all instances, and that in row "Best" of "AVG.(ALL)," called *avg-best-OCR*, shows the average value of the best occupation rates of all instances. Table 33.3 shows the computational environment and the computation time in seconds spent by each algorithm. The rows "Avg." and "#run" show the average CPU time and the number of executions of the algorithms, respectively.

The computational results in Table 33.2 shows that algorithm GCS obtained the best solutions for 14 instances among 15 benchmark instances expect instance "Shirts," for which algorithm BS found the best layout. Algorithms ILSQN, ELS, and CFREFP obtained good solutions; the differences between their avg-best-OCRs and the best values in row "Best" of "AVG.(ALL)" are less than 1.5%. Comparison on the average qualities (i.e., rows "Avg.") shows that GCS algorithm obtained the highest occupation rate for 10 instances among all 15 instances, and for the remaining 5 instances it achieved results with gaps less than

TABLE 33.3   Computation Times of Seven Algorithms for Benchmark Instances

Instance		SAHA [17] P4, 2.4 GHz 512 MB	ILSQN [29] Xeon, 2.8 GHz 1 GB	FITS [31] Xeon, 2.8 GHz 1 GB	BS [24] PD, 2.8 GHz	ELS [33] P4, 2.4 GHz 2 GB	CFREFP [35] i7, 2.8 GHz 4 GB	GCS [34] i7, 2.2 GHz 6 GB
Albano	Avg.	2257	1200	1200	5460	1203	21600	1200
	#run	20	10	10	1	10	7	10
Dagli	Avg.	5110	1200	1200	17331	1205	21600	1200
	#run	20	10	10	1	10	7	10
Dighe1	Avg.	83	600	1200	1.4	601	600	600
	#run	20	10	10	1	10	4	10
Dighe2	Avg.	22	600	1200	0.3	600	600	600
	#run	20	10	10	1	10	4	10
Fu	Avg.	296	600	1200	1192	600	600	600
	#run	20	10	10	1	10	4	10
Jakobs1	Avg.	332	600	1200	2193	603	1800	600
	#run	20	10	10	1	10	4	10
Jakobs2	Avg.	454	600	1200	75	602	5400	600
	#run	20	10	10	1	10	7	10
Mao	Avg.	8245	1200	1200	16757	1204	21600	1200
	#run	20	10	10	1	10	7	10
Marques	Avg.	7507	1200	1200	10692	1204	5400	1200
	#run	20	10	10	1	10	7	10
Shapes0	Avg.	3914	1200	1200	8	1207	21600	1200
	#run	20	10	10	1	10	7	10
Shapes1	Avg.	10314	1200	1200	398	1212	908175	1200
	#run	20	10	10	1	10	7	10
Shapes2	Avg.	2136	1200	1200	5603	1205	5400	1200
	#run	20	10	10	1	10	7	10
Shirts	Avg.	10391	1200	1200	6217	1293	1812654	1200
	#run	20	10	10	1	10	6	10
Swim	Avg.	6937	1200	1200	15721	1246	5993945	1200
	#run	20	10	10	1	10	6	10
Trousers	Avg.	8588	1200	1200	5988	1237	86400	1200
	#run	20	10	10	1	10	7	10

TABLE 33.4   Speed of CPUs

CPU	SPECint2000	SPECint2006	Estimated Speed
Pentium 4, 2.4 GHz	833–1006	–	1
Xeon, 2.8 GHz	1288–1387	–	1.4–1.6
Pentium D, 2.8 GHz	1331–1422	–	1.4–1.6
Pentium 4, 3.8 GHz	1852	12.3	1.9–2.2
Core i7, 2.2 GHz	–	47.7 (32.7–53.5)	5–10
Core i7, 2.8 GHz	–	35.0–69.7	6–12

0.53% from the best average values. For the average case, algorithms FITS and ELS also performed good; the differences between their avg-avg-OCRs and the best values in row "Avg." of "AVG.(ALL)" are less than one percent.

Because different CPUs were used to obtain the results in these tables, we roughly estimated the speed of the CPUs based on the benchmark results on the SPEC website (http://www.spec.org/). We found the results of SPECint2006 for Intel Core i7, 2.2 GHz, and 2.8 GHz, and those of SPECint2000 for Pentium 4, 2.4 GHz, Xeon 2.8 GHz, and Pentium D, 2.8 GHz, which are shown in Table 33.4. For Core i7, 2.2 GHz, we only found an entry with a value of 47.7, but we found more entries for Core i7 with similar clock frequencies, and the value range of such entries is shown in parentheses; we show this because the Core i7 family contains a variety of CPUs and it is not safe to show the entry of only one CPU. To estimate the ratio between SPECint2000 and SPECint2006, we also show the results of a Pentium 4, 3.8 GHz for which both benchmark results are available. The estimated speeds of these CPUs are shown in column "estimated speed," a larger value of which indicates that the CPU is faster. The estimation in this table may not be accurate, but we can at least observe that the speeds of computers used to obtain the results of CFREFP and GCS are much faster than those of SAHA, ILSQN, FITS, BS, and ELS, and even though the time limits of GCS were set to almost the same values as ILSQN and ELS, the actual CPU resource used for GCS must be larger than that for ILSQN and ELS. For this reason, even though algorithm GCS reported higher occupation rates for most of the instances, it does not necessarily indicate that algorithm GCS will outperform the other algorithms if they are run on the same CPU with the same time limit.

33.2.3 Irregular Bin Packing Problem

There are also papers for the irregular bin packing problem, in which containers with a fixed shape (i.e., width and height are fixed) are considered and the objective is to minimize the number of used containers to pack all the given items.

Okano [47] proposed a heuristic algorithm for the irregular two-dimensional bin packing problem using a scanline representation. One of the core ideas is to approximate a two-dimensional item with a set of parallel line segments. He also uses a clustering technique: some items are gathered and packed tightly, and then these items are treated as one new item. Okano designed his algorithm for the irregular two-dimensional bin packing problem; however, his technique is also useful for treating other irregular packing problems. He conducted computational experiments with real instances from a shipbuilding company and reported that the quality of the resulting layouts was sufficiently high for practical use.

López-Camacho et al. [48] also considered the irregular bin packing problem. They proposed a generalization of the *Djang and Finch* (*DJD*) heuristic [49], which was originally designed for the one-dimensional bin packing problem, and they gave it an efficient implementation. The generalized DJD heuristic is used to select the next item to be packed. They also proposed a placement heuristic to pack the selected item. Four variants of the DJD heuristic were considered and compared on benchmark instances with seven other selection heuristics combined with four different placement heuristics. All the combinations of selection and placement heuristics were run, each with a time limit of 10 seconds on a PC with an Intel Core duo 2.33 GHz processor. The results showed that the DJD heuristic required more time for all instances but obtained better solutions than those obtained by other heuristics for most instances before 2013.

Han et al. [50] considered the irregular bin packing problem with guillotine cut constraint. There exist many algorithms [51–53] designed for rectangle cutting and packing problems with guillotine cut constraint. This is the first algorithm for the irregular convex shape packing considering guillotine cut constraints that requires every cut to extend from one edge of the stock sheet to another. Their approach consists of two stages: It first encloses individual or pairs of items into rectangles and then packs a greedily selected subset of the resulting rectangles into bins. They considered two clustering strategies when enclosing items into rectangles and adopted the approach proposed by Charalambous and Fleszar [51] to pack rectangles. They tested their algorithm on real industrial instances and also those created by using the properties of the industrial data. The algorithm was evaluated by the utilization of bins that was defined as the value obtained by dividing the total area of all the items by the sum of the total area of the closed bins (i.e., all the bins except the last bin) and the area of the smallest rectangular area containing all items in the last bin. The results showed that the utilization of bins obtained by their algorithm was about 0.80 on average.

Martinez-Sykora et al. [54] also considered the irregular bin packing problem with guillotine cut constraint. They gave a constructive heuristic that determines the positions of convex items in bins, in which bins are packed one by one and a new bin is opened only when there is no remaining items that can be packed into the current bin. They represented the guillotine cut constraints by a MIP model. The NFP technique takes an important role in their MIP model. They also proposed an improvement procedure to modify the placement of items before closing a bin. The computational results showed that the algorithm improved the solutions obtained in Reference [50] and performed good also on instances of the rectangle bin packing problem with guillotine cut constraint.

33.3 Bitmap Representation

In computer graphics, bitmap representation (also known as raster representation) is a dot matrix data structure representing a grid of pixels. We assume that each pixel is square, and each pair of (horizontally or vertically) adjacent pixels share a common boundary. A *bitmap shape* is defined as the union of the interiors and boundaries of a set of pixels (see Figure 33.1b). Bitmap format is commonly used to represent computer images such as GIF, JPEG, and PNG, which are popular on the World Wide Web. Any arbitrarily shaped object in a scene or image, no matter how complex its shape is, can be approximately represented in bitmap format as shown in Figure 33.1. Bitmap representation has advantages that it is not necessary to identify enclosed area of the shapes and it is easier to detect overlaps among shapes. The resulting bitmap image that represents an irregular shape can be treated as a rectilinear block by considering each dot of a bitmap image as a square. Consequently, methods developed for the rectilinear block packing problem will also work for a variety of irregular packing problems. We explain, in the next section, some representative methods for the rectilinear block packing problem.

In this section, we focus on the problem of packing a set of bitmap shapes into a larger rectangular container with fixed width so as to minimize its height. A bitmap shape is technically characterized by the resolution of the image in pixels, and the computational cost depends on the number of pixels. In the packing problems, when the resolution of the given bitmap shapes becomes high, the computational cost to check the intersection between shapes becomes large.

One popular idea for executing the intersection test efficiently is to represent items (irregular shapes) approximately. Oliveira and Ferreira [55] proposed two approaches to the irregular strip packing problem, and one of them uses this idea. Their approach approximately represents each item in bitmap format and this approximation allows a quick test of overlapping, but suffers from inaccuracy, caused by the approximation inherent in the bitmap representation. Their another approach uses a polygon-based representation that does not use any approximation technique. Both methods allow overlap in the solutions, and the extent of overlap is penalized by an evaluation function. They try to find a good solution via simulated annealing, which aims to reduce the overlap penalty to zero.

Wong et al. [56] proposed a genetic algorithm (GA) for the bitmap shape strip packing problem. Their GA first generates orders of items and then applies a two-stage packing approach to place all the items. In their GA, each individual represents a feasible solution and a chromosome consists of a sequence of integer numbers that indicate the items. It adopts a famous selection scheme called *biased roulette wheel scheme* [57], in which individuals are selected according to their respective probabilities, and it uses the order crossover operator proposed in Reference [58] and the inversion mutation operator in Reference [59]. The placement approach packs items according to the order computed by the GA. It first places an item at the leftmost and infinitely high corner and then moves the item downward toward the placed items as low as possible. They compared on benchmark instances their genetic algorithm with an algorithm that adopted the bottom-left strategy and used random search instead of GA. The results showed that the proposed algorithm outperformed the algorithm using the bottom-left strategy in combination with random search.

Toledo et al. [60] considered representing each shape by a set of dots on a board. They proposed a MIP model in which a 0-1 variable is associated with each pair of dot on the board and type of item shape; the variable is equal to one if an item of the corresponding type of shape is placed at that dot. Their model was tested on 45 instances and obtained optimal solutions for 34 instances having 16 to 56 items.

Shalaby and Kashkoush [61] also used the pixel method to represent irregular shapes. They proposed a sequence-based algorithm that contains two phases: first searching for a good sequence and then packing items compactly. When searching for sequences, they applied a local search algorithm. In the packing phase, they combined a pixel-based placement procedure and a compaction heuristic. Their algorithm was tested on 31 benchmark instances and the results showed that they updated the best-known results of 8 instances and obtained the best-known results for two instances.

Fukatsu et al. [62] proposed an efficient overlap detection method for the bitmap shape packing problem. They proposed an efficient method for checking for overlaps among items based on NFP technique and gave efficient implementations of two construction algorithms, the bottom-left and the best-fit algorithms, which were generalized from the rectangle packing problem to the rectilinear block packing problem in Reference [63]. The analysis on the time complexity of the resulting implementations of the two construction algorithms shows that under a weak assumption, which is satisfied unless the shapes of given items are pathologically complex, the computation time of both algorithms grows almost linearly to the vertical or horizontal resolution of given shapes. They also performed a series of experiments for instances generated from well-known benchmark instances of the irregular packing problem. The occupation rates of the packing layouts obtained by the algorithms were more than 80% for several instances. Even for instances with more than 3000 items, represented with more than 10 million pixels in total, the bottom-left and the best-fit algorithms with the efficient implementation ran in less than 50 seconds. The computational results showed that the proposed algorithms are especially effective for large-scale instances.

Sato et al. [64, 65] also solved the irregular strip packing problem by approximately representing shapes in bitmap format. They transformed the bitmap shape strip packing problem into an overlap minimization problem in which the dimensions of the container are fixed; the height of the container is controlled manually or automatically by a shrinking strategy. In their papers, the technique of NFP is adopted to detect intersections among items. They first considered in Reference [64] a discrete *Voronoi mountain*, which is obtained by extruding all regions from the Voronoi diagram, to evaluate the extent of overlap between two items. They then expanded it to a version of the bitmap method by a multi-resolution approach using the technique of *distance transform*, which is used in image processing to perform skeletonizing and blurring effects. The local search proposed by Elkeran [34] is used to search for the minimum overlap placement for one item in the layout. They also applied a metaheuristic, guided local search, proposed by Umetani et al. [31] to obtain a layout without overlap. The algorithm in Reference [64] was tested on instances with simple shapes to verify that the algorithm was able to obtain good layouts in reasonable time. The algorithm of Reference [65] was tested on four benchmark instances and the results showed that it obtained good packing layouts and running time was significantly improved.

33.4 Rectilinear Block Packing Problem

Rectilinear block packing problem involves packing a set of arbitrarily shaped rectilinear blocks into a larger rectangular container without overlap so as to minimize or maximize a given objective function. A rectilinear block is a polygonal block whose interior angle is either 90° or 270° (see Figure 33.1c). This problem involves many industrial applications, such as VLSI design, timber/glass cutting, and newspaper layout.

It is difficult to represent the relationships among rectilinear blocks than those among rectangles. To represent the relationships among rectilinear blocks, several coding schemes have been designed such as the *bounded-sliceline grid* (*BSG*) [66–68], *sequence-pairs* [69–72], O-tree [73], B*-tree [74], *transitive closure graph* (*TCG*) [75], and *corner block list* (*CBL*) [76]. Several heuristics have been proposed for the rectilinear block packing problem based on these schemes.

We first explain in Sections 33.4.1 and 33.4.2 two representative techniques, the BSG, and the sequence-pair representations. The methods based on these representations perform well for small-scale instances in reasonable running time and are especially effective for instances consisting of rectilinear blocks that have simple shapes (e.g., L- or T-shape). To the best of our knowledge, all existing algorithms based on these representations deal with the rectilinear block area minimization problem, whose objective is to minimize the total area containing all the items. We gave in Section 33.4.3 some construction and exact algorithms for the rectilinear block strip packing problem.

33.4.1 Bounded-Sliceline Grid

Nakatake et al. [67] proposed a *BSG* structure to represent the placement of items. A BSG is a topology defined in the plane and orthogonal line segments, called *BSG-segs*, dissect the plane into square zones, called *BSG-rooms* (see Figure 33.6a as an example). The items are assigned to distinct BSG-rooms and they have the same relationship on the placement with that among the BSG-rooms. A BSG of size $n \times n$ with assignment of rectangles to BSG-rooms can represent the relationship among given n rectangles in a placement of the rectangles. Figure 33.6b shows a placement of four items according to the assignment in a BSG of size 4×4 in Figure 33.6a.

Many approaches for the area minimization problems based on BSG are proposed in the literature. It is not guaranteed that there is a corresponding BSG representation for every placement for instances of the L-shaped block packing problem. Because of this limitation, optimal solutions may not be included in the search space of algorithms based on BSG.

The algorithms for the L-shaped block packing problem were proposed in References [67, 77]. They partition each L-shaped block to two rectangles and assign them into two adjacent BSG-rooms. As adjacent BSG-rooms have a common BSG-seg, it is easy to align the rectangles into the original shapes. These approaches may still miss the optimal solution.

In Reference [66], a method for packing rectilinear blocks and blocks with positional constraints is proposed. The algorithm divides rectilinear blocks into a set of rectangles and packs them based on BSG. Because of the limitation of BSG representation, it may also miss the optimal solution.

Although the above-mentioned algorithms may miss optimal solutions, their performance is often good in practice.

A rectilinear block is *orthogonally convex* if any two points in the block can be connected by a shortest Manhattan path drawn inside the block. Sakanushi et al. [68] proposed an algorithm for packing orthogonally convex rectilinear blocks based on multi-BSG, which is an arrangement of plural BSGs on multiple layers. It also represents a rectilinear block as a set of rectangles. The algorithm guarantees that an optimal placement is contained in its solution space.

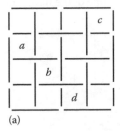
(a)

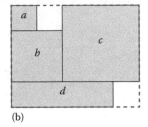
(b)

FIGURE 33.6 An example of the BSG representation.

33.4.2 Sequence-Pair

The *sequence-pair* [78, 79] was proposed in 1995 as a coding scheme to represent a packing layout of rectangles. The basic idea is that any feasible packing layout can be represented by fixing the relative positions of all the rectangles. This technique is also used to represent the solution space of the area minimization problem.

A sequence-pair represents a solution of δ rectangles by a pair of permutations of their names. The number of all possible sequence-pairs for δ rectangles is $(\delta!)^2$. A sequence-pair describes the relative horizontal or vertical positions for each pair of rectangles as follows:

- If $(\cdots a \cdots b \cdots, \cdots a \cdots b \cdots)$ holds, then a is to the left of b (i.e., the right edge of a is left to the left edge of b).
- If $(\cdots b \cdots a \cdots, \cdots a \cdots b \cdots)$ holds, then a is below b (i.e., the top edge of a is below the bottom edge of b).

For a simple example, if we have three rectangles named a, b, and c, (abc, bac) is a sequence-pair that represents their packing layout. The sequence-pair (abc, bac) means that a is to the left of c, b is below a, and b is to the left of c. Figure 33.7 shows a placement with minimum area for sequence-pair (abc, bac).

One-dimensional compaction can be used to construct from a given sequence-pair a placement with minimum area. It greedily pushes every rectangle to the bottom left corner of the container and it can obtain such a placement in $O(\delta^2)$ time.

Fujiyoshi and Murata [69] generalized this technique to solve the rectilinear block packing problem. It deals with the packing problem whose objective is to minimize the area of the container. The algorithm first partitions each rectilinear block as a set of rectangles as a sequence-pair can only deal with rectangles. A procedure called X/Y-alignment was also proposed to align such rectangles. An algorithm to obtain a rectilinear block packing layout from a given sequence-pair (i.e., a decoding algorithm to obtain a feasible packing layout) was proposed. The time complexity of the decoding algorithm is $O(m^3)$, where m is the total number of rectangles representing all the rectilinear blocks.

Machida and Fujiyoshi [80] proposed an improved decoding algorithm that runs in $O(m^2 + n^3)$, where m is the total number of rectangles to represent all the given rectilinear blocks and n is the number of rectilinear blocks.

In a sequence-pair, if rectangles a, b, c, d are assigned in the order of

- $(\cdots a \cdots bc \cdots d \cdots, \cdots c \cdots ad \cdots b \cdots)$ or
- $(\cdots a \cdots bc \cdots d \cdots, \cdots b \cdots da \cdots c \cdots)$,

rectangles a, b, c, d make an *adjacent cross* [81]. There are at most $\lceil (\delta - 2)/2 \rceil \lfloor (\delta - 2)/2 \rfloor$ adjacent crosses in a sequence-pair with δ elements.

An efficient coding scheme for rectangle packing called *selected sequence-pair* [82] is defined as a sequence-pair with k adjacent crosses. Any layout of rectangles can be represented by a selected sequence-pair, and a packing layout with minimum area for a given selected sequence-pair can be obtained in $O(\delta + k)$ time. Takashima and Murata [83] proved that the necessary number of adjacent crosses for representing an arbitrary placement of δ rectangles is at most $\delta - \lfloor \sqrt{4\delta - 1} \rfloor$ (i.e., $k \leq \delta - \lfloor \sqrt{4\delta - 1} \rfloor$). Hence, a given selected sequence-pair can be decoded in linear time of the number of rectangles.

Sequence-pair: (abc, bac)

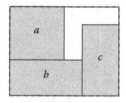

FIGURE 33.7 An example of a minimum area placement for a sequence-pair.

The idea of selected sequence-pair was generalized to the case of rectilinear blocks in Reference [71]. They proposed a decoding algorithm whose running time is $O((p+1)m)$ for a given selected sequence-pair for rectilinear blocks, where m also denotes the total number of rectangles representing all the rectilinear blocks and p is the number of rectilinear blocks that are represented by more than one rectangle (not shaped as rectangles).

Janiak et al. [84] also solved the rectilinear block area minimization problem based on sequence-pair representation. They proposed efficient methods of neighborhood exploration by constructing a heuristic based on *insertion technique* that is proposed for flow-shop scheduling problem in Reference [85]. The results showed that the construction algorithm was efficient compared to those of annealing-based algorithms.

33.4.3 Construction and Exact Algorithms

Most of the methods introduced in Sections 33.4.1 and 33.4.2 deal only with small-scale instances. However, for industrial application purposes, good solutions for large-scale instances are often necessary. The construction heuristics tend to perform well on large-scale instances and it is possible to design efficient implementations to make their running time very short.

Hu et al. [63] proposed two efficient construction algorithms for the rectilinear strip packing problem. Each rectilinear block was regarded as a set of rectangles whose relative positions are fixed. Hence, their algorithms are able to deal with the problem having noncontinuous items, each of which consists of a set of (separate) rectilinear blocks whose relative positions are fixed. They adopted the bottom-left strategy, in which each item is packed at the bottom-left position. Recall that the bottom-left position of a new rectilinear block relative to a packing layout is defined as the leftmost location among the lowest bottom-left stable feasible positions, where a bottom-left stable feasible position is a location at which the new item can be placed without overlap and cannot be moved leftward or downward. They first generalized the bottom-left and the best-fit algorithms for the rectangle packing problem to solve the rectilinear block packing problem and then proposed efficient implementations using an efficient method for finding bottom-left positions that was originally proposed for rectangle packing in Reference [86]. They used sophisticated data structures that keep the information dynamically so that the bottom-left position of each item can be found in sub-linear amortized time. Time complexities of the bottom-left and the best-fit algorithms with the efficient implementations were proven to be $O(mM \log M)$ time, where m is the number of rectangles that represent all distinct shapes of rectilinear blocks and M is the number of rectangles that represent all rectilinear blocks. The algorithms were tested on instances that were generated from nine benchmark instances and ran on a Mac PC with a 2.3 GHz Intel Core i5 processor and 4 GB memory. The results showed that the algorithms are especially efficient for large-scale instances with a relatively small number of distinct shapes. Even for instances with more than 10,000 rectilinear blocks with up to 60 distinct shapes, the proposed algorithms ran in less than 12 seconds. The occupation rate of the packing layouts often reached higher than 95% for large-scale instances. They also created large-scale instances with up to 10,000 distinct shapes to test the performance of the computation time and showed that the running time increased almost linearly with the number of distinct shapes. For instances with 10,000 distinct shapes, the algorithms ran in less than 6000 seconds.

Hu et al. [87] then analyzed the strength and weakness of the bottom-left and the best-fit algorithms from the viewpoint of the quality of the packing results obtained in Reference [63] and summarized the reasons why the best-fit algorithm outperforms the bottom-left algorithm for many instances and the opposite holds for some kinds of instances. Based on those observations, they proposed a new constructive heuristic algorithm, called the *partition-based best-fit* (PBF) algorithm, as a bridge between the bottom-left and the best-fit algorithms. In the PBF algorithm, it is assumed that all the items are partitioned into groups and the algorithm packs items into the container in a group-by-group manner, where the best-fit algorithm is taken as the internal tactics to pack items of each group. They showed that the PBF algorithm runs in the same time complexity $O(mM \log M)$ using implementation similar to the one in Reference [63]. They also gave some effective rules to partition items into groups. The PBF algorithm were also tested on 168 instances generated from benchmark instances and the occupation rate of the packing results was more than 93% on average. The results showed that the improvement of the performance of the occupation rate is remarkable compared with the bottom-left and the best-fit algorithm and significant improvement was observed for small-scale instances. The experiments on instances in which the size of rectilinear blocks are quite different showed that the PBF algorithm performed especially effective for this kind of instances.

Matsushita et al. [88] proposed an exact algorithm for the rectilinear block strip packing problem. They solved the problem by alternately calling two phases: one is to compute a good lower bound on the value of an optimal solution of the currently active search space, and the other is to compute a corresponding upper bound. In the phase of computing lower bounds, they relaxed the packing problem to a parallel processor scheduling problem in which each rectilinear block is cut into rectangles with a constant width and only the horizontal relationship of these rectangles are fixed. By solving the relaxation of the original problem, a lower bound and x-coordinates of the blocks are obtained. Then, in the second phase, according to the x-coordinates of blocks obtained in the first phase, they formulated the packing problem as an IP problem. By solving this problem, an upper bound and a feasible solution of the rectilinear block packing problem are obtained. Their exact algorithm alternately calls these two phases, reducing the search space in each iteration to strengthen the corresponding lower bounds (i.e., to reduce the gap between the lower and upper bounds) until the lower bound on the value of an optimal solution for the current search space is not smaller than the height of the best solution layout obtained during the search by then, which means that an optimal solution of this problem is found. The computational results on benchmark instances showed that the proposed methods to compute lower and upper bounds obtained good bounds, and the exact algorithm obtained exact solutions for some instances with up to 20 rectilinear blocks.

Matsushita et al. [89] further proposed a new solution representation scheme for the rectilinear block packing problem based on an order of items, to decide a layout of rectilinear blocks whose x-coordinates are fixed. When the shapes of given items have certain characteristics, the representation guarantees optimality. They gave decoding algorithm of their representation and generalized the result to the case of general continuous shapes. A branch-and-bound method was also considered to compute an optimal order. They then incorporated the solution representation scheme to their exact algorithm in Reference [88] by replacing the method to compute upper bounds. It iteratively generates x-coordinates of items and finds the corresponding optimal layout. The computational results showed that the new algorithm obtained five exact and one heuristic solutions for six instances and it improved the running time of the exact algorithm in Reference [88].

33.5 Conclusion

In this chapter, we surveyed practical algorithms for the two-dimensional packing problem of general shapes, the irregular packing problem, the bitmap shape packing problem, and the rectilinear block packing problem, all of which have many industrial applications. Irregular shapes are usually represented by a list of vertices (vector graphics) or approximately represented in bitmap format (raster graphics). The resulting bitmap image that represents an irregular shape can also be treated as a rectilinear block. Hence, we focused on these three types of packing problems.

Packing problems with items having complex shapes have been studied extensively in the last decade. The main difference between packing problems of rectangles and those of complex shapes is that the intersection test between items having complex shapes is considerably more complex. To overcome this difficulty, various methodologies have been proposed; no-fit polygon is one of the representative and effective ones. We introduced some coding schemes (in other words, how to represent a solution) such as permutation, bounded-sliceline grid, and sequence-pair and then explained various heuristic, metaheuristic, and some exact algorithms, most of which are based on these coding schemes.

The survey in this chapter is by no means comprehensive, but we hope that this gives valuable information to the readers who are interested in devising practical algorithms for cutting and packing problems.

Acknowledgment

This work was partially supported by KEIT grant funded by MOTIE [Grant No. 10053204], and by JSPS KAKENHI [Grant Nos. 15K16293, 17K12981, 15H02969, 17K00038].

References

1. H. Dyckhoff. A typology of cutting and packing problems. *European Journal of Operational Research*, 44(2):145–159, 1990.
2. G. Wäscher, H. Haußner, and H. Schumann. An improved typology of cutting and packing problems. *European Journal of Operational Research*, 183(3):1109–1130, 2007.
3. E. Hopper and B.C.H. Turton. A review of the application of meta-heuristic algorithms to 2D strip packing problems. *Artificial Intelligence Review*, 16(4):257–300, 2001.
4. J.A. Bennell and J.F. Oliveira. A tutorial in irregular shape packing problems. *Journal of the Operational Research Society*, 60(1):S93–S105, 2009.
5. A.M. Gomes. Irregular packing problems: Industrial applications and new directions using computational geometry. In *Proceedings of the 11th IFAC Workshop on Intelligent Manufacturing Systems (IMS), IFAC Proceedings Volumes*, 46(7):378–383, 2013.
6. B.S. Baker, E.G. Coffman, Jr., and R.L. Rivest. Orthogonal packings in two dimensions. *SIAM Journal on Computing*, 9(4):846–855, 1980.
7. M. Adamowicz and A. Albano. Nesting two-dimensional shapes in rectangular modules. *Computer-Aided Design*, 8(1):27–33, 1976.
8. K.A. Dowsland, S. Vaid, and W.B. Dowsland. An algorithm for polygon placement using a bottom-left strategy. *European Journal of Operational Research*, 141(2):371–381, 2002.
9. J.A. Bennell, G. Scheithauer, Y. Stoyan, T. Romanova, and A. Pankratov. Optimal clustering of a pair of irregular objects. *Journal of Global Optimization*, 61(3):497–524, 2015.
10. R.C. Art, Jr. *An approach to the two dimensional irregular cutting stock problem*. PhD Thesis, Massachusetts Institute of Technology, Cambridge, MA, 1966.
11. E.K. Burke, R. Hellier, G. Kendall, and G. Whitwell. A new bottom-left-fill heuristic algorithm for the two-dimensional irregular packing problem. *Operations Research*, 54(3):587–601, 2006.
12. B. Chazelle. The bottomn-left bin-packing heuristic: An efficient implementation. *IEEE Transactions on Computers*, C-32(8):697–707, 1983.
13. A. Albano and G. Sapuppo. Optimal allocation of two-dimensional irregular shapes using heuristic search methods. *IEEE Transactions on Systems, Man and Cybernetics*, 10(5):242–248, 1980.
14. A.M. Gomes and J.F. Oliveira. A 2-exchange heuristic for nesting problems. *European Journal of Operational Research*, 141(2):359–370, 2002.
15. J.F. Oliveira, A.M. Gomes, and J.S. Ferreira. TOPOS—A new constructive algorithm for nesting problems. *OR Spectrum*, 22(2):263–284, 2000.
16. J.A. Bennell and K.A. Dowsland. Hybridising tabu search with optimisation techniques for irregular stock cutting. *Management Science*, 47(8):1160–1172, 2001.
17. A.M. Gomes and J.F. Oliveira. Solving irregular strip packing problems by hybridising simulated annealing and linear programming. *European Journal of Operational Research*, 171(3):811–829, 2006.
18. Z. Li and V. Milenkovic. Compaction and separation algorithms for non-convex polygons and their applications. *European Journal of Operational Research*, 84(3):539–561, 1995.

19. J.A. Bennell, K.A. Dowsland, and W.B. Dowsland. The irregular cutting-stock problem—a new procedure for deriving the no-fit polygon. *Computers & Operations Research*, 28(3):271–287, 2001.

20. F. Avnaim and J.D. Boissonnat. Polygon placement under translation and rotation. In *Proceedings of the 5th Annual Symposium on Theoretical Aspects of Computer Science (STACS), Lecture Notes in Computer Science*, 294:322–333, 1988.

21. E.K. Burke, R. Hellier, G. Kendall, and G. Whitwell. Complete and robust no-fit polygon generation for the irregular stock cutting problem. *European Journal of Operational Research*, 179(1):27–49, 2007.

22. E.K. Burke, R. Hellier, G. Kendall, and G. Whitwell. Irregular packing using the line and arc no-fit polygon. *Operations Research*, 58(4):948–970, 2010.

23. J. Brazewicz, P. Hawryluk, and R. Walkowiak. Using a tabu search for solving the two-dimensional irregular cutting problem. *Annals of Operations Research*, 41(4):313–325, 1993.

24. J.A. Bennell and X. Song. A beam search implementation for the irregular shape packing problem. *Journal of Heuristics*, 16(2):167–188, 2010.

25. S. Jakobs. On genetic algorithms for the packing of polygons. *European Journal of Operational Research*, 88(1):165–181, 1996.

26. R. Dighe and M.J. Jakiela. Solving pattern nesting problems with genetic algorithms employing task decomposition and contact detection. *Evolutionary Computation*, 3(3):239–266, 1995.

27. T.C. Martins and M.S.G. Tsuzuki. Simulated annealing applied to the irregular rotational placement of shapes over containers with fixed dimensions. *Expert Systems with Applications*, 37(3):1955–1972, 2010.

28. T.C. Martins and M.S.G. Tsuzuki. Simulated annealing applied to the rotational polygon packing. In *Proceedings of the 12th IFAC Symposium on Information Control Problems in Manufacturing (INCOM), IFAC Proceedings Volumes*, 39(3):475–480, 2006.

29. T. Imamichi, M. Yagiura, and H. Nagamochi. An iterated local search algorithm based on nonlinear programming for the irregular strip packing problem. *Discrete Optimization*, 6(4):345–361, 2009.

30. L. Zhang, Y.J. Kim, G. Varadhan, and D. Manocha. Generalized penetration depth computation. *Computer-Aided Design*, 39(8):625–638, 2007.

31. S. Umetani, M. Yagiura, S. Imahori, T. Imamichi, K. Nonobe, and T. Ibaraki. Solving the irregular strip packing problem via guided local search for overlap minimization. *International Transactions in Operational Research*, 16(6):661–683, 2009.

32. J. Egeblad, B.K. Nielsen, and A. Odgaard. Fast neighborhood search for two- and three-dimensional nesting problems. *European Journal of Operational Research*, 183(3):1249–1266, 2007.

33. S.C.H. Leung, Y. Lin, and D. Zhang. Extended local search algorithm based on nonlinear programming for two-dimensional irregular strip packing problem. *Computers & Operations Research*, 39(3):678–686, 2012.

34. A. Elkeran. A new approach for sheet nesting problem using guided cuckoo search and pairwise clustering. *European Journal of Operational Research*, 231(3):757–769, 2013.

35. A.K. Sato, T.C. Martins, and M.S.G. Tsuzuki. An algorithm for the strip packing problem using collision free region and exact fitting placement. *Computer-Aided Design*, 44(8):766–777, 2012.

36. A.K. Sato, T.C. Martins, and M.S.G. Tsuzuki. Placement heuristics for irregular packing to create layouts with exact placements for two moveable items. In *Proceedings of the 11th IFAC Workshop on Intelligent Manufacturing Systems (IMS), IFAC Proceedings Volumes*, 46(7):384–389, 2013.

37. R. Alvarez-Valdes, A. Martinez, and J.M. Tamarit. A branch & bound algorithm for cutting and packing irregularly shaped pieces. *International Journal of Production Economics*, 145(2):463–477, 2013.

38. M. Fischetti and I. Luzzi. Mixed-integer programming models for nesting problems. *Journal of Heuristics*, 15(3):201–226, 2009.

39. M.C. Santoro and F.K. Lemos. Irregular packing: MILP model based on a polygonal enclosure. *Annals of Operations Research*, 235(1):693–707, 2015.

40. L.H. Cherri, L.R. Mundim, M. Andretta, F.M.B. Toledo, J.F. Oliveira, and M.A. Carravilla. Robust mixed-integer linear programming models for the irregular strip packing problem. *European Journal of Operational Research*, 253(3):570–583, 2016.

41. J.A. Bennell and J.F. Oliveira. The geometry of nesting problems: A tutorial. *European Journal of Operational Research*, 184(2):397–415, 2008.

42. T. Imamichi. *Nonlinear Programming Based Algorithms to Cutting and Packing Problems*. PhD Thesis, Kyoto University, Kyoto, Japan, 2009.

43. D.R. Jones. A fully general, exact algorithm for nesting irregular shapes. *Journal of Global Optimization*, 59(2-3):367–404, 2014.

44. V.J. Milenkovic. Rotational polygon containment and minimum enclosure using only robust 2D constructions. *Computational Geometry*, 13(1):3–19, 1999.

45. P. Rocha, R. Rodrigues, A.M. Gomes, F.M.B. Toledo, and M. Andretta. Two-phase approach to the nesting problem with continuous rotations. In *Proceedings of the 15th IFAC Symposium on Information Control Problems in Manufacturing (INCOM)*, IFAC-PapersOnLine, 48(3):501–506, 2015.

46. P. Rocha. *Geometrical Models and Algorithms for Irregular Shapes Placement Problems*. PhD Thesis, Universidade do Porto, Portugal, 2014.

47. H. Okano. A scanline-based algorithm for the 2D free-form bin packing problem. *Journal of the Operations Research Society of Japan*, 45(2):145–161, 2002.

48. E. López-Camacho, G. Ochoa, H. Terashima-Marín, and E.K. Burke. An effective heuristic for the two-dimensional irregular bin packing problem. *Annals of Operations Research*, 206(1):241–264, 2013.

49. P. Ross, S. Schulenburg, J.G. Marín-Blázquez, and E. Hart. Hyper-heuristics: Learning to combine simple heuristics in bin-packing problems. In *Proceedings of the 4th Annual Conference on Genetic and Evolutionary Computation (GECCO)*, pp. 942–948, 2002.

50. W. Han, J.A. Bennell, X. Zhao, and X. Song. Construction heuristics for two-dimensional irregular shape bin packing with guillotine constraints. *European Journal of Operational Research*, 230(3):495–504, 2013.

51. C. Charalambous and K. Fleszar. A constructive bin-oriented heuristic for the two-dimensional bin packing problem with guillotine cuts. *Computers & Operations Research*, 38(19):1443–1451, 2011.

52. K. Fleszar. Three insertion heuristics and a justification improvement heuristic for two-dimensional bin packing with guillotine cuts. *Computers & Operations Research*, 40(1):463–474, 2013.

53. S. Polyakovsky and R. M'Hallah. An agent-based approach to the two-dimensional guillotine bin packing problem. *European Journal of Operational Research*, 192(3):767–781, 2009.

54. A. Martinez-Sykora, R. Alvarez-Valdes, J.A. Bennell, and J.M. Tamarit. Constructive procedures to solve 2-dimensional bin packing problems with irregular pieces and guillotine cuts. *Omega*, 52:15–32, 2015.

55. J.F. Oliveira and J.S. Ferreira. Algorithms for nesting problems. In R.V.V. Vidal, ed., *Applied Simulated Annealing*, pp. 255–273. Springer, Berlin, Germany, 1993.

56. W.K. Wong, X.X. Wang, P.Y. Mok, S.Y.S. Leung, and C.K. Kwong. Solving the two-dimensional irregular objects allocation problems by using a two-stage packing approach. *Expert Systems with Applications*, 36(2):3489–3496, 2009.

57. D.E. Goldberg. *Genetic Algorithms in Search, Optimization and Machine Learning*. Addison-Wesley, Boston, MA, 1989.

58. L. Davis. Job shop scheduling with genetic algorithms. In *Proceedings of the 1st International Conference on Genetic Algorithms (ICGA)*, pp. 136–140, 1985.

59. J.H. Holland. *Adaptation in Natural and Artificial Systems: An Introductory Analysis with Applications to Biology, Control, and Artificial Intelligence*. MIT Press, Cambridge, MA, 1992.

60. F.M.B. Toledo, M.A. Carravilla, C. Ribeiro, José F. Oliveira, and A.M. Gomes. The dotted-board model: A new MIP model for nesting irregular shapes. *International Journal of Production Economics*, 145(2):478–487, 2013.

61. M.A. Shalaby and M. Kashkoush. A particle swarm optimization algorithm for a 2-D irregular strip packing problem. *American Journal of Operations Research*, 3(2):268–278, 2013.

62. S. Fukatsu, Y. Hu, H. Hashimoto, S. Imahori, and M. Yagiura. An efficient method for checking overlaps and construction algorithms for the bitmap shape packing problem. In *Proceedings of the 2014 IEEE International Conference on Industrial Engineering and Engineering Management (IEEM)*, 2014.

63. Y. Hu, H. Hashimoto, S. Imahori, and M. Yagiura. Efficient implementations of construction heuristics for the rectilinear block packing problem. *Computers & Operations Research*, 53: 206–222, 2015.

64. A.K. Sato, M.S.G. Tsuzuki, T.C. Martins, and A.M. Gomes. Irregular packing overlap minimization using discrete voronoi mountain. In *Proceedings of the 19th IFAC World Congress, IFAC Proceedings Volumes*, 47(3): 5241–5246, 2014.

65. A.K. Sato, M.S.G. Tsuzuki, T.C. Martins, and A.M. Gomes. Multiresolution based overlap minimization algorithm for irregular packing problems. In *Proceedings of the 15th IFAC Symposium on Information Control Problems in Manufacturing (INCOM)*, *IFAC-PapersOnLine*, 48(3):484–489, 2015.

66. S. Nakatake, M. Furuya, and Y. Kajitani. Module placement on BSG-structure with pre-placed modules and rectilinear modules. In *Proceedings of the 3rd Asia and South Pacific Design Automation Conference (ASP-DAC)*, pp. 571–576, 1998.

67. S. Nakatake, K. Fujiyoshi, H. Murata, and Y. Kajitani. Module packing based on the BSG-structure and IC layout applications. *IEEE Transactions on Computer-Aided Design of Integrated Circuits and Systems*, 17(6):519–530, 1998.

68. K. Sakanushi, S. Nakatake, and Y. Kajitani. The multi-BSG: Stochastic approach to an optimum packing of convex-rectilinear blocks. In *Proceedings of the 1998 IEEE/ACM International Conference on Computer-Aided Design (ICCAD)*, pp. 267–274, 1998.

69. K. Fujiyoshi and H. Murata. Arbitrary convex and concave rectilinear block packing using sequence-pair. *IEEE Transactions on Computer-Aided Design of Integrated Circuits and Systems*, 19(2):224–233, 2000.

70. M. Kang and W.W.-M. Dai. Arbitrary rectilinear block packing based on sequence pair. In *Proceedings of the 1998 IEEE/ACM International Conference on Computer-Aided Design (ICCAD)*, pp. 259–266, 1998.

71. K. Fujiyoshi, C. Kodama, and A. Ikeda. A fast algorithm for rectilinear block packing based on selected sequence-pair. *Integration, the VLSI Journal*, 40(3):274–284, 2007.

72. J. Xu, P.-N. Guo, and C.-K. Cheng. Rectilinear block placement using sequence-pair. In *Proceedings of the 1998 International Symposium on Physical Design (ISPD)*, pp. 173–178, 1998.

73. Y. Pang, C.-K. Cheng, K. Lampaert, and W. Xie. Rectilinear block packing using O-tree representation. In *Proceedings of the 2001 International Symposium on Physical Design (ISPD)*, pp. 156–161, 2001.

74. G.-M. Wu, Y.-C. Chang, and Y.-W. Chang. Rectilinear block placement using B*-trees. *ACM Transactions on Design Automation of Electronic Systems*, 8(2):188–202, 2003.

75. J.-M. Lin, H.-L. Chen, and Y.-W. Chang. Arbitrarily shaped rectilinear module placement using the transitive closure graph representation. *IEEE Transactions on Very Large Scale Integration Systems*, 10(6):886–901, 2002.

76. Y. Ma, X. Hong, S. Dong, Y. Cai, S. Chen, C.-K. Cheng, and J. Gu. Arbitrary convex and concave rectilinear block packing based on corner block list. In *Proceedings of the 2003 International Symposium on Circuits and Systems (ISCAS)*, pp. V493–496, 2003.

77. M. Kang and W.W.-M Dai. General floorplanning with L-shaped, T-shaped and soft blocks based on bounded slicing grid structure. In *Proceedings of the 1997 Asia and South Pacific Design Automation Conference (ASP-DAC)*, pp. 265–270, 1997.

78. H. Murata, K. Fujiyoshi, S. Nakatake, and Y. Kajitani. Rectangle-packing-based module placement. In *Proceedings of the 1995 IEEE/ACM International Conference on Computer-Aided Design (ICCAD)*, pp. 472–479, 1995.

79. H. Murata, K. Fujiyoshi, S. Nakatake, and Y. Kajitani. VLSI module placement based on rectangle-packing by the sequence-pair. *IEEE Transactions on Computer-Aided Design of Integrated Circuits and Systems*, 15(12):1518–1524, 1996.

80. K. Machida and K. Fujiyoshi. The improvement of rectilinear block packing using sequence-par. IEICE Technical Report, VLD2000-134, (in Japanese), 2001.

81. H. Murata, K. Fujiyoshi, T. Watanabe, and Y. Kajitani. A mapping from sequence-pair to rectangular dissection. In *Proceedings of the 1997 Asia and South Pacific Design Automation Conference (ASP-DAC)*, pp. 625–633, 1997.

82. C. Kodama and K. Fujiyoshi. Selected sequence-pair: An efficient decodable packing representation in linear time using sequence-pair. In *Proceedings of the 2003 Asia and South Pacific Design Automation Conference (ASP-DAC)*, pp. 331–337, 2003.

83. Y. Takashima and H. Murata. The tight upper bound of the empty rooms in floorplan. *The 2001 Workshop on Synthesis and System Integration of Mixed Information Technologies (SASIMI)*, pp. 264–271, 2001.

84. A. Janiak, A. Kozik, and M. Lichtenstein. New perspectives in VLSI design automation: Deterministic packing by sequence pair. *Annals of Operations Research*, 179(1):35–56, 2010.

85. M. Nawaz, E.E. Enscore, and I. Ham. A heuristic algorithm for the m-machine, n-job flow-shop sequencing problem. *Omega*, 11(1):91–95, 1983.

86. S. Imahori, Y. Chien, Y. Tanaka, and M. Yagiura. Enumerating bottom-left stable positions for rectangle placements with overlap. *Journal of the Operations Research Society of Japan*, 57(1):45–61, 2014.

87. Y. Hu, H. Hashimoto, S. Imahori, T. Uno, and M. Yagiura. A partition-based heuristic algorithm for the rectilinear block packing problem. *Journal of the Operations Research Society of Japan*, 59(1):110–129, 2016.

88. K. Matsushita, Y. Hu, H. Hashimoto, S. Imahori, and M. Yagiura. An exact algorithm with successively strengthened lower bounds for the rectilinear block packing problem. In *Proceedings of the 7th International Symposium on Scheduling (ISS)*, pp. 242–247, 2015.

89. K. Matsushita, Y. Hu, H. Hashimoto, S. Imahori, and M. Yagiura. A new solution representation for the rectilinear block packing problem. In *Proceedings of the 2016 IEEE International Conference on Industrial Engineering and Engineering Management (IEEM)*, 2016.

34

Prize Collecting Traveling Salesman and Related Problems

Giorgio Ausiello

Vincenzo Bonifaci

Stefano Leonardi

Alberto
Marchetti-Spaccamela

34.1 Introduction

The most general version of the Prize-Collecting Traveling Salesman Problem (PCTSP) was first introduced by Balas [1]. In this problem, a salesman has to collect a certain amount of prizes (the *quota*) by visiting cities. A known prize can be collected in every city. Furthermore, by not visiting a city, the salesman incurs a pecuniary *penalty*. The goal is to minimize the total travel distance and the total penalty, while starting from a given city and collecting the quota.

The problem generalizes both the Quota TSP, which is obtained when all the penalties are set to zero, and the Penalty TSP (sometimes unfortunately also called PCTSP), in which there is no required quota, only penalties. A special case of the Quota TSP is the k-TSP, in which all prizes are unitary (k is the quota). The k-TSP is strongly tied to the problem of finding a tree of minimum cost spanning any k vertices in a graph, called the k-MST problem.

The k-MST and the k-TSP are NP-hard. They have been the subject of several studies for good approximation algorithms [2–6]. A 2-approximation scheme for both the k-MST and the k-TSP given by Garg [6] is the best known approximation ratio. Interestingly enough, all these algorithms use the primal-dual algorithm of Goemans and Williamson [7] for the Prize-Collecting Steiner Tree as a subroutine.

The Quota TSP was also considered by some researchers. It was considered by Awerbuch et al. [8], who gave an $O(\log^2(\min(Q, n)))$ approximation algorithm for instances with n cities and quota Q. Ausiello et al. [9] give an algorithm with an approximation ratio of 5 for what could be called the Quota MST, by extending some of the ideas of Reference 5.

The Penalty TSP (PTSP) has apparently a better approximation ratio. Goemans and Williamson, as an application of their primal-dual technique, give an approximation algorithm with a ratio of 2 for the PTSP [7]. Their algorithm uses a reduction to the Prize-Collecting Steiner Tree Problem. The running time of the algorithm was reduced from $O(n^2 \log n)$ to $O(n^2)$ by Gabow and Pettie [10]. A refined approximation algorithm with ratio a 1.915 was subsequently proposed by Goemans [11].

We finally remark that both the MST and the TSP also admit a *budget* version; in these cases, a budget B is specified in the input and the objective is to find the largest k-MST, respectively, the k-TSP, whose cost is no more the given budget. The k-TSP with a budget is also called the orienteering problem in the literature [12].

Awerbuch et al., in the aforementioned work, were the first to give an approximation algorithm for the general PCTSP. Their approximation ratio is again $O(\log^2(\min(Q, n)))$. They achieve this ratio by concatenating the tour found by the Quota TSP algorithm to a tour found by the Goemans–Williamson algorithm. As an application of the 5-approximate algorithm for the Quota MST [9], it follows that the approximation ratio of the PCTSP is constant.

In what follows, we introduce formal details and we provide a review of the main results in the area. In Section 34.2, we present the algorithm by Goemans and Williamson for the Prize-Collecting Steiner Tree, which is a basic building block for all the other algorithms in this chapter and has also an application to the Penalty TSP. In Section 34.3, we analyze Garg's technique showing a 5-approximation for both the k-MST and k-TSP and show how to extend it to the Quota TSP. In Section 34.4, we describe an algorithm for the general PCTSP that builds over the algorithms for Quota TSP and Penalty TSP. In Section 34.5, we show some applications of these algorithms to the minimum latency problem and graph searching. Finally, in Section 34.6, we discuss some more recent research developments concerning the PCTSP and related problems.

34.2 The Prize-Collecting Steiner Tree Problem and Penalty TSP

34.2.1 Definitions

Prize-Collecting Steiner Tree: Given an undirected graph $G = (V, E)$ with vertex penalties $\pi : V \to \mathbb{Q}^+$, edge costs $c : E \to \mathbb{Q}^+$ and a root node r, the Prize-Collecting Steiner Tree Problem asks to find a tree $T = (V_T, E_T)$ including r that minimizes

$$c(T) = \sum_{e \in E_T} c_e + \sum_{v \in V \setminus V_T} \pi_v.$$

Penalty Traveling Salesman Problem: Given an undirected graph $G = (V, E)$ with vertex penalties $\pi : V \to \mathbb{Q}^+$, edge costs $c : E \to \mathbb{Q}^+$ satisfying the triangle inequality and a root node r, the Penalty Traveling Salesman Problem asks to find a tour $T = (V_T, E_T)$ including r that minimizes

$$c(T) = \sum_{e \in E_T} c_e + \sum_{v \in V \setminus V_T} \pi_v.$$

34.2.2 History of the Results

The first approximation algorithms for the Prize-Collecting Steiner Tree and Penalty TSP were developed by Bienstock et al. [13]. They gave LP-based algorithms achieving a 3-approximation for the PCST and a 5/2-approximation for the Penalty TSP with triangle inequality. Both bounds were later improved to 2 by Goemans and Williamson [7] with a combinatorial algorithm. The NP-hardness (more precisely, APX-hardness) of the problems follows from that of the Steiner Tree problem [14] and of the TSP [15], respectively. A refined 1.915 approximation algorithm has been proposed by Goemans [11].

34.2.3 The Primal-Dual Algorithm of Goemans and Williamson

We review the algorithm of Goemans and Williamson [7] for the *Prize-Collecting Steiner Tree* and *Penalty Traveling Salesman* problem. We are given an undirected graph $G = (V, E)$, non-negative edge costs c_e, and non-negative vertex penalties π_i. The goal in the Prize-Collecting Steiner Tree Problem is to minimize the total cost of a Steiner tree and the penalties of the vertices that are not spanned by the Steiner tree. In the penalty traveling salesman problem, we aim to minimize the cost of a tour and of the penalties of the vertices that are not included in the tour. In this section, we revise the primal dual algorithm of Reference 7 that provides a 2-approximation for both problems. We first show a 2-approximation for the Steiner tree version and then show how to obtain a 2-approximation for the TSP version. Subsequently, we use GW to refer to this algorithm.

GW is a primal-dual algorithm, that is, the algorithm constructs both a feasible and integral primal and a feasible dual solution for a linear programming formulation of the problem and its dual, respectively. We will consider the version of the problem in which a root vertex r is given and the Steiner tree or the traveling salesman tour will contain r. This is without loss of generality if we can run the algorithm for all possible choices of r.

An integer programming formulation for the Steiner tree problem has a binary variable x_e for all edges $e \in E$: x_e has value 1 if edge e is part of the resulting forest and 0 otherwise. Let us denote by $\mathscr{S}$ all subsets of $V/\{r\}$. The integer programming formulation has a binary variable x_U for each set $U \in \mathscr{S}$. The cost of set U is $\sum_{v \in U} \pi_v$. For a subset $U \in \mathscr{S}$, we define $\delta(U)$ to be the set of all edges that have exactly one endpoint in U. Let T be the set of edges with $x_e = 1$ and let A be the union of sets U for which $x_U = 1$. For any $U \in \mathscr{S}$, any feasible solution must *cross* U at least once, that is, $|\delta(U) \cap T| \geq 1$, or U must be included in the set of vertices that are not spanned, that is, $U \subset A$.

This gives rise to the following integer programming formulation for the Prize-Collecting Steiner Tree Problem:

$$\text{opt}_{\text{IP}} = \min \quad \sum_{e \in E} c_e \cdot x_e + \sum_{U \in \mathscr{S}} x_U \cdot \sum_{i \in U} \pi_i \tag{IP}$$

$$\text{s.t.} \quad \sum_{e \in \delta(U)} x_e + \sum_{U': U \subseteq U'} x_{U'} \geq 1 \qquad \forall U \in \mathscr{S} \tag{34.1}$$

$$x_e, x_U \in \{0, 1\} \qquad \forall e \in E, \quad \forall U \in \mathscr{S}$$

It is easy to observe that any solution to the Prize-Collecting Steiner Tree Problem is an integral solution of this integer linear program: we set $x_e = 1$ for all edges of the Steiner tree T that connects the spanned vertices to the root r. We set $x_U = 1$ for the set U of vertices not spanned by the tree. In the linear programming relaxation of IP, we drop the integrality constraints on variables x_e and x_U.

The dual (D) of the linear programming relaxation (LP) of (IP) has a variable y_U for all sets $U \in \mathscr{S}$. There is a constraint for each edge $e \in E$ that limits the total dual assigned to sets $U \in \mathscr{S}$ that contain exactly one endpoint of e to be at most the cost c_e of the edge. There is a constraint for every $U \in \mathscr{S}$ that limits the total dual from subsets of U by at most the total penalty from vertices in U.

$$\text{opt}_D = \max \quad \sum_{U \in \mathscr{S}} y_U \qquad\qquad (D)$$

$$\text{s.t.} \quad \sum_{U \in \mathscr{S}: e \in \delta(U)} y_U \leq c_e \qquad\qquad \forall e \in E \qquad\qquad (34.2)$$

$$\sum_{U' \subseteq U} y_{U'} \leq \sum_{i \in U} \pi_i \qquad\qquad \forall U \in \mathscr{S} \qquad\qquad (34.3)$$

$$y_U \geq 0 \qquad\qquad \forall U \in \mathscr{S}$$

Algorithm GW constructs a primal solution for (IP) and a dual solution for (D). The algorithm starts with an infeasible primal solution and reduces the degree of infeasibility as it progresses. At the same time, it creates a dual feasible packing of sets of largest possible total value. The algorithm raises dual variables of certain subsets of vertices. The final dual solution is maximal in the sense that no single set can be raised without violating a constraint of type (34.2) or (34.3).

We can think of an execution of GW as a process over time. Let x^τ and y^τ, respectively, be the primal incidence vector and feasible dual solution at time τ. Initially, $x_e^0 = 0$ for all $e \in E$, $x_U^0 = 0$ for all $U \in \mathscr{S}$, and $y_U^0 = 0$ for all $U \in \mathscr{S}$. In the following, we say that an edge $e \in E$ is *tight* if the corresponding constraint (34.2) holds with equality, and that a set U is tight if the corresponding constraint (34.3) holds with equality. We use F^τ to denote the forest formed by the collection of tight edges corresponding to x^τ, and $\mathscr{S}^\tau$ to denote the collection of tight sets corresponding to x^τ. Assume that the solution x^τ at time τ is infeasible. A set $U \in \mathscr{S}$ is *active* at time τ if it is spanned by a connected component in forest F^τ, there is no tight edge $e \in \delta(U)$, and the corresponding constraint (34.3) is not tight. Let $\mathscr{A}^\tau$ be the collection of sets that are active at time τ. GW raises the dual variables for all sets in $\mathscr{A}^\tau$ uniformly at all times $\tau \geq 0$.

Two kind of events are possible: i) An edge $e \in \delta(U)$ becomes tight for a set $U \in \mathscr{A}^\tau$. We set $x_e = 1$ and update $\mathscr{A}^\tau$. (Observe that if the tight edge connects U to a component containing r, the newly formed connected component of F^τ is part of $\mathscr{A}^\tau$. If the tight edge connects set U to a component of F^τ that does not contain the root (either active or inactive), the newly formed component is part of $\mathscr{A}^\tau$.) ii) A constraint (34.3) becomes tight for a set $U \in \mathscr{A}^\tau$. We set $x_U = 1$. GW ends with a reverse pruning phase applied to all sets $U \in \mathscr{S}$ with $x_U = 1$. They are analyzed in order of decrease time at which the corresponding constraint (34.3) became tight. If the vertices of U are actually connected to the root via tight edges in the current solution, we set $x_U = 0$. The algorithm ends with a tree T connecting a set of vertices to the root r and a set A of vertices for which the penalties are paid. Denote by $c(T, A)$ the total cost of the solution, that is, $c(T, A) = \sum_{e \in T} c_e + \sum_{i \in A} \pi_i$.

Theorem 34.1 ([7]) *Suppose that algorithm* GW *outputs a tree T, a set of vertices A and a feasible dual solution $\{y_U\}_{U \in \mathscr{S}}$. Then*

$$c(T, A) \leq 2 \cdot \sum_{U \in \mathscr{S}} y_U \leq 2 \cdot \text{opt},$$

where opt *is the minimum-cost solution for the Prize-Collecting Steiner Tree Problem.*

The proof of the above-mentioned theorem [7] is along the following lines. The total dual of subsets of $U \in \mathscr{S}$ with $x_U = 1$ will pay for the penalties of vertices in A, that is $\sum_{U : x_U = 1} \sum_{U' \subseteq U} y_{U'} \leq \sum_{i \in A} \pi_i$, since constraints (34.3) are tight for these sets. Due to the reverse pruning phase, subsets of $U \in \mathscr{S}$ with $x_U = 1$ do not contribute to make tight any edge in T. The cost of T is then paid by twice the total dual of sets $U \in \mathscr{S}$ loading edges of T, that is those that contribute to making the corresponding constraints (34.2) tight.

GW also provides a 2-approximation for the penalty traveling salesman problem when edge costs obey triangle inequality. First, we run the PCST algorithm with halved penalties. Then, the resulting tree is converted to a tour by doubling every edge and shortcutting the resulting Eulerian tour. For the proof of 2-approximation, we observe that the following integer linear program is a formulation for the problem:

$$\text{opt}_{\text{IP}} = \min \quad \sum_{e \in E} c_e \cdot x_e + \sum_{U \in \mathscr{S}} x_U \cdot \sum_{i \in U} \frac{\pi_i}{2} \tag{IP}$$

$$\text{s.t.} \quad \sum_{e \in \delta(U)} x_e + \sum_{U':U \subseteq U'} x_{U'} \geq 2 \qquad \forall U \in \mathscr{S} \tag{34.4}$$

$$x_e \in \{0,1\}, \qquad \forall e \in E$$

$$x_U \in \{0,2\}, \qquad \forall U \in \mathscr{S}$$

Constraint (34.4) imposes that each subset must me crossed at least twice unless we pay the penalties for all the vertices of the subset. The dual of the corresponding relaxation is

$$\text{opt}_{\text{D}} = \max \quad 2 \cdot \sum_{U \in \mathscr{S}} y_U \tag{D}$$

$$\text{s.t.} \quad \sum_{U \in \mathscr{S}:e \in \delta(U)} y_U \leq c_e \qquad \forall e \in E \tag{34.5}$$

$$\sum_{U' \subseteq U} y_{U'} \leq \sum_{i \in U} \frac{\pi_i}{2} \qquad \forall U \in \mathscr{S} \tag{34.6}$$

$$y_U \geq 0 \qquad \forall U \in \mathscr{S}$$

The cost of the solution provided by GW is given by at most twice the cost of tree T and the total penalty of vertices in A. Since twice the total dual collected by the algorithm is a lower bound to the optimal solution, the following theorem has been proved.

Theorem 34.2 ([7]) *Suppose that algorithm* GW *outputs a cycle C, a set of vertices A and a feasible dual solution* $\{y_U\}_{U \in \mathscr{S}}$. *Then*

$$c(C,A) \leq 2 \cdot \left(2 \cdot \sum_{U \in \mathscr{S}} y_U \right) \leq 2 \cdot \text{opt},$$

where opt *is the minimum-cost solution for the penalty traveling salesman problem.*

34.3 The *k*-Minimum Spanning Tree, *k*-TSP and Quota TSP

34.3.1 Definitions

k-Minimum Spanning Tree Problem: Given an undirected graph $G = (V, E)$, a tree on G spanning exactly k nodes is called a *k-tree*. Given such a graph with edge costs $c : E \to \mathbb{Q}^+$ and a positive integer k, the (*unrooted*) k-minimum spanning tree problem asks to find a k-tree of minimum total cost. In the *rooted* version of the problem, the k-tree has to include a given root node r.

k-Traveling Salesman Problem: Given an undirected graph $G = (V, E)$, a cycle of G spanning exactly k nodes is called a *k-tour*. Given such a graph with edge costs $c : E \to \mathbb{Q}^+$ satisfying the triangle inequality, a positive integer k and a root node r, the k-Traveling Salesman Problem asks to find a k-tour including r of minimum total cost.

Quota Traveling Salesman Problem: Given an undirected graph $G = (V, E)$ with vertex weights $w : V \to \mathbb{Z}^+$ and a non-negative integer Q, a cycle C of G such that $\sum_{v \in C} w(v) \geq Q$ is called a *quota Q-tour*. Given such a graph with edge costs $c : E \rightarrow \mathbb{Q}^+$ satisfying the triangle inequality and a root node r, the Quota Traveling Salesman Problem asks to find a quota Q-tour including r of minimum total cost.

34.3.2 History of the Results

The k-MST problem is known to be an NP-hard problem [16]. Heuristics were given by Cheung and Kumar [17], who studied the problem in the context of communication networks. The first approximation algorithms were considered by Ravi et al. [18], who gave an algorithm achieving an approximation ratio of $O(\sqrt{k})$. Later, this ratio was improved to $O(\log^2 k)$ by Awerbuch et al. [8]. The first constant-ratio algorithm was given by Blum et al. [19]. Subsequently, Garg [5] gave a simple 5-approximation algorithm and a more complicated 3-approximation algorithm, whereas a 2.5-approximation algorithm for the unrooted case was found by Arya and Ramesh [3]. Arora and Karakostas gave a $(2 + \epsilon)$-approximation scheme for the rooted version. A 2-approximation by Garg [6] is the current best bound.

We observe that the rooted and the unrooted versions of the k-MST are equivalent with respect to the approximation ratio. In fact, given a c-approximation algorithm for the rooted case it is sufficient to run n times the k-MST algorithm with all possible choices for the root node r, and return the cheapest k-tree found to obtain a c-approximation algorithm for the rooted case. Garg [6] observed that a c-approximation algorithm for the unrooted case gives a c-approximation algorithm for the rooted case.

Some of these works also addressed the k-TSP and Quota TSP. The algorithms for the k-MST by Garg, as well as the scheme by Arora and Karakostas, extend to the k-TSP, thus giving a 2-approximation algorithm for this problem as the current best bound. Finally, as an application of their $O(\log^2 k)$-approximation algorithm for k-MST, Awerbuch et al. give an $O(\log^2(\min(Q, n)))$-approximation algorithm for the Quota TSP, where n is the number of nodes of the graph.

Finally, we remark that a dual version of the TSP is known as the *orienteering problem* [12]. In this problem, we are given an edge weighted graph and a budget and the goal consists in visiting as many vertices of the graph as possible and return to the origin without incurring a cost greater than the allowed budget. In Reference 20, a 4-approximation algorithm that makes use of min cost path algorithms of Reference 21 is presented for the orienteering problem. In Reference 22 an improved algorithm leading to a 3-approximation bound for the problem is shown in the context of a more general approach to constrained vehicle routing problem. The current best bound is a $(2 + \varepsilon)$-approximation (for any fixed $\varepsilon > 0$) by Chekuri et al. [23]. Other budget versions have been defined also for MST and STP [6,20,24].

34.3.3 A 5-Approximation Algorithm for k-MST and k-TSP

In this section, we present and discuss the algorithm by Garg achieving a 5-approximation for the rooted k-MST and its modification yielding the same approximation for the k-TSP. Our analysis follows Chudak et al. [25].

Several assumptions can be made with no loss of generality. First, we can suppose that the edge costs satisfy the triangle inequality, by using well-known techniques [18]. Also, we will assume that the distance from the root to the farthest vertex is a lower bound on the optimum value. It turns out that this is easy to ensure: we can run the algorithm $n - 1$ times with all possible choices of a "farthest" vertex, every time disregarding nodes farther than the chosen one, and return the best solution found. The last assumption is that $\text{opt} \geq c_0$, where c_0 is the smallest non-zero edge cost. This is not the case only if $\text{opt} = 0$, meaning that the optimal solution is a connected component containing r of size k in the graph of zero-cost edges, and the existence of such a component can be easily checked in a preprocessing phase.

A possible formulation of the rooted k-MST as an integer linear problem is the following:

$$\text{opt} = \min \quad \sum_{e \in E} c_e x_e \tag{IP}$$

$$\text{subject to:} \quad \sum_{e \in \delta(S)} x_e + \sum_{T:T \supseteq S} z_T \geq 1 \qquad \forall S \subseteq V \setminus \{r\} \tag{34.7}$$

$$\sum_{S:S \subseteq V \setminus \{r\}} |S| z_S \leq n - k \tag{34.8}$$

$$x_e \in \{0,1\} \qquad \forall e \in E$$

$$z_S \in \{0,1\} \qquad \forall S \subseteq V \setminus \{r\}$$

In the above-mentioned formulation, r is the root node and $\delta(S)$ is the set of edges with exactly one endpoint in S. The variable x_e indicate whether the edge e is included in the tree; the variable z_S indicates whether the set of vertices S is not spanned by the tree. The set of constraints Equation 34.7 enforces, for each $S \subseteq V \setminus \{r\}$, either some edge of $\delta(S)$ to be in the tree or all the vertices in S to be not spanned by the tree. Thus, every vertex not in any S such that $z_S = 1$ will be connected to the root r. Constraint Equation 34.8 enforces at least k vertices to be spanned. Finally, the LP relaxation of this integer program is obtained by replacing the integrality constraints with non-negativity constraints (in an optimal solution, $x_e \leq 1$ and $z_S \leq 1$ for all e and S).

All the proposed constant approximation algorithms for the k-MST problem use as a subroutine the primal-dual 2-approximation algorithm for the Prize-Collecting Steiner Tree of Goemans and Williamson [7]. This is not by chance, because this problem is essentially the Lagrangean relaxation of the k-MST. Indeed, if we apply Langrangean relaxation to constraint Equation 34.8 of the LP relaxation of the k-MST program, we obtain the following

$$\min \quad \sum_{e \in E} c_e x_e + \lambda \left(\sum_{S \subseteq V \setminus \{r\}} |S| z_S - (n - k) \right) \tag{LR}$$

$$\text{subject to:} \quad \sum_{e \in \delta(S)} x_e + \sum_{T:T \supseteq S} z_T \geq 1 \qquad \forall S \subseteq V \setminus \{r\}$$

$$x_e \geq 0 \qquad \forall e \in E$$

$$z_S \geq 0 \qquad \forall S \subseteq V \setminus \{r\}$$

where $\lambda \geq 0$ is the Lagrangean variable. Apart from the constant term $-\lambda(n-k)$ in the objective function, this is the same as the LP relaxation of the Prize-Collecting Steiner Tree Problem with $\pi_v = \lambda$ for all v. Moreover, any solution feasible for the LP relaxation of k-MST is also feasible for (LR), so the value of this program is a lower bound on the cost of an optimal k-MST.

Before discussing Garg's algorithm, we recall that the primal-dual approximation algorithm for the Prize-Collecting Steiner Tree returns a solution (F, A), where F is a tree including the root r and A is the set of vertices not spanned by F. The algorithm also constructs a feasible solution y for the dual of the LP relaxation of PCST.

Theorem 34.3 ([7]) *The primal solution (F, A) and the dual solution y produced by the prize-collecting algorithm satisfy*

$$\sum_{e \in F} c_e + \left(2 - \frac{1}{n-1} \right) \pi(A) \leq \left(2 - \frac{1}{n-1} \right) \sum_{S \subseteq V \setminus \{r\}} y_S$$

where $\pi(A) = \sum_{v \in A} \pi_v$.

A corollary of the theorem is that the prize-collecting algorithm has an approximation ratio of 2, by weak duality and the feasibility of y.

We would like to use the prize-collecting algorithm to solve the k-MST problem. Thus suppose that we run the algorithm with $\pi_v = \lambda$ for all $v \in V$, for some value $\lambda \geq 0$. Then by Theorem 34.3, we obtain (F, A) and y such that

$$\sum_{e \in F} c_e + 2|A|\lambda \leq 2 \sum_{S \subseteq V \setminus \{r\}} y_S. \qquad (34.9)$$

Consider the dual of the Lagrangean relaxation of the k-MST LP:

$$\begin{aligned}
\max \quad & \sum_{S \subseteq V \setminus \{r\}} y_S - (n - k)\lambda \qquad \text{(LR-D)}\\
\text{subject to:} \quad & \sum_{S: e \in \delta(S)} y_S \leq c_e \qquad \forall e \in E\\
& \sum_{T: T \subseteq S} y_T \leq |S|\lambda \qquad \forall S \subseteq V \setminus \{r\}\\
& y_S \geq 0 \qquad \forall S \subseteq V \setminus \{r\}
\end{aligned}$$

Since this dual is, apart from the objective function, the same as the dual of the LP relaxation of the PCST instance, the solution y is feasible for this dual, and the value of the objective function is a lower bound on the cost of an optimal k-MST, by weak duality. Subtracting $2(n-k)\lambda$ from both sides of Equation 34.9,

$$\sum_{e \in F} c_e + 2\lambda(|A| - (n - k)) \leq 2\left(\sum_{S \subseteq V \setminus \{r\}} y_S - (n - k)\lambda \right) \leq 2\,\mathrm{opt}$$

where opt is the cost of an optimal solution to the k-MST instance.

Now if the term $|A| - (n - k)$ is zero, we can conclude that the tree F is a k-tree and has cost no more than twice optimal. Unfortunately, if $|A| - (n - k)$ is positive then the tree F is not feasible, whereas if it is negative we cannot conclude anything about the approximation ratio. However, it turns out that it is possible to find values of λ such that even if these cases occur, they can be taken care of, although at the cost of resulting in an approximation ratio higher than 2.

What Garg's algorithm does is indeed a binary search for these critical values of λ, through a sequence of calls to the prize-collecting algorithm. Notice that if the prize-collecting algorithm is called with $\lambda = 0$, it will return the empty tree spanning only r as a solution, whereas for $\lambda = \sum_{e \in E} c_e$ it will return a tree spanning all vertices. Thus the initial interval of the binary search will be $[0, \sum_{e \in E} c_e]$, and at every iteration, if the current interval is $[\lambda_1, \lambda_2]$, the prize-collecting algorithm is run with $\lambda = \frac{1}{2}(\lambda_1 + \lambda_2)$. If the returned tree has less than k vertices, we update λ_1 to λ; if it has more than k vertices, we update λ_2 to λ. Notice that in the lucky event that, at any point, a tree with exactly k vertices is returned, we can stop, since by the above-mentioned discussion that must be within a factor 2 of optimal. So assume that this event does not happen. We will stop when we have found two values λ_1, λ_2 such that:

1. $\lambda_2 - \lambda_1 \leq \frac{c_0}{2n(n+1)}$ (recall that c_0 is the smallest non-zero edge cost).
2. for $i = 1, 2$, the prize-collecting algorithm run with λ set to λ_i returns a primal solution (F_i, A_i) spanning k_i vertices and a dual solution $y^{(i)}$, with $k_1 < k < k_2$.

Notice that these two values will be found at most after $O(\log \frac{n^2 \sum_e c_e}{c_0})$ calls to the prize-collecting algorithm. The final step of the algorithm is combining the two solutions (F_1, A_1) and (F_2, A_2) into a single k-tree. Solution (F_1, A_1) is within a factor of 2 of optimal, but infeasible, whereas solution (F_2, A_2) can be easily made feasible but not within a factor of 2 of optimal. More precisely, as a consequence of Theorem 34.3,

$$\sum_{e \in F_1} c_e \le \left(2 - \frac{1}{n}\right)\left(\sum_{S \subseteq V \setminus \{r\}} y_S^{(1)} - |A_1|\lambda_1\right)$$

$$\sum_{e \in F_2} c_e \le \left(2 - \frac{1}{n}\right)\left(\sum_{S \subseteq V \setminus \{r\}} y_S^{(2)} - |A_2|\lambda_2\right).$$

To get a bound on the cost of F_1 and F_2 in terms of opt, let $\alpha_1 = \frac{n-k-|A_2|}{|A_1|-|A_2|}$ and $\alpha_2 = \frac{|A_1|-(n-k)}{|A_1|-|A_2|}$. Then $\alpha_1|A_1| + \alpha_2|A_2| = n - k$ and $\alpha_1 + \alpha_2 = 1$, and after defining, for all $S \subseteq V \setminus \{r\}$, $y_S = \alpha_1 y_S^{(1)} + \alpha_2 y_S^{(2)}$, it is possible to prove the following lemma.

Lemma 34.1 [25]

$$\alpha_1 \sum_{e \in F_1} c_e + \alpha_2 \sum_{e \in F_2} c_e < 2\text{opt}.$$

Proof. We omit the proof for brevity; the reader can find it in the overview by Chudak et al. [25]. ∎

We now show how to obtain a 5-approximation algorithm by choosing one of two solutions. First, if $\alpha_2 \ge \frac{1}{2}$, the tree F_2, besides spanning more than k vertices, satisfies

$$\sum_{e \in F_2} c_e \le 2\alpha_2 \sum_{e \in F_2} c_e \le 4\text{opt}$$

by Lemma 34.1. If instead $\alpha_2 < \frac{1}{2}$, the solution is constructed by extending F_1 with nodes from F_2. Let $\ell \ge k_2 - k_1$ be the number of nodes spanned by F_2 but not by F_1. Then we can obtain a path on $k - k_1$ vertices by doubling the tree F_2, shortcutting the corresponding Eulerian tour to a simple tour of the ℓ nodes spanned only by F_2, and choosing the cheapest path of $k - k_1$ vertices from this tour. The resulting path has cost at most

$$2\frac{k - k_1}{k_2 - k_1} \sum_{e \in F_2} c_e.$$

Notice that this path is disconnected from F_1. However, we can connect it by adding an edge from the root to any node of the set, costing at most opt by one of the assumptions at the beginning of the section. Since

$$\frac{k - k_1}{k_2 - k_1} = \frac{n - k_1 - (n - k)}{n - k_1 - (n - k_2)} = \frac{|A_1| - (n - k)}{|A_1| - |A_2|} = \alpha_2$$

the total cost of the produced solution is bounded by

$$\sum_{e \in F_1} c_e + 2\alpha_2 \sum_{e \in F_2} c_e + \text{opt} \le 2\left(\alpha_1 \sum_{e \in F_1} c_e + \alpha_2 \sum_{e \in F_2} c_e\right) + \text{opt} \le 4\text{opt} + \text{opt}$$

using Lemma 34.1 and the fact that $\alpha_2 < \frac{1}{2}$ implies $\alpha_1 > \frac{1}{2}$.

As for the k-TSP, it suffices to run the prize-collecting subroutine with halved penalties and then short-cut the Eulerian walk obtained after doubling the k-tree found, in the same way as we went from the Prize-Collecting Steiner Tree to the Penalty TSP.

34.3.4 From the k-MST to the Quota TSP

In this section, we will describe a 5-approximation algorithm for the Quota TSP. However, the discussion will be easier if we consider the following problem first.

Quota Minimum Spanning Tree Problem: Given an undirected graph $G = (V, E)$ with vertex weights $w : V \to \mathbb{Z}^+$ and a positive integer Q, a tree F of G such that $\sum_{v \in F} w(v) \geq Q$ is called a *quota Q-tree*. Given such a graph with edge costs $c : E \to \mathbb{Q}^+$ and a root node r, the quota minimum spanning tree problem asks to find a quota Q-tree including r of minimum total cost.

Theorem 34.4 ([9]) *There is a 5-approximation algorithm for the Quota Minimum Spanning Tree Problem.*

The idea behind the theorem is that we can run the 5-approximation algorithm for the k-MST by Garg, but instead of setting the penalties uniformly to λ, we set $\pi_v = \lambda \cdot w_v$ when calling the Prize-Collecting Steiner Tree subroutine. The two solutions obtained at the end of the binary search phase can then be patched essentially as before.

Now, we can obtain an algorithm for the Quota TSP in the same way as we went from the Prize-Collecting Steiner Tree to the penalty TSP in Section 34.2. That is, it is sufficient to run the prize-collecting subroutine with $\pi_v = \frac{1}{2}\lambda w_v$. The analysis remains the same.

34.4 The Prize Collecting Traveling Salesman Problem

34.4.1 Definitions

Prize-Collecting Traveling Salesman Problem. Given an undirected graph $G = (V, E)$ with vertex weights $w : V \to \mathbb{Z}^+$, vertex penalties $\pi : V \to \mathbb{Q}^+$, and a non-negative integer Q, a cycle C of G such that $\sum_{v \in C} w(v) \geq Q$ is called a *quota Q-tour*. Given such a graph with edge costs $c : E \to \mathbb{Q}^+$ satisfying the triangle inequality and a root node r, the PCTSP asks to find a quota Q-tour $T = (V_T, E_T)$ including r that minimizes

$$c(T) = \sum_{e \in E_T} c_e + \sum_{v \in V \setminus V_T} \pi_v.$$

34.4.2 History of the Results

In the general form given here, the PCTSP was first formulated by Balas [1,26], who gave structural properties of the PCTS polytope as well as heuristics. The problem arised during the task of developing daily schedules for a steel rolling mill.

The only results on guaranteed heuristics for the PCTSP are due to Awerbuch et al. [8]. They give polynomial time algorithm with an approximation ratio of $O(\log^2(\min(Q, n)))$, where n is the number of vertices of the graph and Q is the required vertex weight to be visited. However, the PCTSP contains as special cases both the Penalty TSP and the k-TSP, which received more attention in the literature (for the history of results on these problems, the reader can refer to the previous sections). From known results on these problems, we derive a constant-approximation algorithm for the general PCTSP in the following section.

34.4.3 A Constant-Factor Approximation Algorithm for PCTSP

A simple idea exploited by the algorithm of Awerbuch et al. [8] is that, for a given instance I of the PCTSP, the following quantities constitute lower bounds on the cost `opt` of an optimal solution:

1. the cost `optp` of an optimal solution to a Penalty TSP instance I_p defined on the same graph and having the same penalties as in the PCTSP instance (since a feasible solution to I is also feasible for I_p and has the same cost);
2. the cost `optq` of an optimal solution to a Quota TSP instance I_q defined on the same graph and having the same weights and quota as in I (since every feasible solution to I can be turned into a feasible solution for I_q of at most the same cost).

Thus, to approximate an optimal solution to the PCTSP instance I, we can:

1. run an α-approximation algorithm for Penalty TSP on I_p to obtain a tour T_p such that $c(T_p) \leq \alpha \cdot \text{optp}$;
2. run a β-approximation algorithm for Quota TSP on I_q to obtain a tour T_q such that $c(T_q) \leq \beta \cdot \text{optq}$;
3. concatenate T_p and T_q to obtain a tour T feasible for the Prize Collecting TSP instance I of cost

$$c(T) \leq c(T_p) + c(T_q) \leq \alpha \cdot \text{optp} + \beta \cdot \text{optq} \leq (\alpha + \beta)\text{opt}.$$

This means that, by using the best algorithms currently known for Penalty TSP and Quota TSP, one can obtain a constant-factor approximation to the Prize-Collecting TSP.

34.5 The Minimum Latency Problem and Graph Searching

Suppose that a plumber receives calls from various customers and decides to organize a tour of the customers for the subsequent day; for sake of simplicity let us also assume that, at each visit, the time the plumber needs to fix the customer's problem is constant. A selfish plumber would decide to schedule his tour in such way as to minimize the overall time he takes to serve all customers and come back home; such approach would require the solution of an instance of TSP. In alternative, a non-selfish plumber would decide to schedule his tour in such a way to minimize the average time customers have to wait his visit the day after. In this case, he will have to solve an instance of the so-called *traveling repairman problem* (TRP).

The TRP is more frequently known in the literature as *Minimum Latency Problem* (MLP) [27], but it is also known as *school-bus driver problem* [28] and the *delivery man problem* [29,30]. Strictly related to MLP is the so-called *graph searching problem* (GSP) [31]. In such problem we assume that a single prize is hidden in a vertex of an edge weighted graph and the vertices are labeled with the probability that the prize is stored in the vertex. The goal is to minimize the expected cost to find the prize by exploring all vertices of the graph. The relationship between MLP and GSP is discussed in Reference 9.

34.5.1 Definitions

Minimum Latency Problem. Given an undirected graph $G = (V, E)$, with edge costs $c : E \to \mathbb{Q}^+$ satisfying the triangle inequality, let T be a tour that visits the vertices in some order. The *latency* $l_{v_i,T}$ of a vertex $v_i \in T$ is the cost of the prefix of T ending in v_i. The *Minimum Latency Problem* asks to find a tour T such that the sum of the latencies of all vertices along T is minimum.

34.5.2 History of the Results

The problem has been shown to be NP-hard by Sahni and Gonzales [32]. Afrati et al. [33] showed that the problem can be solved in polynomial time on trees with bounded number of leaves. Subsequently, Sitters [34] has shown that the problem is NP-hard also in the case of general weighted trees.

From the approximability point of view, the MLP is, clearly, as the TSP, hard to approximate for any given constant ratio on general graphs (i.e., when the triangle inequality does not hold), whereas in the case of metric spaces it can be shown to be APX-complete, that is it allows approximation algorithms but does not allow approximation schemes.

The first constant factor approximation algorithm for the MLP on general metric spaces has been presented by Blum et al. [27] who show that given a c-approximate algorithm for the k-MST then there exists a $8c$-approximation ratio for the MLP. Subsequently, Goemans and Kleinberg [35] showed that the constant 8 above can be lowered to 3.59, thus implying a 7.18-approximation algorithm for MLP. The best current bound is 3.59, as given by Chaudhuri et al. [21].

Arora and Karakostas [36] showed the existence of a quasi-polynomial time approximation algorithm when the input graph is a tree; to compute a $(1+\epsilon)$-approximation the algorithm requires $n^{O(\log n/\epsilon)}$ time.

We finally remark that the problem has also been extended to the case of k repairmen. Fakcharoenphol, et al. [37] showed the first constant approximation algorithm for the problem. This result has been improved to 8.49-approximation by Chaudhuri et al. [21].

34.5.3 A 3.59-Approximation Algorithm for the Minimum Latency Problem

We first present the algorithm proposed by Goemans and Kleinberg [35], which gives a 7.18-approximation algorithm. The procedure proposed by the authors computes, for every $j = 1, 2, \ldots, n$, the tour T_j of minimum length that visits j vertices. Then we have to concatenate a subsequence of the tours to form the desired tour. Clearly, the goal is to select those values $j_1, \ldots, j_m$ such that the latency of the final tour obtained by stitching together tours $T_{j_1} \ldots T_{j_m}$ is minimized.

Let d_j and p_j be the length of tour T_j and the number of new vertices visited during the same tour, respectively. Since the number of vertices discovered up to the i-th tour is certainly no smaller than j_i, the following claim holds:

$$\sum_{i=1}^{m} p_i d_i \le \sum_{i=1}^{m} (j_i - j_{i-1}) d_i.$$

It follows that for a number of vertices equal to $\sum_{k=1}^{i} p_k - j_i$, we sum a contribution at most d_k on the left side of the equation while a contribution larger than d_k on the right side of the equation. Moreover, each tour T_i is traversed in the direction that minimizes the total latency of the vertices discovered during tour T_i. This allows to rewrite the total latency of the tour obtained by concatenating $T_{j_1}, \ldots, T_{j_m}$ as

$$\sum_{i} (n - \sum_{k=1}^{i} p_k) d_{j_i} + \frac{1}{2} \sum_{i} p_i d_j$$
$$\le \sum_{i} (n - j_i) d_{j_i} + \frac{1}{2} \sum_{i} (j_i - j_{i-1}) d_{j_i}$$
$$= \sum_{i} (n - \frac{j_{i-1} + j_i}{2}) d_{j_i}.$$

The previous formula allows to rewrite the total latency of the algorithm only in terms of the indices j_i and of the length d_j, independently from the number of new vertices discovered during each tour. A complete digraph of $n + 1$ vertices is then constructed in the following way. Arc (i, j) goes from $min(i, j)$ to $max(i, j)$ and has length $(n - \frac{i+j}{2}) d_j$. The algorithm computes a shortest path from node 0 to n. Assume that the path goes through nodes $0 = j_0 < j_1 < \ldots < j_m = n$. The tour is then obtained by concatenating tours $T_{j_1}, \ldots, T_{j_m}$.

The obtained solution is compared against the following lower bound $OPT \ge \sum_{k=1}^{n} \frac{d_k}{2}$. This lower bound follows from the observation that the k-th vertex cannot be visited before $d_k/2$ in any optimum tour. The approximation ratio of the algorithm is determined by bounding the maximum over all the possible set of distances $d_1, \ldots, d_n$ of the ratio between the shortest path in G_n and the lower bound on the optimum solution. This value results to be smaller than 3.59.

Theorem 34.5 ([35]) *Given a c-approximation algorithm for the problem of finding an a tour of minimum length spanning at least k vertices on a specific metric space, then there exists a 3.59c-approximation algorithm for the MLP on the same metric space.*

Again by making use of the 2-approximation algorithm proposed by Garg [6] for k-MST and k-TSP, we may achieve a ratio 7.18 for MLP.

With respect to the results we have seen so far, a remarkable step forward has been achieved by Chaudhuri et al. [21]. Using techniques from Garg [5], Arora and Karakostas [2] and Archer et al. [38], the authors are able to find a k-MST whose cost is no more than $(1 + \epsilon)$ the cost of the minimum path visiting k vertices. Since such a cost is a lower bound on the latency of a k tour, the result implies the following theorem.

Theorem 34.6 [21] *There exists a 3.59-approximation algorithm for the MLP on general metric spaces.*

By the argumentes provided in Reference 9 the same approximation bound also holds for GSP.

34.6 Further Developments

In this section, we review some results from the last decade concerning the PCTSP and related problems: the Penalty TSP, the k-TSP, the orienteering problem, and the Minimum Latency Problem. As the problems considered in this chapter find many applications, it is not surprising that they have been studied in realms other than that of approximation algorithms, such as online optimization [39–41], algorithmic game theory [42], robust optimization [43], and algorithm engineering [44]. In the following, we focus on worst-case approximation results.

34.6.1 Penalty TSP

The natural linear programming relaxation of the Penalty TSP, with an integrality gap of 2, acted as a barrier to further improvements to the approximation ratio of the problem. Fourteen years after Goemans and Williamson's breakthrough 2-approximation algorithm [7], Archer et al. [45,46] were the first to break the barrier and provide an improved performance guarantee of 1.991. For sufficiently small $\varepsilon > 0$, they showed $(2 - \varepsilon)$-approximation algorithms for the penalty versions of the Steiner tree, traveling salesman, and stroll problem. Subsequently, Goemans [11] provided a 1.915-approximation algorithm by combining the rounding algorithm of Bienstock et al. [13] with Archer et al.'s refined analysis of the primal-dual algorithm of Goemans and Williamson.

Bateni et al. [47] study the Penalty TSP and other prize-collecting network design problems on planar graphs, proving the existence of polynomial-time approximation schemes. They also show that a more general prize-collecting Steiner forest problem is APX-hard on planar graphs.

An et al. [48] present a deterministic 1.619-approximation for the s-t path TSP, where two vertices s and t are the specified endpoints of the Hamiltonian path to be constructed. One of the implications of their result is an improved 1.954-approximation algorithm for the version of the problem with penalties, the prize-collecting s-t path problem.

An extension of the Penalty TSP to asymmetric spaces satisfying the triangle inequality is discussed by Nguyen [49], who provides a $\lceil \log n \rceil$-approximation primal-dual algorithm.

34.6.2 k-TSP and k-MST

Despite rather intensive activity on related problems, the best approximability result on the k-TSP and k-MST remains Garg's 2-approximation algorithm [6], which has not been improved or, to our knowledge, simplified.

Chekuri and Pál [50] consider an orienteering problem with time windows on directed graphs satisfying the triangle inequality. One of the applications of their results is to the k-TSP problem on the same class of digraphs; for such a directed k-TSP, their approach yields an $O(\log^2 k)$ approximation running in quasi-polynomial time. The directed k-TSP is also considered by Bateni and Chuzoy [51], who provide an $O(\log^2 n/\log\log n)$-approximation algorithm.

Gupta and Srinivasan [52] study a Covering Steiner problem, where, in addition to an edge-weighted graph, several group of vertices are given, each group with a certain non-negative requirement. The problem is to find a minimum-cost tree which spans at least the required number of vertices from every group. The Covering Steiner problem is a common generalization of the Group Steiner tree and of the k-MST. Gupta and Srinivasan present an algorithm with an approximation ratio matching the best known ratio for the group Steiner problem.

Lau et al. [53] study network design problems with degree or order constraints, and consider the problem of finding a minimum cost λ-edge connected subgraph with at least k vertices, which generalizes the k-MST. When $\lambda = 2$, they present an $O(\log^3 k)$-approximation algorithm, whereas they provide evidence that the problem is hard to approximate when λ is arbitrary.

Gupta et al. [54] consider the k-Forest problem, in which a finite metric space is given together with a set of pairs of nodes. The goal is to find the least-cost subgraph that connects at least k pairs. The k-Forest problem is a common generalization of the k-MST and of the *dense-k-subgraph* problems. The paper gives an $O(\min\{\sqrt{n}\log k, \sqrt{k}\})$-approximation algorithm for k-forest, and also investigates some implications of the result for classic vehicle routing problems.

In a different variation of the k-MST and Quota MST, proposed by Moss and Rabani [55], costs are given on the *vertices* of the graph, as well as prizes on the vertices as in the Quota MST, and the goal is to find a connected subset containing the root that collects the given quota of prizes and minimizes the cost of the selected vertices. Moss and Rabani [55] describe an $O(\log n)$-approximation algorithm for this problem and related variants.

34.6.3 The Orienteering Problem

The best currently known bound for orienteering is proposed by Chekuri et al. [23], who gave a $(2 + \varepsilon)$-approximation algorithm for any fixed $\varepsilon > 0$, thus improving upon the previous 3-approximation of Bansal et al. [22]. The same paper also considers an extension of the problem with time-windows.

Chekuri et al. [23] also consider orienteering in directed graphs with the triangle inequality, for which they provide an $O(\log^2 \mathrm{opt})$-algorithm, where $\mathrm{opt} \le n$ is the optimal value of the orienteering instance. A similar result is independently obtained, with different techniques, by Nagarajan and Ravi [56], who gave an $O(\log^2 n/\log\log n)$-approximation algorithm.

Chen and Har-Peled [57] obtained a polynomial-time approximation scheme for orienteering for points in fixed-dimensional Euclidean space. Angelelli et al. [58] consider the orienteering problem and the Quota TSP in special graph classes such as path, cycle, and star graphs, exhibiting polynomial time algorithms or fully polynomial time approximation schemes for these cases.

Khuller et al. [59] propose a problem called *the gas station problem*, which is somewhat reminiscent of orienteering, but directly incorporates fuel costs and tank capacity in the model. In the input graph, each node represents a gas station, with a certain individual price for gas. The goal is to go from a source node to a destination node in the cheapest possible way, using at most Δ stops to fill gas. The authors show that the problem can be solved exactly in $O(\Delta n^2 \log n)$ time, and consider several harder variations.

34.6.4 The Minimum Latency Problem

Although the bound of 3.59 on the approximability of the MLP by Chaudhuri et al. [21] has not been improved, a number of generalizations and extensions have been obtained.

The MLP is NP-hard on weighted trees, as shown by Sitters [34], but Archer and Blasiak [60] show how to obtain an improved 3.03-approximation algorithm on trees and, in fact, on any class of graphs where one can solve in polynomial time the problem of visiting k vertices from the root, while minimizing the path cost plus the penalty costs for the unvisited vertices. The result was subsequently strengthened by

Sitters [61], who provides the first polynomial-time approximation scheme for MLP in the case of the Euclidean plane, on weighted planar graphs, and on weighted trees.

The paper by Chaudhuri et al. [21] already implied a 6-approximation algorithm for the MLP with k-repairmen. Jothi and Raghavachari [62] consider a generalization of the MLP with k-repairmen where each vertex has an associated "repair time." They present a $(\frac{3}{2}\beta + \frac{1}{2})$-approximation algorithm for arbitrary k, and a $(\beta + 2)$-approximation for fixed k, where β is the best approximation ratio for the MLP with k-repairmen. Post and Swamy [63] investigate the effectiveness of some LP-relaxations for single- and multi-repairmen versions of the MLP.

A version of the MLP on directed graphs (satisfying the triangle inequality) was first considered by Nagarajan and Ravi [64], who gave an $O(n^{1/2+\varepsilon})$-approximation algorithm for such problem. The same problem was revisited by Friggstad et al. [65] who provide an $O(\log n)$-approximation algorithm for directed MLP.

References

1. E. Balas. The prize collecting traveling salesman problem. *Networks*, 19:621–636, 1989.
2. S. Arora and G. Karakostas. A $2 + \epsilon$ approximation algorithm for the k-MST problem. *Mathematical Programming*, 107(3):491–504, 2006.
3. S. Arya and H. Ramesh. A 2.5-factor approximation algorithm for the k-MST problem. *Information Processing Letters*, 65(3):117–118, 1998.
4. A. Blum, R. Ravi, and S. Vempala. A constant-factor approximation algorithm for the k-MST problem. In *Proceeding of the 28th Symposium on Theory of Computing*, ACM, New York, pp. 442–448, 1996.
5. N. Garg. A 3-approximation for the minimum tree spanning k vertices. In *Proceeding of the 37th Symposium on Foundations of Computer Science*, IEEE, Los Alamitos, CA, pp. 302–309, 1996.
6. N. Garg. Saving an epsilon: A 2-approximation for the k-MST problem in graphs. In *Proceeding of the 37th Symposium on Theory of Computing*, ACM, New York, pp. 396–402, 2005.
7. M. Goemans and D. P. Williamson. A general approximation technique for constrained forest problems. *SIAM Journal on Computing*, 24(2):296–317, 1995.
8. B. Awerbuch, Y. Azar, A. Blum, and S. Vempala. New approximation guarantees for minimum-weight k-trees and prize-collecting salesmen. *SIAM Journal on Computing*, 28(1):254–262, 1999.
9. G. Ausiello, S. Leonardi, and A. Marchetti-Spaccamela. On salesmen, repairmen, spiders, and other traveling agents. In G. Bongiovanni, G. Gambosi, and R. Petreschi, Eds., *Proceeding of the 4th Italian Conference on Algorithms and Complexity*, volume 1767 of Lecture Notes in Computer Science, pp. 1–16. Springer-Verlag, Berlin, Germany, 2000.
10. H. N. Gabow and S. Pettie. The dynamic vertex minimum problem and its application to clustering-type approximation algorithms. In M. Penttonen and E. M. Schmidt, Eds., *Proceeding of the 8th Scandinavian Workshop on Algorithm Theory*, volume 2368 of Lecture Notes in Computer Science, pp. 190–199. Springer-Verlag, Berlin, Germany, 2002.
11. M. X. Goemans. Combining approximation algorithms for the prize-collecting TSP. *CoRR*, arXiv: abs/0910.0553, 2009.
12. B. L. Golden, L. Levy, and R. Vohra. The orienteering problem. *Naval Research Logistics*, 34(3):307–318, 1987.
13. D. Bienstock, M. X. Goemans, D. Simchi-Levi, and D. Williamson. A note on the prize collecting traveling salesman problem. *Mathematical Programming*, 59(3):413–420, 1993.
14. M. W. Bern and P. E. Plassmann. The Steiner problem with edge lengths 1 and 2. *Information Processing Letters*, 32(4):171–176, 1989.
15. C. H. Papadimitriou and M. Yannakakis. The traveling salesman problem with distances one and two. *Mathematics of Operations Research*, 18(1):1–11, 1993.

16. M. Fischetti, H. W. Hamacher, K. Jørnsten, and F. Maffioli. Weighted k-cardinality trees: Complexity and polyhedral structure. *Networks*, 24(1):11–21, 1994.

17. S. Y. Cheung and A. Kumar. Efficient quorumcast routing algorithms. In *Proceedings of the 13th IEEE Networking for Global Communications*, IEEE, Los Alamitos, CA, 2:840–847, 1994.

18. R. Ravi, R. Sundaram, M. V. Marathe, D. J. Rosenkrantz, and S. S. Ravi. Spanning trees short or small. *SIAM Journal on Discrete Mathematics*, 9(2):178–200, 1996.

19. A. Blum, R. Ravi, and S. Vempala. A constant-factor approximation algorithm for the k-MST problem. *Journal of Computer and Systems Sciences*, 58(1):101–108, 1999.

20. A. Blum, S. Chawla, D. R. Karger, T. Lane, A. Meyerson, and M. Minkoff. Approximation algorithms for orienteering and discounted-reward TSP. In *Proceedings of the 44th Symposium on Foundations of Computer Science*, IEEE, Los Alamitos, CA, pp. 46–55, 2003.

21. K. Chaudhuri, B. Godfrey, S. Rao, and K. Talwar. Paths, trees, and minimum latency tours. In *Proceedings of the 44th Symposium on Foundations of Computer Science*, IEEE, Los Alamitos, CA, pp. 36–45, 2003.

22. N. Bansal, A. Blum, P. Chalasani, and A. Meyerson. Approximation algorithms for deadline-TSP and vehicle routing with time-windows. In *Proceedings of the 36th Symposium on Theory of Computing*, ACM, New York, NY, pp. 166–174, 2004.

23. C. Chekuri, N. Korula, and M. Pál. Improved algorithms for orienteering and related problems. *ACM Transanctions on Algorithms*, 8(3):23:1–23:27, 2012.

24. D. S. Johnson, M. Minkoff, and S. Phillips. The prize collecting steiner tree problem: Theory and practice. In *Proceedings of the 11th Symposium on Discrete Algorithms*, SIAM, Philadelphia, PA, pp. 760–769, 2000.

25. F. A. Chudak, T. Roughgarden, and D. P. Williamson. Approximate k-MSTs and k-Steiner trees via the primal-dual method and Lagrangean relaxation. *Mathematical Programming*, 100(2):411–421, 2004.

26. E. Balas. The prize collecting traveling salesman problem: II. Polyhedral results. *Networks*, 25:199–216, 1995.

27. A. Blum, P. Chalasani, D. Coppersmith, W. R. Pulleyblank, P. Raghavan, and M. Sudan. The minimum latency problem. In *Proceedings of the 26th Symposium on Theory of Computing*, ACM, New York, pp. 163–171, 1994.

28. T. G. Will. Extremal results and algorithms for degree sequences of graphs. PhD thesis, University of Illinois at Champaign, IL, 1993.

29. E. Minieka. The delivery man problem on a tree network. *Annals of Operations Research*, 18(1–4):261–266, 1989.

30. M. Fischetti, G. Laporte, and S. Martello. The delivery man problem and cumulative matroids. *Operations Research*, 41(6):1055–1064, 1993.

31. E. Koutsoupias, C. H. Papadimitriou, and M. Yannakakis. Searching a fixed graph. In F. Meyer auf der Heide and B. Monien, Eds., *Proceedings of the 23rd International Colloquium on Automata, Languages, and Programming*, volume 1099 of Lecture Notes in Computer Science, pp. 280–289. Springer-Verlag, Berlin, Germany, 1996.

32. S. Sahni and T. F. Gonzalez. P-complete approximation problems. *Journal of the ACM*, 23(3):555–565, 1976.

33. F. N. Afrati, S. S. Cosmadakis, C. H. Papadimitriou, G. Papageorgiou, and N. Papakostantinou. The complexity of the travelling repairman problem. *Informatique Théorique et Applications*, 20(1):79–87, 1986.

34. R. Sitters. The minimum latency problem is NP-hard for weighted trees. In *Proceedings of the 9th Integer Programming and Combinatorial Optimization Conference*, Springer, Berlin, pp. 230–239, 2002.

35. M. Goemans and J. Kleinberg. An improved approximation ratio for the minimum latency problem. *Mathematical Programming*, 82(1):111–124, 1998.

36. S. Arora and G. Karakostas. Approximation schemes for minimum latency problems. *SIAM Journal on Computing*, 32(5):1317–1337, 2003.

37. J. Fakcharoenphol, C. Harrelson, and S. Rao. The *k*-traveling repairman problem. In *Proceedings of the 14th Symposium on Discrete Algorithms*, pp. 655–664, 2003.

38. A. Archer, A. Levin, and D. P. Williamson. A faster, better approximation algorithm for the minimum latency problem. *SIAM Journal on Computing*, 37(5):1472–1498, 2008.

39. S. O. Krumke, W. E. de Paepe, D. Poensgen, and L. Stougie. News from the online traveling repairman. *Theoretical Computer Science*, 295(1):279–294, 2003.

40. G. Ausiello, V. Bonifaci, and L. Laura. The online prize-collecting traveling salesman problem. *Information Processing Letters*, 107(6):199–204, 2008.

41. V. Bonifaci and L. Stougie. Online *k*-server routing problems. *Theory of Computing Systems*, 45(3):470–485, 2009.

42. A. Gupta, J. Könemann, S. Leonardi, R. Ravi, and G. Schäfer. Efficient cost-sharing mechanisms for prize-collecting problems. *Mathematical Programming*, 152(1):147–188, 2015.

43. S. Büttner and S. O. Krumke. Robust optimization for routing problems on trees. *TOP*, 24(2): 338–359, 2016.

44. M. Chimani, M. Kandyba, I. Ljubić, and P. Mutzel. Obtaining optimal *k*-cardinality trees fast. *Journal of Experimental Algorithmics*, 14:5:2.5–5:2.23, January 2010.

45. A. Archer, M. H. Bateni, M. T. Hajiaghayi, and H. J. Karloff. Improved approximation algorithms for prize-collecting Steiner tree and TSP. In *Proceedings IEEE Symposium on Foundations of Computer Science*, pp. 427–436. IEEE, 2009.

46. A. Archer, M. H. Bateni, M. T. Hajiaghayi, and H. Karloff. Improved approximation algorithms for prize-collecting Steiner tree and TSP. *SIAM Journal on Computing*, 40(2):309–332, 2011.

47. M. Bateni, C. Chekuri, A. Ene, M. T. Hajiaghayi, N. Korula, and D. Marx. Prize-collecting Steiner problems on planar graphs. In *Proceedings of the 22nd Annual ACM-SIAM Symposium on Discrete Algorithms*, SIAM, Philadelphia, PA, pp. 1028–1049. SIAM, 2011.

48. H. C. An, R. Kleinberg, and D. B. Shmoys. Improving Christofides' algorithm for the s-t path TSP. *Journal of the ACM*, 62(5):34:1–34:28, 2015.

49. V. H. Nguyen. A primal-dual approximation algorithm for the asymmetric prize-collecting TSP. *Journal of Combinatorial Optimization*, 25(2):265–278, 2013.

50. C. Chekuri and M. Pal. A recursive greedy algorithm for walks in directed graphs. In *Proceedings of the 46th Annual IEEE Symposium on Foundations of Computer Science*, pp. 245–253, October 2005.

51. M. H. Bateni, M. T. Hajiaghayi, and V. Liaghat. Improved approximation algorithms for (budgeted) node-weighted Steiner problems. In F. V. Fomin, R. Freivalds, M. Kwiatkowska, and D. Peleg, Eds., *Proceedings of the 40th International Colloquium on Automata, Languages, and Programming, Part I*, pp. 81–92. Springer, Berlin, Germany, 2013.

52. A. Gupta and A. Srinivasan. An improved approximation ratio for the covering steiner problem. *Theory of Computing*, 2(3):53–64, 2006.

53. L. C. Lau, J. (Seffi) Naor, M. R. Salavatipour, and M. Singh. Survivable network design with degree or order constraints. *SIAM Journal on Computing*, 39(3):1062–1087, 2009.

54. A. Gupta, M. Hajiaghayi, V. Nagarajan, and R. Ravi. Dial a ride from *k*-forest. *The ACM Transactions on Algorithms*, 6(2):41:1–41:21, 2010.

55. A. Moss and Y. Rabani. Approximation algorithms for constrained node weighted Steiner tree problems. *SIAM Journal on Computing*, 37(2):460–481, 2007.

56. V. Nagarajan and R. Ravi. The directed orienteering problem. *Algorithmica*, 60(4):1017–1030, 2011.

57. K. Chen and S. Har-Peled. The Euclidean orienteering problem revisited. *SIAM Journal on Computing*, 38(1):385–397, 2008.

58. E. Angelelli, C. Bazgan, M. G. Speranza, and Z. Tuza. Complexity and approximation for traveling salesman problems with profits. *Theoretical Computer Science*, 531:54–65, 2014.
59. S. Khuller, A. Malekian, and J. Mestre. To fill or not to fill: The gas station problem. *The ACM Transactions on Algorithms*, 7(3):36:1–36:16, 2011.
60. A. Archer and A. Blasiak. Improved approximation algorithms for the minimum latency problem via prize-collecting strolls. In *Proceedings of the 31st Annual ACM-SIAM Symposium on Discrete Algorithms*, Philadelphia, PA, pp. 429–447. SIAM, 2010.
61. R. Sitters. Polynomial time approximation schemes for the traveling repairman and other minimum latency problems. In *Proceedings of the 25th Annual ACM-SIAM Symposium on Discrete Algorithms*, Philadelphia, PA, pp. 604–616. SIAM, 2014.
62. R. Jothi and B. Raghavachari. Approximating the k-traveling repairman problem with repairtimes. *Journal of Discrete Algorithms*, 5(2):293–303, 2007.
63. I. Post and C. Swamy. Linear programming-based approximation algorithms for multi-vehicle minimum latency problems: Extended abstract. In *Proceedings of the 26th Annual ACM-SIAM Symposium on Discrete Algorithms*, Philadelphia, PA, pp. 512–531. SIAM, 2015.
64. V. Nagarajan and R. Ravi. The directed minimum latency problem. In *Approximation, Randomization and Combinatorial Optimization. Algorithms and Techniques*, pp. 193–206. Springer, Berlin, Germany, 2008.
65. Z. Friggstad, M. R. Salavatipour, and Z. Svitkina. Asymmetric traveling salesman path and directed latency problems. *SIAM Journal on Computing*, 42(4):1596–1619, 2013.

35

A Development and Deployment Framework for Distributed Branch-and-Bound

Peter Cappello

Christopher James
Coakley

35.1 Introduction

Branch-and-bound intelligently searches the set of feasible solutions to a combinatorial optimization problem. It, in effect, proves that the optimal solution is found without necessarily examining all feasible solutions. The feasible solutions are not given. They can be generated from the problem description. However, doing so usually is computationally infeasible. The number of feasible solutions typically grows exponentially as a function of the size of the problem input. For example, the set of feasible tours in a symmetric traveling salesman problem (TSP) of a complete graph with 23 nodes is 22!/2 or around 8×10^{14} tours. The space of feasible solutions is progressively partitioned (branching), forming a search *tree*. Each tree node has a partial feasible solution. The node represents the set of feasible solutions that are extensions of its partial solution. For example, in a TSP branch-and-bound, a search tree node has a partial tour, representing the set of all tours that contain that partial tour. As branching continues (progresses down the problem tree), each search node has a more complete partial solution, and thus represents a smaller set of feasible solutions. For example, in a TSP branch-and-bound, a tree node's children each represent an extension of the partial tour to a more complete tour (e.g., one additional city or one additional edge). As one progresses down the search tree, each node represents a larger partial tour. As the size of a partial tour increases, the number of full tours containing the partial tour clearly decreases.

```
activeSet = { originalTask };
u = infinity; // u = the cost of the best solution known
currentBest = null;
while (  ! activeSet.isEmpty() ) {
    k = remove some element of the activeSet;
    children = generate k's children;
    for each element of children {
        if ( element's lower bound <= u )
            if ( element is a complete solution ) {
                u = element's cost;
                currentBest = element;
            }
            else
                add element to activeSet;
    }
}
```

FIGURE 35.1 A sequential algorithm for branch and bound.

In traversing the search tree, we may come to a node that represents a set of feasible solutions, all of which are provably more costly than a feasible solution already found. When this occurs, we *prune* this node of the search tree. We discontinue further exploration of this set of feasible solutions. In the example of the TSP problem, the cost of any feasible tour that has a given partial tour surely can be bounded from below by the cost of the partial tour: the sum of the edge weights for the edges in the partial tour. (In our experiments, we use a Held-Karp lower bound, which is stronger but more computationally complex.) If the lower bound for a node is *higher* than the current upper bound (i.e., best known complete solution), then the cost of all complete solutions (e.g., tours) represented by the node is higher than a complete solution that is already known; the node is pruned. See Papadimitriou and Steiglitz [1] for a more complete discussion of branch-and-bound. Figure 35.1 gives a basic, sequential branch-and-bound algorithm.

Branch-and-bound algorithms may be easily modified to generate suboptimal solutions. The total search time decreases as the desired accuracy decreases.

The framework that we present here is designed for deployment in a distributed setting. Moreover, the framework supports *adaptive parallelism*: During the execution, the set of compute servers can grow (if new compute servers become available) or shrink (if compute servers become unavailable or fail). The branch-and-bound computation thus cannot assume a fixed number of compute servers.

The branch-and-bound computation is decomposed into tasks, each of which is executed on a compute server. Each element of the active set (Figure 35.1) is a task that, in principle, can be scheduled for execution on any compute server. Indeed, parallel efficiency requires load balancing of tasks among compute servers. This distributed setting implies the following compute server requirements:

- Tasks (active set elements) are generated during the computation—they cannot be scheduled *a priori.*
- When a compute server discovers a new best cost, it must be propagated to the other compute servers.
- Detecting termination requires "knowing" when all branches (children) have been either fully examined or pruned. In a distributed setting, the implied communication must not be a bottleneck.

Our goal is to facilitate the development of branch-and-bound computations for deployment as a distributed computation. We provide a development-deployment infrastructure that requires the developer to write code for *only* the particular aspects of the branch-and-bound computation under development, primarily the branching rule, the lower bound computation, and the upper bound computation. We present this framework and some experimental results of its application to a medium complexity TSP code running on a Beowulf cluster.

35.2 Related Work

Held et al. give a short history of the TSP [2]. In this work, they note that, in 1963, Little et al. [3] were the first to use the term "branch-and-bound" to describe their enumerative procedure for solving TSP instances. As we understand it, Little et al. and Land and Doig [4] independently discovered the technique of branch-and-bound. This discovery led to "a decade of enumeration."

Parallel branch-and-bound also has been widely studied [5,6]. Rather early on it was discovered that there are speedup anomalies in parallel branch-and-bound [7]. Completion times are not monotonically nonincreasing as a function of the number of processors. In the discussion that follows, let T denotes the search tree, c^* denotes the cost of a minimum cost leaf in T, $T^* \subseteq T$ denotes the subtree of T whose nodes cost less than or equal to c^*, n denotes the number of nodes in T^*, and h denotes the height of T^*. Karp and Zhang [8] present a universal randomized method called Local Best-First Search for parallelizing sequential branch-and-bound algorithms. When executing on a completely connected, message-passing multiprocessor, the method's computational complexity is asymptotically optimal with high probability; $O(n/p + h)$, where p is the number of processors. The computational complexity of maintaining the local data structure and the communication overhead are ignored in their analysis. When $n > p^2 \log p$, Liu et al. [9] give a method for branch-and-bound that is asymptotically optimal with high probability, assuming that interprocessor communication is controlled by a central FIFO arbiter. Herley et al. [10], give a deterministic parallel algorithm for branch-and-bound based on the parallel heap selection algorithm of Frederickson [11], combined with a parallel priority queue. The complexity of their method is $O(n/p + h \log^2(np))$ on an EREW-PRAM, which is optimal for $h = O(n/(p \log^2(np)))$. This bound includes communication costs on an EREW-PRAM.

Distributed branch-and-bound also has been widely studied. Tschöeke et al. contributed experimental work on distributed branch-and-bound for TSP [12] using over 1,000 processors. When the number of processors gets large, fault tolerance becomes an issue. Yahfoufi and Dowaji [13] present perhaps the first distributed fault-tolerant branch-and-bound algorithm.

There also has been a lot of work on what might be called programming frameworks for distributed branch-and-bound computation. This occurs for two reasons: (1) branch-and-bound is best seen as a meta-algorithm for solving large combinatorial optimization problems: It is a framework that must be completed with problem-specific code; (2) programming a fault tolerant distributed system is sufficiently complex to motivate a specialization of labor: distributed systems research vs. operations research. In 1995, Shinano et al. [14] presented a Parallel Utility for Branch-and-Bound (PUBB), based on the C programming language. They illustrate the use of their utility on TSP and 0/1 ILP. They introduce the notion of a Logical Computing Unit (LCU). Although in parts of their paper, an LCU sounds like a computational task, we are persuaded that it closely resembles a processor, based on their explanation of its use. "The master process maintains in a queue, all the subproblems that are likely to lead to an optimal solution. As long as this queue is not empty and an idle LCU exists, the master process selects subproblems and assigns them to an idle LCU for evaluation one after the other." When discussing their experimental results, they note "The results indicate that, up to using about 10 LCUs, the execution time rapidly decreases as more LCUs are added. When the number of LCUs exceeds about 20, the computing time for one run, remains almost constant." Indeed, from their Figure 9 in [14], we can see that PUBB's parallel efficiency steadily goes down when the number of LCUs is above 10, and is well below 0.5, when the number of LCUs is 55. Aversa et al. [15] report on a the Magda project for mobile agent programming with parallel skeletons. Their divide-and-conquer skeleton is used to implement branch-and-bound, which they provide experimental data for up to 8 processors. Moe [16] reports on great international branch and bound search (GRRIB) and infrastructure for branch-and-bound on the Internet. Experimental results on an SGI Origin 2000 with 32 processors machines show good speedups when the initial bound is tight, and about 67% of ideal speedup, when a simple initial bound is used. Dorta et al. [17] present C++ skeletons for divide-and-conquer and branch-and-bound, where deployment is intended for clusters. In their experiments, using a 0/1 Knapsack problem of size 1000 on a Linux cluster with 7 processors, the average speedup was 2.25. On an Origin 3000 with 16 processors, the average speedup was 4.6. On a Cray T3E with 128 processors, the average speedup was 5.02. They explain "Due to the fine grain of the 0/1 knapsack problem, there is no lineal increase in the speed up when the number of processor increase. For large numbers of processors the speed up is poor."

Neary et al. [18,19] present an infrastructure/framework for distributed computing, including branch-and-bound, that tolerates faulty compute servers, and is in pure Java, allowing application codes to run on a heterogeneous set of machine types and operating systems. They experimentally achieved about 50% of ideal speedup for their TSP code, when running on 1000 processors. Their schemes for termination detection and fault tolerance of a branch-and-bound computation both exploit its *tree-structured* search space. The management of these schemes is centralized. Iamnitchi and Foster [20] build on this idea of exploiting branch-and-bound's tree-structured search space, producing a branch and bound-specific fault tolerance scheme that is distributed, although they provide only simulation results.

35.3 The Deployment Architecture

JICOS, a Java-centric network computing service that supports high-performance parallel computing, is an ongoing project that: virtualizes compute cycles, stores/coordinates *partial* results - supporting fault-tolerance, is partially self-organizing, may use an open grid services architecture [21,22] frontend for service discovery (not presented), is largely independent of hardware/OS, and is intended to scale from a LAN to the Internet. JICOS is designed to: **support scalable, adaptively parallel computation** (i.e., the computation's organization reduces *completion* time, using many *transient* compute servers, called *hosts*, that may join and leave during a computation's execution, with high system efficiency, regardless of how many hosts join/leave the computation); **tolerate basic faults**: JICOS must tolerate host failure and network failure between hosts and other system components; **hide communication latencies**, which may be long, by overlapping communication with computation. JICOS comprises three service component classes.

Hosting Service Provider (HSP): JICOS clients (i.e., processes seeking computation done on their behalf) interact solely with the hosting service provider component. A client logs in, submits computational tasks, requests results, and logs out. When interacting with a client, the hosting service provider thus acts as an agent for the entire network of service components. It also manages the network of task servers described in the following. For example, when a task server wants to join the distributed service, it first contacts the hosting service provider. The HSP tells the task server where it fits in the task server network.

Task Server: This component is a store of Task objects. Each Task object that has been spawned but has not yet been computed, has a representation on some task server. Task servers balance the load of ready tasks among themselves. Each task server has a number of hosts associated with it. When a host requests a task, the task server returns a task that is ready for execution, if any are available. If a host fails, the task server reassigns the host's tasks to other hosts.

Host: A host (aka compute server) joins a particular task server. Once joined, each host repeatedly requests a task for execution, executes the task, returns the results, and requests another task. It is the central service component for virtualizing compute cycles.

When a client logs in, the HSP propagates that login to all task servers, who in turn propagate it to all their hosts. When a client logs out, the HSP propagates that logout to all task servers, which *aggregate* resource consumption information (execution statistics) for each of their hosts. This information, in turn, is aggregated by the HSP for each task server, and returned to the client. Currently, the task server network topology is a torous. However, scatter/gather operations, such as login and logout, are transmitted via a task server *tree*, a subgraph of the torous (Figure 35.2).

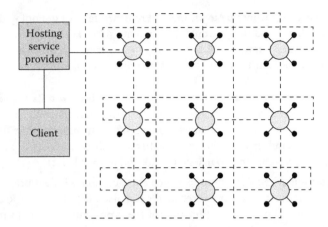

FIGURE 35.2 A JICOS system that has 9 task servers. The task server topology, a 2D torous, is indicated by the dashed lines. Each task server has 4 associated hosts (the little black discs). An actual task server might serve about 40 hosts (although our experiments indicate that 128 hosts/task server is not too much). The client interacts *only* with the HSP.

Task objects *encapsulate* computation. Their inputs and outputs are managed by JICOS. Task execution is idempotent, supporting the requirement for host transience and failure recovery. Communication latencies between task servers and hosts are reduced or hidden via task caching, task pre-fetching, and task execution on task servers for selected task classes.

35.3.1 Tolerating Faulty Hosts

To support self-healing, all proxy objects are leased [23,24]. When a task server's lease manager detects an expired host lease and the offer of renewal fails, the host proxy: (1) returns the host's tasks for reassignment and (2) is deleted from the task server. Because of explicit continuation passing, recomputation is minimized: Systems that support divide-and-conquer but which do not use explicit continuation passing [25], such as Satin [26], need to recompute some task decomposition computations, even if they completed successfully. In some applications, such as sophisticated TSP codes, decomposition can be computationally complex. On Jicos, only the task that was currently being executed needs to be recomputed. This is a substantial improvement. In the TSP problem, instance that we use for our performance experiments, the average task time is 2 s. Thus, the recomputation time for a failed host is, in this instance, a mere 1 s, on average.

35.4 Performance Considerations

JICOS's API includes a simple set of *application-controlled* directives for improving performance by reducing communication latency or overlapping it with task execution.

Task Caching: When a task constructs subtasks, the first constructed subtask is cached on its host, obviating its host's need to ask the Task Server for its next task. *The application programmer thus implicitly controls which subtask is cached.*

Task Pre-fetching: The *application* can help hide communication latency via task pre-fetching:

    Implicit: A task that never constructs subtasks is called *atomic*. The Task class has a Boolean method, *isAtomic*. The default implementation of this method returns true, if and only if the task's class implements the marking interface, *Atomic*. Before invoking a task's *execute* method, a host invokes the task's *isAtomic* method. If it returns true, the host pre-fetches another task via another thread before invoking the task's execute method.

Explicit: When a task object whose *isAtomic* method returned false (it did not know prior to the invocation of its *execute* method that it would not generate subtasks) nonetheless comes to a point in its *execute* method when knows that it is not going to construct any subtasks, it can invoke its environment's *pre-fetch* method. This causes its host to request a task from the task server in a separate thread.

Task pre-fetching overlaps the host's execution of the current task with its request for the next task. Application-directed pre-fetching, both implicit and explicit, thus motivates the programmer to (1) identify atomic task classes and (2) constitute atomic tasks with compute time that is at least as long as a Host–TaskServer round trip (on the order of 10s of milliseconds, depending on the size of the returned task, which affects the time to marshal, send, and unmarshal it).

Task Server Computation: When a task's encapsulated computation is little more complex than reading its inputs, it is faster for the task server to execute the task itself than to send it to a host for execution. This is because the time to marshal and unmarshal the input and the time to marshal and unmarshal the result is less than the time to simply compute the result (not to mention network latency). Binary Boolean operators, such as min, max and sum (typical linear-time gather operations), should execute on the task server. All Task classes have a Boolean method, executeOnServer. The default implementation returns true, if and only if the task's class implements the marking interface, *ExecuteOnServer*. When a task is ready for execution, the task server invokes its executeOnServer method. If it returns true, the task server executes the task itself: *The application programmer controls the use of this important performance feature.*

Taken together, these features reduce or hide much of the delay associated with Host–TaskSever communication.

35.5 The Computational Model

Computation is modeled by an *directed acyclic graph* (DAG) whose nodes represent tasks. An arc from node v to node u represents that the output of the task represented by node v is an input to the task represented by node u. A computation's tasks all have access to an *environment* consisting of an immutable *input* object and a mutable *shared* object. The semantics of "shared" reflects the envisioned computing context—a computer network: The object is shared *asynchronously*. This limited form of sharing is of value in only a limited number of settings. However, branch-and-bound is one such setting, constituting a versatile paradigm for coping with computationally intractable optimization problems.

35.6 The Branch-and-Bound API

Tasks correspond to nodes in the search tree: Each task gives rise to a set of smaller subtasks, until it represents a node in the search tree that is small enough to be explored by a single compute server. We refer to such a task as *atomic*; it does not decompose into subtasks.

35.6.1 The Environment

For branch-and-bound computations, the environment *input* is set to the problem instance. For example, in a TSP, the input can be set to the distance matrix. Doing so materially reduces the amount of information needed to describe a task, which reduces the time spent to marshal and unmarshal such objects.

The cost of the least cost known solution at any point in time is shared among the tasks. It is encapsulated as the branch-and-bound computation's *shared* object see IntUpperBound in the following.

In branch-and-bound, this value is used to decide if a particular subtree of the search tree can be pruned. Thus, sharing the cost of the least cost known solution enhances the pruning ability of concurrently executing tasks that are exploring disjoint parts of the search tree. Indeed, this improvement in pruning is essential to the efficiency of parallel branch-and-bound. When a branch-and-bound task finds a complete solution whose cost is less than the current least cost solution, it sets the shared object to this new value, which implicitly causes JICOS to propagate the new least cost throughout the distributed system.

35.6.2 The JICOS Branch-and-Bound Framework

The classes comprising the JICOS branch-and-bound framework are based on two assumptions:

- The branch-and-bound problem is formulated as a minimization problem. Maximization problems typically can be reformulated as minimization problems.
- The cost can be represented as in int.

If the two assumptions above prove troublesome, we will generalize this framework.

Before giving the framework, we describe the problem-specific class that the *application developer must provide*; A class that implements the *Solution* interface. This class represents nodes in the search tree: A Solution object is a partial feasible solution. For example, in a TSP, it could represent a partial tour. Since it represents a node in the search tree, its children represent more complete partial feasible solutions. For example, in a TSP, a child of a Solution object would represent its parent's partial tour, but including[excluding] one more edge (or including one more city, depending on the branching rule).

The Solution interface has the following methods:

getChildren returns a queue of the Solution objects that are the children of this Solution. The queue's retrieval order represents the application's selection rule, from most promising to least promising. In particular, the first child is cached (Section 35.4 for an explanation of task caching).

getLowerBound returns the lower bound on the cost of any complete Solution that is an extension of this partial Solution.

getUpperBound returns an upper bound on the cost of any complete Solution, and enables an upper bound heuristic for incomplete solutions.

isComplete returns true if and only if the partial Solution is, in fact, complete.

reduce omit loose constraints. For example, in a TSP solution, this method may omit edges whose cost is greater than the current best solution, and therefore cannot be part of any better solution. This method returns void, and can have an empty implementation.

The classes that comprise the branch-and-bound framework—provided by JICOS to the application programmer—are described in the following:

BranchAndBound: This is a Task class, which resides in the jicos.applications.branchandbound package, whose objects represent a search node. A BranchAndBound Task either:
  - Constructs smaller BranchAndBound tasks that correspond to its children search nodes, or
  - Fully searches a subtree, returning:
    - Null, if it does not find a solution that is better than the currently known best solution
    - The best solution it finds, if it is better than the currently known best solution

IntUpperBound: An object that represents the minimum cost among all known complete solutions. This class is in the jicos.services.shared package. It implements the Shared interface (for details about this interface, please see the JICOS API), which defines the shared object. In this case, the shared object is an Integer that holds the current upper bound on the cost of a minimal solution. Consequently, IntUpperBound u "is newer than" IntUpperBound v when $u < v$.

MinSolution: This task is included in the jicos.services.tasks package. It is a composition class whose execute method:

- Receives an array of Solution objects, some of which may be null.
- Returns the one whose cost is minimum, provided it is less than or equal to the current best solution. Equality is included to ensure that the minimum cost *solution* is reported: It is not enough just to know the *cost* of the minimum cost solution.
- From the standpoint of the JICOS system (not a consideration for application programming), the compose tasks form a tree that performs a gather operation, which, in this case, is a min operation on the cost of the Solution objects it receives. Each task in this gather tree is assigned to some task server, distributing the gather operation throughout the network of task servers. (This task is indeed executed on a task server, not a compute server [see Section 35.4].)

Q: A queue of Solution objects.

Using this framework, it is easy to construct a branch and bound computation. The JICOS website tutorial [27] illustrates this, giving a complete code for a simple TSP branch and bound computation.

35.7 Experimental Results

35.7.1 The Test Environment

We ran our experiments on a Linux cluster. The cluster consists of 1 head machine, and 64 compute machines, composed of two processor types. Each machine is a dual 2.6 GHz (or 3.0 GHz) Xeon processor with 3 GB (2 GB) of PC2100 memory, two 36 GB (32 GB) SCSI-320 disks with on-board controller, and an on-board 1 GB Ethernet adapter. The machines are connected via the gigabit link to one of two Asante FX5-2400 switches. Each machine is running CentOS 4 with the Linux smp kernel 2.6.9-5.0.3.ELsmp, and the Java j2sdk1.4.2. Hyperthreading is enabled on most, but not all, machines.

35.7.2 The Test Problem

We ran a branch-and-bound TSP application, using kroB200 from TSPLIB, a 200 city Euclidean instance. In an attempt to ensure that the speedup could not be super-linear, we set the initial upper bound for the minimal-length tour with the optimum tour length. Consequently, each run explored exactly the same search tree: Exactly the same set of nodes is pruned regardless of the number of parallel processors used. Indeed, the problem instance decomposes into exactly 61,295 BranchAndBound tasks whose average execution time was 2.05 s, and exactly 30,647 MinSolution tasks whose average execution time was less than 1 ms.

35.7.3 The Measurement Process

For each experiment, a hosting service provider was launched, followed by a single task server on the same machine. When additional task servers were used, they were started on dedicated machines. Each compute server was started on its own machine. Except for 28 compute servers in the 120 processor case (which were calibrated with a separate base case), each compute server thus had access to two hyper-threaded processors which are presented to the JVM as four processors (we report physical CPUs in our results). After the JICOS system was configured, a client was started on the same machine as the HSP (and task server), which executed the application. The application consists of a *deterministic workload* on a very unbalanced task graph. Measured times were recorded by JICOS's *invoice system*, which reports elapsed time (wall clock, not processor) between submission of the application's source task (aka root task) and receipt of the application's sink task's output. JICOS also automatically computes the critical path using the obvious recursive formulation for a DAG. Each test was run eight times (or more) and averages are reported.

One processor in the OS does not correspond to one physical processor. It therefore is difficult to get meaningful results for one processor. We consequently use one machine, which is two physical CPUs, as our base case. For the 120 processor measurements, we used the speedup formula a heterogeneous processor set [28]. We thus had three separate base cases for computing the 120 processor speedup.

For our fault tolerance test, we launched a JICOS system with 32 processors as compute servers. We issued a kill command to various compute servers after 1500 s, approximately 3/4 through the computation. The completion time for the total computation was recorded, and was compared to the ideal completion time: $1500 + (T_{32} - 1500) \times 32/P_{final}$, where P_{final} denotes the number of compute servers that did *not* fail.

To test the overhead of running a task server on the same machine as a compute server, we ran a 22 processor experiment both with a dedicated task server and with a task server running on the same machine as one of the compute server. We recorded the completion times and report the averages of 8 runs.

35.7.4 The Measurements

T_P denotes the time for P physical processors to run the application. A computation's *critical path time*, denoted T_∞, is a maximum time path from the source task to the sink task. We captured the critical path time for this problem instance; it is 37 s. It is well known [25] that $\max\{T_\infty, T_1/P\} \leq T_P$. Thus, $0 \leq \max\{T_\infty, T_1/P\}/T_P \leq 1$ is a *lower bound* on the fraction of perfect speedup that is actually attained. Figure 35.3 presents speedup data for several experiments. The ordinate in the figure is the *lower bound* of fraction of perfect speedup. As can be seen from the figure, in all cases, the actual fraction of perfect speedup exceeds 0.94; it exceeds 0.96, when using an appropriate number of task servers. Specifically, the 2-processor base case ran in 9 h and 33 min, whereas the 120-processor experiment (2 processors per host) ran in just 11 min!

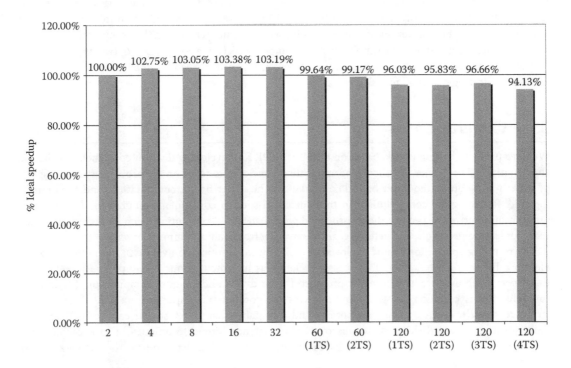

FIGURE 35.3 Number of processors versus % of ideal speedup.

TABLE 35.1   Efficiency of Compute Server Fault Tolerance.

Processors (final)	30	26	12	8	6	4
Theoretical time (s)	2119.43	2214.73	3048.58	3822.87	4597.16	6145.74
Measured time (s)	2194.95	2300.92	2974.35	4182.62	4884.86	6559.91
Percent overhead (%)	3.6	3.9	−2.4	9.4	6.3	6.7

Each experiment started with 32 processors. The experiment in which 30 processors finished had two fail; the experiment in which four finished had 28 fail.

We get superlinear speedups for 4, 8, 16, and 32 processors. The standard deviation was less than 1.6% of the size of the average. As such, the superlinearity cannot be explained by statistical error. However, differences in object placement in the memory hierarchy can have impacts greater than the gap in speedup we observe [29]. Therefore, within experimental factors beyond our control, JICOS performs well.

We are very encouraged by these measurements, especially considering the small average task times. Javelin, for example, was not able to achieve such good speedups for 2 s tasks. Even CX [28,30] is not capable of such fine task granularity.

$P_\infty = T_1/T_\infty$ is a lower bound on the number of processors necessary to extract the *maximum parallelism* from the problem. For this problem instance, $P_\infty = 1857$ processors. Thus, 1857 processors is a lower bound on the number of processors necessary to bring the completion time down to T_∞, namely 37 s.

Our fault tolerance data is summarized in Table 35.1. Overhead is caused by the rescheduling of tasks lost when a compute server failed as well as some time taken by the TaskServer to recognize a faulty compute server. Negative overhead is a consequence of network traffic and thread scheduling preventing a timely transfer of the kill command to the appropriate compute server.

When measuring the overhead of running a task server on a machine shared with a compute server, we received an average of 3115.1 s for a dedicated task server and 3114.8 s for the shared case. Both of these represent 99.7% ideal speedup. This is not too surprising: there is a slight reduction in communication latency having the task server on the same machine as a compute server, and the computational load of the task server is small due to the simplicity of the compose task (it is a comparison of two upper bounds). It therefore appears beneficial to place a compute server on every available computer in a JICOS system without dedicating machines to task servers.

35.8 Conclusion

We have presented a framework, based on the JICOS API, for developing distributed branch-and-bound computations. The framework allows the application developer to focus on the problem-specific aspects of branch-and-bound: the lower bound computation, the upper bound computation, and the branching rule. Reducing the code required to these problem-specific components reduces the likelihood of programming errors, especially those associated with distributed computing, such as threading errors, communication protocols, and detecting, and recovering from, faulty compute servers.

The resulting application can be deployed as a distributed computation via JICOS running, for example, on a Beowulf cluster. JICOS [31] scales efficiently as indicated by our speedup experiments. This, we believe, is because we have carefully provided (1) for divide-and-conquer computation; (2) an environment that is common to all compute servers for computation input (e.g., a TSP distance data structure, thereby reducing task descriptors) and a mutable shared object that can be used to communicate upper bounds as they are discovered; (3) latency hiding techniques of task caching and pre-fetching; and (4) latency reduction by distributing termination detection on the network of task servers.

Faulty compute servers are tolerated with high efficiency, both when faults occur (as indicated by our fault tolerance performance experiments) and when they do not (as indicated by our speedup experiments, in which no faults occur). Finally, the overhead of task servers is shown to be quite small, further confirming the efficiency of JICOS as a distributed system.

The vast majority of the code concerns JICOS, the distributed system of fault tolerant compute servers. The Java classes comprising the branch-and-bound framework are few, and easily enhanced, or added to, by operations researchers; the source code is cleanly designed and freely available for download from the JICOS website [27]. Our branch-and-bound framework may be used for almost any divide-and-conquer computation. JICOS may be easily adapted to solve in a distributed environment any algorithm which can be defined as a computation over directed acyclic graph, where the nodes refer to computations and the edges specify a precedence relation between computations.

References

1. Papadimitriou, C.H. and Steiglitz, K., *Combinatorial Optimization: Algorithms and Complexity*, Prentice-Hall, Englewood Cliffs, NJ, 1982.
2. Held, M., Hoffman, A.J., Johnson, E.L., and Wolfe, P., Aspects of the traveling salesman problem, *IBM J. Res. Develop.*, 28(4), 476, 1984.
3. Little, J.D.C., Murty, K.G., Sweeney, D.W., and Karel, C., An algorithm for the traveling salesman problem, *Oper. Res.*, 11(6), 972, 1963.
4. Land, A.H. and Doig, A.G., An automatic method for solving discrete programming problems, *Econometrica*, 28(3), 497, 1960.
5. Wah, B.W., Li, G.J., and Yu, C.F., Multiprocessing of combinatorial search problems, *IEEE Comput.*, 18(6), 93–108, 1985.
6. Wah, B.W. and Yu, C.F., Stochastic modeling of branch-and-bound algorithms with best-first search, *IEEE Trans. Software Eng.*, SE-11(9), 922, 1985.
7. Lai, T.H. and Sahni, S., Anomalies in parallel branch-and-bound algorithms, *CACM*, 27(6), 594, 1984.
8. Karp, R. and Zhang, Y., A randomized parallel branch-and-bound procedure, in *Proceedings of the 12th Ann. ACM Symposium on Theory of Computing*, 1988, p. 290.
9. Liu, P., Aiello, W., and Bhatt, S., An atomic model for message-passing, in *Proceedings of the Symposium on Parallel Algorithms and Architectures*, 1993, p. 154.
10. Herley, K., Pietracaprina, A., and Pucci, G., Fast deterministic parallel branch-and-bound, *Parallel Proc. Lett.*, 9, 325, 1999.
11. Frederickson, G., An optimal algorithm for selection in a min-heap, *Information and Computation*, 104(2), 197, 1993.
12. Tschöke, S., Lüling, R., and Monien, B., Solving the traveling salesman problem with a distributed branch-and-bound algorithm on a 1024 processor network, in *Proceedings of the 9th International Symposium Parallel Processing Symp.*, IEEE, Santa Barbara, CA, 1995, p. 182.
13. Yahfoufi, N. and Dowaji, S., Self-stabilizing distributed branch-and-bound algorithm, in *Proceedings of the International Phoenix Conference on Computers and Communications*, IEEE, Scottsdale, AZ, 1996, p. 246.
14. Shinano, Y., Higaki, M., and Hirabayashi, R., A generalized utility for parallel branch and bound algorithms, in *Proceedings of the International Symposium on Parallel and Distributed Processing*, IEEE, San Antonio, TX, 1995, p. 392.
15. Aversa, R., Martino, S.D., Mazzocca, N., and Venticinque, S., Mobile agent programming for clusters with parallel skeletons, in *Proceedings of the Conference on High Performance Computing for Computational Science*, LNCS, Springer, Porto, Portugal, 2002, p. 622.

16. Moe, R., GRIBB: Branch-and-bound methods on the Internet, in *Proceedings of the Parallel Processing and Applied Mathematics*, LNCS, Springer, Czestochowa, Poland, 2003, p. 1020.

17. Dorta, I., Leon, C., Rodriguez, C., and Rojas, A., Parallel skeletons for divide-and-conquer and branch-and-bound techniques, in *Proceedings of the Euromicro Conference on Parallel, Distributed and Network-Based Processing*, IEEE, Genova, Italy, 2003.

18. Neary, M.O., Phipps, A., Richman, S., and Cappello, P., Javelin 2.0: Java-based parallel computing on the Internet, in *European Conference on Parallel Processing*, Springer, Munich, Germany, 2000, p. 1231.

19. Neary, M.O. and Cappello, P., Advanced eager scheduling for Java-based adaptively parallel computing, *Concurrency and Computation: Practice and Experience*, ACM, Seattle, WA, 17(7–8), 797, 2005.

20. Iamnitchi, A. and Foster, I., A problem-specific fault-tolerance mechanism for asynchronous, distributed systems, in *Proceedings of the International Conference on Parallel Processing*, IEEE, Toronto, Ontario, 2000.

21. Seed, R. and Sandholm, T., A note on Globus toolket 3 and J2EE, Technical Report, Globus, 2003.

22. Seed, R., Sandholm, T., and Gawor, J., Globus toolkit 3 core—A Grid service container framework, Technical Report, Globus, 2003.

23. Miller, M.S. and Drexler, K.E., Markets and computation: Agoric open systems, in *The Ecology of Computation*, Huberman, B., Ed., Elsevier Science Publishers, New York, 1988.

24. Arnold, K., *The Jini Specifications*, Addison-Wesley, Boston, MA, 2nd ed., 1999.

25. Blumofe, R.D., Joerg, C.F., Kuszmaul, B.C., Leiserson, C.E., Randall, K.H., and Zhou, Y., Cilk: An efficient multithreaded runtime system, in *Proceedings of the Symposium on Principles and Practice of Parallel Programming*, ACM, Santa Barbara, CA, 1995, p. 207.

26. Wrzesinska, G., van Nieuwpoort, R.V., Maasen, J., Kielmann, T., and Bal, H.E., Fault-tolerant scheduling of fine-grained tasks in Grid environments, *International Journal of High Performance Computing Applications*, 20(1), 103, 2006.

27. JICOS: A Java-Centric Network Computing Service, Website, http://cs.ucsb.edu/projects/jicos. Nov. 7, 2017.

28. Cappello, P. and Mourloukos, D., CX: A scalable, robust network for parallel computing, *Scientific Programming*, 10(2), 159, 2001.

29. Krintz, C. and Sherwood, T., private communication, Santa Barbara, CA, 2005.

30. Cappello, P. and Mourloukos, D., A scalable, robust network for parallel computing, in *Proceedings of the Joint ACM Java Grande/ISCOPE Conference*, ACM, Santa Barbara, CA, 2001, p. 78.

31. Cappello, P. and Coakley, C.J., JICOS: A Java-centric network computing service, in *Proceedings of the International Conference on Parallel and Distributed Computing and Systems*, 2005.

36

Approximations for Steiner Minimum Trees

Ding-Zhu Du

Weili Wu

36.1 Introduction

Designing optimal network for certain propose, such as rail road design, telecommunication, computer networks, has been an important research topic in applied mathematics and theoretical computer science for a long time. Many such problems are very difficult. Thus, finding polynomial-time approximation solutions plays an important role.

For example, given a set of points in the Euclidean plane, the shortest network interconnecting all points in the set is called the *Steiner minimum tree*. The Steiner minimum tree may contain some vertices which are not given points. Those vertices are called *Steiner points*, whereas the points given are called *terminals*. The Steiner minimum tree for three terminals was first studied by Fermat (1601–1665). Recent research on mathematical history showed that the Steiner minimum tree for more than three points was first studied by Gauss in a letter of Gauss to Schumacher [1].

Actually, on March 19, 1836, Schumacher wrote a letter to Gauss. In this letter, he mentioned a paradox about Fermat's problem: Consider a convex quadrilateral $ABCD$. It has been known that the solution of Fermat's problem for four points A, B, C, and D is the intersection E of diagonals AC and BD. Suppose extending DA and CB can obtain an intersection F. Now, moving A and B to F. Then E will also be moved to F. However, when the angle at F is less than $120°$, the point F cannot be the solution of Fermat's problem for three given points F, D, and C. What happens? (Figure 36.1.)

On March 21, 1836, Gauss wrote a letter to Schumacher in which he explained that the mistake of Schumacher's paradox occurs at the place where Fermat's problem for four points A, B, C, and D is changed to Fermat's problem for three points F, C, and D. When A and B are identical to F, the total distance from E to the four points A, B, C, and D equals $2EF + EC + ED$, not $EF + EC + ED$. Thus, the point E may not be the solution of Fermat's problem for F, C, and D. More importantly, Gauss proposed

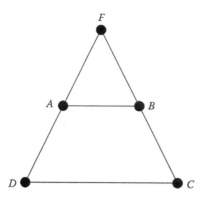

FIGURE 36.1   Schumacher's paradox.

a new problem. He said that it is more interesting to find a shortest network rather than a point. Gauss also presented several possible connections of the shortest network for four given points.

Unfortunately, Gauss' letter was discovered only in 1986. From 1941 to 1986, many publications have followed Crourant and Robbins [2] who in their popular book *What is mathematics?* (published in 1941) referred to Gauss' problem as the Steiner tree problem.

It is well known that the Steiner minimum tree is a NP-hard problem [3,4], that is, the optimal solution is unlikely computable in polynomial-time. Therefore, it is important to concentrate our research efforts seeking for good polynomial-time approximation solutions.

In 1990, Smith and Shor [5] proposed a greedy construction as follows: Start with all n terminals, regarded as a forest of n 1-node trees. At any stage, add the shortest possible line segment to the current forest, which cause two trees to merge. Continue until the forest is completely merged into a single tree. They conjectured that this greedy heuristic has a performance ratio $(3+2\sqrt{3})/6$ which is smaller than $\frac{2}{\sqrt{3}}$. (Here, the performance ratio for an approximation algorithm is defined to be the least upper bound for the ratio of lengths between the approximation solution and the Steiner minimum tree for the same set of terminals.) However, Du [6] disproved their conjecture by proving that this greedy heuristic has performance ratio $\frac{2}{\sqrt{3}}$. This closed the door to finding a better polynomial-time approximations for Steiner minimum trees in the Euclidean plane using this approach.

In fact, it was a long-standing open problem whether there exists a polynomial-time approximation for Steiner minimum trees in the Euclidean plane with performance ratio smaller than $\frac{2}{\sqrt{3}}$. The number $\frac{\sqrt{3}}{2}$ is the value of the Steiner ratio in the Euclidean plane. The Steiner ratio in a metric space is the largest lower bound for the ratio of lengths between the Steiner minimum tree and the minimum spanning tree for the same set of points in the space. In the other words, it is the inverse of the performance ratio of the minimum spanning tree as an approximation of the Steiner minimum tree. In 1968, Gilbert and Pollak [7] conjectured that the Steiner ratio in the Euclidean plane equals $\frac{\sqrt{3}}{2}$. This conjecture received a lots of attention and efforts [8–15], and was finally proved by Du and Hwang [16]; the solution received a lots of publicity.

In general, it was called the *better approximation problem* whether there exists a polynomial-time approximation for Steiner minimum trees in each metric space with performance ratio smaller than the inverse of the Steiner ratio.

Usually, when we talk about the Steiner trees in various metric space, we simply called the Steiner tree problem in the Euclidean plane as *the Euclidean Steiner tree*, the Steiner tree problem in the rectilinear plane as *the rectilinear Steiner tree*, and the Steiner tree problem in weighted graphs as *the network Steiner tree*. These three Steiner tree problems are classical and very well studied in the literature.

Zelikovsky [17] made the first breakthrough. He found a polynomial-time 11/6-approximation for the network Steiner tree which beats the inverse of the Steiner ratio in graphs, $\rho_2^{-1} = 2$. Soon later, Berman and Ramaiyer [18] gave a polynomial-time 92/72 -approximation for the rectilinear Steiner tree and Du et al. [19] showed a general solution for the open problem. They showed that in any metric space, there exists a polynomial-time approximation with performance ratio better than the inverse of the Steiner ratio provided that for any set of a fixed number of points, the Steiner minimum tree is polynomial-time computable. A main part of these works is to establish the lower bound for the k-Steiner ratio. Let us explain it as follows.

A tree interconnecting a terminal set is called a *Steiner tree* if every leaf is a terminal. However, a terminal in a Steiner tree may not be a leaf. A Steiner tree is *full* if every terminal is a leaf. When a terminal is not a leaf, the tree can be decomposed into several small trees at this terminal. In this way, every Steiner tree can be decomposed into smaller trees in each of which every terminal is a leaf. Those smaller trees are called *full components* of the tree. The *size* of a full component is the number of terminals in the full component.

A *k-restricted* Steiner tree is a Steiner tree with all full components of size at most k. The k-restricted Steiner minimum tree is the shortest one among all k-restricted Steiner trees. The 2-restricted Steiner minimum tree is the minimum spanning tree. The k-Steiner ratio in a metric space is the least ratio of lengths between the Steiner minimum tree and the k-restricted Steiner minimum tree for the same set of terminals in the metric space. The 2-Steiner ratio is exactly the Steiner ratio. A better lower bound for the k-Steiner ratio will give a better performance ratio for approximations of Zelikovsky's type. Zelikovsky [17] showed that the 3-Steiner ratio in graphs is at least 3/5. Du et al. [19] showed that the k-Steiner ratio in graphs is at least $\lfloor \log_2 k \rfloor / (1 + \lfloor \log_2 k \rfloor)$.

In 1996, both Arora [20] and Mitchell [21] found that actually, the Euclidean Steiner tree and the rectilinear Steiner tree have polynomial-time approximation scheme (PTAS). However, the network Steiner tree problem is found to be MAX SNP-complete, that is, unlikely to have PTAS [22].

Hwang et al. [23] gave a very useful reference. However, many important results appeared after this book was published. In this survey, we will pay attention to those important results appearing after 1990, and identify some open problems.

36.2 The Steiner Ratio

A *minimum spanning tree* on a set of terminals is the shortest network interconnecting all terminals with all edges between terminals. While the Steiner tree problem is intractable, the minimum spanning tree can be computed pretty fast. As we mentioned in the introduction, the Steiner ratio is a measure of performance of the minimum spanning tree as a polynomial-time approximation of the Steiner tree. Determining the Steiner ratio in each metric space is a traditional problem on Steiner trees. In 1976, Hwang [24] determined that the Steiner ratio in the rectilinear plane is 2/3. However, it took 22 years for completing the story of determining the Steiner ratio in the Euclidean plane. In 1968, Gilbert and Pollak [7] conjectured that the Steiner ratio in the Euclidean plane is $\sqrt{3}/2$. Through efforts made by Pollak [25], Du et al. [8], Friedel and Widmayer [9], Booth [10], Rubinstein and Thomas [11,26], Graham and Hwang [12], Chung and Hwang [13], Du and Hwang [14], and Chung and Graham [15], the conjecture was finally proved by Du and Hwang [16,27] in 1990. The significance of their proof stems also from the potential applications of the new approach included in the proof.

In their approach, the central part is a new minimax theorem for minimizing the maximum value of several concave functions over a polytope as follows:

Minimax Theorem. Let $f(x) = \max_{i \in I} g_i(x)$ where I is a finite set and $g_i(x)$ is a continuous, concave function in a polytope X. Then the minimum value of $f(x)$ over the polytope X is achieved at some critical point, namely, a point satisfying the following property:

(*) There exists an extreme subset Y of X such that $x \in Y$ and the index set $M(x) (= \{i \mid f(x) = g_i(x)\})$ is maximal over Y.

The Steiner ratio problem is first transformed to such a minimax problem [$g_i(x)$ = (the length of a Steiner tree)−(the Steiner ratio) × (the length of a spanning tree with graph structure i) where x is a vector whose components are edge-lengths of the Steiner tree] and the minimax theorem reduces the minimax problem to the problem of finding the minimax value of the concave functions at critical points. Then each critical point is transformed back to an input set of points with special geometric structure; it is a subset of a lattice formed by equilateral triangles. This special structure enables us to verify the conjecture corresponding to the nonnegativeness of minimax value of the concave functions.

Using the same approach, Gao et al. [28] proved that in any Minkowski plane (or 2-dimensional Banach space), the Steiner ratio is at least 2/3. They also proved an upper bound 0.8686 for the Steiner ratio in Minkowski plane. This upper bound is very close to the conjectured upper bound $\sqrt{3}/2 = 0.8666...$ which was made independently by Cieslik [29] and Du et al. [30]. Actually, many interesting open problems have a possible solution with this approach. For example, the above-mentioned upper bound problem can be transformed to the following max-min problem:

$$\max_{\|\cdot\|} \min_{(A,B)} MST_{\|\cdot\|}(A, B, C) \le \sqrt{3}$$

where the maximization is over all norm $\| \cdot \|$ in the plane, the minimization is over all possible directions of edge (A, B) in the equilateral triangle ABC with unit edges, and $MST_{\|\cdot\|}(A, B, C)$ is the length of the minimum Steiner tree for three points A, B, C in the plane with norm $\| \cdot \|$. Clearly, the Euclidean norm is the most symmetric one. However, the above-mentioned minimax theorem cannot be used. One of the reasons is that possible directions of edge (A, B) form an interval which is not a finite discrete set.

The research on this minimax approach has the following three aspects:

- Develop new minimax theorems of the previous type.
- Find new methods to determine critical structures.
- Generate new proof techniques for each open problem on critical structures.

Any research development on this approach would help to solve the following open problems about the Steiner ratio.

Chung–Gilbert's Conjecture on the Steiner Ratio in Euclidean Space: Steiner trees in Euclidean spaces have applications in constructing phylogenetic trees [31]. Gilbert and Pollak [7] also mentioned a possible generalization of their conjecture, that is, in any Euclidean space the Steiner ratio is achieved by the vertex set of a regular simplex. Chung and Gilbert [32] constructed a sequence of Steiner trees on regular simplexes. The lengths of constructed Steiner trees goes decreasingly to $\sqrt{3}/(4 - \sqrt{2})$. Although the constructed trees are not known to be Steiner minimum trees, Chung and Gilbert conjectured that $\sqrt{3}/(4 - \sqrt{2})$ is the best lower bound for Steiner ratios in Euclidean spaces. Clearly, if $\sqrt{3}/(4 - \sqrt{2})$ is the limiting Steiner ratio in d-dimensional Euclidean space as d goes to infinity, then Chung–Gilbert's conjecture is a corollary of Gilbert and Pollak's generalization. However, Smith [33] and Du and Smith [34] disproved Gilbert–Pollak's general conjecture for dimension more than two. This leaves a question mark to Chung and Gilbert's conjecture. It is quite interesting that Chung–Gilbert's conjecture could still be true. In fact, the n-dimensional regular simplex may reaches the smallest Steiner ratio over all sets of $n + 1$ terminals in the Euclidean space. Smith and Smith [35] proved that this is true for $n = 4$.

Graham–Hwang's Conjecture on the Steiner ratio in Rectilinear Space: Rectilinear Steiner trees in high dimensional space can be found in biology [3,31] and optimal traffic multicasting for some communication networks [36,37]. Although the Steiner ratio in rectilinear plane was determined by Hwang [24] many years ago, there is still no progress on the Steiner ratio in rectilinear spaces, which was conjectured by Graham and Hwang [12] to be $d/(2d - 1)$.

The Steiner Ratio in Banach Spaces: A Minkowski space is a Banach space of finite dimension. In the other words, it is a finite dimensional linear space with a norm. Some applications of Steiner minimum trees with L_p norm can be found in Reference 38. Examining the proof of Gilbert–Pollak's conjecture in the Euclidean plane, we observe that the proof has nothing to do with special properties of the Euclidean norm except the last part, verification of the conjecture on point sets with critical structure. This means that using the minimax approach to determine the Steiner ratio in Minkowski planes, we would have no problem on finding a transformation and determining critical structures. We need to work on only point sets with critical structure. Gao et al. [28] showed that the Steiner ratios in Minkowski planes is at least 2/3. Meanwhile, Du et al. [30] showed that the Steiner ratios in Minkowski planes is at most and conjectured that this upper bound is $\sqrt{3}/2$, that is, the upper is achieved at the Euclidean norm. This conjectured is independently made by Cieslik [29].

36.3 Better Approximations

Starting from a minimum spanning tree, improve it by adding Steiner points. This is a natural idea to obtain an approximation solution for the Steiner minimum tree. Every approximation solution obtained in this way would have a performance ratio at most the inverse of the Steiner ratio. The problem is how much better than the inverse of the Steiner ratio one can make.

Over the last twenty years numerous heuristics [36,39–44] for Steiner minimum trees have been proposed for terminals in various metric spaces. Their superiority over minimum spanning trees were often claimed by computation experiments. But no theoretical proof of superiority was ever given. It was a long-standing problem whether there exists a polynomial-time approximation with performance ratio better than the inverse of the Steiner ratio or not. For simplicity, a polynomial-time approximation with performance ratio smaller than the inverse of the Steiner ratio will be called a *better* approximation.

In the following, we study two ideas to add Steiner points greedily. They give approximation algorithms with better performance ratio.

36.3.1 Chang's Idea

Chang [37,40] proposed the following approximation algorithm for Steiner minimum trees in the Euclidean plane: Start from a minimum spanning tree and at each iteration choose a Steiner point such that using this Steiner point to connect three vertices in the current tree could replace two edges in the minimum spanning tree and this replacement achieves the maximum saving among such possible replacements.

Smith et al. [45] also use the idea of the greedy improvement. But, they start with Delaunay triangulation instead of a minimum spanning tree. Since every minimum spanning tree is contained in Delaunay triangulation, the performance ratio of their approximation algorithm can also be bounded by the inverse of the Steiner ratio. The advantage of Smith–Lee–Liebman algorithm is on the running time. While Chang's algorithm runs in $O(n^3)$ time, Smith–Lee–Liebman algorithm runs only in $O(n \log n)$ time.

Kahng and Robins [46] proposed an approximation algorithm for the rectilinear Steiner tree by using the same idea as that of Chang. Approximations obtained with this idea are called *iterated 1-Steiner trees*.

36.3.2 Zelikovsky's Idea

As we mentioned in introduction, Zelikovsky's idea [17] is based on study of k-restricted Steiner trees. Clearly, for any $k \geq 3$, a k-restricted Steiner minimum tree usually has shorter length compared with a minimum spanning tree. It is natural to think about using a minimum k-restricted Steiner tree to approximate the Steiner minimum tree. However, this does not work because computing a k-restricted

Steiner minimum tree is still an intractable problem. Zelikovsky's idea is to approximate the Steiner minimum tree by a 3-restricted Steiner tree generated by a polynomial-time greedy algorithm. The key fact is that the length of such an approximation is smaller than the arithmetic mean of lengths of a minimum spanning tree and a 3-restricted Steiner minimum tree; that is, the performance ratio of his approximation satisfies

$$PR \le \frac{\rho_2^{-1} + \rho_3^{-1}}{2}$$

where ρ_k is the k-Steiner ratio. Thus, if the 3-Steiner ratio ρ_3 is bigger than the Steiner ratio ρ_2, then this greedy algorithm is a better approximation for the Steiner minimum tree. Zelikovsky was able to prove that 3-Steiner ratio in graphs is at least 3/5 which is bigger than 1/2, the Steiner ratio in graphs [41]. So, he solved the better approximation problem in graphs. Zelikovsky's idea has been extensively studied in the literature.

Du et al. [19] generalized Zelikovsky's idea to the k-size Steiner tree. They showed that a generalized Zelikovsky's algorithm has performance ratio

$$PR \le \frac{(k-2)\rho_2^{-1} + \rho_k^{-1}}{k-1}.$$

Berman and Ramaiyer [18] employed a different idea to generalize Zelikovsky's result. They obtained an algorithm with the performance ratio satisfying

$$PR \le \rho_2^{-1} - \sum_{i=3}^{k} \frac{\rho_{i-1}^{-1} - \rho_i^{-1}}{i-1}.$$

They also showed that in the rectilinear plane, the 3-Steiner ratio is at least 72/94 which is bigger than 2/3 [24], the Steiner ratio in rectilinear plane. So, they solved the better heuristic problem in rectilinear plane.

Du et al. [19] proved a lower bound for the k-Steiner ratio in any metric space. This lower bound goes to one as k goes to infinity. So, in any metric space with the Steiner ratio less than one, there exists a k-Steiner ratio bigger than the Steiner ratio. Thus, they proved that the better heuristic exists in any metric space satisfying the following conditions:

1. The Steiner ratio is smaller than one.
2. The Steiner minimum tree on any fixed number of points can be computed in polynomial-time.

These metric spaces include Euclidean plane and Euclidean spaces.

36.3.3 The k-Steiner Ratio ρ_k

As we see in Section 36.3.2 that the determination of the k-Steiner ratio plays an important role in estimation of the performance ratio of several recent better approximations. Borchers and Du [47] completely determined the k-Steiner ratio in graphs that for $k = 2^r + h \ge 2$,

$$\rho_k = \frac{r2^r + h}{(r+1)2^r + h}$$

and Borchers et al. [48] completely determined the k-Steiner ratio in the rectilinear plane that $\rho_2 = 2/3$, $\rho_3 = 4/5$, and for $k \ge 4$, $\rho_k = (2k-1)/(2k)$. However, the k-Steiner ratio in the Euclidean plane for $k \ge 3$ is still an open problem.

36.3.4 Variable Metric Method

Berman and Ramaiyer [18] introduced an interesting approach to generalize Zelikovsky's greedy approximation. Let us call the Steiner minimum tree for a subset of k terminals as a k-*tree*. Their approach consists of two steps. The first step processes all i-trees, $3 \leq i \leq k$, sequentially in the following way: For each i-tree T with positive saving in the current graph, put T in a stack and if two leaves x and y of T are connected by a path p in a minimum spanning tree without passing any other leaf of T, then put an edge between x and y with weight equal to the length of the longest edge in p minus the saving of T. In the second step, it repeatedly pops i-trees from the stack remodifying the original minimum spanning tree for all terminals and keeping only i-trees with the current positive saving. Adding weighted edges to a point set would change the metric on the points set. Let E be an arbitrary set of weighted edges such that adding them to the input metric space makes all i-trees for $3 \leq i \leq k$ have nonpositive saving in the resulting metric space M_E. Denote by $t_k(P)$ a supermum of the length of a minimum spanning tree for the point set P in metric space M_E over all such E's. Then Berman-Ramaiyer's algorithm produces a k-size Steiner tree with total length at most

$$t_2(P) - \sum_{i=3}^{k} \frac{t_{i-1}(P) - t_i(P)}{i-1} = \frac{t_2(P)}{2} + \sum_{i=3}^{k-1} \frac{t_i(P)}{(i-1)i} + \frac{t_k(P)}{k-1}.$$

The bound for the performance ratio of Berman-Ramaiyer's approximation in Section 36.3.2 is obtained from this bound and the fact that $t_k(P) \leq \rho_k^{-1} SMT(P)$ where $SMT(P)$ is the length of the Steiner minimum tree for point set P.

36.3.5 Preprocessing and Loss

Karpinski and Zelikovsky [49] proposed a preprocessing procedure to improve approximations of Zilikovsky type. First, they use this procedure to choose some Steiner points and then run a better approximation algorithm on the union of the set of terminals and the set of chosen Steiner points. This preprocessing can improve the performance ratio for many approximations that we mentioned previously.

The preprocessing procedure is similar to the algorithm of Berman and Ramaiyer. But, it uses a *related gain*, instead of the saving, as the greedy potential function.

Karpinski and Zelikovsky [49] also introduced the loss of a Steiner tree, which the length of a shortest forest connecting all Steiner points to terminals. With the loss, Hougardy and Prömel [50] defined a new greedy potential function and obtain a 1.598-approximation for the network Steiner tree. Robins and Zelikovsky [51] improved the performance ratio to 1.55. This is the best performance ratio for polynomial-time approximations of the network Steiner tree.

36.3.6 Comparison of the Two Ideas

Although Zelikovsky's idea starts from a point different from Chang's one, the two approximations are actually similar. To see this, let us describe Zelikovsky's algorithm as follows: Start from a minimum spanning tree and at each iteration choose a Steiner point such that using this Steiner point to connect three terminals could replace two edges in the minimum spanning tree and this replacement achieves the maximum saving among such possible replacements.

Clearly, they both start from a minimum spanning tree and improve it step by step by using a greedy principal to choose a Steiner point to connect a triple of vertices. The difference is only that this triple in Chang's algorithm may contain some Steiner points while it contains only terminals in Zelikovsky's algorithm. This difference makes Chang's approximation hard to be analyzed. Why? A recent paper of Du et al. [52] indicated that the saving for k-restricted Steiner trees is submodular, whereas the saving for iterated 1-Steiner trees is not. They also indicated that the analysis of iterated 1-Steiner trees in [51] contains a serious mistake and presented a correct one.

36.4 Approximation for Geometric Steiner Minimum Trees

In 1996, *New York Times* reported that Arora [20] obtained a surprising result: many geometric optimization problems, including the Euclidean traveling salesman and the Euclidean Steiner minimum tree, have PTAS. In this context, for any $\varepsilon > 0$, there exists a polynomial-time approximation algorithm which produces an approximation solution within a factor $1 + \varepsilon$ from optimal with run time $n^{O(1/\varepsilon)}$. Several weeks later, Mitchell [53] claimed that a *minor modification* of his earlier work [21] leads to the similar results. One year later, Arora [54] made further progress by improving run time from $n^{O(1/\varepsilon)}$ to $n^3 (\log n)^{O(1/\varepsilon)}$. His new PTAS also runs randomly in time $n(\log n)^{O(1/\varepsilon)}$. Soon after, Mitchell [55] claimed again that his approach provides similar results.

Actually, both Arora's and Mitchell's approaches fall into the same category, the *adaptive partition*, consisting of a sequential partition and dynamic programming. The adaptive partition was first introduced to the design of approximation algorithm by Du et al. [56] in 1986 with guillotine cut. A sequence of guillotine cuts partitions a problem into subproblems in order to find an approximation solution in polynomial time by dynamic programming (see also Chap. 3 and Vol. 2, Chap. 6).

Both Arora and Mitchell found that the cut needs not to be completely guillotine. In other words, the dynamic programming can still runs in polynomial time if subproblems have some relations but the number of relations should be smaller. As the number of relations goes up, the obtained approximation solution approaches to the optimal one, while the run time, of course, goes up.

To do the incomplete guillotine cut, Arora [20] employed *miportals*, whereas Mitchell [21,53] utilized *m-guillotine subdivision*. These two techniques differ in their applicability, that is, there exist many problems for which the portal works, but the *m*-guillotine subdivision does not, and vice versa. Actually, their combination produces the further improvement of Arora [54] and Mitchell [55]. Volume 2, Chapter 2 discusses the application of Arora's approach to the minimum cost k-connectivity problem in geometric graphs.

36.4.1 Guillotine Cut

Roughly speaking, a *guillotine cut* is subdivision with a line segment, which divides the area into two subareas. To explain the technique of adaptive partition. Let us first consider a rectangular partition problem as follows: Given a rectilinear polygon possibly with some rectangular holes, partition it into rectangles with minimum total edge length (MELRP).

Holes in the input rectangular polygon may degenerate into a line segment or a point or partially into a line segment. The existence of holes makes difference in the computational complexity. Although the MELRP in general is NP-hard, the MELRP in the hole-free case can be solved in time $O(n^4)$, where n is the number of vertices in the input rectilinear polygon.

In 1986, Du et al. [56] initiated an idea on adaptive partition with guillotine cut in order to design an approximation for the above-mentioned rectangular partition problem. Today, we know that the idea is applicable to many geometric optimization problems, including the Euclidean Steiner minimum tree and the rectilinear Steiner minimum tree.

A cut is called a *guillotine cut* if it breaks a connected area into two parts. A rectangular partition is called a *guillotine rectangular partition* if it can be performed adaptively by a sequence of guillotine cuts. The guillotine cut features dynamic programming since each guillotine cut breaks a problem into two subproblems. Moreover, Du et al. [56] noted that the minimum length guillotine rectangular partition also satisfies the property that there exists an optimal rectangular guillotine partition in which each maximal line-segment contains a vertex of the boundary. Hence, the minimum length guillotine rectangular partition can be computed by a dynamic programming in $O(n^5)$ time. Therefore, they suggested to use the minimum length guillotine rectangular partition to approximate the MELRP and tried to analyze the performance ratio. Unfortunately, they failed to get a constant ratio in general and obtained a result only in a special case. In this special case, the input is a rectangle with points inside. Those points are holes. It had been showed that even the rectangular partition case is NP-hard, and Gonzalez and Zheng [57] had introduced divide and conquer approximation algorithms for this problem.

In their proof, Du et al. [56] introduced two terms $proj_x(P)$ and $proj_y(P)$ to form a stronger inequality than the one that they want to prove. That is, instead of proving that for each rectangular partition P, by adding some segments, a guillotine rectangular partition P' can be obtained to satisfy that $length(P') \leq 2 \cdot length(P)$, they show that the P' satisfies $length(P') \leq 2 \cdot length(P) - proj_x(P) - proj_y(P)$, where $proj_x(P)$ ($proj_y(P)$) denote the total length of segments on a horizontal (vertical) line covered by vertical (horizontal) projection of the partition P. Volume 2, Chapter 6 discusses a different proof to establish this approximation bound as well as other approximation algorithms for this problem.

The term $proj_x(P)$ ($proj_y(P)$) plays an important role in the induction. When a vertical (horizontal) guillotine cut receives horizontal (vertical) projections from both sides, a portion of $proj_x(P)$ ($proj_y(P)$) will be doubled from the application of the induction hypothesis to the two subproblems resulting from the cut. The increased negative value would cancel the increased positive value caused by the length of the new cut.

For simplicity, let us call any point in such cut a *1-dark point*. That is, a point lying inside of a given area is a horizontal (vertical) 1-dark point if it receives horizontal (vertical) projections from both sides. What Du et al. [56] did was to find various guillotine cuts consisting of 1-dark points. They succeeded only in the mentioned special case.

36.4.2 *m*-Guillotine Subdivision

Mitchell [21] made an important discovery about 1-dark points.

Lemma 36.1 (Mitchell's Lemma) *Let H (V) be the set of all horizontal (vertical) 1-dark points. Then there exists either a horizontal line L such that*

$$length(L \cap H) \leq length(L \cap V)$$

or a vertical line L such that

$$length(L \cap H) \geq length(L \cap V),$$

The first inequality means that if we use all horizontal 1-dark points on the horizontal line L to form an incomplete guillotine cut, then the set of vertical 1-dark points on L will receive enough two-side projection to cancel the length of the cut. Such cut is called a horizontal *1-guillotine cut*. Similarly, the set of all vertical 1-dark points on a vertical line forms a vertical *1-guillotine cut*. The lemma actually says that either an expected horizontal 1-guillotine cut or an expected vertical 1-guillotine cut exists.

A rectangular partition is called a *1-guillotine rectangular partition* if it can be performed by a sequence of 1-guillotine cuts. It has been shown that there exists minimum 1-guillotine rectangular partition such that every maximal segment contains at least a vertex of the boundary.

Now, the question is whether the 1-guillotine cut can also feature the dynamic programming. The answer is yes, although the 1-guillotine cut partitions a rectangular partition problem into two subproblems with some boundary conditions that two open segments may be created on the boundary by a 1-guillotine cut. This boundary condition increases the number of subproblems in the dynamic programming. Since, each subproblem is based on a rectangle with four sides. The condition on each side can be described by two possible open segments at two ends. Hence each side has $O(n^2)$ possible conditions. So, the total number of possible boundary conditions is $O(n^8)$. Thus the total number of possible subproblems is $O(n^{12})$. For each subproblem, there are $O(n^3)$ possible 1-guillotine cuts. Therefore, the minimum 1-guillotine rectangular partition can be computed by dynamic programming in $O(n^{15})$ time. With 1-guillotine cuts, the approximation ratio 2 can be established not only for the special case, but also in general. However, to reduce the number of boundary conditions in the general case, Mitchell [21] initially covered the input by a rectangle and kept performing a *rectangular partition* at each iteration.

Mitchell [21] also presented the proof in a different way. Instead of introducing the terms $proj_x$ and $proj_y$, he did a symmetric charge. For example, in case when a horizontal 1-guillotine cut is used, charge 0.5 to all horizontal segments whose projection directly contribute horizontal 1-dark points on the cut. Since the charge is performed symmetrically, no segment in P can be charged more than twice on the same side. Therefore, each segment in P can be charged with at most value 1 and the total length of charged segments cannot exceed the total length of P. This argument has the same power as that of using projection, however more directly to the point.

The 1-guillotine cut can be easily generalized in the following way: A point p is a horizontal (vertical) *m-dark point* if the horizontal (vertical) line passing through p intersects at least $2m$ vertical (horizontal) segments of the considered rectangular partition P, among which at least m are on the left of p (above p) and at least m are on the right of p (below p). A horizontal (vertical) cut is an *m-guillotine cut* if it consists of all horizontal (vertical) m-dark points on the cut line. In other words, let H_m (V_m) denote the set of all horizontal (vertical) m-dark points. An m-guillotine cut is either a horizontal line L with cut segment

$$L \cap H_m \subseteq L \cap P$$

or a vertical line L with cut segment

$$L \cap V_m \subseteq L \cap P,$$

where P is a considered partition.

A rectangular partition is m-guillotine if it can be realized by a sequence of m-guillotine cuts. The minimum m-guillotine rectangular partition can also be computed by dynamic programming in $O(n^{10m+5})$ time. In fact, at each step, an m-guillotine cut has at most $O(n^{2m+1})$ choices. There are $O(n^4)$ possible rectangles appearing in the algorithm. Each rectangle has $O(n^{8m})$ possible boundary conditions. Mitchell's lemma can also be generalized to the following.

Lemma 36.2 (Mitchell's Lemma) *There exists either a horizontal line L such that*

$$length(L \cap H_m) \le length(L \cap V_m)$$

or a vertical line L such that

$$length(L \cap H_m) \ge length(L \cap V_m).$$

With the m-guillotine subdivision, Mitchell [53] showed that there exists a polynomial-time $(1 + \varepsilon)$-approximation with run time $n^{O(\log 1/\varepsilon)}$ for the rectangular partition.

To apply the m-guillotine subdivision to the Steiner tree problem, we only need to include all terminals in a rectangle at the initial step.

36.4.3 Portals

Arora's PTAS [20] is also based on a sequential rectangular partition. Consider the rectilinear Steiner minimum tree. Initially, use a minimal square to cover n input terminals. Then with a sequence of cuts, partition the square into small rectangles, each containing one terminals. Arora [20] managed the partition to have $O(\log n)$ levels with the following techniques:

1. Equally divide the initial square into $n^2 \times n^2$ grid. Move each input point to the center of the cell containing the input point.
2. Choose a grid line in a range between $1/3$ and $2/3$ of the longest edge to cut.

Let P be the set of n input points and P' the set of n' cell-centers receiving the n input points in (1). (Possibly, $n' < n$.) Then, it is shown that there is a PTAS for P if and only if there exists a PTAS for P'. Therefore, one may work on P' instead of P.

Furthermore, with technique (2), the rectangle at the ith level has area at most $S^2(2/3)^{i-1}$ where S is the edge length of the initial square. Since the ratio between the lengths of longer edge and shorter edge is at most three, the rectangle at the last level, say the sth level, has area at least $(1/3)(S/n^2)^2$. Therefore, $S^2(2/3)^{s-1} \geq (1/3)(S/n^2)^2$. That is, $s = O(\log n)$.

To reduce the number of crosspoints at each cut line, Arora employed a technique, called the *portal*. Portals are points on cut line equally dividing cut segments. For rectilinear Steiner tree, crosspoints on a cut line can be moved to portals by increasing a small amount of the length. This would decrease the number of crosspoints on the cut line. Suppose the number of portals is p and call such portals p-portals. By properly choosing cut line, at each level of the adaptive partition, moving crosspoints to portals would increase the length of the tree within three pth of the total length of the Steiner tree. To see this, consider each rectangle R at a certain level. Suppose its longer edge has length a and shorter edge has length b ($a/3 \leq b \leq a$). Look at every possible cut in a range between $1/3$ and $2/3$ of a longer edge. Choose the cut line to intersect the Steiner tree with the smallest number of crosspoints, say c points. Then the length of the part of the Steiner tree lying in rectangle R is at least $ca/3$. Moving c crosspoints to portals requires to add some segments with total length at most $cb/(p+1) \leq ca/(p+1) \leq (3/p)(ca/3)$.

Since the guillotine rectangular partition has $O(\log n)$ levels, the total length of the resulting Steiner tree is within $(1 + \frac{3}{p})^{O(\log n)}$ times the length of the optimal one. To obtain $(1 + \frac{3}{p})^{O(\log n)} \leq 1 + \varepsilon$, we have to choose $p = O(\frac{\log n}{\varepsilon})$.

Summarizing the earlier, we obtained the structure theorem for the rectilinear Steiner minimum tree.

Theorem 36.1 *For rectilinear SMT, there exists a $(1 + \varepsilon)$-approximation T together with a guillotine rectangular partition P such that each cut intersects T only at p-portals where $p = O(\frac{\log n}{\varepsilon})$.*

Now, one can employ dynamic programming to find the shortest one among $(1 + \varepsilon)$-approximations described in the structure theorem. To estimate the run time of dynamic programming, note that each subproblem is characterized by a rectangle and conditions on the boundary. There are $O(n^8)$ possible rectangles. Each rectangle has four sides. One of them must contain p portals. However, each of other three may contain less than p portals resulting from previous cuts. There are $O(n^4)$ possible previous cuts for each edge. Thus, the number of possible sets of positions for portals on each of these three sides is $O(n^4)$. Hence, the total number of possible sets of portal positions on the boundary is $O(n^{20})$. For each fixed set of portal positions, we need also consider whether a portal is a crosspoint or not and how crosspoints are connected with each other inside the rectangle. It brings us $2^{O(p)}$ possibilities. Therefore, the total number of possible subproblems is $n^{O(1/\varepsilon)}$. Moreover, in each iteration of dynamic programming, the number of all possible cuts is $O(n^2)$. Therefore, the dynamic programming runs in time $n^{O(1/\varepsilon)}$.

36.4.4 Comparison and Combination

Let us compare two techniques, the m-guillotine subdivision and the portal.

For problems in d-dimensional space, the cut line should be replaced by $(d-1)$-dimensional hyperplane. The number of portals would be $O((m\log n)^{d-1})$ where $m = 1/\varepsilon$. With so many possible crosspoints, the dynamic programming runs in time $2^{(m\log n)^{d-1}}$ which is not a polynomial for $d \geq 3$. However, the m-guillotine cut has $O(m)$ crosspoints in each dimension and totally $O(m^{d-1})$ crosspoints in $(d-1)$-dimensional cut plane. Therefore, the dynamic programming runs in time $n^{O(m^{d-1})}$ which is still polynomial for fixed m. Thus, although the m-guillotine works well in high-dimensional space, the portal does not.

The portal technique cannot apply to the above-mentioned rectangular partition problems and the rectilinear Steiner arborescence [58]. In fact, for these three problems, moving crosspoints to portals is sometimes impossible. But, the m-guillotine cut works well in these problems.

On the other hand, the portal preserves the topological structure of original figure, but, the m-guillotine cut does not. Therefore, for problems such as the Euclidean k-medians and the Euclidean facility locations [59], the portal works well, but the m-guillotine cut does not.

In case when the problem is suitable to use both techniques, such as the rectilinear Steiner minimum tree and the Euclidean Steiner tree, a rough idea suggests a combination of two techniques, which may reduce the run time. In fact, one may first move crosspoints to $O(m \log n)$ portals and then choose $2m$ from the $O(m \log n)$ portals to construct an m-guillotine cut ($m = 1/\varepsilon$). This way, the dynamic programming procedure will run in time $n^c (\log n)^{O(m)}$ for some constant c. However, the problem is that in two techniques, two different principles are used for choosing the position of each cut. For the m-guillotine cut, the cut line has to satisfy the inequality in Mitchell's lemma. But, when portals are used, the cut line is chosen to minimize the number of crosspoints.

Again, Arora [54] and Mitchell [55] found two different ways to overcome this problem.

Arora [54] borrowed a shafting technique from nonadaptive partition and employed random argument. He found family of n^2 adaptive partitions obtained by shafting, satisfying property that the total increased length caused by moving crosspoints to portal and m-guillotine cutting in average bounded by $\varepsilon \cdot OPT$.

Mitchell [55] tried another way. First he employed a 1-guillotine cut and put portals on the 1-guillotine cut segment; then constructed an m-guillotine cut with the portals. The purpose of putting portals on the 1-guillotine cut is to bound the length increased by moving crosspoints to portals by the length of the 1-guillotine cut segment. A new lemma of Mitchell's type on the combination of the 1-guillotine cut and the m-guillotine cut would resolve the problem.

It is interesting to mention that combining Arora's and Mitchell's approaches of 1997 again may produce a further improvement in the run time [60].

Rao and Smith [61] reduced time complexity of PTAS to $m^{O(m)} n \log n$ for the rectilinear Steiner tree and the Euclidean Steiner tree. The basic idea is to compute a *banyan* instead of the tree. The banyan contains a $(1 + \varepsilon)$-approximation solution. It has $O(n)$ vertices and the total length is within a constant factor from optimal solution. Moreover, it can be computed in time $O(n \log n)$.

References

1. Gauss, C.F., Briefwechsel Gauss-Schuhmacher, in *Werke Bd. X, 1*, Kgl. Gesellschaft der Wissenschaften, Göttingen, Germany, 1917, p. 459.
2. Crourant, R. and Robbins, H., *What Is Mathematics?* Oxford University Press, New York, 1941.
3. Foulds, L.R. and Graham, R.L., The Steiner problem in Phylogeny is NP-complete, *Advanced Appl. Math.*, 3, 43, 1982.
4. Garey, M.R., Graham, R.L., and Johnson, D.S., The complexity of computing Steiner minimal trees, *SIAM J. Appl. Math.*, 32, 835, 1977.
5. Smith, W.D. and Shor, P.W., Steiner tree problems, *Algorithmica*, 7, 329, 1992.
6. Du, D.-Z., On greedy heuristics for Steiner minimum trees, *Algorithmica*, 13, 381, 1995.
7. Gilbert, E.N. and Pollak, H.O., Steiner minimal trees, *SIAM J. Appl. Math.*, 16, 1, 1968.
8. Du, D.-Z., Hwang, F.K., and Yao, E.Y., The Steiner ratio conjecture is true for five points, *J. Combinatorial Theory, Ser. A*, 38, 230, 1985.
9. Friedel, J. and Widmayer, P., A simple proof of the Steiner ratio conjecture for five points, *SIAM J. Appl. Math.*, 49, 960, 1989.

10. Booth, R.S., The Steiner ratio for five points, *Ann. Oper. Res.*, 33, 419–436, 1991.
11. Rubinstein, J.H. and Thomas, D.A., A variational approach to the Steiner network problem, *Annals of Operations Research*, 33 (6), 481–499, 1991.
12. Graham, R.L. and Hwang, F.K., Remarks on Steiner minimal trees, *Bull. Inst. Math. Acad. Sinica*, 4, 177, 1976.
13. Chung, F.R.K. and Hwang, F.K., A lower bound for the Steiner tree problem, *SIAM J. Appl. Math.*, 34, 27, 1978.
14. Du, D.-Z. and Hwang, F.K., A new bound for the Steiner ratio, *Trans. Amer. Math. Soc.*, 278, 137, 1983.
15. Chung, F.R.K. and Graham, R.L., A new bound for Euclidean Steiner minimum trees, *Ann. N.Y. Acad. Sci.*, 440, 328, 1985.
16. Du, D.-Z. and Hwang, F.K., The Steiner ratio conjecture of Gilbert-Pollak is true, in *Proceedings of National Academy of Sciences*, 87, 1990, p. 9464.
17. Zelikovsky, A.Z., The 11/6-approximation algorithm for the Steiner problem on networks, *Algorithmica*, 9, 463, 1993.
18. Berman, P. and Ramaiyer, V., Improved approximations for the Steiner tree problem, *J. Algorithms*, 17, 381, 1994.
19. Du, D.-Z., Zhang, Y., and Feng, Q., On better heuristic for Euclidean Steiner minimum trees, in *Proceedings of 32nd Annual Symposium on Foundations of Computer Science*, IEEE Press, New York, 1991.
20. Arora, S., Polynomial-time approximation schemes for Euclidean TSP and other geometric problems, in *Proceeding of Foundations of Computer Science*, IEEE Press, New York, 1996, p. 2.
21. Mitchell, J.S.B., Guillotine subdivisions approximate polygonal subdivisions: A simple new method for the geometric k-MST problem, in *Proceedings of the Seventh Annual ACM-SIAM Symposium on Discrete Algorithms*, Society for Industrial and Applied Mathematics Philadelphia, PA, 1996, p. 402.
22. Bern, M. and Plassmann, P., The Steiner problem with edge lengths 1 and 2, *Inf. Proc. Lett.*, 32, 171, 1989.
23. Hwang, F.K., Richards, D.S., and Winter, P., *Steiner Tree Problems*, North-Holland, Amsterdam, the Netherlands, 1992.
24. Hwang, F.K., On Steiner minimal trees with rectilinear distance, *SIAM J. Appl. Math.*, 30, 104, 1976.
25. Pollak, H.O., Some remarks on the Steiner problem, *J. Combinatorial Theory, Ser. A*, 24, 278, 1978.
26. Rubinstein, J.H. and Thomas, D.A., The Steiner ratio conjecture for six points, *J. Combinatorial Theory, Ser. A*, 58, 54, 1991.
27. Du, D.-Z. and Hwang, F.K., An approach for proving lower bounds: Solution of Gilbert-Pollak's conjecture on Steiner ratio, in *Proceedings of 31st Annual Symposium on Foundations of Computer Science*, IEEE Press, New York, 1990, p. 76.
28. Gao, B., Du, D.-Z., and Graham, R.L., A tight lower bound for the Steiner ratio in Minkowski planes, *Disc. Math.*, 142, 49, 1995.
29. Cieslik, D., The Steiner ratio of Banach-Minkowski planes, in *Contemporary Methods in Graph Theory*, Bodendiek, R., Ed., BI-Wissenschatteverlag, Mannheim, Germany, 1990, p. 231.
30. Du, D.-Z., Gao, B., Graham, R.L., Liu, Z.-C., and Wan, P.-J., Minimum Steiner trees in normed planes, *Disc. Comput. Geom.*, 9, 351, 1993.
31. Cavalli-Sforza, L.L. and Edwards, A.W., Phylogenetic analysis: Models and estimation procedures, *Am. J. Human Genetics*, 19, 233, 1967.
32. Chung, F.R.K. and Gilbert, E.N., Steiner trees for the regular simplex, *Bull. Inst. Math. Acad. Sinica*, 4, 313, 1976.
33. Smith, W.D., How to find Steiner minimal trees in Euclidean d-space? *Algorithmica*, 7, 137, 1992.
34. Du, D.-Z. and Smith, W.D., Three disproofs for Gilbert-Pollak conjecture in high dimensional spaces, *J. Comb. Theory*, 74, 115–130, 1996.

35. Smith, W.D. and Smith, J.M., On the Steiner ratio in 3-space, *J. Comb. Theory, Ser. A*, 69, 301, 1995.

36. Bharath-Kumar, K. and Jaffe, J.M., Routing to multiple destinations in computer networks, *IEEE Trans. Comm.*, COM-31, 343, 1983.

37. Chang, S.-K., The design of network configuration with linear or piecewise linear cost functions, in *Proceedings of the Symposium on Computer-Communications Networks and Teletraffic*, Vol. 22, Polytechnic Institute of Brooklyn, John Wiley & Sons, 1972, pp. 363–369.

38. Smith, J.M. and Liebman, J.S., Steiner trees, Steiner circuits and interference problem in building design, *Engineering Opt.*, 4, 15, 1979.

39. Beasley, J.E., A heuristic for the Euclidean and rectilinear Steiner problems, Technical Report of the Management School, Imperial College, London, UK, 1989.

40. Chang, S.-K., The generation of minimal trees with a Steiner topology, *JACM*, 19, 699, 1972.

41. Karp, R.M., Reducibility among combinatorial problems, in *Complexity of Computer Computation*, Miller, R.E. and Tatcher, J.W., Eds., Plenum Press, Berlin, Germany, 1972, p. 85.

42. Kou, L. and Makki, K., An even faster approximation algorithm for the Steiner tree problem in graphs, *Congr. Numer.*, 59, 1987, 147.

43. Kou, L., Markowsky, G., and Berman, L., A fast algorithm for Steiner trees, *Acta Inform.*, 15, 141, 1981.

44. Kuhn, H.W., Steiner's problem revisited, in *Studies in Optimization, Studies in Math.*, Vol. 10, Dantzig, G.B. and Eaves, B.C., Eds., Amer. Math. Assoc., 1975, pp. 53–70.

45. Smith, J.M., Lee, D.T., and Liebman, J.S., An $O(N \log N)$ heuristic for Steiner minimal tree problems in the Euclidean metric, *Networks*, 11, 23, 1981.

46. Kahng, A. and Robins, G., A new family of Steiner tree heuristics with good performance: The iterated 1-Steiner approach, in *Proceedings of International Conference on Computer-Aided Design (ICCAD)*, IEEE Press, New York, 1990, p. 428.

47. Borchers, A. and Du, D.-Z., The k-Steiner ratio in graphs, *SIAM J. Comput.*, 26, 857, 1997.

48. Borchers, A., Du, D.-Z., Gao, B., and Wan, P.-J., The k-Steiner ratio in the rectilinear plane, *J. Algorithms*, 29, 1, 1998.

49. Karpinski, M. and Zelikovsky, A.Z., New approximation algorithms for the Steiner tree problem. *J. Comb. Optim.* 1, 47–65, 1997.

50. Hougardy, S. and Prömel, H.J., A 1.598-approximation algorithm for the Steiner problem in graphs, in *Proceedings of the Tenth Annual ACM-SIAM Symposium on Discrete Algorithms*, Society for Industrial and Applied Mathematics Philadelphia, PA, 1999, p. 448.

51. Robins, G. and Zelikovsky, A., Improved Steiner trees approximation in graphs, in *Proceedings of the Eleventh Annual ACM-SIAM Symposium on Discrete Algorithms*, Society for Industrial and Applied Mathematics Philadelphia, PA, 2000, p. 770.

52. Du, D.-Z., Graham, R.L., Wu, W., Pardalos, P., Wan, P. J, and Zhao, W., Analysis of greedy approximations with nonsubmodular potential functions, in *Proceedings of the Nineteenth Annual ACM-SIAM Symposium on Discrete Algorithms*, Society for Industrial and Applied Mathematics, Philadelphia, PA, 2006, pp. 167–175.

53. Mitchell, J.S.B., Guillotine subdivisions approximate polygonal subdivisions: Part II—A simple polynomial-time approximation scheme for geometric k-MST, TSP, and related problem, *SIAM J. Comput.*, IEEE Press, New York, 29, 515, 1999.

54. Arora, S., Nearly linear time approximation schemes for Euclidean TSP and other geometric problems, in *Proceedings of 38th Annual Symposium on Foundations of Computer Science*, 1997, p. 554.

55. Mitchell, J.S.B., Guillotine subdivisions approximate polygonal subdivisions: Part III—Faster polynomial-time approximation scheme for geometric network optimization, in *Proceedings of Canadian Conference on Computational Geometry*, 1997, p. 229.

56. Du, D.-Z., Pan, L.-Q., and Shing, M.-T., Minimum edge length guillotine rectangular partition, Technical Report 0241886, Mathematical Sciences Research Institute, University of California, Berkeley, CA, 1986.

57. Gonzalez, T.F. and Zheng, S.Q., Bounds for partitioning rectilinear polygons, in *Proceedings of the First Annual Symposium on Computational Geometry*, ACM Press, New York, 1985.

58. Lu, B. and Ruan, L., Polynomial time approximation scheme for the rectilinear Steiner arborescence problem, *J. Comb. Opt.*, 4, 357, 2000.

59. Arora, S., Raghavan, P., and Rao, S., Polynomial time approximation schemes for Euclidean *k*-medians and related problems, in *Proceedings of the Thirtieth Annual ACM Symposium on Theory of Computing*, ACM Press, New York, 1998, p. 106.

60. Arora, S., Polynomial-time approximation schemes for Euclidean TSP and other geometric problems, *JACM*, 45, 753, 1998.

61. Rao, S.B. and Smith, W.D., Approximating geometrical graphs via "spanners" and "banyans", in *Proceedings of the Thirtieth Annual ACM Symposium on Theory of Computing*, ACM Press, New York, 1998, p. 540.

37

Practical Approximations of Steiner Trees in Uniform Orientation Metrics

Andrew B. Kahng

Ion Măndoiu

Alexander Zelikovsky

37.1 Introduction

The Steiner minimum tree problem, which asks for a minimum-length interconnection of a given set of terminals in the plane, is one of the fundamental problems in very large-scale integration (VLSI) physical design. Although advances in VLSI manufacturing technologies have introduced additional routing objectives, minimum length continues to be the primary objective when routing noncritical nets, since the minimum-length interconnection has minimum total capacitance and occupies minimum amount of area.

To simplify design and manufacturing, VLSI interconnect is restricted to a small number of orientations defining the so-called *interconnect architecture*. Until recently, designers have relied almost exclusively on the *Manhattan interconnect architecture*, which allows interconnect routing along two orthogonal directions. However, non-Manhattan interconnect architectures—such as the *Y-architecture*, which allows 0°, 120°, and 240° oriented wires, and the *X-architecture*, which allows 45° diagonal wires in addition to the traditional horizontal and vertical orientations—are becoming increasingly attractive due to the significant potential for reducing interconnect length [1–7]. A common generalization of interconnect architectures of interest in VLSI design is that of *uniform orientation metric*, or λ-*geometry*, in which routing is allowed only along $\lambda \geq 2$ orientations forming consecutive angles of π/λ. The Manhattan, Y-, and X-architectures correspond to $\lambda = 2$, 3, and 4, respectively.

In contrast to the extensive literature on the rectilinear version of the Steiner tree problem, computing Steiner trees in λ-geometries with $\lambda > 2$ has received much less attention in the literature. A hierarchical

Steiner tree construction was proposed by Sarrafzadeh and Wong [8]. Koh [9] showed how to compute the optimal Steiner trees for three terminals under the octilinear metric ($\lambda = 4$) and proposed a heuristic inspired by the iterated 1-Steiner heuristic of Kahng and Robins for computing rectilinear Steiner tree. Li et al. [10] solved the three-terminal case for $\lambda = 3$ and proposed a simulated annealing heuristic. Generalizations of the classical Hanan grid [11] for $\lambda = 3$ and $\lambda = 3$ are also proposed by References 9, 10, but these generalizations lead to multilevel grids typically containing too many points to be of use in designing practical algorithms.

Chiang et al. [12] proposed a simple heuristic for constructing octilinear Steiner trees that computes a rectilinear Steiner tree for the terminals and then iteratively reduces tree length by using diagonal wires to reroute tree edges and by sliding Steiner points. Coulston [13] proposed an exact octilinear Steiner tree algorithm based on a two-phase approach of generating all possible full components for the given set of terminals and then merging them to form an optimal tree. He reports that the proposed algorithm has practical runtime for up to 25 terminals. Nielsen et al. [14] described the extension of the GeoSteiner exact rectilinear/Euclidean Steiner tree package to arbitrary λ-geometries. GeoSteiner uses the same two-phase approach as Coulston's algorithm; however, it incorporates highly effective pruning techniques based on structural properties of full components to greatly improve the running time. With these improvements, GeoSteiner is reported to compute optimal λ-geometry Steiner trees for instances with 100 random terminals in seconds of CPU time.

Recently, there has been a growing interest in practical methods for computing near-optimal Steiner trees for instances with up to *tens of thousands of terminals*. Instances of this size, for example, scan enable nets, are becoming more common in modern VLSI designs due to the increased emphasis on design for test. Such nets are noncritical for chip performance and tend to consume significant routing resources, so minimizing length is the appropriate optimization objective. Furthermore, very large Steiner tree instances are created by reductions that model nonzero terminal dimensions, for example, nets with preroutes. High-quality routing of such instances requires representing each terminal by a set of electrically equivalent points [15], and this results in instances with as much as 100,000 points [16]. Because of combinatorial explosion and/or quadratic memory requirements, instances of this size cannot be solved in practical time with the existing exact methods, such as GeoSteiner, or best-performing heuristics, such as iterated 1-Steiner [17] and Rajagopalan-Vazirani [18].

In this chapter, we present a highly scalable heuristic for computing near-optimal Steiner trees. Our heuristic, referred to as the *batched greedy algorithm* (BGA) in the following, is graph-based and can therefore be easily modified to handle routing in uniform orientation geometries as well as other practical considerations, such as routing obstacles, preferred directions [19], and via costs. Indeed, the results reported in Section 37.5 show only a small factor increase in runtime compared with the rectilinear implementation. The BGA heuristic routes a 34k-terminals net extracted from a real design in less than 25 s compared with over 86 min needed by the $O(n^2)$ edge-based heuristic of Reference 20. More importantly, this dramatic reduction in runtime is achieved with no loss in solution quality. On random instances with more than 100 terminals, our algorithm improves over the rectilinear minimum spanning tree (MST) by an average of 11%, matching in solution quality the edge-based heuristic of Reference 20.

The BGA heuristic derives its efficiency from three key ideas:

- Combining the implementation proposed in Reference 21 for the *greedy triple contraction algorithm* (GTCA) of Zelikovsky [22] with the batched method introduced by the iterated 1-Steiner heuristic of Kahng and Robins [17].
- A new divide-and-conquer method for computing a superset of size $O(n \log n)$ of the set of $O(n)$ triples required by GTCA.
- A new linear size data structure that enables finding a bottleneck (i.e., maximum cost) edge on the tree path between two given nodes in $O(\log n)$ time after $O(n \log n)$ preprocessing, with very low

constants hidden under the big O. This data structure allows computing the gain of a triple (see Section 37.2 for the definition) in $O(\log n)$ time.[*]

The BGA heuristic requires $O(n)$ memory and $O(n \log^2 n)$ time for computing a Steiner tree over n terminals. To the best of our knowledge, this is the first practical subquadratic heuristic with such high performance. The $O(n \log n)$ implementation described in Reference 20 for the edge-based heuristic requires advanced data structures, potentially involving large hidden constants. We are not aware of any implementation demonstrating the practical applicability of this implementation. After the publication of a preliminary version of this work [23], Zhu et al. [24] proposed an $O(n \log n)$ octilinear Steiner tree heuristic based on spanning graphs, which is reported to run faster than BGA at the cost of a small decrease in solution quality.

The rest of the chapter is organized as follows. In Section 37.2, we briefly review the GTCA of Reference 22 and describe our new BGA. In the following two sections, we describe in detail two of the key BGA subroutines: in Section 37.3, we give the new divide-and-conquer method for computing the set of $O(n \log n)$ triples used by BGA, whereas in Section 37.4 we describe the new data structure for computing bottleneck edges. Finally, in Section 37.5, we give experimental results comparing BGA with previous implementations of rectilinear and octilinear Steiner tree heuristics and exact algorithms on test cases both randomly generated and extracted from recent VLSI designs.

37.2 The Greedy Triple Contraction and Batched Greedy Algorithms

We begin this section by introducing the Steiner tree terminology used in the rest of the chapter. A Steiner tree for a set of terminals is a tree spanning the terminals and possibly additional points, called *Steiner points*. A Steiner tree is called a *full Steiner tree* if all terminals are leaves (i.e., have degree 1). Any Steiner tree T can be split into edge-disjoint full Steiner trees called the *full Steiner components* of T [25]. A Steiner tree T is called k-restricted if every full component of T has at most k terminals (an example of a 3-restricted rectilinear Steiner tree is shown in Figure 37.1). The minimum-cost 3-restricted Steiner tree is in general cheaper than the MST of the terminals (note that the MST is the minimum-cost 2-restricted Steiner tree of the terminals).

The GTCA in Reference 22 finds an approximate minimum-cost 3-restricted Steiner tree by greedily choosing 3-restricted full components which reduce the cost of the MST. To describe GTCA, we need to introduce a few more notations. Let G_S be the complete graph on a given set S of terminals, and let $MST(S)$ be a MST of G_S. A *triple* τ is an optimal Steiner tree for a set of three terminals.[†] We denote by $center(\tau)$ the single Steiner point of τ and by $cost(\tau)$ the cost of τ. In the graph $MST(S) \cup \tau$, there are two cycles (Figure 37.2a). To obtain an MST of this graph, we should remove the most expensive edge from each cycle. Let e_1 and e_2 be the two edges that must be removed and let $R(\tau) = \{e_1, e_2\}$. The *gain* of τ is $gain(\tau) = cost(R(\tau)) - cost(\tau)$.

GTCA (Figure 37.3) repeatedly adds a triple τ with the largest gain and *contracts* it, that is, collapses the three terminals of τ into a single new terminal. Contraction of a triple is conveniently implemented by adding two new zero-cost edges $A(\tau) = \{e'_1, e'_2\}$ between the three terminals of τ (Figure 37.2b). It is easy to see that addition of $A(\tau)$ changes the MST of G_S—in the updated MST the two edges in $A(\tau)$ replace

---

[*] Our data structure may be of interest in other applications that require computing bottleneck edges. For example, Zachariasen incorporated it in the beta version of the GeoSteiner 4.0 code for computing optimum geometric Steiner trees. On large instances, computing bottleneck edges with the new data structure was found to be faster than look-up in a precomputed matrix, most likely due to improved memory access locality (Zachariasen, M., Personal communication, 2002.).

[†] The optimum Steiner tree of 3 given terminals can be computed in constant time under the common uniform orientation metrics, including rectilinear [11] and octilinear [3] metrics.

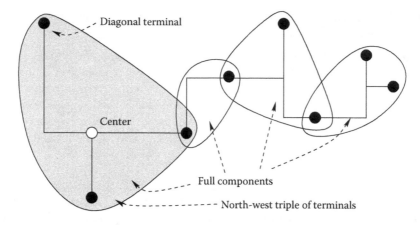

FIGURE 37.1 A 3-restricted rectilinear Steiner tree partitioned into full components. The dark full component is a north-west triple.

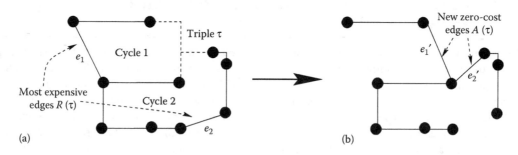

FIGURE 37.2 The MST of G_S (a) before and (b) after contraction of the triple τ. The gain of the dashed triple is the difference between the cost of most expensive edges $R(\tau) = \{e_1, e_2\}$ in each of the two cycles of $T \cup \tau$ and the cost of τ. Two new zero-cost edges $A(\tau) = \{e_1', e_2'\}$ replace e_1 and e_2 in the updated MST.

Input: Set S of terminals
Output: 3-restricted Steiner tree T spanning S

1. $T \leftarrow MST(G_S)$
2. $F \leftarrow \emptyset$
3. Repeat forever

        Find a triple τ with maximum gain

        If $gain(\tau) \leq 0$, then go to Step 4

        $F \leftarrow F \cup \{\tau\}$ // Add τ

        $T \leftarrow T - R(\tau) + A(\tau)$ // Contract τ

4. Output $MST(F \cup MST(G_S))$

FIGURE 37.3 The greedy triple contraction algorithm.

the two edges in $R(\tau)$. Finally, GTCA adds all chosen triples to the original MST of G_S and outputs the MST of this union.

Theorem 37.1 ([26]) *The cost of the rectilinear Steiner tree constructed by GTCA is at most 1.3125 times more than the optimal Steiner tree [26].*

Fößmeier et al. [21] prove that to achieve an approximation ratio of 1.3125 in Theorem 37.1, it is sufficient to consider only *empty tree triples* of terminals. A triple τ is *empty* if the minimum rectangle bounding the triple does not contain any other terminals and is a *tree* triple if the center c of τ is adjacent to all terminals of the triple in $MST(S + c)$ (or, equivalently, if $gain(\tau) > 0$). As shown in Reference 21, there are at most $36n$ empty tree triples. Even so, finding the best triple in Step 3 of GTCA is very time consuming. An efficient $O(n^2 \log n)$ time implementation of GTCA should maintain dynamic MSTs for which, to date, there is no data structure able to handle instances with tens of thousands of nodes in practical running time. Existing data structures are difficult to implement and involve big asymptotic constants, see Cattaneo et al. [27] for a recent empirical study.

Our new heuristic, the BGA (Figure 37.4) adopts the batched method from Reference 17, substantially reducing running time by relaxing the greedy rule used to select triples in GTCA. After contracting a triple we continue by picking the best triple among those with unchanged gain; in general this may not be the best triple overall. Note that a triple τ can change its gain only if one of the edges in $R(\tau)$ is removed when contracting other triple—if none of the contracted triples remove edges from $R(\tau)$ then the gain of τ is unchanged. When done with one such *batched phase* (the body of the while loop in Step 4) it is still possible to have positive gain triples. Therefore, we recompute triple gains and repeat the batched phase selection until no positive gain triples are left. To enable further improvements, we add the centers of triples selected in Step 4 to the terminal set then iterate Steps 2–5 (which constitute a *round* of the algorithm) until no more centers are added to the tree.

In next section, we show how to compute in $O(n \log n)$ time a set of $O(n \log n)$ triples containing all empty tree triples (Theorem 37.3). Then, in Section 37.4, we describe a data structure which enables computing a bottleneck edge on the tree path between any two given nodes in $O(\log n)$ time after $O(n \log n)$

Input: Set S of terminals
Output: Steiner tree T spanning S

1. Compute the minimum spanning tree of S, $MST(S)$
2. Compute a set $Triples$, of size $O(n \log n)$, containing all empty tree triples
3. $SP \leftarrow \emptyset$
4. While $Triples \neq \emptyset$ do

    For each $\tau \in Triples$ compute $R(\tau)$, $A(\tau)$, and $gain(\tau)$, discarding triples with non-positive gain

    Sort $Triples$ in descending order of gain

    Unmark all edges of $MST(S)$

    For each $\tau \in Triples$ do

        If both edges in $R(\tau)$ are unmarked, then mark them and replace them in the MST with the two
            edges in $A(\tau)$, i.e., $MST(S) \leftarrow MST(S) - R(\tau) + A(\tau)$
        $SP \leftarrow SP + center(\tau)$

5. If $SP = \emptyset$ then return the minimum spanning tree of S, else

    $S \leftarrow S + SP$

    Compute the minimum spanning tree of S and discard all Steiner points with degree 1 or 2

    Go to Step 2

FIGURE 37.4 The batched greedy algorithm.

time preprocessing. Since computing the gain of a triple amounts to three bottleneck edge computations, this leads to an $O(n \log^2 n)$ time implementation of the batched phase algorithm. This gives the following:

Theorem 37.2 *The running time of the BGA is $O(Pn \log^2 n)$, where P is the total number of batched phases and n is the number of terminals.*

In practice the total number of phases P is small and can be bounded by a constant. Thus, the runtime of BGA is $O(n \log^2 n)$.

37.3 Generation of Triples

In this section, we show how to compute in $O(n \log n)$ time a set of $O(n \log n)$ triples containing all empty tree triples. For simplicity, we assume that terminals are in general position, that is, no two of them share the same x- or y-coordinate. This assumption is not restrictive since we can always break ties, for example, according to terminal IDs.

In a triple, the terminal which does not share x- and y-coordinates with the center (see the shaded portion of Figure 37.1) is called *diagonal*. There are four types of triples depending on where the diagonal terminal lies with respect to the center: a triple is called *north-west* if the diagonal terminal is in the north-west quadrant of the center (Figure 37.1); north-east, south-west, and south-east triples are defined similarly. We will use the divide and conquer method to find $O(n \log n)$ north-west triples containing all north-west empty tree triples. Triples of the other types are obtained by reflection and application of the same algorithm.

For finding north-west triples, we recursively partition the terminals into (almost) equal halves with a bisector line parallel to line $y = -x$. Let LB (left-bottom) and TR (top-right) be the half-planes defined by the bisector line, and let D, R, and B be the diagonal, right, and bottom terminals of a north-west triple that is intersected by the bisector line (Figure 37.5). We distinguish the following four cases:

Case 1: $D, R \in TR$ **and** $B \in LB$. Figure 37.6a illustrates how to compute for each diagonal terminal D the unique terminal R that can serve as a right terminal in an empty north-west triple with D as the diagonal terminal. All terminals in TR are processed in x-ascending order as follows: (1) if the next terminal has y larger than the current terminal, then a dashed pointer is set from the next to the current terminal, and then the current terminal is advanced to the next terminal; (2) otherwise, a solid pointer is set from the current terminal to the next one, and the current terminal is moved back along the dashed pointer (if it exists, otherwise the current terminal is advanced to the next). Clearly this procedure is linear since the runtime is proportional to the number of pointers established, and each terminal has at most two pointers (one solid and one dashed). When processing of the points in TR is finished, each solid arc connects a terminal D with the leftmost terminal in TR lower than and to the right of D, that is, with the unique terminal R that can serve as a right terminal in an empty north-west triple with D as the diagonal terminal.

    To find all Case 1 north-west triples, we must find for each solid arc (D, R) in TR the node B in LB which can complete the triple, that is, the node B with maximum y-coordinate in the vertical strip defined by D and R. This is done in linear time by one traversal of the terminals in LB in x-ascending order (i.e., strip by strip) while computing the highest point in each strip.

Case 2: $B, R \in TR$ **and** $D \in LB$. For each terminal R, the unique terminal in TR that can serve as the bottom terminal in an empty north-west triple with R as the right terminal (i.e., the highest terminal in TR lower and to the left of R) can be found by a procedure similar to the one in Step 1. An arc (R, B) in TR is completed into a tree north-west triple only when the diagonal node D is the closest to R (and therefore to B) in the octant of LB containing points higher

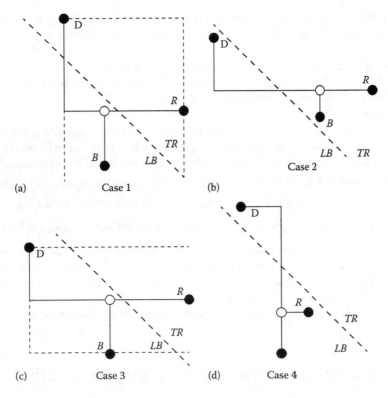

FIGURE 37.5 Four cases of partitioning of a north-west triple: (a) Case 1, (b) Case 2, (c) Case 3, and (d) Case 4.

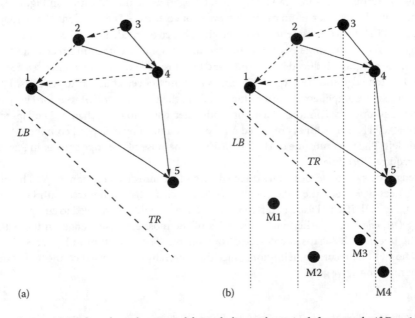

FIGURE 37.6 Case 1: (a) Finding the right terminal for each diagonal terminal, for example, if $D = 1$, then $R = 5$, if $D = 3$, then $R = 4$, and so on. (b) Finding highest terminals in the strips corresponding to consecutive terminals.

than R (cf. [21]). To find the diagonal points D for each arc (R, B) in TR, we simultaneously traverse terminals in TR in y-ascending order and terminals in LB in $(x - y)$-ascending order as follows:

Although there are unprocessed terminals in both TR and LB, do

- Advance in TR until we reach a terminal R which has an arc to the associated B
- Advance in LB until we reach a terminal D higher than R
- Assign D to R

Note that triples found by the above-mentioned procedure are not necessarily empty. With a more careful implementation, it is possible to avoid generating nonempty triples; however, this would not change the asymptotic number of triples generated or the worst-case running time of the algorithm.

Case 3: $R \in TR$ **and** $D, B \in LB$. It is equivalent to the Case 1 after reflection over bisector.

Case 4: $D \in TR$ **and** $B, R \in LB$. It is equivalent to the Case 2 after reflection over bisector.

Theorem 37.3 *A set of size $O(n \log n)$ containing all empty tree triples can be computed in $O(n \log n)$ time.*

Proof. Each north-west empty tree triple crossing the dividing diagonal must fall in one of the four cases considered and finding all crossing triples takes linear time. Thus, the running time is given by the recurrence $T(n) = 2T(n/2) + O(n)$, that is, $T(n) = O(n \log n)$. The number of triples generated by the divide-and-conquer algorithm is also $O(n \log n)$ by the same recurrence; notice that each recursive step generates a linear number of triples. The same argument applies to the other three triple types. ∎

37.4 Computing Maximum Cost Edge on a Tree Path

It is easy to see that computing the gain of a triple τ and the edges in $R(\tau)$ reduces to finding bottleneck (i.e., most expensive) edges on the tree paths between pairs of terminals in τ. The *hierarchical greedy preprocessing* (HGP) algorithm given in Figure 37.7 computes for a given tree on n terminals two auxiliary arrays, *parent* and *edge*, with at most $2n - 1$ elements each. Using these arrays, the bottleneck tree edge between any two terminals u and v can be found in $O(\log n)$ using the algorithm in Figure 37.8.

Assuming that edges are sorted in ascending order of cost, HGP is equivalent to the following recursive procedure. First, for each node u, direct the cheapest edge incident to u, away from u, and save its index in *edge(u)*. As a result some edges remain undirected, some become unidirected, and some become bidirected. In the subgraph induced by the (uni- and bi-) directed edges, each connected component consists of a bidirected edge with two (possibly empty) arborescences attached to its ends. HGP collapses each such connected component K into a single node q, then sets *parent(u)* to q for every $u \in K$. Since each connected component contains at least one bidirected edge, no more than $n/2$ collapsed component nodes are created. The procedure is repeated on the tree induced by collapsed components until there is a single node left. The total runtime of HGP is $O(n \log n)$ because of the edge sorting in Step 1, remaining HGP steps require $O(n)$ time.

Clearly, edge costs decrease along any directed path of a connected component K. Therefore, if u and v are two vertices of K, then the index of the maximum cost edge on the tree path between u and v is $\max\{edge(u), edge(v)\}$. If u and v are in different components K and K', we need compute the maximum between $edge(u)$, $edge(v)$, and the maximum index of the most expensive edge on the path between K and K' in the tree T with collapsed connected components. The algorithm in Figure 37.8 is an iterative implementation of this recursive definition. Since the hierarchy of collapsed connected components has a depth of at most $\log n$, we get:

Theorem 37.4 *The algorithm in Figure 37.8 finds the maximum cost edge on the tree path connecting two given nodes in $O(\log n)$ time after $O(n \log n)$ time for HGP.*

Input: Weighted tree $T = (V, E, cost)$ with $V = \{1, 2, \ldots, n\}$
Output: Arrays $parent(i)$ and $edge(i)$, $i = 1, \ldots, 2n - 1$

---

1. Sort tree edges $e_1, \ldots, e_{n-1}$ in ascending order of cost

2. Initialization:

   $next \leftarrow n$

   For each $i = 1, 2, \ldots, 2n - 1$ do

       $parent(i) \leftarrow NIL$

       $edge(i) \leftarrow NIL$

3. For each edge $e_i = (u, v)$, $i = 1, \ldots, n - 1$, do

   While $u \neq v$ and $parent(u) \neq NIL$ and $parent(v) \neq NIL$ do

       $u \leftarrow parent(u)$

       $v \leftarrow parent(v)$

   If $parent(u) = parent(v) = NIL$, then

       $next \leftarrow next + 1$

       $parent(u) \leftarrow parent(v) \leftarrow next$

       $edge(u) \leftarrow edge(v) \leftarrow i$

   If $parent(u) = NIL$ and $parent(v) \neq NIL$, then

       $parent(u) \leftarrow parent(v)$

       $edge(u) \leftarrow i$

   If $parent(u) \neq NIL$ and $parent(v) = NIL$, then

       $parent(v) \leftarrow parent(u)$

       $edge(v) \leftarrow i$

4. Output the arrays $parent(i)$ and $edge(i)$

FIGURE 37.7   The hierarchical greedy preprocessing algorithm.

Input: Tree edges $e_1, \ldots, e_{n-1}$ in ascending order of cost, arrays $parent(i)$ and $edge(i)$, $i = 1, \ldots, 2n - 1$, and nodes $u, v \in V$
Output: Maximum cost edge on the tree path between u and v

---

1. $index \leftarrow -\infty$,

2. While $u \neq v$ do

   $index \leftarrow \max\{index, edge(u), edge(v)\}$

   $u \leftarrow parent(u)$

   $v \leftarrow parent(v)$

3. Return e_{index}

FIGURE 37.8   Subroutine for computing the maximum cost edge on the tree path between nodes u and v.

37.5 Experimental Results

Comprehensive experimental evaluation indicates that the iterated 1-Steiner heuristic of Kahng and Robins [17] significantly outperforms in solution quality the rectilinear Steiner tree heuristics proposed prior to 1992 [28]. Since then, the edge-based heuristic of Borah et al. [20], and the iterated Rajagopalan-Vazirani (IRV) heuristic [18] have been reported to match or slightly exceed iterated 1-Steiner in solution quality. However, among these best-performing heuristics only the edge-based heuristic can be applied to instances with tens of thousands of terminals, since current implementations of iterated 1-Steiner and IRV require quadratic memory. Besides Borah's $O(n^2)$ implementation of the edge-based heuristic, we compared our $O(n \log^2 n)$ BGA to the recent $O(n \log^2 n)$ Prim-based heuristic of Rohe [29]. For comparison purposes, we also include results from our implementation of the Guibas–Stolfi $O(n \log n)$ rectilinear MST algorithm [30], and, whenever possible, the optimum Steiner trees computed using the beta version of the GeoSteiner 4.0 algorithm recently announced in Reference 14.

All heuristics and MST algorithms were run on a dual 1.4 GHz Pentium III Xeon server with 2 GB of memory running Red Hat Linux 7.1. The GeoSteiner code using the CPLEX 6.6 linear programming solver was run on a 360 MHz SUN Ultra 60 workstation with 2 GB of memory under SunOS 5.7. The test bed for our experiments consisted of two categories of instances: instances drawn uniformly at random from a 1,000,000 × 1,000,000 grid, ranging in size between 100 and 500,000 terminals, and a set of eight test cases extracted from recent industrial designs, ranging in size between 330 and 34,000 terminals.

Table 37.1 gives the percent improvement over the rectilinear MST and running time (in CPU seconds) for experiments on rectilinear instances. On random instances, the batched greedy heuristic matches or slightly exceeds in average solution quality the edge-based heuristic of Reference 20. Both batched greedy and the edge-based heuristic improve the rectilinear MST by an average of 11% in our experiments. This is roughly 1% more than the average improvement achieved by the Prim-based heuristic of Reference 29 and is within 0.7% of the optimum average improvement for the sizes for which the optimum could be computed using GeoSteiner. Results on VLSI instances show that the relative performance of the heuristics is the same to that observed on random instances. However, the improvement over the rectilinear MST and the gaps between heuristics are smaller in this case.

The results in Table 37.1 show that the BGA is highly scalable. Even though batched greedy is not as fast as the MST or the Prim-based heuristic of Reference 29, it can easily handle up to hundreds of thousands of terminals in minutes of CPU time. Compared with Borah's $O(n^2)$ implementation of the edge-based

TABLE 37.1   Percent Improvement Over MST and CPU Time of the Compared Rectilinear Steiner Tree algorithms

#Term.	MST		Prim-Based		Edge-Based		Batched Greedy		GeoSteiner 4.0	
	Len.(μm)	CPU	%Imp.	CPU	%Imp.	CPU	%Imp.	CPU	%Imp.	CPU
Random instances (average results over 10 instances)										
100	85,169.9	0.0005	9.78	0.001	10.97	0.006	10.99	0.003	11.66	0.555
500	184,209.7	0.0036	10.08	0.007	11.12	0.216	11.17	0.081	11.76	15.205
1000	258,926.8	0.0079	10.04	0.014	10.96	0.939	10.99	0.230	11.61	117.916
5000	573,178.8	0.0501	10.02	0.082	11.02	56.348	11.05	1.903	–	–
10,000	809,343.5	0.1268	10.04	0.191	11.01	415.483	11.05	5.192	–	–
50,000	1,808,302.7	1.2330	10.05	1.320	11.01	16,943.777	11.06	69.043	–	–
100,000	2,555,821.9	3.1150	10.08	3.143	11.04	61,771.928	11.08	195.589	–	–
500,000	5,710,906.8	22.9130	10.07	20.570	–	–	11.08	1706.765	–	–
VLSI instances										
337	247.7	0.0020	5.96	0.000	6.50	0.060	6.43	0.040	6.75	16.040
830	675.6	0.0055	3.10	0.010	3.19	0.320	3.20	0.080	3.26	9.480
1944	452.2	0.0165	6.86	0.040	7.77	3.640	7.85	0.400	8.15	1304.270
2437	578.8	0.0217	7.09	0.040	7.96	5.740	7.96	0.680	8.34	13,425.310
2676	887.2	0.0235	8.07	0.040	8.99	5.340	8.93	0.770	9.38	430.800
12,052	2652.7	0.1378	7.65	0.180	8.46	540.840	8.45	5.230	–	–
22,373	13,962.5	0.3419	8.99	0.480	9.83	2263.760	9.85	13.060	–	–
34,728	9900.5	0.5455	8.16	0.690	9.01	5163.060	9.05	24.200	–	–

TABLE 37.2   Percent Improvement Over MST and CPU Time of the Compared Octilinear Steiner Tree Algorithms

#Term.	MST		Edge-Based		Batched Greedy		GeoSteiner 4.0	
	Len.(µm)	CPU	%Imp.	CPU	%Imp.	CPU	%Imp.	CPU
Random instances (average results over 10 instances)								
100	72,375.1	0.0005	4.28	0.530	4.43	0.010	4.75	11.608
500	155,611.7	0.0036	4.12	13.410	4.29	0.118	4.60	311.991
1000	219,030.8	0.0079	4.12	54.641	4.25	0.296	4.59	1321.382
5000	484,650.5	0.0506	4.17	1466.296	4.31	2.820	–	–
10,000	684,409.5	0.1217	4.13	5946.815	4.28	8.362	–	–
50,000	1,528,687.2	1.1940	4.16	147,210.395	4.30	116.419	–	–
100,000	2,160,629.4	3.1060	–	–	4.32	476.307	–	–
500,000	4,826,839.1	23.0610	–	–	4.31	6578.840	–	–
VLSI instances								
337	219.0	0.0020	2.92	5.690	2.99	0.050	3.13	72.960
830	630.4	0.0055	0.93	27.610	0.90	0.120	1.07	195.190
1944	407.2	0.0167	3.33	202.030	3.47	0.750	4.01	5279.870
2437	523.1	0.0218	3.67	345.330	3.77	0.820	4.29	7484.730
2676	780.2	0.0236	3.41	392.340	3.51	1.310	3.89	6080.050
12052	2372.3	0.1417	3.63	7517.680	3.72	10.800	–	–
22373	12,069.8	0.3447	3.65	25,410.340	3.74	21.380	–	–
34728	8724.9	0.5427	3.64	62,971.090	3.74	25.160	–	–

heuristic, batched greedy is two or more orders of magnitude faster as soon as the number of terminals gets into the tens of thousands.

The BGA can be easily adapted to other cost metrics. We have implemented and experimented with an octilinear version of the BGA, modifications to other uniform orientation geometries can be done easily. The only required modifications are in the distance formula and in the procedure for finding the optimum Steiner point of a triple. The octilinear distance between points (x, y) and (x', y') is equal to $\max\{|x - x'|, |y - y'|\} + (\sqrt{2} - 1) \min\{|x - x'|, |y - y'|\}$; this is always smaller than the rectilinear distance, $|x - x'| + |y - y'|$, unless the two points are on the same horizontal or vertical line, in which case the two distances are equal. The computation of the optimum Steiner point of a triple in the octilinear metric can be done in constant time using the method described in Reference 3.

Table 37.2 gives results obtained by the octilinear versions of the Guibas–Stolfi MST, $O(n^2)$ edge-based, batched greedy, and GeoSteiner 4.0 [31] algorithms. Octilinear batched greedy is almost always better than the octilinear edge-based heuristic, and very close to optimum for the instances for which the latter is available. Furthermore, octilinear batched greedy remains highly scalable, with just a small factor increase in runtime compared with the rectilinear version.

37.6 Conclusion

Noncritical nets with tens of thousands of terminals are becoming more common in modern designs due to the increased emphasis on design for test. Even a single net of this size can render quadratic Steiner tree algorithms impractical, given the stringent constraints on routing runtime (e.g., designers expect full chip global and detailed routing to be completed overnight). In this chapter, we have given a high-quality $O(n \log^2 n)$ heuristic that can practically handle these nets without compromising solution quality. Since our heuristic is graph-based, it can be easily modified to handle other practical considerations, such as routing obstacles, preferred directions, and via costs. A C++ implementation of

BGA is freely available as part of the MARCO Gigascale Silicon Research Center VLSI CAD Bookshelf at http://vlsicad.ucsd.edu/GSRC/bookshelf/Slots/RSMT/FastSteiner/.

Acknowledgments

A preliminary version of this work has appeared in Reference 23. The authors wish to thank Manjit Borah, Benny Nielsen, André Rohe, and Martin Zachariasen for giving us access to their implementations, Charles Alpert for providing the VLSI benchmarks, and Jarrod Alexander Roy and Igor Markov for correcting a bug in our original implementation of the BGA algorithm. This work has been supported in part by Cadence Design Systems, Inc., the MARCO Gigascale Silicon Research Center, NSF Grant CCR-9988331, Award No. MM2-3018 of the Moldovan Research and Development Association (MRDA) and the U.S. Civilian Research & Development Foundation for the Independent States of the Former Soviet Union (CRDF), and by the State of Georgia's Yamacraw Initiative.

References

1. Chen, H., Cheng, C.K., Kahng, A.B., Măndoiu, I.I., and Wang, Q., Estimation of wirelength reduction for λ-geometry vs. Manhattan placement and routing, in *Proceedings of ACM International Workshop on System-Level Interconnect Prediction (SLIP)*, ACM Press, New York, 2003, p. 71.
2. Chen, H., Cheng, C.-K., Kahng, A.B., Măndoiu, I.I., Wang, Q., and Yao, B., The Y-architecture for on-chip interconnect: Analysis and methodology, *IEEE Trans. CAD*, 24, IEEE Press, New York, 588, 2005.
3. Koh, C.-K. and Madden, P.H., Manhattan or non-Manhattan? A study of alternative VLSI routing architectures, in *Proceedings of Great Lakes Symposium on VLSI*, 2000, p. 47.
4. Scepanovic, R., Koford, J.S., Kudryavstev, V.B., Andreev, A.E., Aleshin, S.V., and Podkolzin, A.S., Microelectronic Integrated Circuit Structure and Method Using Three Directional Interconnect Routing Based on Hexagonal Geometry, U.S. Patent, No. US5578840, November 1996.
5. Teig, S., The X architecture: Not your father's diagonal wiring, in *Proceedings of ACM/IEEE Workshop on System Level Interconnect Prediction (SLIP)*, 2002, p. 33.
6. Teig, S. and Ganley, J.L., Method and apparatus for considering diagonal wiring in placement, *International Application*, No. WO 02/47165 A2, 2002.
7. http://www.xinitiative.org
8. Sarrafzadeh, M. and Wong, C.K., Hierarchical Steiner tree construction in uniform orientations, *IEEE Trans. CAD*, 11, 1095, 1992.
9. Koh, C.K., Steiner problem in octilinear routing model, Master's Thesis, National University of Singapore, Singapore, 1995.
10. Li, Y.Y., Cheung, S.K., Leung, K.S., and Wong, C.K., Steiner tree constructions in λ_3-metric, *IEEE Trans. Circuits Syst. II, Analog Digit. Signal. Process*, 45, 563, 1998.
11. Hanan, M., On Steiner's problem with rectilinear distance, *SIAM J. Appl. Math.*, 14, 255, 1966.
12. Chiang, C., Su, Q., and Chiang, C.-S., Wirelength reduction by using diagonal wire, in *Proceedings of Great Lakes Symposium on VLSI*, IEEE Press, New York, 2003, p. 104.
13. Coulston, C., Constructing exact octagonal Steiner minimal trees, in *Proceedings of Great Lakes Symposium on VLSI*, IEEE Press, New York, 2003, p. 1.
14. Nielsen, B.K., Winter, P., and Zachariasen, M., An exact algorithm for the uniformly-oriented Steiner tree problem, in *Proceedings of the 10th European Symposium on Algorithms*, LNCS, 2461, Springer-Verlag, Berlin, Heidelberg, 2002, p. 760.
15. Scheffer, L., Rectilinear MST code, available as part of the RMST-Pack at http://www.gigascale.org/bookshelf/slots/RSMT/RMST/

16. Zhou, H., Shenoy, N., and Nicholls, W., Efficient minimum spanning tree construction without Delaunay triangulation, in *Proceedings of Asia-Pacific Design Automation Conference (ASP-DAC)*, IEEE Press, New York, 2001, p. 192.

17. Kahng, A.B. and Robins, G., A new class of iterative Steiner tree heuristics with good performance, *IEEE Trans. CAD*, 11, 1462, 1992.

18. Măndoiu, I.I., Vazirani, V.V., and Ganley, J.L., A new heuristic for rectilinear Steiner trees, *IEEE Trans. CAD*, 19, 1129, 2000.

19. Yildiz, M.C. and Madden, P.H., Preferred direction Steiner trees, in *Proceedings of Great Lakes Symposium on VLSI*, IEEE Press, New York, 2001, p. 56.

20. Borah, M., Owens, R.M., and Irwin, M.J., A fast and simple Steiner routing heuristic, *Disc. Appl. Math.*, 90, 51, 1999.

21. Fößmeier, U., Kaufmann, M., and Zelikovsky, A., Faster approximation algorithms for the rectilinear Steiner tree problem, *Discrete Comput. Geom.*, 18, 93, 1997.

22. Zelikovsky, A., An 11/6-approximation for the network Steiner tree problem, *Algorithmica*, 9, 463, 1993.

23. Kahng, A.B., Măndoiu, I.I., and Zelikovsky, A.Z., Highly scalable algorithms for rectilinear and octilinear Steiner trees, in *Proceedings of Asia and South Pacific Design Automation Conference*, IEEE Press, New York, 2003, p. 827.

24. Zhu, Q., Zhou, H., Jing, T., Hong, X.-L., and Yang, Y., Spanning graph based non-rectilinear Steiner tree algorithms, *IEEE Trans. CAD*, 24, 1066, 2005.

25. Gilbert, E.N. and Pollak, H.O., Steiner minimal trees, *SIAM J. Appl. Math.*, 32, 826, 1977.

26. Berman, P., Fossmeier, U., Kaufmann, M., Karpinski, M., and Zelikovsky, A., Approaching the 5/4-approximation for rectilinear Steiner trees, in *Proceedings of the European Symposium on Algorithms*, LCNS, Springer-Verlag, Berlin Heidelberg, 762, 1994, p. 533.

27. Cattaneo, G., Faruolo, P., Petrillo, U.F., and Italiano, G.F., Maintaining dynamic minimum spanning trees: An experimental study, in *Proceedings of 4th International Workshop on Algorithm Engineering and Experiments (ALENEX)*, LCNS, 2409, Springer-Verlag, 2002, p. 111.

28. Hwang, F.K., Richards, D.S., and Winter, P., *The Steiner Tree Problem*, Volume 53 of *Annals of Discrete Mathematics*, Elsevier, North-Holland, the Netherlands, 1992.

29. Rohe, A., Sequential and parallel algorithms for local routing, PhD Thesis, Bonn University, Bonn, Germany, 2001.

30. Guibas, L.J. and Stolfi, J., On computing all north-east nearest neighbors in the L_1 metric, *Inform. Process. Lett.*, 17, 219, 1983.

31. Warme, D.M., Winter, P., and Zacharisen, M., GeoSteiner 3.1 package, available from http://www.diku.dk/geosteiner/

Algorithms for Chromatic Sums, Multicoloring, and Scheduling Dependent Jobs

Magnús M. Halldórsson

Guy Kortsarz

38.1 Introduction

This survey deals with problems at the intersection of two scientific fields: graph theory and scheduling. They can either be viewed as scheduling *dependent jobs*—jobs with resource conflicts—or as graph coloring optimization involving different objective functions.

Our main aim is to illustrate the various interesting algorithmic techniques that have been brought to bear. We will also survey the state-of-the-art, both in terms of approximation algorithms, lower bounds, and polynomial time solvability.

We first formulate the problems, both from a scheduling and from a graph theory perspective, before setting the stage.

38.1.1 Scheduling Perspective

The input consists of *jobs* that need to be *processed* on *machines*. There are enough machines available (although the case of limited number of machines can also be handled). We view the processing to take place in *time steps* or *rounds*. Each job J has an integral *length* or processing time $x(J)$, so it must be processed by a machine during $x(J)$ of the rounds. The task is to process the jobs on the machines as quickly as possible. What makes the task nontrivial is that each job requires an *exclusive* access to a set of resources. (That means that two jobs that require the same resource cannot be processed simultaneously.) Therefore, the jobs processed during a given round must use disjoint sets of resources.

There are two essential modes of processing the jobs. In the *nonpreemptive* setting, once a job is processed, it must be executed to completion. That means that a job is always processed in consecutive rounds. In the *preemptive* case, there are no constraints of this sort. Finally, in the important special case of *unit-length jobs*, each job is processed only for a single round.

We will consider two objective functions: minimizing the total length of the schedule and minimizing the average completion time. Given a legal schedule, let $f(J)$ be the time J is finished. From the view of the *system*, it is preferable to finish the last job early. This is known as minimizing the *makespan* of the schedule or $\max_J f(J)$. The *users* that "own" the jobs are most concerned that *their* job is finished early. The natural objective to consider for that is to minimize the *sum of completion times*, $\sum_J f(J)$, which is equivalent to minimizing the *average* finish time of a job.

38.1.2 A Graph Theoretic Description

An alternative view of the above-mentioned problems is as *multicoloring* problems, with colors being positive integers. The number $x(J)$ is then called the *color requirement* of J. The jobs naturally define a conflict graph with vertices representing jobs, and with an edge between a pair of jobs that require the same resource. Using graph terminology, a round is *an independent set* (namely a set of vertices no two of which share an edge). Thus, a schedule is an ordered collection of independent sets, or a *multicoloring* (not just a *coloring* since vertices can require multiple colors). Each vertex (job) v is to receive $x(v)$ colors [i.e., belong to $x(v)$ independent sets], so that adjacent vertices receive disjoint sets of colors. In *preemptive multicoloring problems*, there is no restriction on the $x(v)$ colors that each vertex can receive, except that sets assigned to adjacent vertices must be disjoint, as before. A different problem variant occurs when we require the colors assigned to a vertex to form a contiguous interval—we refer to such an assignment as a *nonpreemptive multicoloring*.

We consider two objective functions. We assume that the colors are enumerated as the natural numbers, $1, 2, 3, \ldots$. The *makespan* corresponds to the number of colors used, which equals the largest color used. The *chromatic sum* criteria corresponds to $\sum_v f(v)$, with $f(v)$ being the largest color given to vertex v.

A coloring with few colors does not necessarily imply a low sum coloring. There are examples for which using the minimum $\chi(G)$ number of colors gives a sum that is $\Theta(\sqrt{n})$-factor larger than the minimum sum.

There are also examples in which preemptive coloring leads to a strictly smaller color sum. Consider a path $A - B - C$ with color requirements $x(A) = 1, x(B) = 2, x(C) = 5$. The reader may verify that every nonpreemptive coloring has sum at least 12, whereas there is a preemptive coloring with sum 11.

38.1.3 Notation

The input is an undirected, unweighted graph $G = (V, E)$. For $S \subseteq V$, let $\mathcal{S}(S) = \sum_{v \in S} x(v)$ be the sum of the color requirements of nodes in S, and let $\mathcal{S}(G) = \mathcal{S}(V)$. Let p denotes the largest color requirement. Let $\chi(G)$ be the minimum number of colors required to (multi)color G. Let $N(v)$ be the set of neighbors of vertex v. We may intermix the scheduling and graph theory vocabulary in the remainder.

We denote SC for minimum *sum coloring* problem [where $x(v) = 1$ for every v]; pMC (npMC) for preemptive (nonpreemptive) minimum *makespan multicoloring*; and pSMC (npSMC) for preemptive (nonpreemptive) *sum multicoloring* problem, respectively.

38.1.4 Applications for Sum (Multi)Coloring

38.1.4.1 Wire Minimization in VLSI Design

In wire minimization in VLSI design, [39], terminals lie on a single vertical line, where each terminal is represented by a short interval on this line. To its right are vertical bus lines, separated with unit spacing. Pairs of terminals are to be connected via horizontal wires on each side to a vertical lane, with nonoverlapping pair utilizing the same lane. With the vertical segments fixed, the wire cost corresponds to the total length of horizontal segments. Numbering the lanes in increasing order of distance from the terminal line, lane assignment to a terminal corresponds to coloring the terminal's interval by an integer. The wire-minimization problem then corresponds to sum coloring of an interval graph.

38.1.4.2 Storage Allocation

Storage allocation is a problem of minimizing the total distance traveled by a robot moving in a warehouse [42]. Goods are located in a single corridor from the delivery spot. They are checked in and out at known times. Two robots can move only if their intervals do not intersect. Since the starting point and destinations are on a line, this gives the problem of multicoloring *interval graphs*.

38.1.4.3 Session Scheduling on a Path

In a path network, pairs of vertices need to communicate, for which they need use of the intervening path. If two paths intersect, the corresponding sessions cannot be held simultaneously. In this case, it would be natural to expect the sessions (i.e., jobs) to be of different lengths, leading to the sum multicoloring problem on interval graphs

38.1.4.4 Resource-Constrained Scheduling

Say that we are given a collection of n jobs of integral lengths and a collection of *resources*. We assume that each job requires an exclusive access to particular subset of the resources to execute. The resource-constrained scheduling problem can then be modeled as a multicoloring problem on the conflict graph. We address here the case where each task uses up to k resources. The conflict graph is then $k + 1$-*claw free*, that is, does not contain the star $K_{1,k+1}$ as an induced subgraph, since disjoint vertices must use disjoint sets of resources. A natural example of a limited resource is processors in a multiprocessor system. In the biprocessor task scheduling problem, we are given a set of jobs and a set of processors, and each job requires the exclusive use of two dedicated processors. We are interested in finding a schedule which minimizes the sum of completion times of the jobs. In scheduling terms, this problem is denoted by $P|\text{fix}_j| \sum C_j$, with $|\text{fix}_j| = 2$. In the special case of two resources per task, such as the biprocessor task scheduling problem, the conflict graph is a line graph. Another application of sum multicoloring of $(k+1)$-claw free graphs is scheduling data migration over a network, also known as the *file transfer problem* [10,28]. Suppose that the network is fully connected, and we have a set of files $f_1, \ldots, f_M$; each file f_i needs to migrate from a source s_i to a destination t_i. During migration, f_i requires the exclusive access to s_i and t_i, for a prespecified time interval; thus, f_i and f_j are in conflict if $\{s_i, t_i\} \cap \{s_j, t_j\} \neq \emptyset$. Our goal is to find a migration schedule that minimizes the sum of completion times, where the sum is taken over all files. This translates to the sum multicoloring problem on the conflict graph of the files, which is a line graph.

38.1.5 Approximation Algorithms and Inapproximability

We deal primarily with minimization problems. The goal is to find an s_i with $v(s_i)$ minimum. All the problems we consider are *NP*-hard. Since *NP*-hard problems are unlikely to admit a polynomial time solution, we need to settle for an approximate solutions. Let $opt(I)$ be the value of the best solution for problem instance I. For a minimization (respectively maximization) problem P, an algorithm is called a *ρ-approximation algorithm*, if it runs in time polynomial in the input size, and for every input I, the algorithm outputs a solution of cost at most $\rho \cdot opt(I)$ [respectively, at least $opt(I)/\rho$].

A *lower bound ρ* on the approximability of a problem is a proof that approximating the problem within factor ρ in polynomial time would give a polynomial time algorithm for solving the problem exactly. Since the latter problem is *NP*-hard, so is the former.

A *polynomial-time approximation scheme* (PTAS) for a minimization problem is an algorithm that, given a constant parameter ϵ, produces a $(1 + \epsilon)$-approximate solution for every ϵ. The running time should be polynomial for ϵ constant, but could, for example, be $n^{f(1/\epsilon)}$, for some arbitrary function f, where n is the size of the instance. A *fully polynomial time approximation scheme* (FPTAS) for P is an $1 + \epsilon$-approximation algorithm that runs in time polynomial in n and in $1/\epsilon$.

38.1.6 Graph Classes

We define here the main classes of graphs that we will encounter.

Perfect graphs [20] are those for which the clique number equals the chromatic number, in every induced subgraph. *Chordal graphs* are graphs that contain no cycle of size 4 or larger as an induced subgraph. An *interval graph* is a special case of a chordal graph: vertices correspond to intervals on a line with two vertices adjacent whenever their intervals intersect. *Bipartite graphs* are the 2-colorable graphs.

A *k-tree* is a graph that can be formed by the following process. Starting with a $k + 1$-clique (complete graph on $k + 1$ vertices), and keep adding vertices adjacent to some k-clique. A *partial k-tree* is a subgraph of a k-tree. These graphs are exactly the graphs of *treewidth k* (the definition via treewidth is more complex and outside our scope).

A *line graph* has a vertex for each edge of an underlying graph, with adjacencies between edges that intersect. A graph is $k + 1$-claw free if it does not contain the star $K_{1,k+1}$ as an induced subgraph.

Planar graphs are those that can be drawn on a plane with no two edges intersecting. *Hexagon graphs* are a subclass of planar graph defined in Reference 36, with important applications for cellular networks. A *unit disc graph* [8] has vertices represented by unit circles in the plane, with two vertices being adjacent if the circles intersect.

Outline: In the next section, we overview quickly the state-of-the-art on sum coloring and multicoloring problems. We then devote a section to each of the major approaches used for tackling sum (multi)coloring problems: Greedy approaches in Section 38.3, randomized "doubling" in Section 38.4, partitioning in Section 38.5, and delays in Section 38.6. We then close with a few open problems.

38.2 The State-of-the-Art

We briefly summarize the known results on makespan coloring and sum coloring problems.

38.2.1 Makespan Problems

(Multi)coloring a *line graph* is equivalent to (multi)coloring the edges of a graph, so that adjacent edges are assigned disjoint sets of colors.

For certain graphs, the pMC problem can be reduced to the ordinary coloring problem. A vertex v of color requirement $x(v)$ is replaced by a clique of $x(v)$ vertices (connecting a copy of v to a copy of u if u and v are connected). This reduction is polynomial if p is polynomial in n, but can often be done implicitly for large values of p. This gives an optimum algorithm for pMC on certain perfect graphs, such

TABLE 38.1 Known Results for Multicoloring Problems

	Coloring		Multicoloring	
	u.b.	l.b.	pMC	npMC
General graphs	$O^*(n/\log^3 n)$ [22]	$n^{1-\epsilon}$ [14]	$O^*(n/\log^3 n)$ [22]	
Perfect graphs	1 [20]			
Bipartite graphs	1 Folklore		1 Folklore	1 Folklore
Chordal graphs	1 [19]		1 [20]	
Interval graphs	1 [19]		1 [20]	$2 + \epsilon$ [9]
Partial k-trees	1 [2]		1 [27]	1 [27]
Planar graphs	4/3 [1]	NPC [10,19]	11/6 [32]	
Hexagon graphs	4/3 [36,38]	NPC [36]	4/3 [36]	
Line graphs	Opt+1 [41]	NPC [21]	1.1 [37]	

TABLE 38.2 Known Results for Sum (Multi)Coloring Problems

	SC		SMC	
	u.b.	l.b.	pSMC	npSMC
General graphs*	$n/\log^3 n$ [4,13]	$n^{1-\epsilon}$ [4,14]	$n/\log^3 n$ [6,13]	$n/\log^2 n$ [6]
Perfect graphs	3.591 [16]	APX [7]	5.436 [16]	$O(\log n)$ [6]
Interval graphs	1.796 [16]	APX [40]	5.436 [16]	$11.273 + \epsilon$ [16]
Bipartite graphs	27/26 [34]	APX [7]	1.5 [6]	2.8 [6]
Partial k-trees	1 [27]		PTAS [23]	FPTAS [23]
Planar graphs	PTAS [23]	NPC [23]	PTAS [23]	PTAS [23]
Trees	1 [29]		PTAS [24]	1 [24]
Line graphs	1.8298 [26]	NPC [4]	2 [6]	7.682 [16]
Line graphs of trees	1		PTAS [35]	
$k + 1$-claw free graphs	k	NPC [4]	k [8]	$1.796k^2 + 0.5$
Intersection of k-sets	k [4]	NPC [4]	k [6]	$3.591k + 0.5$ [16]
Unit disc graphs	2 [8]	NPC [8]	2 [8]	

as *chordal graphs*. But large cliques cannot be introduced to *bipartite* (or 2-colorable) graphs, hence the reduction does not work for general perfect graphs. Similarly, one cannot introduce large cliques into a planar graphs.

The above-mentioned reduction also does not work for the nonpreemptive case. For a summary of known results for minimum coloring (both upper and lower bounds), pMC, and npMC see Table 38.1. The $O^*()$ notation hides *poly*(log log n) factors NPC means the problem is NP-complete, while APX indicates that it is hard to approximate within some constant factor greater than 1.

38.2.2 Sum Coloring Problems

The SC problem was first studied explicitly by Kubicka [29]. Efficient algorithms have been given for trees [29], partial k-trees [27], and regular bipartite graphs [33]. See Table 38.2 for a summary of best results known.

38.3 Sum Coloring Fundamentals

We examine first two of the simplest and most natural solution strategies for sum coloring.

38.3.1 Canonical Colorings

The most basic property of a good sum coloring is that it be *minimal*: no vertex can be moved to a smaller color without destroying the coloring property. Namely, each vertex colored c should have neighbors colored $1, 2, \ldots, c - 1$. Such a coloring is sometimes called *canonical*.

We can observe that a canonical coloring has sum at most $n + m$, where n (m) is the number of vertices (edges), respectively. Form an arbitrary order of the vertices that is consistent with the canonical coloring, and let b_i denotes the number of neighbors of vertex i that precede it in the ordering. Observe that each edge (i, j) contributes exactly one to $b_i + b_j$; therefore, the sum equals the number of edges, $\sum_i b_i = m$. We can also note that the color of i is at most $b_i + 1$, and thus the color sum is at most $\sum_i (b_i + 1) = m + n$. This is also equivalent to $\frac{d+2}{2} n$, where d is the average degree of the graph, for an approximation ratio of at most $\frac{d+2}{2}$ [30].

This can be extended to an "everywhere sparse" parameter. A graph has *inductiveness* (or *degeneracy*) D if there is an ordering of the vertices such that each vertex has at most D neighbors that precede it in the ordering. Such an ordering can be found by repeatedly removing the minimum degree vertex. If we then color the vertices in this order, using the smallest available color for each vertex, the resulting chromatic sum is at most $\frac{D+2}{2} n$.

38.3.2 Greedy Colorings

A natural heuristic is to color the vertices one color set at a time, each time finding largest possible independent set in the remaining graph. After all, the more vertices that are colored with the first color, the smaller the sum is likely to be. We call this the *Greedy* algorithm.

More formally, let R_i be the set of vertices that have not been colored by *Greedy* with the first $i - 1$ colors. Initially, clearly $R_0 = V$. In each round $i = 1, 2, \ldots$, *Greedy* finds a maximum independent set X_i in the graph induced by R_i, assigns those vertices the color i, and updates $R_{i+1} = R_i \setminus X_i$.

Theorem 38.1 ([4]) *Greedy achieves 4-approximation for* SC [4].

We give a simpler proof from Reference 15 that was stated for the related *Sum Set Cover* problem.

The proof idea is to compare the areas of geometric figures that represent the two colorings, the greedy and the optimal coloring. We form a *histogram* for both colorings, with a column for each vertex of unit width and height proportional to some measure of the contribution of that vertex to the objective function. The key argument is showing that if we reduce the scale of the greedy histogram by a factor of two, along both the x- and y-dimensions, then it will fit inside the optimal histogram. Since the areas of the original histograms correspond to the color sums, this immediately shows that the sum of the greedy coloring is at most four times that the optimal chromatic sum.

The optimal histogram has a column for each vertex of unit width and height equivalent to the color assigned. It follows immediately that the area equals the color sum. We order the columns of the histogram in nondecreasing height, which corresponds to the order of the vertices in the optimal coloring.

Define the *price* of X_i as $p_i = |R_i|/|X_i|$. The greedy histogram has a unit-width column for each vertex; the height of the column is p_i if the vertex was colored i by *Greedy* (i.e., if contained in X_i). The columns are ordered by the greedy sets X_i, and hence, the heights of rectangles in the greedy histogram are not necessarily monotone.

Let $Gr = \sum_{i=1}^{k} i|X_i|$ be the sum and k be the number of colors of the coloring returned by *Greedy*. Note that $R_i = \cup_{j \geq i} X_j$.

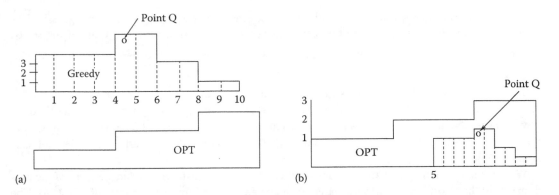

FIGURE 38.1 Optimum and greedy histograms (a); the shrunk greedy histogram aligned with the optimal one (b).

Claim 38.1 *The area of the greedy histogram equals $Gr = \sum_{i=1}^{k} p_i |X_i|$.*

Proof. Since for each vertex in X_i there is a column of height p_i, the area of the greedy histogram is clearly $\sum_{i=1}^{k} p_i |X_i| = \sum_{i=1}^{k} |R_i|$, applying the definition of p_i. On the other hand, the coloring of each X_i by Greedy delays the remaining vertices by R_i; thus, $Gr = \sum_{i=1}^{k} |R_i|$. ∎

Consider Figure 38.1.

We now scale the greedy histogram by halving both the height and width of each column. Thus, in this *shrunk greedy histogram*, the columns corresponding to vertices in X_i have width $1/2$ and height $p_i/2$. We position the shrunk greedy histogram over the optimal histograms so that their lowest right points is collocated. For example, in Figure 38.1, the greedy histogram has 10 rectangles, ranging from $x = 0$ to $x = 10$, so we place the shrunk greedy histogram so that it starts at point $x = 5$.

To establish our claim, we show that the shrunk histogram is completely contained inside the optimal one. To this end, consider an arbitrary point Q inside the shrunk histogram; see Figure 38.1 for an example. Let i be the round in which *Greedy* colored the corresponding vertex (whose rectangle contains Q). Let h be the height of Q from the baseline, and note that h is at most the height of its rectangle in the shrunk greedy solution, or $p_i/2$. Recall the set R_i of vertices that *Greedy* has not colored by the start of round i. Let r be the distance from Q to the right edge of the shrunk greedy histogram (rounded down to the nearest integer). Note that there are $|R_i|$ vertices left to be colored when *Greedy* starts round i, and thus there are at most $|R_i|$ columns to the right of Q in the greedy histogram. In the shrunk histogram, where each is of width $1/2$, that makes for a distance at most $|R_i|/2$; thus, $r \leq |R_i|/2$.

The issue boils down to the height of the rth last column in the optimal histogram, which equals to the number of colors used in the optimal solution for the first $n - r$ vertices. The greedy rule implies that the largest independent set among the vertices in R_i is of size $|X_i|$. To color all but $|R_i|/2 \geq r$ vertices of R_i, the optimal sum coloring uses at least $|R_i|/(2|X_i|) = p_i/2$ colors. Thus, to color all but r vertices of the whole graph, the optimal solution must use at least $p_i/2$ colors. Hence, the height of the column containing Q in the optimal histogram is at least $p_i/2$, and Q is therefore contained inside the optimal histogram. This completes the proof of 4-approximation.

A construction is given in Reference 5 showing that the performance ratio of Greedy is in fact no better than 4.

38.4 Slicing and Dicing Using Randomized Geometric Choices

38.4.1 A Two-Person Game

We will introduce the technique as the following two-person game. Player 1 (the adversary) writes down a secret, a positive real $z \geq 1$, on a piece of paper and folds it. Player 2 (the algorithm) then produces a series of guesses $x_0, x_1, \ldots$ until a guess exceeds z, that is, $x_k \geq z$, for some guess k. The goal of the algorithm is to minimize the total sum of the guesses, *relative* to the value of the secret: minimize $\frac{\sum_{i=0}^{k} x_i}{z}$. As the algorithm must perform well on every z, it effectively seeks to minimize $\max_z \frac{\sum_{i=0}^{k} x_i}{z}$.

The natural deterministic strategy is the classic doubling trick: select $x_i = 2^i$, for $i = 0, 1, \ldots$. This strategy has a worst case *competitive* ratio of 4. The worst case occurs when x_{k-1} just barely undershoots z, in which case the final guess x_k overshoots by a factor of 2; the other factor of 2 comes from the cost of the previous guesses $\sum_{i=0}^{k-1} x_k$ in comparison with x_k. This turns out to be the best possible deterministic strategy [31].

With randomization against an oblivious adversary (one that fixes its choice independent of the random bits of the algorithm), we can do better, obtaining an expected competitive ratio of $e \sim 2.781$. The idea is to average out the cost of overshooting by randomly selecting a starting guess; to do so optimally, we additionally change the base from 2 to e. Choose α uniformly at random from $[0, 1)$ and select guesses $x_i = e^{i+\alpha}$. The last guess is the smallest k for which $x_k = e^{k+\alpha} \geq z$, or $k + \alpha \geq \ln z$. The random variable $k + \alpha - \ln z$ is distributed like α, uniformly in $[0, 1)$, and thus $x_k/z = e^{k+\alpha-\ln z}$ has expected value $E(\alpha) = e - 1$. This means that in expectation, the overshoot is a factor of $e - 1$. The previous guesses contribute a further multiplicative factor of $e/(e - 1)$, since

$$\sum_{i=0}^{k} x_i = \frac{e^{k+\alpha+1} - 1}{e - 1} < \frac{e \cdot e^{k+\alpha}}{e - 1} = \frac{e}{e - 1} \cdot x_k.$$

This technique is often applied when the guesses and the answer must be integers; it is then simple to just round the guesses down.

38.4.2 Using the Game for Sum (Multi)Coloring

An easy application of this game is in solving npSMC on easily colorable graphs. Let us illustrate it on bipartite graphs, using the *color requirement* of each vertex as the value used in the game. We obtain subgraphs $V_0, V_1, \ldots$, where V_i consists of the vertices with color requirement in $(x_{i-1}, x_i]$. We then two-color the underlying subgraph induced by V_i (i.e., compute a bipartition), giving a multicoloring of the subgraph using at most $2x_i$ colors. Finally, the colorings of the subgraphs are concatenated to form the final schedule.

The doubling strategy then ensures that *each vertex v is fully colored within the first $8x(v)$ colors*. This is obviously an 8-approximation, since the vertex cannot be completed before color $x(v)$. Using the randomized strategy, the *expected cost for each vertex is $2e \cdot x(v)$*, resulting in a coloring of expected cost $2e \cdot S(G)$. (It can also be derandomized.) By a finer analysis and by adjusting the multiplier appropriately, the bound can be improved to $2.796S(G)$ [6].

38.4.3 Ways of Applying the Technique

The first application of this technique to sum coloring problems was the above-mentioned application for npSMC of bipartite graphs, using the color requirements. The slicing technique can, perhaps surprisingly, be used in two additional ways.

Another way is to use the fractional values of a linear programming relaxation. This has resulted in $2e$-approximation for pSMC on perfect graphs, a 7.7-approximation for npSMC for line graphs and 11.3-approximation on interval graphs, and 3.6-approximation for SC on perfect graphs [16].

The third way is to use a subroutine for the *maximum induced k-colorable subgraph* problem, letting x_i denote the value of k in the ith call. This problem is polynomially solvable in interval graphs and comparability graphs [25]. That leads to constant-approximation for SC on interval, comparability, co-comparability, trapezoid, and permutation graphs [25].

Interestingly, all three variations of the technique were applied in Reference 12 to the same problem, which is a variant of npSMC where jobs are run in batches and no new job can be started before the whole previous batch completes.

38.5 Approximation Schema via Instance-Sensitive Partitioning

The randomized geometric partitioning technique is oblivious in that the actual distribution of values in the instance does not affect the partitioning. For finer approximation, we may need to construct the partition more carefully. Let us examine when we can hope to obtain a PTAS.

As before, we aim to find breakpoints $b_1, b_2, \ldots$, that define subproblems $V_1, V_2, \ldots$, where V_i consists of vertices with color requirements in the range $(b_{i-1}, b_i]$ (with $b_0 = 0$). We compute a schedule for each V_i, and concatenate them in sequence. That is, if we use c_i colors for subproblem V_i and denote $C_i = c_1 + c_2 + \cdots + c_i$, then we schedule the coloring of V_i during rounds $C_{i-1} + 1, C_{i-1} + 2, \ldots, C_{i-1} + c_i$.

The impact of selecting a particular breakpoint b_i defining a subproblem V_i is that it delays the rest of the vertices (with color requirement larger than b_i). We can compute this *delay cost* as $Delay(b_i) = g(b_i) \cdot C(V_i)$, where $g(b) = |\{v : x(v) > b\}|$ equals the number of vertices with larger color requirement and $C(V_i)$ is the length of the schedule computed for subproblem V_i. The total delay cost of a sequence of breakpoints is then $Delay = \sum_i Delay(b_i) = \sum_{i=1} g(b_i) \cdot C(V_i)$. The key question is how large $Delay$ is compared with the cost of the optimal solution Opt; as a simple bound, we can use that it is at least the sum of the optimal costs of the subproblems, $Opt \geq \sum_i Opt(V_i)$.

To this end, we need a schedule that is *short* (i.e., of small makespan), while at the same time, the coloring on each subproblem must also be efficient, with a good approximation of the chromatic sum. We therefore need a *simultaneous approximation* of the sum and the makespan objectives.

Recall that $S(G) = \sum_{v \in V} x(v)$ denotes the sum of color requirements of a graph G and p denotes the largest color requirement.

The following result achieves the desired partitioning.

Proposition 38.1 (The breakpoint lemma [24, Prop. 2.13]) *Let $X = \{x_1, \ldots, x_m\}$ be a set of nonnegative reals, and let q be a number. Let $g(x)$ be the number of elements in X greater or equal to x. Then, there is a sequence of integral breakpoints b_i, $i = 1, 2, \ldots$, with $\sqrt{q} \leq b_{i+1}/b_i \leq q$, such that $\sum_{i=1}^m g(b_i) \cdot b_i \leq \frac{1}{\ln \sqrt{q}} S(X)$. Additionally, this can be achieved with a polynomial time algorithm.*

With the right algorithm for scheduling the subproblems, we can obtain a good solution for the whole.

We say that an instance is *p-bounded* if all color requirements are at most p. The following result follows easily from the breakpoint lemma, since the delays amount to at most $O(1/\log p)Opt$.

Corollary 38.1 *Suppose there is an efficient algorithm for a sum multicoloring problem on p-bounded graphs in graph class $\mathcal{G}$ that: (1) approximates the sum within $1 + O(1/\log p)$-factor and (2) uses at most $O(p)$ colors. Then, we can approximate the sum multicoloring problem on all graphs in $\mathcal{G}$ within factor $1 + O(1/\log p)$.*

We give examples of this approach on three different graph classes.

38.5.1 Bounded Treewidth Graphs

Graphs of small treewidth are a good example of graphs that can be handled efficiently, by efficient solution of p-bounded graphs and classical *rounding-and-scaling*.

To begin with, p-bounded instances can be handled exactly.

Theorem 38.2 ([23]) *For any integers k and p, the npSMC problem on p-bounded graphs with treewidth k admits an exact algorithm that runs in time $O(n \cdot (k \cdot p \cdot \log n)^{k+1})$ [23].*

This result follows from standard techniques once it is established that the number of colors in optimal sum multicolorings is $O(kp \log n)$.

Suppose we are aiming for a $1 + \epsilon$-approximation, we round the color requirements up to the nearest multiple of $q = \lfloor \epsilon p/n^2 \rfloor$ and produce an n^2/ϵ-bounded instance I', where all lengths are scaled down by a factor of q. We can solve it exactly in $(n/\epsilon)^{O(k)}$ time and scale back the solution to apply to the original instance. The cost formed by the rounding can be shown to increase the sum cost by at most $1 + \epsilon$-factor.

Since the complexity is polynomial in n and $1/\epsilon$, we obtain an FPTAS.

Corollary 38.2 npSMC *admits an* FPTAS *on graphs of constant treewidth.*

38.5.2 Planar Graphs

The starting point of most approximation algorithms for planar graphs is the following classic result.

Theorem 38.3 (Baker's decomposition [3]) *Given a graph $G = (V, E)$ and integer k, we can find in polynomial time a subset $U \subseteq V$ with at most n/k vertices inducing a treewidth-2 graph, such that the graph G' induced by $V \setminus U$ is of treewidth at most k.*

Our approach is to use the breakpoint lemma and apply the following $1 + \epsilon$-approximate algorithm on each subgraph V_i. Let h be such that $2(5 + 8 \lg h)/h = \epsilon/2$.

Step 1: Partition the vertices of V_i into U_i and $V_i \setminus U$, as per Theorem 38.3, with $k = \lceil hp \rceil$.

Step 2: Compute an FPTAS solution of G'_i, the graph induced by $V_i \setminus U$, as per Corollary 38.2.

Step 3: Let W be the set of vertices with color $(1 + 8 \lg h)p$ and larger in the solution of G'_i. Color the subgraph induced by $U \cup W$ using $4p$ colors (using that the underlying graph is planar, and therefore 4-colorable), and append it to the coloring of $V_i \setminus (U \cup W)$.

It can be shown that after using $p(1 + 8 \lg h)$ colors, all but at most $Opt_i/(hp)$ vertices have been fully colored in the coloring of G'_i, where Opt_i is the optimal multichromatic sum of V_i. Thus, W contains at most $Opt_i/(hp)$ vertices. Also, by Theorem 38.3 and the value of k, $|U| \leq Opt_i/(hp)$. The cost of coloring $U \cup W$ is at most $(4 + 1 + 8 \lg h)p \cdot |U \cup W| \leq 2(5 + 8 \lg h)/h \cdot Opt_i = \epsilon Opt_i/2$. Hence, the resulting coloring of V_i is a $1 + \epsilon$-approximation. Combined with the breakpoint lemma, we obtain a PTAS for (unrestricted) planar graphs.

38.5.3 Generalization to Minor-Free Graphs

Let H be a fixed graph (whose size is viewed as a constant). To *contract* an edge (u, v) in an undirected graph is to merge u, v into a single vertex whose neighborhood is $N(u) \cup N(v) \setminus \{u, v\}$. H is a *minor* of G if it can be derived from G by a sequence of vertex and edge deletions and contractions of edges.

Definition 38.1 *A graph is H-minor-free iff H is not a minor of G.*

Theorem 38.4 ([11]) *For every H-minor-free graph and integer k, there is a constant $c = c_H$ such that the vertices can be partitioned to $k + 1$ disjoint sets so that every k of the sets induce a graph with treewidth bounded by $c_h \cdot k$. Furthermore, such a partition can be found in polynomial time [11].*

As graphs with constant treewidth are $O(1)$-colorable, the above is exactly what we need for the proof. Hence:

Theorem 38.5 npSMC *on H-minor-free graph admits a PTAS, for any fixed graph H.*

38.6 Delaying Large Jobs: Paying Your Dues

38.6.1 Sum Coloring $k+1$-Claw-Free Graphs

In nonpreemptive problems, a large job once selected must be processed without preemption and can thereby cause delay for many small jobs. One useful algorithmic strategy is therefore to delay the large jobs. We consider here a delay technique that results in a $2k(2k-1)$-approximation for npSMC on $k+1$-claw free graphs. For simplicity, we assume all color requirements are distinct, resolving ties arbitrarily.

The idea is as follows. We form a new color requirement $y(v) = (\beta+1) \cdot x(v)$ for each vertex v, where β is a parameter to be determined. During the first $\beta x(v)$ rounds that v is selected it *waits*, and only becomes *active* once it is selected for the $\beta x(v) + 1$th time. Note that neighbors cannot both be selected in the same round, and, for example, cannot both be waiting. Note that the rounds of waiting can be nonconsecutive, whereas the active rounds are consecutive. When selecting vertices, our algorithm gives preference to jobs with small length [namely, small $x(v)$].

1. Let $I \leftarrow \emptyset, j \leftarrow 1, I_1 \leftarrow \emptyset$
2. **While** $G \neq \emptyset$ **do:**
   a. Let I_j be the set of active vertices that are not fully processed.
   b. Iteratively add to I_j the vertex of smallest $x(v)$ among vertices with no neighbors in I_j, until I_j is a maximal independent set.
   c. Give color j to the vertices in I_j and update the color requirements y_v.
   d. Delete fully colored vertices.
   e. $j \leftarrow j + 1$.
3. Return the resulting coloring

Analysis: We shall show the following.

Theorem 38.6 *The above-mentioned algorithm yields a $2k \cdot (2k-1)$-approximation for* npSMC *on $k+1$-claw free graphs.*

This implies a 12-approximation for line graphs, which are 3-claw free.

A vertex that is neither active nor waiting in a round (and has not completed its processing) is said to be *delayed*. This must be caused by neighbors that are either active or waiting in that round. It can only be delayed by waiting neighbors that are shorter, due to the rule of preference.

Definition 38.2 *Let $N_s(v)$ be the neighbors of v of smaller color requirement, and let $N_b(v)$ be those neighbors of v that are scheduled to completion before v. Define $d_g(v) = S(N_s(v))$ and $d_b(v) = S(N_b(v))$ to be the total color requirements of these sets.*

Let $f(v)$ be the time (round) when our algorithm finishes scheduling vertex v. A part of $f(v)$ comes from the $(\beta+1)x(v)$ rounds in which v is waiting or active. During other rounds, v is delayed by its neighbors, either the shorter ones in N_s or the longer ones in N_b. We say v experiences a *good delay* if it is delayed by a shorter neighbor; otherwise, the delay is *bad* for v. More formally, if I is the set of vertices active in a round, then v experiences a good delay if $I \cap N_s(v) \neq \emptyset$, and bad delay otherwise [in which case $I \cap N_b(v) \neq \emptyset$].

Our main lower bound on the optimal cost Opt involves the following measure. Let $Q(G) = \sum_{uv \in E} \min\{x(v), x(w)\}$.

Claim 38.2 $Opt \geq S(G) + Q(G)/k$.

Proof. For a vertex v, let $N^-(v)$ be the set of vertices that finish before v in the optimal solution. Since the graph is $k+1$-claw free, at most k vertices in $N^-(v)$ can be simultaneously active in any round. The finish time of v in the optimal solution is then at least $x(v) + \lceil \frac{1}{k} \sum_{u \in N^-(v)} x(u) \rceil$. Summing over the vertices in the graph, the bound follows from the observation that $Q(G) = \sum_v \sum_{u \in N^-(v)} \min(x(v), x(u))$. ∎

We bound the good and bad delays separately. Bounding the former is straightforward.

Claim 38.3 $\sum_v d_g(v) \leq (\beta + 1) \cdot Q(G)$.

Proof. A vertex v can experience at most $(\beta + 1) \sum_{u \in N_s} x(u)$ good delay. Thus, the total good delay is at most $(\beta + 1) \sum_v \sum_{u \in N_s} x(u) = (\beta + 1)Q(G)$. ∎

The key idea is to bound the bad delays in terms of the good ones.

Claim 38.4 $d_b(v) \leq \dfrac{(k-1) \cdot d_g(v)}{\beta - k + 1}$.

Proof. We say that a round is an *event for* v if a neighbor of v in N_b is waiting. The total waiting of nodes in N_b is $\sum_{u \in N_b} \beta \cdot x(u)$, and each such wait occurred during an event for v. Since the graph is $k+1$-claw free, at most k neighbors of v can be active or waiting in the same round. If some neighbor of v is waiting, then v is neither active nor waiting, which means that it was delayed either by a shorter neighbor or an active longer neighbor. Therefore, at most $k - 1$ longer neighbors of v can be waiting in a round, in particular during an event for v. Thus, there are at least

$$\sum_{u \in N_b} \frac{\beta \cdot x(u)}{k - 1} = \beta \cdot \frac{d_b(v)}{k - 1}$$

events for v. Since v is delayed during an event, there are at most $d_g(v) + d_b(v)$ events for v. Combining the bounds on the number of events, we have that $d_g(v) + d_b(v) \geq \beta \cdot d_b(v)/(k-1)$. The claim follows by rearranging the expression. ∎

Proof of Theorem 38.6. Set $\beta = 2(k-1)$. Then, $d_b(v) \leq d_g(v)$, by Claim 38.4. The finish time of vertex v by our algorithm is then $x(v) + d_g(v) + d_b(v) \leq x(v) + 2d_g(v)$. Summing over the vertices, the cost *Alg* of the algorithm's schedule is bounded by

$$Alg - S(G) \leq 2 \cdot (2k-1)Q(G) \leq 2(2k-1) \cdot k(Opt - S(G))$$

by Claims 38.2 and 38.3. ∎

38.7 Open Problems

In addition to improving the results in Table 38.2 consider the following open problems:

1. Does npSMC admits a constant ratio on *chordal graphs*? Maybe even on *perfect graphs*?
2. What is the ratio of npMC on planar graphs?
3. Give a a significant lower bound on approximating Open Shop scheduling.

References

1. Appel, K. and Haken, W., Every planar map is four colourable, *Bull. Amer. Math. Soc.*, 82(5), 711, 1976.
2. Arnborg, S. and Proskurowski, A., Linear time algorithms for NP-hard problems restricted to partial k-trees, *Disc. Appl. Math.*, 23(1), 11, 1989.
3. Baker, B.S., Approximation algorithms for NP-complete problems on planar graphs, *J. ACM*, 41, 153, 1994.
4. Bar-Noy, A., Bellare, M., Halldórsson, M.M., Shachnai, H., and Tamir, T., On chromatic sums and distributed resource allocation, *Inform. Comput.* 140, 183, 1998.
5. Bar-Noy, A., Halldórsson, M.M., and Kortsarz, G., A matched approximation bound for the sum of a greedy coloring, *Inform. Process. Lett.*, 71(3–4), 135, 1998.

6. Bar-Noy, A., Halldórsson, M.M., Kortsarz, G., Shachnai, H., and Salman, R., Sum multicoloring of graphs, *J. Algor.*, 37(2), 422, 2000.
7. Bar-Noy, A. and Kortsarz, G., The minimum color-sum of bipartite graphs, *J. Algo.*, 28, 339, 1998.
8. Borodin, A., Ivan, I., Ye, Y., and Zimny, B., On sum coloring and sum multi-coloring for restricted families of graphs, *Theor. Comput. Sci.*, 418, 1, 2012.
9. Buchsbaum, A.L., Karloff, H., Kenyon, C., Reingold, N., and Thorup, M., OPT versus LOAD in dynamic storage allocation, *SIAM J. Comput.*, 33(3), 632, 2004.
10. Coffman, E.G., Garey, M.R., Johnson, D.S., and LaPaugh, A.S., Scheduling file transfers. *SIAM J. Comput,.* 14(3), 744, 1985.
11. Demaine, E., Hajiaghayi, M., and Kawarabayashi, K., Algorithmic graph minor theory: Decomposition, approximation, and coloring, in *Proceedings of 46th Annual IEEE Symposium on Foundations of Computer Science (FOCS)*, 2005, p. 637.
12. Epstein, L., Halldórsson, M.M., Levin, A., and Shachnai, H., Weighted sum coloring in batch scheduling of conflicting jobs, *Algorithmica*, 55(4), 643, 2009.
13. Feige, U., Approximating maximum clique by removing subgraphs, *SIAM J. Discrete Math.*, 18(2), 219–225, 2004.
14. Feige, U. and Kilian, J., Zero knowledge and the chromatic number, *J. Comput. Syst. Sci.*, 57(2), 187, 1998.
15. Feige, U.L., Lovász, L., and Tetali, P., Approximating min sum set cover, *Algorithmica*, 49(4), 219, 2004.
16. Gandhi, R., Halldórsson, M., Kortsarz, G., and Shachnai, H., Improved bounds for sum multicoloring and weighted completion time of dependent jobs, *ACM Trans. Algor.*, 4(1), 11, 2008.
17. Garey, M.R., Johnson, D.S., and Stockmeyer, L., Some simplified NP-complete graph problems. *Theoretical computer science* 1(3), 237–267, 1976.
18. Garey, M.R. and Johnson, D.S., *Computers and Intractability: A Guide to the Theory of NP-Completeness*. W. H. Freeman, New York, 1979.
19. Gavril, F., Algorithms for minimum coloring, maximum clique, minimum covering by cliques, and maximum independent set of a chordal graph, *SIAM J. Comput.*, 1(2), 180, 1972.
20. Grötschel, M., Lovász, L., and Schrijver, A., *Geometric Algorithms and Combinatorial Optimization*, Springer-Verlag, New York, 1993.
21. Holyer, I., The NP-completeness of edge-coloring, *SIAM J. Comput.*, 10(4), 718–720, 1981.
22. Halldórsson, M.M., A still better performance guarantee for approximate graph coloring. *Inform. Process. Lett.*, 45(1), 19, 1993.
23. Halldórsson, M.M. and Kortsarz, G., Tools for multicoloring with applications to planar graphs and partial k-trees, *J. Algor.*, 42(2), 334, 2002.
24. Halldórsson, M.M., Kortsarz, G., Proskurowski, A., Salman, R., Shachnai, H., and Telle, J.A., Multicoloring trees, *Inform. Comput.*, 180(2), 113, 2003.
25. Halldórsson, M.M., Kortsarz, G., and Shachnai, H., Sum coloring interval and k-claw free graphs with application to scheduling dependent jobs, *Algorithmica*, 37, 187, 2003.
26. Halldórsson, M.M., Kortsarz, G., and Sviridenko, M., Sum edge coloring of multigraphs via configuration LP, *ACM Trans. Algor.*, 7(2), 22, 2011.
27. Jansen, K., The optimum cost chromatic partition problem, in *Proceedings of Third Italian Conference on Algorithms and Complexity (CIAC '97)*, LNCS 1203, 1997, pp. 25–36.
28. Kim, Y., Data migration to minimize average completion time, *Algorithmica*, 63(1–2), 347, 2012.
29. Kubicka, E., The chromatic sum of a graph, PhD Thesis, Western Michigan University, Kalamazoo, MI, 1989.
30. Kubicka, E., Kubicki, G., and Kountanis, D., Approximation algorithms for the chromatic sum, in *Proceedings of First Great Lakes Computer Science Conference*, LNCS 1203, 1989, pp. 15–21.

31. Kortsarts, Y. and Rufinus, J., Teaching the power of randomization using a simple game, in *Proceedings 39th SIGCSE Technical Symposium on Computer Science Education (SIGCSE)*, 2006, pp. 460–463.
32. Kchikech, M. and Togni, O., Approximation algorithms for multicoloring planar graphs and powers of square and triangular meshes, *Discrete Math. Theor. Comput. Sci.*, 8(1), 159, 2006.
33. Malafiejski, M., The complexity of the chromatic sum problem on cubic planar graphs and regular graphs, *Electron. Notes Discrete Math.*, 8, 62, 2001.
34. Malafiejski, M., Giaro, K., Janczewski, R., and Kubale, M., Sum coloring of bipartite graphs with bounded degree, *Algorithmica*, 40(4), 235, 2004.
35. Marx, D., Minimum sum multicoloring on the edges of trees, *Theor. Comput. Sci.*, 361, 133, 2006.
36. McDiarmid, C. and Reed, B., Channel assignment and weighted coloring, *Networks*, 36(2), 114, 2000.
37. Nishizeki, T. and Kashiwagi, K., On the 1.1 edge-coloring of multigraphs, *SIAM J. Discrete Math.*, 3(3), 391, 1990.
38. Narayanan, L. and Shende, S., Static frequency assignment in cellular networks, *Algorithmica*, 29(3), 396, 2001.
39. Nicoloso, S., Sarrafzadeh, M., and Song, X., On the sum coloring problem on interval graphs, *Algorithmica*, 23, 109, 1999.
40. Marx, D., A short proof of the NP-completeness of minimum sum interval coloring, *Oper. Res. Lett.*, 33(4), 382, 2005.
41. Vizing, V.G., On the estimate of the chromatic class of *p*-graphs, *Diskret. Analiz.*, 3, 23, 1964.
42. Woeginger, G., Private communication, 1997.

Approximation Algorithms and Heuristics for Classical Planning

Jeremy Frank

Minh Do

J. Benton

39.1 Introduction

Automated planning has been an active area of research in theoretical computer science and artificial intelligence (AI) for over 45 years. Planning is the study of general purpose algorithms that accept as input an *initial state*, a set of desired *goal states*, and a *planning domain model* that describes how *actions* can transform the state. The problem is to find a sequence of actions that transforms the initial state into one of the goal states. Automated planning is widely applicable to diverse domains, including spacecraft control [1], planetary rover operations [2], automated nursing aides [3], image processing [4], business process generation [5], computer security [6,7], and automated manufacturing [8,9]. Planning is also the subject of continued and lively ongoing research.

This chapter will present an overview of how approximations and related techniques are used in automated planning. The chapter focuses on *classical planning* problems, where states are conjunctions of propositions, all state information is known to the planner, all actions are instantaneous, and their outcomes are deterministic. Classical planning is nonetheless a large problem class that generalizes many combinatorial problems including bin-packing (Chapters 28, 29, and 31), prize collecting TSP (Chapter 34) and scheduling (Chapter 38).

Planning problems include both the *decision* problem, that is, determining whether a plan exists to accomplish the goals, and different *optimization* problems, that is, find a plan optimizing an objective function, such as plan length. While the bulk of this chapter will focus on methods for solving either the decision or optimization problems within the context of classical planning, many techniques naturally generalize to more complex planning settings and their associated objective functions. Examples include explicit models of time and action duration, resources, optimization by choosing subsets of goals, and uncertainty in the world and in action outcomes. Readers interested in learning more about automated planning are referred to [10] and [11].

The rest of this chapter is organized as follows: first, in Section 39.1, we provide some background on the classical planning problem; in Section 39.2, we describe the main algorithmic approaches employed to solve this problem. This short section serves as background material to the main technical contribution in Section 39.3, which covers the ways in which approximation and metaheuristics are used in solving classical planning problems. Section 39.3.1 outlines several modern heuristic families underlying the success of recent classical planners built upon the planning-as-search approach. The key to the success of those heuristic families are rooted in their accuracy and speed, and we shall describe how these heuristics exploit different simplifications of the planning problem. Section 39.3.2 describes incomplete and local search algorithms that sacrifice complete coverage of the search space to reduce computation time. Section 39.3.3 discusses approaches that reformulate the original planning problem into another combinatorial problem. Such techniques exploit advances in solving technologies of the reformulated problem, or use the reformulation to approximate the original problem through different types of compilation. Section 39.3.4 outlines approaches to identify structures that capture either independent or distinct subproblems of the original planning problem. This allows decomposition of the original problem into components that can be individually solved, usually by high-performance specialized solvers. Finally, Section 39.4 outlines approximation and metaheuristics technique for more complex planning setting such as planning with time, resources, and uncertainty. We conclude the chapter with Section 39.5.

39.1.1 Classical Planning

The core of the decision problem in classical planning is *goal achievement*; given a particular state of the world, and actions that can manipulate the state, the problem is to find a sequence of actions that lead to one of a designated set of goal states. More formally, a *planning domain D* consists of a set of world states S and a set of actions A. An action a defines a deterministic mapping $a : S \rightarrow S$. A *planning problem instance* $\langle D, i, G \rangle$ consists of a planning domain $D = \langle A, S \rangle$, an initial state $i \in S$ and a set of goal states $G \subset S$. The set of actions define a directed graph on S; thus, the decision problem is to find a path from i to a state $\gamma \in G$ or to prove that no such path exists.

39.1.1.1 The STRIPS Formalism

For many problems, the size of S and thus the directed graph representing the connections between states in S by actions in A are extremely large. Therefore, S and the associated set of actions A are described *implicitly* by exponentially compressing the descriptions of S and A. The standard form to represent a classical planning problem is by using the Stanford Research Institute Problem Solver (STRIPS) formalism, in which states in S are represented by a set of propositions P and actions are defined by its conditions and effects on propositions in P. Specifically:

- State: Each state $s \in S$ is represented by a complete assignment of True or False values to all propositions $p \in P$. Thus, $|S| = 2^P$. Typically, s is represented *only* by a subset of variables in $P_s \subseteq P$ that are True in s, propositions in $P \setminus P_s$ are implicitly False.
- Goals: The set of goal states G is defined by a conjunction of propositions $P_g = \{g_i\}$ that must be True in the state, that is, $G = s | P_g \subset P_s$.
- Actions: Each action $a \in A$ is represented by three components:
  - Preconditions *pre(a)*: The set of propositions that must be True to enable the application of a. An action a is *applicable* in state s, denoted as *applicable(a, s)* if $s \Rightarrow pre(a)$ (i.e., $pre(a) \subset P_s$).
  - Add effects *add(a)*: The set of propositions that a makes True when applied.
  - Delete effects *del(a)*: The set of propositions that a makes False when applied.

- State transitions: when an action a is applicable in a state s, then applying a in s will lead to a new state $s' = apply(a, s) = P_s \backslash del(a) \bigcup add(a)$.

A *plan* π is a sequence of actions $\pi = \{a_1, a_2, \ldots, a_n\}$. A *solution* is a plan which leads from the initial state i to a goal state $\gamma \in G$. Specifically, (1) the first action a_1 is applicable in the initial state i and $apply(a_1, i) = s_1$; (2) each subsequent action is applicable in the state resulted from applying the previous action: $s_k = apply(a_k, s_{k-1})$; and (3) the last state is a goal state: $s_n = (\gamma \in G)$. When solving the classical planning problem, the decision problem is to find such a plan, or prove one does not exist. Alternatively, we can solve an optimization problem, in which the objective is to minimize the *length* (i.e., number of actions) of the solution plan π. If each action is associated with a *cost* value, then the objective function is to minimize the total cost* of all actions in the plan.

The Planning Domain Definition Language (PDDL) modeling language is a contemporary and well-accepted standard modeling language capable of representing the STRIPS formalism, along with many extensions. A brief discussion of these extensions is provided in Section 39.4; for more information the reader is referred to [12].

39.1.1.2 Complexity

Classical planning using the STRIPS formalism is proven to be $\mathcal{PSPACE}$-complete [13].

39.1.1.3 Example

The Rovers domain description in Figure 39.1 is an example of a domain represented in the STRIPS formalism. In this example, a rover has the goal of taking rock samples and images at varying locations. The foundation of the domain is the locations and different rover subsystems. The propositions in the domain describe the network of locations that the rover can travel between, the operational state for each of the instruments, and the state of possible objectives (images or samples) to be achieved at each location. Finally, we define actions for the rover. It can drive from one location to another, turn its camera on and off, turn a drill on and off, take an picture, and drill for a rock sample. These actions include preconditions (e.g., must be at a location to drill for a sample there) and effects (e.g., change location after driving, have a rock sample after drilling for it).

39.2 Background

In this section, we will provide background on the most popular approaches employed in solving the classical planning problems.

39.2.1 Planning Approaches

This section gives a broad overview of the most common planning methods for classical planning. The approaches outlined here either use approximations and metaheuristics as key algorithmic components (e.g., heuristics for graph search), or are metaheuristic approaches in their own right.

39.2.1.1 Planning as Graph or Tree Search

In this approach, the planner needs to define three things: (1) the root-node, which is the starting point of the search process; (2) a transition function to define the next set of search nodes that can be traversed (i.e., connected) from the search node currently visited; and (3) the termination criterion, which specifies when a search process can stop. Those factors will define the *search space*, which can then be explored by any systematic (e.g., breadth-first, depth-first, best-first, A^*) or local search algorithm. The rationale for

---

* Although action costs are not representable in STRIPS, this common objective generalizes the plan length optimization criteria.

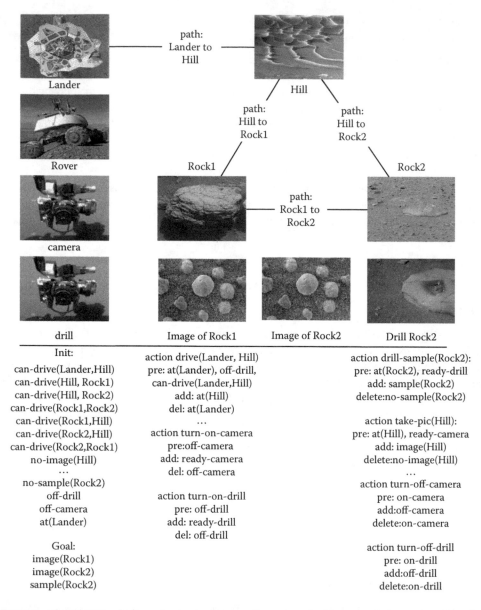

FIGURE 39.1 The STRIPS encoding of a simple planetary Rover domain. The `can-drive` propositions describe the simple road network in the picture. The rover is permitted to drive with the camera on, but cannot drive with the drill on. The drill and camera cannot both be turned on simultaneously. The rover must be `at` a location in order to take a picture of sample. Representative `drive`, `drill-sample`, and `take-picture` actions are shown for a subset of locations.

distinguishing between the state space S and the search space will become clear in the next paragraph. Since the search space is large, the key to the success of using any search algorithm is the heuristic guiding the exploration through the search space.

State-Space Search: One of the most commonly used strategies for planning as graph or tree search is *state-space search*, in which the planner searches for a path through the state space S, introduced in the previous section, between the initial state i and a goal state γ. As noted earlier, the state space and the set of possible transitions is exponential in the number of propositions. To save time and space, states and allowed state transitions are implicitly represented and identified "on the fly" during planning.

This category of planner is further divided as follows:

- *Progression* planners set the root node to be the initial state i, the termination condition to be that a visited state γ is one of the goal states: $\gamma \in G$, the transition function is as defined in the previous section.
- *Regression* planners set the root node to be a "partial" state (a set of states) representing all goal propositions $\{g_i\}$ in any goal state $\gamma \in G$ that need to be achieved, a state s can be regressed through action a if a "contributes" to s, defined by the condition $(\text{del}(a) \cap s = \emptyset) \wedge (\text{add}(a) \cap s \neq \emptyset)$. The transition function is defined through regressing states through actions: $regress(s, a) = P_s \setminus (add(a) \bigcup pre(a))$. The search can be terminated if it visits a state s implied by the initial state: $i \models s$.

Plan-Space Search: Instead of having the search node representing the state that the plan would visit when executing from the initial state to reach the goals, the planner can instead use the possible "partial" (i.e., incomplete) plans to represent the search node. In this framework, a root-node is represented by an empty (i.e., no action) plan. The termination criteria is that we visit a search node represented by a plan that is a solution, as defined previously. The transition function for a current search node n represented by a partial plan π is defined as follows: for each goal proposition g, or a precondition p of a given action a_1 that is "unsupported" (i.e., g or p is not already satisfied by the initial state i or is added by some action a_2 in the current partial plan), then we can expand n by either finding an action a_2 in π that has $g \in add(a_2) \vee p \in add(a_2)$ or adding an action a_3 not already in π that has $g \in add(a_3) \vee p \in add(a_3)$. *Causal links* are added to ensure no other action addition deletes the proposition p or the goal g established by the newly inserted action.

39.2.2 Approximations and Metaheuristics for Planning

This section gives a broad overview of how different approximation and metaheuristic strategies are used for automated planning.

39.2.2.1 Heuristics for Graph and Tree Search

When planning problem is posed as graph or tree search as described earlier, *complete* search algorithms are usually employed. Regardless of the search space representation and choice of progression or regression, the above-mentioned search approaches require a heuristic mechanism to navigate the search space. The challenge is to design cheaply computed but informative heuristics to guide search. A fruitful approach to the construction of these heuristics is to use principled approximations and abstractions of the original problem. Many such families exist and have proven effective at guiding complete search algorithms.

39.2.2.2 Incomplete and Suboptimal Search

Complete algorithms, which are guaranteed to find a solution if one exists, are often very expensive, even when guided by informative heuristics. A search algorithm is *incomplete* if it is not guaranteed to find a solution to a problem instance, even when one exists. A search algorithm is sub-optimal if it is not guaranteed to find an optimal solution. Incomplete and suboptimal algorithms trade coverage and solution quality for reduced run-time. This is often a viable strategy for solving large and complex problems. Algorithms that do not guarantee exploration of the entire search space include local search and sampling methods. Both of these methods can operate on either the state space or plan space, as discussed earlier.

39.2.2.3 Reformulation

Problem reformulation involves mapping a planning problem into another combinatorial problem, to take advantage of powerful solvers or advanced techniques. Satisfiability (SAT), Satisfiability Modulo Theories (SMT), Simplified Action Structures (SAS), Constraint Satisfaction Problem (CSP), or Mixed Integer Linear Program (MILP) are common reformulation targets. A strict reformulation requires each unique solution to a reformulated problem to map directly to a unique solution of the original problem, and vice versa. We can relax this requirement so that solutions exist in the reformulated problem that cannot provably map to a solution in the original problem. These types of relaxations form the basis of many state evaluation heuristics for classical planning problems. We can also reformulate problems so some solutions to the original problem cannot provably map to a unique solution of the reformulated problem. This is the case for some reformulations into other problems with a reduced complexity. In the case of STRIPS planning, for example, we see this when reformulating STRIPS planning problem to propositional satisfiability. Once the problem has been reformulated, numerous algorithms can be used to solve the reformulated problem, including tree or graph search, local search, or sampling, as described previously.

39.2.2.4 Subproblem Identification

The last class of commonly used metaheuristics relies on identifying and exploiting *problem structure*. As noted in the introduction, planning problems often contain *subproblems* that can be solved by specialized combinatorial algorithms, for example, bin-packing, path planning, and scheduling. However, these subproblems are either implicitly encoded in the planning model, hidden in the state space, or both, and must be identified by analyzing the model or state space to be identified. Another common form of structure exploitation is to *partition* the problem, by either ordering or separating goals, thus creating smaller, more easily solved planning problems.

39.3 Approximations and Metaheuristics for Classical Planning

In this section, we will survey different ways approximation and metaheuristic techniques have been used to solve classical planning problems.

39.3.1 Heuristics for Systematic Tree and Graph Search

Planners based on systematic search try to explore the entire search space to guarantee that an (optimal) solution will be found, if one exists. As described in the previous section, the search space for planning problem is normally very large, and thus cannot be represented explicitly. To find a solution within a given time and memory limit, the planners need informative heuristic guidance to navigate the search space. In planning, heuristics estimate the minimal distance from the current search node to a search node in which the plan reaches one of the goal states. If we envision the search space as a tree, this is the distance from the current interior node to the closest descendant that is also a satisfactory leaf node. Heuristic in planning are calculated by solving some simplified version of the original planning problem to make it (sometime significantly) easier to solve than the original problem. Ideally, the cost to compute the heuristic value is tractable, while still retaining enough core aspects of the original planning problem so that its solution is a good approximation of the solution of the original problem.

A heuristic is *admissible* when it under-estimates the true distance. Certain search algorithms guarantee that the first solution found when using an admissible heuristic is optimal, for example, A^* [14]. A^* maintains a list of open nodes ordered by the increasing values of the function $f = g + h$, where g is the cost of the current node (e.g., current plan length) and h is an estimate of the cost to reach the solution. Admissible heuristics used in algorithms such as A^* guarantee optimality, but this comes at the price of high search time, usually due to *inaccuracy* in the heuristic estimate of the distance between the current search state and a solution; this inaccuracy leads to both time and space complexity (since all pending nodes are stored in the open list).

While A^* is one of the most popular search approaches, there are other well-used graph and tree search algorithms that also make use of the g and h values. Greedy Best-First Search (GBFS) [15] orders nodes in the open list using only the h value, prioritizing nodes estimated to be closest to the goals. GBFS can often find a solution quickly, but it might be of poor quality. A^* can be modified to trade solution quality for speed by weighting the heuristic estimate h by a factor $w > 1$. The resulting algorithm, called Weighted A^* (WA^*) orders nodes in the open list minimizing the function $f = g + wh$; as a result, as w increases, WA^* searches more greedily [16]. This approach basically blends GBFS and A^*. Assuming an admissible heuristic, the cost of the solution it finds differs from the optimal by no more than the factor w.

In the remainder of this section, we will outline the most common types of heuristics through planning problem abstraction and approximation. Specifically, we will concentrate on the following categories [17]: delete relaxation, Planning-Graph-based, state-abstraction, critical path, landmarks, and network-flows.

39.3.1.1 Delete Relaxation

This class of heuristics is based on solving a simplified planning problem P^+ where all the action *delete lists* are removed. Ignoring action delete lists makes the problem easier when all goals are described by a set of True propositions, and action preconditions are positive. Thus, a plan π^+ for P^+ starting from a given state s is used as an approximation for the plan starting from s in the original planning problem P. Instead of finding π^+ earlier delete-relaxation heuristics such as h^{max} and h^{sum} only estimate the length of π^+ (i.e., the number of actions in π^+), and use this estimate to differentiate between different states based on how far they are from the goals. Specifically, the heuristic values for all propositions and actions are initially set to ∞, except all facts in the currently visited state have their heuristic values initialized to zero, and the following dynamic programming rules are applied to compute h^{max} and h^{sum} until a fixed-point is reached:

- $\forall p \in P : h(p) = min_{p \in Add(a)} h(a)$
- $h^{max} : \forall a \in A : h(a) = max_{p \in Pre(a)} h(p)$
- $h^{sum} : \forall a \in A : h(a) = sum_{p \in Pre(a)} h(p)$

While h^{max} is proven to underestimate the real length of the plan (i.e., it is an *admissible* heuristic) and thus can be used to find a proven optimal plan, the inadmissible h^{sum} heuristic is more informative and can help guiding planners to find a satisfying plan quicker. HSP and HSPr [18] use this type of heuristic in conjunction with the WA^* search algorithm. HSP is a *progression* planner (Section 39.2.1) whereas HSPr is a *regression* counter-part of HSP that uses the same family of heuristics. Since the *delete relaxation* heuristics measure the distance from a given state to propositions (including goals) that are reachable from that state, regression planners only need to perform the dynamic programming rules once, because the heuristic values are all computed starting from the same initial state. Progression planners need to recompute the heuristic value from every (different) visited state. Despite this advantage, in practice progression planners using delete relaxation heuristics have generally performed better because they avoid unreachable goal states; regression planners may attempt to guide search from an unreachable goal state to the initial state, which is obviously doomed to fail.

FF [19] goes beyond HSP by not only estimating the length of π^+, but also extracts a plan π^+ for different purposes in guiding the planning search. While finding an optimal plan for P^+ is $\mathcal{NP}$ [13], finding a satisfying finding a plan for P^+ is tractable; we will describe how FF extracts a relaxed plan π^+ from the (relaxed) Planning-Graph later. Similar to the HSP planners, FF uses the length of π^+ to estimate how far each state is from the goal; and it also uses the actions in π^+ to limit the number of actions to consider next when extending the current state.

Recently, more advanced variations of the delete-relaxation heuristics have been introduced, such as pairwise max, set-additive, Steiner-tree, and landmark-cut (discussed later) [20]. Some of those provide the backbone of the planners derived from Fast Downward [21]. Lastly, besides finding different ways to improve the heuristics gathered from solving the delete-relaxation problem, the delete-relaxation heuristic quality can be improved in an orthogonal way by incorporating the negative action interactions caused by the delete lists. Keyder et al. [22] has advanced the heuristic by augmenting the delete-relaxed actions with the limited amount of delete information and the h^m family of heuristics [23] have formally introduce a systematic way to utilize *k-ary* mutual-exclusion relations between sets of propositions of size k to improve upon the delete-relaxation heuristics. h^{max} is the special case h^1 where all mutual-exclusion relations are ignored.

39.3.1.2 Planning-Graph-Based Heuristics

The Planning-Graph is a structure built for conducting *reachability* analysis. The original Planning-Graph is part of an algorithm called GRAPHPLAN [24], but the structure has subsequently seen other uses for heuristic search such as checking if goals can be reached from a given state, and as a basis for extracting heuristic guidance.

Building the Planning-Graph: The Planning-Graph is a compact representation of a superset of the valid plans of a fixed plan length. It is represented as a multi-level graph, where each level alternates between a layer L_i containing literals that may become True and a layer A_i containing actions that may be included in a plan at a given step i. The algorithm for building the Planning-Graph:

1. Begins with a fact level L_1 containing literals that are True in the initial state
2. Follow each fact layer L_i with an action layer A_i containing all actions with preconditions satisfied by facts in L_i
3. The following fact layer L_{i+1} is the union of all facts in L_i and add effects of all actions in A_i

With this definition, each fact level contains the previous literal level with the addition of all add effects from the actions in the previous action layer. We stop the process whenever $L_i = L_{i+1}$ and $A_i = A_{i+1}$ or when all goals become *reachable* (i.e., a proposition layer contains all of the goals). An example of the Planning-Graph for the simple Rover domain can be seen in Figure 39.2.

Planning-Graphs also offer an effective approach to propagate *mutual exclusions* (mutex) between actions and propositions. To do this, we must augment the Planning-Graph to include delete effects, and use them to infer mutual exclusion relations between actions. Actions are mutually exclusive if one deletes either a precondition or add effect of another. The Planning-Graph also allows identification of two actions with preconditions that are mutually exclusive with each other; these actions can be marked as mutually exclusive. This, in turn, enables propagation of mutual exclusions to further propositional layers as well. This changes the termination criteria of Planning-Graph expansion, adding the requirement that mutual exclusions must also remain unchanged between layers before halting. Mutual exclusions in the Planning-Graph have applications in both the original GRAPHPLAN algorithm and as part of improving the quality of heuristics extracted from the Planning-Graph. Using these augmentations, also aid in reformulation techniques discussed in Section 39.3.3 [25].

Planning-Graph Based Heuristics: The fact level i at which all goals appear in the Planning-Graph can be used as an admissible heuristic estimating the distance from the state from which the Planning-Graph is built to the goals. When mutexes are not propagated during the Planning-Graph building process (i.e., the action delete lists are ignored), then i is equal to the h^{max} heuristic. When mutexes are propagated and i is the level at which all goals appear nonmutex with each other, then i is equal to the h^2 heuristic within the h^m heuristic family. Both h^{max} (or h^1) and h^m were discussed earlier in this section. To create more informative heuristics, Nguyen et al. [26] introduced several inadmissible heuristics that utilize the information on the levels at which set of facts appear nonmutex in the Planning-Graph. The *partition-k* heuristic partitions the set of goals into multiple set G_i, each of size k, and then use as the heuristic value the sum of the levels i at which each G_i appears nonmutex. Another heuristic, *adjusted-sum* takes the *partition-1* heuristic and adjust it with mutex information on the full goal set.

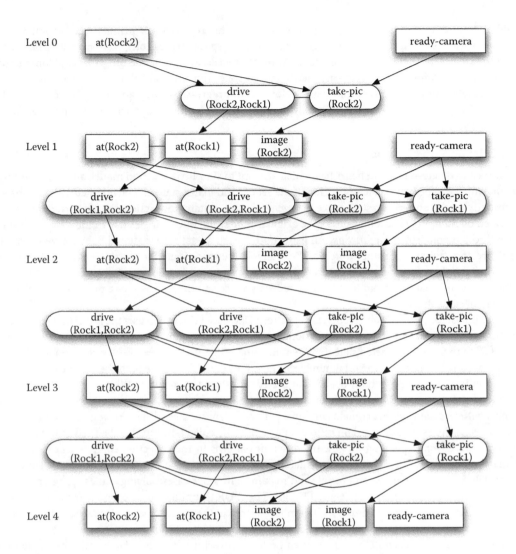

FIGURE 39.2 Part of the Planning-Graph for the Rover domain. Boxes represent propositions, ovals represent actions. Arrows represent action preconditions and effects, while arcs represent mutual exclusions. The Planning-Graph assumes the rover is at (Rock2) and the camera is ready and includes only the actions drive and take-pic for simplicity. Notice that the set of mutual exclusions does not change after level four of this Planning-Graph.

FF [19] is the first planner that introduced the relaxed Planning-Graph heuristic, which is a cost estimate extracted from the mutex-free Planning-Graph. The regression search procedure for plan extraction from the Planning-Graph is done as follows:

1. Begins by identifying goals in the last fact level L_n as subgoals.
2. For every subgoal g at level L_i, chooses an action $a \in A_{i-1}$ that supports g and add a_{i-1} to the relaxed plan π^+.
3. Identifies the set of subgoals at L_i as all action preconditions from the actions in $\Pi_i, p \in pre(a_i \in \Pi_i)$.

This process ends when reaching L_1. We count the number of actions in the relaxed plan π^+ and use it as the heuristic value estimating how many actions are needed to advance the current state to achieve all goals. This relaxed-plan heuristic has been used in many different types of planner (see Section 39.2.1) such as forward state-space (progression), backward state-space (regression), plan-space (partial-order causal-link), and local-search on action graph. As mentioned earlier, given that progression planner needs to build a new planning graph for each newly visited state, simpler—and thus less time-consuming— mutex-free Planning-Graph is normally used. However, for regression and plan-space planner, where the Planning-Graph is only need to be built only once, the full Planning-Graph with mutex propagation is normally used.

39.3.1.3 State Abstraction

In contrast to delete-relaxation heuristics, which attempt to simplify the action model, and the Planning-Graph, which can be seen as an abstraction of the set of valid plans, state-abstraction heuristics are based on simplifying the underlying state-space. In a general form of state-abstraction, the *abstracted state space* is created by mapping subset of states in the original state space into a single state in the abstracted state space through an *abstraction function*. All paths in the original state space should be preserved through the abstraction function. In other word, for the original state space S and the abstraction function α, every state $s \in S$ should map to a unique state $w^{\alpha} \in W^{\alpha}$, and every path in S should map to a (simplified) path in the W^{α}. Plans that are found in the abstracted state space can then be used as a proxy for the plans in the original state-space.

The simplest form of state abstraction is the Pattern Database [27], which uses the *projection* abstraction function α^{p}. This is accomplished by a function mapping two states that agree on the values of all variables in a set S^{p} to the same abstracted state s^{p}. When the remaining number $n = |S^{p}|$ of state variables that are not removed is low enough, the total number of possible states 2^{n} and the paths between them through action applications can potentially be fully computed and stored in memory. While early PDB heuristics use one abstraction to compute the heuristic, recent developments lead to more complex heuristics in which a *pattern collection* that contains multiple pattern databases is used. The longest distance (i.e., *max* rule) between the estimates based on different pattern in the collection can then be used as the heuristic estimate. Moreover, depending on how the different patterns in the collection complement each other, estimates from different PDB can be combined in a more sophisticated way. Examples include summing up the values, summing up when possible and maximize in other cases, solving a linear program to find the weighted sum of individual estimates [28–30]. The iPDB technique goes beyond using a pre-defined pattern collection but instead employs hill-climbing search in the space of pattern collections to select the best one to seed the multiple-pattern heuristic computation process [28].

Due to the fact that each PDB estimate is an under-estimation of the real plan length, the single PDB and max PDB heuristics are admissible and thus guarantee that the planner will find optimal plan. For this reason, PDB heuristics are predominantly used in optimal planners while delete-relaxation heuristics are dominant in satisfying planners (even though some delete-relaxation heuristics are admissible).

Some keys issues in getting state-abstraction heuristic to work well are (1) defining an effective abstraction function; (2) how to represent the abstracted state-space; (3) is one big abstraction better than having multiple smaller abstractions; (4) how to combine results from multiple interdependent abstractions, if we decide to do so; (5) should we build the abstractions up-front or in a online fashion during the planning search. Answering those questions lead to multiple family of state-abstraction heuristics such as Pattern Database [27], Symbolic Pattern Database [31], constrained Pattern Database, merge-and-shrink, structural patterns, or Cartesian abstraction [32–34].

39.3.1.4 Critical Paths

While the pattern database and delete relaxation either simplify the action model or the underlying state-space, respectively, the critical path heuristic is based on simplifying the goal state. The critical path heuristic finds a plan to a "hardest to achieve" goal subset. By only requiring the planner to satisfy a subset $G' \subset G$ of the original goals, the planner can more quickly estimate the length of the plan achieving G', or find the relaxed or valid plan achieving G'. Similar to previous heuristics, the length of the plan, or the plan itself, for the simplified problem is then used as an estimate of the plan for the original problem. The key to this approach is to find solutions for multiple subsets G' and the *critical path* that achieves the hardest G' estimates the difficulty in achieving the original set of goals.

The decisions that planners using this type of heuristic need to make include (1) which subsets of goals to be abstracted out from the original problem; (2) how to compute the plan for the relaxed problems; (3) if we only do a single or multiple abstractions (Similar to multiple pattern databases for the previous heuristic), and if multiple abstractions are used then how to divide, and combine the results later on, in a most beneficial way.

Popular heuristics in this family includes h^m, in which for a predetermined m value, the planner will *concurrently*, through dynamic programming rules, estimate the length of the shortest plans to achieve each subset G' of size m. The longest plan-length estimate among all that achieve a subset G' is used as the heuristic measuring how far a given state s from achieving all goals in G. The higher the value of m, the harder it is to compute the heuristic value (when $m = |G|$ then it's as hard as solving the original problem) but better the quality of h^m. In practice, only $m = 1$ and $m = 2$ are shown to be practical and used by existing planners.

In more advanced version of h^m such as additive h^m [35], and additive-disjunctive heuristic graphs [36], the planners not only use the simple *max* rule to get the longest path to achieve one of the goal subset, but also by analyzing the relationship between different goals subset G' and combining the estimates of plans achieving them, improve the overall informedness of the final heuristic value.

39.3.1.5 Landmarks

While the three previously described heuristics are based on solving the simplified planning problems P' and use the solutions to P' as heuristic estimates of the solution for the original problem, *landmarks* heuristics work quite differently. By definition, a landmark is an invariant; either a state that needs to be visited by any plan on the way to achieving goals, or an action that must be performed in every plan.

Fact landmark: The most obvious fact landmarks are the goals: all plans need to achieve them.[*] The first fact-landmark based work [37] tried to find a "reasonable" ordering between achievement of the goal propositions so that planners can work on achieving them in sequence rather than in parallel. This would allow the planner to *decompose* the original planning task (of achieving all goals) into a sequence of smaller planning tasks that achieve each individual goal, or smaller subset of goals, in sequence. Intuitively, the pair of landmarks l_1 and l_2 has a "reasonable order" between them (i.e., l_1 needs to be achieved before l_2) if it is not possible to reach a state $s_{1,2}$ in which both l_1 and l_2 are True from a state s_2 where only l_2 is True, without having to temporary making l_2 False. Therefore, it is reasonable to achieve l_1 before l_2 to avoid unnecessary effort of having to delete l_2 if it's achieved before l_1. The planning task of achieving both l_1 and l_2 is then broken down into a smaller planning problem P_1 of achieving l_1 first, then when P_1 is solved we can try to find a plan to achieve l_2 from the end state visited when l_1 is achieved. Given that the planning task is generally exponentially harder when the length of the plan increases, being able to decompose the original task into a sequence of smaller planning task can help planners tremendously.

Extending the notion of landmarks beyond goal-ordering is accomplished by identifying other propositions $p \in P$, and the ordering between them, that need to be achieved [38]. These landmark orderings can then be used to guide any state-space planner. To identify landmarks other than goals, Hoffmann et al. [38] first defines the "necessary" ordering between them, which is different from the "reasonable" ordering described earlier: l_1 is *necessarily* ordered before l_2 if all valid solution plans that achieve l_1 before achieve l_2. Referring to Figure 39.1, the goals of imaging Rock1 and Rock2 from the initial state of rover at Lander both require being at the Hill at some point; thus, at(Hill) is a fact landmark of this new class. Utilizing this definition, starting from the goals G as landmarks, the landmark-finding

---

[*] Technically all facts in the initial states are also fact-landmarks, but they are less interesting since there is no ordering between them.

algorithm first considers actions achieving found landmarks, and then identifies shared preconditions between those actions as the next candidate landmarks. In essence, it finds a set of facts l' that are necessarily True before already identified landmarks l need to be achieved through causal relations of actions needing l' as preconditions and achieving l as their effects. This backward reasoning process can continue until it runs out of candidates. Since they are heuristically found, the algorithm also goes through the second step to verify the legitimacy of all found landmark candidates. This validation process is done by evaluating a sufficient (but not necessary) condition on landmark candidates and throwing away those candidates where the condition fails. Specifically, for a given landmark l, a new delete-relaxation planning model D^+ is built where all action delete effects are ignored, and l is removed from all add effect list of all actions. If D^+ is unsolvable, then l is decided to be a fact landmark. When the process terminates, then the algorithm will return a *landmark generation graph*, which is an acyclic directed graph (more specifically a polytree) with nodes are landmarks and the directed edges showing the ordering relationships between them.* The traditional fact-landmarks, can be generalized to "disjunctive" fact-landmark, which is the set of fact in which at least one of them need to be achieved. Later on, we will also see this notion of disjunctive landmark reappear in action landmarks.

LAMA [39] is the first planner that goes beyond using fact-landmarks for decomposition and instead uses them as heuristics guiding the planning search. To do so, it first extends the notion of landmarks to propositional formula over facts represented as multi-valued variables.† Since the planner knows that all landmarks should be achieved, it approximates the goal distance of a state s, reached by a partial plan π_p, by counting the number of landmarks that either are still *not* yet achieved by π_p, or achieved by π_p but are reasoned to be needed again on the way from s to achieving all goals. While the heuristic will assign a non-zero value for any state that does not satisfy all goals (since goals are landmarks), it can assign value non-zero even when s satisfies all goals. This is because the heuristically found landmark generation graph may indicate that some landmarks that were already achieved may still need to be reacquired. Therefore, the goal-checking routine is needed for every visited state.

Action landmark: An action landmark is a disjunctive set of actions l_a for which any valid plan in P should contain at least one action in l_a. This heuristic is used by the Fast Downward family of planners [21]. For any given state s, the distance from s to the goal state is estimated by counting the number of actions that can satisfy all landmarks that are still not satisfied by the current plan leading to s. Note that there is a difference between counting the number of unsatisfied action landmarks and counting the number of actions that can satisfy the action landmarks; this is due to the fact that action landmarks are not independent and they can share actions. For example, $l_1 = \{a_1, a_2\}$ is one landmark while $l_2 = \{a_2, a_3\}$ is another and both of them are not currently satisfied by the plan leading to a state s. While the remaining plan leading from s to the goals need to satisfy both landmarks l_1 and l_2, adding one action a_2 would do so. The above-mentioned simple example demonstrates two main technical issues faced by planners using action landmark-based heuristics: how to find the landmark set, and given the landmark set L that contains multiple landmarks, how to exploit L to compute the most accurate heuristic.

Given that an action landmark is, by definition, the set of actions that are included in any plan that achieves the goals, landmarks are found starting from the *justification graph* that is built backward from the goal to identify actions that are justified to achieve goals and actions that are (definitely) needed to achieve the goals. To build the justification graph, a *precondition choice function* (PCF) that maps each action into one of its preconditions is first defined. For a given PCF, the justification graph is built, with vertices representing facts, and an edge $p \xrightarrow{a} e$ labeled with action a connects the precondition p of a (selected according to the PCF) to the effect e of a. Note that due to the mappings made in the PCF,

---

* Note that computing the set of landmarks exactly is $\mathcal{PSPACE}$-hard [38], which is as hard as solving the original planning problem. Therefore, approximation algorithms are used to compute the landmark generation graph.

† The first version of LAMA [40] still only uses the single-fact and disjunctive landmarks introduced by Hoffmann et al. [38].

there can be multiple edges between two nodes in the justification graph, each one labeled with a different action. Since each goal or needed action's precondition can be achieved by any action that has it as one of the add effect, the set of actions that are relevant (i.e., justified) to a given goal or needed-fact is a disjunctive set. Thus, we only need one of them to be included in the final plan to be able to achieve that goal or needed-fact. After the justification graph is built starting from the goal backward until it reaches the initial state, *cut c* of this graph is a subset of the justification graph's arcs such that all paths from the initial state to goals must use some arc in *c*. It is easily to see that actions representing the arcs in *c* constitutes a landmark as defined earlier. Since actions in general have multiple preconditions, multiple PCFs exist for a given planning problem, and thus multiple justification graphs can be built. Cuts for different graphs constitute different landmark sets.

As outlined previously, the set L of action landmarks is collected by (1) defining multiples PCFs; (2) building justification graphs corresponding to the defined PCFs; (3) conduct the relevant cuts through all justification graphs; and (4) put the set of landmarks representing all cuts through all built justification graph in L. The planner needs to decide on the best way to compute heuristic values from the landmark set L. Some simple ways to utilize landmarks in L, include taking the maximum landmark size in L (admissible), or summing the sizes of all landmark set of L (inadmissible), turned out to be ineffective. The current best approach is to find the smallest set of actions A_{lm} such that A_{lm} contains at least one action from each landmark set l_i, with each landmark $l_i \in L$ is a disjunctive set of actions $\{a_1, a_2, ..., a_n\}$. This is the *NP-complete hitting set* problem that can be solved either by using an exact algorithm, or approximated by posing as a Linear Programming problem. Note that in the most basic classical planning version where actions are not associated with a cost value, then hitting set approach would normally work well. However, landmark-cut heuristics have been used extensively targeting the variation where actions are associated with different cost values. In this scenario, when the number of landmarks are high, then Linear Programming can be a very effective approximation approach compared to the exact hitting-set approaches.

39.3.1.6 Network Flows

The flow-based heuristic [41] is quite different from the other types of heuristics described previously, and is based on the key concept in the network-flow problem: the total amount of flow into the network should be equal to the amount of flow out of the network. By treating each proposition p as a commodity that is either *produced* (i.e., p appears in the action's add list) or *consumed* (p appears in the action's precondition list), the flow-based heuristics try to balance between: (1) the start "quantity" of each state variable available in the initial state; (2) the end quantity of the same variable in the goal state; and (3) the amount of production and consumption of that variables by actions included in the plan, based on the above-mentioned rules outline.

The flow-based heuristic is computed by setting up a Linear Programming (LP) encoding that counts the number of appearance of each action. For example, for a given proposition p that does not appear in the initial state (i.e., initially being False) and appears as one of the goals (i.e., need to be True at the end), assume that there are two action a_1 and a_2 that can produce p while there is a single action a_3 that can consume p. Let $n_1 \geq 0$, $n_2 \geq 0$, and $n_3 \geq 0$ be the number of occurrences of a_1, a_2 and a_3 in a plan, then the network-flow LP constraints should ensure to balance the production and consumption of p by enforcing: $n_1 + n_2 - n_3 \geq 1$.

By minimizing the number of actions in the plan while at the same time balancing the production and consumption of all variables, the flow-based heuristic will be able to estimate the number of times each action may appear in the plan. The key difference between the flow-based heuristic and all other types of heuristics is that it does not try to estimate the real plan by extracting a set of action instances solving a simplified problem, it only estimates *how many times* each action may appear in the final plan. In a more advanced version of the flow-based heuristic, ideas from other heuristics such as landmarks are used to modify the LP formula computing the flow-based heuristic, thereby improving accuracy.

39.3.2 Incomplete Search Techniques: Local Search and Sampling

Incomplete search algorithms trade coverage of the search space and, therefore, either coverage (likelihood to solve decision problems) or solution quality, in return for reduced run-time. In this section, we discuss the two main classes of such incomplete algorithms used in planning, local search and sampling methods.

39.3.2.1 Local Search

Local search algorithms start with a candidate solution, which can be either a complete or partial plan π. It then tries to gradually *refine* π through a sequence of local changes with the most elementary operators involve adding to, removing from, or reordering actions in π. At each step, a local search algorithm will:

1. First, define the set N of eligible local changes to π. This set is called the search *neighborhood* for this search step.
2. Second, heuristically evaluate the local search neighborhood $n \in N$ and choose a move $n^* \in N$.
3. Finally, apply n^* to π to get the new search node: $\pi \leftarrow Apply(n^*, \pi)$.

This process is repeated until a stopping condition is reached, typically either after a pre-defined number of repetitions or if π has been successfully modified from a partial plan to be a complete plan.

The most common heuristic to choose the next search node n^* from the immediate neighborhood is a "greedy" one: the lowest cost neighbor is selected. Lowest cost can either mean a state closest to a valid plan (i.e., requires the fewest number of local moves to become a valid plan), or a state that can be refined into a complete plan with shortest length. While the first cost measurement is designed to minimize algorithm running time, the second one tries to steer search towards a high-quality solution. Many local search algorithms use a cost function that balances between those two cost measures, for example a weighted sum. Moreover, to avoid over-dependence on a node-selection strategy that is heuristic by nature, the local search algorithm may randomly choose the next node using a probability distribution with the costs as input.

To make progress, greedy local search assumes that the node in the neighborhood to move to is *better* than the current node. However, greedy search can be trapped in either *local minimas* where all immediate neighbors are of worse quality than the current node or *plateaus* where all immediate neighbors are of the same quality as the current node. Numerous strategies for avoiding, detecting, and escaping local minima and plateaus have been studied, from (1) injecting a sequence of random moves, random restarts into search (Chapter 17) to (2) adding either temporary or permanent constraints to influence subsequent steps (Chapters 19 and 21), or (3) switch to systematic search algorithm such as breadth-first-search until a descendant node with lower cost outside of the local minima or plateau is reached [19]. All of these strategies are applicable to local search for planning.

One potential problem facing local search algorithms is that of *large neighborhoods* at each search step (Chapter 18), which can be expensive to enumerate and evaluate. One solution to this problem is to bound the neighborhood size, that is, reducing the size of the neighborhood that can be considered at each step to improve the rate at which local moves are made. The possible cost of this strategy is a slow improvement in the optimization criteria.

When feasible plans are easy to find and optimization is required, local search over the space of complete plans is a sensible approach. Planning By Rewriting (PBR) [42] is one such technique. The *plan rewrite rules* are domain-specific local moves that are designed to improve the quality of the current plan. Rules come in four classes: action reordering, collapsing (which reduces the number of actions), expanding (inverse of collapsing), and paralleling (which eliminates ordering constraints). A sample reordering rule from the Rover problem (Figure 39.1) is "if the plan π moves the Rover to satisfy a goal proposition g, then moves back to the current position to accomplish another goal g', reorder g and g' and discard the extra movements from π." Since PBR assumes that the set of plans in the neighborhood is mostly feasible, the local neighborhood can simply be evaluated by the function that determines the cost of each feasible plan in the search neighborhood.

LPG [43] is one of the most well-known local search planners. LPG operates on Linear Action Graphs (LAGs). Similar to the Planning-Graph, described in Section 39.3.1, an action-graph A is a graph interleaving between action and fact *levels*. Each level contains nodes representing either actions or facts. If an action a is in A, then $p \in (pre(a) \cup add(a))$ are also in A and there are connections between a and p. Actiongraphs can have inconsistencies, either in the form of mutual exclusions or propositions not supported by actions. This leads to a natural local search neighborhood: actions can be added to establish support, or removed to eliminate unsupported actions or to resolve mutual exclusions. A LAG τ is an action-graph with at most one action per each action level. When an action is added, τ grows by one level. Conversely, when an action with unsupported preconditions is removed, τ shrinks by one level. Figure 39.3 shows a linear action graph neighborhood for the Rover example where there is only one way to legally insert the `turn-on-drill` action. In general, however, there may be many immediate neighbors to evaluate. LPG only manipulates actions to fix unsupported precondition inconsistencies, which leads to a simpler and smaller search neighborhood.

Since LPG's search neighborhood mostly consists of infeasible plans, the planner estimates the amount of work needed to fix the "flaws" in each plan, and uses this as the heuristic. Given that LPG represents plans using LAGs, it is equal to estimating the work needed to turns an inconsistent LAG τ into a fully consistent one. Let $threats(a, \tau)$ be the set of supporting preconditions of actions in τ that become unsupported if a is added to τ (due to $del(a)$). Let $nopre(a, \tau)$ be the set of unsupported preconditions of a in τ. Also, let $unsup(a', \tau)$ be the set of supporting preconditions of actions in τ that become unsupported if

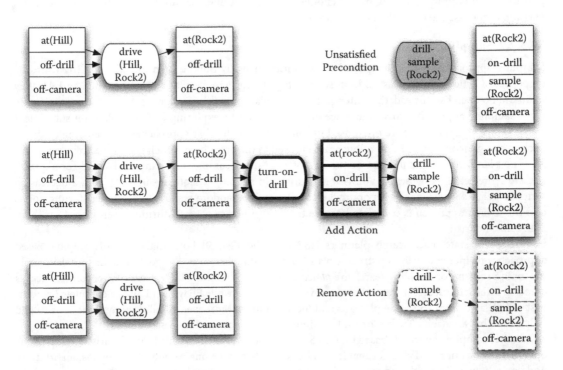

FIGURE 39.3 An example of the search neighborhood for LPG in the Rover domain using Linear Action Graphs. The preconditions for `drill-sample` are unsatisfied, inducing a search neighborhood of size two. There is one legal place for adding `turn-on-drill`; alternately, `drill-sample` can be removed.

another action a' is removed from τ. A simple neighborhood evaluation function for the two local moves corresponding to adding a to τ and removing a' from τ is:

$$E^{add}(a, \tau) = |threats(a, \tau)| + |nopre(a, \tau)|$$
$$E^{remove}(a', \tau) = |unsup(a', \tau)|$$

A more accurate heuristic recursively adds actions to support unsupported facts or remove not-fully-supported actions to reduce the set of total unsupported facts and use $E^{add}(a, \tau)$ and $E^{remove}(a', \tau)$ to heuristically evaluate these moves; an even more advanced heuristic adapt ideas from the delete-relaxation heuristics used in systematic search (Section 39.3.1). Specifically, starting with the unsupported conditions in τ as the initial goal set, a regression planner that ignores action delete effects can extract a solution to the delete-free planning problem π^+. We then count the actions in the π^+ and conflicts between π^+ and τ as the heuristic value estimate the hardness in fixing τ to get a consistent LAG.

In Monte Carlo Random Walk planning [44], local search combined with quick random walks are performed to explore the neighborhood of the current search state. Each transition action is randomly selected from among the applicable actions and several statistical techniques are used for biasing the random selection. Only the end state of each random walk session is evaluated by a heuristic function such as the delete relaxation heuristic. After a number of random walks from the current start state, the search greedily updates to an end state with the minimum heuristic value. A subsequent variant, RWLS [45], performs a local Greedy Best-First Search starting from each state visited during exploration. It also uses well-known enhancements such as delayed evaluation and a second open list containing only states reached via preferred operators. RWLS evaluates the best state s retrieved from an open list, and also performs a random walk starting from s ending in a state r. A linear combination of the heuristic values $h(s)$ and $h(r)$ is used to order the nodes in the open list.

39.3.2.2 Sampling

Sampling approaches use easily applied rules to incrementally build a plan quickly, starting from the empty set of actions. In contrast to local search, they do not remove actions; unlike GBFS and A^*, they maintain no open list. Instead, they attempt to build a plan, and if they fail, restart search from the empty plan. Sampling approaches cover the space of possible plans by exploring different plans on subsequent trials. Sampling approaches are often used in conjunction with other approaches to overcome its drawback of being incomplete. FFs sampling search is used as the first phase, which is followed by a variant of A^* if no plan is found and this can be thought of as a "single sample" mechanism. To find good quality plans, Richter et al. [46] uses an admissible heuristic in running a sequence of WA^* with a decreasing series of weights w to converge on optimal plans. It also uses random restarts to get better starting points and the overall approach is complete because the final run uses $w = 1$, turning itself into A^* with an admissible heuristic.

Variants of state-space search planners HSP [18] and FF [19] both make use of sampling-based approaches to incrementally construct a plan from the initial state to the goal. Similar to local search, the main problems of sampling search are *plateaus* and *local minimas*. HSP restarts if it becomes stuck, as detected by lack of successor states that improve the heuristic, while FF while FF uses a complete breadth-first-search algorithm to escape plateaus and local minimas. Moreover, FF also analyzes actions to add only those that are deemed to be "helpful" to limit the set of potential successor nodes.

Another sampling approach based on GBFS [47] changes mode if the minimum heuristic value on the open list is not improved after a tunable number of node expansions. When this happens, an arbitrary node on the open list is selected for GBFS to expand from and the process repeats. This approach works well using many heuristics on a variety of planning problems.

A recent innovation in sampling approaches to planning [48] uses a large variety of techniques to construct *probes*. The actions in a probe are selected to achieve ordered sequences of landmarks (see Section 39.3.1) that are yet to be achieved. More specifically, the probes use a version of the h^{sum} heuristic that incorporates the notion of *commitment* to improve the estimate. A commitment is an action a_1 with $p \in add(a_1)$ and a set of fluents B; the intent is that p will ensure some fluent in B is added by some subsequent action a_2, that is, $p \in pre(a_2)|b \in (add(a_2) \cap B)$. Let C in the following formula be a set of commitments, and let $O(p)$ be the set of actions a with $p \in add(a)$. The modified heuristic function is

$$\delta(a, s, C) = \max_i \min_{q \in B_i} h(q|s, C)$$
$$h(a|s_i, C) = \delta(a, s_i, C) + h(Pre(a), s_i, C)$$
$$h(p|s, C) = \min_{a \in O(p)} cost(a) + h(a|s, C)$$

The probe is constructed as follows. The landmarks are extracted, and orderings among them are computed. An earliest unachieved goal is selected using the orderings. Once the goal is selected, the action selection routine first regresses from the goal, using minimum cost actions $h(a|s_i, C)$ to build a relaxed plan. The *helpful* actions for a subgoal g in a node $n = s$, C are the actions a with heuristic estimate $h(a|s, C) = 0$ that add a precondition or goal in the relaxed plan; one such action is added to the probe. The action a consumes a commitment (a_i, p_i, B_i) if a adds a fluent in B_i. Similarly, a makes the commitments if p is a fluent added by a, and B is the set of fluents added by actions in the relaxed plan in n that have p as a precondition. Probes can be used as a lookahead mechanism from each expanded state in a standard GBFS.

39.3.3 Reformulation

Reformulation maps a planning problem $\langle D, i, G \rangle$ to a different representation. These transformations offer several potential advantages: (1) we can use "off-the-shelf" solvers of many different types for the new problem in the new representation, then map the solutions back to plans; (2) we can exploit problem structure that is more easily discernible in the new representation; and (3) we can use problem relaxations normally not afforded to us when solving the original problem.

Making use of a reformulation often requires some careful thought. For example, if we use a reformulation that loses no information, this eases the proof that a solution to the reformulated problem can directly map to a plan, and that no solution to the new problem means there is no valid plan. On the other hand, if we intend to use the reformulation as a source for heuristic calculation, we can instead use an abstract reformulation that relaxes certain constraints from the original planning problem.

We will initially look at two popular reformulation targets used for classical planning: SAS+ (Simplified Action Structures+) and propositional satisfiability (SAT). We then discuss other formulations approaches such as to Constraint Satisfaction Problems (CSP), and Mixed Integer-Linear Programming (MILP) problems.

39.3.3.1 SAS+

The Simplified Action Structures+ (SAS+) [49] formalism differs from STRIPS in that it allows multi-valued variables rather than only propositions. Whereas in our simple Rovers example (Figure 39.1), multiple STRIPS at propositions are needed to define the location of the rover, only a single SAS+ variable is required, with values represent all possible locations a rover may go. In other words, action effects provide value assignments to variables. This differs from STRIPS, which has *add* and *delete* effects on binary variables, making them True or False respectively. Preconditions and goals become checks on assigned values. When an action has a precondition on a variable value, and then changes that value as an effect, we can view that action as causing a value transitioning of that given variable from one value to the next. In this case, these variable value checks are called *preconditions*. Variables that need to remain unchanged at action execution are called *prevail conditions*. SAS+ provides a natural way of viewing actions as transitions on atomic graphs, represented as types of state machines. Each precondition-effect pair provides a transition on the state machine, and each prevail condition defines conditions on the states of other machines. The resulting structure is called a *domain transition graph* [21]. This provides a more compact representation of the original planning problem and multiple advantages exist for SAS+ compared to STRIPS, including (1) a reduced number of impractical (though valid) states, (2) more obvious problem decomposition, and (3) explicit inclusion of implicit STRIPS mutual exclusions (e.g., a rover cannot be in two places at once). The planner Fast Downward effectively uses all of these benefits to some degree. In its original incarnation, the Fast Downward *causal graph* heuristic took advantage of reduced state space, problem decomposition and inclusion of mutual exclusions splitting each variable into separate but coupled problems [21].

A translation between STRIPS and SAS+ is described in [50]. To provide intuition about the translation approach, consider the at literals in our simple rover domain (Figure 39.1). Assuming the initial state declares only one at true, then the actions in the domain will always ensure that one at literal will remain true throughout a plan. To prove this, we need to only look for "balanced" effects: whenever an action deletes (makes false) an at literal, it also adds (makes true) another at literal. In this case, that happens for all instances of the drive action. Similar analysis can be done to other literals . For instance, the translation algorithm can notice that image and no-image, as well as sample and no-sample have a mutual exclusion relationship. For the SAS+ formulation of our sample, see Figure 39.4. A general STRIPS/PDDL to SAS+ translation beyond the simple example that we outlined previously requires additional more complex domain analysis; the reader is referred to [51] for more details.

39.3.3.2 Propositional Satisfiability

Extensive research has gone into encoding planning problems as propositional satisfiability (SAT) problems. Given a propositional formula, a SAT problem asks whether there exists a set of True and False values for each variable in the formula such that the propositional formula will return True [52]. SAT is $\mathcal{NP}$-complete problem, and since $\mathcal{NP} \subseteq \mathcal{PSPACE}$, SAT cannot fully represent the complete space of STRIPS planning, which is $\mathcal{PSPACE}$-hard. Therefore, we cannot in general map every planning problem to a SAT problem. In practice, to encode a planning problem in SAT, we need to limit the *plan horizon*,

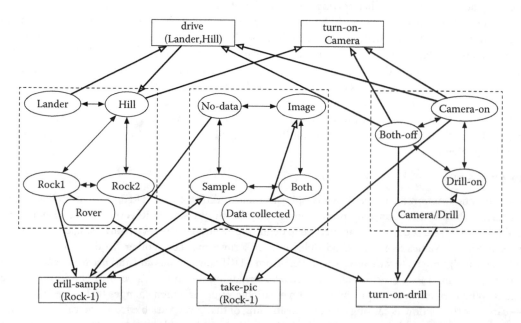

FIGURE 39.4 The Rover domain in the SAS+ formalism. Only one instance of each action is provided for simplicity. Ovals are values of state variables, boxes represent actions. State variables are enclosed in a dashed box. State variable value transitions are shown with black arrows, illustrating each domain transition graph, action preconditions and effects are shown with white arrows.

or number of expected steps a solution may have. If the reformulated SAT problem returns no solution, the *plan horizon* can be increased and another attempt can be made to solve the new SAT encoding representing the newly increased horizon.

One early encoding of planning as SAT from the planner SATPlan [53] works like this: (1) actions imply their preconditions at the previous time-step and their effects at the current time-step; (2) only one action occurs per time-step; (3) facts in the initial state are true in the time-step 0 (i.e., first time-step); and (4) the *frame* axiom holds (facts that are true in one time-step and are not affected by actions executed at that time step will hold in the next time-step). An example of this SAT encoding for our Rover domain is shown in Figure 39.5. Once the translation is done, techniques to manipulate the SAT encoding include using unit propagation (identifying clauses of a single literal, assigning the value that satisfies this literal, and reducing the formula accordingly) and posting single- and double-literal assertions to generate newly inferred clauses.

A later SAT transformation [54] allows concurrent actions and eliminates the SAT variables representing actions from the encoding; all that remains are logical clauses expressing the possible transformations of propositions truth values between consecutive time-steps if a set of actions is invoked. Other encodings have used the Planning-Graph as a basis, hence the plan horizon is based on the size of the Planning-Graph. This reduces the size of the SAT encoding, and thus can improve SAT solver performance.

More efficient encodings have focused on using a *relaxed* notion of parallelism between actions. Specifically, it is possible to find a correct plan where actions can appear simultaneously provided they can be totally ordered so as not to interfere with another set of actions. These are called ∃-step plans. When combined with parallel-search strategies, these encodings lead to very efficient algorithms [55].

39.3.3.3 Other Transformations

Several other reformulation targets and algorithms have been used, most of which have similar plan horizon restrictions as SAT reformulations. For instance, the mixed integer linear program (MILP) planner OptiPlan [56] uses the Planning-Graph structure to define constraints in much the same way as the SAT encodings described earlier, but it uses a integer program as the solving substrate and uses a

Actions imply preconditions and effects	Only one action at a time	Frame axioms (summarized)
(drive-Hill-Rock1–i) => ((at-Hill–i)∧ (can-drive-Hill-Rock1-i) ∧ (at-Rock1-i+1) ∧ (at-Hill-i+1))	((drive-Hill-Rock1–i) ∧ (turn-on-Camera–i) ∧ ... (Sample-Rock1–i)) ∧	((take-pic-Rock–i) ∧ (true-prop–i)) => (true-prop–i+1)
(turn-on-Camera–i) => ((off-Camera–i) ∧ (off-Drill-i) ∧ (ready-Camera-i+1) ∧ (off-Camera-i+1))	((drive-Hill-Rock1-i ∧ (turn-on-Camera–i) ∧ ... (Sample-Rock1–i)) ∧	((drive-Hill-Rock1–i) ∧ true-prop–i)) => (true-prop-i+1)
(take-pic-Rock1–i) => ((at-Rover-Rock1-i) ∧ (ready-Camera–i)∧ (image-Rock1-i+1)) ...	((drive-Hill-Rock1–i) ∧ (turn-on-Camera–i) ∧ ... (Sample-Rock1–i))	((turn-on-Camera–i) ∧ true-prop–i)) => (true-prop-i+1)) ...

FIGURE 39.5 A partial SATPlan encoding of the example Rover problem. This represents a partial encoding. The first column contains constraints related to implications of actions being placed in a plan, the second column contains constraints enforcing serial, nonparallel plans, and the last column the frame axioms that ensure facts that are True remain True and that are False remain False between steps unless an action has an effect on them.

notion of *state change* on variables (from True to False) similar to the domain transition graphs of SAS+. Further innovations in using integer programming as heuristics for search are described in Reference 57. Some of these directly use the SAS+ reformulation of a STRIPS problem, then use linear program relaxations. This means that variable domains typically defined within the set $\{0, 1\}$ are relaxed to be within the interval $[0, 1]$. These formulations define planning heuristics [58,59] for search rather than directly solving the problem within an integer program. These reformulations recast the planning problem as a network flow, where the objective is to find the optimal flow through the network, with some additional constraints added to handle actions with multiple effects (i.e., which can be seen as a mutual flow).

The Planning-Graph structure can be transformed into a Constraint Satisfaction Problem (CSP) [60, 61]. A CSP problems consists of a set of variables, a (usually) finite domain of values it can take on, and a set of constraints that restricts the joint values of variables. In CSP encodings derived from the Planning-Graph, each proposition at each level of the Planning-Graph becomes a variable, and the actions that support it (i.e., actions that make the corresponding proposition True at that level) become values it may take on. Constraints involve ensuring each proposition acquires support from an action, and that mutually exclusive actions are not added to the plan. These include *no-op* actions that mean the proposition persists from the previous planning step.

39.3.4 Subproblem Identification and Decomposition

Structure and subproblem identification is another powerful tool for guiding search. One class of techniques involves identifying common combinatorial problems as subcomponents of planning problems. Such problems can then be solved with specialized, efficient solvers. The other class of techniques identifies structures like partitions and ordering restrictions, which are then used to split the planning problem into smaller, nearly independent planning problems that can be solved quickly with general-purpose planners, after which the resulting plans can easily be merged.

The best known technique for combinatorial subproblem identification uses *type extraction* [62] to identify sets of related propositions and the relations between them. This is done by building up state-machines whose nodes are propositions, and whose transitions are actions. For example, the characteristics of a transportation domain are a set of location propositions, at(x), and a mobility action move(x, y) that enables a change in location. These techniques can identify interesting and nonobvious subproblems. An example in Reference 62 shows how a wall-painting domain leads to a transportation subproblem due to constraints on what colors walls can be painted; that is, the walls are mobile along a map of colors. Type extraction is also extensible to "dependent" maps (e.g., travelers can only travel places with passports), transported objects (e.g., packages carried in trucks), and multiple move propositions (e.g., drive and fly operators). Furthermore, this approach can be extended to identify other types of subproblem structure such as packing or scheduling. A significant disadvantage of these techniques is that humans must write the subproblem descriptions. Further work is needed when multiple subproblems can be extracted from the same parent problem instance.

A more general approach to identifying problem structures relies on translating planning problems into SAS+ [49], using the transformation algorithm discussed in Section 39.3.3.1. In the SAS+ representation, a planning problem consists of variables with finite domains, and actions cause subsets of these variables to change values. A SAS+ domain's *causal graph* is a directed graph over the variables of the problem; there is an edge $u \rightarrow v$ if either a prevail condition on u causes v to change value, or an action causes both u and v to change value. The causal graph can be viewed as superimposed over the set of domain transition graphs to represent the dependencies between each of them. A subclass of SAS+ problem instances restricts goals to variables with zero out-degree in the causal graph. A polynomial time approximation algorithm for these problems enumerates paths of value transitions from shortest to longest [21]; each path is checked to ensure that the transitions satisfy the goal, and that the action descriptions for the low-level goals are satisfied. These plans are converted into distance estimates by summing the number of value transitions for the high-level variable plan and the shortest paths between the initial and final values for the low-level variable. This approach sums transitions and resembles the h^{sum} heuristic, described in Section 39.3.1; unifying these approaches has been explored with notable success [63].

Another class of structure identification and exploitation techniques relies on identifying subproblems that are solved with a planner instead of special-purpose combinatorial solvers. This approach relies on identifying small subproblems and guaranteeing that an overall solution can be constructed from the subproblems. As described in Section 39.3.1, fact landmarks can be used to divide and conquer a problem in such a way that the planner itself can be used to tackle smaller, easier to achieve problems, thereby improving solving time. This is an elegant approach because no special subsolver is needed.

39.4 Advanced Topics

While we have limited the discussion to classical planning problems, real world planning applications frequently require more expressive planning models. STRIPS cannot easily represent some very common, important types of constraints such as time (e.g., to complete a task, an action needs to spend non-zero duration executing) [12,64–66], metric resources (e.g., actions consume fuel or battery power), goal selection (a variation of optimization in which the planner must decide what goals can be achieved), and uncertainty (e.g., action failure). Handling these types of constraints requires extensions to the methods discussed here.

When it comes to planning problems involving transportation and optimization, a common subproblem is the Orienteering problem or Prize-collecting TSP (Chapter 34). The Orienteering problem is quite simple compared to the an optimal planning problem, consisting of a graph with weighted nodes and edges, where the node weights correspond to edge traversal costs, while node weights correspond to rewards for visiting each edge. Nevertheless, this problem generalizes classical planning, in that goals are *disjunctive*; the optimization problem is not to minimize plan length or action cost, but maximize the number of goals achieved.

Methods for automatically constructing and using the Orienteering problem for goal selection-based optimal planning are described in Reference 67. The extraction of the Orienteering problem consists of three phases. The first phase involves a sensitivity analysis step, in which the ground propositions are analyzed for the variable cost of goal achievement. This is done by constructing a relaxed plan for each goal g and calculating costs. For each proposition p in the final plan for g, assume p was true initially, and that propositions mutually exclusive with p are infinitely expensive, then recalculate the cost; a node of the Orienteering problem comprises the set of all propositions for which the cost of goal achievement varies significantly when this proposition is assumed to be true (using a parameter passed to the algorithm). Edge costs are computed by applying each action to each possible state which is both reachable from the initial state, and consists only of propositions in the seed set; this is calculable using a normal Planning-Graph. The Planning-Graph is used to estimate the costs of achieving goals at each state of the resulting problem using the same cost function; this process adds new nodes (states derived from the Planning-Graph) and edges. Approximate solutions to the resulting Orienteering problem then provide heuristics to a classical planner, which can either be an approximate or complete planner.

Table 39.1 provides an overview for further reading on approximation and metaheuristic techniques handling models beyond classical planning.

TABLE 39.1 Further Reading for Advanced Topics.

	Time	Resources	Uncertainty	Oversubscription
Graph plan	[64]	[68]	[69,70]	[67]
Heuristic search	[65,66,71]	[65,71]	[11]	[72,73]
Local search	[74]	[75]	[76]	
SAT/SMT	[77]	[78]	[79]	[80]

Each column represents an extension to STRIPS planning, and each row a methodology to finding solutions.

39.5 Conclusion

In this chapter, we reviewed diverse approximation and metaheuristic techniques that have found use in modern planners. They range from finding informative heuristics through solving a simplified version of the original problem to identification of goal and landmark orderings that allow the planners to decompose the original problem into smaller components that can be solved in sequence or parallel, and extraction of subproblem structures that can be solved by specialized solvers. The various heuristics extracted from solving simplified (or approximated) problems have been used effectively in a wide range of state-of-the-art planners following different planning frameworks (e.g., state-space, plan-space, or reformulation) that utilize different search algorithms (e.g., systematic, local search, or sampling). While we have focussed our discussion on the most fundamental form of classical planning, we have also outlined briefly how the same techniques have been expanded and used effectively in more advanced planning setting such as planning with temporal and numeric resource constraints or stochastic and probabilistic planning. In summary, approximation and metaheuristics approaches have been the key driver behind some of the most effective recent development in planning (e.g., informative heuristic) and they will continue to play an even more important role in both theoretical and application-oriented research in automated planning.

Acknowledgment

The authors gratefully acknowledge Ari Jónsson for his role in writing the first version of this chapter. This work was funded by the NASA Advanced Exploration Systems Program.

References

1. Muscettola, N., Nayak, P., Pell, B., and Williams, B., Remote agent: To boldly go where no AI system has gone before, *Artificial Intelligence*, 103(1–2), 5–47, 1998.
2. Bresina, J., Jonsson, A., Morris, P., and Rajan, K., Activity planning for the mars exploration rovers, in *Proceedings of the 15th International Conference on Automated Planning and Scheduling*, 2005, pp. 40–49.
3. McCarthy, C.E. and Pollack, M.E., A plan-based personalized cognitive orthotic, in *Proceedings of the 6th Conference on Artificial Intelligence Planning Systems*, 2002, pp. 243–252.
4. Golden, K., Pang, W., Nemani, R., and Votava, P., Automating the processing of earth observation data, in *Proceedings of the 7th International Symposium on Artificial Intelligence, Robotics and Space*, 2003.
5. Hoffmann, J., Weber, I., and Kraft, F.M., SAP speaks PDDL: Exploiting a software-engineering model for planning in business process management, *Journal of Artificial Intelligence Research*, 44(1), 587–632, 2012.
6. Boddy, M., Gohde, J., Haigh, T., and Harp, S., Course of action generation for cyber security using classical planning, in *Proceedings of the 15th International Conference on Automated Planning and Scheduling*, 2005, pp. 12–21.
7. Hoffmann, J., Simulated penetration testing: From "Dijkstra" to "Turing Test++", in *Proceedings of the 25th International Conference on Automated Planning and Scheduling*, 2015, pp. 364–372.
8. Ruml, W., Do, M.B., and Fromhertz, M., On-line planning and scheduling for high speed manufacturing, in *Proceedings of the 22nd International Conference on Automated Planning and Scheduling*, 2005, pp. 30–39.
9. Do, M.B., Okajima, K., Uckun, S., Hasegawa, F., Kawano, Y., Tanaka, K., Crawford, L.S., Zhang, Y., and Ohashi, A., Online planning for a material control system for liquid crystal display manufacturing, in *Proceedings of the 25th International Conference on Artificial Intelligence Planning Systems*, 2011, pp. 50–57.

10. Traverso, P., Ghallab, M., and Nau, D., *Automated Planning: Theory and Practice*, Morgan Kauffman, San Francisco, CA, 2004.

11. Mausam A.K., *Planning with Markov Decision Processes: An AI Perspective*, Morgan & Claypool, San Rafael, CA, 2012.

12. Fox, M. and Long, D., PDDL2.1: An extension to PDDL for expressing temporal planning domains, *Journal of Artificial Intelligence Research*, 20(1), 61–124, 2003.

13. Bylander, T., The computational complexity of propositional STRIPS planning, *Artificial Intelligence*, 69(1–2), 165–204, 1994.

14. Hart, P.E., Nilsson, N.J., Raphael, B., A formal basis for the heuristic determination of minimum cost paths, *Transactions on Systems Science and Cybernetics*, 4(2), 100–107, 1968.

15. Dechter, R. and Pearl, J., Generalized best-first search strategies and the optimality of A*, *Journal of the ACM*, 32(3), 505–536, 1985.

16. Pohl, I., Heuristic search viewed as path finding in a graph, *Artificial Intelligence*, 1(3–4), 193–204, 1970.

17. Helmert, M. and Röger, G., A beginner's introduction to heuristic search planning, in *Tutorial at The 29th AAAI Conference on Artificial Intelligence (AAAI-15)*, 2015.

18. Bonet, B. and Geffner, H., Planning as heuristic search, *Artificial Intelligence*, 129(1–2), 5–33, 2001.

19. Hoffmann, J. and Nebel, B., The FF planning system: Fast plan generation through heuristic search, *Journal of Artificial Intelligence Research*, 14(1), 253–302, 2001.

20. Helmert, M. and Domshlak, C., Landmarks, critical paths and abstractions: What's the difference anyway? in *Proceedings of the 19th International Conference on Automated Planning and Scheduling*, 2009.

21. Helmert, M., The fast downward planning system, *Journal of Artificial Intelligence Research*, 26(1), 191–246, 2006.

22. Keyder, E., Hoffmann, J., and Haslum, P., Semi-relaxed plan heuristics, in *Proceedings of the 15th International Conference on Automated Planning and Scheduling*, 2012, pp. 2126–2128.

23. Haslum, P. and Geffner, H., Admissible heuristics for optimal planning, in *Proceedings of the 5th International Conference on Artificial Intelligence Planning Systems*, 2000, pp. 140–149.

24. Blum A. and Furst, M., Fast planning through planning graph analysis, in *Proceedings of the 24th International Joint Conference on Artificial Intelligence*, 1995, pp. 1636–1642.

25. Nguyen, X. and Kambhampati, S., Extracting effective and admissible state-space heuristics from the planning graph, in *Proceedings of the 17th National Association for Artificial Intelligence*, 2000, pp. 798–805.

26. Nguyen, X., Kambhampati, S., and Nigenda, R.S., Planning graph as the basis for deriving heuristics for plan synthesis by state space and CSP search, *Artificial Intelligence*, 135(1–2), 73, 2002.

27. Edelkamp, S., Planning with pattern databases, in *Proceedings of European Conference on Planning (ECP-01)*, 2001, pp. 13–34.

28. Haslum, P., Helmert, M., Bonet, B., Botea, A., and Koenig, S., Domain-independent construction of pattern database heuristics for cost-optimal planning, in *Proceedings of the 21st AAAI Conference on Artificial Intelligence (AAAI-07)*, 2007.

29. Pommerening, F., Röger, G., and Helmert, M., Getting the most out of pattern databases for classical planning, in *Proceedings of the 23rd International Joint Conference on Artificial Intelligence (IJCAI 2013)*, 2013, pp. 2357–2364.

30. Katz, M. and Domshlak, C., Optimal admissible composition of abstraction heuristics, *Artificial Intelligence*, 174(12–13), 2010, pp. 767–798.

31. Edelkamp, S., Symbolic pattern databases in heuristic search planning, in *Proceedings of the 6th International Conference on Artificial Intelligence Planning Systems*, 2002, pp. 274–283.

32. Helmert, M., Haslum, P., and Hoffmann, J., Flexible abstraction heuristics for optimal sequential planning, in *Proceedings of the Seventeenth International Conference on Automated Planning and Scheduling (ICAPS 2007)*, 2007, pp. 176–183.

33. Seipp J. and Helmert, M., Counterexample-guided Cartesian abstraction refinement, in *Proceedings of the 23rd International Conference on Automated Planning and Scheduling (ICAPS 2013)*, 2013, pp. 347–351.

34. Katz, M. and Domshlak, C., Structural pattern heuristics via fork decomposition, in *Proceedings of the 18th International Conference on Automated Planning and Scheduling (ICAPS 2008)*, 2008, pp. 182–189.

35. Haslum, P., Bonet, B., and Geffner, H., New admissible heuristics for domain-independent planning, in *Proceedings of the 19th AAAI Conference on Artificial Intelligence (AAAI-05)*, 2005, pp. 9–13.

36. Coles, A., Fox, M., Long, D., and Smith, A., Additive-disjunctive heuristics for optimal planning, in *Proceedings of the 18th International Conference on Automated Planning and Scheduling (ICAPS 08)*, 2008, pp. 44–51.

37. Koehler, J. and Hoffmann, J., On reasonable and forced Goal orderings and their use in an agenda-driven planning algorithm, *Artificial Intelligence*, 12(1), 338–386, 2000.

38. Hoffmann, J., Porteous, J., and Sebastia, L., Ordered landmarks in planning, *Journal of Artificial Intelligence Research*, 22(1), 215–278, 2004.

39. Richter, S. and Westphal, M., The LAMA planner: Guiding cost-based anytime planning with landmarks, *Journal of Artificial Intelligence Research*, 39(1), 127–177, 2010.

40. Richter, S., Helmert, M., and Westphal, M., Landmarks revisited, in *Proceedings of the 22nd AAAI Conference on Artificial Intelligence (AAAI-08)*, 2008, pp. 975–982.

41. Bonet, B. and Van Den Briel, M., Flow-based heuristics for optimal planning: Landmarks and merges, in *Proceedings of the 24th International Conference on Automated Planning and Scheduling*, 2014, pp. 47–55.

42. Ambite, J.L. and Knoblock, C.A., Planning by rewriting, *Journal of Artificial Intelligence Research*, 15, 207–261, 2001.

43. Gerevini, A., Saetti, L., and Serina, I., Planning through stochastic local search and temporal action graphs, *Journal of Artificial Intelligence Research*, 20(1), 239–290, 2003.

44. Nakhost, H. and Müller, M., Monte-Carlo exploration for deterministic planning, in *Proceedings of the 21st International Joint Conference on Artificial Intelligence*, 2009, pp. 1766–1771.

45. Xie, F., Nakhost, H., and Müller, M., Adding local exploration to Greedy best-first search in satisficing planning, in *Proceedings of the 28th National Conference on Artificial Intelligence*, 2012, pp. 2388–2394.

46. Richter, S., Thayer, J., and Ruml, W., The joy of forgetting: Faster anytime search via restarting, in *Proceedings of the 20th International Conference on Automated Planning and Scheduling*, 2010.

47. Xie, F., Muller, M., and Holte, R., Understanding and improving local exploration for GBFS, in *Proceedings of the 25th International Conference on Automated Planning and Scheduling*, 2015, pp. 244–248.

48. Lipovetzky, N. and Geffner, H., Searching for plans with carefully designed probes, in *Proceedings of the 21st International Conference on Automated Planning and Scheduling*, 2011, pp. 154–161.

49. Bäckström, C. and Nebel, B., Complexity results for SAS$^+$ planning, *Computational Intelligence*, 11(4), 625–655, 1995.

50. Edelkamp, S. and Helmert, M., Exhibiting knowledge in planning problems to minimize state encoding, in *Proceedings of the 5th European Conference on Planning*, 1999, pp. 135–147.

51. Helmert, M., Concise finite-domain representations for PDDL planning tasks, *Artificial Intelligence*, 173(5), 503–535, 2009.

52. Cook, S. A., The complexity of theorem-proving procedures, in *Proceedings of the 3rd Annual ACM Symposium on Theory of Computing*, 1971, pp. 151–158.

53. Kautz, H. and Selman, B., Planning as satisfiability, in *Proceedings of the 10th European Conference on Artificial Intelligence*, 1992, pp. 359–363.

54. Kautz, H. and Selman, B., Unifying SAT-based and graph-based planning, in *Proceedings of the 16th International Joint Conference on Artificial Intelligence*, 1999, pp. 318–325.

55. Rintanen, J., Engineering efficient planners with SAT, in *Proceedings of the European Conference on AI*, 2012, pp. 684–689.

56. van den Briel, M. and Kambhampati, S., OptiPlan: Unifying IP-based and graph-based planning, *Journal of Artificial Intelligence Research*, 24, 919–931, 2006.

57. Bylander, T., A linear programming heuristic for optimal planning, in *Proceedings of the 14th National Conference on Artificial Intelligence*, 1997, pp. 694–699.

58. van den Briel, M., Benton, J., Kambhampati, S., and Vossen, T., An LP-based heuristic for optimal planning, in *Proceedings of the 13th International Conference on Principles and Practice of Constraint Programming*, 2007, pp. 651–665.

59. Bonet, B., An admissible heuristic for SAS+ planning obtained from the state equation, in *Proceedings of the 23rd International Joint Conference on Artificial Intelligence*, 2013, pp. 2268–2274.

60. Do, M.B. and Kambhampati, S., Solving planning-graph by compiling it into CSP, in *Proceedings of the 5th International Conference on Artificial Intelligence Planning Systems*, 2000, p. 82.

61. Lopez, A. and Bacchus, F., Generalizing graphplan by formulating planning as a CSP, in *Proceedings of the 18th International Joint Conference on Artificial Intelligence*, 2003, pp. 954–960.

62. Long, D. and Fox, M., Automatic synthesis and use of generic types in planning, in *Proceedings of the 5th Artificial Intelligence Planning Systems*, 2000, pp. 196–205.

63. Helmert, M. and Geffner, H., Unifying the causal graph and additive heuristics, in *Proceedings of the 18th International Conference on Automated Planning and Scheduling*, 2008, pp. 140–147.

64. Smith, D. and Weld, D., Temporal planning with mutual exclusion reasoning, in *Proceedings of the 16th International Joint Conference on Artificial Intelligence*, 1999, pp. 326–333.

65. Do, M. and Kambhampati, S., Sapa: A multi-objective metric temporal planner, *Journal of Artificial Intelligence Research*, 20, 155–194, 2003.

66. Cushing, W., Kambhampati, S., and Weld, D., When is temporal planning really temporal? in *Proceedings of the 20th International Joint Conference on Artificial Intelligence*, 2007, pp. 1852–1859.

67. Smith, D., Choosing objectives in over-subscription planning, in *Proceedings of the 14th International Conference on Automated Planning and Scheduling*, 2004, pp. 393–401.

68. Koehler, J., Planning under resource constraints, in *Proceedings of the 13th European Conference on Artificial Intelligence*, 1998, pp. 489–493.

69. Weld, D., Anderson, C., and Smith, D., Extending graphplan to handle uncertainty & sensing actions, in *Proceedings of the 15th National Conference on Artificial Intelligence*, 1998, pp. 897–904.

70. Weber, C. and Bryce, D., Planning and acting in incomplete domains, in *Proceedings of the 21th International Conference on Automated Planning and Scheduling*, 2011, pp. 274–281.

71. Coles, A.J., Coles, A.I., Fox, M., and Long, D., COLIN: Planning with continuous linear numeric change, *Journal of Artificial Intelligence Research*, 44, 1–96, 2012.

72. Benton, J., Do, M., and Kambhampati, S., Anytime heuristic search for partial satisfaction planning, *Artificial Intelligence*, 173(5), 562–592, 2009.

73. Domshlak, C. and Mirkis, V., Deterministic oversubscription planning as heuristic search: Abstractions and Reformulations, *Journal of Artificial Intelligence Research*, 52, 97–169, 2015.

74. Gerevini, A., Saetti, A., and Serina, I., Temporal planning with problems requiring concurrency through action graphs and local search, in *Proceedings of the 20th International Conference on Automated Planning and Scheduling*, 2010, pp. 226–229.

75. Hoffmann, J., The Metric-FF planning system: Translating "Ignoring Delete Lists" to numeric state variables, *Journal of Artificial Intelligence Research*, 20(1), 291, 2003.

76. Domshlak, C. and Hoffmann, J., Probabilistic planning via heuristic forward search and weighed model counting, *Journal of Artificial Intelligence Research*, 30, 565–620, 2007.

77. Shin, J.A. and Davis, E., Processes and continuous change in a SAT-based planner, *Artificial Intelligence*, 166(1), 194–253, 2005.

78. Hoffmann, J., Gomes, C.P., Selman, B. and Kautz, H.A., SAT encodings of state-space reachability problems in numeric domains, in *Proceedings of the 20th International Joint Conference on Artificial Intelligence*, 2007, pp. 1918–1923.

79. Palacios, H. and Geffner, H., Mapping conformant planning into SAT through compilation and projection, in *Proceedings of the Conference of the Spanish Association for Artificial Intelligence*, 2005, pp. 311–320.

80. Juma, F., Hsu, E.I., and McIlraith, S.A., Preference-based planning via MaxSAT, in *Proceedings of the 25th Canadian Conference on Artificial Intelligence*, 2012, pp. 109–120.

81. Fikes, R. and Nilsson, N., STRIPS: A new approach to the application of theorem proving to problem solving, *Artificial Intelligence*, 2(3/4), 189–208, 1971.

82. Kautz, H., McAllester, D., and Selman, B., Encoding plans in propositional logic, in *Proceedings of the 5th International Conference on the Principle of Knowledge Representation and Reasoning*, 1996, pp. 374–384.

83. Nguyen, X. and Kambhampati, S., Reviving partial ordering planning, in *Proceedings of the 17th International Joint Conference on Artificial Intelligence*, 2001, pp. 459–464.

84. van den Briel, M., Sanchez Nigenda, R., Do, M.B., and Kambhampati, S., Effective approaches for partial satisfaction (oversubscription) planning, in *Proceedings of the 19th National Conference on Artificial Intelligence*, 2004, pp. 562–569.

85. Vossen, T., Ball, M., Lotem, A., and Nau, D., On the use of integer programming models in AI planning, in *Proceedings of the 16th International Joint Conference on Artificial Intelligence*, 1999, pp. 304–309.

86. Sanchez Nigenda, R. and Kambhampati, S., Planning graph heuristics for selecting objectives in over-subscription planning problems, in *Proceedings of the 15th International Conference on Automated Planning and Scheduling*, 2005, pp. 192–201.

87. Do, M.B. and Kambhampati, S., Improving temporal flexibility of position constrained metric temporal plans, in *Proceedings of the 13th International Conference on Automated Planning and Scheduling*, 2003, pp. 42–51.

88. Bäckström, C., Computational aspects of reordering plans, *Journal of Artificial Intelligence Research*, 9(1), 99–137, 1998.

89. Buttner, M. and Rintanen, J., Improving parallel planning with constraints on the number of actions, in *Proceedings of the 15th International Conference on Automated Planning and Scheduling*, 292, 2005.

90. Wah, B. and Chen, Y., Subgoal partitioning and global search for solving temporal planning problems in mixed space, *International Journal on Artificial Intelligence Tools*, 13(4), 767, 2004.

91. Kautz, H. and Walser, J.P., State-space planning by integer optimization, in *Proceedings of the 16th National Conference on Artificial Intelligence*, 1999, pp. 526–533.

92. Kautz, H. and Selman, B., Pushing the envelope: Planning, propositional logic and stochastic search, *Proceedings of the 13th National Conference on Artificial Intelligence*, 1996, pp. 1194–1201.

93. Smith, D., Frank, J., and Jónsson, A., Bridging the gap between planning and scheduling, *Knowledge Engineering Review*, 15(1), 47–83, 2000.

94. Srivastava, B., Realplan: Decoupling causal and resource reasoning in planning, in *Proceedings of the 17th National Conference on Artificial Intelligence*, 2000, pp. 812–818.

95. Tran, D., Chien, S., Sherwood, R., Castaño, R., Cichy, B., Davies, A., and Rabbideau, G., The Autonomous sciencecraft experiment onboard the EO-1 spacecraft, in *Proceedings of the 19th National Conference on Artificial Intelligence*, 2004, pp. 1040–1041.
96. Benton, J., van den Briel, M., and Kambhampati, S., A hybrid linear programming and relaxed plan heuristic for partial satisfaction planning problems, in *Proceedings of the 17th International Conference on Automated Planning and Scheduling*, 2007, pp. 34–41.

40

Generalized Assignment Problem

Wei Wu

Mutsunori Yagiura

Toshihide Ibaraki

40.1 Introduction

The *generalized assignment problem* (GAP) is one of the representative combinatorial optimization problems known to be NP-hard. Given n jobs and m agents, we undertake to determine a minimum cost assignment such that every job is assigned to exactly one agent and the resource constraint for each agent is satisfied. This problem is a natural generalization of such representative combinatorial optimization problems as bipartite matching, knapsack, and bin packing problems, and has many important applications, including flexible manufacturing systems [1], facility location [2], and vehicle routing problems [3]. Consequently, designing good exact and/or heuristic algorithms for GAP has significant practical as well as theoretical value. Among various heuristic algorithms developed for GAP are: a combination of the greedy method and the local search by Martello and Toth [4,5]; a tabu search and simulated annealing by Osman [6]; a genetic algorithm by Chu and Beasley [7]; a scalable parallel genetic algorithm by Liu and Wang [8]; variable depth search algorithms by Racer and Amini [9,10]; a tabu search based on ejection chain approach by Laguna et al. [11] (which is proposed for a generalization of GAP); another tabu search by Díaz and Fernández [12]; a hybrid approach of tabu search and branch-and-bound by Woodcock and Wilson [13]; a Lagrangian heuristic algorithm by Haddadi and Ouzia [14]; Lagrangian relaxation guided problem space search heuristics by Jeet and Kutanoglu [15]; a MAX-MIN ant system combined with local search and tabu search by Lourenço and Serra [16]; a path-relinking algorithm by Alfandari et al. [17]; ejection chain approaches by Yagiura et al. [18,19]; a very large-scale variable neighborhood search heuristic by Mitrović-Minić and Punnena [20]; a bees algorithm by Özbakir et al. [21], and so on. Research for exact algorithms also has long history since early papers by Ross and Soland [22], Martello and Toth [4], and Fisher et al. [23]. Among more recent exact algorithms successful for GAP are the branch-and-bound methods by Savelsbergh [24], Nauss [25], Haddadi and Ouzia [26], Avella et al. [27], Munapo et al. [28] and Posta et al. [29], and an ascending hyper-plane approach through network flows by Munapo et al. [30].

In this chapter, we review heuristic and metaheuristic algorithms for GAP. We first define the problem formally and introduce some related problems and their complexities in Section 40.2. We then explain basic strategies of greedy method and local search in Sections 40.3 and 40.4. As relaxation problems often provide useful information to improve the search, we introduce Lagrangian relaxation problems in Section 40.5 and then explain some basic ideas of Lagrangian heuristics in Section 40.6. Although branch-and-bound is usually used to design exact algorithms, it can also be used for approximation algorithms; hence we describe the basics of branch-and-bound in Section 40.8. We then report some computational results of various metaheuristic algorithms and branch-and-bound methods in Section 40.9. A brief survey on polynomial-time approximation algorithms with performance guarantees is given in Section 40.10.

40.2 Generalized Assignment Problem

In this section, we first give a formal definition of the GAP and then introduce some equivalent formulations, special cases, generalizations, and their complexities.

40.2.1 Definition of the Problem

Given n jobs $J = \{1, 2, \ldots, n\}$ and m agents $I = \{1, 2, \ldots, m\}$, we determine a minimum cost assignment subject to the constraints of assigning each job to exactly one agent and satisfying a resource constraint for each agent. Assigning job j to agent i incurs a cost c_{ij} and consumes an amount a_{ij} (> 0) of the resource, whereas the total amount of the resource available at agent i is b_i (> 0). An assignment is a mapping σ: $J \to I$, where $\sigma(j) = i$ means that job j is assigned to agent i. For convenience, we define a 0–1 variable x_{ij} for each pair of $i \in I$ and $j \in J$ by

$$x_{ij} = 1 \iff \sigma(j) = i.$$

Then, the GAP is formulated as follows:

$$\text{GAP:} \qquad \text{minimize} \qquad cost(x) = \sum_{i \in I} \sum_{j \in J} c_{ij} x_{ij} \tag{40.1}$$

$$\text{subject to} \qquad \sum_{j \in J} a_{ij} x_{ij} \le b_i, \qquad \forall i \in I \tag{40.2}$$

$$\sum_{i \in I} x_{ij} = 1, \qquad \forall j \in J \tag{40.3}$$

$$x_{ij} \in \{0, 1\}, \qquad \forall i \in I \text{ and } \forall j \in J. \tag{40.4}$$

The first condition (40.2) signifies that the amount of resource required by the jobs assigned to each agent must not exceed the available amount of resource at the agent, which is called the resource constraint, and the second condition (40.3) signifies that each job must be assigned to exactly one agent, which is called the assignment constraint [6] or semi-assignment constraint [5]. The objective value $cost(x)$ is also referred to as $cost(\sigma)$, as it represents the cost of assignment $\sigma : J \to I$.

40.2.2 Related Problems and Complexity

The GAP is sometimes defined as a maximization problem whose objective is to maximize $\sum_{i\in I}\sum_{j\in J}\rho_{ij}x_{ij}$, where ρ_{ij} are profits of assigning job j to agent i, subject to constraints (40.2)–(40.4). This is equivalent to the above-mentioned minimization formulation; simply let $c_{ij} := -\rho_{ij}$, or let $c_{ij} := \bar{\rho} - \rho_{ij}$ for a constant $\bar{\rho}$ with $\bar{\rho} \geq \max_{i\in I, j\in J}\rho_{ij}$, if negative costs are not preferable. In the latter case, $cost(x) = \sum_{i\in I}\sum_{j\in J}(\bar{\rho} - \rho_{ij})x_{ij} = n\bar{\rho} - \sum_{i\in I}\sum_{j\in J}\rho_{ij}x_{ij}$ holds for any feasible x by constraints (40.3). This is also a natural generalization of the knapsack problem [5], where $|I| = 1$ holds in the definition of GAP.

GAP is known to be NP-hard, and the (supposedly) simpler problem of determining the existence of a feasible solution for GAP is NP-complete in the strong sense, since (the decision version of) the bin packing problem [31] can be reduced to this problem without introducing large numbers. This feasibility problem of GAP remains to be NP-complete even if m is a fixed constant, since the partition problem [31] can be reduced to this problem with $m = 2$.

On the other hand, we can consider the problem GAP′, which is GAP with constraints (40.3) being replaced by

$$\sum_{i\in I} x_{ij} \leq 1, \; \forall j \in J. \tag{40.5}$$

This GAP′ always has a feasible solution. (Note that, in this case, a job j will not be assigned to an agent i unless $c_{ij} \leq 0$, and hence the formulation of profit maximization seems more natural. Such maximization formulation is suitable when performance guarantees are considered [32,33] and is called by various names such as LLGAP [5] and max-profit GAP [33].) The GAP′ is however equivalent to GAP as shown in the following. For any instance of GAP′, we can construct an equivalent instance of GAP by adding a dummy agent "$m + 1$" with $b_{m+1} = n$, $c_{m+1,j} = 0$, and $a_{m+1,j} = 1$ for all $j \in J$. That is, for a solution x' of GAP′, the corresponding GAP instance has a feasible solution x with the same cost as x', which is obtained by setting $x_{m+1,j} = 1$ (resp., 0) for all j with $\sum_{i=1}^{m} x'_{ij} = 0$ (resp., 1) and $x_{ij} = x'_{ij}$ for all $i \in \{1, 2, \ldots, m\}$ and $j \in J$. Conversely, for any instance of GAP, we can consider an instance of GAP′ having costs $c_{ij} - \bar{c}$ for all $i \in I$ and $j \in J$, where $\bar{c}$ is a sufficiently large constant (e.g., $\bar{c} > \max_{j\in J}\max_{i\in I} c_{ij}$). Then, if an optimal solution to the transformed instance of GAP′ satisfies constraints (40.3), the original instance also has an optimal feasible solution whose cost is larger by $n\bar{c}$; otherwise, we can conclude that there is no feasible solution to the instance of GAP.

If m is a fixed constant, GAP admits a pseudo-polynomial time exact algorithm, which is based on dynamic programming. More formally, let $f^*(j, k_1, k_2, \ldots, k_m)$ denote the optimal solution value for the subproblem of GAP consisting of job set $\{1, 2, \ldots, j\}(\subseteq J)$ and resource amount $k_i(\leq b_i)$ for each agent i. The optimal solution value $f^*(n, b_1, b_2, \ldots, b_m)$ for the original GAP can be computed by the following recurrence formula:

$$
f^*(j, k_1, k_2, \ldots, k_m)
$$
$$
= \begin{cases}
+\infty & \text{if } \exists i \in M, k_i < 0; \\
0 & \text{if } j = 0 \text{ and } \forall i \in M, k_i \geq 0; \\
\min_{i\in M}\left\{c_{ij} + f^*(j-1, k_1, \ldots, k_{i-1}, k_i - a_{ij}, k_{i+1}, \ldots, k_m)\right\} & \text{otherwise.}
\end{cases}
$$

This yields an overall running time of $O(mnb_1 b_2 \cdots b_m)$.

The GAP with $a_{ij} = 1$ for all $i \in I$ and $j \in J$ becomes a special case of the minimum cost flow problem and is solvable in polynomial time. (Actually, it is not hard to show that this special case is equivalent to the bipartite maximum weight matching problem.) A more restricted case with $m = n$, $a_{ij} = 1$, and $b_i = 1$ for all $i \in I$ and $j \in J$ is known as the *assignment problem* and has efficient special algorithms. Efficient algorithms for network flow, matching and assignment problems are found in References [34–36].

The GAP with the 0-1 constraint $x_{ij} \in \{0, 1\}$ of (40.4) relaxed to $x_{ij} \geq 0$ for all $i \in I$ and $j \in J$ is formulated as follows:

$$\text{minimize} \qquad cost(x) = \sum_{i \in I} \sum_{j \in J} c_{ij} x_{ij}$$

$$\text{subject to} \qquad \sum_{j \in J} a_{ij} x_{ij} \leq b_i, \qquad \forall i \in I$$

$$\sum_{i \in I} x_{ij} = 1, \qquad \forall j \in J \qquad\qquad (40.6)$$

$$x_{ij} \geq 0, \qquad \forall i \in I \text{ and } \forall j \in J.$$

(Though $0 \leq x_{ij} \leq 1$ seems more natural than the last constraint $x_{ij} \geq 0$, $x_{ij} \leq 1$ is redundant because of the assignment constraints (40.3).) Problem (40.6) is a linear programming (LP) problem, and hence solvable in polynomial time.

The GAP can also be formulated as a set partitioning problem. For each agent $i \in I$, suppose that all possible subsets $J' \subseteq J$ (including the empty set) satisfying $\sum_{j \in J'} a_{ij} \leq b_i$ are numbered and let C_i be the set of such indices. A subset $J_i^k \subseteq J$ with $k \in C_i$ represents a set of jobs assigned to agent i. For each J_i^k, define $d_i^k = \sum_{j \in J_i^k} c_{ij}$ and $\delta_{ij}^k = 1$ (resp., 0) if $j \in J_i^k$ (resp., $j \notin J_i^k$). Then the set partitioning formulation of GAP is given as follows:

$$\text{minimize} \qquad \sum_{i \in I} \sum_{k \in C_i} d_i^k y_i^k \qquad\qquad (40.7)$$

$$\text{subject to} \qquad \sum_{i \in I} \sum_{k \in C_i} \delta_{ij}^k y_i^k = 1, \qquad \forall j \in J \qquad\qquad (40.8)$$

$$\sum_{k \in C_i} y_i^k = 1, \qquad \forall i \in I \qquad\qquad (40.9)$$

$$y_i^k \in \{0, 1\}, \qquad \forall k \in C_i \text{ and } \forall i \in I. \qquad\qquad (40.10)$$

It is not practical to generate all elements in C_i ($i \in I$), and hence *column generation* approach is usually used to generate only promising subsets [24,37].

Motivated by practical applications, various generalizations of GAP have been proposed. The *multi-resource generalized assignment problem* (MRGAP), in which more than one resource constraint is considered for each agent, is a natural generalization of GAP. For MRGAP, Gavish and Pirkul proposed a branch-and-bound algorithm and two simple Lagrangian heuristics [38], Rego et al. [39] applied a meta-heuristic approach called RAMP (relaxation adaptive memory programming) [40], and Yagiura et al. [41] devised a very large-scale neighborhood search algorithm. There are many other generalizations of GAP such as the multi-level GAP [11], the multi-resource generalized quadratic assignment problem [42], the dynamic MRGAP [43], the generalized multi-assignment problem [44], the MRGAP with additional constraints [45], and the min-max regret GAP with interval data [46]. The survey by Cattrysse and van Wassenhove [47] discusses some other generalizations of GAP and summarizes existing results before early 1990s.

40.3 Greedy Methods

There are several useful tools used to design approximation algorithms. The most common one is perhaps the *greedy method*, which directly constructs approximate solutions by successively determining the values of variables on the basis of some local information. Another important tool is the *local search* (LS), which will be discussed in Section 40.4. In this section, we will describe simple examples of greedy methods for GAP.

The basic framework of greedy methods for GAP is as follows: Start with an empty assignment (i.e., $x_{ij} = 0$ for all $i \in I$ and $j \in J$), and in each iteration, choose a pair of unassigned job j (i.e., with $x_{ij} = 0$ for all $i \in I$) and agent i that minimizes an evaluation function $f_{ij}(x)$ among those having sufficient space at agent i (i.e., $\sum_{j' \in J} a_{ij'} x_{ij'} + a_{ij} \leq b_i$), and assign the job j to the agent i (i.e., let $x_{ij} := 1$). This is repeated until a feasible solution is found or it is concluded that the algorithm failed to find a feasible solution. The function $f_{ij}(x)$ measures the desirability of assigning job j to agent i when the current (incomplete) solution is x, where various definitions of $f_{ij}(x)$ are possible, as will be described later. For convenience, let

$$\tilde{b}_i(x) = b_i - \sum_{j' \in J} a_{ij'} x_{ij'}$$

be the currently available amount of resource at agent i and

$$F_j(x) = \left\{ i \in I \mid a_{ij} \leq \tilde{b}_i(x) \right\} \tag{40.11}$$

be the set of agents to which job j can be assigned without violating the resource constraints (40.2). Then the framework of the greedy method is formally described as follows, where $J' \subseteq J$ denotes the set of jobs not assigned yet.

Algorithm GREEDY

Step 1. Let $x_{ij} := 0$ for all $i \in I$ and $j \in J$, and let $J' := J$.

Step 2. If $F_j(x) = \emptyset$ holds for some $j \in J'$, then output "failed" and stop. If $J' = \emptyset$ holds, output the current solution x and stop.

Step 3. Choose a pair (i, j) that minimizes $f_{ij}(x)$ among those satisfying $j \in J'$ and $i \in F_j(x)$. Then let $x_{ij} := 1, J' := J' \setminus \{j\}$, and return to Step 2.

Among conceivable functions for $f_{ij}(x)$ are: (1) c_{ij}, (2) $(c_{ij} - \bar{c}_j)/a_{ij}$, (3) a_{ij}, and (4) a_{ij}/b_i, where $\bar{c}_j$ is a parameter satisfying $\bar{c}_j \geq \max_{i \in I} c_{ij}$. The evaluation function (2) comes from the profit maximization formulation of GAP, where the counterpart is ρ_{ij}/a_{ij}, the profit per unit resource (larger values are preferable). It should be noted that this measure ρ_{ij}/a_{ij} is meaningful only if $\rho_{ij} \geq 0$ for all $i \in I$ and $j \in J$.

In what follows is a more sophisticated way of defining $f_{ij}(x)$, which is based on the concept of "regret" measures. Let f' be a simple evaluation function such as those listed earlier. For each job $j \in J'$, let i_1 be the agent i in $F_j(x)$ that minimizes $f'_{ij}(x)$. Moreover, if $|F_j(x)| \geq 2$ holds, let i_2 be the agent i that minimizes $f'_{ij}(x)$ among those in $F_j(x) \setminus \{i_1\}$. Then, $f_{ij}(x)$ is defined as $f_{i_1 j}(x) = -\infty$ if $|F_j(x)| = 1$, and $f_{i_1 j}(x) = f'_{i_1 j}(x) - f'_{i_2 j}(x)$ otherwise, while $f_{ij}(x) = +\infty$ for $i \neq i_1$. Of course, in the iteration of GREEDY, the x_{ij} with the minimum $f_{ij}(x)$ is chosen and set to $x_{ij} := 1$. This regret measure gives priority to the job having the largest difference in the values of $f'_{ij}(x)$ between the best and the second best agents. Although simple original measures are myopic, the regret measures consider one step forward, and hence lead to better performance in many cases. An efficient implementation of such a greedy method is discussed in Reference [5].

40.4 Local Search

Local search is one of the basic tools for designing approximation algorithms. It starts from an initial solution x and repeatedly replaces x with a better solution in its *neighborhood* $N(x)$ until no better solution is found in $N(x)$, where a neighborhood $N(x)$ is a set of solutions obtainable from x with slight perturbations. The resulting solution x is *locally optimal* in the sense that no better solution exists in its neighborhood. Local search is sometimes terminated before a locally optimal solution is reached according to some stopping criterion; for example, when it becomes computationally expensive to continue the search.

Shift and swap neighborhoods, denoted N_{shift} and N_{swap} respectively, are commonly used in local search methods for GAP, where

$$N_{shift}(x) = \{x' \mid x' \text{ is obtainable from } x \text{ by changing the assignment of one job}\},$$
$$N_{swap}(x) = \{x' \mid x' \text{ is obtainable from } x \text{ by exchanging the assignments of two jobs}\}.$$

The size of these neighborhoods are $O(mn)$ and $O(n^2)$, respectively. Note that the number of jobs assigned to each agent will not be changed by swap moves; hence it is not appropriate to use the swap neighborhood alone.

The definition of the search space is also an important issue in designing local search. For GAP, it is not effective to search only within the feasible region unless the resource constraints (40.2) are loose, since the problem of determining the existence of a feasible solution is already NP-complete as mentioned in Section 40.2. The following two search spaces are commonly used:

SS1: The set of all possible x that satisfy constraints (40.3) and (40.4) (i.e., the set of all possible assignments $\sigma : J \to I$), implying that the resource constraints (40.2) can be violated during the search.

SS2: The set of all possible x that satisfy constraints (40.2), (40.4) and (40.5) (i.e., the constraints (40.3) are relaxed to (40.5)). This is equivalent to adding a dummy agent "$m + 1$" and considering all possible assignments $\sigma : J \to I \cup \{m + 1\}$ that satisfy the resource constraints (40.2).

Then, we need to evaluate solutions taking the degree of constraint violation into consideration. An objective function penalized by infeasibility is usually used for this purpose. For example, if the above-mentioned search space SS1 is used, a natural penalized cost function is as follows:

$$pcost(x) = cost(x) + \sum_{i \in I} \alpha_i p_i(x), \tag{40.12}$$

where

$$p_i(x) = \max \left\{ 0, \sum_{j \in J} a_{ij} x_{ij} - b_i \right\}$$

denotes the amount of infeasibility at agent $i \in I$, and $\alpha_i \ (> 0)$ are parameters called *penalty weights*. For the search space SS2, a natural evaluation function is

$$cost(x) + \sum_{j \in J} \beta_j \left(1 - \sum_{i \in I} x_{ij} \right)$$

with appropriate penalty weights $\beta_j \ (> 0)$. (Equivalently, we can add a dummy agent "$m + 1$" and let $c_{m+1,j} = \beta_j$, $a_{m+1,j} = 1$ for all $j \in J$ and $b_{m+1} = n$, and let the search space to be the feasible region with respect to the modified instance.) The parameters α_i and β_j can be given as fixed constants as in References [48,49], or can be adaptively controlled during the search by using such an algorithm as in Reference [18] (see also Section 40.7.2). A locally optimal solution under the above-mentioned penalized costs may not always be feasible; however, we can increase the probability of obtaining feasible solutions by (1) keeping the best feasible solution obtained during the search as an incumbent solution and (2) using large values for the penalty weights.

Various algorithms can be realized within the framework of local search by specifying detailed rules in the algorithm. For example, local search with the shift neighborhood in search space SS2 includes the greedy method in Section 40.3, where all jobs are initially assigned to the dummy agent $m + 1$, and shifted to new agents i with $1 \leq i \leq m$ one by one.

As a concrete example of a simple local search algorithm, we introduce the algorithm MTHG proposed by Martello and Toth [4,5] in the following. This algorithm is often used as a subprocedure for other approximation or exact algorithms. It starts with the greedy method in Section 40.3 using a regret measure as the evaluation function f and then improves the solution by shift moves, where the feasible shift move with the largest improvement is chosen for each job, and a job will not be shifted again once its assignment is changed; hence the number of shift moves is limited to n. The algorithm is formally described as follows. Recall that $F_j(x)$ defined in (40.11) is the set of agents to which job j can be shifted without violating the resource constraint.

Algorithm MTHG

Step 1. Call algorithm GREEDY using a regret measure. If it failed to find a feasible solution, output "failed" and stop; otherwise, let x be the solution output by GREEDY and $j := 1$.

Step 2. Let i' be the agent to which the job j is currently assigned (i.e., $x_{i'j} = 1$). If $F_j(x) \setminus \{i'\} = \emptyset$ holds, proceed to Step 3; otherwise, let i'' be the agent that minimizes c_{ij} among those in $F_j(x) \setminus \{i'\}$. If $c_{i''j} < c_{i'j}$ holds, change the assignment of job j from i' to i''; that is, let $x_{i'j} := 0$ and $x_{i''j} := 1$.

Step 3. If $j = n$ holds, output the current solution x and stop; otherwise, let $j := j + 1$ and return to Step 2.

For simplicity, in the above-mentioned algorithm, shift moves are tested according to the order of their job indices (as described in the same way as in Reference [5]); however, it is often preferable to shuffle the indices beforehand to avoid undesirable bias. If appropriately implemented, algorithm MTHG runs in $O(nm \log m + n^2)$ time, provided that the simple evaluation f' to be used to compute the regret measure f can be evaluated independently of x as in the case of the four examples discussed in Section 40.3.

40.5 Lagrangian Relaxation

Relaxation problems often provide useful information for designing efficient algorithms to solve the original problem. In this section, we give two types of *Lagrangian relaxation problems* for GAP: One relaxes the resource constraints (40.2), and another relaxes the assignment constraints (40.3). The first type is formally defined as follows:

$$L^{\text{rec}}(u) = \min \sum_{i \in I} \sum_{j \in J} (c_{ij} + u_i a_{ij}) x_{ij} - \sum_{i \in I} u_i b_i$$

$$\text{s.t.} \quad \sum_{i \in I} x_{ij} = 1, \quad \forall j \in J \tag{40.13}$$

$$x_{ij} \in \{0, 1\}, \quad \forall i \in I \text{ and } \forall j \in J,$$

where $u = (u_1, u_2, \ldots, u_m) \in R^m_+$ (R_+ is the set of nonnegative reals) is a Lagrangian multiplier vector given to the resource constraints (40.2). An optimal solution x for this relaxation problem is obtained by the following simple rule: For each job j, let i_j^* be an agent i that minimizes $c_{ij} + u_i a_{ij}$, and let $x_{ij} := 1$ for $i = i_j^*$ and $x_{ij} := 0$ for $i \neq i_j^*$.

For any $u \in R_+^m$, $L^{rec}(u)$ gives a lower bound on the objective value of problem GAP, and the Lagrangian dual problem is to find an $u \in R_+^m$ that maximizes the lower bound $L^{rec}(u)$. Any optimal solution $u = u^*$ to the dual of the LP relaxation (40.6) of GAP,

$$\text{LP} = \max \quad \sum_{j \in J} v_j - \sum_{i \in I} b_i u_i$$

$$\text{s.t.} \quad v_j - a_{ij} u_i \leq c_{ij}, \quad \forall i \in I \text{ and } \forall j \in J, \tag{40.14}$$

$$u_i \geq 0, \quad \forall i \in I,$$

is an optimal solution to the Lagrangian dual [50]. However, computing such u^* by solving problem (40.14) is expensive for large-scale instances. Hence the *subgradient method* [50,51] explained in the following is often used for finding a near-optimal u.

Define the subgradient vector $s(u) \in R^m$ by

$$s_i(u) = \sum_{j \in J} a_{ij} x_{ij}(u) - b_i$$

for all $i \in M$, where $x(u)$ is an optimal solution to $L^{rec}(u)$. The method generates a sequence of Lagrangian multiplier vectors $u^{(0)}, u^{(1)}, \ldots$, where $u^{(0)}$ is an initial vector, and $u^{(k+1)}$ is updated from $u^{(k)}$ by the following formula:

$$u_i^{(k+1)} := \max \left\{ u_i^{(k)} + t_k s_i(u^{(k)}), \ 0 \right\}, \quad \forall i \in I,$$

where $t_k > 0$ is a scalar called step size. One of the most common rules of determining step sizes is

$$t_k = \lambda_k \frac{UB - L^{rec}(u^{(k)})}{||s(u^{(k)})||^2},$$

where UB is an upper bound on $cost(x)$ (e.g., obtained by a greedy method or a local search) and $\lambda_k \in (0, 2]$ is a scalar. The sequence of λ_k is often determined by setting $\lambda_0 := 2$ and halving λ_k whenever $\max_{k \geq 0} L^{rec}(u^{(k)})$ has failed to improve during some number of consecutive iterations.

Another type of Lagrangian relaxation problem is obtained by relaxing the assignment constraints (40.3):

$$L^{asgn}(v) = \min \sum_{i \in I} \sum_{j \in J} (c_{ij} - v_j) x_{ij} + \sum_{j \in J} v_j$$

$$\text{s.t.} \quad \sum_{j \in J} a_{ij} x_{ij} \leq b_i, \quad \forall i \in I \tag{40.15}$$

$$x_{ij} \in \{0, 1\}, \quad \forall i \in I \text{ and } \forall j \in J,$$

where $v = (v_1, v_2, \ldots, v_n) \in R^n$ (R is the set of reals) is a Lagrangian multiplier vector given to the assignment constraint. Problem (40.15) decomposes into m 0-1 knapsack problems, which is also known to be NP-hard [31] but is much easier to solve exactly compared to GAP [5,52]. The Lagrangian dual to problem (40.15) is similarly defined as in the case of problem (40.13), and the subgradient method is useful also in this case.

For an optimal solution (u^*, v^*) to the linear programming problem (40.14),

$$L^{asgn}(v^*) \geq L^{rec}(u^*) = \text{LP}$$

holds, and $L^{asgn}(v^*)$ often provides a tighter lower bound on the optimal value of GAP than $L^{rec}(u^*) = \text{LP}$.

40.6 Lagrangian Heuristics

If a good Lagrangian multiplier vector u (resp., v) is given, an optimal solution x to problem (40.13) (resp., (40.15)) is often close to be feasible. (If an optimal x to problem (40.15) happens to be feasible, it is optimal for the original GAP, but this property does not hold for problem (40.13).) Hence it is often advantageous to obtain feasible solutions by slightly modifying them. Moreover, the Lagrangian *relative costs* (or *reduced costs*) $c_{ij} + u_i a_{ij}$ and $c_{ij} - v_j$ with respect to the Lagrangian relaxation problems (40.13) and (40.15), respectively, often provide more accurate estimation on the desirability of assigning a job j to an agent i than the original cost c_{ij}. (The relative cost $c_{ij} + u_i a_{ij}$ is suitable for comparing the desirability of agents i for a specified job j, whereas the relative cost $c_{ij} - v_j$ is suitable for comparing the desirability of jobs j for a specified agent i.) A heuristic algorithm based on such information is called a *Lagrangian heuristic algorithm* [50]. For GAP, many such algorithms have been proposed and are confirmed to be quite useful [14,25,26,37,53–55]. Some are based on different relaxation problems from those illustrated in Section 40.5 [37,53–55]. Among them, Cattrysse et al. [37] proposed an algorithm based on the set partitioning formulation of GAP. See Reference [47], for various types of relaxation problems of GAP. Lagrangian heuristics is also known to be very efficient to solve large-scale instances of other representative combinatorial optimization problems such as location and set covering problems. We briefly illustrate basic ideas for such heuristic algorithms in the following.

A standard framework of Lagrangian heuristics is as follows:

> **Lagrangian Heuristic Algorithm**
> For each Lagrangian multiplier vector $u = u^{(k)}$ generated by a subgradient method, if u (or $L^{\text{rec}}(u)$ or $x(u)$) satisfies certain prespecified conditions, construct a feasible solution (by a method as described in the following) based on the information from the Lagrangian relaxation problem (40.13). (The algorithm for the Lagrangian relaxation problem (40.15) is similar.)

As the construction of feasible solutions is tried many times in the above-mentioned iteration, quick methods are preferable. Moreover, it is not fruitful to construct a feasible solution by using a poor multiplier vector u (e.g., those generated in the early stage of the subgradient method), and hence some conditions are applied to decide whether the construction is executed or not. A simple rule may be to execute only if $(UB - L^{\text{rec}}(u))/L^{\text{rec}}(u)$ is below a prespecified constant.

For the Lagrangian relaxation problem (40.13) with a given u, an optimal solution x to it satisfies the assignment constraints (40.3). It is therefore natural to use such x as the initial solution for a local search algorithm with search space SS1, and improve its feasibility with the shift neighborhood (or the combination of the shift and swap neighborhoods) using sufficiently large penalty weights α_i in $pcost(x)$ of (40.12). Another reasonable method is to call algorithm MTHG in which $f'_{ij}(x) = c_{ij} + u_i a_{ij}$ is used to compute the regret measure.

For the Lagrangian relaxation problem (40.15) with a given v, an optimal solution x to it satisfies the resource constraints (40.2), but may not satisfy the assignment constraints (40.3). A job j may be assigned to two or more agents, or may not be assigned to any agent. Let

$$J_0 = \left\{ j \in J \mid \sum_{i \in I} x_{ij} = 0 \right\},$$
$$J_1 = \left\{ j \in J \mid \sum_{i \in I} x_{ij} = 1 \right\},$$
$$J_2 = \left\{ j \in J \mid \sum_{i \in I} x_{ij} \geq 2 \right\}$$

be the set of jobs assigned to no agent, exactly one agent, and more than one agent, respectively. Then, it is often the case that $|J_0 \cup J_2|$ is small. Hence it is usually effective to fix the assignment of jobs in J_1

to the currently assigned agents, and try to modify the assignments of jobs in J_0 or J_2 to obtain a feasible solution. A standard method is to remove each job in J_2 from all but one agent, and then assign jobs in J_0 to agents. For choosing an agent for a job j in J_2 among those i with $x_{ij} = 1$, reasonable criteria are (1) c_{ij}, (2) $(c_{ij} - v_j)/a_{ij}$, and so forth, where a smaller value is more preferable. Note that $c_{ij} - v_j < 0$ holds for all $i \in I$ and $j \in J$ satisfying $x_{ij} = 1$ in an optimal solution x to problem (40.15). To assign jobs in J_0 to agents, we can use the greedy method in Section 40.3 or algorithm MTHG to the restricted instances whose free variables are x_{ij} with $j \in J_0$. Once a feasible solution is found, it is often effective to improve it by a local search algorithm. The efficiency of the above-mentioned method may degrade if $|J_0 \cup J_2|$ is large; hence it is advantageous to try the construction only when $|J_0 \cup J_2|$ is below a prespecified threshold [26].

40.7 Metaheuristics

Metaheuristic algorithms are widely recognized as one of the most practical approaches for hard combinatorial optimization problems. Metaheuristics is a set of comprehensive guidelines and/or frameworks useful to devise efficient algorithms by combining basic strategies such as greedy methods and local search in a unified manner. Among representative metaheuristic algorithms are genetic algorithms, simulated annealing, tabu search, and so on. See other chapters of this handbook (e.g., Chapters 17, 18, 19, 21, 22, and 23) for details of metaheuristic algorithms. Although basic strategies such as greedy methods, local search, and Lagrangian heuristics usually find good solutions quickly, metaheuristic algorithms aim at obtaining better solutions by investing more computation time.

For GAP, many metaheuristic algorithms have been proposed. Among them are a tabu search and a simulated annealing by Osman [6]; a genetic algorithm by Chu and Beasley [7]; variable depth search algorithms by Racer and Amini [9,10]; a tabu search based on ejection chain approach by Laguna et al. [11] (which is proposed for a generalization of GAP); another tabu search by Díaz and Fernández [12]; a MAX-MIN ant system combined with local search and tabu search by Lourenço and Serra [16]; a path-relinking algorithm by Alfandari et al. [17]; variable depth search algorithms [48,49] and ejection chain approaches by Yagiura et al. [18,19]; RAMP (relaxation adaptive memory programming) and primal-dual RAMP algorithms by Rego et al. [39], and so forth. The algorithm by Rego et al. is based on a general framework called the RAMP approach proposed in Reference [40]. We will illustrate basic ideas of the ejection chain and path-relinking approaches by Yagiura et al. [18,19] in the following, which are among the most efficient metaheuristic algorithms in those listed above.

It utilizes a powerful neighborhood search algorithm [18], called ejection chains (EC) probe, based on the idea of *ejection chains* [56]. An ejection chain is an embedded neighborhood construction that compounds simple moves to create more complex and powerful moves. In Reference [18], EC probe was first incorporated in the framework of tabu search, and promising results were obtained. We call this algorithm *TSEC* (tabu search with ejection chains). In Reference [19], EC probe was then incorporated with the *path relinking* approach, which provides an "evolutionary" mechanism for generating new solutions by combining two or more reference solutions. We call the resulting algorithm path relinking with ejection chains (PREC).

As mentioned in Section 40.4, it is difficult to conduct the search only within the feasible region. Algorithms TSEC and PREC are both based on the search space SS1, and evaluate solutions by the penalized cost (40.12). As the search is sensitive to penalty weights α_i, the algorithms employ a sophisticated adaptive control mechanism of the penalty weights, thereby introducing strategic oscillation mechanism in the algorithms. This adaptive control method improves the performance and robustness of the algorithms as well as their usability.

In the following subsections, we briefly explain the idea of EC probe and adaptive control of penalty weights and then summarize the frameworks of TSEC and PREC. For convenience, in the rest of this section, we use an assignment σ instead of x to represent a solution and use notations such as $pcost(\sigma)$ instead of $pcost(x)$, as their meanings are clear.

40.7.1 EC Probe

An ejection chain move in algorithms TSEC and PREC is a sequence of shift moves such that a solution σ' is obtained from the current solution σ by shifting l (where l satisfies $2 \leq l \leq n$) jobs $j_1, j_2, \ldots, j_l$ simultaneously in such a way that satisfies $\sigma'(j_r) = \sigma(j_{r-1})$ for $r = 2, 3, \ldots, l$, where $\sigma'(j_1)$ is arbitrary. In other words, for $r = 2, 3, \ldots, l$, job j_r is shifted from agent $\sigma(j_r)$ to agent $\sigma(j_{r-1})$ after ejecting job j_{r-1}. An ejection chain move is called *cyclic* if $\sigma'(j_1) = \sigma(j_l)$ holds, that is, the first job j_1 is inserted into the agent from which the last job j_l is ejected. The *length* of an ejection chain move is the number of shift moves l in the sequence. The ejection chain neighborhood is the set of solutions obtainable by such ejection chain moves. Both shift and swap neighborhoods are subsets of the ejection chain neighborhood, since a shift move is an ejection chain move of length one, and a swap move is a cyclic ejection chain move of length two. However, the size of the ejection chain neighborhood can become exponential unless intelligently controlled. For this purpose, three subsets of the neighborhood called *shift*, *double shift*, and *long chain* are considered, where a double shift move is an ejection chain move of length two, and a long chain move is an ejection chain of any length. For the double shift neighborhood, we restrict the number of candidates for j_2 to $\max\{m, \log n\}$ when j_1 is fixed, and for the long chain neighborhood, we consider only one candidate for j_r when $j_1, j_2, \ldots, j_{r-1}$ are fixed, where $r = 2, 3, \ldots, l$. Such candidates are chosen by using heuristic rules based on the observation that, given a good multiplier vector v, the Lagrangian relative cost $c_{ij} - v_j$ for Lagrangian relaxation (40.15) tends to represent desirability of assigning job j to agent i as discussed in Section 40.6. More details of such rules are found in Reference [18].

As a result, the sizes of shift, double shift, and long chain neighborhoods become $O(mn)$, $O(n \max\{m, \log n\})$ and $O(n^2)$, respectively. Yagiura et al. [18] also showed that the expected size of the long chain neighborhood was $O(n^{(3/2)+\varepsilon})$ for an arbitrarily small positive ε under a simplified random model. These three neighborhoods are used alternately to form an improving phase of local search, which is called *EC probe*.

40.7.2 Adaptive Control of the Penalty Weights

In this section, we briefly explain the basic idea of the adaptive control of the penalty weights of (40.12) as given in [18,19]. The initial values of α_i are decided by solving a quadratic programming problem (abbreviated as QP), whose main aim is to balance the estimated change in the cost and penalty after shift moves. The definition of the QP in Reference [18] is slightly complicated and is omitted here. Then, whenever an EC probe stops at a locally optimal solution σ_{lopt}, α_i values are updated by the following rule. If no feasible solution is found during the previous EC probe after the last update of α_i, the penalty weights are increased by

$$\alpha_i := \alpha_i \left(1 + \delta_{\text{inc}} \cdot \frac{p_i(\sigma_{\text{lopt}})/b_i}{\max_{i' \in I}(p_{i'}(\sigma_{\text{lopt}})/b_{i'})}\right)$$

for all $i \in I$, where δ_{inc} is a parameter. Otherwise (i.e., if at least one feasible solution is found during the previous EC probe after the last update of α_i), the penalty weights are decreased by

$$\alpha_i := \alpha_i(1 - \delta_{\text{dec}})$$

for all $i \in I$ that satisfy $p_i(\sigma_{\text{lopt}}) = 0$, where δ_{dec} is a parameter. In the computational experiments in Section 40.9, δ_{inc} and δ_{dec} are set to 0.01 and 0.1, respectively, as in [18,19].

40.7.3 Algorithms TSEC and PREC

Algorithm TSEC applies an EC probe to a solution obtained by perturbing an available good solution σ_{seed}. Solution σ_{seed} is initially generated randomly and is replaced with the locally optimal solution σ_{lopt} obtained by the previous EC probe whenever $pcost(\sigma_{\text{lopt}}) \leq pcost(\sigma_{\text{seed}})$ holds. The perturbed solution to which the next EC probe is applied is determined by first generating solutions by applying shift moves to σ_{seed} and then choosing the one that minimizes *pcost* among those not tested yet from the current σ_{seed}. Note that the perturbed solution σ generated by applying a shift move to σ_{seed} is improved first by the cyclic double shift neighborhood in EC probe. This strategy was motivated by the success of SSS probe [48] and was confirmed to be effective to avoid short cycling. The idea is based on the fact that a cyclic move does not change the number of jobs assigned to each agent, while a shift move changes those of relevant agents. See [18] for more discussion.

Algorithm PREC applies EC probes to solutions generated by path relinking, which is a methodology to generate solutions from two or more solutions. It keeps a reference set R ($|R| = \pi$ is a parameter) of good solutions. Initially R is prepared by applying EC probes to randomly generated solutions. Then, it is updated by reflecting the outcomes of local search. The incumbent solution (i.e., the best feasible solution) is always stored as a member of R. Other solutions in R are maintained as follows. Whenever an EC probe stops, the locally optimal solution σ_{lopt} is exchanged with the worst (with respect to *pcost*) solution σ_{worst} in R (excluding the incumbent solution), provided that σ_{lopt} is not worse than σ_{worst} and is different from all solutions in R. Path relinking is applied to two solutions σ_A (initiating solution) and σ_B (guiding solution) randomly chosen from R, where a random shift is applied to σ_B with probability 1/2 (no shift with the remaining probability 1/2) before applying the path relinking (for the purpose of keeping the diversity of the search), and the resulting solution is redefined to be σ_B. Let the distance between two solutions σ and σ' be $dist(\sigma, \sigma') = \left| \{ j \in J \mid \sigma(j) \neq \sigma'(j) \} \right|$, that is, the number of jobs assigned to different agents, and let $d = dist(\sigma_A, \sigma_B)$ be the distance between solutions σ_A and σ_B. The algorithm generates a sequence $\sigma_0, \sigma_1, \ldots, \sigma_d$ of solutions from two solutions σ_A and σ_B as follows. Starting from $\sigma_0 := \sigma_A$, for $k = 1, 2, \ldots, d$, let σ_k be the solution in $N_{\text{shift}}(\sigma_{k-1})$ with the best *pcost* among those whose distances to σ_B are smaller than that from σ_{k-1}. Let S be the best γ (a parameter) solutions with respect to *pcost* from $\{ \sigma_2, \sigma_3, \ldots, \sigma_{d-1} \}$.* EC probes are then applied to all solutions in S, which completes this phase of path relinking. For each starting solution $\sigma \in S$, it is improved first by the cyclic double shift neighborhood in EC probe to avoid short cycling, as in algorithm TSEC.

Yagiura et al. [19] also tested other rules, for example, perturbing guiding solutions σ_B or not, using more than one guiding solution for generating a path, restricting the distance between solutions in the reference set, rules to choose initiating and guiding solutions, rules to define the set S, and so on, and they observed that the algorithm was not sensitive to such changes. In the computational experiment in Section 40.9, the parameters in PREC were set to $\pi = 20$ and $\gamma = 10$ as in Reference [19].

40.8 Branch-and-Bound Algorithms

Branch-and-bound is one of the major enumerative methods to solve combinatorial optimization problems. It is based on the idea that the original problem can be equivalently solved as a result of solving all partial problems decomposed from the original problem. The decomposition can be then applied to the generated partial problems. A problem instance of GAP can be decomposed into two partial problems by fixing a variable x_{ij} to 0 and 1, where such a variable is called a *branching variable*, and the corresponding operation is called a *branching operation*. We denote the original problem by P_0 and the kth partial problem generated during computation by P_k. Starting from P_0, if we apply branching operations repeatedly until all partial problems become trivial, we can enumerate all possible solutions. During this process, we usually examine only a small portion of all partial problems, based on the following idea. If an optimal solution to a partial problem P_k is found or it is concluded for some reason that an optimal solution to the original problem can be found even without solving P_k, then it is not necessary to consider P_k further. The operation of removing such a P_k from the list of partial problems to be solved is called a *bounding operation*, and we say that this operation *terminates* P_k. For example, denoting the incumbent value by

---

* A part of their neighbors are also considered as candidates for S.

UB, and a lower bound on the objective value for P_k by $LB(P_k)$, we can terminate P_k if $LB(P_k) \geq UB$ holds. This is called a *lower bound test*. The process of branching operations can be expressed by a search tree, whose root corresponds to P_0, and the children of a node correspond to the partial problems generated by a branching operation applied to the node. A partial problem is called *active* if it has not been terminated nor decomposed into partial problems. A branch-and-bound algorithm obtains an exact optimal solution if no active partial problem remains when it stops. The rule to choose an active partial problem to test is called the *search strategy*, which affects the efficiency of the search.

A common way for obtaining a lower bound *LB* is to solve relaxation problems of GAP such as the Lagrangian relaxations (40.13) and (40.15) in Section 40.5, and the LP relaxation (40.6) in Section 40.2.2. As mentioned in Section 40.5, the lower bound $L^{\mathrm{asgn}}(v)$ from problem (40.15) tends to be better than the bound $L^{\mathrm{rec}}(u)$ from problem (40.13) or the lower bound from the LP relaxation. Computing better lower bounds will result in reducing the number of partial problems generated during the search, and hence the bound $L^{\mathrm{asgn}}(v)$ is often used in branch-and-bound algorithms. Another common approach to obtain better lower bounds is the use of *valid inequalities*, where a linear inequality constraint is called valid if all feasible solutions to the original GAP satisfy it (but solutions to the LP relaxation problem may not satisfy it). The optimal values of LP relaxations with such additional constraints will usually give better lower bounds. Branch-and-bound algorithms using such valid inequalities are called *branch-and-cut* methods. We give a simple example of a valid inequality in the following. Let $\bar{x}$ be an optimal solution to the LP relaxation of GAP, and let $i \in I$ be an agent having at least one $\bar{x}_{ij}$ fractional (i.e., $0 < \bar{x}_{ij} < 1$). Sort the jobs j with $\bar{x}_{ij} > 0$ in the nonincreasing order of a_{ij}, and let $j_1, j_2, \ldots$ be the resulting order (i.e., $a_{ij_1} \geq a_{ij_2} \geq \cdots$). Let k be the index that satisfies $\sum_{l=1}^{k-1} a_{ij_l} \leq b_i < \sum_{l=1}^{k} a_{ij_l}$, and let $J_i = \{j_1, j_2, \ldots, j_k\}$. Then $\sum_{j \in J_i} x_{ij} \leq |J_i| - 1$ is a valid inequality. Therefore, if the solution $\bar{x}$ violates it (i.e., $\sum_{j \in J_i} \bar{x}_{ij} > |J_i| - 1$), it will be useful to improve the lower bound.

The objective value of any feasible assignment is an upper bound on the optimal value. We can use approximation algorithms to obtain good bounds *UB*; for example, greedy methods, local search, Lagrangian heuristics or metaheuristic algorithms. Having a good *UB* at an early stage is useful for reducing computation time, because many partial problems may be terminated by lower bound tests. It is often effective to apply approximation algorithms to partial problems with small lower bounds. Lagrangian heuristic algorithms are often used for this purpose when Lagrangian relaxation is used to compute lower bounds.

While branch-and-bound algorithms are usually designed to obtain exact optimal solutions, they can also be used as approximation algorithms if the computation is cut off before normal termination. Approximation algorithms of this type are sometimes effective, especially when the size of the problem instance is not very large as will be observed in Section 40.9. One of the merits of such an approximation branch-and-bound is that the minimum of $LB(P_k)$ among all active P_k's gives a lower bound that tends to be better than the lower bound to the original problem $LB(P_0)$. Search strategy to find good upper bound earlier may be different from the one that makes the exact branch-and-bound algorithm efficient. Hence it will be worth trying to design a good search strategy for the purpose of obtaining a good approximation branch-and-bound algorithm.

Among early papers on branch-and-bound algorithms for GAP are [4,22,23]. Cattrysse and van Wassenhove [47] surveyed various relaxation problems used to compute lower bounds for branch-and-bound methods, such as Lagrangian relaxation, surrogate relaxation and so on. Savelsbergh [24] specialized a general purpose solver, MINTO, to solve GAP based on a branch-and-price approach to the set partitioning formulation of GAP. Nauss [25] incorporated various ideas to improve lower bounds and fixing variables, and obtained impressive results. Haddadi and Ouzia [26] also gave good results with a relatively simple branch-and-bound algorithm. Avella et al. [27] considered a cutting plane algorithm based on a branch-and-cut framework. Munapo et al. [28] presented a branch-and-bound algorithm in which partial problems are relaxed to the transportation problems. Posta et al. [29] used variable fixing in a simple depth-first Lagrangian branch-and-bound method. See a survey by Nauss [57] for more information on branch-and-bound algorithms for GAP.

40.9 Computational Results

In this section, we show computational results of some metaheuristic algorithms and branch-and-bound algorithms. There are five types of benchmark instances called types A, B, C, D, and E [4,7,11]. Out of these, we use three types C, D, and E, since types A and B are relatively easy. Instances of these types are generated as follows:

Type C: a_{ij} are random integers from interval $[5, 25]$, c_{ij} are random integers from interval $[10, 50]$, and $b_i = 0.8 \sum_{j \in J} a_{ij}/m$.

Type D: a_{ij} are random integers from interval $[1, 100]$, $c_{ij} = 111 - a_{ij} + e_{ij}$, where e_{ij} are random integers from interval $[-10, 10]$, and $b_i = 0.8 \sum_{j \in J} a_{ij}/m$.

Type E: $a_{ij} = 1 - 10 \ln e_{ij}$, where e_{ij} are random numbers from interval $(0, 1]$, $c_{ij} = 1000/a_{ij} - 10\hat{e}_{ij}$, where $\hat{e}_{ij}$ are random numbers from interval $[0, 1]$, and $b_i = 0.8 \sum_{j \in J} a_{ij}/m$.

Types D and E are somewhat harder than other types, since c_{ij} and a_{ij} are inversely correlated. Note that types D and E instances should be solved as minimization problems, since otherwise they are trivial. Among the three size categories of problem instances SMALL, MEDIUM, and LARGE used in Reference [18], we use MEDIUM in Section 40.9.1 and LARGE in Section 40.9.2.

MEDIUM: Total of 18 instances of types C, D, and E with n up to 200, each type consisting of six instances. Among them, types C and D instances were taken from OR-Library[*], and type E instances were generated by us, which are available at our site.[†]

LARGE: Total of 27 instances of types C, D, and E with n up to 1600. All of them were generated by us, which are available at our site.

40.9.1 Comparison for MEDIUM Instances

We first compare various metaheuristic algorithms: (1) variable depth search by Yagiura et al. [49] (denoted VDS), (2) two algorithms of branching variable depth search by Yagiura et al. [48] (denoted BVDS-l and BVDS-j), (3) tabu search based on ejection chains by Yagiura et al. [18] (denoted TSEC), (4) path relinking approach based on ejection chains by Yagiura et al. [19] (denoted PREC), (5) variable depth search by Racer and Amini [10] (denoted RA), (6) tabu search by Laguna et al. [11] (denoted LKGG), (7) tabu search for the general purpose constraint satisfaction problem by Nonobe and Ibaraki [58] (denoted NI), (8) a MAX-MIN ant system combined with local search and tabu search by Lourenço and Serra [16] (denoted LoSe), (9) the genetic algorithm by Chu and Beasley [7] (denoted CB), and (10) the tabu search by Díaz and Fernández [12] (denoted DF). The results of these algorithms for MEDIUM instances are shown in Table 40.1. The results of a commercial solver CPLEX 6.5[‡] are also shown in column CPLEX. All the data were taken from Table 2 of Reference [19]. Unless otherwise stated, algorithms were run on a workstation Sun Ultra 2 Model 2300 (two UltraSPARC II 300 MHz processors with 1 GB memory), where the computation was executed on a single processor. SPECint95 of this workstation is 12.3 according to the SPEC site[*] (Standard Performance Evaluation Corporation). The table shows the best costs obtained by the algorithms within 150 seconds for $n = 100$, and 300 seconds for $n = 200$, respectively, unless otherwise stated below the table. The computation time of LoSe, CB and DF is longer than this time limit as discussed in Reference [18] and [19]. Table 40.1 also shows the lower bounds (denoted LB) obtained by solving the Lagrangian relaxation (40.15) of GAP. Each "⋆" mark indicates that the best cost among all the

---

[*] URL of OR-Library: http://people.brunel.ac.uk/˜mastjjb/jeb/orlib/gapinfo.html.

[†] URL of our site: http://www.co.mi.i.nagoya-u.ac.jp/˜yagiura/gap/.

[‡] CPLEX 8.1.0 was also tested, but the results of CPLEX 6.5 were slightly better on average.

[*] URL of SPEC site: http://www.spec.org/.

TABLE 40.1 The Best Costs Obtained by the Tested Algorithms (with Time Limits 150 and 300 seconds for $n = 100$ and 200, Respectively; One Execution per Instance Except LoSe, CB and DF)

Type	n	m	LB	PREC	TSEC	BVDS-1	BVDS-j	VDS	RA	LKGG	NI	[d]LoSe	[e]CB	[f]DF	CPLEX
C	100	5	1930	*1931	*1931	*1931	*1931	*1931	1938	*1931	*1931	1942	*1931	*1931	*1931
C	100	10	1400	*1402	*1402	*1402	1403	*1402	1405	1403	1403	1407	1403	*1402	*1402
C	100	20	1242	*1243	*1243	1244	1244	1246	1250	1245	1245	1247	1244	*1243	*1243
C	200	5	3455	*3456	*3456	*3456	3457	3457	3469	3457	3465	3467	3458	3457	*3456
C	200	10	2804	2807	*2806	2809	2808	2809	2835	2812	2817	2818	2814	2807	*2806
C	200	20	2391	*2391	2392	2401	2400	2405	2419	2396	2407	2405	2397	*2391	*2391
D	100	5	6350	*6353	6357	6358	6362	6365	–	6386	6415	6476	6373	6357	6358
D	100	10	6342	6356	6358	6367	6370	6380	6532	6406	6487	6469	6379	*6355	6381
D	100	20	6177	*6211	6221	6275	6245	6284	6428	6297	6368	6358	6269	6220	6280
D	200	5	12,741	*12,744	12,746	12,755	12,755	12,778	–	12788	12973	12923	12,796	12,747	12,750
D	200	10	12,426	*12,438	12,446	12,480	12,473	[a]12,496	12,799	12,537	12,889	12,746	12,601	12,457	12,457
D	200	20	12,230	*12,269	12,284	12,440	12,318	[a]12,335	12,665	12,436	12,793	12,617	12,452	12,351	12,393
E	100	5	12,673	*12,681	12,682	*12,681	12,682	12,685	12,917	[b]12,687	[c]12,686	12,836	N.A.	*12,681	*12,681
E	100	10	11,568	*11,577	*11,577	11,585	11,599	11,585	12,047	[b]11,641	[c]11,590	11,780	N.A.	11,581	11,593
E	100	20	8431	8444	*8443	8499	8484	8490	9004	[b]8522	[c]8509	8717	N.A.	8460	8565
E	200	5	24,927	*24,930	*24,930	24,942	24,933	24,948	25,649	[b]25,147	[c]24,958	25,317	N.A.	24,931	*24,930
E	200	10	23,302	23,310	*23,307	23,346	23,348	23,340	24,717	[b]23,567	[c]23,396	23,620	N.A.	23,318	23,321
E	200	20	22,377	*22,379	22,391	22,475	22,437	†22,452	24,117	[b]22,659	[c]22,551	22,779	N.A.	22,422	22,457

[a] Results after 1000 seconds on Sun Ultra 2 Model 2300.
[b] Results after 20,000 seconds on Sun Ultra 2 Model 2300.
[c] Results after 5000 seconds on Sun Ultra 2 Model 2300.
[d] Computation time is reported in Reference [18].
[e] Results in Reference [7].
[f] Results in Reference [12].

tested algorithms is attained, and "—" means that no feasible solution was found. The results of algorithm CB for type E instances are not available (i.e., not reported in their paper) and are denoted "N.A." in the table. Note that all algorithms except CB and DF were run only once for each instance in the computational results in this section in order to make the comparison fair. From the table, we can observe that PREC and TSEC are highly effective, especially for the type D and E instances. They obtained the best solutions for most of the tested instances.

We next show the results of branch-and-bound algorithms by Nauss [25] (denoted NaussBB) and Haddadi and Ouzia [26] (denoted HOBB). The results for MEDIUM instances are shown in Table 40.2. The results of NaussBB were taken from Tables 1 and 5 of Reference [25], and those of HOBB were taken from Table 4 of Reference [26]. (Table 4 of Reference [26] also includes the results by NaussBB, but they seem to be taken from Table 4 of Reference [18], which are slightly different from those in Reference [25] because the experiments were conducted under different settings as mentioned in Reference [18].) NaussBB was coded in FORTRAN 77 and run on a Dell XPS D300 (Pentium II 300 MHz with 64 MB memory), whereas HOBB was coded in Turbo-Pascal 7.0 and run on an IBM compatible PC (Pentium MMX 200 MHz). SPECint95 of these computers are 11.9 and 6.4, respectively, according to the SPEC site; hence the speed of the Dell XPS D300 is approximately the same as the Sun Ultra 2, while the IBM compatible PC is slower approximately by a factor of two. The time limit of NaussBB was 3000 seconds, whereas that of HOBB was 2400 seconds. Column "best" shows the best cost obtained by the algorithm within the time limit. Column "time to best" shows the computation time used to obtain the solutions reported in column "best." Column "total time" shows the whole computation time needed for the algorithm until it stops after confirming optimality, where "T.O." signifies that the algorithm stopped when the prespecified time limit was reached before confirming optimality. If the algorithm stopped after confirming optimality, column "opt?" is "yes"; otherwise the column is "no." Column "best known" shows the best known values reported in Table 6 of Reference [19].

From the table, we can observe the following: (1) type C instances with up to $n = 200$ can be solved exactly within a reasonable amount of computation time; (2) type D instances are much harder to solve exactly, but the best results obtained by the branch-and-bound methods are of high quality. Moreover, the computation time of HOBB to obtain such solutions are reasonably small in most cases compared to the time limit of Table 40.1, which implies the effectiveness of the algorithm when it is used as heuristics.

TABLE 40.2 Results of Branch-and-Bound Algorithms by Nauss [25] (NaussBB) and Haddadi and Ouzia [26] (HOBB), NaussBB with Time Limit of 3000 seconds on a Pentium II 300 MHz, and HOBB with Time Limit 2400 seconds on a Pentium MMX 200 MHz

				NaussBB				HOBB			
Type	n	m	Best known	Best	Time to best	Total time	Opt?	Best	Time to best	Total time	Opt?
C	100	5	1931	1931	0.1	3.7	yes	1931	0.8	7.5	yes
C	100	10	1402	1402	7.3	15.1	yes	1402	1.2	20.6	yes
C	100	20	1243	1243	90.2	115.2	yes	1243	22.2	26.8	yes
C	200	5	3456	3456	30.4	40.6	yes	3456	27.7	34.9	yes
C	200	10	2806	2806	312.6	490.4	yes	2806	177.0	193.3	yes
C	200	20	2391	2391	968.7	1028.3	yes	2391	94.4	96.0	yes
D	100	5	6353	6353	349.9	362.2	yes	6353	65.5	471.0	yes
D	100	10	6347	6349	2831.5	T.O.	no	6349	371.6	T.O.	no
D	100	20	6190	6200	2829.4	T.O.	no	6196	17.8	T.O.	no
D	200	5	12742	12745	2937.0	T.O.	no	12742	535.7	1481.5	yes
D	200	10	12431	12447	1896.8	T.O.	no	12436	2068.1	T.O.	no
D	200	20	12241	12263	2375.4	T.O.	no	12250	55.6	T.O.	no

40.9.2 Comparison for LARGE Instances

We then report the computational results of the following algorithms for LARGE instances in Table 40.3: (1) tabu search based on ejection chains by Yagiura et al. [18] (denoted TSEC), (2) path relinking approach based on ejection chains by Yagiura et al. [19] (denoted PREC). The results of a commercial solver CPLEX 12.6 are also shown in column CPLEX. All the experiments were carried out on a PC with i7-5820K at 3.30 GHz and 32 GB RAM memory, where the computation was always conducted on a single processor. The table shows the best costs obtained by the algorithms within 20,000 seconds.

Column "opt (LB)" shows the optimal cost, except for those values inside parentheses. The latter case indicates that optimal values were not obtained by the tested algorithms, and the best lower bounds obtained by running CPLEX 20,000 seconds are shown. The notations "time to best," "best," "total time," and "T.O." are the same as those in Table 40.2.

For 15 instances, exact optimal solutions were obtained by CPLEX, and for two instances (the type C instance with $n = 1600$ and $m = 20$ and the type D instance with $n = 1600$ and $m = 80$), the best

TABLE 40.3 The Results for LARGE Instances obtained by IBM ILOG CPLEX 12.6 (CPLEX) and Algorithms by Yagiura et al. [18] (TSEC) and [19] (PREC) (One Execution per Instance with Time Limit 20,000 seconds on i7-5820k 3.3 GHz)

Type	n	m	Opt (LB)	PREC		TSEC		CPLEX		
				Best	Time to best	Best	Time to best	Best	Time to best	Total time
C	400	10	5597	5597	1.7	5597	1.2	5597	0.4	1.3
C	400	20	4782	4782	6.9	4782	6.9	4782	4.5	38.6
C	400	40	4244	4244	83.3	4244	14.3	4244	9.0	9.0
C	900	15	11,340	11,340	527.8	11,340	5505.7	11340	255.5	255.5
C	900	30	9982	9982	1087.3	9983	6287.7	9982	1662.9	1662.9
C	900	60	9326	9327	637.9	9327	8512.3	9326	4902.5	7016.4
C	1600	20	18,802	18,802	337.6	18,803	633.7	18,803	64.4	T.O.
C	1600	40	(17,144)	17,145	1184.6	17,147	2318.0	17,146	418.4	T.O.
C	1600	80	(16,283)	16,286	9884.3	16,288	15,724.3	16,287	314.8	T.O.
D	400	10	(24,959)	24,963	1842.7	24,968	2658.6	24,969	6667.8	T.O.
D	400	20	(24,558)	24,569	11,168.0	24,605	3379.5	24,591	18,209.8	T.O.
D	400	40	(24,349)	24,373	2379.3	24,450	3008.6	24,447	18,119.1	T.O.
D	900	15	(55,402)	55,408	16,862.6	55,426	15,684.2	55,417	11,293.6	T.O.
D	900	30	(54,831)	54,852	16,089.9	54,913	9018.5	54,891	19,499.2	T.O.
D	900	60	(54,551)	54,569	17,037.0	54,652	38.2	54,738	19,332.2	T.O.
D	1600	20	(97,823)	97,837	659.3	97,864	16,359.8	97,850	19,499.9	T.O.
D	1600	40	(97,105)	97,109	4473.1	97,164	25.7	97,296	18,158.9	T.O.
D	1600	80	97,034	97,034	18,025.3	97,092	111.1	97,288	18,386.8	T.O.
E	400	10	45,746	45,746	2.2	45,746	1.2	45,746	7.3	7.3
E	400	20	44,877	44,877	1909.9	44,878	6025.0	44,877	30.6	30.6
E	400	40	44,561	44,565	5125.9	44,569	5351.8	44,561	624.4	764.7
E	900	15	102,421	102,421	808.8	102,421	1186.5	102,421	7.1	7.1
E	900	30	100,427	100,428	19,533.8	100,435	1268.7	100,427	19.7	19.7
E	900	60	100,149	100,159	3189.9	100,163	19,152.9	100,149	2348.8	2348.8
E	1600	20	180,645	180,645	493.9	180,647	4668.7	180,645	255.4	265.9
E	1600	40	178,293	178,300	2501.2	178,310	12,200.1	178,293	455.4	455.4
E	1600	80	176,820	176,851	10,850.5	176,852	18,413.6	176,820	866.1	866.1

costs obtained by PREC coincide with the lower bounds provided by CPLEX and hence are optimal. Thus PREC could solve to optimality 17 out of 27 instances. Between the two metaheuristic algorithms, PREC dominates TSEC for most instances, and both PREC and TSEC provided good solutions with optimality gap less than 1% for all instances. By checking more details, we observe the following results: (1) type C instances with up to $n = 900$ were solved exactly within a reasonable amount of computation time by CPLEX; (2) for the three type C instances with $n = 1600$ for which CPLEX failed to obtain optimal solutions, PREC were more effective and obtained better costs than CPLEX; (3) type D instances were much harder to solve exactly, and for all type D instances in this table, CPLEX failed to find optimal solutions; (4) for all type D instances in this table, PREC obtained better solutions than the other two algorithms, and the differences between the costs obtained by PREC and those by CPLEX were larger for the instances with $n = 1600$ and $m \geq 40$; (5) type E instances were solved to optimality within 1 hour by CPLEX.

We next show the comparison among metaheuristics by Mitrović-Minić and Punnen [20] (denoted VLSVNS), Yagiura et al. [18] (denoted TSEC) and [19] (denoted PREC). The results for LARGE instances are shown in Table 40.4. The results of VLSVNS were taken from Tables 3 of Reference [20]. Algorithm VLSVNS was coded in C++ and tested on a Intel Xeon 2.0 GHz processor with 512 MB memory, while TSEC and PREC were coded in C and run on a i7-5820k 3.3 GHz professor with 32 GB memory. The time limits for VLSVNS in Reference [20] are 3000, 10,000, and 50,000 seconds for $n = 400, 900$, and 1600, respectively. We compared the performance between two computers by using SPECint scores. The computer used for TSEC and PREC is scored 53.7 under SPECint2006, while the score of the computer used in Reference [20] is 837 under SPECint2000. As there seems no formula for converting SPECint2000 scores to SPECint2006 scores and vice versa, we adopted the rate by referring to the computers whose SPECint2000 and SPECint2006 are both reported on SPEC site[*] (e.g., NovaScale B260 Intel Xeon processor E5335, ProLiant DL380 G5 Intel Xeon processor 5150, and IBM BladeCenter LS41 AMD Opteron 8220). From such rates, we estimated that the speed of Intel Xeon 2.0 GHz is slower than i7-5820k by a factor of μ that ranges from 8.2 to 10.4. For convenience, we set $\mu = 10$; based on this, the time limits for TSEC and PREC were set to 300, 1000 and 5000 seconds for $n = 400, 900$ and 1600, respectively. In Table 40.4, the notations "opt (LB)," "time to best," "best," "total time," and "T.O." are the same as those in Table 40.3. We compare the costs obtained by the three algorithms, and the best of them is marked by "*". We observe that PREC dominated TSEC and VLSVNS, obtaining the best results for most instances, and VLSVNS obtained better results than TSEC for 16 instances out of 27. This tendency remains unchanged (except for a small number of changes in "*" marks) when factor μ is set to any value between 8 and 11.

40.10 Performance Guarantee

In this section, we briefly summarize theoretical results on the performance guarantees of approximation algorithms for GAP. As mentioned in Section 40.2, determining the existence of a feasible solution is already NP-complete; hence no polynomial-time approximation algorithm is guaranteed to obtain a feasible solution unless some assumptions are made or a different formulation is considered. For terminologies and more details on approximation algorithms, we refer the interested reader to Vazirani [59], Ausiello et al. [60], and other chapters of this handbook (e.g., Chapters 8, 10 through 12, 14).

We first consider the case where the resource constraints (40.2) can be violated. Without loss of generality, we assume in this section that the available amount of resource at agents satisfy $b_1 = b_2 = \cdots = b_m = \bar{b}$ by scaling the amount of resource. Lin and Vitter [61] gave a polynomial-time algorithm that, given a cost value $\bar{c}$, amount of resource $\bar{b}$ and any constant $\varepsilon > 0$, finds a solution of cost at most $(1 + \varepsilon)\bar{c}$ where the required amount of resource at each agent is at most $(2 + 1/\varepsilon)\bar{b}$, if there exists a feasible solution

---

[*] URL of SPEC site: http://www.spec.org/.

TABLE 40.4 The Results for LARGE Instances Obtained by the Algorithms by Mitrović-Minić and Punnen [20] (VLSVNS), Yagiura et al. [18] (TSEC) and [19] (PREC) (for VLSVNS, One Execution per Instance with Time Limits 3000, 10,000, and 50,000 seconds for $n = 400, 900$, and 1600, Respectively, on Intel Xeon 2.0 GHz, and for TSEC and PREC, One Execution per Instance with Time Limits 300, 1000, and 5000 seconds for $n = 400, 900$, and 1600, Respectively, on i7-5820k 3.3 GHz)

Type	n	m	Opt (LB)	PREC		TSEC		VLSVNS†	
				Best	Time to best	Best	Time to best	Best	Time to best
C	400	10	5597	*5597	2	*5597	1	*5597	90
C	400	20	4782	*4782	7	*4782	7	*4782	444
C	400	40	4244	*4244	83	*4244	14	*4244	1012
C	900	15	11,340	*11,340	528	11,341	28	11341	1225
C	900	30	9982	9983	900	9984	304	*9982	2125
C	900	60	9326	*9327	638	9328	257	*9327	7988
C	1600	20	18,802	*18,802	338	18,803	634	18,803	5740
C	1600	40	(17,144)	*17,145	1185	17,147	2318	17,146	35,696
C	1600	80	(16,283)	*16,287	667	16,290	1268	*16,287	11426
D	400	10	(24,959)	*24,965	147	24,975	18	24,973	1348
D	400	20	(24,558)	*24,583	208	24,611	110	24,610	1286
D	400	40	(24,349)	*24,402	150	24,456	282	24,436	2783
D	900	15	(55,402)	*55,413	389	55,431	803	55,423	9886
D	900	30	(54,831)	*54,859	934	54,915	8	54,905	6178
D	900	60	(54,551)	*54,581	969	54,652	38	54,835	3803
D	1600	20	(97,823)	*97,837	659	97,872	6	97,863	10,661
D	1600	40	(97,105)	*97,109	4473	97,164	26	97,244	30,954
D	1600	80	97,034	*97,039	3653	97,092	111	97,374	32,585
E	400	10	45,746	*45,746	2	*45,746	1	*45,746	64
E	400	20	44,877	44,879	46	44,882	21	*44,877	2509
E	400	40	44,561	*44,572	30	44,581	23	44,581	2933
E	900	15	102,421	*102,421	809	102,423	83	102,423	779
E	900	30	100,427	*100,430	79	100,439	194	100,431	3732
E	900	60	100,149	*100,162	893	100,184	840	100,212	9574
E	1600	20	180,645	*180,645	494	180,647	4669	180,646	6593
E	1600	40	178,293	*178,300	2501	178,312	2280	178,301	39159
E	1600	80	176,820	176,853	4139	176,854	4156	*176,845	49620

†Results in Reference [20].

of cost at most $\bar{c}$. Shmoys and Tardos [62] showed that there is a polynomial-time algorithm that, given a value $\bar{c}$, either proves that no feasible solution of cost at most $\bar{c}$ exists, or else finds a solution of cost at most $\bar{c}$ where the resource requirement at each agent is at most $2\bar{b}$. On the negative side, Lenstra et al. [63] considered the problem of minimizing $\bar{b}$ (i.e., the assignment cost is not considered and $\bar{b}$ becomes the objective value instead of being a given constant), and showed that, for any $\alpha < 3/2$, no polynomial-time α-approximation algorithm exists unless P = NP, where an α-approximation algorithm guarantees to produce a solution with objective function value at most (resp., at least) α times the optimum for a minimization (resp., maximization) problem. Shmoys and Tardos also considered the problem of minimizing the weighted sum of the assignment cost $\sum_{i \in I} \sum_{j \in J} c_{ij} x_{ij}$ and the required amount of resource $\bar{b}$ at each agent, and they gave a polynomial-time 2-approximation algorithm.

We next consider another variant of GAP, in which the objective is to maximize the total profit $\sum_{i \in I} \sum_{j \in J} \rho_{ij} x_{ij}$ and the assignment constraints (40.3) are relaxed to (40.5), that is, $\sum_{i \in I} x_{ij} \leq 1$ ($\forall j \in J$), and call it the max-profit GAP (or MPGAP in short). As discussed in Section 40.2, this problem is equivalent to the original GAP when the objective is to find an exact optimal solution; however, MPGAP is suitable for considering approximation algorithms with performance guarantees because it always has a feasible solution while the original GAP may not. Chekuri and Khanna [32] and Nutov et al. [33] showed that highly restricted cases of MPGAP are APX-hard.

On the positive side, Chekuri and Khanna showed that a polynomial-time 1/2-approximation algorithm exists, which is based on the above-mentioned result by Shmoys and Tardos and is explained as follows. The theorem by Shmoys and Tardos can be restated as follows. (Though they did not state some of the following facts explicitly, they are immediate from the proof in Reference [62] as pointed out in Reference [32].)

Theorem 40.1 (Shmoys and Tardos [62]) *Given an instance of GAP that has a feasible solution of cost $\bar{c}$, there is a polynomial-time algorithm that produces a solution x that satisfies the following three conditions.*

1. *The cost of the solution x is at most $\bar{c}$.*
2. *Each job j assigned to an agent i satisfies $a_{ij} \leq b_i$.*
3. *If $\sum_{j \in J} a_{ij} x_{ij} > b_i$ holds for an agent $i \in I$, there exists a job j' that satisfies $\sum_{j \in J \setminus \{j'\}} a_{ij} x_{ij} \leq b_i$.*

Then the algorithm Chekuri and Khanna works as follows. Given an instance of MPGAP whose optimal profit is ρ^*, obtain an equivalent instance of GAP as explained in Section 40.2, and call the algorithm of Shmoys and Tardos. Then we have a solution x that has profit at least ρ^* and satisfies the above-mentioned conditions 2 and 3. For each i with $\sum_{j \in J} a_{ij} x_{ij} > b_i$, let j' be a job that satisfies $\sum_{j \in J \setminus \{j'\}} a_{ij} x_{ij} \leq b_i$, and let $x_{ij'} := 0$ if $\rho_{ij'} < \sum_{j \in J} \rho_{ij} x_{ij}/2$, or let $x_{ij} := 0$ for all $j \neq j'$ otherwise. Then the obtained solution is feasible and has profit at least $\rho^*/2$.

Fleischer et al. [64] and Cohen et al. [65] also provided 1/2-approximation algorithms by using a simple polynomial-time local search and by applying the local-ratio technique, respectively.

Nutov et al. [33] considered the case with $\rho_{ij} = \rho_j$ ($\forall j \in J$), and showed that a polynomial-time $(1 - 1/e)$-approximation algorithm exists, where e is the base of the natural logarithm function. When an instance of MPGAP satisfies $\rho_{ij} = \rho_j$ and $a_{ij} = a_j$ ($\forall j \in J$), the problem is called the *multiple knapsack problem* (MKP). Chekuri and Khanna proposed a PTAS for MKP and showed that there is no FPTAS for MKP even with $m = 2$ unless P = NP.

For the MPGAP in general, Fleischer et al. [64] proposed a polynomial-time $(1-1/e)$-approximation by an LP-based algorithm. Calinescu et al. [66] achieved a $(1-1/e-O(1))$-approximation with a conceptually simpler algorithm, which is obtained as a consequence of a result for a more general problem. Feige and Vondrák [67] improved the approximation ratio to $(1 - 1/e + \varepsilon)$, where ε is a small constant ($\varepsilon > 0$) whose definition is given in Reference [67]. On the other hand, Chakrabarty and Goel [68] obtain the hardness result, proving that it is NP-hard to approximate MPGAP to any factor better than 10/11.

40.11 Conclusion

In this chapter, we reviewed various algorithms for the GAP, such as greedy methods, local search, Lagrangian heuristics, metaheuristics and branch-and-bound approaches. As examples of efficient approximation methods, basic components of metaheuristic algorithms by Yagiura et al. [18,19] were explained in Section 40.7. We then gave some computational comparisons of representative metaheuristic algorithms as well as some branch-and-bound methods. We also surveyed performance guarantees of some polynomial-time approximation algorithms. The survey in this chapter is by no means comprehensive, but we hope this article gives useful information for people who are interested in devising efficient algorithms to solve this basic problem of GAP, which is of practical as well as theoretical importance.

References

1. Mazzola, J.B., Neebe, A.W., and Dunn, C.V.R., Production planning of a flexible manufacturing system in a material requirements planning environment, *Intl. J. Flex. Manuf Sys.*, 1/2, 115, 1989.
2. Ross, G.T. and Soland, R.M., Modeling facility location problems as generalized assignment problems, *Manag. Sci.*, 24, 345, 1977.
3. Fisher, M.L. and Jaikumar, R., A generalized assignment heuristic for vehicle routing, *Networks*, 11, 109, 1981.
4. Martello, S. and Toth, P., An algorithm for the generalized assignment problem, In *Proceedings of IFORS International Conference on Operational Research*, Brans, J.P., Ed., North-Holland, Amsterdam, the Netherlands 1981, p. 589.
5. Martello, S. and Toth, P., *Knapsack Problems: Algorithms and Computer Implementations*, John Wiley & Sons, Chichester, West Sussex, UK, 1990.
6. Osman, I.H., Heuristics for the generalized assignment problem: Simulated annealing and tabu search approaches, *OR Spektrum*, 17, 211, 1995.
7. Chu, P.C. and Beasley, J.E., A genetic algorithm for the generalized assignment problem, *Comput. Oper. Res.*, 24, 17, 1997.
8. Liu, Y.Y. and Wang, S., A scalable parallel genetic algorithm for the generalized assignment problem, *Parallel Comput.*, 46, 98, 2015.
9. Amini, M.M. and Racer, M., A hybrid heuristic for the generalized assignment problem, *Eur. J. Oper. Res.*, 87, 343, 1995.
10. Racer, M. and Amini, M.M., A robust heuristic for the generalized assignment problem, *Ann. Oper. Res.*, 50, 487, 1994.
11. Laguna, M., Kelly, J.P., González-Velarde, J.L., and Glover, F., Tabu search for the multilevel generalized assignment problem, *Eur. J. Oper. Res.*, 82, 176, 1995.
12. Díaz, J.A. and Fernández, E., A tabu search heuristic for the generalized assignment problem, *Eur. J. Oper. Res.*, 132, 22, 2001.
13. Woodcock, A.J. and Wilson, J.M., A hybrid tabu search/branch & bound approach to solving the generalized assignment problem, *Eur. J. Oper. Res.*, 207, 566, 2010.
14. Haddadi, S. and Ouzia, H., An effective Lagrangian heuristic for the generalized assignment problem, *INFOR*, 39, 351, 2001.
15. Jeet, V. and Kutanoglu, E., Lagrangian relaxation guided problem space search heuristics for generalized assignment problems, *Eur. J. Oper. Res.*, 182, 1039, 2007.
16. Lourenço, H.R. and Serra, D., Adaptive search heuristics for the generalized assignment problem, *Mathware Soft Comput.*, 9, 209, 2002.
17. Alfandari, L., Plateau, A., and Tolla, P., A path relinking algorithm for the generalized assignment problem, in *Metaheuristics: Computer Decision-Making*, Resende, M.G.C., and de Sousa, J.P., Eds., Kluwer Academic Publishers, Boston, MA, 2003, p. 1.
18. Yagiura, M., Ibaraki, T., and Glover, F., An ejection chain approach for the generalized assignment problem, *INFORMS J. Comput.*, 16, 133, 2004.
19. Yagiura, M., Ibaraki, T., and Glover, F., A path relinking approach with ejection chains for the generalized assignment problem, *Eur. J. Oper. Res.*, 169, 548, 2006.
20. Mitrović-Minić, S. and Punnen, A.P., Very large-scale variable neighborhood search for the generalized assignment problem, *J. Interdiscip. Math.*, 11, 653, 2008.
21. Özbakir, L., Baykasoğluand, A., and Tapkan, P., Bees algorithm for generalized assignment problem, *Appl. Math. Comput.*, 215, 3782, 2010.
22. Ross, G.T. and Soland, R.M., A branch and bound algorithm for the generalized assignment problem, *Math. Prog.*, 8, 91, 1975.

23. Fisher, M.L., Jaikumar, R., and van Wassenhove, L.N., A multiplier adjustment method for the generalized assignment problem, *Manag. Sci.*, 32, 1095, 1986.
24. Savelsbergh, M., A branch-and-price algorithm for the generalized assignment problem, *Oper. Res.*, 45, 831, 1997.
25. Nauss, R.M., Solving the generalized assignment problem: An optimizing and heuristic approach, *INFORMS J. Comput.*, 15, 249, 2003.
26. Haddadi, S. and Ouzia, H., Effective algorithm and heuristic for the generalized assignment problem, *Eur. J. Oper. Res.*, 153, 184, 2004.
27. Avella, P., Boccia, M., and Vasilyev, I., A computational study of exact knapsack separation for the generalized assignment problem, *Comput. Optim. Appl.*, 45, 543, 2010.
28. Munapo, E., Lesaoana, M., Nyamugure, P., and Kumar S., A transportation branch and bound algorithm for solving the generalized assignment problem, *Int. J. Syst. Assur. Eng. Manag.*, 6, 217, 2015.
29. Posta, M., Ferland J.A., and Michelon, P., An exact method with variable fixing for solving the generalized assignment problem, *Comput. Optim. Appl.*, 52, 629, 2012.
30. Munapo, M., Kumar S., and Musekwa, S.D., A revisit to the generalized assignment problem: An ascending hyper-plane approach through network flows, *Int. J. Manag. Prud.*, 1(2), 7, 2010.
31. Garey, M.R. and Johnson, D.S., *Computers and Intractability: A Guide to the Theory of NP-Completeness*, Freeman, New York, 1979.
32. Chekuri, C. and Khanna, S., A polynomial time approximation scheme for the multiple knapsack problem, *SIAM J. Comput.*, 35, 713, 2006.
33. Nutov, Z., Beniaminy, I., and Yuster, R., A $(1 - 1/e)$-approximation algorithm for the generalized assignment problem, *Oper. Res. Lett.*, 34, 283, 2006.
34. Ahuja, R.K., Magnanti, T.L., and Orlin, J.B., *Network Flows: Theory, Algorithms, and Applications*, Prentice Hall, Upper Saddle River, NJ, 1993.
35. Korte, B. and Vygen, J., *Combinatorial Optimization: Theory and Algorithms*, 3rd ed., Springer, Berlin, Germany, 2006.
36. Schrijiver, A., *Combinatorial Optimization: Polyhedra and Efficiency*, Springer, Berlin, Germany, 2003.
37. Cattrysse, D.G., Salomon, M., and van Wassenhove, L.N., A set partitioning heuristic for the generalized assignment problem, *Eur. J. Oper. Res.*, 72, 167, 1994.
38. Gavish, B. and Pirkul, H., Algorithms for the multi-resource generalized assignment problem, *Manag. Sci.*, 37, 695, 1991.
39. Rego, C., Sagbansua, L., Alidaee, B., and Glover, F., RAMP and primal-dual RAMP algorithms for the multi-resource generalized assignment problem, Research Report, Hearin Center for Enterprise Science, School of Business Administration, University of Mississippi, MS, 2005.
40. Rego, C., RAMP: A new metaheuristic framework for combinatorial optimization, in *Metaheuristic Optimization via Memory and Evolution: Tabu Search and Scatter Search*, Rego, C. and Alidaee, S., Eds., Kluwer Academic Publishers, Boston, MA, 2005, p. 441.
41. Yagiura, M., Iwasaki, S., Ibaraki, T., and Glover, F., A very large-scale neighborhood search algorithm for the multi-resource generalized assignment problem, *Disc. Opt.*, 1, 87, 2004.
42. Yagiura, M., Komiya, A., Kojima, K., Nonobe, K., Nagamochi, H., Ibaraki, T., and Glover, F., A path relinking approach for the multi-resource generalized quadratic assignment problem, In *Proceedings of the Engineering Stochastic Local Search Algorithms—Designing, Implementing and Analyzing Effective Heuristics (SLS 2007)*, Brussels, Belgium, September 6–8, 2007, Lecture Notes in Computer Science, 4638, 2007, p. 121.
43. Shtub, A. and Kogan, K., Capacity planning by the dynamic multi-resource generalized assignment problem (DMRGAP), *Eur. J. Oper. Res.*, 105, 91, 1998.
44. Park, J.S., Lim, B.H., and Lee, Y., A Lagrangian dual-based branch-and-bound algorithm for the generalized multi-assignment problem, *Manag. Sci.*, 44, 271, 1998.

45. Privault, C. and Herault, L., Solving a real world assignment problem with a metaheuristic, *J. Heuristics*, 4, 383, 1998.
46. Wu, W., Iori, M., Martello, S., and Yagiura, M., Algorithms for the min-max regret generalized assignment problem with interval data, In *Proceedings of the IEEE International Conference on Industrial Engineering and Engineering Management (IEEM 2014)*, Selangor, Malaysia, December 9–12, 2014, p. 734.
47. Cattrysse, D.G. and van Wassenhove, L.N., A survey of algorithms for the generalized assignment problem, *Eur. J. Oper. Res.*, 60, 260, 1992.
48. Yagiura, M., Yamaguchi, T., and Ibaraki, T., A variable depth search algorithm with branching search for the generalized assignment problem, *Optim. Methods Softw.*, 10, 419, 1998.
49. Yagiura, M., Yamaguchi, T., and Ibaraki, T., A variable depth search algorithm for the generalized assignment problem, in *Meta-Heuristics: Advances and Trends in Local Search Paradigms for Optimization*, Voß, S., Martello, S., Osman, I.H., and Roucairol, C., Eds., Kluwer Academic Publishers, Boston, MA, 1999, p. 459.
50. Fisher, M.L., The Lagrangian relaxation method for solving integer programming problems, *Manag. Sci.*, 27, 1, 1981.
51. Held, M. and Karp, R.M., The traveling salesman problem and minimum spanning trees: Part II, *Math. Prog.*, 1, 6, 1971.
52. Martello, S., Pisinger, D., and Toth, P., Dynamic programming and strong bounds for the 0-1 knapsack problem, *Manag. Sci.*, 45, 414, 1999.
53. Haddadi, S., Lagrangian decomposition based heuristic for the generalized assignment problem, *INFOR*, 37, 392, 1999.
54. Lorena, L.A.N. and Narciso, M.G., Relaxation heuristics for a generalized assignment problem, *Eur. J. Oper. Res.*, 91, 600, 1996.
55. Narciso, M.G. and Lorena, L.A.N., Lagrangian/surrogate relaxation for generalized assignment problems, *Eur. J. Oper. Res.*, 114, 165, 1999.
56. Glover, F., Ejection chains, reference structures and alternating path methods for traveling salesman problems, *Disc. Appl. Math.*, 65, 223, 1996.
57. Nauss, R.M., The generalized assignment problem, in *Integer Programming: Theory and Practice*, Karlof, J.K., Ed., CRC Press, Boca Raton, FL, 2006, p. 39.
58. Nonobe, K. and Ibaraki, T., A tabu search approach to the CSP (constraint satisfaction problem) as a general problem solver, *Eur. J. Oper. Res.*, 106, 599, 1998.
59. Vazirani, V., *Approximation Algorithms*, Springer, Berlin, Germany, 2003.
60. Ausiello, G., Crescenzi, P., Gambosi, G., Kann, V., Marchetti-Spaccamela, A., and Protasi, M., *Complexity and Approximation: Combinatorial Optimization Problems and Their Approximability*, Springer, Berlin, Germany, 1999.
61. Lin, J.-H. and Vitter, J.S., ε-Approximations with minimum packing constraint violation, In *Proceedings of the 24th Annual ACM Symposium on the Theory of Computing (STOC 1992)*, Victoria, BC, Canada, May 4–5, 1992, p. 771.
62. Shmoys, D.B. and Tardos, É., An approximation algorithm for the generalized assignment problem, *Math. Prog.*, 62, 461, 1993.
63. Lenstra, J.K., Shmoys, D.B., and Tardos, É., Approximation algorithms for scheduling unrelated parallel machines, *Math. Prog.*, 46, 259, 1990.
64. Fleischer, L., Goemans, M.X., Mirrokni, V.S., and Sviridenko, M., Tight approximation algorithms for maximum general assignment problems, *Math. Oper. Res.*, 36, 611, 2006,
65. Cohen, R., Katzir, L., and Raz, D., An efficient approximation for the generalized assignment problem, *Inform. Process. Lett.*, 100, 162, 2006.
66. Calinescu, G., Chekuri C., Pál, M. and Vondrák, J., Maximizing a monotone submodular function subject to a matroid constraint, *SIAM J. Comput.*, 40, 1740, 2011.

67. Feige, U. and Vondrák, J., Approximation algorithms for allocation problems: Improving the factor of $1 - 1/e$, In *Proceedings of the 47th Annual IEEE Symposium on Foundations of Computer Science* (*FOCS 2006*), Berkeley, CA, USA, October 21–24, 2006, p. 667.
68. Chakrabarty, D. and Goel, G., On the approximability of budgeted allocations and improved lower bounds for submodular welfare maximization and GAP, *SIAM J. Comput.*, 39, 2189, 2010.

41

Linear Ordering Problem

Celso S. Sakuraba

Mutsunori Yagiura

41.1 Introduction

Given an $n \times n$ matrix containing a value c_{ij} in each pair of row i and column j, the linear ordering problem (LOP) consists of finding a permutation π of n indices that maximizes the sum of the values in the upper triangle (i.e., above the diagonal) of the matrix obtained by permuting both rows and columns of that matrix by π. The first appearance of the LOP is usually credited to Reference 1, where the authors analyze the stability of an economy based on the monetary flow among its sectors given by the input/output matrix [2]. Through the analysis of the matrix arranged by the permutation, it is possible to define supplier/consumer relationships among sectors on that economy.

Charon and Hudry [3,4] presented the LOP as a generalization of two problems, the Kemeny's problem [5] and the Slater's problem [6]. The first one is a problem of aggregating individual preferences and can be stated as follows: given a set of individual preferences each of which can be expressed by a permutation, aggregate the preferences of all individuals in a permutation π that minimizes the number of disagreements between π and the individual preferences. The equivalence between the problems is shown by generating a score matrix whose values c_{ij} represent the number of individuals that prefer a candidate i to candidate j, placing i in a position before j in his/her permutation. The Slater's problem consists of determining the best fitting permutation of the vertices of a tournament. The best fitting permutation can be defined as the one that generates the smallest number of reverse arcs, that is, arcs leaving a vertex in a position after the vertex it enters.

As a ranking method based on the relationship between pairs of objects, the LOP has a variety of practical applications, which include determination of ancestry relationships [7], scheduling [8], assessment of corruption perception [9], minimizing crossing numbers in graph drawing [10], and sports teams ranking [11].

In this chapter, we review several methods developed for the LOP. The next section defines the problem formally, discussing its complexity and some approximation algorithms. The following sections present exact, greedy, and local search methods developed for the LOP, followed by a summary of metaheuristics for the problem. Inside the sections of local search and metaheuristics, we give special attention to methods specially tailored to deal with large instances of the LOP.

41.2 Formulation and Complexity

The sum of the elements in the upper triangle of a permuted matrix can be calculated directly from a permutation π using the c_{ij} values, which represent the relation between the sectors in row i and column j. Only when sector j is placed after sector i in π, c_{ij} is placed in the upper triangle of the permuted matrix. Thus, if we define binary variables x_{ij}, where $x_{ij} = 1$ means that sector i precedes sector j in the permutation π (and therefore, $x_{ij} = 0$ means that sector j precedes sector i in π), the LOP can be formulated as follows:

$$\text{maximize} \qquad value(\pi) = \sum_{i \in V} \sum_{\substack{j \in V \\ j \neq i}} c_{ij} x_{ij} \qquad\qquad (41.1)$$

$$\text{subject to} \qquad x_{ij} + x_{ji} = 1 \qquad\qquad \forall i, j \in V, i \neq j \qquad\qquad (41.2)$$

$$x_{ij} + x_{jk} + x_{ki} \leq 2 \qquad\qquad \forall i, j, k \in V, i \neq j \neq k \neq i \qquad\qquad (41.3)$$

$$x_{ij} \in \{0, 1\} \qquad\qquad \forall i, j \in V. \qquad\qquad (41.4)$$

The objective function (41.1) is the maximization of the values in the upper triangle of the permuted matrix. Equation (41.2) signifies that we must choose an order between any pair of sectors i and j, that is, whether sector i is placed either before or after sector j in permutation π. Equation (41.3) guarantees that when a sector i is placed before sector j and j is placed before another sector k in π, k cannot be placed before i, which suffices to guarantee the same condition for any subset with four or more sectors [12,13]. Therefore, the constraints assure that any x satisfying Equations (41.2) through (41.4) corresponds to a valid permutation π.

The LOP can also be modeled over a digraph $G = (V, A)$, where V is the set of vertices ($|V| = n$) and A is the set of arcs ($|A| = m$), if we regard each value c_{ij} in the matrix as the weight of the arc from vertex i to vertex j, as shown in Figure 41.1. If $c_{ij} = 0$, we can either assume that vertices i and j are not directly connected by an arc, or consider that they are connected by an arc with weight 0. It is possible to see that by arranging the order of the rows and columns of the matrix (a) in Figure 41.1 by a permutation $\pi = (4, 1, 3, 2)$, we obtain the matrix (a) in Figure 41.2, which corresponds to ordering the vertices of the corresponding input graph in Figure 41.1(b) in a single row by the same permutation as shown in Figure 41.2(b). The maximization of values in the upper triangle of the matrix, which corresponds to the sum of the weights on the *forward arcs* (arcs going from the left to the right in Figure 41.2(b)), implies the minimization of the sum of the weights on *reverse arcs* (arcs going from the right to the left) or

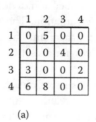

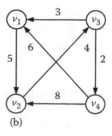

(a)   (b)

FIGURE 41.1   (a) Matrix and (b) graph representations.

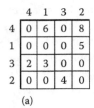
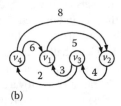

(a)
(b)

FIGURE 41.2 Relationship between (a) matrix and (b) graph representations of the LOP.

values in the lower triangle of the matrix. Thus, the LOP can also be defined as the problem of finding a permutation of the vertices of a digraph with weights on the arcs to minimize the sum of the weights on reverse arcs. When considering such minimization, weights are often called costs. Since various works use different notations to deal with the LOP, in this paper we use the digraph notation with the objective function of minimizing the cost unless otherwise stated to avoid confusion. Using a similar definition of the variables x_{ij} to model this problem (i.e., $x_{ij} = 1$ when vertex i is placed before vertex j in the permutation and 0 otherwise), the only change is the objective function, which becomes the minimization of $cost(\pi) = \sum_{i \in V} \sum_{\substack{j \in V \\ j \neq i}} c_{ij}(1 - x_{ij})$.

This problem can also be regarded as the problem of making the graph acyclic by deleting some arcs (an arc (i, j) is deleted if and only if $x_{ij} = 0$) so as to minimize the sum of the weights of the deleted arcs. This problem is known as the *feedback arc set problem* (FASP). When we consider the maximization of total weight of remaining arcs in the resulting acyclic graph, the problem is called the *maximum acyclic subgraph problem*. When all arc weights are identical (i.e., $c_{ij} = 1$ for all $(i, j) \in A$) and the minimization of the total weight on reverse arcs is considered, the problem is called Slater's problem [6], and its solution is often called Slater's ranking. Even if such a restricted problem is considered, the LOP is known to be NP-hard (the FASP on graphs with identical arc weights was shown to be NP-hard in 1972 [14]).

We assume without loss of generality that if we regard G as an undirected graph, it is connected, which implies $m \geq n - 1$. We also define Δ as the maximum degree of G, that is, the maximum number of vertices connected to one of the vertices of G. Note that there are at most two arcs connecting each pair of vertices, and we can remove one of them without changing the essential nature of the problem by the following transformation: If $c_{ij} \geq c_{ji}$ holds, keep only arc (i, j) and set its weight to $c_{ij} - c_{ji}$; the opposite case is similar. In particular, we can remove both of them when $c_{ij} = c_{ji}$. For these reasons, we can assume without loss of generality that $c_{ij} > 0$ for all arcs (i, j). For the same reason, the graph can also be transformed without loss of generality, to a tournament having nonnegative arc weights (if we allow arcs with weight 0), where a tournament is a digraph that has exactly one arc (i, j) or (j, i) for every pair of vertices $i, j \in V$. Moreover, when the given digraph is general but has identical arc weights, we can use the same idea to transform it into a tournament whose arc weights are 0 or 1. However, when arcs with weight 0 are not allowed in the resulting tournament, such transformation is not immediate. Indeed, the complexity of LOP when restricted to tournaments with identical arc weights had been open for more than a decade since its NP-hardness was conjectured in 1992 in Reference 15; it was in 2006 when a positive proof of this conjecture was provided [16].

The LOP is shown to be APX-Hard [17], which means that assuming P $\neq$ NP, there is a constant α for which there is no polynomial-time approximation algorithm that can guarantee to find a solution under α times the value of the optimal solution. There is also a proof that such α is not less than 1.3606 [18]. Polynomial-time algorithms that can guarantee a solution under $O(\log n \log \log n)$ times the optimal solution value are also presented in References 19, 20.

For the problem of maximizing the total weight of forward arcs, there is a simple method to obtain a solution that has at least half the value of the optimal one. Taking an arbitrary permutation π and its inverse permutation π', the sets of forward arcs for π and for π' are mutually complementary, and the sum of the weights on the arcs in both sets is the total weight of all arcs. Since the value of optimal solution has to be under the sum of the weights of all arcs, if we take the solution with greater value between π and π', its value is at least half of the value of the optimal solution. For graphs such that all arc weights are the same and there are no cycles of length 2, Berger and Shor [21] proposed an algorithm that obtains a permutation with the number of forward arcs at least $(1/2 + \Omega(1/\sqrt{\Delta}))m$. Other results on the approximation guarantees can be found in Reference 4.

There are also several polynomial-time algorithms proposed for graphs having special structures. We present two results for cases where all weights are the same. Conitzer [22] proposed a linear-time algorithm for graphs with a special layered structure (the definition of this structure is quite complex; see [22] for more details). Laslier [23] showed that it is rather simple to obtain an optimal solution when the tournament has a regular structure called "cyclic."

41.3 Exact Algorithms

Researches based on one of the most well-known exact methods, the branch-and-bound, date from long ago [24]. The branch-and-bound method divides a problem into several smaller subproblems, called partial problems, and is based on the idea that solving all these partial problems is equivalent to solving the original one. There are cases where we can conclude without exactly solving a partial problem that it will not give an optimal solution to the original problem. One of the most common technologies for this is the lower bound test. We define the incumbent solution π^* as the one having the smallest cost among solutions π found by constructive methods or during the search of a branch-and-bound method. Then, when the lower bound on the cost of solutions for a partial problem is larger than or equal to $cost(\pi^*)$, we can conclude that it is impossible to obtain a solution better than π^* for that partial problem. Thus, there is no need to investigate it any further, and we can stop the search on that partial problem, reducing the computational effort necessary to find an optimal solution.

We present an example of methods to calculate a lower bound for the LOP. Consider a set of disjoint dicycles, which is a set of dicycles such that any two dicycles among them do not have a common arc. Then, a lower bound on the optimal solution value is given by summing up the weights of the arcs with the smallest weight of each dicycle (Theorem 37 of Reference 4). It is not easy to compute the best lower bound among all combinations of disjoint dicycles, but a relatively good lower bound can be obtained by combining dicycles whose minimum arc weights are large. Lower bounds based on such cycles have also been proved to be effective for the solution of other problems such as the MAX-2-SAT [25].

A relatively recent branch-and-bound algorithm for the LOP is given in Reference 26. The authors used a metaheuristic called noising method [27] to obtain the initial solution, and proposed a method of obtaining a lower bound using Lagrangian relaxation. Although the relaxation problem can be solved much faster than the original one, as the frequency of lower bound calculations for the branch-and-bound is quite high, it is expensive to solve the relaxation problem for every partial problem. To alleviate this, they developed a method to terminate the partial problems without solving the relaxation problem. The computational results using randomly generated instances with up to 100 vertices on a Sun SPARC workstation showed that most of the instances were solved within 1 hour. They also reported that the difficulty of the instances is related to the number of vertices, and also is highly influenced by the way they were generated.

There are also algorithms for the LOP based on the cutting plane method. Valid inequalities are a set of linear constraints satisfied by all feasible solutions of an integer programming problem, and the cutting plane method consists of adding such inequalities one after another and repeatedly solving the LP relaxation problem while there are valid inequalities not satisfied by the best solution of the LP relaxation problem. For instance, Mitchell and Borchers [28] presented a cutting plane method based on the primal-dual interior point method. They also proposed a method using cutting planes based on the primal-dual interior point method in the first half of iterations, changing to the simplex method in the latter half [29]. Improved results were reported when compared to the two cases where only the first or latter half of the two phases is used separately. Computational experiments using randomly generated instances with 100–250 vertices showed that the algorithm could solve every instance within 30 minutes on a Sun SPARC 20/71.

The branch-and-cut, a method resulting from the combination of the branch-and-bound and the cutting plane methods, has its efficiency shown for many problems, including the LOP. Grötschel et al. [12] proposed an approach combining the branch-and-bound with a cutting plane method using various valid inequalities, which solved instances with up to 60 vertices in less than 20 minutes on an IBM 370/168. Oswald et al. [30] presented some results on a branch-and-cut approach using valid inequalities called mod-k cut.

Branch-and-bound and branch-and-cut methods are often used also as heuristics by stopping their execution before normal termination with the guarantee of optimality. One of the merits of such an approach is that it also outputs a lower bound (for minimization problems, and an upper bound for maximization), which can be used to evaluate the quality of the obtained solution value. Commercial integer programming solvers are often based on the branch-and-cut framework, and their performance with respect to exact methods as well as heuristics has been greatly improved during the last few decades.

It might therefore be worth trying to obtain optimal or heuristic solutions for the LOP using the integer programming formulation presented in Section 41.2 on a commercial integer programming solver. We conducted a preliminary test on this approach for some instances in http://www.optsicom. es/lolib/#instances. Using the solver CPLEX 12.6.2 on an Intel Core i7-5820K 3.3 GHz with 32 GB of RAM, it was possible to obtain optimal solutions for all instances called SGB with 75 vertices on an average time of 1 minute. We also tested larger instances, and the results are shown in Table 41.1. The table presents the values of the objective functions found in 1 hour of processing time by CPLEX on the same environment for larger instances, for which no optimal solution was found (the presented values refer to the maximization version of the LOP). The instances used in the experiments belong to the RandomAI set, which contains 25 instances for each value of $n = 100, 150$, and 200. In the table, the first column refers to the instance number and each following set of four columns refers to a different value of n, that is, the values in the first row of the columns under $n = 100$ correspond to values found for the instance t1d100.1. For each instance, four values are presented: "CPLEX" and "UB" refers to the values of the objective function of the best feasible solution and the upper bound found by CPLEX after 1 hour of processing time; "Best" corresponds to the values of the objective function of the best known solutions reported in Reference 31, all of which coincide with the best of those obtained in [31] by several metaheuristics, each with a time limit of 10 minutes on a Dual Intel Xeon 3.06 GHz with 3.2 GB of RAM; and "dif. (%)" is the relative difference in % given by 100("Best"−"CPLEX")/"Best."

It is possible to see from Table 41.1 that solutions obtained by CPLEX have an average gap of approximately 28% from their upper bounds for instances with $n = 100$ or 150, and of about 50% for those with $n = 200$. These large gaps might signify that it will be difficult for CPLEX to obtain exact optimal solutions to these instances even if much longer computation time is allowed. We can also observe that the relative differences between the best known solutions (Best) and obtained solution values (CPLEX) are around 23%, 20%, and 18% for $n = 100, 150$, and 200, respectively, which are quite large. In contrast, it is possible to obtain solutions with a relative error of up to 0.5% from the best known solution values just by running a local search algorithm multiple times from random solutions with a time limit of a few minutes (we confirmed this with a local search algorithm implemented with a data structure of TREE described later in Section 41.5 on a PC with Intel Xeon 3.0 GHz, much slower than the one used for CPLEX). Thus, for instances with $n = 100$ or larger, it is highly recommended to use heuristic methods instead of optimization software to obtain good quality solutions in a reasonable time.

TABLE 41.1 Best Solution Values and Upper Bounds

Instance no.	n = 100				n = 150				n = 200			
	CPLEX	Best	Dif. (%)	UB	CPLEX	Best	Dif. (%)	UB	CPLEX	Best	Dif. (%)	UB
1	81,262	106,852	24	114,142	183,371	235,056	22	261,413	333,135	410,871	19	672,568
2	81,396	105,947	23	113,801	184,514	234,421	21	259,050	334,677	407,729	18	666,887
3	80,966	109,819	26	117,534	189,909	236,319	20	262,076	334,158	407,420	18	667,017
4	85,018	109,252	22	117,408	187,616	234,506	20	259,619	331,048	410,101	19	670,817
5	83,564	108,859	23	117,228	187,475	234,601	20	259,863	337,481	411,522	18	675,232
6	82,448	108,201	24	116,721	190,781	234,465	19	260,035	333,205	406,451	18	664,988
7	84,881	108,803	22	116,785	192,532	235,283	18	260,922	338,940	412,482	18	675,871
8	83,794	107,480	22	115,449	185,722	237,230	22	260,956	333,477	408,850	18	669,243
9	84,405	108,549	22	116,267	183,663	237,253	23	261,817	333,475	409,308	19	668,719
10	84,611	108,771	22	117,218	185,524	234,821	21	260,359	336,375	406,453	17	665,237
11	83,613	107,920	23	116,343	188,705	234,157	19	259,316	336,310	410,239	18	671,789
12	83,583	108,389	23	116,323	184,705	236,318	22	262,105	333,233	412,831	19	672,490
13	82,055	108,702	25	116,516	184,032	237,116	22	262,728	331,870	409,234	19	673,189
14	82,239	105,583	22	113,884	186,049	234,453	21	259,553	329,079	408,879	20	667,017
15	84,726	108,667	22	116,927	184,383	232,065	21	257,699	340,476	409,061	17	669,391
16	84,290	107,426	22	115,442	188,974	232,948	19	258,627	337,091	408,059	17	668,958
17	82,117	105,612	22	113,543	188,830	236,656	20	262,929	334,647	410,280	18	671,323
18	81,421	107,861	25	115,675	186,567	234,348	20	259,865	333,901	407,709	18	668,122
19	84,582	108,026	22	116,664	187,658	234,994	20	260,286	340,759	412,947	17	674,808
20	85,313	109,968	22	118,675	190,949	235,411	19	260,653	331,828	406,418	18	665,495
21	81,976	107,255	24	116,000	185,739	233,980	21	259,087	326,807	408,037	20	664,821
22	83,442	108,250	23	116,691	189,141	235,415	20	260,134	332,688	407,333	18	669,855
23	80,615	106,146	24	112,983	188,535	233,492	19	258,077	332,344	408,552	19	667,462
24	81,714	108,782	25	116,611	182,314	236,016	23	261,152	332,419	410,583	19	668,498
25	84,598	106,933	21	115,024	188,241	236,428	20	261,299	333,608	406,356	18	664,812

41.4 Constructive Heuristics

In this section, we present some methods that build solutions for the LOP by inserting and changing the position of vertices in the permutation using different rules and measures to evaluate the vertices in relation to the position they should be allocated.

Borda [32] suggested a greedy algorithm based on the idea that vertices with large costs on the outgoing arcs should be ranked first. He defines the attractiveness of each vertex i to be $a_i = \sum_{j=1}^{n} c_{ij}$, and builds a solution by ordering the vertices in the nonincreasing order of attractiveness.

Another measure used to rank vertices is defined by Becker [33]. He defines a quotient given by $q_i = \sum_{k \neq i} c_{ik} / \sum_{k \neq i} c_{ki}$, and ranks the vertices in decreasing order of their quotients. He also proposed an alternative that recalculates all quotients each time the vertex with the highest quotient is inserted in the permutation using only the costs of the remaining vertices.

Constructive procedures not based on any measure associated with the vertices are presented in References 34–36. Chanas and Kobylanski [34] proposed a procedure based on two operations, *sort* and *reverse*. The first one utilizes a criterion to rearrange the permutation obtaining an improved solution, whereas the latter one returns the reversed permutation that usually results in a bad solution. The authors showed that the application of reverse and then sort operations to a solution results in a permutation that is as good as or better than the original one, and their algorithm consists of applying both operations to an initial solution alternately.

Masson [35] presented a procedure developed to solve a small instance of the French economy with 28 sectors that evaluates the vertices based directly on the relation between pairs or vertices. The main idea of their heuristic can be explained as follows. The proposed heuristic maintains a partition of V and a partial order $P \subseteq V \times V$, where $(i, j) \in P$ signifies that i precedes j in the resulting permutation. The algorithm starts from $P := \emptyset$ and the partition consisting of a single block V, and it gradually increases elements in P by partitioning one of the blocks and adding into P some ordered pairs between new blocks until it becomes impossible to add more elements in P. This is done by repeatedly applying the following rule. The algorithm randomly chooses a vertex i with a high degree (among those that have not been chosen as i yet) and divide the block B that contains i into the following four blocks: (1) $L(i) = \{j \in B \mid c_{ij} < c_{ji}\}$, (2) $R(i) = \{j \in B \mid c_{ij} > c_{ji}\}$, (3) $\{i\}$, and (4) $I(i) = B \setminus (L(i) \cup R(i) \cup \{i\})$. Then, it computes $P' := P \cup \{(j, i) \mid j \in L(i)\} \cup \{(i, j) \mid j \in R(i)\}$ and replaces P with the transitive closure of P'. During this process, the algorithm also tries to add into P' some of the ordered pairs between the vertex i and vertices j in $I(i')$ with $c_{ij} \neq c_{ji}$ for some of vertices i' that have been chosen before, where care should be taken in the choice of i' and j to avoid making cycles in P'. When the process of increasing P has stopped, the algorithm outputs one of the linear extensions of P.

Guo et al. [36] proposed an algorithm based on the idea that for some networks, vertices can be classified as leaders or followers based on the costs of the arcs entering them. Vertices are divided into two groups according to this classification, and the group of leaders, that is, vertices with large costs on their incoming arcs, is placed after the group of followers in the permutation. Then, leaders and followers are redefined inside each of the groups, repeating the procedure until all groups contain only one vertex. They explore some real Internet applications of ranking vertices to predict the directions of information flows, allowing structural investigations of networks.

There is a method that regards a deleted arc set with respect to the FASP formulation as the set of arcs to change the direction. It successively changes the direction of arcs until there is no dicycle in the graph. As mentioned before, the problem can be defined on tournaments without loss of generality. We define $\delta_+(i)$ to be the outdegree of a vertex i, which is the number of arcs leaving vertex i. When a tournament has an arc (i, j) with $\delta_+(i) \leq \delta_+(j)$, there is at least one dicycle of length 3 containing the arc (i, j) [37]. Moreover, by changing the direction of (i, j), the number of dycycles of length 3 is reduced by $\delta_+(j) - \delta_+(i) + 1$. Therefore, by repeatedly changing the direction of such arcs, at the end a tournament with no cycles is obtained.

There are several measures to choose arcs to be reversed. For the case when all arcs have the same weight, a method that chooses the arc (i, j) with the largest value of $\delta_+(j) - \delta_+(i) + 1$ is proposed in

Reference 38. When arc weights are general, Berthélemy et al. [39] proposed a method that chooses the arc (i, j) with the largest value of $(\delta_+(j) - \delta_+(i) + 1)/c_{ij}$, and another that chooses the arc (i, j) whose inversion generates the largest difference between the sum of the weights on the dicycles of length 3 containing (i, j) and that of dicycles of length 3 that would be generated by the reversed arc (j, i).

41.5 Local Search Algorithms

A local search algorithm starts from an initial solution and repeatedly replaces the current solution with a better solution in its neighborhood until no such solution is found, where a neighborhood is a set of solutions obtainable from the current solution by slight perturbations. The two neighborhoods most commonly used for local search algorithms for the LOP are *insert* and *exchange*. The insert neighborhood consists of solutions obtained by removing one vertex from its original position and inserting it into a different one. The *shift* neighborhood is a subset of the insert neighborhood, where the insertion is restricted to adjacent positions. The exchange neighborhood consists of solutions obtained by switching the positions of two vertices in a solution. It is known that any solution improved by exchange can be improved by insert, although the opposite is not always true [40]. It is also empirically observed that local search algorithms based on the insert neighborhood show better results than those based on exchange [41], being used in the majority of metaheuristics for the LOP as in References 42–44.

Ceberio et al. [45] proposed a method that calculates a set of positions k for each vertex i, for which any permutation π having i assigned to position k would never be a locally optimal solution in the sense that there are better positions to insert i, and we can exclude k from the list of positions to insert i into. By analyzing the structure of a solution of the LOP, they observed that each vertex i has a contribution to the objective function determined only by the position it occupies in the permutation π and the sets of vertices before and after i, regardless of the order inside each set. They calculate vectors of differences for each vertex i containing the changes in the value of the objective function when i is inserted in different positions. By sorting this vector properly, it is possible to show that when some inequalities regarding the vertex i, a position k and the weights on the arcs entering and leaving i do not hold, any solution having i assigned to position k can always be improved by inserting i in a different position. Based on this property, they proposed an $O(n^3)$ preprocessing algorithm that calculates a matrix of positions k for each vertex i in which i should not be inserted. Such algorithm is ran once before the use of any method that uses the insert neighborhood, and positions k are excluded from the list of possible positions to insert each vertex i during the execution of any local search independently of the current permutation π, creating a reduced search space called *restricted insert neighborhood*. The authors showed that including this method in two state-of-the-art algorithms for the LOP improved their results.

To compare the efficiency of local search methods, Yagiura and Ibaraki [46] defined the one-round time as the time necessary to find an improved solution in a neighborhood or to conclude that such solution does not exist, including the time necessary to update the data structures maintained for such purposes. Schiavinotto and Stützle [47] showed that it is possible to achieve a one-round time of $O(n^2)$ for the insert neighborhood by conducting the search in an ordered manner. This complexity is achieved by sequentially calculating the objective values of solutions obtained by consecutive shift operations performed in the permutation, each in constant time. Two efficient local search algorithms for the insert neighborhood are presented in Reference 48. Their algorithms, called LIST and TREE, are designed to deal with sparse and large instances of the LOP, respectively, and are described in the following subsections.

41.5.1 LIST Algorithm

The LIST algorithm is based on a set of linked lists containing one list per vertex, represented in its head. Each element of a list corresponds to a vertex u connected to the head vertex v, containing the index of u and the weight of the connecting arc (u, v) or (v, u), where we assume without loss of generality that only one of the two arcs exists and its weight is nonnegative as discussed before. Elements are kept in the same order of the current solution, with forward and reverse arcs having positive and negative values, respectively. The search for an improving insert position is made by scanning a list from its head to one direction at a time and accumulating the values in the elements of the list, eliminating the calculation of insertions that do not change the value of the cost (i.e., inserting a vertex v at a position before or after a vertex not connected to v will not affect the cost).

Once an improving insertion is found, it is necessary to update the lists of the inserted vertex and of all vertices connected to it to reflect the change and make the list correspond to the new solution. This is done in two steps: For the list of the inserted vertex, we remove the head from its current position and insert it in its new position in the list, multiplying the values of the elements between its current and new positions by -1. For the list of each vertex connected to the inserted one, the element corresponding to the inserted vertex is removed from its current position and inserted in its new position, multiplying its own value by -1 if it crosses the list's head during this operation. This algorithm has a one-round time of $O(m)$, obtaining good results for sparse instances of the LOP.

41.5.2 TREE Algorithm

The TREE algorithm has a one-round time of $O(n + \Delta \log \Delta)$, which is achieved by using a set of trees containing one tree for each vertex v of the permutation. Leaves of the trees represent gaps between vertices connected to v, so that the set of reverse arcs connected to v when v is inserted in each of these gaps is different. The internal nodes keep information used to find the positions in which v can be inserted to reduce the total cost of the permutation.

The search for an improved solution in the neighborhood is achieved by implementing the following operations:

- Find a vertex v that, when inserted in an appropriate position, generates a solution π' with a cost smaller than the cost of the current solution π, or conclude that no such vertex exists, in $O(n)$ time.
- Find the position where v should be inserted to generate π' in $O(\log d_v)$ time, where d_v is the degree of v.
- Update the data structure to make the trees correspond to the new solution π' in $O(d_v \log \Delta)$ time.

To process these operations, a tree with $O(d_v)$ leaves such as the one illustrated in Figure 41.3 is built for each vertex v. An index $v_{\text{name}}(z)$ represented on the left side of each node z of the tree in the figure is

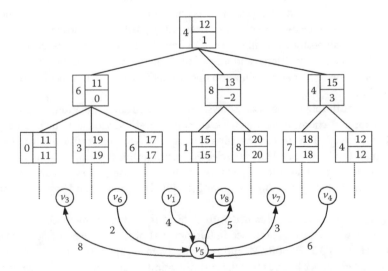

FIGURE 41.3 Example of a tree for vertex v_5 and the vertices connected to v_5 with their costs. The value 12 in the root represents the minimum cost v_5 can have if inserted in an appropriate position of the current permutation π.

used to find the position of each leaf. Note that the leftmost and rightmost leaves correspond to positions where v could be inserted before the first and after the last vertex connected to v. We define the cost of a vertex v in a permutation π, denoted $cost(v, \pi)$, as the sum of the reverse arcs in π connected to v. The value $\gamma(z)$ represented in the bottom part of each node z, when summed up from a leaf l through the path to the root, corresponds to the cost of vertex v in a solution when v is inserted in a position corresponding to l. For instance, consider the leaf between vertices v_1 and v_8 in Figure 41.3; if we sum up its bottom values with those of its ancestors, we obtain a value of $15 + (-2) + 1 = 14$, which is the sum of the reverse arcs connected to v_5 if we insert v_5 in a position between v_1 and v_8. The value in the top part of each node z, $\gamma_{\min}(z)$, represents the minimum of the values of γ on a path among all paths between z and the leaves in the subtree that has z as its root. The value $\gamma_{\min}(r_v)$, where r_v is the root of the tree for vertex v, represents the minimum cost v can have if inserted in an appropriate position of the permutation π. From the above-mentioned definitions, we can obtain a solution with a cost smaller than the current solution π if and only if there is a vertex v with $\gamma_{\min}(r_v) < cost(v, \pi)$.

41.6 Metaheuristics

There are several metaheuristic algorithms proposed for the LOP. Martí et al. [31] presented a comparison of some major ones developed by them, including hybrid genetic algorithm (HGA) [40], memetic algorithm (MA) [41], iterated local search (ILS) [41], tabu search (TS) [42], variable neighborhood search (VNS) [43], scatter search (SS) [44], and two multi-start procedures based on [49] and [50]. Among these methods, they concluded that MA outperformed the others. This section gives a description of it along with some other methods.

In the MA proposed in Reference 41, an initial population is generated randomly and local search is applied to each individual. Then, at each iteration (also called generation), new individuals (permutations) are generated by crossover and mutation from randomly selected individuals of the current generation. The crossover operators that presented the best results were the ones called CX and OB. The CX operator starts by copying to the offspring all the vertices that occupies the same position in both parent solutions, keeping them in the same position as their parents. Then it randomly chooses a parent π^A from the two parents and an empty position i_1, and it assigns the element a in position i_1 of the parent π^A to position i_1 in the offspring, and element b_1 in position i_1 of the other parent π^B to position i_2 where b_1 belongs in π^A. Element b_2 in position i_2 in π^B is then assigned to position i_3 where b_2 belongs in π^A, and this assignment procedure is repeated until element a is chosen again or all positions are filled. If there are still empty positions, one of the two parents and a new empty position are chosen randomly and the same process is iterated until all positions are filled. The OB operator generates an offspring as a copy of one of the parents, and then selects k positions randomly, reordering the vertices in these positions according to their relative order in the other parent. Mutation operations consist of a number of exchange operations applied to a solution.

The iterative part of the algorithm is as follows: form the current population, a number of crossovers and mutations is applied, choosing parents to be combined and individuals to be mutated using an uniform distribution. Each offspring is improved by applying local search to it after any crossover or mutation is performed. Then, the best individuals are selected, maintaining the same population size. When the average solution quality is the same for a long time, a diversification process occurs by selecting the best individual of the population, generating the other individuals randomly and applying local search to all of them.

Martí et al. [31] reported results obtained with state-of-the-art exact methods and metaheuristics in the literature, among which the MA proposed in Reference 41 obtained the best results. They used a Dual Intel Xeon 3.06 GHz with 3.2 GB of RAM, and reported average gaps between MA solutions obtained in 10 minutes and the best upper bounds (for the maximization version of the problem) obtained through different methods in 2 hours. The average gaps are ranging between 7% and 12% for instances with $100 \leq n \leq 200$ and around 42% for instances with $n = 500$ (these gaps seem large, but this is probably because the upper bounds are not tight). Moreover, the MA and ILS proposed in Reference 41 were able to obtain, within 10 seconds of processing time, optimal solutions for 228 instances among the 229 instances used in the experiment for which the optimal solution values are known.

Ingram et al. [51] proposed an exploration method to increase the efficiency of the insertion neighborhood under TS framework. Their main idea consists of improving the calculation of the objective function based on the fact that exploring insertions in consecutive positions has a much lower computational cost. The resulting algorithm obtained solutions of similar quality to the original tabu search proposed in Reference 42 in smaller computation time.

41.6.1 Metaheuristics for Large Instances

Metaheuristics designed so that they are efficient especially for large instances of the LOP are proposed in Reference 52. The authors proposed an ILS and a great deluge algorithm (GDA) based on the data structure of TREE [48] to solve instances with up to 8000 vertices. These algorithms keep one solution at a time, instead of maintaining a population consisting of several solutions, which is essential for a good performance of the TREE data structure because the initialization of TREE requires a certain amount of time. For both metaheuristics, the constructive heuristic presented in Reference 33 is used to generate the initial solution. Moreover, the characteristics of TREE are exploited in designing the criteria used to perform the search to improve its results. They also showed that the performance of these metaheuristics designed to deal with large instances is comparable to the state-of-the-art algorithms in Reference 31 for smaller instances.

The ILS consists of, from an initial solution, a repeated use of local search until a locally optimal solution is found, each starting from a solution obtained by perturbing a good solution obtained by then. The ILS proposed in Reference 52 takes advantage of the structure of TREE for both local search and perturbation procedures. The local search is performed likewise the one explained in Subsection 41.5.2, and the perturbation consists of k insert operations on randomly chosen vertices v, whose insert positions are decided by walking down the trees of v looking for the insert positions that *increase* $cost(v, \pi)$ the most, generating diversification.

The GDA starts from an initial solution and an initial water level WL. Then, the following procedure is iterated until a stopping criterion is reached: a randomly chosen neighborhood operation is applied to the current solution, and the obtained solution replaces the current one if its cost is smaller than WL; if the current solution is updated, WL is decreased. In the proposed GDA, each iteration is executed efficiently using the TREE structure to always find a new solution with cost smaller than WL. In this process, randomness is incorporated in going down the trees, instead of finding the best insert positions, to randomize the choice of neighbors. The value of WL is sometimes reset if a prespecified number of iterations are executed without a solution update.

41.7 Computational Results

In this section, we present some results on the performance of the local search algorithms described in this chapter. Several results obtained with the algorithms presented in this work can be found in the cited references, among which we selected the ones related to local search using the insert neighborhood since they are the majority of the metaheuristic methods. We compare the procedure presented in Reference 47, referred to as SCST, with LIST and TREE algorithms presented in Reference 48. The instances used in the experiments were randomly generated and are divided into classes based on their number of vertices n and density p (in %), which represents the probability of having an arc connecting each pair of vertices in the graph ($m \approx n(n - 1)/2 \times p/100$). A full description of the instances is available online at http://www.co.cm.is.nagoya-u.ac.jp/~yagiura/lop/ for benchmark purposes. Numerical experiments in this section were run on an Intel Xeon 3.0 GHz processor with 8 GB of RAM memory.

TABLE 41.2   Time to Reach a Locally Optimal Solution

n	Algorithm	Move strategy	Density p				
			1%	5%	10%	50%	100%
500	SCST	Best	0.9069	1.4504	1.6432	2.0189	2.1747
		First	0.3697	0.8009	0.9705	1.4496	1.6392
	LIST	Best	0.0174	0.4383	0.9751	10.1641	31.8788
		First	0.0016	0.0220	0.0952	19.0473	92.7353
	TREE	Best	0.0026	0.0126	0.0356	0.5183	1.2052
1000	SCST	Best	27.4378	41.2255	47.4462	59.5248	66.6577
		First	6.6118	11.9438	15.2223	21.3865	24.3017
	LIST	Best	0.1776	1.3428	11.0109	138.4430	365.5362
		First	0.0106	0.2212	2.6592	265.2471	1,521.3867
	TREE	Best	0.0130	0.1084	0.3172	3.2533	6.8180
2000	SCST	Best	351.9283	496.3623	567.0452	698.2716	756.9867
		First	80.2860	131.8742	155.7777	220.4153	243.8055
	LIST	Best	1.9439	64.8015	172.2008	1,382.6318	3,603.7983
		First	0.1038	9.4876	61.5922	3,729.2617	21,018.6371
	TREE	Best	0.0866	0.7513	1.9749	17.1772	35.4242
3000	SCST	Best	1,395.6812	1,894.9275	2,121.5547	2,673.6799	2,883.9076
		First	330.3462	519.0211	607.3933	825.4955	912.4719
	LIST	Best	23.4644	275.1180	657.6628	5,358.4320	14,201.3319
		First	0.6265	48.3467	281.6440	16,540.8074	T.O.
	TREE	Best	0.2632	2.1393	5.6573	45.2669	92.8771
4000	SCST	Best	3,631.5215	5,016.5952	5,465.2006	6,712.5637	7,129.9031
		First	965.4084	1,456.2072	1,737.1501	2,308.0524	2,548.0524
	LIST	Best	86.4585	738.4881	1,720.6672	13,585.8200	37,759.4419
		First	2.2847	140.4271	812.2275	T.O.	T.O.
	TREE	Best	0.5585	4.3577	11.4467	88.6669	179.2260
8000	SCST	Best	38,321.0989	T.O.	T.O.	T.O.	T.O.
		First	20,642.0433	24,213.2550	28,582.3740	(41,490.6257)	T.O.
	LIST	Best	963.1642	7,139.7846	17,717.8765	T.O.	T.O.
		First	31.1471	1,750.9126	10,232.6312	T.O.	T.O.
	TREE	Best	3.4837	25.2932	62.5775	457.9856	941.0565

Note: The value in parentheses represents the average excluding an instance for which the time limit was reached without confirming local optimality.

Table 41.2 shows the experimental results, corresponding to computation times spent to obtain locally optimal solutions from the beginning of the search, that is, without considering the times to read instances and build data structures. In the table, rows with label "best" (respectively, "first") show the results when the best (respectively, first) admissible move strategy is used. The values represent the average amount of time in seconds spent until the algorithms stopped after finding locally optimal solutions, starting from a randomly generated solution. In the table, (time over) means that the algorithm ran for 12 hours without finding a locally optimal solution for all the instances of that class. All data were taken from Table 4 of Reference 48.

From the results in Table 41.2, we can see that the TREE algorithm is the only one that was able to reach locally optimal solutions in less than 12 hours for all of the tested instances; TREE reached a locally optimal solution in less than 20 minutes for every instance. We can also observe that LIST reaches locally optimal solutions faster than SCST for instances with density smaller than 50%, and TREE reaches locally optimal solutions faster than the other two algorithms for almost all the instance classes.

41.8 Conclusion

The linear ordering problem (LOP) has attracted the interest of many researchers due to its theoretical and practical importance and complexity. In this chapter, we gave an overview of contributions to this problem mainly in terms of nonexact methods, dividing them into three parts: constructive heuristics, local search methods, and metaheuristics. Given the increase in the amount of data that needs to be processed nowadays, some methods specially tailored to solve large instances of the LOP were also described. Besides presenting results similar to the state-of-the-art algorithms for instances with hundreds of vertices, such methods allow us to deal with instances with up to several thousand vertices.

We presented a summary of well-known existing methods collected from various works. Although this chapter by itself may not suffice to have a deep understanding of the LOP, our aim was to provide a variety of useful information on this problem, which has been extensively studied in the literature and has many practical applications. We tried to organize the chapter to be useful for readers without preliminary knowledge on the LOP, and for those with further interest on this problem, we recommend reviews such as [4,31] and [41], which present a more mathematical and computational perspective, and an analysis of the search space, respectively.

Acknowledgments

The authors would like to thank André Luis Azevedo Ramanho and Wei Wu for their contributions on the bibliographical search and on the CPLEX experiments, respectively.

References

1. Chenery, H.B. and Watanabe, T., International comparisons of the structure of production, *Econometrica*, 26, 487, 1958.
2. Leontief, W.W., Quantitative input-output relations in the economic system of the United States, *The Review of Economics and Statistics*, 18, 105, 1936.
3. Charon, I. and Hudry, O., A survey on the linear ordering problem for weighted or unweighted tournaments, *4OR: A Quarterly Journal of Operations Research*, 5, 5, 2007.
4. Charon, I. and Hudry, O., An updated survey on the linear ordering problem for weighted or unweighted tournaments, *Annals of Operations Research*, 175, 107, 2010.
5. Kemeny, J.G., Mathematics without numbers, *Daedalus*, 88, 577, 1959.
6. Slater, P., Inconsistencies in a schedule of paired comparisons, *Biometrika*, 48, 303, 1961.
7. Glover, F., Klastorin, T., and Klingman, D., Optimal weighted ancestry relationships, *Management Science*, 20, 1190, 1974.
8. Jünger, M. and Mallach, S., An integer programming approach to optimal basic block instruction scheduling for single-issue processors, *Discrete Optimization*, 22-A, 37, 2016.
9. Achatz, H., Kleinschmidt, P., and Lambsdorff, J., Der corruption perceptions index und das linear ordering problem, *ORNews*, 26, 10, 2006.
10. Jünger, M. and Mutzel, P., 2-layer straightline crossing minimization: Performance of exact and heuristic algorithms, *Journal of Graph Algorithms and Applications*, 1, 1, 1997.
11. Sukegawara, N., Yamamoto, Y., and Zhang, L., Lagrangian relaxation and pegging test for linear ordering problems, *Journal of the Operations Research Society of Japan*, 54, 142, 2011.
12. Grötschel, M., Jünger, M., and Reinelt, G., A cutting plane algorithm for the linear ordering problem, *Operations Research*, 32, 1195, 1984.

13. Grötschel, M., Jünger, M., and Reinelt, G., Facets of the linear ordering polytope, *Mathematical Programming*, 33, 43, 1985.

14. Karp, R.M., Reducibility among combinatorial problems, in: Miller, R.E. and Thatcher, J.W., Eds., *Complexity of Computer Computations*, Plenum Press, New York, 1972, p. 85.

15. Bang-Jensen, J. and Thomassen, C., A polynomial algorithm for the 2-path problem for semicomplete digraphs, *SIAM Journal on Discrete Mathematics*, 5, 366, 1992.

16. Alon, N., Ranking tournaments, *SIAM Journal on Discrete Mathematics*, 20, 137, 2006.

17. Kann, V., On the approximability of NP-complete optimization problems, PhD Thesis, KTH Royal Institute of Technology, Sweden, 1992.

18. Dinur, I. and Safra, S., On the hardness of approximating minimum vertex cover, *Annals of Mathematics*, 162, 439, 2005.

19. Even, G., Naor, J.S., Sudan, M., and Schieber, B., Approximating minimum feedback sets and multicuts in directed graphs, *Algorithmica*, 20, 151, 1998.

20. Seymour, P.D., Packing directed circuits fractionally, *Combinatorica*, 15, 281, 1995.

21. Berger, B. and Shor, P., Approximation algorithms for the maximum acyclic subgraph problem, *Proceedings of the 1st ACM-SIAM Symposium on Discrete Algorithms—SODA'90*, San Francisco, CA, 1990, p. 236.

22. Conitzer, V., Computing Slater rankings using similarities among candidates, *Proceedings of the 21st National Conference on Artificial Intelligence-AAAI2006*, Boston, MA, 2006, p. 613.

23. Laslier, J.F., *Tournament Solutions and Majority Voting*, Springer, Berlin, Germany, 1997.

24. Kaas, R., A branch and bound algorithm for the acyclic subgraph problem, *European Journal of Operational Research*, 8, 355, 1981.

25. Ibaraki, T., Imamichi, T., Koga, Y., Nagamochi, H., Nonobe, K., and Yagiura, M., Efficient branch-and-bound algorithms for weighted MAX-2-SAT, *Mathematical Programming*, 127, 297, 2011.

26. Charon, I. and Hudry, O., A branch-and-bound algorithm to solve the linear ordering problem for weighted tournaments, *Discrete Applied Mathematics*, 154, 2097, 2006.

27. Charon, I. and Hudry, O., The noising methods: A generalization of some metaheuristics, *European Journal of Operational Research*, 135, 86, 2001.

28. Mitchell, J.E. and Borchers, B., Solving real world linear ordering problems using a primal-dual interior point cutting plane method, *Annals of Operations Research*, 62, 253, 1996.

29. Mitchell, J.E. and Borchers, B., Solving linear ordering problems with a combined interior point/simplex cutting plane algorithm, *High Performance Optimization*, Kluwer Academic Publishers, Dordrecht, the Netherlands, 2000.

30. Oswald, M., Reinelt, G., and Seitz, H., Applying mod-k-cuts for solving linear ordering problems, *TOP*, 17, 158, 2009.

31. Martí, R., Reinelt, G., and Duarte A., A benchmark library and a comparison of heuristic methods for the linear ordering problem, *Computational Optimization and Applications*, 51, 1297, 2012.

32. Borda, J.C., Mémoire sur les élections au scrutin, *Histoire de l'Académie Royale des Sciences pour 1781*, Imprimerie Royale, Paris, France, 1784.

33. Becker, O., Das Helmstädtersche Reihenfolgeproblem—die Effizienz verschiedener Näherungsverfahren, *Computer Uses in the Social Science*, Berichteiner Working Conference, Vienna, Austria, 1967.

34. Chanas, S. and Kobylanski, P., A new heuristic algorithm solving the linear ordering problem, *Computational Optimization and Applications*, 6, 191, 1996.

35. Masson, D., Méthode de triangulation du tableau européen des échanges interindustriels, *Revue Économique*, 11, 239, 1960.

36. Guo, F., Yang, Z., and Zhou, T., Predicting link directions via a recursive subgraph-based ranking, *Physica A*, 392, 3402, 2013.

37. Moon, J.W., *Topics on Tournaments*, Holt, Rinehart and Winston, New York, 1968.

38. Smith, A.F.M. and Payne, C.D., An algorithm for determining Slater's *i* and all nearest adjoining orders, *British Journal of Mathematical and Statistical Psychology*, 27, 49, 1974.

39. Berthélemy, J.P., Guénoche, A., and Hudry, O., Median linear orders: Heuristics and a branch-and-bound algorithm, *European Journal of Operational Research*, 42, 313, 1989.

40. Huang, G. and Lim, A., Designing a hybrid genetic algorithm for the linear ordering problem, *Proceedings of Genetic and Evolutionary Computation Conference—GECCO 2003*, Chicago, IL, USA (LNCS 2723/2724, Springer, Berlin), 2003, p.1053.

41. Schiavinotto T. and Stützle, T., The linear ordering problem: Instances, search space analysis and algorithms, *Journal of Mathematical Modelling and Algorithms*, 3, 367, 2004.

42. Laguna, M., Martí, R., and Campos, V., Intensification and diversification with elite tabu search solutions for the linear ordering problem, *Computers & Operations Research*, 26, 1217, 1999.

43. Garcia, C.G., Pérez-Brito, D., Campos, V., and Martí, R., Variable neighborhood search for the linear ordering problem, *Computers & Operations Research*, 33, 3549, 2006.

44. Campos, V., Glover, F., Laguna, M., and Martí, R., An experimental evaluation of a scatter search for the linear ordering problem, *Journal of Global Optimization*, 21, 397, 2001.

45. Ceberio, J., Mendiburu, A., and Lozano, J.A., The linear ordering problem revisited, *European Journal of Operational Research*, 241, 686, 2015.

46. Yagiura, M. and Ibaraki, T., Analyses on the 2 and 3-flip neighborhoods for the MAX SAT, *Journal of Combinatorial Optimization*, 3, 95, 1999.

47. Schiavinotto T. and Stützle, T., Search space analysis of the linear ordering problem, in: Raidl, G.R., Ed., *Applications of Evolutionary Computing*, Springer, Berlin, Germany, 2003, p. 322.

48. Sakuraba, C.S. and Yagiura, M., Efficient local search algorithms for the linear ordering problem, *International Transactions in Operational Research*, 17, 711, 2010.

49. Kernighan, B.W. and Lin, S., An efficient heuristic procedure for partitioning graphs, *Bell System Technical Journal*, 49, 291, 1970.

50. Chanas, S. and Kobylański, P., A new heuristic algorithm solving the linear ordering problem, *Computational Optimization and Applications*, 6, 191, 1996.

51. Ingram, F.E.G., Valdez, G.C., and Huacuja, H.J.F., Using consecutive swaps to explore the insertion neighborhood in tabu search solution of the linear ordering problem, in: Melin, P., Kacprzyk, J., and Pedrycz, W., Eds., *Soft Computing for Recognition Based on Biometrics*, Springer, Berlin, Germany, 2010, p. 267.

52. Sakuraba, C.S., Ronconi, D.P., Birgin, E.G., and Yagiura, M., Metaheuristics for large-scale instances of the linear ordering problem, *Expert Systems with Applications*, 42, 4432, 2015.

42

Submodular Functions Maximization Problems

Niv Buchbinder

Moran Feldman

42.1 Introduction

In this chapter, we study fundamental results on maximizing a special class of functions called *submodular functions* under various combinatorial constraints. The study of submodular functions is motivated both by their many real-world applications and by their frequent occurrence in more theoretical fields, such as economy and algorithmic game theory. In particular, submodular functions and submodular maximization play a major role in combinatorial optimization as several well-known combinatorial functions turn out to be submodular. A few examples of such functions include cuts functions of graphs and hypergraphs, rank functions of matroids and covering functions. We discuss some of these examples further in the following.

Let us begin by providing basic notation used throughout the chapter. We then give two definitions of submodular functions and prove that they are equivalent. Let $\mathcal{N} = \{u_1, u_2, \ldots, u_n\}$ be a ground set of elements. For a set A and an element $u \in \mathcal{N}$, we denote the union $A \cup \{u\}$ by $A + u$. Similarly, we denote $A \setminus \{u\}$ as $A - u$. The following is the first definition of submodular functions.

Definition 42.1 (Submodular Function: Definition 1) *A function* $f : 2^{\mathcal{N}} \to \mathbb{R}_{\geq 0}$ *is submodular if:*

$$f(A) + f(B) \geq f(A \cap B) + f(A \cup B) \qquad \forall A, B \subseteq \mathcal{N}. \tag{42.1}$$

In many cases it is more convenient to use the following (equivalent) definition.

Definition 42.2 (Submodular Function: Definition 2) *A function $f : 2^{\mathcal{N}} \to \mathbb{R}_{\geq 0}$ is submodular if:*

$$f(A + u) - f(A) \geq f(B + u) - f(B) \qquad \forall A \subseteq B \subseteq \mathcal{N}, u \in \mathcal{N} \setminus B. \tag{42.2}$$

The second definition illustrates an important property of submodular functions known as diminishing returns. Informally, Definition 42.2 states that adding an element to a larger set results in smaller marginal increase in the value of f (compared to adding the element to a smaller set). This property makes submodular functions natural candidates for modeling production profit (returns), utility functions, accuracy of a learning algorithm, and so on.

We now prove that the above-mentioned two definitions are indeed equivalent.

Lemma 42.1 *Definition 42.1 and Definition 42.2 are equivalent.*

Proof. Assume f satisfies Definition 42.1, and consider arbitrary sets $A \subseteq B \subseteq \mathcal{N}$ and element $u \in \mathcal{N} \setminus B$. By applying Definition 42.1 with $A + u$ and B, and observing that $(A+u) \cup B = B+u$ and $(A+u) \cap B = A$, we get

$$f(A + u) + f(B) \geq f((A + u) \cup B) + f((A + u) \cap B) = f(B + u) + f(A).$$

Rearranging the terms yields that f satisfies Definition 42.2.

On the other hand, assume that f satisfies Definition 42.2, and consider arbitrary sets $A, B \subseteq \mathcal{N}$. Our objective is to show that $f(A) + f(B) \geq f(A \cup B) + f(A \cap B)$. We prove this claim by induction on $|A \cup B| - |A \cap B|$. If $|A \cup B| - |A \cap B| = 0$ (which implies $A = B$), then the claim is trivial. Next, suppose that $|A \cup B| - |A \cap B| = k$, and let us assume without loss of generality that there exists an element $u \in A \setminus B$. Then,

$$
\begin{aligned}
f(A) + f(B) &= \big[f(A) - f(A - u)\big] + \big[f(A - u) + f(B)\big] \\
&\geq \big[f(A) - f(A - u)\big] + \big[f((A \cup B) - u) + f(A \cap B)\big] \tag{42.3} \\
&\geq \big[f(A \cup B) - f((A \cup B) - u)\big] + \big[f((A \cup B) - u) + f(A \cap B)\big] \tag{42.4} \\
&= f(A \cup B) + f(A \cap B).
\end{aligned}
$$

where Inequality (42.3) follows by the induction hypothesis, and Inequality (42.4) follows by applying Definition 42.2. ∎

Recall that this chapter deals with algorithms maximizing submodular functions subject to combinatorial constraints. In addition to being submodular, the function to be maximized often has additional natural properties that can be used by the algorithm and often lead to an improved performance.

Definition 42.3 *Consider a function $f : 2^{\mathcal{N}} \to \mathbb{R}_{\geq 0}$.*

- *f is monotone if $f(A) \leq f(B)$ for every two sets $A \subseteq B \subseteq \mathcal{N}$.*
- *f is symmetric if $f(A) = f(\mathcal{N} \setminus A)$ for every set $A \subseteq \mathcal{N}$.*
- *f is normalized if $f(\varnothing) = 0$.*

The following lemma provides basic properties of submodular functions used throughout the chapter. These properties follow quite easily from the definition of submodularity, and thus, we leave the proof of most of them to the reader. In this lemma, and in the rest of the chapter, we use the shorthand $f(u \mid A)$ to denote $f(A + u) - f(A)$ (i.e., the marginal increase in $f(A)$ following the addition of u to A).

Lemma 42.2 (Basic Properties) *Let $f, g : 2^{\mathcal{N}} \to \mathbb{R}_{\geq 0}$ be submodular functions, and let $B \subseteq \mathcal{N}$ be a subset of $\mathcal{N}$. Then,*

1. *For every $c \in \mathbb{R}_{\geq 0}$, $h(A) = c \cdot f(A)$ is a submodular function.*
2. *$h(A) = f(A) + g(A)$ is a submodular function.*

3. $h(A) = f(A \cap B)$ *is a submodular function.*
4. $h(A) = f(\mathcal{N} \setminus A)$ *is a submodular function.*
5. *If f is monotone and $c \in \mathbb{R}_{\geq 0}$, then $h(A) = \min\{f(A), c\}$ is a monotone submodular function.*
6. $\sum_{u \in B} f(u \mid A) \geq f(B \cup A) - f(A)$.

Proof. We only prove here Property 6. The proof relies on the frequently used idea of a telescopic sum. Let $B = \{u_1, \ldots, u_k\}$.

$$f(B \cup A) - f(A) = \sum_{i=1}^{k} f(u_i \mid A \cup \{u_1, \ldots, u_{i-1}\}) \leq \sum_{i=1}^{k} f(u_i \mid A),$$

where the equality follows by a telescopic sum, and the inequality follows by the submodularity of f. ∎

42.1.1 Examples of Submodular Functions

In this section we survey briefly several important submodular functions.

42.1.1.1 Additive Function, Budget Additive

Suppose each $u_i \in \mathcal{N}$ is associated with a weight $w_i \geq 0$. Then, it is easy to see that the function $f(A) = \sum_{u_i \in A} w_i$ is a normalized monotone submodular function. A slight generalization of this idea is captured by the following function, called budget additive.

$$f(A) = \min \left\{ \sum_{u_i \in A} w_i, b \right\} \qquad \text{(budget additive function)} \qquad (42.5)$$

Budget additive functions are used, for example, to model the amount of money a buyer (with budget $b \geq 0$) is willing to pay for a set of items. Lemma 42.2 shows that these functions are also normalized, monotone, and submodular.

42.1.1.2 Coverage Function

Let $E = \{e_1, \ldots, e_m\}$ be a set of elements. Let $\mathcal{S} = \{s_1, \ldots, s_n\}$ be a collection of subsets of E (i.e., $s_i \subseteq E$). We define $f : 2^{\mathcal{S}} \to \mathbb{Z}_{\geq 0}$ to be a function assigning for every subcollection of subsets the total number of elements covered by subsets of this collection. Formally,

$$f(A) = \left| \bigcup_{s_i \in A} s_i \right| \qquad \text{(coverage function)} \qquad (42.6)$$

It is easy to see that f is normalized and monotone. Additionally, f is also submodular, since for every $A \subseteq B \subseteq \mathcal{S}$ and $s \in \mathcal{S} \setminus B$,

$$f(A + s) - f(A) = \left| \left\{ e \in E \mid e \in s \setminus \left(\bigcup_{s_i \in A} s_i \right) \right\} \right|$$

$$\geq \left| \left\{ e \in E \mid e \in s \setminus \left(\bigcup_{s_i \in B} s_i \right) \right\} \right| = f(B + s) - f(B).$$

42.1.1.3 Cut Function

Let $G = (V, E)$ be a directed graph with capacities $c_e \geq 0$ on the edges. For every subset of vertices $A \subseteq V$, let $\delta(A) = \{e = uv \mid u \in A, v \in V \setminus A\}$. The *cut capacity function* is defined as the total capacity of edges that cross the cut $(A, V \setminus A)$. Formally,

$$f(A) = \sum_{e \in \delta(A)} c_e \qquad \text{(cut function)} \qquad (42.7)$$

Clearly f is normalized, and for undirected graphs it is also symmetric. Additionally, f is submodular for every graph, since for every $A \subseteq B \subseteq V$ and $r \in V \setminus B$,

$$
\begin{aligned}
f(A + r) - f(A) &= \sum_{e \in \delta(A+r)} c_e \quad - \sum_{e \in \delta(A)} c_e \\
&= \sum_{e = rv \mid v \in V \setminus A} c_e \quad - \sum_{e = ur \mid u \in A} c_e \\
&\geq \sum_{e = rv \mid v \in V \setminus B} c_e \quad - \sum_{e = ur \mid u \in B} c_e \qquad (42.8) \\
&= \sum_{e \in \delta(B+r)} c_e \quad - \sum_{e \in \delta(B)} c_e = f(B + r) - f(B),
\end{aligned}
$$

where Inequality (42.8) follows since we have both $\{e = rv \mid v \in V \setminus B\} \subseteq \{e = rv \mid v \in V \setminus A\}$ and $\{e = ur \mid u \in A\} \subseteq \{e = ur \mid u \in B\}$.

42.1.1.4 Rank Function

Let $\{a_1, \ldots, a_m\}$ be a set of vectors in $\mathbb{R}^n$. For a subset $A \subseteq \{a_1, \ldots, a_m\}$ we define the *rank* of A as the dimension of the vector space spanned by the vectors in A. Let,

$$f(A) = \text{rank}(A) \qquad \text{(independent set rank function)} \qquad (42.9)$$

The function f is a normalized monotone submodular function, where submodularity follows by basic properties of linear algebra. In fact, this function is a special case of a more general family of matroid rank functions (which is outside the scope of this chapter). Informally, matroids are important combinatorial objects generalizing of the notion of linear independence of vector spaces. Matroids and submodular functions are closely related. In particular, each matroid is associated with a submodular rank function.

42.1.2 Maximizing Submodular Functions

Let $f: 2^{\mathcal{N}} \to \mathbb{R}_{\geq 0}$ be a nonnegative submodular function. The basic problem we discuss in this chapter is maximizing the function f subject to a (possibly empty) set of constraints. Formally, let $\mathcal{I}$ be the set of subsets of $\mathcal{N}$ obeying the constraint. We are interested in the following problem:

$$
\begin{aligned}
\max \quad & f(A) \\
\text{s.t.} \quad & A \in \mathcal{I} \subseteq 2^{\mathcal{N}}
\end{aligned}
\qquad (42.10)
$$

We denote by $OPT \in \mathcal{I}$ a feasible subset with maximum value, and look for algorithms that can find a set whose value approximates the optimal value $f(OPT)$. In the following subsections we survey briefly important special cases of this problem, however, we first consider two technical issues with the above-mentioned problem. First, we note that we restrict ourselves to nonnegative functions. This restriction is necessary since we will be looking for algorithms with a multiplicative approximation ratio for this problem. Moreover, as seen by the above-mentioned examples, many important submodular functions are normalized and monotone, which implies that they are also nonnegative.

The second technical issue is that the representation of a general submodular function might be exponential in the size $|\mathcal{N}|$ of the ground set (we reserve n throughout the chapter to denote this size). To deal with this issue, the algorithms we present assume access to a procedure that given any subset $A \subseteq \mathcal{N}$ returns the value $f(A)$. Such a procedure is called *value oracle*. In many cases, we measure the complexity of algorithms by the number of oracle queries they performs. To obtain the real-time complexity one needs to take into account the time complexity of the value oracle itself, as well as the time required for additional calculations performed by the algorithm. However, counting the number of value oracle queries is a clean measure that in many cases represents (and dominates up to poly-logarithmic terms) the actual time complexity of the algorithm.

As promised, we now survey special cases of the above-mentioned basic problem.

42.1.2.1 Maximum Coverage Problem

Consider the coverage function given by (42.6). The *maximum coverage problem* is the problem of maximizing $f(A)$ subject to the constraint $|A| \leq k$ when f is a coverage function and k is a parameter. In other words, given a set of elements and a collection of subsets of these elements, we are looking for a subcollection of k subsets maximizing the number of covered elements. The constraint that allows A to be of size at most k is called a *cardinality constraint*. Section 42.2.1 discusses in details algorithms for maximizing general nonnegative submodular functions subject to a cardinality constraint.

42.1.2.2 Maximum Cut Problem

Consider the cut function given by (42.6). The *maximum cut problem* is the problem of maximizing $f(A)$ when f is a cut function. In other words, given a (possibly directed) graph $G = (V, E)$ with capacities on its edges we are interested in finding a subset A of the vertices that maximizes the total capacity of the edges crossing the cut $(A, V \setminus A)$. The maximum cut problem is a special case of the more general unconstrained submodular maximization problem in which one looks for an arbitrary set maximizing a given nonnegative submodular function. We discuss this problem in details in Section 42.2.2.

42.1.2.3 Submodular Welfare

In the Submodular Welfare problem there is a set of items, $\mathcal{N}^{SW}$, that we would like to allocate to a set of k buyers. Each buyer i is associated with a monotone submodular function $f_i : 2^{\mathcal{N}^{SW}} \to \mathbb{R}_{\geq 0}$ representing his/her utility from each subset of items. The problem is to partition the items between the buyers so as to maximize the total welfare measured as $\sum_{i=1}^{k} f_i(A_i)$, where A_i is the subset of items allocated to buyer i. At first sight, it looks like this problem is not a special case of the general problem (42.10). To see that SW can, in fact, be presented as a special case of this general problem, consider an instance of SW, and let us define a new ground set $\mathcal{N} = \mathcal{N}^{SW} \times \{1, 2, \ldots, k\}$, where $\mathcal{N}^{SW}$ is the ground set of the SW instance and k is the number of buyers in this instance. Intuitively, each element (u, i) of $\mathcal{N}$ should be understood as corresponding to assigning item u to buyer i. Using this interpretation, the objective function of the SW instance can be formulated as

$$f(A) = \sum_{i=1}^{k} f_i(\{u \in \mathcal{N}^{SW} : (u, i) \in A\}) \quad \forall A \subseteq \mathcal{N}.$$

It is not difficult to verify that the previous function f is nonnegative, monotone and submodular whenever the utility functions of the buyers have these properties. Additionally, the requirement of SW that each element is assigned to at most one buyer can be captured by the general problem (42.10) by choosing $\mathcal{I} = \{A \subseteq \mathcal{N} : |A \cap (\{u\} \times \{1, 2, \ldots, k\})| \leq 1 \quad \forall u \in \mathcal{N}^{SW}\}$, which means that a solution is feasible if and only if it contains at most one pair (u, i) involving every given element $u \in \mathcal{N}^{SW}$. We discuss the Submodular Welfare problem further in Sections 42.2.3, 42.3.2, and 42.3.3.

42.1.3 Organization

In Section 42.2, we discuss basic discrete (combinatorial) algorithms for maximizing submodular functions subject to various constraints. In Section 42.2.1, we discuss maximization subject to a cardinality constraint. In Section 42.2.2, we discuss the unconstrained submodular maximization problem. Finally, in Section 42.2.3, we discuss a discrete algorithm for the submodular welfare problem.

In Section 42.3, we discuss continuous methods that are based on extensions of submodular functions to the n-dimensional cube. We discuss the basic properties of these extensions in Section 42.3.1. Sections 42.3.2 and 42.3.3 describe basic algorithms obtaining fractional solutions which approximately maximize these extensions. Rounding these fractional solutions to obtain integral solutions is discussed both in Sections 42.3.2 and 42.3.4. Finally, Section 42.3.5 describes a method for obtaining inapproximability results for submodular maximization problems which is based on similar techniques.

42.2 Discrete Greedy Algorithms

In this section, we present discrete (combinatorial) algorithms for three submodular maximization problems: maximizing a submodular function subject to a cardinality constraint, unconstrained submodular maximization, and submodular welfare.

42.2.1 Cardinality Constraint

One of the first submodular maximization problems that was considered is the problem of maximizing a nonnegative monotone submodular function subject to a cardinality constraint. More specifically, in this problem, the input is a nonnegative monotone submodular function $f: \mathcal{N} \to \mathbb{R}_{\geq 0}$ and an integer k between 1 and n. The objective of the problem is to find a set A of size at most k maximizing f.

A natural greedy algorithm for this problem starts with the empty set, and then adds elements to it in k iterations. In each iteration the algorithm adds to its current solution the single element increasing the value of this solution by the most (i.e., the element with the largest marginal value with respect to the current solution). In the context of submodular maximization this simple algorithm is usually referred to simply as "the greedy algorithm." A formal statement of the greedy algorithm is given as Algorithm 42.1. It is important to note that Algorithm 42.1 denotes by A_i the solution produced at the end of iteration i. The use of a different variable for the solution produced at the end of each iteration is not necessary for the algorithm, but is convenient for the analysis.

The approximation ratio of the greedy algorithm is $1 - 1/e \approx 0.632$ [1], and this turns out to be the best approximation ratio possible for the problem [2]. Here we give only the analysis of the approximation ratio of the greedy algorithm. More information about the matching hardness result can be found in Section 42.3.5.

Theorem 42.1 (Due to [1]) *The greedy algorithm is a $(1 - 1/e)$-approximation algorithm for maximizing a nonnegative monotone submodular function subject to a cardinality constraint.*

ALGORITHM 42.1   THE GREEDY ALGORITHM (f, k)

1 Let $A_0 \leftarrow \varnothing$.
2 **for** $i = 1$ **to** k **do**
3   Let u_i be the element of $\mathcal{N}$ maximizing $f(u_i \mid A_{i-1})$.
4   Let $A_i \leftarrow A_{i-1} + u_i$.
5 **return** A_k

Proof. Observe that, for every $1 \leq i \leq k$,

$$f(u_i \mid A_{i-1}) \geq \max_{u \in OPT} f(u \mid A_{i-1}) \geq \frac{\sum_{u \in OPT} f(u \mid A_{i-1})}{k}$$

$$\geq \frac{f(OPT \cup A_{i-1}) - f(A_{i-1})}{k} \geq \frac{f(OPT) - f(A_{i-1})}{k},$$

where the first inequality follows since the element u_i is chosen as the element with the largest marginal contribution with respect to A_{i-1}, the second inequality holds by an averaging argument since OPT contains at most k elements and the third inequality follows from the submodularity of f (Lemma 42.2, Property 6). Finally, the last inequality follows from the monotonicity of f.

Recall that $A_i = A_{i-1} + u_i$, and hence, $f(u_i \mid A_{i-1}) = f(A_i) - f(A_{i-1})$. Plugging this equality into the previous inequality gives

$$f(A_i) - f(A_{i-1}) \geq \frac{f(OPT) - f(A_{i-1})}{k}.$$

The last inequality states that $f(A_i)$ improves over $f(A_{i-1})$ by at least $1/k$ of the gap between $f(A_{i-1})$ and OPT. Intuitively, it is clear that combining this inequality for all $1 \leq i \leq k$ should give a lower bound on $f(A_k)$ (and thus, also on the approximation ratio of the greedy algorithm). To derive this bound formally, we rearrange the last inequality as follows.

$$f(OPT) - f(A_i) \leq (1 - 1/k) \cdot [f(OPT) - f(A_{i-1})].$$

Combining the last inequality for every $1 \leq i \leq k$ gives

$$f(OPT) - f(A_k) \leq (1 - 1/k)^k \cdot [f(OPT) - f(A_0)]$$

Finally, we rearrange again and get:

$$f(A_k) \geq f(OPT) - (1 - 1/k)^k \cdot [f(OPT) - f(A_0)] \geq (1 - 1/e) \cdot f(OPT),$$

where the second inequality follows from the nonnegativity of f and since $(1 - 1/k)^k \leq 1/e$. ∎

The greedy algorithm is often used in practice also when the submodular objective function f is not monotone. However, from a theoretical point of view the greedy algorithm has no constant approximation guarantee in this case. This is demonstrated, for example, by a nonnegative submodular function $f : 2^N \to \mathbb{R}_{\geq 0}$ defined as follows:

$$f(A) = \begin{cases} |A| & \text{if } u_n \notin A, \\ 2 & \text{otherwise.} \end{cases}$$

It can be verified that f is a nonnegative (nonmonotone) submodular function. The value of f for every set is $|A|$ unless the set contains u_n. If the set contains u_n then its value is 2, regardless of the presence of other elements in the set.

Assume the greedy algorithm is faced with the above-mentioned function f and some value $2 \leq k < n$. In the first iteration of the greedy algorithm the element u_n has a marginal gain of 2, which is larger than the marginal gains of the other elements. Hence, the greedy algorithm takes the "bait" and adds u_n to its solution. This means that u_n remains in the solution of the greedy algorithm until the end, and thus, the value of this solution is 2 (notice that this is true even if one tries to improve the greedy algorithm by allowing it to stop as soon as no element has a positive marginal contribution). On the other hand, the optimal solution can take any k elements other than u_n and get a value of k. Thus, the approximation ratio of greedy is at most $2/k$.

ALGORITHM 42.2   RANDOM GREEDY (f, k)

1 Let $A_0 \leftarrow \varnothing$.
2 **for** $i = 1$ to k **do**
3 | Let $M_i = \arg\max_{B:|B|\leq k} \left\{ \sum_{u\in B} f(u \mid A_{i-1}) \right\}$.
4 | **with** *probability* $(1 - |M_i|/k)$ **do** $A_i \leftarrow A_{i-1}$.
5 | **otherwise** Let u_i be a uniformly random element of M_i, and set $A_i \leftarrow A_{i-1} + u_i$.

6 **return** A_k

The nonconstant approximation ratio of the greedy algorithm for nonmonotone functions motivated [3] to suggest a randomized version of the greedy algorithm called *Random Greedy*. Intuitively, Random Greedy tries to avoid "baits" by introducing randomization into the elements selection logic. A formal statement of Random Greedy appears as Algorithm 42.2.

Random Greedy finds in every iteration the set M of size at most k maximizing the total marginal contribution of the elements in the set with respect to the current solution. Then, the algorithm adds at most one element of M to its current solution. The algorithm picks the element to be added in such a way that every element of M has a probability of exactly $1/k$ to be added to the solution. Notice that this implies that, when $|M| < k$, there is a positive probability that the solution of the algorithm is unchanged by the iteration. This might be considered wasteful, however, avoiding this wastefulness does not improve the theoretical guarantee of the algorithm as far as we know.

Lemma 42.3 *For every* $0 \leq i \leq k$, $\mathbb{E}[f(A_i \cup OPT)] \geq (1 - 1/k)^i \cdot f(OPT)$.

Proof. Consider any $1 \leq j \leq k$, and let E_{j-1} be an arbitrary possible choice for the random decisions of Random Greedy during its first $j - 1$ iterations. In the first part of this proof, we implicit condition on E_{j-1}, all the expectations and random quantities (i.e., we fix the random choices of the algorithm in the first $j - 1$ iterations, and consider the distribution induced by its remaining random choices). Since A_{j-1} is now deterministic, we get

$$\mathbb{E}[f(A_j \cup OPT)] - f(A_{j-1} \cup OPT) = \frac{1}{k} \sum_{u \in M_j} f(u \mid A_{j-1} \cup OPT)$$

$$\geq \frac{1}{k} \left[f(M_j \cup A_{j-1} \cup OPT) - f(A_{j-1} \cup OPT) \right]$$

$$\geq -\frac{1}{k} f(A_{j-1} \cup OPT),$$

where the equality follows since each element of M_j is added to the solution with probability $1/k$, the first inequality follows by submodularity and the last inequality follows by the nonnegativity of f. Rearranging the last inequality yields

$$\mathbb{E}[f(A_j \cup OPT)] \geq \left(1 - \frac{1}{k}\right) \cdot f(A_{j-1} \cup OPT).$$

We now remove the implicit conditioning on E_{j-1}. Taking an expectation over all the possible ways to choose E_{j-1}, the last inequality yields

$$\mathbb{E}[f(A_j \cup OPT)] \geq \left(1 - \frac{1}{k}\right) \cdot \mathbb{E}[f(A_{j-1} \cup OPT)].$$

Combining the above-mentioned inequality for every $1 \leq j \leq i$ implies the lemma since $A_0 \cup OPT = \varnothing \cup OPT = OPT$. ∎

We are now ready to analyze the approximation ratio of Random Greedy.

Theorem 42.2 (Due to [3]) *Random greedy is a $1/e$-approximation algorithm for maximizing a nonnegative submodular function subject to a cardinality constraint. If f is monotone the approximation ratio of random greedy is $1 - 1/e$.*

Proof. As in the proof of Lemma 42.3, consider any $1 \leq i \leq k$, and let E_{i-1} be an arbitrary possible choice for the random decisions of Random Greedy during its first $i - 1$ iterations. The following inequality implicitly condition on E_{i-1} in the expectation and random quantities. Observe that

$$\mathbb{E}[f(u_i \mid A_{i-1})] = \frac{\sum_{u \in M_i} f(u \mid A_{i-1})}{k} \geq \frac{\sum_{u \in OPT} f(u \mid A_{i-1})}{k} \geq \frac{f(A_{i-1} \cup OPT) - f(A_{i-1})}{k},$$

where the first inequality follows from the definition of M_i as the set that maximizes the total marginal contribution, and the second from the submodularity of f. We now remove the implicit conditioning on E_{i-1}. Taking an expectation over all the possible ways to choose E_{i-1}, the last inequality yields

$$\mathbb{E}[f(u_i \mid A_{i-1})] \geq \frac{\mathbb{E}[f(A_{i-1} \cup OPT)] - \mathbb{E}[f(A_{i-1})]}{k} \geq \frac{(1 - 1/k)^{i-1} \cdot f(OPT) - \mathbb{E}[f(A_{i-1})]}{k},$$

where the second inequality is due to Lemma 42.3. Note that if f is monotone then $f(A_{i-1} \cup OPT) \geq f(OPT)$, and thus, in this case we get the same bound as in Theorem 42.1. Hence, if we continued the analysis for monotone f, we would have got the same bound of $1 - 1/e$ as in Theorem 42.1.

For nonmonotone functions, we observe that $\mathbb{E}[f(u_i \mid A_{i-1})]$ is the expected increase in $f(A_i)$ compared to $f(A_{i-1})$, and thus, the last inequality gives a lower bound on the improvement in the expected value of the algorithm's solution in each iteration. This bound is given in terms of $f(OPT)$ and the expected value of the current solution. The reminder of the proof "translates" this lower bound into an explicit lower bound on the expected value of the algorithm's solution after any given number of iterations. More specifically, we prove by induction that $\mathbb{E}[f(A_i)] \geq \frac{i}{k} \cdot (1 - \frac{1}{k})^{i-1} \cdot f(OPT)$. For $i = 0$, this is true since $f(A_0) \geq 0 = \frac{0}{k} \cdot (1 - \frac{1}{k})^{-1} \cdot f(OPT)$.* Assume now that the claim holds for every $0 \leq i' < i$, and let us prove it for $i > 0$.

$$
\begin{aligned}
\mathbb{E}[f(A_i)] &= \mathbb{E}[f(A_{i-1})] + \mathbb{E}[f(u_i \mid A_{i-1})] \geq \mathbb{E}[f(A_{i-1})] + \frac{\left(1 - 1/k\right)^{i-1} \cdot f(OPT) - \mathbb{E}[f(A_{i-1})]}{k} \\
&= \left(1 - \frac{1}{k}\right) \cdot \mathbb{E}[f(A_{i-1})] + \frac{(1 - 1/k)^{i-1}}{k} \cdot f(OPT) \\
&\geq \left(1 - \frac{1}{k}\right) \cdot \left[\frac{i-1}{k} \cdot \left(1 - \frac{1}{k}\right)^{i-2} \cdot f(OPT)\right] + \frac{(1 - 1/k)^{i-1}}{k} \cdot f(OPT) \\
&= \frac{i}{k} \cdot \left(1 - \frac{1}{k}\right)^{i-1} \cdot f(OPT).
\end{aligned}
$$

Plugging $i = k$ into the earlier claim yields the theorem.

$$\mathbb{E}[f(A_k)] \geq \frac{k}{k} \cdot (1 - 1/k)^{k-1} \cdot f(OPT) \geq 1/e \cdot f(OPT). \quad \blacksquare$$

---

* For $k = 1$ the term $(1 - 1/k)^{-1}$ is undefined. However, one can verify that Random Greedy is in fact optimal for $k = 1$.

ALGORITHM 42.3   SAMPLE GREEDY (f, k, ε)

1 Let $A_0 \leftarrow \varnothing$.
2 **for** $i = 1$ **to** k **do**
3 $\quad$ Let B_i be a uniformly random subset of $\mathcal{N}$ of size $\left\lceil \frac{n \cdot \ln(1/\varepsilon)}{k} \right\rceil$.
4 $\quad$ Let u_i be the element of B_i maximizing $f(u_i \mid A_i)$.
5 $\quad$ Let $A_i \leftarrow A_{i-1} + u_i$.
6 **return** A_k

When the function f is monotone both greedy and random greedy achieve an approximation ratio of $1 - 1/e$. It can also be verified that both algorithms require $O(n)$ value oracle queries in each one of their k iterations, and thus, require $O(nk)$ queries in total. In the case of Random Greedy using $O(n)$ value, oracle queries in each iteration is a bit wasteful. More specifically, Random Greedy queries the marginal contributions of all the elements to construct all of M, and then chooses a uniformly random element out of M (note that when f is monotone, we can assume M always contains exactly k elements). Instead, one can query the marginal contributions of a random subset of the elements, and use it to recover, with high enough probability, at least one element of M (i.e., to find an element whose marginal contribution, with respect to the current solution, is among the top k marginal contributions). This is the idea behind Algorithm 42.3, which we term here *Sample Greedy*, originally suggested independently by References 4, 5.

Sample Greedy has a parameter $\varepsilon \in [e^{-k}, 1 - 1/e]$ controlling a tradeoff between the number of value oracle queries used by the algorithm and the quality of its output. This tradeoff is captured by the following theorem.

Theorem 42.3 (Due to [4,5]) *Sample Greedy is a $(1 - 1/e - \varepsilon)$-approximation algorithm for maximizing a nonnegative monotone submodular function subject to a cardinality constraint which uses $O(n \ln(1/\varepsilon))$ value oracle queries.*[*]

The analysis of the number of value oracle queries used by Sample Greedy is quite straightforward, and thus, we concentrate on proving its approximation ratio. However, we would like to direct the reader's attention to the fact that the number of queries used by Sample Greedy is independent of k, which is quite surprising. The following lemma lower bounds the expected improvement in the value of the solution of Sample Greedy after every given iteration.

Lemma 42.4 *For every $1 \leq i \leq k$, $\mathbb{E}[f(u_i \mid A_{i-1})] \geq (1 - \varepsilon) \cdot \frac{f(OPT) - \mathbb{E}[f(A_{i-1})]}{k}$.*

Proof. Let E_{i-1} be an event specifying the random decisions of Sample Greedy up to iteration $i-1$ (including). If the lemma holds conditioned on every given event E_{i-1}, then it holds also unconditionally. Hence, in the rest of the proof we fix an event E_{i-1} and prove the lemma conditioned on this event. For simplicity, all the probabilities, expectations and random quantities in the proof are implicitly conditioned on E_{i-1}. In particular, note that the set A_{i-1} is now deterministic since it is fully determined by E_{i-1}.

Let $v_1, v_2, \ldots, v_k$ be the k elements with the largest marginal contributions with respect to A_{i-1}, sorted in a nonincreasing marginal contribution order. Additionally, let X_j be an indicator for the event that

---

[*] Observe that the restriction of ε to the range $[e^{-k}, 1 - 1/e]$ is not an issue because Theorem 42.3 gives no approximation guarantee for $\varepsilon > 1 - e^{-1}$, and its guarantee for the number of value oracle queries is no better than that of Random Greedy for $\varepsilon < e^{-k}$.

Sample Greedy selects v_j in its ith iteration (i.e., $u_i = v_j$). Using this notation and the linearity of the expectation, it is possible to lower bound $\mathbb{E}[f(u_i \mid A_{i-1})]$ as follows:

$$\mathbb{E}[f(u_i \mid A_{i-1})] \geq \mathbb{E}\left[\sum_{j=1}^{k} X_j \cdot f(v_j \mid A_{i-1})\right] = \sum_{j=1}^{k} \mathbb{E}[X_j] \cdot f(v_j \mid A_{i-1}).$$

Recall that u_i is selected as the element with the highest marginal contribution in B_i; hence, when multiple elements of $v_1, v_2, \ldots, v_k$ belong to B_i, Sample Greedy selects the element with the lowest index among them. This implies that $\mathbb{E}[X_j] = \Pr[u_i = v_j]$ is a nonincreasing function of j. Consider now the sum on the right-hand side of the above-mentioned inequality. Every term of this sum is a multiplication of two nonincreasing functions of j: $\mathbb{E}[X_j]$ and $f(v_j \mid A_{i-1})$. Thus, by Chebyshev's sum inequality, this rightmost hand side can be bounded by

$$\mathbb{E}[f(u_i \mid A_{i-1})] \geq \sum_{j=1}^{k} \mathbb{E}[X_j] \cdot \frac{\sum_{j=1}^{k} f(v_j \mid A_{i-1})}{k}. \tag{42.11}$$

Next, observe that

$$\sum_{j=1}^{k} \mathbb{E}[X_j] = \Pr[\{v_1, v_2, \ldots, v_k\} \cap B_i \neq \varnothing] \geq 1 - \left(1 - \frac{k}{n}\right)^{\left\lceil \frac{n \cdot \ln(1/\varepsilon)}{k} \right\rceil} \geq 1 - \varepsilon.$$

The first equality follows since one of variables X_j is 1 if and only if $\{v_1, v_2, \ldots, v_k\} \cap B_i \neq \varnothing$, and the first inequality follows since the probability of choosing an element from $\{v_1, v_2, \ldots, v_k\}$ when the sample set B_i is chosen with repetitions is smaller than this probability when the set is chosen with no repetitions. Plugging the last inequality into Inequality (42.11) yields

$$\mathbb{E}[f(u_i \mid A_{i-1})] \geq (1 - \varepsilon) \cdot \frac{\sum_{j=1}^{k} f(v_j \mid A_{i-1})}{k}.$$

The lemma now follows by observing that the definition of the v_j's and the submodularity and monotonicity of f imply together

$$\sum_{j=1}^{k} f(v_j \mid A_{i-1}) \geq \sum_{u \in OPT} f(u \mid A_{i-1}) \geq f(OPT \cup A_{i-1}) - f(A_{i-1}) \geq f(OPT) - f(A_{i-1}). \quad \blacksquare$$

We are now ready to prove the approximation ratio of Sample Greedy.

Proof of the Approximation Ratio of Sample Greedy: Let us denote $\alpha = (1 - \varepsilon)/k$. Then, by Lemma 42.4, for every $1 \leq i \leq k$,

$$\mathbb{E}[f(A_i) - f(A_{i-1})] = \mathbb{E}[f(u_i \mid A_{i-1})] \geq \alpha[f(OPT) - \mathbb{E}[f(A_{i-1})]].$$

Rearranging gives

$$f(OPT) - \mathbb{E}[f(A_i)] \leq (1 - \alpha) \cdot [f(OPT) - \mathbb{E}[f(A_{i-1})]].$$

Combining the above-mentioned inequality for every $1 \leq i \leq k$ yields

$$f(OPT) - \mathbb{E}[f(A_k)] \leq (1 - \alpha)^k \cdot [f(OPT) - \mathbb{E}[f(A_0)]] \leq (1 - \alpha)^k \cdot f(OPT),$$

where the last inequality follows from the nonnegativity of f. Rearranging once more, we get

$$\mathbb{E}[f(A_k)] \geq \left[1 - (1 - \alpha)^k\right] \cdot f(OPT) \geq (1 - e^{-k \cdot \alpha}) \cdot f(OPT)$$
$$= (1 - e^{\varepsilon - 1}) \cdot f(OPT) \geq (1 - 1/e - \varepsilon) \cdot f(OPT)$$

where the last inequality holds for every $\varepsilon \in [0, 1]$.   ∎

42.2.2 Unconstrained Maximization

Perhaps the simplest variant of a submodular maximization problem is *unconstrained submodular maximization*, which is a problem asking to maximize a submodular function subject to no constraints on the feasible subset we may choose. When the function f is monotone the solution is trivial—simply take all elements to the solution. However, the problem is nontrivial when the function is nonmonotone. We present in this section two algorithms for the problem. Amazingly, the first algorithm does not even evaluate the function. It simply takes each element to its solution with probability $1/2$. We prove that this algorithm achieves an approximation ratio of $1/4$. This result is due to Reference 6, but the proof we present here for this result is different from the original proof of Reference 6. The advantage of the proof we present is that it provides us with some basic tools that are used later for analyzing the second algorithm. The second algorithm itself is a greedy algorithm which was shown to have an approximation ratio of $1/2$ by Reference 7. A matching inapproximability result, showing that no polynomial time algorithm can achieve an approximation ratio of $1/2 + \varepsilon$ for any constant $\varepsilon > 0$, was proved by Reference 6.

Let us begin with the analysis of the first algorithm. Recall that this algorithm simply takes every element to the solution with probability $1/2$, independently. For the analysis's purpose we give a more verbose description of this algorithm as Algorithm 42.4. While reading Algorithm 42.4, it is important to recall that $u_1, u_2, \ldots, u_n$ is some (arbitrary) ordering of the elements of $\mathcal{N}$.

Notice that the algorithm maintains two random solutions. The values of these solutions after i iterations are denoted by X_i and Y_i. It can be easily seen that X_i and Y_i always agree on the first i elements, and thus, at the end of the execution $X_n = Y_n$. To analyze the performance of the algorithm we need some additional notation. Let $OPT_i \triangleq (OPT \cup X_i) \cap Y_i$. By the above-mentioned discussion, we have $OPT_0 = OPT$ and $OPT_n = X_n = Y_n$. Similarly, let $\overline{OPT} = \mathcal{N} \setminus OPT$ and let $\overline{OPT}_i \triangleq (\overline{OPT} \cup X_i) \cap Y_i$, then $\overline{OPT}_0 = \overline{OPT}$ and $\overline{OPT}_n = X_n = Y_n$.

One can view $f(OPT_i)$, $f(\overline{OPT}_i)$, $f(X_i)$, and $f(Y_i)$ as functions of i. The following lemma takes this point of view and bounds the expected decrease of value in OPT_i and $\overline{OPT}_i$ when i increases by the total expected increase in the values of X_i and Y_i.

Lemma 42.5 *For every* $1 \leq i \leq n$,

$$\mathbb{E}[f(OPT_{i-1}) - f(OPT_i) + f(\overline{OPT}_{i-1}) - f(\overline{OPT}_i)] \leq \mathbb{E}[f(X_i) - f(X_{i-1}) + f(Y_i) - f(Y_{i-1})], \quad (42.12)$$

where the expectations are taken over the random choices of the algorithm.

ALGORITHM 42.4   RANDOM SAMPLE (f)

1   Let $X_0 \leftarrow \varnothing, Y_0 \leftarrow \mathcal{N}$.
2   **for** $i = 1$ **to** n **do**
3   **with** *probability* $1/2$ **do** $X_i \leftarrow X_{i-1} + u_i, Y_i \leftarrow Y_{i-1}$.
4   **otherwise** $X_i \leftarrow X_{i-1}, Y_i \leftarrow Y_{i-1} - u_i$.   // Also done with probability $1/2$.
5   **return** X_n *(or equivalently Y_n).*

Proof. Notice that it suffices to prove Inequality (42.12) conditioned on any event of the form $X_{i-1} = A_{i-1}$ for which the probability that $X_{i-1} = A_{i-1}$ is non-zero (notice that this can only happen when $A_{i-1} \subseteq \{u_1, u_2, \ldots, u_{i-1}\}$). Fix such an event. The rest of the proof implicitly assumes everything is conditioned on this event. Note that, due to the conditioning, the random variables $Y_{i-1}, OPT_{i-1},$ and $\overline{OPT}_{i-1}$ become deterministic.

If $u_i \notin OPT$ (and, hence, $u_i \in \overline{OPT}$), then

$$\mathbb{E}[f(X_i) - f(X_{i-1}) + f(Y_i) - f(Y_{i-1})]$$
$$= \frac{1}{2} \left[f(X_{i-1} + u_i) - f(X_{i-1}) + f(Y_{i-1} - u_i) - f(Y_{i-1}) \right]$$
$$\geq \frac{1}{2} \left[f(\overline{OPT}_{i-1}) - f(\overline{OPT}_{i-1} - u_i) + f(OPT_{i-1}) - f(OPT_{i-1} + u_i) \right]$$
$$= \mathbb{E} \left[f(OPT_{i-1}) - f(OPT_i) + f(\overline{OPT}_{i-1}) - f(\overline{OPT}_i) \right].$$

To see that the inequality follows by submodularity notice that $X_{i-1} \subseteq \left[(\overline{OPT} \cup X_{i-1}) \cap Y_{i-1} \right] - u_i = \overline{OPT}_{i-1} - u_i$ and $OPT_{i-1} = (OPT \cup X_{i-1}) \cap Y_{i-1} \subseteq Y_{i-1} - u_i$ (since $u_i \notin OPT$).

The proof that the lemma holds also when $u_i \in OPT$ is similar, and is, therefore, omitted. ∎

The performance of the algorithm easily follows from Lemma 42.5.

Theorem 42.4 (Due to [6]) *Algorithm 42.4 is a randomized linear time 1/4-approximation algorithm for the unconstrained submodular maximization problem. If the function f is symmetric, in addition to being submodular, then the algorithm is a 1/2-approximation algorithm.*

Proof. Summing up the inequality in Lemma 42.5 for every $1 \leq i \leq n$ we get

$$\sum_{i=1}^{n} \mathbb{E} \left[f(OPT_{i-1}) - f(OPT_i) + f(\overline{OPT}_{i-1}) - f(\overline{OPT}_i) \right] \leq \sum_{i=1}^{n} \mathbb{E} \left[f(X_i) - f(X_{i-1}) + f(Y_i) - f(Y_{i-1}) \right].$$

The previous sum is telescopic. Collapsing it yields

$$\mathbb{E} \left[f(OPT_0) - f(OPT_n) + f(\overline{OPT}_0) - f(\overline{OPT}_n) \right] \leq \mathbb{E} \left[f(X_n) - f(X_0) + f(Y_n) - f(Y_0) \right]$$

Using the nonnegativity of f, and recalling the values of $OPT_0, OPT_n, \overline{OPT}_0,$ and $\overline{OPT}_n$, we obtain

$$\mathbb{E}[f(X_n)] = \mathbb{E}[f(Y_n)] \geq \frac{1}{4} \left(f(OPT) + f(\overline{OPT}) + f(\varnothing) + f(\mathcal{N}) \right) \geq \frac{1}{4} \cdot f(OPT).$$

If f is also symmetric then $f(\overline{OPT}) = f(OPT)$, and the right hand side of the above-mentioned inequality can be improved to $1/2 \cdot f(OPT)$. ∎

We next present a 1/2-approximation algorithm for unconstrained submodular maximization. The algorithm is formally described in Algorithm 42.5.

Similar to the previous algorithm, X_i and Y_i are random variables denoting the two solutions maintained by the algorithm after i iterations. The behavior of these random variables is very similar to their behavior in Algorithm 42.4, except that now the decision whether to add the element u_i to X_{i-1} or to remove it from Y_{i-1} is done with probabilities that depend on the function f. The analysis of the algorithm requires the random variable OPT_i, which is defined exactly like in the analysis of Algorithm 42.4.

The following lemma plays the same role as Lemma 42.5.

ALGORITHM 42.5   DOUBLE GREEDY (f)

1 Let $X_0 \leftarrow \varnothing, Y_0 \leftarrow \mathcal{N}$.
2 **for** $i = 1$ **to** n **do**
3   Let $a_i \leftarrow \max\{f(X_{i-1} + u_i) - f(X_{i-1}), 0\}$.
4   Let $b_i \leftarrow \max\{f(Y_{i-1} - u_i) - f(Y_{i-1}), 0\}$.
5   **with** *probability* $a_i/(a_i + b_i)^*$ **do** $X_i \leftarrow X_{i-1} + u_i, Y_i \leftarrow Y_{i-1}$.
6   **otherwise** $X_i \leftarrow X_{i-1}, Y_i \leftarrow Y_{i-1} - u_i$.   // Done with probability $b_i/(a_i + b_i)$.
7 **return** X_n *(or equivalently Y_n).*
  * *If* $a_i = b_i = 0$, *we assume* $a_i/(a_i + b_i) = 1$.

Lemma 42.6 *For every $1 \leq i \leq n$,*

$$\mathbb{E}[f(OPT_{i-1}) - f(OPT_i)] \leq \frac{1}{2} \cdot \mathbb{E}\left[f(X_i) - f(X_{i-1}) + f(Y_i) - f(Y_{i-1})\right], \qquad (42.13)$$

where the expectations are taken over the random choices of the algorithm.

Proof. Again, notice that it suffices to prove Inequality (42.13) conditioned on any event of the form $X_{i-1} = A_{i-1}$ for which the probability that $X_{i-1} = A_{i-1}$ is non-zero (recall that this can only happen when $A_{i-1} \subseteq \{u_1, u_2, \ldots, u_{i-1}\}$). Fix such an event. The rest of the proof implicitly assumes everything is conditioned on this event. Notice that, as before, due to the conditioning, the random variables Y_{i-1} and OPT_{i-1} become deterministic once we fix the event. Moreover, the values a_i, b_i also become deterministic when we do it. We claim that

$$\mathbb{E}\left[f(X_i) - f(X_{i-1}) + f(Y_i) - f(Y_{i-1})\right] \qquad (42.14)$$

$$= \frac{a_i}{a_i + b_i}\left(f(X_{i-1} + u_i) - f(X_{i-1})\right) + \frac{b_i}{a_i + b_i}\left(f(Y_{i-1} - u_i) - f(Y_{i-1})\right) = \frac{a_i^2 + b_i^2}{a_i + b_i}.$$

The first equality follows by definition. To see why the second equality holds observe that, by submodularity, $f(X_{i-1} + u_i) - f(X_{i-1}) + f(Y_{i-1} - u_i) - f(Y_{i-1}) \geq 0$ since $X_{i-1} \subseteq Y_{i-1} - u_i$. This means that $f(X_{i-1} + u_i) - f(X_{i-1}) < 0$ implies $a_i/(a_i + b_i) = 0$ and $f(Y_{i-1} - u_i) - f(Y_i) < 0$ implies $b_i/(a_i + b_i) = 0$.

Let us now analyze the expected change in OPT_i. If $u_i \notin OPT$, then $OPT_{i-1} - u_i = OPT_{i-1}$. Thus,

$$\mathbb{E}[f(OPT_{i-1}) - f(OPT_i)] = \frac{a_i}{a_i + b_i} \cdot [f(OPT_{i-1}) - f(OPT_{i-1} + u_i)]$$

$$\leq \frac{a_i}{a_i + b_i} \cdot [f(Y_{i-1} - u_i) - f(Y_{i-1})] \leq \frac{a_i b_i}{a_i + b_i},$$

where the first inequality follows by the submodularity of f as $OPT_{i-1} = (OPT \cup X_{i-1}) \cap Y_{i-1} \subseteq Y_{i-1} - u_i$ (since $u_i \notin OPT$). On the other hand, if $u_i \in OPT$ then $OPT_{i-1} + u_i = OPT_{i-1}$. Thus,

$$\mathbb{E}[f(OPT_{i-1}) - f(OPT_i)] = \frac{b_i}{a_i + b_i} \cdot [f(OPT_{i-1}) - f(OPT_{i-1} - u_i)]$$

$$\leq \frac{b_i}{a_i + b_i} \cdot [f(X_{i-1} + u_i) - f(X_{i-1})] \leq \frac{a_i b_i}{a_i + b_i},$$

where the first inequality follows by again the submodularity of f as $X_{i-1} \subseteq (OPT \cup X_{i-1}) \cap Y_{i-1} - u_i = OPT_{i-1} - u_i$.

The lemma follows by combining the last two inequalities with Inequality (42.14) since, for every $a, b \in \mathbb{R}$, $2ab \leq a^2 + b^2$. ∎

Using Lemma 42.6 the next theorem follows easily.

Theorem 42.5 (Due to [7]) *Algorithm 42.5 is a randomized linear time* $1/2$-*approximation algorithm for the* unconstrained submodular maximization *problem.*

Proof. Summing up Lemma 42.6 for every $1 \leq i \leq n$ gives

$$\sum_{i=1}^{n} \mathbb{E}[f(OPT_{i-1}) - f(OPT_i)] \leq \frac{1}{2} \cdot \sum_{i=1}^{n} \mathbb{E}[f(X_i) - f(X_{i-1}) + f(Y_i) - f(Y_{i-1})].$$

The previous sum is telescopic. Collapsing it, we get

$$\mathbb{E}[f(OPT_0) - f(OPT_n)] \leq \frac{1}{2} \cdot \mathbb{E}[f(X_n) - f(X_0) + f(Y_n) - f(Y_0)] \leq \frac{\mathbb{E}[f(X_n) + f(Y_n)]}{2}.$$

Recalling that $OPT_0 = OPT$ and $OPT_n = X_n = Y_n$, we obtain $\mathbb{E}[f(X_n)] = \mathbb{E}[f(Y_n)] \geq f(OPT)/2$. ∎

42.2.3 The Submodular Welfare Problem

In this section, we design a simple $1/2$-approximation algorithm for the `Submodular Welfare` problem. The algorithm is very natural. It begins with the empty assignment. Then, in each iteration, it assigns a yet unassigned item to the buyer for whom the item has the largest marginal contribution. A formal statement of this algorithm is given as Algorithm 42.6.

Theorem 42.6 (Due to [8]) *Algorithm 42.6 is a* $1/2$-*approximation algorithm for the* `Submodular Welfare` *problem.*

Proof. Let A_j^i be the set of items allocated by the algorithm to buyer j out of items $u_1, \ldots, u_i$. Let OPT_j^i be the set of items allocated by the optimal solution to buyer j out of items $u_{i+1}, u_{i+2}, \ldots, u_n$. For every $1 \leq i \leq k$, we claim that the following inequality holds:

$$\sum_{j=1}^{k} \left[f_j\left(A_j^i\right) - f_j\left(A_j^{i-1}\right) \right] \geq \sum_{j=1}^{k} \left[f_j\left(A_j^{i-1} \cup OPT_j^{i-1}\right) - f_j\left(A_j^i \cup OPT_j^i\right) \right]. \tag{42.15}$$

Let us explain why this is the case. Assume that the algorithm allocated item i to buyer j, whereas OPT allocated item i to buyer j^*. Note that this means that on the left hand side of the last inequality only the jth term may be non-zero, while on the right hand side only terms j and j^* may be non-zero. Moreover, the

---

ALGORITHM 42.6 THE GREEDY ALGORITHM FOR SUBMODULAR WELFARE ($\{f_i\}_{i=1}^{k}$)

1   $A_i \leftarrow \varnothing$, for each $i = 1, \ldots, k$.
2   **for** $i = 1$ **to** n **do**
3 | Let j be the player maximizing $f_j(u_i \mid A_j)$.
4 | Let $A_j \leftarrow A_j + u_i$
5   **return** $\{A_j\}_{j=1}^{k}$

monotonicity of f_j guarantees that the jth term on the right-hand side is nonpositive. Thus, it is enough to prove the following.

$$f_j\left(A_j^i\right) - f_j\left(A_j^{i-1}\right) = f_j\left(u_i \mid A_j^{i-1}\right) \geq f_{j^*}\left(u_i \mid A_{j^*}^{i-1}\right)$$

$$\geq f_{j^*}\left(u_i \mid A_{j^*}^{i-1} \cup OPT_{j^*}^i\right) \geq f_{j^*}\left(A_{j^*}^{i-1} \cup OPT_{j^*}^{i-1}\right) - f_{j^*}\left(A_{j^*}^i \cup OPT_{j^*}^i\right),$$

where the first inequality follows since j maximizes $f_j(u_i \mid A_j)$, and the second inequality follows by submodularity. Finally, the last inequality holds as an equality when $j \neq j^*$ because in this case $A_{j^*}^{i-1} = A_{j^*}^i$ and $OPT_{j^*}^{i-1} = OPT_{j^*}^i + u_i$. When $j = j^*$ the last inequality still holds because its left hand side is nonnegative by the monotonicity, and its right-hand side is 0.

Summing up Inequality (42.15) for $1 \leq i \leq n$, and using the nonnegativity of each f_j, we get:

$$\sum_{j=1}^{k} f_j\left(A_j\right) \geq \sum_{j=1}^{k}\left[f_j\left(A_j^n\right) - f_j\left(A_j^0\right)\right]$$

$$\geq \sum_{j=1}^{k} f_j\left(OPT_j^0 \cup A_j^0\right) - \sum_{j=1}^{k} f_j\left(OPT_j^n \cup A_j^n\right) = \sum_{j=1}^{k} f_j\left(OPT_j\right) - \sum_{j=1}^{k} f_j\left(A_j\right),$$

which concludes the proof. ∎

The following observation shows that the above-mentioned analysis of Algorithm 42.6 cannot be improved.

Observation 42.1 *The approximation ratio of Algorithm 42.6 is exactly $1/2$ even when there are only two buyers.*

Proof. Consider a ground set with only two elements, $\mathcal{N} = \{u_1, u_2\}$. The utility of buyer 1 is equal to 1 if he is assigned at least one of the elements, and 0 if he is assigned no elements. The utility of buyer 2 is 1 if he is assigned u_1, and 0 otherwise. Formally, the utility functions of both players are:

$$f_1(A) = \min\{|A|, 1\} \quad \forall A \subseteq \mathcal{N} \qquad \text{and} \qquad f_2(A) = |A \cap \{u_1\}| \quad \forall A \subseteq \mathcal{N}.$$

One can observe that, given the empty assignment, the marginal contribution of adding u_1 to buyer 1 is 1, and no other assignment of a single element to any user has a larger marginal contribution. Hence, Algorithm 42.6 might assign u_1 to buyer 1 in its first iteration;[*] in which case, regardless of the assignment of u_2, buyer 1 ends up with a utility of 1 and buyer 2 ends up with no utility, leading to a total utility of 1. On the other hand, the optimal assignment assigns u_1 to buyer 2 and u_2 to buyer 1, which leads to a total utility of 2. ∎

42.2.4 Notes and Open Questions

42.2.4.1 Cardinality Constraint

The greedy algorithm (Algorithm 42.1) and its analysis are due to Reference 1. Random Greedy (Algorithm 42.2) and its analysis are due to Reference 3. Finally, the faster Sample Greedy (Algorithm 42.3) is due to Reference 5 and independently [4].

In contrast to the case of the greedy algorithm, the approximation ratio of Random Greedy is not optimal. A much more involved algorithm by Reference 3 achieves an improved approximation ratio of

---

[*] The exact behavior of the algorithm in this case depends on the tie breaking rules used to implement it. If one wants to avoid this, f_1 should be replaced with $f_1(A) = \min\{|A|, 1\} + \varepsilon \cdot |A \cap \{u_1\}|$ for an arbitrary small constant $\varepsilon > 0$.

$1/e + 0.004$ for the same problem. On the other hand, a hardness result by Reference 9 shows that no algorithm can achieve an approximation ratio better than 0.491 for the problem using only a polynomial number of value oracle queries. Arguably, one of the most important open questions related to the results discussed in this chapter is closing this gap. The unnatural state of the art approximation ratio of $1/e + 0.004$ due to Reference 3 can probably be improved. On the other hand, it is unclear whether the corresponding inapproximability result of 0.491 of Reference 9 is tight or not.

An additional interesting question is whether one can get deterministic algorithms achieving the same bounds. Buchbinder and Feldman [10] designed a deterministic variant of Random Greedy achieving the same approximation ratio of $1/e$. A different technique was used by Badanidiyuru and Vondrák [11] to get a *deterministic* $(1 - 1/e - \varepsilon)$-approximation algorithm for maximizing a nonnegative monotone submodular function subject to a cardinality constraint which uses only $O((n/\varepsilon)\log(n/\varepsilon))$ value oracle queries. Later, Buchbinder et al. [4] developed two incomparable randomized $(1/e - \varepsilon)$-approximation algorithms for objective functions that are not necessarily monotone. One algorithm is based on the technique of Reference 11 and uses $O(k\sqrt{(n/\varepsilon)\ln(k/\varepsilon)} + (n/\varepsilon)\ln(k/\varepsilon))$ value oracle queries, whereas the other is a slight variation of Sample Greedy requiring $O(\frac{n}{\varepsilon^2}\ln(1/\varepsilon))$ value oracle queries.

The cardinality constraint is also related to the *strict cardinality* constraint which allows only sets of size *exactly k*. For monotone objective functions the two constraints are equivalent since one can increase to k the size of any set whose size is smaller than k by adding to it arbitrary elements, and this never decreases the value of the set. For nonmonotone objectives, however, the strict cardinality constraint seems to be more difficult. The 0.491 inapproximability result of Reference 9 applies also to maximizing a nonnegative submodular function subject to a strict cardinality constraint, but the best approximation ratio known for this problem is only $1/e - o(1)$ [12].

42.2.4.2 Unconstrained

The random sample algorithm (Algorithm 42.4) is due to Reference 6. The double greedy algorithm (Algorithm 42.5) is due to Reference 7. It is known that no polynomial time algorithm for the unconstrained submodular maximization can have an approximation ratio of $1/2 + \varepsilon$ for any constant $\varepsilon > 0$ [6], and a deterministic algorithm achieving the optimal approximation ratio of $1/2$ for the problem was presented in Reference 10.

Algorithm 42.4 is nonadaptive in the sense that the list of the sets on which it queries the value oracle is chosen before the first query is made. Feige et al. [6] showed that the approximation ratio of $1/4$ achieved by this algorithm is optimal for nonadaptive algorithms that are also required to return one of the sets whose value they queried. If the algorithm is allowed to return a different set, then an improved $1/3$-approximation nonadaptive algorithm exists [6].

42.2.4.3 Submodular Welfare

The greedy algorithm presented for Submodular Welfare (Algorithm 42.6) is due to Reference 8. In fact, [8] designed a greedy algorithm for the much more general problem of maximizing a monotone submodular function with matroid constraints. The best inapproximability result for SW shows that no polynomial time algorithm can achieve an approximation ratio of $1 - (1 - 1/k)^k + \varepsilon$ for any constant $\varepsilon > 0$ [13], and thus, for general k, no polynomial time algorithm can achieve an approximation ratio of $1 - 1/e + \varepsilon$. The large gap between this inapproximability result and the approximation ratio of the greedy algorithm motivated Călinescu et al. [14] to develop the Continuous Greedy algorithm, which yields $(1 - 1/e)$-approximation for SW. We discuss this in more details in Sections 42.3.2 and 42.3.3.

42.3 Continuous Methods

Many results for submodular maximization problems are based on a continuous relaxation of these problems known as the *multilinear* relaxation. This section surveys a few of the most important results based on this relaxation, and is organized as follows. Sections 42.3.1 gives definitions and preliminary results necessary for the other sections. Sections 42.3.2 and 42.3.3 describe algorithms for obtaining approximately optimal fractional solutions for the multilinear relaxation. Rounding of these fractional solutions is discussed both in Sections 42.3.2 and 42.3.4. Finally, Section 42.3.5 describes a method for obtaining inapproximability results for submodular maximization problems which is based on similar techniques.

42.3.1 Definitions and Preliminaries

The sets of $2^{\mathcal{N}}$ (or $\{0, 1\}^{\mathcal{N}}$) can be naturally identified with points of the $[0, 1]^{\mathcal{N}}$ cube by identifying each set $A \subseteq \mathcal{N}$ with its characteristic vector (which we denote here by $\mathbf{1}_A$). For linear functions this gives a natural way to extend any function from $\{0, 1\}^{\mathcal{N}}$ to the cube $[0, 1]^{\mathcal{N}}$. However, finding the right extension for submodular functions is more difficult. One useful extension of submodular functions is the Lovász extension defined as follows.

Definition 42.4 (The Lovász Extension) *Given a vector $x \in [0, 1]^{\mathcal{N}}$ and a scalar $\lambda \in [0, 1]$, let $T_\lambda(x) = \{u \in \mathcal{N} \mid x_u \geq \lambda\}$ be the set of elements in $\mathcal{N}$ whose coordinate in x is at least λ. Then,*

$$\hat{f}(x) = \int_{\lambda=0}^{1} f(T_\lambda(x)) d\lambda.$$

Definition 42.4 can also be interpreted in probabilistic terms as the expected value of f over the set $T_\lambda(x)$, where λ is selected uniformly at random from the range $[0, 1]$.

The following lemma proves an important property of the Lovász extension. One way to prove this lemma is via the equality, which follows from the work of Reference 15, between the Lovász extension of a submodular function and another extension known as *the convex closure*. However, the proof we give here is based on a proof originally given by Buchbinder et al. [3].

Lemma 42.7 *Let $f : 2^{\mathcal{N}} \to \mathbb{R}$ be a submodular function, and let $\hat{f}$ be its Lovász extension. For every $x \in [0, 1]^{\mathcal{N}}$, let D_x denote an arbitrary distribution over $\{0, 1\}^{\mathcal{N}}$ such that $\Pr[u \in D_x] = x_u$ for every $u \in \mathcal{N}$ (i.e., its marginals agree with x). Then, $\hat{f}(x) \leq \mathbb{E}_{A \sim D_x}[f(A)]$.*

Proof. Let us denote by x_i the coordinate of x corresponding to element u_i. Assume without loss of generality that $x_1 \geq x_2 \geq \ldots \geq x_n$ is a nonincreasing series (this can always be guaranteed by choosing the right order for the elements), and let $B_i = \{u_1, u_2, \ldots, u_i\}$. Let A be a random set distributed according to the distribution D_x, and let us denote by Y_i an indicator for the event that $u_i \in A$. Then,

$$\mathbb{E}_{A \sim D_x}[f(A)] = \mathbb{E}\left[f(\varnothing) + \sum_{i=1}^{n} Y_i \cdot f(u_i \mid A \cap B_{i-1})\right] \geq \mathbb{E}\left[f(\varnothing) + \sum_{i=1}^{n} Y_i \cdot f(u_i \mid B_{i-1})\right]$$

$$= f(\varnothing) + \sum_{i=1}^{n} \mathbb{E}[Y_i] \cdot f(u_i \mid B_{i-1}) = f(\varnothing) + \sum_{i=1}^{n} x_i \cdot f(u_i \mid B_{i-1})$$

$$= (1 - x_1) \cdot f(\varnothing) + \sum_{i=1}^{n-1}(x_i - x_{i+1}) \cdot f(B_i) + x_n \cdot f(\mathcal{N}) = \hat{f}(A),$$

where the inequality follows from submodularity and the last equality is true since our assumption that $x_1 \geq x_2 \geq \cdots \geq x_n$ implies that $T_\lambda(x) = B_i$ for every $\lambda \in (x_{i+1}, x_i]$ (except for the cases $i = 0$ and $i = n$ for which the correct ranges are $(x_1, 1]$ and $[0, x_n]$, respectively). ∎

Remark: Recall that $\hat{f}(x)$ was defined as the expected value of f over the distribution of the set $T_\lambda(x)$ when λ is chosen uniformly at random from the range $[0, 1]$. One can observe that the marginals of this distribution agree with x. Hence, Lemma 42.7 can be interpreted as claiming that, among all the distributions D_x whose marginals agree with x, the distribution defining $\hat{f}(x)$ is the one with the minimum expected value.

Another important extension of submodular functions is the multilinear extension.

Definition 42.5 (The Multilinear Extension) *Given a vector* $x \in [0,1]^{\mathcal{N}}$, *let* $R(x)$ *denote a random subset of* $\mathcal{N}$ *containing every element* $u \in \mathcal{N}$ *independently with probability* x_u. *Then,*

$$F(x) = \mathbb{E}[R(x)] = \sum_{A \subseteq \mathcal{N}} (f(A) \cdot \Pr[R(x) = A]) = \sum_{A \subseteq \mathcal{N}} \left(f(A) \cdot \prod_{u \in A} x_u \cdot \prod_{u \notin A} (1 - x_u) \right).$$

The explicit formula for the multilinear extension makes it clear that, as suggested by its name, the multilinear extension is a multilinear function.[*] The multilinear extension is central to all the results presented in the remaining sections of this chapter. By definition $F(x)$ is the expected value of f over the distribution of the random set $R(x)$. Since the marginals of this distribution agree with x, we get by Lemma 42.7 the following immediate corollary.

Corollary 42.1 *Let* $f: 2^{\mathcal{N}} \to \mathbb{R}$ *be a submodular function, and let* $\hat{f}$ *and* F *be its Lovász and multilinear extensions, respectively. Then, for every* $x \in [0,1]^{\mathcal{N}}$, $F(x) \geq \hat{f}(x)$.

We conclude this section with an additional notation which is used heavily in the next sections. For two vectors $x, y \in [0,1]^{\mathcal{N}}$, we use $x \vee y$ to denote the coordinate-wise maximum of x and y (formally, $(x \vee y)_u = \max\{x_u, y_u\}$ for every $u \in \mathcal{N}$).

42.3.2 Continuous Greedy

Recall that the problems we are interested in fall into the framework of maximizing a given non-negative submodular function $f: 2^{\mathcal{N}} \to \mathbb{R}_{\geq 0}$ subject to a (possibly trivial) constraint. Denoting by $\mathcal{I}$ the collection of subsets of $\mathcal{N}$ obeying the constraint, this class of problems can be formulated as follows:

$$
\begin{aligned}
\max \quad & f(A) \\
\text{s.t.} \quad & A \in \mathcal{I} \subseteq \{0, 1\}^{\mathcal{N}}
\end{aligned}
\tag{42.16}
$$

An important relaxation of the general problem (42.16), called the multilinear relaxation, is obtained by replacing f with its multilinear extension F and $\mathcal{I}$ with a convex body P which is contained in the cube $[0,1]^{\mathcal{N}}$ on the one hand, and contains the characteristic vectors of the sets in $\mathcal{I}$ on the other hand. Formally, the relaxation we get is

$$
\begin{aligned}
\max \quad & F(x) \\
\text{s.t.} \quad & x \in P \subseteq [0, 1]^{\mathcal{N}}
\end{aligned}
\tag{42.17}
$$

Remark: Usually the natural relaxation P of the constraint is a polytope. However, this is not necessary for the results we present, and thus, we allow P to be an arbitrary convex body.

Given a relaxation in the form of (42.17), one can attempt to approximate the original (integral) problem by first finding an approximate solution for the relaxation, and then rounding this solution without losing too much in the objective. Note that this approach is similar to the standard approach for linear optimization problems in which the algorithm first solves a linear programming (LP) relaxation of the problem, and then rounds the fractional solution obtained.

---

[*] A multilinear function is a polynomial function in which the degree of every variable is at most 1.

The Continuous Greedy algorithm is analogous to the LP solver in the above-mentioned approach. In other words, given a relaxation in the form of (42.17), Continuous Greedy finds a fractional solution whose value approximates the value of the optimal integral solution. This fractional solution can then be rounded by various methods discussed later in this section and in Section 42.3.4. It is interesting to note that unlike LP solvers, the output of Continuous Greedy only approximates the value of the optimal integral solution. This is unavoidable, as is shown later in this section. Continuous Greedy assumes that P is *solvable* meaning that one can optimize linear functions over it.[*]

We are now ready to describe Continuous Greedy. Continuous Greedy maintains a solution that evolves during the time interval $[0, 1]$. The solution starts at time 0 as the empty solution, that is, $\mathbf{1}_\varnothing$. At every time point $t \in [0, 1)$ the algorithm calculates a weight vector $w(t) \in \mathbb{R}^n$ whose value for each element $u \in N$ is $w_u(t) = F(x \vee \mathbf{1}_{\{u\}}) - F(x)$. In other words, $w_u(t)$ is the gain that the algorithm can get by increasing the coordinate of u in its solution to be 1. The algorithm then finds a vector $x(t)$ in P maximizing the linear objective function $w(t) \cdot x(t)$ defined by these weights. Intuitively, $x(t)$ tells Continuous Greedy which coordinates should be increased at the current time. The algorithm uses this information by adding an infinitesimal fraction of $x(t)$ to its current solution. A formal statement of Continuous Greedy is given as Algorithm 42.7.

Naturally, the above-mentioned continuous formulation of Continuous Greedy cannot be implemented in polynomial time. However, there is a discretized version of Continuous Greedy that has the same guarantee (up to low order terms that can be removed by an appropriate preprocessing step) and can be implemented efficiently. The discretization involves two main steps. First, instead of having a continuous time, the algorithm advances time in small steps. Second, instead of calculating $w_u(t)$ by evaluating the multilinear extension F exactly (which might require exponential time), the algorithm approximates $w_u(t) = F(y(t) \vee \mathbf{1}_{\{u\}}) - F(y(t)) = \mathbb{E}[f(R(y(t) \vee \mathbf{1}_{\{u\}})) - f(R(y))]$ by averaging a polynomial number of samples from the distribution of the random quantity $f(R(y(t) \vee \mathbf{1}_{\{u\}})) - f(R(y))$.

The continuous version of Continuous Greedy is cleaner, and thus, we give here only its analysis and leave out the technical details of how to apply this analysis to the discretized version of the algorithm. Unfortunately, the analysis of the continuous version is not completely formal since it involves integration of various functions, and we implicitly assume that these functions are integrable. The reader should keep in mind, however, that this issue goes away when the analysis is applied to the discretized version since the above-mentioned integrations are then replaced by finite sums. Călinescu et al. [14] proved the following guarantee for Continuous Greedy.

Theorem 42.7 (Due to [14]) *Let f be a nonnegative monotone submodular function, and let P is a solvable convex body. Then, Continuous Greedy outputs a vector $y(1) \in P$ such that $F(y(1)) \geq (1 - 1/e) \cdot f(OPT)$, where OPT is a solution in $\mathcal{I}$ maximizing $f(OPT)$.*

ALGORITHM 42.7   CONTINUOUS GREEDY (f, P)

1 Let $y(0) \leftarrow \mathbf{1}_\varnothing$.
2 **foreach** $t \in [0, 1)$ **do**
3   For each $u \in \mathcal{N}$, let $w_u(t) \leftarrow F(y(t) \vee \mathbf{1}_{\{u\}}) - F(y(t))$.
4      $x(t) \leftarrow \arg\max_{x \in P}\{w(t) \cdot x\}$.
5   Increase $y(t)$ at a rate of $\frac{dy(t)}{dt} = x(t)$.
6 **return** $y(1)$

---

[*] Continuous Greedy can also be used when one can only approximately maximize linear functions over P, but this translates into a poorer guarantee on its output.

Proof. Recall that $x(t)$ is a vector inside P for every time $t \in [0, 1)$. Hence, $y(1) = \int_{t=0}^{1} x(t)dt$ is a convex combination of vectors in P, and thus, belongs to P by the convexity of P. Additionally, by the chain rule,

$$\frac{dF(y(t))}{dt} = \sum_{u \in \mathcal{N}} \left(\frac{dy_u(t)}{dt} \cdot \frac{\partial F(y)}{\partial y_u} \bigg|_{y=y(t)} \right) = \sum_{u \in \mathcal{N}} \left(x_u(t) \cdot \frac{\partial F(y)}{\partial y_u} \bigg|_{y=y(t)} \right).$$

Recall, now, that F is multilinear. Thus, its partial derivative with respect to a single coordinate is equal to the difference between the value of the function for two different values of this coordinate over the difference between the values. Plugging this observation into the previous inequality yields (for $t < 1$)

$$\frac{dF(y(t))}{dt} = \sum_{u \in \mathcal{N}} \left(x_u(t) \cdot \frac{\partial F(y)}{\partial y_u} \bigg|_{y=y(t)} \right)$$

$$= \sum_{u \in \mathcal{N}} \left(x_u(t) \cdot \frac{F(y(t) \vee \mathbf{1}_{\{u\}}) - F(y(t))}{1 - y_u(t)} \right) \geq \sum_{u \in \mathcal{N}} x_u(t) \cdot w_u(t) = x(t) \cdot w(t).$$

Next, note that one possible candidate to be $x(t)$ is $\mathbf{1}_{OPT}$ since P contains the characteristic vectors of all the feasible sets. Hence, $x(t) \cdot w(t) \geq \mathbf{1}_{OPT} \cdot w(t)$. Plugging this into the previous inequality gives

$$\frac{dF(y(t))}{dt} \geq x(t) \cdot w(t) \geq \mathbf{1}_{OPT} \cdot w(t) = \sum_{u \in OPT} [F(y(t) \vee \mathbf{1}_{\{u\}}) - F(y(t))] \tag{42.18}$$

$$= \sum_{u \in OPT} \mathbb{E}[f(R(y(t) \vee \mathbf{1}_{\{u\}})) - f(R(y(t)))] \geq \mathbb{E}[f(R(y(t) \vee \mathbf{1}_{OPT})) - f(R(y(t)))]$$

$$= F(y(t) \vee \mathbf{1}_{OPT}) - F(y(t)) \geq f(OPT) - F(y(t)),$$

where the penultimate inequality holds by the submodularity of f and the last inequality holds by its monotonicity.

Inequality Equation 42.18 is a differential inequality with respect to the function $F(y(t))$. Solving this inequality for the initial condition $F(y(0)) \geq 0$ gives

$$F(y(t)) \geq (1 - e^{-t}) \cdot f(OPT).$$

The theorem now follows by plugging $t = 1$ into the last inequality. $\blacksquare$

42.3.2.1 Rounding the Output of Continuous Greedy

We consider next the problem of rounding the output of Continuous Greedy. In particular, as an illustrative example, let us consider rounding for Submodular Welfare (SW). Recall that we have already shown (in Section 42.1.2) an embedding for SW into the general problem (42.16). The natural multilinear relaxation for the embedded form of SW is as follows:

$$\begin{aligned}
\max \quad & F(x) = \mathbb{E}[\sum_{i=1}^{k} f_i(\{u \in \mathcal{N}^{SW} : (u, i) \in R(x)\})] \\
\text{s.t.} \quad & \sum_{i=1}^{k} x_{(u,i)} \leq 1 && \forall u \in \mathcal{N}^{SW} \\
& x_{(u,i)} \geq 0 && \forall u \in \mathcal{N}^{SW}, 1 \leq i \leq k
\end{aligned}$$

Notice that every fractional solution for this relaxation splits each item fractionally among the k buyers, so that the total fractions of all the buyers add up to at most 1. This means that one can round such a solution x by assigning each item $u \in \mathcal{N}^{SW}$, independently, to at most one user where the probability that u is assigned to user $1 \leq i \leq k$ is exactly $x_{(u,i)}$. Let us denote the random output of this rounding process by $R_d(x)$ (the subscript d standards for "dependent" because, unlike in $R(x)$, here there is a dependency between the rounded values of all the coordinates corresponding to the same item u). The following lemma shows that this rounding does not loose in expectation.

Lemma 42.8 *For every vector* $x \in [0,1]^{\mathcal{N}}$ *obeying* $\sum_{i=1}^{k} x_{(u,i)} \leq 1$ *for every item* $u \in \mathcal{N}^{SW}$, $F(x) = \mathbb{E}[\sum_{i=1}^{k} f_i(\{u \in \mathcal{N}^{SW} : (u,i) \in R_d(x)\})]$.

Proof. For every given buyer $1 \leq i \leq k$, every item $u \in \mathcal{N}^{SW}$ is assigned to i with probability $x_{(u,i)}$ both by $R_d(x)$ and by $R(x)$. Moreover, in both cases this assignment is independent of the assignment of the other items of $\mathcal{N}^{SW}$. Hence, for every buyer i, $\{u \in \mathcal{N}^{SW} : (u,i) \in R_d(x)\}$ has the same distribution as $\{u \in \mathcal{N}^{SW} : (u,i) \in R(x)\}$. The observation now follows by the linearity of the expectation since:

$$F(x) = \mathbb{E}\left[\sum_{i=1}^{k} f_i(\{u \in \mathcal{N}^{SW} : (u,i) \in R(x)\})\right] = \sum_{i=1}^{k} \mathbb{E}[f_i(\{u \in \mathcal{N}^{SW} : (u,i) \in R(x)\})]$$

$$= \sum_{i=1}^{k} \mathbb{E}[f_i(\{u \in \mathcal{N}^{SW} : (u,i) \in R_d(x)\})] = \mathbb{E}\left[\sum_{i=1}^{k} f_i(\{u \in \mathcal{N}^{SW} : (u,i) \in R_d(x)\})\right]. \qquad \blacksquare$$

In conclusion, one can get a $(1 - 1/e)$-approximation algorithm for SW, which improves on the greedy approach described in Section 42.2.3, using the following two steps. First, use Continuous Greedy to get a fractional solution y for the multilinear relaxation of SW such that $F(y)$ is at least $(1 - 1/e)$ times the value of the best assignment. Then, round y randomly and output $R_d(y)$. By the last lemma, the expected value of this random assignment is equal to $F(y)$; and thus, the approximation ratio of this algorithm is indeed $(1 - 1/e)$.

Remark: The approximation ratio of the above-mentioned algorithm matches the inapproximability result of [13] for SW with general k values. Thus, the guarantee of Continuous Greedy cannot be improved in general since any such improvement (except for a low order terms improvement) will violate the result of [13].

42.3.3 Measured Continuous Greedy

This section presents a variant of continuous greedy, called Measured Continuous Greedy and originally suggested by Feldman et al. [16], that has several advantages over the original algorithm. The first advantage is the ability of Measured Continuous Greedy to achieve an improved approximation in some cases (although, as discussed earlier, the approximation ratio of Continuous Greedy cannot be improved for the general case). Intuitively, the improved approximation of Measured Continuous Greedy is based on the following observation. The analysis of Continuous Greedy shows that the solution that it maintains at time t has value of at least $(1 - e^{-t}) \cdot f(OPT)$. This guarantee improves as t grows, and thus, it is best to let the algorithm run for as long as possible. Unfortunately, Continuous Greedy cannot run after time $t = 1$ since this might result in an infeasible solution. Measured Continuous Greedy tries to avoid this problem by increasing the coordinates of its solution slower, which can sometimes allow it to reach larger values of t. Algorithm 42.8 is a formal statement of Measured Continuous Greedy. The parameter T is the amount of time we let the algorithm run.

ALGORITHM 42.8　MEASURED CONTINUOUS GREEDY (f, P, T)

1　Let $y(0) \leftarrow 1_{\varnothing}$.
2　**foreach** $t \in [0, T)$ **do**
3　　　For each $u \in \mathcal{N}$, let $w_u(t) \leftarrow F(y(t) \vee 1_{\{u\}}) - F(y(t))$.
4　　　$x(t) \leftarrow \arg\max_{x \in P}\{w(t) \cdot x\}$.
5　　　Increase $y(t)$ at a rate of $\frac{dy(t)}{dt} = (1 - y(t)) \cdot x(t)$.
6　**return** $y(T)$

It is not obvious from the description of Measured Continuous Greedy that $y(t)$ is always within $[0, 1]^{\mathcal{N}}$. The following lemma proves a stronger claim which we use later on.

Lemma 42.9 *For every time $t \in [0, T]$ and element $u \in \mathcal{N}$, $y_u(t) \leq 1 - e^{-t} < 1$.*

Proof. Since $x(t)$ is always in $P \subseteq [0, 1]^{\mathcal{N}}$, $y_u(t)$ obeys the differential inequality

$$\frac{dy_u(t)}{dt} = (1 - y_u(t)) \cdot x_u(t) \leq 1 - y_u(t).$$

Using the initial condition $y_u(0) = 0$, the solution for this differential inequality is $y_u(t) \leq 1 - e^{-t}$. ∎

Let us now analyze the approximation guarantee of Measured Continuous Greedy.

Lemma 42.10 *Let f be a nonnegative monotone submodular function, and let P be a solvable convex body. Measured Continuous Greedy outputs a vector $y(T)$ such that $F(y(T)) \geq (1 - e^{-T}) \cdot f(OPT)$, where OPT is a solution in $\mathcal{I}$ maximizing $f(OPT)$.*

Proof. By the chain rule,

$$\frac{dF(y(t))}{dt} = \sum_{u \in \mathcal{N}} \left(\frac{dy_u(t)}{dt} \cdot \frac{\partial F(y)}{\partial y_u} \bigg|_{y=y(t)} \right) = \sum_{u \in \mathcal{N}} \left((1 - y_u(t)) \cdot x_u(t) \cdot \frac{\partial F(y)}{\partial y_u} \bigg|_{y=y(t)} \right).$$

As F is multilinear, its partial derivative with respect to a single coordinate is the difference between the value of the function for two different values of this coordinate over the difference between the values. Plugging this observation into the previous inequality yields

$$\frac{dF(y(t))}{dt} = \sum_{u \in \mathcal{N}} \left((1 - y_u(t)) \cdot x_u(t) \cdot \frac{\partial F(y)}{\partial y_u} \bigg|_{y=y(t)} \right)$$

$$= \sum_{u \in \mathcal{N}} \left((1 - y_u(t)) \cdot x_u(t) \cdot \frac{F(y(t) \vee \mathbf{1}_{\{u\}}) - F(y(t))}{1 - y_u(t)} \right)$$

$$= \sum_{u \in \mathcal{N}} x_u(t) \cdot w_u(t) = x(t) \cdot w(t)$$

The rest of the proof is completely analogous to the corresponding part in the proof of Theorem 42.7, and is, thus, omitted. ∎

The values of T for which the output of Measured Continuous Greedy is feasible depend on the properties of the convex body P. In fact the output of Measured Continuous Greedy might not be feasible even for $T = 1$. We demonstrate this issue by studying the values of T for which Measured Continuous Greedy outputs a feasible solution when given the multilinear relaxation of SW. As a consequence, we get an improved approximation algorithm for SW.

Lemma 42.11 *Given the multilinear relaxation of SW and $T = -k \ln(1 - 1/k)$, Measured Continuous Greedy outputs a feasible solution of value at least $[1 - (1 - 1/k)^k] \cdot f(OPT)$.*

Proof. Fix an arbitrary item $u \in \mathcal{N}^{\text{SW}}$. We need to prove that $\sum_{i=1}^{k} y_{(u,i)}(T) \leq 1$. For every $1 \leq i \leq k$ and time $t \in [0, T)$ we have

$$\frac{dy_{(u,i)}(t)}{dt} = (1 - y_{(u,i)}(t)) \cdot x_{(u,i)}(t).$$

Solving this differential equation for $y_{(u,i)}(t)$ and plugging the initial condition $y_{(u,i)}(0) = 0$ yield

$$y_{(u,i)}(t) = 1 - e^{-\int_{\tau=0}^{t} x_{(u,i)}(\tau) d\tau}. \tag{42.19}$$

Notice that the sum $\sum_{i=1}^{k} \int_{0}^{T} x_{(u,i)}(\tau)d\tau$ is at most T since $\sum_{i=1}^{k} x_{(u,i)}(t) \leq 1$ for every time $t \in [0, T)$. Additionally, the function $1 - e^{-x}$ is concave. Combining these observations with Equation (42.19) gives

$$\sum_{i=1}^{k} y_{(u,i)}(T) = \sum_{i=1}^{k} \left[1 - e^{-\int_{\tau=0}^{T} x_{(u,i)}(\tau)d\tau} \right] \leq k \left[1 - e^{-k^{-1} \cdot \sum_{i=1}^{k} \int_{\tau=0}^{T} x_{(u,i)}(\tau)d\tau} \right]$$

$$\leq k \left[1 - e^{-k^{-1} \cdot T} \right] = k \left[1 - e^{\ln(1-1/k)} \right] = 1,$$

where the penultimate equality follows by plugging in the value of T. It remains to analyze the quality of the solution $y(T)$ outputted by Measured Continuous Greedy. By Lemma 42.10,

$$F(y(T)) \geq (1 - e^{-T}) \cdot f(OPT) = \left[1 - e^{k \ln(1-1/k)} \right] \cdot f(OPT) = \left[1 - (1 - 1/k)^{k} \right] \cdot f(OPT). \quad \blacksquare$$

Lemma 42.8 shows that the fractional solution obtained by Lemma 42.11 can be efficiently rounded without any loss in the objective, yielding a $(1-(1-1/k)^{k})$-approximation algorithm for SW. The approximation ratio of this algorithm matches (up to low order terms) the inapproximability result of [13] for any given value of k. For general values of k this result does not improve over the $(1 - 1/e)$-approximation algorithm presented in Section 42.3.2, however, for small values of k the improvement is significant. For example, for $k = 2$ the improved approximation ratio is $1 - (1 - 1/2)^{2} = 0.75$, whereas the algorithm presented previously achieves, even for this case, an approximation ratio of only $(1 - 1/e) \cong 0.632$.

An additional advantage of Measured Continuous Greedy is its ability to handle well non-monotone functions. A convex body $P \subseteq [0, 1]^{\mathcal{N}}$ is called *down-closed* if decreasing the coordinates of a vector inside it (while keeping them nonnegative) can never take the vector outside of P. Formally, P is down-closed if, for every two vectors $x, y \in [0, 1]^{\mathcal{N}}$, $x \leq y$ and $y \in P$ imply $x \in P$. We prove the following result for down-closed convex bodies.

Theorem 42.8 (Due to [16]) *Let f be a nonnegative submodular function, and let $\mathcal{P}$ be a solvable down-closed convex body. Then, for $T = 1$ Measured Continuous Greedy outputs a vector $y(1) \in \mathcal{P}$ such that $F(y(1)) \geq 1/e \cdot f(OPT)$, where OPT is a solution in $\mathcal{I}$ maximizing $f(OPT)$.*

Proof. Recall that $x(t)$ is a vector inside P for every time $t \in [0, 1]$, and thus, since P is down-closed, $(1 - y(t)) \cdot x(t)$ also belongs to P. Hence, $y(1) = \int_{t=0}^{1}(1 - y(t) \cdot x(t))dt$ is a convex combination of vectors in P, and therefore, belongs to P by the convexity of P.

Let us now analyze the value of $F(y(1))$. Repeating the proof of Lemma 42.10 we get

$$\frac{dF(y(t))}{dt} = x(t) \cdot w(t).$$

As before, one possible candidate to be $x(t)$ is $\mathbf{1}_{OPT}$ since P contains the characteristic vectors of all the feasible sets. Hence, $x(t) \cdot w(t) \geq \mathbf{1}_{OPT} \cdot w(t)$. Plugging this into the previous inequality gives

$$\frac{dF(y(t))}{dt} = x(t) \cdot w(t) \geq \mathbf{1}_{OPT} \cdot w(t) = \sum_{u \in OPT} \left[F(y(t) \vee \mathbf{1}_{\{u\}}) - F(y(t)) \right]$$

$$\geq F(y(t) \vee \mathbf{1}_{OPT}) - F(y(t)) \geq \hat{f}(y(t) \vee \mathbf{1}_{OPT}) - F(y(t)),$$

where $\hat{f}$ is the Lovász extension of f, the penultimate inequality holds by the submodularity of f and the last inequality holds by Corollary 42.1. To lower bound the term $\hat{f}(y(t) \vee \mathbf{1}_{OPT})$, we recall that by

Lemma 42.9 every coordinate of $y(t)$ is upper bounded by $1 - e^{-t}$. Hence,

$$\hat{f}(y(t) \vee 1_{OPT}) = \int_{\lambda=0}^{1} f(T_\lambda(y(t) \vee 1_{OPT}))d\lambda \geq \int_{\lambda=1-e^{-t}}^{1} f(T_\lambda(y(t) \vee 1_{OPT}))d\lambda = e^{-t} \cdot f(OPT).$$

Plugging the last inequality into the previous one gives

$$\frac{dF(y(t))}{dt} \geq e^{-t} \cdot f(OPT) - F(y(t)),$$

which is a differential inequality with respect to the function $F(y(t))$. Solving this inequality for the initial condition $F(y(0)) \geq 0$ gives

$$F(y(t)) \geq te^{-t} \cdot f(OPT).$$

Finally, the theorem follows by plugging $t = T = 1$ into the last inequality. ∎

Theorem 42.8 sets T to 1. Similar to the monotone case, it is often possible to get a feasible output also for other values of T. However, this does not seem to be beneficial for nonmonotone functions as the formula te^{-t} used by the proof of Theorem 42.8 to determine the approximation ratio of Measured Continuous Greedy attains its maximal value at $t = 1$.

Implementation of Measured Continuous Greedy requires discretization, which reduces the approximation ratio by a low order term. Similar to the case of Continuous Greedy, this loss in the approximation ratio can be fixed by an appropriate preprocessing for monotone objective functions. However, this is not known to be the case for nonmonotone functions. Hence, for polynomial time implementations of Measured Continuous Greedy, the guarantee of Theorem 42.8 should be reduced from $1/e \cdot f(OPT)$ to $(1/e - o(1)) \cdot f(OPT)$, where $o(1)$ is a term which diminishes when $n = |\mathcal{N}|$ increases.

42.3.4 Contention Resolution Schemes

Contention resolution schemes, originally proposed by Reference 17, are a framework for rounding fractional solutions under submodular and linear objective functions. As such, contention resolution schemes play important role in many submodular maximization algorithms that are based on the standard algorithmic scheme of first finding an approximately optimal fractional solution and then rounding it. Moreover, contention resolution schemes found uses also outside the world of submodular maximization (see, Reference 18–20).

We begin the presentation of the contention resolution schemes framework with some definitions.

Definition 42.6 (Contention Resolution Scheme) *A contention resolution scheme (CRS) for a given down-closed body* $P \subseteq [0,1]^{\mathcal{N}}$ is a (possibly random) function $\pi_x : 2^{\mathcal{N}} \to 2^{\mathcal{N}}$, where $x \in P$ and $A \subseteq \mathcal{N}$ that satisfies that for every $x \in P$ and $A \subseteq \mathcal{N}$:*

1. *$\pi_x(A) \subseteq A$.*
2. *The characteristic vector of $\pi_x(A)$ always belongs to P.*

Intuitively, we think of $\pi_x(R(x))$ as a random rounding for the fractional point $x \in P$. Notice that this is indeed a rounding in the sense that it gets as input a possibly fractional point $x \in P$ and outputs a set corresponding to an integral point in P. Given this point of view, we can think of the parameter x of

---

* It is usually assumed that P is a down-closed polytope, however, this assumption is not essential for the results we present in this section.

Instance		CRS
$\mathcal{N}$	$= \{u, v\}$	$\pi_x(\emptyset)$ $= \emptyset$
$f(A)$	$= \min\{|A|, 1\}\ \forall\ A \subseteq \mathcal{N}$	$\pi_x(\{u\})$ $= \emptyset$
P	$= [0, 1]^{\mathcal{N}}$	$\pi_x(\{v\})$ $= \emptyset$
x	$= (\frac{1}{2}, \frac{1}{2})$	$\pi_x(\{u, v\}) = \{u,v\}$

FIGURE 42.1 An example of a c-balanced CRS for which $\mathbb{E}[f(\pi_x(\mathrm{R}(x)))] \geq c \cdot F(x)$ is not true.

the CRS as informing the CRS about the distribution of its other parameter, and allowing it to choose a response for every given input set which yields a good (in some sense) rounding overall. To get some guarantee on the quality of the rounding induced by a CRS, we first need to define additional properties of CRSs.

Definition 42.7 *Let $P \subseteq [0,1]^{\mathcal{N}}$ be a down-closed body and let π be a CRS for P.*

- *π is c-balanced (for $c \in [0, 1]$) if $\Pr[u \in \pi_x(\mathrm{R}(x))] \geq c \cdot x_u$ for every element $u \in \mathcal{N}$ and vector $x \in P$.*
- *π is monotone if $\Pr[u \in \pi_x(A)] \geq \Pr[u \in \pi_x(B)]$ whenever $x \in P$ and $A \subseteq B \subseteq \mathcal{N}$.*

It is immediate that a c-balanced CRSs can be used for rounding solutions under a linear function objective as follows. Let $f \colon 2^{\mathcal{N}} \to \mathbb{R}_{\geq 0}$ be a nonnegative linear function, and let F be its natural extension. Then, $\mathbb{E}[f(\pi_x(\mathrm{R}(x)))] \geq c \cdot F(x)$ for every $x \in P$. We are interested in proving a similar claim for a nonnegative submodular function f and its multilinear extension F. Unfortunately, requiring the CRS to be c-balanced is not enough even when f is a monotone function. This is demonstrated by the example described in Figure 42.1. Notice that in this example π is $(1/2)$-balanced (at least for the x given in this figure) since $\Pr[u \in \pi_x(\mathrm{R}(x))] = \Pr[v \in \pi_x(\mathrm{R}(x))] = 1/4 = x_u/2 = x_v/2$. However, $\mathbb{E}[f(\pi_x(\mathrm{R}(x)))] = 1/4 \not\geq F(x)/2 = 3/8$.

The problem, intuitively, is that the CRS π described by Figure 42.1 is an unnatural choice for a rounding scheme. We expect a rounding scheme to be more liberal in removing elements when there are more elements in its input set (and thus, the set is more likely to violate down-closed constraints). However, the behavior of π is quite the opposite. Namely, π keeps elements only when its input set contains both u and v. This intuitive idea is formalized by the monotonicity property in Definition 42.7 (which is violated in this example). Assuming both properties of Definition 42.7 (c-balance and monotonicity) allow us to obtain the desired result for submodular functions.

Theorem 42.9 (Due to [17]) *Let $f \colon 2^{\mathcal{N}} \to \mathbb{R}_{\geq 0}$ be a nonnegative monotone submodular function. Let π be a monotone c-balanced CRS for a down-closed body P. Then, $\mathbb{E}[f(\pi_x(\mathrm{R}(x)))] \geq c \cdot F(x)$ for every $x \in P$.*

Proof. Let $A \sim \mathrm{R}(x)$ be a random set distributed like $\mathrm{R}(x)$, and for every $1 \leq i \leq n$ let X_i be an indicator for the event that $u_i \in \pi_x(A)$, and let $B_i = \{u_1, \ldots, u_i\}$. Using this notation we get for every $1 \leq i \leq n$:

$$\mathbb{E}_{A \sim \mathrm{R}(x)}[f(\pi_x(A) \cap B_i) - f(\pi_x(A) \cap B_{i-1})] = \mathbb{E}[X_i \cdot f(u_i \mid \pi_x(A) \cap B_{i-1})]$$
$$\geq \mathbb{E}[X_i \cdot f(u_i \mid A \cap B_{i-1})] = \mathbb{E}[\Pr[u_i \in \pi_x(A)] \cdot f(u_i \mid A \cap B_{i-1})]$$
$$\geq \mathbb{E}[\Pr[u_i \in \pi_x(A)]] \cdot \mathbb{E}[f(u_i \mid A \cap B_{i-1})]$$
$$\geq c x_{u_i} \cdot \mathbb{E}[f(u_i \mid A \cap B_{i-1})].$$

The first inequality follows by submodularity since $\pi_x(A) \subseteq A$. The second inequality follows by the FKG inequality since $\Pr[u_i \in \pi_x(A)]$ is a nonincreasing function of A due to the monotonicity of π and

$f(u_i \mid A \cap B_{i-1})$ is also a nonincreasing function of A due to the submodularity of f. The final inequality follows since π is c-balanced. Summing up the above-mentioned inequality for all $1 \leq i \leq n$ we get

$$\mathbb{E}_{A \sim R(x)}[f(\pi_x(A))] = \mathbb{E}[f(\pi_x(A) \cap B_n)] = f(\varnothing) + \sum_{i=1}^{n} \left(\mathbb{E}_{A \sim R(x)}[f(\pi_x(A) \cap B_i) - f(\pi_x(A) \cap B_{i-1})] \right)$$

$$\geq f(\varnothing) + \sum_{i=1}^{n} cx_{u_i} \cdot \mathbb{E}[f(u_i \mid A \cap B_{i-1})] \geq c \cdot F(x),$$

where the last inequality holds since the nonnegativity of f guarantees that $f(\varnothing) \geq c \cdot f(\varnothing)$. ∎

There is an extension of Theorem 42.9 for nonmonotone submodular functions. The only difference between this extension and the original theorem is that it requires the output of the CRS to go through an additional post-processing step which might remove from it additional elements.

Theorem 42.10 (Due to [17]) *Let $f: 2^{\mathcal{N}} \to \mathbb{R}_{\geq 0}$ be a nonnegative submodular function. Let π be a monotone c-balanced CRS for a down-closed body P. Then, there exists a CRS π' which can be efficiently computed (given that π can) such that $\mathbb{E}[f(\pi'_x(R(x)))] \geq c \cdot F(x)$ for every $x \in P$.*

We do not prove Theorem 42.10 in this survey. However, we would like to point out that π' is very simple. It first apply π, and then it simply scans the elements left in some predetermined order, and removes every element whose marginal contribution to the subset of the elements scanned so far is negative.

42.3.4.1 Contention Resolution Scheme for Matchings

At this point we would like to give an example of a CRS and analyze its properties. The down-closed body of our example CRS is the natural relaxation of matching. More formally, given a graph $G = (V, E)$, we define the polytope P to be

$$(P) \quad \begin{array}{ll} \sum_{u \in \delta(a)} x_u \leq 1 & \forall\, a \in V \\ x_u \geq 0 & \forall\, u \in E \end{array},$$

where $\delta(a)$ stands for the set of edges emanating from node a. Notice that P is indeed a down-closed body. The CRS we define for P is given by Algorithm 42.9. The technique used by this CRS was first given by Feldman et al. [16] (see also [21] for more details). It is important to keep in mind while reading Algorithm 42.9 and its analysis that the ground set here is the set of edges E, and thus, we use the term "edge" to denote the elements of this ground set.

Note, for example, that if two edges in B share a node, then both of them are not added to C, although it is possible that one of them could be added without violating the matching constraint. This unnatural behavior is necessary to guarantee the monotonicity of the CRS (Lemma 42.12). One can verify that Algorithm 42.9 is indeed a CRS for P, that is, it always outputs a subset of the edges in its input set, and the edges in the output subset form a legal matching. The next lemma analyzes the properties of this CRS.

Lemma 42.12 *Algorithm 42.9 is a monotone e^{-2}-balanced CRS.*

ALGORITHM 42.9   CRS FOR MATCHING(x, A)

1 Let B be a subset of A containing every edge u of A with probability $(1 - e^{-x_u})/x_u$, independently.
2 Let $C \leftarrow \varnothing$.
3 For each $u \in B$: Add u to C only if no other edge of B shares a node with u.
4 **return** C.

Proof. Observe that every edge of A is copied to B with a probability independent of the membership of other edges in A. Hence, given that an edge u belongs to A, adding other edges to A can only increase the probability that some edge adjacent to u ends up in B. Since u belongs to output set C if and only if it is in B and no other edge of B shares a node with it, we get that adding other edges to A can only decrease the probability of u to belong to the output set. Thus, by definition, the CRS defined by Algorithm 42.9 is monotone.

It remains to analyze the balance of this CRS. Consider an arbitrary edge $u \in E$. We assume in the rest of the proof that $A \sim R(x)$, and our objective is to show that $\Pr[u \in C] \geq e^{-2} \cdot x_u$. The probability that u ends up in B is

$$\Pr[u \in B] = \Pr[u \in A] \cdot \frac{1 - e^{-x_u}}{x_u} = 1 - e^{-x_u}.$$

Let a and b denote the two end nodes of u. Since the probability of every edge to get to B is independent, the probability that no other edge hitting a is in B is

$$\prod_{v \in \delta(a) - u} \Pr[v \notin B] = \prod_{v \in \delta(a) - u} (1 - \Pr[v \in B]) = \prod_{v \in \delta(a) - u} e^{-x_v} = e^{-\sum_{v \in \delta(a) - u} x_v} \geq e^{x_u - 1},$$

where the inequality holds since x belongs to the polytope P. Similarly, we get that with probability at least $e^{x_u - 1}$ no edge hitting b belongs to B. As these events are independent (no edge other than u hits both a and b), we get

$$\Pr[u \in C] = \Pr[e \in B] \cdot \left(\prod_{v \in \delta(a) - u} \Pr[v \notin B] \right) \cdot \left(\prod_{v \in \delta(b) - u} \Pr[v \notin B] \right)$$

$$\geq (1 - e^{-x_u}) \cdot (e^{x_u - 1})^2 \geq (e^{x_u} - 1) \cdot e^{-2} \geq x_u \cdot e^{-2}. \qquad \blacksquare$$

In the last proof, we used the fact that $\sum_{u \in \delta(a)} x_u \leq 1$ for every vector $x \in P$ and node $a \in V$. Observe that we would have gotten a better balance guarantee if we had $\sum_{u \in \delta(a)} x_u \leq b$ for some $b \in (0, 1)$. In other words, the balance of the CRS improves when $x \in bP$. This common phenomenon motivates the following definition.

Definition 42.8 ((b, c)-Balanced CRS) *A CRS π for a down-closed body $P \subseteq [0, 1]^{\mathcal{N}}$ is (b, c)-balanced (for $b, c \in [0, 1]$) if $\Pr[u \in \pi_x(R(x))] \geq c \cdot x_u$ for every element $u \in \mathcal{N}$ and vector $x \in bP$.*

A slight modification of the proof of Lemma 42.12 shows that the CRS given as Algorithm 42.9 is in fact (b, e^{-2b})-balanced for every $b \in [0, 1]$. The parameters of a few other known CRSs are given by Table 42.1. One can derive from these CRSs new CRSs for more involved constraints using the following important lemma.

Lemma 42.13 (Combining CRSs) *Let π^1 be a monotone (b, c_1)-balanced CRS for a down-closed body P_1, and let π^2 be a monotone (b, c_2)-balanced CRS for a down-closed body P_2. Then, there is a monotone $(b, c_1 c_2)$-balanced CRS π for $P_1 \cap P_2$. Moreover, π can be computed efficiently whenever π^1 and π^2 can be computed efficiently.*

Proof. For every vector $x \in P_1 \cap P_2$, we define $\pi_x(A) = \pi_x^1(A) \cap \pi_x^2(A)$. One can verify that π is indeed a CRS whenever π^1 and π^2 are CRSs. To see that π is monotone, consider two sets $A \subseteq B \subseteq \mathcal{N}$, an element

TABLE 42.1 The parameters of a few known CRSs. All the CRSs in this table are monotone and can be calculated efficiently.

Polytope	Balance	Reference
Matching polytope	(b, e^{-2b})-balanced	[16]
Matroid polytope	$(b, (1 - e^{-b})/b)$-balanced	[17]
The natural relaxation of a knapsack constraint	$(1 - \varepsilon, 1 - \varepsilon)$-balanced*	[17]
The natural relaxation of "Independent set in an interval graph"	(b, e^{-b})-balanced	[21]

*ε can be any constant greater than 0. The use of this CRS requires a pre-processing step whose time complexity depends on ε.

$u \in A$ and some vector $x \in P_1 \cap P_2$. Since $\pi_x^1(A)$ and $\pi_x^2(A)$ are independent when A is deterministic, the monotonicity of π^1 and π^2 implies

$$\Pr[u \in \pi_x(A)] = \Pr[u \in \pi_x^1(A) \cap \pi_x^2(A)] = \Pr[u \in \pi_x^1(A)] \cdot \Pr[u \in \pi_x^2(A)]$$

$$\geq \Pr[u \in \pi_x^1(B)] \cdot \Pr[u \in \pi_x^2(B)] = \Pr[u \in \pi_x^1(B) \cap \pi_x^2(B)] = \Pr[u \in \pi_x(B)].$$

It remains to analyze the balance of π. Let A be a set distributed like R(x). By the monotonicity of π^1 and π^2, the probabilities of the two events $u \in \pi_x^1(A)$ and $u \in \pi_x^2(A)$ are both nonincreasing in A. Hence, by the FKG inequality, for every vector $x \in bP$ and element u such that $x_u > 0$,

$$\Pr[u \in \pi_x(R(x))] = x_u \cdot \Pr[u \in \pi_x(A) \mid u \in A] = x_u \cdot \Pr[u \in \pi_x^1(A) \wedge u \in \pi_x^2(A) \mid u \in A]$$

$$\geq x_u \cdot \Pr[u \in \pi_x^1(A) \mid u \in A] \cdot \Pr[u \in \pi_x^2(A) \mid u \in A]$$

$$= \frac{1}{x_u} \cdot \Pr[u \in \pi_x^1(R(x))] \cdot \Pr[u \in \pi_x^2(R(x))] \geq x_u \cdot c_1 c_2.$$

To complete the proof that π is $(b, c_1 c_2)$-balanced we observe that the inequality $\Pr[u \in \pi_x(R(x))] \geq x_u \cdot c_1 c_2$ is trivial for an element u such that $x_u = 0$. ∎

Using Lemma 42.13 it is easy to derive CRSs for relaxations of very involved constraints. For example, plugging CRSs from Table 42.1 into Lemma 42.13 can yield a monotone $(b, e^{-4b} - \varepsilon)$-balanced CRS for a relaxation of a constraint allowing a set of edges only if it is a legal matching in two different graphs (over the same set of edges) and also obeys some knapsack constraint.

Finally, once we have a monotone (b, c)-balanced CRS for a solvable down-closed convex body P which is a relaxation of some constraint, it is possible to get an approximation algorithm for this constraint as summarized by the following theorem.

Theorem 42.11 *Let P be a solvable down-closed convex body, and let π be a monotone (b, c)-balanced CRS for it. Then,*

- *There is a cb-approximation algorithm for maximizing a linear function over the integral points of P.*
- *There is a (cbe^{-b})-approximation algorithm for maximizing a submodular function over the integral points of P.*
- *There is a $c(1 - e^{-b})$-approximation algorithm for maximizing a **monotone** submodular function over the integral points of P.*

Proof. Consider first a linear objective function f. Since we assumed P is solvable, it is possible to find a fractional solution $x \in P$ maximizing f. Since $bx \in bP$, we can feed bx into the CRS to get an integral solution whose value is, in expectation, at least $c \cdot f(bx) = bc \cdot f(x)$. Hence, the approximation ratio of the algorithm is bc.

For a submodular objective function it is possible to find an approximately optimal fractional solution $x \in P$ using Measured Continuous Greedy. We can then use bx as our vector in bP, and the approximation ratio of this vector is be^{-1} for nonmonotone functions and $b(1 - e^{-1})$ for monotone functions. However, there is a better way. If one stops Measured Continuous Greedy at time $T = b$, then its output is a vector which already belongs to bP. The approximation ratio of this vector is be^{-b} for nonmonotone functions and $1 - e^{-b}$ for monotone functions. Feeding this vector into the monotone (b, c)-balanced CRS we have yields an approximation algorithm whose approximation ratio is $c \cdot be^{-b}$ for nonmonotone functions and $c(1 - e^{-b})$ for monotone functions. ∎

42.3.5 Inapproximability Results

In this section, we describe the symmetry gap technique (originally proposed by Vondrák [22]) which can be used to prove many of the state of the art inapproximability results in the field of submodular maximization. We begin by studying a very simple instance of unconstrained submodular maximization induced by the cut function of the graph depicted in Figure 42.2. Formally, our instance contains a ground set of two elements $\mathcal{N} = \{u, v\}$. The function defined over this ground set is

$$f(A) = \begin{cases} 1 & \text{if } |A| = 1, \\ 0 & \text{otherwise.} \end{cases}$$

It is easy to verify that this function is indeed nonnegative and submodular. One can also observe that u and v play completely symmetric roles in this instance. Let us assume, for the moment, that an algorithm for unconstrained submodular maximization must output a vector $x \in [0, 1]^{\mathcal{N}}$ which is symmetric with respect to every symmetry that its input instance have. In particular, for the instance we consider, it must be that $x_u = x_v = y$ for some value $y \in [0, 1]$. The value of any such symmetric vector, with respect to the multilinear extension F of f, is

$$F(x) = F(y \cdot \mathbf{1}_{\mathcal{N}}) = 2y(1 - y) \leq \frac{1}{2}.$$

On the other hand, there is an integral solution for this instance (e.g., the vector $(1, 0)$) whose value is 1. Hence, there is a multiplicative gap of $(1/2) : 1 = 1/2$ between the best symmetric and nonsymmetric (feasible) fractional solutions of this instance. This gap is called the *symmetry gap*.

On the face of it, the symmetry gap does not seem to be related to inapproximability. After all, real algorithms output integral solutions, and these solutions are allowed to be asymmetric. However, it turns out that there is a way to force polynomial time algorithms to output a close to symmetric fractional solution. As a first step in that direction, let us "blow" each element of the above-mentioned instance into t elements. This yields a new ground set that we denote by $\mathcal{N}_t = \{u_i, v_i\}_{i=1}^{t}$. The objective function that we associate with this ground set is

$$f_t(A) = F(t^{-1} \cdot |A \cap \{u_i\}_{i=1}^{t}|, t^{-1} \cdot |A \cap \{v_i\}_{i=1}^{t}|).$$

Informally, every set $A \subseteq \mathcal{N}_t$ is interpreted as a fractional solution over the ground set $\mathcal{N}$, where the fraction of u is proportional to the number of $\{u_i\}_{i=1}^{t}$ elements in A, and the fraction of v is proportional to the number of $\{v_i\}_{i=1}^{t}$ elements in A. It can be verified that the resulting objective function

FIGURE 42.2   A graph inducing a simple unconstrained submodular maximization instance.

f_t is nonnegative and submodular whenever the objective function f of the original instance has these properties (this follows, e.g., from the more general results of [22]).

Assume that the elements of $\mathcal{N}_t$ are assigned names, uniformly at random, from the list $1, 2, \ldots, 2t$, and the algorithm can access the elements only via these names (i.e., the algorithm has no access to the original names of the elements). The randomness of the names assignment means that the algorithm does not know which of the elements descend from u and which descend from v. If the algorithm fails to find out more information about the origin of the different elements, then its output set will necessarily be chosen independently of this information. Thus, assuming t is large, the output set is likely to contain a similar number of elements descending from u and v. In other words, if the algorithm cannot find information about the origin of the different elements, then its output is close to being a symmetric fractional solution for the original (nonblown) instance; and thus, its approximation ratio cannot be significantly better than the symmetry gap.

The only way in which the algorithm can gather information is through the value oracle. To query this oracle the algorithm constructs a set A and forward it to the oracle. Note that when the number of element in A originating from u and v is equal, then the response of the oracle is

$$F(t^{-1} \cdot |A|/2, t^{-1} \cdot |A|/2) = 2 \cdot \frac{|A|}{2t} \left(1 - \frac{|A|}{2t}\right) = \frac{|A|}{t}\left(1 - \frac{|A|}{2t}\right),$$

which is a function of the size of A alone. Hence, queries about sets with equal number of elements originating from u and v does not give the algorithm any information about the origin of the different elements. Thus, getting information about the origin of the elements requires the algorithm to construct a set which is unbalanced in terms of the number elements in it originating from u and v.

By a slight modification of f_t, one can further guarantee that getting information about the origin of the elements requires the algorithm to construct a set which is *significantly* unbalanced in terms of the number elements in it originating from u and v. However, such a set cannot be constructed without prior knowledge about the origin of elements because any set constructed without such knowledge is likely, for large values of t, to contain a similar number of elements descending from u and v.

This concludes our intuitive explanation about how to convert the symmetry gap of unconstrained submodular maximization into an inapproximability result. The next theorem, due to Reference 6, gives a formal proof, based on this idea, of this inapproximability.

Theorem 42.12 (Due to [6]) *For every constant $\varepsilon > 0$, there is no polynomial time $(1/2+\varepsilon)$-approximation algorithm for unconstrained submodular maximization.*

Proof. For simplicity, we prove the equivalent claim that there is no polynomial time $(1/2 + 3\varepsilon)$-approximation algorithm for unconstrained submodular maximization. This claim is meaningless for $\varepsilon > 1/6$, thus, we assume $\varepsilon \in (0, 1/6]$. Additionally, we also assume $1/\varepsilon$ is integral (otherwise, one can replace ε with some value from the range $[\varepsilon/2, \varepsilon]$ having this property).

Let t be an arbitrary large positive integer dividable by $1/\varepsilon$ (meaning that εt is integral), and let $\mathcal{N}_t = \{1, 2, \ldots, 2t\}$. Let $U \subseteq \mathcal{N}_t$ be an arbitrary subset of size t. Intuitively, the elements of U are the elements originating from u in the above-mentioned general description of the proof. For every set $A \subseteq \mathcal{N}_t$ let $u_A = |A \cap U|$ and $v_A = |A \setminus U|$. Using this notation we define a function $f_t : 2^{\mathcal{N}_t} \to \mathbb{R}_{\geq 0}$ as follows.

$$f_t(A) = \begin{cases} \dfrac{u_A + v_A}{t}\left(1 - \dfrac{u_A + v_A}{2t}\right) = \dfrac{|A|}{t}\left(1 - \dfrac{|A|}{2t}\right) & \text{when } |u_A - v_A| \leq \varepsilon t, \\[3mm] \dfrac{u_A}{t}\left(1 - \dfrac{v_A}{t}\right) + \dfrac{v_A}{t}\left(1 - \dfrac{u_A}{t}\right) + \dfrac{\varepsilon^2}{2} - \dfrac{\varepsilon|u_A - v_A|}{t} & \text{when } |u_A - v_A| \geq \varepsilon t. \end{cases}$$

The definition of this function in the case $|u_A - v_A| \geq \varepsilon t$ is identical to the definition of the function f_t from the intuitive description above up to the additional terms involving ε. These additional terms change the maximum of the function very little, but allow for the two cases of the definition to combine into one submodular function. For ease of the reading, we omit the technical verification of the following claim.

Claim 42.1 f_t *is a nonnegative submodular function. Its maximum value* $1 + \varepsilon^2/2 - \varepsilon \in (1 - \varepsilon, 1)$ *is attained both when* $u_A = t$ *and* $v_A = 0$ *and when* $u_A = 0$ *and* $v_A = t$.

The definition of f_t depends on the set U which we assumed so far to be fixed. From now on we assume it is a uniformly random subset of $\mathcal{N}_t$ of size t. Hence, f_t is now a sample out of a distribution of nonnegative submodular functions. All the functions in the support of this distribution share the same maximum value. We would like to show that, given access to a function f_t sampled from this distribution, no polynomial time algorithm can output, in expectation, a solution of value at least $1/2 + \varepsilon$ times this optimal value (where the expectation is over the randomness of U and the random coins of the algorithm). By Yao's principal, it is enough to prove this for deterministic algorithms (if there is a single distribution over instances which is hard for any given polynomial time deterministic algorithm, then Yao's principal guarantees that this distribution is hard also for polynomial time randomized algorithms); and thus, we restrict ourselves to deterministic algorithms in the rest of the proof.

Consider an arbitrary polynomial time deterministic algorithm ALG for unconstrained submodular maximization, and consider an execution of this algorithm on the instance induced by the following nonnegative submodular function:

$$g_t(A) = \frac{|A|}{t}\left(1 - \frac{|A|}{2t}\right).$$

Let $Q_1, Q_2, \ldots, Q_r$ be the list of sets that ALG passes as queries to its value oracle given this instance. Since ALG is polynomial, r (the number of queries) is bounded by a polynomial function of t. Additionally, we can assume, without loss of generality, that the output set of ALG is one of the sets in this list. For every set Q_i, $u_{Q_i} = |Q_i \cap U|$ has a hypergeometric distribution, and thus, by bounds given in Reference 23 (which are based on results of References 24, 25) we get

$$\Pr[|u_{Q_i} - v_{Q_i}| \geq \varepsilon t] = \Pr[|2|Q_i \cap U| - |Q_i|| \geq \varepsilon t] = \Pr\left[||Q_i \cap U| - \mathbb{E}[|Q_i \cap U|]| \geq |Q_i| \cdot \frac{\varepsilon t}{2|Q_i|}\right]$$

$$\leq 2e^{-2 \cdot \left(\frac{\varepsilon t}{2|Q_i|}\right)^2 \cdot |Q_i|} = 2e^{-\frac{\varepsilon^2 t^2}{2|Q_i|}} \leq 2e^{-\varepsilon^2 t/4},$$

where the last inequality holds since $|Q_i| \leq |\mathcal{N}_t| = 2t$. Notice that the event $|u_{Q_i} - v_{Q_i}| \leq \varepsilon t$ implies $f_t(Q_i) = g_t(Q_i)$. Hence, what we proved is in fact that $f_t(Q_i) = g_t(Q_i)$ with probability at least $1 - 2e^{-\varepsilon^2 t/4}$ for every given set Q_i. By the union bound, we get that, with probability at least $1 - 2re^{-\varepsilon^2 t/4}$, $f_t(Q_i) = g_t(Q_i)$ for every $1 \leq i \leq r$. Let us denote by $\mathcal{E}$ the last event (i.e., the event that $f_t(Q_i) = g_t(Q_i)$ for every $1 \leq i \leq r$).

The main observation of the proof is that when $\mathcal{E}$ happens ALG has an identical execution given either f_t or g_t. Thus, $\mathcal{E}$ guarantees that given both inputs ALG outputs the same output set, and this set has the same value under both functions because we assumed that it is one of the sets in the list $Q_1, Q_2, \ldots, Q_r$. Since the maximum value of g_t is $1/2$, this implies that ALG outputs a set of value at most $1/2$ whenever $\mathcal{E}$ occurs. We can now upper bound the expected value of the output of ALG given f_t (where the expectation is over the randomness of U) by

$$\frac{1}{2} \cdot \Pr[\mathcal{E}] + \max_{A \subseteq \mathcal{N}_t} f_t(A) \cdot (1 - \Pr[\mathcal{E}]) \leq \frac{1}{2} \cdot 1 + 1 \cdot (1 - \Pr[\mathcal{E}]) \leq \frac{1}{2} + 2re^{-\varepsilon^2 t/4} \leq \frac{1}{2} + \varepsilon,$$

where the last inequality holds for large enough t since r is polynomial in t. On the other hand, by the claim the maximum value of f_t is at least $1 - \varepsilon$, and thus, the approximation ratio of ALG is worse than

$$\frac{1/2 + \varepsilon}{1 - \varepsilon} \leq \frac{1}{2} + 3\varepsilon.$$

This completes the proof of the theorem by the above-mentioned discussion. ∎

In general, the work of Reference 22 shows that the symmetry gap can be converted into an inapproximability result in many cases. We do not state the results of Reference 22 explicitly here, however, intuitively, the conversion of the symmetry gap into an inapproximability result can be done whenever the following two conditions hold.

- The symmetries of the objective function that one uses in order to calculate the symmetry gap cannot be broken by the algorithm using some bypass. In particular, the constraint of the problem should obey these symmetries as well.
- As seen in the proof of Theorem 42.12, the hardness result applies in fact to a distribution over blown up instances. Hence, it should be possible to somehow present blown up instances as new instances of the original problem for which the inapproximability result should apply.

To better understand these conditions, we now give an informal proof based on the symmetry gap technique for another inapproximability result—the problem of maximizing a monotone submodular function subject to a cardinality constraint. Consider an instance of this problem in which one is allowed to pick up to one element out of some ground set $\mathcal{N}$ of size n. The objective function of the instance is the monotone submodular function $f(A) = \min\{|A|, 1\}$. Clearly, all the elements of the ground set are symmetric with respect to both this objective function and the constraint. Hence, any symmetric fractional solution must assign an identical value to each one of the elements. To keep this fractional solution feasible, this identical value must be at most $1/n$. Thus, the best feasible symmetric fractional solution is $n^{-1} \cdot 1_{\mathcal{N}}$, and its value with respect to the multilinear extension F of f is

$$F(n^{-1} \cdot 1_{\mathcal{N}}) = 1 - (1 - 1/n)^n.$$

One the other hand, any set containing a single element of $\mathcal{N}$ is feasible and has a value of 1. Hence, the symmetry gap of this instance of maximizing a monotone submodular function subject to a cardinality constraint is $[1 - (1 - 1/n)^n]/1 = 1 - (1 - 1/n)^n$.

When blowing up the above-mentioned instance by a factor of t one gets a new instance with a ground set of size nt in which a feasible solution contains at most t elements. Clearly, this blown up instance is a valid instance of maximizing a monotone submodular function subject to a cardinality constraint, and thus, the two intuitive conditions given above hold. This means that the symmetry gap we have shown translates into an inapproximability result for this problem. In conclusion, we got that no polynomial time algorithm can achieve an approximation ratio of $1 - (1 - 1/n)^n + \varepsilon$ for any constant $\varepsilon > 0$ and integer n. As $(1 - 1/n)^n$ approaches $1/e$ when n grows, this implies that no polynomial time algorithm can achieve an approximation ratio of $1 - 1/e + \varepsilon$ for any constant $\varepsilon > 0$. This inapproximability result is optimal (as shown in Section 42.2.1). It was first proved by Nemhauser and Wolsey [2] using a different technique.

It is interesting to note that the inapproximability result proved by Theorem 42.12, like all the other inapproximability results discussed in this chapter, is unconditional. In particular, this inapproximability result does not rely on complexity assumptions such as $P \neq NP$. Instead, its proof argues that a polynomial number of value oracle queries does not suffice in order to distinguish between two functions f_t and g_t having very different maximum values—which is an information theoretic argument. The use of this information theoretic argument means that our assumption that the objective function can be accessed only via a value oracle is essential for the proof. More specifically, the proof does not hold if the objective function is specified to the algorithm using some succinct representation. Dobzinski and Vondrák [26] showed that inapproximability results based on the symmetry gap, such as the inapproximability result proved by Theorem 42.12, can usually be modified to allow some succinct representation of the input functions. This modification, however, comes at the cost of losing the unconditionality of the result and relaying on the widely believed complexity assumption $RP \neq NP$.

42.3.6 Notes and Open Questions

42.3.6.1 Measured Continuous Greedy

The $1/e$-approximation guarantee of Measured Continuous Greedy for nonmonotone functions on down-closed convex body P is not known to be the best possible. For this problem it is only known that no polynomial time algorithm can output a fractional solution of value at least $0.478 \cdot f(OPT)$ [9]. Closing the gap between the guarantee of Measured Continuous Greedy and this impossibility result is one of the most important open problems related to the topics presented in this section.

42.3.6.2 Rounding Techniques

The fractional solutions produced by Continuous Greedy and Measured Continuous Greedy are of little use without rounding algorithms that convert their fractional solutions into integral solutions. In Section 42.3.4 we presented a powerful technique which yields rounding algorithms for many types of constraints. We would like to mention a few rounding algorithms which are based on other techniques.

The problem of maximizing a submodular function subject to a matroid constraint* was one of the first problems studied in the context of submodular functions maximization [8]. Matroid constraint generalizes, for example, both the cardinality constraint and the constraints in SW. Two rounding algorithms called Pipage Rounding and Swap Rounding were suggested by References 14 and 27, respectively, for the multilinear relaxation of this problem in which the convex body P is the matroid polytope. Both rounding techniques loose nothing in the objective, and thus, yield an optimal $(1 - 1/e)$-approximation algorithm for maximizing a nonnegative monotone submodular function subject to a matroid constraint as well as a $(1/e - o(1))$-approximation for the nonmonotone counterpart of this problem. Swap Rounding, suggested later, is faster than Pipage Rounding and also enjoys additional concentration bounds, which make it more suitable for some applications.

Kulik et al. [28] designed a rounding algorithm for the natural multilinear relaxation of the problem of maximizing a nonnegative monotone submodular function subject to a *constant* number of knapsack constraints. This rounding algorithm looses only a factor of $1 - \varepsilon$ (for any fixed $\varepsilon > 0$), and thus, combined with Continuous Greedy, implies $(1 - 1/e - \varepsilon)$-approximation for this problem. Notice that this result is optimal up to the term ε since the problem generalizes maximization of a submodular function subject to a cardinality constraint. In a later work Kulik et al. [29] extended their rounding algorithm also to nonmonotone functions, which, combined with Measured Continuous Greedy, yields a $(1/e - \varepsilon)$-approximation algorithm for the nonmonotone counterpart of the above-mentioned problem.

42.3.6.3 Inapproximability

Despite its origin in the field of submodular maximization, the symmetry gap technique found some applications also in the study of submodular minimization problems [30].

References

1. G. L. Nemhauser, L. A. Wolsey, and M. L. Fisher. An analysis of approximations for maximizing submodular set functions—I. *Mathematical Programming*, 14:265–294, 1978.

---

* Unfortunately, the definition of matroids is outside the scope of this survey; however, we mention a few results concerning matroid constraints for the sake of more advanced readers.

2. G. L. Nemhauser and L. A. Wolsey. Best algorithms for approximating the maximum of a submodular set function. *Mathematics of Operations Research*, 3(3):177–188, 1978.

3. N. Buchbinder, M. Feldman, J. (Seffi) Naor, and R. Schwartz. Submodular maximization with cardinality constraints. In *Proceedings of the 25th Annual ACM-SIAM Symposium on Discrete Algorithms*, SIAM, Portland, OR, pp. 1433–1452, 2014.

4. N. Buchbinder, M. Feldman, and R. Schwartz. Comparing apples and oranges: Query tradeoff in submodular maximization. In *Mathematics of Operations Research*, 42(2):308–329, 2015.

5. B. Mirzasoleiman, A. Badanidiyuru, A. Karbasi, J. Vondrák, and A. Krause. Lazier than lazy greedy. In *Proceedings of the 29th AAAI Conference on Artificial Intelligence*, AAAI Press, Austin, TX, pp. 1812–1818, 2015.

6. U. Feige, V. S. Mirrokni, and J. Vondrák. Maximizing non-monotone submodular functions. *SIAM Journal on Computing*, 40(4):1133–1153, 2011.

7. N. Buchbinder, M. Feldman, J. Naor, and R. Schwartz. A tight linear time (1/2)-approximation for unconstrained submodular maximization. *SIAM Journal on Computing*, 44(5):1384–1402, 2015.

8. M. L. Fisher, G. L. Nemhauser, and L. A. Wolsey. An analysis of approximations for maximizing submodular set functions—II. *Mathematical Programming Study*, 8:73–87, 1978.

9. S. O. Gharan and J. Vondrák. Submodular maximization by simulated annealing. In *Proceedings of the 22nd annual ACM-SIAM Symposium on Discrete Algorithms*, SIAM, San Francisco, CA, pp. 1098–1116, 2011.

10. N. Buchbinder and M. Feldman. Deterministic algorithms for submodular maximization problems. In *Proceedings of the 27th Annual ACM-SIAM Symposium on Discrete Algorithms*, SIAM, Arlington, VA, pp. 392–403, 2016.

11. A. Badanidiyuru and J. Vondrák. Fast algorithms for maximizing submodular functions. In *Proceedings of the 25th Annual ACM-SIAM Symposium on Discrete Algorithms*, SIAM, Portland, OR, pp. 1497–1514, 2014.

12. M. Feldman. Maximizing symmetric submodular functions. In *ACM Transactions on Algorithms*, 13(3):39:1–39:36, 2015.

13. V.S. Mirrokni, M. Schapira, and J. Vondrák. Tight information-theoretic lower bounds for welfare maximization in combinatorial auctions. In *Proceedings of the 9th ACM Conference on Electronic Commerce*, ACM, Chicago, IL, pp. 70–77, 2008.

14. G. Călinescu, C. Chekuri, M. Pál, and J. Vondrák. Maximizing a monotone submodular function subject to a matroid constraint. *SIAM Journal on Computing*, 40(6):1740–1766, 2011.

15. L. Lovász. Submodular functions and convexity. In A. Bachem, M. Grötschel, and B. Korte, Eds., *Mathematical Programming: The State of the Art*, pp. 235–257. Springer, New York, 1983.

16. M. Feldman, J. Naor, and R. Schwartz. A unified continuous greedy algorithm for submodular maximization. In *IEEE 52nd Annual Symposium on Foundations of Computer Science*, IEEE, Palm Springs, CA, pp. 570–579, 2011.

17. C. Chekuri, J. Vondrák, and R. Zenklusen. Submodular function maximization via the multilinear relaxation and contention resolution schemes. *SIAM Journal on Computing*, 43(6):1831–1879, 2014.

18. A. Gupta and V. Nagarajan. A stochastic probing problem with applications. In *International Conference on Integer Programming and Combinatorial Optimization*, Springer, Valparaiso, Chile, pp. 205–216, 2013.

19. M. Adamczyk. Non-negative submodular stochastic probing via stochastic contention resolution schemes. *CoRR*, abs/1508.07771, 2015.

20. M. Feldman, O. Svensson, and R. Zenklusen. Online contention resolution schemes. In *Proceedings of the 27th Annual ACM-SIAM Symposium on Discrete Algorithms*, SIAM, Arlington, VA, pp. 1014–1033, 2016.

21. M. Feldman. Maximization problems with submodular objective functions. PhD Thesis, Computer Science Department, Technion - Israel Institute of Technology, Israel, 2013.

22. J. Vondrák. Symmetry and approximability of submodular maximization problems. *SIAM Journal on Computing*, 42(1):265–304, 2013.
23. M. Skala. Hypergeometric tail inequalities: Ending the insanity. *CoRR*, abs/1311.5939, 2013.
24. V. Chvátal. The tail of the hypergeometric distribution. *Discrete Mathematics*, 25(3):285–287, 1979.
25. W. Hoeffding. Probability inequalities for sums of bounded random variables. *Journal of the American Statistical Association*, 58(301):13–30, 1963.
26. S. Dobzinski and J. Vondrák. From query complexity to computational complexity. In *Proceedings of the 44th Annual ACM Symposium on Theory of Computing*, ACM, New York, pp. 1107–1116, 2012.
27. C. Chekuri, J. Vondrák, and R. Zenklusen. Dependent randomized rounding via exchange properties of combinatorial structures. In *Proceedings of the IEEE 51st Annual Symposium on Foundations of Computer Science*, IEEE, Las Vegas, NV, pp. 575–584, 2010.
28. A. Kulik, H. Shachnai, and T. Tamir. Maximizing submodular set functions subject to multiple linear constraints. In *Proceedings of the 20th Annual ACM-SIAM Symposium on Discrete Algorithms*, SIAM, New York, pp. 545–554, 2009.
29. A. Kulik, H. Shachnai, and T. Tamir. Approximations for monotone and nonmonotone submodular maximization with knapsack constraints. *Mathematics of Operations Research*, 38(4):729–739, 2013.
30. A. Ene, J. Vondrák, and Y. Wu. Local distribution and the symmetry gap: Approximability of multiway partitioning problems. In *Proceedings of the 24th Annual ACM-SIAM Symposium on Discrete Algorithms*, SIAM, New Orleans, LA, pp. 306–325, 2013.

Index